Springer Collected Works in Mathematics

For further volumes:
http://www.springer.com/series/11104

New York, 1961

Harish-Chandra

Collected Papers III

1959 – 1968

Editor

Veeravalli Seshadri Varadarajan

Reprint of the 1984 Edition

 Springer

Author
Harish-Chandra (1923 Kanpur,
  India – 1983 Princeton, USA)

Editor
Veeravalli Seshadri Varadarajan
University of California
Los Angeles, USA

ISSN 2194-9875
ISBN 978-3-662-45452-7   (Softcover)
     978-0-387-90782-6   (Hardcover)
DOI 10.1007/978-3-662-45453-4
Springer Heidelberg New York Dordrecht London

Library of Congress Control Number: 2012954381

Harish-Chandra
Collected Papers

Volume III
(1959–1968)

Edited by V. S. Varadarajan

Springer-Verlag Berlin Heidelberg GmbH

Harish-Chandra
1923–1983

Editor

V. S. Varadarajan
Department of Mathematics
University of California
Los Angeles, CA 90024
U.S.A.

AMS Subject Classifications: 01A75, 22-XX

Library of Congress Cataloging in Publication Data
Harish-Chandra.
  Harish-Chandra collected papers.
  Contents: v. 1. 1944–1954—v. 2. 1955–1958—
[etc.]—v. 4. 1970–present.
  1. Mathematics—Collected works. I. Varadarajan,
V. S. II. Title.
QA3.H294 1983   510   83-4828

9 8 7 6 5 4 3 2 1

ISBN 978-1-4899-7409-9   ISBN 978-1-4899-7407-5 (eBook)
DOI 10.1007/978-1-4899-7407-5

Table of Contents
Volume III

Bibliography of Harish-Chandra

[1944a]   (with Bhabha, H. J.) On the theory of point-particles. *Proc. Royal Soc. A.* **183**, 134–141.

[1944b]   On the removal of the infinite self-energies of point-particles. *Proc. Royal Soc. A.* **183**, 142–167.

[1945a]   On the scattering of scalar mesons. *Proc. Indian Acad. Sci. Sect. A.* **21**, 135–146.

[1945b]   Algebra of the Dirac-matrices. *Proc. Indian Acad. Sci. Sect. A.* **22**, 30–41.

[1946a]   (with Bhabha, H. J.) On the fields and equations of motion of point particles. *Proc. Royal Soc. A.* **185**, 250–268.

[1946b]   On the equations of motion of point particles. *Proc. Royal Soc. A.* **185**, 269–287.

[1946c]   A note on the σ-symbols. *Proc. Indian Acad. Sci. Sect. A.* **23**, 152–163.

[1946d]   The correspondence between the particle and the wave aspects of the meson and the photon. *Proc. Royal Soc. A.* **186**, 502–525.

[1947a]   On the algebra of the meson matrices. *Proc. Camb. Phil. Soc.* **43**, 414–421.

[1947b]   On relativistic wave equations. *Phys. Rev.* **71**, 793–805.

[1947c]   Equations for particles of higher spin. Report of an International Conference on Fundamental particles and Low temperatures held at the Cavendish Laboratory, Cambridge (1946). Vol. I, Fundamental Particles, 185–188. The Physical Society, London.

[1947d]   Infinite irreducible representations of the Lorentz group. *Proc. Royal Soc. A.* **189**, 372–401.

[1948a]   Relativistic equations for elementary particles. *Proc. Royal Soc. A.* **192**, 195–218.

[1948b]   Motion of an electron in the field of a magnetic pole. *Phys. Rev.* **74**, 883–887.

[1949a]   Faithful representations of Lie algebras. *Ann. of Math.* **50**, 68–76.

[1949b]   On representations of Lie algebras. *Ann. of Math.* **50**, 900–915.

[1950a]   On the radical of a Lie algebra. *Proc. Amer. Math. Soc.* **1**, 14–17.

[1950b]   On faithful representations of Lie groups. *Proc. Amer. Math. Soc.* **1**, 205–210.

[1950c]   Lie algebras and the Tannaka duality theorem. *Ann. of Math.* **51**, 299–330.

[1951a]   On some applications of the universal enveloping algebra of a semisimple Lie algebra. *Trans. Amer. Math. Soc.* **70**, 28–96.

[1951b]   Representations of semisimple Lie groups on a Banach space. *Proc. Nat. Acad. Sci. U.S.A.* **37**, 170–173.

[1951c] Representations of semisimple Lie groups. II. *Proc. Nat. Acad. Sci. U.S.A.* **37**, 362–365.

[1951d] Representations of semisimple Lie groups. III. Characters. *Proc. Nat. Acad. Sci. U.S.A.* **37**, 366–369.

[1951e] Representations of semisimple Lie groups. IV. *Proc. Nat. Acad. Sci. U.S.A.* **37**, 691–694.

[1951f] Plancherel formula for complex semisimple Lie groups. *Proc. Nat. Acad. Sci. U.S.A.* **37**, 813–818.

[1952] Plancherel formula for the 2×2 real unimodular group. *Proc. Nat. Acad. Sci. U.S.A.* **38**, 337–342.

[1953] Representations of a semisimple Lie group on a Banach space. I. *Trans. Amer. Math. Soc.* **75**, 185–243.

[1954a] Representations of semisimple Lie groups. II. *Trans. Amer. Math. Soc.* **76**, 26–65.

[1954b] Representations of semisimple Lie groups. III. *Trans. Amer. Math. Soc.* **76**, 234–253.

[1954c] The Plancherel formula for complex semisimple Lie groups. *Trans. Amer. Math. Soc.* **76**, 485–528.

[1954d] On the Plancherel formula for the right K-invariant functions on a semisimple Lie group. *Proc. Nat. Acad. Sci. U.S.A.* **40**, 200–204.

[1954e] Representations of semisimple Lie groups. V. *Proc. Nat. Acad. Sci. U.S.A.* **40**, 1076–1077.

[1954f] Representations of semisimple Lie groups. VI. *Proc. Nat. Acad. Sci. U.S.A.* **40**, 1078–1080.

[1955a] Integrable and square-integrable representations of a semi-simple Lie group. *Proc. Nat. Acad. Sci. U.S.A.* **41**, 314–317.

[1955b] On the characters of a semisimple Lie group. *Bull. Amer. Math. Soc.* **61**, 389–396.

[1955c] Representations of semisimple Lie groups. IV. *Amer. J. of Math.* **77**, 743–777.

[1956a] Representations of semisimple Lie groups. V. *Amer. J. of Math.* **78**, 1–41.

[1956b] Representations of semisimple Lie groups. VI. Integrable and square-integrable representations. *Amer. J. of Math.* **78**, 564–628.

[1956c] The characters of semisimple Lie groups. *Trans. Amer. Math. Soc.* **83**, 98–163.

[1956d] On a lemma of F. Bruhat. *J. Math. Pures. Appl.* (9) **35**, 203–210.

[1956e] Invariant differential operators on a semisimple Lie algebra. *Proc. Nat. Acad. Sci. U.S.A.* **42**, 252–253.

[1956f] A formula for semisimple Lie groups. *Proc. Nat. Acad. Sci. U.S.A.* **42**, 538–540.

[1957a] Representations of semisimple Lie groups. *Proceedings of the International Congress of Mathematicians*, Amsterdam (1954). Vol. 1, 299–304. Erven P. Noordhoff N. V., Groningen, North Holland Publishing Company.

[1957b] Differential operators on a semisimple Lie algebra. *Amer. J. of Math.* **79**, 87–120.

[1957c] Fourier transforms on a semisimple Lie algebra. I. *Amer. J. of Math.* **79**, 193–257.

[1957d] Fourier transforms on a semisimple Lie algebra. II. *Amer. J. of Math.* **79**, 653–686.

[1957e] A formula for semisimple Lie groups. *Amer. J. of Math.* **79**, 733–760.

[1957f] Spherical functions on a semisimple Lie group. *Proc. Nat. Acad. Sci. U.S.A.* **43**, 408–409.

[1958a] Spherical functions on a semisimple Lie group. I. *Amer. J. of Math.* **80**, 241–310.

[1958b] Spherical functions on a semisimple Lie group. II. *Amer. J. of Math.* **80**, 553–613.

[1959a] Automorphic forms on a semisimple Lie group. *Proc. Nat. Acad. Sci. U.S.A.* **45**, 570–573.

[1959b] Some results on differential equations and their applications. *Proc. Nat. Acad. Sci. U.S.A.* **45**, 1763–1764.

BIBLIOGRAPHY OF HARISH-CHANDRA

[1960a]   Some results on differential equations.
[1960a'] Supplement to "Some results on differential equations"
[1960b]   Differential equations and semisimple Lie groups.
[1961]   (with Borel, A.) Arithmetic subgroups of algebraic groups. *Bull. Amer. Math. Soc.* **67**, 579–583.
[1962]   (with Borel, A.) Arithmetic subgroups of algebraic groups. *Ann. of Math.* **75**, 485–535.
[1963]   Invariant eigendistributions on semisimple Lie groups. *Bull. Amer. Math. Soc.* **69**, 117–123.
[1964a]   Invariant distributions on Lie algebras. *Amer. J. of Math.* **86**, 271–309.
[1964b]   Invariant differential operators and distributions on a semi-simple Lie algebra. *Amer. J. of Math.* **86**, 534–564.
[1964c]   Some results on an invariant integral on a semisimple Lie algebra. *Ann. of Math.* **80**, 551–593.
[1965a]   Invariant eigendistributions on a semisimple Lie algebra. *Publ. Math. IHES* No. **27**, 5–54.
[1965b]   Invariant eigendistributions on a semisimple Lie group. *Trans. Amer. Math. Soc.* **119**, 457–508.
[1965c]   Discrete series for semisimple Lie groups. I. Construction of invariant eigendistributions. *Acta Math.* **113**, 241–318.
[1966a]   Two theorems on semisimple Lie groups. *Ann. of Math.* **83**, 74–128.
[1966b]   Discrete series for semisimple Lie groups. II. Explicit determination of the characters. *Acta. Math.* **116**, 1–111.
[1966c]   Harmonic analysis on semisimple Lie groups. Some recent advances in the basic sciences. Vol. 1 (1962, 1963, 1964), 35–40. *Belfer Graduate School of Science Annual Science Conference Proceedings*. Edited by A. Gelbart. Belfer Graduate School of Science, Yeshiva University, New York, New York.
[1967]   Characters of semisimple Lie groups. *Symposia on Theoretical Physics*, **4**, 137–142. (Lectures presented at the 1965 Third Anniversary Symposium of the Institute of Mathematical Sciences, Madras, India. Edited by A. Ramakrishnan.) Plenum Press, New York.
[1968a]   Harmonic analysis on semisimple Lie groups. *Proceedings of the International Congress of Mathematicians*, Moscow (1966), 89–94. "Mir", Moscow.
[1968b]   Automorphic forms on semisimple Lie groups. Notes by J. G. M. Mars. *Lecture Notes in Mathematics*, No. 62, Springer-Verlag, Berlin–Heidelberg–New York.
[1970a]   Some applications of the Schwartz space of a semisimple Lie group. *Lectures in Modern Analysis and Applications*. *II*, 1–7. Edited by C. T. Tamm. *Lecture Notes in Mathematics*. No. **140**, Springer-Verlag, Berlin–Heidelberg–New York.
[1970b]   Eisenstein series over finite fields. Functional analysis and related fields, 76–88. *Proceedings of a Conference in honor of Professor Marshall Stone held at the University of Chicago, May 1968*. Edited by F. E. Browder. Springer-Verlag, Berlin–Heidelberg–New York.
[1970c]   Harmonic analysis on semisimple Lie groups. *Bull. Amer. Math. Soc.* **76**, 529–551.
[1970d]   Harmonic analysis on reductive p-adic groups. Notes by G. van Dijk. *Lecture Notes in Mathematics*, No. **162**, Springer-Verlag, Berlin–Heidelberg–New York.
[1972]   On the theory of the Eisenstein integral. Conference on Harmonic Analysis, College Park, Maryland (1971), 123–149. Edited by D. Gulick and R. L. Lipsman. *Lecture Notes in Mathematics*, No. **266**, Springer-Verlag, Berlin–Heidelberg–New York.
[1973]   Harmonic analysis on reductive p-adic groups. Harmonic analysis on homogeneous spaces, 167–192. Edited by C. C. Moore. *Proceedings of Symposia in Pure Mathe-*

matics, Vol. **XXVI**, Amer. Math. Soc., Providence, R. I., U.S.A.

[1975]   Harmonic analysis on real reductive groups. I. The theory of the constant term. *J. of Functional Analysis* **19**, 104–204.

[1976a]   Harmonic analysis on real reductive groups. II. Wave packets in the Schwartz space. *Inventiones Math*. **36**, 1–55.

[1976b]   Harmonic analysis on real reductive groups. III. The Maass–Selberg relations and the Plancherel formula. *Ann. of Math*. **104**, 117–201.

[1977a]   The characters of reductive *p*-adic groups. *Contributions to Algebra*, 175–182. A collection of papers dedicated to Ellis Kolchin. Edited by H. Bass, P. J. Cassidy, and J. Kovacic. Academic Press, New York–San Francisco–London.

[1977b]   The Plancherel formula for reductive *p*-adic groups.

[1977b'] Corrections to "The Plancherel formula for reductive *p*-adic groups"

[1978]   Admissible invariant distributions on reductive *p*-adic groups. Lie theories and their applications, 281–347. *Proceedings of the 1977 annual seminar of the Canadian Mathematical Congress*. Edited by W. Rossmann. Queen's papers in Pure and Applied Math. No. **48**. (Editors: A. J. Coleman and P. Ribenboim). Kingston, Ontario.

[1980]   A submersion principle and its applications. Geometry and Analysis, 95–102. Papers dedicated to the memory of V. K. Patodi, Indian Academy of Sciences, Bangalore, and the Tate Institute of Fundamental Research, Bombay.

[1983]   Supertempered distributions on real reductive groups. *Studies in Applied Mathematics, Advances in Mathematics, Supplementary Studies Series*, Vol. 8, Academic Press Inc., pp. 139–153, edited by Victor Guillemin.

Abbreviations Used in the Bibliography

Proc. Royal Soc. A. ...	Proceedings of the Royal Society, A...
Proc. Indian Acad. Sci. Sect. A	Proceedings of the Indian Academy of Sciences, Section A
Proc. Camb. Phil. Soc.	Proceedings of the Cambridge Philosophical Society
Phys. Rev.	The Physical Review
Ann. of Math.	Annals of Mathematics
Proc. Amer. Math. Soc.	Proceedings of the American Mathematical Society
Trans. Amer. Math. Soc.	Transactions of the American Mathematical Society
Proc. Nat. Acad. Sci. U.S.A.	Proceedings of the National Academy of Sciences of the United States of America
Bull. Amer. Math. Soc.	Bulletin of the American Mathematical Society
Amer. J. of Math.	American Journal of Mathematics
J. Math. Pures Appl.	Journal de Mathématiques Pures et Appliquées
Publ. Math. IHES	Publications Mathématiques, Institut des Hautes Études Scientifiques
Acta Math.	Acta Mathematica
J. Functional Analysis	Journal of Functional Analysis
Inventiones Math.	Inventiones Mathematicae

Preface

These volumes of the Collected Papers of Harish-Chandra are being brought out in response to a widespread feeling in the mathematical community that they would immensely benefit scholars and researchworkers, especially those in analysis, representation theory, arithmetic, mathematical physics, and other related areas. It is hoped that in addition to making his contributions more accessible by collecting them in one place, these volumes would help focus renewed attention on his ideas and methods as well as lend additional perspective to them.

The papers are arranged chronologically. Harish-Chandra is still an extraordinarily productive mathematician, and so I feel that there is no need to impose any structure on his output (past and current) other than that coming from the evolution of his own thinking. It was also decided to divide the papers into four volumes so that each volume would have a reasonable size. However, I have attempted to make the lines of division in such a way that each volume has a certain unity within itself.

It is impossible to mention all the individuals and organisations that helped me with this project. Nevertheless, I do want to record my special thanks to Phyllis Parris, for typing the two handwritten manuscripts on differential equations; to Walter Kaufmann-Bühler, for his constant enthusiasm and encouragement; and to Elise Oranges for her tireless work in producing the final copy. In addition I am extremely grateful to Roger Howe and to Nolan Wallach for agreeing to write commentaries on selected parts of Harish-Chandra's work, and doing it within the time frame that I gave them. Finally, I want to thank Mrs. Lily Harish-Chandra for providing me with the photographs of Harish-Chandra that appear in these volumes.

Pacific Palisades, California
May, 1983

V. S. VARADARAJAN

Biographical Note

Harish-Chandra was born on October 11, 1923 in Kanpur, India. He received the M.Sc degree from Allahabad University in 1943 and spent the years 1943–1945 at the Indian Institute of Science, Bangalore, India. He went to Cambridge University in England as a graduate student of Dirac in 1945 and received his Ph.D from there in 1947. He was at Columbia University, New York, from 1950 to 1963. In 1963 he became a permanent member of the Institute for Advanced Study at Princeton, New Jersey, where he has been the IBM von Neumann Professor from 1970 onwards.

He was a Guggenheim Fellow from 1957 to 1958 and a Sloan Fellow from 1961 to 1963. He was elected a Fellow of the Royal Society in 1973 and a member of the National Academy of Sciences of the U.S.A. in 1981. He was awarded honorary doctorates by Delhi University in 1973 and by Yale University in 1981.

He and his wife live in Princeton. They have two daughters. Although he has not returned to India except for brief visits, he has always been profoundly concerned with India and Indian science. He remains a figure of great inspiration in his native land.

Postscript

Harish-Chandra is no more. He suffered a fatal heart attack while out for a walk on the evening of Sunday, October 16, 1983. He survives in his work, which is a faithful reflection of his personality—lofty, intense, uncompromising.

Princeton   V. S. VARADARAJAN
October 19, 1983

Introduction

V. S. VARADARAJAN

> *Bottom*: Masters, I am to discourse wonders:
> but ask me not what; for if I tell you, I am
> no true Athenian. I will tell you every-
> thing, right as it fell out.
>
> —Shakespeare, *A Midsummer Night's
> Dream*, Act Four, Scene II.

§0. The first scientific papers of Harish-Chandra were in theoretical physics. However, his interests shifted to mathematics rather early in his career. His prolonged and intense preoccupation with group representations began in the late 1940's when, at the suggestion of Dirac, he started investigating the infinite dimensional unitary representations of the Lorentz group [1947d]. Since then he has erected, almost singlehandedly, a monumental theory of harmonic analysis on reductive groups and their homogeneous spaces. His work is a profound synthesis of algebra, geometry, and analysis. The great force and continued resonance of his ideas have inspired a generation of mathematicians. The singlemindedness and courage with which he has pursued his goals, the beauty of his results, the power and originality of his methods, and the ultimate simplicity of his philosophy, as well as the conviction that sustains it, compel our admiration. There can be no doubt that his achievement is one of the greatest in mathematics in our time.

This introduction attempts to give an overview of this work. I have sketched, in an impressionistic way, the main lines of it in §2; the remaining sections contain additional detail on selected parts of his work so that one of his great achievements —the harmonic analysis of $L^2(G)$ for a real reductive group G—may come into clearer focus. I hope that this account, although incomplete and imperfect, may still be of value as a rough guide to the prospective reader.

§1. The early sources of motivation for representation theory and harmonic analysis were the classical theory of Fourier series and integrals and the theory of linear representations of finite groups. After the rise of topology and functional analysis in the early decades of this century, far-reaching generalizations of these classical themes began to emerge.

For a locally compact abelian group G let $\hat{G}$ be the group of characters of G. $\hat{G}$ is also locally compact abelian and the Pontryagin–van Kampen theory pairs G and $\hat{G}$ in a canonical duality. Given any function $f \in L^1(G)$ its Fourier transform is the function $\hat{f}$ on $\hat{G}$ given by

$$\hat{f}(\hat{x}) = \int_G f(x)\langle x, \hat{x}\rangle \, dx \qquad (\hat{x} \in \hat{G}, \, dx \text{ a Haar measure})$$

and $f \mapsto \hat{f}$ is a homomorphism of the convolution algebra $L^1(G)$ into the algebra of continuous functions on $\hat{G}$ that vanish at infinity. The Fourier transform can also be introduced on $L^2(G)$; it is then a unitary isomorphism of $L^2(G)$ with $L^2(\hat{G})$ where the Haar measure $d\hat{x}$ on $\hat{G}$ is uniquely determined by dx (the measure dual to dx). In particular we have the *Plancherel formula*

$$\int_G |f(x)|^2 \, dx = \int_{\hat{G}} |\hat{f}(\hat{x})|^2 \, d\hat{x} \qquad (f \in L^2(G))$$

and the *inversion formula*

$$f(x) = \int_{\hat{G}} \hat{f}(\hat{x})\langle x, \hat{x}\rangle^{\text{conj}} \, d\hat{x} \qquad (f \in L^1(G), \hat{f} \in L^1(\hat{G}))$$

which generalize the corresponding formulae from classical Fourier analysis on $\mathbb{R}^n$ ([W 1] [Lo] [Bl]).

The classical Poisson Summation formula can also be generalized to this setting. Let $\Gamma \subset G$ be a lattice in G, i.e., a discrete subgroup of G such that G/Γ is compact. Then $\Gamma^{\vee}$, the annihilator of Γ, is also a lattice in $\hat{G}$. If Fourier transforms are defined using the Haar measure that gives the measure 1 to G/Γ, then the Poisson formula takes the form

$$\sum_{\gamma \in \Gamma} f(\gamma) = \sum_{\gamma^{\vee} \in \Gamma^{\vee}} \hat{f}(\gamma^{\vee})$$

where f and $\hat{f}$ are suitably restricted ([W 2]).

The Poisson formula has always been a powerful tool in number theory. The importance of Fourier analysis on general locally compact abelian groups became very clear with the work of Tate who, at the suggestion of Artin, used the general Poisson formula in the setting provided by the adele group of a number field to derive the analytic theory of the Hecke L-series attached to a Grössencharakter ([T] [W 2]).

Suppose now that G is compact. Let dx be the normalized Haar measure on G so that $\int_G dx = 1$. Let $\hat{G}$ be the set of equivalence classes of irreducible (finite dimensional) unitary representations of G. The general facts concerning linear representations of finite groups extend essentially unchanged to G. Thus if $\pi \in \omega \in \hat{G}$ and ϕ, ψ are unit vectors in the representation space of π,

$$\int_G |(\pi(x)\phi, \psi)|^2 \, dx = \frac{1}{d(\omega)}$$

where $d(\omega)$ is the degree of ω. Moreover, for $\omega \in \hat{G}$ let Θ_{ω} be the corresponding character; then

$$\int_G \Theta_{\omega}\Theta_{\omega'}^{\text{conj}} \, dx = \delta_{\omega\omega'} \qquad (\omega, \omega' \in \hat{G}).$$

The fundamental result is the Peter–Weyl theorem. It asserts the completeness of the irreducible representations and can be stated as the completeness formula

$$\int_G |f|^2 \, dx = \sum_{\omega \in \hat{G}} d(\omega) \|\pi_\omega(f)\|_2^2$$

where $f \in L^2(G)$, $\pi_\omega \in \omega$, and $\|\cdot\|_2$ is the Hilbert–Schmidt norm. The measure on $\hat{G}$, which assigns the mass $d(\omega)$ to ω, may thus be viewed as the *Plancherel measure* of G.

Let G be a compact Lie group. For $f \in L^2(G)$ let $\hat{f}$ be its Fourier transform, defined as the function on $\hat{G}$ given by

$$\hat{f}(\omega) = \int_G f \Theta_\omega^{\text{conj}} \, dx \qquad (\omega \in \hat{G}).$$

The completeness formula can then be restated as

$$f(e) = \sum_{\omega \in \hat{G}} d(\omega) \hat{f}(\omega) \qquad (f \in C^\infty(G)).$$

In other words, we have the relation

$$\delta = \sum_{\omega \in \hat{G}} d(\omega) \Theta_\omega$$

in the space of *distributions* on the manifold G, δ being the delta function at the identity element of G.

It is natural to ask for an explicit determination of the irreducible representations of special groups such as the simple Lie groups which had been classified in the late 1890's by Killing and Élie Cartan. The work of Cartan and Hermann Weyl in the 1920's gave a complete solution of this problem. For Cartan it was a question of determining the irreducible finite dimensional representations of the (complex) simple *Lie algebras*; he obtained them as highest weight modules. Weyl had a more global view and worked with the *compact simple Lie groups*. He discovered beautiful formulae for the characters and dimensions of the irreducible representations. Moreover his discovery of the existence of a compact form of a complex simple Lie algebra ("unitarian trick") showed that the representation theory of a complex simple Lie algebra is essentially the same as that of the (simply connected) compact form associated to it, thus unifying his theory with that of Cartan ([We 2]; see also [V 1]).

The notion of distributions, which came into widespread use with the work of Laurent Schwartz [S], led to a deeper understanding of the notion of Fourier transform and to a great extension of the domain of its applicability. In the language of distributions the Plancherel formula for $\mathbb{R}^n$ became the "expansion" of the delta function at the origin in terms of the characters:

$$\delta = (2\pi)^{-n/2} \int_{\mathbb{R}^n} e(\lambda) \, d\lambda \qquad (e(\lambda)(x) = e^{i\lambda \cdot x}).$$

Suppose now that G is a locally compact group, unimodular and second countable, which is neither compact nor abelian. Let $\hat{G}$ be the set of equivalence classes of its irreducible unitary representations (finite or infinite dimensional). There was no systematic theory of representations of such groups in the 1930's. But in Quantum Physics unitary representations of the various symmetry groups had already begun to play a prominent role. This is because the covariance properties of a system with a (connected) group of symmetries are expressed by means of a unitary representation of the group in question or an extension of it. In relativistic quantum physics this meant the inhomogeneous Lorentz group; its physically significant irreducible unitary representations (both single and double valued in the physicists' terminology) were classified by Wigner in a famous memoir in 1939 [Wi].

The situation changed completely in the 1940's. For, by then, the epoch-making papers of von Neumann (by himself and with Murray) on operator algebras in Hilbert spaces had appeared [MvN] [vN]. This theory provided the framework for the study of infinite dimensional unitary representations; in addition it revealed that new phenomena such as type II and type III algebras, direct integral decompositions, and so on, would have to be taken into account.

The systematic beginnings of unitary representation theory of general locally compact groups go back to 1943 when Gel'fand and Raikov proved that every such group had enough irreducible unitary representations to separate the points of the group [GR]. The subject developed rapidly in the late 1940's and early 1950's mainly at the hands of Godement, Mackey, Mautner, and Segal. Their point of view was functional analytic, and their main tool was the theory of operator algebras. Their work led, among other things, to a general decomposition theory of unitary representations, abstract Plancherel theorems, and to the recognition that a reasonable harmonic analysis can be expected only for type I groups, namely groups whose unitary representations were all of type I [Ma 1] [Ma 2] [Ma 3]. It was also in this period that the papers of Mackey on systems of imprimitivity and induced representations appeared, in which he extended the classical Frobenius theory to all separable locally compact groups. Mackey's work, which went far beyond what is suggested in these remarks, has proved to be very influential in the representation theory of nilpotent and solvable groups (cf. [AM] [Mo]).

These developments led to the following formulation of the central problem of harmonic analysis: given a separable locally compact group G acting transitively on a space X with an invariant measure μ, and an irreducible unitary representation τ of the stabilizer in G of a given point of X, decompose into irreducible constituents the representation of G induced by τ. If τ is the trivial representation, the induced representation is just the natural representation of G in $L^2(X, \mu)$; if $G = X$ and the action is by left translations, the problem is the decomposition of the left regular representation of G. The decomposition may not be entirely discrete so that it has to be understood as a direct integral.

In this generality this problem is hopeless because the category of all separable locally compact groups is simply too vast for classical harmonic analysis to be anything more than a very uncertain guide. New examples were clearly needed, and it was natural to look to the simple Lie groups once again for this purpose. The papers of Gel'fand and Neumark [GN 1] and Bargmann [B], dealing with SL(2, C) and SL(2, R), and published independently of each other in 1947, mark the beginning of the subject of infinite dimensional representations of real semisimple Lie groups.

In [GN 1] Gel'fand and Neumark determined all the irreducible unitary representations of $\mathrm{SL}(2,\mathbb{C})$ upto equivalence, discovered an explicit Plancherel formula, and proved that any unitary representation of $\mathrm{SL}(2,\mathbb{C})$ could be decomposed as a direct integral of irreducible representations in a natural way. They discovered the surprising fact that this group had *additional* irreducible unitary representations which played no role in the harmonic analysis of $L^2(G)$. They called these the representations of the *supplementary series*. They followed this up with a remarkable and seminal series of papers in which they extended their theory to $\mathrm{SL}(n,\mathbb{C})$ and, to a lesser extent, to the other *complex classical groups*. This work, which was presented in detail in their beautiful monograph [GN 2], completely changed the perspective of the theory.

The group $\mathrm{SL}(n,\mathbb{C})$ acts transitively on a number of projective varieties such as the Grassmannians and the various flag spaces. The spaces of *holomorphic* sections of the complex analytic line bundles on these homogeneous spaces carry the finite dimensional complex analytic representations of $\mathrm{SL}(n,\mathbb{C})$. With remarkable insight Gel'fand and Neumark realized that the *smooth* sections of the *real analytic* line bundles on the same homogeneous spaces would transform (roughly speaking) according to the *infinite dimensional* irreducible representations of the group. These representations would be unitary under appropriate conditions. In this way, one had a single scheme for *all* irreducible representations, which contained the finite dimensional as well as the unitary ones as special cases. For example, $G = \mathrm{SL}(n,\mathbb{C})$ acts transitively on the space of all flags (e_m) $(1 \leqslant m \leqslant n,\ e_1 \subset e_2 \subset \cdots \subset e_n,\ e_m$ is an m-dimensional linear subspace of $\mathbb{C}^n$). Then $X \approx G/B$ where B is the subgroup of the upper triangular matrices. Gel'fand and Neumark proved that the unitary representations of G induced by the unitary one dimensional representations of B, which they called the *principal series* of representations, are all irreducible, and that the regular representation of the group is a direct integral of them. They extended to $\mathrm{SL}(n,\mathbb{C})$ (as well as to the other complex classical groups) the construction of the supplementary series.

Although these representations are all infinite dimensional, Gel'fand and Neumark succeeded in associating *characters* to them. If ξ is any one dimensional representation of B and π_ξ the representation of G induced by ξ, the operator

$$\pi_\xi(x) = \int_G x(g)\pi_\xi(g)\,dg$$

is of trace class if x is suitably restricted, and there is a (unique) class function θ_ξ on G such that

$$\mathrm{tr}\big(\pi_\xi(x)\big) = \int_G x(g)\theta_\xi(g)\,dg.$$

It is natural to call θ_ξ the *character* of π_ξ. They evaluated θ_ξ on the subgroup D of diagonal matrices and found for it a formula of the same general structure as Weyl's character formula. In terms of these characters they obtained the *Plancherel formula* for $SL(n, \mathbb{C})$ in the form

$$x(e) = \int_{\hat{D}} \hat{x}(\xi)\omega(\xi)\,d\xi \qquad (x \in C_c^\infty(G))$$

where

$$\hat{x}(\xi) = \operatorname{trace}(\pi_\xi(x)) \qquad \text{(Fourier transform!)}$$

and $\omega(\xi)$ is an explicitly determined function $\geqslant 0$. If we rewrite this in the form

$$\delta = \int \omega(\xi)\theta_\xi\,d\xi,$$

its analogy with the Plancherel formula for $\mathbb{R}^n$ and compact Lie groups becomes clear; $\omega(\xi)\,d\xi$ is the Plancherel measure.

Let me now turn to the work of Bargmann which treated both $SL(2, \mathbb{C})$ and $SL(2, \mathbb{R})$. So far as $SL(2, \mathbb{C})$ was concerned the results were essentially the same as those in [GN 1]; however $SL(2, \mathbb{R})$ presented new phenomena. In addition to the principal and supplementary series, defined essentially as for the complex unimodular group, there now appeared a discretely parametrized series of irreducible unitary representations characterized by the remarkable fact that their matrix elements were already square integrable on the group manifold. This property meant that these representations would occur as *discrete direct summands* of the regular representation. The property of having square integrable matrix coefficients makes sense for irreducible unitary representations of arbitrary separable locally compact unimodular groups and is equivalent to their occurring as direct summands of the regular representation; they form the so-called *discrete series* of the group, and the set of the corresponding equivalence classes is denoted by $\hat{G}_d$. Their appearance in the case of $SL(2, \mathbb{R})$ meant that the decomposition of its regular representation had both discrete and continuous parts.

A highlight of the paper of Bargmann is the emergence of another new theme in our subject. He viewed the matrix elements of the irreducible representations as eigenfunctions of the Casimir differential operator defined on the group manifold (originally introduced by the physicist Casimir for $SU(2)$). In this way the problem of harmonic analysis in $L^2(G)$ became one of eigenfunction expansions for the Casimir operator. The matrix elements have simple transformation properties relative to the action (from both left and right) of the rotation subgroup of $SL(2, \mathbb{R})$; and so they are essentially determined by their values at the elements $\begin{pmatrix} e^t & 0 \\ 0 & e^{-t} \end{pmatrix}$ $(t > 0)$ of the group. Thus they may be viewed as functions on the half-line of positive reals; moreover the action of the Casimir operator coincides with that of a second order differential operator on this half-line. The spectral theory of such differential operators goes back to Hermann Weyl [We 1]. Bargmann used Weyl's theory to prove the completeness theorem, i.e., the fact that the linear space, spanned by the matrix elements of the discrete series and the "wave packets" of the matrix elements of the principal series, is dense in $L^2(G)$.

For the matrix elements of the discrete series Bargmann discovered remarkable generalizations of the Schur orthogonality relations. For the principal series the corresponding problem is the explicit determination of the Plancherel measure.

Bargmann found a beautiful expression for it in terms of the leading coefficients of the asymptotic expansions of the matrix elements.

The papers of Gel'fand and Neumark and Bargmann form a great watershed in our subject. Their work showed clearly, at least in retrospect, that to do harmonic analysis on real semisimple Lie groups it is absolutely necessary to operate at a level that allowed for a true interaction among the algebraic, geometric, and analytic properties of these groups.

This was the state of affairs in the early 1950's when Harish-Chandra entered the field. His methods combined in a deep and original manner the infinitesimal and global aspects of these problems, and, through an astonishing sequence of papers spanning almost three decades, led to a profound understanding of harmonic analysis of real reductive groups. Sometime in the 1960's he realized that his work contained deep analogies with the spectral theory of automorphic forms that was then being built through the efforts of Selberg, Gel'fand (and his collaborators), and Langlands. His exploration of these analogies has led in recent years to significant results in the harmonic analysis of reductive groups over local, global, and even finite fields, and has revealed the essential unity of the structure of Fourier analysis in these diverse contexts.

§2. Before starting with a description of Harish-Chandra's work we introduce a few definitions and some notation. G is a real semisimple Lie group, connected and with finite center; K is a maximal compact subgroup of G; and θ, the involution that fixes K pointwise; $\mathfrak{g} = \mathrm{Lie}(G)$, $\mathfrak{k} = \mathrm{Lie}(K)$. For any real Lie algebra $\mathfrak{m}$, $U(\mathfrak{m})$ is the universal enveloping algebra of the complexification $\mathfrak{m}_c$ of $\mathfrak{m}$.

Manifolds are smooth and second countable. They need not be connected but all the connected components will have the same dimension. For any manifold M, $C^\infty(M)$ (resp. $C_c^\infty(M)$) is the space of complex valued functions on M that are smooth (resp. smooth with compact support). If ω is a volume element (= exterior differential form of maximal degree which is nowhere zero) on M, then given any smooth differential operator D on M its transpose D^t is the differential operator such that $\int_M Df \cdot g\omega = \int_M f \cdot D^t g\omega$ for all $f, g \in C_c^\infty(M)$. If T is a distribution on M the distribution DT is defined by $\langle DT, f \rangle = \langle T, D^t f \rangle$ ($f \in C_c^\infty(M)$). Any function F on M that is locally integrable with respect to ω will be usually identified with the distribution $f \mapsto \int_M Ff\omega$. The actions of differential operators on smooth functions and distributions are compatible under this identification. If M is a Lie group we shall always choose ω to be a left invariant volume element. In this case $U(\mathfrak{m}) = U(\mathfrak{m})^t$ is canonically isomorphic to the algebra of all left invariant differential operators on M; $X^t = -X$ for $X \in \mathfrak{m}$.

We write $\mathfrak{Z}$ for the center of $U(\mathfrak{g})$. Then $\mathfrak{Z}$ is canonically isomorphic to the algebra of all differential operators on G that are both left and right invariant. $\mathfrak{Z}$ is a polynomial algebra in l indeterminates where l is the rank of G.

The centralizer in G of a Cartan subalgebra of $\mathfrak{g}$ is called a Cartan subgroup. Cartan subgroups are parameter spaces for conjugacy classes, and so are absolutely fundamental objects in character theory which deals with class functions and distributions. They are abelian if G is linear and are maximal tori if G is compact. A compact Cartan subgroup is always connected and is in fact a torus. An element $x \in G$ is regular if its centralizer in $\mathfrak{g}$ is a Cartan subalgebra. If $l = rk(G)$, t is an indeterminate and $D(x)$ is the coefficient of t^l in $\det(\mathrm{Ad}(x) - 1 + t)$, x is regular if and only if $D(x) \neq 0$. The set of regular points is denoted by G'; it is a dense open subset of G. D is often called the discriminant of G.

Parabolic subgroups (psgrps) are normalizers in G of parabolic subalgebras of $\mathfrak{g}$; a subalgebra $\mathfrak{p}$ of $\mathfrak{g}$ is parabolic if $\mathfrak{p}_c$ contains a Borel subalgebra of $\mathfrak{g}_c$. Any psgrp P has a canonical decomposition $P = MAN \simeq M \times A \times N$. Here N is the unipotent radical of P, namely the maximal normal subgroup of P that consists entirely of unipotent elements; $M_1 = MA$ is reductive and is defined as $P \cap \theta(P)$, so that $P = M_1 N$ is essentially the Levi decomposition of P, while A is the maximal $\mathbb{R}$-split* subgroup lying in the center of M_1. The subgroup M is the intersection of the kernels of all the continuous homomorphisms of M_1 into the positive reals. The decomposition $P = MAN$ is called the Langlands decomposition of P. The parabolic subgroups are needed for defining the various series of irreducible unitary representations of G; moreover the formulation of many of the asymptotic questions in harmonic analysis depend in a fundamental way on the geometry of the set of all parabolic subgroups.

Harish-Chandra's early work was largely devoted to the development of algebraic methods for studying infinite dimensional representations of real semisimple Lie groups. These methods led him to discover the fundamental facts about unitary representations of G. Among these are the results that assert that G is a type I group and that in every irreducible unitary representation of G the irreducible representations of K occur only with finite multiplicities.

Let us define a $(\mathfrak{g}, K)$-module to be a module V for $\mathfrak{g}$ which has a compatible structure as a K-module such that V is the algebraic sum of finite dimensional K-modules. V is unitary if there is a positive definite scalar product $(\ ,\ )$ for V such that $(Xu, v) = -(u, Xv)$ for all $u, v \in V$, $X \in \mathfrak{g}$. One of Harish-Chandra's major discoveries was that the space $V(\pi)$ of K-finite vectors in an irreducible unitary representation π is an irreducible $(\mathfrak{g}, K)$-module and that the assignment $\pi \mapsto V(\pi)$ induces a bijection of $\hat{G}$ with the set of isomorphism classes of irreducible $(\mathfrak{g}, K)$-modules that are unitary. He also proved that any irreducible $(\mathfrak{g}, K)$-module is of the form $V(\pi)$ for some irreducible representation π acting on a Hilbert space and possessing an infinitesimal character, thus showing that in the notion of a $(\mathfrak{g}, K)$-module which is irreducible we have the correct formulation of the vague idea of an "arbitrary" irreducible representation of G.

In spite of the promise and power of the algebraic method he never pursued it beyond his initial investigations and was content to use his early work as the point of departure for the analytical theory.

The basic step in the analytical theory is the introduction of characters. The algebraic treatment had already shown that the multiplicities of the irreducible representations of K that occur in an irreducible unitary representation of G do not exceed a constant multiple of the degrees of the representations of K. It follows from this that for any function $f \in C_c^\infty(G)$ the operator

$$\pi(f) = \int_G f(x)\pi(x)\,dx$$

---

*This means that $\mathrm{Ad}(a)$ is diagonalizable over $\mathbb{R}$ for all $a \in A$.

is of trace class, and the linear functional

$$\Theta_\pi : f \mapsto \operatorname{tr}(\pi(f))$$

is a distribution on the group G. Θ_π depends only on the equivalence class ω of π; it is denoted by Θ_ω and is called the character of π (or ω). It is an invariant distribution, i.e., a distribution that is invariant under all inner automorphisms of G. It determines ω completely. If $(\phi_i)_{i \geqslant 1}$ is an orthonormal basis of K-finite vectors for π and f_i is the matrix element defined by ϕ_i, i.e., $f_i(x) = (\pi(x)\phi_i, \phi_i)$ $(x \in G)$,

$$\Theta_\omega = \sum_{i \geqslant 1} f_i,$$

this equation being understood in the sense of distributions, i.e., for all $u \in C_c^\infty(G)$,

$$\Theta_\omega(u) = \sum_{i \geqslant 1} \int_G f_i u \, dx.$$

Since any matrix element satisfies the differential equation

$$zf = \chi_\omega(z)f \qquad (z \in \mathfrak{Z})$$

where χ_ω is the infinitesimal character of ω, it follows that

$$z\Theta_\omega = \chi_\omega(z)\Theta_\omega \qquad (z \in \mathfrak{Z}).$$

Here we recall that the infinitesimal character of ω is the homomorphism according to which $\mathfrak{Z}$ acts (by Schur's lemma) on the space $V(\pi)$. In other words, Θ_ω is an *invariant eigendistribution* on G.

Using the notion of characters one may formulate the problem of harmonic analysis on G as follows (cf. the introductory remarks to [1957e] [1970c]): given any invariant distribution T on G, express T as a "linear combination" of the irreducible characters. For $f \in C_c^\infty(G)$ define its *Fourier transform* $\hat{f}$ to be the function on $\hat{G}$ given by

$$\hat{f}(\omega) = \Theta_\omega(f) \qquad (\omega \in \hat{G}).$$

The problem is then to define a "distribution" $\hat{T}$ on $\hat{G}$ such that

$$T(f) = \hat{T}(\hat{f}) \qquad (f \in C_c^\infty(G)).$$

$\hat{T}$ will be the "Fourier transform" of T. If $T = \delta$, the delta function at the identity element of G, the determination of $\hat{\delta}$ is the problem of finding the explicit Plancherel formula for G; $\hat{\delta}$ will have to be a nonnegative measure, the Plancherel measure. The derivation by Gel'fand and Neumark of the Plancherel formula for $SL(n,\mathbb{C})$, Harish-Chandra's generalization of this to all complex semisimple Lie groups [1951f] [1954c], and his treatment of the case of $SL(2,\mathbb{R})$ [1952], and the work of Gel'fand and Graev on $SL(n,\mathbb{R})$ [GG 1] [GG 2], may be regarded as initial evidence for this view.

A second approach to harmonic analysis is through *eigenfunction expansions*; this goes back to Bargmann as we saw in §1. Let $H \subset G$ be a closed unimodular subgroup, $X = G/H$, and let $\mathcal{H} = L^2(X)$ be the Hilbert space of functions square integrable with respect to the invariant measure on X. Let λ be the natural representation of G in $\mathcal{H}$. If $\mathcal{H}^\infty$ is the space of differentiable vectors in $\mathcal{H}$ for λ, we have a representation $a \mapsto \lambda(a)$ of $U(\mathfrak{g})$ in $\mathfrak{K}^{\infty}$. If $z \in \mathfrak{Z}$ is Hermitian, i.e., $(z^{\mathrm{conj}})^t = z$, the operator $\lambda(z)$ is essentially self adjoint on $\mathfrak{K}^{\infty}$ and is invariant under G. So, in the first approximation, one may view the problem of decomposition of λ as an eigenfunction expansion problem for the commuting algebra of differential operators $\lambda(z)$, $z \in \mathfrak{Z}$ (see the introductory remarks to [1968b]). The eigenspaces are parametrized by the points of the spectrum of the commutative algebra $\mathfrak{Z}$, i.e., by the homomorphisms $\chi \colon \mathfrak{Z} \to \mathbf{C}$. For a given χ, the corresponding eigenspace would be a G-module which, typically, would be infinite dimensional. In view of the relationship between the eigenvalue problem and the representation theoretic question, we should expect that only those eigenfunctions will contribute to the spectral expansion in $L^2(X)$ whose χ are the infinitesimal characters of irreducible *unitary* representations of G (not necessarily all of them either).

The basic example in this approach is the case $X = G$, viewed as a homogeneous space for G (left action) or $G \times G$ (two-sided action). A variant of this in which the eigenspaces are finite dimensional is obtained as follows. Let U be a finite dimensional Hilbert space which is a bimodule for K, i.e., which carries a unitary representation τ of $K \times K^{\vee}$, $K^{\vee}$ being the group opposite to K. The Hilbert space is now the space $L^2(G, \tau)$ of functions $f \colon G \to U$ which are square integrable and satisfy

$$f(k_1 x k_2) = k_1 f(x) k_2$$

for all $k_1, k_2 \in K$ and almost all $x \in G$. Such functions are called *τ-spherical*; if $U = \mathbf{C}$ and $K \times K^{\vee}$ acts trivially, one speaks of *spherical* or *strictly spherical* functions. The terminology reminds us that we have a generalization of the classical theory of spherical harmonics. If χ is a homomorphism of $\mathfrak{Z}$ into $\mathbf{C}$ and $f \in L^2(G, \tau)$ is such that $zf = \chi(z)f$ for all $z \in \mathfrak{Z}$, f is C^{∞} (even analytic), and the space of all such f is of finite dimension (cf. [1960b]). One of Harish-Chandra's basic results is that the eigenfunctions needed for the L^2-expansion are precisely those that satisfy the so-called weak inequality ([1970c], p. 539).

The theory of automorphic forms gives another example of this point of view. The classical automorphic forms on the Poincaré half-plane may be interpreted as eigenfunctions for the Laplace–Beltrami operator. They can then be generalized to the context of nonanalytic forms and higher dimensional symmetric spaces, as was done by Maass [M] and Selberg [Se 1]. Following the work of Borel and Harish-Chandra [1962] on the structure of arithmetic subgroups of reductive groups and the associated homogeneous spaces, Gel'fand [G], Selberg [Se 1] [Se 2], and Langlands [L 1] constructed a general theory of automorphic forms. They approached the fundamental problem as the spectral theory of the invariant differential operators in $L^2(G/\Gamma)$, G being the group of real points of a reductive group defined over $\mathbf{Q}$, and Γ is the associated arithmetic subgroup of G. The classical Eisenstein series and its generalizations are examples of eigenfunctions of these differential operators. It is interesting to note that the homomorphisms χ in the spectrum of $\mathfrak{Z}$ that correspond to these eigenfunctions do not come from unitary representations, and so one must make an analytic continuation in χ in order to get the eigenfunctions needed for the spectral theory (cf. [1968b]).

Whichever point of view is adopted, it is not possible to go far without a good supply of irreducible unitary representations. The known results for complex groups and $SL(n, \mathbb{R})$ seemed to suggest that each conjugacy class of Cartan subgroups of G gives rise to a "series" of irreducible unitary representations, and that the Plancherel measure is a *disjoint sum* of the measures contributed by each of these series (see [1957a] [1954e]). As a partial justification of this principle Harish-Chandra sketched in [1954e] a method of constructing a series of irreducible unitary representations associated to any Cartan subgroup L. Given L, one can select a parabolic subgroup $P = MAN$ such that A is the maximal $\mathbb{R}$-split* component of L. If η is an irreducible unitary representation of M and $\nu \in \hat{A}$, the unitary representation $\pi_{\eta, \nu}$ of G is defined as the representation induced by the representation $m \cdot a \cdot n \mapsto \nu(a)\eta(m)$ of P. The $\pi_{\eta, \nu}$ depend only on the conjugacy class of L and are discretely decomposable into irreducible subrepresentations. If L is chosen so that $\dim(A)$ is maximum, P is a minimal parabolic subgroup, M is compact, and what we have is called the *principal series*; for complex G it *is* the principal series of Gel'fand and Neumark. Bruhat's work [Br 1] had shown that for minimal P the $\pi_{\eta, \nu}$ are irreducible for ν in "general position." So one could hope that this would be so in the case of arbitrary L also.

If P is not minimal, M is not compact and so we are still faced with the problem of determining $\hat{M}$. Harish-Chandra's remarkable idea, which is implicit but very clear in [1957a] and [1954e], was that if we restrict η to vary always over the *discrete series* of M, the various series arising from a complete system of mutually nonconjugate Cartan subgroups would determine essentially disjoint parts of $\hat{G}$; and these representations would suffice to obtain the Plancherel formula for G. The discrete series is thus associated to the (unique) conjugacy class of compact Cartan subgroups, and the general discussion makes it clear that these may have to be constructed a priori and by quite different methods.

Discrete series: beginnings. Already in 1947 Godement had proved that the matrix elements of the discrete series satisfy orthogonality relations [Go]. In particular, one can associate to each $\omega \in \hat{G}_d$ a positive number $d(\omega)$, the *formal degree of* ω, such that

$$\int_G |(\pi(x)\phi, \psi)|^2 \, dx = \frac{1}{d(\omega)}$$

where $\pi \in \omega$ and ϕ, ψ are arbitrary unit vectors in the representation space of π. Moreover if $^0L^2(G)$ is the closed linear span of the irreducible subspaces of $L^2(G)$ (= the "discrete part" of $L^2(G)$), and E is the orthogonal projection $L^2(G) \to {}^0L^2(G)$, then, for all $f \in L^1(G) \cap L^2(G)$,

$$\|Ef\|^2 = \sum_{\omega \in \hat{G}_d} d(\omega)\|\pi_\omega(f)\|_2^2$$

where $\pi_\omega \in \omega$ and $\|\cdot\|_2$ is the Hilbert-Schmidt norm; this shows that the Plancherel

---

*This means that A is the maximal subgroup lying in the connected component of the identity of L such that $\mathrm{Ad}(a)$ is diagonalizable over $\mathbb{R}$ for all $a \in A$.

measure assigns the mass $d(\omega)$ to the class ω [1956b]. These results are actually true for all second countable locally compact groups ([Ma 2], p. 640).

In a succession of beautiful papers Harish-Chandra added new evidence to the picture of harmonic analysis of $L^2(G)$ suggested above [1954f], [1955a], [1955c], [1956a], [1956b]. Assuming not only that $rk(G) = rk(K)$ but further that G/K is Hermitian symmetric (this means that G/K has a G-invariant complex structure) he constructed a whole family of discrete series representations. They were realized on certain Hilbert spaces of holomorphic functions on which G acts naturally; they are the *holomorphic discrete series*. When G/K is not Hermitian symmetric this method would give nothing; even in the Hermitian symmetric case it will not give the entire discrete series. Nevertheless the introductory remarks to [1955c] reveal his conviction that in these cases different methods would have to be invented.

These papers revealed profound analogies between the (holomorphic) discrete series and the finite dimensional representations. These representations were *highest weight modules* with respect to positive systems coming from the complex structure of G/K. Their formal degrees were given *exactly by Weyl's dimension formula* applied to their highest weights. Finally, their *characters made sense as point functions* on the set of regular points of a compact Cartan subgroup and *could be described there by precise analogues of the Weyl character formulae.*

These results were the first serious indication that the discrete series would be parametrized by the characters of the compact Cartan subgroup, and hence that the representations of the series associated to an arbitrary Cartan subgroup would be parametrized by the characters of that Cartan subgroup. One would thus have a natural map

$$\coprod_{1 \leqslant i \leqslant r} \hat{A}_i \to \hat{G}$$

where $A_1, A_2, \ldots, A_r$, is a complete system of mutually nonconjugate Cartan subgroups of G (cf. the introductory remarks to [1957e]).

It is thus clear that by 1956 Harish-Chandra had essentially visualized the lines along which the Fourier analysis of $L^2(G)$ would have to be constructed. In retrospect it is really remarkable that he saw so clearly and so far; the clues available to him were certainly very meager.

Discrete series: construction. The determination of the discrete series turned out to be a prodigious undertaking ([1957b]–[1957e], [1964a]–[1966b]). His method, which was entirely novel, was roughly in three parts.

The first part is a direct and explicit construction of the invariant eigendistributions Θ_ξ, parametrized by the (regular) characters ξ of a compact Cartan subgroup $B \subset K$; these would ultimately be the characters of the representations of the discrete series. In the second part the question is that of showing that the Fourier coefficients (with respect to K) of the Θ_ξ, which are clearly eigenfunctions for $\mathfrak{Z}$ that are K-finite, are in $L^2(G)$. The third part is concerned with the completeness of the Θ_ξ; and the main result here is that these eigenfunctions and their translates span a dense subspace of $^0L^2(G)$.

Before the invariant eigendistributions could be constructed it is absolutely essential to understand the nature of their *singularities*. An initial exploration [1956c] had already shown that any invariant eigendistribution Θ on G is an analytic function F_Θ on G', the set of regular points of G. This function is locally integrable around the singular points also and so defines an invariant distribution $f \mapsto \int F_\Theta f \, dx$ ($f \in C_c^\infty(G)$) on G, also denoted by F_Θ. Experience suggested that $\Theta = F_\Theta$ (cf. [1955b], although the statement there is somewhat guarded), and this actually turned out to be true in complete generality. This result, known as the *Regularity theorem*, lies at a great depth, and is the corner stone for the entire structure of harmonic analysis.

The function F_Θ will in general blow up in the neighborhood of a singular point of the group. This behavior however cannot be completely arbitrary because the differential equations satisfied by the characters remain valid in the neighborhood of the singular points also. Harish-Chandra's work revealed the true nature of the limitations on the behaviour of F_Θ, namely that F_Θ should satisfy conditions of the following form: for suitable (explicitly determined) invariant differential operators ∇ on G', ∇F_Θ should extend continuously to all of G.

The starting point of the construction of the Θ_ξ is now quite clear; one defines Θ_ξ at the regular points on the compact Cartan subgroup by the analogue of the Weyl character formula. But the problem of defining Θ_ξ on the other Cartan subgroups is nontrivial because an invariant eigendistribution is not uniquely determined by its values on $B' = B \cap G'$. Nevertheless Harish-Chandra discovered that there is uniqueness *if the distribution is suitably restricted at infinity on the group*. The condition is that the distributions should be tempered; this however is to anticipate later ideas. At this stage the condition had to be formulated in a more direct manner, and he did it by requiring that $|D|^{-1/2}$ be a *global majorant* for Θ_ξ, D being the discriminant of G. The existence problem, i.e., the construction of the Θ_ξ, is however much deeper and depends in a fundamental way on *Fourier transforms on the Lie algebra of G*. Harish-Chandra proved that the Θ_ξ can be obtained by exponentiating from the Lie algebra the Fourier transforms of the invariant measures on appropriately chosen regular semisimple orbits [1965c].

The regularity theorem is the key that opens all the doors. If we look at the smooth, almost magical manner in which the theory flows once the regularity theorem is established, and also remember that there was very little in the known history of the subject to guide him in his prolonged struggle or to reassure him during moments of doubt, we have no choice except to place this among his greatest creative achievements.

The idea now is to use the method of integration over the conjugacy classes ($=$ orbits) to reduce the harmonic analysis of the discrete series to that on the compact Cartan subgroup. This is of course exactly the method used by Weyl in his famous completeness proof for the compact case. However the present situation is much more complicated, for one must take into account the contributions of the other Cartan subgroups. Before explaining Harish-Chandra's method of doing this I shall discuss in some detail the method of orbital integrals.

It was Gel'fand and Neumark who first realized that the theory of orbital integrals can be made the basis for a proof of the Plancherel formula [GN 2]. They treated $\mathrm{SL}(n, \mathbb{C})$ by this method. In [1951f], [1954c] Harish-Chandra pushed this method through for all *complex* semisimple Lie groups. To motivate this let us first assume that G is compact and let $T \subset G$ be a maximal torus. Using Weyl's formulae we can rewrite the Plancherel formula as a sum over $\hat{T}$:

$$f(e) = \sum_{\xi \in \hat{T}} \Omega(\xi) \int_G \Theta_\xi f \, dx \qquad (f \in C^\infty(G)).$$

Here $\Omega(\xi)$ is a polynomial function of ξ, namely a constant multiple of the product of the roots of a positive system; Θ_ξ is the Weyl character, given on T by

$$\Delta(t)\Theta_\xi(t) = \sum_{s \in W} \varepsilon(s)(s\xi)(t) \qquad (t \in T)$$

where W is the Weyl group and Δ is the Weyl denominator function. Integrating over the orbits and setting

$$F_f(t) = \Delta(t) \int_G f(xtx^{-1}) \, dx \qquad (t \in T)$$

we have

$$\int_G \Theta_\xi f \, dx = (-1)^p \hat{F}_f(\xi)$$

where p is the number of positive roots and $\hat{F}_f$ is the Fourier transform of F_f (F_f is obviously in $C^\infty(T)$). So the Plancherel formula is equivalent to the assertion

$$f(e) = (-1)^p \sum_{\xi \in \hat{T}} \Omega(\xi) \hat{F}_f(\xi).$$

But, by Fourier transform theory on T, $(-1)^p\Omega(\xi)\hat{F}_f(\xi)$ is $\widehat{\Omega F_f}(\xi)$ where Ω *is now viewed as a constant coefficient differential operator on* T. So the Plancherel formula is equivalent to the relation

(L.F.) $$f(e) = (\Omega F_f)(e) \qquad (f \in C^\infty(G)).$$

We shall refer to (L.F.) as the *limit formula for the orbital integral.*

This derivation suggests a method for proving the Plancherel formula: first prove (L.F.) and then apply Fourier transform to it. It is remarkable that this method continues to work when G is complex. The maximal torus is now replaced by a Cartan subgroup L. For any $f \in C_c^\infty(G)$, $F_{f,L}(h)$ will denote the integral of f over the orbit of h:

$$F_{f,L}(h) = \Delta(h) \int_{G/L} f(xhx^{-1}) \, d\dot{x}$$

where $d\dot{x}$ is the invariant measure on G/L. Initially this makes sense only for h regular, and in the regular subset of L, $F_{f,L}$ will be a smooth function. It can then be proved that $F_{f,L}$ extends to all of L as an element of $C_c^\infty(L)$. The character formulae for the principal series now show that, with appropriate normalization,

$$\langle T_\xi, f \rangle = \hat{F}_{f,L}(\xi) \qquad (\xi \in \hat{L})$$

where T_ξ is the distribution character of the principal series representation corresponding to the character ξ of L. But now the limit formula

(L.F.) $$f(e) = (\Omega F_{f,L})(e) \qquad (f \in C_c^\infty(G))$$

holds, with same Ω as before, and the Fourier transformation of this relation gives the Plancherel formula

$$f(e) = \int_L \hat{\Omega}(\xi)\langle T_\xi, f \rangle \, d\xi \qquad (f \in C_c^\infty(G))$$

where $\hat{\Omega}$ is the polynomial function which is the Fourier transform of the differential operator Ω.

The set $\{e\}$ is a very complicated singular point in the space of conjugacy classes. In particular it is not the limit of orbits in general position, and so $f(e)$ is not the limit of $F_{f,L}(h)$ as $h \to e$. It is therefore striking that one can still recover $f(e)$ as the limit as $h \to e$ of *suitable derivatives* of $F_{f,L}(h)$.

It is natural to try to extend this method to all cases by establishing the limit formula and relating the Fourier transforms of the orbital integrals to the irreducible characters. However for arbitrary G the orbital integrals and their derivatives have jumps at the singular points. Moreover there is more than one orbital integral to be considered since there is in general more than one conjugacy class of Cartan subgroups. Nevertheless calculations in special cases [1952] seemed to suggest that the Fourier transform of the orbital integral is very closely related to the irreducible characters, and so a limit formula should still represent a primitive version of the Plancherel formula (see the remarks at the beginning of [1957e]). In a series of papers in the 1950's ([1957b]–[1957e] and [1964c]) Harish-Chandra established the limit formula for the orbital integral associated to a *fundamental* Cartan subgroup, namely a Cartan subgroup whose maximal compact subgroup has the largest possible dimension (these form a single conjugacy class, and when $rk(G) = rk(K)$, fundamental means compact). In the limit formula

$$(\text{L.F.}) \qquad\qquad f(e) = (\Omega F_{f,L})(e) \qquad (L \text{ fundamental})$$

$F_{f,L}$ may still have jumps at singular points of L; *nevertheless the differential operator Ω, which is still the product of roots of a positive system, kills all the jumps, i.e., $\Omega F_{f,L}$ extends continuously to all of L, so that the right side makes sense.*

The jumps of $F_{f,L}$ make the Fourier analysis of the relation (L.F.) a very delicate matter (see [1952]). These jumps are in fact the orbital integrals associated to Cartan subgroups which are "less compact" than L, suggesting that the jumps are caused by the contributions to f from the representations associated to these less compact Cartan subgroups. Thus when L is compact, one must work with matrix elements of the discrete series representations in order to get rid of the jumps. This can only be done by working with a space of rapidly decreasing functions instead of $C_c^\infty(G)$.

For this purpose Harish-Chandra introduced in [1966b] the Schwartz space $\mathcal{C}(G)$ of G. Overcoming many technical difficulties he proved in [1966b] that the theory of orbital integrals can be extended to the Schwartz space, that the limit formula (L.F.) is true for all f in this space, and that the distributions Θ_ξ constructed earlier are *tempered*, with

$$\langle \Theta_\xi, f \rangle = \int_G \Theta_\xi f \, dx,$$

the integrals converging absolutely. Moreover he proved that *if f is an eigenfunction* *for β in the Schwartz space, the integrals of f on the regular non-elliptic orbits are all zero*; here we recall that an orbit is said to be *elliptic* if it meets a compact Cartan subgroup. If G has no compact Cartan subgroup this means that all the orbital integrals of f are zero; the limit formula would then imply that $f = 0$. We shall see presently that eigenfunctions of β in $L^2(G)$ are automatically in Schwartz space; so this argument already shows that the condition $rk(G) = rk(K)$ is necessary for the existence of the discrete series. If $rk(G) = rk(K)$, the same argument would still allow us to conclude that the orbital integral $F_{f,B}$ ($B \subset K$) has no jumps, i.e., it lies in $C^\infty(B)$. The Fourier transformation of the limit formula coupled with the formula for Θ_ξ on B would show that the harmonic analysis of f is entirely controlled by the Θ_ξ. This is substantially Harish-Chandra's argument for the completeness of the Θ_ξ.

To round out this picture it is necessary to know a method of determining when a given eigenfunction of β belongs to the Schwartz space. In [1966b] Harish-Chandra developed a powerful theory to describe the asymptotic behaviour of τ-spherical eigenfunctions for β that satisfy the weak inequality. If f is any τ-spherical eigenfunction satisfying the weak inequality, Harish-Chandra was able to associate to f, at each point at infinity of the group, a so called *constant term*, that was the dominant term in the asymptotic development of f there. These dominant terms are parametrized by the parabolic subgroups of G. If $P = MAN$ is a parabolic subgroup and f_P is the associated constant term, f_P is essentially a function of the same nature as f, but lives on the group MA; the difference between f and f_P has very rapid decay at infinity "in the direction of P." An important consequence of this theory is the result that f lies in $L^2(G, \tau)$ *if and only if all its constant terms are zero; and that in this case, f is in the Schwartz space.*

In order to apply this theory successfully two things have to be done. First, it must be shown that the Fourier coefficients of the Θ_ξ satisfy the weak inequality. This is technically rather subtle and is done in [1966a]. The second point is to show that all the constant terms of these Fourier coefficients are zero. Harish-Chandra's proof of this made use of a beautiful property of these constant terms, namely that they are *transitive* in a natural sense with respect to the partial ordering among the parabolic subgroups. If f is a τ-spherical eigenfunction built out of the Fourier coefficients of the Θ_ξ, and $P \neq G$ is chosen so that $f_P \neq 0$ and $\dim(M)$ is minimal, the transitivity mentioned above would show that the restrictions of f_P to M are eigenfunctions in the Schwartz space of M (actually linear combinations of such); so M has to have a compact Cartan subgroup, implying that P is associated to a Cartan subgroup of G itself. This Cartan subgroup cannot be compact, and hence as the eigenhomomorphism to which f belongs is regular and is associated to the compact Cartan subgroup, we have a contradiction.

These brief remarks cannot give an adequate idea of the pathbreaking nature of the papers [1965c] [1966b]. The audacity of their conceptions as well as the delicacy and craftsmanship that were needed in executing them, compel one to place them among the greatest works in Fourier analysis.

The continuous spectrum: the theory of the Eisenstein integral and the explicit Plancherel formula. With the discrete series completely determined Harish-Chandra turned to the problem of explicitly decomposing the regular representation of G. He went much farther than a mere L^2-theory and obtained a Fourier analysis of the

Schwartz space of G. The results were worked out in three long papers [1975], [1976a], and [1976b]; but the lectures [1970a], [1970c], and [1972] containing the announcements of these and many other results are interesting in themselves, for they give a bird's eye view of his entire approach.

Let L be a Cartan subgroup written (canonically) as $T \cdot A$ where T is compact and A is connected and split over $\mathbb{R}$. Let $\mathscr{P}(A)$ be the finite set of parabolic subgroups $P = MAN$. Since $rk(M) = \dim(T)$, $\hat{M}_d \neq \phi$. For $\eta \in \omega \in \hat{M}_d$ and $\nu \in \mathfrak{a}^*$ $(\mathfrak{a} = \mathrm{Lie}(A))$, $\pi_{\eta,\nu}$ is the unitary representation of G induced by the representation $m \cdot a \cdot n \mapsto e^{i\nu(\log a)} \cdot \eta(m)$ of P. The character $\Theta_{\omega,\nu}$ of $\pi_{\eta,\nu}$ is independent of P (M and A are the same for all $P \in \mathscr{P}(A)$). Then (see [1976b])

(a)　$\pi_{\eta,\nu}$ is finitely decomposable; it is irreducible for regular ν.

(b)　Let $W(A)$ be the Weyl group of A, i.e., the normalizer of A modulo the centralizer of A; then $W(A)$ acts on $\hat{M}_d \times \mathfrak{a}^*$ and

$$\Theta_{s\omega,s\nu} = \Theta_{\omega,\nu} \qquad (s \in W(A)).$$

Actually results much stronger than (a) were proved by Harish-Chandra. For example, if L is fundamental, *all* the $\pi_{\eta,\nu}$ are irreducible ([1976b]); special cases of this were known earlier, going all the way back to the irreducibility of the principal series of $SL(n, \mathbb{C})$ due to Gel'fand and Neumark.

One can now associate to L the subspace $\mathcal{C}_L$ of $\mathcal{C}$ consisting of all functions that are orthogonal to the matrix elements of the representations $\pi_{\eta,\nu}$ attached to the Cartan subgroups that are not conjugate to L. $\mathcal{C}_L$ is closed in $\mathcal{C}$ and is invariant under translations. Central to Harish-Chandra's theory is the decomposition (Theorem 12, [1970c])

$$\mathcal{C} = \bigoplus_L \mathcal{C}_L \qquad\qquad (*)$$

where the sum is over a complete set of mutually nonconjugate Cartan subgroups; the sum is orthogonal and smooth (i.e., the projections are continuous in $\mathcal{C}$). It will turn out that $\mathcal{C}_L$ is exactly the space of "wave packets" of the matrix elements of the $\pi_{\eta,\nu}$ associated to L (this can be formulated more precisely). Thus $(*)$ is a beautiful and definitive formulation of the heuristic principle that "each Cartan subgroup makes a separate contribution to the Fourier analysis of G."

The splitting $(*)$ allows one to work on each $\mathcal{C}_L$ separately. Let $\Omega_L = \hat{M}_d \times \mathfrak{a}^*$ and let us define, for $f \in \mathcal{C}$, its Fourier transform $\hat{f} = (\hat{f}_L)$ by

$$\hat{f}_L(\omega, \nu) = \int_G \Theta_{\omega,\nu}^{\mathrm{conj}} f \, dx \qquad ((\omega, \nu) \in \Omega_L).$$

Then $\hat{f}_L$ is $W(A)$-invariant, and we have an initial formulation of the Plancherel formula ([1972], Theorem 11): there is a unique $W(A)$-invariant continuous function μ_L on Ω_L ($\hat{M}_d$ is given the discrete topology) such that for all $f \in \mathcal{C}_L$,

$$f(e) = \sum_{\omega \in \hat{M}_d} d(\omega) \int_{\mathfrak{a}^*} \hat{f}_L(\omega : \nu) \mu_L(\omega : \nu) \, d\nu$$

the series converging absolutely. The function μ_L is $\geqslant 0$ and has other nice properties. For fixed ω, it extends to a meromorphic function on $\mathfrak{a}_c^*$ that is holomorphic on $\mathfrak{a}^*$ and > 0 at the regular points of $\mathfrak{a}^*$; moreover it is of at most polynomial growth on $\mathfrak{a}^*$.

The basic question now is of course the explicit calculation of μ_L. Harish-Chandra's method for doing this was based on a remarkable *product formula* for μ_L. To describe this, fix a parabolic subgroup $P \in \mathscr{P}(A)$; then one can associate in a canonical manner a set of pairs (M_i, Q_i) $(1 \le i \le m)$ where M_i is a reductive subgroup of G with no split component and $Q_i = MA_iN_i$ is a parabolic subgroup of M_i with $A_i \subset A$ and $\dim(A_i) = 1$. If $\mathfrak{a}_i = \mathrm{Lie}(A_i)$ and $\nu_i = \nu|_{\mathfrak{a}_i}$, then

$$\mu_L = c \prod_{1 \le i \le m} \mu_{L,i}, \qquad \mu_{L,i}(\omega:\nu) = \mu_{L,i}(\omega:\nu_i)$$

where $\mu_{L,i}$ is the μ-function associated to (M_i, Q_i) (see [1972], pp. 144–145 for a description of the (M_i, Q_i)); c is a constant > 0 independent of ω and ν and can be explicitly evaluated. This reduces the problem of explicit determination of μ_L to the case (G, P) where G has no split component, $P = MAN$, and $\dim(A) = 1$. If $rk(G) > rk(K)$, then L is fundamental and the limit formula, by Fourier transformation on L, leads to the determination of μ_L as a *polynomial function* on $\hat{L}$; here we must remember that we are operating not on the whole Schwartz space but only on $\mathcal{C}_L$ so that the contributions of the other Cartan subgroups of G do not enter the limit formula. Thus the analysis proceeds as if G has a single conjugacy class of Cartan subgroups (see §24, [1976b]). If $rk(G) = rk(K)$, the treatment follows the model where $rk(G/K) = 1$ (cf. [1966a] and §36, [1976b]).

It is impossible to describe except in bare outline the architecture of this beautiful theory. I shall however try to compensate by discussing two important sources of motivation behind Harish-Chandra's treatment. The first is the harmonic analysis of *spherical functions* [1958a] [1958b]. Many of the crucial features that are typical of the continuous spectrum were discovered first in this context. The second source is the spectral theory of $L^2(G/\Gamma)$, due to Selberg [Se 2], Gel'fand and Piatetsky-Shapiro [G] [GP], and Langlands [L 1], Γ being an arithmetic subgroup of G. Harish-Chandra's treatment of this in [1968b] is especially useful for this purpose.

The theory of spherical functions is essentially the harmonic analysis of $L^2(G/K)$. The fundamental objects are the elementary spherical functions $\varphi(\nu: x) = \varphi_\nu(x)$ and the transform that is defined by them. In [1958a] and [1958b] Harish-Chandra made the remarkable discovery that the φ_ν, regarded as functions on A (of the Iwasawa decomposition), look like a sum of plane waves at infinity on A; and that all of these have the same amplitude, in terms of which the Plancherel measure can be determined rather simply. More precisely,

$$d(a)\varphi(\nu: a) \sim \sum_{s \in W(A)} c_s(\nu) e^{is\nu(\log a)},$$

d being a certain homomorphism of A into the positive reals; the amplitudes satisfy

$$|c_s(\nu)| = |c(\nu)| \qquad (c = c_1)$$

and

$$|c(\nu)|^{-2} d\nu$$

is the Plancherel measure. Actually we have the relations

$$c_s(\nu) = c(s\nu)$$

which mirror the *functional equations*

$$\varphi_{s\nu} = \varphi_\nu$$

satisfied by the elementary spherical functions.

The computation of the spherical Plancherel measure thus reduces to determining the function c, which is a special case of the famous c-function of Harish-Chandra. It turned out that this function is meromorphic on $\mathfrak{a}_c^*$ and that it can be expressed as a product of the analogous c-functions associated to the real rank one subgroups of G defined by the root system of (G, A) (the Gindikin–Karpelevič product formula). Since the c-function can be explicitly evaluated for groups of real rank one, the story is complete.

For the harmonic analysis of arbitrary (not necessarily spherical) functions in the Schwartz space Harish-Chandra introduced the eigenfunctions that generalized the elementary spherical functions. Let U be a finite dimensional Hilbert space carrying a unitary double representation τ of K. Let $\tau_M = \tau|_M$ and let $\mathfrak{L}$ be the space of τ_M-spherical maps of M into U whose components (with respect to some basis of U) are in $\mathcal{C}(M) \cap {}^0L^2(M)$. Then $\dim(\mathfrak{L})$ is finite and for each $\psi \in \mathfrak{L}$, $\nu \in \mathfrak{a}^*$, and a given choice of P, a parabolic subgroup in $\mathcal{P}(A)$, Harish-Chandra defined an eigenfunction for $\mathfrak{Z}$ in $C^\infty(G, \tau)$, denoted by $E_{\psi,\nu}$ or $E_{P,\psi,\nu}$, called the *Eisenstein integral*. If P is minimal, $U = \mathbf{C}$, τ is the trivial representation, and ψ is 1, this is just the elementary spherical function (see [1972], p. 133).

The Eisenstein integrals satisfy the weak inequality, and so one can determine their behaviour at infinity on the group by applying the asymptotic theory of tempered eigenfunctions developed in [1966b] [1975]. If Q is a parabolic subgroup in $\mathcal{P}(A)$ which may be different from P, then one finds that the constant term of the Eisenstein integral $E_{P,\psi,\nu}$ associated to Q is of the form

$$\sum_{s \in W(A)} (\psi_{s,\nu})(m) e^{is\nu(\log a)}$$

where ν is a regular element of $\mathfrak{a}^*$ and $\psi_{s,\nu} \in \mathfrak{L}$. The maps

$$c_{Q|P}(s:\nu): \psi \mapsto \psi_{s,\nu}$$

are uniquely determined endomorphisms of $\mathfrak{L}$ and the relation

$$E_{P,\psi,\nu} \underset{Q}{\sim} \sum_{s \in W(A)} c_{Q|P}(s:\nu)\psi e^{is\nu}$$

expresses the fact that $E_{P,\psi,\nu}$ looks like a sum of plane waves at infinity in the direction of Q. This is clearly analogous to what happens in the spherical case.

Let us now fix ω and choose τ so that the subspace $\mathfrak{L}(\omega)$ of $\mathfrak{L}$ that "transforms according to ω" is nonzero; this is always possible. It can then be shown that for regular $\nu c_{Q|P}(1:\nu)$ is a bijection of $\mathfrak{L}(\omega)$ onto itself. Let $c_{Q|P,\omega}(1:\nu)$ be the restriction of $c_{Q|P}(1:\nu)$ to $\mathfrak{L}(\omega)$. Then the fundamental result expressing the relationship between the Plancherel formula and the asymptotics of the Eisenstein integrals can be stated as follows (see [1976b], §13):

$$\mu_L(\omega:\nu)c_{Q|P,\omega}(1:\nu)^{\dagger}c_{Q|P,\omega}(1:\nu) = c \cdot id_{\omega}$$

where id_{ω} is the identity endomorphism of $\mathfrak{L}(\omega)$, $c > 0$ is a constant independent of ω and ν which can be explicitly calculated. One must also note that as $\mathfrak{L} \subset {}^0L(M, \tau_M)$, $\mathfrak{L}$ has a natural structure as a Hilbert space, and $\dagger$ is the adjoint operation in this structure.

The c-functions are closely related to the intertwining operators between the induced representations. The latter have natural product formulae which then lead to product formulae for the c-functions and thence to the μ-functions. This is the origin of the product formula for the Plancherel measure.

A central role in Harish-Chandra's treatment of these questions is played by the functional equations of the Eisenstein integrals. The analogy with the theory of automorphic forms is most significant here, and so it may be useful to make a few remarks comparing the Eisenstein integral with the Eisenstein series. The starting point of this comparison is the following characterization of the "discrete part" of the Schwartz space of G, obtained by Harish-Chandra: if ${}^0\mathcal{C} = \mathcal{C}(G) \cap {}^0L^2(G)$, then an element f in $\mathcal{C}(G)$ is in ${}^0\mathcal{C}$ if and only if

$$\int_N f(xn)\, dn = 0 \qquad (x \in G)$$

for all parabolic subgroups $P = MAN \neq G$ (Theorems 14 and 15, [1970c]). This is obviously analogous to the condition that defines a cusp form on G/Γ, so that it is natural to speak of the elements of ${}^0\mathcal{C}$ as *cusp forms* on G. Thus $\mathfrak{L}$ may be viewed as the space of τ_M-spherical cusp forms on M; and the definition of the Eisenstein integral imitates that of the Eisenstein series with integration over K replacing the summation over Γ (cf. [1968b], p. 29). Both the characterization of ${}^0\mathcal{C}$ and the finite dimensionality of $\mathfrak{L}$ are deep lying consequences of the theory of the discrete spectrum.

The functional equations of the Eisenstein integrals involve the comparison of $E_{P,\psi,\nu}$ and $E_{Q,\psi',s\nu}$, and Harish-Chandra obtained them as a consequence of the remarkable principle asserting that *if just one summand is common to the constant terms of two such functions, they must be identical*. This principle follows from what he called the *Maass – Selberg relations* which may be described as follows: if f is a τ-spherical eigenfunction satisfying the weak inequality and suitable additional conditions (which are satisfied by the Eisenstein integrals), the intensities (= norms in ${}^0L^2(M, \tau_M)$) of the plane waves that occur in its asymptotic forms along the parabolic subgroups in $\mathscr{P}(A)$ are all equal (cf. Theorem 14.1 and the results of §17, especially Lemmas 17.2 and 17.3 of [1976b]). The Maass–Selberg relations originally arose in the theory of automorphic forms ([1968b], Chap. IV, §2); the discovery and proof of their analogues in the present context is a highlight of [1976b].

Representation theory of $\mathfrak{p}$-adic groups. The Lefschetz principle and the philosophy of cusp forms. The beginnings of the representation theory of reductive $\mathfrak{p}$-adic groups go back to the late 1950's when Mautner's paper [Mau 1] on the representations of $GL(2, \Omega)$ and $PGL(2, \Omega)$ appeared, Ω being a local field (also called a $\mathfrak{p}$-adic field), i.e., a locally compact non-discrete field. It developed rapidly in the early 1960's at the hands of Bruhat, Gel'fand–Graev, Satake and others. Harish-Chandra became actively interested in this area in the mid 1960's; his ideas and results since then have proved to be of fundamental importance for the subject. I shall first try to motivate these developments.

Let k be an algebraic number field. Then one can attach to k the local fields k_v which are its completions at the various places v, and the adele ring $\mathbf{A}$ which is a "restricted" direct product of the $k_v \cdot \mathbf{A}$ is an abelian ring which is locally compact in a natural topology; k is diagonally imbedded in it as a discrete subring and $\mathbf{A}/k$ is compact. By the 1950's it had become a well understood principle in number theory that many arithmetic questions on k could be studied over each k_v, and then by working over $\mathbf{A}$, the local data coming from the various k_v could often be put together to reach global results. This is so in classfield theory, in the theory of simple algebras and Brauer groups, and in the arithmetic theory of quadratic forms [W 2].

After the development of the theory of linear algebraic groups in the 1950's it was realized that the formalism of adeles could be applied to any algebraic group defined over k. Let $\mathbf{G}$ be such a group, and for any k-algebra R, let $\mathbf{G}(R)$ be the group of R-points of $\mathbf{G}$. We then have the groups $G_v = \mathbf{G}(k_v)$ for each v and the adele group $G_\mathbf{A} = \mathbf{G}(\mathbf{A})$. G_v and $G_\mathbf{A}$ are locally compact, separable, and $G_k = \mathbf{G}(k)$ is (diagonally) imbedded naturally as a discrete subgroup of $G_\mathbf{A}$. With the availability of the Chevalley theory of reductive algebraic groups and the work of Borel and Harish-Chandra [1962] (see also [Bo 1]) on the structure of $G_\mathbf{A}/G_k$ it was natural to hope that one could begin studying the arithmetic aspects of the theory of reductive groups. A succession of closely related ideas and papers, especially those of Weil [W 3] [W 4] [W 5] and Langlands [L 1] [JL] (with Jacquet) [L 2] [L 3] led to an explosive development of this point of view in the 1960's. As a consequence, the supreme importance of the harmonic analysis of $L^2(G_\mathbf{A}/G_k)$ for arithmetic was recognized. For example, if $\mathbf{G} = GL(1)$, $G_\mathbf{A} = \mathbb{I}$ the idele group and $G_\mathbf{A}/G_k = \mathbb{I}/k^\times$ is the idele class group; a character of $\mathbb{I}/k^\times$ is essentially a Grössencharakter of Hecke. Tate's work (cf. §1) in 1947 had derived the Hecke theory of the associated L-series from the Poisson formula for $\mathbf{A}/k$. Also one knows from classfield theory that the characters of $\mathbb{I}/k^\times$ of finite order define (through Artin's reciprocity law) cyclic extensions of k. If $\mathbf{G} = GL(2)$, the modular forms on the Poincaré half plane which are eigenfunctions for the Hecke operators define (under suitable spectral conditions) irreducible representations of $G_\mathbf{A}$ that occur in $L^2(G_\mathbf{A}/G_k)$, showing that in the notion of an irreducible representation occurring in $L^2(G_\mathbf{A}/G_k)$ we have a far reaching generalization of the classical automorphic forms and Eisenstein series. The work of [L 3] suggested that for any integer $n \geqslant 1$ and any n-dimensional representation of the Galois group $\mathrm{Gal}(\bar{k}/k)$ ($\bar{k} = $ an algebraic closure of k) there would be associated an irreducible unitary representation occurring in $L^2(G_\mathbf{A}/G_k)$.

The irreducible unitary representations of $G_\mathbf{A}$ are tensor products of irreducible unitary representations of the local groups G_v. It is therefore natural to try to understand the representation theory of the groups G_v as a first step. Of course these representations would reflect the arithmetic of the local fields k_v. For example, following [L 3], one would expect any n-dimensional representations of $\mathrm{Gal}(\bar{k}_v/k_v)$ to determine an irreducible unitary representation of $GL(n, k_v)$. This is a very rough sketch of some of the ideas that led to the great interest in the 1960's for constructing a theory of representations of groups $G = G(\Omega)$ where Ω is a p-adic field and G is a reductive group defined over Ω.

From the very beginning Harish-Chandra emphasized the striking similarities between the theory for real groups and the newly developing theory for the p-adic groups [1968a]. In [1970d] he formulated this as a broad heuristic principle, which he called the *Lefschetz principle*, asserting that whatever is true for real groups is also true for the p-adic ones. Its thrust was to use the rich theory of Fourier analysis on real reductive groups as a guide in the search for discovering the basic facts to be proved in the representation theory of the p-adic groups. In [1970d], as a first illustration of this principle, Harish-Chandra explored the theory of characters and orbital integrals on $G = G(\Omega)$ as above. These investigations were completed in [1978] where he showed, assuming $\operatorname{char}\Omega = 0$, that the theory of characters of G resembles the theory for real groups to a remarkably close extent. Thus, characters of irreducible admissible representations are locally summable functions on G which are locally constant on G'; and they can be obtained by exponentiating the Fourier transforms of the invariant measures on the (nilpotent) orbits in the Lie algebra. One should recall that in the real case such theorems came out of the profound study of the differential equations satisfied by the characters. In the present case there are no differential equations, and so it is quite surprising that such results are true. Harish-Chandra's work uses a beautiful theorem of Howe on invariant distributions on the Lie algebra of G as a substitute for the differential equations.

The philosophy of cusp forms, which he had already introduced in [1970b] as the unifying principle for Fourier analysis on reductive groups is an even more striking illustration of the Lefschetz principle. In [1970b] he had illustrated it for reductive groups over finite fields, and it was the driving force in his work on the Fourier analysis of the Schwartz space of a real reductive group. He showed that the harmonic analysis of these groups can be worked out quite explicitly, based on the concept of cusp forms and parabolic induction. In [1973] and [1977a] he succeeded in pushing this method through for reductive p-adic groups. The cusp forms are once again (essentially) the matrix elements of the discrete series and this method led him to the theory of the Eisenstein integral, in particular, to the Maass–Selberg relations and Plancherel formula, in virtually complete analogy with the real case.

For groups over p-adic fields the discrete series is still not completely determined. Nevertheless, by taking Fourier analysis of reductive groups to this stage, Harish-Chandra has shown that its inner structure can be understood in terms of a small number of simple and yet general principles. This is a truly extraordinary accomplishment.

§3. *Orbital Integrals and Limit Formula.* I now wish to supplement this general survey with a little more detailed discussion of selected parts of Harish-Chandra's work. I shall begin with the theory of orbital integrals and the limit formula. The bulk of this work is contained in [1957b]–[1957e] and [1964c]; the extension of the theory to the Schwartz space of G is treated in [1965b], [1966a], and [1966b].

We take G to be the real form of a complex simply connected semi simple group. Let $L \subset G$ be a Cartan subgroup with $\mathfrak{h} = \mathrm{Lie}(L)$; P, a positive system of roots of $(\mathfrak{g}_c, \mathfrak{h}_c)$; and Δ, the Weyl denominator on L. Define ϖ and π by $\varpi = \prod_{\alpha \in P} H_\alpha$, $\pi = \prod_{\alpha \in P} \alpha$. The orbital integral associated to L is then defined for $f \in C_c^\infty(G)$, $h \in L' = L \cap G'$ by

$$F_{f,L}(h) = \varepsilon_R(h)\Delta(h)\int_{G/L = G^*} f(xhx^{-1})\, dx^* \tag{1}$$

where $\varepsilon_R(h)$ is a locally constant sign factor (§22, [1965b]). The analogous definition on $\mathfrak{g}$ is (see §5, [1964c])

$$\psi_{f,\mathfrak{h}}(H) = \varepsilon_R(H)\pi(H)\int_{G^*} f(H^x)\, dx^* \qquad (f \in C_c^\infty(\mathfrak{g})). \tag{2}$$

These are well defined because the regular orbits are closed, and it is easy to show that $F_{f,L}$ and $\psi_{f,\mathfrak{h}}$ are C^∞ on L' and $\mathfrak{h}'$, respectively.

The key to the entire theory is the system of differential equations satisfied by F_f and ψ_f:

$$F_{zf,L} = \mu_{\mathfrak{g}/\mathfrak{h}}(z)F_{f,L} \qquad (z \in \mathfrak{Z}) \tag{3G}$$

$$\psi_{\partial(p)f,\mathfrak{h}} = \partial(p_\mathfrak{h})\psi_{f,\mathfrak{h}} \qquad (p \in I(\mathfrak{g})). \tag{3g}$$

Here, $I(\mathfrak{g})$ is the algebra of G-invariant elements in $S(\mathfrak{g}_c)$ and $p_\mathfrak{h}$ is the "restriction" of p to $\mathfrak{h}$ while $\mu_{\mathfrak{g}/\mathfrak{h}}$ is as in §12 of [1965b]. To illustrate this let us examine how they control the growth of the orbital integrals near a singular point. Consider for instance $\psi_{f,\mathfrak{h}}$. The equations (3g) and the formula for integration on $\mathfrak{g}$ give the estimate

$$\int_{\mathfrak{h}'} |\pi||\partial(p_\mathfrak{h})\psi_{f,\mathfrak{h}}|\, d\mathfrak{h} \leqslant \int_{(\mathfrak{h}')^G} |\partial(p)f|\, d\mathfrak{g}. \tag{4}$$

From this one can conclude that all derivatives of $\psi_{f,\mathfrak{h}}$ are locally bounded, and in fact even more. The question is the often encountered one of obtaining point estimates in terms of L^1-norms of a function and its derivatives; but there is a technical complication because of the weight function $|\pi|$ which vanishes on $\mathfrak{h} \setminus \mathfrak{h}'$. In [1965b] (cf. Theorem 3) Harish-Chandra obtained a global estimate of the Sobolev type that can be used in the present situation. This can be formulated as follows (see [V 3], Proposition 7, p. 157). Let V be a real finite dimensional vector space; $w = \prod_{1 \leqslant j \leqslant q}|\lambda_j|^{a_j}$ where $a_j > 0$, $\lambda_j \in V_c^*$, and $V' =$ the set where $w \neq 0$; let S_0 be a subalgebra of $S = S(V_c)$ such that S is a finite module over S_0. Write $\mathfrak{U}^1(S_0: w)$ (resp. $\mathfrak{U}^\infty$) for the Frechet space of all $g \in C^\infty(V')$ such that $\int|\partial(u)g|w\, dV < \infty$ (resp. $\sup|\partial(u)g| < \infty$) for all $u \in S_0$ (resp. $u \in S$). Then

$$\mathfrak{U}^1(S_0: w) \subset \mathfrak{U}^\infty$$

and the inclusion is continuous. Since $\psi_{f,\mathfrak{h}} \in \mathfrak{U}^1(I(\mathfrak{h}):|\pi|)$ by (4), we must have $\psi_{f,\mathfrak{h}} \in \mathfrak{U}^\infty$. Actually, replacing f by qf where q is an invariant polynomial on $\mathfrak{g}$, we find the much stronger result that $\psi_{f,\mathfrak{h}}$ lies in the Schwartz space $\mathcal{C}(\mathfrak{h}')$ of $\mathfrak{h}'$ and that the map $f \mapsto \psi_{f,\mathfrak{h}}$ is continuous with respect to the collection of seminorms of the form

$$f \mapsto \int_{(\mathfrak{h}')^G} |\partial(p)(qf)|\, d\mathfrak{g} \qquad (f \in C_c^\infty(\mathfrak{g})).$$

Since *these are all continuous on* $\mathcal{C}(\mathfrak{g})$, the orbital integral extends to a continuous map $\mathcal{C}(\mathfrak{g}) \to \mathcal{C}(\mathfrak{h}')$; in particular, the invariant measures on the orbits are all tempered. The original proof in [1957c] was essentially a variant of this type of argument.

The same argument works (up to a point) on the group also. The crucial case is when $L = B$ is compact and let us make this assumption. Let $\mathcal{U}^\infty$ now be the space of all $g \in C^\infty(B')$ such that $\sup|ug| < \infty$ for all $u \in S(\mathfrak{b}_c)(\mathfrak{b} = \mathrm{Lie}(B))$, regarded as a Frechet space in the usual way. Then the above argument will show that $F_{f,B} \in \mathcal{U}^\infty$ and that $f \mapsto F_{f,B}$ will be continuous with respect to the topology induced on $C_c^\infty(G)$ by the seminorms

$$\nu_z : f \mapsto \int_{G_e} |zf|\, dx \qquad (z \in \mathfrak{Z})$$

where $G_e = (B')^G$ is the elliptic set. If one can show that

$$V(f) = \int_{G_e} |f|\, dx < \infty$$

for any $f \in \mathcal{C}(G)$ and that V is a continuous seminorm on $\mathcal{C}(G)$, the continuity of the ν_z will follow, leading to the extension of the theory of orbital integrals to $\mathcal{C}(G)$. Let φ_B be the characteristic function of the set G_e and let

$$b(x) = \int_K \varphi_B(xk)\, dk \qquad (x \in G).$$

Since

$$\int_{G_e} |f|\, dx \leqslant \int_G |f| b\, dx$$

it is enough to prove that b satisfies the weak inequality.

Harish-Chandra does this in [1966a] (Theorem 5). It is obvious that $b(x)$ is the measure of the set of all k in K for which xk is elliptic. One can use the boundedness of the finite dimensional characters on G_e to conclude that $b(x) \to 0$ when x goes to infinity on G. For the sharper result needed here, one needs class functions with good growth properties on *all* Cartan subgroups. Harish-Chandra's proof depends on the existence of a locally summable class function Θ such that

(i) Θ is a nonzero constant on G_e.

(ii)   If $L = T \cdot A$ is any Cartan subgroup, and $\mathfrak{z}$ is the centralizer of A in $\mathfrak{g}$,

$$|\Theta(x)| \leqslant \mathrm{const.}|\det(1 - \mathrm{Ad}(x^{-1}))_{\mathfrak{g}/\mathfrak{z}}|^{-1/2} \qquad (x \in L').$$

(iii) $\int_K \Theta(xk)\, dk = 0 \qquad (x \in G).$

Using his theory of the distributions $\Theta_\xi(\xi \in \hat{B})$ he shows that $\Theta = \sum_{s \in W} \varepsilon(s)\Theta_{s\rho}$ ($W =$ the Weyl group of $(\mathfrak{g}_c, \mathfrak{b}_c)$ and ρ has the usual meaning), will have the required properties. Θ is essentially the character of the sum of the discrete series representations which have the same infinitesimal character as the trivial representation. It is interesting to observe that the same argument works in the $\mathfrak{p}$-adic case also; Θ may then be taken as the Steinberg character ([1973]).

The analysis of the jumps of the orbital integrals at the singular points comes down, through general arguments, to the case when the singular point is semi regular, and hence, to a calculation in $\mathfrak{sl}(2,\mathbb{R})$. Here there are two Cartan subalgebras $\mathfrak{h} = \mathbb{R} \cdot H$, $\mathfrak{b} = \mathbb{R} \cdot (X - Y)$ where

$$H = \begin{pmatrix} 1 & 0 \\ 0 & -1 \end{pmatrix}, \qquad X = \begin{pmatrix} 0 & 1 \\ 0 & 0 \end{pmatrix}, \qquad Y = \begin{pmatrix} 0 & 0 \\ 1 & 0 \end{pmatrix}.$$

The orbit E_θ of $\theta(X - Y)$ is a two-sheeted elliptic hyperboloid, and the orbit H_t of tH is a single sheeted hyperbolic hyperboloid; these are separated by C, the cone of nilpotents (light cone!), composed of two halves $C_\pm$ and the vertex 0. $\psi_{f,\mathfrak{h}}$ is C^∞ while $\psi_{f,\mathfrak{b}}$ is not. To see why, let $\theta \to 0+$, $t \to 0$. Then the $E_{\pm\theta}$ tend to $C_\pm$ while H_t tends to C (draw a figure). Thus

$$\lim_{\theta \to \pm 0} \psi_{f,\mathfrak{b}}(\pm\theta(X - Y)) = \pm \int_{C_\pm} f \, dC$$

$$\lim_{t \to 0+} \psi_{f,\mathfrak{b}}(tH) = \int_C f \, dC$$

giving the "jump formula"

$$\psi_{f,\mathfrak{b}}(0+) - \psi_{f,\mathfrak{b}}(0-) = \psi_{f,\mathfrak{h}}(0). \tag{5}$$

Generalized to arbitrary $\mathfrak{g}$, this is the basic relation that links the orbital integrals associated to different Cartan subalgebras ([1964c] Theorem 2); there is of course a corresponding version on the group. It shows that jumps arise only when *crossing root hyperplanes corresponding to singular imaginary roots*, and that for derivatives that are skew with respect to these roots, the jumps are 0 (Theorem 1, [1964c]). The continuity of $\partial(\varpi)\psi_{f,\mathfrak{h}}$ and $\varpi F_{f,L}$ follow from this, as well as the vanishing of their values at 0 and e respectively if there are real roots, i.e., if L is not fundamental. The formula (5) or rather its generalization, namely Theorem 2 of [1964c], implies that if $F_{f,\bar{L}} = 0$ for Cartan subgroups $\bar{L}$ "more split" than L, then $F_{f,L}$ is of class C^∞ on L. We have already seen in §2 how this principle was decisive in the determination of the discrete spectrum. In particular, if G is complex, more generally, if L is "Iwasawa", $F_{f,L}$ is of class C^∞; of course one can prove this by a direct evaluation of $F_{f,L}(h)$:

$$F_{f,L}(h) = d_P(h) \iint_{K \times N} f(khnk^{-1}) \, dk \, dn$$

$$\left(d_P(h) = |\det(\mathrm{Ad}(h)_\mathfrak{n})|^{1/2} \right).$$

Let us now turn to the limit formula. It is enough to prove it on the Lie algebra. Write

$$T(f) = \left(\partial(\varpi)\psi_{f,\mathfrak{h}} \right)(0) \qquad (f \in \mathcal{C}(\mathfrak{g})). \tag{6}$$

T is a *tempered* invariant distribution on $\mathfrak{g}$ and the limit formula is the assertion that $T = c\delta$ where $c = c(\mathfrak{h})$ is a real constant which, for fundamental $\mathfrak{h}$, is $(-1)^q$ times a positive number, q being the integer $1/2\{\dim(G/K) - rk(G) + rk(K)\}$.

If $\mathfrak{g}$ is complex or if it has a single conjugacy class of Cartan subalgebras, one can show that (cf. [1975], §§35–36) that

$$\psi_{\hat{f},\mathfrak{h}} = c\hat{\psi}_{f,\mathfrak{h}} \qquad (f \in \mathcal{C}(\mathfrak{g})) \tag{7}$$

where c is a constant $\neq 0$ (recall that $\psi_{f,\mathfrak{h}}$ is smooth and so lies in $\mathcal{C}(\mathfrak{h})$). The integration formula

$$\int_{\mathfrak{g}} u \, d\mathfrak{g} = \text{const.} \int_{\mathfrak{h}} \pi\psi_{u,\mathfrak{h}} \, d\mathfrak{h} \qquad (u \in \mathcal{C}(\mathfrak{g})),$$

on Fourier transformation, gives the limit formula ([1954c]). For arbitrary $\mathfrak{g}$ the relation (7) has to be suitably modified, and this requires a deeper study of the Fourier transforms $\hat{\mu}_H$ of the orbital measures μ_H: $f \mapsto \int_{G^*} f(H^x) \, dx^*$. Since the invariant polynomials are constant on the orbits, the $\hat{\mu}_H$ are invariant eigendistributions:

$$\partial(p)\hat{\mu}_H = \tilde{p}(iH)\hat{\mu}_H \qquad (p \in I(\mathfrak{g})) \tag{8}$$

where $\tilde{p}$ is the polynomial corresponding to p. From the theory of invariant eigendistributions (see §4 below) it follows that $\hat{\mu}_H$ is an analytic function on $\mathfrak{g}'$, say $\hat{\mu}(H: \cdot)$, which can be studied in detail. If $\bar{\mathfrak{h}}$ is a Cartan subalgebra, $\bar{\mathfrak{h}}^+$ a connected component of $\bar{\mathfrak{h}}'$, $\overline{W}$ the Weyl group of $(\mathfrak{g}_c, \bar{\mathfrak{h}}_c)$, and y is an element of the complex adjoint group taking $\mathfrak{h}_c$ to $\bar{\mathfrak{h}}_c$,

$$\pi(H)\bar{\pi}(\overline{H})\hat{\mu}(H: \overline{H}) = \sum_{s \in \overline{W}} c_s(H) e^{i\langle sH^y, \overline{H}\rangle} \tag{9}$$

for $H \in \mathfrak{h}'$, $\overline{H} \in \bar{\mathfrak{h}}^+$; the equations (3$\mathfrak{g}$) now imply that *the c_s are locally constant on* $\mathfrak{h}'$ (Lemma 24, [1957c]). With this proviso, (9) is a generalization of (7) and leads directly to the conclusion that $\hat{T}$ is locally constant on $\mathfrak{g}'$. But $\text{supp}(T) \subset$ the set of nilpotents of $\mathfrak{g}$ so that $\hat{T}$ is $I(\mathfrak{g})$-finite. Since an invariant $I(\mathfrak{g})$-finite distribution on $\mathfrak{g}$, which is locally constant on $\mathfrak{g}'$, is constant on $\mathfrak{g}$, we have $\hat{T} = $ a constant. This argument (see §10, [1965a]) makes use of the regularity theorem on the Lie algebra; the earlier proof in [1957d] was more direct but was also more involved.

One should think of the $\hat{\mu}_H (H \in \mathfrak{h}')$ as the analogues of the irreducible characters of G associated to L by the inducing process. For example one can prove that if $\bar{\mathfrak{h}}$ is a Cartan subalgebra, $\hat{\mu}_H$ is 0 on $(\bar{\mathfrak{h}}')^G$ if $\bar{\mathfrak{h}}$ is not conjugate to a Cartan subalgebra of the centralizer of the split component $\mathfrak{h}_{\mathbf{R}}$ of $\mathfrak{h}$; in particular this will be the case if $\dim(\bar{\mathfrak{h}}_{\mathbf{R}}) < \dim(\mathfrak{h}_{\mathbf{R}})$, i.e., $\bar{\mathfrak{h}}$ is "more compact" than $\mathfrak{h}$.

It remains to look at the case when $rk(G) = rk(K)$ and $\mathfrak{h} \subset \mathfrak{k}$. Harish-Chandra uses a fundamental solution Ξ, due to de Rham, of $\partial(\Omega)^{[n/2]}$ ($\Omega = $ the Casimir element, $n = \dim(\mathfrak{g})$), whose restriction to $\mathfrak{h}$ is a fundamental solution to $(-1)^q \partial(\Omega_{\mathfrak{h}})^{l/2}$, $l = \dim(\mathfrak{h})$. He now chooses $f \in \mathcal{C}(\mathfrak{g})$ such that

$$f(0) = 1, \qquad \psi_{f,\bar{\mathfrak{h}}} = 0 \quad \text{if } \bar{\mathfrak{h}} \text{ is not conjugate to } \mathfrak{h}. \tag{10}$$

Then in view of earlier remarks on the jumps of ψ_f, $\psi_{f,\mathfrak{h}}$ will be smooth and so will lie in $\mathcal{C}(\mathfrak{h})$; and the relation

$$1 = \int_{\mathfrak{g}} \Xi \cdot \left(\partial(\Omega)^{[n/2]} f \right) d\mathfrak{g}$$

will reduce to an integral on $\mathfrak{h}$ that gives the limit formula ([1964c]).

Thus everything comes down to the choice of f. Harish-Chandra takes $f = \hat{g}$ where $g \in C_c^\infty((\mathfrak{h}')^G)$ and $\int_{\mathfrak{g}} g \, d\mathfrak{g} = 1$; by our earlier remarks on $\hat{\mu}_H$, $\psi_{f,\bar{\mathfrak{h}}} = 0$ for $\bar{\mathfrak{h}}$ not conjugate to $\mathfrak{h}$.

The functions such as f in the above proof are truly remarkable because their Fourier analysis is controlled completely by $\mathfrak{h}$. Their analogues on the group are clearly cusp forms. Since we do not have access to these until a substantial amount of Fourier analysis on the group is done, it is now clear why the proof of the limit formula takes place on the Lie algebra. Nevertheless the reader should note how closely this proof resembles the proof, on the group, of the separation of the discrete spectrum. I think the essential originality of Harish-Chandra's treatment of the limit formula was to have conceived of the transition to $\mathfrak{g}$, and to have realized that the theory of Fourier transforms on the Lie algebra could provide him with the "additive" analogues of the cusp forms that held the key to the solution of the problem.

§4. *The Regularity Theorem.* Announced in [1963], this is the central result in the long series of papers [1964a]–[1965b] on the structure of invariant eigendistributions. Harish-Chandra's approach was to prove it first on $\mathfrak{g}$ and then carry the result over to G. For mostly technical reasons he worked on the group with invariant distributions that were $\mathfrak{Z}$-finite (resp. $I(\mathfrak{g})$-finite on $\mathfrak{g}$). The essential problem is to investigate the structure of invariant distributions around semi simple points.

His main technique for doing this was the *method of descent*. Let M be a manifold on which a Lie group L acts, $m_0 \in M$, and let E be a submanifold of M which contains m_0 and is transversal to the orbit of m_0 at m_0. Then the descent mechanism is a method of reducing questions involving invariant distributions and differential operators defined on M around m_0, to similar questions on E around m_0. The classical example is the reduction of rotationally symmetric problems to problems on a half-line.

To set up the descent machinery (see [1964a]) we need volume elements dE, dM on E and M respectively, with dM invariant under L. Assume $\pi: L \times E \to M$ is submersive and let $M' = \pi(L \times E)$. Integrating on the fibers of π with respect to the volume element $dL \, dE/dM$ will then give a "pull-back" $T \mapsto \beta_T$, taking distributions T on M' to distributions β_T on $L \times E$. If T is invariant, β_T will be of the form $1 \otimes \sigma_T$, and σ_T will be the *transfer* of T; the map $T \mapsto \sigma_T$ is injective. If F is any L-invariant function and $F_E = F|_E$, F is locally integrable on M' if and only if F_E is so on E; and $\sigma_F = F_E$. If L is unimodular and L_0 is a closed subgroup of L such that E is L_0-stable and dE is L_0-invariant, σ_T is L_0-invariant. Suppose now that D is an analytic invariant differential operator on M'. If E is sufficiently small, we can choose an analytic differential operator D' on $L \times E$, invariant under the action x, $(x', m) \mapsto (xx', m)$ of L, such that D' and D are π-related. The analytic differential operator

$\Delta(D)$ on E, obtained by retaining only the terms in D' not involving differentiations with respect to L, is called a *radial component of* D. It is in general not unique because D' is not unique; but it always satisfies

$$\sigma_{DT} = \Delta(D)\sigma_T. \tag{1}$$

If, however, E is a local section at m_0 for the action of L, then $\Delta(D)$ is uniquely determined by the requirement

$$(DF)_E = \Delta(D)F_E \qquad \left(F \in (C^\infty(M'))^L\right) \tag{2}$$

and $D \mapsto \Delta(D)$ will be a homomorphism.

In the present context it is necessary to compute explicitly the radial components of $\mathfrak{Z}$ and $\partial(I(\mathfrak{g}))$ for the actions of G on itself and on $\mathfrak{g}$, around semisimple points. Let a (resp. X) be a semisimple element of G (resp. $\mathfrak{g}$); then its centralizer Z(resp. $\mathfrak{z}$) in G (resp. $\mathfrak{g}$) is transversal to the orbit of a (resp. X). The radial components and distributions on Z (resp. $\mathfrak{z}$) are then defined around a (resp. X) and invariant under $\Xi = Z^0$. If a or X is not central, $Z(X)$ will have lower dimension than $G(\mathfrak{g})$ and so induction on dimension is possible.

Harish-Chandra's main calculations may be summarized as follows.

(i) Here G acts on $\mathfrak{g}$, X is regular, and $\mathfrak{z} = \mathfrak{h}$ is a Cartan subalgebra; $M = \mathfrak{g}'$. We take $E = \mathfrak{h}'$ and find ([1957b] Lemma 8)

$$\Delta(\partial(p)) = \pi^{-1} \circ \partial(p_\mathfrak{h}) \circ \pi \qquad (p \in I(\mathfrak{g})). \tag{3}$$

Let $\mathfrak{D}(\mathfrak{g})$ (resp. $\mathfrak{D}(\mathfrak{h})$) be the algebra of differential operators with polynomial coefficients on $\mathfrak{g}$ (resp. $\mathfrak{h}$). In [1964b] Harish-Chandra proved, as a partial extension of (3), that for $D \in \mathfrak{D}(\mathfrak{g})$, $\delta(D) = \pi \circ \Delta(D) \circ \pi^{-1}$ is in $\mathfrak{D}(\mathfrak{h})$, invariant under the Weyl group, and

$$\delta(\hat{D}) = \widehat{\delta(D)} \quad (\wedge = \text{Fourier transform}). \tag{4}$$

(ii) G acts on itself by conjugacy, $E = L'$ where L is a Cartan subgroup, and $M = G'$. Then the radial components are unique, and ([1956c] Theorem 2)

$$\Delta(z) = \Delta^{-1} \circ \mu_{\mathfrak{g}/\mathfrak{h}}(z) \circ \Delta \qquad (z \in \mathfrak{Z}) \tag{5}$$

where $\Delta = \Delta_L$ is the Weyl denominator and $\mathfrak{h} = \text{Lie}(L)$.

(iii) G acts on $\mathfrak{g}$; $X \in \mathfrak{g}$ semisimple. Let $\zeta(Z) = \det(\text{ad } Z)_{\mathfrak{g}/\mathfrak{z}}$; $\mathfrak{z}'$ is the subset of $\mathfrak{z}$ where ζ is not zero. The radial components are not unique; nevertheless, if Ω is an invariant open neighbourhood of X in $\mathfrak{g}$ and T an invariant distribution on Ω,

$$\Delta(\partial(p))\sigma_t = |\zeta|^{-1/2} \circ \partial(p_\mathfrak{z}) \circ |\zeta|^{1/2}\sigma_T \qquad (p \in I(\mathfrak{g})) \tag{6}$$

thus generalizing (3) ([1964b] Theorem 2). In particular, T is $I(\mathfrak{g})$-finite on $(\Omega \cap \mathfrak{z}')^G$ if and only if $|\zeta|^{1/2}\sigma_T$ is $I(\mathfrak{z})$-finite on $\Omega \cap \mathfrak{z}'$.

(iv) G acts on itself; $a \in G$ is semisimple. Let $\nu(y) = \det (\text{Ad}((ay))^{-1} - 1)_{\mathfrak{g}/\mathfrak{z}})$ where $\mathfrak{z} = \text{Lie}(Z)$. Let Ξ' be the subset of Ξ where ν is nonzero. If Ω is a completely invariant (cf. [1963]) open neighbourhood of a in G,

$$\Delta(z)\theta = \left(|\nu|^{-1/2} \circ \mu_{\mathfrak{g}/\mathfrak{z}}(z) \circ |\nu|^{1/2}\right)\theta \tag{7}$$

for all Ξ-invariant distributions θ on $\Omega_\Xi = \Xi' \cap (a^{-1}\Omega)$. In particular, an invariant distribution T on $(a\Omega_\Xi)^G$ is $\mathfrak{Z}$-finite if and only if $|\nu|^{1/2}\sigma_T$ is $\mathfrak{Z}_a$-finite on Ω_Ξ; $\mathfrak{Z}_a = $ center of $U(\mathfrak{z})$ ([1965b], Lemma 22).

(v) Transfer from $\mathfrak{g}$ to G via the exponential map. Let $z \mapsto p_z$ be the isomorphism (§12, [1965b]) of $\mathfrak{Z}$ with $I(\mathfrak{g})$ such that $\mu_{\mathfrak{g}/\mathfrak{h}}(z) = (p_z)_{\mathfrak{h}}$ for all Cartan subalgebras $\mathfrak{h} \subset \mathfrak{g}$. Define the invariant function $\xi = \xi_\mathfrak{g}$ on $\mathfrak{g}$ by

$$\xi(X) = |\det\{(e^{\mathrm{ad}\, X/2} - e^{-\mathrm{ad}\, X/2})/\mathrm{ad}\, X\}|^{1/2} \qquad (X \in \mathfrak{g}).$$

Then $dx = \xi^2 dX$ ($x = \exp X$) and $|\pi|\xi = |\Delta \circ \exp|$ on any Cartan subalgebra $\mathfrak{h}$. Let U be a completely invariant open neighbourhood of 0 in $\mathfrak{g}$ such that $U_G = \exp U$ is completely invariant open in G and $\exp : U \to U_G$ is a diffeomorphism. We then have a G-equivariant pull back isomorphism $T \mapsto \tilde{T}$ of distributions from U_G to U such that at the level of locally summable functions

$$\tilde{F} = (F \circ \exp)\xi. \tag{8}$$

The main formula here is

$$(zT)^{\tilde{}} = \partial(p_z)\tilde{T} \tag{9}$$

for all invariant distributions T (§14, [1965b]). Formula (9) implies that T is $\mathfrak{Z}$-finite if and only if $\tilde{T}$ is $I(\mathfrak{g})$-finite.

The formulae (3) and (6) on the Lie algebra are proved essentially by direct calculation. But such a method cannot handle (7) or (9). It is easy in both cases to determine how the transferred differential operators act on smooth invariant functions; but their actions on invariant distributions are quite difficult to determine directly. Harish-Chandra reduced everything to calculations with invariant *smooth* functions by first proving the following remarkably general and beautiful theorem (Theorem 1, [1965b]): if D is an invariant analytic differential operator on Ω such that $Df = 0$ for all invariant $f \in C^\infty(\Omega)$, then $DT = 0$ for all invariant distributions T on Ω. Its proof comes down, via descent, to an analogous result on $\mathfrak{g}$; the proof of this however uses the regularity theorem on $\mathfrak{g}$ ([1965a] Theorems 4 and 5; for an illuminating sketch of the argument see [1963]).

Fix a completely invariant open set Ω of $\mathfrak{g}$ and an invariant $I(\mathfrak{g})$-finite distribution T on Ω. Using (3) one finds that T is an analytic function F on $\Omega' = \Omega \cap \mathfrak{g}'$; F as well as $\partial(p)F(p \in I(\mathfrak{g}))$ are locally summable on Ω. The first step in proving the regularity theorem is to show that T_F, the distribution defined by F on Ω, is $I(\mathfrak{g})$-finite. If this is done, $\mathfrak{U} = T - T_F$ is invariant, $I(\mathfrak{g})$-finite, and supported on the singular set; the second step is to show that 0 is the only such distribution.

For the first step it is a question of proving that $\partial(p)T_F = T_{\partial(p)F}$ for all $p \in I(\mathfrak{g})$. A formal argument (cf. [1957b], p. 99) reduces this to $p = \omega$, the Casimir operator. Let $J_0 = \partial(\omega)T_F - T_{\partial(\omega)F}$. J_0 can be expressed rather explicitly in terms of boundary values of the orbital integrals ([1965a] p. 15). Theorem 4 of [1964c] guarantees that such a distribution is zero on Ω if it is zero around the semi regular points of the noncompact type in Ω. For J_0, this condition is equivalent, by descent, to the regularity theorem on $\mathfrak{sl}(2,\mathbb{R})$, which can be verified by direct calculation.

The second step, by (6) and induction on $\dim(\mathfrak{g})$, reduces to showing that 0 is the only invariant $I(\mathfrak{g})$-finite distribution on $\mathfrak{g}$ supported by the set $\mathfrak{N}$ of nilpotents. The key to Harish-Chandra's proof of this (Theorems 4 and 5, [1964a]) is to view the space $\mathfrak{I}$ of invariant distributions supported by $\mathfrak{N}$ as a module for a Lie subalgebra

$\mathcal{D} \subset \mathcal{D}(\mathfrak{g})$ which is isomorphic to $\mathfrak{sl}(2, \mathbb{C})$ with a standard basis $\{H', X', Y'\}$ where $2Y' = \partial(\omega)$, $2X' = $ multiplication by $\tilde{\omega}$, and $H' = D + \frac{1}{2}\dim(\mathfrak{g})$ where D is the Euler vector field. Using descent theory and the finiteness of $G \setminus \mathfrak{N}$ it is proved that $\mathfrak{T}$ is a weight module for $\mathcal{D}$ none of whose weights (= eigenvalues of H') is an integer $\geqslant 0$. This implies at once the much stronger result that 0 is the only $\partial(\omega)$-finite element of $\mathfrak{T}$. This completes the sketch of the proof of the regularity theorem on $\mathfrak{g}$; the formulae (7) and (9) then lead to the theorem on the group.

The behaviour around the singular points of the analytic functions defined by the invariant $\mathfrak{Z}$-finite (or $I(\mathfrak{g})$-finite) distributions can be studied in detail. Theorems 2, 3 and Lemma 25 of [1965a] do this on the Lie algebra while Lemmas 31, 34–37 of [1965b] treat the group situation. Let L be a Cartan subgroup and F the invariant locally summable function on Ω that defines a $\mathfrak{Z}$-finite distribution. If $\Phi_L = \Delta_L \cdot (F|_{\Omega \cap L'})$, then Φ_L extends analytically to $L'(R)$, the subset of L where no real (global) root takes the value 1. Moreover, $\varpi_L \Phi_L = \Psi_L$ extends continuously on all of $\Omega \cap L$ and the system of functions (Ψ_L) are compatible in the sense that

$$\Psi_{L_1} = \Psi_{L_2} \quad \text{on } L_1 \cap L_2 \cap \Omega. \tag{10}$$

These are generally known as the Harish-Chandra *matching conditions*; They are equivalent to the statement that $\nabla_G F$ extends to a continuous function on Ω (for the definition of ∇_G see Lemma 36 of [1965b]). Entirely analogous results are true on $\mathfrak{g}$.

§5. *Construction of the Invariant Eigendistributions* Θ_ξ. In [1965c] the regularity theorem and the theory of Fourier transforms are used to construct the distributions Θ_ξ that will be the characters of the discrete series. Let G be as in §3 with $rk(G) = rk(K)$; let $B \subset K$ be a Cartan subgroup, with $\mathfrak{b} = \mathrm{Lie}(B)$. We select a compact form U of G_c such that $K = U \cap G$ and $B \subset U$. W and W_c denote the respective Weyl groups of (K, B) and (U, B). W_c operates on B and the $\xi \in \hat{B}$ that have trivial stabilizer in W_c are called regular. Theorem 3 of [1965c] then asserts that for any regular $\xi \in \hat{B}$ there is a unique invariant eigendistribution Θ_ξ on G such that

(a) $\quad |\Theta_\xi| \leqslant \mathrm{const.} |D|^{-1/2} \quad$ on G'

(b) $\quad \Theta_\xi = \Delta^{-1} \Sigma_{s \in W} \varepsilon(s)(s\xi) \quad$ on $B' = B \cap G'$. $\hfill (1)$

The Θ_ξ are obtained by exponentiating suitable invariant eigendistributions on the Lie algebra. The condition (a) is essentially equivalent to the *temperedness* of these distributions on the Lie algebra; it thus allows Fourier transforms to be used for constructing them. In fact, if η is the analogue of D on $\mathfrak{g}(\eta = \pi^2$ on any Cartan subalgebra), the formula (8) of §4 shows that the conditions $|F| \leqslant \mathrm{const.} |D|^{-1/2}$ and $|\tilde{F}| \leqslant \mathrm{const.} |\eta|^{-1/2}$ are equivalent; the theory of orbital integrals can be used to show that the latter condition, for invariant eigendistributions with regular eigenvalues, is equivalent to temperedness. The fundamental result concerning the distributions on the Lie algebra is Theorem 2 of [1965c]: let $\lambda \in i\mathfrak{b}^*$ be regular; then there is a unique invariant distribution $T_\lambda = T_{\mathfrak{g},\lambda}$ on $\mathfrak{g}$ such that

(a) $\quad T_\lambda$ is tempered, $\partial(p)T_\lambda = p_\mathfrak{b}(\lambda)T_\lambda \quad (p \in I(\mathfrak{g}))$

(b) $\quad T_\lambda = \pi^{-1} \Sigma_{s \in W} \varepsilon(s)e^{s\lambda} \quad$ on $\mathfrak{b}' = \mathfrak{b} \cap \mathfrak{g}'$.

Let $\mathfrak{T}_\lambda$ be the space of all tempered invariant distributions T on $\mathfrak{g}$ such that

$$\partial(p)T = p_\mathfrak{h}(\lambda)T \qquad (p \in I(\mathfrak{g})).$$

Let A_λ be the space of all functions on $\mathfrak{h}$ of the form $\Sigma_{s \in W_c} c_s e^{s\lambda}(c_s \in \mathbb{C})$ which are W-skew. If $T \in \mathfrak{T}_\lambda$, $\pi \cdot (T|_{\mathfrak{h}'})$ belongs to A_λ (Theorem 2, [1965a]), and the uniqueness part of the theorem above is the statement that the restriction map $R_\mathfrak{h}: T \mapsto \pi \cdot (T|_{\mathfrak{h}'})$ is injective. On the other hand, $-iH_\lambda$ is in $\mathfrak{h}$ and we have

$$(-iH_\lambda)^{G_c} \cap \mathfrak{g} = \coprod_{s \in W \setminus W_c} (-isH_\lambda)^G$$

showing that the Fourier transforms $\hat{\mu}_s (s \in W \setminus W_c)$ of the invariant measures on the G-orbits $(-isH_\lambda)^G$ are linearly independent elements of $\mathfrak{T}_\lambda$ (cf. (8) of §3). Since $\dim(A_\lambda) = [W_c: W]$, the injectivity of $R_\mathfrak{h}$ will imply that $\mathfrak{T}_\lambda$ is *spanned* by the $\hat{\mu}_s$ and that $R_\mathfrak{h}$ is an isomorphism of $\mathfrak{T}_\lambda$ with A_λ. This of course will prove Theorem 2 of [1965c].

The key result is thus the injectivity of $R_\mathfrak{h}$. To get a feeling for it consider $\mathfrak{g} = \mathfrak{sl}(2,\mathbb{R})$ with $\lambda \neq 0$ in $i\mathbb{R}$, $T \in \mathfrak{T}_\lambda$, $\partial(\omega)T = \lambda^2 T$ where $\omega = H^2 + 4YX$. Let $i\lambda > 0$. Then

$$i\theta T(\theta(X - Y)) = c_\mathfrak{h}^+ e^{\lambda\theta} - c_\mathfrak{h}^- e^{-\lambda\theta} \qquad (0 \neq \theta \in \mathbb{R})$$

$$tT(tH) = a_\mathfrak{h}^+ e^{-i\lambda t} - a_\mathfrak{h}^- e^{i\lambda t} \qquad (t > 0).$$

The matching conditions give

$$c_\mathfrak{h}^+ + c_\mathfrak{h}^- = a_\mathfrak{h}^+ + a_\mathfrak{h}^- \tag{2}$$

while the temperedness of T implies that only bounded exponentials can appear in $tT(tH)$, so that $a_\mathfrak{h}^- = 0$. But then $a_\mathfrak{h}^+ = c_\mathfrak{h}^+ + c_\mathfrak{h}^-$, showing that T is completely determined by $T|_{\mathfrak{h}'}$.

Such an argument works in general. In Lemma 28 of [1965c] Harish-Chandra proves that if $\mathfrak{h}$ is a Cartan subalgebra and $\mathfrak{h}^+$ is a component of $\mathfrak{h}'(R)$, one can choose Cartan subalgebras $\tilde{\mathfrak{h}}$ "adjacent" to $\mathfrak{h}$ (i.e., $\dim(\tilde{\mathfrak{h}}_\mathbf{R}) = \dim(\mathfrak{h}_\mathbf{R}) - 1$) and components $\tilde{\mathfrak{h}}^+$ of $\tilde{\mathfrak{h}}'(R)$ such that (a) $Cl(\tilde{\mathfrak{h}}^+)$ and $Cl(\mathfrak{h}^+)$ interface along $\tilde{\mathfrak{h}} \cap \mathfrak{h}$ (which is a hyperplane in both $\mathfrak{h}$ and $\tilde{\mathfrak{h}}$); (b) the coefficients of the exponentials occurring in the formula for T on $\mathfrak{h}^+$ and those coming from $\tilde{\mathfrak{h}}^+$ are related in a manner similar to (2); (c) only bounded exponentials occur. The proof that $T = 0$ on $\mathfrak{h}' \Rightarrow T = 0$ then goes by induction on $\dim(\mathfrak{h}_\mathbf{R})$. The regularity of λ is decisive here.

For technical reasons it is necessary to prove the uniqueness even when T is not everywhere defined. The domains Ω however cannot be arbitrary; roughly speaking $\Omega \cap \mathfrak{h}$ must be "unbounded in real directions." To make this precise note first that any semisimple $X \in \mathfrak{g}$ can be uniquely written as $X = X_e + X_h$ where (i) $X_h \in [\mathfrak{g}, \mathfrak{g}]$ (to allow for cases where $\mathfrak{g}$ is only reductive), X_e, X_h are semisimple and $[X_e, X_h] = 0$; (ii) $\mathrm{ad}\, X_e$(resp. $\mathrm{ad}\, X_h$) has only pure imaginary (resp. real) eigenvalues; X_e is called the *elliptic component* of X. The uniqueness theorem is then true for the class $\mathfrak{S}(\mathfrak{g})$ of open sets Ω with the following properties:

(a) Ω is completely invariant; if $X \in \Omega$, then $tX \in \Omega$ for $0 \leqslant t \leqslant 1$;
(b)   if $X \in \mathfrak{g}$ is semisimple, then Ω contains all semisimple elements X' of $\mathfrak{g}$ such that $X'_e = X_e$ (in particular $X_e \in \Omega$).

Typical examples of sets in $\mathcal{E}(\mathfrak{g})$ are $\mathfrak{g}(\varepsilon)$ ($\varepsilon > 0$), the subset of all X in $\mathfrak{g}$ such that $|\mathrm{Im}\,\lambda| < \varepsilon$ for any eigenvalue λ of ad X. If $\mathfrak{z} \subset \mathfrak{g}$ is a reductive subalgebra containing $\mathfrak{b}$, then $\mathfrak{z} \cap \mathfrak{g}(\varepsilon) \in \mathcal{E}(\mathfrak{z})$. These are the domains that occur in [1965c].

In lifting results to G one can use the exponentiation only around *elliptic* elements of G; for, only these have centralizers in $\mathfrak{g}$ that admit Cartan subalgebras of compact type. For any $b \in B$ let $\mathfrak{z}_b$ be the centralizer of b in $\mathfrak{g}$; and for any $\varepsilon > 0$, let

$$G_\varepsilon^{(b)} = \left(b\exp\left(\mathfrak{z}_b \cap \mathfrak{g}(\varepsilon)\right)\right)^G. \tag{4}$$

The sets $G_\varepsilon^{(b)}$, which are used by Harish-Chandra in the lifting process, have remarkable properties. They are of course completely invariant, open if $\varepsilon > 0$ is sufficiently small. Moreover, if $\varepsilon_b > 0$ are arbitrary numbers, $(b \in B)$,

$$G = \bigcup_{b \in B} G_{\varepsilon_b}^{(b)} \tag{5}$$

while for $b', b'' \in B$ and $\varepsilon', \varepsilon'' > 0$ (small enough),

$$G_{\varepsilon'}^{(b')} \cap G_{\varepsilon''}^{(b'')}$$

is the union of sets of the form $G_\varepsilon^{(b)}$. The uniqueness principle on the Lie algebras $\mathfrak{z}_b$ (in their localized version) will then give, in view of these properties, the uniqueness, not only on the whole group, but also for invariant $\mathfrak{Z}$-finite distributions satisfying (a) and (b) of (1), but defined only on $G_\varepsilon^{(b)}$. For the existence, a suitable linear combination of $T_{\mathfrak{z}_b, s\lambda}$ ($s \in W$) will lift to a distribution $\Theta_\xi^{(b)}$ on $G_{\varepsilon_b}^{(b)}$ satisfying (a) and (b) of (1) on it if $\varepsilon_b > 0$ is sufficiently small; the compatibility of these on the overlaps of the $G_{\varepsilon_b}^{(b)}$ is a consequence of the strengthened form of the uniqueness mentioned above, and so, in view of (5), we would have constructed Θ_ξ on all of G (cf. [V 3], Part II, §§2, 5).

The technique of using the matching conditions on the interfaces of adjacent Cartan subalgebras, which was used in the proof of uniqueness, has other applications. As an illustration, let us consider the distribution

$$\Theta_\xi^* = \sum_{s \in W_c} \varepsilon(s)\Theta_{s\xi}. \tag{6}$$

On B, Θ_ξ^* coincides, up to a constant factor, with the character of an irreducible finite dimensional representation of U. In particular, it is bounded and W_c-invariant. If L is any other Cartan subgroup, and $W_{L,I}$ is the subgroup of the Weyl group of (G_c, L_c) generated by the reflexions associated to the imaginary roots, the technique mentioned above can be used to prove by induction on $\dim(A)$ (where A is the $\mathbb{R}$-split component of L), that $\Theta_\xi^*|_{L'}$ is $W_{L,I}$-invariant.[†] This will imply that

$$\sup_{L'} \left(|D_L|^{1/2}|\Theta_\xi^*|\right) < \infty \tag{7}$$

where

$$D_L(x) = \det\left(\left(Ad(x^{-1})-1\right)_{\mathfrak{g}/\mathfrak{z}}\right)$$

[†] It can be shown that $W_{L,I}$ operates on L and stabilizes each component of $L'(R)$; see p. 307, [1965c].

($\mathfrak{z}$ = centralizer of A in $\mathfrak{g}$) (§24, [1965c]). We have already seen in §3 the decisive role of the distributions Θ_ξ^* and the estimate (7) in the problem of extending the theory of orbital integrals to $\mathcal{C}(G)$.

§6. *Eigenfunction Asymptotics, Weak Inequality, and Analysis in the Schwartz Space.* I shall conclude this introduction with a few comments on the Schwartz space and Harish-Chandra's treatment of the asymptotic behaviour of the matrix elements of the irreducible unitary representations and their wave packets [1958a] [1958b] [1966b] [1975] [1976a]. Let $G = KA_0N_0$ be an Iwasawa decomposition with associated (minimal) parabolic subgroup $P_0 = M_0A_0N_0$; A_0^+, the positive chamber of A_0; $\mathfrak{a}_0 = \mathrm{Lie}(A_0)$ and $\rho_0 \in \mathfrak{a}_0^*$ defined as usual by $\rho_0(H) = \frac{1}{2}\mathrm{tr}(\mathrm{ad}\,H)_{\mathfrak{n}_0}$ where $\mathfrak{n}_0 = \mathrm{Lie}(N_0)$. We write σ for the spherical function on G such that $\sigma(a) = \|\log a\|(a \in A_0)$; $\sigma(x)$ is then the distance from K to xK in the Riemannian space G/K.

The concern in [1958a] [1958b] was with the elementary spherical functions φ_ν. The basic estimate for them is (Theorem 3, [1958a]) the following

$$|\varphi_\nu(a)| \leqslant Ce^{-\rho_0(\log a)}(1 + \sigma(a))^m \qquad (a \in A_0^+, \nu \in \mathfrak{a}_0^*)$$

where $C > 0, m \geqslant 0$ are constants. In [1958b] it was proved that wave packets f of the φ_ν satisfy, for each $m \geqslant 0$,

$$|f(a)| \leqslant C_m e^{-\rho_0(\log a)}(1 + \sigma(a))^{-m} \qquad (a \in A_0^+)$$

$C_m > 0$ being a constant. Since the Jacobian for the polar decomposition $G = K\,Cl(A_0^+)K$ behaves roughly as $e^{2\rho_0}$ on A_0^+, these estimates show that the φ_ν are in $L^{2+\varepsilon}(G)$ for every $\varepsilon > 0$ and that the wave packets are actually in $L^2(G)$. They motivate to some extent the following definition ([1970c], p. 539): a continuous function f on G is said to satisfy the *weak inequality* if there are constants $C > 0, m \geqslant 0$, such that

$$|f(k_1ak_2)| \leqslant Ce^{-\rho_0(\log a)}(1 + \sigma(a))^m \qquad (k_1, k_2 \in K,\ a \in A_0^+). \qquad (1)$$

The Schwartz space $\mathcal{C}(G)$ is then defined as the space of all $f \in C^\infty(G)$ such that for $u, v \in U(\mathfrak{g})$ and any $m \geqslant 0$, the two-sided derivative ufv of f satisfies

$$|(ufv)(k_1ak_2)| \leqslant Ce^{-\rho_0(\log a)}(1 + \sigma(a))^{-m} \qquad (k_1, k_2 \in K,\ a \in A_0^+) \qquad (2)$$

for some constant $C > 0$.

Actually, if we write $\varphi_0 = \Xi$, then

$$e^{-\rho_0(\log a)} \leqslant \Xi(a) \leqslant 1 \qquad (a \in A_0^+) \qquad (3a)$$

so that in (1) and (2) we can replace $e^{-\rho_0(\log a)}$ by $\Xi(a)$. Since Ξ has good properties from the point of view of harmonic analysis, this change improves the formal aspects of the theory also (cf. the proof that $\mathcal{C}(G)$ is a topological algebra under convolution, in §14, [1975]). In particular

$$\int_G \Xi^2(1 + \sigma)^{-m}\,dx < \infty \qquad (3b)$$

if $m \gg 0$.

The space $\mathcal{C}(G)$ gives rise to a natural notion of tempered distributions on G. The fundamental nature of the weak inequality is then revealed by Theorem 14.1 of [1975]: for a K-finite $\mathfrak{Z}$-finite function on G, weak inequality is equivalent to temperedness. We also mention the corresponding results for class functions. For invariant $\mathfrak{Z}$-finite distributions Θ temperedness is equivalent to the estimate

$$|\Theta| \leqslant \text{const.}|D|^{-1/2}(1+\sigma)^m \qquad \text{(pointwise on } G') \tag{4}$$

for some $m \geqslant 0$, and then

$$\Theta(f) = \int_G \Theta f \, dx \qquad (f \in \mathcal{C}(G)) \tag{5}$$

the integral converging absolutely; the counterpart to (3b) is

$$\int_G |D|^{-1/2}\Xi(1+\sigma)^{-m} dx < \infty \tag{6}$$

for $m \gg 0$ (cf. [1975], §§10–14; see also the introductory remarks to [1970a]). In (4) the factor $(1+\sigma)^m$ can be dropped if Θ is an eigenfunction for a *regular* infinitesimal character, thereby elucidating the meaning of the condition $\sup|D|^{1/2}|\Theta_\xi| < \infty$ in the construction of the distributions Θ_ξ. Temperedness (of the characters as well as the matrix elements) is the characteristic property of the series of representations $(\pi_{\eta,\nu})$ associated to the various Cartan subgroups of G.

In [1958b] Harish-Chandra discovered that the differential equations satisfied by the φ_ν become (roughly speaking), at infinity on A_0, differential equations with *constant coefficients*. This fact led to the result that φ_ν is asymptotically just a sum of plane waves. In [1966b] Harish-Chandra extended this method to handle all tempered K-finite $\mathfrak{Z}$-finite functions. Let τ be a finite dimensional double unitary representation of K and let $\mathcal{C}(G, \tau)$ be the space of all τ-spherical tempered $\mathfrak{Z}$-finite functions on G. Let $f \in \mathcal{C}(G, \tau)$. Fix a parabolic subgroup $P = MAN$. Then $G = K(MA)K$, and one can use the τ-sphericity of f to reduce the differential equations satisfied by f on G to a system of differential equations on MA; in the limit on MA, when $a \in A$ and $a \to {}_P\infty$ (this means that a goes to infinity in such a way that for some $\varepsilon > 0, \alpha(\log a) \geqslant \varepsilon\sigma(a)$ for all roots α of (P, A)), the latter differential equations become the ones that define elements of $\mathcal{C}(MA, \tau_M)$. This suggests that f may be approximated by an element of $\mathcal{C}(MA, \tau_M)$ at infinity on MA. That this is true is the content of Theorem 21.1 of [1975]; $f_P \in \mathcal{C}(MA, \tau_M)$ defined there is the constant term of f along P. The fact that f_P is a good approximation to f in a suitable regime is contained in the following estimate (Lemma 23.4 [1975]):

$$|e^{\rho_0(\log a)}f(a) - e^{\rho_\infty(\log a)}f_P(a)| \leqslant C(1+\sigma(a))^m e^{-\kappa\rho_0(\log a)}; \tag{7}$$

here ρ_{00} is the "ρ_0" of MA, $\kappa > 0$, $C > 0$, $m \geqslant 0$ are constants, and a is to be restricted to regions of the following form ($\varepsilon > 0$ can be arbitrary)

$$A_0^+(P: \varepsilon) = \{a \in Cl(A_0^+)|\alpha(\log a) \geqslant \varepsilon\rho_0(\log a) \quad \text{for all roots } \alpha \text{ of } (P, A)\}. \tag{8}$$

We have already seen in §2 how decisive the theory of the constant term is in the treatment of the discrete spectrum. In particular the estimates (7) on sets $A_0^+(P: \varepsilon)$

show that for an $f \in \mathcal{C}(G, \tau)$,

$$f \in L^2(G, \tau) \Leftrightarrow f \in \mathcal{C}(G, \tau) \Leftrightarrow f_P = 0 \quad \text{for all } P.$$

Let me mention now another illustration of the method of differential equations. Let g be in $\mathcal{C}(G) \cap {}^0L^2(G)$. We can write $g = \Sigma_j f_j$ where the f_j are K-finite eigenfunctions, the sum being convergent in $L^2(G)$. The f_j of course have vanishing constant terms and so decay exponentially in view of (7); by being careful in the analysis that led to (7) one can keep track of the dependence of (7) on the parameters of f_j (the class $\omega \in \hat{G}_d$ to which f_j belongs, and the K-types according to which f_j transforms). The result is that the series $\Sigma_j f_j$ is convergent *in the topology of the Schwartz space*. On the other hand, the f_j, being $\mathfrak{Z}$-finite functions in $\mathcal{C}(G)$, are cusp forms (Theorem 18.1, [1975]) so that g is a cusp form, showing that $\mathcal{C}(G) \cap {}^0L^2(G)$ is precisely the space of cusp forms ([1970c], §8; [V 3], Part II, §16).

In the theory of the continuous spectrum one considers families of eigenfunctions, for instance the Eisenstein integrals. Harish-Chandra introduces the concept of (τ-spherical) eigenfunctions (ϕ_ν) of type II(λ) associated to a Cartan subgroup $L = T \cdot A$ (§8, [1976a]) here $\lambda \in it^*$ ($t = \mathrm{Lie}(T)$) is regular, ν varies in $\mathfrak{a}^*$ and ϕ_ν is smooth in ν, satisfies the weak inequality in a uniform way, and satisfies the differential equation

$$z\phi_\nu = \mu_{\mathfrak{g}/\mathfrak{h}}(z)(\lambda + i\nu)\phi\nu \quad (z \in \mathfrak{Z}).$$

The fundamental idea now is to form wave packets

$$\phi_\alpha = \int_{\mathfrak{a}^*} \alpha(\nu)\phi_\nu d\nu \quad (\alpha \in \mathcal{C}(\mathfrak{a}^*)) \tag{9}$$

and to try to prove that $\phi_\alpha \in \mathcal{C}(G, \tau)$. This may not always be true even if $\alpha \in C_c^\infty(\mathfrak{a}^*)$. The point is that the constant terms $(\phi_{P, \nu})(P \in \mathcal{P}(A))$ are not eigenfunctions on MA but only linear combinations of such (this is because the asymptotic form on MA of the differential operator $z \in \mathfrak{Z}$ is $\mu_{\mathfrak{g}/\mathfrak{h}}(z)$, and $\mu_{\mathfrak{g}/\mathfrak{h}}(\mathfrak{Z}) \neq \mathfrak{Z}_{MA}$, the center of $U(\mathfrak{m} \oplus \mathfrak{a})$). For regular $\nu \in \mathfrak{a}^*$ one can write $\phi_{P, \nu}$ as $\Sigma_{s \in W(A)}\phi_{P, s, \nu}$; the $\phi_{P, s, \nu}$ will now be eigenfunctions (these statements have to be formulated with some care; see §§8–9 of [1976a]) but they may become singular in ν if ν approaches root hyperplanes in $\mathfrak{a}^*$ (for instance, in the spherical case, they are $c(s\nu)e^{is\nu}$). To overcome this, Harish-Chandra introduces the concept of eigenfunctions (ϕ_ν) of type II$'(\lambda)$ in §9, loc.cit, and proves that if (ϕ_ν) is of type II$'(\lambda)$, $\phi_\alpha \in \mathcal{C}(G, \tau)$ for any $\alpha \in \mathcal{C}(\mathfrak{a}^*)$, and $\alpha \mapsto \phi_\alpha$ is a continuous map (Theorem 13.1, loc.cit). The usefulness of this concept becomes clear from Theorem 9.1 of loc.cit asserting that if (ϕ_ν) is of type II(λ), $(\varpi(\lambda + i\nu)\phi_\nu)$ is of type II$'(\lambda)$. Theorem 11.1 of loc.cit actually gives a necessary and sufficient condition that (ϕ_ν) which is of type II(λ), is actually of type II$'(\lambda)$.

The wave packet theorem is proved by induction on $\dim(G)$. The basic point of course is to check that if (ϕ_ν) is of type II$'(\lambda)$ on G, the $(\phi_{P, s, \nu})$ lead to eigenfunctions of type II$'(\lambda)$ on MA (Lemma 9.1, [1976a]). If one had simply restricted oneself to Eisenstein integrals, the inductive step would have become much more complicated.

The wave packets of $E_{P,\psi,\nu}$ define *essentially* the inverse of the Fourier transform. Essentially, because we are using only $d\nu$ as the weight and not the Plancherel measure; or (equivalently), because we have not yet determined the true normalization of the $E_{P,\psi,\nu}$. To correct this one has to compute the Fourier transform of the wave packet ϕ_α and relate it to α. Theorem 13.2 of [1976a] does this. It is actually a generalization of the corresponding theorem 4 of [1958b] for spherical functions, and is based on the same circle of ideas. What it says is that if $\bar{P}$ is the parabolic subgroup opposite to P, the integral

$$\phi_\alpha^{\bar{P}}(m) = d_P(m) \int_{\bar{N}} \phi_\alpha(\bar{n}m)\, d\bar{n} \qquad (m \in MA)$$

can be evaluated by substituting for ϕ_α the "truncated wave packets"

$$\phi_{P,\alpha}(m) = \int_{\mathfrak{a}*} \alpha(\nu)\phi_{P,\nu}(m)\, d\nu \qquad (m \in MA)$$

(which are obtained by replacing ϕ_ν by its constant term in the formation of the wave packet). If (ϕ_ν) is the Eisenstein integral associated to a $Q \in \mathcal{P}(A)$, $\phi_{P,\nu}$ is expressible as a sum of plane waves $c_{P|Q}(s:\nu)e^{is\nu}$, and the integral of the truncated wave packet has an explicit formula in which the c-functions enter directly (Theorem 19.2, [1976b]). The theory of the c-functions and the functional equations of the Eisenstein integrals are now combined to derive the explicit Plancherel formula. I should mention however one technical difficulty in forming wave packets with the properly normalized weight function, namely, the control of the growth of the Plancherel measure at infinity on $\mathfrak{a}*$; to overcome this one already needs the product formula and the explicit calculations in the case when $\dim(A)$ is equal to 1. I refer the reader to [1976b] for details.

References

[AM]   Auslander, L., and Moore, C. C., Unitary representations of solvable Lie groups. *Memoirs of the Amer. Math. Soc.* **62** (1966), 1–199.

[B]   Bargmann, V., Irreducible unitary representations of the Lorentz group, *Ann. of Math.* **48** (1947), 568–640.

[Bl]   Blattner, R. J., *General Background. Harmonic Analysis and Representations of Semisimple Lie Groups.* Edited by J. A. Wolf, M. Cahen, and M. De wilde. D. Reidel Publishing Company, Holland, 1980, pp. 1–67.

[Bo 1]   Borel, A., *Introduction aux groupes arithmétiques.* Hermann, Paris, 1969.

[Bo 2]   —, *Formes automorphes et séries de Dirichlet.* Springer Lecture Notes **514** (1976), pp. 183–222.

[Br 1]   Bruhat, F., Sur les représentations induites des groupes de Lie. *Bull. Soc. Math. France*, **84** (1956), 97–205.

[Br 2]   —, Sur les représentations des groupes classiques p-adiques, I, II. *Amer. J. of Math.* **83** (1961), 321–338, 343–368.

[G]   Gel'fand, I. M., Automorphic functions and the theory of representations. *Proc. Int. Cong. of Mathematicians, Stockholm, 1962*, pp. 74–85.

[GG 1]   Gel'fand, I. M., and Graev, M. I., On a general method of decomposition of the regular representation of a Lie group into irreducible representations. *Dokl. Akad. Nauk. SSSR*, **92** (1953), 221–224.

[GG 2]   —, The analogue of Plancherel's theorem for real unimodular groups. *Dokl. Akad. Nauk, SSSR* **92** (1953), 461–464.

[GN 1]   Gel'fand, I. M., and Neumark, M. A., Unitary representations of the Lorentz group. *Izvestiya. Akad. Nauk. SSSR* **11** (1947), 411–504.

[GN 2]   —, Unitary representations of the classical groups. *Trudy Mat. Inst. Steklova*, **36** (1950), 1–288.

[GP]   Gel'fand, I. M., and Piatetsky-Shapiro, I. I., Unitary representations in the G/Γ space, where G is the group of real n^{th} order matrices and Γ is the subgroup of integral matrices. *Dokl. Akad. Nauk. SSSR* **147** (1962), 275–278.

[GR]   Gel'fand, I. M., and Raikov, D. A., Irreducible unitary representations of arbitrary locally bicompact groups. *Mat. Sbornik* (N.S) **13** (55) (1943), 301–316.

[Go]   Godement, R., Sur les relations d'orthogonalité de V. Bargmann. I.II. I: Résultats préliminaires, *C.R. Acad. Sci. Paris*. **225** (1947), 521–523; II: Démonstration génerale, *C.R. Acad. Sci. Paris*, **225** (1947), 657–659.

[JL]   Jacquet, H., and Langlands, R. P., *Automorphic Forms on* GL (2), Springer Lecture Notes **114** (1970).

[L 1]   Langlands, R. P., *On the Functional Equations Satisfied by Eisenstein Series*. Springer Lecture Notes **544** (1976).

[L 2]   —, *Euler Products*. Yale University Press, 1967.

[L 3]   —, *Problems in the Theory of Automorphic Forms. Lectures in Modern Analysis and Applications*, Springer Lecture Notes, **170** (1970), 18–86.

[Lo]   Loomis, L. H., *An Introduction to Abstract Harmonic Analysis*. Van Nostrand, New York, 1953.

[M]   Maass, H., Über eine neue Art von nichtanalytischen automorphen Funktionen. *Math. Ann.* **121** (1949), 141–183.

[Mau 1] Mautner, F. I., Spherical functions over $\mathfrak{p}$-adic fields, I. *Amer. J. of Math.*, **80** (1958), 441–457.

[Mau 2] —, Spherical functions over $\mathfrak{p}$-adic fields, II. *Amer. J. of Math.* **86** (1964), 171–200.

[Ma 1]   Mackey, G. W., *The Theory of Unitary Group Representations. Chicago Lectures in Mathematics*. The Chicago University Press, Chicago and London, 1976.

[Ma 2]   —, Infinite dimensional group representations. *Bull. Amer. Math. Soc.* **69** (1963), 628–686.

[Ma 3]   —, *Unitary Group Representations in Physics, Probability, and Number Theory*. Benjamin, 1978.

[Mo]   Moore, C. C., Representations of Solvable and nilpotent groups and harmonic analysis on nil and solvmanifolds. Harmonic analysis on homogeneous spaces. *Proceedings of Symposia in Pure Mathematics*, **XXVI**. Edited by C. C. Moore. *Amer. Math. Soc.* 1973, pp. 3–44.

[MvN]   Murray, F. J., and von Neumann, J., On rings of operators. I, II, IV. I: *Ann. of Math.* **37** (1936), 116–229; II: *Trans. Amer. Math. Soc.* **41** (1937), 208–248; IV: *Ann. of Math.* **44** (1943), 716–808.

[Se 1]   Selberg, A., Harmonic analysis and discontinuous groups in weakly symmetric Riemannian spaces with applications to Dirichlet series. *J. Indian Math. Soc.* **20** (1956), 47–87.

[Se 2]   —, Discontinuous groups and harmonic analysis. *Proc. Int. Cong. of Mathematicians, Stockholm* (1962), 177–189.

[S]   Schwartz, L., *Théorie des Distributions*, Hermann, Paris, 1973 (Nouvelle édition).

[T]   Tate, J., Fourier analysis in number fields and Hecke's Zeta functions. Thesis (Princeton), 1950. Reproduced in *Algebraic Number Theory*, Edited by J. W. S. Cassels and A. Frölich, Thompson Book Company Inc., Washington D.C., 1967.

[vN]   von Neumann, J., (a) On rings of operators III. *Ann. of Math.* **41** (1940), 94–161. (b) On rings of operators. Reduction theory. *Ann. of Math.* **50** (1949), 401–485.

[V 1]   Varadarajan, V. S., *Lie Groups, Lie Algebras, and Their Representations*. Prentice Hall, Englewood Cliffs, N.J., 1974.

[V 2]   —, The theory of characters and the discrete series for semisimple groups. Harmonic analysis on homogeneous spaces, *Proceedings of Symposia in Pure Mathematics*, **XXVI**. Edited by C. C. Moore. *Amer. Math. Soc.* 1973, pp. 45–99.

[V 3]   —, *Harmonic Analysis on Real Reductive Groups*. Springer Lecture Notes **576** (1976).

[W 1]   Weil, A., *L'integration dans les groupes topologiques et ses applications*. Hermann, Paris, 1940.

[W 2]   —, *Basic Number Theory*. Springer-Verlag, New York, 1967.

[W 3]   —, (a) Sur certains groupes d'opérateurs unitaires *Acta Math.* **111** (1964), 143–211. (b) Sur la formule de Siegel dans la théorie des groupes classiques. *Acta Math.* **113** (1965), 1–87.

[W 4]   —, Über die bestimmung Dirichletscher Reihen durch Funktionalgleichungen. *Math. Ann.* **168** (1967), 149–156.

[W 5]   —, *Automorphic Forms and Dirichlet Series*. Springer Lecture Notes **189** (1971).

[We 1] Weyl, H., Über gewöhnliche Differentialgleichungen mit singularitäten und die zugehorigen Entwicklungen willkürlicher Funktionen. *Math. Ann.* **68** (1910), 220–269.

[We 2] —, Theorie der Darstellung kontinuierlicher halbeinfacher Gruppen durch lineare Transformationen. I, II, III, und Nachtrag. I: *Math. Zeist.* **23** (1925), 271–309; II: *Math. Zeist.* **24** (1926), 328–376; III: *Math. Zeist.* **24** (1926), 377–395; Nachtrag: *Math. Zeist.* **24** (1926), 789–791.

[Wi]   Wigner, E. P., On unitary representations of the inhomogeneous Lorentz group. *Ann. Math.* **40** (1939), 149–204.

Some Additional Aspects of Harish-Chandra's Work on Real Reductive Groups

Nolan R. Wallach

The main emphasis in the introduction to these volumes is the work of Harish-Chandra which led to the Plancherel theorem for reductive groups over local fields. This work did not historically follow a straight line and it included important results that are not directly related to the main theme. We include here a partial guide to some of these theorems and ideas. We label each paper to be discussed using the scheme of the Bibliography, e.g. [1951a].

[1951a] In this paper Harish-Chandra shows how to simultaneously construct a semisimple Lie algebra and all of its irreducible finite dimensional representations from its Cartan matrix. Recall that a Cartan matrix is an integral $l \times l$ matrix

$$A = (a_{ij})$$

with

(1) $a_{ii} = 2$, $a_{ij} \leqslant 0$, $i \neq j$;
(2)   if $a_{ij} = 0$, then $a_{ji} = 0$;
(3) $\det(A) \neq 0$;
(4)   the group generated by the transformations

$$s_i: x_j \mapsto x_j - a_{ji} x_i$$

     is finite.

Starting with such a matrix A he studies the algebra (either associative or Lie) with generators X_i, Y_i, H_i, $1 \leqslant i \leqslant l$, and the relations

(R)
$$[H_i, H_j] = 0, \qquad [X_i, Y_j] = \delta_{ij} H_i$$
$$[H_i, X_j] = a_{ji} X_j, \qquad [H_i, Y_j] = -a_{ji} Y_j.$$

At first Harish-Chandra studies the associative algebra subject to (R). He later looks at the Lie algebra subject to (R) (the Lie subalgebra generated by the X_i, Y_i, H_i in the associative algebra). Given a linear functional Λ on the span of the H_i with $\Lambda(H_i) = n_i$, $n_i \in \mathbb{Z}$, $n_i \geq 0$, he constructs a module for the algebra by taking a suitable quotient of what has come to be called a Verma module. He then takes its unique irreducible quotient (which we denote), $L(\Lambda)$. If $\mathfrak{g}'$ is the Lie algebra determined by the relations (R) and if $I = \{X \in \mathfrak{g}' \,|\, X$ acts by zero on every $L(\Lambda)\}$, then $\mathfrak{g} = \mathfrak{g}'/I$ defines the semisimple Lie algebra. The $L(\Lambda)$ are the finite dimensional irreducible representations of $\mathfrak{g}$. Parts of this argument are attributed by Harish-Chandra to Chevalley.

It is of some interest that the conditions (3), (4) are used only to guarantee that $\dim \mathfrak{g} < \infty$ and $\dim L(\Lambda) < \infty$. If the conditions (3) and (4) are dropped, the construction of Harish-Chandra gives the Kac-Moody Lie algebras and their standard modules.

[1953] In the first part of this paper Harish-Chandra initiates the algebraic theory of $(\mathfrak{g}, \mathfrak{k})$-modules (or Harish-Chandra modules). Here $\mathfrak{g}$ is a semisimple Lie algebra over the real numbers, θ is a Cartan involution of $\mathfrak{g}$, and $\mathfrak{k}$ is the fixed point algebra of θ. A $(\mathfrak{g}, \mathfrak{k})$-module is a $\mathfrak{g}$-module, M, over $\mathbb{C}$ that splits into a (not necessarily finite) direct sum of irreducible $\mathfrak{k}$-modules. Harish-Chandra calls such modules quasi-semisimple for $\mathfrak{k}$. The basic theorem in the first part is Theorem 1 which may be stated as follows. Let $U(\mathfrak{g}_\mathbb{C}) \supset U(\mathfrak{k}_\mathbb{C})$ be the universal enveloping algebras of $\mathfrak{g}_\mathbb{C}$ and $\mathfrak{k}_\mathbb{C}$ respectively. Let $Z(\mathfrak{g}_\mathbb{C})$ be the center of $U(\mathfrak{g}_\mathbb{C})$. The theorem then says: if W_1, W_2 are finite dimensional semisimple $\mathfrak{k}_\mathbb{C}$-modules then

$$\mathrm{Hom}_\mathfrak{k}\left(W_1, U(\mathfrak{g}_\mathbb{C}) \underset{U(\mathfrak{k}_\mathbb{C})}{\otimes} W_2\right)$$

is finitely generated as a $Z(\mathfrak{g}_\mathbb{C})$-module (the action of $Z(\mathfrak{g}_\mathbb{C})$ is $(z \cdot T)(v) = z \cdot T(v)$).

This theorem in particular implies that a $(\mathfrak{g}, \mathfrak{k})$-module M that is finitely generated and is $Z(\mathfrak{g}_\mathbb{C})$-finite ($\dim Z(\mathfrak{g}_\mathbb{C}) \cdot m < \infty$, $m \in M$) has finite $\mathfrak{k}$-multiplicities, i.e., is admissible. This implication is the starting point for the theory of admissible $(\mathfrak{g}, \mathfrak{k})$-modules. For the special case when $\mathfrak{g}$ is a complex Lie algebra looked upon as a real Lie algebra this result (as well as the basic idea of its proof) is in [1951a]. In [1951a] Harish-Chandra indicates that the result in this special case was suggested to him by Mautner.

In the second part of this paper Harish-Chandra initiates the theory of analytic vectors for Banach space representations of Lie groups. These vectors are called well-behaved in this paper. Let (π, H) be a Banach representation of a connected Lie group G. Let H^ω be the space of analytic vectors. His main results are that H^ω is dense in H and is $\mathfrak{g}$-invariant, and that the closure of a $\mathfrak{g}$-invariant subspace of H^ω is G-invariant.

Let G be connected semisimple with finite center and let $K \subset G$ be the analytic subgroup of G corresponding to $\mathfrak{k}$. Let (π, H) be a Banach representation of G such that if $v \in H^\omega$ $\dim Z(\mathfrak{g}_\mathbb{C})v < \infty$, and suppose that there exist $v_i \in H^\omega \cap H_K$, $1 \leq i \leq N$ such that H is the closed linear span of $\pi(x)v_i$, $x \in G$, $1 \leq i \leq N$; here $H_K = \{v \in H \,|\, \pi(K) \cdot v$ spans a finite dimensional space$\}$. Then Theorem 1 and the theory of analytic vectors combine to prove

(1) $H_K \subset H^\omega$; all the K-multiplicities in H are finite.
(2) $V \mapsto \mathrm{Cl}(V)$ is a bijection between the lattice of $\mathfrak{g}$-stable subspaces of H_K and the lattice of G-stable closed subspaces of H.

These results are then used by him to prove that a connected semisimple Lie group is of type I in the sense of Murray and von Neumann.

[1954a] This paper contains the proof of his celebrated subquotient theorem. Let G be linear, connected and semisimple. The theorem asserts that any irreducible $(\mathfrak{g}, \mathfrak{k})$-module that integrates to a K-module is isomorphic to a subquotient of the space of K-finite vectors of a representation (in general nonunitarily) induced from a one dimensional representation of an Iwasawa subgroup of G. He was later able to drop the linearity condition by using his results on differential equations [1960a].

[1954c] The main result of this paper is the Plancherel theorem for semisimple Lie groups defined over the field of complex numbers. Theorems 1 and 2 give the formula for the characters of the principal series representations for arbitrary connected semisimple Lie groups over $\mathbb{R}$, with finite center. The Plancherel theorem is based on this character computation and an integro-differential formula (Lemma 14) which generalizes earlier work of Gel'fand and Naimark. He generalized Lemma 14 to arbitrary semisimple Lie algebras in [1957c, d]; this provided one of the main steps in his proof of the Plancherel theorem for real reductive groups (see Varadarajan's introduction). Lemma 10 and its Corollary are important ingredients in the proofs of Lemma 14 and Theorem 2. He later used this result in his proof [1956d] of the Bruhat lemma.

[1956b] Section 7 of this paper is titled "A Digression on a Theorem of Cartan". Harish-Chandra considers the case $\mathfrak{g}_\mathbb{C}$ simple, and $\mathfrak{g}_\mathbb{C} = \mathfrak{k}_\mathbb{C} \oplus \mathfrak{p}_\mathbb{C}$ (complexified Cartan decomposition) with $\mathfrak{p}_\mathbb{C} = \mathfrak{p}^+ \oplus \mathfrak{p}^-$ a direct sum of two Ad K-invariant nonzero subspaces. He had earlier shown [1955c] that there is $J \in \mathfrak{k}$ with ad $J|_{\mathfrak{p}^+} = iI$, ad $J|_{\mathfrak{p}^-} = -iI$, ad $J|_\mathfrak{k} = 0$. Thus $[\mathfrak{p}^+, \mathfrak{p}^+] = [\mathfrak{p}^-, \mathfrak{p}^-] = 0$. Let G be linear; then G is contained in $G_\mathbb{C}$, a connected Lie group with Lie algebra $\mathfrak{g}_\mathbb{C}$. By going to a covering one may assume that $G_\mathbb{C}$ is simply connected. In $G_\mathbb{C}$ there is the connected parabolic subgroup

$$K_\mathbb{C} \exp \mathfrak{p}^+.$$

In [1956a] it is shown that

$$GK_\mathbb{C} \exp \mathfrak{p}^+ \subset (\exp \mathfrak{p}^-) K_\mathbb{C} (\exp \mathfrak{p}^+)$$

is an open subset, and that

$$G \cap (K_\mathbb{C} \exp \mathfrak{p}^+) = K.$$

One therefore has

$$GK_\mathbb{C} \exp \mathfrak{p}^+ = \exp(\Omega) K_\mathbb{C} \exp \mathfrak{p}^+$$

with $\Omega \subset \mathfrak{p}^-$ an open subset.

This gives rise to a diffeomorphism

$$\psi: G/K \to \Omega$$

given by

$$(\exp \psi(g)) K_{\mathbb{C}} \exp \mathfrak{p}^+ = g K_{\mathbb{C}} \exp \mathfrak{p}^+$$

Using ad J (defined above) one shows that G/K has a G-invariant complex structure and it is easy to see that $\psi: G/K \to \Omega$ is complex analytic. Using his theory of strongly orthogonal roots and Lemma 20 Harish-Chandra shows that Ω is a bounded, symmetric domain in $\mathfrak{p}^-$. This gives a proof without going through classification of Cartan's basic theorem that every Hermitian symmetric domain is biholomorphic with a bounded symmetric domain. The map ψ is now called the Harish-Chandra imbedding and is the starting point in the study of Hermitian symmetric domains.

In this paper there is also a very interesting argument which transfers the calculation of an integral on G to the calculation of a corresponding integral on a compact form of $G_{\mathbb{C}}$ (see Lemma 28).

[1960a] This previously unpublished paper extends work in [1958a,b]. In this paper Harish-Chandra generalizes the classical theory of regular singular points and Frobenius's method to several variables. Most notable is his handling of what is now called the asymptotic expansion along the walls. These results were basic to the Langlands classification of irreducible $(\mathfrak{g}, \mathfrak{k})$-modules.

[1962] In this paper with A. Borel a general reduction theory of arithmetic subgroups of semisimple Lie groups over $\mathbb{R}$ is developed. Theorem 9.4 says that the volume of a fundamental domain for such a discrete subgroup is finite. Theorem 11.8 (which gives a positive answer to Godement's conjecture) is now called the Borel–Harish-Chandra criterion for compactness of a fundamental domain. It says that if $\Gamma \subset G$ is arithmetic then G/Γ is compact if Γ has no unipotent elements. This criterion is now the main method of proving compactness for arithmetic quotients.

[1976b] This paper is the culmination of Harish-Chandra's work on the Plancherel formula for real reductive groups. In addition to the completion of the proof of the Plancherel theorem this paper contains the completeness theorem for intertwining operators for cuspidal principal series representations (Theorem 37.1). This theorem is the basic ingredient in the analysis of the irreducibility of the cuspidal principal series representations.

[1983] The purpose of this paper is to study the basic ingredients that will be necessary for the spectral analysis of tempered, invariant distributions (for example the orbital integrals) on a real reductive group G. As in the case of the Plancherel theorem (which is concerned with the spectral analysis of the Dirac delta function at the identity of G) it is necessary to have the correct notion of the constant term. The theory of the constant term of a tempered, $Z(\mathfrak{g})$-finite, central distribution on G is one of the main themes of this article.

Let Θ be a tempered, invariant, $Z(\mathfrak{g})$-finite distribution on G. Let $K \subset G$ be a maximal compact subgroup of G. Let P be a proper parabolic subgroup of G with standard Levi decomposition $P = MN$. The constant term of Θ along P, Θ_P, should be a $Z(\mathfrak{m})$-finite tempered, central distribution on M which is the limit in a suitable sense of Θ along the positive chamber relative to P in A (= the standard split component of M). If (π, H) is an irreducible tempered representation of G and if Θ is the character of π then there is a natural notion of Θ_P which we now describe. If H_K is the space of K-finite vectors of H then

$$H_K / \mathfrak{n} \cdot H_K$$

is an admissible, finitely generated, $(\mathfrak{m}, K \cap M)$-module. As a module for $\mathfrak{a}$, it splits into a direct sum of generalized weight spaces for $\mathfrak{a}$:

$$(H_K / \mathfrak{n} \cdot H_K)_\lambda.$$

Let ρ be as usual the differential of the square root of the modular function, δ, of P. Let $\bar{P} = \theta(P)$. Then $\Theta_{\bar{P}}$ is $\delta^{1/2}$ times the sum of the characters of the $(\mathfrak{m}, K \cap M)$-modules

$$(H_K / \mathfrak{n} \cdot H_K)_\lambda$$

with

$$\operatorname{Re}(\lambda - \rho, \alpha) = 0$$

for α a root of (P, A). Harish-Chandra's definition in the general case is consistent with this definition for characters.

A tempered, invariant, $Z(\mathfrak{g})$-finite distribution Θ is called *super tempered* if $\Theta_P = 0$ for all proper parabolic subgroups P of G. Thus the super tempered distributions are the analogues of the cusp forms of G. Using the results in [1960a] it is not difficult to see that if Θ is the character of an irreducible representation π of G which is tempered and unitary, then Θ is super tempered if and only if π is square integrable. However, there are super tempered, invariant, eigendistributions that are not linear combinations of characters of discrete series representations. For example, if (π, H) is the reducible unitary principle series representation of SL $(2,\mathbb{R})$, if π^+ and π^- are the irreducible constituents of π, and if $\Theta^\pm$ are the characters of $\pi^\pm$, then $\Theta^+ - \Theta^-$ is super tempered, although neither Θ^+ nor Θ^- is super tempered. One of the main results of this paper is that all super tempered eigendistributions that are not linear combinations of discrete series characters are given in terms of distribu-. tions generalizing $\Theta^+ - \Theta^-$ for SL $(2,\mathbb{R})$.

The Work of Harish-Chandra on Reductive p-adic Groups

ROGER HOWE

Harish-Chandra's work on p-adic groups is broadly based on his experience with Lie groups. (In this survey, the term *Lie group* means an analytic group over $\mathbf{R}$.) He himself made the connection quite explicit, formulating what he called the Lefschetz Principle and the Philosophy of Cusp Forms. His Lefschetz Principle [1970d] is a paraphrase for harmonic analysis of the rule of thumb of the same name from algebraic geometry. It asserts that the phenomena of representation theory for groups over p-adic fields, finite fields or adele rings are, suitably interpreted, essentially the same as for groups over $\mathbf{R}$, i.e. Lie groups. In applying this principle the "suitable interpretation" is sometimes less than obvious; there are many respects in which p-adic harmonic analysis might seem rather different from real harmonic analysis. Nevertheless, it is a valuable principle, and there have been many examples where one theory will enlighten the other, or an attempt to reconcile apparent differences will lead to insight in both theories. For example, the cleanness of the theory of the constant term in the p-adic case led to a reworking of it for real groups [CM], [Wa]. It is a tribute to the robustness of Harish-Chandra's methods and to his tenacity that, despite serious obstacles including a lack of detailed understanding of the discrete series, characters and orbital integrals, he was able to reach a main goal, the Plancherel Formula [1977b].

The Philosophy of Cusp Forms is much more specific to harmonic analysis than is the Lefschetz Principle; it might be regarded as the primary concrete embodiment of that Principle in representation theory of reductive groups. It is most explicitly stated in [1970b] where it is traced back as far as a talk of Gel'fand at the Stockholm International Congress. It has become a basic feature of reductive harmonic analysis. In particular it is a unifying feature of Harish-Chandra's work.

The gist of the Philosophy of Cusp Forms is that the collection of all representations of a reductive group G should be partitioned into disjoint classes, with each class being attached to a certain parabolic subgroup. (Or more precisely, to a class of associated parabolic subgroups. Recall that two parabolic subgroups are *associated* if their Levi components (the reductive components in their Levi decomposions) are conjugate.) The representations attached to G itself are called *cuspidal* representations and the matrix coefficients of cuspidal representations are *cusp forms*. The representations associated to a proper parabolic subgroup P are further associated to Weyl group orbits of cuspidal representations of M, the Levi component of P, and this further association is finite-to-one. Moreover, the representations of G associated to a given cuspidal representation σ of M may be recovered as components of the induced representation

$$\mathrm{ind}_P^G\left(\sigma \otimes \delta_P^{1/2}\right)$$

where σ is extended from M to P by letting it be trivial on the unipotent radical of P, and δ_P is the modular function of P. Finally, the method of passing from a representation of G to its associated P and representation σ of M is explicit; it is based on what Harish-Chandra calls the "theory of the constant term".

Thus the Philosophy of Cusp Forms provides an inductive strategy for understanding the representations of G. The total problem is divided into two parts: to determine the cusp forms and to analyze induced representations. Both problems are quite difficult and neither is solved in all cases, but the division has provided a fruitful method of attack and substantial progress has been made on each partial problem.

It should perhaps be pointed out that what one means by "cusp form" varies slightly according to context. For purposes of the Plancherel formula for real or p-adic groups, "cusp form" has an analytic meaning: a representation is cuspidal if and only if it is in the discrete series (modulo the center). For purposes of the theory of admissible representations of p-adic groups only supercuspidal representations are cuspidal, while in the Langlands classification [BW], [L2], [Vo] all representations which are tempered on the commutator subgroup play the part of cuspidal representations. The discrepancy between the cusp forms and the discrete spectrum is perhaps most vexed in the case of automorphic forms. All cusp forms in $L^2(G_\mathbf{A}/G_k)$ (for notation see [Bo]) belong to the discrete spectrum, but the non-cuspidal or "residual discrete spectrum" is very rich and almost as poorly understood as the cusp forms. The variability of the notion of cusp form of course complicates the theory, but pays off amply in increased flexibility and applicability.

The determination of the discrete series of a semisimple Lie group was one of Harish-Chandra's great achievements. For semisimple p-adic groups this problem is still unsolved, except in some cases, and in fact it still seems quite far from solution. (See [MS] for some discussion of its current status.) An explicit description even of the representations of the compact group of norm units in a p-adic division algebra has not yet been given.

There are several reasons for the greater mystery of the discrete series, in the p-adic case. Recall that if G is a connected semisimple Lie group, then [V]:

1) Up to conjugacy G contains at most one compact Cartan subgroup T.
2) The discrete series of G are parametrized in a natural way by the regular characters of T.

Some groups, including complex groups and $SL_n(\mathbf{R})$, $n \geqslant 3$, have no compact Cartan subgroup and therefore no discrete series. By contrast, all p-adic groups have compact Cartan subgroups and discrete series in abundance. Further, while it is roughly true that characters of the compact Cartan subgroups parametrize discrete series representations, there can be no such clean correspondence as for real groups. Some characters (even "regular" characters, although the definition of that term is itself problematic) will not correspond to representations. Further, some representations would seem to correspond to characters of several tori. And finally, there are some discrete series, analogous to the unipotent cuspidal representations constructed by Lusztig [Lu1] for groups over finite fields, which seem not to correspond to a character of any torus. Langlands [Bo], [L1] has conjectured that the discrete series should be parametrized by certain homomorphisms of the Weil group (of the local field over which the group is defined) into an appropriate group, called the L-group. This has been verified in some cases [K], [M], [He], [KM], [T]. But even the conjectures on how to describe the discrete series of p-adic groups are still undergoing refinement [Ar], [Lu2].

Because of the relative intractability of the determination of cuspidal representations for p-adic groups, Harish-Chandra along with most other workers in this area concentrated mainly on the second part of the cusp form program, namely, the connection between representations of a reductive group G and its parabolic subgroups P. He also established what qualitative results he could about discrete series, and these were sufficient to yield a version of the Plancherel formula [1977b] quite analogous to, though somewhat less precise than, the real case. We will describe below his progress toward this goal in more detail.

Harish-Chandra's papers on p-adic groups are [1970d], [1973], [1977a], [1977b], [1978], [1980]. However, it should be noted that [1973] is only a summary of results, essentially an extended research announcement. The full proofs of the results in [1973] have not been published by Harish-Chandra. Instead they appear in Allan Silberger's book [Si], which is largely based on lectures of Harish-Chandra at the Institute for Advanced Study in the academic years 1971–1972 and 1972–1973. For understanding Harish-Chandra's work on p-adic groups, this book is an essential supplement to the papers appearing under his own name. Neither have the full proofs of some results stated in [1977b], [1978], or [1980] yet appeared.

The main topics treated in these papers are ones familiar from Harish-Chandra's work on real groups: the theory of characters, orbital integrals, the constant term, discrete series and the Schwartz space. All of these topics are mutually related via the Philosophy of Cusp Forms. In the real case all were subservient to the Plancherel Formula, but in the p-adic case neither character theory nor orbital integrals has been made as precise as in the real case, and eventually were bypassed by Harish-Chandra in his proof of the p-adic Plancherel Formula. We will discuss these topics in turn.

As a prelude to the more detailed discussion, a remark on technique is in order. Throughout his work on real groups, Harish-Chandra relies on the differential equations supplied by the universal enveloping algebra, especially its center. This fundamental avenue of approach is of course not available in the study of p-adic groups and replacing it is a major technical problem. For some purposes, the place of the differential equations is taken by certain finiteness theorems. Thus a result of Jacquet [J] forms the basis for the theory of the constant term, a result of Howe [Ho] is useful in character theory, and a finiteness result in [1977b] provides the final step in the Plancherel Theorem. We will discuss these in more detail below. But in fact, the lack of the differential equations has not been completely made good, and as was already noted, complete information about discrete series, character formulas, and orbital integrals is not yet available.

Character theory is the primary topic of [1977a], [1978], and [1980], and occupies the bulk of [1970d]. The paper [1977a] is an announcement of the results in [1978], which contains the most precise results on characters, and largely supersedes [1970d]. However, the paper [1978] is restricted to groups over fields of characteristic zero; in [1980] some of the results are extended to groups over fields of positive characteristic.

Before one can describe characters one must know in what sense they exist. For this, and much of the rest of representation theory for p-adic groups, a basic notion is that of *admissible representation*, first formalized in [JL]. Let G be a locally compact topological group which is totally disconnected as topological space—a *t.d. group* in Harish-Chandra's terminology [1973]. This is the same as to say G has a basis of neighborhoods of the identity consisting of open compact subgroups. All p-adic algebraic groups are t.d. groups. Let ρ be a representation of G on a vector space V. A vector $v \in V$ is called a *smooth vector* for ρ if there is an open subgroup $K \subseteq G$ such that v is invariant under $\rho(K)$, i.e., $\rho(k)v = v$ for $k \in K$. The set of smooth vectors for ρ form a subspace $V^\infty \subseteq V$; this subspace V^∞ is invariant by G. If V is a complete locally convex topological vector space, and ρ is a strongly continuous representation of G, then V^∞ is a dense subspace of V.

For a given compact open group $K \subseteq G$, let V^K denote the subspace of vectors in V fixed by $\rho(K)$. Then $V^K \subseteq V^\infty$, and V^∞ is a union of the V^K as K varies over all compact open subgroups of G. If $V = V^\infty$, then the representation ρ is called *smooth*. If in addition all the spaces V^K are finite dimensional, then ρ is called *admissible*. A continuous representation of ρ on a locally convex space whose associated smooth subrepresentation is admissible is also called admissible.

An admissible representation has a character in the following sense. Let $C_c^\infty(G)$ denote the space of locally constant, compactly supported, complex-valued functions on G. Then $C_c^\infty(G)$ is a convolution algebra under the usual definition of convolution on G. If ρ is a smooth representation of G on V, then we can define in an obvious and standard manner a representation, also denoted ρ, of $C_c^\infty(G)$ on V. If ρ is an admissible representation, then $\rho(f)$ will be a finite rank operator. (If f is invariant under left translation by the open subgroup $K \subseteq G$, then $\rho(f)(V) \subseteq V^K$.) Hence the trace of $\rho(f)$ will be well defined, and will depend linearly on f. Thus

$$\theta_\rho(f) = \operatorname{trace} \rho(f)$$

defines a linear functional on $C_c^\infty(G)$. By definition a linear functional on $C_c^\infty(G)$ is called a *distribution* on G. Thus we may say that an admissible representation has a character defined by a distribution on G—its "distribution character".

Thus the first question to ask concerning character theory of representations of G is, are all reasonable (say unitary, though one can be more general) irreducible representations of G admissible? The analogous question for Lie groups was one of the first issues resolved by Harish-Chandra [1953]. In the case of a reductive p-adic group the answer was much longer in coming. That the answer was "yes" was first established by Bernstein [Be]. A stronger result was proved for groups over fields of characteristic zero by Harish-Chandra in [1978], generalizing and extending [Ho]. Both results were based on [1970d] and in particular were resolutions of conjectures made in [1970d]. We will explain these conjectures.

The conjectures concern the supercuspidal representations, which are the most characteristic and the most mysterious phenomenon of representation theory on reductive p-adic groups.

Let G be a locally compact group. Let f be a function on G. Define the left and right translate $\lambda_g f$ and $\rho_g f$ of f by g in the usual way:

$$\lambda_g(f)(x) = f(g^{-1}x) \qquad \rho_g(f)(x) = f(xg) \qquad x, g \in G. \tag{1}$$

Let $H \subseteq G$ be a subgroup. We say f is left (right) H-*invariant* if $\lambda_h(f) = f(\rho_h(f) = f)$ for $h \in H$. If f is complex- or vector-valued we say f is left (right) H-*finite* if the left translates $\lambda_h(f)$ (right translates $\rho_h(f)$) span a finite dimensional space. We say f is left (right) compactly supported mod H if there is a compact set $C \subseteq G$ such that $f = 0$ outside HC (outside CH). If H is central in G, we may omit the adjectives "left" and "right" in the definitions above.

Let G be a reductive p-adic group—the rational points of a connected reductive algebraic group $\mathbf{G}$ defined over a p-adic field Ω. Let Z be the center of G. Let P be a parabolic subgroup of G and let N be the unipotent radical of P. Let δ_P be the modular function of P. Let f be a smooth complex- or vector-valued function on G, compactly supported mod Z. Then in analogy with the real case we can define a function $f^{(P)}$ on P/N by the formula

$$f^{(P)}(m) = \delta_P(m)^{1/2} \int_N f(mn)\, dn \qquad m \in P.$$

Here dn is Haar measure on N.

We say a function f on G is a *supercusp form* if

   i) f is smooth;
  ii) f is compactly supported mod Z;
 iii) f is Z-finite; and
 iv) $(\lambda_g f)^{(P)} = 0$ for all proper parabolics $P \subseteq G$ and $g \in G$.

A *supercuspidal representation* is a representation whose matrix coefficients are supercusp forms. Let π be a smooth representation of G on the space V. Consider $T \in \text{End } V$. We say T is a *smooth operator* if there is an open subgroup $K \subseteq G$ such that

$$\pi(k)T = T = T_\pi(k) \qquad k \in K.$$

Let $\text{End}^\circ(V)$ denote the space of smooth operators which also have finite rank. (Observe that if π is admissible, then a smooth operator automatically has finite rank.) Then $\text{End}^\circ(V)$ is a subalgebra of V, and is invariant by right and left multiplication by $\pi(g)$, for $g \in G$. Set

$$f_T(g) = \operatorname{trace} T\pi(g).$$

The function f_T is called a *matrix coefficient* of ρ. It is easy to see that $f_T \in C^\infty(G)$, the space of locally constant functions on G. The space spanned by all matrix coefficients of all admissible representations is denoted $\mathcal{C}(G)$. If π is admissible and all matrix coefficients of π are supercusp forms, then π is called supercuspidal. It turns out ([1973] §6) that for π to be supercuspidal, it is sufficient that its matrix coefficients be compactly supported mod Z.

Supercuspidal representations are obviously discrete series representations (mod Z). A general argument ([1970d], p. 6) using the Schur orthogonality relations for discrete series shows that all discrete series representations are admissible. In fact if π is an irreducible square integrable representation (mod Z) of G on a space V, and V^K is the space of $\pi(K)$-invariant vectors in V for an open subgroup $K \subseteq G$, one has

$$\dim V^K \leqslant \left(d(\pi)\mu(K) \right)^{-1}$$

where $d(\pi)$ is the formal degree of π and $\mu(K)$ is the Haar measure of K. If 1_K denotes the trivial representation of K, then $\dim V^K$ is the multiplicity of the trivial representation of K in π. It will be denoted $[\pi:1_K]$.

Harish-Chandra shows in [1970d], Part II, that the question of admissibility hinges on the possibility of bounding $[\pi:1_K]$ independently of π. He considers three possible facts that would do this. Observe that if π is an irreducible admissible representation of G, then $\pi|Z$ is a multiple of some character $\chi_\pi = \chi$, called the *central character* of π. Consider the following statements.

   i)   For each compact open subgroup K there is a number $\delta(K)$ such that $[\pi:1_K] \leqslant \delta(K)$ for all supercuspidal irreducible representations π of G.

   ii) For each compact open subgroup K there is a number $m(K)$ such that, for any character χ of Z,

$$\sum_\pi [\pi:1_K] \leqslant m(K)$$

        where π ranges over all (equivalence classes of) irreducible supercuspidal representations of G with central character χ.

   iii) There is $\varepsilon > 0$ such that $d(\pi) > \varepsilon$ for all irreducible supercuspidal representations π of G.

Obviously statement ii) implies i). The Schur orthogonality relations in fact imply that iii) implies ii). In [1970d], Part II, Harish-Chandra shows that statement i) implies that all irreducible unitary representations of G are admissible. In [Be] Bernstein shows by a simple argument based on the structure of $C^\infty(G//K)$, the K bi-invariant functions of compact support, that statement i) is true. In [1978] Harish-Chandra, as a consequence of a study of characters of supercuspidal representations, shows that if the ground field Ω has characteristic 0, then statement iii) is true. In fact, he shows that if Haar measure on G is appropriately normalized, then the formal degrees $d(\pi)$ of supercuspidal representations are integers.

Once the issue of existence of characters is settled, one would like to know more precisely what they look like. For supercuspidal characters and when the ground field is of characteristic 0, Harish-Chandra already establishes an important qualitative result in [1970d]. Let dg denote Haar measure on G, and let φ be a complex-valued function on G which is locally L^1, i.e. the integral

$$\int_C |\varphi|(g)\, dg$$

where $|\varphi|$ indicates the absolute value of φ and C is any open compact subset of G, is finite. Then φ defines a distribution D_φ on G by the obvious formula

$$D_\varphi(f) = \int_G \varphi(g)f(g)\, dg \qquad f \in C_c^\infty(G).$$

If a distribution D is equal to D_φ for some φ as above, we say D is *locally L^1*. If the function φ may be taken to be locally constant on some open set $S \subseteq G$, we say D is *locally constant* on S. In [1970d], Part V, Harish-Chandra shows that the character of an irreducible supercuspidal representation is a locally constant function on the set G' of regular elements of G. In Parts VII and VIII of [1970d] he goes on to show that if the ground field Ω has characteristic zero, then in fact the character of an irreducible supercuspidal representation is locally L^1 on G. These results are based on an integral formula ([1970d], Theorem 9) for the character of a supercuspidal representation and on a detailed analysis of the geometry of conjugacy classes, analogous to his work on real groups.

In [1978] Harish-Chandra extends these basic facts to all irreducible admissible representations. Furthermore he gives more precise information about the behavior of characters near singular points. These results make use of the exponential map from the Lie algebra $\mathfrak{g}$ of G to G, and so are valid only when the ground field Ω has characteristic zero. They depend on a finiteness result, proved first for GL_n in [Ho], and extended and generalized by Harish-Chandra in [1978]. This may be stated as follows. Let $L \subseteq \mathfrak{g}$ be a lattice, and let $C_c^\infty(\mathfrak{g}/L)$ be the functions on $\mathfrak{g}$ which are compactly supported and constant on cosets of L. Let $\mathrm{Ad}\,G$ denote the adjoint action of G on $\mathfrak{g}$. Let $\omega \subseteq \mathfrak{g}$ be a compact set, let $\mathrm{Ad}\,G(\omega)$ denote the subset of $\mathfrak{g}$ swept out by the action of $\mathrm{Ad}\,G$ on ω, and let $J(\omega)$ denote the space of distributions which are supported on $\mathrm{Ad}\,G(\omega)$ and which are invariant under $\mathrm{Ad}\,G$. Let $j_L J(\omega)$ denote the space of linear functionals on $C_c^\infty(\mathfrak{g}/L)$ obtained by restricting elements of $J(\omega)$ to $C_c^\infty(\mathfrak{g}/L)$. Then $j_L J(\omega)$ has finite dimension.

This finiteness result has two main consequences for characters. First, it gives a description of the possible singularities of characters near any point. It is in terms of Fourier transforms of invariant measures on nilpotent conjugacy classes. Let $x \in \mathfrak{g}$ be a nilpotent element, and let $\mathcal{O} = \mathcal{O}_x = \mathrm{Ad}\,G(x)$ denote the G-conjugacy class of x. There are only finitely many nilpotent conjugacy classes. On $\mathcal{O}$ there is supported a G-invariant measure ν_x, unique up to multiples. Furthermore, a result of Deligne and Rao [Ra] says that the measure ν_x assigns finite mass to bounded subsets of $\mathcal{O}$; consequently ν_x defines a distribution on $\mathfrak{g}$.

We want to consider the Fourier transforms of the measures ν_x. These may be defined in the usual way. If $f \in C_c^\infty(\mathfrak{g})$, its Fourier transform $\hat{f}$ is defined by

$$\hat{f}(x) = \int f(x') \chi(B(x', x))\, dx'$$

where $B(\ ,\ )$ is an Ad G-invariant, non-degenerate symmetric bilinear form on $\mathfrak{g}$ (the Killing form if $\mathfrak{g}$ is semisimple), χ is a unitary character of the (additive group of the) base field Ω, and dx' is a Haar measure on $\mathfrak{g}$, normalized to make the map $f \to \hat{f}$ unitary. We can then extend $\hat{\ }$ to distributions by the recipe

$$\hat{D}(f) = D(\hat{f}) \qquad f \in C_c^\infty(\mathfrak{g}),\ D \in C_c^\infty(\mathfrak{g})^*. \tag{2}$$

Recall that $J(\omega)$ is the space of Ad G-invariant distributions supported on Ad $G(\omega)$ where $\omega \subseteq \mathfrak{g}$ is some compact set. Harish-Chandra in [1978] has described the Fourier transforms of elements of $J(\omega)$. To state his result we need the functions $\eta_\mathfrak{g}$ and D_G, familiar from the real theory, which measure how regular elements of $\mathfrak{g}$ or G are.

Consider the polynomial $\det(t - \operatorname{ad} x)$ on $\mathfrak{g} \times \Omega$. Regard this as a polynomial in t with coefficients depending on x. Then if l is the rank of $\mathfrak{g}$ (i.e., the dimension of any Cartan subalgebra), the coefficient of t^b is identically zero if $b < l$. The function $\eta_\mathfrak{g}$ is the coefficient of t^l. Thus

$$\det(t - \operatorname{ad} x) = t^l \eta_\mathfrak{g}(x) + t^{l+1} R(x, t). \tag{3}$$

The function $\eta_\mathfrak{g}$ may be described in another way, as follows. Let $x \in \mathfrak{g}$ be regular, let $\mathfrak{t} \subseteq \mathfrak{g}$ be the kernel of $\operatorname{ad} x$ and let $\mathfrak{t}^\perp$ be the orthogonal complement of $\mathfrak{t}$ with respect to the bilinear form B. Then

$$\eta_\mathfrak{g}(x) = \det\left(\operatorname{ad} x | \mathfrak{t}^\perp\right).$$

The function D_G is the analogue on G of $\eta_\mathfrak{g}$. The polynomial $\det(1 + t - \operatorname{Ad} g)$ also begins with the term t^l and one writes

$$\det(1 + t - \operatorname{Ad} g) = t^l D_G(g) + t^{l+1} R'(g, t).$$

The functions $\eta_\mathfrak{g}$ and D_G take values in Ω, the ground field. Recall [W] that on Ω there is defined a natural absolute value

$$|\ \ |: \Omega \to \mathbf{R}.$$

We will need the real-valued functions gotten by taking absolute values of $\eta_\mathfrak{g}(x)$ or $D_G(x)$. Thus we write

$$|\eta_\mathfrak{g}|(x) = |\eta_\mathfrak{g}(x)| \qquad |D_G|(x) = |D_G(x)|.$$

Consider a distribution $D \in J(\omega)$. Harish-Chandra ([1978], Theorem 3) shows that $\hat{D}$ is locally L^1, locally constant on $\mathfrak{g}'$, the set of regular elements in $\mathfrak{g}$, and that $|\eta_\mathfrak{g}|^{1/2}\hat{D}$ is locally bounded on $\mathfrak{g}$. These results apply in particular to the Fourier transforms $\hat{\nu}_x$ of the invariant measures on nilpotent orbits.

Finally we can describe characters. Let θ_π be the character of an irreducible admissible representation π of G. Then θ_π is a locally L^1 function, locally constant on G', and the function $|D_G|^{1/2}\theta_\pi$ is locally bounded. Furthermore, if $\gamma \in G$ is a semisimple element, then the behavior of θ_π near γ is as follows. Let M be the centralizer of γ in G, and $\mathfrak{m}$ the Lie algebra of M. Let C be a small neighborhood of 0 in $\mathfrak{m}$, on which the exponential map exp is defined. Then $\gamma \exp C$ is a neighborhood of γ in M, and $\operatorname{Ad} G(\gamma \exp C)$ is an $\operatorname{Ad} G$-invariant neighborhood of γ in G. If C is chosen sufficiently small, then we may write

$$\theta_\pi(\gamma \exp Y) = \sum_\xi c_\xi \hat{\nu}_\xi(Y) \qquad Y \in C$$

where ξ runs over the nilpotent conjugacy classes *in* $\mathfrak{m}$, and the c_ξ are complex numbers depending on π and γ. Roughly, we may say that locally θ_π is a linear combination of Fourier transforms of nilpotent orbits.

This result is analogous to the knowledge that an invariant eigendistribution on a real reductive group is a locally L^1 function, and is given locally on each Cartan subgroup by $D_G^{-1/2}$ times a linear combination of certain exponentials. What is still missing, except in special cases, are results as precise as Harish-Chandra's character formula for discrete series of real groups.

As was already stated, the above results apply when the ground field Ω has characteristic zero. When Ω is of positive characteristic, much less is known. However, in [1980] it is shown that in all characteristics characters are locally constant on the regular set. The techniques of this paper are different from those of [1973] or [1978]. The basic fact used is the submersiveness of a certain map ([1980], Theorem 1). Also, the following interesting fact is established. Let K be a "good" maximal compact subgroup of G in the sense of Bruhat and Tits [BT]. For an admissible representation π of G on a vector space V and a regular element x in G, set

$$T_x = \int_K \pi(kxk^{-1})\, dk.$$

Then $T_x \in \operatorname{End}^\circ(V)$, and in particular T_x has finite rank. Then evidently $\theta_\pi(x) = \operatorname{tr} T_x$.

The study of the map F_f or, in common current parlance, orbital integrals, is entwined with and in some sense dual to the study of characters, and currently is in roughly the same state. One has substantial knowledge of the qualitative behavior of F_f, but precise control analogous to the "jump formula" for real groups is still lacking. This would be provided by an explicit description of the "Shalika germs" which will be defined and discussed below.

First recall the definition of F_f. Fix a Cartan subgroup $A \subseteq G$, and fix an invariant measure dg^* on G/A. Then for $f \in C_c^\infty(G)$ and $a \in A'$, where $A' = G' \cap A$ one sets

$$F_f(a) = |D_G(a)|^{1/2} \int_{G/A} f(gag^{-1})\, dg^*. \tag{4}$$

There is a close analogue of F_f for the Lie algebra. If $\mathfrak{a}$ is the Lie algebra of A and $x \in \mathfrak{a}$, put

$$\phi_f(x) = |\eta_\mathfrak{a}(x)|^{1/2} \int_{G/A} f(\operatorname{Ad} g(x))\, dg^* \qquad f \in C_c^\infty(\mathfrak{g}). \tag{5}$$

In formulas (4) and (5) the functions D_G and $\eta_\mathfrak{a}$ are the ones defined above in (2) and (3).

Formulas (4) and (5) are very near parallels of the definitions of F_f and ϕ_f for real groups. However, we should remark on one difference that has not been important up to now, but which needs clarification before these maps can be thoroughly understood. For real groups the function D_G is a smooth function. The function $|D_G|^{1/2}$ is not smooth. However, on any Cartan subgroup A, there is a function $D_G^{A\ 1/2}$ such that $(D_G^{A\ 1/2})^2 = D_G$ on A and which yields an optimal theory of orbital integrals. In other words, if one multiplies $|D_G|^{1/2}$ by appropriate phase factors, one obtains a better theory, in which the functions F_f are as smooth as possible. It is the function $D_G^{A\ 1/2}$ rather than the positive $|D_G|^{1/2}$ which is used in Harish-Chandra's definition of F_f for real groups. It would seem that some modification of $|D_G|^{1/2}$ by appropriate phases is also preferable in the p-adic case; but precisely how to do this has not yet been made clear.

The parts of Harish-Chandra's work devoted to F_f on p-adic groups are [1970d], Part VI; [1973], §16; [1978] §§3, 4, 8, 9; and [1980], §5. In [1970d] it is already established that when the ground field Ω has characteristic 0, the function F_f is bounded for any $f \in C_c^\infty(G)$. However, it is not shown that F_f is locally constant. This is done in [1973], §16 (for the proofs see [Si]) by the same technique that establishes local constancy of characters. In fact, local constancy of F_f is even established for f belonging to Harish-Chandra's Schwartz space $\mathcal{C}(G)$ (see below for the definition of $\mathcal{C}(G)$); however local boundedness of F_f for $f \in \mathcal{C}(G)$ is not proven. Local constancy of F_f for Ω of positive characteristic, and $f \in \mathcal{C}(G)$ is proved in [1980], but boundedness not.

For real groups the behavior of $F_f(x)$ as x approaches singular points is of much interest. This is also true for p-adic groups. The standard approach to the study of this limiting behavior is provided by an observation of Shalika [Sh]. Harish-Chandra's version of this result is Theorem 14 of [1978], which applies to the Lie algebra. As above, we let ν_x denote the invariant measures on the nilpotent G-conjugacy classes $\mathcal{O}_x$ in $\mathfrak{g}$. Fix a Cartan subalgebra $\mathfrak{a} \subseteq \mathfrak{g}$. Then there are functions $\Gamma_\mathcal{O}^\mathfrak{a} = \Gamma_\mathcal{O}$, for each nilpotent conjugacy class $\mathcal{O} = \mathcal{O}_x$ such that

$$\phi_f(a) - \sum \nu_x(f)\Gamma_\mathcal{O}(a)$$

vanishes in some neighborhood of the origin in $\mathfrak{a}$. The functions $\Gamma_\mathcal{O}$ are strictly speaking, only germs of functions, an equivalence class of functions equal in some neighborhood of 0; they are generally called "Shalika germs". However, they may be defined uniquely on all of $\mathfrak{a}$ by requiring them to satisfy an appropriate condition of homogeneity, namely

$$\Gamma_\mathcal{O}(t^2a) = |t|^m \Gamma_\mathcal{O}(a) \qquad t \in \Omega^\times,\ a \in \mathfrak{a}$$

where m is an integer depending on $\mathcal{O}$. There are analogous germ expansions around non-zero singular points of $\mathfrak{a}$, and for F_f.

Clearly the functions $\Gamma_\mathcal{O}$ control the singularities of ϕ_f. Considerable effort consequently has been devoted to determining them [Vi], [Ko], [Re1], [Re2], [Ro2]. However, they remain mysterious. Harish-Chandra shows in [1978], §9 that the $\Gamma_\mathcal{O}^\mathfrak{a}$ separate the nilpotent orbits $\mathcal{O}$ in the sense that the only linear combination $\sum c_\mathcal{O} \Gamma_\mathcal{O}^\mathfrak{a}$ which vanishes for all Cartan subalgebras $\mathfrak{a}$ is the trivial combination with all $c_\mathcal{O} = 0$. This is implied by his result ([1978], Theorem 10) that all invariant distributions annihilate any function $f \in C_c^\infty(\mathfrak{g})$ such that $\phi_f = 0$ for all Cartan subalgebras $\mathfrak{a}$. He also shows that $\Gamma_0^\mathfrak{a}$, the germ associated to the origin, is zero if $\mathfrak{a}$ is not an elliptic Cartan subalgebra, and on an elliptic Cartan is a constant times $|\eta_\mathfrak{g}|^{1/2}$. The value of the constant was conjectured by Harish-Chandra and determined by Rogawski [Ro1].

The third major topic in Harish-Chandra's work is the Plancherel Formula, and the associated analysis on G: the constant term, induced series of representations, wave packets, Eisenstein integrals, c-functions. For real groups, character theory and orbital integrals also contribute to the Plancherel Formula, but as explained above, for p-adic groups, they have not reached the state necessary for use in the Plancherel Formula. However certain technical aspects of the p-adic situation allow Harish-Chandra to complete his program without knowledge of the discrete series. The resulting Plancherel Formula is not as explicit as the one for real groups. But except for the lack of explicit knowledge of the discrete series, the Plancherel Formula for p-adic groups is quite parallel to that for real groups. Hence our discussion of it will be relatively brief, and will focus on the aspects which are particular to p-adic groups.

Since the theory of the constant term for real groups depends heavily on the differential equations supplied by the center of the enveloping algebra, its transference to p-adic groups requires new techniques. Harish-Chandra made a preliminary essay in [1970d], establishing the existence of the constant term contingent on a conjecture. However a firmer basis was provided shortly after by a result of Jacquet [J].

Let π be an admissible representation of G on the vector space V. Let $P \subseteq G$ be a parabolic subgroup with unipotent radical N. Let $V(N)$ be the subspace of V spanned by vectors of the form $\pi(n)v - v$. Since N is normalized by P, one can easily see that $V(N)$ is stable under P. Hence there is defined a representation of P on $V_N = V/V(N)$. From the definition of $V(N)$ it is clear that N acts trivially on V_N. Hence the action of P on V_N factors to an action π_M of $P/N \simeq M$ on V_N. Jacquet showed (for $G = GL_n(\Omega)$, but with a general proof) that V_N is also an admissible module for M. It turns out that π is supercuspidal if and only if $V_N = \{0\}$.

Recall that $\mathcal{A}(G)$ is the space of all matrix coefficients of all admissible representations of G. Using Jacquet's theorem, Harish-Chandra establishes the existence of the constant term in a very strong form. Let M now denote a Levi component of P, so that $M \simeq P/N$. Let A denote the center of M. Let $\{\alpha_i\}$ be the simple roots of A acting on N by conjugation. Each α_i is a rational homomorphism

$$\alpha_i : A \to \Omega^\times.$$

For a given $t > 0$, set

$$A^+(t) = \{a \in A : |\alpha_i(a)| > t \text{ for each } \alpha_i\}.$$

As above let δ_P denote the modular function of P. Then Harish-Chandra shows ([1973], §6; see [Si], Chapter 2 for proofs) that given $f \in \mathcal{A}(G)$, there is a unique $f_P \in \mathcal{A}(M)$ such that, for any compact set $\omega \subseteq M$, one has

$$f(ma) = \delta_P(ma)^{1/2} f_P(ma) \qquad m \in \omega, \ a \in A^+(t)$$

for t sufficiently large. Thus in the p-adic case one eventually has actual equality between f and f_P, its *constant term along* P, and not merely an asymptotic relation.

Since f_P is in $\mathcal{Q}(M)$, it is in particular A-finite. The finite dimensional space spanned by the A-translates of f_P will be spanned by generalized eigenspaces for A. If Y is such an eigenspace, then there is a quasicharacter ψ of A such that $\lambda_a - \psi(a)$ is nilpotent on Y. (Here λ_a is the left action of A, as in (1).) Harish-Chandra calls these characters ψ the *exponents* of f relative to the pair (P, A). The A-finiteness of f_P takes the place in many situations of the differential equations governing the constant term in the real case, and the exponents take the place of infinitesimal characters. To place limits on the elements of $\mathcal{Q}(M)$ which can be constant terms of matrix coefficients of a fixed irreducible representation of G, Harish-Chandra develops a sharp form of the theory of intertwining operators for induced representations, pioneered by Bruhat for real groups [Br1].

With the constant term, Harish-Chandra can proceed toward the Plancherel Formula along very much the same road used in the real case. The theory of induced representations, of the Eisenstein integral, the c-functions, the Maass–Selberg relations, of wave packets and the Schwartz space proceed very much as in the real case. Already in [1973] he is able to state the Plancherel Formula for the wave packets constructed from a given series of induced representations (Theorem 34). In this Plancherel Formula, the Plancherel measure is determined by a product formula, as it is for real groups. However, it is not explicitly known. But luckily, it is not necessary to know it explicitly in the p-adic case. This is because any p-adic torus A, modulo an open compact subgroup, is discrete. Hence $\hat{A}^0$, the identity component of the Pontrjagin dual of A, is compact. It is a real torus—a product of circles. If M is the Levi component of a parabolic subgroup P, and A the center of M, then a series from which one builds wave packets consists of representations of the form $\pi \otimes \chi$ where π is a square integrable (mod A) unitary representation of M and $\chi \in \hat{A}^0$. In the real case, $\hat{A}^0$ is a vector group, and it is necessary to know that Plancherel measure on $\hat{A}^0$ grows only polynomially at ∞ in order to be able to control the Fourier transform. In fact, for real groups, Harish-Chandra gives an explicit expression for the Plancherel measure, and one can see directly that this expression has moderate growth. But since $\hat{A}^0$ is compact for p-adic groups, the growth of Plancherel measure is not an issue, and no explicit knowledge of it is necessary.

We will close this account by describing what is involved in the passage from the "local" Plancherel Formula of [1973] to the full formula announced in [1977b].

First the notion of constant term described above must be modified so that matrix coefficients of representations which are square integrable but not supercuspidal have zero constant term. This leads to the notion of the weak constant term.

Let K be a good maximal compact subgroup of G. Let Ξ be, as for real groups, the K-spherical matrix coefficient of the representation unitarily induced from the trivial representation of a minimal parabolic subgroup of G. Let σ be the measure of slow growth on G; roughly σ is the logarithm of Ξ (see [1973]; §14). For each compact open $K_1 \subseteq K$, one defines $\mathcal{C}(G//K_1)$ to be the space of K_1 bi-invariant functions f on G such that

$$|f| \leqslant c(f,k)\,\Xi(1+\sigma)^{-k}$$

for an appropriate positive number $c(f, k)$ and for each positive number k. Then $\mathcal{C}(G)$, the *Schwartz space* of G is the union over compact open K_1 of the spaces $\mathcal{C}(G//K_1)$.

Consider $\phi \in \mathcal{C}(G)$. One says ϕ satisfies the *weak inequality* if for some constant c and some integer k one has the estimate

$$|\phi| \leqslant c\Xi(1+\sigma)^k.$$

Such a ϕ is called *tempered*, and the subspace of $\mathcal{C}(G)$ consisting of tempered matrix coefficients is denoted $\mathcal{C}^w(G)$.

Let f be in $\mathcal{C}^w(G)$, and let f_P be its constant term along the parabolic subgroup P. Write

$$f_P = \sum f_{P,\psi}$$

where ψ runs over the exponents of f, and $f_{P,\psi}$ is the component of f in the generalized ψ-eigenspace of the A translates of f, A being the center of a Levi component M of P. The sum of the $f_{P,\psi}$ over those ψ which are unitary is called the *weak constant term* of f along P, and is denoted f_P^w. It is in $\mathcal{C}^w(M)$. A *cusp form* on G is an element of $\mathcal{C}^w(G)$ such that the weak constant terms $(\lambda(g)f)_P^w$ of all left translates of f are zero for all proper parabolic subgroups P of G. The space of cusp forms on G is denoted ${}^0\mathcal{C}^w(G)$.

There is a notion of constant term for $\mathcal{C}(G)$ that is dual to that for $\mathcal{C}^w(G)$. For $f \in \mathcal{C}(G)$ and P a parabolic subgroup of G, set

$$f^{(P)}(m) = \delta_P(m)^{1/2} \int_N g(mn)\, dn \qquad m \in P/N$$

where N is the unipotent radical of P. Let ${}^0\mathcal{C}(G)$ denote the subspace of $\mathcal{C}(G)$ consisting of those f such that $(\lambda(g)f)^{(P)} = 0$ for all proper parabolic subgroups.

If G were semisimple rather than reductive, then it is easy to see that ${}^0\mathcal{C}^w(G) \subseteq {}^0\mathcal{C}(G)$. If G possesses a non-compact center, then one must formulate the relation between ${}^0\mathcal{C}^w$ and ${}^0\mathcal{C}$ in a more complicated way, but the situation is in essence the same. So to simplify the discussion we take G semisimple. Then ${}^0\mathcal{C}^w(G)$ consists of matrix coefficients of the discrete series, while ${}^0\mathcal{C}(G)$ is the subspace of $\mathcal{C}(G)$ which cannot be obtained by inducing tempered representations from proper parabolic subgroups. Hence to pass from the Plancherel Theorem of [1973], §17, to the Plancherel Formula for G, one needs to show that in fact ${}^0\mathcal{C}^w(G) = {}^0\mathcal{C}(G)$. This essentially amounts to the statement that (for G semisimple), the representation of G on the span of the left translates of $f \in {}^0\mathcal{C}(G)$ is admissible, i.e., consists of finitely many irreducible summands. This is a consequence of Lemma 4 of [1977b]. A crucial step in the proof of Lemma 4 is Theorem 11, which states that if $\theta \in {}^0\mathcal{C}(G)$, then the map

$$f \to \int_{G/Z} dy^* \int_G f(x)\theta(yxy^{-1})\, dx$$

is well defined and yields a tempered distribution on G. Although the details of this result are not yet published, in flavor it reminds one of the analysis of supercuspidal characters in [1970d].

The form of Harish-Chandra's p-adic Plancherel Theorem is this. The space $\mathcal{C}(G)$ is the orthogonal direct sum of wave packets formed from series of representations induced unitarily from discrete series of (the Levi components of) parabolic subgroups of P. Moreover if two such series of induced representations yield the same subspace of $\mathcal{C}(G)$, then the parabolics from which they are induced are associate, and the representations of the Levi components are conjugate. To complete the analogy with real groups, one needs only to explicitly determine the discrete series; this is the outstanding problem left in p-adic representation theory.

REFERENCES

[Ar] J. Arthur, On some problems suggested by the trace formula, Special Year in Representation Theory, U. of Maryland, Nov. 1982, Springer Lecture Notes, to appear.

[Be] I. N. Bernstein, All reductive p-adic groups are tame, *Fun. Anal. and App.* **8** (1974), 91–93.

[Bo] A. Borel, Automorphic L-functions, in *Automorphic Forms, Representations, and L-functions*, *Proc. Sym. Pure Math.*, **XXXIII**, Part 2, Amer. Math. Soc. Providence, R.I., 1974, 27–63.

[BW] A. Borel and N. Wallach, Continuous cohomology, discrete subgroups, and representations of reductive groups, *Ann. of Math. Studies* **94**, Princeton University Press, Princeton, 1980.

[Br1] F. Bruhat, Sur les representations induites des groupes de Lie, *Bull. Soc. Math. France*, **84** (1956), 97–205.

[BT] F. Bruhat and J. Tits, Groupes reductifs sur un corps local, I, Donnees radicielles valuees, *Pub. Math. I.H.E.S.* **41** (1972), 5–251.

[CM] W. Casselman and D. Milicic, Asymptotic behavior of matrix coefficients of admissible representations, *Duke Math. J.* **49** (1982), 869–930.

[He] G. Henniart, La conjecture de Langlands locale pour GL(3), *I.H.E.S. Notes* (1982).

[Ho] R. Howe, The Fourier transform and germs of characters, *Math. Ann.* **208** (1974), 305–322.

[H] H. Jacquet, Representations des groupes lineaires p-adiques, Theory of Group Representations and Harmonic Analysis (C.I.M.E., II, Ciclo, Montecatini terme, 1970) Edizione Cremonese, Roma, 1971, 119–220.

[JL] H. Jacquet and R. Langlands, Automorphic forms on GL(2). *Lec. Notes Math.*, **260**, Springer-Verlag, Berlin, New York, 1972.

[Ko] R. E. Kottwitz, Orbital integrals on GL(3), *Am. J. Math*, **102** (1980), 327–384.

[K] P. Kutzko, The Langlands conjecture for GL_2 of a local field, *Ann. Math.* **112** (1980), 381–412.

[KM] P. Kutzko and A. Moy, On the local Langlands Conjecture in prime dimension, preprint.

[L1] R. Langlands, Problems in the theory of automorphic forms, *Lectures in Modern Analysis and Applications, Lecture Notes in Math.*, **170**, Springer-Verlag, New York, 1973, 18–86.

[L2] R. Langlands, On the classification of irreducible representations of real reductive groups, *Notes, I.A.S.*, Princeton, 1973.

[Lu1] G. Lusztig, Irreducible representations of finite classical groups, *Inv. Math.* **43** (1977), 125–175.

[Lu2] G. Lusztig, Some examples of square integrable representations of semisimple p-adic groups, preprint, *I.H.E.S.*, March 1982.

[M] A. Moy, Local constants and the tame Langlands correspondence, Thesis, University of Chicago, 1982.

[MS] A. Moy and P. Sally, Supercuspidal representations of SL_n over a p-adic field: the tame case, preprint.

[Ra] R. Rao, Orbital integrals in reductive groups, *Ann. of Math.* **96** (1972), 505–510.

[Re1] J. Repka, Shalika's germs for p-adic $GL(n)$: the leading term, preprint.

[Re2] J. Repka, Shalika's germs for p-adic $GL(n)$, II: the subregular term, preprint.

[Ro1] J. Rogawski, An application of the building to orbital integrals, *Comp. Math.* **42** (1981), 417–423.

[Ro2] J. Rogawski, Some remarks on Shalika germs, preprint.

[Sh] J. Shalika, A theorem on semisimple p-adic groups, *Ann. Math.* **95** (1972), 226–242.

[Si] A. Silberger, Introduction to Harmonic Analysis on Reductive p-adic Groups, *Math. Notes*, Princeton University Press, Princeton, N.J., 1979.

[T] J. Tunnell, Report on the local Langlands conjecture for GL_2, *Proc. Symp. Pure Math.*, **XXXIII**, Part 2, Amer. Math. Soc., Providence, R.I., 1979, 135–138.

[V] V. Varadarajan, Harmonic Analysis on Real Reductive Groups, *Lecture Notes in Math.*, **576**, Springer-Verlag, New York, 1976.

[Vi] M. F. Vigneras, Caracterisation des integrales orbitales sur un groupe reductif p-adique, *J. Fac. Sci. Tokyo*, Sec. IA, **28**, (1982), 945–961.

[Vo] D. Vogan, Representation of Real Reductive Lie Groups, *Progress in Math.*, **15**, Birkhauser, Boston, Basel, Stuttgart, 1981.

[Wa] N. Wallach, Asymptotic expansions of generalized matrix entries, special year on representation theory, University of Maryland, Vol. I, *Lecture Notes in Mathematics*. Springer-Verlag, Berlin, Heidelberg, New York, Tokyo, 1983.

[W] A. Weil, *Basic Number Theory*, 2nd ed., Grund. Math. Wiss., **144**, Springer-Verlag, Berlin, Heidelberg, New York, 1973.

Permissions

Springer-Verlag would like to thank the original publishers of Harish-Chandra's papers for granting permissions to reprint specific papers in this collection. The following list contains the credit lines for those articles.

[1944a] Reprinted from *Proc. Royal Soc. A.* **183**, ©1944 by The Royal Society of London.
[1944b] Reprinted from *Proc. Royal Soc. A.* **183**, ©1944 by The Royal Society of London.
[1945a] Reprinted from *Proc. Indian Acad. Sci. Sect. A.* **21**, ©1945 by The Indian Academy of Sciences.
[1945b] Reprinted from *Proc. Indian Acad. Sci. Sect. A.* **22**, ©1945 by The Indian Academy of Sciences.
[1946a] Reprinted from *Proc. Royal Soc. A.* **185**, ©1946 by The Royal Society of London.
[1946b] Reprinted from *Proc. Royal Soc. A.* **185**, ©1946 by The Royal Society of London.
[1946c] Reprinted from *Proc. Indian Acad. Sci. Sect. A.* **23**, ©1946 by The Indian Academy of Sciences.
[1946d] Reprinted from *Proc. Royal Soc. A.* **186**, ©1946 by The Royal Society of London.
[1947a] Reprinted from *Proc. Camb. Phil. Soc.* **43**, ©1947 by Cambridge University Press.
[1947b] Reprinted from *Phys. Rev.*, **71**, ©1947 by The American Physical Society.
[1947c] Reprinted from Report of an International Conf. on Fundamental Particles and Low Temperatures held at the Cavendish Laboratory, Cambridge, (1946). Vol I, Fundamental Particles, ©1947 by The Institute of Physics.
[1947d] Reprinted from *Proc. Royal Soc. A.* **189**, ©1947 by The Royal Society of London.
[1948a] Reprinted from *Proc. Royal Soc. A.* **192**, ©1948 by The Royal Society of London.
[1948b] Reprinted from *Phys. Rev.* **74**, ©1948 by The American Physical Society.
[1949a] Reprinted from *Ann. of Math.* **50**, ©1949 by Princeton University Press.
[1949b] Reprinted from *Ann. of Math.* **50**, ©1949 by Princeton University Press.
[1950a] Reprinted from *Proc. Amer. Math. Soc.* **1**, ©1950 by The American Mathematical Society.
[1950b] Reprinted from *Proc. Amer. Math. Soc.* **1**, ©1950 by The American Mathematical Society.
[1950c] Reprinted from *Ann. of Math.* **51**, ©1950 by Princeton University Press.
[1951a] Reprinted from *Trans. Amer. Math. Soc.* **70**, ©1951 by The American Mathematical Society.

PERMISSIONS

Automorphic Forms on a Semisimple Lie Group*

HARISH-CHANDRA

DEPARTMENT OF MATHEMATICS, COLUMBIA UNIVERSITY
Communicated by Paul A. Smith, February 19, 1959

Let G be a connected semisimple Lie group with finite center and K a maximal compact subgroup of G. We denote the corresponding Lie algebras by $\mathfrak{g}$ and $\mathfrak{k}$ respectively. Put $B(X, Y) = tr(ad\,X\,ad\,Y)$ $(X, Y \in \mathfrak{g})$ and let $\mathfrak{p}$ denote the space of all $Y \in \mathfrak{g}$ which are orthogonal to $\mathfrak{k}$ under the bilinear form B. Then $\mathfrak{g}$ is the direct sum of $\mathfrak{k}$ and $\mathfrak{p}$. Put $\theta(X + Y) = X - Y$ $(X \in \mathfrak{k}, Y \in \mathfrak{p})$. Then θ is an automorphism of $\mathfrak{g}$ and the quadratic form $\|Z\|^2 = -B(Z, \theta(Z))$ $(Z \in \mathfrak{g})$ is positive definite. We regard $\mathfrak{g}$ as a (real) Hilbert space under this quadratic form and put $\|x\|^2 = tr(Ad(x)Ad(x)^*)$ $(x \in G)$ where $Ad(x)^*$ is the adjoint of the linear transformation $Ad(x)$.

Let $\mathfrak{Z}$ be the algebra of all differential operators on G which are invariant under both left- and right-translations of G. Fix a homomorphism χ of $\mathfrak{Z}$ into the field of complex numbers. Let Γ be a discrete subgroup of G and σ and μ unitary representations of K and Γ respectively on a finite-dimensional complex Hilbert space U. We assume that σ is continuous and U is a left K-module (under σ) and a right Γ-module (under μ). We shall denote the norm of u by $|u|$ $(u \in U)$. An automorphic form on G of type (σ, μ, χ) is a C^∞ function f on G with values in U such that (1) $f(kx\gamma) = \sigma(k)f(x)\mu(\gamma)$ $(k \in K, x \in G, \gamma \in \Gamma)$ and (2) $zf = \chi(z)f$ $(z \in \mathfrak{Z})$. It is not difficult to show that such a function is always analytic. Let $\mathfrak{F} = \mathfrak{F}(\sigma, \mu, \chi)$ denote the space of all such automorphic forms. If G/Γ is compact one proves easily that $\dim \mathfrak{F} < \infty$. The main problem here is to establish a similar result under some weaker hypothesis on G/Γ.

A form $f \in \mathfrak{F}$ is called normal at infinity if there exists an integer $m \geqslant 0$ and a real number $c > 0$ such that $|f(x)| \leqslant c\|x\|^m$ for all $x \in G$. Let $\mathfrak{F}_0 = \mathfrak{F}_0(\sigma, \mu, \chi)$ be the space of all $f \in \mathfrak{F}$ which are normal at infinity. Also let dx denote the Haar measure on G.

Lemma 1. *Let f be a function in $\mathfrak{F}$ such that $\int_{G/\Gamma} |f(x)|\,dx < \infty$. Then $f \in \mathfrak{F}_0$.*

Let $\mathfrak{a}$ be a maximal abelian subspace of $\mathfrak{p}$. We introduce an arbitrary but fixed lexicographic order in the space $\mathfrak{a}'$ of (real) linear functions on $\mathfrak{a}$. For any $\alpha \in \mathfrak{a}'$, let $\mathfrak{n}_\alpha$ denote the space of all $X \in \mathfrak{g}$ such that $[H, X] = \alpha(H)X$ for every $H \in \mathfrak{a}$. Consider the set Σ of those $\alpha > 0$ for which $\mathfrak{n}_\alpha \neq \{0\}$. Then Σ is a finite set. A subset

*

Σ' of Σ is called closed if $\alpha + \beta \in \Sigma'$ whenever $\alpha, \beta \in \Sigma'$ and $\alpha + \beta \in \Sigma$. Let $\Sigma' \supset \Sigma''$ be two subsets of Σ such that Σ' is closed. We say that Σ'' is an ideal in Σ' if $\alpha + \beta \in \Sigma''$ whenever $\alpha \in \Sigma''$ and $\beta, \alpha + \beta \in \Sigma'$. For any closed Σ' put $\mathfrak{n}(\Sigma') = \Sigma_{\alpha \in \Sigma'} \mathfrak{n}_\alpha$. Then $\mathfrak{n}(\Sigma')$ is a subalgebra of the nilpotent Lie algebra $\mathfrak{n} = \mathfrak{n}(\Sigma)$. Moreover if Σ' is an ideal in Σ, $\mathfrak{n}(\Sigma')$ is an ideal in $\mathfrak{n}$. Let $N(\Sigma')$, N and A denote the analytic subgroups of G corresponding to $\mathfrak{n}(\Sigma')$, $\mathfrak{n}$ and $\mathfrak{a}$ respectively.

Put $l = \dim \mathfrak{a}$. Then one can select l linearly independent elements $\alpha_1, \ldots, \alpha_l$ in Σ such that every $\alpha \in \Sigma$ is of the form $\alpha = m_1 \alpha_1 + \cdots + m_l \alpha_l$ where $m_1, \ldots, m_l$ are nonnegative integers. Let $\mathfrak{a}^+$ denote the set of all points $H \in \mathfrak{a}$ where $\alpha(H) > 0$ for every $\alpha \in \Sigma$ and put $A^+ = \exp \mathfrak{a}^+$.

Γ being a discrete subgroup of G, we shall say that Γ is of the type I if $\int_{G/\Gamma} dx < \infty$. Moreover we say that Γ is of type II if the following conditions hold:

(1) For every ideal Σ' in Σ, $N(\Sigma')/N(\Sigma') \cap \Gamma$ is compact.
(2) There exists an element $a_0 \in A$ such that $G = Ka_0(A^+)^{-1}N\Gamma$.

It is easy to show that if Γ is of type II then it is also of type I.

Let S be the class of all subsets of the set $(\alpha_1, \ldots, \alpha_l)$. For every $Q \in S$ let Σ_Q denote the smallest closed set in Σ containing Q and let Σ_Q' be the complement of Σ_Q in Σ. Then Σ_Q' is an ideal in Σ. Put $N_Q = N(\Sigma_Q)$ and $V_Q = N(\Sigma_Q')$. Select an element $H_0 \in \mathfrak{a}$ such that $\alpha(H_0) = 0$ $(\alpha \in \Sigma)$ if and only if $\alpha \in \Sigma_Q$. This is always possible. Let $\mathfrak{c}_Q$ denote the centralizer of H_0 in $\mathfrak{g}$. Then $\mathfrak{c}_Q$ (which depends only on Q and not on the choice of H_0) is reductive in $\mathfrak{g}$. Put $\mathfrak{g}_Q = [\mathfrak{c}_Q, \mathfrak{c}_Q]$ and $\mathfrak{a}_Q = \mathfrak{a} \cap \mathfrak{g}_Q$. Then $\mathfrak{g}_Q$ is semisimple. Let G_Q and A_Q denote the analytic subgroups of G corresponding to $\mathfrak{g}_Q$ and $\mathfrak{a}_Q$ respectively. Put $A_Q^+ = \exp \mathfrak{a}_Q^+$ where $\mathfrak{a}_Q^+$ is the set of those $H \in \mathfrak{a}_Q$ where $\alpha(H) > 0$ for every $\alpha \in Q$. Also let $\Gamma_Q = G_Q \cap ((G_Q V_Q \cap \Gamma) V_Q)$. Then Γ_Q is a discrete subgroup of G_Q. We shall say that Γ is of type III, if Γ_Q is of type II in G_Q for every $Q \in S$.

Theorem 1. *Let Γ_μ denote the kernel of μ in Γ. Suppose $N \cap \Gamma / N \cap \Gamma_\mu$ is finite and Γ is of type* III. *Then* $\dim \mathfrak{F}_0(\sigma, \mu, \chi) < \infty$.

The proof, which proceeds the induction on l, depends on a lemma of Godement.[1]

Put $\langle H, H' \rangle = B(H, H')$ $(H, H' \in \mathfrak{a})$ and, for any $\lambda \in \mathfrak{a}'$, define the dual element $H_\lambda \in \mathfrak{a}$ by the condition that $\langle H, H_\lambda \rangle = \lambda(H)$ for every $H \in \mathfrak{a}$. For any $\alpha \in \Sigma$ we define the Weyl reflexion s_α in $\mathfrak{a}$ by $s_\alpha H = H - 2\{\alpha(H)/\alpha(H_\alpha)\}H_\alpha$ $(H \in \mathfrak{a})$. Then s_α can be "extended" to an automorphism $a \to a^{s_\alpha}$ of A. Also s_α operates on $\mathfrak{a}'$ by duality. Select a base $H_1, \ldots, H_l$ for $\mathfrak{a}$ such that $\alpha_i(H_j) = \delta_{ij}$ $1 \leqslant i, j \leqslant l$. By a theorem of Iwasawa,[2] corresponding to any $x \in G$, there exists a unique element $H(x) \in \mathfrak{a}$ such that $x \in K \exp H(x) N$.

Let π be a representation of G on a finite-dimensional complex vector space $U \neq \{0\}$. We denote the corresponding representation of $\mathfrak{g}$ also by π. One can always introduce a Hilbert-space structure in U in such a way that the adjoint of $\pi(X)$ is $-\pi(\theta(X))$ $(X \in \mathfrak{g})$. We shall always tacitly assume that such a structure on U is given. For any $\lambda \in \mathfrak{a}'$, let U_λ denote the set of all $u \in U$ such that $\pi(H)u = \lambda(H)u$ for every $H \in \mathfrak{a}$. λ is called a weight of π if $U_\lambda \neq \{0\}$. Λ being the highest weight of π, we denote the orthogonal projection of U on U_Λ by E_Λ. A vector $u \in U$ is said to belong to a weight λ if $u \in U_\lambda$.

Put $G_i = G_Q$ where $Q = \{\alpha_i\}$ and let $s_i = s_{\alpha_i}$ $(1 \leqslant i \leqslant l)$. We shall say that Γ is of type IV if the following conditions hold:

(1) $N(\Sigma')/N(\Sigma') \cap \Gamma$ is compact for every ideal Σ' in Σ.
(2) $\inf_{\gamma \in \Gamma} \langle H_i, H(\gamma) \rangle > -\infty$ $(1 \leqslant i \leqslant l)$.[3]
(3)   We can choose a compact set N_0 in N, elements $\gamma_i \in \Gamma \cap G_i$ $(1 \leqslant i \leqslant l)$, a representation π of G on a finite-dimensional complex vector space $U \neq \{0\}$ and a unit vector ψ belonging to the highest weight Λ of π such that:
(a) $N_0(N \cap \Gamma) = N$,
(b) $\gamma_i a \gamma_i^{-1} = a^{s_i}$ $(a \in A, 1 \leqslant i \leqslant l)$,
(c) $s_i \Lambda < \Lambda$ $(1 \leqslant i \leqslant l)$,
(d) $|E_\Lambda \pi(n\gamma_i)\psi| < 1$ for $n \in N_0$ $(1 \leqslant i \leqslant l)$.

One can prove that if Γ is of type IV then it is also of type III.

Let R and C be the fields of real and complex numbers respectively, Z the ring of rational integers and $Z' = Z[(-1)^{1/2}]$ the ring of Gaussian integers. Then it is easy to check that Γ is of type IV in G in the following cases: (1) $G = SL(n, R)$, $\Gamma = SL(n, Z)$, (2) $G = Sp(n, R)$, $\Gamma = Sp(n, Z)$, (3) $G = SL(n, C)$, $\Gamma = SL(n, Z')$, (4) $G = Sp(n, C)$, $\Gamma = Sp(n, Z')$. (Here n is any positive integer.) Hence our results are applicable to these four cases.

Now Γ' being any subgroup of Γ, the Haar measure dx on G defines an invariant measure on G/Γ'. Consider the Hilbert space $\mathfrak{H} = L_2(G/\Gamma')$ corresponding to this measure. Since G operates on G/Γ' on the left in the obvious way, we get a unitary representation λ of G on $\mathfrak{H}$. Let π be an irreducible unitary representation of G on some Hilbert space. Given an integer $m \geqslant 0$, we say that π occurs as a discrete component of λ at least m times, if we can find m mutually orthogonal closed invariant subspaces $\mathfrak{H}_i$ $(1 \leqslant i \leqslant m)$ of $\mathfrak{H}$ such that the representation of G defined on $\mathfrak{H}_i$ under λ is equivalent to π. If π occurs at least once we say π is a discrete component of λ. Moreover we say that π occurs only a finite number of times if there exists an integer $m \geqslant 1$ such that it is impossible to choose $\mathfrak{H}_i$ $(1 \leqslant i \leqslant m)$ with the above properties.

The following result is an immediate consequence of Lemma 1 and Theorem 1.

Theorem 2. *Let Γ be a discrete subgroup of G of type* III *and Γ' a subgroup of finite index in Γ. Put $\Gamma_0 = \cap_{\gamma \in \Gamma} \gamma \Gamma' \gamma^{-1}$ and suppose that $N \cap \Gamma / N \cap \Gamma_0$ is finite. Then every discrete irreducible component of the representation λ of G on $L_2(G/\Gamma')$ occurs only a finite number of times in λ.*

*This work was supported in part by a grant from the National Science Foundation.
[1]See Séminaire H. Cartan, 1957/58, *Exposé 8*, pp. 8–10.
[2]*Ann. of Math.*, **50**, 525 (1949).
[3]This condition was suggested by Godement.

Reprinted from the Proceedings of the National Academy of Sciences
Vol. 45, No. 12, pp. 1763–1764. December, 1959.

SOME RESULTS ON DIFFERENTIAL EQUATIONS AND THEIR APPLICATIONS*

By Harish-Chandra

DEPARTMENT OF MATHEMATICS, COLUMBIA UNIVERSITY

Communicated by S. Eilenberg, October 19, 1959

Let $\mathbf{R}$ and $\mathbf{C}$ be the fields of real and complex numbers respectively and $\mathbf{Z}_+$ the set of all nonnegative integers. Fix an integer $l \geq 1$ and denote by $\mathbf{C}^l$ the Cartesian product of $\mathbf{C}$ with itself l times. Similarly for $\mathbf{R}^l$ and $\mathbf{Z}_+^l$. Put $\langle \mathbf{t}, \mathbf{t}' \rangle = t_1 t_1' + \ldots + t_l t_l'$ for $\mathbf{t} = (t_1, \ldots, t_l)$, $\mathbf{t}' = (t_1', \ldots, t_l')$ in $\mathbf{C}^l$. Also let[1] $s(\mathbf{t}) = \mathrm{Re}(t_1 + \ldots + t_l)$, $|\mathbf{t}| = \sup_j |t_j|$ and $\mathbf{t}^{\mathbf{k}} = t_1^{k_1} t_2^{k_2} \ldots t_l^{k_l}$ ($\mathbf{t} \in \mathbf{C}^l$, $\mathbf{k} \in \mathbf{Z}_+^l$).

Let U be a complex Banach space of finite dimension. By a U-polynomial we mean a polynomial function on $\mathbf{C}^l$ with values in U. Consider the space $\mathcal{E}(U)$ of all endomorphisms of U and put $|T| = \sup_{|u| \leq 1} |Tu|$ ($u \in U$) for $T \in \mathcal{E}(U)$. In this way $\mathcal{E}(U)$ also becomes a Banach space. If $p(\mathbf{t}) = \sum_{\mathbf{k}} u(\mathbf{k}) \mathbf{t}^{\mathbf{k}}$ ($u(\mathbf{k}) \in U$, $\mathbf{k} \in \mathbf{Z}_+^l$) is a U-polynomial, we define $\|p\| = \sum_{\mathbf{k}} |u(\mathbf{k})|$.

Let $\mathbf{C}_+^l$ denote the set of all points $\mathbf{t} \in \mathbf{C}^l$ such that $\mathrm{Re}\, t_i > 0$ ($1 \leq i \leq l$). Put $\mathbf{R}_+^l \doteq \mathbf{C}_+^l \cap \mathbf{R}^l$.

THEOREM 1. *Let* Γ_i *($1 \leq i \leq l$) be holomorphic functions on* $\mathbf{C}_+^l$ *with values in* $\mathcal{E}(U)$. *Suppose V is a nonempty connected open set in* $\mathbf{R}_+^l$ *and ϕ a function on V with values in U, which satisfies the system of differential equations $\partial\phi/\partial t_i = \Gamma_i(\mathbf{t})\phi$ ($1 \leq i \leq l$) on V. Then there exists exactly one holomorphic function Φ on* $\mathbf{C}_+^l$ *such that 1) $\Phi = \phi$ on V and 2) $\partial\Phi/\partial t_i = \Gamma_i(\mathbf{t})\Phi$ ($1 \leq i \leq l$) on* $\mathbf{C}_+^l$.

The proof depends on the fact that $\mathbf{C}_+^l$ is a simply connected subdomain in $\mathbf{C}$.

Now let $c_i(\mathbf{k})$ ($1 \leq i \leq l$, $\mathbf{k} \in \mathbf{Z}_+^l$) be elements in $\mathcal{E}(U)$ such that $\sum_i \sum_{\mathbf{k}} |c_i(\mathbf{k})| e^{-s(\mathbf{k})\eta} < \infty$ for every $\eta > 0$. Put

$$C_i(\mathbf{t}) = \sum_{\mathbf{k}} c_i(\mathbf{k}) e^{-\langle \mathbf{k},\, \mathbf{t} \rangle} \qquad (\mathbf{t} \in \mathbf{C}_+^l).$$

Then obviously C_i is a holomorphic function on $\mathbf{C}_+^l$ with values in $\mathcal{E}(U)$. Let $W(\varepsilon, \eta)$ ($0 < \varepsilon < 1$, $\eta > 0$) denote the set of all $\mathbf{t} \in \mathbf{C}^l$ such that $\mathrm{Re}\, t_i > \max(\varepsilon |\mathbf{t}|, \eta)$ ($1 \leq i \leq l$).

THEOREM 2. *Let Φ be a holomorphic function on* $\mathbf{C}_+^l$ *such that $\partial\Phi/\partial t_i = C_i(\mathbf{t})\Phi$ ($1 \leq i \leq l$). Then we can choose a set Λ of points in* $\mathbf{C}^l$ *and, for each $\lambda \in \Lambda$, a U-polynomial p_λ such that the following conditions are fulfilled.*

(1) $p_\lambda \neq 0$ *for every* $\lambda \in \Lambda$.

(2) *For any* ε, η *($0 < \varepsilon < 1$, $\eta > 0$) and $\mu \in \mathbf{C}^l$, the series* $\sum_\lambda |p_\lambda(\mathbf{t}) e^{\langle \lambda + \mu,\, \mathbf{t} \rangle}|$

converges uniformly on $W(\varepsilon, \eta)$.

(3) $\Phi(\mathbf{t}) = \sum_\lambda p_\lambda(\mathbf{t}) e^{\langle \lambda,\, \mathbf{t} \rangle}$ *($\mathbf{t} \in \mathbf{C}_+^l$).*

Moreover the set[2] Λ and the polynomials p_λ ($\lambda \in \Lambda$) are uniquely determined by these conditions.

Let us call an element $\lambda \in \Lambda$ maximal if $\lambda + \mathbf{k}$ does not lie in Λ for any $\mathbf{k} \neq 0$ in $\mathbf{Z}_+^l$. Let Λ_0 denote the set of all maximal elements in Λ.

THEOREM 3. Λ_0 *is a finite set and every element in Λ is of the form $\lambda_0 - \mathbf{k}$ where*

5

$\lambda_0 \epsilon \Lambda_0$ *and* $\mathbf{k} \epsilon \mathbf{Z}_+{}'$. *Moreover the degrees of the U-polynomials* $p_\lambda (\lambda \epsilon \Lambda)$ *remain bounded and we can choose positive numbers* a, b *such that* $\|p_\lambda\| \leq ab^{-s(\lambda)}$ *for every* $\lambda \epsilon \Lambda$.

We call the function $\Phi_0(t) = \sum_{\lambda \epsilon \Lambda_0} p_\lambda(t) e^{\langle \lambda, t \rangle}$ the principal part of Φ. Notice that $\Phi_0 \neq 0$ unless $\Phi = 0$.

We shall now present some applications of these results to the theory of spherical functions on a connected semisimple Lie group G. Suppose the center of G is finite and K is a maximal compact subgroup of G. Let $\sigma = (\sigma_1, \sigma_2)$ be a (continuous) double representation of K on a finite-dimensional (complex) vector space U and let V be an open set in G such that $KVK = V$. By a spherical function ϕ on V of type σ, we mean a continuous function with values in U such that $\phi(k_1 x k_2) = \sigma_1(k_1)\phi(x)\sigma_2(k_2)$ for $x \epsilon V$ and $k_1, k_2 \epsilon K$. Put[3] $G^+ = KA_\mathfrak{p}{}^+ K$. Then G^+ is open in G.

LEMMA 1. *Suppose* $V \subseteq G^+$ *and* ϕ *is a spherical function on* V *of type* σ *and class* C^∞ *such that* ϕ *is an eigenfunction of every differential operator in*[3] $\mathfrak{Z}$. *Then if* V *is connected and nonempty,* ϕ *can be extended to an analytic function on* G^+ *which is also spherical of type* σ.

This is a consequence of Theorem 1.

Let π be a quasi-simple[4] irreducible representation of G on a Banach space B. Let $\mathfrak{b}$ be an equivalence class of finite-dimensional irreducible representations of K and $E_\mathfrak{b}$ the canonical projection[4] in B corresponding to $\mathfrak{b}$. We assume that $\mathfrak{b}$ is so chosen that $B_\mathfrak{b} = E_\mathfrak{b} B \neq \{0\}$. Let $\pi_\mathfrak{b}(k)$ $(k \epsilon K)$ denote the restriction of $\pi(k)$ on $B_\mathfrak{b}$. Then we have a double representation $\sigma = (\sigma_1, \sigma_2)$ of K on $\mathcal{E}(B_\mathfrak{b})$ given by $\sigma_1(k_1)T\sigma_2(k_2) = \pi_\mathfrak{b}(k_1)T\pi_\mathfrak{b}(k_2)(T \epsilon \mathcal{E}(B_\mathfrak{b}); k_1, k_2 \epsilon K)$. Put $\phi(x) = E_\mathfrak{b}\pi(x)E_\mathfrak{b}(x \epsilon G)$. Then ϕ can be regarded as a spherical function on G of type σ. Let M denote the centralizer of[3] $A_\mathfrak{p}$ in K and $d(\mathfrak{b})$ the degree of any representation in $\mathfrak{b}$.

THEOREM 4. Dim $B_\mathfrak{b} \leq d(\mathfrak{b})^2$ *and we can choose a linear function* λ *on*[3] $\mathfrak{h}_p$ *and an element* $S \epsilon \mathcal{E}(B_\mathfrak{b})$ *such that* S *commutes with* $\pi_\mathfrak{b}(m)$ $(m \epsilon M)$ *and*[3]

$$\phi(x) = \int_K \pi_\mathfrak{b}(k_x)S\pi_\mathfrak{b}(k^{-1})e^{\lambda(H(xk))}dk \qquad (x \epsilon G).$$

Here dk *is the normalized Haar measure on* K.

This is an extension of an earlier result.[5] The proof depends on the notion of the principal part introduced above.

Let τ be a representation of K on a finite-dimensional space U. Then we can define a double representation $\sigma = (\sigma_1, \sigma_2)$ of K on $\mathcal{E}(U)$ as follows: $\sigma_1(k_1)T\sigma_2(k_2) = \tau(k_1)T\tau(k_2)(k_1, k_2 \epsilon K, T \epsilon \mathcal{E}(U))$. Let $\mathcal{E}_0$ denote the set of all elements in $\mathcal{E}(U)$ which commute with $\tau(m)$ for every $m \epsilon M$. Fix $T \epsilon \mathcal{E}_0$ and consider the function[3]

$$\phi_{H'}(x) = \int_K \tau(k_x)T\tau(k^{-1}) \exp\left\{(-1)^{1/2}\langle H', H(xk)\rangle - \rho(H(xk))\right\}dk$$

for $H'\epsilon\mathfrak{h}_p$ and $x\epsilon G$. Then for a fixed H', $\phi_{H'}$ is a spherical function of type σ. Theorem 2 enables us to obtain a formula[6] for $\phi_{H'}$ which expresses its asymptotic behaviour very clearly.

* This work was supported by a grant from the National Science Foundation and a contract with the U. S. Air Force.

[1] Rec denotes the real part of a complex number c.

[2] Note that Λ is empty in case $\Phi = 0$.

[3] See *Amer. Jour. Math.*, **80**, 241–310 (1958), for the meaning of undefined symbols.

[4] See *Trans. Amer. Math. Soc.*, **75**, 185–243 (1953).

[5] See *Trans. Amer. Math. Soc.*, **76**, 63 (1954).

[6] This is a generalization of Lemma 37 of the paper in reference 3.

Some Results on Differential Equations[†]

§1.   Introduction

Our object here is to obtain some results on certain systems of differential equations, which will then be applied in another paper to the theory of spherical functions on a semisimple Lie group. Although these results are not difficult to derive, they seem to be new. This is perhaps due to the fact that unless one had a definite application in mind, it would not have been easy to know what to look for.

The problem which we consider here can be described in the simplest case as follows. Let U be a finite-dimensional complex Banach space and $\mathcal{E}(U)$ the algebra of all endomorphisms of U. Let ϕ and $\mathcal{C}$ be functions of a real variable $t > 0$ with values in U and $\mathcal{E}(U)$ respectively and suppose $d\phi/dt = \mathcal{C}(t)\phi(t)$. We assume that $\mathcal{C}$ has an expansion of the form

$$\mathcal{C}(t) = \sum_{k \geqslant 0} c_k e^{-kt} \qquad \left(c_k \in \mathcal{E}(U) \right)$$

where k runs over all nonnegative integers and the series is absolutely convergent for $t > 0$. Then it is not difficult to prove that ϕ can be expanded in the form

$$\phi(t) = \sum_{1 \leqslant i \leqslant r} \sum_{k \geqslant 0} p_{ik}(t) e^{(\lambda_i - k)t}$$

and this relation holds for all sufficiently large positive values of t. (Actually we shall prove that it is valid for all $t > 0$.) Here $(\lambda_1, \ldots, \lambda_r)$ is a finite set of complex numbers and p_{ik} are polynomials in t with values in U.

We shall extend the above result to the case of several variables. Our method is straightforward and uses induction on the number of variables. The precise form in which we choose to state our result (see Theorem 1) is, of course, largely determined by the application mentioned above.

The contents of this paper are as follows. In Section 2 we introduce two concepts which play an important role throughout this paper, namely the notions of decent convergence and of a normal sequence of polynomials. Lemma 7 contains a simple and well-known result which reduces a differential equation to an integral equation. This lemma is the key to our entire method. It allows us to prove (see Lemma 10)

[†] This paper and the following one were written in 1960. Due to various circumstances their publication was postponed and eventually abandoned altogether. They are being reproduced here without any additions or revisions. —V.

that solutions of certain differential equations, which are defined on a small connected open set, can always be extended by analytic continuation to a much larger set. Our induction argument is based on Lemma 11 and the main results of the paper are contained in Theorems 1 and 2. The proof of Lemma 11 is long and requires considerable preparation. This proof is given in Section 5.

This work was supported at various stages by grants from the Guggenheim Memorial Foundation, the National Science Foundation and a contract with the U.S. Air Force. A short account of the results of this paper has been published in [2].

§2. Decent Convergence

Let $\mathbf{R}$ and $\mathbf{C}$ be the fields of real and complex numbers respectively and $\mathbf{Z}_+$ the set of nonnegative integers. For any integer $l \geqslant 1$ we denote by $\mathbf{C}^l$ the Cartesian product of $\mathbf{C}$ with itself l times. Similarly for $\mathbf{R}^l$ and $\mathbf{Z}_+^l$. For $\mathbf{t} = (t_1, \ldots, t_l)$ and $\mathbf{t}' = (t_1', \ldots, t_l')$ in $\mathbf{C}^l$, put $\langle \mathbf{t}, \mathbf{t}' \rangle = t_1 t_1' + \cdots + t_l t_l'$ and $|\mathbf{t}| = \sup_j |t_j|$. With this norm $\mathbf{C}^l$ becomes a complex Banach space. We set[1] $\sigma(\mathbf{t}) = \mathrm{Re}(t_1 + \cdots + t_l)$.

Let W be an open set in $\mathbf{C}^l$ and $\{f_j\}_{j \in J}$ an (indexed) family of holomorphic functions on W with values in a complex Banach space U of finite dimension. We shall say that the series $\sum_{j \in J} f_j$ converges *decently* on W if the following condition holds. Put $D_{\mathbf{k}} = \partial^{k_1 + \cdots + k_l}/\partial t_1^{k_1} \cdots \partial t_l^{k_l}$ for $\mathbf{k} \in \mathbf{Z}_+^l$. Then for any $\lambda \in \mathbf{C}^l$ and $\mathbf{k} \in \mathbf{Z}_+^l$, the series $\sum_{j \in J} |e^{\langle \lambda, \mathbf{t} \rangle} D_{\mathbf{k}} f_j|$ should converge uniformly on W.

Let ω be any subset of $\mathbf{C}^l$ and $\{f_j\}_{j \in J}$ a family of functions on ω with values in U. We say that the series $\sum_{j \in J} f_j$ converges decently on ω, if we can choose an open neighborhood W of ω in $\mathbf{C}^l$ and for each $j \in J$, a holomorphic function F_j on W such that 1) $F_j = f_j$ on ω and 2) the series $\sum_{j \in J} F_j$ converges decently on W.

Lemma 1. *Let W and W' be open sets in $\mathbf{C}^l$ such that $W \subset W'$. Let d denote the distance between W and the complement of W' in $\mathbf{C}^l$. Let $\{f_j\}_{j \in J}$ be a family of holomorphic functions on W' with values in U such that for every $\lambda \in \mathbf{C}^l$, the series $\sum_{j \in J} |e^{\langle \lambda, \mathbf{t} \rangle} f_j|$ converges uniformly on W'. Then if $d > 0$, the series $\sum_{j \in J} f_j$ converges decently on W.*

This is an easy consequence of Lemma 58 of [1].

Lemma 2. *Let S be a set and $\{f_i\}_{i \in I}$, $\{g_j\}_{j \in J}$ two families of functions on S with nonnegative real values. Suppose the series $\sum_i f_i, \sum_j g_j$ both converge uniformly on S and their sums remain bounded on S. Then the series $\sum_{i,j} f_i g_j$ also converges uniformly on S and its sum remains bounded on S.*

Let A, B be two positive numbers such that $\sum_i f_i \leqslant A$, $\sum_j g_j \leqslant B$ on S. For a given $\varepsilon > 0$, choose finite subsets F and G of I and J respectively, such that[2] $\sum_{i \in {}^cF} f_i \leqslant \varepsilon$

---

[1] Re c denotes the real part of a complex number c.

[2] ${}^cF$ denotes the complement of F in I. Similarly for the others.

and $\sum_{j \in {}^cG} g_j \leqslant \varepsilon$. Then $H = F \times G$ is a finite subset of $I \times J$ and

$$\sum_{(i,j) \in {}^cH} f_i g_j \leqslant \sum_i f_i \sum_{j \in {}^cG} g_j + \sum_{i \in {}^cF} f_i \sum_j g_j \leqslant \varepsilon(A + B).$$

Also, $\sum_{i,j} f_i g_j \leqslant AB$.

Lemma 3. *Let W be a set in $\mathbf{C}^l$ and $\{f_i\}_{i \in I}$, $\{g_j\}_{j \in J}$ two families of complex-valued functions on W such that*

(1) *the series $\sum_i |f_i e^{\langle \lambda, \mathbf{t} \rangle}|$ and $\sum_j |g_j e^{\langle \lambda, \mathbf{t} \rangle}|$ converge uniformly on W for every $\lambda \in \mathbf{C}^l$;*

(2) *there exist elements μ, ν in $\mathbf{C}^l$ and positive constants a, b such that $\sum_i |f_i| \leqslant a|e^{\langle \mu, \mathbf{t} \rangle}|$ and $\sum_j |g_j| \leqslant b|e^{\langle \nu, \mathbf{t} \rangle}|$ on W.*

Then for every $\lambda \in \mathbf{C}^l$, the series $\sum_{i,j} |f_i g_j e^{\langle \lambda, \mathbf{t} \rangle}|$ converges uniformly on W.

Fix λ and for a given $\varepsilon > 0$, choose finite subsets F and G of I and J respectively such that

$$\sum_{i \in {}^cF} |f_i e^{\langle \lambda + \nu, \mathbf{t} \rangle}| \leqslant \varepsilon, \quad \sum_{j \in {}^cG} |g_j e^{\langle \lambda + \mu, \mathbf{t} \rangle}| \leqslant \varepsilon$$

on W. Then if $H = F \times G$,

$$\sum_{(i,j) \in {}^cH} |f_i g_j e^{\langle \lambda, \mathbf{t} \rangle}| \leqslant \sum_{i \in {}^cF} |f_i e^{\langle \lambda + \nu, \mathbf{t} \rangle}| \sum_j |g_j e^{-\langle \nu, \mathbf{t} \rangle}|$$

$$+ \sum_{i \in F} |f_i e^{-\langle \mu, \mathbf{t} \rangle}| \sum_{j \in {}^cG} |g_j e^{\langle \lambda + \mu, \mathbf{t} \rangle}| \leqslant \varepsilon(b + a).$$

This proves our assertion.

By a *U-polynomial* (on $\mathbf{C}^l$) we mean a polynomial function on $\mathbf{C}^l$ with values in U. Put $\mathbf{t}^{\mathbf{k}} = t_1^{k_1} t_2^{k_2} \cdots t_l^{k_l}$ for $\mathbf{t} \in \mathbf{C}^l$ and $\mathbf{k} \in \mathbf{Z}_+^l$. Let $p = \sum_{\mathbf{k}} u(\mathbf{k}) \mathbf{t}^{\mathbf{k}}$ $(u(\mathbf{k}) \in U)$ be a U-polynomial. We set $\|p\| = \sum_{\mathbf{k}} |u(\mathbf{k})|$. If $p \neq 0$, the degree of p, which we denote by $d^0 p$, is defined, as usual, to be the least integer $d \geqslant 0$ such that $u(\mathbf{k}) = 0$ for all $\mathbf{k}$ with $\sigma(\mathbf{k}) > d$. Also it is convenient to put $d^0 p = -\infty$ in case $p = 0$.

Suppose for each $\mathbf{k} \in \mathbf{Z}_+^l$, we are given a U-polynomial $v_{\mathbf{k}}$. We say that $v_{\mathbf{k}}$ is a *normal sequence* (of U-polynomials) if the following two conditions hold.

(1) $\sup_{\mathbf{k}} d^0 v_{\mathbf{k}} < \infty$.

(2) There exists a positive number a such that

$$\|v_{\mathbf{k}}\| \leqslant a^{\sigma(\mathbf{k}) + 1}$$

for all $\mathbf{k} \in \mathbf{Z}_+^l$.

Note that $|p(\mathbf{t})| \leqslant \|p\|(1 + |\mathbf{t}|)^d$ for a U-polynomial p if $d \geqslant d^0 p$.

For any $M \geqslant 0$ and $0 < \varepsilon < 1$, let $W_l(\varepsilon, M)$ denote the set of all points $\mathbf{t} \in \mathbf{C}^l$ such that $\mathrm{Re}\, t_i > \max(\varepsilon|\mathbf{t}|, M)$ $(1 \leqslant i \leqslant l)$. Then $W_l(\varepsilon, M)$ is a nonempty open set in $\mathbf{C}^l$. Moreover if $\mathbf{t}$ is in $W_l(\varepsilon, M)$, the same holds for $\alpha \mathbf{t}$ for any number $\alpha \geqslant 1$.

Lemma 4. *Let v_k ($k \in \mathbb{Z}_+^l$) be a normal sequence of U-polynomials on $\mathbb{C}^l$. Then for any ε ($0 < \varepsilon < 1$) we can choose $M \geqslant 0$ such that the series*

$$\sum_k v_k e^{-\langle k, t \rangle}$$

converges decently on $W_l(\varepsilon, M)$.

Choose $M > 1 + 2\varepsilon^{-1}$ and put $W = W_l(\varepsilon, M)$, $W' = W_l(\varepsilon/2, M-1)$. Then $W' \supset W$. Fix t in W and suppose t' is a point in $\mathbb{C}^l$ such that $|t - t'| \leqslant 1$. We claim $t' \in W'$. For $\operatorname{Re} t'_j \geqslant \operatorname{Re} t_j - 1 > M - 1$. Moreover $\operatorname{Re} t'_i > (\varepsilon/2)|t'_j|$ ($1 \leqslant i, j \leqslant l$). This is seen as follows. Obviously $\operatorname{Re} t'_i \geqslant \operatorname{Re} t_i - 1$ and $|t'_j| \leqslant |t_j| + 1$. Hence it is sufficient to verify that $\operatorname{Re} t_i - 1 > (\varepsilon/2)(|t_j| + 1)$. But $\operatorname{Re} t_i - 1 > \varepsilon |t_j| - 1$ since $t \in W$. So it is enough to show that $\varepsilon |t_j| - 1 > (\varepsilon/2)(|t_j| + 1)$ or $(\varepsilon/2)|t_j| > 1 + (\varepsilon/2)$. But this is obvious since $|t_j| > M > 1 + 2\varepsilon^{-1}$. Therefore, in view of Lemma 1, it is sufficient to prove that the series $\sum_k |v_k e^{\langle \lambda - k, t \rangle}|$ converges uniformly on W' for every $\lambda \in \mathbb{C}^l$, provided M is sufficiently large.

Now $|\langle \lambda, t \rangle| \leqslant l|\lambda||t|$. On the other hand if $t \in W'$,

$$\operatorname{Re}\langle k, t \rangle = \sum_{1 \leqslant i \leqslant l} k_i \operatorname{Re} t_i \geqslant \frac{\varepsilon}{2} \sigma(k)|t|.$$

Therefore

$$\operatorname{Re}\langle \lambda - k, t \rangle \leqslant \left\{ l|\lambda| - \frac{\varepsilon}{2} \sigma(k) \right\} |t|.$$

Now $l|\lambda| - 4^{-1}\varepsilon \sigma(k) < 0$ for all $k \in \mathbb{Z}_+^l$ except a finite number. So it would be enough to select M in such a way that the series

$$\sum_k |v_k(t)| e^{-\varepsilon \sigma(k)|t|/4}$$

converges uniformly on W'. Since v_k is a normal sequence, we can choose an integer $d \geqslant 0$ and a number $a \geqslant 1$ such that $d^0 v_k \leqslant d$ and $\|v_k\| \leqslant a^{\sigma(k)+1}$ for all $k \in \mathbb{Z}_+^l$. Then

$$|v_k(t)| \leqslant \|v_k\|(1 + |t|)^d \leqslant a^{2\sigma(k)}(1 + |t|)^{d\sigma(k)}$$

if $\sigma(k) \geqslant 1$. Now choose $M > 1 + 2\varepsilon^{-1}$ so large that $a^2(1 + s)^d e^{-\varepsilon s/4} \leqslant \frac{1}{2}$ for $s \geqslant M - 1$. Then $|v_k| e^{-\varepsilon \sigma(k)|t|/4} \leqslant 2^{-\sigma(k)}$ if $\sigma(k) \geqslant 1$ and $|t| \geqslant M - 1$. Since $\sum_k 2^{-\sigma(k)} = 2^l$, it follows that the series

$$\sum_k |v_k(t)| e^{-\varepsilon \sigma(k)|t|/4}$$

converges uniformly on W'. This proves the lemma.

Put $\lambda = 0$ in the above proof. Then

$$\sum_k |v_k(t) e^{-\langle k, t \rangle}| = |v_0(t)| + \sum_{\sigma(k) \geqslant 1} |v_k(t) e^{-\langle k, t \rangle}|$$

$$\leqslant \|v_0\|(1 + |t|)^d + \sum_{\sigma(k) \geqslant 1} 2^{-\sigma(k)} \leqslant e^{\varepsilon|t|/4} + (2^l - 1) \leqslant 2^l e^{\varepsilon|t|/4}$$

$$\leqslant 2^l |e^{t_1/4}| \quad \text{for } t \in W'.$$

Let η be a positive number. Then we can choose a compact subset Ω of $\mathbb{C}_+^l$ such that $W_l(\varepsilon, \eta) \subset \Omega \cup W_l(\varepsilon, M)$. Therefore if we put $\mathbf{1} = (1, 1, \ldots, 1) \in \mathbb{C}^l$, we get the following result.

Lemma 5. *For any $\eta > 0$, we can choose a number $b > 0$ such that*

$$\sum_{\mathbf{k}} |v_{\mathbf{k}} e^{-\langle \mathbf{k}, \mathbf{t}\rangle}| \leqslant b |e^{\langle \mathbf{l}, \mathbf{t}\rangle}| = b e^{\sigma(\mathbf{t})}$$

for $\mathbf{t} \in W_l(\varepsilon, \eta)$.

Let $\mathcal{E}(U)$ denote the algebra of all endomorphisms of U. We put $|T| = \sup_{|u| \leqslant 1} |Tu|$ $(u \in U)$ for $T \in \mathcal{E}(U)$. In this way $\mathcal{E}(U)$ also becomes a Banach space.

Lemma 6. *Let r be an integer such that $1 \leqslant r \leqslant l$. Suppose $P_{\mathbf{k}}$ $(\mathbf{k} \in \mathbb{Z}_{+}^l)$ is a normal sequence of $\mathcal{E}(U)$-polynomials on $\mathbb{C}^l$ and $q_{\mathbf{m}}$ $(\mathbf{m} \in \mathbb{Z}_{+}^r)$ a normal sequence of U-polynomials on $\mathbb{C}^r$. We regard $q_{\mathbf{m}}$ as a polynomial on $\mathbb{C}^l$ by setting $q_{\mathbf{m}}(\mathbf{t}) = q_{\mathbf{m}}(t_1, \ldots, t_r)$ $(\mathbf{t} \in \mathbb{C}^l)$. Also if $\mathbf{k} \in \mathbb{Z}_{+}^l$ and $\mathbf{m} \in \mathbb{Z}_{+}^r$, we denote by $\mathbf{k} + \mathbf{m}$ the element $(k_1 + m_1, \ldots, k_r + m_r, k_{r+1}, \ldots, k_l)$ in $\mathbb{Z}_{+}^l$. Put*

$$p_{\mathbf{k}}(\mathbf{t}) = \sum P_{\mathbf{k}}(\mathbf{t}) q_{\mathbf{m}}(\mathbf{t}) \qquad \left(\mathbf{t} \in \mathbb{C}^l,\ \mathbf{k} \in \mathbb{Z}_{+}^l\right)$$

where the sum is over all pairs $(\mathbf{k}', \mathbf{m})$ $(\mathbf{k}' \in \mathbb{Z}_{+}^l,\ \mathbf{m} \in \mathbb{Z}_{+}^r)$ such that $\mathbf{k}' + \mathbf{m} = \mathbf{k}$. Then $p_{\mathbf{k}}$ is a normal sequence of U-polynomials on $\mathbb{C}^l$.

Choose an integer $d \geqslant 0$ and a number $a \geqslant 1$ such that $d^0 P_{\mathbf{k}} \leqslant d$, $\|P_{\mathbf{k}}\| \leqslant a^{\sigma(\mathbf{k})+1}$, $d^0 q_{\mathbf{m}} \leqslant d$ and $\|q_{\mathbf{m}}\| \leqslant a^{\sigma(\mathbf{m})+1}$ for all $\mathbf{k} \in \mathbb{Z}_{+}^l$ and $\mathbf{m} \in \mathbb{Z}_{+}^r$. Then it is obvious that $d^0 p_{\mathbf{k}} \leqslant 2d$ and

$$\|p_{\mathbf{k}}\| \leqslant a^2 \sum a^{\sigma(\mathbf{k}') + \sigma(\mathbf{m})}$$

$$\leqslant a^2 (2a)^{\sigma(\mathbf{k})} \sum 2^{-\sigma(\mathbf{k}') - \sigma(\mathbf{m})}$$

$$\leqslant a^2 (2a)^{\sigma(\mathbf{k})} 2^{l+r}$$

where the sum is over all pairs $(\mathbf{k}', \mathbf{m})$ with $\mathbf{k}' + \mathbf{m} = \mathbf{k}$. Hence if we put $A = 2^{l+r} a^2$, $\|p_{\mathbf{k}}\| \leqslant (A)^{\sigma(\mathbf{k})+1}$. This proves our assertion.

Following is a simple and well-known result in the theory of differential equations which we shall use frequently.

Lemma 7. *Let $\mathcal{T}$ denote the compact interval $a \leqslant t \leqslant b$ in $\mathbb{R}$ and let $\mathcal{C}$ be a continuous function from $\mathcal{T}$ to $\mathcal{E}(U)$. Suppose ϕ is a function from $\mathcal{T}$ to U such that*

$$\frac{d\phi}{dt} = \mathcal{C}(t)\phi$$

on $\mathcal{T}$. Put $K_0 = 1 \in \mathcal{E}(U)$ and

$$K_q = \int_{b \geqslant t_1 \geqslant \cdots \geqslant t_q \geqslant a} \mathcal{C}(t_1)\mathcal{C}(t_2)\cdots \mathcal{C}(t_q)\, dt_1 \cdots dt_q \qquad (q \geqslant 1).$$

Then $\sum_{q \geqslant 0} |K_q| < \infty$ and if $K = \sum_{q \geqslant 0} K_q$, $\phi(b) = K\phi(a)$.

We give a proof for completeness. Put

$$K_q(t) = \int_{t \geqslant t_1 \geqslant \cdots \geqslant t_q \geqslant a} \mathcal{C}(t_1)\cdots \mathcal{C}(t_q)\, dt_1 \cdots dt_q \qquad (q \geqslant 1,\ t \in \mathcal{T}).$$

It is obvious from the above differential equation that ϕ is continuous and

$$\phi(t) = \phi(a) + \int_a^t \mathcal{C}(t_1)\phi(t_1)\,dt_1.$$

We claim that

$$\phi(t) = \phi(a) + \sum_{1 \leqslant r < q} K_r(t)\phi(a) + \int_{t \geqslant t_1 \geqslant \cdots \geqslant t_q \geqslant a} \mathcal{C}(t_1)\cdots \mathcal{C}(t_q)\phi(t_q)\,dt_1\cdots dt_q$$

for any $q \geqslant 1$ and $t \in \mathcal{T}$. For $q = 1$ this is the relation just obtained above. Assuming it for q one proves it for $q+1$ by replacing $\phi(t_q)$ by the expression

$$\phi(a) + \int_a^{t_q} \mathcal{C}(t_{q+1})\phi(t_{q+1})\,dt_{q+1}.$$

Let $M = \sup_{a \leqslant t \leqslant b}|\mathcal{C}(t)|$ and $N = \sup_{a \leqslant t \leqslant b}|\phi(t)|$. Then

$$|K_q| \leqslant M^q \int_{b \geqslant t_1 \geqslant \cdots \geqslant t_q \geqslant a} dt_1 \cdots dt_q = M^q \frac{(b-a)^q}{q!}$$

and

$$\left|\int_{b \geqslant t_1 \geqslant \cdots \geqslant t_q \geqslant a} \mathcal{C}(t_1)\cdots\mathcal{C}(t_q)\phi(t_q)\,dt_1\cdots dt_q\right| \leqslant NM^q \frac{(b-a)^q}{q!}$$

Hence it is clear that

$$\left|\phi(b) - \sum_{0 \leqslant r < q} K_r\phi(a)\right| \to 0$$

as $q \to \infty$. Also $\sum_{q \geqslant 0}|K_q| \leqslant e^{M(b-a)} < \infty$. The statement of this lemma is now obvious.

Remark. One proves similarly that

$$\phi(a) = \phi(b) + \sum_{q \geqslant 1} (-1)^q \int_{a \leqslant t_1 \leqslant t_2 \leqslant \cdots \leqslant t_q \leqslant b} \mathcal{C}(t_1)\mathcal{C}(t_2)\cdots\mathcal{C}(t_q)\phi(b)\,dt_1\cdots dt_q.$$

§3. Analytic Continuation of Solutions

By a domain in $\mathbf{C}^l$ we mean, as usual, a nonempty open connected subset of $\mathbf{C}^l$.

Lemma 8. *Let D be a simply connected domain in $\mathbf{C}$ and let $\mathcal{C}$ be a holomorphic function on D with values in $\mathcal{E}(U)$. Fix a point $z_0 \in D$. Then there exists exactly one holomorphic function Γ on D with values in $\mathcal{E}(U)$ such that*[3] *1) $\Gamma(z_0) = 1$ and 2) $d\Gamma(z)/dz = \mathcal{C}(z)\Gamma(z)\,(z \in D)$.*

Let f be a holomorphic function from D to $\mathcal{E}(U)$. Since D is simply connected, it is clear that there exists exactly one holomorphic function F from D to $\mathcal{E}(U)$ such that 1) $F(z_0) = 0$ and 2) $dF/dz = f$ on D. For every integer $q \geqslant 0$, we now define a

---

[3] Here 1 denotes the identity mapping of U which is the unit element of the algebra $\mathcal{E}(U)$.

holomorphic function γ_q on D (with values in $\mathcal{E}(U)$) by induction on q as follows. $\gamma_0 = 1$ and γ_q $(q \geqslant 1)$ is defined by the conditions that $\gamma_q(z_0) = 0$ and $d\gamma_q/dz = \mathcal{C}(z)\gamma_{q-1}(z)$ $(z \in D)$. We shall now show that the series $\Sigma_{q \geqslant 0}|\gamma_q(z)|$ converges uniformly on every compact subset of D. Fix a point z_1 in D and a positive number ε. Let V be the set of all points $z \in \mathbf{C}$ such that $|z - z_1| \leqslant \varepsilon$. We assume that ε is so small that $V \subset D$. It would be sufficient to prove that the above series converges uniformly on V. Let $\zeta(t)$ $(a \leqslant t \leqslant b)$ be a piecewise differentiable curve in D such that $\zeta(a) = z_0$. Put $\zeta(b) = z$. Then one proves by an easy induction on q that

$$\gamma_q(z) = \int_a^b \mathcal{C}(\zeta(t))\gamma_{q-1}(\zeta(t))\zeta'(t)\, dt$$

$$= \int_{b \geqslant t_1 \geqslant t_2 \geqslant \cdots \geqslant t_q \geqslant a} \mathcal{C}(\zeta(t_1))\mathcal{C}(\zeta(t_2))\cdots \mathcal{C}(\zeta(t_q))\zeta'(t_1)\cdots \zeta'(t_q)\, dt_1 \cdots dt_q$$

where $\zeta'(t) = d\zeta(t)/dt$. Let $M_\zeta = \sup_{a \leqslant t \leqslant b}|\mathcal{C}(\zeta(t))|$. Then

$$|\gamma_q(z)| \leqslant M_\zeta^q \int_{b \geqslant t_1 \geqslant \cdots \geqslant t_q \geqslant a} |\zeta'(t_1)\cdots \zeta'(t_q)|\, dt_1 \cdots dt_q.$$

Let s denote the arc-length along ζ so that $ds/dt = |\zeta'(t)|$. Then if L is the total length of ζ, it is clear that

$$\int_{b \geqslant t_1 \geqslant \cdots \geqslant t_q \geqslant a} |\zeta'(t_1)\cdots \zeta'(t_q)|dt_1 \cdots dt_q = \int_{L \geqslant s_1 \geqslant \cdots \geqslant s_q \geqslant 0} ds_1 \cdots ds_q = \frac{L^q}{q!}.$$

This shows that

$$|\gamma_q(z)| \leqslant (M_\zeta L)^q/q!$$

Now fix a curve $\zeta(t)$ $(a \leqslant t \leqslant b)$ in D such that $\zeta(a) = z_0$, $\zeta(b) = z_1$ and let $M = \sup_{a \leqslant t \leqslant b}|\mathcal{C}(\zeta(t))|$. z being any point in V we define a curve $\eta(t)$ $(a \leqslant t \leqslant b+1)$ as follows: $\eta(t) = \zeta(t)$ $(a \leqslant t \leqslant b)$ and $\eta(t) = (1 + b - t)z_1 + (t - b)z$ for $b \leqslant t \leqslant b + 1$. Then η is obtained by following the curve ζ by the straight line segment joining z_1 to z. Clearly η is also a piecewise differentiable curve in D and if L is the length of ζ, the length of η is $L + |z - z_1| \leqslant L + \varepsilon$. Moreover $M_\eta \leqslant \max(M, \sup_{z \in V}|\mathcal{C}(z)|) = N$ (say). Therefore

$$|\gamma_q(z)| \leqslant N^q(L + \varepsilon)^q/q! \qquad (q \geqslant 0).$$

This proves that the series $\Sigma_{q \geqslant 0}|\gamma_q(z)|$ converges uniformly on V.

Now put $\Gamma(z) = \Sigma_{q \geqslant 0}\gamma_q(z)$ $(z \in D)$. Then in view of the uniform convergence proved above, Γ is a holomorphic function on D and

$$\frac{d\Gamma(z)}{dz} = \sum_{q \geqslant 0} \frac{d\gamma_q(z)}{dz} = \sum_{q \geqslant 1} \mathcal{C}(z)\gamma_{q-1}(z) = \mathcal{C}(z)\Gamma(z).$$

Also $\Gamma(z_0) = \gamma_0(z_0) = 1$.

Now we come to the uniqueness of Γ. Fix a point $z_1 \in D$ and let $\zeta(t)$ $(a \leqslant t \leqslant b)$ be a differentiable curve in D such that $\zeta(a) = z_0$ and $\zeta(b) = z_1$. Let Γ be any holomorphic function on D such that $d\Gamma/dz = \mathcal{C}(z)\Gamma(z)$ on D. Then if we put $\gamma(t) = \Gamma(\zeta(t))$, it is clear that $d\gamma(t)/dt = \mathcal{C}(\zeta(t))\gamma(t)\zeta'(t)$. Hence we can conclude from Lemma 7 that

$$\Gamma(z_1) = \gamma(b) = 1 + \sum_{q \geqslant 1} \int_{b \geqslant t_1 \geqslant \cdots \geqslant t_q \geqslant a} \mathcal{C}\big(\zeta(t_1)\big)\mathcal{C}\big(\zeta(t_2)\big)$$

$$\cdots \mathcal{C}\big(\zeta(t_q)\big)\zeta'(t_1)\cdots\zeta'(t_q)\, dt_1 \cdots dt_q$$

since $\gamma(a) = \Gamma(z_0) = 1$. This proves that Γ is unique.

Lemma 9. *Suppose D and z_0 have the same meaning as in Lemma 8 and Λ is an open set in $\mathbf{C}^r$ $(r \geqslant 1)$. Assume that $\mathcal{C}(z, \lambda)$ $(z \in D, \lambda \in \Lambda)$ is a holomorphic function on $D \times \Lambda$ with values in $\mathcal{E}(U)$. For a fixed $\lambda \in \Lambda$, let $\Gamma(z, \lambda)$ denote the function Γ of Lemma 8 corresponding to $\mathcal{C}(z, \lambda)$. Then $\Gamma(z, \lambda)$ is a holomorphic function of (z, λ) on $D \times \Lambda$.*

Define $\gamma_q(z, \lambda)$ $(q \geqslant 0)$ corresponding to the function γ_q of the proof of Lemma 8. One proves without difficulty by induction on q, that $\gamma_q(z, \lambda)$ is a holomorphic function on $D \times \Lambda$. Moreover a repetition of the above argument shows that the series $\sum_q |\gamma_q(z, \lambda)|$ converges uniformly on every compact subset of $D \times \Lambda$. Hence $\Gamma(z, \lambda) = \sum_q \gamma_q(z, \lambda)$ is holomorphic on $D \times \Lambda$.

For any $M \geqslant 0$, we denote by $\mathbf{C}^l(M)$ the set of all $\mathbf{t} \in \mathbf{C}^l$ such that $\operatorname{Re} t_i > M$ $(1 \leqslant i \leqslant l)$. Also we set $\mathbf{R}^l(M) = \mathbf{R}^l \cap \mathbf{C}^l(M)$. Fix $M_0 \geqslant 0$ and let $\mathcal{C}_i$ $(1 \leqslant i \leqslant l)$ be given holomorphic functions on $\mathbf{C}^l(M_0)$ with values in $\mathcal{E}(U)$.

Lemma 10. *Let V be a nonempty open connected set in $\mathbf{R}^l(M_0)$ and ϕ a function from V to U such that $\partial\phi/\partial t_i = \mathcal{C}_i(\mathbf{t})\phi$ $(1 \leqslant i \leqslant l)$ on V. Then there exists a unique holomorphic function Φ on $\mathbf{C}^l(M_0)$ such that 1) $\Phi = \phi$ on V and 2) $\partial\Phi/\partial t_i = \mathcal{C}_i(\mathbf{t})\Phi$ $(1 \leqslant i \leqslant l)$ on $\mathbf{C}^l(M_0)$.*

Fix a point $\mathbf{t}^0 = (t_1^0, \ldots, t_l^0)$ in V. Then by Lemma 8, for each i $(1 \leqslant i \leqslant l)$ there exists a holomorphic function $K_i(\mathbf{t})$ from $\mathbf{C}^l(M_0)$ to $\mathcal{E}(U)$ such that $\partial K_i(\mathbf{t})/\partial t_i = \mathcal{C}_i(\mathbf{t})K_i(\mathbf{t})$ and $K_i(\mathbf{t}) = 1$ if $t_i = t_i^0$. For any $\mathbf{t} \in \mathbf{C}^l(M_0)$, let $\mathbf{t}^r$ denote the point $(t_1, \ldots, t_r, t_{r+1}^0, \ldots, t_l^0)$ $(1 \leqslant r \leqslant l)$. Put

$$\Phi(\mathbf{t}) = K_l(\mathbf{t}^l)K_{l-1}(\mathbf{t}^{l-1})\cdots K_1(\mathbf{t}^1)\phi(\mathbf{t}^0) \qquad \big(\mathbf{t} \in \mathbf{C}^l(M_0)\big).$$

Obviously Φ is holomorphic on $\mathbf{C}^l(M_0)$ and $\Phi(\mathbf{t}^0) = \phi(\mathbf{t}^0)$. Fix a number $\varepsilon > 0$ and let A be the set of all points $\mathbf{t} \in \mathbf{R}^l$ such that $|\mathbf{t} - \mathbf{t}^0| < \varepsilon$. We assume that ε is so small that $A \subset V$. It is obvious that if $\mathbf{t} \in A$, then $\mathbf{t}^r$ is also in A $(1 \leqslant r \leqslant l)$. Now fix r and choose real numbers t_j $(1 \leqslant j < r)$ such that $|t_j - t_j^0| < \varepsilon$. Put $\phi_r(t) = \phi(t_1, \ldots, t_{r-1}, t, t_{r+1}^0, \ldots, t_l^0)$ $(|t - t_r^0| < \varepsilon, t \in \mathbf{R})$. Then it is clear that

$$\frac{d\phi_r(t)}{dt} = \mathcal{C}_r\big(t_1, \ldots, t_{r-1}, t, t_{r+1}^0, \ldots, t_l^0\big)\phi_r(t).$$

On the other hand let

$$\psi_r(t) = K_r\big(t_1, \ldots, t_{r-1}, t, t_{r+1}^0, \ldots, t_l^0\big)\phi\big(t_1, \ldots, t_{r-1}, t_r^0, t_{r+1}^0, \ldots, t_l^0\big),$$

when t is real and $|t - t_r^0| < \varepsilon$. Then it follows from the definition of K_r that

$$\frac{d\psi_r(t)}{dt} = \mathcal{C}_r\big(t_1,\ldots,t_{r-1}, t, t_{r+1}^0,\ldots,t_l^0\big)\psi_r(t).$$

Since $\psi_r(t_r^0) = \phi_r(t_r^0) = \phi(t_1,\ldots,t_{r-1}, t_r^0, t_{r+1}^0,\ldots,t_l^0)$, we conclude from Lemma 7 that $\phi_r = \psi_r$. This proves that

$$\phi(\mathbf{t}^r) = K_r(\mathbf{t}^r)\phi(\mathbf{t}^{r-1}) \qquad (1 \leqslant r \leqslant l)$$

for $\mathbf{t} \in A$. Hence $\phi(\mathbf{t}) = \Phi(\mathbf{t})$ for $\mathbf{t} \in A$. Therefore in particular $\partial\Phi/\partial t_i = \mathcal{C}_i(\mathbf{t})\Phi$ $(1 \leqslant i \leqslant l)$ on A. But since Φ is holomorphic on $\mathbf{C}^l(M_0)$ this implies that these equations actually hold on $\mathbf{C}^l(M_0)$.

Now let τ be a point in V. Choose a differentiable path $\mathbf{x}(t)$ $(a \leqslant t \leqslant b)$ in V such that $\mathbf{x}(a) = \mathbf{t}^0$ and $\mathbf{x}(b) = \tau$. Put $f(t) = \phi(\mathbf{x}(t))$ and $F(t) = \Phi(\mathbf{x}(t))$. Then we have $df/dt = \mathcal{C}(t)f$ and $dF/dt = \mathcal{C}(t)F$ where $\mathcal{C}(t) = \sum_{1 \leqslant i \leqslant l}\mathcal{C}_i(\mathbf{x}(t))(dx_i(t)/dt)$. Since $f(a) = \phi(\mathbf{t}^0) = \Phi(\mathbf{t}^0) = F(a)$, it follows from Lemma 7 that $f(b) = F(b)$ and so $\phi(\tau) = \Phi(\tau)$. This proves that $\phi = \Phi$ on V. Since a holomorphic function on $\mathbf{C}^l(M_0)$ which is zero on V must vanish identically, the uniqueness of Φ is obvious.

§4. The Main Theorems

Suppose we are given a number $M_0 \geqslant 0$ and for each $\mathbf{k} \in \mathbf{Z}_+^l$ an element $c(\mathbf{k}) \in \mathcal{E}(U)$ such that

1) $\displaystyle\sum_{\mathbf{k}} |c(\mathbf{k})| e^{-\sigma(\mathbf{k})M_0} < \infty,$

2) $c(\mathbf{k}) = 0$ if $k_l = 0$ and $\mathbf{k} \neq 0$.

Put $\mathcal{C}(\mathbf{t}) = \sum_{\mathbf{k}} c(\mathbf{k}) e^{-\langle \mathbf{k}, \mathbf{t}\rangle}$ $(\mathbf{t} \in \mathbf{C}^l(M_0))$. Then obviously $\mathcal{C}$ is a holomorphic function on $\mathbf{C}^l(M_0)$. Fix a number $a > M_0$ and let K denote the holomorphic function from $\mathbf{C}^l(M_0)$ to $\mathcal{E}(U)$ such that $K(\mathbf{t}) = 1$ if $t_l = a$ and $\partial K(\mathbf{t})/\partial t_l = \mathcal{C}(\mathbf{t})K(\mathbf{t})$. We know from Lemmas 8 and 9 that such a function exists and is unique.

Lemma 11. *We can choose 1) a finite number of points $\beta_1,\ldots,\beta_N$ in $\mathbf{C}^l$ and 2) for each j $(1 \leqslant j \leqslant N)$ a normal sequence $P_{j,\mathbf{k}}$ of $\mathcal{E}(U)$-polynomials on $\mathbf{C}^l$ such that the following condition holds. Fix ε $(0 < \varepsilon < 1)$ and choose $M \geqslant M_0$ so large that the series*

$$\sum_{1 \leqslant j \leqslant N} \sum_{\mathbf{k}} P_{j,\mathbf{k}}(\mathbf{t}) e^{\langle \beta_j - \mathbf{k}, \mathbf{t}\rangle}$$

converges decently on $W_l(\varepsilon, M)$. Then its sum coincides with $K(\mathbf{t})$ on $W_l(\varepsilon, M)$.

We shall give a proof of this lemma later in §5.

Lemma 12. *Suppose $l \geqslant 2$, $q_{\mathbf{m}}$ $(\mathbf{m} \in \mathbf{Z}_+^{l-1})$ is a normal sequence of U-polynomials on $\mathbf{C}^{l-1}$ and μ is a given point in $\mathbf{C}^{l-1}$. Then we can choose 1) a finite number of points $\lambda_1,\ldots,\lambda_N$ in $\mathbf{C}^l$ and 2) for each j $(1 \leqslant j \leqslant N)$ a normal sequence $p_{j,\mathbf{k}}$ $(\mathbf{k} \in \mathbf{Z}^l)$ of U-polynomials on $\mathbf{C}^l$, such that the following condition is fulfilled. For any ε $(0 < \varepsilon < 1)$,* *there exists a number $M \geqslant M_0$ such that:*

(1) *The series $\sum_m q_m(s)e^{-\langle m,s\rangle}$ and $\sum_k p_{j,k}(t)e^{-\langle k,t\rangle}$ $(1 \leqslant j \leqslant N)$ converge decently for $s \in W_{l-1}(\varepsilon, M)$ and $t \in W_l(\varepsilon, M)$;*

(2) *Put $\psi(s) = \sum_m q_m(s)e^{-\langle m,s\rangle}$ $(s \in W_{l-1}(\varepsilon, M))$. Then*

$$K(t)\psi(t_1,\ldots,t_{l-1}) = \sum_{1 \leqslant j \leqslant N} \sum_k p_{j,k}(t)e^{\langle \lambda_j - k, t\rangle}$$

for $t \in W_l(\varepsilon, M)$.

This is an immediate consequence of Lemmas 11, 6 and 4.
Put $\mathbf{C}_+^l = \mathbf{C}^l(0)$ and $\mathbb{R}_+^l = \mathbb{R}^l(0)$.

Lemma 13. *Let $c_i(k)$ $(1 \leqslant i \leqslant l, \, k \in \mathbf{Z}_+^l)$ be given elements in $\mathcal{E}(U)$ such that:*

a) $\displaystyle\sum_k |c_i(k)|e^{-\eta\sigma(k)} < \infty \qquad (1 \leqslant i \leqslant l) \quad$ *for every $\eta > 0$,*

b) $c_i(k) = 0 \quad$ *if $k_i = 0$ and $k \neq 0$* $\quad (1 \leqslant i \leqslant l)$.

Put $\mathcal{C}_i(t) = \sum_k c_i(k)e^{-\langle k,t\rangle}$ $(t \in \mathbf{C}_+^l)$ and let Φ be a holomorphic function from $\mathbf{C}_+^l$ to U such that $\partial\Phi/\partial t_i = \mathcal{C}_i(t)\Phi$ $(1 \leqslant i \leqslant l)$ on $\mathbf{C}_+^l$. Then we can select 1) a finite number of elements $\lambda_j \in \mathbf{C}^l$ $(1 \leqslant j \leqslant N)$ and 2) for each j, a normal sequence $p_{j,k}(k \in \mathbf{Z}_+^l)$ of U-polynomials on $\mathbf{C}^l$ such that the following condition holds. For any ε $(0 < \varepsilon < 1)$, there exists a number $M \geqslant 0$ such that:

(1) *the series $\displaystyle\sum_k p_{j,k}(t)e^{-\langle k,t\rangle}(1 \leqslant j \leqslant N)$ converge decently on $W_l(\varepsilon, M)$,*

(2) $\displaystyle\Phi(t) = \sum_{1 \leqslant j \leqslant N} \sum_k p_{j,k}(t)e^{\langle \lambda_j - k, t\rangle} \qquad (t \in W_l(\varepsilon, M))$.

Fix a number $a > 0$ and let K_l be the holomorphic function from $\mathbf{C}_+^l$ to $\mathcal{E}(U)$ such that $\partial K_l(t)/\partial t_l = \mathcal{C}_l(t)K_l(t)$ and $K_l(t) = 1$ if $t_l = a$. If $l = 1$, $\Phi(t) = K_1(t)\Phi(a)$ $(t \in \mathbf{C}_+^1)$ and our assertion follows from Lemma 11. So now suppose $l \geqslant 2$ and use induction. Put $\psi(s) = \Phi(s, a)$ $(s \in \mathbf{C}_+^{l-1})$. Then $\Phi(t) = K_l(t)\psi(t_1,\ldots,t_{l-1})$ $(t \in \mathbf{C}_+^l)$ from Lemma 7. Therefore if we apply the induction hypothesis to ψ, our assertion follows from Lemma 12.

Now we want to remove the restriction b) of Lemma 13 on the elements $c_i(k)$. For this we proceed as follows. Let τ denote the endomorphism of $\mathbf{C}^l$ given by $\tau(t) = t + s(t)\mathbf{1}$ $(t \in \mathbf{C}^l)$ where $s(t) = t_1 + \cdots + t_l$ and $\mathbf{1} = (1, 1, \ldots, 1)$. Then τ is non-singular and it maps $\mathbf{C}^l(M)$ into itself for $M \geqslant 0$. Moreover $\tau(\mathbf{Z}_+^l) \subset \mathbf{Z}_+^l$ and if $k \in \tau(\mathbf{Z}_+^l)$ and $k \neq 0$ then $k_i \geqslant 1$ $(1 \leqslant i \leqslant l)$. Now suppose Φ and $\mathcal{C}_i$ have the same meaning as in Lemma 13 except that we drop the condition b) on $c_i(k)$. Put

$$\Phi'(t) = \Phi(\tau(t)),$$

$$\mathcal{C}_i'(t) = \mathcal{C}_i(\tau(t)) + \sum_{1 \leqslant j \leqslant l} \mathcal{C}_j(\tau(t)) \qquad (t \in \mathbf{C}_+^l, 1 \leqslant i \leqslant l).$$

Then Φ' and $\mathcal{C}_i'$ are holomorphic functions on $\mathbf{C}_+^l$ and

$$\partial\Phi'/\partial t_i = \mathcal{C}_i'(t)\Phi'(t) \qquad (1 \leqslant i \leqslant l).$$

For any $\mathbf{k} \in \mathbf{Z}_+^l$ define $c_i'(\mathbf{k}) \in \mathcal{E}(U)$ as follows: $c_i'(\mathbf{k}) = 0$ if $\mathbf{k} \notin \tau(\mathbf{Z}_+^l)$ and $c_i'(\mathbf{k}) = c_i(\mathbf{k}') + \sum_{1 \leqslant j \leqslant l} c_j(\mathbf{k}')$ if $\mathbf{k}' = \tau^{-1}(\mathbf{k})$ is in $\mathbf{Z}_+^l$. Then it is clear that the elements $c_i'(\mathbf{k})$ $(1 \leqslant i \leqslant l, \mathbf{k} \in \mathbf{Z}_+^l)$ satisfy the conditions a) and b) of Lemma 13 and

$$\mathcal{C}_i'(\mathbf{t}) = \sum_{\mathbf{k}} c_i'(\mathbf{k}) e^{-\langle \mathbf{k}, \mathbf{t} \rangle} \qquad \left(\mathbf{t} \in \mathbf{C}_+^l \right).$$

Therefore from Lemma 13 (applied to Φ'), we can choose a finite number of points $\mu_1, \ldots, \mu_N \in \mathbf{C}^l$ and for each j $(1 \leqslant j \leqslant N)$ a normal sequence $q_{j,\mathbf{k}}$ of U-polynomials on $\mathbf{C}^l$ such that the following condition holds. For any ε $(0 < \varepsilon < 1)$ choose $M \geqslant 0$ so large that the series $\sum_{\mathbf{k}} q_{j,\mathbf{k}} e^{-\langle \mathbf{k}, \mathbf{t} \rangle}$ converges decently on $W_l(\varepsilon, M)$ for every j. Then

$$\Phi'(\mathbf{t}) = \sum_{1 \leqslant j \leqslant N} \sum_{\mathbf{k}} q_{j,\mathbf{k}} e^{\langle \mu_j - \mathbf{k}, \mathbf{t} \rangle}$$

on $W_l(\varepsilon, M)$. Let Λ' be the set of all points in $\mathbf{C}^l$ of the form $\mu_j - \mathbf{k}$ $(1 \leqslant j \leqslant N, \mathbf{k} \in \mathbf{Z}_+^l)$. For any $\lambda \in \Lambda'$, define $Q_\lambda = \sum q_{j,\mathbf{k}}$ where the sum is over pairs $(j, \mathbf{k})$ such that $\mu_j - \mathbf{k} = \lambda$. Let Λ be the set of those elements $\lambda \in \Lambda'$ for which $Q_\lambda \neq 0$. Then it is obvious that

$$\Phi'(\mathbf{t}) = \sum_{\lambda \in \Lambda} Q_\lambda(\mathbf{t}) e^{\langle \lambda, \mathbf{t} \rangle} \qquad (\mathbf{t} \in W_l(\varepsilon, M)),$$

the series converging decently on $W_l(\varepsilon, M)$. Since for every j, the sequence $q_{j,\mathbf{k}}$ is normal, it is clear that we can choose an integer $d \geqslant 0$ and two positive numbers a and b such that $d^0 Q_\lambda \leqslant d$ and $\|Q_\lambda\| \leqslant a b^{-\sigma(\lambda)}$ for every $\lambda \in \Lambda$.

Let Λ_0 be the set of all $\lambda \in \Lambda$ such that $\lambda + \tau(\mathbf{k}) \notin \Lambda$ for any $\mathbf{k} \neq 0$ in $\mathbf{Z}_+^l$.

Lemma 14. Λ_0 *is a finite set and every element in* Λ *is of the form* $\lambda - \tau(\mathbf{k})$ *with* $\lambda \in \Lambda_0$ *and* $\mathbf{k} \in \mathbf{Z}_+^l$.

Let μ be an element in Λ. Then it is obvious that $\mu = \mu_j - \mathbf{k}$ for some j $(1 \leqslant j \leqslant N)$ and $\mathbf{k} \in \mathbf{Z}_+^l$. Hence $\sigma(\mu) \leqslant \sup_j \sigma(\mu_j) = A$ (say). Let $S(\mu)$ be the set of all $\mathbf{k} \in \mathbf{Z}_+^l$ such that $\mu + \tau(\mathbf{k}) \in \Lambda$. Then if $\mathbf{k} \in S(\mu)$, $\sigma(\mu) + \sigma(\tau(\mathbf{k})) = \sigma(\mu) + (l+1)\sigma(\mathbf{k}) \leqslant A$. This means that $S(\mu)$ is a finite set which is not empty since it contains zero. Choose $\mathbf{k} \in S(\mu)$ with the largest possible value of $\sigma(\mathbf{k})$. Then it is obvious that $\lambda = \mu + \tau(\mathbf{k}) \in \Lambda_0$ and therefore $\mu = \lambda - \tau(\mathbf{k})$.

Now in order to prove that Λ_0 is a finite set, we may obviously assume that it is not empty. It follows from Lemma 5 that we can choose $M_1 \geqslant M$ and $a, b > 0$ such that

$$\sum_{\lambda \in \Lambda} |Q_\lambda(\mathbf{t}) e^{\langle \lambda, \mathbf{t} \rangle}| \leqslant \sum_{1 \leqslant j \leqslant N} \sum_{\mathbf{k}} |q_{j,\mathbf{k}}(\mathbf{t}) e^{\langle \mu_j - \mathbf{k}, \mathbf{t} \rangle}| \leqslant a |e^{b\langle \mathbf{1}, \mathbf{t} \rangle}|$$

for $\mathbf{t} \in W_l(\varepsilon, M_1)$. Put

$$A = \sum_{1 \leqslant i \leqslant l} \sum_{\mathbf{k}} |c_i'(\mathbf{k})| e^{-\sigma(\mathbf{k})}$$

Then it is obvious that $A < \infty$ and $|c_i'(\mathbf{k})| \leqslant A e^{\sigma(\mathbf{k})}$. Hence for each i, we can regard $c_i'(\mathbf{k})$ $(\mathbf{k} \in \mathbf{Z}_+^l)$ as a normal sequence of $\mathcal{E}(U)$-polynomials (which are actually constants). Hence by choosing M_1 sufficiently large, we can assume that the series

$$\sum_{\mathbf{k}} c_i'(\mathbf{k}) e^{-\langle \mathbf{k}, \mathbf{t} \rangle} \qquad (1 \leqslant i \leqslant l)$$

converge decently on $W_l(\varepsilon, M_1)$. Put $\gamma_i(\mathbf{k}) = c_i(\mathbf{k}) + \Sigma_{1 \le j \le l} c_j(\mathbf{k})$. Then it follows from the definition of $c_i'(\mathbf{k})$ that

$$\mathcal{C}_i'(\mathbf{t}) = \sum_{\mathbf{k}} c_i'(\mathbf{k}) e^{-\langle \mathbf{k}, \mathbf{t} \rangle} = \sum_{\mathbf{k}} \gamma_i(\mathbf{k}) e^{-\langle \tau(\mathbf{k}), \mathbf{t} \rangle} \qquad (1 \le i \le l),$$

both the series converging decently on $W_l(\varepsilon, M_1)$. On the other hand $\partial \Phi'/\partial t_i = \mathcal{C}_i'(\mathbf{t}) \Phi'$ and

$$\Phi'(\mathbf{t}) = \sum_{\lambda \in \Lambda} Q_\lambda(\mathbf{t}) e^{\langle \lambda, \mathbf{t} \rangle}$$

where the series converges decently on $W_l(\varepsilon, M_1)$. Hence we can differentiate it term by term and obtain

$$\sum_{\lambda \in \Lambda} \left(\lambda_i Q_\lambda + \frac{\partial Q_\lambda}{\partial t_i} \right) e^{\langle \lambda, \mathbf{t} \rangle} = \sum_{\mathbf{k}} \sum_{\lambda \in \Lambda} \gamma_i(\mathbf{k}) Q_\lambda(\mathbf{t}) e^{\langle \lambda - \tau(\mathbf{k}), \mathbf{t} \rangle}$$

for $\mathbf{t} \in W_l(\varepsilon, M_1)$. (Here λ_i denotes the ith coordinate of the point λ.) Now

$$\sum_{\mathbf{k}} |\gamma_i(\mathbf{k}) e^{-\langle \tau(\mathbf{k}), \mathbf{t} \rangle}| \le 2 \sum_{1 \le j \le l} \sum_{\mathbf{k}} |c_i(\mathbf{k})| e^{-(l+1) M_1 \sigma(\mathbf{k})} < \infty$$

for $\mathbf{t} \in W_l(\varepsilon, M_1)$. Hence it follows from Lemma 3 that the series

$$\sum_{\mathbf{k}} \sum_{\lambda \in \Lambda} |\gamma_i(\mathbf{k}) Q_\lambda(\mathbf{t}) e^{\langle \lambda + \mu - \tau(\mathbf{k}), \mathbf{t} \rangle}|$$

converges uniformly on $W_l(\varepsilon, M_1)$ for any $\mu \in \mathbf{C}^l$. Now fix an element $\nu \in \Lambda_0$. Then by the Corollary of Lemma 57 of [1], the coefficients of $e^{\langle \nu, \mathbf{t} \rangle}$ on the two sides of our equation above must be equal. Since $\nu \in \Lambda_0$, $\lambda - \tau(\mathbf{k}) \ne \nu$ for any $\lambda \in \Lambda$ and $\mathbf{k} \in \mathbf{Z}_+^l$, unless $\lambda = \nu$ and $\mathbf{k} = 0$. Hence we get

$$\nu_i Q_\nu + \frac{\partial Q_\nu}{\partial t_i} = \gamma_i(0) Q_\nu.$$

This means that $(\partial Q_\nu/\partial t_i) = (\gamma_i(0) - \nu_i) Q_\nu$ and therefore by repeated differentiations,

$$\frac{\partial^r Q_\nu}{\partial t_i^r} = (\gamma_i(0) - \nu_i)^r Q_\nu \qquad (r \ge 1).$$

If $r > d^0 Q_\nu$, the left side is zero. Hence $(\gamma_i(0) - \nu_i)^r Q_\nu = 0$ for r sufficiently large. But since $\nu \in \Lambda_0$, $Q_\nu \ne 0$ and so we conclude that ν_i is an eigenvalue of $\gamma_i(0)$. This being true for every i, it follows that Λ_0 can have at most $(\dim U)^l$ elements. This proves that Λ_0 is a finite set.

Let $\nu_1, \ldots, \nu_N$ be all the distinct elements in Λ_0. For any $\mathbf{k} \in \mathbf{Z}_+^l$, define the U-polynomials $p_{j,\mathbf{k}}'$ $(1 \le j \le N)$ on $\mathbf{C}^l$ as follows. $p_{j,\mathbf{k}}' = Q_{\nu_j - \tau(\mathbf{k})}$ provided $\nu_j - \tau(\mathbf{k}) \in \Lambda$ and $\nu_j - \tau(\mathbf{k}) \ne \nu_i - \tau(\mathbf{k}')$ for any $i < j$ and $\mathbf{k}' \in \mathbf{Z}_+^l$. Otherwise $p_{j,\mathbf{k}}' = 0$. Then it is clear that

$$\Phi'(\mathbf{t}) = \sum_\lambda Q_\lambda(\mathbf{t}) e^{\langle \lambda, \mathbf{t} \rangle} = \sum_{1 \le j \le N} \sum_{\mathbf{k}} p_{j,\mathbf{k}}'(\mathbf{t}) e^{\langle \nu_j - \tau(\mathbf{k}), \mathbf{t} \rangle}$$

on $W_l(\varepsilon, M)$. Put $\lambda_j = \tau^{-1}(\nu_j)$ and $p_{j,\mathbf{k}}(\mathbf{t}) = p_{j,\mathbf{k}}'(\tau^{-1}(\mathbf{t}))$. Then since $\langle \tau(\lambda), \mu \rangle =$

$\langle \lambda, \tau(\mu) \rangle$ $(\lambda, \mu \in \mathbb{C}^l)$, we get

$$\Phi(\tau(\mathbf{t})) = \Phi'(\mathbf{t}) = \sum_{1 \leqslant j \leqslant N} \sum_{\mathbf{k}} p_{j,\mathbf{k}}(\tau(\mathbf{t})) e^{\langle \lambda_j - \mathbf{k}, \tau(\mathbf{t}) \rangle}$$

for $\mathbf{t} \in W_l(\varepsilon, M)$. Now we claim that $p_{j,\mathbf{k}}$ is a normal sequence for every j. Choose an integer $d \geqslant 0$ such that $d^0 Q_\lambda \leqslant d$ for all $\lambda \in \Lambda$. Let P_d denote the space of all U-polynomials on $\mathbb{C}^l$ of degree $\leqslant d$. Then P_d is a Banach space of finite dimension with the norm $\| p \|$ ($p \in P_d$). Let E denote the linear transformation in P_d defined as follows. If $q \in P_d$, Eq is the function $\mathbf{t} \to q(\tau^{-1}(\mathbf{t}))$ $(\mathbf{t} \in \mathbb{C}^l)$. We can choose a number $A \geqslant 0$ such that $\| Eq \| \leqslant A \| q \|$ for every $q \in P_d$. It is obvious that $d^0 p_{j,\mathbf{k}} \leqslant d$ and if $p_{j,\mathbf{k}} \neq 0$, $p_{j,\mathbf{k}} = E Q_{\nu_j - \tau(\mathbf{k})}$. Hence

$$\| p_{j,k} \| \leqslant A \| Q_{\nu_j - \tau(\mathbf{k})} \|.$$

But as we have seen, there exist positive numbers a, b such that $\| Q_\lambda \| \leqslant a b^{-\sigma(\lambda)}$ for every $\lambda \in \Lambda$. Choose a number B so large that $B \geqslant \sup_{1 \leqslant j \leqslant N} a b^{-\sigma(\nu_j)}$ and $B \geqslant b^{-(l+1)}$. Then $\| p_{j,\mathbf{k}} \| \leqslant B^{\sigma(\mathbf{k})+1}$. This proves that the sequence $p_{j,\mathbf{k}}$ is normal.

Now for any ε $(0 < \varepsilon < 1)$ choose $M \geqslant 0$ such that the series $\sum_{\mathbf{k}} p_{j,\mathbf{k}}(\mathbf{t}) e^{-\langle \mathbf{k}, \mathbf{t} \rangle}$ $(1 \leqslant j \leqslant N)$ converge decently on $W_l(\varepsilon, M)$. We claim that

$$\Phi(\mathbf{t}) = \sum_{1 \leqslant j \leqslant N} \sum_{\mathbf{k}} p_{j,\mathbf{k}}(\mathbf{t}) e^{\langle \lambda_j - \mathbf{k}, \mathbf{t} \rangle} \qquad (\mathbf{t} \in W_l(\varepsilon, M)).$$

Since both sides are holomorphic functions on $W_l(\varepsilon, M)$, it is enough to verify this equality on some nonempty open subset of $V = \mathbb{R}^l \cap W_l(\varepsilon, M)$. However, if M' is sufficiently large $(M' \geqslant M)$, we have seen above that

$$\Phi(\tau(\mathbf{t})) = \sum_j \sum_{\mathbf{k}} p_{j,\mathbf{k}}(\tau(\mathbf{t})) e^{\langle \lambda_j - \mathbf{k}, \tau(\mathbf{t}) \rangle} \qquad (\mathbf{t} \in W_l(\varepsilon, M')).$$

This shows that the required equation holds on $\tau(V')$ where $V' = \mathbb{R}^l \cap W_l(\varepsilon, M')$. But obviously V' is a nonempty open subset of V and $\tau(V') \subset \tau(V) \subset V$. Since τ is nonsingular, $\tau(V')$ is a nonempty open subset of V, and so our assertion follows.

For later use we make the following two observations which are immediate consequences of the definition of λ_j and $p_{j,\mathbf{k}}$.

(1) If $i \neq j$ then $\lambda_i \neq \lambda_j - \mathbf{k}$ for any $\mathbf{k} \in \mathbb{Z}_+^l$.
(2) $p_{j,0} \neq 0$ $(1 \leqslant j \leqslant N)$.

Lemma 15. *Let ε, η be any two real numbers such that $0 < \varepsilon < 1$ and $\eta > 0$. Then the series $\sum_{1 \leqslant j \leqslant N} \sum_{\mathbf{k}} | p_{j,\mathbf{k}}(\mathbf{t}) e^{\langle \lambda_j - \mathbf{k}, \mathbf{t} \rangle} |$ converges uniformly on $W_l(\varepsilon, \eta)$. Moreover*

$$\Phi(\mathbf{t}) = \sum_{1 \leqslant j \leqslant N} \sum_{\mathbf{k}} p_{j,\mathbf{k}}(\mathbf{t}) e^{\langle \lambda_j - \mathbf{k}, \mathbf{t} \rangle} \qquad \left(\mathbf{t} \in \mathbb{C}_+^l \right).$$

Choose $M \geqslant \eta$ so large that the series $\sum_{\mathbf{k}} p_{j,\mathbf{k}}(\mathbf{t}) e^{-\langle \mathbf{k}, \mathbf{t} \rangle}$ $(1 \leqslant j \leqslant N)$ converge decently on $W_l(\varepsilon, M)$. It is obvious that there exists a compact subset Ω of $\mathbb{C}_+^l$ such that $W_l(\varepsilon, \eta) \subset W_l(\varepsilon, M) \cup \Omega$. Hence it is sufficient to prove that the series $\sum_{1 \leqslant j \leqslant N} \sum_{\mathbf{k}} | p_{j,\mathbf{k}}(\mathbf{t}) e^{\langle \lambda_j - \mathbf{k}, \mathbf{t} \rangle} |$ converges uniformly on a compact subset of $\mathbb{C}_+^l$. But any compact subset $\mathbb{C}_+^l$ is contained in $W_l(\varepsilon, \eta)$ for suitable ε, η. Hence it would be enough to verify that this series converges uniformly on every compact subset of $W_l(\varepsilon, \eta)$ for any given values of ε and η.

So let us fix ε, η and choose M as above. It is obvious that $W_l(\varepsilon, \eta) + M\mathbf{1} \subset W_l(\varepsilon, M)$. Put $c(\mathbf{k}) = \Sigma_{1 \leqslant i \leqslant l} c_i(\mathbf{k})$ ($\mathbf{k} \in \mathbf{Z}_+^l$) and $\mathcal{C}(\mathbf{t}) = \Sigma_{1 \leqslant i \leqslant l} \mathcal{C}_i(\mathbf{t})$ ($\mathbf{t} \in \mathbf{C}_+^l$). It is obvious that

$$d\Phi(\mathbf{t} + s\mathbf{1})/ds = \mathcal{C}(\mathbf{t} + s\mathbf{1})\Phi(\mathbf{t} + s\mathbf{1}) \qquad (\mathbf{t} \in \mathbf{C}_+^l,\ s \geqslant 0).$$

Put $K_0(\mathbf{t}) = 1 \in \mathcal{E}(U)$ and

$$K_q(\mathbf{t}) = (-1)^q \int_{0 \leqslant s_1 \leqslant s_2 \leqslant \cdots \leqslant s_q \leqslant M} \mathcal{C}(\mathbf{t} + s_1\mathbf{1})\mathcal{C}(\mathbf{t} + s_2\mathbf{1}) \cdots \mathcal{C}(\mathbf{t} + s_q\mathbf{1})\, ds_1 \cdots ds_q$$

for $q \geqslant 1$ and $\mathbf{t} \in \mathbf{C}_+^l$. Let Ω be a compact subset of $\mathbf{C}_+^l$. It is obvious that we can choose a positive number m such that $|\mathcal{C}(\mathbf{t} + s\mathbf{1})| \leqslant m$ for $\mathbf{t} \in \Omega$ and $0 \leqslant s \leqslant M$. Then $|K_q(\mathbf{t})| \leqslant m^q M^q/q!$ for $\mathbf{t} \in \Omega$. Hence the series $\Sigma_{q \geqslant 0} |K_q(\mathbf{t})|$ converges uniformly for $\mathbf{t} \in \Omega$. Put

$$K(\mathbf{t}) = \sum_{q \geqslant 0} K_q(\mathbf{t}) \qquad (\mathbf{t} \in \mathbf{C}_+^l).$$

Then as we have seen in §1, $\Phi(\mathbf{t}) = K(\mathbf{t})\Phi(\mathbf{t} + M\mathbf{1})$ ($\mathbf{t} \in \mathbf{C}_+^l$). Now put $A = \Sigma_{\mathbf{k}} |c(\mathbf{k})| e^{-\sigma(\mathbf{k})\eta/2}$. Then $A < \infty$ and

$$\mathcal{C}(\mathbf{t} + s\mathbf{1}) = \sum_{\mathbf{k}} c(\mathbf{k}:s) e^{-\langle \mathbf{k}, \mathbf{t} \rangle} \qquad (\mathbf{t} \in \mathbf{C}_+^l,\ s \geqslant 0)$$

where $c(\mathbf{k}:s) = c(\mathbf{k})e^{-\sigma(\mathbf{k})s}$. For any $\mathbf{k} \in \mathbf{Z}_+^l$ and $q \geqslant 0$, define $\kappa_q(\mathbf{k}) \in \mathcal{E}(U)$ as follows. $\kappa_0(\mathbf{k}) = 1$ or 0 according as $\mathbf{k} = 0$ or not and if $q \geqslant 1$,

$$\kappa_q(\mathbf{k}) = (-1)^q \sum_{\mathbf{k}_1 + \cdots + \mathbf{k}_q = \mathbf{k}} \int_{0 \leqslant s_1 \leqslant \cdots \leqslant s_q \leqslant M} c(\mathbf{k}_1:s_1)c(\mathbf{k}_2:s_2) \cdots$$

$$\times c(\mathbf{k}_q:s_q)\, ds_1 \cdots ds_q$$

where the sum is over all sequences $(\mathbf{k}_1, \ldots, \mathbf{k}_q)$ of q elements in $\mathbf{Z}_+^l$ such that $\mathbf{k}_1 + \cdots + \mathbf{k}_q = \mathbf{k}$. Put

$$c(\mathbf{k}:s_1, \ldots, s_q) = \sum_{\mathbf{k}_1 + \cdots + \mathbf{k}_q = \mathbf{k}} c(\mathbf{k}_1:s_1) \cdots c(\mathbf{k}_q:s_q) \qquad (s_1, \ldots, s_q \geqslant 0).$$

Since $\Sigma_{\mathbf{k}} |c(\mathbf{k}:s)| e^{-\sigma(\mathbf{k})\eta/2} \leqslant A$, it is obvious that

$$|c(\mathbf{k}:s_1, \ldots, s_q)| \leqslant A^q e^{\sigma(\mathbf{k})\eta/2}.$$

On the other hand

$$\mathcal{C}(\mathbf{t} + s_1\mathbf{1}) \cdots \mathcal{C}(\mathbf{t} + s_q\mathbf{1}) = \sum_{\mathbf{k}} c(\mathbf{k}:s_1, \ldots, s_q) e^{-\langle \mathbf{k}, \mathbf{t} \rangle}$$

and

$$\sum_{\mathbf{k}} |e^{-\langle \mathbf{k}, \mathbf{t} \rangle}| \int_{0 \leqslant s_1 \leqslant \cdots \leqslant s_q \leqslant M} |c(\mathbf{k}:s_1, \ldots, s_q)|\, ds_1 \ldots ds_q$$

$$\leqslant \frac{A^q M^q}{q!} \sum_{\mathbf{k}} e^{-\sigma(\mathbf{k})\eta/2} = \frac{A^q M^q}{q!} (1 - e^{-\eta/2})^{-l}$$

for $\mathbf{t} \in \mathbf{C}^l(\eta)$. Hence it follows immediately that

$$K_q(\mathbf{t}) = \sum_{\mathbf{k}} \kappa_q(\mathbf{k}) e^{-\langle \mathbf{k}, \mathbf{t} \rangle} \qquad (\mathbf{t} \in \mathbf{C}^l(\eta)).$$

Moreover $|\kappa_q(\mathbf{k})| \leqslant (M^q A^q/q!)e^{\sigma(\mathbf{k})\eta/2}$ and therefore

$$\sum_{q \geqslant 0} |\kappa_q(\mathbf{k})| \leqslant \exp\left(\sigma(\mathbf{k})\frac{\eta}{2} + MA\right).$$

Put $\kappa(\mathbf{k}) = \sum_{q \geqslant 0}\kappa_q(\mathbf{k})$. Then

$$K(\mathbf{t}) = \sum_{q \geqslant 0} K_q(\mathbf{t}) = \sum_{q \geqslant 0}\sum_{\mathbf{k}}\kappa_q(\mathbf{k})e^{-\langle \mathbf{k},\mathbf{t}\rangle} \qquad (\mathbf{t} \in \mathbf{C}'(\eta)).$$

But $\sum_{q \geqslant 0}\sum_{\mathbf{k}}|\kappa_q(\mathbf{k})e^{-\langle \mathbf{k},\mathbf{t}\rangle}| \leqslant e^{MA}\sum_{\mathbf{k}}e^{-\sigma(\mathbf{k})\eta/2} = e^{MA}(1 - e^{-\eta/2})^{-l}$. This proves that

$$K(\mathbf{t}) = \sum_{\mathbf{k}}\kappa(\mathbf{k})e^{-\langle \mathbf{k},\mathbf{t}\rangle} \qquad (\mathbf{t} \in \mathbf{C}'(\eta)).$$

Now consider the equation

$$\Phi(\mathbf{t}) = K(\mathbf{t})\Phi(\mathbf{t} + M\mathbf{l}) \qquad (\mathbf{t} \in W_l(\varepsilon, \eta)).$$

Since $W_l(\varepsilon, \eta) + M\mathbf{l} \subset W_l(\varepsilon, M)$,

$$\Phi(\mathbf{t} + M\mathbf{l}) = \sum_{1 \leqslant j \leqslant N}\sum_{\mathbf{k}} p_{j,\mathbf{k}}(\mathbf{t} + M\mathbf{l})e^{\langle \lambda_j - \mathbf{k}, \mathbf{t} + M\mathbf{l}\rangle}.$$

Put

$$q_{j,\mathbf{k}}(\mathbf{t}) = \sum_{\mathbf{k}_1 + \mathbf{k}_2 = \mathbf{k}} \kappa(\mathbf{k}_1) p_{j,\mathbf{k}_2}(\mathbf{t} + M\mathbf{l})e^{M\langle \lambda_j - \mathbf{k}_2, \mathbf{l}\rangle}.$$

Then $q_{j,\mathbf{k}}$ are U-polynomials on $\mathbf{C}^l$. We have seen that the series $\sum_{\mathbf{k}}|\kappa(\mathbf{k})e^{-\langle \mathbf{k},\mathbf{t}\rangle}|$ converges uniformly on $\mathbf{C}'(\eta)$. Let Ω be a compact subset of $W_l(\varepsilon, \eta)$. Then $\Omega + M\mathbf{l} \subset W_l(\varepsilon, M)$ and therefore the series

$$\sum_{1 \leqslant j \leqslant N}\sum_{\mathbf{k}} |p_{j,\mathbf{k}}(\mathbf{t} + M\mathbf{l})e^{\langle \lambda_j - \mathbf{k}, \mathbf{t} + M\mathbf{l}\rangle}|$$

converges uniformly for $\mathbf{t} \in \Omega$. Hence it follows from Lemma 2 that the series

$$\sum_{1 \leqslant j \leqslant N}\sum_{\mathbf{k}} |q_{j,\mathbf{k}}(\mathbf{t})e^{\langle \lambda_j - \mathbf{k},\mathbf{t}\rangle}|$$

converges uniformly on Ω and

$$\Phi(\mathbf{t}) = K(\mathbf{t})\Phi(\mathbf{t} + M\mathbf{l})$$
$$= \sum_{1 \leqslant j \leqslant N}\sum_{\mathbf{k}} q_{j,\mathbf{k}}(\mathbf{t})e^{\langle \lambda_j - \mathbf{k},\mathbf{t}\rangle} \qquad (\mathbf{t} \in \Omega).$$

We claim that for each j, $q_{j,\mathbf{k}}$ is a normal sequence of U-polynomials. Choose an integer $d \geqslant 0$ such that $d^0 p_{j,\mathbf{k}} \leqslant d$ for all j and $\mathbf{k}$. For any U-polynomial p let Ep denote the polynomial function $\mathbf{t} \to p(\mathbf{t} + M\mathbf{l})$. Then as we have seen earlier, we can choose a number $\alpha \geqslant 0$ such that $|Ep| \leqslant \alpha|p|$ for every U-polynomial p with degree $\leqslant d$. Hence

$$\|q_{j,\mathbf{k}}\| \leqslant e^{M\sigma(\lambda_j) + \sigma(\mathbf{k})\eta} \sum_{\mathbf{k}_1 + \mathbf{k}_2 = \mathbf{k}} |\kappa(\mathbf{k}_1)e^{-\sigma(\mathbf{k}_1)\eta}|\alpha\|p_{j,\mathbf{k}_2}\|.$$

But we have seen above that

$$a = \sum_{\mathbf{k}}|\kappa(\mathbf{k})|e^{-\sigma(\mathbf{k})\eta} < \infty.$$

21

Also we can choose $b \geqslant 1$ such that

$$\| p_{j,\mathbf{k}} \| \leqslant b^{\sigma(\mathbf{k})+1}$$

for all j and $\mathbf{k}$. Hence

$$\| q_{j,\mathbf{k}} \| \leqslant \alpha a e^{M\sigma(\lambda_j)} b \left(b e^{\eta} \right)^{\sigma(\mathbf{k})} \leqslant B^{\sigma(\mathbf{k})+1}$$

if $B \geqslant b e^{\eta}$ and $B \geqslant \alpha a b \sup_j e^{M\sigma(\lambda_j)}$. This proves our assertion.

Define U-polynomials $Q_{j,\mathbf{k}}$ ($1 \leqslant j \leqslant N$, $\mathbf{k} \in \mathbf{Z}^l_+$) as follows. $Q_{j,\mathbf{k}} = 0$ if there exists an $i < j$ such that $\lambda_j - \mathbf{k} = \lambda_i - \mathbf{k}'$ for some $\mathbf{k}' \in \mathbf{Z}^l_+$. Otherwise

$$Q_{j,\mathbf{k}} = \sum q_{i,\mathbf{k}'}$$

where the sum is over all pairs $(i,\mathbf{k}')$ ($1 \leqslant i \leqslant N$, $\mathbf{k}' \in \mathbf{Z}^l_+$) such that $\lambda_j - \mathbf{k} = \lambda_i - \mathbf{k}'$. Let μ be a point in $\mathbf{C}^l$ and W a subset of $\mathbf{C}^l_+$ such that the series $\Sigma_j \Sigma_{\mathbf{k}} | q_{j,\mathbf{k}} e^{\langle \lambda_j + \mu - \mathbf{k}, t \rangle} |$ converges uniformly on W. Then it is obvious that the series $\Sigma_j \Sigma_{\mathbf{k}} | Q_{j,\mathbf{k}} e^{\langle \lambda_j + \mu - \mathbf{k}, t \rangle} |$ also converges uniformly on W and

$$\sum_j \sum_{\mathbf{k}} q_{j,\mathbf{k}} e^{\langle \lambda_j - \mathbf{k}, t \rangle} = \sum_j \sum_{\mathbf{k}} Q_{j,\mathbf{k}} e^{\langle \lambda_j - \mathbf{k}, t \rangle}$$

on W. This holds in particular if W is a compact subset of $W_l(\varepsilon, \eta)$.

We shall now show that $Q_{j,\mathbf{k}} = p_{j,\mathbf{k}}$. This will prove the required result that the series $\Sigma_j \Sigma_{\mathbf{k}} | p_{j,\mathbf{k}} e^{\langle \lambda_j - \mathbf{k}, t \rangle} |$ converges uniformly on any compact subset of $W_l(\varepsilon, \eta)$. Since $q_{j,\mathbf{k}}$ is a normal sequence for every j, we can choose $M' \geqslant M$ so large that the series $\Sigma_{\mathbf{k}} q_{j,\mathbf{k}} e^{-\langle \mathbf{k}, t \rangle}$ ($1 \leqslant j \leqslant N$) converge decently on $W_l(\varepsilon, M')$. Then, in view of what we have said above, the series

$$\sum_j \sum_{\mathbf{k}} | Q_{j,\mathbf{k}}(t) e^{-\langle \lambda_j + \mu - \mathbf{k}, t \rangle} |$$

also converges uniformly on $W_l(\varepsilon, M')$ for every $\mu \in \mathbf{C}^l_+$. Also as we have seen above,

$$\Phi(t) = \sum_j \sum_{\mathbf{k}} q_{j,\mathbf{k}}(t) e^{\langle \lambda_j - \mathbf{k}, t \rangle} = \sum_j \sum_{\mathbf{k}} Q_{j,\mathbf{k}}(t) e^{\langle \lambda_j - \mathbf{k}, t \rangle}$$

for $t \in W_l(\varepsilon, \eta)$. But

$$\Phi(t) = \sum_j \sum_{\mathbf{k}} p_{j,\mathbf{k}}(t) e^{\langle \lambda_j - \mathbf{k}, t \rangle} \qquad (t \in W_l(\varepsilon, M))$$

the series on the right converging decently on $W_l(\varepsilon, M)$. Hence

$$\sum_j \sum_{\mathbf{k}} p_{j,\mathbf{k}}(t) e^{\langle \lambda_j - \mathbf{k}, t \rangle} = \sum_j \sum_{\mathbf{k}} Q_{j,\mathbf{k}}(t) e^{\langle \lambda_j - \mathbf{k}, t \rangle}$$

on $W_l(\varepsilon, M')$. We now apply the Corollary of Lemma 57 of [1]. By equating corresponding coefficients on the two sides and taking into account the definitions of $p_{j,\mathbf{k}}$ and $Q_{j,\mathbf{k}}$ we get $p_{j,\mathbf{k}} = Q_{j,\mathbf{k}}$ ($1 \leqslant j \leqslant N$, $\mathbf{k} \in \mathbf{Z}^l_+$).

This proves that the series $\Sigma_j \Sigma_{\mathbf{k}} | p_{j,\mathbf{k}} e^{\langle \lambda_j - \mathbf{k}, t \rangle} |$ converges uniformly on every compact subset of $\mathbf{C}^l_+$. Now

$$\Phi(t) = \sum_{j=1}^{N} \sum_{\mathbf{k}} p_{j,\mathbf{k}}(t) e^{\langle \lambda_j - \mathbf{k}, t \rangle}$$

on $W_l(\varepsilon, M)$. However, it is now obvious that both sides of this equation are holomorphic functions on $\mathbf{C}^l_+$. Hence this equation holds everywhere on $\mathbf{C}^l_+$ and so our lemma is proved.

Corollary. *The series* $\sum_{1 \leqslant j \leqslant N} \sum_{\mathbf{k}} p_{j,\mathbf{k}}(\mathbf{t}) e^{\langle \lambda_j - \mathbf{k}, \mathbf{t} \rangle}$ *converges decently on* $W_l(\varepsilon, \eta)$.

Choose a compact subset Ω of $\mathbf{C}^l_+$ such that $W_l(\varepsilon, \eta) \subset W_l(\varepsilon, M) \cup \Omega$. By Lemma 1, it would be enough to show that the series $\sum_j \sum_{\mathbf{k}} |p_{j,\mathbf{k}} e^{\langle \lambda_j - \mathbf{k}, \mathbf{t} \rangle}|$ converges uniformly on some compact neighborhood of Ω in $\mathbf{C}^l_+$. But this we have already done above.

Let Λ be the set of all points λ in $\mathbf{C}^l$ such that we can select j and $\mathbf{k}$ ($1 \leqslant j \leqslant N$, $\mathbf{k} \in \mathbf{Z}^l_+$) in such a way that $\lambda = \lambda_j - \mathbf{k}$ and $p_{j,\mathbf{k}} \neq 0$. It follows from the definition of the polynomials $p_{j,\mathbf{k}}$, that for any $\lambda \in \Lambda$, j and $\mathbf{k}$ are uniquely determined by the conditions that $\lambda = \lambda_j - \mathbf{k}$ and $p_{j,\mathbf{k}} \neq 0$. We define $p_\lambda = p_{j,\mathbf{k}}$. Then it is clear that the series $\sum_{\lambda \in \Lambda} p_\lambda e^{\langle \lambda, \mathbf{t} \rangle}$ is obtained from $\sum_j \sum_{\mathbf{k}} p_{j,\mathbf{k}} e^{\langle \lambda_j - \mathbf{k}, \mathbf{t} \rangle}$ by just omitting those terms which are zero. We have seen that there exists a number $B > 0$ such that $\|p_{j,\mathbf{k}}\| \leqslant B^{\sigma(\mathbf{k})+1}$ for all $j, \mathbf{k}$. Suppose λ is in Λ and $\lambda = \lambda_j - \mathbf{k}$ with $p_{j,\mathbf{k}} \neq 0$. Then $\|p_\lambda\| = \|p_{j,\mathbf{k}}\| \leqslant B^{\sigma(\mathbf{k})+1} = B^{\sigma(\lambda_j)+1-\sigma(\lambda)}$. Therefore if $A = \sup_j B^{\sigma(\lambda_j)+1}$, $\|p_\lambda\| \leqslant A B^{-\sigma(\lambda)}$ for every $\lambda \in \Lambda$. Also $\sup_{\lambda \in \Lambda} d^0 p_\lambda \leqslant \sup_{j,\mathbf{k}} d^0 p_{j,\mathbf{k}} < \infty$. Thus we get the following theorem.

Theorem 1. *Let $c_i(\mathbf{k})$ ($1 \leqslant i \leqslant l$, $\mathbf{k} \in \mathbf{Z}^l_+$) be given elements in $\mathcal{E}(U)$ such that*

$$\sum_{1 \leqslant i \leqslant l} \sum_{\mathbf{k}} |c_i(\mathbf{k})| e^{-\sigma(\mathbf{k})\eta} < \infty$$

for every $\eta > 0$. Put

$$\mathcal{C}_i(\mathbf{t}) = \sum_{\mathbf{k}} c_i(\mathbf{k}) e^{-\langle \mathbf{k}, \mathbf{t} \rangle} \qquad \left(1 \leqslant i \leqslant l, \ \mathbf{t} \in \mathbf{C}^l_+ \right).$$

Suppose Φ is a holomorphic function on $\mathbf{C}^l_+$ with values in U such that $\partial \Phi / \partial t_i = \mathcal{C}_i(\mathbf{t}) \Phi$ ($1 \leqslant i \leqslant l$) on $\mathbf{C}^l_+$. Then we can choose a finite or countable subset Λ of $\mathbf{C}^l$ and for each $\lambda \in \Lambda$ a U-polynomial, p_λ on $\mathbf{C}^l$ such that the following conditions hold.

(1) $p_\lambda \neq 0$ *for any $\lambda \in \Lambda$.*

(2) *The series $\sum_{\lambda \in \Lambda} p_\lambda(\mathbf{t}) e^{\langle \lambda, \mathbf{t} \rangle}$ converges decently on $W_l(\varepsilon, \eta)$ for any two real numbers ε, η such that $0 < \varepsilon < 1$, $\eta > 0$.*

(3) $\Phi(\mathbf{t}) = \sum_{\lambda \in \Lambda} p_\lambda(\mathbf{t}) e^{\langle \lambda, \mathbf{t} \rangle} \qquad \left(\mathbf{t} \in \mathbf{C}^l_+ \right).$

The set Λ and the polynomial p_λ, corresponding to any $\lambda \in \Lambda$, are uniquely determined by these conditions. Moreover there exists an integer $d \geqslant 0$ and two positive numbers a, b such that $d^0 p_\lambda \leqslant d$ and $\|p_\lambda\| \leqslant a b^{-\sigma(\lambda)}$ for any $\lambda \in \Lambda$.

The uniqueness of Λ and the polynomial p_λ ($\lambda \in \Lambda$) is an immediate consequence of the Corollary of Lemma 57 of [1].

Call an element $\lambda \in \Lambda$ *maximal* if $\lambda + \mathbf{k} \notin \Lambda$ for any $\mathbf{k} \neq 0$ in $\mathbf{Z}^l_+$. Since $p_{\lambda_j} = p_{j,0} \neq 0$ ($1 \leqslant j \leqslant N$), it follows from our remark before Lemma 15 that λ_j ($1 \leqslant j \leqslant N$) are distinct maximal elements in Λ. Since every element in Λ is of the form $\lambda_j - \mathbf{k}$ for some j and $\mathbf{k}$, these are all the maximal elements in Λ. Hence we get the following result.

Theorem 2. *Let Λ_0 denote the set of all maximal elements in Λ. Then Λ_0 is a finite set and every element of Λ is of the form $\lambda - \mathbf{k}$ where $\lambda \in \Lambda_0$ and $\mathbf{k} \in \mathbf{Z}_+^l$.*

By an *exponent* of Φ, we mean an element in Λ. An exponent λ will be called *principal* if it lies in Λ_0. Moreover the function

$$\Phi_P(\mathbf{t}) = \sum_{\lambda \in \Lambda_0} p_\lambda(\mathbf{t}) e^{\langle \lambda, \mathbf{t} \rangle}$$

will be called *the principal part* of Φ.

We note the following fact for later use.

Lemma 16. *Fix ε $(0 < \varepsilon < 1)$ and $\eta > 0$. Then we can choose positive numbers α and β such that*

$$\sum_{\lambda \in \Lambda} |p_\lambda(\mathbf{t}) e^{\langle \lambda, \mathbf{t} \rangle}| \leqslant \alpha |e^{\beta \langle \mathbf{t}, 1 \rangle}|$$

for $\mathbf{t} \in W_l(\varepsilon, \eta)$.

Since the sequence $p_{j,\mathbf{k}}$ is normal for each j $(1 \leqslant j \leqslant N)$, this is an immediate consequence of Lemma 5.

§5.   Proof of Lemma 11

Let $\mathscr{P}^l(U)$ denote the space of all U-polynomials on $\mathbf{C}^l$. Put $D_i = \partial / \partial t_i$ $(1 \leqslant i \leqslant l)$. For any complex number λ, we denote by $(D_i + \lambda)^{-1}$ the linear mapping of $\mathscr{P}^l(U)$ into itself defined as follows. If $\lambda = 0$

$$D_i^{-1}(u\mathbf{t}^{\mathbf{k}}) = u\mathbf{t}^{\mathbf{k}} \frac{t_i}{k_i + 1} \qquad (u \in U, \ \mathbf{k} \in \mathbf{Z}_+^l)$$

and if $\lambda \neq 0$,

$$(D_i + \lambda)^{-1} f = \sum_{r \geqslant 0} (-1)^r \lambda^{-r-1} D_i^r f \qquad (f \in \mathscr{P}^l(U)).$$

It is easy to verify that $(D_i + \lambda)(D_i + \lambda)^{-1} f = f$ in every case. However in general $(D_i + \lambda)^{-1}(D_i + \lambda) f \neq f$.

Let A be a linear transformation of U. We wish to define a linear mapping $(D_i + A)^{-1}$ of $\mathscr{P}^l(U)$ into itself such that $(D_i + A)(D_i + A)^{-1} f = f$ for all $f \in \mathscr{P}^l(U)$. First suppose A has only one eigenvalue a. Then $A - a$ is nilpotent and so $e^{(A-a)t_i}$ is an $\mathscr{E}(U)$-polynomial on $\mathbf{C}^l$. Put

$$(D_i + A)^{-1} f = e^{-(A-a)t_i}(D_i + a)^{-1}(e^{(A-a)t_i} f) \qquad (f \in \mathscr{P}^l(U)).$$

Since $D_i e^{-(A-a)t_i} = -(A-a) e^{-(A-a)t_i}$, it is clear that

$$(D_i + a)(D_i + A)^{-1} f = -(A-a)(D_i + A)^{-1} f$$

$$+ e^{-(A-a)t_i}(D_i + a)\{(D_i + a)^{-1}(e^{(A-a)t_i} f)\}$$

$$= -(A-a)(D_i + A)^{-1} f + f.$$

Therefore $(D_i + A)(D_i + A)^{-1} f = f$.

Now we come to the general case. Let $a_1,\ldots,a_r$ be all the distinct eigenvalues of A. Consider the space U_j of all $u \in U$ such that $(A - a_j)^k u = 0$ for some integer $k \geq 1$. Then U is the direct sum of $U_1,\ldots,U_r$. Let E_j denote the projection of U on U_j corresponding to this direct sum. Then $E_j^2 = E_j$, $E_j E_k = 0$ ($j \neq k$), A commutes with E_i and $E_1 + \cdots + E_r = 1$. Put

$$(D_i + A)^{-1}f = \sum_{1 \leqslant j \leqslant r} e^{-(A - a_j)t_i}(D_i + a_j)^{-1}\left(e^{(A - a_j)t_i}E_j f\right)$$

for $f \in \mathcal{P}^l(U)$. Since A has only the eigenvalue a_j on U_j, $e^{(A - a_j)t_i}E_j$ is an $\mathcal{E}(U)$-polynomial. Hence the right side of the above equation lies in $\mathcal{P}^l(U)$. Moreover $E_j f$ is a U_j-polynomial and a_j is the only eigenvalue of A on U_j. Hence it follows from what we have seen above that

$$(D_i + A)\left\{e^{-(A - a_j)t_i}(D_i + a_j)^{-1}\left(e^{(A - a_j)t_i}E_j f\right)\right\} = E_j f.$$

Therefore

$$(D_i + A)(D_i + A)^{-1}f = \sum_{1 \leqslant j \leqslant r} E_j f = f.$$

By the minimal polynomial of A, we mean, as usual, a polynomial $F(X)$ in an indeterminate X with complex coefficients such that 1) $F \neq 0$ and the highest coefficient of F is 1, 2) $F(A) = 0$ and 3) F has the lowest possible degree consistent with the two preceding conditions. It is well known that $F(X) = (X - a_1)^{m_1} \cdots (X - a_r)^{m_r}$ where $m_1,\ldots m_r$ are certain positive integers.

Lemma 17. *Let X^m be the highest power of X which divides the minimal polynomial of A. Then*

$$d^0\left((D_i + A)^{-1}f\right) \leqslant d^0 f + 2m$$

for $f \in \mathcal{P}^l(U)$.

First we need the following result.

Lemma 18. *Suppose A is nonsingular and f and g are two elements in $\mathcal{P}^l(U)$ such that $(D_i + A)g = f$. Then*

$$g = \sum_{k \geqslant 0} (-1)^k A^{-k-1}D_i^k f.$$

Put $g_0 = \sum_{k \geqslant 0}(-1)^k A^{-k-1}D_i^k f$. It is easy to check directly that $(D_i + A)g_0 = f$. Hence if $h = g - g_0$, $(D_i + A)h = 0$. We claim $h = 0$. For otherwise let d be the degree of h in t_i. Then the degree of $D_i h$ in t_i is at most $d - 1$. On the other hand since A is nonsingular, the degree of Ah in t_i is d. But this is impossible since $D_i h = - Ah$. Therefore $h = 0$ and $g = g_0$.

Now we come to Lemma 17. For any j ($1 \leqslant j \leqslant r$), $g_j = (D_i + A)^{-1}E_j f$ is a U_j-polynomial such that $(D_i + A)g_j = E_j f$. If $a_j \neq 0$, A is nonsingular on U_j and so by Lemma 18, $d^0 g_j \leqslant d^0 E_j f$. On the other hand if $a_j = 0$,

$$g_j = e^{-At_i}D_i^{-1}\left(e^{At_i}E_j f\right).$$

Since $e^{At_i} = \sum_{0 \leqslant k < m} A^k t_i^k / k!$, it is clear that

$$d^0 g_j \leqslant 2(m-1) + 1 + d^0 E_j f \leqslant 2m + d^0 E_j f.$$

Hence

$$d^0 (D_i + A)^{-1} f \leqslant \sup_j d^0 (D_i + A)^{-1} E_j f \leqslant 2m + d^0 f.$$

Lemma 19. *Let d be a nonnegative integer. Then if $d^0 f \leqslant d$ ($f \in \mathcal{P}^l(U)$), and $\lambda \neq 0$ is a complex number, we have the inequalities*

1) $\|D_i^{-1} f\| \leqslant \|f\|$

2) $\|(D_i + \lambda)^{-1} f\| \leqslant (d+1)! \|f\| \max(|\lambda|^{-1}, |\lambda|^{-(d+1)})$.

Suppose $f = u t^k$ where $u \in U$ and $\sigma(k) \leqslant d$. Then $\|D_i^{-1} f\| = (k_i + 1)^{-1} |u| \leqslant \|f\|$. Moreover,

$$\|(D_i + \lambda)^{-1} f\| = \sum_{0 \leqslant m \leqslant k_i} k_i (k_i - 1) \cdots (k_i - m + 1) |\lambda|^{-(m+1)} \|f\|$$

$$\leqslant d! \|f\| \sum_{0 \leqslant m \leqslant d} |\lambda|^{-(m+1)} \leqslant (d+1)! \|f\| \sup_{0 \leqslant m \leqslant d} |\lambda|^{-(m+1)}.$$

The required inequalities are now obvious.

Lemma 20. *Fix an element A in $\mathcal{E}(U)$. Then we can choose numbers $b > 0$ and $c \geqslant 1$ with the following property. Let d be a nonnegative integer. Then*

$$\|(D_i + A)^{-1} f\| \leqslant d! b c^d \|f\|$$

for any $f \in \mathcal{P}^l(U)$ with $d^0 f \leqslant d$.

We keep to our earlier notation. Then

$$\|(D_i + A)^{-1} f\| \leqslant \sum_{1 \leqslant j \leqslant r} \|e^{-(A-a_j)t_i} (D_i + a_j)^{-1} \left(e^{(A-a_j)t_i} E_j f \right)\|.$$

It is obvious that if $p \in \mathcal{P}^l(\mathcal{E}(U))$ and $g \in \mathcal{P}^l(U)$, then $\|pg\| \leqslant \|p\| \|g\|$. Hence

$$\|(D_i + A)^{-1} f\| \leqslant \sum_j \|e^{-(A-a_j)t_i} E_j\| \|(D_i + a_j)^{-1} \left(e^{(A-a_j)t_i} E_j f \right)\|.$$

But

$$\|e^{-(A-a_j)t_i} E_j\| = \sum_{k \geqslant 0} |(A - a_j)^k E_j| (k!)^{-1} \leqslant |E_j| e^{|A - a_j|}.$$

Also from Lemma 19 we can choose $c \geqslant 1$ such that

$$\|(D_i + a_j)^{-1} g\| \leqslant (m+1)! c^{m+1} \|g\| \qquad (1 \leqslant j \leqslant r)$$

for any integer $m \geqslant 0$ and any $g \in \mathcal{P}^l(U)$ with $d^0 g \leqslant m$. Since $d^0 f \leqslant d$, it is clear that $d^0 (e^{(A-a_j)t_i} E_j f) \leqslant d + m_j - 1$ where m_j is the least positive integer such that

$(A - a_j)^{m_j} E_j = 0$. Therefore

$$\|(D_i + a_j)^{-1}(e^{(A - a_j)t_i} E_j f)\| \leqslant (d + m_j)! c^{(d + m_j)} \|e^{(A - a_j)t_i} E_j f\|$$

$$\leqslant (d + m_j)! c^{d + m_j} e^{|A - a_j|} \|E_j\| \|f\|$$

$$\leqslant d! m_j! (2c)^{d + m_j} e^{|A - a_j|} \|E_j\| \|f\|$$

since $(d + m_j)!/d! m_j! \leqslant 2^{d + m_j}$. Therefore

$$\|(D_i + A)^{-1} f\| \leqslant d! \sum_j m_j! (2c)^{d + m_j} e^{2|A - a_j|} |E_j|^2 \|f\| \leqslant d! b c^d \|f\|$$

if

$$b = r \sup_j m_j! (2c)^{m_j} e^{2|A - a_j|} |E_j|^2.$$

Now we shall begin the proof of Lemma 11. We recall that the elements $c(\mathbf{k}) \in \mathcal{E}(U)$ satisfy the conditions

1) $\displaystyle\sum_{\mathbf{k}} |c(\mathbf{k})| e^{-\sigma(\mathbf{k}) M_0} < \infty,$

2) $c(\mathbf{k}) = 0$ if $k_l = 0$ and $\mathbf{k} \neq 0$.

Let us write $\mathbf{k} \cdot > \mathbf{k}'$ (or $\mathbf{k}' < \cdot \mathbf{k}$) if $\mathbf{k} - \mathbf{k}' \in \mathbf{Z}_+^l$ and $k_l > k_l'$.

Put $\Xi = c(0)$ and fix an eigenvalue ξ of Ξ. Let u be a vector in U such that $(\Xi - \xi)^k u = 0$ for some $k \geqslant 1$. Put

$$p_0 = u + \sum_{1 \leqslant k < m} t_l^k \frac{(\Xi - \xi)^k}{k!} u$$

where m is the least positive integer such that $(\Xi - \xi)^m u = 0$. Then $p_0 \in \mathcal{P}^l(U)$ and one verifies by direct computation that

$$(D_l + \xi - \Xi) p_0 = 0.$$

Now we define an element $p_{\mathbf{k}} \in \mathcal{P}^l(U)$ for $\mathbf{k} \in \mathbf{Z}_+^l$ by induction on k_l as follows. If $k_l = 0$ and $\mathbf{k} \neq 0$, put $p_{\mathbf{k}} = 0$. If $k_l > 0$, put

$$p_{\mathbf{k}} = (D_l + \xi - k_l - \Xi)^{-1} q_{\mathbf{k}}$$

where

$$q_{\mathbf{k}} = \sum_{\mathbf{k}' < \cdot \mathbf{k}} c(\mathbf{k} - \mathbf{k}') p_{\mathbf{k}'}.$$

Let $n = \dim U$. It follows from Lemma 17 that

$$d^0 p_{\mathbf{k}} \leqslant d^0 q_{\mathbf{k}} + 2n \leqslant \sup_{\mathbf{k}' < \cdot \mathbf{k}} d^0 p_{\mathbf{k}'} + 2n.$$

Since $d^0 p_0 \leqslant n$, one proves easily by induction on k_l that $d^0 p_{\mathbf{k}} \leqslant (2k_l + 1)n$. Now if $k \geqslant 1$,

$$\xi - k - \Xi = -k\left(1 + \frac{\Xi - \xi}{k}\right).$$

Choose an integer $N \geqslant 2$ such that $k^{-1}|\Xi - \xi| \leqslant 1/2$ for $k \geqslant N$. Then $1 + (\Xi - \xi/k)$ is nonsingular and $|(1 + (\Xi - \xi/k))^{-1}| \leqslant 1 + 1/2 + 1/2^2 + \cdots = 2$ $(k \geqslant N)$. Therefore

$$|(\xi - k - \Xi)^{-1}| \leqslant \frac{2}{k} \leqslant 1$$

for $k \geqslant N$ and so it follows from Lemma 18 that $d^0 p_{\mathbf{k}} \leqslant d^0 q_{\mathbf{k}} \leqslant \sup_{\mathbf{k}' < \cdot \mathbf{k}} d^0 p_{\mathbf{k}'}$ if $k_l \geqslant N$. Therefore it is obvious that $d^0 p_{\mathbf{k}} \leqslant (2N+1)n$ for all $\mathbf{k}$. Put $d = \sup_{\mathbf{k}} d^0 p_{\mathbf{k}}$. By Lemma 20, we can choose $b > 0$ such that

$$\|(D_l + \xi - k - \Xi)^{-1} f\| \leqslant b \|f\|$$

if $1 \leqslant k < N$ and $d^0 f \leqslant d$ $(f \in \mathscr{P}^l(U))$. We may assume that $b \geqslant (d+1)!$. Then from Lemma 18,

$$\|(D_l + \xi - k - \Xi)^{-1} f\| \leqslant \sum_{r \geqslant 0} \|D_l^r f\| \leqslant (d+1)! \|f\| \leqslant b \|f\|$$

for $k \geqslant N$ under the same assumptions on f. Hence

$$\|p_{\mathbf{k}}\| \leqslant b \|q_{\mathbf{k}}\| \qquad (k_l \geqslant 1).$$

Put $A = \sum_{\mathbf{k}} |c(\mathbf{k})|^{-\sigma(\mathbf{k}) M_0} < \infty$. Then

$$\|q_{\mathbf{k}}\| e^{-\sigma(\mathbf{k}) M_0} \leqslant e^{-\sigma(\mathbf{k}) M_0} \sum_{\mathbf{k}' < \cdot \mathbf{k}} |c(\mathbf{k} - \mathbf{k}')| \|p_{\mathbf{k}'}\|$$

$$\leqslant A \sup_{\mathbf{k}' < \cdot \mathbf{k}} \|p_{\mathbf{k}'}\| e^{-\sigma(\mathbf{k}') M_0} \qquad (k_l \geqslant 1).$$

Choose $\eta > 0$ such that $e^\eta \geqslant Ab$. Then

$$\|p_{\mathbf{k}}\| e^{-\sigma(\mathbf{k}) M_0} \leqslant e^\eta \sup_{\mathbf{k}' < \cdot \mathbf{k}} \|p_{\mathbf{k}'}\| e^{-\sigma(\mathbf{k}') M_0} \qquad (k_l \geqslant 1).$$

Now $k_l' < k_l$ if $\mathbf{k}' < \cdot \mathbf{k}$. Hence

$$\|p_{\mathbf{k}}\| e^{-\sigma(\mathbf{k}) M_0 - k_l \eta} \leqslant \sup_{\mathbf{k}' < \cdot \mathbf{k}} \|p_{\mathbf{k}'}\| e^{-\sigma(\mathbf{k}') M_0 - k_l' \eta} \qquad (k_l \geqslant 1)$$

and this proves that

$$\|p_{\mathbf{k}}\| e^{-\sigma(\mathbf{k}) M_0 - k_l \eta} \leqslant \|p_0\|.$$

Thus we have obtained the following result.

Lemma 21. *Let ξ be an eigenvalue of $\Xi = c(0)$ and u a vector in U such that $(\Xi - \xi)^k u = 0$ for some $k \geqslant 1$. Then we can select for every $\mathbf{k} \in \mathbb{Z}_+^l$ a u-polynomial $p_{\mathbf{k}}$ on $\mathbb{C}^l$ with the following properties.*

1) $p_{\mathbf{k}}(\mathbf{t})$ *depends only on* t_l.

2) $\sup_{\mathbf{k}} d^0 p_{\mathbf{k}} < \infty$.

3) $(D_l + \xi - k_l - \Xi) p_{\mathbf{k}} = \sum_{\mathbf{k}' < \cdot \mathbf{k}} c(\mathbf{k} - \mathbf{k}') p_{\mathbf{k}'}$ *for* $k_l \geqslant 1$.

4) $p_{\mathbf{k}} = 0$ *if* $k_l = 0$ *and* $\mathbf{k} \neq 0$. *Also the constant term of* p_0 *is* u.

5)   *There exists a positive number η such that*

$$\| p_{\mathbf{k}} \| \leq \| p_0 \| e^{\sigma(\mathbf{k}) M_0 + k_l \eta}$$

for all $\mathbf{k} \in \mathbf{Z}^l_+$.

Choose $M_1 \geq M_0 + \eta$ and let $\mathbf{C}^l(M_0 : M_1)$ denote the set of all points $\mathbf{t} \in \mathbf{C}^l$ such that $\operatorname{Re} t_i > M_0$ ($1 \leq i < l$) and $\operatorname{Re} t_l > M_1$. Then if $d = \sup_{\mathbf{k}} d^0 p_{\mathbf{k}}$,

$$| p_{\mathbf{k}}(\mathbf{t}) | \leq \| p_{\mathbf{k}} \| (1 + |t_l|)^d$$

and therefore the series

$$\sum_{\mathbf{k}} | p_{\mathbf{k}}(\mathbf{t}) e^{-\langle \mathbf{k}, \mathbf{t} \rangle} |$$

converges uniformly on every compact subset of $\mathbf{C}^l(M_0 : M_1)$. Put

$$\phi(\mathbf{t}) = \sum_{\mathbf{k}} p_{\mathbf{k}}(\mathbf{t}) e^{\xi t_l - \langle \mathbf{k}, \mathbf{t} \rangle} \qquad \left(\mathbf{t} \in \mathbf{C}^l(M_0 : M_1) \right).$$

Then ϕ is a holomorphic function from $\mathbf{C}^l(M_0 : M_1)$ to U. Also by Lemma 1, the above series converges decently on every compact subset of $\mathbf{C}^l(M_0 : M_1)$. Since the same evidently holds for the series $\sum_{\mathbf{k}} c(\mathbf{k}) e^{-\langle \mathbf{k}, \mathbf{t} \rangle}$, it follows immediately from condition 3) of Lemma 21 that

$$\frac{\partial \phi}{\partial t_l} = \mathcal{C}(\mathbf{t}) \phi$$

on $\mathbf{C}^l(M_0 : M_1)$. Here $\mathcal{C}(\mathbf{t}) = \sum_{\mathbf{k}} c(\mathbf{k}) e^{-\langle \mathbf{k}, \mathbf{t} \rangle}$ ($\mathbf{t} \in \mathbf{C}^l(M_0)$). Now for any number $a > M_0$ let $K(\mathbf{t} : a)$ denote the holomorphic function from $\mathbf{C}^l(M_0)$ to $\mathcal{E}(U)$ such that $K(\mathbf{t} : a) = 1$ if $t_l = a$ and $\partial K(\mathbf{t} : a) / \partial t_l = \mathcal{C}(\mathbf{t}) K(\mathbf{t} : a)$. Now choose $b > M_1$ and put

$$\Phi(\mathbf{t}) = K(\mathbf{t} : b) \phi(t_1, \ldots, t_{l-1}, b) \qquad \left(\mathbf{t} \in \mathbf{C}^l(M_0) \right).$$

Then Φ is a holomorphic function on $\mathbf{C}^l(M_0)$ such that $\partial \Phi / \partial t_l = \mathcal{C}(\mathbf{t}) \Phi$ and $\Phi(t_1, \ldots, t_{l-1}, b) = \phi(t_1, \ldots, t_{l-1}, b)$ if $\operatorname{Re} t_i > M_0$ ($1 \leq i < l$). It then follows from Lemma 7 that $\Phi = \phi$ on $\mathbf{C}^l(M_0 : M_1)$. We denote this holomorphic extension of ϕ on $\mathbf{C}^l(M_0)$ again by ϕ. Thus we have obtained the following result.

Lemma 22. *There exists a holomorphic function ϕ on $\mathbf{C}^l(M_0)$ with values in U such that*

1) $\partial \phi / \partial t_l = \mathcal{C}(\mathbf{t}) \phi.$

2) $\phi(\mathbf{t}) = \sum_{\mathbf{k}} p_{\mathbf{k}}(\mathbf{t}) e^{\xi t_l - \langle \mathbf{k}, \mathbf{t} \rangle}$ *on* $\mathbf{C}^l(M_0 : M_1).$

Now choose a base $u_1, \ldots, u_n$ for U with the following property. For each i, there exists an eigenvalue ξ_i of Ξ such that $(\Xi - \xi_i)^n u_i = 0$ ($1 \leq i \leq n$). Obviously this is possible. Let $p_{i,\mathbf{k}}$ denote the sequence of polynomials of Lemma 21 corresponding to $u = u_i$ and let ϕ_i denote the corresponding functions of Lemma 22. We may assume that M_1 is so large that

$$\| p_{i, \mathbf{k}} \| \leq \| p_{i, 0} \| e^{\sigma(\mathbf{k}) M_0 + k_l (M_1 - M_0)}$$

for $1 \leqslant i \leqslant n$ and $\mathbf{k} \in \mathbb{Z}_+^l$. Then

$$\phi_i(\mathbf{t}) = \sum_{\mathbf{k}} p_{i,\mathbf{k}}(\mathbf{t}) e^{\xi_i t_l - \langle \mathbf{k}, \mathbf{t} \rangle}$$

on $\mathbb{C}^l(M_0 : M_1)$.

Now suppose $l \geqslant 2$. Then

$$\phi_i(\mathbf{s}, t) = \sum_{k \geqslant 0} \sum_{\mathbf{m}} p_{i,\mathbf{m},k}(t) e^{(\xi_i - k)t - \langle \mathbf{m}, \mathbf{s} \rangle}$$

for $\mathbf{s} \in \mathbb{C}^{l-1}(M_0)$ and $t \in \mathbb{C}^1(M_1)$. Here $\mathbf{m}$ runs over $\mathbb{Z}_+^{l-1}$ and $(\mathbf{m}, k)$ is to be considered as an element in $\mathbb{Z}_+^l$. We can choose an integer $d \geqslant 0$ and a number $A \geqslant 0$ such that $\| p_{i,\mathbf{m},k} \| \leqslant A e^{\sigma(\mathbf{m}) M_0 + k M_1}$ and $d^0 p_{i,\mathbf{m},k} \leqslant d$ for $1 \leqslant i \leqslant n$, $\mathbf{m} \in \mathbb{Z}_+^{l-1}$ and $k \geqslant 0$. Therefore the series

$$\sum_{k \geqslant 0} | p_{i,\mathbf{m},k}(t) e^{(\xi_i - k)t} |$$

converges uniformly on every compact subset of $\mathbb{C}^1(M_1)$. Put

$$v_i(\mathbf{m} : t) = \sum_{k \geqslant 0} p_{i,\mathbf{m},k}(t) e^{(\xi_i - k)t} \quad (\mathbf{m} \in \mathbb{Z}_+^{l-1}, \ t \in \mathbb{C}^1(M_1)).$$

Then $v_i(\mathbf{m}, t)$ is a holomorphic function of t on $\mathbb{C}^1(M_1)$ with values in U. Also

$$\phi_i(\mathbf{s}, t) = \sum_{\mathbf{m}} v_i(\mathbf{m} : t) e^{-\langle \mathbf{m}, \mathbf{s} \rangle} \quad \left(\mathbf{s} \in \mathbb{C}^{l-1}(M_0), \ t \in \mathbb{C}^1(M_1) \right).$$

Since the series $\sum_{\mathbf{m}} | v_i(\mathbf{m} : t) e^{-\langle \mathbf{m}, \mathbf{s} \rangle} |$ converges uniformly with respect to $(\mathbf{s}, t)$ on every compact subset of $\mathbb{C}^{l-1}(M_0) \times \mathbb{C}^1(M_1)$, it follows that

$$\partial \phi_i(\mathbf{s}, t) / \partial t = \sum_{\mathbf{m}} \{ \partial v_i(\mathbf{m} : t) / \partial t \} e^{-\langle \mathbf{m}, \mathbf{s} \rangle}.$$

On the other hand

$$\partial \phi_i(\mathbf{s}, t) / \partial t = \mathcal{C}(\mathbf{s}, t) \phi_i(\mathbf{s}, t).$$

Now put

$$c(\mathbf{m} : t) = \sum_{k \geqslant 0} c(\mathbf{m}, k) e^{-kt} \quad \left(\mathbf{m} \in \mathbb{Z}_+^{l-1}, \ t \in \mathbb{C}^1(M_0) \right).$$

Then $\mathcal{C}(\mathbf{s}, t) = \sum_{\mathbf{m}} c(\mathbf{m} : t) e^{-\langle \mathbf{m}, \mathbf{s} \rangle}$. Now put $z_j = e^{-s_j}$ $(1 \leqslant j < l)$. Then $\mathbf{z} = (z_1, \ldots, z_{l-1})$ is a point in $\mathbb{C}^{l-1}$ and for a fixed $t \in \mathbb{C}^1(M_1)$, $\phi_i(\mathbf{s}, t)$ and $\mathcal{C}(\mathbf{s}, t)$ can be regarded as power series in $\mathbf{z} = (z_1, \ldots, z_{l-1})$ which converge absolutely for $|\mathbf{z}| < e^{-M_0}$. Hence we can equate the coefficients of $\mathbf{z}^{\mathbf{m}}$ in the two sides of the equation

$$\frac{\partial \phi_i}{\partial t}(\mathbf{s}, t) = \mathcal{C}(\mathbf{s}, t) \phi_i(\mathbf{s}, t).$$

In particular, by considering the case $\mathbf{m} = 0$, we get

$$\frac{\partial v_i}{\partial t}(0 : t) = c(0 : t) v_i(0 : t) \quad \left(t \in \mathbb{C}^1(M_1) \right).$$

Lemma 23. *Suppose $b > M_1$. Then the vectors $v_i(0 : b)$ $(1 \leqslant i \leqslant n)$ are linearly independent.*

For otherwise we can choose complex numbers a_i $(1 \leqslant i \leqslant n)$ not all zero such that $\Sigma_i a_i v_i(0 : b) = 0$. Put $v(t) = \Sigma_i a_i v_i(0 : t)$ $(t \in \mathbf{C}^1(M_1))$. Then $(\partial v(t)/\partial t) = c(0 : t)v(t)$ and $v(b) = 0$. Hence we conclude from Lemma 7 that $v(t) = 0$ for all real $t \geqslant b$. Now $d^0 p_{i,0,k} \leqslant d$ and $\|p_{i,0,k}\| \leqslant Ae^{kM_1}$. Hence if M is a sufficiently large number $(M \geqslant b)$, the series

$$\sum_{k \geqslant 0} p_{i,0,k}(t)e^{(\xi_i - k)t} \qquad (1 \leqslant i \leqslant n)$$

converge decently on $W_1(\tfrac{1}{2}, M)$. Without loss of generality we may assume that $\operatorname{Re}\xi_1 \geqslant \operatorname{Re}\xi_2 \geqslant \cdots \geqslant \operatorname{Re}\xi_n$. Let j be the least index $(1 \leqslant j \leqslant n)$ such that $a_j \neq 0$. Then since

$$\sum_i a_i \sum_{k \geqslant 0} p_{i,0,k}(t)e^{(\xi_i - k)t} = v(t) = 0$$

for $t \geqslant b$, we can (by the Corollary of Lemma 57 of [1]) equate the coefficient of $e^{\xi_j t}$ in the left hand side of this equation to zero. Let J be the set of all indices i such that $\xi_i = \xi_j$. Then we get

$$\sum_{i \in J} a_i p_{i,0}(t) = 0.$$

But since the constant term of $p_{i,0}$ is u_i, this implies that $\Sigma_{i \in J} a_i u_i = 0$. However this contradicts the linear independence of $u_1, \ldots, u_n$ since $a_j \neq 0$. So the lemma is proved.

Now fix $b > M_1$ and put $v_i(\mathbf{m}) = v_i(\mathbf{m} : b)$ $(1 \leqslant i \leqslant n, \mathbf{m} \in \mathbf{Z}_+^{l-1})$. Then

$$|v_i(\mathbf{m})| \leqslant (1 + b)^d \sum_{k \geqslant 0} \|p_{i,\mathbf{m},k}\| |e^{(\xi_i - k)b}|$$

$$\leqslant A(1 + b)^d |e^{\xi_i b}| e^{\sigma(\mathbf{m}) M_0} \sum_{k \geqslant 0} e^{-k(b - M_1)} = Be^{\sigma(\mathbf{m}) M_0}$$

where

$$B = A(1 + b)^d |e^{\xi_i b}| \sum_{k \geqslant 0} e^{-k(b - M_1)} < \infty.$$

Then

$$\phi_i(\mathbf{s}, b) = \sum_{\mathbf{m}} v_i(\mathbf{m})e^{-\langle \mathbf{m}, \mathbf{s} \rangle} \qquad \left(\mathbf{s} \in \mathbf{C}^{l-1}(M_0)\right).$$

Let α_i be the linear function on U such that[4] $\alpha_i(u_j) = \delta_{ij}$ $(1 \leqslant i, j \leqslant n)$. Then

$$\alpha_j(\phi_i(\mathbf{s}, b)) = \sum_{\mathbf{m}} \alpha_j(v_i(\mathbf{m}))\mathbf{z}^{\mathbf{m}},$$

the power series being absolutely convergent for $|\mathbf{z}| \leqslant e^{-M_0}$. Put $\Delta(\mathbf{s}) =$

---

[4] $\delta_{ij} = 1$ or 0 according as $i = j$ or not.

$\det(\alpha_j(\phi_i(s, b)))_{1 \leqslant j, i \leqslant n}$. Then it is obvious that we have an expansion of the form

$$\Delta(s) = \sum_{\mathbf{m}} \delta(\mathbf{m}) e^{-\langle \mathbf{m}, s \rangle} \qquad \left(s \in \mathbf{C}^{l-1}(M_0) \right)$$

where $\delta(\mathbf{m})$ are complex numbers and the series converges absolutely for $s \in \mathbf{C}^{l-1}(M_0)$. Now $\delta(0) = \det(\alpha_j(v_i(0))_{1 \leqslant j, i \leqslant n} \neq 0$ from Lemma 23. Hence $\Delta(s)^{-1}$ can also be expanded as a power series in $\mathbf{z}$ for sufficiently small values of $|\mathbf{z}|$. This means that we can choose $M_2 \geqslant M_0$ and $\gamma_{ij}(\mathbf{m}) \in \mathbf{C}$ such that the series

$$\Gamma_{ij}(s) = \sum_{\mathbf{m}} \gamma_{ij}(\mathbf{m}) e^{-\langle \mathbf{m}, s \rangle} \qquad (1 \leqslant i, j \leqslant n)$$

converge absolutely for $s \in \mathbf{C}^{l-1}(M_2)$ and $(\Gamma_{ij}(s))_{1 \leqslant i, j \leqslant n}$ is the inverse of the matrix $\{\alpha_i(\phi_j(s, b))\}_{1 \leqslant i, j \leqslant n}$. Then

$$u_i = \sum_{1 \leqslant j \leqslant n} \Gamma_{ij}(s) \phi_j(s, b) \qquad \left(s \in \mathbf{C}^{l-1}(M_2) \right).$$

Put $\psi_j(t) = \phi_j(t) - K(t : b) \phi_j(t_1, \ldots, t_{l-1}, b)$. It is clear that $\partial \psi_j(t) / \partial t_l = \mathcal{C}(t) \psi_j(t)$ and $\psi_j(t) = 0$ if $t_l = b$. Since ψ_j is a holomorphic function on $\mathbf{C}^l(M_0)$, it follows from Lemma 7 that $\psi_j = 0$. Therefore $\phi_j(t) = K(t : b) \phi_j(t_1, \ldots, t_{l-1}, b)$ for $t \in \mathbf{C}^l(M_0)$ and

$$K(t : b) u_i = \sum_{1 \leqslant j \leqslant n} \Gamma_{ij}(t_1, \ldots, t_{l-1}) K(t : b) \phi_j(t_1, \ldots, t_{l-1} : b)$$

$$= \sum_{1 \leqslant j \leqslant n} \Gamma_{ij}(t_1, \ldots, t_{l-1}) \phi_j(t) \qquad \left(t \in \mathbf{C}^l(M_2) \right).$$

Put

$$q_{ij, \mathbf{m}, k}(t) = \sum_{\mathbf{m}_1 + \mathbf{m}_2 = \mathbf{m}} \gamma_{ij}(\mathbf{m}_1) p_{j, \mathbf{m}_2, k}(t_l) \qquad \left(\mathbf{m} \in \mathbf{Z}_+^{l-1}, \ k \geqslant 0 \right).$$

Since the series for $\Gamma_{ij}(s)$ converges absolutely for $s = (M_2 + 1)\mathbf{l}$, it is clear that $A_1 = \sup_{i, j} \sum_{\mathbf{m}} |\gamma_{ij}(\mathbf{m})| e^{-\sigma(\mathbf{m})(M_2 + 1)} < \infty$. Also

$$\| p_{j, \mathbf{m}, k} \| \leqslant A e^{\sigma(\mathbf{m}) M_0 + k M_1}.$$

Hence

$$\| q_{ij, \mathbf{m}, k} \| \leqslant A_1 A e^{\sigma(\mathbf{m})(M_2 + 1) + M_k j}.$$

Also, $d^0 q_{ij, \mathbf{m}, k} \leqslant d$ and therefore $q_{ij, \mathbf{k}} \ (\mathbf{k} \in \mathbf{Z}_+^l)$ is a normal sequence of U-polynomials and

$$K(t : b) u_i = \sum_{1 \leqslant j \leqslant n} \sum_{\mathbf{k}} q_{ij, \mathbf{k}}(t) e^{t_j t_l - \langle \mathbf{k}, t \rangle} \qquad \left(t \in \mathbf{C}^l(M_2) \right).$$

Let $Q_{j, \mathbf{k}}$ denote the $\mathcal{E}(U)$-polynomial defined by the condition that $Q_{j, \mathbf{k}}(t) u_i = q_{ij, \mathbf{k}}(t)$. Then $d^0 Q_{j, \mathbf{k}} \leqslant d$. Since $u_1, \ldots, u_n$ is a base for U, it is obvious that we can choose a number $B > 0$ such that $|T| \leqslant B \sum_i |T u_i|$ for any $T \in \mathcal{E}(U)$. Let $Q = \sum_{\mathbf{k}} T(\mathbf{k}) t^{\mathbf{k}}$ $(T(\mathbf{k}) \in \mathcal{E}(U))$ be a $\mathcal{E}(U)$-polynomial. Then

$$\| Q \| = \sum_{\mathbf{k}} |T(\mathbf{k})| \leqslant B \sum_i \sum_{\mathbf{k}} |T(\mathbf{k}) u_i| = B \sum_i \| q_i \|$$

where q_i is the U-polynomial given by $q_i(\mathbf{t}) = Q(\mathbf{t})u_i$. Hence, in particular,

$$\|Q_{j,\mathbf{k}}\| \leqslant B\sum_i \|q_{ij,\mathbf{k}}\| \leqslant nBA_1 Ae^{\sigma(\mathbf{k})M}$$

when $M = \max(M_2 + 1, M_1)$. This proves that $Q_{j,\mathbf{k}}$ $(\mathbf{k} \in \mathbf{Z}_+^l)$ is a normal sequence of $\mathcal{E}(U)$-polynomials. It is obvious that

$$K(\mathbf{t}:b) = \sum_{1 \leqslant j \leqslant n} \sum_{\mathbf{k}} Q_{j,\mathbf{k}}(\mathbf{t}) e^{\xi_j t_l - \langle \mathbf{k},\mathbf{t}\rangle} \qquad \big(\mathbf{t} \in \mathbf{C}^l(M)\big).$$

Now fix a number $a > M_0$. The number b above was arbitrary except for the condition that $b > M_1$. Hence we may assume that $b > a$. Put $K_0(\mathbf{t}) = K(\mathbf{t}:a) - K(\mathbf{t}:b)K(t_1,\ldots,t_{l-1},b:a)$ $(\mathbf{t} \in \mathbf{C}^l(M_0))$. Then $\partial K_0(\mathbf{t})/\partial t_l = \mathcal{C}(\mathbf{t})K_0(\mathbf{t})$ and $K_0(\mathbf{t}) = 0$ if $t_l = b$. Hence it follows from Lemma 7 that $K_0(\mathbf{t}) = 0$ if t_l is real and $\geqslant b$. But since K_0 is a holomorphic function, this implies that $K_0 = 0$ and so

$$K(\mathbf{s}, t:a) = K(\mathbf{s}, t:b)K(\mathbf{s}, b:a) \qquad \big(\mathbf{s} \in \mathbf{C}^{l-1}(M_0),\ t \in \mathbf{C}^l(M_0)\big).$$

Put $K_0(\mathbf{s}, b:a) = 1 \in \mathcal{E}(U)$ and

$$K_q(\mathbf{s}, b:a) = \int_{b \geqslant t_1 \geqslant \cdots \geqslant t_q \geqslant a} \mathcal{C}(\mathbf{s}, t_1)\mathcal{C}(\mathbf{s}, t_2)\cdots \mathcal{C}(\mathbf{s}, t_q)\, dt_1 \cdots dt_q \qquad (q \geqslant 1)$$

for $\mathbf{s} \in \mathbf{C}^{l-1}(M_0)$. Then, by Lemma 7, $K(\mathbf{s}, b:a) = \sum_{q \geqslant 0} K_q(\mathbf{s}, b:a)$. Now $\mathcal{C}(\mathbf{s}, t) = \sum_{\mathbf{m}} c(\mathbf{m}:t) e^{-\langle \mathbf{m},\mathbf{s}\rangle}$ and if $t > M_0$,

$$\sum_{\mathbf{m}} |c(\mathbf{m}:t)| e^{-\sigma(\mathbf{m})M_0} \leqslant \sum_{k \geqslant 0} \sum_{\mathbf{m}} |c(\mathbf{m}, k)| e^{-\sigma(\mathbf{m})M_0 - kM_0} = A \quad \text{(say)}.$$

Then $A < \infty$ and $|c(\mathbf{m}:t)| \leqslant Ae^{\sigma(\mathbf{m})M_0}$. Put $\kappa_0(0:b, a) = 1 \in \mathcal{E}(U)$ and $\kappa_0(\mathbf{m}:b, a) = 0$ if $\mathbf{m} \neq 0$ $(\mathbf{m} \in \mathbf{Z}_+^{l-1})$. Moreover, let

$$\kappa_q(\mathbf{m}:b, a) = \sum_{\mathbf{m}_1 + \cdots + \mathbf{m}_q = \mathbf{m}} \int_{b \geqslant t_1 \geqslant \cdots \geqslant t_q \geqslant a} c(\mathbf{m}_1:t_1)c(\mathbf{m}_2:t_2)\cdots$$

$$\times c(\mathbf{m}_q:t_q)\, dt_1 \cdots dt_q$$

for $q \geqslant 1$. Then

$$|\kappa_q(\mathbf{m}:b, a)| \leqslant A^q e^{\sigma(\mathbf{m})M_0} \frac{(b-a)^q}{q!}.$$

Therefore $\sum_{\mathbf{m}} \sum_q |\kappa_q(\mathbf{m}:b, a)| e^{-\sigma(\mathbf{m})M_0} \leqslant e^{A(b-a)} < \infty$. Put

$$\kappa(\mathbf{m}:b, a) = \sum_{q \geqslant 0} \kappa_q(\mathbf{m}:b, a).$$

Then $|\kappa(\mathbf{m}:b, a)| \leqslant \sum_{q \geqslant 0} |\kappa_q(\mathbf{m}:b, a)| \leqslant e^{A(b-a)+\sigma(\mathbf{m})M_0}$. It is clear that

$$K_q(\mathbf{s}:b, a) = \sum_{\mathbf{m}} \kappa_q(\mathbf{m}:b, a) e^{-\langle \mathbf{m},\mathbf{s}\rangle}$$

for $\mathbf{s} \in \mathbf{C}^{l-1}(M_0)$. Hence

$$K(\mathbf{s}:b, a) = \sum_{q \geqslant 0} K_q(\mathbf{s}:b, a) = \sum_{\mathbf{m}} \kappa(\mathbf{m}:b, a) e^{-\langle \mathbf{m},\mathbf{s}\rangle} \qquad \big(\mathbf{s} \in \mathbf{C}^{l-1}(M_0)\big).$$

Now put

$$P_{j,\mathbf{k}}(\mathbf{t}) = \sum_{\mathbf{k}'+\mathbf{m}=\mathbf{k}} Q_{j,\mathbf{k}'}(\mathbf{t})\kappa(\mathbf{m}:b,a) \qquad (\mathbf{k}\in\mathbf{Z}_+^l)$$

where the sum is over all pairs $(\mathbf{k}',\mathbf{m})$ $(\mathbf{k}'\in\mathbf{Z}_+^l, \mathbf{m}\in\mathbf{Z}_+^{l-1})$ such that $\mathbf{k}'+\mathbf{m} = (k_1'+m_1, k_2'+m_2,\ldots,k_{l-1}'+m_{l-1}, k_l') = \mathbf{k}$. Clearly $d^0 P_{j,\mathbf{k}}\leqslant d$ and

$$\|P_{j,\mathbf{k}}\| \leqslant \sum_{\mathbf{k}'+\mathbf{m}=\mathbf{k}} \|Q_{j,\mathbf{k}'}\|\,|\kappa(\mathbf{m}:b,a)|.$$

We have seen that we can choose numbers $M\geqslant M_0+1$ and $B\geqslant 0$ such that $\|Q_{j,\mathbf{k}}\|\leqslant Be^{\sigma(\mathbf{k})M}$ for $1\leqslant j\leqslant n$, $\mathbf{k}\in\mathbf{Z}_+^l$. Then

$$\|P_{j,\mathbf{k}}\|e^{-\sigma(\mathbf{k})M} \leqslant Be^{A(b-a)}\sum_{\mathbf{m}} e^{-\sigma(\mathbf{m})} < \infty.$$

This proves that $P_{j,\mathbf{k}}$ $(\mathbf{k}\in\mathbf{Z}_+^l)$ is a normal sequence of $\mathfrak{E}(U)$-polynomials. Moreover it is now obvious that

$$K(\mathbf{t}:a) = K(\mathbf{t}:b)K(t_1,\ldots,t_{l-1},b:a)$$

$$= \sum_{1\leqslant j\leqslant n}\sum_{\mathbf{k}} P_{j,\mathbf{k}}(\mathbf{t})e^{\xi_j t_l - \langle\mathbf{k},\mathbf{t}\rangle} \qquad (\mathbf{t}\in\mathbf{C}^l(M)).$$

The statement of Lemma 11 now follows immediately from Lemma 4.

We have assumed above that $l\geqslant 2$. The case $l=1$ is treated in the same way except for some minor changes. Corresponding to Lemma 23, we first show that the vectors $\phi_1(b),\ldots,\phi_n(b)$ are linearly independent. Hence

$$u_i = \sum_j \Gamma_{ij}\phi_j(b) \qquad (1\leqslant i\leqslant n)$$

when Γ_{ij} are certain complex numbers. Therefore

$$K(t:b)u_i = \sum_j \Gamma_{ij}\phi_j(t) \qquad \left(t\in\mathbf{C}^1(M_0)\right).$$

Since $K(t:a) = K(t:b)K(b:a)$, we now define the U-polynomials $P_{j,k}$ $(1\leqslant j\leqslant n, k\geqslant 0)$ by $P_{j,k}(t)=Q_{j,k}(t)K(b:a)$ where $Q_{j,k}(t)u_i = p_{j,k}(t)u_i$ $(1\leqslant i\leqslant n)$. Then $P_{j,k}$ $(k\geqslant 0)$ is a normal sequence of $\mathfrak{E}(U)$-polynomials and

$$K(t:a) = \sum_{1\leqslant j\leqslant n}\sum_{k\geqslant 0} P_{j,k}(t)e^{(\xi_j-k)t}$$

for $t\in\mathbf{C}^1(M)$, where M is some number $\geqslant M_0$. Taking into account Lemma 4, this proves Lemma 11 in this case.

§ 6.

Let $\mathfrak{R}$ be the ring of meromorphic functions on $\mathbf{C}^l$ generated over $\mathbf{C}$ by the functions 1, $(e^{\langle\mathbf{k},\mathbf{t}\rangle}+1)^{-1}$ and $(e^{\langle\mathbf{k},\mathbf{t}\rangle}-1)^{-1}$ $(\mathbf{k}\in\mathbf{Z}_+^l, \mathbf{k}\neq 0)$.

Suppose $l=l_1+l_2$ where l_1, l_2 are two positive integers. Then $\mathbf{C}^l = \mathbf{C}^{l_1}\times\mathbf{C}^{l_2}$. For any $\mathbf{t}\in\mathbf{C}^l$, we denote by $\mathbf{t}_1,\mathbf{t}_2$ the projections of $\mathbf{t}$ in $\mathbf{C}^{l_1}$ and $\mathbf{C}^{l_2}$ respectively so that $\mathbf{t}=(\mathbf{t}_1,\mathbf{t}_2)$.

Lemma 24. *Let $f \in \mathcal{R}$. Then we can choose complex numbers $a(\mathbf{k})$ $(\mathbf{k} \in \mathbb{Z}_+^l)$ such that*

$$\sum_{\mathbf{k} \in \mathbb{Z}_+^l} |a(\mathbf{k})| e^{-\sigma(\mathbf{k})\alpha} < \infty$$

for every $\alpha > 0$ and

$$f(\mathbf{t}) = \sum_{\mathbf{k}} a(\mathbf{k}) e^{-\langle \mathbf{k}, \mathbf{t} \rangle} \qquad \left(\mathbf{t} \in \mathbb{C}_+^l \right).$$

It is clear that if our lemma holds for two elements $f_1, f_2 \in \mathcal{R}$, then it also holds for $f_1 + f_2, f_1 f_2$ and cf_1 $(c \in \mathbb{C})$. Hence it is enough to verify it for $f = \langle e^{\langle \mathbf{k}, \mathbf{t} \rangle} \pm 1 \rangle^{-1}$ $(\mathbf{k} \neq 0)$. But in this case, it is obvious.

Lemma 25. *Let Ω be a bounded domain in $\mathbb{C}^{l_1}$ and let $f \in \mathcal{R}$. Then we can choose a finite number of hyperplanes $H_1, \ldots, H_r$ in $\mathbb{C}^{l_1}$ and a number $M > 0$ such that:*

1) $H_i \cap \mathbb{C}_+^{l_1}$ *is empty* $(1 \leqslant i \leqslant r)$.
2) *Let Ω' denote the complement of $\cup_{1 \leqslant i \leqslant r} H_i$ in Ω. Then f is holomorphic on*

$$\Omega' \times \mathbb{C}^{l_2}(M).$$

3) *For each $\mathbf{m} \in \mathbb{Z}_+^{l_2}$, there exists a holomorphic function $a(\mathbf{z}:\mathbf{m})$ $(\mathbf{z} \in \Omega')$ on Ω' such that the series*

$$\sum_{\mathbf{m}} |a(\mathbf{z}:\mathbf{m})| e^{-\sigma(\mathbf{m})M}$$

converges uniformly with respect to $\mathbf{z}$ on every compact subset of Ω' and

$$f(\mathbf{t}) = \sum_{\mathbf{m}} a(\mathbf{t}_1:\mathbf{m}) e^{-\langle \mathbf{m}, \mathbf{t}_2 \rangle}$$

on $\Omega' \times \mathbb{C}^{l_2}(M)$.

Again it follows without difficulty (from Lemma 2) that if the above lemma holds for two elements f_1, f_2 in $\mathcal{R}$, then it also holds for cf_1 $(c \in \mathbb{C})$, $f_1 + f_2$ and $f_1 f_2$. Hence it is sufficient to consider the case when $f = \langle e^{\langle \mathbf{k}, \mathbf{t} \rangle} + \varepsilon \rangle^{-1}$ where $\varepsilon = +1$ or -1 and $\mathbf{k} \neq 0$ $(\mathbf{k} \in \mathbb{Z}_+^l)$. Now $\mathbf{k} = (\mathbf{k}_1, \mathbf{k}_2)$ where $\mathbf{k}_1 \in \mathbb{Z}_+^{l_1}$, $\mathbf{k}_2 \in \mathbb{Z}_+^{l_2}$. First suppose $\mathbf{k}_2 \neq 0$. Choose $M > 0$ such that $|\langle \mathbf{k}_1, \mathbf{z} \rangle| \leqslant M/2$ for $\mathbf{z} \in \Omega$. Then if $\mathbf{t} \in \Omega \times \mathbb{C}^{l_2}(M)$, $|e^{-\langle \mathbf{k}, \mathbf{t} \rangle}| \leqslant e^{M(-1/2 - \sigma(\mathbf{k}_2))} \leqslant e^{-M/2}$ and therefore

$$\left(e^{\langle \mathbf{k}, \mathbf{t} \rangle} + \varepsilon \right)^{-1} = e^{-\langle \mathbf{k}, \mathbf{t} \rangle} \left(1 + \varepsilon e^{-\langle \mathbf{k}, \mathbf{t} \rangle} \right)^{-1}$$

$$= \sum_{q \geqslant 0} \varepsilon^q e^{-(q+1)\langle \mathbf{k}, \mathbf{t} \rangle}$$

for $\mathbf{t} \in \Omega \times \mathbb{C}^{l_2}(M)$. Hence if we choose $r = 0$ and

$$a(\mathbf{z}:\mathbf{m}) = \begin{cases} 0 & \text{if } \mathbf{m} \neq q\mathbf{k}_1 \\ \varepsilon^q e^{-(q+1)\langle \mathbf{k}_1, \mathbf{z} \rangle} & \text{if } \mathbf{m} = q\mathbf{k}_1 \ (q \geqslant 0) \end{cases}$$

all our conditions are satisfied.

Now suppose $\mathbf{k}_2 = 0$. Put $\eta = 1$ or 0 according as $\varepsilon = 1$ or -1. For every integer m, consider the hyperplane $H(m)$ in $\mathbf{C}^{l_1}$ given by the equation $\langle \mathbf{k}_1, \mathbf{z} \rangle = (-1)^{1/2}\pi\{2m + \eta\}$. The function $F(\mathbf{z}) = (e^{\langle \mathbf{k}_1, \mathbf{z} \rangle} + \varepsilon)^{-1}$ is holomorphic on $\mathbf{C}^{l_1}$ outside the set $\cup_m H(m)$. Note that $H(m) \cap \mathbf{C}^{l_1}_+$ is empty. Since Ω is bounded, $H(m)$ meets Ω only for a finite number of integers m. Hence we can choose a finite set of hyperplanes $H_1, \ldots, H_r$ among $H(m)$, such that F is holomorphic on the complement Ω' of $\cup_{1 \leqslant i \leqslant r} H_i$ in Ω. Now put

$$a(\mathbf{z}:\mathbf{m}) = \begin{cases} F(\mathbf{z}) & \text{if } \mathbf{m} = 0 \\ 0 & \text{if } \mathbf{m} \neq 0 \end{cases} \quad (\mathbf{m} \in \mathbf{Z}^{l_2}_+)$$

and let M be any positive number. Then all our conditions are again fulfilled.

Let D_1, D_2 be two domains in $\mathbf{C}^{l_1}$ and $\mathbf{C}^{l_2}$ respectively and let p be a holomorphic function on $D_1 \times D_2$ with values in U. We say that $p(\mathbf{t})$ is a polynomial function of $\mathbf{t}_2$, if there exist holomorphic functions $g_\mathbf{m}$ $(\mathbf{m} \in \mathbf{Z}^{l_2}_+)$ from D_1 to U such that the following conditions hold.

1) $g_\mathbf{m} = 0$ for all $\mathbf{m} \in \mathbf{Z}^{l_2}_+$ except a finite number.
2) $p(\mathbf{t}) = \sum_\mathbf{m} g_\mathbf{m}(\mathbf{t}_1)\mathbf{t}_2^\mathbf{m}$ for $\mathbf{t} \in D_1 \times D_2$.

Obviously if p is a polynomial function on $\mathbf{t}_2$, we can regard it as a holomorphic function on $D_1 \times \mathbf{C}^{l_2}$.

Obviously the functions $g_\mathbf{m}$, if they exist, are unique. Let d be a nonnegative integer. We say that p is of degree $\leqslant d$ in $\mathbf{t}_2$ if $g_\mathbf{m} = 0$ for all $\mathbf{m}$ with $\sigma(\mathbf{m}) > d$. It follows from Taylor's formula that in order that p be a polynomial function in $\mathbf{t}_2$ of degree $\leqslant d$, it is necessary and sufficient that all derivatives of p with respect to $\mathbf{t}_2$ of order $> d$ be zero.

V being a finite-dimensional complex Banach space, let $\mathcal{R}_V$ denote the tensor product $\mathcal{R} \otimes V$. Obviously an element of $\mathcal{R}_V$ can be regarded as a meromorphic function on $\mathbf{C}^l$ with values in V. Let D be a domain in $\mathbf{C}^{l_1}$ such that $D \supset \mathbf{C}^{l_1}_+$.

Theorem 3. *Let* $\mathcal{C}_i$ *$(1 \leqslant i \leqslant l)$ be given functions in* $\mathcal{R}_{\mathcal{E}(U)}$ *and* ϕ *a holomorphic function on* $D \times \mathbf{C}^{l_1}_+$ *with values in* U *such that*

$$\frac{\partial \phi}{\partial t_i} = \mathcal{C}_i(\mathbf{t})\phi \qquad (1 \leqslant i \leqslant l).$$

Then we can choose a finite or countable set Λ_2 *of points in* $\mathbf{C}^{l_2}$ *and for each* $\lambda \in \Lambda_2$, *a holomorphic function* ϕ_λ *from* $D \times \mathbf{C}^{l_2}$ *to* U *such that the following conditions are fulfilled.*

1) $\phi_\lambda(\mathbf{t})$ *is a polynomial function of* $\mathbf{t}_2$.
2) *Let* ω *be an open subset of* D *whose closure in* D *is compact. Then we can choose* $M > 0$ *such that for any* ε *$(0 < \varepsilon < 1)$, the series*

$$\sum_{\lambda \in \Lambda_2} \phi_\lambda(\mathbf{t})e^{\langle \lambda, \mathbf{t}_2 \rangle}$$

converges decently on $\omega \times W_{l_2}(\varepsilon, M)$ *and its sum is equal to* $\phi(\mathbf{t})$.

Moreover if we require that $\phi_\lambda \neq 0$ *for* $\lambda \in \Lambda_2$, *then the set* Λ_2 *and the functions* ϕ_λ *$(\lambda \in \Lambda_2)$ are uniquely determined by the above conditions.*

The last statement of the theorem is an immediate consequence of the Corollary of Lemma 57 of [1]. For the rest we may obviously assume that $\phi \neq 0$. It follows from Lemma 24 that Theorem 1 is applicable. Therefore

$$\phi(\mathbf{t}) = \sum_{\lambda \in \Lambda} p_\lambda(\mathbf{t}) e^{\langle \lambda, \mathbf{t} \rangle} \qquad \left(\mathbf{t} \in \mathbf{C}_+^l \right)$$

in the notation of Theorem 1. Let Λ_0 be the set of all maximal elements in Λ and let Λ' be the set of all elements in $\mathbf{C}^l$ of the form $\lambda - \mathbf{k}$ ($\lambda \in \Lambda_0$, $\mathbf{k} \in \mathbf{Z}_+^l$). Define $p_\mu = 0$ for any $\mu \in \Lambda'$ if μ is not in Λ. Then from Theorem 2, Λ_0 is a finite set and

$$\phi(\mathbf{t}) = \sum_{\lambda \in \Lambda'} p_\lambda(\mathbf{t}) e^{\langle \lambda, \mathbf{t} \rangle} \qquad \left(\mathbf{t} \in \mathbf{C}_+^l \right).$$

Let Ω be a bounded subdomain of D such that $\Omega \cap \mathbf{C}_+^{l_1}$ is not empty and Ω contains the closure of ω. By Lemma 25 we can choose a number $M_0 > 0$ and a finite number of hyperplanes $H_1, \ldots, H_r$ in $\mathbf{C}^{l_1}$ such that

1) H_i does not meet $\mathbf{C}_+^{l_1}$ ($1 \leqslant i \leqslant r$).
2) Let Ω' denote the complement of $\cup_{1 \leqslant i \leqslant r} H_i$ in Ω. Then the functions $\mathcal{C}_i(\mathbf{t})$ ($1 \leqslant i \leqslant l$) are holomorphic on $\Omega' \times \mathbf{C}^{l_2}(M_0)$.
3) There exist holomorphic functions $c_i(\mathbf{z} : \mathbf{m})$ ($1 \leqslant i \leqslant l$, $\mathbf{m} \in \mathbf{Z}_+^{l_2}$, $\mathbf{z} \in \Omega'$) from Ω' to $\mathcal{E}(U)$ such that the series

$$\sum_{1 \leqslant i \leqslant l} \sum_{\mathbf{m}} |c_i(\mathbf{z} : \mathbf{m})| e^{-\sigma(\mathbf{m}) M_0}$$

converges uniformly with respect to $\mathbf{z}$ on every compact subset of Ω' and

$$\mathcal{C}_i(\mathbf{t}) = \sum_{\mathbf{m}} c_i(\mathbf{t}_1 : \mathbf{m}) e^{-\langle \mathbf{m}, \mathbf{t}_2 \rangle} \qquad \left(\mathbf{t} \in \Omega' \times \mathbf{C}^{l_2}(M_0),\ 1 \leqslant i \leqslant l \right).$$

Let $\mathfrak{J}$ denote the closed interval $0 \leqslant s \leqslant 1$ in $\mathbf{R}$. By a path ζ in Ω, we mean a continuously differentiable function ζ on $\mathfrak{J}$ with values in Ω'. For such a path ζ, put

$$\mathcal{C}_\zeta(s : \mathbf{u}) = \sum_{1 \leqslant i \leqslant l_1} \mathcal{C}_i(\zeta(s), \mathbf{u}) \frac{d\zeta_i(s)}{ds} \qquad (s \in \mathfrak{J})$$

and

$$K_\zeta(\mathbf{u}) = 1 + \sum_{q \geqslant 1} \int_{1 \geqslant s_1 \geqslant s_2 \geqslant \cdots \geqslant s_q \geqslant 0} \mathcal{C}_\zeta(s_1, \mathbf{u}) \mathcal{C}_\zeta(s_2 : \mathbf{u}) \cdots \mathcal{C}_\zeta(s_q : \mathbf{u})\, ds_1 \cdots ds_q$$

for $\mathbf{u} \in \mathbf{C}^{l_2}(M_0)$.

Lemma 26. *Let ζ be a path in Ω'. Then K_ζ is a holomorphic function on $\mathbf{C}^{l_2}(M_0)$ with values in $\mathcal{E}(U)$. Moreover we can choose elements $\kappa_\zeta(\mathbf{m}) \in \mathcal{E}(U)$ ($\mathbf{m} \in \mathbf{Z}_+^{l_2}$) such that*

$$\sum_{\mathbf{m}} |\kappa_\zeta(\mathbf{m})| e^{-\sigma(\mathbf{m}) M_0} < \infty$$

and

$$K_\zeta(\mathbf{u}) = \sum_{\mathbf{m}} \kappa_\zeta(\mathbf{m}) e^{-\langle \mathbf{m}, \mathbf{u} \rangle} \qquad \left(\mathbf{u} \in \mathbf{C}^{l_2}(M_0) \right).$$

Put

$$c_\zeta(s:\mathbf{m}) = \sum_{1 \leqslant i \leqslant l_1} c_i(\zeta(s):\mathbf{m})\frac{d\zeta(s)}{ds} \qquad (s \in \mathfrak{T}, \; \mathbf{m} \in \mathbf{Z}_+^{l_2}).$$

Then

$$\mathcal{C}_\zeta(s:\mathbf{u}) = \sum_{\mathbf{m}} c_\zeta(s:\mathbf{m}) e^{-\langle \mathbf{m}, \mathbf{u} \rangle} \qquad (s \in \mathfrak{T}, \; \mathbf{u} \in \mathbf{C}^{l_2}(M_0)).$$

Let

$$\kappa_\zeta(q:\mathbf{m}) = \sum_{\mathbf{m}_1 + \cdots + \mathbf{m}_q = \mathbf{m}} \int_{1 \geqslant s_1 \geqslant \cdots \geqslant s_q \geqslant 0} c_\zeta(s_1:\mathbf{m}_1) c_\zeta(s_2:\mathbf{m}_2) \cdots$$
$$\times c_\zeta(s_q:\mathbf{m}_q)\, ds_1 \cdots ds_q$$

for $\mathbf{m} \in \mathbf{Z}_+^{l_2}$ and $q \geqslant 1$. Here this sum is over all sequences $(\mathbf{m}_1, \ldots, \mathbf{m}_q)$ of q elements in $\mathbf{Z}_+^{l_2}$ such that $\mathbf{m}_1 + \cdots + \mathbf{m}_q = \mathbf{m}$. Since the image of $\mathfrak{T}$ under ζ is compact, we can choose a number $A \geqslant 0$ such that $\sum_{\mathbf{m}} |c_\zeta(s:\mathbf{m})| e^{-\sigma(\mathbf{m})M_0} \leqslant A$ for $s \in \mathfrak{T}$. Then it is clear that

$$\sum_{\mathbf{m}} |\kappa_\zeta(q:\mathbf{m})| e^{-\sigma(\mathbf{m})M_0} \leqslant \frac{A^q}{q!}.$$

Now define $\kappa_\zeta(0:\mathbf{m}) = 1$ or 0 according as $\mathbf{m} = 0$ or not $(\mathbf{m} \in \mathbf{Z}_+^{l_2})$. Then

$$\sum_{q \geqslant 0} \sum_{\mathbf{m}} |\kappa_\zeta(q:\mathbf{m})| e^{-\sigma(\mathbf{m})M_0} \leqslant e^A.$$

Also, one proves without difficulty that

$$\int_{1 \geqslant s_1 \geqslant \cdots \geqslant s_q \geqslant 0} \mathcal{C}_\zeta(s_1:\mathbf{u}) \mathcal{C}_\zeta(s_2:\mathbf{u}) \cdots \mathcal{C}_\zeta(s_q:\mathbf{u})\, ds_1 \cdots ds_q = \sum_{\mathbf{m}} \kappa_\zeta(q:\mathbf{m}) e^{-\langle \mathbf{m}, \mathbf{u} \rangle}$$

for $\mathbf{u} \in \mathbf{C}^l(M_0)$ and $q \geqslant 1$. Put $\kappa_\zeta(\mathbf{m}) = \sum_{q \geqslant 0} \kappa_\zeta(q:\mathbf{m})$. Then

$$\sum_{\mathbf{m}} |\kappa_\zeta(\mathbf{m})| e^{-\sigma(\mathbf{m})M_0} \leqslant \sum_{q \geqslant 0} \sum_{\mathbf{m}} |\kappa_\zeta(q:\mathbf{m})| e^{-\sigma(\mathbf{m})M_0} \leqslant e^A < \infty$$

and

$$K_\zeta(\mathbf{u}) = \sum_{\mathbf{m}} \kappa_\zeta(\mathbf{m}) e^{-\langle \mathbf{m}, \mathbf{u} \rangle} \qquad (\mathbf{u} \in \mathbf{C}^{l_2}(M_0)).$$

Since the series $\sum_{\mathbf{m}} |\kappa_\zeta(\mathbf{m}) e^{-\langle \mathbf{m}, \mathbf{u} \rangle}|$ obviously converges uniformly on every compact subset of $\mathbf{C}^{l_2}(M_0)$, K_ζ is a holomorphic function on $\mathbf{C}^{l_2}(M_0)$.

Fix a point $\mathbf{z}_0 \in \Omega'$ and let V be an open and convex neighborhood of $\mathbf{z}_0$ in Ω'. We assume moreover that the closure of V in Ω' is compact. Fix a point $\mathbf{w} \in \Omega \cap \mathbf{C}_+^{l_1} = \Omega' \cap \mathbf{C}_+^{l_1}$. Since Ω' is connected, we can select a path ζ in Ω' such that $\zeta(0) = \mathbf{w}$ and $\zeta(1) = \mathbf{z}_0$. Let $\mathbf{z}$ be a point in V. Put $\xi(s) = (1-s)\mathbf{z}_0 + s\mathbf{z}$ $(0 \leqslant s \leqslant 1)$. Since V is convex, ξ is a path in V. Put $K(\mathbf{z}:\mathbf{u}) = K_\xi(\mathbf{u}) K_\zeta(\mathbf{u})$ $(\mathbf{u} \in \mathbf{C}^l(M_0))$.

Lemma 27. *For each $\mathbf{m} \in \mathbf{Z}_+^{l_2}$, we can choose a holomorphic function $\kappa(\mathbf{z}:\mathbf{m})$ on V such that:*

1) $\sum_{\mathbf{m}} |\kappa(\mathbf{z}:\mathbf{m})| e^{-\sigma(\mathbf{m})M_0}$ *remains bounded for $\mathbf{z}$ in V.*

2) *For any $M > M_0$, the series $\sum_{\mathbf{m}}|\kappa(\mathbf{z}:\mathbf{m})|e^{-\sigma(\mathbf{m})M}$ converges uniformly with respect to $\mathbf{z} \in V$.*

3) $K(\mathbf{z}:\mathbf{u}) = \sum_{\mathbf{m}} \kappa(\mathbf{z}:\mathbf{m})e^{-\langle \mathbf{m},\mathbf{u}\rangle} \qquad (\mathbf{u} \in \mathbb{C}^l(M_0),\ \mathbf{z} \in V).$

For any $\mathbf{z} \in V$, define ξ as above and put $K_0(\mathbf{z}:\mathbf{u}) = K_\xi(\mathbf{u})$. Then since $\xi(s) = (1-s)\mathbf{z}_0 + s\mathbf{z}$ $(0 \leqslant s \leqslant 1)$,

$$c_\xi(\mathbf{m}:s) = c(\mathbf{z}:\mathbf{m}:s)$$

where

$$c(\mathbf{z}:\mathbf{m}:s) = \sum_{1 \leqslant i \leqslant l_1} c_i\big(\mathbf{m}:(1-s)\mathbf{z}_0 + s\mathbf{z}\big)(z_i - z_{0i}) \qquad (s \in \mathfrak{I}).$$

Now put

$$\kappa_0(\mathbf{z}:q:\mathbf{m}) = \sum_{\mathbf{m}_1 + \cdots + \mathbf{m}_q = \mathbf{m}} \int_{1 \geqslant s_1 \geqslant \cdots \geqslant s_q \geqslant 0} c(\mathbf{z}:\mathbf{m}_1:s_1)c(\mathbf{z}:\mathbf{m}_2:s_2)\cdots$$
$$\times c(\mathbf{z}:\mathbf{m}_q:s_q)\,ds_1 \cdots ds_q$$

for $q \geqslant 1$. It is obvious that $\kappa_0(\mathbf{z}:q:\mathbf{m})$ is a holomorphic function of $\mathbf{z}$ on V with values in $\mathcal{E}(U)$. Since the closure of V in Ω' is compact, we can choose $A \geqslant 0$ such that

$$\sum_{\mathbf{m}} |c(\mathbf{z}:\mathbf{m}:s)|e^{-\sigma(\mathbf{m})M_0} \leqslant A$$

for $\mathbf{z} \in V$ and $s \in \mathfrak{I}$. Then

$$\sum_{\mathbf{m}} |\kappa_0(\mathbf{z}:q:\mathbf{m})|e^{-\sigma(\mathbf{m})M_0} \leqslant \frac{A^q}{q!}.$$

Now define $\kappa_0(\mathbf{z}:0:\mathbf{m}) = 1$ or 0 according as $\mathbf{m} = 0$ or not. Then

$$\sum_{q \geqslant 0} \sum_{\mathbf{m}} |\kappa_0(\mathbf{z}:q:\mathbf{m})|e^{-\sigma(\mathbf{m})M_0} \leqslant e^A < \infty.$$

Put

$$\kappa_0(\mathbf{z}:\mathbf{m}) = \sum_{q \geqslant 0} \kappa_0(\mathbf{z}:q:\mathbf{m}).$$

Since $|\kappa_0(\mathbf{z}:q:\mathbf{m})| \leqslant e^{\sigma(\mathbf{m})M_0}(A^q/q!)$, the series $\sum_{q \geqslant 0}|\kappa_0(\mathbf{z}:q:\mathbf{m})|$ converges uniformly with respect to $\mathbf{z}$ on V. Hence $\kappa_0(\mathbf{z}:\mathbf{m})$ is a holomorphic function of $\mathbf{z}$ on V. Also, as we have seen during the proof of Lemma 26,

$$K_0(\mathbf{z}:\mathbf{u}) = \sum_{\mathbf{m}} \kappa_0(\mathbf{z}:\mathbf{m})e^{-\langle \mathbf{m},\mathbf{u}\rangle} \qquad \big(\mathbf{u} \in \mathbb{C}^{l_2}(M_0)\big).$$

On the other hand, by Lemma 26,

$$K_\xi(\mathbf{u}) = \sum_{\mathbf{m}} \kappa_\xi(\mathbf{m})e^{-\langle \mathbf{m},\mathbf{u}\rangle} \qquad \big(\mathbf{u} \in \mathbb{C}^{l_2}(M_0)\big)$$

and

$$\sum_{\mathbf{m}} |\kappa_\xi(\mathbf{m})|e^{-\sigma(\mathbf{m})M_0} = B < \infty.$$

Put

$$\kappa(\mathbf{z}:\mathbf{m}) = \sum_{\mathbf{m}_1 + \mathbf{m}_2 = \mathbf{m}} \kappa_0(\mathbf{z}:\mathbf{m}_1)\kappa_\zeta(\mathbf{m}_2) \qquad (\mathbf{z} \in V,\ \mathbf{m} \in \mathbf{Z}_+^{l_2}).$$

Then it is obvious that

$$K(\mathbf{z}:\mathbf{u}) = K_0(\mathbf{z}:\mathbf{u})K_\zeta(\mathbf{u}) = \sum_{\mathbf{m}} \kappa(\mathbf{z}:\mathbf{m})e^{-\langle \mathbf{m},\mathbf{u}\rangle} \qquad (\mathbf{z} \in V,\ \mathbf{u} \in \mathbf{C}^{l_2}(M_0)).$$

Now

$$\sum_{\mathbf{m}} |\kappa(\mathbf{z}:\mathbf{m})|e^{-\sigma(\mathbf{m})M_0} \leqslant e^A B \qquad (\mathbf{z} \in V).$$

Therefore if $M > M_0$,

$$|\kappa(\mathbf{z}:\mathbf{m})|e^{-\sigma(\mathbf{m})M} \leqslant Be^{A - \sigma(\mathbf{m})(M - M_0)}.$$

Since $\sum_{\mathbf{m}} e^{-\sigma(\mathbf{m})(M - M_0)} < \infty$, it follows that the series

$$\sum_{\mathbf{m}} |\kappa(\mathbf{z}:\mathbf{m})|e^{-\sigma(\mathbf{m})M}$$

converges uniformly with respect to $\mathbf{z} \in V$.

Put $d = \sup_{\lambda \in \Lambda} d^0 p_\lambda$.

Lemma 28. *Let Λ_2 denote the projection of Λ' on $\mathbf{C}^{l_2}$. For each $\lambda \in \Lambda_2$, we can choose a holomorphic function ψ_λ on $V \times \mathbf{C}^{l_2}(M_0)$ such that*

1) $\psi_\lambda(\mathbf{t})$ *is a polynomial function in* $\mathbf{t}_2$ *of degree* $\leqslant d$.
2) *For any* $\lambda_0 \in \mathbf{C}^{l_2}$ *the series*

$$\sum_{\lambda \in \Lambda_2} |\psi_\lambda(\mathbf{t})e^{\langle \lambda_0 + \lambda, \mathbf{t}_2\rangle}|$$

converges uniformly on $V \times W_{l_2}(\varepsilon, M)$ *if* $0 < \varepsilon < 1$ *and* $M > M_0$.

3) $\phi(\mathbf{t}) = \sum_{\lambda \in \Lambda_2} \psi_\lambda(\mathbf{t})e^{\langle \lambda, \mathbf{t}_2\rangle} \qquad (\mathbf{t} \in V \times \mathbf{C}^{l_2}(M_0))$.

For $\lambda \in \Lambda_2$, put

$$\phi_\lambda(\mathbf{t}) = \sum_{\mu_2 = \lambda} p_\mu(\mathbf{t})e^{\langle \mu_1, \mathbf{t}_1\rangle} \qquad (\mathbf{t} \in \mathbf{C}_+^l)$$

where the sum is over all elements $\mu = (\mu_1, \mu_2)$ in Λ' such that $\mu_2 = \lambda$. By Theorem 1 the series $\sum_{\mu \in \Lambda'} |p_\mu(\mathbf{t})e^{\langle \mu, \mathbf{t}\rangle}|$ converges uniformly on $W_l(\varepsilon, \eta)$ for any ε, η ($0 < \varepsilon < 1$, $\eta > 0$). Hence it is obvious that ϕ_λ is a holomorphic function on $\mathbf{C}_+^l$, and the above series can be differentiated term by term. Since $d^0 p_\mu \leqslant d$, it follows that all derivatives of ϕ_λ with respect to $\mathbf{t}_2$ of order $> d$ are zero. Hence ϕ_λ is a polynomial in $\mathbf{t}_2$ of degree $\leqslant d$. It is obvious that

$$\phi(\mathbf{w}, \mathbf{u}) = \sum_{\lambda \in \Lambda_2} \phi_\lambda(\mathbf{w}, \mathbf{u})e^{\langle \lambda, \mathbf{u}\rangle} \qquad (\mathbf{u} \in \mathbf{C}^{l_2}(M_0))$$

and

$$\phi(\mathbf{z}, \mathbf{u}) = K(\mathbf{z}:\mathbf{u})\phi(\mathbf{w}, \mathbf{u}) \qquad (\mathbf{z} \in V).$$

Put

$$\psi_\lambda(\mathbf{z},\mathbf{u}) = \sum_{\mathbf{m}} \kappa(\mathbf{z}:\mathbf{m})\phi_{\lambda+\mathbf{m}}(\mathbf{w},\mathbf{u}) \qquad (\mathbf{z}\in V,\ \lambda\in\Lambda_2)$$

where the sum is over all $\mathbf{m}\in\mathbf{Z}_+^{l_2}$ such that $\lambda+\mathbf{m}\in\Lambda_2$. It is obvious that if $\sigma(\mathbf{m})$ is sufficiently large, $\lambda+\mathbf{m}\notin\Lambda_2$ and therefore the above sum is actually finite. Then $\psi_\lambda(\mathbf{z},\mathbf{u})$ is a holomorphic function on $V\times\mathbf{C}^{l_2}$ which is a polynomial of degree $\leqslant d$ in $\mathbf{u}$. Now fix ε, M ($0<\varepsilon<1$, $M>M_0$). Then if $\lambda_0\in\mathbf{C}^{l_2}$ and $\mathbf{u}\in W_{l_2}(\varepsilon,M)$, $|\langle\lambda_0,\mathbf{u}\rangle|\leqslant l_2|\lambda_0||\mathbf{u}|\leqslant aM$ when $a=l_2|\lambda_0|\varepsilon^{-1}$. Hence if $\alpha=\frac{1}{2}(M-M_0)$,

$$|\kappa(\mathbf{z}:\mathbf{m})e^{\langle\lambda_0-\mathbf{m},\mathbf{u}\rangle}|\leqslant|\kappa(\mathbf{z}:\mathbf{m})|e^{aM-\sigma(\mathbf{m})M}\leqslant|\kappa(\mathbf{z}:\mathbf{m})|e^{-\sigma(\mathbf{m})(M_0+\alpha)}$$

if $\sigma(\mathbf{m})\alpha\geqslant aM$. By Lemma 27 the series $\sum_{\mathbf{m}}|\kappa(\mathbf{z}:\mathbf{m})|e^{-\sigma(\mathbf{m})(M_0+\alpha)}$ converges uniformly with respect to $\mathbf{z}$ in V. Therefore the series

$$\sum_{\mathbf{m}}|\kappa(\mathbf{z}:\mathbf{m})e^{\langle\lambda_0-\mathbf{m},\mathbf{u}\rangle}|$$

converges uniformly for $(\mathbf{z},\mathbf{u})\in V\times W_{l_2}(\varepsilon,M)$. Also $\sum_{\mathbf{m}}|\kappa(\mathbf{z}:\mathbf{m})|e^{-\sigma(\mathbf{m})M_0}$ remains uniformly bounded for $\mathbf{z}\in V$. Therefore $\sum_{\mathbf{m}}|\kappa(\mathbf{z}:\mathbf{m})e^{-\langle\mathbf{m},\mathbf{u}\rangle}|$ remains bounded on $V\times W_{l_2}(\varepsilon,M)$.

On the other hand

$$\sum_{\lambda\in\Lambda_2}|\phi_\lambda(\mathbf{w},\mathbf{u})e^{\langle\lambda,\mathbf{u}\rangle}|\leqslant\sum_{\mu\in\Lambda}|p_\mu(\mathbf{w},\mathbf{u})e^{\langle\mu_1,\mathbf{w}\rangle+\langle\mu_2,\mathbf{u}\rangle}|.$$

Therefore it follows from Lemma 16 that we can choose positive numbers A and ν such that

$$\sum_{\lambda\in\Lambda_2}|\phi_\lambda(\mathbf{w},\mathbf{u})e^{\langle\lambda,\mathbf{u}\rangle}|\leqslant Ae^{\nu\sigma(\mathbf{u})}$$

for $\mathbf{u}\in W_{l_2}(\varepsilon,M)$. Now let $\lambda_0\in\mathbf{C}^{l_2}$. Thus for $\lambda\in\Lambda_2$,

$$|\phi_\lambda(\mathbf{w},\mathbf{u})e^{\langle\lambda+\lambda_0,\mathbf{u}\rangle}|\leqslant\sum_{\mu_2=\lambda}|p_\mu(\mathbf{w},\mathbf{u})e^{\langle\mu_1,\mathbf{w}\rangle+\langle\lambda_0+\mu_2,\mathbf{u}\rangle}|$$

where the sum is over all $\mu\in\Lambda$ such that $\mu_2=\lambda$. By Theorem 1, the series

$$\sum_{\mu\in\Lambda}|p_\mu(\mathbf{w},\mathbf{u})e^{\langle\mu_1,\mathbf{w}\rangle+\langle\lambda_0+\mu_2,\mathbf{u}\rangle}|$$

converges uniformly with respect to $\mathbf{u}\in W_{l_2}(\varepsilon,M)$. Hence the same holds for the series

$$\sum_{\lambda\in\Lambda_2}|\phi_\lambda(\mathbf{w},\mathbf{u})e^{\langle\lambda_0+\lambda,\mathbf{u}\rangle}|.$$

Therefore it follows (see the proof of Lemma 3) that the series

$$\sum_{\lambda\in\Lambda_2}\sum_{\mathbf{m}}|\kappa(\mathbf{z}:\mathbf{m})||\phi_\lambda(\mathbf{w},\mathbf{u})e^{\langle\lambda_0+\lambda-\mathbf{m},\mathbf{u}\rangle}|$$

converges uniformly on $V\times W_{l_2}(\varepsilon,M)$. From this the 2nd statement of the lemma follows immediately.

Finally,

$$\phi(\mathbf{z},\mathbf{u}) = K(\mathbf{z}:\mathbf{u})\phi(\mathbf{w},\mathbf{u}) = \sum_{\mathbf{m}} \sum_{\lambda \in \Lambda_2} \kappa(\mathbf{z}:\mathbf{m})\phi_\lambda(\mathbf{w},\mathbf{u})e^{\langle \lambda - \mathbf{m}, \mathbf{u}\rangle} = \sum_{\lambda \in \Lambda} \psi_\lambda(\mathbf{z},\mathbf{u})e^{\langle \lambda, \mathbf{u}\rangle}$$

for $\mathbf{z} \in V$ and $\mathbf{u} \in \mathbb{C}^{l_2}(M_0)$. This proves the 3rd statement.

Lemma 29. *Fix a point $\mathbf{z} \in \Omega'$. Then there exist U-polynomials $\psi_{\lambda,\mathbf{z}}$ $(\lambda \in \Lambda_2)$ on $\mathbb{C}^{l_2}$ which are uniquely characterized by the following two properties.*

1) *If $M > M_0$ and $0 < \varepsilon < 1$, the series*

$$\sum_{\lambda \in \Lambda_2} |\psi_{\lambda,\mathbf{z}}(\mathbf{u})e^{\langle \lambda_0 + \lambda, \mathbf{u}\rangle}|$$

converges uniformly on $W_{l_2}(\varepsilon, M)$ for any $\lambda_0 \in \mathbb{C}^{l_2}$.

2) $\phi(\mathbf{z},\mathbf{u}) = \displaystyle\sum_{\lambda \in \Lambda_2} \psi_{\lambda,\mathbf{z}}(\mathbf{u})e^{\langle \lambda, \mathbf{u}\rangle} \qquad (\mathbf{u} \in \mathbb{C}^{l_2}(M_0)).$

Put $\psi_\lambda(\mathbf{z},\mathbf{u}) = \psi_{\lambda,\mathbf{z}}(\mathbf{u})$ $(\mathbf{z} \in \Omega', \ \mathbf{u} \in \mathbb{C}^{l_2})$. Then ψ_λ is a holomorphic function on $\Omega' \times \mathbb{C}^{l_2}$ which is a polynomial function of $\mathbf{u}$. Moreover, if K is a compact set in Ω' and λ_0 a point in $\mathbb{C}^{l_2}$, the series

$$\sum_{\lambda \in \Lambda_2} |\psi_\lambda(\mathbf{z},\mathbf{u})e^{\langle \lambda_0 + \lambda, \mathbf{u}\rangle}|$$

converges uniformly for $\mathbf{z} \in K$ and $\mathbf{u} \in W_{l_2}(\varepsilon, M)$ if $0 < \varepsilon < 1$ and $M > M_0$.

The existence of $\phi_{\lambda,\mathbf{z}}$ follows from Lemma 28 and their uniqueness from the Corollary of Lemma 57 of [1]. The rest is an immediate consequence of Lemma 28.

We now intend to prove that for any $\lambda \in \Lambda_2$, ψ_λ can be extended to a holomorphic function on $\Omega \times \mathbb{C}^{l_2}$. Fix a point $\mathbf{z}_0 \in \Omega$. It is enough to show that such an extension is possible on $V_0 \times \mathbb{C}^{l_2}$ where V_0 is some open neighborhood of $\mathbf{z}_0$ in Ω. If $\mathbf{z}_0 \in \Omega'$, there is nothing to prove. So we may assume that $\mathbf{z}_0 \in \cup_{1 \leqslant i \leqslant r} H_i$. By the Lemma of the Appendix, we can choose positive numbers η_i $(1 \leqslant i \leqslant l_1)$ with the following properties.

1) If $\mathbf{z} = (z_1,\ldots,z_{l_1})$ is any point in $\mathbb{C}^{l_1}$ such that $|z_i - z_{0i}| < 2\eta_i$ $(1 \leqslant i \leqslant l_1)$, then $\mathbf{z} \in \Omega$ and
2) if $|z_i - z_{0i}| = \eta_i$ $(1 \leqslant i \leqslant l_1)$ then $\mathbf{z} \in \Omega'$.

Let V be the open neighborhood of $\mathbf{z}_0$ in Ω consisting of all points $\mathbf{z} \subset \mathbb{C}^{l_1}$ such that $|z_i - z_{0i}| < \eta_i$ $(1 \leqslant i \leqslant l_1)$. Then if $\mathbf{z} \in V$, $\mathbf{u} \in \mathbb{C}^l(M_0)$,

$$\phi(\mathbf{z},\mathbf{u}) = \left(2\pi(-1)^{1/2}\right)^{-l_1} \oint_1 \cdots \oint_{l_1} \phi(\mathbf{w},\mathbf{u}) \prod_{1 \leqslant i \leqslant l_1} (w_i - z_i)^{-1} dw_1 \cdots dw_{l_1}$$

when $\oint_k$ denotes complex integration around the circle $|w_k - z_{0k}| = \eta_k$ $(1 \leqslant k \leqslant l_1)$. Put

$$\phi_\lambda(\mathbf{z},\mathbf{u}) = \left(2\pi(-1)^{1/2}\right)^{-l_1} \oint_1 \cdots \oint_{l_1} \psi_\lambda(\mathbf{w},\mathbf{u}) \prod_{1 \leqslant i \leqslant l_1} (w_i - z_i)^{-1} dw_1 \cdots dw_{l_1}$$

for $\lambda \in \Lambda_2$. Then $\phi_\lambda(\mathbf{z}, \mathbf{u})$ is a holomorphic function on $V \times \mathbf{C}^{l_2}$ which is a polynomial function in $\mathbf{u}$ of degree $\leqslant d$. Let K be the set of points $\mathbf{z} \in V$ such that $|z_i - z_{0i}| \leqslant \frac{1}{2}\eta_i$ $(1 \leqslant i \leqslant l_1)$. Then K is compact. Put $2\delta = \min_i \eta_i$ and let I denote the interval $0 \leqslant t \leqslant 2\pi$ in $\mathbf{R}$. Put $J = I^{l_1} \subset \mathbf{R}^{l_1}$. For any point $\theta = (\theta_1, \ldots, \theta_{l_1})$ in J, define $\mathbf{w}(\theta)$ to be the point $(w_1, \ldots, w_{l_1})$ given by $w_i = z_{0i} + \eta_i e(-1)^{1/2\theta_i}$ $(1 \leqslant i \leqslant l_1)$. Then if $\mathbf{z} \in K$, $|z_i - w_i(\theta)| \geqslant \frac{1}{2}\eta_i \geqslant \delta$. Hence

$$|\phi_\lambda(\mathbf{z}, \mathbf{u})| \leqslant (\eta/2\pi\delta)^{-l_1} \int_J |\psi_\lambda(\mathbf{w}(\theta), \mathbf{u})| d\theta_1 \cdots d\theta_{l_1}$$

where $\eta = \max_i \eta_i$. Let $\lambda_0 \in \mathbf{C}^{l_2}$. Then

$$|\phi_\lambda(\mathbf{z}, \mathbf{u}) e^{\langle \lambda_0 + \lambda, \mathbf{u} \rangle}| \leqslant (\eta/2\pi\delta)^{-l_1} \int_J |\psi_\lambda(\mathbf{w}(\theta), \mathbf{u}) e^{\langle \lambda_0 + \lambda, \mathbf{u} \rangle}| d\theta_1 \cdots d\theta_{l_1}.$$

Let F be the set of all points $\mathbf{w}(\theta)$ with $\theta \in J$. Then F is a compact subset of Ω'. Hence the series $\Sigma_{\lambda \in \Lambda_2} |\psi_\lambda(\mathbf{w}, \mathbf{u}) e^{\langle \lambda_0 + \lambda, \mathbf{u} \rangle}|$ converges uniformly on $F \times W_{l_2}(\varepsilon, M)$ if $0 < \varepsilon < 1$ and $M > M_0$. But this means that the series $\Sigma_{\lambda \in \Lambda_2} |\phi_\lambda(\mathbf{z}, \mathbf{u}) e^{\langle \lambda_0 + \lambda, \mathbf{u} \rangle}|$ also converges uniformly on $K \times W_{l_2}(\varepsilon, M)$. Now

$$\phi(\mathbf{w}, \mathbf{u}) = \sum_{\lambda \in \Lambda_2} \psi_\lambda(\mathbf{w}, \mathbf{u}) e^{\langle \lambda, \mathbf{u} \rangle}$$

for $\mathbf{w} \in F$ and $\mathbf{u} \in \mathbf{C}^{l_2}(M_0)$. Hence it is clear that

$$\phi(\mathbf{z}, \mathbf{u}) = \sum_{\lambda \in \Lambda_2} \phi_\lambda(\mathbf{z}, \mathbf{u}) e^{\langle \lambda, \mathbf{u} \rangle}$$

on $K \times \mathbf{C}^{l_2}(M_0)$. But then it follows from Lemma 29 that $\phi_\lambda(\mathbf{z}, \mathbf{u}) = \psi_\lambda(\mathbf{z}, \mathbf{u})$ for $\mathbf{z} \in K \cap \Omega'$ and $\mathbf{u} \in \mathbf{C}^{l_2}(M_0)$. Since both $\phi_\lambda(\mathbf{z}, \mathbf{u})$ and $\psi_\lambda(\mathbf{z}, \mathbf{u})$ are polynomial functions in $\mathbf{u}$, this proves that $\phi_\lambda = \psi_\lambda$ on $(K \cap \Omega') \times \mathbf{C}^{l_2}$. We can now take V_0 to be the interior of K in $\mathbf{C}^{l_1}$.

Let ϕ_λ denote the holomorphic extension of ψ_λ on $\Omega \times \mathbf{C}^{l_2}$. Then $\phi_\lambda(\mathbf{t})$ is a polynomial function of $\mathbf{t}_2$ of degree $\leqslant d$. Now any point $\mathbf{z}_0 \in \Omega$ has a compact neighborhood K in Ω such that the series

$$\sum_{\lambda \in \Lambda_2} \phi_\lambda(\mathbf{z}, \mathbf{u}) e^{\langle \lambda, \mathbf{u} \rangle}$$

converges decently on $K \times W_{l_2}(\varepsilon, M)$ for every ε, M $(0 \leqslant \varepsilon < 1, M > M_0)$. In view of Lemma 4 this follows from Lemma 29 if $\mathbf{z}_0 \in \Omega'$ and from what we have seen above in case $\mathbf{z}_0 \notin \Omega'$. The statements of Theorem 3 are now obvious.

The lemma below follows immediately from the Corollary of Lemma 57 of [1].

Lemma 30. *If we assume that $\phi_\lambda \neq 0$ for $\lambda \in \Lambda_2$, in Theorem 3, then Λ_2 is exactly the projection of Λ in $\mathbf{C}^{l_2}$ and for any $\lambda \in \Lambda_2$,*

$$\phi_\lambda(\mathbf{t}) = \sum_{\mu_2 = \lambda} p_\mu(\mathbf{t}) e^{\langle \mu_1, \mathbf{t}_1 \rangle} \qquad \left(\mathbf{t} \in \mathbf{C}_+^l \right)$$

where the sum is over all $\mu \in \Lambda$ such that $\mu_2 = \lambda$.

Call an element $\lambda \in \Lambda_2$ *maximal* if $\phi_{\lambda + \mathbf{m}} = 0$ for every $\mathbf{m} \neq 0$ in $\mathbf{Z}_+^{l_2}$ such that $\lambda + \mathbf{m} \in \Lambda_2$. Let $\Lambda_2(0)$ denote the set of all maximal elements in Λ_2. Define Λ_0 as in

Theorem 1. Then $\Lambda_2(0)$ is the projection of Λ_0 in $\mathbf{C}^{l_2}$. Define $c_i(\mathbf{z}:\mathbf{m})$ as in the beginning of the proof of Theorem 3.

Lemma 31. *Let* $\lambda_0 \in \Lambda_2(0)$. *Then*

$$\frac{\partial \phi_{\lambda_0}}{\partial z_i}(\mathbf{z},\mathbf{u}) = c_i(\mathbf{z}:0)\phi_{\lambda_0}(\mathbf{z},\mathbf{u}) \qquad (1 \leqslant i \leqslant l_1)$$

for $\mathbf{z} \in \Omega'$ *and* $\mathbf{u} \in \mathbf{C}^{l_2}$.

Clearly $(\partial \phi / \partial z_i)(\mathbf{z},\mathbf{u}) = \mathcal{C}_i(\mathbf{z},\mathbf{u})\phi(\mathbf{z},\mathbf{u})$. Now fix a point $\mathbf{z}_0 \in \Omega'$ and use the notation of the proof of Lemma 27. Then

$$\sum_{\lambda \in \Lambda_2} |\phi_\lambda(\mathbf{z},\mathbf{u})e^{\langle \lambda,\mathbf{u}\rangle}| \leqslant \sum_{\mathbf{m}} |\kappa(\mathbf{z}:\mathbf{m})|e^{-\sigma(\mathbf{m})M_0} \sum_{\lambda \in \Lambda_2} |\phi_\lambda(\mathbf{w},\mathbf{u})e^{\langle \lambda,\mathbf{u}\rangle}|$$

$$\leqslant Be^{\nu\sigma(\mathbf{u})}$$

for $\mathbf{u} \in W_{l_2}(\varepsilon, M)$ and $\mathbf{z} \in V$ where B is a suitable constant. Also

$$\sum |c_i(\mathbf{z}:\mathbf{m})|e^{-\sigma(\mathbf{m})M_0} < \infty.$$

Hence it follows from Lemma 3 that the double series

$$\sum_{\lambda \in \Lambda_2} \sum_{\mathbf{m}} |c_i(\mathbf{z}:\mathbf{m})\phi_\lambda(\mathbf{z},\mathbf{u})e^{\langle \lambda+\alpha-\mathbf{m},\mathbf{u}\rangle}|$$

converges uniformly with respect to $\mathbf{u} \in W_{l_2}(\varepsilon, M)$ for any $\alpha \in \mathbf{C}^{l_2}$ and $\mathbf{z} \in V$. Since the series $\sum_{\lambda \in \Lambda_2}\phi_\lambda(\mathbf{z},\mathbf{u})e^{\langle \lambda,\mathbf{u}\rangle}$ converges decently on $V \times W_{l_2}(\varepsilon, M)$, we can (by the Corollary of Lemma 57 of [1]) compare the coefficients of $e^{\langle \lambda_0,\mathbf{u}\rangle}$ on the two sides of the equation

$$\sum_{\lambda \in \Lambda_2} \frac{\partial \phi_\lambda}{\partial z_i}(\mathbf{z},\mathbf{u})e^{\langle \lambda,\mathbf{u}\rangle} = \sum_{\lambda \in \Lambda_2} \sum_{\mathbf{m}} c_i(\mathbf{z}:\mathbf{m})\phi_\lambda(\mathbf{z},\mathbf{u})e^{\langle \lambda-\mathbf{m},\mathbf{u}\rangle}$$

and obtain

$$\frac{\partial \phi_{\lambda_0}}{\partial z_i}(\mathbf{z},\mathbf{u}) = c_i(\mathbf{z}:0)\phi_{\lambda_0}(\mathbf{z},\mathbf{u}).$$

§ 7.

We now consider some applications of Theorem 1 in the case $l = 1$. Put $\mathbf{C}_+ = \mathbf{C}^1_+$ and let $\mathcal{C}(t) = \sum_{k \geqslant 0} c(k)e^{-kt}$ $(c(k) \in \mathcal{E}(U), t \in \mathbf{C}_+)$ where, we assume, as before, that $\sum_{k \geqslant 0}|c(k)|e^{-k\eta} < \infty$ for every $\eta > 0$. Let F be the vector space consisting of all holomorphic functions ϕ from $\mathbf{C}_+$ to U which satisfy the differential equation $d\phi/dt = \mathcal{C}(t)\phi$ on $\mathbf{C}_+$. Fix a point $t_0 \in \mathbf{C}_+$. By Lemma 8 there exists a holomorphic function K from $\mathbf{C}_+$ to $\mathcal{E}(U)$ such that $dK/dt = \mathcal{C}(t)K$ and $K(t_0) = 1$. Also by Lemma 7, $\phi(t) = K(t)\phi(t_0)$ for any $\phi \in F$. Conversely for any given $u \in U$, the function $\phi(t) = K(t)u$ obviously lies in F. This shows that the mapping $\phi \to \phi(t_0)$ is a linear isomorphism of F on U. Hence $\dim F = \dim U = n$ (say).

Note that $\mathcal{C}(t) = \mathcal{C}(t + 2\pi(-1)^{1/2})$ $(t \in \mathbf{C}_+)$. Therefore it is obvious that if $\phi \in F$, then the function $t \to \phi(t + 2\pi(-1)^{1/2})$ is also in F. Let $L\phi$ denote this function. Then L is an endomorphism of F. Our object is to prove the following result about L.

Theorem 4. *Let $\Xi = c(0)$. Then the characteristic polynomials of L and $e^{2\pi(-1)^{1/2}\Xi}$ are the same. Suppose ξ is a semisimple eigenvalue of Ξ such that $\xi + k$ is not an eigenvalue of Ξ for any integer $k \neq 0$. Then $e^{2\pi(-1)^{1/2}\xi}$ is a semisimple eigenvalue of L.*

Choose a base $u_1, \ldots, u_n$ for U such that the following conditions hold. (1) For each i, there exists an eigenvalue ξ_i of Ξ such that $(\Xi - \xi_i)^n u_i = 0$; (2) $(\Xi - \xi_i)u_i$ is a linear combination of u_k $(1 \leqslant k < i)$ and (3) $\operatorname{Re} \xi_1 \geqslant \operatorname{Re} \xi_2 \geqslant \cdots \geqslant \operatorname{Re} \xi_n$. Obviously this is possible. Construct the polynomials $p_{i,k}$ $(k \geqslant 0)$ of Lemma 21 corresponding to $u = u_i$ and put

$$\phi_i(t) = \sum_{k \geqslant 0} p_{ik}(t) e^{(\xi_i - k)t} \qquad (t \in \mathbf{C}_+).$$

It follows from the results of §5 and Theorem 1 that the series $\sum_{k \geqslant 0} |p_{ik}(t) e^{(\xi_i - k)t}|$ converges uniformly on every compact subset of $\mathbf{C}_+$ and the functions $\phi_1, \ldots, \phi_n$ lie in F and are linearly independent. Hence they form a base for F.

Now let $\phi \in F$. Then $\phi = \sum_{1 \leqslant i \leqslant n} a_i \phi_i$ $(a_i \in \mathbf{C})$. Let λ be the highest index such that $a_\lambda \neq 0$. Since the constant term of $p_{i,0}$ is u_i, the polynomials p_{i0} $(1 \leqslant i \leqslant n)$ are linearly independent. Hence it is obvious that ξ_λ is a principal exponent of ϕ.

Now fix j and consider $L\phi_j$. Then

$$\phi_j\left(t + 2\pi(-1)^{1/2}\right) = e^{2\pi(-1)^{1/2}\xi_j} \sum_{k \geqslant 0} p_{j,k}\left(t + 2\pi(-1)^{1/2}\right) e^{(\xi_j - k)t}$$

$$= e^{2\pi(-1)^{1/2}\xi_j}\left\{\phi_j + \sum_{k \geqslant 0} q_{j,k}(t) e^{(\xi_j - k)t}\right\}$$

where $q_{jk}(t) = p_{jk}(t + 2\pi(-1)^{1/2}) - p_{jk}(t)$. Put $\psi_j = L\phi_j - e^{2\pi(-1)^{1/2}\xi_j}\phi_j$. Then $\psi_j \in F$ and if it is not zero, it is clear that it has exactly one principal exponent which is of the form $\xi_j - k$ $(k \in \mathbf{Z}_+)$. Put $F_0 = \{0\}$ and $F_k = \sum_{1 \leqslant i \leqslant k} \mathbf{C}\phi_i$ $(1 \leqslant k \leqslant n)$. Similarly let $U_0 = \{0\}$ and $U_k = \sum_{1 \leqslant i \leqslant k} \mathbf{C} u_i$. We claim that $\psi_j \in F_{j-1}$. In order to prove this we may assume that $\psi_j \neq 0$. Let $\psi_j = \sum_{1 \leqslant i \leqslant n} a_i \phi_i$ $(a_i \in \mathbf{C})$ and let λ be the greatest index such that $a_\lambda \neq 0$. Then, as we have seen above, ξ_λ is a principal exponent of ψ_j and therefore $\xi_\lambda = \xi_j - k$ where $k \in \mathbf{Z}_+$. If $k \geqslant 1$, it is obvious that $\lambda < j$ and therefore $\psi_j \in F_{j-1}$. Therefore we may suppose that $k = 0$. Then $\xi_\lambda = \xi_j$ is the only principal exponent of ψ_j. Let J be the set of all indices i $(1 \leqslant i \leqslant \lambda)$ such that $\xi_i = \xi_j$. Then by comparing the coefficients of $e^{\xi_j t}$ on the two sides of the equation $\psi_j = \sum_{1 \leqslant i \leqslant \lambda} a_i \phi_i$, we get

$$e^{2\pi(-1)^{1/2}\xi_j} q_{j,0}(t) = \sum_{i \in J} a_i p_{i0}(t).$$

Since $(\Xi - \xi_i)^r u_i \in U_{i-1}$ for $r \geqslant 1$ and since $p_{i0}(t) = e^{(\Xi - \xi_i)t} u_i$ (see the definition of p_0 in §5), it is clear that $p_{i0}(t) \equiv u_i \bmod U_{i-1}$ for any $t \in \mathbf{C}$ $(1 \leqslant i \leqslant n)$. Hence

$$q_{j0}(t) = p_{j0}\left(t + 2\pi(-1)^{1/2}\right) - p_{j0}(t) \equiv 0 \bmod U_{j-1}.$$

Now our assertion would follow if $\lambda < j$. So let us assume that $\lambda \geqslant j$. Then $q_{j0}(t) \equiv 0 \mod U_{\lambda-1}$ and $\sum_{i \in J} a_i p_{i0}(t) \equiv a_\lambda u_\lambda \mod U_{\lambda-1}$. Since $a_\lambda \neq 0$, this implies that $u_\lambda \in U_{\lambda-1}$, which is of course false. Therefore we conclude that $\lambda < j$ and hence $\psi_j \in F_{j-1}$.

This proves that $L\phi_j \equiv e^{2\pi(-1)^{1/2}\xi_j}\phi_j \mod F_{j-1}$ ($1 \leqslant j \leqslant n$) and from this the first assertion of the theorem is obvious.

Now we come to the second statement of the theorem. Let J be the set of all indices j such that $\xi_j = \xi$. Then if $j \in J$, it is clear from the proof of Lemma 21 that the polynomials p_{jk} are actually all constants which we shall denote by a_{jk}. Then

$$\phi_j(t) = \sum_{k \geqslant 0} a_{jk} e^{(\xi-k)t}$$

and therefore $L\phi_j = e^{2\pi(-1)^{1/2}\xi}\phi_j$. Now if $i \notin J$, $\xi_i - \xi$ is not an integer and therefore $e^{2\pi(-1)^{1/2}\xi_i} \neq e^{2\pi(-1)^{1/2}\xi}$. So it is obvious that $e^{2\pi(-1)^{1/2}\xi}$ is a semi-simple eigenvalue of L.

In view of its applications to the theory of semisimple Lie groups which will be made in another paper, we consider the following problem. Let V be a complex Banach space of finite dimension and z a complex variable moving within a domain ω in $\mathbb{C}$. Let f_1, f_2 be two holomorphic functions from ω to $\mathcal{E}(V)$. Consider a holomorphic function ϕ on some subdomain ω_1 of ω with values in V, which satisfies the differential equation $D\phi = 0$ where

$$D = \frac{d^2}{dz^2} + f_1(z)\frac{d}{dz} + f_2(z).$$

We wish to consider the question of analytic continuation (see [3]) of ϕ along any path in ω whose initial end-point lies in ω_1. Let Ω be the simply connected complex covering manifold of ω. Then we can regard z as a local coordinate everywhere on Ω. Hence d/dz and D can still be considered as differential operators on Ω. Let F be the space of all holomorphic functions ϕ from Ω to V such that $D\phi = 0$.

Lemma 32. *Put $\phi' = d\phi/dz$ for any $\phi \in F$ and fix a point $p_0 \in \Omega$. Then the mapping $\phi \to (\phi(p_0), \phi'(p_0))$ ($\phi \in F$) is a linear isomorphism of F onto $V \times V$.*

Put $U = V \times V$ and for $z \in \omega$, let $\mathcal{C}(z)$ denote the endomorphism

$$(v_1, v_2) \to (v_2, -f_2(z)v_1 - f_1(z)v_2) \qquad (v_1, v_2 \in V)$$

of U. Let $\mathcal{F}$ denote the space of all holomorphic functions Φ from Ω to U such that $d\Phi/dz = \mathcal{C}(z)\Phi$. We define a linear mapping α of F into $\mathcal{F}$ as follows. For any $\phi \in F$, $\alpha\phi = (\phi, d\phi/dz)$. It is easy to check that $\alpha\phi \in \mathcal{F}$ and α is a linear isomorphism of F with $\mathcal{F}$. Hence it would be enough to verify that the mapping $\Phi \to \Phi(p_0)$ is a linear isomorphism of $\mathcal{F}$ with U. First suppose $\Phi(p_0) = 0$ for some $\Phi \in F$. Since $d\Phi/dz = \mathcal{C}(z)\Phi$, it follows by induction on k, that $(d^k\Phi/dz^k)_{p_0} = 0$ for $k \geqslant 0$. Since Φ is holomorphic, this proves that $\Phi = 0$. Now fix an element $u \in U$. We show exactly as in Lemma 8, that there exists a holomorphic function K from Ω to $\mathcal{E}(U)$ such that $K(p_0) = 1$ and $dK/dz = \mathcal{C}(z)K$. Put $\Phi = Ku$. Then obviously $\Phi \in F$ and $\Phi(p_0) = u$. This proves the lemma.

Corollary. *Let z_0 be a point in ω and ω_1 a subdomain of ω containing z_0. Let $\zeta(\lambda)$ $(0 \leqslant \lambda \leqslant 1)$ be a path in ω such that $\zeta(0) = z_0$ and ϕ a holomorphic function from ω_1 to V such that $D\phi = 0$. Then ϕ can be continued analytically along the path ζ.*

This is an immediate consequence of Lemma 32.

Now let us assume that ω is the region $0 < |z| < e^{-M}$ where M is some real number. Also let $f_1(z) = z^{-1}f(z), f_2(z) = z^{-2}g(z)$ where f and g are holomorphic on the open disc $|z| < e^{-M}$. Then f, g have Taylor expansions of the form

$$f(z) = \sum_{k \geqslant 0} a_k z^k, \qquad g(z) = \sum_{k \geqslant 0} b_k z^k \qquad (a_k, b_k \in \mathcal{E}(U))$$

and the series

$$\sum_{k \geqslant 0} |a_k||z|^k, \qquad \sum_{k \geqslant 0} |b_k||z|^k$$

converge for $|z| < e^{-M}$. Now

$$D = \frac{d^2}{dz^2} + \frac{f(z)}{z}\frac{d}{dz} + \frac{g(z)}{z^2}.$$

As usual let $\mathbf{C}(M)$ be the set of all points $t \in \mathbf{C}$ such that $\operatorname{Re} t > M$. Then under the mapping $z = e^{-t}$ $(t \in \mathbf{C}(M))$, the region $\mathbf{C}(M)$ is a simply connected complex covering manifold of ω and a simple calculation shows that $D = e^{2t}\Delta$ on $\mathbf{C}(M)$, where

$$\Delta = \frac{d^2}{dt^2} + \left(1 - f(e^{-t})\right)\frac{d}{dt} + g(e^{-t}).$$

Therefore the space F of Lemma 32 consists of all holomorphic functions ϕ from $\mathbf{C}(M)$ to V such that $\Delta\phi = 0$. Put $U = V \times V$ as before and let $\mathcal{C}(t)$ denote the endomorphism $(v_1, v_2) \to (v_2, - g(e^{-t})v_1 + (f(e^{-t}) - 1)v_2)$ $(v_1, v_2 \in V)$ of U. Consider the space $\mathcal{F}$ of all holomorphic functions Φ from $\mathbf{C}(M)$ to U such that $d\Phi/dt = \mathcal{C}(t)\Phi$. Then we verify as before that the mapping $\alpha : \phi \to (\phi, d\phi/dt)$ is a linear isomorphism of F onto $\mathcal{F}$.

Now define two elements L and Λ in $\mathcal{E}(F)$ and $\mathcal{E}(\mathcal{F})$ respectively as follows. For any $\phi \in F$, $L\phi$ is the function $t \to \phi(t + 2\pi(-1)^{1/2})$. Similarly $\Lambda\Phi$ is the function $t \to \Phi(t + 2\pi(-1)^{1/2})$ $(\Phi \in \mathcal{F})$. It is obvious that $\alpha L = \Lambda\alpha$ and therefore the eigenvalues of L and Λ are the same. Now clearly $\mathcal{C}(t)$ has an expansion of the form

$$\mathcal{C}(t) = \sum_{k \geqslant 0} c_k e^{-kt} \qquad (t \in \mathbf{C}(M))$$

where $c_k \in \mathcal{E}(U)$. Moreover $\Xi = c_0$ is the mapping

$$(v_1, v_2) \to (v_2, - b_0 v_1 + (a_0 - 1)v_2) \qquad (v_1, v_2 \in V)$$

of U. Now suppose $a_0 \in \mathbf{C}$. Then if $-\mu$ is an eigenvalue of Ξ, it is clear that $-\mu(\mu + a_0 - 1)$ is an eigenvalue of b_0. Therefore we conclude from Theorem 4, that every eigenvalue of L is of the form $e^{-2\pi(-1)^{1/2}\mu}$ where μ is some complex number such that $-\mu(\mu + a_0 - 1)$ is an eigenvalue of b_0.

Theorem 5. *Let ρ be a positive number and ω the region $0 < |z| < \rho$ in $\mathbf{C}$. Suppose f and g are two holomorphic functions on the open disc $|z| < \rho$ with values in $\mathcal{E}(V)$. Let ω_1 be a subdomain of ω and $\phi \neq 0$ a holomorphic function from ω_1 to V such that*

$$\frac{d^2\phi}{dz^2} + \frac{f(z)}{z}\frac{d\phi}{dz} + \frac{g(z)}{z^2}\phi = 0.$$

Let z_0 be a point in ω_1. Suppose by analytic continuation along the circle $\zeta(s) = z_0 e^{2\pi(-1)^{1/2}s}$ ($0 \leqslant s \leqslant 1$), ϕ is transformed into $c\phi$ where c is a complex number. Then if $f(0) \in \mathbf{C}$, we can choose a complex number μ such that $c = e^{2\pi(-1)^{1/2}\mu}$ and $-\mu(\mu + f(0) - 1)$ is an eigenvalue of $g(0)$.

Put $M = -\log\rho$ where the logarithm is real and choose $t_0 \in C(M)$ such that $z_0 = e^{-t_0}$. By Lemma 32, there exists a holomorphic function ψ on $C(M)$ such that $D\psi = e^{2t}\Delta\psi = 0$ and $\psi(t) = \phi(e^{-t})$ if t is sufficiently near t_0. Now it follows from our hypothesis that $\psi(t - 2\pi(-1)^{1/2}) = c\psi(t)$ and therefore $cL\psi = \psi$. Evidently $\psi \neq 0$ since $\phi \neq 0$ and therefore c^{-1} is an eigenvalue of L. But then in view of what we have seen above, our assertion is obvious.

Appendix

Lemma. *Let $\lambda_i(z) = \sum_{j=1}^n c_{ij}z_j$ ($1 \leqslant i \leqslant q$) be a finite number of linear functions on $\mathbf{C}^n$, none of which is identically zero. Then, for any $\varepsilon > 0$, we can choose positive numbers $r_1, \ldots, r_n$ such that 1) $0 < r_i < \varepsilon$ ($1 \leqslant i \leqslant n$) and 2) if $z = (z_1, \ldots, z_n)$ is any point such that $|z_i| = r_i$ ($1 \leqslant i \leqslant n$) then $\lambda_j(z) \neq 0$ ($1 \leqslant j \leqslant q$).*

If $n = 1$, this is obvious. So assume $n \geqslant 2$ and use induction. Put $\bar{\lambda}_j = \sum_{1 \leqslant j < n} c_{ij}z_j$ ($1 \leqslant i \leqslant q$). Then by induction hypothesis we can choose $r_1, \ldots r_{n-1}$ such that 1) $0 < r_i < \varepsilon$ ($1 \leqslant i < n$) and 2) if $\bar{z} = (z_1, \ldots, z_{n-1})$ is any point of $\mathbf{C}^{n-1}$ such that $|z_i| = r_i$ ($1 \leqslant i < n$), then $\bar{\lambda}_j(\bar{z}) \neq 0$ provided $\bar{\lambda}_j \neq 0$ ($1 \leqslant j \leqslant q$). Let ω denote the set of all points $(z_1, \ldots, z_{n-1})$ in $\mathbf{C}^{n-1}$ such that $|z_i| = r_i$ ($i \leqslant i < n$). Then ω is a compact set. If $\bar{\lambda}_j \neq 0$, $\bar{\lambda}_j$ never takes the value zero on ω. Let $\rho_j = \inf_{\bar{z} \in \omega}|\bar{\lambda}_j(\bar{z})|$ ($1 \leqslant j \leqslant q$). Then $\rho_j > 0$ unless $\bar{\lambda}_j = 0$. Now $\lambda_j = \bar{\lambda}_j + c_j z_n$ ($c_j \in \mathbf{C}$). Choose r_n ($0 < r_n < \varepsilon$) such that $|c_j|r_n < \rho_j$ if $\rho_j \neq 0$ ($1 \leqslant j \leqslant q$). Then if $|z_i| = r_i$ ($1 \leqslant i \leqslant n$),

$$|\lambda_j(z)| \geqslant \rho_j - |c_j|r_n > 0$$

if $\rho_j \neq 0$. However, if $\rho_j = 0$, $\bar{\lambda}_j = 0$ and so $\lambda_j(z) = c_j z_n$. But since $\lambda_j \neq 0$, $c_j \neq 0$ and so $|\lambda_j(z)| = |c_j|r_n > 0$. $\hfill$ Q.E.D.

References

1. Harish-Chandra, Spherical functions on a semisimple Lie group, *I. Amer. J. Math.* **80** (1958), 241–310.
2. —, Some results on differential equations and their applications. *Proc. Nat. Acad. Sci. U.S.A.*, 45 (1959), 1763–1764.
3. Weyl, H., *The Concept of a Riemann Surface* (translated from the German), Addison-Wesley Publishing Company, Inc., Massachusetts. Third edition, 1955.

Supplement[†] to "Some Results on Differential Equations"

§1.

Let D be a domain (i.e. nonempty open connected set) in $\mathbb{R}^{l_1}$ such that $D \supset \mathbb{R}^{l_1}_+$.

Theorem 3. *Let $\mathcal{C}_i$ $(1 \leqslant i \leqslant l)$ be given functions in $\mathcal{R}_{\mathcal{S}(U)}$ and ϕ an analytic function on $D \times \mathbb{R}^{l_2}_+$ with values in U such that*

$$\frac{\partial \phi}{\partial t_i} = \mathcal{C}_i(t) \phi \qquad (1 \leqslant i \leqslant l).$$

Then we can choose a finite or countable set Λ_2 of points in $\mathbb{C}^{l_2}$ and for each $\lambda \in \Lambda_2$ an analytic function ϕ_λ from $D \times \mathbb{R}^{l_2}$ to U such that

1) *$\phi_\lambda(t)$ is a polynomial function of t_2.*
2) *Let ω_0 be a subdomain of D whose closure in D is compact. Then we can choose a domain $\omega \supset \omega_0$ in $\mathbb{C}^{l_1}$ and a number $M > 0$ such that:*

   (a) *For every $\lambda \in \Lambda_2$, ϕ_λ can be extended to a holomorphic function on $\omega \times \mathbb{C}^{l_2}$.*

   (b) *For any ε $(0 < \varepsilon < 1)$ the series*

   $$\sum_{\lambda \in \Lambda_2} \phi_\lambda(t) e^{\langle \lambda, t_2 \rangle}$$

   converges decently on $\omega \times W_{l_2}(\varepsilon, M)$.

   (c) *The sum of this series is equal to $\phi(t)$ for $t_1 \in \omega_0$ and $t_2 \in \mathbb{R}^{l_2} \cap W_{l_2}(\varepsilon, M)$.*

Moreover if we require that $\phi_\lambda \neq 0$ for $\lambda \in \Lambda_2$, then the set V_2 and the functions ϕ_λ $(\lambda \in \Lambda_2)$ are uniquely determined by the above conditions.

We know

$$\phi(t) = \sum_{\lambda \in \Lambda'} p_\lambda(t) e^{\langle \lambda, t \rangle} \qquad \left(t \in \mathbb{R}^l_+ \right).$$

Let Ω_0 be a subdomain of D such that 1) $\Omega_0 \cap \mathbb{R}^l_+$ is not empty 2) $\Omega_0 \supset Cl(\omega_0)$ and 3) the closure of Ω_0 in D is compact. Fix a bounded domain Ω_1 in $\mathbb{C}^{l_1}$ such that

---

[†] The reader should note the identical numbering of some of the theorems and lemmas of the supplement with those in the paper—V.

$\Omega_1 \supset \Omega_0$. By Lemma 25, we can choose a number $M_0 > 0$ and a finite number of hyperplanes $H_1, \ldots, H_r$ in $\mathbb{C}^{l_1}$ such that:

1) H_i does not meet $\mathbb{C}^{l_1}_+$ $(1 \leqslant i \leqslant r)$.
2) Let Ω'_1 denote the complement of $\cup_{1 \leqslant i \leqslant r} H_i$ in Ω_1. Then the functions $\mathcal{C}_i(\mathbf{t})$ $(1 \leqslant i \leqslant l)$ are holomorphic on $\Omega'_1 \times \mathbb{C}^{l_2}(M_0)$.
3) There exist holomorphic functions $c_i(\mathbf{z}:\mathbf{m})$ $(1 \leqslant i \leqslant l, \ \mathbf{m} \in \mathbb{Z}^{l_2}_+, \ z \in \Omega'_1)$ from Ω'_1 to $\mathcal{E}(U)$ such that the series

$$\sum_{1 \leqslant i \leqslant l} \sum_{\mathbf{m}} |c_i(\mathbf{z}:\mathbf{m})| e^{-\sigma(\mathbf{m}) M_0}$$

converges uniformly with respect to $\mathbf{z}$ on every compact subset of Ω'_1 and

$$\mathcal{C}_i(\mathbf{t}) = \sum_{\mathbf{m}} c_i(\mathbf{t}_1:\mathbf{m}) e^{-\langle \mathbf{m}, \mathbf{t}_2 \rangle} \qquad \left(\mathbf{t} \in \Omega'_1 \times \mathbb{C}^{l_2}(M_0), \ 1 \leqslant i \leqslant l \right).$$

Fix a nonempty domain $\mathcal{D}_0$ in $\mathbb{R}^{l_2}(M_0)$ such that its closure in $\mathbb{R}^{l_2}(M_0)$ is compact. Then ϕ being analytic on $\Omega_0 \times \mathcal{D}_0$ and the closure of $\Omega_0 \times \mathcal{D}_0$ in $D \times \mathbb{R}^{l_2}(M_0)$ being compact, it is clear that we can choose a domain Ω in $\mathbb{C}^{l_1}$ and a domain $\mathcal{D} \subset \mathbb{C}^{l_2}(M_0)$ such that $\Omega_0 \subset \Omega \subset \Omega_1$, $\mathcal{D} \supset \mathcal{D}_0$ and such that ϕ can be extended to a holomorphic function on $\Omega \times \mathcal{D}$. Put $\Omega' = \Omega \cap \Omega'_1$.

Let J denote the closed interval $0 \leqslant s \leqslant 1$ in $\mathbb{R}$. By a path in Ω', we mean a continuously differentiable function ζ on J with values in Ω'. For such a path ζ, put

$$\mathcal{C}_\zeta(s:\mathbf{u}) = \sum_{1 \leqslant i \leqslant l_1} \mathcal{C}_i(\zeta(s), \mathbf{u}) \frac{d\zeta_i(s)}{ds} \qquad (s \in J)$$

$$K_\zeta(\mathbf{u}) = 1 + \sum_{q \geqslant 1} \int_{1 \geqslant s_1 \geqslant \cdots \geqslant s_q \geqslant 0} \mathcal{C}_\zeta(s_1:\mathbf{u}) \mathcal{C}_\zeta(s_2:\mathbf{u}) \cdots \mathcal{C}_\zeta(s_q:\mathbf{u}) \, ds_1 \, ds_2 \cdots ds_q$$

for $\mathbf{u} \in \mathbb{C}^{l_2}(M_0)$.

Lemma 26. *Let ζ be a path in Ω'. Then K_ζ is a holomorphic function on $\mathbb{C}^{l_2}(M_0)$ with values in $\mathcal{E}(U)$. Moreover we can choose elements $\kappa_\zeta(\mathbf{m}) \in \mathcal{E}(U)$ $(\mathbf{m} \in \mathbb{Z}^{l_2}_+)$ such that*

$$\sum_{\mathbf{m}} |\kappa_\zeta(\mathbf{m})| e^{-\sigma(\mathbf{m}) M_0} < \infty$$

and

$$K_\zeta(\mathbf{u}) = \sum_{\mathbf{m}} \kappa_\zeta(\mathbf{m}) e^{-\langle \mathbf{m}, \mathbf{u} \rangle} \qquad \left(\mathbf{u} \in \mathbb{C}^{l_2}(M_0) \right).$$

The proof is the same as before (see Lemma 26 of the paper).

Fix a point $\mathbf{z}_0 \in \Omega'$ and let V be an open convex neighborhood of $\mathbf{z}_0$ in Ω'. We assume moreover that the closure of V in Ω' is compact. Fix a point $\mathbf{w} \in \Omega_0 \cap \mathbb{R}^{l_1}_+ \subset \Omega'$. Since Ω' is connected, we can select a path ζ in Ω' such that $\zeta(0) = \mathbf{w}$ and $\zeta(1) = \mathbf{z}_0$. Let $\mathbf{z}$ be any point in V. Put $K(\mathbf{z}:\mathbf{u}) = K_\xi(\mathbf{u}) K_\zeta(\mathbf{u})$ $(\mathbf{u} \in \mathbb{C}^l(M_0))$ where ξ is the path $\xi(s) = (1-s)\mathbf{z}_0 + s\mathbf{z}$ $(0 \leqslant s \leqslant 1)$.

Lemma 27. *For each* $\mathbf{m} \in \mathbb{Z}_+^{l_2}$ *we can choose a holomorphic function* $\kappa(\mathbf{z}:\mathbf{m})$ *on* V *such that*

1) $\sum_{\mathbf{m}} |\kappa(\mathbf{z}:\mathbf{m})| e^{-\sigma(\mathbf{m})M_0}$ *remains bounded for* $\mathbf{z} \in V$.
2) *For any* $M > M_0$, *the series* $\sum_{\mathbf{m}} |\kappa(\mathbf{z}:\mathbf{m})| e^{-\sigma(\mathbf{m})M}$ *converges uniformly with respect to* $\mathbf{z} \in V$.
3) $K(\mathbf{z}:\mathbf{u}) = \sum_{\mathbf{m}} \kappa(\mathbf{z}:\mathbf{m}) e^{-\langle \mathbf{m}, \mathbf{u} \rangle}$ $\quad \left(\mathbf{u} \in \mathbb{C}^{l_2}(M_0),\ \mathbf{z} \in V \right).$

Proof same as before (see Lemma 27 of the paper.)

Put $d = \sup_{\lambda \in \Lambda} d^0 p_\lambda$.

Lemma 28. *Let* Λ_2 *denote the projection of* Λ *on* $\mathbb{C}^{l_2}$. *Suppose* $\mathbf{z}_0 \in \Omega_0' = \Omega_0 \cap \Omega'$. *For each* $\lambda \in \Lambda_2$, *we can choose a holomorphic function* ψ_λ *on* $V \times \mathbb{C}^{l_2}$ *with values in* U *such that:*

1) $\psi_\lambda(\mathbf{t})$ *is a polynomial function in* $\mathbf{t}_2$ *of degree* $\leqslant d$.
2) *For any* $\lambda_0 \in \mathbb{C}^{l_2}$ *the series*

$$\sum_{\lambda \in \Lambda_2} \left| \psi_\lambda(\mathbf{t}) e^{\langle \lambda_0 + \lambda, \mathbf{t}_2 \rangle} \right|$$

converges uniformly on $V \times W_{l_2}(\varepsilon, M)$ *if* $0 < \varepsilon < 1$ *and* $M > M_0$.
3) $\phi(\mathbf{t}) = \sum_{\lambda \in \Lambda_2} \psi_\lambda(\mathbf{t}) e^{\langle \lambda, \mathbf{t}_2 \rangle}$ $\quad \left(\mathbf{t} \in (\Omega_0 \cap V) \times \mathbb{R}^{l_2}(M_0) \right).$

For $\lambda \in \Lambda_2$, put

$$\phi_\lambda(\mathbf{t}) = \sum_{\mu_2 = \lambda} p_\mu(\mathbf{t}) e^{\langle \mu_1, \mathbf{t} \rangle} \qquad \left(\mathbf{t} \subset \mathbb{C}_+^l \right).$$

Then ϕ_λ is a holomorphic function on $\mathbb{C}_+^l$ which is a polynomial function in $\mathbf{t}_2$ of degree $\leqslant d$. Obviously

$$\phi(\mathbf{w}, \mathbf{u}) = \sum_{\lambda \in \Lambda_2} \phi_\lambda(\mathbf{w}, \mathbf{u}) e^{\langle \lambda, \mathbf{u} \rangle} \qquad \left(\mathbf{u} \in \mathbb{R}^l(M_0) \right)$$

and also if $\mathbf{u} \in \mathcal{D}$. Now

$$\phi(\mathbf{z}, \mathbf{u}) = K(\mathbf{z}:\mathbf{u}) \phi(\mathbf{w}, \mathbf{u}) \qquad (\mathbf{z} \in V,\ \mathbf{u} \in \mathcal{D}).$$

Fix $\mathbf{z} \in V \cap \Omega_0$. Then both sides being analytic on $\mathbb{R}^{l_2}(M_0)$ in $\mathbf{u}$, we conclude that they are equal for $\mathbf{z} \in V \cap \Omega_0$, $\mathbf{u} \in \mathbb{R}^{l_2}(M_0)$. Put

$$\psi_\lambda(\mathbf{z}, \mathbf{u}) = \sum_{\mathbf{m}} \kappa(\mathbf{z}:\mathbf{m}) \phi_{\lambda + \mathbf{m}}(\mathbf{w}, \mathbf{u}) \qquad (\mathbf{z} \in V,\ \lambda \in \Lambda_2),$$

the sum on the right being obviously finite. Then ψ_λ is a holomorphic function on $V \times \mathbb{C}^{l_2}$, which is a polynomial function in $\mathbf{u}$ of degree $\leqslant d$. Now fix ε, M $(0 < \varepsilon < 1$, $M > M_0)$. Then if $\lambda_0 \in \mathbb{C}^{l_2}$ and $\mathbf{u} \in W_{l_2}(\varepsilon, M)$, $|\langle \lambda, \mathbf{u} \rangle| \leqslant l_2 |\lambda_0| |\mathbf{u}| \leqslant aM$ where $a = l_2 |\lambda_0| \varepsilon^{-1}$. Hence if $\alpha = \frac{1}{2}(M - M_0)$,

$$|\kappa(\mathbf{z}:\mathbf{m}) e^{\langle \lambda_0 - \mathbf{m}, \mathbf{u} \rangle}| \leqslant |\kappa(\mathbf{z}:\mathbf{m})| e^{aM - \sigma(\mathbf{m})M} \leqslant |\kappa(\mathbf{z}:\mathbf{m})| e^{-\sigma(\mathbf{m})(M_0 + \alpha)}$$

if $\sigma(\mathbf{m})\alpha \geqslant aM$. By Lemma 27, the series

$$\sum_{\mathbf{m}} |\kappa(\mathbf{z}:\mathbf{m})| e^{-\sigma(\mathbf{m})(M_0+\alpha)}$$

converges uniformly with respect to $\mathbf{z} \in V$ and therefore the series

$$\sum_{\mathbf{m}} |\kappa(\mathbf{z}:\mathbf{m}) e^{\langle \lambda_0 - \mathbf{m}, \mathbf{u} \rangle}|$$

converges uniformly for $(\mathbf{z},\mathbf{u}) \in V \times W_{l_2}(\varepsilon, M)$. Also $\sum_{\mathbf{m}} |\kappa(\mathbf{z}:\mathbf{m})| e^{-\sigma(\mathbf{m})M_0}$ remains uniformly bounded for $\mathbf{z} \in V$. So $\sum_{\mathbf{m}} |\kappa(\mathbf{z}:\mathbf{m}) e^{-\langle \mathbf{m},\mathbf{u} \rangle}|$ remains bounded on $V \times W_{l_2}(\varepsilon, M)$.

On the other hand

$$\sum_{\lambda \in \Lambda_2} |\phi_\lambda(\mathbf{w},\mathbf{u}) e^{\langle \lambda, \mathbf{u} \rangle}| \leqslant \sum_{\mu \in \Lambda} |p_\mu(\mathbf{w},\mathbf{u}) e^{\langle \mu_1 \cdot \mathbf{w} \rangle + \langle \mu_2, \mathbf{u} \rangle}|.$$

Therefore it follows from Lemma 16 that we can choose positive numbers A and ν such that

$$\sum_{\lambda \in \Lambda_2} |\phi_\lambda(\mathbf{w},\mathbf{u}) e^{\langle \lambda, \mathbf{u} \rangle}| \leqslant A e^{\nu \sigma(\mathbf{u})}$$

for $\mathbf{u} \in W_{l_2}(\varepsilon, M)$. Now let $\lambda_0 \in \mathbb{C}^{l_2}$. Then for any $\lambda \in \Lambda_2$,

$$|\phi_\lambda(\mathbf{w},\mathbf{u}) e^{\langle \lambda + \lambda_0, \mathbf{u} \rangle}| \leqslant \sum_{\mu_2 = \lambda} |p_\mu(\mathbf{w},\mathbf{u}) e^{\langle \mu_1, \mathbf{w} \rangle + \langle \lambda_0 + \mu_2, \mathbf{u} \rangle}|$$

where the sum is over all $\mu \in \Lambda$ such that $\mu_2 = \lambda$. By Theorem 1, the series

$$\sum_{\mu \in \Lambda} |p_\mu(\mathbf{w},\mathbf{u}) e^{\langle \mu_1, \mathbf{w} \rangle + \langle \lambda_0 + \mu_2, \mathbf{u} \rangle}|$$

converges uniformly with respect to $\mathbf{u} \in W_{l_2}(\varepsilon, M)$. Therefore the same holds for the series

$$\sum_{\lambda \in \Lambda_2} |\phi_\lambda(\mathbf{w},\mathbf{u}) e^{\langle \lambda + \lambda_0, \mathbf{u} \rangle}|.$$

From this we conclude (see Lemma 3) that the series

$$\sum_{\lambda \in \Lambda_2} \sum_{\mathbf{m}} |\kappa(\mathbf{z}:\mathbf{m})| |\phi_\lambda(\mathbf{w},\mathbf{u}) e^{\langle \lambda_0 + \lambda - \mathbf{m}, \mathbf{u} \rangle}|$$

converges uniformly on $V \times W_{l_2}(\varepsilon, M)$. From this the 2nd statement of the lemma follows immediately.

Finally

$$\phi(\mathbf{z},\mathbf{u}) = K(\mathbf{z},\mathbf{u})\phi(\mathbf{w},\mathbf{u}) = \sum_{\mathbf{m}} \sum_{\lambda \in \Lambda_2} \kappa(\mathbf{z}:\mathbf{m})\phi_\lambda(\mathbf{w},\mathbf{u}) e^{\langle \lambda - \mathbf{m}, \mathbf{u} \rangle}$$

$$= \sum_{\lambda \in \Lambda_2} \psi_\lambda(\mathbf{z},\mathbf{u}) e^{\langle \lambda, \mathbf{u} \rangle} \qquad ((\mathbf{z},\mathbf{u}) \in (V \cap \Omega_0) \times \mathbb{R}^{l_2}(M_0)).$$

This proves the 3rd statement.

Lemma 29. *Fix a point* $z \in \Omega_0'$. *Then there exist U-polynomials* $\psi_{\lambda,z}$ $(\lambda \in \Lambda_2)$ *on* $\mathbb{C}^{l_2}$ *which are characterized by the following two properties.*

1) *If* $M > M_0$ *and* $0 < \varepsilon < 1$, *the series*

$$\sum_{\lambda \in \Lambda_2} \left| \psi_{\lambda,z}(\mathbf{u}) e^{\langle \lambda_0 + \lambda, \mathbf{u} \rangle} \right|$$

converges uniformly on $W_{l_2}(\varepsilon, M)$ *for any* $\lambda_0 \in \mathbb{C}^{l_2}$.

2) $\quad \phi(\mathbf{z}, \mathbf{u}) = \sum_{\lambda \in \Lambda_2} \psi_{\lambda,z}(\mathbf{u}) e^{\langle \lambda, \mathbf{u} \rangle} \qquad \left(\mathbf{u} \in \mathbb{R}^{l_2}(M_0) \right).$

Now put $\psi_\lambda(\mathbf{z}, \mathbf{u}) = \psi_{\lambda,z}(\mathbf{u}) e^{\langle \lambda, \mathbf{u} \rangle}$ $(\mathbf{z} \in \Omega_0', \mathbf{u} \in \mathbb{C}^{l_2})$. *Then we can choose a subdomain* $\Omega_1' \subset \Omega'$ *such that* $\Omega_0' \subset \Omega_1'$ *and* ψ_λ *can be extended to a holomorphic function on* $\Omega_1' \times \mathbb{C}^{l_2}$ *which is a polynomial function of* $\mathbf{u}$. *Moreover if* K *is a compact set in* Ω_1' *and* λ_0 *a point in* $\mathbb{C}^{l_2}$, *the series*

$$\sum_{\lambda \in \Lambda_2} \left| \psi_\lambda(\mathbf{z}, \mathbf{u}) e^{\langle \lambda_0 + \lambda, \mathbf{u} \rangle} \right|$$

converges uniformly for $\mathbf{z} \in K$ *and* $\mathbf{u} \in W_{l_2}(\varepsilon, M)$ *if* $0 < \varepsilon < 1$ *and* $M > M_0$.

Proof the same as before (see Lemma 29 of the paper).

Lemma 30. *Fix a point* $z_0 \in \Omega_0$. *Then we can choose an open neighborhood* V_0 *of* z_0 *in* Ω *such that* ψ_λ *can be extended to a holomorphic function on* $V_0 \times \mathbb{C}^{l_2}$ *for every* $\lambda \in \Lambda_2$.

Proof the same as before (see Lemma 29 of the paper).

This essentially completes the proof of Theorem 3.

Let $\mathbf{s}, \mathbf{t}$ be two points in $\mathbb{C}^l$. We say $\mathbf{s} > \mathbf{t}$ (or $\mathbf{t} \prec \mathbf{s}$) if $s_i - t_i$ is real and ≥ 0 for every i $(1 \leq i \leq l)$.

Lemma 31. (We use the notations of Theorem 1 and Theorem 2.) *Let* α *be a point in* $\mathbb{C}^l$ *such that* $\lambda + \alpha \prec 0$ *for every* $\lambda \in \Lambda_0$. *Then we can select positive numbers* M_0 *and* A *such that*

$$\sum_{\lambda \in \Lambda} \left| p_\lambda(\mathbf{t}) e^{\langle \lambda, \mathbf{t} \rangle} \right| \leq A (1 + |\mathbf{t}|)^d |e^{-\langle \alpha, \mathbf{t} \rangle}|$$

for all $\mathbf{t} \in \mathbb{C}^l(M_0)$.

Note that

$$|p_\lambda(\mathbf{t})| \leq (1 + |\mathbf{t}|)^d \| p_\lambda \| \leq a b^{-\sigma(\lambda)} (1 + |\mathbf{t}|)^d.$$

Now clearly $\lambda + \alpha \prec 0$ for every $\lambda \in \Lambda$. Hence if $\mathbf{t} \in \mathbb{C}^l(M_0)$,

$$\left| e^{\langle \lambda, \mathbf{t} \rangle} \right| = \left| e^{\langle \lambda + \alpha, \mathbf{t} \rangle} \right| \left| e^{-\langle \alpha, \mathbf{t} \rangle} \right|$$

$$\leq e^{\sigma(\lambda + \alpha) M_0} \left| e^{-\langle \alpha, \mathbf{t} \rangle} \right|$$

since $\lambda + \alpha \prec 0$. Therefore

$$\sum_{\lambda \in \Lambda} \left| p_\lambda(\mathbf{t}) e^{\langle \lambda, \mathbf{t} \rangle} \right| \leq (1 + |\mathbf{t}|)^d e^{\sigma(\alpha) M_0} |e^{-\langle \alpha, \mathbf{t} \rangle}| \sum_{\lambda \in \Lambda} a (b^{-1} e^{M_0})^{\sigma(\lambda)}.$$

Now if M_0 is sufficiently large, $b^{-1}e^{M_0} > 1$ and the series $\Sigma_{\lambda \in \Lambda}(b^{-1}e^{M_0})^{\sigma(\lambda)}$ obviously converges. Hence our assertion follows.

Lemma 32. *We use the notation of Theorem 3. Then under the assumptions of Lemma 31, we can choose $M_1 \geqslant M_0$ and $B \geqslant 0$ such that*

$$\sum_{\lambda \in \Lambda_2} \left| \phi_\lambda(\mathbf{t})\, e^{\langle \lambda, t_2 \rangle} \right| \leqslant B(1 + |\mathbf{t}|)^d \left| e^{-\langle \alpha, t \rangle} \right|$$

for all $\mathbf{t} = (t_1, t_2) \in \omega_0 \times \mathbb{R}^{l_2}(M_1)$.

Since $Cl(\omega_0)$ is compact, it is enough to prove the following result.

Lemma 33. *Fix a point $\mathbf{z}_0 \in \omega$. Then we can choose an open neighborhood V of $\mathbf{z}_0$ in ω and numbers $M_1 \geqslant M_0$, $B \geqslant 0$ such that*

$$\sum_{\lambda \in \Lambda_2} \left| \phi_\lambda(\mathbf{t})\, e^{\langle \lambda, t_2 \rangle} \right| \leqslant B(1 + |\mathbf{t}|)^d \left| e^{-\langle \alpha, t \rangle} \right|$$

for $\mathbf{t} \in V \times \mathbb{C}^{l_2}(M_1)$.

We use the notation of the proof of Lemma 28. First assume $\mathbf{z}_0 \in \omega \cap \Omega'$. Select V as in Lemma 28. Choose $M_1 \geqslant M_0$ sufficiently large. Then we can choose $A_1 \geqslant 0$ such that

$$\sum_{\mathbf{m}} |\kappa(\mathbf{z}, \mathbf{m})| e^{-\sigma(\mathbf{m})M_1} \leqslant A_1$$

for all $\mathbf{z} \in V$. Now

$$\phi_\lambda(\mathbf{z}, \mathbf{u}) = \sum_{\mathbf{m}} \kappa(\mathbf{z}, \mathbf{m}) \phi_{\lambda+\mathbf{m}}(\mathbf{w}, \mathbf{u}) \qquad (\lambda \in \Lambda_2)$$

$\mathbf{z} \in V$, $\mathbf{u} \in \mathbb{C}^{l_2}$. Hence

$$\sum_{\lambda \in \Lambda_2} \left| \phi_\lambda(\mathbf{z}, \mathbf{u}) e^{\langle \lambda, u \rangle} \right| \leqslant \sum_{\lambda \in \Lambda_2} \sum_{\mathbf{m}} |\kappa(\mathbf{z}, \mathbf{m})| \left| \phi_{\lambda+\mathbf{m}}(\mathbf{w}, \mathbf{u}) e^{\langle \lambda, u \rangle} \right|$$

$$\leqslant \sum_{\mathbf{m}} \left| \kappa(\mathbf{z}, \mathbf{m}) e^{-\langle \mathbf{m}, u \rangle} \right| \sum_{\lambda \in \Lambda_2} \left| \phi_\lambda(\mathbf{w}, \mathbf{u}) e^{\langle \lambda, u \rangle} \right|.$$

Now if $\mathbf{u} \in \mathbb{C}^l(M_1)$, $\left| e^{-\langle \mathbf{m}, u \rangle} \right| \leqslant e^{-\sigma(\mathbf{m})M_1}$. So

$$\sum_{\lambda \in \Lambda_2} \left| \phi_\lambda(\mathbf{z}, \mathbf{u}) e^{\langle \lambda, u \rangle} \right| \leqslant A_1 \sum_{\lambda \in \Lambda_2} \left| \phi_\lambda(\mathbf{w}, \mathbf{u}) e^{\langle \lambda, u \rangle} \right|$$

for $\mathbf{z} \in V$ and $\mathbf{u} \in \mathbb{C}^{l_2}(M_1)$. But

$$\sum_{\lambda \in \Lambda_2} \left| \phi_\lambda(\mathbf{w}, \mathbf{u}) e^{\langle \lambda, u \rangle} \right| \leqslant \sum_{\mu \in \Lambda} \left| p_\mu(\mathbf{w}, \mathbf{u}) e^{\langle \mu_1, w \rangle + \langle \mu_2, u \rangle} \right|$$

$$\leqslant A(1 + |\mathbf{w} + \mathbf{u}|)^{d'} \left| e^{-\langle \alpha, w+u \rangle} \right|$$

if $\mathbf{u} \in \mathbb{C}^{l_2}(M_1)$. (Here we have used Lemma 32 and the fact that $(\mathbf{w}, \mathbf{u}) \in \mathbb{C}^{l_2}(M_0)$, if

w had been suitably chosen.) Now

$$(1+|\mathbf{w}+\mathbf{u}|) \leqslant (1+|\mathbf{w}|)(1+|\mathbf{u}|)$$

$$\left|e^{-\langle \alpha, \mathbf{w}+\mathbf{u}\rangle}\right| = \left|e^{-\langle \alpha_1, \mathbf{w}\rangle}\right|\left|e^{-\langle \alpha_2, \mathbf{u}\rangle}\right|.$$

Therefore we can choose A_2 such that

$$\sum_{\lambda \in \Lambda_2} \left|\phi_\lambda(\mathbf{z},\mathbf{u})e^{\langle \lambda,\mathbf{u}\rangle}\right| \leqslant A_2(1+|\mathbf{u}|)^d\left|e^{-\langle \alpha_2,\mathbf{u}\rangle}\right|$$

for $\mathbf{z} \in V$ and $\mathbf{u} \in \mathbf{C}^{l_2}(M_1)$. Since V is a bounded set, $|\langle \alpha_1,\mathbf{z}\rangle|$ remains bounded for $\mathbf{z} \in V$. Also $|\mathbf{u}+\mathbf{z}| \geqslant |\mathbf{u}|$. Hence the assertion of Lemma 33 is now obvious in this case.

Now suppose $\mathbf{z}_0 \notin \Omega_0'$. Then we have seen (see discussion following Lemma 29) that

$$\left|\phi_\lambda(\mathbf{z},\mathbf{u})e^{\langle \lambda,\mathbf{u}\rangle}\right| \leqslant \left(\frac{\eta}{2\pi\delta}\right)^{-l_1}\int_J\left|\phi_\lambda(\mathbf{w}(\theta),\mathbf{u})e^{\langle \lambda,\mathbf{u}\rangle}\right|d\theta_1\cdots d\theta_l$$

for $\mathbf{z} \in K$, $\mathbf{u} \in \mathbf{C}^{l_2}$. Let F be the set of all points $\mathbf{w}(\theta)$ with $\theta \in J$. Then F is a compact subset of $\omega \cap \Omega'$. Hence in view of what has been proved above, there exist numbers $B_1 \geqslant 0$, $M_1 \geqslant M_0$ such that

$$\sum_{\lambda \in \Lambda_2} \left|\phi_\lambda(\mathbf{w},\mathbf{u})e^{\langle \lambda,\mathbf{u}\rangle}\right| \leqslant B_1(1+|\mathbf{w}+\mathbf{u}|)^d\left|e^{\langle \alpha,\mathbf{w}+\mathbf{u}\rangle}\right|$$

for $\mathbf{w} \in F$, $\mathbf{u} \in \mathbf{C}^{l_2}(M_1)$. Hence

$$\sum_{\lambda \in \Lambda_2} \left|\phi_\lambda(\mathbf{z},\mathbf{u})e^{\langle \lambda,\mathbf{u}\rangle}\right| \leqslant B_2(1+|\mathbf{u}|)^d\left|e^{-\langle \alpha_2,\mathbf{u}\rangle}\right|$$

for $\mathbf{z} \in K$ and $\mathbf{u} \in \mathbf{C}^{l_2}(M_1)$ where B_2 is a suitable constant. Since Int K is an open neighborhood of $\mathbf{z}_0$ our assertion follows.

Theorem 4. *Suppose ϕ is analytic on some domain in $\mathbf{R}^l$ containing $Cl(\mathbf{R}^l_+)$. Then if $\lambda + \alpha \prec 0$ for every $\lambda \in \Lambda_0$, we can choose a number $A \geqslant 0$ such that*

$$|\phi(\mathbf{t})| \leqslant A(1+|\mathbf{t}|)^d\left|e^{-\langle \alpha,\mathbf{t}\rangle}\right|$$

for all $\mathbf{t} \in Cl(\mathbf{R}^l_+)$.

Put $\Omega = Cl(\mathbf{R}^l_+)$ and let F be any subset of Ω. We say Theorem 4 holds on F, if we can choose a number A_F such that

$$|\phi(\mathbf{t})| \leqslant A_F(1+|\mathbf{t}|)^d\left|e^{-\langle \alpha,\mathbf{t}\rangle}\right|$$

for all $\mathbf{t} \in F$. If $F_1,\ldots,F_r$ is a finite number of subsets of Ω and Theorem 4 holds on F_i for every i, then it is clear that it also holds for $\cup_i F_i$.

For any two numbers M_1, M_2, $(0 \leqslant M_1 < M_2, M_2$ can be $\infty)$, let $E(M_1, M_2)$ denote the characteristic function of the interval $M_1 \leqslant x < M_2$ on the set of all nonnegative reals. Let $E_i(M_1, M_2)$ denote the function on Ω such that $E_i(M_1, M_2:\mathbf{t}) = E(M_1, M_2, t_i)$. If J is any subset of $I = (1,2,\ldots,l)$, we denote by $\Omega_J(M_1, M_2)$ the set of all points $\mathbf{t} \in \Omega$ where $E_J(M_1, M_2) = \prod_{i \in J}E_i(M_1, M_2)$ is $\neq 0$.

Lemma 34. *Let $J \subset I$. Then for any M $(0 \leqslant M < \infty)$ we can choose $M_1 > M$ such that Theorem 4 holds on $\Omega_J(0, M) \cap \Omega_{c_J}(M_1, \infty)$.*

If $J = I$, $\Omega_I(0, M)$ is a bounded subset of Ω and so Theorem 4 obviously holds on it. Let $l_J =$ number of elements in J. We may assume $l_J < l$. Also if $J = \varnothing$, the set of our lemma is $\mathbb{R}^l_+(M_1)$. This case is covered by Lemma 31. So now we can assume $0 < l_J < l$. Without loss of generality we may assume $J = (1, 2, \ldots, l_1)$. Then since the projection of $\Omega_J(0, M)$ on $\mathbb{R}^{l_1}$ is a bounded set, our assertion follows from Lemma 32.

Since $1 = E(0, M) + E(M, \infty)$ on $Cl(\mathbb{R}^+)$ for any $M > 0$, it is clear that

$$1 = \sum_{J \subset I} E_{c_J}(0, M) E_J(M, \infty)$$

on Ω. Hence

$$\Omega = \bigcup_{J \subset I} \left(\Omega_{c_J}(0, M) \cap \Omega_J(M, \infty) \right).$$

So in order to prove Theorem 4 it is enough to prove the following lemma.

Lemma 35. *For any $M > 0$, Theorem 4 holds on $\Omega_{c_J}(0, M) \cap \Omega_J(M, \infty)$.*

We shall prove this by induction on l_J. If $l_J = 0$, this set is bounded in Ω and so Theorem 4 holds on it. Now suppose $l_J > 0$. By Lemma 34, we can choose $M_1 > M$ such that Theorem 4 holds on $\Omega_{c_J}(0, M) \cap \Omega_J(M_1, \infty)$. Since

$$E(M, \infty) = E(M, M_1) + E(M_1, \infty),$$

it is clear that

$$E_J(M, \infty) = \sum_{K \subset J} E_{c_{K \cap J}}(M, M_1) E_K(M_1, \infty).$$

Hence

$$E_J(M, \infty) - E_J(M_1, \infty) = \sum_{\substack{K \subset J \\ K \neq J}} E_{c_{K \cap J}}(M, M_1) E_K(M_1, \infty).$$

This proves that

$$\Omega_J(M, \infty) \cap {}^c\Omega_J(M_1, \infty) = \bigcup_{\substack{K \subset J \\ K \neq J}} \Omega_{c_{K \cap J}}(M, M_1) \cap \Omega_K(M_1, \infty).$$

Since Theorem 4 holds on $\Omega_{c_J}(0, M) \cap \Omega_J(M_1, \infty)$, it is enough to prove that it holds on

$$\Omega_{c_J}(0, M) \cap \Omega_{c_{K \cap J}}(M, M_1) \cap \Omega_K(M_1, \infty)$$

for every $K \subset J$ with $l_K < l_J$. But the above set is contained in

$$\Omega_{c_K}(0, M_1) \cap \Omega_K(M_1, \infty)$$

and since $l_K < l_J$, Theorem 4 holds on it by induction hypothesis.

Differential Equations and Semisimple Lie Groups[†]

§1. Introduction

Let G be a connected semisimple Lie group with finite center and K a maximal compact subgroup of G. Let $\mathfrak{Z}$ be the algebra of all differential operators on G which are invariant under both left and right translations of G. It is an important problem, in the theory of harmonic analysis on G, to obtain those C^∞-functions ϕ on G which satisfy the following two conditions:

(1) ϕ is an eigenfunction of every differential operator in $\mathfrak{Z}$.

(2) ϕ has a prescribed behaviour at infinity.

For example, we might ask that ϕ should lie in $L_1(G)$ or $L_2(G)$ (with respect to the Haar measure on G). Since K is compact, one can, without essential loss of generality, assume the following additional condition.

(3) The left and right translates of ϕ by elements of K span a vector space of finite dimension.

The main point then is to investigate the asymptotic behaviour of functions satisfying (1) and (3). We can restate this problem in a slightly different form as follows. Let V be a finite-dimensional complex Hilbert space and $\sigma = (\sigma_1, \sigma_2)$ a double unitary representation of K on V. By a *spherical function ϕ of type σ* we mean a C^∞-function from G to V such that $\phi(k_1 x k_2) = \sigma_1(k_1)\phi(x)\sigma_2(k_2)$ for all $x \in G$ and $k_1, k_2 \in K$. Let $S(\sigma)$ denote the space of all such ϕ. If χ is a homomorphism of $\mathfrak{Z}$ into the field $\mathbb{C}$ of complex numbers, we denote by $S(\sigma, \chi)$ the subspace consisting of those $\phi \in S(\sigma)$ which satisfy the system of differential equations $z\phi = \chi(z)\phi$ $(z \in \mathfrak{Z})$. Our main problem then is to find an asymptotic expansion for a given ϕ in $S(\sigma, \chi)$ near infinity.

We can choose a closed, connected and simply connected abelian subgroup A of G such that $G = KAK$. If $\phi \in S(\sigma)$, then $\phi(k_1 a k_2) = \sigma_1(k_1)\phi(a)\sigma_2(k_2)$ $(k_1, k_2 \in K, a \in A)$. Hence ϕ is completely determined by its restriction $\bar\phi$ on A. It is easy to see that if $\phi \in S(\sigma)$ and $z \in \mathfrak{Z}$, then $z\phi$ is also in $S(\sigma)$. We wish to compare $\overline{z\phi}$ and $\bar\phi$.

Let M be the centralizer and M^* the normalizer of A in K. Then M is a closed normal subgroup of M^* and $\mathfrak{w} = M^*/M$ is a finite group, which operates on A as

---

[†]See footnote to previous paper—V.

follows. If $s \in \mathfrak{w}$ and $a \in A$, then $a^s = mam^{-1}$ where m is any element of M^* lying in the coset s. An element $a \in A$ is called *singular* if $a^s = a$ for some $s \neq 1$ in $\mathfrak{w}$. Otherwise it is called *regular*. A, being a simply connected abelian Lie group, is essentially a real Euclidean space of dimension l (say). The singular set consists of a finite number of hyperplanes passing through the origin. Let A' be the set of regular elements of A. Then A' is open and dense in A. Fix a connected component A_+ of A'. Then $A' = \cup_{s \in \mathfrak{w}} A^s_+$. If $\phi \in S(\sigma)$ then $\phi(a^s) = \phi(mam^{-1}) = \sigma_1(m)\phi(a)\sigma_2(m^{-1})$ ($a \in A$, $s \in \mathfrak{w}$) where m lies in the coset s. Hence ϕ is determined completely by its restriction $\bar{\phi}$ on A_+. Let $\mathfrak{E}(V)$ denote the associative algebra of all endomorphisms of V. Then we have the following result.

Lemma. *For each $z \in \mathfrak{Z}$, one can define a differential operator $\delta'(z)$ on A_+, whose coefficients are analytic functions from A_+ to $\mathfrak{E}(V)$, such that $\overline{z\phi} = \delta'(z)\bar{\phi}$ for every $\phi \in S(\sigma)$.*

Now let L be the rank of G. It is known that $\mathfrak{Z}$ is abelian and one can select L elements $z_1,\ldots,z_L \in \mathfrak{Z}$ such that $\mathfrak{Z} = \mathbb{C}[z_1,\ldots,z_L]$. Hence the system of differential equations $z\phi = \chi(z)\phi$ ($z \in \mathfrak{Z}$), reduces to the finite set $\delta'(z_i)\bar{\phi} = \chi(z_i)\bar{\phi}$ ($1 \leqslant i \leqslant L$).

Let us consider the following example. Let G be the group of 2×2 real matrices with determinant 1 and K the subgroup of all orthogonal matrices in G. Finally let A be the subgroup of diagonal matrices with positive diagonal coefficients. Then the mapping

$$t \to \begin{pmatrix} e^{t/2} & 0 \\ 0 & e^{-t/2} \end{pmatrix}$$

is an isomorphism of the additive group of the real line onto A. In this case $L = l = 1$ and so $\mathfrak{Z}$ is generated by a single element. The group $\mathfrak{w}$ consists of two elements 1 and s where s corresponds to the reflexion $t \to -t$. The only singular point in A is $t = 0$. So A_+ can be chosen as the set of all points where $t > 0$. The Lie algebra $\mathfrak{g}$ of G can be thought of as consisting of all real 2×2 matrices with trace zero. Put

$$H = \begin{pmatrix} 1 & 0 \\ 0 & -1 \end{pmatrix}, \qquad X = \begin{pmatrix} 0 & 1 \\ 0 & 0 \end{pmatrix}, \qquad Y = \begin{pmatrix} 0 & 0 \\ 1 & 0 \end{pmatrix}$$

and let $\mathbb{R}$ denote the field of real numbers. Then (H, X, Y) is a base for $\mathfrak{g}$ over $\mathbb{R}$. Put $Z = \frac{1}{2}(X - Y) = \begin{pmatrix} 0 & 1/2 \\ -1/2 & 0 \end{pmatrix}$. Then $\mathbb{R} Z$ is the Lie algebra of K. Now we regard elements of $\mathfrak{g}$ as left-invariant differential operators on G in the usual way. Then $\omega = \frac{1}{2}H^2 + XY + YX$ becomes a differential operator in $\mathfrak{Z}$ such that $\mathfrak{Z} = \mathbb{C}[\omega]$. Since σ_1, σ_2 are representations of K, they define representations of $\mathbb{R} Z$ which we again denote by σ_1, σ_2. The operator $\delta'(\omega)$ is then given by the formula

$$\tfrac{1}{2}\delta'(\omega)\bar{\phi} = d^2\bar{\phi}/dt^2 + \coth t \, d\bar{\phi}/dt + (\sinh t)^{-2}\left(\sigma_1(z)^2\bar{\phi} + \bar{\phi}\sigma_2(z)^2\right)$$

$$- 2\cosh t (\sinh t)^{-2}\sigma_1(z)\bar{\phi}\sigma_2(z) \qquad (\phi \in S(\sigma)).$$

The remarkable thing about this operator is that for $t > 0$, its coefficients can be expanded in powers of e^{-t}.

Now this situation is quite general. In fact we can always choose a Cartesian coordinate system $(t_1,\ldots,t_l)$ in the Euclidean space A such that:

1) A_+ is precisely the set of points where $t_i > 0$ $(1 \leqslant i \leqslant l)$.
2) For any $z \in \mathfrak{Z}$, the coefficients of $\delta'(z)$ (with respect to this coordinate system) are functions f from A_+ to $\mathcal{E}(V)$ of the form

$$f(t) = \sum_{m_1,\ldots,m_l \geqslant 0} c(m_1,\ldots,m_l)e^{-(m_1 t_1 + \cdots + m_l t_l)} \qquad (t_1,\ldots,t_l > 0),$$

where $c(m_1,\ldots,m_l) \in \mathcal{E}(V)$ and the series[1] $\sum_m |c(m_1,\ldots,m_l)|e^{-(m_1 + \cdots + m_l)\varepsilon}$ converges for every $\varepsilon > 0$.

Let $\mathfrak{A}$ be the algebra of all differential operators on A with constant coefficients. Then it is possible to find a finite number of elements $u_1 = 1, u_2,\ldots,u_r$ in $\mathfrak{A}$ such that any $u \in \mathfrak{A}$ can be written in the form[2]

$$u = \sum_j \sum_i g_{ij} u_i \circ \delta'(z_{ij})$$

on A_+. Here $z_{ij} \in \mathfrak{Z}$ and g_{ij} are a finite number of functions from A_+ to $\mathcal{E}(V)$ with expansions of the above form. Such relations hold in particular for $u = (\partial/\partial t_j) \circ u_i$ $(1 \leqslant j \leqslant l,\ 1 \leqslant i \leqslant r)$. Let V^r denote the direct sum of V with itself r times and put

$$\Phi = \left(u_1 \bar{\phi}, \ldots, u_r \bar{\phi} \right)$$

for a fixed $\phi \in S(\sigma, \chi)$. Then Φ is a function from A_+ to V^r. Since $\delta'(z)\bar{\phi} = \chi(z)\bar{\phi}$ $(z \in \mathfrak{Z})$, we get the relation

$$u\bar{\phi} = \sum_j \sum_i g_{ij} \chi(z_{ij}) u_i \bar{\phi}$$

from the above equation. But this obviously implies equations of the form

$$\partial\Phi/\partial t_i = \Gamma_i(t)\Phi \qquad (1 \leqslant i \leqslant l).$$

Here Γ_i are functions from A_+ to the algebra $\mathcal{E}(V^r)$ of endomorphisms of V^r, which have expansions of the form

$$\Gamma_i(t) = \sum_{k_1,\ldots,k_l \geqslant 0} c_i(k_1,\ldots,k_l)e^{-(k_1 t_1 + \cdots + k_l t_l)} \qquad (t_1,\ldots,t_l > 0)$$

$c_i(k_1,\ldots,k_l) \in \mathcal{E}(V^r))$. We can now apply the results of [11] to this system of equations for Φ and thus obtain an expansion for Φ which describes the behaviour of Φ when $t_1,\ldots,t_l$ are all large and positive. Then since $u_1 = 1$, this also gives the required expansion for $\bar{\phi}$.

We have given above a brief summary of the principal ideas of this paper. Now we shall give a short description of its contents. The main results of §2 are contained in Lemma 7 and its corollaries. In §3 we apply them to spherical functions. With the help of the results of [11] we introduce the notion of the principal exponents and the

---

[1] $|u|$ denotes the bound of the linear transformation $u \in \mathcal{E}(V)$.
[2] $\circ$ denotes product for two differential operators.

leading term of a function $\phi \in S(\sigma, \chi)$. These concepts play an important role later in this paper.

In §4 we define the mapping δ' indicated above. The main results here are contained in Lemmas 10 and 12 and their corollaries. They are applied in §6A to the theory of irreducible representations of G on a Banach space. Theorem 2 is the principal result of this section. The earlier results in this direction (see [2, p. 63, Theorem 4]) had been obtained only under additional assumptions on G, for example that it should have a finite-dimensional faithful representation.

Section 6B is devoted to the proof of Lemma 19 and in §7 we compute $\delta'(\omega)$ for the Casimir operator ω. In §8 we give some heuristic considerations which are then made precise in §9. (This part can be regarded as a generalization of the results of [8, §8].) They enable us to construct (see Theorem 3) certain analytic functions ψ from A_+ to V such that $\delta'(z)\psi = \chi(z)\psi$ for all $z \in \mathfrak{z}$. The behaviour of ψ at infinity is quite obvious from its construction. The rest of the paper is then devoted to showing that the spherical functions θ arising in the theory of representations (see Theorem 2) can be expressed on A_+ as a linear combination of the special functions ψ constructed above. This gives us the required asymptotic expansion of θ (see Theorem 4). In Theorem 4 certain meromorphic functions on $\mathfrak{h}_\mathfrak{p}$ appear as coefficients of the main terms of the expansion. The results of [8, 9] make it plausible that these functions should play an important role in the problem of the explicit determination of the Plancherel measure for G. There are also some indications that these functions are likely to appear in the theory of automorphic forms on G with respect to a suitable discrete subgroup Γ of G. Therefore it seems important that they should be studied in some detail. It is possible to obtain for them an integral formula which generalizes Theorem 4 of [8].

This work was begun in Paris in 1958 and it has been supported at various periods by grants from the Guggenheim Memorial Foundation, the National Science Foundation and a contract with the U.S. Air Force. A brief sketch of some of the results of this paper has appeared in [10].

§2. Proof of Lemma 7

Let $\mathbb{R}$ and $\mathbb{C}$ be the fields of real and complex numbers respectively and G a connected semisimple Lie group and $\mathfrak{g}_0$ its Lie algebra over $\mathbb{R}$. Define $\mathfrak{k}_0, \mathfrak{p}_0, \mathfrak{h}_{\mathfrak{p}_0}, \mathfrak{h}_{\mathfrak{k}_0}$ and $\mathfrak{h}_0$ as in [8, §2]. We complexify $\mathfrak{g}_0, \mathfrak{k}_0, \mathfrak{p}_0, \mathfrak{h}_{\mathfrak{p}_0}, \mathfrak{h}_{\mathfrak{k}_0}, \mathfrak{h}_0$ to $\mathfrak{g}, \mathfrak{k}, \mathfrak{p}, \mathfrak{h}_\mathfrak{p}, \mathfrak{h}_\mathfrak{k}$ and $\mathfrak{h}$ respectively. Introduce compatible orders (see [7, p. 195]) in the spaces of real-valued linear functions on[3] $\mathfrak{h}_{\mathfrak{p}_0} + (-1)^{1/2}\mathfrak{h}_{\mathfrak{k}_0}$ and $\mathfrak{h}_{\mathfrak{p}_0}$ respectively. Let P denote the set of all positive roots of $\mathfrak{g}$ (with respect to $\mathfrak{h}$) under this order and P_+ the subset of those $\alpha \in P$ which do not vanish identically on $\mathfrak{h}_\mathfrak{p}$. For any root α define X_α as usual (see [1, p. 188]) and put $\mathfrak{n}_+ = \Sigma_{\alpha \in P_+} \mathbb{C} X_\alpha$ and $\mathfrak{n}_- = \Sigma_{\alpha \in P_+} \mathbb{C} X_{-\alpha}$. Then $\mathfrak{n}_+$ and $\mathfrak{n}_-$ are nilpotent subalgebras of $\mathfrak{g}$. Put $\mathfrak{n}_0 = \mathfrak{n}_+ \cap \mathfrak{g}_0$ and let K, A and N be the analytic

---

[3] We fix a square-root of -1 and denote it throughout by $(-1)^{1/2}$.

subgroups of G corresponding to $\mathfrak{k}_0$, $\mathfrak{h}_{\mathfrak{p}_0}$ and $\mathfrak{n}_0$ respectively. Then we have the Iwasawa decomposition $G = KAN$.

Let $\mathfrak{G}$ be the universal enveloping algebra of $\mathfrak{g}$ and $S(\mathfrak{g})$ the symmetric algebra over $\mathfrak{g}$. We denote by λ the canonical mapping [1, p. 192] of $S(\mathfrak{g})$ into $\mathfrak{G}$. For any integer d let $S_d(\mathfrak{g})$ denote the space of homogeneous elements in $S(\mathfrak{g})$ of degree d. (If $d < 0$, $S_d(\mathfrak{g}) = \{0\}$.) Then an element $g \in \mathfrak{G}$ is said to be homogeneous of degree d, if it lies in $\lambda(S_d(\mathfrak{g}))$. We denote by $\mathfrak{G}_d$ the subspace $\Sigma_{0 \leqslant e \leqslant d}\lambda(S_e(\mathfrak{g}))$ of $\mathfrak{G}$. For any subspace $\mathfrak{q}$ of $\mathfrak{g}$, we can regard the symmetric algebra $S(\mathfrak{q})$ over $\mathfrak{q}$ as a subalgebra of $S(\mathfrak{g})$. Put $S_d(\mathfrak{q}) = S_d(\mathfrak{g}) \cap S(\mathfrak{q})$. The following result is well known.

Lemma 1. *Let $p \in S_d(\mathfrak{g})$ and $q \in S_e(\mathfrak{g})$. Then*

$$\lambda(p)\lambda(q) \equiv \lambda(pq) \bmod \mathfrak{G}_{d+e-1}.$$

For any element $g \neq 0$ of $\mathfrak{G}$ or $S(\mathfrak{g})$, we denote by $d^0 g$ the degree of g. If $g = 0$, we put $d^0 g = -\infty$.

Let $\mathfrak{m}$ be the centralizer of $\mathfrak{h}_{\mathfrak{p}}$ in $\mathfrak{g}$. Then $\mathfrak{m} = \mathfrak{m} \cap \mathfrak{k} + \mathfrak{h}_{\mathfrak{p}}$. Denote by $\mathfrak{H}_{\mathfrak{p}}$, $\mathfrak{M}$ and $\mathfrak{R}$ the enveloping algebras[4] of $\mathfrak{h}_{\mathfrak{p}}$, $\mathfrak{m}$ and $\mathfrak{k}$ respectively in $\mathfrak{G}$. Put $U_d = U \cap \mathfrak{G}_d$ for $U = \mathfrak{H}_{\mathfrak{p}}$, $\mathfrak{R}$ or $\mathfrak{M}$.

Lemma 2. *Let g be an element in $\mathfrak{G}$ such that[5] $[H, g] = 0$ for every $H \in \mathfrak{h}_{\mathfrak{p}}$. Then there exists a unique element $m \in \mathfrak{M}$ such that $g - m \in \mathfrak{G}\mathfrak{n}_+$. If $g \in \mathfrak{G}_d$, then $m \in \mathfrak{M}_d$ and $g - m \in \mathfrak{n}_- \mathfrak{G}_{d-2}\mathfrak{n}_+$.*

Let $\mathfrak{R}_+$ and $\mathfrak{R}_-$ be the enveloping algebras of $\mathfrak{n}_+$ and $\mathfrak{n}_-$ respectively in $\mathfrak{G}$. Since[6] $\mathfrak{g} = \mathfrak{n}_- + \mathfrak{m} + \mathfrak{n}$, we conclude from Lemma 12 of [1] that $\mathfrak{G} = \mathfrak{R}_- \mathfrak{M} \mathfrak{R}_+ = \mathfrak{R}_- \mathfrak{M} + \mathfrak{G}\mathfrak{n}_+$. Therefore m, if it exists, is clearly unique. Now suppose $g \in \mathfrak{G}_d$. Then since $\mathfrak{G}_d = \mathfrak{G}_d \cap (\mathfrak{R}_- \mathfrak{M}) + \mathfrak{G}_d \cap \mathfrak{G}\mathfrak{n}_+$ from the same lemma, we can choose $m \in \mathfrak{G}_d \cap (\mathfrak{R}_- \mathfrak{M})$ such that $g - m \in \mathfrak{G}_d \cap \mathfrak{G}\mathfrak{n}_+$. We note that both $\mathfrak{R}_- \mathfrak{M}$ and $\mathfrak{G}\mathfrak{n}_+$ are stable under the mapping $a \to [H, a]$ ($a \in \mathfrak{G}$) corresponding to a fixed $H \in \mathfrak{h}_{\mathfrak{p}}$. Hence it follows from the directness of the sum $\mathfrak{R}_- \mathfrak{M} + \mathfrak{G}\mathfrak{n}_+$ that m commutes with H. Since the centralizer of $\mathfrak{h}_{\mathfrak{p}}$ in $\mathfrak{R}_- \mathfrak{M}$ is obviously $\mathfrak{M}$, it follows that $m \in \mathfrak{M}$. Now $g - m \in \mathfrak{G}_d \cap \mathfrak{G}\mathfrak{n}_+$ and $g - m$ commutes with $\mathfrak{h}_{\mathfrak{p}}$. Since $\mathfrak{G} = \mathfrak{R}_- \mathfrak{M}\mathfrak{R}_+$, the centralizer of $\mathfrak{h}_{\mathfrak{p}}$ in $\mathfrak{G}\mathfrak{n}_+$ is obviously contained in $\mathfrak{n}_- \mathfrak{G}\mathfrak{n}_+$. Therefore $g - m \in \mathfrak{G}_d \cap (\mathfrak{n}_- \mathfrak{G}\mathfrak{n}_+) = \mathfrak{n}_- \mathfrak{G}_{d-2}\mathfrak{n}_+$.

Let $\mathfrak{Z}_{\mathfrak{m}}$ denote the center of $\mathfrak{M}$ and $\mathfrak{Z}$ the center of $\mathfrak{G}$.

Corollary. *There exists a homomorphism μ' of $\mathfrak{Z}$ into $\mathfrak{Z}_{\mathfrak{m}}$ such that $z - \mu'(z) \in \mathfrak{G}\mathfrak{n}_+$ for $z \in \mathfrak{Z}$.*

---

[4]This means that for example $\mathfrak{H}_{\mathfrak{p}}$ is the subalgebra of $\mathfrak{G}$ generated by $(1, \mathfrak{h}_{\mathfrak{p}})$.

[5]$[a, b] = ab - ba$ for any two elements a, b of an associative algebra.

[6]$\dot{+}$ denotes direct sum.

Let $z \in \mathfrak{Z}$ and choose $m \in \mathfrak{M}$ such that $z - m \in \mathfrak{G}\mathfrak{n}_+$. If $X \in \mathfrak{m}$, it is obvious that $[X, \mathfrak{n}_+] \subset \mathfrak{n}_+$. Therefore $[X, z - m] = -[X, m] \in \mathfrak{G}\mathfrak{n}_+ \cap \mathfrak{M} = \{0\}$. Hence $m \in \mathfrak{Z}_{\mathfrak{m}}$. For two elements, $z_1, z_2 \in \mathfrak{Z}$, let m_1, m_2 denote the corresponding elements in $\mathfrak{Z}_{\mathfrak{m}}$. Then $z_1 z_2 - m_1 m_2 = z_1(z_2 - m_2) + m_2(z_1 - m_1) \in \mathfrak{G}\mathfrak{n}_+$. Hence μ' is a homomorphism.

Put $\rho = \frac{1}{2}\sum_{\alpha \in P_+} \bar{\alpha}$ where $\bar{\alpha}$ denotes the restriction of α on $\mathfrak{h}_\mathfrak{p}$. Define an automorphism $m \to {}'m$ of $\mathfrak{M}$ as follows. ${}'H = H - \rho(H)$ $(H \in \mathfrak{h}_\mathfrak{p})$ and ${}'X = X$ for $X \in \mathfrak{m} \cap \mathfrak{k}$. Put $\mu(z) = {}'(\mu'(z))$ $(z \in \mathfrak{Z})$. Since $\mathfrak{h}_\mathfrak{p}$ lies in the center of $\mathfrak{m}$, it is obvious that $\mu(z) \in \mathfrak{Z}_{\mathfrak{m}}$.

Let W be the Weyl group of $\mathfrak{g}$ with respect to $\mathfrak{h}$ and W_- the subgroup of W generated by the Weyl reflexions s_α corresponding to the roots $\alpha \in P_-$. (P_- is the complement of P_+ in P.) W operates on the symmetric algebra $S(\mathfrak{h})$ in the obvious way. Let $I(\mathfrak{h})$ and $I_-(\mathfrak{h})$ be the subalgebras consisting of those elements of $S(\mathfrak{h})$ which are invariant under W and W_- respectively. Let $\mathfrak{H}$ be the enveloping algebra of $\mathfrak{h}$ in $\mathfrak{G}$. We identify $S(\mathfrak{h})$ with $\mathfrak{H}$ under the isomorphism which leaves every point of $\mathfrak{h}$ fixed. Then $I(\mathfrak{h})$ and $I_-(\mathfrak{h})$ become subalgebras of $\mathfrak{H}$. Let γ and γ_- be the canonical isomorphisms (see [4, Lemma 19]) of $\mathfrak{Z}$ and $\mathfrak{Z}_{\mathfrak{m}}$ onto $I(\mathfrak{h})$ and $I_-(\mathfrak{h})$ respectively.

Lemma 3. $\gamma_-(\mu(z)) = \gamma(z)$ *for* $z \in \mathfrak{Z}$. *Therefore* μ *is injective.*

This is an immediate consequence of the definitions of γ and γ_- respectively.

Lemma 4. *Let* $z \in \mathfrak{Z}$. *Then* $d^0\gamma(z) = d^0 z$. *Similarly if* $v \in \mathfrak{Z}_{\mathfrak{m}}$ *then* $d^0\gamma_-(v) = d^0 v$.

We give the proof only for γ since the other case is handled in the same way. We shall use induction on $d = d^0 z$. One can obviously assume that z is homogeneous and $\neq 0$. Choose $q \in S_d(\mathfrak{g})$ such that $\lambda(q) = z$. For any $X \in \mathfrak{g}$, extend adX to a derivation d_X of $S(\mathfrak{g})$ and let $I(\mathfrak{g})$ be the subalgebra of those elements $p \in S(\mathfrak{g})$, for which $d_X p = 0$ for every $X \in \mathfrak{g}$. Then $q \in I(\mathfrak{g})$ (see the Corollary of Lemma 11 of [1]). We regard elements of $S(\mathfrak{g})$ and $S(\mathfrak{h})$ as polynomial functions on $\mathfrak{g}$ and $\mathfrak{h}$ in the usual way [6, §2]. For any $p \in I(\mathfrak{g})$, let $\bar{p}$ denote the restriction of the polynomial function p on $\mathfrak{h}$. Then by a theorem of Chevalley [6, Lemma 9], $\bar{p} \in I(\mathfrak{h})$ and $p \to \bar{p}$ is an isomorphism of $I(\mathfrak{g})$ onto $I(\mathfrak{h})$ which obviously preserves degrees. Hence $\bar{q}$ is a homogeneous element in $I(\mathfrak{h})$ of degree d and it is not zero. Put $\check{\mathfrak{n}}_+ = \sum_{\alpha \in P} \mathbf{C} X_\alpha$, $\check{\mathfrak{n}}_- = \sum_{\alpha \in P} \mathbf{C} X_{-\alpha}$. Then $\mathfrak{g} = \check{\mathfrak{n}}_- + \mathfrak{h} + \check{\mathfrak{n}}_+$. Also $d_H(q - \bar{q}) = 0$ for every $H \in \mathfrak{h}$ and $q - \bar{q}$ vanishes identically on $\mathfrak{h}$. Hence we conclude without difficulty that

$$q - \bar{q} \in \sum_{\substack{e_1 + e_2 + e_3 = d \\ e_1 > 0,\, e_3 > 0}} S_{e_1}(\check{\mathfrak{n}}_-) S_{e_2}(\mathfrak{h}) S_{e_3}(\check{\mathfrak{n}}_+).$$

Moreover $\bar{q} \in I(\mathfrak{h}) \subset S(\mathfrak{h})$ and in view of our identification of $S(\mathfrak{h})$ with $\mathfrak{H}$, $\lambda(\bar{q}) = \bar{q}$. Hence from Lemma 1,

$$z - \bar{q} = \lambda(q - \bar{q}) \in \mathfrak{G}\check{\mathfrak{n}}_+ + \mathfrak{G}_{d-1}.$$

But then it is clear from the definition of γ that $\gamma(z) \equiv \bar{q} \bmod(\mathfrak{H} \cap \mathfrak{G}_{d-1})$. Since $\bar{q} \neq 0$, this proves that $d^0\gamma(z) = d$.

Let $r = [W : W_-]$.[7] Since W and W_- are both finite groups generated by reflexions, Lemma 8 of [8] is applicable. Therefore in view of Lemmas 3 and 4 above, we get the following result.

Lemma 5. *We can choose r elements $v_1 = 1, v_2, \ldots, v_r \in \mathfrak{Z}_m$ with the following properties.*

1) $\gamma_-(v_j)$ $(1 \leqslant j \leqslant r)$ *are homogeneous.*
2) *Every element $v \in \mathfrak{Z}_m$ can be written uniquely in the form*

$$v = \sum_{1 \leqslant i \leqslant r} \mu(z_i) v_i$$

with $z_i \in \mathfrak{Z}$. Moreover $d^0 v = d^0 z_i + d^0 v_i$ $(1 \leqslant i \leqslant r)$.

Let θ denote the involution of $\mathfrak{g}$ defined by $\theta(X + Y) = X - Y$ $(X \in \mathfrak{k}, Y \in \mathfrak{p})$. Also put $\tilde{\theta}(X + (-1)^{1/2} Y) = \theta(X) - (-1)^{1/2} \theta(Y)$ $(X, Y \in \mathfrak{g}_0)$ and $B(X, Y) = \mathrm{tr}(adX adY)$ $(X, Y \in \mathfrak{g})$. Then

$$\|X\|^2 = - B(X, \tilde{\theta}(X)) \qquad (X \in \mathfrak{g})$$

is a positive definite Hermitian form which defines the structure of a Hilbert space on $\mathfrak{g}$. Let $\mathfrak{q}$ be the orthogonal complement of $\mathfrak{m} \cap \mathfrak{k}$ in $\mathfrak{k}$ with respect to this form. Let $\mathfrak{h}'_{\mathfrak{p}_0}$ denote the set of all points $H \in \mathfrak{h}_{\mathfrak{p}_0}$ where $\alpha(H) \neq 0$ for every $\alpha \in P_+$. Then $A' = \exp \mathfrak{h}'_{\mathfrak{p}_0}$ is an open dense subset of A. We know [8, Lemma 21] that[8] $\mathfrak{g} = \mathfrak{q}^{h^{-1}} + \mathfrak{h}_\mathfrak{p} + \mathfrak{k}$ for any $h \in A'$. Let $\mathfrak{Q} = \lambda(S(\mathfrak{q}))$ and put

$$f_\alpha^\pm (h) = \left(e^{\alpha(\log h)} \pm 1 \right)^{-1} \qquad (h \in A')$$

for $\alpha \in P_+$. Then $f_\alpha^\pm$ are analytic functions on the open submanifold A' of A. Let $X_\alpha = Y_\alpha + Z_\alpha$ where $Y_\alpha \in \mathfrak{p}$ and $Z_\alpha \in \mathfrak{k}$.

Lemma 6. $\theta(X_\alpha) = 2(f_\alpha^+ (h) + f_\alpha^+ (h) f_\alpha^- (h)) Z_\alpha^{h^{-1}} - 2 f_\alpha^+ (h) f_\alpha^- (h) Z_\alpha$ *for $h \in A'$ and $\alpha \in P_+$.*

For

$$Z_\alpha = \tfrac{1}{2}(X_\alpha + \theta(X_\alpha)) \quad \text{and if } H = \log h,$$

$$Z_\alpha^{h^{-1}} = \tfrac{1}{2}\left(e^{-\alpha(H)} X_\alpha + e^{\alpha(H)} \theta(X_\alpha) \right)$$

$$= -(\sinh \alpha(H)) Y_\alpha + (\cosh \alpha(H)) Z_\alpha$$

$$= \sinh \alpha(H) \theta(X_\alpha) + e^{-\alpha(H)} Z_\alpha.$$

Therefore

$$\theta(X_\alpha) = 2 e^{\alpha(H)} \left(e^{2\alpha(H)} - 1 \right)^{-1} Z_\alpha^{h^{-1}} - 2 \left(e^{2\alpha(H)} - 1 \right)^{-1} Z_\alpha$$

and this is equivalent to the statement of the lemma.

---

[7] $[W : W_-]$ denotes the index of W_- in W.

[8] It is often convenient to write $x^y = y x y^{-1}$ and $X^y = Ad(y) X$ for $x, y \in G$ and $X \in \mathfrak{g}$. Also $g \to g^y$ denotes the corresponding automorphism of $\mathfrak{G}$.

Let $\mathscr{R}_0$ be the algebra generated over $\mathbb{C}$ by the functions f_α^+, f_α^- ($\alpha \in P_+$). (Note that the constant function 1 is *not* included among the generators of $\mathscr{R}_0$.) Let $m \to m'$ ($m \in \mathfrak{M}$) denote the automorphism of $\mathfrak{M}$ which is the inverse of the automorphism $m \to {}'m$ defined earlier. Select $v_1, \ldots, v_r$ as in Lemma 5.

Lemma 7. *Let $g \in \mathfrak{G}$. Then we can select a finite number of elements $g_0, g_1, \ldots, g_s$ in $(\Sigma_{1 \leqslant i \leqslant r} \mathfrak{Z} v_i')\mathscr{R}$ and $f_1, \ldots, f_s \in \mathscr{R}_0$, $q_1, \ldots, q_s \in \mathfrak{Q}$ such that the following conditions hold.*

1) $g = g_0 + \Sigma_{1 \leqslant i \leqslant s} f_i(h) q_i^{h^{-1}} g_i$ *for all* $h \in A'$.
2) $d^0 g_0 \leqslant d^0 g$ *and* $d^0 q_i + d^0 g_i \leqslant d^0 g$ $(1 \leqslant i \leqslant s)$.
3) $g \equiv g_0 \bmod \mathfrak{n}_- \mathfrak{G}$.

We can assume that $g \neq 0$ and use induction on $d = d^0 g$. Since $\mathfrak{g} = \mathfrak{n}_- + \mathfrak{h}_\mathfrak{p} + \mathfrak{k}$, we can choose a unique element $g' \in \mathfrak{H}_\mathfrak{p} \mathscr{R}$ such that $g - g' \in \mathfrak{n}_- \mathfrak{G}_{d-1}$. Then $d^0 g' \leqslant d$ and since $\mathfrak{H}_\mathfrak{p} \subset \mathfrak{Z}_\mathfrak{m}$, we can, from Lemma 5, choose $z_{ij} \in \mathfrak{Z}$, $k_{ij} \in \mathscr{R}$ $(1 \leqslant i \leqslant r, 1 \leqslant j \leqslant t)$ such that

$$g' = \sum_{i,j} \mu'(z_{ij}) v_i' k_{ij}$$

and $d^0 z_{ij} + d^0 v_i' + d^0 k_{ij} \leqslant d^0 g' \leqslant d$. Put $\nu(z) = z - \mu'(z)$ for $z \in \mathfrak{G}$. Then if $e = d^0 z$ ($z \neq 0$), it follows from Lemma 2 that $\nu(z) \in \mathfrak{n}_- \mathfrak{G}_{e-1}$. Hence

$$g \equiv g' \equiv g_0 \bmod \mathfrak{n}_- \mathfrak{G}_{d-1}$$

where $g_0 = \Sigma_{i,j} z_{ij} v_i' k_{ij}$. Since $\mathfrak{n}_- = \theta(\mathfrak{n}_+)$, it is clear that

$$g - g_0 = \sum_{\alpha \in P_+} \theta(X_\alpha) g_\alpha$$

where $g_\alpha \in \mathfrak{G}_{d-1}$. On the other hand from Lemma 6,

$$\theta(X_\alpha) = F_\alpha(h) Z_\alpha^{h^{-1}} + F_\alpha'(h) Z_\alpha \qquad (h \in A', \alpha \in P_+)$$

where F_α, F_α' are suitable functions in $\mathscr{R}_0$. Therefore

$$g - g_0 = \sum_{\alpha \in P_+} \left(F_\alpha(h) Z_\alpha^{h^{-1}} g_\alpha + F_\alpha'(h) Z_\alpha g_\alpha \right) \qquad (h \in A').$$

Now $d^0 g_\alpha \leqslant d - 1$. Let $\mathscr{R}$ be the algebra generated by $\mathscr{R}_0$ and the constant function 1 on A'. Then by induction hypothesis (applied to g_α), we can choose a finite set J of indices and elements $g_{\alpha j} \in \Sigma_{1 \leqslant i \leqslant r}(\mathfrak{Z} v_i')\mathscr{R}$, $q_{\alpha j} \in \mathfrak{Q}$ and $f_{\alpha j} \in \mathscr{R}$ $(j \in J)$ such that

1) $g_\alpha = \Sigma_{j \in J} f_{\alpha j}(h) q_{\alpha j}^{h^{-1}} g_{\alpha j}$ *for all* $h \in A'$,
2) $d^0 q_{\alpha j} + d^0 g_{\alpha j} \leqslant d^0 g_\alpha$ $(\alpha \in P_+, j \in J)$.

Since $\mathfrak{k} = \mathfrak{q} + \mathfrak{m} \cap \mathfrak{k}$, $\mathscr{R} = \mathfrak{Q} \mathfrak{M}_\mathfrak{k}$ where $\mathfrak{M}_\mathfrak{k} = \lambda(S(\mathfrak{m} \cap \mathfrak{k}))$. Therefore

$$Z_\alpha q_{\alpha j} = \sum_{l \in L} q_{\alpha j l} m_{\alpha j l} \qquad (\alpha \in P_+, j \in J)$$

where L is a finite set of indices, $m_{\alpha j l} \in \mathfrak{M}_\mathfrak{k}$, $q_{\alpha j l} \in \mathfrak{Q}$ and $d^0 q_{\alpha j l} + d^0 m_{\alpha j l} \leqslant$

$d^0 q_{\alpha j} + 1$. Hence

$$Z_\alpha^{h^{-1}} g_\alpha = \sum_{j \in J} \sum_{l \in L} f_{\alpha j}(h)(q_{\alpha jl})^{h^{-1}} g_{\alpha jl}$$

where $g_{\alpha jl} = m_{\alpha jl} g_{\alpha j}$ and $d^0 q_{\alpha jl} + d^0 g_{\alpha jl} \leqslant d^0 q_{\alpha j} + d^0 g_{\alpha j} + 1 \leqslant d^0 g_\alpha + 1 \leqslant d$. Now put $g_\alpha' = [Z_\alpha, g_\alpha]$. Then $d^0 g_\alpha' \leqslant d^0 g_\alpha$ and $Z_\alpha g_\alpha = g_\alpha Z_\alpha + g_\alpha'$. Therefore

$$Z_\alpha g_\alpha = \sum_{j \in J} f_{\alpha j}(h) q_{\alpha j}^{h^{-1}} g_{\alpha j} Z_\alpha + f_\alpha(h) g_\alpha'$$

for $\alpha \in P_+$ and $h \in A'$. Now put $F_{\alpha j} = F_\alpha f_{\alpha j}$, $F_{\alpha j}' = F_\alpha' f_{\alpha j}$ and $g_{\alpha j}' = g_{\alpha j} Z_\alpha$. Then

$$g - g_0 = \sum_{\alpha, j, l} F_{\alpha j}(h)(q_{\alpha jl})^{h^{-1}} g_{\alpha jl}$$

$$+ \sum_{\alpha, j} F_{\alpha j}'(h) q_{\alpha j}^{h^{-1}} g_{\alpha j}' + \sum_\alpha F_\alpha'(h) g_\alpha'.$$

If we now apply the induction hypothesis once again to g_α' and note that $\mathcal{R}_0$ is an ideal in $\mathcal{R}$, we get the required result immediately.

Corollary 1. *For any* $h \in A'$, $\mathfrak{G} = \mathfrak{Q}^{h^{-1}}(\sum_{1 \leqslant i \leqslant r} \mathfrak{Z} v_i')\mathfrak{R}$.

This is obvious from Lemma 7.

Let $\mathfrak{v}_1$, $\mathfrak{v}_2$ be two equivalence classes of irreducible finite-dimensional (continuous) representations of K. Let ϕ be a continuous complex-valued function on G. We say that ϕ is of *type* $(\mathfrak{v}_1, \mathfrak{v}_2)$ if the following conditions hold. Let V_ϕ denote the vector space of all functions of the form $x \to \phi(k_1 x k_2)$ $(x \in G)$ where k_1, k_2 are any two fixed elements in K. Then we require first of all that $\dim V_\phi < \infty$. Define two representations r and l of K on V_ϕ as follows. If $\psi \in V_\phi$ and $k \in K$, $r(k)\psi$ is the function $x \to \psi(xk)$ and $l(k)\psi$ the function $x \to \psi(k^{-1}x)$ $(x \in G)$. Then we further require that r and l are fully reducible and every irreducible component of l lies in $\mathfrak{v}_1$ and of r in $\mathfrak{v}_2$.

We recall that elements of $\mathfrak{G}$ may be regarded as left-invariant differential operators on G.

Corollary 2. *Let* χ *be a homomorphism of* $\mathfrak{Z}$ *into* $\mathbf{C}$ *and let* $E(\chi, \mathfrak{v}_1, \mathfrak{v}_2)$ *denote the space of all complex-valued* C^∞-*functions* ϕ *on* G *of type* $(\mathfrak{v}_1, \mathfrak{v}_2)$ *which satisfy the system of differential equations* $z\phi = \chi(z)\phi$ $(z \in \mathfrak{Z})$. *Then*[9]

$$\dim E(\chi, \mathfrak{v}_1, \mathfrak{v}_2) \leqslant [W : W_-] \, d(\mathfrak{v}_1)^2 d(\mathfrak{v}_2)^2.$$

We shall now use the notation concerning differential operators which has been introduced in [6, p. 89, Footnote 1]. Then $\phi(x; g)$ $(x \in G, g \in \mathfrak{G})$ denotes the value of $g\phi$ at x. Also we define

$$\phi(g; x) = \phi\left(x; g^{x^{-1}}\right)$$

---

[9] For any equivalence class $\mathfrak{v}$ of finite-dimensional representations of K, $d(\mathfrak{v})$ denotes the degree of any representation in $\mathfrak{v}$.

and

$$\phi(g_1; x; g_2) = \phi\left(x; g_1^{x^{-1}} g_2\right) \qquad (g_1, g_2 \in \mathfrak{G}).$$

Put $E = E(\chi, \mathfrak{d}_1, \mathfrak{d}_2)$ and let $\mathfrak{K}_1$ be the ideal consisting of all $k \in \mathfrak{K}$ such that $\phi(k; x) = 0$ for all $\phi \in E$ and $x \in G$. Similarly let $\mathfrak{K}_2$ be the ideal of all $k \in \mathfrak{K}$ such that $k\phi = 0$ for every $\phi \in E$. Then if $E \neq \{0\}$, $\dim \mathfrak{K}/\mathfrak{K}_i = d_i^2$ where $d_i = d(\mathfrak{d}_i)$ $(i = 1, 2)$. Let ω denote the Casimir operator of $\mathfrak{g}$. Choose a base $X_1, \dots, X_n$ for $\mathfrak{g}_0$ over $\mathbb{R}$ such that $(X_1, \dots, X_p)$ is a base for $\mathfrak{p}_0$, $(X_{p+1}, \dots, X_n)$ is a base for $\mathfrak{k}_0$, $B(X_i, X_i) = 1$ $(1 \leqslant i \leqslant p)$ and $B(X_i, X_i) = -1$ for $p < i \leqslant n$. Put

$$\omega_1 = \sum_{1 \leqslant i \leqslant p} X_i^2, \qquad \omega_2 = \sum_{p < i \leqslant n} X_i^2.$$

Then $\omega = \omega_1 - \omega_2$ and it is obvious that $\Omega = \omega_1 + \omega_2$ is an elliptic differential operator on G. Moreover ω_2 lies in the center of $\mathfrak{K}$ and hence by Schur's lemma there exists a complex number c such that $\sigma(\omega_2) = c\sigma(1)$ for any[10] $\sigma \in \mathfrak{d}_2$. Now if $\phi \in E$,

$$\Omega\phi = (\omega + 2\omega_2)\phi = (\chi(\omega) + 2c)\phi.$$

This shows that ϕ satisfies an elliptic differential equation and therefore we conclude from well-known results that ϕ is an analytic function on G.

Now fix a point $h \in A'$. Suppose $\phi(k_1; h; v_i'k_2) = 0$ for all $k_1, k_2 \in \mathfrak{K}$ and $1 \leqslant i \leqslant r$. Then we claim $\phi = 0$. Since ϕ is analytic it would be enough to verify that $\phi(h; g) = 0$ for every $g \in \mathfrak{G}$. But this is an immediate consequence of Corollary 1 above. Let k_{ij} $(1 \leqslant j \leqslant d_i^2)$ be a base for $\mathfrak{K} \bmod \mathfrak{K}_i$ $(1 = 1, 2)$. Then if $\phi(k_{1j}; h; v_i'k_{2l})$ $= 0$ $(1 \leqslant j \leqslant d_1^2, \ 1 \leqslant i \leqslant r, \ 1 \leqslant l \leqslant d_2^2)$ it follows from the above result that $\phi = 0$. This means that $rd_1^2 d_2^2$ linear conditions on a function $\phi \in E$ are sufficient to ensure that $\phi = 0$. This obviously implies that $\dim E \leqslant rd_1^2 d_2^2$.

The result of Corollary 2 above can be used to obtain a direct proof of Theorem 4 (p. 236) of [3]. This had previously been derived from an unpublished result of Chevalley (see [2, p. 37]).

§3.   Spherical Functions

Let $\sigma = (\sigma_1, \sigma_2)$ be a (continuous) double representation of K on a finite-dimensional (complex) vector space V and let χ be a homomorphism of $\mathfrak{Z}$ into[11] $\mathfrak{E}(V)$ such that $\chi(1) = 1$. By a *spherical function*[12] *of type* (σ, χ) we mean a C^∞-function ϕ from G to V such that:

1) $\phi(k_1 x k_2) = \sigma_1(k_1)\phi(x)\sigma_2(k_2)$ $(k_1, k_2 \in K, \ x \in G)$,
2) $z\phi = \chi(z)\phi$ $(z \in \mathfrak{Z})$.

We denote by $S(\sigma, \chi)$ the space of all spherical functions of type (σ, χ).

---

[10] We identify corresponding representations of K and $\mathfrak{K}$.

[11] For any complex vector space V of finite dimension, we denote by $\mathfrak{E}(V)$ the associative algebra of all endomorphisms of V. Let i be the identity mapping of V. Then we identify $\mathbb{C}$ with the center $\mathbb{C}i$ of $\mathfrak{E}(V)$ under the mapping $c \to ci$ $(c \in \mathbb{C})$.

[12] This definition of spherical functions is not quite the same as that of Godement [13].

Lemma 8. *Every function in $S(\sigma, \chi)$ is analytic and*

$$\dim(S(\sigma, \chi)) \leqslant [W : W_-]\dim V.$$

Define ω, Ω and ω_2 as during the proof of Corollary 2 of Lemma 7. Then if $\phi \in S(\sigma, \chi)$,

$$\phi(x; \Omega) = E\phi(x) \qquad (x \in G)$$

where E is the endomorphism of V given by the relation[10]

$$Ev = \chi(\omega)v - 2v\sigma_2(\omega_2) \qquad (v \in V).$$

Now E satisfies an equation of the form

$$E^g + c_1 E^{g-1} + \cdots + c_g = 0$$

for suitable complex numbers $c_1, \ldots, c_g$. Then

$$\left(\Omega^g + c_1 \Omega^{g-1} + \cdots + c_g \right)\phi = 0$$

and therefore ϕ satisfies an elliptic differential equation. This implies that ϕ is analytic. Now fix a point $h \in A'$. Suppose $\phi(h; v_i') = 0$ $(1 \leqslant i \leqslant r)$ for some ϕ in $S(\sigma, \chi)$. Then it follows from Corollary 1 of Lemma 7 that $\phi(h; g) = 0$ for every $g \in \mathfrak{G}$. Since ϕ is analytic, this implies that $\phi = 0$. Thus the conditions $\phi(h; v_i') = 0$ $(1 \leqslant i \leqslant r)$ for an element $\phi \in S(\sigma, \chi)$ imply that $\phi = 0$. This shows that $\dim S(\sigma, \chi) \leqslant r \dim V$.

Let Σ denote the set of all linear functions $\bar{\beta}$ on $\mathfrak{h}_\mathfrak{p}$ $(\beta \in P_+)$ where $\bar{\beta}$ stands for the restriction of β on $\mathfrak{h}_\mathfrak{p}$. If $l = \dim_\mathbf{R} \mathfrak{h}_{\mathfrak{p}_0}$, we can choose (see [7, Lemma 1]) l elements $\alpha_1, \ldots, \alpha_l \in \Sigma$ such that 1) $\alpha_1, \ldots, \alpha_l$ are linearly independent over $\mathbf{R}$ and 2) every $\alpha \in \Sigma$ can be written in the form $\alpha = m_1 \alpha_1 + \cdots + m_l \alpha_l$ where $m_1, \ldots, m_l$ are nonnegative integers. We choose a base $H_1, \ldots, H_l$ for $\mathfrak{h}_{\mathfrak{p}_0}$ over $\mathbf{R}$ such that[13] $\alpha_i(H_j) = \delta_{ij}$. Let $\mathfrak{h}_\mathfrak{p}^+$ be the set of all points $H \in \mathfrak{h}_\mathfrak{p}$ such that[14] $\operatorname{Re} \alpha(H) > 0$ for every $\alpha \in \Sigma$. Also put $\mathfrak{h}_{\mathfrak{p}_0}^+ = \mathfrak{h}_\mathfrak{p}^+ \cap \mathfrak{h}_{\mathfrak{p}_0}$. Then $H = t_1 H_1 + \cdots + t_l H_l$ $(t_i \in \mathbf{C})$ lies in $\mathfrak{h}_\mathfrak{p}^+$ if and only if $\operatorname{Re} t_i > 0$ $(1 \leqslant i \leqslant l)$.

Now for any $\phi \in S(\sigma, \chi)$, put $\phi_i = v_i'\phi$ $(1 \leqslant i \leqslant r)$. Fix an element $H \in \mathfrak{h}_{\mathfrak{p}_0}$ and put $\mathfrak{U} = (\Sigma_{1 \leqslant i \leqslant r} \mathfrak{Z} v_i')\mathfrak{R}$. Then from Lemma 7 we can choose elements $g_{ij} \in \mathfrak{U}$ $(0 \leqslant j \leqslant s)$ and $f_{ij} \in \mathfrak{R}_0$, $q_{ij} \in \mathfrak{Q}$ $(1 \leqslant j \leqslant s)$ such that

$$Hv_i' = g_{i0} + \sum_{1 \leqslant j \leqslant s} f_{ij}(h)q_{ij}^{h-1}g_{ij}$$

for $h \in A'$ and $1 \leqslant i \leqslant r$. Then

$$\phi_i(h; H) = \phi(h; Hv_i')$$

$$= \phi(h; g_{i0}) + \sum_{1 \leqslant j \leqslant s} f_{ij}(h)\sigma_1(q_{ij})\phi(h; g_{ij}).$$

---

[13]As usual the Kronecker symbol δ_{ij} is defined to be 1 if $i = j$ and 0 otherwise.

[14]For any complex number c, $\operatorname{Re} c$ and $\operatorname{Im} c$ stand for the real parts of c and $-(-1)^{1/2}c$ respectively.

Now suppose $z \in \mathfrak{Z}$ and $k \in \mathfrak{R}$. Then for any $x \in G$,

$$\phi(x; zv_i'k) = \chi(z)\left(\phi(x; v_i')\sigma_2(k)\right).$$

Hence it is obvious that there exist linear mappings ξ_i $(1 \leqslant i \leqslant r)$ of $\mathfrak{U}$ into $\mathfrak{E}(V)$ such that

$$\phi(x; g) = \sum_{1 \leqslant i \leqslant r} \xi_i(g)\phi_i(x)$$

for all $\phi \in S(\sigma, \chi)$, $g \in \mathfrak{U}$ and $x \in G$. Therefore

$$\phi_i(h; H) = \sum_{1 \leqslant p \leqslant r} \xi_p(g_{i0})\phi_p(h) + \sum_{1 \leqslant j \leqslant s} \sum_{1 \leqslant p \leqslant r} f_{ij}(h)\sigma_1(q_{ij})\left(\xi_p(g_{ij})\phi_p(h)\right)$$

for $1 \leqslant i \leqslant r$ and $h \in A'$.

Let V^r denote the direct sum of V with itself r times. For any $\phi \in S(\sigma, \chi)$, let Φ_ϕ denote the function from G to V^r defined by

$$\Phi_\phi(x) = \left(\phi_1(x), \phi_2(x), \ldots, \phi_r(x)\right) \qquad (x \in G).$$

Also let us recall that $\mathfrak{R}$ is the algebra of functions on A' generated over $\mathbf{C}$ by $\mathfrak{R}_0$ and 1. Then the above result can be expressed in the following form.

Lemma 9. *Let E_p $(1 \leqslant p \leqslant a)$ be a base for $\mathfrak{E}(V^r)$. Then we can choose functions $f_{jp} \in \mathfrak{R}$ $(1 \leqslant j \leqslant l,\ 1 \leqslant p \leqslant a)$ with the following property. Set $\Xi_j(h) = \sum_p f_{jp}(h)E_p$ $(h \in A')$. Then*

$$\Phi_\phi(h; H_j) = \Xi_j(h)\Phi_\phi(h)$$

for $\phi \in S(\sigma, \chi)$, $h \in A'$ and $1 \leqslant j \leqslant l$.

Let L be the set of all linear functions λ on $\mathfrak{h}_\mathfrak{p}$ of the form $\lambda = m_1\alpha_1 + \cdots + m_l\alpha_l$ where $m_1, \ldots, m_l$ are nonnegative integers. Define $m(\lambda) = m_1 + \cdots + m_l$. Then it follows from the definition of $\mathfrak{R}$ that for any given $f \in \mathfrak{R}$, we can choose complex numbers $c(\lambda)$ $(\lambda \in L)$ such that the following conditions hold.

1) $\sum_{\lambda \in L} |c(\lambda)| e^{-m(\lambda)\varepsilon} < \infty$ for $\varepsilon > 0$.
2) $f(\exp H) = \sum_{\lambda \in L} c(\lambda) e^{-\lambda(H)}$ for $H \in \mathfrak{h}_{\mathfrak{p}_0}^+$.

Therefore it is clear from Lemma 9 above that Lemma 10 and Theorem 1 of [11] are applicable to the function $H \to \Phi_\phi(\exp H)$ $(H \in \mathfrak{h}_{\mathfrak{p}_0}^+)$. Let M be the centralizer of A in K and V_0 the subspace of those elements $u \in V$ for which $\sigma_1(m)u = u\sigma_2(m)$ $(m \in M)$. Since $h^m = h$ $(h \in A,\ m \in M)$, it is clear that $\phi(h) \in V_0$ for any $\phi \in S(\sigma, \chi)$. Let F be any projection of V onto V_0. Since $v_1' = 1$ and $F\phi(h) = \phi(h)$ $(h \in A)$, we get the following result from Lemma 10 and Theorem 1 of [11].

Theorem 1. *Fix an element $\phi \in S(\sigma, \chi)$. Then we can select an at most enumerable set $E(\phi)$ of linear functions on $\mathfrak{h}_\mathfrak{p}$ and for each $\lambda \in E(\phi)$, a polynomial function p_λ from $\mathfrak{h}_\mathfrak{p}$ to V_0 such that the following conditions are fulfilled.*

(1) *$p_\lambda \neq 0$ for any $\lambda \in E(\phi)$.*

(2) *For any two real numbers ε, ν $(0 < \varepsilon < 1, \nu > 0)$ let $\mathfrak{h}_\mathfrak{p}(\varepsilon, \nu)$ denote the set of all points $H \in \mathfrak{h}_\mathfrak{p}$ where $\operatorname{Re}\alpha_i(H) > \max(\nu, \varepsilon|\alpha_1(H)|, \ldots, \varepsilon|\alpha_l(H)|)$ $(1 \leqslant i \leqslant l)$. Then the series*

$$\sum_{\lambda \in E(\phi)} p_\lambda(H) e^{\lambda(H)}$$

converges decently for $H \in \mathfrak{h}_\mathfrak{p}(\varepsilon, \nu)$.

(3) $\displaystyle \phi(\exp H) = \sum_{\lambda \in E(\phi)} p_\lambda(H) e^{\lambda(H)} \qquad \left(H \in \mathfrak{h}_{\mathfrak{p}_0}^+ \right).$

Moreover the set $E(\phi)$ and the polynomials p_λ $(\lambda \in E(\phi))$ are uniquely determined by these conditions.

We shall call the elements of $E(\phi)$, the *exponents* of ϕ.

Let λ be a linear function on $\mathfrak{h}_\mathfrak{p}$. We say λ is *real* if it takes only real values on $\mathfrak{h}_{\mathfrak{p}_0}$. Any linear function λ can be written uniquely in the form $\lambda = \lambda_R + (-1)^{1/2}\lambda_I$ where λ_R and λ_I are real linear functions which are called, respectively, the *real and imaginary parts* of λ. Given two linear functions λ and μ we write $\lambda > \mu$ (or $\mu < \lambda$) if $\lambda_R > \mu_R$ or $\lambda_R = \mu_R$ and $\lambda_I > \mu_I$. Also we write $\lambda \gg \mu$ (or $\mu \ll \lambda$) if $\lambda \neq \mu$ and $\lambda - \mu \in L$. Clearly $\lambda \gg \mu$ implies that $\lambda > \mu$. Let $E_0(\phi)$ be all the elements of $E(\phi)$ which are maximal with respect to the relation $\gg$. Then we know from Theorem 2 of [11] that $E_0(\phi)$ is a finite set and if λ is an element of $E(\phi)$ which is not in $E_0(\phi)$ then $\lambda \ll \lambda_0$ for some $\lambda_0 \in E_0(\phi)$. Hence $E_0(\phi)$ is not empty unless $\phi = 0$. Elements of $E_0(\phi)$ will be called the *principal exponents* of ϕ and the sum $\sum_{\lambda \in E_0(\phi)} p_\lambda e^\lambda$ the *principal part* of ϕ. Let λ be the highest principal exponent under the relation $>$. Then λ is called the *highest* or the *leading exponent* of ϕ and the corresponding term $p_\lambda e^\lambda$ will be called the *highest term* of ϕ. Let d be the degree of p_λ and let q be the homogeneous component of degree d of p_λ. Then qe^λ will be called the *leading term* of ϕ and we shall denote it by ϕ^L. By the *degree* of ϕ^L we mean the degree d of q.

§4. The Mapping δ'

Let $\mathfrak{D}$ be the algebra of all differential operators on A' (regarded as an open submanifold of the Lie group A). Consider the tensor product $\mathfrak{C} = \mathfrak{R} \otimes \mathfrak{D} \otimes \mathfrak{R}$. We turn $\mathfrak{C}$ into an associative algebra by defining multiplication as follows.

$$(k_1 \otimes D_1 \otimes k_1') \cdot (k_2 \otimes D_2 \otimes k_2') = k_1 k_2 \otimes D_1 D_2 \otimes k_2' k_1'$$

for k_1, k_1', k_2, k_2' in $\mathfrak{R}$ and $D_1, D_2 \in \mathfrak{D}$. (Note the inversion of the order of factors in $k_2' k_1'$.) For any integer d, let $\mathfrak{D}_d$ denote the subspace of all elements of $\mathfrak{D}$ of degree $\leqslant d$. Put $\mathfrak{C}_d = \mathfrak{R} \otimes \mathfrak{D}_d \otimes \mathfrak{R}$. If Δ is an element of $\mathfrak{C}$ and $\Delta \neq 0$, then $d^0\Delta$ is the smallest integer d such that $\Delta \in \mathfrak{C}_d$. If $\Delta = 0$, we put $d^0\Delta = -\infty$.

Let U be an open set in A'. We denote by $C^\infty(U:V)$ the space of all C^∞-functions from U to V. Then $\mathfrak{C}$ operates on $C^\infty(U:V)$ as follows. If $F \in C^\infty(U:V)$,

$$(k_1 \otimes D \otimes k_2)F = \sigma_1(k_1)(DF)\sigma_2(k_2) \qquad (k_1, k_2 \in \mathfrak{R}, D \in \mathfrak{D}).$$

It follows from the definition of multiplication in $\mathcal{C}$, that in this way $C^\infty(U:V)$ becomes a left module over $\mathcal{C}$. Whenever convenient, we shall write $F(h; \Delta)$ to denote the value of ΔF at h ($\Delta \in \mathcal{C}$, $h \in U$).

In particular $\mathfrak{H}_\mathfrak{p}$ is a subalgebra of $\mathfrak{D}$ and therefore $\mathfrak{U} = \mathfrak{R} \otimes \mathfrak{H}_\mathfrak{p} \otimes \mathfrak{R} \subset \mathcal{C}$. Fix a point $h \in A'$ and for any $D \in \mathfrak{D}$ let D_h denote the local expression of D at h (see [4, p. 112]). Then $D \to D_h$ is a linear mapping of $\mathfrak{D}$ into $\mathfrak{H}_\mathfrak{p}$ which obviously has a unique extension to a linear mapping $\Delta \to \Delta_h$ ($\Delta \in \mathcal{C}$) of $\mathcal{C}$ into $\mathfrak{U}$ such that $(k_1 \otimes D \otimes k_2)_h = k_1 \otimes D_h \otimes k_2$ ($k_1, k_2 \in \mathfrak{R}$, $D \in \mathfrak{D}$). We shall call Δ_h the *local expression* of Δ at h. It is clear that $F(h; \Delta) = F(h; \Delta_h)$ for $F \in C^\infty(U:V)$.

Let $\mathfrak{H}_\mathfrak{p}^{(d)}$ be the set of all homogeneous elements in $\mathfrak{H}_\mathfrak{p}$ of degree d. Put $\mathfrak{U}^{(d)} = \mathfrak{R} \otimes \mathfrak{H}_\mathfrak{p}^{(d)} \otimes \mathfrak{R}$. We say that an element in $\mathfrak{U}$ is homogeneous of degree d, if it lies in $\mathfrak{U}^{(d)}$. On the other hand put

$$\mathfrak{U}^{[d]} = \sum_{d_1 + d_2 + d_3 = d} \mathfrak{R}^{(d_1)} \otimes \mathfrak{H}_\mathfrak{p}^{(d_2)} \otimes \mathfrak{R}^{(d_3)}$$

where $\mathfrak{R}^{(e)} = \lambda(S_e(\mathfrak{k}))$. (Here λ is the canonical mapping of $S(\mathfrak{g})$ into $\mathfrak{G}$.) Then an element in $\mathfrak{U}$ is said to be homogeneous of total degree d, if it lies in $\mathfrak{U}^{[d]}$. For $\Delta \in \mathfrak{U}$, we denote by $d^0\Delta$ the degree of Δ and by $d_t^0\Delta$ the total degree of Δ. (As usual $d^0\Delta = d_t^0\Delta = -\infty$ if $\Delta = 0$.)

For any $h \in A'$, let Γ_h denote the linear mapping of $\mathfrak{R} \otimes \mathfrak{H}_\mathfrak{p} \otimes \mathfrak{R}$ into $\mathfrak{G}$ given by $\Gamma_h(k_1 \otimes u \otimes k_2) = k_1^{h^{-1}} u k_2$ ($k_1, k_2 \in \mathfrak{R}$, $u \in \mathfrak{H}_\mathfrak{p}$). We know (see Corollary 1 of Lemma 21 of [8]) that Γ_h defines a one-one linear mapping of $\mathfrak{Q} \otimes \mathfrak{H}_\mathfrak{p} \otimes \mathfrak{R}$ onto $\mathfrak{G}$. Let δ_h' denote its inverse so that $\delta_h' : \mathfrak{G} \to \mathfrak{Q} \otimes \mathfrak{H}_\mathfrak{p} \otimes \mathfrak{R}$.

Lemma 10. *For any $g \in \mathfrak{G}_d$, we can select a finite number of elements $f_1, \ldots, f_r \in \mathfrak{R}_0$ and $\Delta_0, \ldots, \Delta_r \in \mathfrak{Q} \otimes \mathfrak{H}_\mathfrak{p} \otimes \mathfrak{R}$ such that $d_t^0\Delta_j \leqslant d$ ($0 \leqslant j \leqslant r$), $d^0\Delta_i \leqslant d-1$ ($1 \leqslant i \leqslant r$) and*

$$\delta_h'(g) = \Delta_0 + \sum_{1 \leqslant i \leqslant r} f_i(h)\Delta_i$$

for every $h \in A'$.

We shall use induction on d. Therefore we may assume that $g \neq 0$ and $d^0 g = d \geqslant 1$. Since $\mathfrak{g} = \mathfrak{n}_- + \mathfrak{h}_\mathfrak{p} + \mathfrak{k}$, we can choose $g_0 \in \mathfrak{H}_\mathfrak{p}\mathfrak{R}$ such that $g \equiv g_0 \bmod \mathfrak{n}_- \mathfrak{G}$. Then $d^0 g_0 \leqslant d$ and $g - g_0 \in \mathfrak{n}_- \mathfrak{G}_{d-1}$. Put $\Delta_0 = \delta_h'(g_0)$ for any $h \in A'$. It is clear that Δ_0 is actually independent of h and it lies in $1 \otimes \mathfrak{H}_\mathfrak{p} \otimes \mathfrak{R}$. Now select $g_\alpha \in \mathfrak{G}_{d-1}$ such that

$$g - g_0 = \sum_{\alpha \in P_+} \theta(X_\alpha) g_\alpha.$$

Then from Lemma 6, we can choose F_α and F_α' in $\mathfrak{R}_0$ such that

$$g - g_0 = \sum_{\alpha \in P_+} \left(F_\alpha(h) Z_\alpha^{h^{-1}} + F_\alpha'(h) Z_\alpha \right) g_\alpha$$

$$= \sum_{\alpha \in P_+} F_\alpha(h) Z_\alpha^{h^{-1}} g_\alpha + \sum_{\alpha \in P_+} F_\alpha'(h)\{[Z_\alpha, g_\alpha] + g_\alpha Z_\alpha\}$$

for all $h \in A'$. Then by applying the induction hypothesis to $[Z_\alpha, g_\alpha]$ and g_α, it follows immediately that we can choose a finite number of elements $\Delta'_1, \ldots, \Delta'_r$ in $\Re \otimes \mathfrak{H}_\mathfrak{p} \otimes \Re$ and $f_1, \ldots, f_r \in \Re_0$ such that $d_t^0 \Delta'_i \leqslant d$ and $d^0 \Delta'_j \leqslant d - 1$ $(1 \leqslant j \leqslant r)$ and

$$\sum_{1 \leqslant i \leqslant r} f_i(h) \Gamma_h(\Delta'_i) = g - g_0 = g - \Gamma_h(\Delta_0) \qquad (h \in A').$$

Now $\mathfrak{k} = \mathfrak{q} + \mathfrak{m} \cap \mathfrak{k}$ and so $\Re = \mathfrak{Q}\mathfrak{M}_\mathfrak{k}$. Moreover $\Gamma_h(qm \otimes u \otimes k) = \Gamma_h(q \otimes u \otimes mk)$ $(q \in \mathfrak{Q}, m \in \mathfrak{M}_\mathfrak{k}, u \in \mathfrak{H}_\mathfrak{p}, k \in \Re)$. Hence it follows easily that we can select elements $\Delta_j \in \mathfrak{Q} \otimes \mathfrak{H}_\mathfrak{p} \otimes \Re$ such that $d_t^0 \Delta_j \leqslant d_t^0 \Delta'_j$, $d^0 \Delta_j \leqslant d^0 \Delta'_j$ and $\Gamma_h(\Delta_j) = \Gamma_h(\Delta'_j)$ $(1 \leqslant j \leqslant r)$ for every $h \in A'$. Therefore

$$g = \Gamma_h(\Delta_0) + \sum_{1 \leqslant i \leqslant r} f_i(h) \Gamma_h(\Delta_i)$$

and so

$$\delta'_h(g) = \Delta_0 + \sum_{1 \leqslant i \leqslant r} f_i(h) \Delta_i \qquad (h \in A').$$

It follows from the above proof that we can take $\Delta_0 = \delta'_h(g_0)$. We shall now show that this is necessarily so.

Corollary 1. *Let g be an element in $\mathfrak{G}$ and $\Delta_0, \Delta_1, \ldots, \Delta_r$ a finite number of elements in $\mathfrak{Q} \otimes \mathfrak{H}_\mathfrak{p} \otimes \Re$ and $f_1, \ldots, f_r \in \Re_0$ such that*

$$\delta'_h(g) = \Delta_0 + \sum_{1 \leqslant i \leqslant r} f_i(h) \Delta_i$$

for all $h \in A'$. Let g_0 be the unique element in $\mathfrak{H}_\mathfrak{p}\Re$ such that $g - g_0 \in \mathfrak{n}_- \mathfrak{G}$. Then $\Delta_0 = \delta'_h(g_0)$ and $d^0 \delta'_h(g - g_0) \leqslant d^0 g - 1$ for $h \in A'$.

It follows from the proof of the above lemma that

$$\delta'_h(g) = \Delta'_0 + \sum_{1 \leqslant i \leqslant s} f'_i(h) \Delta'_i \qquad (h \in A')$$

where $\Delta'_0 = \delta'_h(g_0)$, $\Delta'_i \in \mathfrak{Q} \otimes \mathfrak{H}_\mathfrak{p} \otimes \Re$, $f'_i \in \Re_0$ and $d^0 \Delta'_i \leqslant d^0 - 1$ $(1 \leqslant i \leqslant s)$. Hence we get

$$\Delta_0 - \Delta'_0 = \sum_{j \in J} F_j(h) D_j \qquad (h \in A')$$

where J is a finite set of indices, $F_j \in \Re_0$ and $D_j \in \mathfrak{Q} \otimes \mathfrak{H}_\mathfrak{p} \otimes \Re$. Fix $H \in \mathfrak{h}^+_{\mathfrak{p}_0}$. Then it is easy to see that $F(\exp tH) \to 0$ as $t \to +\infty$ for any $F \in \Re_0$. Hence by setting $h = \exp tH$ and making t tend to $+\infty$, we get $\Delta_0 - \Delta'_0 = 0$. This proves that $\Delta_0 = \Delta'_0 = \delta'_h(g_0)$ $(h \in A')$. Since

$$\delta'_h(g - g_0) = \sum_{1 \leqslant i \leqslant s} f'_i(h) \Delta'_i,$$

it follows that $d^0 \delta'_h(g - g_0) \leqslant d^0 g - 1$.

As before let M denote the centralizer of A in K. For any $m \in M$, we define an automorphism $\Delta \to \Delta^m$ of $\Re \otimes \mathfrak{H}_\mathfrak{p} \otimes \Re$ such that $(k_1 \otimes u \otimes k_2)^m = k_1^m \otimes u \otimes k_2^m$ $(k_1, k_2 \in \Re, u \in \mathfrak{H}_\mathfrak{p})$. Note that $\mathfrak{q}^m = \mathfrak{q}$ and so $(\mathfrak{Q} \otimes \mathfrak{H}_\mathfrak{p} \otimes \Re)^m = \mathfrak{Q} \otimes \mathfrak{H}_\mathfrak{p} \otimes \Re$.

Corollary 2. *Suppose that, in Lemma* 10, $g^m = g$ *for all* $m \in M$. *Then we can select* Δ_j *in such a way that* $\Delta_j^m = \Delta_j$ $(0 \leqslant j \leqslant r)$.

Without loss of generality we may assume that $f_1, \ldots, f_r$ are linearly independent over **C**. Since $\lim_{t \to +\infty} f(\exp tH) = 0$ for any $f \in \mathfrak{R}_0$ and $H \in \mathfrak{h}_{\mathfrak{p}_0}^+$, it follows that $(1, f_1, \ldots, f_r)$ are all linearly independent. It is obvious that $\Gamma_h(\Delta^m) = (\Gamma_h(\Delta))^m$ for $\Delta \in \mathfrak{Q} \otimes \mathfrak{H}_\mathfrak{p} \otimes \mathfrak{R}$, $h \in A'$ and $m \in M$. Therefore $\delta_h'(g_1^m) = (\delta_h'(g_1))^m$ for $g_1 \in \mathfrak{G}$. This means that

$$\Delta_0 - \Delta_0^m + \sum_{1 \leqslant j \leqslant r} f_j(h)\big(\Delta_j - \Delta_j^m\big) = 0 \qquad (h \in A').$$

Since $(1, f_1, \ldots, f_r)$ are linearly independent, we conclude that $\Delta_j = \Delta_j^m$ $(0 \leqslant j \leqslant r)$.

It is obvious from Lemma 10 that, for each $g \in \mathfrak{G}$, there exists an element $\delta'(g) \in \mathfrak{Q}$ such that $\delta_h'(g)$ is the local expression of $\delta'(g)$ for any point $h \in A'$. We identify $\mathfrak{D}$ with a subalgebra of $\mathfrak{Q}$ under the mapping $D \to 1 \otimes D \otimes 1$ $(D \in \mathfrak{D})$. Then $\mathfrak{Q}$ becomes a two-sided module over $\mathfrak{D}$. Since $\mathfrak{D}$ contains all C^∞-functions on A', the symbols $f\Delta \circ g$ $(f, g \in C^\infty(A'), \Delta \in \mathfrak{Q})$ have a well-defined meaning in $\mathfrak{Q}$. (Here the circle $\circ$ denotes multiplication in $\mathfrak{Q}$.) Let λ be a linear function on $\mathfrak{h}_\mathfrak{p}$. Then we agree to denote by e^λ the function $h \to e^{\lambda(\log h)}$ $(h \in A)$ on A. Put $\delta(g) = e^\rho \delta'(g) \circ e^{-\rho}$ $(g \in \mathfrak{G})$ where ρ is defined as in §2.

Let $I_\mathfrak{g}$ denote the centralizer of $\mathfrak{k}$ in $\mathfrak{G}$ and $\mathfrak{w}$ the "little Weyl group" (see [8, §3]) of $\mathfrak{g}_0$ with respect to $\mathfrak{h}_{\mathfrak{p}_0}$. We denote by $I(\mathfrak{h}_\mathfrak{p})$ the subalgebra of those elements of $S(\mathfrak{h}_\mathfrak{p})$ which are invariant under $\mathfrak{w}$. Let γ denote the homomorphism of $I_\mathfrak{g}$ onto $I(\mathfrak{h}_\mathfrak{p})$ given by Theorem 1 (p. 260) of [8]. Put $\gamma'(q) = e^{-\rho}\gamma(q) \circ e^\rho$ $(q \in I_\mathfrak{g})$ where $\gamma(q)$ is to be regarded as an element of $\mathfrak{H}_\mathfrak{p}$. Then it follows from the definition of γ that $\gamma'(q)$ is the unique element in $\mathfrak{H}_\mathfrak{p}$ such that $q - \gamma'(q) \in \mathfrak{k}\mathfrak{G} + \mathfrak{G}\mathfrak{n}_+$. Now we extend θ to an automorphism of $\mathfrak{G}$ and denoted by $g \to g^*$ the anti-automorphism of $\mathfrak{G}$ such that $X^* = -X$ for $X \in \mathfrak{g}$. Since $\mathfrak{H}_\mathfrak{p}$ is abelian and $\theta(H^*) = H$ $(H \in \mathfrak{h}_\mathfrak{p})$, it is obvious that $\theta(u^*) = u$ for $u \in \mathfrak{H}_\mathfrak{p}$. Now g and u be two elements in $\mathfrak{G}$ and $\mathfrak{H}_\mathfrak{p}$ respectively such that $g - u \in \mathfrak{k}\mathfrak{G} + \mathfrak{G}\mathfrak{n}_+$. Since $\theta(\mathfrak{n}_+) = \mathfrak{n}_-$, it is clear that

$$\theta(g^*) - u \in \mathfrak{G}\mathfrak{k} + \mathfrak{n}_- \mathfrak{G}.$$

This shows that $\theta(q^*) - \gamma'(q) \in \mathfrak{G}\mathfrak{k} + \mathfrak{n}_- \mathfrak{G}$ for $q \in I_\mathfrak{g}$. Replacing q by $\theta(q^*)$, we get $q - \gamma'(\theta(q^*)) \in \mathfrak{G}\mathfrak{k} + \mathfrak{n}_- \mathfrak{G}$. But by Lemma 47 of the Appendix (§15), $\gamma(\theta(q^*)) = \gamma(q)$. Therefore since $\mathfrak{g} = \mathfrak{n}_- + \mathfrak{h}_\mathfrak{p} + \mathfrak{k}$, it is clear that $\gamma'(q)$ is also the unique element in $\mathfrak{h}_\mathfrak{p}$ such that

$$q - \gamma'(q) \in \mathfrak{G}\mathfrak{k} + \mathfrak{n}_- \mathfrak{G}.$$

Moreover we know (see [8, Lemma 3]) that $d^0\gamma'(q) \leqslant d^0 q$.

Lemma 11. *Let* $q \in I_\mathfrak{g}$. *Then* $d^0(\delta(q) - \gamma(q)) \leqslant d^0 q - 1$.

$\mathfrak{G} = \mathfrak{N}_- \mathfrak{H}_\mathfrak{p} \mathfrak{R}$ where $\mathfrak{N}_-$ is, as before, the enveloping algebra of $\mathfrak{n}_-$. Let q_0 be the unique element in $\mathfrak{H}_\mathfrak{p}\mathfrak{R}$ such that $q - q_0 \in \mathfrak{n}_- \mathfrak{G}$. Then $d^0 q_0 \leqslant d^0 q$ and $q_0 - \gamma'(q) \in \mathfrak{H}_\mathfrak{p}\mathfrak{R}\mathfrak{k}$. Therefore it is obvious that $d^0\delta_h'(q_0 - \gamma'(q)) \leqslant d^0(q_0 - \gamma'(q)) - 1 \leqslant d^0 q - 1$ for $h \in A'$. On the other hand $d^0\delta_h'(q - q_0) \leqslant d^0 q - 1$ from Corollary 1 of Lemma

10. Therefore $d^0\delta'_h(q - \gamma'(q)) \leqslant d^0 q - 1$ and from this our assertion follows immediately.

σ and V being defined as in §3, we denote by $S(\sigma)$ the set of all C^∞ functions ϕ from G to V such that $\phi(k_1 x k_2) = \sigma_1(k_1)\phi(x)\sigma_2(k_2)$ $(k_1, k_2 \in K, \; x \in G)$. The significance of the mapping δ' is expressed by the following result (cf. [8, Lemma 23]).

Lemma 12. *For any function ψ on G let $\bar{\psi}$ denote its restriction on A'. Then if $\phi \in S(\sigma)$,*

$$\overline{g\phi} = \delta'(g)\bar{\phi} \qquad (g \in \mathfrak{G}).$$

Fix a point $h \in A'$ and suppose

$$\delta'(g) = \sum_{1 \leqslant i \leqslant r} f_i(q_i \otimes u_i \otimes k_i)$$

where $f_i \in \mathfrak{R}$, $q_i \in \mathfrak{Q}$, $u_i \in \mathfrak{H}_\mathfrak{p}$ and $k_i \in \mathfrak{R}$. Then

$$g = \Gamma_h(\delta'_h(g)) = \sum_i f_i(h) q_i^{h^{-1}} u_i k_i.$$

Hence

$$\phi(h; g) = \sum_i f_i(h)\phi\left(h; q_i^{h^{-1}} u_i k_i\right)$$

$$= \sum_i f_i(h)\sigma_1(q_i)\phi(h; u_i)\sigma_2(k_i) = \bar{\phi}(h; \delta'(g)).$$

This proves the lemma.

Corollary. *If $q_1, q_2 \in I_\mathfrak{g}$ and $\phi \in S(\sigma)$, then*

$$\delta'(q_1 q_2)\bar{\phi} = \delta'(q_1)(\delta'(q_2)\bar{\phi}).$$

For $q_1 q_2 \phi = q_1 \phi_2$ where $\phi_2 = q_2 \phi$. Since $q_2 \in I_\mathfrak{g}$, it is clear that $\phi_2 \in S(\sigma)$. Hence it follows from Lemma 12, that

$$\delta'(q_1 q_2)\bar{\phi} = \overline{q_1 \phi_2} = \delta'(q_1)\bar{\phi}_2 = \delta'(q_1)(\delta'(q_2)\bar{\phi}).$$

§5. Application to Spherical Functions

We now define two linear mappings ξ' and Y of $I_\mathfrak{g}$ into the subalgebra $1 \otimes \mathfrak{H}_\mathfrak{p} \otimes \mathfrak{R}$ of $\mathfrak{C}$ as follows. Since $\mathfrak{g} = \mathfrak{n}_- + \mathfrak{h}_\mathfrak{p} + \mathfrak{k} = \mathfrak{n}_+ + \mathfrak{h}_\mathfrak{p} + \mathfrak{k}$, there exist, for any $q \in I_\mathfrak{g}$, unique elements $g_1, g_2 \in \mathfrak{H}_\mathfrak{p}\mathfrak{R}$ such that $q - g_1 \in \mathfrak{n}_- \mathfrak{G}$ and $q - g_2 \in \mathfrak{n}_+ \mathfrak{G}$. We put $\xi'(q) = \delta'(g_1)$, $\mathsf{Y}(q) = \delta'(g_2)$. Also let $\xi(q) = e^\rho \xi'(q) \circ e^{-\rho}$ and $\zeta(q) = e^{-\rho}\mathsf{Y}(q) \circ e^\rho$. For any linear function λ on $\mathfrak{h}_\mathfrak{p}$, let η_λ denote the anti-homomorphism of $1 \otimes \mathfrak{H}_\mathfrak{p} \otimes \mathfrak{R}$ into $\mathfrak{R}$ such that $\eta_\lambda(1 \otimes H \otimes k) = \lambda(H)k$ $(H \in \mathfrak{h}_\mathfrak{p}, \; k \in \mathfrak{R})$. We put $\xi'_\lambda(q) = \eta_\lambda(\xi'(q))$, $\xi_\lambda(q) = \eta_\lambda(\xi(q))$, $\mathsf{Y}_\lambda(q) = \eta_\lambda(\mathsf{Y}(q))$, $\zeta_\lambda(q) = \eta_\lambda(\zeta(q))$ for $q \in I_\mathfrak{g}$.

Lemma 13. *For any linear function λ, ξ'_λ, ξ_λ, Y_λ and ζ_λ are anti-homomorphisms of $I_\mathfrak{g}$ into $\mathfrak{R}$.*

Let $q_1, q_2 \in I_{\mathfrak{g}}$ and choose $g_1, g_2 \in \mathfrak{H}_{\mathfrak{p}}\mathfrak{K}$ such that $q_i - g_i \in \mathfrak{n}_- \mathfrak{G}$ $(k = 1, 2)$. Suppose

$$g_i = \sum_{1 \leqslant j \leqslant s} u_{ij} k_{ij} \qquad (i = 1, 2)$$

where $u_{ij} \in \mathfrak{H}_{\mathfrak{p}}$ and $k_{ij} \in \mathfrak{K}$. Then $(q_1 - g_1)q_2 \in \mathfrak{n}_- \mathfrak{G}$ and therefore since $q_2 \in I_{\mathfrak{g}}$,

$$q_1 q_2 \equiv g_1 g_2 \equiv \sum_j u_{1j} q_2 k_{1j} \bmod \mathfrak{n}_- \mathfrak{G}.$$

However

$$u_{1j}(q_2 - g_2) \in u_{1j}\mathfrak{n}_- \mathfrak{G} \subset \mathfrak{n}_- \mathfrak{G}$$

since $[\mathfrak{h}_{\mathfrak{p}}, \mathfrak{n}_-] \subset \mathfrak{n}_-$. Therefore

$$q_1 q_2 \equiv \sum_j u_{1j} g_2 k_{1j} \equiv \sum_{j,l} u_{1j} u_{2l} k_{2l} k_{1j} \bmod \mathfrak{n}_- \mathfrak{G}$$

and so

$$\xi'_\lambda(q_1 q_2) = \xi'_\lambda(q_2)\xi'_\lambda(q_1).$$

Since $\xi_\lambda = \xi'_{\lambda - \rho}$, this proves the ξ'_λ and ξ_λ are anti-homomorphisms. The proof for Υ_λ and ζ_λ is similar.

We now return to the space $S(\sigma, \chi)$ of §3. We have seen that it is a finite-dimensional space which is obviously stable under the differential operators in $I_{\mathfrak{g}}$.

Lemma 14. *Let Λ be the leading exponent of an element $\phi \neq 0$ in $S(\sigma, \chi)$ and let $q \in I_{\mathfrak{g}}$. Then if $\phi^L \sigma_2(\xi'_\Lambda(q)) \neq 0$, we can conclude that $q\phi \neq 0$ and $(q\phi)^L = \phi^L \sigma_2(\xi'_\Lambda(q))$. On the other hand suppose $\phi^L \sigma_2(\xi'_\Lambda(q)) = 0$ and $q\phi \neq 0$. Let Λ' be the leading exponent of $q\phi$ and d and d' the degrees of ϕ^L and $(q\phi)^L$ respectively. Then $\Lambda' \leqslant \Lambda$ and, in case $\Lambda' = \Lambda$, we must have $d' < d$.*

We know from Lemma 12 that $\overline{q\phi} = \delta'(q)\bar{\phi}$. Moreover by Corollary 1 of Lemma 10, we can choose elements $f_i \in \mathfrak{R}_0$, $\Delta_i \in \mathfrak{Q} \otimes \mathfrak{H}_{\mathfrak{p}} \otimes \mathfrak{K}$ $(1 \leqslant i \leqslant r)$ such that

$$\delta'(q) = \xi'(q) + \sum_i f_i \Delta_i.$$

However every f in $\mathfrak{R}_0$ can be written in the form

$$f(\exp H) = \sum_{\lambda \in L} c_\lambda e^{-\lambda(H)} \qquad \left(H \in \mathfrak{h}_{\mathfrak{p}_0}^+\right)$$

where $c_\lambda \in \mathbf{C}$, $c_0 = 0$ and the series converges decently on $\mathfrak{h}_{\mathfrak{p}}(\varepsilon, \nu)$ for any ε, ν $(0 < \varepsilon < 1, \nu > 0)$. Therefore it follows from Theorem 1 together with the results of §2 of [11] and Lemma 16 of [11], that if Δ is any element in $\mathfrak{K} \otimes \mathfrak{H}_{\mathfrak{p}} \otimes \mathfrak{K}$, the series obtained by applying Δ to the series of $\bar{\phi}$ term by term and by multiplying the resulting series with the above series for f, still converges decently on $\mathfrak{h}_{\mathfrak{p}}(\varepsilon, \nu)$ and its sum is equal to $f\Delta\bar{\phi}$. We can therefore compare coefficients of the various exponents in the two sides of the equation

$$\overline{q\phi} = \xi'(q)\bar{\phi} + \sum_{1 \leqslant i \leqslant r} f_i \Delta_i \bar{\phi}.$$

Let ϕ^H denote the highest term of ϕ and $(q\phi)^H$ the highest term of $q\phi$ in case $q\phi \neq 0$. Since every exponent of f_i is negative $(1 \leqslant i \leqslant r)$, the term on the right side corresponding to the exponent Λ is $\xi'(q)\phi^H$ and every other term has a lower exponent. Hence $\xi'(q)\phi^H \neq 0$ implies that $q\phi \neq 0$ and $(q\phi)^H = \xi'(q)\phi^H$. It is clear that $\xi'(q)\phi^H = \phi^L \sigma_2(\xi'_\Lambda(q)) + pe^\Lambda$ where p is a polynomial of degree $< d$. Hence $\phi^L \sigma_2(\xi'_\Lambda(q)) \neq 0$ implies that $q\phi \neq 0$ and $(q\phi)^L = \phi^L \sigma_2(\xi'_\Lambda(q))$. Now suppose $\phi^L \sigma_2(\xi'_\Lambda(q)) = 0$. Then $\xi'(q)\phi^H = pe^\Lambda$. If $p \neq 0$, we get $(q\phi)^H = pe^\Lambda$ and so $\Lambda' = \Lambda$ and $d' < d$. On the other hand suppose $q\phi \neq 0$ while $p = 0$. Then $\xi'(q)\phi^H = 0$ and so all the exponents on the right side of our equation are lower than Λ. Hence $\Lambda' < \Lambda$ in this case. This completes the proof of the lemma.

Lemma 15. *Let E be a vector subspace of $S(\sigma, \chi)$ which is stable under $I_\mathfrak{g}$ and which is irreducible as an $I_\mathfrak{g}$-module. Let ϕ_1, ϕ_2 be two nonzero elements in E. Then the leading exponents and the degrees of the leading terms of ϕ_1 and ϕ_2 are the same.*

Let Λ_1, Λ_2 be the leading exponents and d_1, d_2 the degrees of the leading terms of ϕ_1, ϕ_2 respectively. Since E is irreducible under $I_\mathfrak{g}$, we can choose $q \in I_\mathfrak{g}$ such that $\phi_2 = q\phi_1$. Then it follows from Lemma 14 that $\Lambda_2 \leqslant \Lambda_1$. We prove similarly that $\Lambda_1 \leqslant \Lambda_2$ and therefore $\Lambda_1 = \Lambda_2$. But then again by Lemma 14, $d_2 \leqslant d_1$ and therefore by symmetry $d_1 = d_2$.

Corollary 1. *Let Λ denote the leading exponent of any nonzero element in E. Suppose q is an element in $I_\mathfrak{g}$ such that $\sigma_2(\xi'_\Lambda(q)) = 0$. Then $q\phi = 0$ for every $\phi \in E$.*

Suppose $q\phi \neq 0$ for some $\phi \in E$. Then since ϕ and $q\phi$ have the same leading exponent, it follows that $d' < d$ (in the notation of Lemma 14). But this contradicts Lemma 15.

Corollary 2. *There exists a linear function λ on $\mathfrak{h}_\mathfrak{p}$ with the following property. If q is any element in $I_\mathfrak{g}$ such that $\sigma_2(\zeta_\lambda(q)) = 0$, then $q\phi = 0$ for every $\phi \in E$.*

This obtained from Corollary 1 above simply by reversing the order in the space of real linear functions on $\mathfrak{h}_\mathfrak{p}$.

§6A. Application to Representations of G

Let π be an irreducible representation of G on a Banach space B. We shall say that it is *quasi-simple* if the following conditions hold.

1) Let Z denote the center of G. Then there exists a homomorphism η of Z into the multiplicative group of complex numbers such that $\pi(z) = \eta(z)I$ $(z \in Z)$ where I is the identity mapping of B.

2) Let B^0 denote the Gårding subspace [1] of B with respect to π and let us denote the corresponding representation of $\mathfrak{G}$ on B^0 again by π. Then there should

exist a homomorphism χ of $\mathfrak{Z}$ into $\mathbf{C}$ such that $\chi(1) = 1$ and $\pi(z) = \chi(z)I_0$ $(z \in \mathfrak{Z})$ where I_0 is the identity mapping of B^0.

η is called the *central character* and χ the *infinitesimal character* of π.

Lemma 16. *Let π be an irreducible quasi-simple representation of G on a Banach space B. Let Ω be the set of all equivalence classes of finite-dimensional irreducible representations of K. Fix $\mathfrak{v} \in \Omega$ and let $E_{\mathfrak{v}}$ denote the corresponding canonical projection (see [1, p. 225]) in B. Then there exists a linear function λ on $\mathfrak{h}_{\mathfrak{p}}$ with the following property. If q is an element in $I_{\mathfrak{g}}$ such that $\sigma_0(\zeta_\lambda(q)) = 0$ for $\sigma_0 \in \mathfrak{v}$, then $E_{\mathfrak{v}}\pi(q)E_{\mathfrak{v}} = 0$.*

Put $B_{\mathfrak{v}} = E_{\mathfrak{v}}B$. Then $\dim B_{\mathfrak{v}} < \infty$ and $B_{\mathfrak{v}} \subset B^0$ (see [1]). Put $\mathfrak{A} = I_{\mathfrak{g}}\mathfrak{R}$. Then $\mathfrak{A}$ is a subalgebra of $\mathfrak{G}$ and $B_{\mathfrak{v}}$ is stable and irreducible under $\pi(\mathfrak{A})$ (see [2, Theorem 2, p. 32]). Let τ be the representation of $\mathfrak{A}$ on $B_{\mathfrak{v}}$ defined under π. First we claim that the associative algebra $\tau(I_{\mathfrak{g}})$ is semisimple. Let q_0 be any element in $I_{\mathfrak{g}}$ such that $\operatorname{tr}\tau(q_0 q) = 0$ for $q \in I_{\mathfrak{g}}$. Then in order to establish our assertion, it is enough to prove that $\tau(q_0) = 0$. Let $a \in \mathfrak{A}$. Then it follows from Lemma 7 of [1] that a can be written in the form $a = q + \sum_{1 \leqslant i \leqslant r}[X_i, a_i]$ where $q \in I_{\mathfrak{g}}$, $a_i \in \mathfrak{A}$, $X_i \in \mathfrak{k}$. Hence $q_0 a = q_0 q + \sum_i [X_i, q_0 a_i]$ and therefore $\operatorname{tr}\tau(q_0 a) = \operatorname{tr}\tau(q_0 q) = 0$. But since τ is irreducible, $\tau(\mathfrak{A})$ is a simple algebra and therefore $\tau(q_0) = 0$. Now we shall prove that $\tau(I_{\mathfrak{g}})$ is a simple algebra. Let q_0 be an element in $I_{\mathfrak{g}}$ such that $\tau(q_0)$ lies in the center of $\tau(I_{\mathfrak{g}})$. Then since q_0 commutes with every element of $\mathfrak{R}$, it is obvious that $\tau(q_0)$ lies in the center of $\tau(\mathfrak{A})$. But τ being irreducible, it follows from Schur's lemma that $\tau(q_0) = c\tau(1)$ where $c \in \mathbf{C}$. This shows that $\tau(I_{\mathfrak{g}})$ is simple.

Put[11] $V = \mathcal{E}(B_{\mathfrak{v}})$ and let χ denote the infinitesimal character of π. Put

$$\sigma_1(k_1)F\sigma_2(k_2) = \pi_{\mathfrak{v}}(k_1)F\pi_{\mathfrak{v}}(k_2) \qquad (k_1, k_2 \in K, \ F \in V)$$

where $\pi_{\mathfrak{v}}(k)$ denotes the restriction of $\pi(k)$ $(k \in K)$ on $B_{\mathfrak{v}}$. Then we have a double representation $\sigma = (\sigma_1, \sigma_2)$ of K on V. Put

$$\phi_0(x) = E_{\mathfrak{v}}\pi(x)E_{\mathfrak{v}} \qquad (x \in G).$$

Then it is obvious that $\phi_0 \in S(\sigma, \chi)$. Let $\mathcal{F}$ be the subspace of $S(\sigma, \chi)$ consisting of all elements of the form $q\phi_0$ $(q \in I_{\mathfrak{g}})$. Note that $\phi_0(x; q) = E_{\mathfrak{v}}\pi(x)\pi(q)E_{\mathfrak{v}} = \phi_0(x)\tau(q)$ for $x \in G$ and $q \in I_{\mathfrak{g}}$. Let $\tau'(q)$ denote the mapping $\phi \to q\phi$ $(\phi \in \mathcal{F})$ of $\mathcal{F}$. Then τ' is a representation of $I_{\mathfrak{g}}$ on $\mathcal{F}$ and $\dim \mathcal{F} \leqslant \dim S(\sigma, \chi) < \infty$ from Lemma 8. Moreover if $\tau(q_0) = 0$ for some $q_0 \in I_{\mathfrak{g}}$, then $q_0 q\phi_0 = \phi_0\tau(q_0 q) = 0$ for all $q \in I_{\mathfrak{g}}$ and therefore $\tau'(q_0) = 0$. This shows that τ' can also be regarded as a representation of the simple algebra $\tau(I_{\mathfrak{g}})$. Therefore τ' is fully reducible and all its irreducible components are equivalent.

Now for the proof of Lemma 16, we may obviously assume that $E_{\mathfrak{v}} \neq 0$. Then $\phi_0 \neq 0$ and therefore $\mathcal{F} \neq \{0\}$. Let $E \neq \{0\}$ be a subspace of $\mathcal{F}$ which is invariant and irreducible under τ'. By Corollary 2 of Lemma 15 there exists a linear function λ on $\mathfrak{h}_{\mathfrak{p}}$ with the following property. If q is an element in $I_{\mathfrak{g}}$ such that $\sigma_2(\zeta_\lambda(q)) = 0$ then $q\phi = 0$ for $\phi \in E$. Since all irreducible components of τ' are equivalent, this implies that $\tau'(q) = 0$. This shows in particular that $q\phi_0 = 0$. But $\phi_0(x; q) = \phi_0(x)\tau(q)$. Therefore putting $x = 1$, we get $\tau(q) = 0$. This proves Lemma 16.

It is now convenient to introduce the assumption that the center of G is finite or, what is equivalent, that K is compact. (This is done in order to avoid certain technical complications, which, although they are rather cumbersome to handle, present no essential difficulty. The main point to note here is that we *do not* assume that G has a faithful finite-dimensional representation.)

Consider the space $L_2(K)$ (regarded as a Hilbert space in the usual way) defined with respect to the Haar measure on K. For any $x \in G$, let $\kappa(x)$ and $H(x)$ denote the unique elements k and H in K and $\mathfrak{h}_{\mathfrak{p}_0}$ respectively such that $x \in (k \exp H)N$. We shall also write $k_x = \kappa(xk)$ ($x \in G$, $k \in K$) whenever it is convenient to do so. Fix $\mathfrak{v}$ and λ as in Lemma 16 and consider the representation T of G on $L_2(K)$ defined as follows. If $x \in G$ and $f \in L_2(K)$ then[15]

$$T(x)f(k) = \exp\left(-\lambda_+\left(H(x^{-1}k)\right)\right)f\left(\kappa(x^{-1}k)\right) \qquad (k \in K)$$

where $\lambda_+ = \lambda + \rho$. Then[8]

$$T\left((nau)^k\right)f(k) = f(ku^{-1})e^{\lambda_+(\log a)} \qquad (k, u \in K, \ a \in A, \ n \in N).$$

Select a unitary matrix representation $\sigma_0 \in \mathfrak{v}$ and put[9] $f_0(k) = d(\mathfrak{v}) \mathrm{tr}\, \sigma_0(k^{-1})$ ($k \in K$). Let U be the subspace of $L_2(K)$ spanned by all functions of the form $T(u)f_0$ ($u \in K$). If g_{ij} ($1 \leqslant i, j \leqslant d(\mathfrak{v})$) are the matrix coefficients of σ_0, then the functions[16] $k \to g_{ij}(k^{-1}) = \mathrm{conj}\, g_{ji}(k)$ form a base for U and U consists of all functions $f \in L_2(K)$ which transform under $T(K)$ according to the class $\mathfrak{v}$. Let $F_{\mathfrak{v}}$ denote the orthogonal projection of $L_2(K)$ on U and V the Gårding subspace of $L_2(K)$ with respect to T. Then $V_{\mathfrak{v}} = F_{\mathfrak{v}}V$ is dense in U and therefore since U is of finite dimension, $V_{\mathfrak{v}} = U$. Since $V_{\mathfrak{v}} \subset V$, it follows that $f_0 \in V$. Now

$$T\left((nau)^k\right)f_0(k) = d(\mathfrak{v}) \mathrm{tr}\, \sigma_0(k^{-1}u)e^{(\lambda+\rho)(\log a)}$$

as we have seen above. Since $f_0 \in V$, we conclude immediately that

$$T(q)f_0(k) = T(q^k)f_0(k) = d(\mathfrak{v})\mathrm{tr}\left(\sigma_0(k^{-1})\sigma_0(\mathfrak{z}_\lambda(q))\right) \qquad \left(q \in I_{\mathfrak{g}}\right)$$

where the corresponding representation of $\mathfrak{G}$ on V is also denoted by T. Now suppose $\mathrm{tr}(\sigma_0(k^{-1})\sigma_0(\mathfrak{z}_\lambda(q))) = 0$ for all $k \in K$. Since σ_0 is irreducible, this implies that $\sigma_0(\mathfrak{z}_\lambda(q)) = 0$. Therefore, from Lemma 16, $T(q)f_0 = 0$ implies that $E_{\mathfrak{v}}\pi(q)E_{\mathfrak{v}} = 0$ ($q \in I_{\mathfrak{g}}$). Moreover we have the following lemma.

Lemma 17. *Let* π_1, π_2 *be two finite-dimensional representations of a subalgebra* $\mathfrak{B}$ *of* $\mathfrak{G}$ *such that* $\mathfrak{B} \supset \mathfrak{A} = I_{\mathfrak{g}}\mathfrak{R}$. *We assume that* π_2 *is semisimple and* $\pi_2(q) = 0$ *whenever* $\pi_1(q) = 0$ $(q \in I_{\mathfrak{g}})$. *Then* $\pi_2(b) = 0$ *whenever* $\pi_1(b) = 0$ $(b \in \mathfrak{B})$.

Fix b in $\mathfrak{B}$ such that $\pi_1(b) = 0$. We have to show that $\pi_2(b) = 0$. Define a representation t of $\mathfrak{R}$ on $\mathfrak{B}$ as follows: $t(X)c = [X, c]$ ($X \in \mathfrak{k}$, $c \in \mathfrak{B}$). Fix c in $\mathfrak{B}$. It is clear that the degree of any element in $t(\mathfrak{R})c$ cannot exceed $d^0 c$. Hence $\dim t(\mathfrak{R})c$

---

[15]$T(x)f(k)$ denotes the value of $T(x)f$ at a point k in K.

[16]conj c denotes the conjugate of a complex number c.

$< \infty$ and we conclude from Lemmas 3 and 7 of [1] that c can be written in the form

$$c = q + \sum_{1 \leqslant i \leqslant r} [X_i, c_i]$$

where $q \in I_\mathfrak{g} \cap (t(\Re)c)$, $X_i \in \mathfrak{k}$ and $c_i \in t(\Re)c$ $(1 \leqslant i \leqslant r)$. Now a being any element in $\mathfrak{B}$, we apply this result to $c = ab$. Let $\mathfrak{B}_1$ denote the kernel of π_1 in $\mathfrak{B}$. Then $c \in \mathfrak{B}_1$ and since $\mathfrak{B}_1$ is an ideal in $\mathfrak{B}$, it follows that $t(\Re)c \subset \mathfrak{B}_1$. Hence $\pi_1(q) = 0$. But then $\pi_2(q) = 0$ from our hypothesis. On the other hand it is obvious that $\operatorname{tr} \pi_2([X_i, c_i]) = 0$ and therefore $\operatorname{tr} \pi_2(c) = \operatorname{tr} \pi_2(q) = 0$. This shows that $\operatorname{tr} \pi_2(ab) = 0$ for every $a \in \mathfrak{B}$. Since π_2 is a semisimple representation of $\mathfrak{B}$, this implies that $\pi_2(b) = 0$.

Now let us assume that $B_\mathfrak{v} = E_\mathfrak{v}B \neq \{0\}$ in Lemma 16. Let $\mathfrak{B}$ be the algebra of all elements $b \in \mathfrak{G}$ such that $T(b)U \subset U$ and $\pi(b)B_\mathfrak{v} \subset B_\mathfrak{v}$. Then $\mathfrak{B} \supset \mathfrak{A}$. Let π_1, π_2 be the representations of $\mathfrak{B}$ defined on U and $B_\mathfrak{v}$ under T and π respectively. Since $B_\mathfrak{v}$ is irreducible under $\pi(\mathfrak{A})$, π_2 is irreducible. Also if $\pi_1(q) = 0$ for some $q \in I_\mathfrak{g}$, then $T(q)f_0 = 0$ and therefore, as we have already seen, $\pi_2(q) = 0$. Hence we conclude from Lemma 17 that $\pi_2(b) = 0$ whenever $\pi_1(b) = 0$ $(b \in \mathfrak{B})$ and so we can choose two subspaces U_1, U_2 of U with the following properties:

(1) $U_1 \supset U_2$ and U_1, U_2 are both stable under $T(\mathfrak{B})$, (2) the representation of $\mathfrak{B}$ defined on the factor space U_1/U_2 under T is equivalent to the irreducible representation π_2 of $\mathfrak{B}$ on $B_\mathfrak{v}$. Choose an element $f_1 \in U_1$ such that $f_1 \notin U_2$. Let $\mathfrak{B}_1$ be the set of all $b \in \mathfrak{B}$ such that $T(b)f_1 \in U_2$. Then $\mathfrak{B}_1$ is a maximal left ideal in $\mathfrak{B}$. We claim $\mathfrak{G}\mathfrak{B}_1 \neq \mathfrak{G}$. For otherwise we could choose $g_i \in \mathfrak{G}$ and $b_i \in \mathfrak{B}_1$ $(1 \leqslant i \leqslant r)$ such that $f_1 = \sum_{1 \leqslant i \leqslant r} T(g_i b_i)f_1$. Let $\mathfrak{v}_1, \ldots, \mathfrak{v}_s$ be a finite number of distinct elements in Ω none of which is equal to $\mathfrak{v}$. Let F_j and E_j respectively denote the canonical projections in $L_2(K)$ and B corresponding to $\mathfrak{v}_j$ $(1 \leqslant j \leqslant s)$. Then we can choose (see [2, Lemma 4, p. 32]) an element $e \in \Re$ such that $T(e)F_\mathfrak{v} = F_\mathfrak{v}$ and $T(e)F_j = 0$ $(1 \leqslant j \leqslant s)$. It is then clear that the corresponding relations $\pi(e)E_\mathfrak{v} = E_\mathfrak{v}$, $\pi(e)E_j = 0$ $(1 \leqslant j \leqslant s)$ also hold. Moreover we can obviously choose the set $(\mathfrak{v}_1, \ldots, \mathfrak{v}_s)$ so large that $T(g_i)U \subset FL_2(K)$ and $\pi(g_i)B_\mathfrak{v} \subset EB$ $(1 \leqslant i \leqslant r)$ where $F = F_\mathfrak{v} + F_1 + \cdots + F_r$, $E = E_\mathfrak{v} + E_1 + \cdots + E_r$. Then $T(eg_i)U \subset T(e)FL_2(K) = F_\mathfrak{v}L_2(K) \subset U$ and similarly $\pi(eg_i)B_\mathfrak{v} \subset B_\mathfrak{v}$. This proves that $eg_i \in \mathfrak{B}$ and therefore $eg_i b_i \in \mathfrak{B}_1$ $(1 \leqslant i \leqslant r)$. On the other hand $f_1 \in U$ and so

$$f_1 = T(e)f_1 = \sum_{1 \leqslant i \leqslant r} T(eg_i b_i)f_1 \in T(\mathfrak{B}_1)f_1 \subset U_2 .$$

As this contradicts the definition of f_1, we conclude that $\mathfrak{G}\mathfrak{B}_1 \neq \mathfrak{G}$.

Now by Zorn's lemma we can choose a maximal left ideal $\mathfrak{G}_1$ in $\mathfrak{G}$ containing $\mathfrak{G}\mathfrak{B}_1$. Then $\mathfrak{G}_1 \cap \mathfrak{B} \supset \mathfrak{B}_1$ and $1 \notin \mathfrak{G}_1 \cap \mathfrak{B}$. Since $\mathfrak{B}_1$ is a maximal left ideal in $\mathfrak{B}$, this implies that $\mathfrak{G}_1 \cap \mathfrak{B} = \mathfrak{B}_1$. Let V_1, V_2 be the closures in $L_2(K)$ of the spaces $T(\mathfrak{G})f_1$ and $T(\mathfrak{G}_1)f_1$ respectively. Then it follows from well-known facts (see [1, Theorem 5, p. 228]) that V_1, V_2 are both invariant under $T(G)$ and $U_i = V_i \cap U$ $(i = 1, 2)$. Let V' denote the orthogonal complement of V_2 in V_1 and F' the orthogonal projection of V_1 on V'. For any $x \in G$ let $T'(x)$ denote the linear transformation of V' given by $T'(x)v = F'T(x)v$ $(v \in V')$. Then T' is a representation of G on the Hilbert space V' which is irreducible and quasi-simple since $\mathfrak{G}_1$ is a maximal left ideal in $\mathfrak{G}$

containing $\mathfrak{B}_1$. Also it follows from Theorem 2 of [2] that T' and π are infinitesimally equivalent.

Let dk denote the normalized Haar measure on K so that $\int_K dk = 1$. We denote by $\langle f, g \rangle$ the scalar product of two elements $f, g \in L_2(K)$ so that

$$\langle f, g \rangle = \int (\operatorname{conj} f(k)) g(k)\, dk.$$

Lemma 18. *Let S denote the linear transformation in U given by*

$$Sg = f_0 \langle f_0, g \rangle \qquad (g \in U).$$

Then for any $x \in G$,

$$F_{\mathfrak{v}} T(x) F_{\mathfrak{v}} = \int_K T(k_x) ST(k^{-1}) \exp \lambda_- \left(H(xk) \right) dk$$

where $\lambda_- = \lambda - \rho$.

Put $d_0 = d(\mathfrak{v})$. Then if $u \in K$,

$$\langle f_0, T(u) f_0 \rangle = d_0^2 \int \operatorname{conj}\!\left(tr\sigma_0(k^{-1}) \right) tr\sigma_0(k^{-1}u)\, dk.$$

Let g_{ij} $(1 \le i, j \le d_0)$ denote as before the coefficients of the unitary matrix representation σ_0. Then it follows from the Schur orthogonality relations that

$$\langle f_0, T(u) f_0 \rangle = d_0^2 \sum_{i, j, m} \int (\operatorname{conj} g_{ii}(k)) g_{mj}(k) g_{jm}(u)\, dk$$

$$= d_0 \sum_i g_{ii}(u) = f_0(u^{-1}) = T(u) f_0(1).$$

Since U is spanned by vectors of the form $T(u) f_0$ $(u \in K)$, we conclude that $\langle f_0, g \rangle = g(1)$ for every $g \in U$. Hence

$$g(u) = \langle f_0, T(u^{-1}) g \rangle = \langle T(u) f_0, g \rangle$$

for $u \in K$ and $g \in U$. Now suppose $g_1, g_2 \in U$. Then

$$\langle g_1, T(x) g_2 \rangle = \int (\operatorname{conj} g_1(k)) g_2(k_{x^{-1}}) e^{-\lambda_+(H(x^{-1}k))} dk.$$

Since $k \to k_x$ is a homeomorphism of K and since $dk_x = e^{-2\rho(H(x^{-1}k))}\, dk$ (see [1, p. 241]) we find by replacing k by k_x in the above integral that

$$\langle g_1, T(x) g_2 \rangle = \int (\operatorname{conj} g_1(k_x)) g_2(k) e^{\lambda_+(H(xk))}\, dk_x$$

$$= \int (\operatorname{conj} g_1(k_x)) g_2(k) e^{\lambda_-(H(xk))}\, dk$$

since $H(x^{-1}k_x) = -H(xk)$. But

$$(\operatorname{conj} g_1(k_x)) g_2(k) = \langle g_1, T(k_x) f_0 \rangle \langle f_0, T(k^{-1}) g_2 \rangle$$

$$= \langle g_1, T(k_x) ST(k^{-1}) g_2 \rangle$$

and therefore

$$\langle g_1, T(x)g_2 \rangle = \int \langle g_1, T(k_x)ST(k^{-1})g_2 \rangle e^{\lambda_-(H(xk))} \, dk.$$

This being true for any two elements $g_1, g_2 \in U = F_{\mathfrak{v}} L_2(K)$, the assertion of the lemma follows immediately.

Now let $F_{\mathfrak{v}}'$ denote the orthogonal projection of $L_2(K)$ on $V' \cap U = V_{\mathfrak{v}}'$. Then if $v \in V_{\mathfrak{v}}'$,

$$F_{\mathfrak{v}}'T'(x)v = F_{\mathfrak{v}}'T(x)v = \int T(k_x)S'T(k^{-1})v e^{\lambda_-(H(xk))} dk$$

where $S' = F_{\mathfrak{v}}'SF_{\mathfrak{v}}'$. (Here we make use of the obvious fact that $F_{\mathfrak{v}}'$ commutes with $T(k)$ for any $k \in K$). Since $T(k)$ coincides with $T'(k)$ on V', we get

$$F_{\mathfrak{v}}'T'(x)F_{\mathfrak{v}}' = \int T'(k_x)S'T'(k^{-1})e^{\lambda_-(H(xk))} dk.$$

Moreover $\dim V_{\mathfrak{v}}' \leqslant \dim U = d(\mathfrak{v})^2$. However since T' and π are infinitesimally equivalent, we have obtained the following result.

Theorem 2. *Suppose the center of G is finite and π is a quasi-simple irreducible representation of G on a Banach space B. Fix a class $\mathfrak{v} \in \Omega$ which occurs in π and put $B_{\mathfrak{v}} = E_{\mathfrak{v}} B$ where $E_{\mathfrak{v}}$ is the canonical projection corresponding to π. Let $\pi_{\mathfrak{v}}(x)$ denote the restriction of $E_{\mathfrak{v}}\pi(x)E_{\mathfrak{v}}$ $(x \in G)$ on $B_{\mathfrak{v}}$ and similarly for $\pi_{\mathfrak{v}}(g)$ $(g \in \mathfrak{G})$. Then there exists a linear function λ on $\mathfrak{h}_p$ with the following property. If q is an element in $I_{\mathfrak{g}}$, then $\pi_{\mathfrak{v}}(q) = 0$ whenever $\pi_{\mathfrak{v}}(\zeta_\lambda(q)) = 0$. Fix one such λ and let M denote the centralizer of A in K. Then we can choose a linear transformation S_λ in $B_{\mathfrak{v}}$ with the following properties.*

1) $\pi_{\mathfrak{v}}(m)S_\lambda = S_\lambda \pi_{\mathfrak{v}}(m)$ *for all $m \in M$.*

2) $\pi_{\mathfrak{v}}(x) = \displaystyle\int_K \pi_{\mathfrak{v}}(k_x)S_\lambda \pi_{\mathfrak{v}}(k^{-1})e^{\lambda_-(H(xk))} dk \qquad (x \in G)$

   where $\lambda_- = \lambda - \rho$.

Finally $\dim B_{\mathfrak{v}} \leqslant d(\mathfrak{v})^2$.

We have seen above that there exists a linear transformation S in $B_{\mathfrak{v}}$ for which condition (2) is fulfilled. Put

$$S_\lambda = \int_M \pi_{\mathfrak{v}}(m)S\pi_{\mathfrak{v}}(m^{-1}) \, dm$$

where dm is the normalized Haar measure on M. We claim S_λ also satisfies condition (2). This is seen as follows.

$$\pi_{\mathfrak{v}}(x) = \int \pi_{\mathfrak{v}}(\kappa(xk))S\pi_{\mathfrak{v}}(k^{-1})e^{\lambda_-(H(xk))} dk$$

$$= \int \pi_{\mathfrak{v}}(\kappa(xkm))S\pi_{\mathfrak{v}}(m^{-1}k^{-1})e^{\lambda_-(H(xkm))} \, dk \, dm.$$

But $\kappa(xkm) = \kappa(xk)m$ and $H(xkm) = H(xk)$ $(m \in M)$. Hence the right side is equal to

$$\int \pi_{\mathfrak{v}}(\kappa(xk)) S_{\lambda} \pi_{\mathfrak{v}}(k^{-1}) e^{\lambda_-(H(xk))} dk$$

and this proves our assertion.

In case G has a faithful finite-dimensional representation the above result could have been obtained from Theorem 4 (p. 63) of [2].

§6B. Proof of Lemma 19

Let $\mathfrak{M}$ be the subspace of $\mathfrak{C}$ spanned by elements of the form

$$k_1 m \otimes D \otimes k_2 - k_1 \otimes D \otimes mk_2 \qquad (k_1, k_2 \in \mathfrak{K}, \; D \in \mathfrak{D}, \; m \in \mathfrak{M}_{\mathfrak{k}}).$$

(We recall that $\mathfrak{M}_{\mathfrak{k}}$ is the enveloping algebra of $\mathfrak{m} \cap \mathfrak{k}$ in $\mathfrak{G}$.) It is obvious that $\mathfrak{M}$ is a left ideal in $\mathfrak{C}$. We now intend to prove the following result.

Lemma 19. $\delta'(q_1 q_2) - \delta'(q_1) \circ \delta'(q_2) \in \mathfrak{M}$ *for any two elements* q_1, q_2 *in* $I_{\mathfrak{g}}$.

Put $\mathfrak{M}_0 = \mathfrak{M} \cap (\mathfrak{K} \otimes \mathfrak{H}_p \otimes \mathfrak{K})$ and fix a point $h \in A'$. Then if $\Delta = \delta'(q_1 q_2) - \delta'(q_1) \circ \delta'(q_2)$, we have to prove that $\Delta_h \in \mathfrak{M}_0$. Let Z denote the center of G. Then $Z \subset K$ and under the natural mapping of G on $G^* = G/Z$, A is mapped isomorphically on a subgroup of G^*. Therefore if we identify A with its image under this mapping, it is obvious that Δ_h has the same value whether we consider A as a subgroup of G or of G^*. This shows that, for the proof of Lemma 19, it is permissible to replace G by any group locally isomorphic to it. Hence we can assume that the center of G is finite and K is compact. This assumption will be in force for the rest of this section.

Lemma 20. *Fix* $h_0 \in A'$ *and suppose* a *is an element in* $\mathfrak{Q} \otimes \mathfrak{H}_p \otimes \mathfrak{K}$ *with the following property. If* σ *is any double representation of* K *on a finite-dimensional vector space* V, *then* $\phi(h_0; a) = 0$ *for* $\phi \in S(\sigma)$. *Under this condition we can conclude that* $a = 0$.

For otherwise suppose $a \neq 0$. Put $g = \Gamma_{h_0}(a) \in \mathfrak{G}$. Since Γ_{h_0} is injective on $\mathfrak{Q} \otimes \mathfrak{H}_p \otimes \mathfrak{K}$, $g \neq 0$. We now regard $C^{\infty}(G)$ as a double K-module as follows. If $\phi \in C^{\infty}(G)$ and $k_1, k_2 \in K$, then $k_1 \phi k_2$ is the function $x \to \phi(k_1^{-1} x k_2^{-1})$ $(x \in G)$. Let $\mathcal{L}(G)$ be the subspace consisting of those $\phi \in C^{\infty}(G)$ whose transforms $k_1 \phi k_2$ $(k_1, k_2, \in K)$ span a finite-dimensional subspace of $C^{\infty}(G)$.

Similarly we regard $C^{\infty}(K \times K)$ as a double K-module as follows. If $k_1, k_2 \in K$ and $f \in C^{\infty}(K \times K)$, then $k_1 f k_2$ is the function $(k, k') \to f(k_1^{-1} k : k' k_2^{-1})$. Let $\mathcal{L}(K \times K)$ denote the subspace of those $f \in C^{\infty}(K \times K)$ whose transforms $k_1 f k_2$ $(k_1, k_2 \in K)$ span a finite-dimensional subspace of $C^{\infty}(K \times K)$.

For any $\phi \in \mathcal{L}(G)$, we define a mapping $T_{\phi} : G \to \mathcal{L}(K \times K)$ as follows. If $x \in G$, $T_{\phi}(x) = f$ where $f(k_1 : k_2) = \phi(k_1 x k_2)$ $(k_1, k_2 \in K)$. Note that $T_{k \phi k'}(x) = k T_{\phi}(x) k'$ and therefore since $\phi \in \mathcal{L}(G)$, it is obvious that $T_{\phi}(x) \in \mathcal{L}(K \times K)$. Let L_{ϕ} be the subspace of $\mathcal{L}(G)$ spanned by $k_1 \phi k_2$ $(k_1, k_2 \in K)$. Then $\dim L_\phi < \infty$. Suppose $\phi \neq 0$ and $\phi = \phi_1, \phi_2, \ldots, \phi_r$ is a base for L_ϕ. Then it is obvious that there exist elements $f_{ij} \in \mathcal{L}(K \times K)$ $(1 \leqslant i, j \leqslant r)$ such that

$$k_1 \phi_i k_2 = \sum_{1 \leqslant j \leqslant r} f_{ji}(k_1^{-1} : k_2^{-1}) \phi_j \qquad (k_1, k_2 \in K).$$

Also it is clear that[13] $f_{ji}(1:1) = \delta_{ji}$ and

$$T_{\phi_i}(x) = \sum_j \phi_j(x) f_{ji} \qquad (1 \leqslant i \leqslant r).$$

Let V_ϕ be the subspace of $\mathcal{L}(K \times K)$ spanned by $T_{k_1 \phi k_2}(x)$ for all $k_1, k_2 \in K$ and $x \in G$. Since $f_{ji} \in \mathcal{L}(K \times K)$, it is obvious from the above relation that $\dim V_\phi < \infty$. Moreover, being a submodule of $\mathcal{L}(K \times K)$, V_ϕ is also a double K-module. Therefore it is clear that we have a (continuous) double representation σ of K on V_ϕ. Put $\Phi(x) = T_\phi(x)$ $(x \in G)$. Then $\Phi \in S(\sigma)$ and therefore $\Phi(h_0; a) = 0$. But since $g = \Gamma_{h_0}(a)$, it is obvious that $\Phi(h_0; a) = \Phi(h_0; g)$. Hence $\Phi(h_0; g) = 0$. Now put $f_j = f_{j1}$ $(1 \leqslant j \leqslant r)$. Then

$$\Phi(x) = T_\phi(x) = \sum_j \phi_j(x) f_j$$

and therefore

$$\Phi(x; g) = \sum_j \phi_j(x; g) f_j.$$

Hence

$$0 = \Phi(h_0; g) = \sum_j \phi_j(h_0; g) f_j.$$

But $f_j(1:1) = \delta_{j1}$ and so

$$\phi(h_0; g) = \sum_j \phi_j(h_0; g) f_j(1:1) = 0.$$

This proves that $\phi(h_0; g) = 0$ for every $\phi \in \mathcal{L}(G)$. But now we have the following lemma.

Lemma 21. *Let g be an element in $\mathfrak{G}$ and x_0 an element in G. Suppose $\phi(x_0; g) = 0$ for every $\phi \in \mathcal{L}(G)$. Then $g = 0$.*

Assuming this for a moment, it follows from our result above that $g = \Gamma_{h_0}(a) = 0$ which gives a contradiction. This proves Lemma 20.

Now we shall prove Lemma 21. Suppose $g \neq 0$. Then we can choose $\phi_0 \in C^\infty(G)$ such that $\phi_0(x_0; g) \neq 0$ (see Lemma 13 of [4]). Choose a neighborhood ω of 1 in K such that $\omega = \omega^{-1}$ and

$$\left| \phi_0(x_0; g) - \phi_0(k_1 x_0 k_2^{-1}; g^{k_2}) \right| \leqslant \tfrac{1}{2} c$$

for k_1, k_2 in ω. Here $c = |\phi_0(x_0; g)| > 0$. Also select a nonnegative continuous function f_0 on K such that $\int f_0(k)\, dk = 1$ and $f_0 = 0$ outside ω. (As usual dk is the normalized Haar measure on K). Fix $\varepsilon > 0$ and let $\mathcal{L}(K)$ denote the subspace of $C^\infty(K)$ spanned by the matrix coefficients of all finite-dimensional representations of K. By the Peter–Weyl Theorem, we can select $f \in \mathcal{L}(K)$ such that $|f(k) - f_0(k)| \leq \varepsilon$ for $k \in K$. Put

$$\phi(x) = \int_{K \times K} f(k_1)\phi_0\big(k_1^{-1}xk_2^{-1}\big)f(k_2)\,dk_1\,dk_2.$$

Then obviously $\phi \in \mathcal{L}(G)$ and

$$\phi(x_0; g) = \int f(k_1)\phi_0\big(k_1^{-1}x_0k_2^{-1}; g^{k_2}\big)f(k_2)\,dk_1\,dk_2.$$

Let $\alpha = \sup_{k_1, k_2 \in K}|\phi_0(k_1^{-1}x_0k_2^{-1}; g^{k_2})|$. Then

$$\left|\phi_0(x_0; g) - \int f_0(k_1)\phi_0\big(k_1^{-1}x_0k_2^{-1}; g^{k_2}\big)f_0(k_2)\,dk_1\,dk_2\right|$$

$$\leq \alpha\varepsilon\int|f(k_2)|\,dk_2 + \alpha\varepsilon\int|f_0(k_1)|\,dk_1$$

$$\leq \alpha\varepsilon(1+\varepsilon) + \alpha\varepsilon(2+\varepsilon).$$

But since f_0 is zero outside ω,

$$\left|\int f_0(k_1)\phi_0\big(k_1^{-1}x_0k_2^{-1}; g^{k_2}\big)f_0(k_2)\,dk_1\,dk_2 - \phi_0(x_0; g)\right| \leq \tfrac{1}{2}c.$$

Therefore

$$|\phi(x_0; g) - \phi_0(x_0; g)| \leq \tfrac{1}{2}c + \alpha\varepsilon(2+\varepsilon) \leq \tfrac{2}{3}c$$

if ε is chosen sufficiently small. This proves that $|\phi(x_0; g)| \geq \tfrac{1}{3}c > 0$. Since $\phi \in \mathcal{L}(G)$, we get a contradiction with our hypothesis. Hence the lemma.

Now we come to Lemma 19. Let $\Delta = \delta'(q_1q_2) - \delta'(q_1)\circ\delta'(q_2)$. Since $\mathfrak{R} = \mathfrak{Q}\mathfrak{M}_\mathfrak{k}$, it is obvious that we can choose elements $p_i \in \mathfrak{Q}$, $\kappa_j \in \mathfrak{R}$ and $D_{ij} \in \mathfrak{D}$ ($1 \leq i \leq r$, $1 \leq j \leq s$) such that the sets $(p_1,\ldots,p_r)$ and $(\kappa_1,\ldots,\kappa_s)$ are separately linearly independent and

$$\Delta \equiv \sum_{i,j} p_i \otimes D_{ij} \otimes \kappa_j \bmod \mathfrak{M}.$$

Fix a point $h \in A'$. We know from the corollary of Lemma 12 that $\Delta\bar{\phi} = 0$ for any $\phi \in S(\sigma)$. On the other hand it is easy to verify that $\mu\bar{\phi} = 0$ for $\mu \in \mathfrak{M}$. Hence if u_{ij} is the local expression of D_{ij} at h and

$$\Delta_0 = \sum_{i,j} p_i \otimes u_{ij} \otimes \kappa_j \in \mathfrak{Q} \otimes \mathfrak{H}_p \otimes \mathfrak{R},$$

we have $\phi(h; \Delta_0) = 0$ for every $\phi \in S(\sigma)$. This being true for every σ, we conclude from Lemma 20 that $\Delta_0 = 0$ and therefore $u_{ij} = 0$ ($1 \leq i \leq r$, $1 \leq j \leq s$). This shows that the local expression of D_{ij} at h is zero. But since h was an arbitrary element in A', we conclude that $D_{ij} = 0$ and so $\Delta \in \mathfrak{M}$. This concludes the proof of Lemma 19.

§7. Computation of $\delta'(\omega)$

Let ω be the Casimir operator of $\mathfrak{g}$ so that $\omega \in \mathfrak{Z}$. We propose to compute $\delta'(\omega)$. Put $B(X, Y) = \operatorname{tr}(adX adY)$ $(X, Y \in \mathfrak{g})$ as before and let $X_1, \ldots, X_n$ be a base for $\mathfrak{g}$. Then if $g_{ij} = B(X_i, X_j)$, ω is given by the formula $\omega = \Sigma_{1 \leqslant i, j \leqslant n} g^{ij} X_i X_j$ where $(g^{ij})_{1 \leqslant i, j \leqslant n}$ is the inverse of the matrix $(g_{ij})_{1 \leqslant i, j \leqslant n}$. For every $\alpha \in P_+$, choose the elements X_α and $X_{-\alpha}$ of §2 in such a way that $B(X_\alpha, X_{-\alpha}) = 1$. It is clear that the spaces $\mathfrak{h}_\mathfrak{p}$, $\mathfrak{m} \cap \mathfrak{k}$ and $\Sigma_{\alpha \in P_+}(\mathbf{C} X_\alpha + \mathbf{C} X_{-\alpha})$ are mutually orthogonal under the bilinear form B. Hence it follows by an easy computation that

$$\omega = \omega_\mathfrak{m} + \sum_{\alpha \in P_+} (X_\alpha X_{-\alpha} + X_{-\alpha} X_\alpha).$$

Here $\omega_\mathfrak{m}$ is defined as follows. Let $Z_1, \ldots, Z_p$ be a base for $\mathfrak{m} \cap \mathfrak{k}$. Put $g'_{ij} = B(Z_i, Z_j)$ $(1 \leqslant i, j \leqslant p)$. If $(g^{ij})_{1 \leqslant i, j \leqslant l}$ and $(g''^{ij})_{1 \leqslant i, j \leqslant p}$ respectively are the inverses of the matrices[17] $(B(H_i, H_j))_{1 \leqslant i, j \leqslant l}$ and $(g'_{ij})_{1 \leqslant i, j \leqslant p}$,

$$\omega_\mathfrak{m} = \sum_{1 \leqslant i, j \leqslant l} g^{ij} H_i H_j + \omega_{\mathfrak{m} \cap \mathfrak{k}}$$

where $\omega_{\mathfrak{m} \cap \mathfrak{k}} = \Sigma_{1 \leqslant i, j \leqslant p} g''^{ij} Z_i Z_j$. For any root β, whose restriction on $\mathfrak{h}_\mathfrak{p}$ is not zero, we write $X_\beta = Y_\beta + Z_\beta$ $(Y_\beta \in \mathfrak{p}, Z_\beta \in \mathfrak{k})$. Also for any linear function μ on $\mathfrak{h}_\mathfrak{p}$, let H_μ denote the unique element in $\mathfrak{h}_\mathfrak{p}$ such that $B(H, H_\mu) = \mu(H)$ for all $H \in \mathfrak{h}_\mathfrak{p}$. We have already identified $\mathfrak{D}$ with a subalgebra of $\mathfrak{C}$ in §4.

Lemma 22. *For any linear function μ on $\mathfrak{h}$ let $\bar{\mu}$ denote its restriction on $\mathfrak{h}_\mathfrak{p}$. Also let us agree to denote the function $h \to \mu(\log h)$ $(h \in A)$ on A again by μ. Then*

$$\delta'(\omega) = \delta'(\omega_{\mathfrak{m} \cap \mathfrak{k}}) + \sum_{1 \leqslant i, j \leqslant l} g^{ij} H_i H_j + \sum_{\alpha \in P_+} \coth \alpha H_{\bar{\alpha}}$$

$$- 2 \sum_{\alpha \in P_+} (\sinh \alpha)^{-2} \{(1 \otimes 1 \otimes Z_\alpha Z_{-\alpha}) + (Z_\alpha Z_{-\alpha} \otimes 1 \otimes 1)\}$$

$$+ 4 \sum_{\alpha \in P_+} (\sinh \alpha)^{-1} \coth \alpha (Z_\alpha \otimes 1 \otimes Z_{-\alpha}).$$

Fix $h \in A'$ and let α be a root such that $\bar{\alpha} \neq 0$. Then if $c = \alpha(\log h)$,

$$Ad(h^{-1}) Z_\alpha = \tfrac{1}{2} Ad(h^{-1})(X_\alpha + \theta(X_\alpha)) = \tfrac{1}{2}(e^{-c} X_\alpha + e^c \theta(X_\alpha))$$

$$= \tfrac{1}{2}(e^c + e^{-c}) Z_\alpha - \tfrac{1}{2}(e^c - e^{-c}) Y_\alpha.$$

Hence

$$Y_\alpha = \coth c Z_\alpha - (\sinh c)^{-1} Ad(h^{-1}) Z_\alpha,$$

$$X_\alpha = 2(e^c - e^{-c})^{-1} \{e^c Z_\alpha - Z_\alpha^{h^{-1}}\}.$$

Therefore

$$- X_\alpha X_{-\alpha} = 4(e^c - e^{-c})^{-2} \{ Z_\alpha Z_{-\alpha} + (Z_\alpha Z_{-\alpha})^{h^{-1}} - e^c Z_\alpha Z_{-\alpha}^{h^{-1}} - e^{-c} Z_\alpha^{h^{-1}} Z_{-\alpha}\}.$$

---

[17] We recall that $H_1, \ldots, H_l$ is the base of $\mathfrak{h}_\mathfrak{p}$ chosen in §3.

But

$$Z_\alpha Z_{-\alpha}^{h^{-1}} = Z_{-\alpha}^{h^{-1}} Z_\alpha + \left[Z_\alpha, Z_{-\alpha}^{h^{-1}} \right]$$

$$= Z_{-\alpha}^{h^{-1}} Z_\alpha + \cosh c [Z_\alpha, Z_{-\alpha}] + \sinh c [Z_\alpha, Y_{-\alpha}].$$

Therefore

$$- X_\alpha X_{-\alpha} = 4(e^c - e^{-c})^{-2} \Big\{ Z_\alpha Z_{-\alpha} + (Z_\alpha Z_{-\alpha})^{h^{-1}} - e^c Z_{-\alpha}^{h^{-1}} Z_\alpha - e^{-c} Z_\alpha^{h^{-1}} Z_{-\alpha}$$

$$- \tfrac{1}{2}(e^{2c} + 1)[Z_\alpha, Z_{-\alpha}] - \tfrac{1}{2}(e^{2c} - 1)[Z_\alpha, Y_{-\alpha}] \Big\}.$$

Applying θ to this equation, we get

$$- \theta(X_\alpha X_{-\alpha}) = 4(e^c - e^{-c})^{-2} \Big\{ Z_\alpha Z_{-\alpha} + (Z_\alpha Z_{-\alpha})^h - e^c Z_{-\alpha}^h Z_\alpha - e^{-c} Z_\alpha^h Z_{-\alpha}$$

$$- \tfrac{1}{2}(e^{2c} + 1)[Z_\alpha, Z_{-\alpha}] + \tfrac{1}{2}(e^{2c} - 1)[Z_\alpha, Y_{-\alpha}] \Big\}.$$

Replacing α and h by $-\alpha$ and h^{-1} respectively in this equation, we obtain

$$- \theta(X_{-\alpha} X_\alpha) = 4(e^c - e^{-c})^{-2} \Big\{ Z_{-\alpha} Z_\alpha + (Z_{-\alpha} Z_\alpha)^{h^{-1}} - e^c Z_\alpha^{h^{-1}} Z_{-\alpha} - e^{-c} Z_{-\alpha}^{h^{-1}} Z_\alpha$$

$$+ \tfrac{1}{2}(e^{2c} + 1)[Z_\alpha, Z_{-\alpha}] + \tfrac{1}{2}(e^{2c} - 1)[Z_{-\alpha}, Y_\alpha] \Big\}.$$

Therefore

$$- (X_\alpha X_{-\alpha} + \theta(X_{-\alpha} X_\alpha)) = 4(e^c - e^{-c})^{-2} \Big\{ (Z_\alpha Z_{-\alpha} + Z_{-\alpha} Z_\alpha)$$

$$+ (Z_\alpha Z_{-\alpha} + Z_{-\alpha} Z_\alpha)^{h^{-1}} - 2e^c Z_{-\alpha}^{h^{-1}} Z_\alpha - 2e^{-c} Z_\alpha^{h^{-1}} Z_{-\alpha} \Big\}$$

$$+ 2e^c (e^c - e^{-c})^{-1} \{ [Z_{-\alpha}, Y_\alpha] - [Z_\alpha, Y_{-\alpha}] \}.$$

Put $\omega_+ = \sum_{\alpha \in P_+} \{ X_\alpha X_{-\alpha} + X_{-\alpha} X_\alpha \}$. Since $\omega = \omega_m + \omega_+$, $\theta(\omega) = \omega$ and $\theta(\omega_m) = \omega_m$, it is clear that $\theta(\omega_+) = \omega_+$. Hence

$$- \omega_+ = - \frac{1}{2}(\omega_+ + \theta(\omega_+)) = - \frac{1}{2} \sum_{\alpha \in P_+} \{ X_\alpha X_{-\alpha} + X_{-\alpha} X_\alpha + \theta(X_\alpha X_{-\alpha} + X_{-\alpha} X_\alpha) \}$$

$$= \sum_{\alpha \in P_+} 4(e^{\alpha(H)} - e^{-\alpha(H)})^{-2} \Big\{ (Z_\alpha Z_{-\alpha} + Z_{-\alpha} Z_\alpha) + (Z_\alpha Z_{-\alpha} + Z_{-\alpha} Z_\alpha)^{h^{-1}}$$

$$- 2e^{\alpha(H)} Z_{-\alpha}^{h^{-1}} Z_\alpha - 2e^{-\alpha(H)} Z_\alpha^{h^{-1}} Z_{-\alpha} \Big\}$$

$$+ \sum_{\alpha \in P_+} \coth \alpha(H) \{ [Z_{-\alpha}, Y_\alpha] - [Z_\alpha, Y_{-\alpha}] \}$$

where $H = \log h$. But $[X_\alpha, X_{-\alpha}] = H_\alpha$ where H_α is the element in $\mathfrak{h}$ such that

$B(H_\alpha, H') = \alpha(H')$ for every $H' \in \mathfrak{h}$. Therefore $[Z_\alpha, Y_{-\alpha}] + [Y_\alpha, Z_{-\alpha}] = H_{\bar\alpha}$. Hence

$$\omega_+ = \sum_{\alpha \in P_+} \coth \alpha(H) H_{\bar\alpha} - \sum_{\alpha \in P_+} 4\left(e^{\alpha(H)} - e^{-\alpha(H)}\right)^{-2}$$

$$\times \left\{ (Z_\alpha Z_{-\alpha} + Z_{-\alpha} Z_\alpha) + (Z_\alpha Z_{-\alpha} + Z_{-\alpha} Z_\alpha)^{h^{-1}} \right.$$

$$\left. -2e^{\alpha(H)} Z_{-\alpha}^{h^{-1}} Z_\alpha - 2e^{-\alpha(H)} Z_\alpha^{h^{-1}} Z_{-\alpha} \right\}.$$

Since X_α and $X_{-\alpha}$ ($\alpha \in P_+$) are orthogonal to $\mathfrak{m}$, it is obvious that $Z_\alpha, Z_{-\alpha}$ are in $\mathfrak{q}$. Put $\alpha^* = -\theta\alpha$ for $\alpha \in P_+$. Then $\alpha \to \alpha^*$ is a permutation of the set P_+. Hence $X_{\alpha^*} = c_\alpha \theta(X_{-\alpha})$ and $X_{-\alpha^*} = c_{-\alpha} \theta(X_\alpha)$ ($c_\alpha, c_{-\alpha} \in \mathbf{C}$) for $\alpha \in P_+$. But since $B(X_\alpha, X_{-\alpha}) = B(\theta(X_\alpha), \theta(X_{-\alpha})) = B(X_{\alpha^*}, X_{-\alpha^*}) = 1$, we conclude that $c_\alpha c_{-\alpha} = 1$. Therefore $Z_{\alpha^*} Z_{-\alpha^*} = c_\alpha c_{-\alpha} Z_{-\alpha} Z_\alpha = Z_{-\alpha} Z_\alpha$. Since $\alpha = \alpha^*$ on $\mathfrak{h}_\mathfrak{p}$ this shows that

$$\delta'(\omega_+) = \sum_{\alpha \in P_+} \coth \alpha H_{\bar\alpha} - 2 \sum_{\alpha \in P_+} (\sinh \alpha)^{-2} \{ 1 \otimes 1 \otimes Z_\alpha Z_{-\alpha} + Z_\alpha Z_{-\alpha} \otimes 1 \otimes 1 \}$$

$$+ 4 \sum_{\alpha \in P_+} (\sinh \alpha)^{-1} \coth \alpha (Z_\alpha \otimes 1 \otimes Z_{-\alpha})$$

and from this the assertion of the lemma is obvious.

Corollary 1. *Define the mapping γ' as in Section 4. Then*

$$\delta'(\omega) = \delta'(\omega_{\mathfrak{m} \cap \mathfrak{k}}) + \gamma'(\omega) + 2 \sum_{\alpha \in P_+} e^{-2\alpha} (1 - e^{-2\alpha})^{-1} H_{\bar\alpha}$$

$$-8 \sum_{\alpha \in P_+} e^{-2\alpha} (1 - e^{-2\alpha})^{-2} \{ (1 \otimes 1 \otimes Z_\alpha Z_{-\alpha}) + (Z_\alpha Z_{-\alpha} \otimes 1 \otimes 1) \}$$

$$+8 \sum_{\alpha \in P_+} e^{-\alpha} (1 + e^{-2\alpha})(1 - e^{-2\alpha})^{-2} (Z_\alpha \otimes 1 \otimes Z_{-\alpha}).$$

We know (see [8, p. 269]) that $\gamma'(\omega) = \sum_{1 \leqslant i, j \leqslant l} g^{ij} H_i H_j + 2H_\rho$. Since

$$\coth \alpha = 1 + 2e^{-2\alpha}(1 - e^{-2\alpha})^{-1}$$

$$(\sinh \alpha)^{-2} = 4e^{-2\alpha}(1 - e^{-2\alpha})^{-2}$$

$$(\sinh \alpha)^{-1} \coth \alpha = 2e^{-\alpha}(1 + e^{-2\alpha})(1 - e^{-2\alpha})^{-2},$$

we get the corollary immediately from Lemma 22.

Corollary 2. *Let M be the centralizer of A in K and Σ the set of all linear functions on $\mathfrak{h}_\mathfrak{p}$ of the form $\bar\alpha$ ($\alpha \in P_+$). For a given $\beta \in \Sigma$, let P_β denote the set of all $\alpha \in P_+$ such that $\bar\alpha = \beta$. Then if $m \in M$, we have $(\omega_{\mathfrak{m} \cap \mathfrak{k}})^m = \omega_{\mathfrak{m} \cap \mathfrak{k}}$ and*

$$\sum_{\alpha \in P_\beta} Z_\alpha^m \otimes 1 \otimes Z_{-\alpha}^m = \sum_{\alpha \in P_\beta} (Z_\alpha \otimes 1 \otimes Z_{-\alpha})$$

for any $\beta \in \Sigma$.

Put $F_\beta = e^{-\beta}(1 + e^{-2\beta})(1 - e^{-2\beta})^{-2}$ $(\beta \in \Sigma)$. Then F_β is an analytic function on A'. We claim that the functions F_β $(\beta \in \Sigma)$ are linearly independent. For otherwise suppose $\Sigma_\beta c_\beta F_\beta = 0$ where $c_\beta \in \mathbf{C}$ and not all c_β are zero. Let β_0 be the lowest element in Σ such that $c_{\beta_0} \neq 0$. Define $\alpha_1, \ldots, \alpha_l$, $\mathfrak{h}_\mathfrak{p}^+$ and $\mathfrak{h}_{\mathfrak{p}_0}^+$ as in §3. The expansion

$$F_\beta = \sum_{k \geqslant 0} (2k + 1) e^{-(2k+1)\beta}$$

is valid on $A^+ = \exp \mathfrak{h}_{\mathfrak{p}_0}^+$. Moreover the above series can be regarded as a power series in $e^{-\alpha_1}, \ldots, e^{-\alpha_l}$ which converges on $\mathfrak{h}_\mathfrak{p}^+$. By equating the coefficient of $e^{-\beta_0}$ to zero in the corresponding series expansion of $\Sigma_\beta c_\beta F_\beta$, we get $c_{\beta_0} = 0$ contrary to our hypothesis.

Fix $H \in \mathfrak{h}_{\mathfrak{p}_0}^+$. Then obviously $\lim_{t \to +\infty} F_\beta(\exp tH) = 0$. Hence the functions 1 and F_β $(\beta \in \Sigma)$ are also linearly independent. On the other hand $\omega \in \mathfrak{Z}$ and so $\omega^m = \omega$. Therefore it follows from Corollary 2 of Lemma 10 that $\delta_h'(\omega) = (\delta_h'(\omega))^m$ for any $h \in A'$. In view of the linear independence of 1 and F_β $(\beta \in \Sigma)$, the required result now follows immediately from Corollary 1 above.

§8. Some Heuristic Considerations

Define $\alpha_1, \ldots, \alpha_l$, $\mathfrak{h}_{\mathfrak{p}_0}^+$ and L as in §3 and let $\sigma = (\sigma_1, \sigma_2)$ be a double representation of K on a finite-dimensional complex vector space V. Let V_0 be the subspace of all elements $v \in V$ such that $\sigma_1(m)v = v\sigma_2(m)$ for $m \in M$. (As before, M is the centralizer of A in K.) Let U be an open set in $A^+ = \exp \mathfrak{h}_{\mathfrak{p}_0}^+$. Then $G_1 = KUK$ is an open set in G. Let $S_1 = S(\sigma, G_1)$ denote the set of all C^∞ functions ϕ from G_1 to V such that $\phi(k_1 x k_2) = \sigma_1(k_1)\phi(x)\sigma_2(k_2)$ $(k_1, k_2 \in K, x \in G_1)$. Let $\phi \in S_1$. Then if $h \in U$ and $m \in M$, $\phi(h) = \phi(h^m) = \sigma_1(m)\phi(h)\sigma_2(m^{-1})$. This shows that $\phi(h) \in V_0$. Therefore if $\bar{\phi}$ denotes the restriction of ϕ on U, $\bar{\phi}$ is a C^∞ function from U to V_0. It is obvious that $\bar{\phi} = 0$ implies that $\phi = 0$ and in fact it is not difficult to show that $\phi \to \bar{\phi}$ is a bijective linear mapping of S_1 on the space of all C^∞ functions from U to V_0. We may therefore identify S_1 with its image under this mapping. Let Λ be a linear function on $\mathfrak{h}_\mathfrak{p}$. We regard e^Λ as a function on A in the usual way. Now fix Λ. Then for each $\lambda \in L$, we propose to define an element[11] $\beta_\lambda(\Lambda) \in \mathcal{E}(V_0)$ in such a way that the following conditions hold.

(1) $\beta_0(\Lambda) = 1$ and the series $\Sigma_{\lambda \in L} e^{\Lambda - \lambda}\beta_\lambda(\Lambda)$ converges on U in a suitable sense so that its sum Φ_Λ is an analytic function from U to $\mathcal{E}(V_0)$.

Fix an element $v \in V_0$. Then $\Phi_\Lambda v$ is a function in S_1. Our second condition is as follows.[18, 19]

(2) $\omega(\Phi_\Lambda v) = \Phi_\Lambda\{\langle \Lambda, \Lambda + 2\rho \rangle v + v\sigma_2(\omega_{\mathfrak{m} \cap \mathfrak{k}})\}$ for every $v \in V_0$.

For any $k \in K$ (or in $\mathfrak{K}$) let $\sigma_2'(k)$ denote the element of $\mathcal{E}(V)$ defined by $\sigma_2'(k)v = v\sigma_2(k)$ $(v \in V)$. It is clear from the argument of Lemma 12, that $\omega(\Phi_\Lambda v)$

---

[18] $\langle \Lambda, \Lambda' \rangle = \Lambda'(H_\Lambda)$ for two linear functions Λ and Λ' on $\mathfrak{h}_\mathfrak{p}$.

[19] It follows from Corollary 2 of Lemma 22 that if $v \in V_0$ then $v\sigma_2(\omega_{\mathfrak{m} \cap \mathfrak{k}})$ is also in V_0.

$= \delta'(\omega)(\Phi_\Lambda v)$ on U ($v \in V_0$). Hence our second condition above may be expressed as follows.

(2)′ $\delta'(\omega)(\Phi_\Lambda v) = \Phi_\Lambda\{\langle \Lambda, \Lambda + 2\rho \rangle + \sigma_2'(\omega_{\mathfrak{m} \cap \mathfrak{k}})\}v$ on U for every $v \in V_0$.

Now if we use term by term differentiation for our series and make use of the expansions

$$e^{-2\alpha}\left(1 - e^{-2\alpha}\right)^{-1} = \sum_{k \geqslant 1} e^{-2k\alpha}$$

$$e^{-2\alpha}\left(1 - e^{-2\alpha}\right)^{-2} = \sum_{k \geqslant 1} ke^{-2k\alpha}$$

$$e^{-\alpha}\left(1 + e^{-2\alpha}\right)\left(1 - e^{-2\alpha}\right)^{-2} = \sum_{k \geqslant 0} (2k+1)e^{-(2k+1)\alpha},$$

which are valid on A^+ for $\alpha \in P_+$, we get the following recurrence relation for $\beta_\lambda(\Lambda)$ ($\lambda \in L$) from Corollary 1 of Lemma 22:

$$\langle \lambda, 2\Lambda - 2\rho - \lambda \rangle \beta_\lambda(\Lambda) + \left[\beta_\lambda(\Lambda), \sigma_2'(\omega_{\mathfrak{m} \cap \mathfrak{k}})\right]$$

$$= 2 \sum_{\alpha \in P_+} \sum_{k \geqslant 1} \langle \Lambda - \lambda + 2k\bar{\alpha}, \bar{\alpha} \rangle \beta_{\lambda - 2k\bar{\alpha}}(\Lambda)$$

$$-8 \sum_{\alpha \in P_+} \sum_{k \geqslant 1} k\left(\sigma_1(Z_\alpha Z_{-\alpha}) + \sigma_2'(Z_\alpha Z_{-\alpha})\right)\beta_{\lambda - 2k\bar{\alpha}}(\Lambda)$$

$$+8 \sum_{\alpha \in P_+} \sum_{k \geqslant 0} (2k+1)\sigma_1(Z_\alpha)\sigma_2'(Z_{-\alpha})\beta_{\lambda - (2k+1)\bar{\alpha}}(\Lambda).$$

Here it is understood that $\beta_0(\Lambda) = 1$ and $\beta_\mu(\Lambda) = 0$ for any linear function μ on $\mathfrak{h}_\mathfrak{p}$ which does not lie in L, so that the sums on the right hand side are all finite. Now assume for a moment that the linear mapping

$$T \to \langle \lambda, 2\Lambda - 2\rho - \lambda \rangle T + \left[T, \sigma_2'(\omega_{\mathfrak{m} \cap \mathfrak{k}})\right] \qquad (T \in \mathscr{E}(V_0))$$

of $\mathscr{E}(V_0)$ into itself is nonsingular for every $\lambda \neq 0$ in L. Then the above relation determines $\beta_\lambda(\Lambda)$ completely by recurrence.

In the next section we shall make the above considerations precise.

§9. Some Consequences of Lemma 22[20]

Let U be an open set in $\mathfrak{h}_\mathfrak{p}$ whose closure is compact. For any real number $\nu \geqslant 0$, let $\mathfrak{h}_\mathfrak{p}(\nu)$ denote the set of all points $H \in \mathfrak{h}_\mathfrak{p}$ such that[14] $\operatorname{Re}\alpha_i(H) > \nu$ ($1 \leqslant i \leqslant l$). Moreover if $0 < \varepsilon < 1$, let $\mathfrak{h}_\mathfrak{p}(\varepsilon, \nu)$ denote as before the set of all $H \in \mathfrak{h}_\mathfrak{p}(\nu)$ such that $\operatorname{Re}\alpha_i(H) > \varepsilon\max_{1 \leqslant j \leqslant l}|\alpha_j(H)|$.

Let V be a complex Banach space of finite dimension. We denote by $|v|$ the norm of any element $v \in V$. For any $T \in \mathscr{E}(V)$, put $|T| = \sup_{|v| \leqslant 1}|Tv|$ ($v \in V$). Then with this norm $\mathscr{E}(V)$ also becomes a Banach space. Define $m(\lambda)$ ($\lambda \in L$) as in §3.

---

[20] The reader should compare the methods and results of this section with those of [8, §8].

Lemma 23. *Suppose for each $\lambda \in L$, we are given a holomorphic function β_λ from U to V. Moreover suppose there exists a positive number c such that $|\beta_\lambda(H')| \leq c^{m(\lambda)+1}$ for all $H' \in U$ and $\lambda \in L$. Then we can choose $\nu \geq 0$ with the following property. Let U' be an open subset of U whose closure in U is compact, a a complex number and ε a real number $(0 < \varepsilon < 1)$. Then the series* [21]*

$$\sum_{\lambda \in L} \beta_\lambda(H') e^{a\langle H', H \rangle - \lambda(H)}$$

converges decently for $(H, H') \in U' \times \mathfrak{h}_\mathfrak{p}(\varepsilon, \nu)$.

By Lemma 1 of [11], it would be sufficient to choose $\nu > 0$ in such a way that the following condition holds. For any real linear function (see §3) μ on $\mathfrak{h}_\mathfrak{p}$, a complex number a and any ε $(0 < \varepsilon < 1)$, the series

$$\sum_{\lambda \in L} |\beta_\lambda(H') e^{a\langle H', H \rangle - \lambda(H) + \mu(H)}|$$

should converge uniformly for $(H', H) \in U \times \mathfrak{h}_\mathfrak{p}(\varepsilon, \nu/2)$. Fix a, μ and ε. Since U is bounded, we can choose $b > 0$ such that

$$|a\langle H', H \rangle + \mu(H)| \leq b \sum_{1 \leq i \leq l} |\alpha_i(H)| \leq b l \varepsilon^{-1} \gamma(H)$$

for $H' \in U$ and $H \in \mathfrak{h}_\mathfrak{p}(\varepsilon, 0)$. Here $\gamma(H) = \min_{1 \leq i \leq l} \mathrm{Re}\, \alpha_i(H)$. So it would be enough to choose ν in such a way that the series

$$\sum_{\lambda \in L} c^{m(\lambda)+1} |e^{k\gamma(H) - \lambda(H)}|$$

converges uniformly on $\mathfrak{h}_\mathfrak{p}(\nu/2)$ for any number $k \geq 0$. But if $m(\lambda) \geq k$, $|e^{k\gamma(H) - \lambda(H)}| \leq e^{(k - m(\lambda))\nu/2}$ for $H \in \mathfrak{h}_\mathfrak{p}(\nu/2)$. Hence our condition is fulfilled if $c < e^{\nu/2}$. The lemma is therefore proved.

Let σ be a double representation of K on a finite-dimensional Banach space V. As before let V_0 be the subspace of those elements $v \in V$ for which $\sigma_1(m)v = v\sigma_2(m)$ $(m \in M)$. Let $Q = \mathbb{C}(S(\mathfrak{h}_\mathfrak{p}))$ denote the quotient field of $S(\mathfrak{h}_\mathfrak{p})$ and put $\mathcal{E}_Q(V_0) = Q \otimes \mathcal{E}(V_0)$. For any $k \in K$ (or in $\mathfrak{K}$) we denote by $\sigma_2'(k)$ the linear transformation $v \to v\sigma_2(k)$ $(v \in V)$ in V. Put $\omega_0 = \omega_{\mathfrak{m} \cap \mathfrak{t}}$ for convenience. It follows from Corollary 2 of Lemma 22 that V_0 is stable under $\sigma_2'(\omega_0)$. Let E be the restriction of $\sigma_2'(\omega_0)$ on V_0. Put [5] $DT = [E, T]$ for any $T \in \mathcal{E}(V_0)$. Then D is a linear mapping of $\mathcal{E}(V_0)$ into itself. We extend it by linearity over Q to an endomorphism of $\mathcal{E}_Q^0 = \mathcal{E}_Q(V_0)$. Now for any $\lambda \in L$, we define an element $`\Gamma_\lambda \in \mathcal{E}_Q^0$ by induction on $m(\lambda)$ as follows. $`\Gamma_0 = 1$ and if $\lambda \neq 0$, $`\Gamma_\lambda$ is given by the relation

$$\{2\lambda - \langle \lambda, \lambda - 2\rho \rangle\} `\Gamma_\lambda - D`\Gamma_\lambda = 2 \sum_{\alpha \in P_+} \sum_{k \geq 1} \{\bar{\alpha} - \langle \bar{\alpha}, \lambda - 2k\bar{\alpha} \rangle\} `\Gamma_{\lambda - 2k\bar{\alpha}}$$

$$+ 8 \sum_{\alpha \in P_+} \sum_{k \geq 1} (2k - 1)\sigma_1(Z_\alpha)\sigma_2'(Z_{-\alpha}) `\Gamma_{\lambda - (2k-1)\bar{\alpha}}$$

$$- 8 \sum_{\alpha \in P_+} \sum_{k \geq 1} k\{\sigma_1(Z_\alpha Z_{-\alpha}) + \sigma_2'(Z_\alpha Z_{-\alpha})\} `\Gamma_{\lambda - 2k\bar{\alpha}}.$$

---

[21] As usual $\langle H', H \rangle = \mathrm{tr}(\mathrm{ad}H' \mathrm{ad}H)$ for $H, H' \in \mathfrak{h}_\mathfrak{p}$.

Here k runs over all positive integers and $`\Gamma_\mu$ is defined to be zero in case $\mu \notin L$, so that all the sums are actually finite. It follows from Corollary 2 of Lemma 22 that V_0 is stable under $\sum_{\alpha \in P_\beta} \sigma_1(Z_\alpha)\sigma_2'(Z_{-\alpha})$, $\sum_{\alpha \in P_\beta} \sigma_1(Z_\alpha Z_{-\alpha})$ and $\sum_{\alpha \in P_\beta} \sigma_2'(Z_\alpha Z_{-\alpha})$ for any $\beta \in \Sigma$. Hence in view of our induction hypothesis, the right side of the above equation is in $\mathscr{E}_Q^0$. On the other hand let T_λ denote the endomorphism of the vector space $\mathscr{E}_Q^0$ over Q defined by $T_\lambda F = \{2\lambda - \langle \lambda, \lambda - 2\rho \rangle\}F - DF$ $(F \in \mathscr{E}_Q^0)$. We claim T_λ is nonsingular. For suppose $T_\lambda F = 0$ $(F \in \mathscr{E}_Q^0)$. We have to prove that $F = 0$. By multiplying by a nonzero element in $S(\mathfrak{h}_\mathfrak{p})$, we may obviously assume that F is a polynomial function on $\mathfrak{h}_\mathfrak{p}$ with values in $\mathscr{E}(V_0)$. Then if H is any point in $\mathfrak{h}_\mathfrak{p}$,

$$\{2\lambda(H) - \langle \lambda, \lambda - 2\rho \rangle\}F(H) - DF(H) = 0$$

where $F(H)$ is the value of F at H. Let $d_1, \ldots, d_r$ be all the distinct eigenvalues of the endomorphism D of $\mathscr{E}(V_0)$. Put

$$f = \prod_{1 \leqslant i \leqslant r} \{2\lambda - \langle \lambda, \lambda - 2\rho \rangle - d_i\}.$$

Since $\lambda \neq 0$, f is a nonzero element in $S(\mathfrak{h}_\mathfrak{p})$ and it is obvious that if H is a point in $\mathfrak{h}_\mathfrak{p}$ where $f(H) \neq 0$, then the linear transformation $2\lambda(H) - \langle \lambda, \lambda - 2\rho \rangle - D$ of $\mathscr{E}(V_0)$ is nonsingular. Hence $F(H) = 0$ if $f(H) \neq 0$. Since F is a polynomial function and $f \neq 0$, this implies that $F = 0$. Therefore T_λ is nonsingular and so the above relation defines uniquely an element $`\Gamma_\lambda \in \mathscr{E}_Q^0$.

Let L' be the set of all elements $\lambda \neq 0$ in L. Define $d_1, \ldots, d_r$ as above and for any $\lambda \in L'$ and i $(1 \leqslant i \leqslant r)$, let $\tau_{\lambda i}'$ denote the hyperplane consisting of all points H in $\mathfrak{h}_\mathfrak{p}$ such that $2\lambda(H) = \langle \lambda, \lambda - 2\rho \rangle + d_i$. Clearly any compact subset of $\mathfrak{h}_\mathfrak{p}$ meets $\cup_{1 \leqslant i \leqslant r} \tau_{\lambda i}'$ for only a finite number of $\lambda \in L'$. Let $\mathfrak{h}_\mathfrak{p}''$ denote the complement of $\cup_{\lambda \in L'} \cup_i \tau_{\lambda i}'$ in $\mathfrak{h}_\mathfrak{p}$. Then $\mathfrak{h}_\mathfrak{p}''$ is an open, connected and dense subset of $\mathfrak{h}_\mathfrak{p}$ and the rational function $`\Gamma_\lambda$ (from $\mathfrak{h}_\mathfrak{p}$ to $\mathscr{E}(V_0)$) is well defined on $\mathfrak{h}_\mathfrak{p}''$ for every $\lambda \in L$. We denote by $\beta(H)$ the value of a rational function β at a point $H \in \mathfrak{h}_\mathfrak{p}$ where it is defined.

Put $f_0 = 1$ and

$$f_\lambda = \prod_{1 \leqslant i \leqslant r} \{2\lambda - \langle \lambda, \lambda - 2\rho \rangle - d_i\}^{m_i} \qquad (\lambda \in L')$$

where m_i is the multiplicity of the eigenvalue d_i of D. Put $F_\lambda = f_\lambda \prod_{\mu \ll \lambda} f_\mu$ where μ runs over all elements in L such that $\mu \ll \lambda$. It follows from the definition of $`\Gamma_\lambda$ by an easy induction on $m(\lambda))$ that $F_\lambda \, `\Gamma_\lambda$ is a polynomial function for every $\lambda \in L$.

Lemma 24. *Let U be a compact set in $\mathfrak{h}_\mathfrak{p}$ and let L_U denote the set of those $\lambda \in L'$ for which U meets $\cup_{1 \leqslant i \leqslant r} \tau_{\lambda i}'$. Put $F_U = \prod_{\lambda \in L_U} F_\lambda$. $(F_U = 1$ if L_U is empty.) Then for every $\lambda \in L$, the rational function $\beta_\lambda = F_U \, `\Gamma_\lambda$ is defined everywhere on U. Also there exists a positive number c such that*

$$|\beta_\lambda(H)| \leqslant c^{m(\lambda)+1}$$

for all $\lambda \in L$ and $H \in U$.

Fix an element λ in L. Then we know that $F_\lambda \, `\Gamma_\lambda$ is a polynomial function and therefore is defined everywhere on $\mathfrak{h}_\mathfrak{p}$. Let $L_U(\lambda)$ denote the set of all $\mu \in L_U$ such that $\lambda - \mu \in L$. Then if $F'_\lambda = \prod_{\mu \in L_U(\lambda)} f_\mu$, it is clear from the definition of F_λ that $F''_\lambda = F_\lambda / F'_\lambda$ is a polynomial function, which is nowhere zero on U. Hence $F'^\wedge_\lambda F_\lambda = (F_\lambda `\Gamma_\lambda)/F''_\lambda$ is defined everywhere on $U$. But obviously $F'_\lambda$ divides $F_U$ in $S(\mathfrak{h}_\mathfrak{p})$ since $L_U(\lambda) \subset L_U$. Hence $\beta_\lambda = F_U `\Gamma_\lambda$ is defined everywhere on U.

Put $m_0 = \sup_{\lambda \in L_U} m(\lambda)$. (In case L_U is empty we put $m_0 = 0$.) Since U is compact, we can obviously choose a number $c_0 \geqslant 1$ such that $|\beta_\lambda(H)| \leqslant c_0^{m(\lambda)+1}$ for all $H \in U$ if $m(\lambda) \leqslant m_0$ $(\lambda \in L)$. Let $L(m_0)$ denote the set of all $\lambda \in L$ for which $m(\lambda) > m_0$. For any $\lambda \in L(m_0)$, the mapping

$$T_{\lambda, H} : u \to \{2\lambda(H) - \langle \lambda, \lambda - 2\rho \rangle\} u - Du \qquad (u \in \mathcal{E}(V_0))$$

is nonsingular for every $H \in U$ since $\lambda \notin L_U$. Moreover

$$|T_{\lambda, H} u| \geqslant \langle \lambda, \lambda \rangle |u| - |Du| - 2|\lambda(H) + \langle \lambda, \rho \rangle| |u|$$

$$\geqslant \{\langle \lambda, \lambda \rangle - 2|\lambda(H) + \langle \lambda, \rho \rangle| - |D|\} |u|.$$

Also we can obviously choose positive numbers a_1, a_2 such that $\langle \lambda, \lambda \rangle \geqslant a_1 m(\lambda)^2$ and $2|\lambda(H) + \langle \lambda, \rho \rangle| \leqslant a_2 m(\lambda)$ for all $\lambda \in L$ and $H \in U$. Hence we can select an integer $m_1 \geqslant m_0$ such that

$$|T_{\lambda, H} u| \geqslant \left(a_1 m(\lambda)^2 - a_2 m(\lambda) - |D| \right) |u| \geqslant \tfrac{1}{2} a_1 m(\lambda)^2 |u|$$

for $H \in U$, $u \in \mathcal{E}(V_0)$ and $\lambda \in L$ if $m(\lambda) > m_1$. On the other hand there are only a finite number of elements $\lambda \in L(m_0)$ with $m(\lambda) \leqslant m_1$.

Let $\Delta_\lambda(H) = \det T_{\lambda, H}$. Then Δ_λ is a polynomial function on $\mathfrak{h}_\mathfrak{p}$ which is nowhere zero in U if $\lambda \in L(m_0)$. Let $b_\lambda = \inf_{H \in U} |\Delta_\lambda(H)|$. Since U is compact, $b_\lambda > 0$ if $\lambda \in L(m_0)$. On the other hand it is clear that there exists a polynomial function S_λ from $\mathfrak{h}_\mathfrak{p}$ to $\mathcal{E}^2(V_0) = \mathcal{E}(\mathcal{E}(V_0))$ such that[11]

$$T_{\lambda, H} S_\lambda(H) = S_\lambda(H) T_{\lambda, H} = \Delta_\lambda(H)$$

for $H \in \mathfrak{h}_\mathfrak{p}$. Then if $\lambda \in L(m_0)$ and $H \in U$,

$$|u| = |\Delta_\lambda(H)|^{-1} |S_\lambda(H) T_{\lambda, H} u| \leqslant b_\lambda^{-1} s_\lambda |T_{\lambda, H} u| \qquad (u \in \mathcal{E}(V_0))$$

where $s_\lambda = \sup_{H \in U} |S_\lambda(H)| < \infty$. This shows that for each $\lambda \in L(m_0)$, we can choose a positive number t_λ such that $|T_{\lambda, H} u| \geqslant t_\lambda m(\lambda)^2 |u|$ for $H \in U$ and $u \in \mathcal{E}(V_0)$. Let c_1 denote the minimum among the numbers $\tfrac{1}{2} a_1$ and t_λ $(\lambda \in L(m_0), m(\lambda) \leqslant m_1)$. Then $c_1 > 0$ and it is obvious that

$$|T_{\lambda, H}(u)| \geqslant c_1 m(\lambda)^2 |u|$$

for all $H \in U$, $u \in \mathcal{E}(V_0)$ and $\lambda \in L(m_0)$. Also we can choose a positive number c_2 such that

$$|\alpha(H) - \langle \lambda, \bar{\alpha} \rangle| \leqslant c_2(m(\lambda) + 1)$$

for $\lambda \in L$, $\alpha \in P_+$ and $H \in U$. Then if $\lambda \in L(m_0)$ and $\lambda' \ll \lambda$ $(\lambda' \in L)$,

$$|\alpha(H) - \langle \lambda', \bar{\alpha} \rangle| |T_{\lambda, H}^{-1} u| \leqslant c_2(m(\lambda') + 1) c_1^{-1} m(\lambda)^{-2} |u|$$

$$\leqslant c_2 c_1^{-1} m(\lambda)^{-1} |u|$$

for $H \in U$ and $u \in \mathfrak{G}(V_0)$. Put $c_3 = c_2 c_1^{-1}$ and select c_4 such that

$$|\sigma_1(Z_\alpha)\sigma_2'(Z_{-\alpha})| + |\sigma_1(Z_\alpha Z_{-\alpha}) + \sigma_2'(Z_\alpha Z_{-\alpha})| \leqslant c_4/8$$

for every $\alpha \in P_+$. (Here the norms are to be taken in $\mathfrak{G}(V)$.) Then if $\lambda \in L(m_0)$, it follows from our recursion formula that

$$\begin{aligned}
|\beta_\lambda(H)| &\leqslant 2c_3 m(\lambda)^{-1} \sum_{\alpha \in P_+} \sum_{k \geqslant 1} |\beta_{\lambda - 2k\bar\alpha}(H)| \\
&\quad + c_4 c_1^{-1} m(\lambda)^{-2} \sum_{\alpha \in P_+} \sum_{k \geqslant 1} |\beta_{\lambda - (2k-1)\bar\alpha}(H)| \\
&\quad + c_4 c_1^{-1} m(\lambda)^{-2} \sum_{\alpha \in P_+} \sum_{k \geqslant 1} k|\beta_{\lambda - 2k\bar\alpha}(H)| \\
&\leqslant \left\{ 2c_3 m(\lambda)^{-1} + 2c_4 c_1^{-1} m(\lambda)^{-1} \right\} \sum_{\alpha \in P_+} \sum_{k \geqslant 1} |\beta_{\lambda - k\bar\alpha}(H)|
\end{aligned}$$

since $k \leqslant m(\lambda)$ if $\lambda - 2k\bar\alpha \in L$. Let p be the number of roots in P_+. Put

$$c = \max\{c_0, 2p(c_3 + c_4 c_1^{-1})\}.$$

We shall prove by induction on $m(\lambda)$ that $|\beta_\lambda(H)| \leqslant c^{m(\lambda)+1}$ for $\lambda \in L$ and $H \in U$. If $m(\lambda) \leqslant m_0$, this is obvious since $c \geqslant c_0$. So now suppose that $m(\lambda) > m_0$. Then it follows from the above result and the induction hypothesis that

$$|\beta_\lambda(H)| \leqslant (2c_3 + 2c_4 c_1^{-1}) m(\lambda)^{-1} pm(\lambda) c^{m(\lambda)} \leqslant c^{m(\lambda)+1}$$

since $c \geqslant c_0 \geqslant 1$. This proves Lemma 24.

Corollary 1. *Let U' be the interior of U. Then we can choose $v \geqslant 0$ such that for any complex number a and any ε $(0 < \varepsilon < 1)$, the series*

$$\sum_{\lambda \in L} \beta_\lambda(H') e^{a\langle H', H \rangle - \lambda(H)}$$

converges decently for $(H', H) \in U' \times \mathfrak{h}_\mathfrak{p}(\varepsilon, v)$.

Obviously we can choose a compact neighborhood U_1 of U in $\mathfrak{h}_\mathfrak{p}$ such that $L_{U_1} = L_U$. By above lemma (applied to U_1), we can choose $c \geqslant 0$ such that $|\beta_\lambda(H')| \leqslant c^{m(\lambda)+1}$ for $\lambda \in L$, $H' \in U_1$. Our assertion now follows from Lemma 23.

Corollary 2. *Put*

$$\phi'(H' : H) = e^{\langle H', H \rangle} \sum_{\lambda \in L} \beta_\lambda(H') e^{-\lambda(H)} \qquad \left(H' \in U', \; H \in \mathfrak{h}_\mathfrak{p}(v) \right).$$

Then ϕ' is a holomorphic function from $U' \times \mathfrak{h}_\mathfrak{p}(v)$ to $\mathfrak{G}(V_0)$.

This is an immediate consequence of Corollary 1 above.

Let $\mathfrak{F}$ be the space of all linear mappings of V_0 in V. Then $\mathfrak{F}$ is a left-module over the algebra $\mathfrak{G}(V)$ and a right-module over $\mathfrak{G}(V_0)$. Define a double representation $S = (S_1, S_2)$ of K on $\mathfrak{F}$ as follows: $S_1(k)F = \sigma_1(k)F$, $FS_2(k) = \sigma_2'(k)F$ for $F \in \mathfrak{F}$ and

$k \in K$. Also put $|F| = \sup_{|v| \leqslant 1} |Fv|$ $(v \in V_0)$. Under this norm $\mathcal{F}$ becomes a Banach space. Note that $\mathcal{E}(V_0) \subset \mathcal{F}$.

Let $\mathfrak{B}$ denote the set of all functions b from L to $\mathcal{F}$ such that $\sup_{\lambda \in L} (m(\lambda) + 1)^{-1} \log|b(\lambda)| < +\infty$. For any $b \in \mathfrak{B}$ and $E_0 \in \mathcal{E}(V_0)$, we denote by bE_0 the function $\lambda \to b(\lambda)E_0$ $(\lambda \in L)$. It is obvious that bE_0 is also in $\mathfrak{B}$. In this way $\mathfrak{B}$ becomes a right-module over $\mathcal{E}(V_0)$. Similarly it is a left module over $\mathcal{E}(V)$.

It follows from Lemma 23 that for any $b \in \mathfrak{B}$, we can choose a number $v(b) \geqslant 0$ such that (1) the series $\sum_{\lambda \in L} |b(\lambda)e^{-\lambda(H)}|$ converges uniformly for $H \in \mathfrak{h}_\mathfrak{p}(v(b))$ and (2) the series $\sum_{\lambda \in L} b(\lambda)e^{-\lambda(H)}$ converges decently on $\mathfrak{h}_\mathfrak{p}(\frac{1}{2}, v(b))$. For any $v \geqslant 0$, let $A(v)$ denote the set of all points $h \in A$ such that $\log h \in \mathfrak{h}_\mathfrak{p}(v)$.

Lemma 25. *Fix $b \in \mathfrak{B}$, $g \in \mathfrak{G}$ and let Λ be a linear function on $\mathfrak{h}_\mathfrak{p}$. Put*

$$\Phi(h) = e^{\Lambda(H)} \sum_{\lambda \in L} b(\lambda)e^{-\lambda(H)} \qquad \left(h \in A(v(b)) \right)$$

where $H = \log h$. Then we can choose an element $B \in \mathfrak{B}$ with the following property. Let $v = \max(v(b), v(B))$. Then

$$\Phi(h; \delta'(g)) = e^{\Lambda(\log h)} \sum_{\lambda \in L} B(\lambda)e^{-\lambda(\log h)}$$

for $h \in A(v)$.

We note that since $A(v(b))$ is an open subset of A', since Φ is an analytic function from $A(v(b))$ to $\mathcal{F}$ and $\mathcal{F}$ is a double module over K, the function $\delta'(g)\Phi$ is well defined.

Fix $v_0 > v(b)$ and define $\mathfrak{R}$ as in §2. Then $v_0 > 0$ and we have seen in §4 that every function $f \in \mathfrak{R}$ has an expansion of the form

$$f(\exp H) = \sum_{\lambda \in L} a_\lambda e^{-\lambda(H)} \qquad \left(H \in \mathfrak{h}_{\mathfrak{p}_0} \cap \mathfrak{h}_\mathfrak{p}(v_0) \right)$$

where $a_\lambda \in \mathbf{C}$ and the series $\sum_{\lambda \in L} |a_\lambda e^{-\lambda(H)}|$ converges uniformly for $H \in \mathfrak{h}_\mathfrak{p}(v_0)$. Put

$$\phi(H) = e^{\Lambda(H)} \sum_{\lambda \in L} b(\lambda)e^{-\lambda(H)} \qquad \left(H \in \mathfrak{h}_\mathfrak{p}(v_0) \right).$$

For any $u \in S(\mathfrak{h}_\mathfrak{p})$, define the differential operator $\partial(u)$ on $\mathfrak{h}_\mathfrak{p}$ as in [6, §2]. Then it is clear that

$$\phi(H; \partial(u)) = e^{\Lambda(H)} \sum_{\lambda \in L} b(\lambda)u(\Lambda - \lambda)e^{-\lambda(H)} \qquad \left(H \in \mathfrak{h}_\mathfrak{p}(v_0) \right)$$

where $u(\Lambda - \lambda)$ is the value of the polynomial at the point $\Lambda - \lambda$ (regarded as a point in $\mathfrak{h}_\mathfrak{p}$). Now put

$$b'(\lambda) = \sum_{\lambda_1 + \lambda_2 = \lambda} a_{\lambda_1} u(\Lambda - \lambda_2)b(\lambda_2) \qquad (\lambda \in L)$$

where the sum is over all pairs (λ_1, λ_2) of elements in L such that $\lambda_1 + \lambda_2 = \lambda$. Then it is obvious that

$$\sum_{\lambda \in L} |b'(\lambda) e^{-\lambda(H)}| \leqslant \sum_{\lambda \in L} |a_\lambda e^{-\lambda(H)}| \sum_{\lambda \in L} |b(\lambda) u(\Lambda - \lambda) e^{-\lambda(H)}| < \infty$$

for $H \in \mathfrak{h}_\mathfrak{p}(\nu_0)$. Fix a point $H_0 \in \mathfrak{h}_\mathfrak{p}(\nu_0)$. Then we can choose a positive number c such that (1) $|e^{\lambda(H_0)}| \leqslant c^{m(\lambda)}$ for all $\lambda \in L$ and (2) $\sum_{\lambda \in L} |b'(\lambda) e^{-\lambda(H_0)}| \leqslant c$. Then $|b'(\lambda)| \leqslant |e^{\lambda(H_0)}| c \leqslant c^{m(\lambda)+1}$ $(\lambda \in L)$ and therefore $b' \in \mathfrak{B}$. Now from Lemma 10,

$$\delta'(g) = \Delta_0 + \sum_{1 \leqslant i \leqslant r} f_i \Delta_i$$

where $f_i \in \mathfrak{R}_0$ $(1 \leqslant i \leqslant r)$ and Δ_j $(0 \leqslant j \leqslant r)$ are in $\mathfrak{D} \otimes \mathfrak{H}_\mathfrak{p} \otimes \mathfrak{R}$. Hence it follows from the above remarks that there exists an element $B \in \mathfrak{B}$ such that

$$\Phi(h; \delta'(g)) = e^{\Lambda(\log h)} \sum_{\lambda \in L} B(\lambda) e^{-\lambda(\log h)}$$

for $h \in A(\nu_1)$ if ν_1 is a sufficiently large positive number. However the left and right sides of this equation are analytic functions on $A(\nu(b))$ and $A(\nu(B))$ respectively. So it is obvious that they must coincide on $A(\nu)$.

Corollary. *The element B of the above lemma is uniquely determined. In particular if $g \in I_\mathfrak{g}$, $B(0) = \sigma_2'(\xi_\Lambda'(g))b(0)$.*

The uniqueness of B is an immediate consequence of the Corollary of Lemma 57 of [8]. If $g \in I_\mathfrak{g}$, we know from Corollary 1 of Lemma 10 that $\Delta_0 = \xi'(g)$. Hence the coefficient of $e^{\Lambda(\log h)}$ in $\Phi(h; \Delta_0)$ is clearly $\sigma_2'(\xi_\Lambda'(g))b(0)$. This proves the corollary.

The element B defined above depends on Λ, g and b. Hence we write $B(\Lambda : g : b)$ instead of B and $B(\Lambda : g : b : \lambda)$ instead of $B(\lambda)$ $(\lambda \in L)$.

Fix $g \in \mathfrak{G}$. Then it follows from Lemma 10 that we can choose $u_i \in \mathfrak{H}_\mathfrak{p}$, $f_i \in \mathfrak{R}$ and $E_i \in \mathfrak{E}(V)$ $(i \leqslant i \leqslant r)$ such that

$$\phi(h; \delta'(g)) = \sum_{1 \leqslant i \leqslant r} f_i(h) E_i \phi(h; u_i) \qquad (h \in A')$$

for any function ϕ which is defined and of class C^∞ on some open neighborhood in A' of the point h and whose values are in $\mathfrak{F}$. Moreover f_i has an expansion of the form

$$f_i(h) = \sum_{\lambda \in L} a_{i\lambda} e^{-\lambda(\log h)} \qquad (h \in A^+)$$

where $a_{i\lambda} \in \mathbf{C}$ and the series $\sum_{\lambda \in L} |a_{i\lambda} e^{-\lambda(H)}|$ converges uniformly for $H \in \mathfrak{h}_\mathfrak{p}(a)$ for any positive number a. Under our identification of $S(\mathfrak{h}_\mathfrak{p})$ with $\mathfrak{H}_\mathfrak{p}$, we can regard u_i as polynomial functions on $\mathfrak{h}_\mathfrak{p}$. Also we recall that $\mathfrak{h}_\mathfrak{p}$ has been identified with its dual and $u(\mu)$ denotes the value of a polynomial $u \in \mathfrak{H}_\mathfrak{p}$ at $\mu \in \mathfrak{h}_\mathfrak{p}$.

Lemma 26. $B(\Lambda : g : b : \lambda) = \sum_{\lambda_1 + \lambda_2 = \lambda} \sum_{1 \leqslant i \leqslant r} a_{i\lambda_1} u_i(\Lambda - \lambda_2) E_i b(\lambda_2)$ *where the sum is over all pairs (λ_1, λ_2) of elements in L such that $\lambda_1 + \lambda_2 = \lambda$.*

This follows immediately from the proof of Lemma 25.

Corollary. *Fix g, b and λ. Then $B(\Lambda:g:b:\lambda)$ is a polynomial function of Λ. Also if $g \in I_{\mathfrak{g}}$ and $b(\mu) \in \mathscr{E}(V_0)$ for every $\mu \in L$, then $B(\Lambda:g:b:\lambda) \in \mathscr{E}(V_0)$.*

The first part is obvious from Lemma 26. Now suppose q_j and k_j $(1 \leqslant j \leqslant s)$ are elements in $\mathfrak{Q}$ and $\mathfrak{R}$ respectively such that for any $m \in M$, $\Sigma_j q_j^m \otimes k_j^m = \Sigma_j q_j \otimes k_j$ in the tensor product $\mathfrak{Q} \otimes \mathfrak{R}$. Put $E = \Sigma_j \sigma_1(q_j)\sigma_2'(k_j) \in \mathscr{E}(V)$. Then if $F \in \mathscr{E}(V_0)$ and $v \in V_0$, we have

$$EFv_0 = \sum_j \sigma_1(q_j^m)(Fv_0)\sigma_2(k_j^m)$$

$$= \sigma_1(m)(EFv_0)\sigma_2(m^{-1}) \qquad (m \in M)$$

since $Fv_0 \in V_0$. This shows that $EFv_0 \in V_0$ and therefore $EF \in \mathscr{E}(V_0)$. This shows that $\mathscr{E}(V_0)$ is stable under the mapping $F \to EF$ $(F \in \mathscr{F})$ of $\mathscr{F}$.

Now if $g \in I_{\mathfrak{g}}$, it follows from Corollary 2 of Lemma 10 that we can assume that $\mathscr{E}(V_0)$ is stable under multiplication by E_i $(1 \leqslant i \leqslant r)$ on the left. The required result then follows immediately from Lemma 26.

Fix $E_0 \in \mathscr{E}(V_0)$. Then the linear mapping $F \to FE_0$ $(F \in \mathscr{F})$ commutes with the operations of K on $\mathscr{F}$. Hence it follows that $B(\Lambda:g:bE_0) = B(\Lambda:g:b)E_0$ for $b \in \mathfrak{B}$ and $g \in \mathfrak{G}$.

Let $\mathfrak{B}_0$ be the set of those elements $b \in \mathfrak{B}$ for which $b(\lambda) \in \mathscr{E}(V_0)$ for every $\lambda \in L$.

Lemma 27. *Let $q_1, q_2 \in I_{\mathfrak{g}}$ and $b \in \mathfrak{B}_0$. Then*

$$B(\Lambda:q_1 q_2:b) = B(\Lambda:q_1:b_2)$$

where $b_2 = B(\Lambda:q_2:b)$. Moreover $b_2 \in \mathfrak{B}_0$.

Define $\mathfrak{M}$ as in Lemma 19. It is easy to check that if $F \in \mathscr{E}(V_0)$ then $S_1(m)F = FS_2(m)$ for $m \in M$. Ω being an open set in A' and ϕ a C^∞ function from Ω to $\mathscr{E}(V_0)$, it follows without difficulty that $\mu\phi = 0$ for $\mu \in \mathfrak{M}$. Let

$$\Phi(h) = e^{\Lambda(\log h)} \sum_{\lambda \in L} b(\lambda) e^{-\lambda(\log h)} \qquad (h \in A(\nu(b))).$$

Then we conclude from Lemma 19 that

$$\delta'(q_1 q_2)\Phi = \delta'(q_1)\Phi_2$$

where $\Phi_2 = \delta'(q_2)\Phi$. But if $\nu = \max(\nu(b), \nu(b_2))$,

$$\Phi_2(h) = \Phi(h; \delta'(q_2)) = e^{\Lambda(\log h)} \sum_{\lambda \in L} b_2(\lambda) e^{-\lambda(\log h)} \qquad (h \in A(\nu))$$

from the definition of b_2. Therefore we conclude from Lemma 25 and its corollary that $B(\Lambda:q_1 q_2:b) = B(\Lambda:q_1:b_2)$. Also it follows from the corollary of Lemma 26 that $b_2 \in \mathfrak{B}_0$.

Lemma 28. *As before let ω be the Casimir operator of $\mathfrak{g}$. Then if $b \in \mathfrak{B}$ and $H' \in \mathfrak{h}_\mathfrak{p}$, we have*[22]

$$B(H': \omega: b: \lambda)$$

$$= \{\langle H', H' \rangle + 2\rho(H') - 2\lambda(H') + \langle \lambda, \lambda - 2\rho \rangle\} b(\lambda) + \sigma_2'(\omega_{\mathfrak{m} \cap \mathfrak{t}}) b(\lambda)$$

$$+ 2 \sum_{\alpha \in P_+} \sum_{k \geqslant 1} \{\alpha(H') - \langle \bar{\alpha}, \lambda - 2k\bar{\alpha} \rangle\} b(\lambda - 2k\bar{\alpha})$$

$$+ 8 \sum_{\alpha \in P_+} \sum_{k \geqslant 1} (2k - 1)\sigma_1(Z_\alpha)\sigma_2'(Z_{-\alpha}) b(\lambda - (2k - 1)\bar{\alpha})$$

$$- 8 \sum_{\alpha \in P_+} \sum_{k \geqslant 1} k\{\sigma_1(Z_\alpha Z_{-\alpha}) + \sigma_2'(Z_\alpha Z_{-\alpha})\} b(\lambda - 2k\bar{\alpha}) \qquad (\lambda \in L)$$

where $b(\mu)$ is defined to be zero if $\mu \notin L$.

This follows from Lemma 26 and Corollary 1 of Lemma 22 by a direct computation.

Let E be an element in $\mathfrak{E}(V)$ which maps V_0 into itself. Let E_0 denote the restriction of E on V_0. Then we define $FE = FE_0$ and $bE = bE_0$ for $F \in \mathfrak{F}$ and $b \in \mathfrak{B}$. We now return to the notation of Corollary 1 of Lemma 24. For $H' \in U'$ let $\beta(H')$ denote the function $\lambda \to \beta_\lambda(H')$ on L. Then $\beta(H') \in \mathfrak{B}_0$. Put $B_{H'}(q) = B(H': q: \beta(H'))$ for $H' \in U'$ and $q \in I_\mathfrak{g}$.

Lemma 29. $B_{H'}(\omega) = \{\langle H', H' \rangle + 2\rho(H')\}\beta(H') + \beta(H')\sigma_2'(\omega_{\mathfrak{m} \cap \mathfrak{t}})$ *for $H' \in U'$.*

From Corollary 2 of Lemma 22, $\sigma_2'(\omega_{\mathfrak{m} \cap \mathfrak{t}})$ maps V_0 into itself. Hence $\beta(H')\sigma_2'(\omega_{\mathfrak{m} \cap \mathfrak{t}})$ is well defined. Lemma 29 is an immediate consequence of Lemma 28 and the definitions of $`\Gamma_\lambda$ and β_λ.

Corollary 1. *Let $q \in I_\mathfrak{g}$ and $H' \in U'$. Then*

$$B_{H'}(q) = \beta(H')\sigma_2'(\xi_{H'}'(q)).$$

It is easy to verify that $(\xi'(q))^m = \xi'(q)$ for $m \in M$. Hence $\sigma_2'(\xi_{H'}'(q))$ maps V_0 into itself. Since $\omega q = q\omega$, it follows from Lemma 27 that

$$B_{H'}(\omega q) = B(H': q\omega: \beta(H')) = B(H': q: B_{H'}(\omega)).$$

On the other hand

$$B_{H'}(\omega q) = B(H': \omega: B_{H'}(q))$$

again by Lemma 27. Therefore

$$B(H': \omega: B_{H'}(q)) = B(H': q: B_{H'}(\omega)).$$

---

[22]As usual we identify $\mathfrak{h}_\mathfrak{p}$ with its dual.

But then we conclude from Lemmas 28 and 29 that

$$\{2\lambda(H') - \langle \lambda, \lambda - 2\rho \rangle\} B_{H'}(q:\lambda) - \sigma_2'(\omega_{m \cap t}) B_{H'}(q:\lambda) + B_{H'}(q:\lambda)\sigma_2'(\omega_{m \cap t})$$

$$= 2 \sum_{\alpha \in P_+} \sum_{k \geqslant 1} \{\alpha(H') - \langle \bar{\alpha}, \lambda - 2k\bar{\alpha} \rangle\} B_{H'}(q:\lambda - 2k\bar{\alpha})$$

$$+ 8 \sum_{\alpha \in P_+} \sum_{k \geqslant 1} (2k-1)\sigma_1(Z_\alpha)\sigma_2'(Z_{-\alpha}) B_{H'}(q:\lambda - (2k-1)\bar{\alpha})$$

$$- 8 \sum_{\alpha \in P_+} \sum_{k \geqslant 1} k\{\sigma_1(Z_\alpha Z_{-\alpha}) + \sigma_2'(Z_\alpha Z_{-\alpha})\} B_{H'}(q:\lambda - 2k\bar{\alpha}) \qquad (\lambda \in L).$$

On the other hand $B_{H'}(q:0) = \sigma_2'(\xi'_{H'}(q))\beta_0(H')$ from the corollary of Lemma 25. However '$\Gamma_0 = 1$ and therefore $\beta_0(H') \in \mathbf{C}$. This shows that $B_{H'}(q:0) = \beta_0(H')\sigma_2'(\xi'_{H'}(q))$. Now the above recurrence relation for $B_{H'}(q:\lambda)$ is exactly the same as that satisfies by $\beta_\lambda(H')$. Moreover $B_{H'}(q) \in \mathfrak{B}_0$ from Lemma 27. Therefore if $H' \in \mathfrak{h}''_\mathfrak{p}$, it follows immediately by induction on $m(\lambda)$ that $B_{H'}(q:\lambda) = \beta_\lambda(H')\sigma_2'(\xi'_{H'}(q))$. However $\mathfrak{h}''_\mathfrak{p} \cap U'$ is dense in U'. Hence in order to complete our proof, it is enough to verify that for fixed q and λ, $B_{H'}(q:\lambda) = B(H':q:\beta(H'):\lambda)$ is a continuous function of H' on U'. But this is obvious from Lemma 26.

Corollary 2. *Put*

$$\Phi'_{H'}(h) = e^{\langle H', H \rangle} \sum_{\lambda \in L} \beta_\lambda(H') e^{-\lambda(H)} \qquad (h \in A(\nu),\ H' \in U')$$

where $H = \log h$ and ν is a real number chosen in accordance with Lemmas 23 and 24. Then if $q \in I_\mathfrak{g}$, we have

$$\delta'(q)\Phi'_{H'} = \Phi'_{H'}\sigma_2'(\xi'_{H'}(q)).$$

Fix H' and q. Then it follows from Corollary 1 above that we can choose a number $\nu_0 \geqslant \nu$ such that

$$\Phi'_{H'}(h;\delta'(q)) = \Phi'_{H'}(h)\sigma_2'(\xi'_{H'}(q))$$

for $h \in A(\nu_0)$. But since $\Phi'_{H'}$ is an analytic function on $A(\nu)$, this relation must hold for all $h \in A(\nu)$.

§10.   Proof of Lemma 31

Let η denote the mapping of $\mathfrak{h}_\mathfrak{p}$ given by $\eta(H) = -(-1)^{1/2}(H + H_\rho)$. Then $\eta^{-1}(H) = (-1)^{1/2}H - H_\rho$. For any $p \in S(\mathfrak{h}_\mathfrak{p})$, let p^η denote the polynomial function given by $p^\eta(H) = p(\eta^{-1}(H))$ $(H \in \mathfrak{h}_\mathfrak{p})$. Then $p \to p^\eta$ is an automorphism of $S(\mathfrak{h}_\mathfrak{p})$ over $\mathbf{C}$ which can be extended uniquely to its quotient field Q. Define $D \in \mathcal{E}^2(V_0)$ as in §9 (see the definition of 'Γ_λ). We recall that $d_1, d_2, \ldots, d_r$ are all the distinct eigenvalues of D. We shall assume from now on that V is a Hilbert space and the representations σ_1, σ_2 of K on V are unitary. Then every eigenvalue of the transformation $\sigma_2'(\omega_{m \cap t})$ is real. This shows that $d_1, d_2, \ldots, d_r$ are real.

Put $\tau_{\lambda i} = \eta(\tau'_{\lambda i})$ ($\lambda \in L'$, $1 \leqslant i \leqslant r$). Then $\tau_{\lambda i}$ is the set of all points $H \in \mathfrak{h}_{\mathfrak{p}}$ where $2(-1)^{1/2}\lambda(H) = \langle \lambda, \lambda \rangle + d_i$. Put $\Gamma_\lambda = ({}^{\prime}\Gamma_\lambda)^\eta$. Then the rational function Γ_λ is defined everywhere on ${}^{\prime}\mathfrak{h}_{\mathfrak{p}} = \eta(\mathfrak{h}''_{\mathfrak{p}})$. Suppose $H \in \tau_{\lambda i} \cap \mathfrak{h}_{\mathfrak{p}_0}$. Then since d_i is real, $\lambda(H) = \langle \lambda, \lambda \rangle + d_i = 0$. Clearly there are only a finite number of $\lambda \in L'$ such that $\Pi_{1 \leqslant i \leqslant r}(\langle \lambda, \lambda \rangle + d_i) = 0$. Hence ${}^{\prime}\mathfrak{h}_{\mathfrak{p}_0} = {}^{\prime}\mathfrak{h}_{\mathfrak{p}} \cap \mathfrak{h}_{\mathfrak{p}_0}$ is the complement in $\mathfrak{h}_{\mathfrak{p}_0}$ of a finite number of hyperplanes passing through the origin.

Put $\xi(q:H) = \xi_H(q) = \xi'_{H-H_\rho}(q)$ for $H \in \mathfrak{h}_{\mathfrak{p}}$ and $q \in I_{\mathfrak{g}}$. Also define $\delta(q) = e^\rho \delta'(q) \circ e^{-\rho}$ as before.

Theorem 3. *Let U be an open relatively compact[23] subset of $\mathfrak{h}_{\mathfrak{p}}$. Then we can choose an element $p \in S(\mathfrak{h}_{\mathfrak{p}})$ and a number $v \geqslant 0$ such that the following conditions hold.*

1) *p is nowhere zero on ${}^{\prime}\mathfrak{h}_{\mathfrak{p}}$.*
2) *The rational functions $p_\lambda = p\Gamma_\lambda$ ($\lambda \in L$) are all everywhere defined on U.*
3) *The series $\sum_{\lambda \in L}|p_\lambda(H_0)e^{-\lambda(H)}|$ converges uniformly for $H_0 \in U$ and $H \in \mathfrak{h}_{\mathfrak{p}}(v)$.*
4) *Put $P_\lambda(H_0 : H) = p_\lambda(H_0)\exp\{(-1)^{1/2}\langle H_0, H \rangle - \lambda(H)\}$. Then for any ε ($0 < \varepsilon < 1$) the series $\sum_{\lambda \in L}P_\lambda(H_0 : H)$ converges decently on $U \times \mathfrak{h}_{\mathfrak{p}}(\varepsilon, v)$.*
5) *Put $\Psi(H_0 : h) = \sum_{\lambda \in L}P_\lambda(H_0 : \log h)$ for $H_0 \in U$ and $h \in A_{\mathfrak{p}}(v)$. Then*

$$\Psi(H_0 : h; \delta(q)) = \Psi(H_0 : h)\sigma'_2\big(\xi\big(q : (-1)^{1/2}H_0\big)\big)$$

for $h \in A(v)$, $q \in I_{\mathfrak{g}}$ and $H_0 \in U$.

This is merely a restatement of the results of §9.

Define γ as in Lemma 11 and let $\mathfrak{w}$ denote the little Weyl group of $\mathfrak{g}_0$ with respect to $\mathfrak{h}_{\mathfrak{p}_0}$ (see [8, §3]). Let w be the order of $\mathfrak{w}$. Put $S = S(\mathfrak{h}_{\mathfrak{p}})$ and $J = I(\mathfrak{h}_{\mathfrak{p}})$. By the Corollary of Lemma 8 of [8], we can choose homogeneous elements $u_1 = 1, u_2, \ldots, u_w$ in S such that $S = \sum_{1 \leqslant i \leqslant w}Ju_i$.

Lemma 30. *Let u be a homogeneous element in $\mathfrak{S}_{\mathfrak{p}}$ of degree d. Choose elements $q_i \in I_{\mathfrak{g}}$ such that $u = \sum_{1 \leqslant i \leqslant w}\gamma(q_i)u_i$ and $d^0 q_i + d^0 u_i \leqslant d$. Then*

$$d^0\bigg(u - \sum_{1 \leqslant i \leqslant w} u_i \circ \delta(q_i)\bigg) \leqslant d - 1.$$

Note that it follows from Lemmas 19 and 20 of [8] that elements q_i with the required properties can always be found. Moreover

$$u - \sum_{1 \leqslant i \leqslant w} u_i \circ \delta(q_i) = \sum_i u_i \circ (\gamma(q_i) - \delta(q_i))$$

and so our assertion is an immediate consequence of Lemma 11.

---

[23]This means that the closure of U in $\mathfrak{h}_{\mathfrak{p}}$ is compact.

Corollary. *For any $u \in \mathfrak{H}_\mathfrak{p}$, we can select a finite number of elements $f_{ij} \in \mathcal{R}$, $k_{ij}, k'_{ij} \in \mathfrak{R}$, $q_{ij} \in I_\mathfrak{g}$ ($1 \leqslant i \leqslant w$, $1 \leqslant j \leqslant r$) such that*

$$u = \sum_{1 \leqslant j \leqslant r} \sum_{1 \leqslant i \leqslant w} f_{ij}(k_{ij} \otimes u_i \otimes k'_{ij}) \circ \delta(q_{ij}).$$

Let $d = d^0 u$. We shall use induction on d. Obviously we can assume that u is homogeneous and not zero and $d \geqslant 1$. Choose q_i as in Lemma 30. Then $d^0(u - \sum_{1 \leqslant i \leqslant w} u_i \circ \delta(q_i)) < d$ and therefore from Lemma 10,

$$u - \sum_{1 \leqslant i \leqslant w} u_i \circ \delta(q_i) = \sum_{1 \leqslant j \leqslant s} f_j(k_j \otimes v_j \otimes k'_j)$$

where $f_j \in \mathcal{R}$, $k_j, k'_j \in \mathfrak{R}$, $v_j \in \mathfrak{H}_\mathfrak{p}$ and $d^0 v_j \leqslant d - 1$. Now

$$k_j \otimes v_j \otimes k'_j = (k_j \otimes 1 \otimes k'_j) \circ v_j.$$

Hence if we apply the induction hypothesis to v_j, our assertion follows immediately.

For any $u \in \mathfrak{H}_\mathfrak{p}$, let $\partial(u)$ denote the corresponding differential operator on $\mathfrak{h}_\mathfrak{p}$ (see [6, §2]). Let f be an element in $\mathcal{R}$. Then it is obvious that there exists a unique holomorphic function f^* on[24] $\mathfrak{h}'_\mathfrak{p}$ such that $f^*(H) = f(\exp H)$ for $H \in \mathfrak{h}'_{\mathfrak{p}_0}$. Clearly $f \to f^*$ ($f \in \mathcal{R}$) is an isomorphism onto a ring $\mathcal{R}^*$ of holomorphic functions on $\mathfrak{h}'_\mathfrak{p}$. As before let $\mathcal{D}$ denote the algebra of all differential operators on A' and let $\mathcal{D}_0$ be the subalgebra of $\mathcal{D}$ generated by $\mathfrak{H}_\mathfrak{p}$ and $\mathcal{R}$. Also let $\mathcal{D}^*$ be the subalgebra of all holomorphic differential operators on $\mathfrak{h}'_\mathfrak{p}$ generated by $\partial(\mathfrak{H}_\mathfrak{p})$ and $\mathcal{R}^*$. Then it is easy to verify that there exists a unique isomorphism $D \to D^*$ of $\mathcal{D}_0$ onto $\mathcal{D}^*$ such that $(fu)^* = f^* \partial(u)$ for $f \in \mathcal{R}$ and $u \in \mathfrak{H}_\mathfrak{p}$. Put $\mathcal{C}^* = \mathfrak{R} \otimes \mathcal{D}^* \otimes \mathfrak{R}$. We turn $\mathcal{C}^*$ into an algebra in a way similar to that used for $\mathcal{C}$. Let Ω be a nonempty open set in $\mathfrak{h}'_\mathfrak{p}$. Then the space of C^∞ functions from Ω to V (or to $\mathcal{F}$) becomes a left-module over $\mathcal{C}^*$ in an analogous way. Extend the above isomorphism of $\mathcal{D}_0$ onto $\mathcal{D}^*$ to an isomorphism of $\mathcal{C}_0 = \mathfrak{R} \otimes \mathcal{D}_0 \otimes \mathfrak{R}$ onto $\mathcal{C}^*$ by setting

$$(k_1 \otimes D \otimes k_2)^* = k_1 \otimes D^* \otimes k_2 \qquad (k_1, k_2 \in \mathfrak{R},\ D \in \mathcal{D}_0).$$

We know from Lemma 10 that $\delta(g) \in \mathcal{C}_0$ for $g \in \mathfrak{G}$. Put $\delta^*(g) = (\delta(g))^*$.

We now return to the notation of Theorem 3. For any $H_0 \in U$, put

$$\psi(H_0 : H) = \sum_{\lambda \in L} P_\lambda(H_0 : H) \qquad (H \in \mathfrak{h}_\mathfrak{p}(\nu)).$$

We intend to prove that if Ω is any compact subset of $\mathfrak{h}_\mathfrak{p}^+$, the series $\sum_\lambda |P_\lambda(H_0 : H)|$ converges uniformly for $(H_0, H) \in U \times \Omega$. In any case, it follows from the 4th statement of Theorem 3 that, for a fixed H_0, $\psi(H_0 : H)$ is a holomorphic function of H on $\mathfrak{h}_\mathfrak{p}(\nu)$. Therefore it is obvious from the 5th statement that

$$\psi(H_0 : H; \delta^*(q)) = \psi(H_0 : H)\sigma'_2\left(\xi\left(q : (-1)^{1/2} H_0\right)\right)$$

for $H \in \mathfrak{h}_\mathfrak{p}(\nu)$.

---

[24] We recall that $\mathfrak{h}'_\mathfrak{p}$ is the set of all points H in $\mathfrak{h}_\mathfrak{p}$ such that $\alpha(H) \neq 0$ for every $\alpha \in P_+$. Also $\mathfrak{h}'_{\mathfrak{p}_0} = \mathfrak{h}_{\mathfrak{p}_0} \cap \mathfrak{h}'_\mathfrak{p}$.

Let E be a complex vector space of finite dimension. Consider the tensor products $\mathcal{R}^* \otimes E$ and $S(\mathfrak{h}_\mathfrak{p}) \otimes \mathcal{R}^* \otimes E$. An element f of $\mathcal{R}^* \otimes E$ is a holomorphic function from $\mathfrak{h}'_\mathfrak{p}$ to E. We denote by $f(H)$ $(H \in \mathfrak{h}'_\mathfrak{p})$ the value (in E) of this function at H. Similarly an element Ξ of $S(\mathfrak{h}_\mathfrak{p}) \otimes \mathcal{R}^* \otimes E$ is a polynomial function from $\mathfrak{h}_\mathfrak{p}$ to $\mathcal{R}^* \otimes E$. Suppose $H \in \mathfrak{h}_\mathfrak{p}$ and $H' \in \mathfrak{h}'_\mathfrak{p}$. Then $\Xi(H)$ is the value in $\mathcal{R}^* \otimes E$ of this function at H. But $\Xi(H)$ being itself a holomorphic function from $\mathfrak{h}'_\mathfrak{p}$ to E, we denote by $\Xi(H : H')$ the value in E of the function $\Xi(H)$ at H'.

Let u be an element in $\mathfrak{H}_\mathfrak{p}$. Then from the corollary of Lemma 30 we can choose a finite set of elements $\Xi_{ij}(u) \in \mathcal{R}^* \otimes \mathcal{E}(\mathcal{F})$ and $q_{ij} \in I_\mathfrak{g}$ $(1 \leqslant i \leqslant w,\ 1 \leqslant j \leqslant r)$ such that[25]

$$\psi\big(H_0 : H;\ \partial(u)\big) = \sum_{i,j} \Xi_{ij}(u : H)\psi\big(H_0 : H;\ \partial(u_i)\big)\sigma'_2\Big(\xi\big(q_{ij} : (-1)^{1/2}H_0\big)\Big)$$

for $H_0 \in U$ and $H \in \mathfrak{h}_\mathfrak{p}(\nu)$. Here $\Xi_{ij}(u : H)$ is the value of $\Xi_{ij}(u)$ in $\mathcal{E}(\mathcal{F})$ at H. Let $H_1, \ldots, H_l$ be the base of $\mathfrak{h}_{\mathfrak{p}_0}$ over $\mathbb{R}$ introduced in §3. Put $\psi_i(H_0 : H) = \psi(H_0 : H;\ \partial(u_i))$, $1 \leqslant i \leqslant w$. Then by applying the above result successively to $u = u_i H_p$ $(1 \leqslant i \leqslant w,\ 1 \leqslant p \leqslant l)$, we get the following result. There exist elements $\Xi_{ijpk} \in \mathcal{R}^* \otimes \mathcal{E}(\mathcal{F})$ and $q_{ijpk} \in I_\mathfrak{g}$ $(1 \leqslant i, j \leqslant w,\ 1 \leqslant p \leqslant l,\ 1 \leqslant k \leqslant r)$ such that

$$\psi\big(H_0 : H : \partial(u_i H_p)\big) = \sum_{j,k} \Xi_{ijpk}(H)\psi\big(H_0 : H;\ \partial(u_j)\big)\sigma'_2\Big(\xi\big(q_{ijpk} : (-1)^{1/2}H_0\big)\Big)$$

for $1 \leqslant i \leqslant w,\ 1 \leqslant p \leqslant l$. Let $\mathcal{F}^w$ denote the direct sum of $\mathcal{F}$ with itself w times. Put

$$\Psi(H_0 : H) = \big(\psi_1(H_0 : H), \psi_2(H_0 : H), \ldots, \psi_w(H_0 : H)\big).$$

Then Ψ is a holomorphic function from $U \times \mathfrak{h}_\mathfrak{p}(\nu)$ to $\mathcal{F}^w$. The above result shows that there exist elements $\Xi_p \in S(\mathfrak{h}_\mathfrak{p}) \otimes \mathcal{R}^* \otimes \mathcal{E}(\mathcal{F}^w)$ $(1 \leqslant p \leqslant l)$ such that

$$\Psi\big(H_0 : H;\ \partial(H_p)\big) = \Xi_p(H_0 : H)\Psi(H_0 : H) \qquad (1 \leqslant p \leqslant l).$$

Put $|\mathbf{F}| = |F_1| + \cdots + |F_w|$ for $\mathbf{F} = (F_1, \ldots, F_w) \in \mathcal{F}^w$. Then $\mathcal{F}^w$ becomes a Banach space and we can apply Lemma 10 and Theorem 1 of [11] to the function Ψ for a fixed $H_0 \in U$. Put

$$\mathbf{P}_\lambda(H_0 : H) = \big(P_{\lambda 1}(H_0 : H), P_{\lambda 2}(H_0 : H), \ldots, P_{\lambda w}(H_0 : H)\big)$$

where $P_{\lambda i}(H_0 : H) = P_\lambda(H_0 : H;\ \partial(u_i))$. Then it follows from Theorem 1 and Lemma 16 of [11], that, for any fixed $H_0 \in U$, the series

$$\sum_{\lambda \in L} |\mathbf{P}_\lambda(H_0 : H)|$$

converges decently for $H \in \mathfrak{h}_\mathfrak{p}(\varepsilon, \eta)$ for any ε, η $(0 < \varepsilon < 1,\ \eta > 0)$ and

$$\Psi(H_0 : H) = \sum_{\lambda \in L} \mathbf{P}_\lambda(H_0 : H) \qquad \big(H \in \mathfrak{h}_\mathfrak{p}(\nu)\big).$$

This shows that Ψ can be extended on $U \times \mathfrak{h}_\mathfrak{p}^+$ in such a way that, for fixed $H_0 \in U$, it is a holomorphic function of H on $\mathfrak{h}_\mathfrak{p}^+$. Naturally the extended function (which we

---

[25] As usual we follow the mode of writing introduced in [6, §2].

again denote by Ψ) continues to satisfy the differential equations above. This follows by the principle of analytic continuation.

Fix a point $H' = \Sigma_{1 \leqslant p \leqslant l} c_p H_p$ $(c_p > 0)$ in $\mathfrak{h}_{\mathfrak{p}_0}^+$. Then if T is a sufficiently large positive number, $\mathfrak{h}_{\mathfrak{p}}^+ + TH' \subset \mathfrak{h}_{\mathfrak{p}}(\nu)$. Put $\Xi = \Sigma_p c_p \Xi_p$ and

$$\Theta_k(H_0 : H) = (-1)^k \int_{0 \leqslant t_1 \leqslant \cdots \leqslant t_k \leqslant T} \Xi(H_0 : H : t_1) \Xi(H_0 : H : t_2) \cdots$$

$$\times \Xi(H_0 : H : t_k) \, dt_1 \cdots dt_k$$

where $\Xi(H_0 : H : t) = \Xi(H_0 : H + tH')$ $(H_0 \in \mathfrak{h}_{\mathfrak{p}}, H \in \mathfrak{h}_{\mathfrak{p}}^+, k \geqslant 1)$. Also define $\Theta_0(H_0 : H) = 1 \in \mathcal{E}(\mathcal{F}^w)$ and put $\Psi(H_0 : H : t) = \Psi(H_0 : H + tH')$ $(t \geqslant 0)$. Then

$$d\Psi(H_0 : H : t)/dt = \Xi(H_0 : H : t)\Psi(H_0 : H : t).$$

Now Θ_k is a holomorphic function from $U \times \mathfrak{h}_{\mathfrak{p}}^+$ to $\mathcal{E}(\mathcal{F}^w)$. Let Ω be a compact subset of $\mathfrak{h}_{\mathfrak{p}}^+$. Then there obviously exists a number a such that $|\Xi(H_0 : H : t)| \leqslant a$ for $H_0 \in U$, $H \in \Omega$ and $0 \leqslant t \leqslant T$. Hence $|\Theta_k(H_0 : H)| \leqslant a^k T^k / k!$ and therefore the series $\Sigma_{k \geqslant 0} |\Theta_k(H_0 : H)|$ converges uniformly on $U \times \Omega$. Therefore

$$\Theta(H_0 : H) = \sum_{k \geqslant 0} \Theta_K(H_0 : H)$$

is a holomorphic function from $U \times \mathfrak{h}_{\mathfrak{p}}^+$ to $\mathcal{E}(\mathcal{F}^w)$ and it follows from our differential equation that $\Psi(H_0 : H) = \Theta(H_0 : H)\Psi(H_0 : H + TH')$. But Ψ is holomorphic on $U \times \mathfrak{h}_{\mathfrak{p}}(\nu)$ from Theorem 3 and $\mathfrak{h}_{\mathfrak{p}}^+ + TH' \subset \mathfrak{h}_{\mathfrak{p}}(\nu)$. Hence we conclude from this relation that Ψ is holomorphic on $U \times \mathfrak{h}_{\mathfrak{p}}^+$.

Now it is obvious from the definition of Ξ that we have an expansion of the form

$$\Xi(H_0 : H) = \sum_{\lambda \in L} \Xi_\lambda(H_0) e^{-\lambda(H)} \qquad (H_0 \in U, \ H \in \mathfrak{h}_{\mathfrak{p}}^{\downarrow})$$

where Ξ_λ are polynomial functions from U to $\mathcal{E}(\mathcal{F}^w)$ of bounded degree and the series

$$\sum_{\lambda \in L} |\Xi_\lambda(H_0)| e^{-\nu_0 m(\lambda)}$$

converges uniformly for $H_0 \in U$ for every $\nu_0 > 0$. Fix $\nu_0 > 0$ and choose $b > 0$ such that

$$\sum_{\lambda \in L} |\Xi_\lambda(H_0) e^{-\lambda(H)}| \leqslant \sum_{\lambda \in L} |\Xi_\lambda(H_0)| e^{-\nu_0 m(\lambda)} \leqslant b$$

for $H_0 \in U$ and $H \in \mathfrak{h}_{\mathfrak{p}}(\nu_0)$. Put $\Xi_\lambda(H_0 : t) = e^{-t\lambda(H')}\Xi_\lambda(H_0)$. Then

$$\Xi(H_0 : H : t) = \sum_{\lambda \in L} \Xi_\lambda(H_0 : t) e^{-\lambda(H)}$$

and

$$\sum_\lambda |\Xi_\lambda(H_0 : t) e^{-\lambda(H)}| \leqslant \sum_\lambda |\Xi_\lambda(H_0) e^{-\lambda(H_0)}| \leqslant b$$

for $H_0 \in U$, $H \in \mathfrak{h}_{\mathfrak{p}}(\nu_0)$ and $t \geq 0$. Put

$$\theta_{k,\lambda}(H_0) = (-1)^k \sum_{\lambda_1 + \cdots + \lambda_k = \lambda} \int_{0 \leq t_1 \leq \cdots \leq t_k \leq T} \Xi_{\lambda_1}(H_0: t_1) \cdots$$

$$\times \Xi_{\lambda_k}(H_0: t_k) \, dt_1 \cdots dt_k$$

where the sum is over sequences $(\lambda_1, \ldots, \lambda_k)$ of k elements in L such that $\lambda_1 + \cdots + \lambda_k = \lambda$. (If $k = 0$, we put $\theta_{0,\lambda}(H_0) = 0$ for $\lambda \neq 0$ and $\theta_{00}(H_0) = 1$.) Then $\theta_{k,\lambda}$ is a polynomial function from $\mathfrak{h}_{\mathfrak{p}}$ to $\mathcal{E}(\mathcal{F}^w)$ and

$$\sum_{\lambda \in L} |\theta_{k,\lambda}(H_0) e^{-\lambda(H)}| \leq b^k T^k / k! \qquad \left(H_0 \in U, \ H \in \mathfrak{h}_{\mathfrak{p}}(\nu_0) \right).$$

This shows that the series $\sum_{k \geq 0} \sum_{\lambda \in L} |\theta_{k,\lambda}(H_0) e^{-\lambda(H)}|$ converges uniformly for $(H_0, H) \in U \times \mathfrak{h}_{\mathfrak{p}}(\nu_0)$ and it is obvious that

$$\Theta(H_0: H) = \sum_{k \geq 0} \sum_{\lambda \in L} \theta_{k,\lambda}(H_0) e^{-\lambda(H)}$$

on $U \times \mathfrak{h}_{\mathfrak{p}}^+$. Now

$$\Psi(H_0: H) = \Theta(H_0: H) \Psi(H_0: H + TH')$$

and

$$\Psi(H_0: H + TH') = \sum_{\lambda \in L} \mathbf{P}_\lambda(H_0: H + TH') \qquad \left(H_0 \in U, \ H \in \mathfrak{h}_{\mathfrak{p}}^+ \right).$$

Put

$$\mathbf{Q}_\lambda(H_0: H) = \sum_{k \geq 0} \sum_{\lambda_1 + \lambda_2 = \lambda} \theta_{k,\lambda_1}(H_0) e^{-\lambda_1(H)} \mathbf{P}_{\lambda_2}(H_0: H + TH')$$

$(H_0 \in U, H \in \mathfrak{h}_{\mathfrak{p}}^+)$. Then the series $\sum_{\lambda \in L} |\mathbf{Q}_\lambda(H_0: H)|$ converges uniformly on $U \times \Omega$ for any compact subset Ω of $\mathfrak{h}_{\mathfrak{p}}^+$. But now we claim that $\mathbf{Q}_\lambda = \mathbf{P}_\lambda$. Assuming this for a moment, we obtain

$$\Psi(H_0: H) = \sum_{\lambda \in L} \mathbf{Q}_\lambda(H_0: H) = \sum_{\lambda \in L} \mathbf{P}_\lambda(H_0: H) \qquad \left(H_0 \in U, \ H \in \mathfrak{h}_{\mathfrak{p}}^+ \right),$$

the series $\sum_{\lambda \in L} |\mathbf{P}_\lambda(H_0: H)|$ converging uniformly on $U \times \Omega$.

In order to verify that $\mathbf{Q}_\lambda = \mathbf{P}_\lambda$ ($\lambda \in L$), we proceed as follows. It is obvious from the definition of $\mathbf{P}_\lambda$ that there exists a holomorphic function $\mathbf{p}_\lambda$ from U to $\mathcal{F}^w$ such that

$$\mathbf{P}_\lambda(H_0: H) = \mathbf{p}_\lambda(H_0) \exp\{(-1)^{1/2} \langle H_0, H \rangle - \lambda(H)\} \qquad \left(H_0 \in U, \ H \in \mathfrak{h}_{\mathfrak{p}} \right).$$

Put

$$\mathbf{q}_\lambda(H_0) = e^{(-1)^{1/2} T \langle H_0, H' \rangle} \sum_{k \geq 0} \sum_{\lambda_1 + \lambda_2 = \lambda} \theta_{k,\lambda_1}(H_0) \mathbf{p}_{\lambda_2}(H_0) e^{-T\lambda_2(H')}$$

for $H_0 \in U$. In order to prove that $\mathbf{Q}_\lambda = \mathbf{P}_\lambda$, it would be sufficient to show that

$\mathbf{q}_\lambda = \mathbf{p}_\lambda$ $(\lambda \in L)$. Fix $H_0 \in U$ and put

$$\theta_\lambda(H_0) = \sum_{k \geqslant 0} \theta_{k,\lambda}(H_0) \qquad (\lambda \in L).$$

Then

$$\Theta(H_0 : H) = \sum_{\lambda \in L} \theta_\lambda(H_0) e^{-\lambda(H)}$$

and, as we have seen, the series $\sum_{\lambda \in L}|\theta_\lambda(H_0) e^{-\lambda(H)}|$ converges uniformly for $H \in \mathfrak{h}_\mathfrak{p}(\nu_0)$ for every $\nu_0 > 0$. On the other hand

$$\Psi(H_0 : H) = \Theta(H_0 : H)\Psi(H_0 : H + TH')$$

and the expansion

$$\Psi(H_0 : H) = e^{(-1)^{1/2}\langle H_0, H \rangle} \sum_{\lambda \in L} \mathbf{p}_\lambda(H_0) e^{-\lambda(H)}$$

is valid for $H \in \mathfrak{h}_\mathfrak{p}(\nu)$, the series being decently convergent. Hence we can compare the coefficients of $e^{-\lambda}$ on both sides of the above relation and obtain

$$\mathbf{p}_\lambda(H_0) = e^{(-1)^{1/2}T\langle H_0, H' \rangle} \sum_{\lambda_1 + \lambda_2 = \lambda} \theta_{\lambda_1}(H_0)\mathbf{p}_{\lambda_2}(H_0) e^{-T\lambda_2(H')} = \mathbf{q}_\lambda(H_0).$$

Thus we have shown that the series $\sum_\lambda |\mathbf{P}_\lambda(H_0 : H)|$ converges uniformly on $U \times \mathfrak{h}_\mathfrak{p}(\nu_0)$ for any $\nu_0 > 0$ and

$$\Psi(H_0 : H) = \sum_{\lambda \in L} \mathbf{P}_\lambda(H_0 : H) \qquad \left(H_0 \in U, \ H \in \mathfrak{h}_\mathfrak{p}^+\right).$$

Since $u_1 = 1$, we have obtained the following result.

Lemma 31. *Define P_λ as in Theorem 3. Then for any compact subset Ω of $\mathfrak{h}_\mathfrak{p}^+$, the series $\sum_{\lambda \in L}|P_\lambda(H_0 : H)|$ converges uniformly on $U \times \Omega$. Put*

$$\psi(H_0 : H) = \sum_{\lambda \in L} P_\lambda(H_0 : H) \qquad \left(H_0 \in U, \ H \in \mathfrak{h}_\mathfrak{p}^+\right).$$

Then ψ is a holomorphic function on $U \times \mathfrak{h}_\mathfrak{p}^+$ and for a fixed $H_0 \in U$, the above series converges decently on $\mathfrak{h}_\mathfrak{p}(\varepsilon, \eta)$ for any ε, η $(0 < \varepsilon < 1, \eta > 0)$. Finally

$$\psi(H_0 : H; \delta^*(q)) = \psi(H_0 : H)\sigma_2'\!\left(\xi\!\left(q : (-1)^{1/2} H_0\right)\right)$$

for $q \in I_\mathfrak{g}$.

§11.   Statement of Lemma 33

For any $H_0 \in \mathfrak{'h}_\mathfrak{p}$ put

$$\Phi(H_0 : h) = e^{(-1)^{1/2}\langle H_0, H \rangle} \sum_{\lambda \in L} \Gamma_\lambda(H_0) e^{-\lambda(H)} \qquad (h \in A^+)$$

where $H = \log h$. Then it follows from Theorem 3 and Lemma 31 that for a fixed H_0, Φ is an analytic function from A^+ to $\mathscr{E}(V_0)$.

Lemma 32. *Let* $\Lambda_1, \ldots, \Lambda_r$ *be distinct elements in* $`\mathfrak{h}_{\mathfrak{p}_0}$ *and suppose* $v_1, \ldots, v_r$ *are elements in* V_0 *such that*

$$\sum_{1 \leqslant i \leqslant r} \Phi(\Lambda_i : h) v_i = 0$$

for all $h \in A^+$. *Then* $v_1 = v_2 = \cdots = v_r = 0$.

Let J be the set of all indices i such that $v_i \neq 0$. Suppose that contrary to our assertion, J is not empty. Let $(-1)^{1/2}\Lambda_j$ be the highest element among[22] $(-1)^{1/2}\Lambda_i$ $(i \in J)$. Then by Theorem 3, we can equate the coefficient of $e^{(-1)^{1/2}\Lambda_j}$ to zero in the equation

$$\sum_{i \in J} \sum_{\lambda \in L} e^{(-1)^{1/2}\Lambda_i(H) - \lambda(H)} \Gamma_\lambda(\Lambda_i) v_i = 0.$$

This gives $v_j = 0$ since $\Gamma_0(\Lambda_j) = 1$. But $j \in J$ and therefore $v_j \neq 0$. This contradiction proves the lemma.

From now on we shall assume that the center of G is finite and therefore K is compact. As before let $\mathfrak{w}$ denote the little Weyl group of $\mathfrak{g}_0$ with respect to $\mathfrak{h}_{\mathfrak{p}_0}$ (see [8, p. 249]). We use the notation of Theorem 2.

Lemma 33. *Fix an element* $H_0 \in \mathfrak{h}'_\mathfrak{p}$ *such that* $sH_0 \in `\mathfrak{h}_\mathfrak{p}$ *for every* $s \in \mathfrak{w}$. *Then for any* $v \in V$ *there exist unique elements* $v_s \in V_0$ $(s \in \mathfrak{w})$ *such that*

$$e^{\rho(\log h)} \int \sigma_1(k_h) v \sigma_2(k^{-1}) \exp\{(-1)^{1/2}\langle H_0, H(hk)\rangle - \rho(H(hk))\} \, dk$$

$$= \sum_{s \in \mathfrak{w}} \Phi(sH_0 : h) v_s$$

for $h \in A^+$.

Since $H_0 \in \mathfrak{h}'_\mathfrak{p}$, the elements sH_0 $(s \in \mathfrak{w})$ are all distinct. Hence the uniqueness of the elements v_s is obvious from Lemma 32 and therefore only their existence remains to be proved. This, however, requires considerable preparation which will be undertaken in the next section.

§12. Preparation for the Proof of Lemma 33

We recall that $\mathfrak{h} = \mathfrak{h}_\mathfrak{p} + \mathfrak{h}_\mathfrak{t}$ and W is the Weyl group of $\mathfrak{g}$ with respect to $\mathfrak{h}$ (see §2.). Fix an element $H_0 \in \mathfrak{h}$ and let W_{H_0} be the subgroup of those elements s in W for which $sH_0 = H_0$. W operates on $S = S(\mathfrak{h})$ in the obvious way. Let $J = I(\mathfrak{h})$ be the subalgebra consisting of those elements of S which are left fixed by all operations of W. Similarly let J_{H_0} denote the algebra of invariants of W_{H_0} in S. We denote by $J_{H_0}^+$ the subspace spanned by homogeneous elements of J_{H_0} of degree $\geqslant 1$. Then $J_{H_0}^+$ is an ideal in J_{H_0}. For $p, q \in S$, define the scalar product $\langle p, q \rangle$ as in [6, p. 90]. Let $\mathfrak{X}(H_0)$ be the subspace of all $p \in S$ such that $\langle p, q \rangle = 0$ for every $q \in SJ_{H_0}^+$. It is obvious that an element p of S lies in $\mathfrak{X}(H_0)$ if and only if $\partial(q)p = 0$ for all $q \in J_{H_0}^+$.

Lemma 34. *Let p_0 be an element in S and let f denote the function $H \to p_0(H)e^{\langle H, H_0 \rangle}$ $(H \in \mathfrak{h})$ on $\mathfrak{h}$. Then the following statements are equivalent.*

   (1) *p_0 lies in $\mathfrak{X}(H_0)$.*
   (2) *f is an eigenfunction of $\partial(q)$ for every $q \in J_{H_0}$.*
   (3) *f is an eigenfunction of $\partial(q)$ for every $q \in J$.*

Moreover[26] $\dim \mathfrak{X}(H_0) = [W_{H_0}]$.

Define an automorphism $p \to p^\alpha$ of S as follows: $p^\alpha(H) = p(H + H_0)$ $(H \in \mathfrak{h})$. Let g denote the function $H \to e^{\langle H, H_0 \rangle}$ on $\mathfrak{h}$. We claim $g^{-1}\partial(p) \circ g = \partial(p^\alpha)$ for $p \in S$. The mapping $p \to g\partial(p^\alpha) \circ g^{-1}$ is obviously a homomorphism of S into $\partial(S)$. So it would be enough to verify that $g\partial(p^\alpha) \circ g^{-1} = \partial(p)$ when $d^0 p \leq 1$. But this follows immediately by a direct computation. Note that since H_0 is left fixed by W_{H_0}, it is obvious that $(J_{H_0})^\alpha = J_{H_0}$.

We shall first show that (2) is equivalent to (1). Let $q \in J_{H_0}$. Then $\partial(q)f = g\partial(q^\alpha)p_0$. But $\partial(q^\alpha)p_0 = q^\alpha(0)p_0 + \partial(q^\alpha - q^\alpha(0))p_0$ and we may obviously assume for our proof that $p_0 \neq 0$. Then it is obvious that

$$d^0\big(\partial(q^\alpha - q^\alpha(0))p_0\big) < d^0 p_0.$$

Hence f is an eigenfunction of $\partial(q)$ if and only if $\partial(q^\alpha - q^\alpha(0))p_0 = 0$. Since $(J_{H_0})^\alpha = J_{H_0}$ this shows that (2) is equivalent to the condition that $\partial(q)p = 0$ for all $q \in J_{H_0}^+$. But this latter condition is equivalent to (1).

Now we prove that $\dim \mathfrak{X}(H_0) = [W_{H_0}]$. For any $p \in S$, define a polynomial function $p_* \in S$ by[16] $p_*(H) = \mathrm{conj}\, p(H)$ for $H \in \mathfrak{h}_{\mathfrak{p}_0} + (-1)^{1/2}\mathfrak{h}_{t_0}$. Then $\langle p, p_* \rangle$ is a positive definite Hermitian form on S (see [6, p. 110]) and since $\mathfrak{h}_{\mathfrak{p}_0} + (-1)^{1/2}\mathfrak{h}_{t_0}$ is stable under W, it is clear that $J_{H_0}^+$ is stable under the mapping $p \to p_*$, and $\mathfrak{X}(H_0)$ is the orthogonal complement in S of $SJ_{H_0}^+$ under the above positive definite form. It now follows easily (see the proof of Lemma 32 of [8]) that

$$\dim \mathfrak{X}(H_0) = \dim S/SJ_{H_0}^+ = \left[W_{H_0}\right].$$

Here one has to make use of the result of Chevalley (see Lemma 49 of the Appendix) that W_{H_0} is generated by the Weyl reflexions s_α corresponding to those roots α which vanish at H_0.

Since $J \subset J_{H_0}$, it remains to show that (3) implies (2). Let $r = [W : W_{H_0}]$. Choose elements s_i $(0 \leq i < r, s_0 = 1)$ in W such that $W = \cup_{0 \leq i < r} s_i W_{H_0}$ and put $H_i = s_i H_0$. Let E be the space of all holomorphic functions F on $\mathfrak{h}$ such that $\partial(q)F = q(H_0)F$ for all $q \in J$. We know from [6, Lemma 13] that $\dim E \leq [W]$. Let E_i be the space of all functions F on $\mathfrak{h}$ of the form $F(H) = p(H)e^{\langle H, H_i \rangle}$ $(H \in \mathfrak{h})$ where $p \in \mathfrak{X}(H_i)$ $(0 \leq i < r)$. Then since $J_{H_i} \supset J$, it follows from the equivalence of (1) and (2) proved

---

[26] $[W_{H_0}]$ denotes the order of the group W_{H_0}.

above that $E_i \subset E$. Also $\dim E_i = \dim \mathfrak{X}(H_i) = [W_{H_i}] = [W_{H_0}]$. Moreover the elements H_i $(0 \leqslant i < r)$ being distinct, the functions $g_i : H \to e^{\langle H, H_i \rangle}$ $(0 \leqslant i < r)$ are linearly independent over S. Therefore the sum $\Sigma_{0 \leqslant i < r} E_i$ is direct. Since $\Sigma_{0 \leqslant i < r} \dim E_i = r[W_{H_0}] = [W] \geqslant \dim E$, this shows that $E = \Sigma_{0 \leqslant i < r} E_i$.

Now suppose condition (3) holds for f. Then $\partial(q^\alpha - q^\alpha(0)) p_0 = 0$ for every $q \in J$. But $q^\alpha(0) = q(H_0)$ and so it is clear that $f \in E$. Since the functions g_i $(0 \leqslant i < r)$ are linearly independent over S and since $E = \Sigma_{0 \leqslant i < r} E_i$, it follows that $f \in E_0$. This shows that (3) implies (1) and therefore also (2). The proof of Lemma 34 is now complete.

Corollary. *Suppose* $W_{H_0} = \{1\}$. *Then* $\mathfrak{X}(H_0) = \mathbf{C}$.

For it follows from Lemma 34 that $\dim \mathfrak{X}(H_0) = 1$. On the other hand it is obvious that $\mathbf{C} \subset \mathfrak{X}(H_0)$. Therefore $\mathfrak{X}(H_0) = \mathbf{C}$.

Put $\mathfrak{X}_{\mathfrak{p}}(H_0) = \mathfrak{X}(H_0) \cap S(\mathfrak{h}_{\mathfrak{p}})$. U being a complex vector space of finite dimension, we denote by $\mathfrak{X}_{\mathfrak{p}}(H_0 : U)$ the space of all polynomial mappings p from $\mathfrak{h}_{\mathfrak{p}}$ to U such that $\beta \circ p \in \mathfrak{X}_{\mathfrak{p}}(H_0)$ for every linear function β on U. Clearly $\mathfrak{X}_{\mathfrak{p}}(H_0 : U)$ can be identified with the tensor product $\mathfrak{X}_{\mathfrak{p}}(H_0) \otimes U$.

Now as before consider the double representation $\sigma = (\sigma_1, \sigma_2)$ of K on V. Let M be the centralizer of A in K and Ω_M the set of all equivalence classes of finite-dimensional irreducible representations of M. Let $\mathfrak{v}_1, \ldots, \mathfrak{v}_r$ be all the distinct classes of Ω_M which occur in the reduction of $\sigma_1(M)$. We denote by V_i^* the subspace of those elements in V which transform under $\sigma_1(M)$ according to $\mathfrak{v}_i$ $(1 \leqslant i \leqslant r)$. Let E_i denote the projection of V on V_i^* corresponding to the direct sum $V = \Sigma_{1 \leqslant i \leqslant r} V_i^*$. Then $\sigma_1(m)$ $(m \in M)$ commutes with E_i for every i and $E_1 + E_2 + \cdots + E_r = 1 \in \mathfrak{E}(V)$. Hence $V_0 = \Sigma_{1 \leqslant i \leqslant r} V_i$ where $V_i = E_i V_0$.

We identify the linear space $\mathfrak{h}$ with its dual in the usual way. Similarly for $\mathfrak{h}_{\mathfrak{p}}$ and $\mathfrak{h}_{\mathfrak{k}}$. Then a linear function λ on $\mathfrak{h}_{\mathfrak{k}}$ (or $\mathfrak{h}_{\mathfrak{p}}$) can also be regarded as a linear function on $\mathfrak{h}$. Let ${}^{\backprime}\lambda_i$ denote the highest among all weights (with respect to $\mathfrak{h}_{\mathfrak{k}}$) of the representation of $\mathfrak{m} \cap \mathfrak{k}$ defined on $E_i V$ under σ_1. Put $\lambda_i = {}^{\backprime}\lambda_i + \tfrac{1}{2}\Sigma_{\alpha \in P_-} \alpha$ where P_- is the set of those positive roots of $\mathfrak{g}$ which vanish identically on $\mathfrak{h}_{\mathfrak{p}}$. Let $\mathfrak{H}$ denote the subalgebra of $\mathfrak{G}$ generated by $(1, \mathfrak{h})$. For any linear function Λ on $\mathfrak{h}$, we denote by η_Λ the homomorphism of $\mathfrak{H}$ into $\mathbf{C}$ given by $\eta_\Lambda(1) = 1$ and $\eta_\Lambda(H) = \Lambda(H)$ $(H \in \mathfrak{h})$.

Lemma 35. *Let* Λ *be a linear function on* $\mathfrak{h}_{\mathfrak{p}}$ *and* p *a polynomial mapping of* $\mathfrak{h}_{\mathfrak{p}}$ *into* V_i. *Suppose the function* $\phi : h \to p(\log h)\exp\{\Lambda(\log h) - \rho(\log h)\}$ $(h \in A^+)$ *is an eigenfunction of the operator* $\delta'(\mu'(z))$ *for every* $z \in \mathfrak{Z}$. *Then* $p \in \mathfrak{X}_{\mathfrak{p}}(\Lambda + \lambda_i : V_i)$.

(We recall that the mapping μ' has been defined in §2.) Suppose for a fixed $z \in \mathfrak{Z}$, $\mu'(z) = \Sigma_{1 \leqslant j \leqslant s} u_j v_j$ where $u_j \in \mathfrak{H}$, $v_j \in \mathfrak{M}_{\mathfrak{k}}$ and $u_1, \ldots, u_s$ are linearly independent. ($\mathfrak{M}_{\mathfrak{k}}$ is, as before, the enveloping algebra of $\mathfrak{m} \cap \mathfrak{k}$ in $\mathfrak{G}$.) Then $\delta'(\mu'(z)) = \Sigma_j 1 \otimes u_j \otimes v_j$ and hence

$$\phi(h; \delta'(\mu'(z))) = \sum_j \phi(h; u_j)\sigma_2(v_j) = \sum_j \sigma_1(v_j)\phi(h; u_j).$$

Define γ and γ_- as in Lemma 3 (§2). Since $z^m = z$ $(m \in M)$, it follows immediately from the definition of μ' that $(\mu'(z))^m = \mu'(z)$ and therefore $v_j^m = v_j$ $(1 \le j \le s)$. Hence it follows from Schur's lemma and the definition of γ_- that $\sigma_1(v_j) E_i = \eta_{\lambda_i}(\gamma_-(v_j)) E_i$ $(1 \le j \le s)$. On the other hand

$$\mu(z) = \sum_j {}^{\scriptscriptstyle \backprime}u_j v_j$$

where ${}^{\scriptscriptstyle \backprime}u_j = e^\rho u_j \circ e^{-\rho}$ and

$$\gamma(z) = \sum_j {}^{\scriptscriptstyle \backprime}u_j \gamma_-(v_j)$$

from Lemma 3. Therefore it follows from the definition of ϕ, that

$$\phi(h; \delta'(\mu'(z))) = \sum_j \eta_{\lambda_i}(\gamma_-(v_j)) \phi(h; u_j)$$

$$= \{\eta_{\Lambda_i}(\gamma(z)) p(\log h) + q(\log h)\} e^{(\Lambda - \rho)(\log h)}$$

where $\Lambda_i = \Lambda + \lambda_i$ and q is a polynomial function from $\mathfrak{h}_\mathfrak{p}$ to V_i with $d^0 q \le d^0 p - 1$. But ϕ being an eigenfunction of $\delta'(\mu'(z))$, we conclude that $q = 0$ and therefore

$$\sum_j \eta_{\lambda_i}(\gamma_-(v_j)) u_j \phi = \eta_{\Lambda_i}(\gamma(z)) \phi.$$

Now extend p to a polynomial function on $\mathfrak{h}$ by setting $p(H + H') = p(H)$ for $H \in \mathfrak{h}_\mathfrak{p}$ and $H' \in \mathfrak{h}_\mathfrak{t}$. Also put

$$F(H) = p(H) e^{\Lambda_i(H)}, \quad F_0(H) = p(H) e^{\Lambda(H)} \quad (H \in \mathfrak{h}),$$

so that $F = F_0 e^{\lambda_i}$. Moreover identity $\mathfrak{S}$ with $S(\mathfrak{h})$ in the obvious way and for any $u \in \mathfrak{S}$ denote by $\partial(u)$ the corresponding differential operator on $\mathfrak{h}$ (see [6, p. 89]). Then

$$\partial(\gamma(z)) F = \sum \eta_{\lambda_i}(\gamma_-(v_j)) e^{\lambda_i} \partial({}^{\scriptscriptstyle \backprime}u_j) F_0.$$

But since $e^{\rho(\log h)} \phi(h) = F_0(\log h)$ $(h \in A^+)$, it follows from our result above that

$$\sum_j \eta_{\lambda_i}(\gamma_-(v_j)) \partial({}^{\scriptscriptstyle \backprime}u_j) F_0 = \eta_{\Lambda_i}(\gamma(z)) F_0.$$

Hence

$$\partial(\gamma(z)) F = \eta_{\Lambda_i}(\gamma(z)) F.$$

Since γ maps $\mathfrak{Z}$ onto $J = I(\mathfrak{h})$, we conclude that F is an eigenfunction of $\partial(q)$ for every $q \in J$ and so $p \in \mathfrak{X}_\mathfrak{p}(\Lambda_i : V_i)$ from Lemma 34.

Corollary. *Suppose Λ takes only pure imaginary values in $\mathfrak{h}_{\mathfrak{p}_0}$ and $s\Lambda \ne \Lambda$ for any $s \ne 1$ in $\mathfrak{w}$. Then p is a constant.*

Put $\langle H, H' \rangle = \operatorname{trad} H \operatorname{ad} H'$ for $H, H' \in \mathfrak{h}$. In view of the Corollary of Lemma 34 and Chevalley's result (Lemma 49), it is enough to verify that $\langle \Lambda_i, \alpha \rangle \ne 0$ for any positive root α of $\mathfrak{g}$. Suppose then that α is a positive root such that $\langle \Lambda_i, \alpha \rangle = 0$. Since $\langle \Lambda_i, \alpha \rangle$ is pure imaginary while $\langle \lambda_i, \alpha \rangle$ is real, this implies that $\langle \Lambda, \alpha \rangle =$

$\langle \lambda_i, \alpha \rangle = 0$. It follows from our hypothesis on Λ that $\langle \Lambda, \beta \rangle \neq 0$ for any $\beta \in P_+$. Hence $\alpha \in P_-$. Now $\lambda_i = {}^{\backprime}\lambda_i + \rho_-$ where $\rho_- = \frac{1}{2} \Sigma_{\alpha \in P_-} \alpha$. Since ${}^{\backprime}\lambda_i$ is the highest weight of a finite-dimensional representation of $\mathfrak{m} \cap \mathfrak{k}$, it follows from known facts (see Weyl [12]) that $\langle {}^{\backprime}\lambda_i, \alpha \rangle \geqslant 0$ and $\langle \rho_-, \alpha \rangle > 0$. Hence $\langle \lambda_i, \alpha \rangle > 0$ and so we get a contradiction. This proves the corollary.

Lemma 36. *Let χ be a homomorphism of $\mathfrak{Z}$ into $\mathbb{C}$ such that $\chi(1) = 1$. Define*[27] *$v'_1, \ldots, v'_r$ and $H_1, \ldots, H_l$ as in Section 3 and let V^r denote the direct sum of V with itself r times. For any function ϕ of class C^∞ from A^+ to V, let Φ_ϕ denote the function from A^+ to V^r given by*

$$\Phi_\phi(h) = \big(\phi_1(h), \ldots, \phi_r(h)\big) \qquad (h \in A^+)$$

where $\phi_i(h) = \phi(h; v'_i)$ $(1 \leqslant i \leqslant r)$. Then there exist elements Ξ_j in $\mathfrak{R} \otimes \mathfrak{E}(V^r)$ such that

$$\Phi_\phi(h; H_j) = \Xi_j(h)\Phi_\phi(h) \qquad (1 \leqslant j \leqslant l,\ h \in A^+)$$

for any C^∞ function ϕ from A^+ to V_0 such that $\delta'(z)\phi = \chi(z)\phi$.

Put $G_+ = KA^+K$ and $K^* = K/M$. For any $k^* \in K^*$ and $h \in A$ define $h^{k^*} = h^k$ where k is some element in the coset k^*. We shall prove that G_+ is open in G and the mapping $\beta : (k, h, k^*) \to kh^{k^*}$ $(k \in K, h \in A^+, k^* \in K^*)$ defines an analytic isomorphism of the manifold $K \times A^+ \times K^*$ with G_+. An elementary computation (see [5, p. 617]) shows that this mapping is everywhere regular and therefore G_+ is open. Thus it only remains to show that β is univalent.

Suppose $k_1 h k_2 = k'_1 h' k'_2$ for some elements $h, h' \in A^+$ and $k_i, k'_i \in K$ $(i = 1, 2)$. Since $(k, X) \to k \exp X$ $(k \in K, X \in \mathfrak{p}_0)$ is a topological mapping of $K \times \mathfrak{p}_0$ onto G, it is clear that $k_1 k_2 = k'_1 k'_2$ and $h' = khk^{-1}$ where $k = (k'_1)^{-1} k_1$. Since h and h' are in A^+, the centralizer in $\mathfrak{p}_0$ of either one of these elements is $\mathfrak{h}_{\mathfrak{p}_0}$. Hence $\mathfrak{h}_{\mathfrak{p}_0}$ is stable under $Ad(k)$. This shows that $k \in M^*$ where M^* is the normalizer of A^+ in K. Let s be the element of $\mathfrak{w}$ induced by k. Then $h' = h^s$. But h and h' are both in A^+ and so it follows from known facts that $s = 1$. This proves that $k \in M$ and from this one concludes without difficulty that β is univalent.

Put $v^{k^*} = \sigma_1(k) v \sigma_2(k^{-1})$ for $v \in V_0$ and $k^* \in K^*$ where k is any point in the coset k^*. Now fix a C^∞ function ϕ from A^+ to V_0 and define a function F_1 from $K \times A^+ \times K^*$ to V as follows:

$$F_1(k, h, k^*) = \sigma_1(k)\big(\phi(h)\big)^{k^*} \qquad (k \in K,\ h \in A^+,\ k^* \in K^*).$$

Then obviously F_1 is also of class C^∞. Now put $F = F_1 \circ \beta^{-1}$. Then F is a C^∞ function on G_+ such that

$$F(k_1 h k_2) = \sigma_1(k_1)\phi(h)\sigma_2(k_2) \qquad (k_1, k_2 \in K,\ h \in A^+).$$

This shows that every C^∞ function ϕ from A^+ to V_0 can be extended to a C^∞ function from G_+ to V by setting $\phi(k_1 h k_2) = \sigma_1(k_1)\phi(h)\sigma_2(k_2)$ $(k_1, k_2 \in K, h \in A^+)$. Let $S_+(\sigma, \chi)$ denote the set of all C^∞ functions F from G_+ to V such that

---

[27]The number r here has of course no connection with that occurring in the decomposition $V_0 = V_1 + V_2 + \cdots + V_r$.

$F(k_1 x k_2) = \sigma_1(k_1) F(x) \sigma_2(k_2)$ $(k_1, k_2 \in K, x \in G_+)$ and $zF = \chi(z)F$ $(z \in \mathfrak{Z})$. Also let $S_0(\sigma, \chi)$ denote the set of C^∞ functions ϕ from A^+ to V_0 such that $\delta'(z)\phi = \chi(z)\phi$ $(z \in \mathfrak{Z})$. For any $F \in S_+(\sigma, \chi)$, let $\bar{F}$ denote the restriction of F on A^+. Then the above result shows that $F \to \bar{F}$ is a bijective mapping of $S_+(\sigma, \chi)$ onto $S_0(\sigma, \chi)$. Now the proof of Lemma 9 is applicable to $S_+(\sigma, \chi)$ (instead of $S(\sigma, \chi)$) without any change whatsoever and this gives Lemma 36 immediately.

It follows from the above lemma that Theorem 1 is applicable to functions in $S_0(\sigma, \chi)$. Hence if $\phi \in S_0(\sigma, \chi)$ and λ is a linear function on $\mathfrak{h}_\mathfrak{p}$, we can speak of the principal exponents of $e^\lambda \phi$.

For any linear function μ on $\mathfrak{h}$, put $\chi_\mu(z) = \eta_\mu(\gamma(z))$ $(z \in \mathfrak{Z})$.

Lemma 37. *Let ϕ be a C^∞ function from A^+ to V_0 which is an eigenfunction of $\delta(z)$ for every $z \in \mathfrak{Z}$. Let Λ be a principal exponent of ϕ. Then we can choose an index i $(1 \leqslant i \leqslant r)$ such that $\delta(z)\phi = \chi_{\Lambda_i}(z)\phi$ $(z \in \mathfrak{Z})$ where $\Lambda_i = \Lambda + \lambda_i$.*

Let ϕ_0 denote the principal part of ϕ. Then $\phi_0 \neq 0$ and therefore $\phi \neq 0$. Let χ be the homomorphism of $\mathfrak{Z}$ into $\mathbf{C}$ such that $\delta(z)\phi = \chi(z)\phi$ $(z \in \mathfrak{Z})$. It follows from Corollary 1 of Lemma 10 that $\delta(\mu'(z))\phi_0 = \chi(z)\phi_0$ $(z \in \mathfrak{Z})$. Let $p_0 e^\Lambda$ be the term in ϕ_0 corresponding to the exponent Λ. Then p_0 is a polynomial function from $\mathfrak{h}_\mathfrak{p}$ to V_0 and $p_0 \neq 0$. Hence we can choose i such that $E_i p_0 \neq 0$. Put $\phi_i(h) = p(\log h) e^{\Lambda(\log h)}$ $(h \in A^+)$ where $p = E_i p_0$. Since E_i commutes with $\sigma_1(m)$ for every $m \in \mathfrak{M}_\mathfrak{k}$, we again have $\delta(\mu'(z))\phi_i = \chi(z)\phi_i$ $(z \in \mathfrak{Z})$. But we have seen during the proof of Lemma 35 that $\delta(\mu'(z))\phi_i = \chi_{\Lambda_i}(z)\phi_i$. Therefore since $\phi_i \neq 0$, we conclude that $\chi = \chi_{\Lambda_i}$.

Corollary. *Let Λ_1, Λ_2 be two principal exponents of ϕ. Then we can choose indices i, j $(1 \leqslant i, j \leqslant r)$ and an element $s \in W$ such that $\Lambda_2 + \lambda_j = s(\Lambda_1 + \lambda_i)$.*

For by the above lemma, we can choose i and j such that $\chi_{\Lambda_1 + \lambda_i} = \chi = \chi_{\Lambda_2 + \lambda_j}$. But then $\eta_{\Lambda_1 + \lambda_i}(u) = \eta_{\Lambda_2 + \lambda_j}(u)$ for every $u \in J = I(\mathfrak{h})$ and this implies that $\Lambda_2 + \lambda_j = s(\Lambda_1 + \lambda_i)$ for some $s \in W$.

Let W_1 be the subgroup of all $s \in W$ such that $s\mathfrak{h}_\mathfrak{p} = \mathfrak{h}_\mathfrak{p}$ and let W_2 denote the complement of W_1 in W. Since the operations of W preserve the fundamental bilinear form on $\mathfrak{h}$, it follows that an element s of W lies in W_1 if and only if $s\mathfrak{h}_\mathfrak{k} = \mathfrak{h}_\mathfrak{k}$. Therefore $(-1)^{1/2}(s\mathfrak{h}_{\mathfrak{k}_0} + \mathfrak{h}_{\mathfrak{k}_0}) \cap \mathfrak{h}_{\mathfrak{p}_0} \neq \{0\}$ for every $s \in W_2$. For every $s \in W_2$, fix an element $H(s) \neq 0$ in $\mathfrak{h}_{\mathfrak{p}_0}$ such that $H(s) \equiv sH \bmod(-1)^{1/2}\mathfrak{h}_{\mathfrak{k}_0}$ for some $H \in (-1)^{1/2}\mathfrak{h}_{\mathfrak{k}_0}$. For a given $s \in W_1$, let τ_s denote the restriction of s on $\mathfrak{h}_\mathfrak{p}$. Then it is known (see Lemma 52 of the Appendix) that $\tau_s \in \mathfrak{w}$ and $s \to \tau_s$ is a homomorphism of W_1 onto $\mathfrak{w}$.

Lemma 38. *Let Λ be a linear function on $\mathfrak{h}_\mathfrak{p}$ satisfying the following two conditions. (1) Λ takes only pure imaginary values on $\mathfrak{h}_{\mathfrak{p}_0}$ and (2) $\Lambda(H(s)) \neq 0$ for every $s \in W_2$. Fix an index i $(1 \leqslant i \leqslant r)$ and let ϕ be a C^∞ function from A^+ to V_0 such that $\delta(z)\phi = \chi_{\Lambda_i}(z)\phi$ $(z \in \mathfrak{Z})$ where $\Lambda_i = \Lambda + \lambda_i$. Then every principal exponent of ϕ is of the form $\tau\Lambda$ $(\tau \in \mathfrak{w})$.*

Let Λ_0 be any principal exponent of ϕ. Then by Lemma 37 we can choose j such that $\chi_{\Lambda_i} = \chi_{\Lambda_0 + \lambda_j}$. This implies that $\Lambda_0 + \lambda_j = s\Lambda_i$ for some $s \in W$. We claim that $s \in W_1$. For otherwise since W_1 is a group, $s^{-1} \in W_2$. Choose H and H' in $(-1)^{1/2}\mathfrak{h}_{t_0}$ such that $H(s^{-1}) = s^{-1}H + H'$. Then $\lambda_j(H) = (\Lambda_0 + \lambda_j)(H) = \Lambda_i(s^{-1}H) = \Lambda(H(s^{-1})) - \lambda_i(H')$. But since λ_i and λ_j take real values on $(-1)^{1/2}\mathfrak{h}_{t_0}$, it follows that $\Lambda(H(s^{-1}))$ is real. Therefore from condition (1) on Λ, we conclude that $\Lambda(H(s^{-1})) = 0$ which however contradicts condition (2). Hence $s \in W_1$ and so both $\mathfrak{h}_{\mathfrak{p}}$ and $\mathfrak{h}_{\mathfrak{t}}$ are stable under s. But then it is obvious that $\Lambda_0 = s\Lambda = \tau\Lambda$ where $\tau = \tau_s$.

§13.   Proof of Lemma 33

We have only to prove the existence of the elements v_s ($s \in \mathfrak{w}$). First note that the left side (in the statement of Lemma 33) remains unchanged if we replace v by $\sigma_1(m)v\sigma_2(m^{-1})$ ($m \in M$). Therefore it is permissible to replace v by

$$v_0 = \int_M \sigma_1(m)v\sigma_2(m^{-1})\, dm$$

where dm is the normalized Haar measure of M. Hence we can assume that $v \in V_0$. Since $V_0 = V_1 + V_2 + \cdots + V_r$ (in the notation of §12), it would be enough to consider the case $v \in V_i$ for some i. So without loss of generality we can suppose that $v \in V_1$.

Let Λ denote the linear function $H \to (-1)^{1/2}\langle H_0, H\rangle$ ($H \in \mathfrak{h}_{\mathfrak{p}}$) on $\mathfrak{h}_{\mathfrak{p}}$. For any $v_0 \in V$, put

$$\phi(x:v_0) = \int_K \sigma_1(k_x)v_0\sigma_2(k^{-1})\exp\{\Lambda(H(xk)) - \rho(H(xk))\}\, dk \qquad (x \in G).$$

Lemma 39. *Let $x, y \in G$ and $v_0 \in V$. Then*

$$\int_K \phi(xy^k:v_0)\, dx = \phi\left(x: \int_K v_0\sigma_2(\kappa(y^k))\exp\{\Lambda(H(y^k)) - \rho(H(y^k))\}\, dk\right).$$

Note that $\kappa(xy) = \kappa(x\kappa(y))$ and $H(xy) = H(x\kappa(y)) + H(y)$. Hence if $\Lambda_- = \Lambda - \rho$, it follows that

$$\int_K \phi(xy^k:v_0)\, dx = \int_{K \times K} \sigma_1(\kappa(xy^k u))v_0\sigma_2(u^{-1})\exp\Lambda_-(H(xy^k u))\, dk\, du$$

$$= \int_{K \times K} \sigma_1(\kappa(xuy^k))v_0\sigma_2(u^{-1})\exp\Lambda_-(H(xuy^k))\, dk\, du.$$

Now $\kappa(xuy^k) = \kappa(xu\kappa(y^k))$ and $H(xuy^k) = H(xu\kappa(y^k)) + H(y^k)$. Therefore if we put $k_1 = u\kappa(y^k)$, we get

$$\int_K \phi(xy^k:v_0)\, dx = \int_{K \times K} \sigma_1(\kappa(xk_1))v_0\sigma_2(\kappa(y^k)k_1^{-1})$$

$$\times \exp\{\Lambda_-(H(xk_1) + \Lambda_-(H(y^k))\}\, dk\, dk_1$$

and this is equivalent to the statement of the lemma.

Corollary. *Let $q \in I_{\mathfrak{g}}$ and suppose $q \equiv \Sigma_{1 \leqslant j \leqslant t} k_j u_j \bmod \mathfrak{G}\mathfrak{n}_+$ where $k_j \in \mathfrak{R}$ and $u_j \in \mathfrak{H}_{\mathfrak{p}}$. Then*

$$\phi(x; q: v_0) = \sum_j \eta_{\Lambda - \rho}(u_j)\phi\big(x: v_0\sigma_2(k_j)\big) \qquad (x \in G).$$

This is an immediate consequence of the above lemma.

Now let $z \in \mathfrak{Z}$ and suppose $\mu'(z) = \Sigma_{1 \leqslant j \leqslant t} u_j v_j$ where $v_j \in \mathfrak{M}_{\mathfrak{k}}$ and $u_1, u_2, \ldots, u_t$ are linearly independent elements of $\mathfrak{H}_{\mathfrak{p}}$. Then as we saw during the proof of Lemma 35, $\gamma(z) = \Sigma_j \mathop{}^{\backprime}\! u_j \gamma_-(v_j)$ where $\mathop{}^{\backprime}\! u_j = e^\rho u_j \circ e^{-\rho}$ and

$$v\sigma_2(v_j) = \sigma_1(v_j)v = \eta_{\lambda_1}\big(\gamma_-(v_j)\big)v$$

since $v \in V_1$. Therefore from the corollary of Lemma 39 we find that

$$\phi(x; z: v) = \sum_j \eta_\Lambda(\mathop{}^{\backprime}\! u_j)\eta_{\lambda_1}\big(\gamma_-(v_j)\big)\phi(x: v)$$

$$= \chi_{\Lambda + \lambda_1}(z)\phi(x: v).$$

Hence if we put $\phi_0(h) = e^{\rho(\log h)}\phi(h: v)$ $(h \in A^+)$, it is obvious that

$$\delta(z)\phi_0 = \chi_{\Lambda + \lambda_1}(z)\phi_0 \qquad (z \in \mathfrak{Z}).$$

Now first suppose $H_0 \in \mathfrak{h}'_{\mathfrak{p}_0}$ and $\langle H_0, H(s)\rangle \neq 0$ for any $s \in W_2$. Then Λ satisfies the conditions of Lemma 38. Hence every principal exponent of ϕ_0 is of the form $s\Lambda$ $(s \in \mathfrak{w})$. We may obviously assume for our proof that $\phi_0 \neq 0$. Let $\Lambda_1, \ldots, \Lambda_k$ be all the distinct principal exponents of ϕ_0. Let p_j $(1 \leqslant j \leqslant k)$ be the polynomial mappings of $\mathfrak{h}_{\mathfrak{p}}$ into V_0 such that $\Sigma_{1 \leqslant j \leqslant k}\phi_j$ is the principal part of ϕ_0 where $\phi_j(h) = p_j(\log h)e^{\Lambda_j(\log h)}$. Then it follows from Corollary 1 of Lemma 10 that

$$\delta(\mu'(z))\phi_j = \chi_{\Lambda + \lambda_1}(z)\phi_j \qquad (z \in \mathfrak{Z}, \ 1 \leqslant j \leqslant k)$$

and so we conclude from the corollary of Lemma 35 that $E_i p_j$ is a constant for every i $(1 \leqslant i \leqslant r)$. This implies that $p_j = \Sigma_{1 \leqslant i \leqslant r} E_i p_j$ is also a constant. So we can choose elements $v_s \in V_0$ such that $\Sigma_{s \in \mathfrak{w}} e^{s\Lambda} v_s$ is the principal part of ϕ_0. Now put

$$\psi(h) = \phi_0(h) - \sum_{s \in \mathfrak{w}} \Phi(sH_0: h)v_s \qquad (h \in A^+),$$

where Φ is defined as in §11. We claim that $\psi = 0$. This is seen as follows. We know from Theorem 3 that

$$\Phi(sH_0: h; \delta(z))v_s = \Phi(sH_0: h)\Big(v_s\sigma_2\big(\xi(z:(-1)^{1/2}sH_0)\big)\Big)$$

for $z \in \mathfrak{Z}$. Let $\mu'(z) = \Sigma_{1 \leqslant j \leqslant t} u_j v_j$ as before. Then since $z - \mu'(z) \in \mathfrak{n}_-\mathfrak{G}\mathfrak{n}_+$ from Lemma 2, we conclude that

$$\xi\big(z:(-1)^{1/2}sH_0\big) = \sum_j \eta_{s\Lambda - \rho}(u_j)v_j.$$

Therefore

$$v_s\sigma_2\big(\xi(z:(-1)^{1/2}sH_0)\big) = \sigma_1\big(\xi(z:(-1)^{1/2}sH_0)\big)v_s$$

$$= \sum_j \eta_{s\Lambda - \rho}(u_j)\sigma_1(v_j)v_s$$

since $v_s \in V_0$. But we have seen above that the function $\phi_s : h \to e^{(-1)^{1/2}\langle sH_0, \log H\rangle} v_s$ satisfies the differential equation $\delta(\mu'(z))\phi_s = \chi_{\Lambda+\lambda_i}(z)\phi_s$. Hence

$$\sum_j \eta_{s\Lambda-\rho}(u_j)\sigma_1(v_j)v_s = \chi_{\Lambda+\lambda_i}(z)v_s$$

and therefore

$$v_s\sigma_2\Big(\xi\big(z : (-1)^{1/2}sH_0\big)\Big) = \chi_{\Lambda+\lambda_i}(z)v_s.$$

This proves that $\delta(z)\psi = \chi_{\Lambda+\lambda_i}(z)\psi$.

Now suppose that contrary to our claim $\psi \neq 0$. Then by Lemma 38 every principal exponent of ψ is of the form $s\Lambda$ ($s \in \mathfrak{w}$). On the other hand since $\sum_{s\in\mathfrak{w}} e^{s\Lambda} v_s$ is the principal part of both ϕ_0 and $\sum_{s\in\mathfrak{w}} \Phi(sH_0 : h)v_s$, it is obvious that no principal exponent of ψ can be of the form $s\Lambda$ ($s \in \mathfrak{w}$). This contradiction proves that $\psi = 0$ and therefore

$$\phi_0(h) = \sum_{s\in\mathfrak{w}} \Phi(sH_0 : h)v_s \qquad (h \in A^+)$$

in this case.

In order to prove Lemma 33 in the general case we need some more work. It follows from what we have seen above that

$$\sigma_1(\xi(z : \Lambda))E_i = \chi_{\Lambda+\lambda_i}(z)E_i \qquad (1 \leqslant i \leqslant r, \ z \in \mathfrak{Z})$$

for any linear function Λ on $\mathfrak{h}_\mathfrak{p}$. Fix i and s ($1 \leqslant i \leqslant r, \ s \in \mathfrak{w}$) and let $V(i, s)$ denote the set of all $v \in V_0$ such that

$$\sigma_1(\xi(z : s\Lambda))\sigma = \chi_{\Lambda+\lambda_i}(z)E_i$$

for every $z \in \mathfrak{Z}$ and every Λ. Let $J(i, s)$ denote the set of all indices j ($1 \leqslant j \leqslant r$) such that $\chi_{s\Lambda+\lambda_j} = \chi_{\Lambda+\lambda_i}$ for every Λ. Since $\sigma_1(\xi(z : s\Lambda))E_j = \chi_{s\Lambda+\lambda_j}(z)E_j$, it follows that $V(i, s) = \sum_{j \in J(i,s)} V_j$.

Let W_- be the subgroup of W generated by the Weyl reflexions s_α corresponding to $\alpha \in P_-$. Then (see Lemma 52 of the Appendix) we can identify $\mathfrak{w}$ with W_1/W_-. Let $I_-(\mathfrak{h})$ denote the algebra of all invariants of W_- in $\mathfrak{S} = S(\mathfrak{h})$. We have the isomorphism γ_- of $\mathfrak{Z}_\mathfrak{m}$ onto $I_-(\mathfrak{h})$ (see §2). Now W operates on $\mathfrak{S}$ in the obvious way. It is clear that if $s \in W_1$ and $u \in I_-(\mathfrak{h})$, then u^s depends only on the coset sW_-. For any $\tau \in \mathfrak{w}$ and $u \in I_-(\mathfrak{h})$, put $u^\tau = u^s$ where s is any element in W_1 whose image in $\mathfrak{w}$ is τ. In this way $\mathfrak{w}$ operates on $I_-(\mathfrak{h})$. Since γ_- is an isomorphism, we can make $\mathfrak{w}$ operate on $\mathfrak{Z}_\mathfrak{m}$ by defining z^τ ($z \in \mathfrak{Z}_\mathfrak{m}, \tau \in \mathfrak{w}$) by the rule $\gamma_-(z^\tau) = (\gamma_-(z))^\tau$. For any linear function λ on $\mathfrak{h}_\mathfrak{f}$, put $\beta_\lambda(z) = \eta_\lambda(\gamma_-(z))$ ($z \in \mathfrak{Z}_\mathfrak{m}$).

Lemma 40. *Fix i, s ($1 \leqslant i \leqslant r, \ s \in \mathfrak{w}$). Then the following three statements for any index j ($1 \leqslant j \leqslant r$) are equivalent.*

(1) $j \in J(i, s)$.

(2) *There exists a linear function Λ on $\mathfrak{h}_\mathfrak{p}$ such that (a) $t\Lambda \neq \Lambda$ for any $t \neq 1$ in $\mathfrak{w}$, (b) Λ satisfies the conditions of Lemma 38 and (c) $\chi_{s\Lambda+\lambda_j} = \chi_{\Lambda+\lambda_i}$.*

(3) $\beta_{\lambda_j}(z^s) = \beta_{\lambda_i}(z)$ *for all* $z \in \mathfrak{Z}_\mathfrak{m}$.

Obviously (1) implies (2). Now suppose (2) holds. Then $s\Lambda + \lambda_j = t(\Lambda + \lambda_i)$ for some $t \in W$. Then we conclude, as in the proof of Lemma 38, that $t \in W_1$, $s\Lambda = t\Lambda$ and $\lambda_j = t\lambda_i$. Let τ be the image of t in $\mathfrak{w}$. Then $s\Lambda = \tau\Lambda$ and therefore $s = \tau$ in view of our hypothesis (a) on Λ. Now if $z \in \mathfrak{Z}_m$,

$$\beta_{\lambda_j}(z^s) = \eta_{\lambda_j}\big((\gamma_-(z))^t\big) = \eta_{\lambda_i}(\gamma_-(z)) = \beta_{\lambda_i}(z)$$

since $\lambda_i = t^{-1}\lambda_j$. This proves that (2) implies (3). Now suppose (3) holds. Fix $z \in \mathfrak{Z}$ and suppose

$$\mu(z) = \sum_{1 \le k \le q} u_k v_k$$

where $u_k \in \mathfrak{H}_\mathfrak{p}$ and $v_k \in \mathfrak{Z}_m \cap \mathfrak{M}_\mathfrak{k}$. Then $\gamma(z) = \sum_k u_k \gamma_-(v_k)$. Therefore since $\gamma(z) = (\gamma(z))^s = \sum u_k^s(\gamma_-(v_k))^s$, we get for any Λ,

$$\chi_{s\Lambda+\lambda_j}(z) = \sum_k \eta_\Lambda(u_k)\beta_{\lambda_j}(v_k^s) = \sum_k \eta_\Lambda(u_k)\beta_{\lambda_i}(v_k) = \chi_{\Lambda+\lambda_i}(z).$$

This shows that (3) implies (1) and so the lemma is proved.

Now put

$$\theta(\Lambda:h:v) = e^{\rho(\log h)}\int_K \sigma_1(k_h)v\sigma_2(k^{-1})\exp\{(-1)^{1/2}\Lambda(H(hk)) - \rho(H(hk))\}\,dk$$

for $h \in A$, $v \in V_0$ and any linear function Λ on $\mathfrak{h}_\mathfrak{p}$. By Corollary 1 of Lemma 7, we can select a finite number of elements $u_1 = 1$, $u_2, \ldots, u_q \in \mathfrak{H}_\mathfrak{p}$ such that

$$\mathfrak{G} = \mathfrak{K}^{h^{-1}}\bigg(\sum_{1 \le j \le q} \mathfrak{Z}u_j'\bigg)\mathfrak{K}$$

for $h \in A^+$. (As usual $u' = c^{-\rho}u \circ c^\rho$ for $u \in \mathfrak{H}_\mathfrak{p}$). Fix a point $h_0 \in A^+$.

Lemma 41. *Let ϕ be a C^∞ function from A^+ to V_0 such that ϕ is an eigenfunction of $\delta'(z)$ for every $z \in \mathfrak{Z}$. Then if $\phi(h_0; u_j') = 0$ $(1 \le j \le q)$, we can conclude that $\phi = 0$.*

Put $G_+ = KA^+K$. Then as we have seen during the proof of Lemma 36, ϕ can be extended to a C^∞ function from G_+ to V such that $\phi(k_1 xk_2) = \sigma_1(k_1)\phi(x)\sigma_2(k_2)$ $(x \in G_+, k_1, k_2 \in K)$. Moreover there exists a homomorphism χ of $\mathfrak{Z}$ into $\mathbf{C}$ such that $z\phi = \chi(z)\phi$ $(z \in \mathfrak{Z})$. Hence we conclude as before (see the proof of Lemma 8) that ϕ is analytic.

Now if $k_1, k_2 \in \mathfrak{K}$ and $z \in \mathfrak{Z}$, it is clear that

$$\phi\Big(h_0; (k_1)^{h_0^{-1}}zu_j'k_2\Big) = \chi(z)\sigma_1(k_1)\phi(h_0; u_j')\sigma_2(k_2) = 0.$$

Since $\mathfrak{G} = \mathfrak{K}^{h_0^{-1}}(\sum_{1 \le j \le q}\mathfrak{Z}u_j')\mathfrak{K}$, this proves that $\phi(h_0; g) = 0$ for every $g \in \mathfrak{G}$. But ϕ being analytic, this implies that $\phi = 0$.

Let V^q denote the direct sum of V with itself q times. For any $\Lambda \in {}^\backprime\mathfrak{h}_\mathfrak{p}$ and $v \in V_0$ consider the vector $\Phi(\Lambda:v)$ in V^q given by

$$\Phi(\Lambda:v) = \big(\Phi(\Lambda:h_0;u_1)v, \ldots, \Phi(\Lambda:h_0;u_q)v\big).$$

Similarly for $\Lambda \in \mathfrak{h}_{\mathfrak{p}}$ and $v \in V_0$ define the vector $\Theta(\Lambda : v)$ in V^q by

$$\Theta(\Lambda : v) = \big(\theta(\Lambda : h_0; u_1 : v),\ldots,\theta(\Lambda : h_0; u_q : v)\big).$$

For $s \in \mathfrak{w}$, select a base v_{si} $(1 \leqslant i \leqslant N_s)$ for $V(1, s)$. Let $*\mathfrak{h}_{\mathfrak{p}}$ be the set of all points $\Lambda \in \mathfrak{h}_{\mathfrak{p}}$ such that $s\Lambda \in {}^\backprime\mathfrak{h}_{\mathfrak{p}}$ for every $s \in \mathfrak{w}$.

Lemma 42. *Let* $N = \sum_{s \in \mathfrak{w}} N_s$. *Then for any* $\Lambda \in *\mathfrak{h}_{\mathfrak{p}}$ *and* $v \in V_1$, *the* $N+1$ *vectors* $\Theta(\Lambda : v)$, $\Phi(s\Lambda : v_{si})$ $(1 \leqslant i \leqslant N_s, s \in \mathfrak{w})$ *are linearly dependent.*

Put $\Phi_{si}(\Lambda) = \Phi(s\Lambda : v_{si})$. Then by Lemma 31, Φ_{si} is a holomorphic function from $*\mathfrak{h}_{\mathfrak{p}}$ to V^q. We arrange these N functions in a sequence Φ_j $(1 \leqslant j \leqslant N)$. Put $\Phi_0(\Lambda) = \Theta(\Lambda : v)$. Then Φ_i $(0 \leqslant i \leqslant N)$ are $N+1$ holomorphic functions from $*\mathfrak{h}_{\mathfrak{p}}$ to V^q. Choose a base β_j $(1 \leqslant j \leqslant q\nu)$ for the space of linear functions on V^q $(\nu = \dim V)$ and put $D_{ji}(\Lambda) = \beta_j(\Phi_i(\Lambda))$ $(1 \leqslant j \leqslant q\nu, 0 \leqslant i \leqslant N)$. Then $D = (D_{ji})$ is a $q\nu \times (N+1)$ matrix whose coefficients are holomorphic functions on $*\mathfrak{h}_{\mathfrak{p}}$. It is enough to prove that for any $\Lambda \in *\mathfrak{h}_{\mathfrak{p}}$, rank $D(\Lambda) \leqslant N$. For this we may obviously assume that $N \geqslant q\nu$. Let Δ be the determinant of an $N \times N$ submatrix of D. We have to prove that $\Delta = 0$. Now it is obvious that there exists a nonempty open set U in $\mathfrak{h}'_{\mathfrak{p}_0}$ such that $\langle H_0, H(s)) \rangle \neq 0$ for $H_0 \in U$ and $s \in W_2$ and $U \subset *\mathfrak{h}_{\mathfrak{p}}$. Then we have seen above that for any $\Lambda \in U$, we can choose elements $v_s \in V_0$ such that

$$\theta(\Lambda : h : v) = \sum_{s \in \mathfrak{w}} \Phi(s\Lambda : h) v_s \qquad (h \in A^+).$$

Moreover since $v \in V_1$, $\theta(\Lambda : h; \delta(z) : v) = \chi_{\Lambda^* + \lambda_1}(z)\theta(\Lambda : h : v)$ $(z \in \mathfrak{Z})$ where $\Lambda^* = (-1)^{1/2}\Lambda$. On the other hand

$$\Phi(s\Lambda : h; \delta(z))E_i = \chi_{s\Lambda^* + \lambda_0}(z)\Phi(s\Lambda : h)$$

for any i $(1 \leqslant i \leqslant r)$ since $\sigma_1(\xi(z : s\Lambda^*))E_i = \chi_{s\Lambda^* + \lambda_i}(z)E_i$. Therefore it is obvious from Lemmas 32 and 40 that $E_i v_s = 0$ unless $i \in J(1, s)$. This shows that $v_s = \sum_i E_i v_s \in V(1, s)$ $(s \in \mathfrak{w})$. Since

$$\theta(\Lambda : h_0; u_j : v) = \sum_{s \in \mathfrak{w}} \Phi(s\Lambda : h_0; u_j) v_s \qquad (1 \leqslant j \leqslant q),$$

it is now obvious that $\Phi_0(\Lambda)$ is a linear combination of $\Phi_j(\Lambda)$ $(1 \leqslant j \leqslant N)$. Hence $\Delta(\Lambda) = 0$ for $\Lambda \in U$. But since Δ is a holomorphic function on $*\mathfrak{h}_{\mathfrak{p}}$ and $*\mathfrak{h}_{\mathfrak{p}}$ is connected[28], it is now obvious that $\Delta = 0$. Thus Lemma 42 is proved.

Lemma 43. *Put* $*\mathfrak{h}'_{\mathfrak{p}} = *\mathfrak{h}_{\mathfrak{p}} \cap \mathfrak{h}'_{\mathfrak{p}}$. *Then the vectors* $\Phi_j(\Lambda)$ $(1 \leqslant j \leqslant N)$ *are linearly independent for every* $\Lambda \subset *\mathfrak{h}'_{\mathfrak{p}}$.

For otherwise fix Λ in $*\mathfrak{h}'_{\mathfrak{p}}$ such that $\Phi_j(\Lambda)$ $(1 \leqslant j \leqslant N)$ are linearly dependent. Then we can choose elements $v_s \in V(1, s)$ $(s \in \mathfrak{w})$ not all zero such that

$$\sum_{s \in \mathfrak{w}} \Phi(s\Lambda : v_s) = 0.$$

---

[28] This follows from the fact that $*\mathfrak{h}_{\mathfrak{p}}$ is the complement in $\mathfrak{h}_{\mathfrak{p}}$ of a countable number of complex hyperplanes such that any compact subset of $\mathfrak{h}_{\mathfrak{p}}$ meets only a finite number of these.

Put

$$\phi(h) = e^{-\rho(\log h)} \sum_{s \in \mathfrak{w}} \Phi(s\Lambda : h) v_s \qquad (h \in A^+).$$

Then if $z \in \mathfrak{Z}$, it is clear that

$$\phi(h; \delta'(z)) = e^{-\rho(\log h)} \sum_{s \in \mathfrak{w}} \Phi(s\Lambda : h; \delta(z)) v_s$$

$$= e^{-\rho(\log h)} \sum_{s \in \mathfrak{w}} \Phi(s\Lambda : h) \sigma_1(\xi(z : s\Lambda^*)) v_s$$

$$= \chi_{\Lambda^* + \lambda_1}(z) \phi(h)$$

since $v_s \in V(1, s)$. $(\Lambda^* = (-1)^{1/2}\Lambda$ as before.) On the other hand

$$\phi(h_0; u_j') = e^{-\rho(\log h_0)} \sum_{s \in \mathfrak{w}} \Phi(s\Lambda : h_0; u_j) v_s = 0$$

since $\sum_{s \in \mathfrak{w}} \Phi(s\Lambda : v_s) = 0$. Therefore $\phi = 0$ from Lemma 41. Now the elements $s\Lambda$ $(s \in \mathfrak{w})$ are all distinct since $\Lambda \in \mathfrak{h}'_\mathfrak{p}$. Hence we conclude from Lemma 32 that $v_s = 0$ $(s \in \mathfrak{w})$. Since this contradicts the definition of v_s, we get the lemma.

It follows from Lemmas 42 and 43 that for any $\Lambda \in {}^*\mathfrak{h}'_\mathfrak{p}$, $\Phi_0(\Lambda)$ is a linear combination of $\Phi_j(\Lambda)$ $(1 \leqslant j \leqslant N)$. Fix a point $\Lambda_0 \in {}^*\mathfrak{h}'_\mathfrak{p}$. Then by Lemma 43, the matrix $(D_{ij}(\Lambda_0))_{1 \leqslant i \leqslant qv, \ 1 \leqslant j \leqslant N}$ is of rank N. Hence we can select an $N \times N$ submatrix, say $(D_{ij}(\Lambda_0))_{1 \leqslant i, j \leqslant N}$ such that $\Delta(\Lambda_0) \neq 0$ where $\Delta = \det(D_{ij})_{1 \leqslant i, j \leqslant N}$. Therefore we can choose an open connected neighborhood U of Λ_0 in ${}^*\mathfrak{h}'_\mathfrak{p}$ such that Δ is nowhere zero on U. From this it is obvious that there exist holomorphic functions c_j $(1 \leqslant j \leqslant N)$ on U such that

$$\Phi_0(\Lambda) = \sum_{1 \leqslant j \leqslant N} c_j(\Lambda) \Phi_j(\Lambda)$$

for any $\Lambda \in U$. Thus we have obtained the following result.

Lemma 44. *For any point* $\Lambda \in {}^*\mathfrak{h}'_\mathfrak{p}$, *there exist unique elements* $v_s(\Lambda) \in V(1, s)$ $(s \in \mathfrak{w})$ *such that*

$$\Phi_0(\Lambda) = \sum_{s \in \mathfrak{w}} \Phi(s\Lambda : v_s(\Lambda)).$$

Moreover $v_s(\Lambda)$ *is a holomorphic function of* Λ *on* ${}^*\mathfrak{h}'_\mathfrak{p}$.

Now for a fixed $\Lambda \in {}^*\mathfrak{h}'_\mathfrak{p}$, put

$$\phi(h) = \theta(\Lambda : h : v) - \sum_{s \in \mathfrak{w}} \Phi(s\Lambda : h) v_s(\Lambda) \qquad (h \in A^+).$$

Then since $v \in V_1$, we can show, as in the proof of Lemma 43 that

$$\delta(z)\phi = \chi_{\Lambda^* + \lambda_1}(z)\phi \qquad (z \in \mathfrak{Z}).$$

Moreover $\phi(h_0; u_j) = 0$ $(1 \leqslant j \leqslant q)$ since

$$\Phi_0(\Lambda) = \sum_{s \in \mathfrak{w}} \Phi(s\Lambda : v_s(\Lambda)).$$

Therefore $\phi = 0$ from Lemma 41. This proves that

$$\theta(\Lambda : h : v) = \sum_{s \in \mathfrak{w}} \Phi(s\Lambda : h) v_s(\Lambda) \qquad (h \in A^+).$$

The proof of Lemma 33 is now complete.

Theorem 4. *Let* $*\mathfrak{h}'_{\mathfrak{p}}$ *denote the set of all* $H_0 \in \mathfrak{h}'_{\mathfrak{p}}$ *such that* $sH_0 \in {}^\backprime\mathfrak{h}_{\mathfrak{p}}$ *for all* $s \in \mathfrak{w}$. *Then for any* $H_0 \in *\mathfrak{h}'_{\mathfrak{p}}$ *there exist uniquely determined elements* $\mathbf{c}_s(H_0) \in \mathcal{E}(V_0)$ $(s \in \mathfrak{w})$ *such that*

$$e^{\rho(\log h)} \int_K \sigma_1(k_h) v \sigma_2(k^{-1}) \exp\{(-1)^{1/2}\langle H_0, H(hk) \rangle - \rho(H(hk))\} \, dk$$

$$= \sum_{s \in \mathfrak{w}} \Phi(sH_0 : h) \mathbf{c}_s(H_0) v$$

for $h \in A^+$ *and* $v \in V_0$. *Moreover the* $\mathbf{c}_s$ *are holomorphic functions from* $*\mathfrak{h}'_{\mathfrak{p}}$ *to* $\mathcal{E}(V_0)$.

This is merely a restatement of the above results.

§14.   The functions $\mathbf{c}_s$

We now intend to prove that the $\mathbf{c}_s$ are actually meromorphic functions on $\mathfrak{h}_{\mathfrak{p}}$.

Lemma 45. *Let* U *be a nonempty, open, connected and relatively compact set in* $\mathfrak{h}_{\mathfrak{p}}$. *Then we can choose a holomorphic function* Δ *on* U *such that* $\Delta \neq 0$ *and for every* $s \in \mathfrak{w}$, $\Delta \mathbf{c}_s$ *can be extended to a holomorphic function* $\mathbf{b}_s$ *from* U *to* $\mathcal{E}(V_0)$.

By enlarging U, if necessary, we can assume that $sU = U$ for $s \in \mathfrak{w}$. Choose an element $p \in S(\mathfrak{h}_{\mathfrak{p}})$ corresponding to Theorem 3.

Lemma 46. *Let* v *be an element in* V_i *for some* i $(1 \leqslant i \leqslant r)$. *Then there exists a holomorphic function* Δ *on* U *such that the following conditions hold.*

(1) $\Delta \neq 0$.

(2)   *For each* $s \in \mathfrak{w}$, *the function* $\Lambda \to \Delta(\Lambda) \mathbf{c}_s(\Lambda) v$ $(\Lambda \in U \cap *\mathfrak{h}'_{\mathfrak{p}})$ *can be extended to a holomorphic function from* U *to* V_0.

Assuming this for a moment, we can prove Lemma 45 as follows. Choose a base v_j $(1 \leqslant j \leqslant \nu_0)$ for V_0 such that every v_j lies in some V_i $(1 \leqslant i \leqslant r)$. For each j choose a holomorphic function Δ_j corresponding to Lemma 46. Then it is obvious that $\Delta = \Pi_{1 \leqslant j \leqslant \nu_0} \Delta_j$ fulfills the requirements of Lemma 45.

For the proof of Lemma 46, we may assume without loss of generality that $i = 1$ so that $v \in V_1$. Define $\Psi(\Lambda : h)$ $(\Lambda \in U, h \in A^+)$ as in Theorem 3 and for any $w \in V_0$

and $\Lambda \in U$, let $\Psi(\Lambda:w)$ denote the vector in V^q given by

$$\Psi(\Lambda:w) = \big(\Psi(\Lambda:h_0;u_1)w,\ldots,\Psi(\Lambda:h_0;u_q)w\big).$$

(We keep to the notation of §13.) Put $\Psi_{si}(\Lambda) = \Psi(s\Lambda:v_{si})$ $(1 \leqslant i \leqslant N_s, s \in \mathfrak{w})$. Then Ψ_{si} are holomorphic functions from U to V^q. Again we arrange them in a sequence Ψ_j $(1 \leqslant j \leqslant N)$. Define β_i $(1 \leqslant i \leqslant q\nu)$ as during the proof of Lemma 42 and put $D_{ij}(\Lambda) = \beta_i(\Psi_j(\Lambda))$. Then D_{ij} are holomorphic functions on U. Since p is not identically zero, it follows from Lemma 43 that we can choose a point $\Lambda_0 \in U$ such that the rank of the matrix $(D_{ij}(\Lambda_0))_{1 \leqslant i \leqslant q\nu, 1 \leqslant j \leqslant N}$ is N. Therefore $q\nu \geqslant N$ and we can choose a $N \times N$ submatrix, say $(D_{ij})_{1 \leqslant i, j \leqslant N} = \mathbf{D}$, such that $\Delta(\Lambda_0) \neq 0$ where $\Delta = \det \mathbf{D}$. Obviously there exists a matrix $\mathbf{D}' = (D_{ij}')_{1 \leqslant i, j \leqslant N}$ whose elements are holomorphic functions on U such that $\mathbf{DD}' = \mathbf{D}'\mathbf{D} = \Delta 1$. (Here 1 stands for the unit matrix of degree N.)

Now put $U' = U \cap {}^*\mathfrak{h}_\mathfrak{p}'$. Then by Theorem 4

$$\Theta(\Lambda:v) = \sum_{s \in \mathfrak{w}} \Psi(s\Lambda:\mathbf{c}_s(\Lambda)v)p(s\Lambda)^{-1}$$

for $\Lambda \in U'$. (We recall that p is nowhere zero on $U \cap {}^{\backprime}\mathfrak{h}_\mathfrak{p}$.) Also, $\mathbf{c}_s(\Lambda)v \in V(1, s)$, as we saw during the proof of Theorem 4. Therefore there exist holomorphic functions c_j $(1 \leqslant j \leqslant N)$ on U' such that

$$\Theta(\Lambda:v) = \sum_{1 \leqslant j \leqslant N} \Psi_j(\Lambda)c_j(\Lambda) \qquad (\Lambda \in U').$$

But then

$$\Delta(\Lambda)c_j(\Lambda) = \sum_{1 \leqslant i \leqslant N} D_{ji}'(\Lambda)\beta_i(\Theta(\Lambda:v)) \qquad (\Lambda \in U').$$

The right side of this equation is obviously defined and holomorphic on U and so this shows that Δc_j can be extended to a holomorphic function on U. This proves that for each $s \in \mathfrak{w}$, there exists a holomorphic function v_s from U to $V(1, s)$ such that

$$\Delta(\Lambda)\Theta(\Lambda:v) = \sum_{s \in \mathfrak{w}} \Psi(s\Lambda:v_s(\Lambda))p(s\Lambda)^{-1}$$

for $\Lambda \in U'$. We note that $\Psi(s\Lambda:h)p(s\Lambda)^{-1} = \Phi(s\Lambda:h)$ $(\Lambda \in U', h \in A^+)$. Therefore by applying Lemma 41, we can conclude as before that

$$\Delta(\Lambda)\theta(\Lambda:h:v) = \sum_{s \in \mathfrak{w}} \Phi(s\Lambda:h)v_s(\Lambda) \qquad (\Lambda \in U').$$

But then it follows from Theorem 4 and Lemma 31 that $v_s(\Lambda) = \Delta(\Lambda)\mathbf{c}_s(\Lambda)v$ $(\Lambda \in U')$. Since $\Delta(\Lambda_0) \neq 0$ and v_s is holomorphic on U, the lemma is proved.

Now assume U is convex and regard it as an open complex submanifold of $\mathfrak{h}_\mathfrak{p}$. For a fixed $s \in \mathfrak{w}$, let $\mathbf{d}_s$ denote the divisor of poles of the meromorphic function $\mathbf{c}_s$ on U. Then since $\mathbf{c}_s$ is regular on U', it is clear that $\mathbf{d}_s$ is an integral linear combination of a finite number of hyperplane sections on U. Therefore it is obvious that the function Δ in Lemma 45 may be chosen to be a finite product of (inhomogeneous) linear functions.

117

§15. Appendix

Put $\mathfrak{B} = \lambda(S(\mathfrak{p}))$ where λ is the canonical mapping of $S(\mathfrak{g})$ onto $\mathfrak{G}$. Let γ be the homomorphism of $I_\mathfrak{g}$ into $\mathfrak{H}_\mathfrak{p}$ defined in [8, §4]. Let $g \to g^*$ denote the anti-automorphism of $\mathfrak{G}$ such that $X^* = -X$ for $X \in \mathfrak{g}$. Then we have the following result.

Lemma 47. $\gamma(\theta(q^*)) = \gamma(q)$ *for* $q \in I_\mathfrak{g}$.

We know from [8, §4] that $I_\mathfrak{g} = I_\mathfrak{g} \cap \mathfrak{B} + I_\mathfrak{g} \cap \mathfrak{G}\mathfrak{k}$ and $I_\mathfrak{g} \cap \mathfrak{G}\mathfrak{k} = I_\mathfrak{g} \cap \mathfrak{k}\mathfrak{G}$ is the kernel of γ. Hence it is clear that this kernel is stable under the transformation $q \to \theta(q^*)$ of $I_\mathfrak{g}$. Therefore it would be sufficient to prove the lemma in case $q \in I_\mathfrak{g} \cap \mathfrak{B}$. But since $X = -\theta(X)$ for $X \in \mathfrak{p}$, it is obvious that $\theta(p^*) = p$ for $p \in \mathfrak{B}$. Hence $\theta(q^*) = q$ in this case and so our assertion is true.

Although the following lemma is undoubtedly known, we give a proof for the sake of completeness.

Lemma 48. *Let L be a compact connected Lie group with Lie algebra $\mathfrak{l}$ and let $\mathfrak{b}$ be an abelian subalgebra of $\mathfrak{l}$. Then the centralizer of $\mathfrak{b}$ in L is connected.*

Extend $\mathfrak{b}$ to a maximal abelian subalgebra $\mathfrak{a}$ of $\mathfrak{l}$. Let A be the analytic subgroup of L corresponding to $\mathfrak{a}$. Then since L is compact, $L = \cup_{x \in L} xAx^{-1}$. Let Z be the centralizer of $\mathfrak{b}$ in L and fix an element x in Z. Then $x \in yAy^{-1}$ for some $y \in L$. Let $\mathfrak{l}_x$ denote the centralizer of x in $\mathfrak{l}$ and Z_x the analytic subgroup of L corresponding to $\mathfrak{l}_x$. Then Z_x is compact and since $yAy^{-1} \subset Z_x$, it follows that rank $\mathfrak{l}_x = $ rank $\mathfrak{l}$. Extend $\mathfrak{b}$ to a maximal abelian subalgebra $\mathfrak{a}_1$ of $\mathfrak{l}_x$. Let A_1 be the analytic subgroup of Z_x corresponding to $\mathfrak{a}_1$. Now $x \in yAy^{-1} \subset Z_x$. Therefore since Z_x is compact, $zxz^{-1} \in A_1$ for some $z \in Z_x$. But x being in the center of Z_x, this shows that $x \in A_1$. Choose $X \in \mathfrak{a}_1$ such that $x = \exp X$ and let $\mathfrak{b}_1 = \mathfrak{b} + \mathbb{R} X$. Then $\mathfrak{b}_1$ is an abelian subalgebra of $\mathfrak{l}$ and if $B_1 = \exp \mathfrak{b}_1$, it is obvious that $x \in B_1 \subset Z$. Since B_1 is connected, this shows that x lies in the connected component of 1 in Z. But x being an arbitrary element of Z, this proves that Z is connected.

The following lemma is due to Chevalley (although his proof is different).

Lemma 49. (Chevalley). *Let $\mathfrak{a}$ be a subset of $\mathfrak{h}$ and W the Weyl group of $\mathfrak{g}$ with respect to $\mathfrak{h}$. Let $W_\mathfrak{a}$ denote the subgroup consisting of those elements of W which leave $\mathfrak{a}$ pointwise fixed. Then $W_\mathfrak{a}$ is generated by the Weyl reflexions s_α corresponding to those roots α which vanish everywhere on $\mathfrak{a}$.*

We can obviously replace $\mathfrak{a}$ by the vector subspace of $\mathfrak{h}$ spanned by it. Hence we may assume that $\mathfrak{a}$ is a subspace. Let $P_\mathfrak{a}$ denote the set of all positive roots which vanish everywhere on $\mathfrak{a}$. If $\alpha \in P_\mathfrak{a}$, it is clear that $s_\alpha \in W_\mathfrak{a}$. Let W' be the subgroup of $W_\mathfrak{a}$ generated by s_α for all $\alpha \in P_\mathfrak{a}$. We have to prove $W' = W_\mathfrak{a}$.

Put $\mathfrak{h}_* = (-1)^{1/2}\mathfrak{h}_{\mathfrak{p}_0} + \mathfrak{h}_{\mathfrak{t}_0}$ and $\mathfrak{g}_* = \mathfrak{k}_0 + (-1)^{1/2}\mathfrak{p}_0$. Then $\mathfrak{g}_*$ is a compact real form of $\mathfrak{g}$. Let U denote the connected adjoint group of $\mathfrak{g}_*$. Then U is compact. We denote by η the conjugation of $\mathfrak{g}$ with respect to $\mathfrak{g}_*$ so that $\eta(X + (-1)^{1/2}Y) = X - (-1)^{1/2}Y$ $(X, Y \in \mathfrak{g}_*)$. Let $\mathfrak{a}_*$ denote the subspace of $\mathfrak{h}_*$ spanned over $\mathbb{R}$ by all elements of the form $H + \eta(H)$ and $(-1)^{1/2}(H - \eta(H))$ $(H \in \mathfrak{a})$. Since a root α takes only pure imaginary values on $\mathfrak{h}_*$, it is obvious that α vanishes identically on α if and only if it does the same on $\mathfrak{a}_*$.

Now fix $s \in W_\alpha$ and choose $u \in U$ such that $uH = sH$ for all $H \in \mathfrak{h}_*$. Let Z denote the centralizer of $\mathfrak{a}_*$ in U. Then by Lemma 48, Z is connected. Let $\mathfrak{z}_0$ be the Lie algebra of Z and $\mathfrak{z}$ its complexification in $\mathfrak{g}$. It is obvious that $\mathfrak{z}$ is the centralizer of $\mathfrak{a}$ (or $\mathfrak{a}_*$) in $\mathfrak{g}$ and therefore $\mathfrak{z} = \mathfrak{h} + \sum_{\alpha \in P_\alpha}(\mathbb{C} X_\alpha + \mathbb{C} X_{-\alpha})$. Now $\mathfrak{z}$ is reductive and its Weyl group (with respect to $\mathfrak{h}$) is generated by s_α $(\alpha \in P_\alpha)$. Since $u \in Z$, we conclude that $s \in W'$. This proves that $W' = W_\alpha$.

Let K be the analytic subgroup of U corresponding to $\mathfrak{k}_0$. If M is the centralizer and M^* the normalizer of $\mathfrak{h}_\mathfrak{p}$ in K, then $\mathfrak{w} = M^*/M$ is the little Weyl group of $\mathfrak{g}_0$ with respect to $\mathfrak{h}_{\mathfrak{p}_0}$.

Lemma 50. *Let s be an element in $\mathfrak{w}$. Then we can choose an element $m \in M^*$ lying in the coset s, such that $m\mathfrak{h} = \mathfrak{h}$.*

Choose $m_1 \in M^*$ such that $m_1 H = sH$ for all $H \in \mathfrak{h}_\mathfrak{p}$. Then $\mathfrak{h}_{\mathfrak{k}_0}$ and $m_1 \mathfrak{h}_{\mathfrak{k}_0}$ are two maximal abelian subalgebras of $\mathfrak{m}_0 \cap \mathfrak{k}_0$. Since M is compact we can choose $m_2 \in M$ such that $m_1 \mathfrak{h}_{\mathfrak{k}_0} = m_2 \mathfrak{h}_{\mathfrak{k}_0}$. Put $m = m_2^{-1} m_1$. Then $mH = m_1 H = sH$ for $H \in \mathfrak{h}_\mathfrak{p}$ and $m\mathfrak{h} = \mathfrak{h}$.

Let W_1 denote the subgroup of those elements of W which map $\mathfrak{h}_\mathfrak{p}$ into itself.

Lemma 51. *Let s be an element in W_1. We can choose $m \in M^*$ such that $sH = mH$ for every $H \in \mathfrak{h}_\mathfrak{p}$.*

Let Q be the set of those roots which do not vanish identically on $\mathfrak{h}_\mathfrak{p}$. It is obvious that s permutes the roots in Q among themselves. Hence $s\mathfrak{h}'_{\mathfrak{p}_0} = \mathfrak{h}'_{\mathfrak{p}_0}$. Therefore $s\mathfrak{h}^+_{\mathfrak{p}_0}$ is a connected component of $\mathfrak{h}'_{\mathfrak{p}_0}$. But we know (see [5, Lemma 37]) that every connected component of $\mathfrak{h}'_{\mathfrak{p}_0}$ is of the form $\sigma\mathfrak{h}^+_{\mathfrak{p}_0}$ for some $\sigma \in \mathfrak{w}$. Hence from Lemma 50, we can choose $m \in M^*$ such that $m\mathfrak{h} = \mathfrak{h}$ and $s\mathfrak{h}^+_{\mathfrak{p}_0} = m\mathfrak{h}^+_{\mathfrak{p}_0}$. Choose $u \in U$ such that $sH = uH$ for $H \in \mathfrak{h}$. Then if $v = m^{-1}u$, $v\mathfrak{h}^+_{\mathfrak{p}_0} = \mathfrak{h}^+_{\mathfrak{p}_0}$. Since $v\mathfrak{h} = \mathfrak{h}$, v defines an element t of W. Let g be the order of t in W. Fix an element $H_0 \in \mathfrak{h}^+_{\mathfrak{p}_0}$ and put $H_1 = g^{-1}\sum_{1 \leqslant q \leqslant g} t^q H_0$. Then since $\mathfrak{h}^+_{\mathfrak{p}_0}$ is convex, $H_1 \in \mathfrak{h}^+_{\mathfrak{p}_0}$ and $tH_1 = H_1$. As before let P_- be the set of all positive roots of $\mathfrak{g}$ which vanish everywhere on $\mathfrak{h}_\mathfrak{p}$ and W_- the subgroup of W generated by the Weyl reflexions s_α corresponding to $\alpha \in P_-$. Since $H_1 \in \mathfrak{h}'_{\mathfrak{p}_0}$, it is obvious that P_- is exactly the set of those roots which vanish at H_1 and therefore from Lemma 49, t lies in W_-. Hence $tH = H$ for all $H \in \mathfrak{h}_\mathfrak{p}$ and this shows that $sH = uH = mH$ $(H \in \mathfrak{h}_\mathfrak{p})$.

Define W_1 and W_- as above. Then obviously $W_- \subset W_1$.

Lemma 52. *For any $s \in W_1$, let τ_s denote the restriction of the operation s on $\mathfrak{h}_\mathfrak{p}$. Then $\tau_s \in \mathfrak{w}$. The mapping $s \to \tau_s$ is a homomorphism of W_1 onto $\mathfrak{w}$ whose kernel is W_-.*

It follows from Lemma 51 that $\tau_s \in \mathfrak{w}$ and from Lemma 50 that the mapping is onto $\mathfrak{w}$. Now $\tau_s = 1$ if and only if $sH = H$ for all $H \in \mathfrak{h}_\mathfrak{p}$. But by Lemma 48 this happens if and only if $s \in W_-$.

119

Harish-Chandra

References

1. Harish-Chandra, Representations of a semisimple Lie group on a Banach space. I. *Trans. Amer. Math. Soc.* **75** (1953), 185–243.
2. —, Representations of semisimple Lie groups. II. *Trans. Amer. Math. Soc.* **76** (1954), 26–65.
3. —, Representations of semisimple Lie groups. III. *Trans. Amer. Math. Soc.* **76** (1954), 234—253.
4. —, The characters of semisimple Lie groups. *Trans. Amer. Math. Soc.* **83**, (1956), 98–163.
5. —, Representations of semisimple Lie groups. VI. Integrable and square-integrable representations. *Amer. J. Math.* **78** (1956), 564–628.
6. —, Differential operators on a semisimple Lie algebra. *Amer. J. Math.* **79** (1957), 87–120.
7. —, Fourier transforms on a semisimple Lie algebra. I. *Amer. J. Math.* **79** (1957), 193–257.
8. —, Spherical functions on a semisimple Lie group. I. *Amer. J. Math.* **80** (1958), 241–310.
9. —, Spherical functions on a semisimple Lie group. II. *Amer. J. Math.* **80** (1958), 553–613.
10. —, Some results on differential equations and their applications. *Proc. Nat. Acad. Sci. U.S.A.* **45** (1959), 1763–1764.
11. —, Some results on differential equations. 1960.
12. Weyl, H., Theorie der Darstellung kontinuierlicher halbeinfacher Gruppen durch lineare Transformationen. I, II, III und Nachtrag. *Math. Zeit.* **23** (1925), 271–309; **24** (1926), 328–376; **24** (1926), 377–395; **24** (1926), 789–791.
13. Godement, R., A theory of spherical functions, I. *Trans. Amer. Math. Soc.* **73** (1952), 496–556.
14. Schwartz, L., Théorie des distributions I, Paris, Hermann, 1950.
15. John, F., General properties of solutions of linear elliptic partial differential equations. *Proc. Symp. on Spectral Theory and Differential Problems*, Stillwater, Okla, 1951, 113–175.

ARITHMETIC SUBGROUPS OF ALGEBRAIC GROUPS

BY ARMAND BOREL AND HARISH-CHANDRA

Communicated by Deane Montgomery, July 22, 1961.

A complex algebraic group G is in this note a subgroup of $GL(n, C)$, the elements of which are all invertible matrices whose coefficients annihilate some set of polynomials $\{P_\mu[X_{11}, \cdots, X_{nn}]\}$ in n^2 indeterminates. It is said to be defined over a field $K \subset C$ if the polynomials can be chosen so as to have coefficients in K. Given a subring B of C, we denote by G_B the subgroup of elements of G which have coefficients in B, and whose determinant is a unit of B. Assume in particular G to be defined over Q. Then G_Z is an "arithmetically defined discrete subgroup" of G_R, or, more briefly, an *arithmetic subgroup* of G_R. A typical example is the group of units of a nondegenerate integral quadratic form, and as a matter of fact, the main results stated below generalize facts known in this case from reduction theory. The proofs will be published elsewhere.

1. **Reductive groups.** A complex algebraic group G is an *algebraic torus* (a torus in the terminology of [1]) if it is connected and can be diagonalized or, equivalently, if it is birationally isomorphic to a product of groups C^* [1, Chapter II]. The group G is *reductive* if its identity component G^0 may be written as $G^0 = T \cdot G'$, where T is a central algebraic torus, and G' is an invariant connected semi-simple group, or, equivalently, if all rational representations of G are fully reducible.

LEMMA 1. *Let $G_1 \supset \cdots \supset G_m$ be reductive algebraic subgroups of $GL(n, C)$, defined over R. Then there exists $a \in SL(n, R)$ such that the groups $a \cdot G_{iR} \cdot a^{-1}$ are stable under $x \to {}^t x$ ($i = 1, \cdots, m$).*

This lemma, formulated in a somewhat different terminology, is due to G. D. Mostow [4]. Lemma 1, for $m = 1$, implies easily that the usual properties of maximal compact subgroups and of the Iwasawa decompositions (see [7] for instance) are valid for real algebraic reductive groups.

LEMMA 2. *Let G be a connected reductive complex algebraic group, H an algebraic subgroup. Then G/H is an affine variety if and only if H is reductive. If G and H are defined over Q, and H is reductive, there exists a rational representation $\pi: G \to GL(m, C)$, defined over Q, such*

579

that there is a point $v \in Z^m$ whose orbit under G is closed and whose isotropy group is H.

The fact that if G/H is an affine variety (or more generally a Stein manifold) then H is reductive, is due to Matsushima [3] (whose proof can be simplified using ordinary cohomology with complex coefficients). The converse is stated in [3] (and attributed to Iwahori and Sigiura); it can be proved by realizing G/H as a closed orbit in a suitable linear representation.

2. **Siegel domains.** Let $G \subset GL(n, C;)$ be an algebraic reductive group defined over R, $G_R = K \cdot A \cdot N$ an Iwasawa decomposition of G_R, and $\mathfrak{g}$, $\mathfrak{k}$, $\mathfrak{a}$, $\mathfrak{n}$ the Lie algebras of G_R, K, A, N respectively. K is a maximal compact subgroup, A is real diagonalizable, connected, N is unipotent, $\mathfrak{a}$ is orthogonal to $\mathfrak{k}$ with respect to the Killing form. Let further $\psi \subset \mathfrak{a}^*$ be the set of roots of $\mathfrak{g}$ with respect to $\mathfrak{a}$ (the restricted roots) and, for $\alpha \in \psi$, let $\mathfrak{g}_\alpha = \{x \in \mathfrak{g}, [h, x] = \alpha(h) \cdot x, h \in \mathfrak{a}\}$. Then, for a suitable ordering on $\mathfrak{a}^*$, we have $\mathfrak{n} = \sum_{\alpha > 0} \mathfrak{g}_\alpha$. Let $A_t = \{a \in A, \alpha(\log a) \leqq t, \alpha \in \psi, \alpha > 0\}$, $(t > 0)$. A *Siegel domain* of G_R, with respect the given Iwasawa decomposition, is a subset $\mathfrak{S}_{t,\omega} = K \cdot A_t \cdot \omega$, where ω is a compact set in N. The $\mathfrak{S}_{t,\omega}$'s, ordered by inclusion, form a filtered set, and their union is G_R. One would obtain equivalent families by letting α run only through the *simple* restricted roots in the definition of A_t, or by replacing A_t by $a_0 \cdot A^-(a_0 \in A)$, A^- being the exponential of the negative Weyl chamber. It is easily seen that if G is semi-simple, a Siegel domain has finite Haar measure.

A *standard Siegel domain* $\mathfrak{S}$ in $GL(n, R)$ is a Siegel domain with respect to the usual Iwasawa decomposition (where $K = O(n)$, A is the group of diagonal matrices with strictly positive entries, N the group of upper triangular unipotent matrices) such that $GL(n, R) = \mathfrak{S} \cdot SL(n, Z)$. The existence of such domains is classical. By a well known theorem of Siegel [8], given $x \in GL(n, Q)$, the set of $y \in SL(n, Z)$ such that $\mathfrak{S} \cap \mathfrak{S} \cdot y \cdot x \neq \varnothing$ (resp. $\mathfrak{S} \cap \mathfrak{S} \cdot x \cdot y \neq \varnothing$) is finite. $\mathfrak{S} \cap SL(n, R)$ will be called a standard Siegel domain of $SL(n, R)$.

LEMMA 3. *Let $\pi: SL(n, C) \to GL(m, C)$ be a right rational representation, defined over Q, $v \in R^m$ be a point whose orbit under $SL(n, R)$ is closed and whose isotropy group in $SL(n, R)$ is stable under $x \to {}^t x$, and let $\mathfrak{S}$ be a standard Siegel domain of $SL(n, R)$. Then $v \cdot \pi(\mathfrak{S}) \cap Z^m$ is finite.*

This lemma can be proved more generally for a rational representation of a reductive group, a suitable Iwasawa decomposition and a Cartan involution compatible with it, but the above special case suffices for the applications. Applied to the natural representation of $SL(n, C)$ in the space of quadratic forms, it yields the finiteness of the number of reduced integral forms with a given nonzero determinant, stated first by Hermite [2].

3. Fundamental sets for arithmetic subgroups. As is well known, Hermite has used the result just mentioned to construct a fundamental set for the group of units of a nondegenerate quadratic form F in the space of majorizing forms of F [2]. The construction of the set U below is in a sense a generalization of his procedure.

THEOREM 1. *Let G be a connected complex algebraic group defined over Q. Then there exists an open set U in G_R with the following properties*: (i) $G_R = U \cdot G_Z$; (ii) $K \cdot U = U$ *for a suitable maximal compact subgroup of G_R*; (iii) *For any $x \in G_Q$, $U^{-1} \cdot U \cap (x \cdot G_Z \cup G_Z \cdot x)$ is finite*; (iv) *if G has no nontrivial rational character defined over Q, U has finite Haar measure.*

The group G is the semi-direct product of a reductive group and an invariant unipotent group N, both defined over Q. Since N_R/N_Z is compact, the proof of Theorem 1 is easily reduced to the case where G is reductive. Assuming moreover, as we may, $G \subset SL(n, C)$, we take a right rational representation $\pi: SL(n, C) \to GL(m, C)$ such that $\pi(SL(n, Z)) \subset SL(m, Z)$, and that there exists $v \in Z^m$ whose orbit is closed and whose isotropy group is G. The existence of π follows mainly from Lemma 2. Let $a \in SL(n, R)$ be such that $a \cdot G_R \cdot a^{-1}$ is stable under $x \to {}^t x$ (Lemma 1) and $\mathfrak{S}$ be an open neighborhood of a standard Siegel domain of $SL(n, R)$ contained in a standard Siegel domain. By Lemma 3, there exists a finite number of elements $b_1, \cdots, b_m \in SL(n, Z)$ such that

$$(1) \qquad v \cdot \pi(SL(n, Z)) \cap v \cdot \pi(a^{-1}\mathfrak{S}) = \left\{ v \cdot \pi(b_1^{-1}), \cdots, v \cdot \pi(b_m^{-1}) \right\}.$$

We have $G_R = a^{-1} \cdot H$, where H is the set of elements in $SL(n, R)$ which map $v \cdot \pi(a^{-1})$ onto v. From this, (1) and the equality $SL(n, R) = \mathfrak{S} \cdot SL(n, Z)$ it follows easily that $U = \bigcup_i (G_R \cap a^{-1} \cdot \mathfrak{S} \cdot b_i)$ satisfies (i), (ii). Property (iii) is then a consequence of the theorem of Siegel recalled in §2. When G is semi-simple, property (iv) follows from the following lemma:

LEMMA 4. *Let $G_1 \subset G$ be algebraic semi-simple groups. Assume that there are Iwasawa decompositions $G_{1R} = K_1 \cdot A_1 \cdot N_1$, $G_R = K \cdot A \cdot N$ of G_{1R} and G_R such that $K_1 \subset K$, $A_1 \subset A$, $N_1 \subset N$, and that a positive root of the Lie algebra of G_R with respect to the Lie algebra of A, for the ordering defined by N, restricts to a positive root of the Lie algebra of G_{1R}, for* *the ordering defined by N_1, Let $\mathfrak{S}$ be a Siegel domain of G_R with respect to the Iwasawa decomposition $K \cdot A \cdot N$, and $x \in G_R$. Then $\mathfrak{S} \cdot x \cap G_{1R}$ is contained in a finite number of translates of a Siegel domain of G_{1R}.*

By an elementary argument, property (iii) implies the

COROLLARY. *Let G be a complex algebraic group defined over $\boldsymbol{Q}$. Then G_Z is finitely generated.*

The result on reduced forms stated at the end of §2 implies the finiteness of the number of classes of integral forms with a given non-zero determinant. The latter has the following generalization.

THEOREM 2. *Let G be a connected reductive algebraic group defined over $\boldsymbol{Q}$, and $\pi: G \to \boldsymbol{GL}(m, \boldsymbol{C})$ a rational representation defined over $\boldsymbol{Q}$, and $H = G_Z \cap \pi^{-1}(\boldsymbol{GL}(m, \boldsymbol{Z}))$. Then for any closed orbit X of G in $\boldsymbol{C}^m$, the integral points of X form a finite number of orbits of H.*

COROLLARY. *Let G, G' be connected algebraic groups, defined over $\boldsymbol{Q}$, and $\mu: G \to G'$ a rational surjective homomorphism with finite kernel, defined over $\boldsymbol{Q}$ (an isogeny). Then $\mu(G_Z)$ and G'_Z are commensurable.*

4. Arithmetic subgroups with compact fundamental sets. The quotient G_R/G_Z is compact for instance when G is the orthogonal group of a form which does not represent zero. This fact has the following generalization, which had been conjectured by R. Godement:

THEOREM 3. *Let G be a complex algebraic group defined over Q. Then the following conditions are equivalent: (i) G_R/G_Z is compact: (ii) the identity component of G has no nontrivial rational character defined over $\boldsymbol{Q}$, and every unipotent element of $G_\boldsymbol{Q}$ (or, equivalently, of G_Z) belongs to the radical of $G_\boldsymbol{Q}$.*[1]

REMARKS. (1) Theorem 1, its corollary and Theorem 3 were known essentially for the classical groups (see [6; 8; 9]). Theorem 3 for algebraic tori is proved in [5]. As is known, it is easy to derive from them similar results on groups of matrices with coefficients in the ring of integers of a number field K, which belong to an algebraic group defined over K.

(2) Theorem 3 and known properties of semi-simple Lie algebras imply easily that a connected semi-simple Lie group G always has discrete subgroups H such that G/H is compact.

---

[1] We understand that another proof of Theorem 3 has since been given by G. D. Mostow and T. Tamagawa.

BIBLIOGRAPHY

1. A. Borel, *Groupes linéaires algébriques*, Ann. of Math. vol. 64(1956) pp. 20–80.

2. C. Hermite, *Oeuvres complètes*, Vol. 1, Paris, Gauthier-Villars, 1905.

3. Y. Matsushima, *Espaces homogènes de Stein des groupes de Lie complexes*, Nagoya Math. J. vol. 16 (1960) pp. 205–218.

4. G. D. Mostow, *Self-adjoint group*, Ann. of Math. vol. 62 (1955), pp. 44–55.

5. T. Ono, *Sur une propriété arithmétique des groupes algébriques commutatifs*, Bull. Soc. Math. France vol. 85 (1957) pp. 307–323.

6. K. G. Ramanathan, *Unit of fixed points in involutorial algebras*, Proceedings of the International Symposium on Algebraic Number Theory, Tokyo, 1955.

7. Séminaire S. Lie, *Théorie des algèbres de Lie, Topologie des groupes de Lie*, Paris, 1954–1955.

8. C. L. Siegel, *Einheiten quadratischer Formen*, Abh. Math. Sem. Univ. Hamburg vol. 13 (1939) pp. 209–239.

9. A. Weil, *Discontinuous subgroups of classical groups*, Notes, University of Chicago, 1958.

THE INSTITUTE FOR ADVANCED STUDY AND
   COLUMBIA UNIVERSITY

Reprinted from
Bull. Amer. Math. Soc.
67 (1961), 579–583

ARITHMETIC SUBGROUPS OF ALGEBRAIC GROUPS

By Armand Borel and Harish-Chandra

(Received October 18, 1961)

Table of Contents

Introduction

A complex matric group $G \subset \mathrm{GL}(n, \mathbf{C})$ is algebraic, defined over $\mathbf{Q}$, if it consists of all invertible matrices whose coefficients annihilate some set of polynomials on $\mathbf{M}(n, \mathbf{C})$ with rational coefficients. In this case, let $G_\mathbf{Z}$ be the subgroup of elements of G which have integral coefficients, determinant ± 1, and $G_\mathbf{R} = G \cap \mathrm{GL}(n, \mathbf{R})$. Then $G_\mathbf{Z}$ is an arithmetically defined subgroup of $G_\mathbf{R}$, or more briefly, an *arithmetic subgroup* of $G_\mathbf{R}$. Typical examples are $\mathrm{SL}(n, \mathbf{Z}) \subset \mathrm{SL}(n, \mathbf{R})$, Siegel's modular group, or the group of units of a non-degenerate rational quadratic form; and the main purpose of this paper is to generalize facts known in these and other cases involving classical groups, chiefly from reduction theory. In particular, we shall prove that $G_\mathbf{Z}$ is finitely generated, and give necessary and sufficient conditions under which $G_\mathbf{R}/G_\mathbf{Z}$ is compact, or of finite invariant measure. In analogy with the terminology used in the classical cases, we shall also call $G_\mathbf{Z}$ the *group of units* of $G_\mathbf{R}$.

In view of known facts about algebraic groups and algebraic tori, the main case to investigate is that of semi-simple groups, and this paper is mainly concerned with the latter. However, it turns out that the reductive groups (that is fully reducible groups, or groups whose identity component is isogenous to the product of an algebraic torus by a semi-simple group) form the natural domain of validity for some results, and part of the discussion will be carried out directly for reductive groups. The two

485

following properties will be particularly useful for our purposes:

(A) Let $G_1 \supset \cdots \supset G_m$ be reductive real algebraic subgroups of $\mathbf{GL}(n, \mathbf{R})$. Then there exists $a \in \mathbf{SL}(n, \mathbf{R})$ such that the groups $a \cdot G_i \cdot a^{-1}$ are all stable under $x \to {}^t x$ (1.9).

(B) Let G be a connected complex algebraic reductive group defined over $\mathbf{Q}$, H a closed subgroup defined over $\mathbf{Q}$. Then H is reductive if and only if G/H can be realized as the closed orbit of a rational point in a rational representation defined over $\mathbf{Q}$ (3.8).

The statement (A) is known [21, § 7], and (B) is a slight sharpening of known facts. Proofs have been included for the sake of completeness. However, the techniques used in them will seldom occur elsewhere in the paper, so that the reader who wishes to proceed as directly as possible to the main part of the paper may skip 1.5 to 1.9, the proofs of 1.10, 1.11, and 3.4 to 3.8 without serious inconvenience. Section 2 is also preliminary, and collects some basic facts and notions about algebraic groups.

Section 4 introduces Siegel domains. Let G be a real algebraic semi-simple, or reductive group, and $G = K \cdot A \cdot N$ an Iwasawa decomposition of G (see 1.11). A Siegel domain $\mathfrak{S}_{t,\omega}(t > 0; \omega$ a compact set in $N)$ is the set of elements $K \cdot A_t \cdot \omega$, where A_t is the set of exponentials of elements on which the simple restricted roots have values smaller than t. Their chief properties are:

(i) the set of elements $a \cdot n \cdot a^{-1}(a \in A_t, n \in \omega)$ is relatively compact (4.2);

(ii) if G is semi-simple, $\mathfrak{S}_{t,\omega}$ has finite Haar measure (4.3).

When $G = \mathbf{GL}(n, \mathbf{R})$ and $K \cdot A \cdot N$ is the standard Iwasawa decomposition, it is classical that $\mathfrak{S}_{t,\omega}$ meets only a finite number of its right translates under $x \cdot G_{\mathbf{Z}} \cup G_{\mathbf{Z}} \cdot x$ $(x \in G_{\mathbf{Q}})$, and that $G = \mathfrak{S}_{t,\omega} \cdot \mathbf{SL}(n, \mathbf{Z})$ for t, ω big enough (see 2.5 for references). These facts will be used in the present paper.

Section 5 is devoted to a finiteness lemma (5.3, 5.4), which generalizes the finiteness of the number of integral reduced forms with a given non-zero determinant. This lemma, together with (A), (B), is used in 6.5 to show that if G is reductive, defined over $\mathbf{Q}$, there exist open sets U in $G_{\mathbf{R}}$ such that $G_{\mathbf{R}} = U \cdot G_{\mathbf{Z}}$, $K \cdot U = U$ for a suitable maximal compact subgroup K; and $U^{-1} \cdot U \cap (x \cdot G_{\mathbf{Z}} \cdot y)$ is finite if $x, y \in G_{\mathbf{Q}}$. The construction of U is analogous to Hermite's procedure to obtain a fundamental domain for the group of units of an indefinite rational form in the space of majorizing forms; however, we shall operate directly in the group, instead of using the symmetric space G/K. The finite generation of $G_{\mathbf{Z}}$ follows immediately. It is also shown (6.9) that in a rational representation of G defined over $\mathbf{Q}$, the integral points contained in a closed orbit form a finite number of orbits of $G_{\mathbf{Z}}$. Applied to the natural representation of $\mathbf{SL}(n, \mathbf{Z})$

in the space of quadratic forms, this yields the finiteness of the number of classes of integral forms with a given non-zero determinant. Theorem 6.5 is extended to general algebraic groups in 6.12. In 6.11 it is proved that if $\mu\colon G \to G'$ is an isogeny, defined over $\mathbf{Q}$, then $\mu(G_\mathbf{Z})$ and $G'_\mathbf{Z}$ are commensurable.

The finiteness of the measure of $G_\mathbf{R}/G_\mathbf{Z}$ is proved in § 7 when G is semi-simple, in § 9 when the identity component G^0 of G has no non-trivial rational character defined over $\mathbf{Q}$ (see 7.8, 9.4). In § 7, the basic lemma is 7.5, which says roughly that if $\mathfrak{S}$ is a Siegel domain of a real algebraic semi-simple Lie group, and G_1 is a suitably embedded semi-simple sub-group of G, then $\mathfrak{S} \cdot x \cap G_1$ is contained in the union of a finite number of translates of a Siegel domain of G_1. Theorem 9.4 follows easily from 7.8, a result of Ono [22] on algebraic tori, and some remarks on rational characters made in § 8.

Section 10 is a preliminary to § 11, and discusses closed conjugacy classes. It is shown that if G is an algebraic group, the conjugacy class of an element x in G, (or in the Lie algebra of G), is closed if x is semi-simple, and not closed if G is reductive and x not semi-simple (10.1).

Section 11 gives a necessary and sufficient condition for $G_\mathbf{R}/G_\mathbf{Z}$ to be compact. When G is semi-simple, the condition is that $G_\mathbf{Q}$ consists of semi-simple elements. It generalizes the compactness of $G_\mathbf{R}/G_\mathbf{Z}$ when $G_\mathbf{Q}$ is the multiplicative group of elements of norm 1 in a division algebra over $\mathbf{Q}$, or when G is the orthogonal group of a rational form which does not represent zero. This condition had been conjectured by R. Godement. Its necessity follows easily from 10.1. Conversely, if $G_\mathbf{Q}$ consists of semi-simple elements, then 10.1 implies the existence of a locally faithful rational representation defined over $\mathbf{Q}$ (the adjoint representation) in which all rational points have closed orbits. The main part of the proof starts from that fact, and is a suitable adaptation of a known argument used in the classical case.

The definition of groups of units given above can be generalized by replacing $\mathbf{Q}$ and $\mathbf{Z}$ by a number field K and the ring of algebraic integers of K respectively. However this case is reduced to the previous one by the well-known operation of "restriction of the ground field", and the main results of the paper extend automatically to groups over number fields, as will be seen in § 12.

In § 13, we have relegated some remarks on algebraic groups, not used in the present paper, but which may be viewed as natural complements to some auxiliary results proved in §§ 1, 8.

The main results of this paper have been summarized in [3]. The applications to adele groups announced in [2] will be published elsewhere.

Notation. Linear representations will always be right representations. $H\backslash G$ (resp. G/H) is the space of cosets Hg (resp. gH) of a group G modulo a subgroup H. The identity component of a topological group G is denoted G^0. If a group G operates on a space M, the subgroup of G leaving a given point $m \in M$ fixed, (the *isotropy group* of m), is denoted by G_m. diag $(a_1, \cdots, a_n)$ is a diagonal $n \times n$ matrix with diagonal coefficients $a_1, \cdots, a_n$. The Lie algebra of a Lie group $G, H, M, \cdots$ is denoted by the corresponding German letter.

1. Reductive real algebraic groups

1.1. Let $\mathfrak{g}$ be a real semi-simple Lie algebra, $\mathfrak{g} = \mathfrak{k} + \mathfrak{p}$ a Cartan decomposition of $\mathfrak{g}$ and $\theta: k + p \to k - p$ the corresponding Cartan involution. It is convenient to allow $\mathfrak{g}$ to be compact, then $\mathfrak{p} = 0$ and θ is the identity. The Cartan involutions and Cartan decompositions are conjugate under Ad$\mathfrak{g}$ (see [20], for instance). The Cartan involutions of $\mathfrak{sl}(n, \mathbf{R})$ are the transformations $x \to -x^*$, where x^* is the adjoint of x with respect to some positive non-degenerate quadratic form on $\mathbf{R}^n$. They are also involutive automorphisms of $\mathfrak{gl}(n, \mathbf{R})$, to be called the Cartan involutions of $\mathfrak{gl}(n, \mathbf{R})$. Similarly, the automorphisms $x \to (x^{-1})^*$ of $SL(n, \mathbf{R})$ or $GL(n, \mathbf{R})$ will be called the Cartan involutions of these groups. The restriction of a Cartan involution of $\mathfrak{gl}(n, \mathbf{R})$ to a semi-simple subalgebra $\mathfrak{g}$, which is stable under it, is a Cartan involution of $\mathfrak{g}$.

Given a subspace $\mathfrak{m}$ of $\mathfrak{g}$ or of $\mathfrak{gl}(n, \mathbf{R})$, and a Cartan involution θ of $\mathfrak{g}$ or $\mathfrak{gl}(n, \mathbf{R})$, we put

$$\mathfrak{m}_\mathfrak{k} = \mathfrak{m} \cap \mathfrak{k}, \qquad \mathfrak{m}_\mathfrak{p} = \mathfrak{m} \cap \mathfrak{p}.$$

If $\mathfrak{m} = \theta(\mathfrak{m})$, then $\mathfrak{m} = \mathfrak{m}_\mathfrak{k} + \mathfrak{m}_\mathfrak{p}$ (and conversely), and $\mathfrak{m}$ is spanned by semi-simple elements.

It is standard that, given a real or complex representation of $\mathfrak{g}$ in a finite dimensional vector space V, there exists a Hilbert space structure on V with respect to which the elements of $\rho(\mathfrak{k})$ (resp. $\rho(\mathfrak{p})$) are skew-hermitian (resp. hermitian). The algebra $\rho(\mathfrak{g})$ and the analytic group it generates are then self-adjoint; this also implies that $\rho(x)$ is a semi-simple endomorphism of V with purely imaginary (resp. real) eigenvalues whenever $x \in \mathfrak{k}$ (resp. $x \in \mathfrak{p}$). If ρ is the adjoint representation, such a scalar product is given by $B^*(x, y) = -B(x, \theta(y))(x, y \in \mathfrak{g})$, where B is the Killing form of $\mathfrak{g}$.

If $\mathfrak{g} \subset \mathfrak{gl}(n, \mathbf{R})$, and ρ is the identical representation, the preceding remarks show that $\mathfrak{g}$ is stable under some Cartan involution of $\mathfrak{gl}(n, \mathbf{R})$. By the conjugacy of Cartan involutions, it follows that the Cartan involutions of $\mathfrak{g}$ are the restrictions to $\mathfrak{g}$ of the Cartan involutions of $\mathfrak{gl}(n, \mathbf{R})$ leaving $\mathfrak{g}$ invariant.

1.2. A subalgebra $\mathfrak{g}$ of a Lie algebra $\mathfrak{m}$ is *reductive in* $\mathfrak{m}$ if $\mathrm{ad}_\mathfrak{m}\mathfrak{g}$, (the image of $\mathfrak{g}$ in the adjoint representation of $\mathfrak{m}$), is completely reducible. A Lie algebra $\mathfrak{g}$ is *reductive* if it is so in itself. This is the case if and only if $\mathfrak{g}$ is the direct product of its center by a semi-simple ideal, which is then necessarily equal to the derived algebra $\mathscr{D}\mathfrak{g}$ of $\mathfrak{g}$ [5, § 6, No. 4].

LEMMA. *Let G be a closed subgroup of $\mathbf{GL}(n, \mathbf{R})$ with a finite number of connected components. Then the following conditions are equivalent:*

(i) *G is completely reducible;*

(ii) *$\mathfrak{g}$ is completely reducible;*

(iii) *$\mathfrak{g}$ is reductive in $\mathfrak{gl}(n, R)$;*

(iv) *$\mathfrak{g}$ is reductive, and its center consists of semi-simple endomorphisms.*

The identity component G^0 of G is invariant in G, of finite index. By an elementary fact (see e.g. G. D. Mostow, Amer. J. Math. 78 (1956), 200–221, Lemma 3.1), G is completely reducible if and only if G^0 is, hence (i) is equivalent to (ii). For the other equivalences, see [5, § 6, Nos. 3–6].

1.3. *A real algebraic group* is a subgroup of $\mathbf{GL}(n, \mathbf{R})$ which consists of all invertible matrices whose coefficients annihilate some set of polynomials with real coefficients, in n^2 indeterminates. A subalgebra of $\mathfrak{gl}(n, \mathbf{R})$ is algebraic if it is the Lie algebra of a real algebraic group. A real algebraic group is *reductive* if it is a completely reducible linear group.

1.4. LEMMA. *Let $\mathfrak{m}$ be an algebraic commutative fully reducible subalgebra of $\mathfrak{gl}(n, \mathbf{R})$. Then $\mathfrak{m} = \mathfrak{m}' + \mathfrak{m}''$, where $\mathfrak{m}'$ (resp. $\mathfrak{m}''$) consists of all elements of $\mathfrak{m}$ with purely imaginary (resp. real) eigenvalues, and is an algebraic subalgebra. $\mathfrak{m}$ is invariant under a Cartan involution of $\mathfrak{gl}(n, \mathbf{R})$. If θ is a Cartan involution of $\mathfrak{gl}(n, \mathbf{R})$ leaving $\mathfrak{m}$ invriant, then $\mathfrak{m}_\mathfrak{t} = \mathfrak{m}'$, $\mathfrak{m}_\mathfrak{p} = \mathfrak{m}''$, and θ leaves invariant every algebraic subalgebra of $\mathfrak{m}$.*

The algebra $\mathfrak{m}$ consists of semi-simple elements, and is therefore contained in a Cartan subalgebra $\mathfrak{c}$ of $\mathfrak{gl}(n, \mathbf{R})$ [4, 2.7]. It is known (see e.g. [11, p. 107]) that a Cartan subalgebra of a semi-simple Lie algebra $\mathfrak{g}$ is invariant under a Cartan involution of $\mathfrak{g}$. If we apply this to $\mathfrak{sl}(n, \mathbf{R})$, and recall that $\mathfrak{c}$ is the product of the center of $\mathfrak{gl}(n, \mathbf{R})$ by a Cartan subalgebra of $\mathfrak{sl}(n, \mathbf{R})$, we see that $\mathfrak{c}$ is invariant under a Cartan involution θ' of $\mathfrak{gl}(n, \mathbf{R})$. Using 1.1, we have then $\mathfrak{c} = \mathfrak{c}_\mathfrak{t} + \mathfrak{c}_\mathfrak{p}$ where $\mathfrak{c}_\mathfrak{t}$ (resp. $\mathfrak{c}_\mathfrak{p}$) is the set of elements of $\mathfrak{c}$ with purely imaginary (resp. real) eigenvalues.

Let $x \in \mathfrak{m}$. We have $x = k + p$ $(k \in \mathfrak{c}_\mathfrak{t}, p \in \mathfrak{c}_\mathfrak{p})$, and in order to prove our first contention, it is enough to show that k, $p \in \mathfrak{m}$ or also, since k and p are real matrices, that k, $p \in \mathfrak{m}_c$. After a suitable complex change of coordinates, we may assume $\mathfrak{c}$ to be diagonal, and write

$$k = \operatorname{diag}(i\mu_1, \cdots, i\mu_n) \qquad p = \operatorname{diag}(\lambda_1, \cdots, \lambda_n)$$

$$(\lambda_i, \mu_i \in \mathbf{R}, i = 1, \cdots, n) \ .$$

But [7a, p. 160] the smallest algebraic algebra in $\mathfrak{gl}(n, \mathbf{C})$ containing x is the set of matrices $\operatorname{diag}(a_1, \cdots, a_n)$, where $(a_1, \cdots, a_n)$ annihilates all linear forms with integral coefficients which are zero on $(\lambda_1 + i\mu_1, \cdots, \lambda_n + i\mu_n)$. It contains therefore k and p; a fortiori k, $p \in \mathfrak{m}_c$.

If now θ is a Cartan involution of $\mathfrak{gl}(n, \mathbf{R})$ leaving $\mathfrak{m}$ invariant, and $\mathfrak{a}$ is an algebraic subalgebra of $\mathfrak{m}$, then we have $\mathfrak{m}' = \mathfrak{m}_\mathfrak{t}$, $\mathfrak{m}'' = \mathfrak{m}_\mathfrak{p}$ by 1.1, hence also $\mathfrak{a}' \subset \mathfrak{m}_\mathfrak{t}$, $\mathfrak{a}'' \subset \mathfrak{m}_\mathfrak{p}$, which shows that $\theta(\mathfrak{a}) = \mathfrak{a}$.

1.5. LEMMA. *Let $\mathfrak{g}$ be either semi-simple or equal to $\mathfrak{gl}(n, \mathbf{R})$ and $\mathfrak{m}$ a subalgebra of $\mathfrak{g}$. Then if $\mathfrak{m}$ is stable under a Cartan involution θ of $\mathfrak{g}$, it is reductive in $\mathfrak{g}$, and the restriction of θ to $\mathscr{D}\mathfrak{m}$ is a Cartan involution of $\mathscr{D}\mathfrak{m}$. Conversely, if $\mathfrak{m}$ is reductive and algebraic, it is stable under some Cartan involution of $\mathfrak{g}$.*

Using the last assertion of 1.1, and identifying $\mathfrak{g}$ with an algebraic subalgebra of $\mathfrak{gl}(n, \mathbf{R})$ $(n = \dim \mathfrak{g})$ by means of the adjoint representation, we see first that it is enough to prove the first part when $\mathfrak{g} = \mathfrak{gl}(n, \mathbf{R})$.

Let $\mathfrak{m}$ be stable under a Cartan involution θ of $\mathfrak{g}$. The ideal $\mathfrak{n}$ formed by the nilpotent matrices of the radical of $\mathfrak{m}$ is then also stable under θ, hence (1.1) spanned by semi-simple matrices; thus $\mathfrak{n} = 0$ and $\mathfrak{m}$ is fully reducible [5; § 6, Théorème 4].

Let now $\mathfrak{m}$ be fully reducible, and algebraic. The existence of a Cartan involution θ of $\mathfrak{g}$ leaving $\mathfrak{m}$ invariant is known if $\mathfrak{m}$ is semi-simple [20], or if $\mathfrak{m}$ is commutative (1.4). Let now $\mathfrak{m}$ be neither semi-simple nor commutative. Its center $\mathfrak{c}$ is fully reducible, algebraic, therefore invariant under a Cartan involution θ of $\mathfrak{g}$. The centralizer $\mathfrak{z}(\mathfrak{c})$ of $\mathfrak{c}$ in $\mathfrak{g}$ is reductive in $\mathfrak{g}$ [4, §§ 3, 4], stable under θ, and θ induces a Cartan involution of $\mathscr{D}\mathfrak{z}(\mathfrak{c})$. By [20] there exists a Cartan involution θ' of $\mathfrak{g}$ which leaves $\mathscr{D}\mathfrak{z}(\mathfrak{c})$ and $\mathscr{D}\mathfrak{m}$ invariant. The restriction of θ and θ' to $\mathscr{D}\mathfrak{z}(\mathfrak{c})$ are Cartan involutions of $\mathscr{D}\mathfrak{z}(\mathfrak{c})$, and the analytic group generated by $\mathscr{D}\mathfrak{z}(\mathfrak{c})$ contains an element g such that $\theta'' = \operatorname{Ad} g \circ \theta \circ \operatorname{Ad} g^{-1}$ has the same restriction to $\mathscr{D}\mathfrak{z}(\mathfrak{c})$ as θ' (1.1). The Cartan involution θ'' leaves then invariant $\mathfrak{c}$, $\mathscr{D}\mathfrak{m}$, hence also $\mathfrak{m}$.

1.6. LEMMA. *Let $\mathfrak{g} \supset \mathfrak{g}'$ be algebraic subalgebras reductive in $\mathfrak{gl}(n, \mathbf{R})$. Then there exists a Cartan involution θ of $\mathfrak{gl}(n, \mathbf{R})$ leaving $\mathfrak{g}$ invariant. Any such Cartan involution is conjugate under an element of the analytic* *group G generated by $\mathfrak{g}$ to a Cartan involution leaving also $\mathfrak{g}'$ invariant.*

Let $\mathfrak{c}$ and $\mathfrak{c}'$ be the centers of $\mathfrak{g}$ and $\mathfrak{g}'$, and θ a Cartan involution of $\mathfrak{gl}(n, \mathbf{R})$ leaving $\mathfrak{g}$ invariant. The algebra $\mathfrak{c} + \mathfrak{c}'$ is commutative, algebraic, reductive in $\mathfrak{gl}(n, \mathbf{R})$, hence so is $\mathfrak{c}'' = \mathscr{D}\mathfrak{g} \cap (\mathfrak{c} + \mathfrak{c}')$, and $\mathfrak{c}'' + \mathscr{D}\mathfrak{g}'$ is algebraic, reductive in $\mathscr{D}\mathfrak{g}$. By 1.5, $\mathfrak{c}'' + \mathscr{D}\mathfrak{g}'$ is stable under a Cartan involution of $\mathscr{D}\mathfrak{g}$. Using 1.1, this shows the existence of $g \in G$ such that $\theta' = \mathrm{Ad}g \circ \theta \circ \mathrm{Ad}g^{-1}$ leaves $\mathfrak{c}'' + \mathscr{D}\mathfrak{g}'$ invariant; since $\mathfrak{g}$ centralizes $\mathfrak{c}$, we also have $\theta'(\mathfrak{c}) = \mathfrak{c}$, hence $\theta'(\mathfrak{c} + \mathfrak{c}'') = \mathfrak{c} + \mathfrak{c}''$, and, by 1.4, $\theta'(\mathfrak{c}') = \mathfrak{c}'$; therefore $\theta'(\mathfrak{g}') = \mathfrak{g}'$.

This proves the second assertion of 1.6. The first one is the special case where $\mathfrak{g} = \mathfrak{gl}(n, \mathbf{R})$ and $\mathfrak{g}' = \mathfrak{g}$.

1.7. Lemma. *Let M be reductive algebraic subgroup of $\mathbf{GL}(n, \mathbf{R})$, θ a Cartan involution of $\mathbf{GL}(n, \mathbf{R})$ leaving the Lie algebra $\mathfrak{m}$ of M invariant, $\mathfrak{gl}(n, \mathbf{R}) = \mathfrak{k} + \mathfrak{p}$ the corresponding Cartan decomposition. Then $M = L \cdot \exp(\mathfrak{m}_{\mathfrak{p}})$ where L is a compact subgroup with Lie algebra $\mathfrak{m}_{\mathfrak{k}}$, which is connected if M is so.*

Let $\mathfrak{c}$ be the center of $\mathfrak{m}$, and Q the normalizer of $(\mathscr{D}\mathfrak{m})_{\mathfrak{k}}$ in M. The group Q normalizes $\mathfrak{c}$, hence also $\mathfrak{c}_{\mathfrak{k}}$ and $\mathfrak{c}_{\mathfrak{p}}$ (1.4), and $(\mathscr{D}\mathfrak{m})_{\mathfrak{p}}$, which is the orthogonal complement of $(\mathscr{D}\mathfrak{m})_{\mathfrak{k}}$ in $\mathscr{D}\mathfrak{m}$ with respect to the Killing form. By a standard result, $(\mathscr{D}\mathfrak{m})_{\mathfrak{k}}$ is equal to its normalizer in $\mathscr{D}\mathfrak{m}$; therefore the Lie algebra $\mathfrak{q}$ of Q is equal to $\mathfrak{c} + (\mathscr{D}\mathfrak{m})_{\mathfrak{k}}$. The algebras $\mathfrak{c}_{\mathfrak{k}} = \mathfrak{c} \cap \mathfrak{k}$ and $(\mathscr{D}\mathfrak{m})_{\mathfrak{k}} = \mathfrak{k} \cap \mathscr{D}\mathfrak{m}$, being algebraic, generate closed, hence compact, subgroups A, B of Q^0, with A central. $\mathfrak{c}_{\mathfrak{p}}$ is the Lie algebra of a vector subgroup V, which is invariant in Q. It follows immediately that $Q^0 = A \cdot B \cdot V$, and hence Q^0/V is compact. Since Q^0 has finite index in Q, the quotient Q/V is also compact. By Iwasawa's theorem (see e.g., [27, Exp. 22, Théorème 1]), there exists a compact subgroup L of Q, containing $A \cdot B$, such that Q is the semi-direct product of L and V. The Lie algebra of L is therefore equal to $\mathfrak{m}_{\mathfrak{k}}$. By the conjugacy of Cartan decompositions of $\mathscr{D}\mathfrak{m}$ under $\mathrm{Ad}\mathscr{D}\mathfrak{m}$, the group Q meets every connected component of M, therefore $M = Q \cdot M^0$. By a classical result of E. Cartan (see [20], [12, Lemma 31] for instance), the analytic subgroup generated by $\mathscr{D}\mathfrak{m}$ is equal to $B \cdot \exp((\mathscr{D}\mathfrak{m})_{\mathfrak{p}})$, therefore $M^0 = A \cdot B \cdot V \cdot \exp((\mathscr{D}\mathfrak{m})_{\mathfrak{p}}) = A \cdot B \cdot \exp \mathfrak{m}_{\mathfrak{p}}$, and $M = L \cdot V \cdot A \cdot B \cdot \exp(\mathfrak{m}_{\mathfrak{p}}) = L \cdot \exp(\mathfrak{m}_{\mathfrak{p}})$. Moreover $L = A \cdot B$ if $M = M^0$.

1.8. Lemma. *Let $G \supset G'$ be reductive algebraic subgroups of $\mathbf{GL}(n, \mathbf{R})$. Then there exists a Cartan involution of $\mathbf{GL}(n, \mathbf{R})$ leaving G invariant. Every such Cartan involution is conjugate by an element $\mathrm{Ad}g\,(g \in G^0)$ to a Cartan involution leaving also G' invariant.*

As in 1.6, the first assertion is a special case of the second one. By 1.6, a Cartan involution leaving G invariant is conjugate by an automorphism

$\mathrm{Ad}x(x \in G^0)$ to a Cartan involution θ leaving also the Lie algebra $\mathfrak{g}'$ of G' invariant. Let $\mathfrak{gl}(n, \mathbf{R}) = \mathfrak{k} + \mathfrak{p}$ and $\mathbf{GL}(n, \mathbf{R}) = K \cdot P$ be the corresponding decompositions of $\mathfrak{gl}(n, \mathbf{R})$ and $\mathbf{GL}(n, \mathbf{R})$. By 1.7, $G' = L \cdot \exp(\mathfrak{g}'_\mathfrak{p})$, where L is a compact subgroup with Lie algebra $\mathfrak{g}'_\mathfrak{k}$. The involution θ clearly leaves invariant G^0, L^0, but not necessarily L. Assume that we have found $y \in G^0$ such that

$$(1) \qquad K' = y \cdot K \cdot y^{-1} \supset L \qquad y \cdot \exp(\mathfrak{g}'_\mathfrak{k}) \cdot y^{-1} = \exp(\mathfrak{g}'_\mathfrak{p}) \ .$$

Then $\theta' = \mathrm{Ad}y \circ \theta \circ \mathrm{Ad}y^{-1}$ will do. In fact it is conjugate under $y \cdot x$ to the initial Cartan involution, acts trivially on L, and acts by $p \to p^{-1}$ on $\exp(\mathfrak{g}'_\mathfrak{p})$, hence it leaves G' invariant.

There remains to find $y \in G^0$ satisfying (1). Let M be the normalizer of $\mathfrak{g}'_\mathfrak{p}$ in G. It contains L, and is stable under θ. We view P as usual as a Riemannian symmetric space of $\mathbf{GL}(n, \mathbf{R})$ under the operations $p \to x \cdot p \cdot x^*$, where $x^* = \theta(x^{-1})$. Then $M_\mathfrak{p} = \exp \mathfrak{m}_\mathfrak{p}$ is a totally geodesic submanifold, and, by E. Cartan's fixed point argument, every compact subgroup of $\mathbf{GL}(n, \mathbf{R})$ leaving $M_\mathfrak{p}$ invariant has a fixed point on $M_\mathfrak{p}$ (for all this, see for instance [20]). In particular L has a fixed point a on $M_\mathfrak{p}$. Let y be the square root of a contained in $M_\mathfrak{p}$. Then, for any $x \in L$,

$$(y^{-1} \cdot x \cdot y)(y^{-1} \cdot x \cdot y)^* = y^{-1} \cdot x \cdot y \cdot y \cdot x^* \cdot y^{-1}$$
$$= y^{-1}x \cdot a \cdot x^* \cdot y^{-1} = y^{-1} \cdot y^2 \cdot y^{-1} = e \ ,$$

and therefore $y^{-1} \cdot L \cdot y \subset K$, which is the first part of (1). Since M normalizes $\mathfrak{g}'_\mathfrak{p}$, we have $[\mathfrak{m}_\mathfrak{p}, \mathfrak{g}'_\mathfrak{p}] \subset \mathfrak{g}'_\mathfrak{p}$; but $[\mathfrak{m}_\mathfrak{p}, \mathfrak{g}'_\mathfrak{k}] \subset [\mathfrak{p}, \mathfrak{p}] \subset \mathfrak{k}$, hence $[\mathfrak{m}_\mathfrak{p}, \mathfrak{g}'_\mathfrak{p}] = 0$, and y commutes elementwise with $\exp \mathfrak{g}'_\mathfrak{p}$. Thus y also verifies the second equality of (1).

1.9. THEOREM (Mostow [21]). *Let $G_1 \supset \cdots \supset G_m$ be reductive real algebraic subgroups of $\mathbf{GL}(n, \mathbf{R})$. Then there exists $a \in \mathbf{SL}(n, \mathbf{R})$ such that the groups $a \cdot G_i \cdot a^{-1} (i = 1, \cdots, m)$ are self-adjoint.*

By 1.1, this theorem is equivalent to the existence of a Cartan involution of $\mathbf{GL}(n, \mathbf{R})$ leaving the G_i's invariant. By 1.8, there exists a Cartan involution leaving G_1 invariant. Assume $\theta(G_i) = G_i$ for $1 \leq i \leq k < m$. By 1.8, θ is conjugate under an element $\mathrm{Ad}x(x \in G_k^0)$ to a Cartan involution θ' leaving G_{k+1} stable. Since the G_i's $(i \leq k)$ are still invariant under θ', the theorem follows by induction.

The following proposition strengthens 1.7, and extends to reductive real algebraic groups well-known properties of semi-simple Lie groups.

1.10. PROPOSITION. *Let G be a reductive algebraic subgroup of $\mathbf{GL}(n, \mathbf{R})$, θ a Cartan involution of $\mathbf{GL}(n, \mathbf{R})$ leaving G invariant (1.6), $\mathfrak{gl}(n, \mathbf{R}) = \mathfrak{k} + \mathfrak{p}$, and $\mathbf{GL}(n, \mathbf{R}) = K \cdot P$ the corresponding decompositions* *of* $\mathfrak{gl}(n, \mathbf{R})$ *and* $\mathbf{GL}(n, \mathbf{R})$. *Then the set* L *of fixed points of* θ *in* G *is a maximal compact subgroup with Lie algebra* $\mathfrak{g}_{\mathfrak{t}}$. *Every compact subgroup of* G *is conjugate by an element* $\mathrm{Ad}x(x \in \exp(\mathfrak{g}_{\mathfrak{p}}))$ *to a subgroup of* L. *The map* $(x, y) \to x \cdot \exp(y)$ *of* $L \times \mathfrak{g}_{\mathfrak{p}}$ *into* G *is an analytic homeomorphism.*

Let $g = k \cdot p (k \in K, p \in P)$ be an element of G. Since $\theta(g^{-1}) \cdot g = p^2$, we have $p^2 \in G$, hence also $p^{2m} \in G, (m \in \mathbf{Z})$. In a suitable coordinate system, $\log p = \mathrm{diag}(\lambda_1, \cdots, \lambda_n)$, $(\lambda_i \text{ real})$, and $p^{2m} = \mathrm{diag}(\exp 2m\lambda_1, \cdots, \exp 2m\lambda_n)$. It is then an elementary fact that every polynomial with real coefficients over the space of $n \times n$ matrices which is annihilated by the elements $p^{2m}(m \in \mathbf{Z})$ is also annihilated by the diagonal matrices $\mathrm{diag}(\exp t\lambda_1, \cdots, \exp t\lambda_n)$ (t real). Since G is algebraic, this shows that G contains the 1-parameter group generated by $\log p$. Therefore $\log p \in \mathfrak{g}_{\mathfrak{p}}, p \in \exp\mathfrak{g}_{\mathfrak{p}}$, and $k \in G \cap K = L$. Thus $G = L \cdot \exp\mathfrak{g}_{\mathfrak{p}}$. That every compact subgroup of G is conjugate under $\exp\mathfrak{g}_{\mathfrak{p}}$ to a subgroup of L is proved by the fixed point argument of E. Cartan used in 1.8. The proof of the last statement for semi-simple Lie groups given in [12, Lemma 31] is valid without change here.

1.11. We now extend Iwasawa's decomposition to reductive algebraic groups. Let $G \subset \mathbf{GL}(n, \mathbf{R})$ be reductive, algebraic, θ a Cartan involution of $\mathbf{GL}(n, \mathbf{R})$ leaving G invariant, K its fixed point set, $\mathfrak{a}$ a maximal subalgebra of $\mathfrak{g}_{\mathfrak{p}}$. For $\lambda \in \mathfrak{a}^*$, let $\mathfrak{g}_\lambda = \{x \in \mathfrak{g}, [a, x] = \lambda(a)x, a \in \mathfrak{a}\}$. Choose some order on $\mathfrak{a}^*$ and let $\mathfrak{n} = \sum_{\lambda > 0} \mathfrak{g}_\lambda$.

PROPOSITION. *We keep the previous notation. Then* $\mathfrak{n}$ *generates a closed unipotent subgroup* N *invariant under* $A = \exp\mathfrak{a}$, *and* $G = K \cdot A \cdot N$. *The map* $(k, a, n) \to k \cdot a \cdot n$ *of* $K \times A \times N$ *onto* G *is an analytic homeomorphism.*

Let $\mathfrak{c}$ be the center of $\mathfrak{g}$. Then $\mathfrak{g} = \mathfrak{c} \times \mathscr{D}\mathfrak{g}$, and $\mathfrak{a} = \mathfrak{c}_{\mathfrak{p}} \times \mathfrak{a}_1$, where $\mathfrak{a}_1$ is a maximal subalgebra of $(\mathscr{D}\mathfrak{g})_{\mathfrak{p}}$. Clearly, for $\lambda \neq 0, \mathfrak{g}_\lambda = \{x \in \mathscr{D}\mathfrak{g}, [a_1, x] = \lambda(a_1)x, a_1 \in \mathfrak{a}_1\}$. By the usual Iwasawa decomposition, the analytic group M generated by $\mathscr{D}\mathfrak{g}$ is of the form $M = K_1 \cdot A_1 \cdot N$ where $K_1 = K \cap M$, $A_1 = \exp\mathfrak{a}_1$, and N is closed, unipotent, with Lie algebra $\mathfrak{n}$. As was remarked in 1.7, $\mathfrak{c}_{\mathfrak{t}}$ and $\mathfrak{c}_{\mathfrak{p}}$ generate respectively a torus T and a vector subgroup V, therefore

$$G^0 = T \cdot K_1 \cdot V \cdot A_1 \cdot N = T \cdot K_1 \cdot A \cdot N.$$

But $G = K \cdot G^0$, with $K \supset T \cdot K_1$, hence $G = K \cdot A \cdot N$. Since V is central, and N is stable under conjugation by elements of A_1, the group N is also invariant under A. The second assertion is proved by standard arguments, as in the usual case [27; Exp. 11].

An Iwasawa decomposition $G = K \cdot A \cdot N$ and a Cartan involution θ

whose fixed point set is K, and which induces $x \to x^{-1}$ on A, will be called *compatible*.

Let G' be an open subgroup of G. Since $G = K \cdot G^{\circ}$ and θ leaves K pointwise fixed, we have $\theta(G') = G'$. The automorphism of G' induced by θ and the decomposition $G' = (K \cap G') \cdot A \cdot N$ will also be called a Cartan involution and an Iwasawa decomposition of G' respectively. Clearly, 1.8 to 1.11 are also valid for open subgroups of reductive real algebraic groups.

2. Algebraic groups

In this paragraph and the next we assume some familiarity with the elementary theory of affine algebraic sets and algebraic groups, and shall recall only some of the relevant definitions and facts. For more details, see e.g. [1, 8, 17].

2.1. A *complex algebraic group*, or simply, an *algebraic group*, is a subgroup G of $\mathbf{GL}(n, \mathbf{C})$, which consists of all invertible matrices $g = (g_{ij})$ whose coefficients annihilate some set of polynomials $\{P_{\alpha}[X_{11}, \cdots, X_{nn}]\}$ with complex coefficients. In other words, G is the intersection of $\mathbf{GL}(n, \mathbf{C})$ with an affine algebraic set in the space $\mathbf{M}(n, \mathbf{C})$ of $n \times n$ complex matrices. The group is said to be defined over a subfield k of $\mathbf{C}$ if the P_{α} may be chosen so as to have coefficients in k.[1] The intersection of all the fields of definition is also a field of definition. An algebraic group is also a complex Lie group; it is connected as a manifold if and only if it is irreducible as an algebraic set, or if and only if it is connected in the Zariski topology.

Let G be an algebraic group, and A a subring of $\mathbf{C}$. Then G_A will denote the subgroup of elements of G whose coefficients are in A and whose determinant is a unit of A. If A is a field of definition, then G_A is an algebraic A-group in the terminology of [2], an algebraic group over A in [7], and if G is connected, G_A is Zariski dense in G [26, p. 44]. Conversely, if H is an algebraic group over A, in the sense of [7], then $H = G_A$, where G is the smallest complex algebraic group containing H.

An algebraic group G is not necessarily a closed subset of $\mathbf{M}(n, \mathbf{C})$; however, if we add one coordinate and put $g_{n+1,n+1} = (\det g)^{-1}$, $g_{n+1,i} = g_{i,n+1} = 0$ $(i = 1, \cdots, n)$, then G becomes an algebraic subgroup of $\mathbf{SL}(n + 1, \mathbf{C})$, which is closed in $\mathbf{M}(n + 1, \mathbf{C})$. This operation clearly does not change G_A.

---

[1] This is the definition of [1], with the universal field specialized to $\mathbf{C}$. Being in characteristic zero, we need not distinguish between Zariski k-closed and defined over k (or between defined and quasi-defined over k [1]).

2.2. A *rational representation* $\pi\colon G \to \mathrm{GL}(V)$ is a homomorphism of G into $\mathrm{GL}(V)$ (V finite dimensional complex vector space) whose restriction to each connected component of G is a rational map of G into the space of endomorphisms $E(V)$ of V. The coefficients of $\pi(g)$ with respect to a basis of V are regular functions on each connected component of G. If G is a closed set in $\mathbf{M}(n, \mathbf{C})$, they are therefore polynomials in the coefficients of g, whose coefficients are constant on each connected component of G. From this and the end remark in 2.1, it follows that in general the coefficients of $\pi(g)$ are polynomials in those of g and in $(\det g)^{-1}$.

The rational representation π will be said to be defined over k if each component of G is, and if there exists a basis (v_i) of v such that the coefficients of $\pi(g)$ with respect to that basis are regular functions defined over k on each connected component of G. In that case, we shall denote by V_A the set of linear combinations of the v_i's with coefficients in $A \supset k$.

Given a rational representation $\pi\colon G \to \mathrm{GL}(V)$, we shall most often write $v \cdot g$ for $v \cdot \pi(g)$ ($v \in V$, $g \in G$). An orbit $v \cdot G$ is always open and everywhere dense in its Zariski closure (smallest affine algebraic set containing it), therefore the latter coincides with the closure in the ordinary topology. If G is connected, and π is defined over k, then $v \cdot G$, its closure and the isotropy group G_v of v are defined over $k(v)$.

Let G be a connected algebraic group, H an algebraic subgroup, both defined over k. Then $H\backslash G$ is in a canonical way an algebraic variety defined over k. Given a rational representation $\pi\colon G \to \mathrm{GL}(V)$ and a point $v \in V$ for which $G_v = H$, the map $g \to v \cdot \pi(g)$ induces a regular bijective map of $H\backslash G$ onto $v \cdot G$, defined over k if π is. Since we are in characteristic zero and these varieties are non-singular, this map is in fact birational and biregular.

2.3. PROPOSITION. *Let G be a connected algebraic group defined over* $\mathbf{R}$, $\pi\colon G \to \mathrm{GL}(V)$ *a rational representation of G defined over* $\mathbf{R}$, *and X an orbit of G in V. Then $X_{\mathbf{R}} = X \cap V_{\mathbf{R}}$ is the union of a finite number of orbits of $(G_{\mathbf{R}})^0$, which are closed if X is so.*

$\dim_{\mathbf{R}}$ and $\dim_{\mathbf{C}}$ will denote the topological and the complex dimension, respectively. Let $s = \dim_{\mathbf{C}} G$, $t = \dim_{\mathbf{C}} X$. The $s - t = \dim_{\mathbf{C}} G_x (x \in X)$. We assume $X_{\mathbf{R}}$ to be non-empty, hence X and its closure $\bar{X}$ are defined over $\mathbf{R}$. The latter is an irreducible algebraic set of complex dimension t. The group $G_{\mathbf{R}}$ being Zariski-dense in G, the set $\bar{X}$ is the smallest algebraic set containing $(\bar{X})_{\mathbf{R}}$, hence $(\bar{X})_{\mathbf{R}}$ is a real algebraic irreducible set of dimension t, and $(\bar{X})_{\mathbf{R}} = A \cup B$, where A is a manifold of dimension t, the set of real simple points of $\bar{X}$, and B a real algebraic set of dimension $<t$ [31, §§ 10, 11]. In view of their characterizations, A and B are both in-

variant under $G_\mathbf{R}$. Let now $x \in X_\mathbf{R}$. It is a simple point of $\bar{X}$, hence $x \in A$, $x \cdot G_\mathbf{R}$ is an open submanifold of A, and $x \cdot G_\mathbf{R}^0$ is a connected component of A. But A, being the complement of a real algebraic set in a real algebraic set, has only a finite number of components [31, Theorem 4], which proves the first part of our assertion. If moreover $X = \bar{X}$, then $(\bar{X})_\mathbf{R} = X_\mathbf{R} \subset A$, hence B is empty, A is a closed submanifold with a finite number of components, which must then also be closed.

2.4. PROPOSITION. *Let G be a connected algebraic group, H an algebraic subgroup, k a field of definition for G and H, and assume $H \backslash G$ to be an affine algebraic set. Then there exists a rational representation $\pi \colon G \to \mathbf{GL}(V)$ defined over k and a point $v \in V_k$ such that $G_v = H$, $v \cdot G$ is closed, and $g \to v \cdot g$ induces a biregular birational map of $H \backslash G$ onto $v \cdot G$.*[2]

Let A and B be rings of regular (i.e. rational, everywhere defined) functions, defined over k, of G and $H \backslash G$ respectively. The variety $H \backslash G$ is defined over k, non-singular and affinely imbeddable. It is therefore also biregularly homeomorphic over k to an affine algebraic set defined over k (see A. Weil, Amer. J. Math. 78 (1950), p. 509–524, Theorem 7). Therefore

(i) B is a finitely generated k-algebra and separates the points of $H \backslash G$.[3]

We let G operate on $A \otimes \mathbf{C}$ on the right by $(f \cdot g)(x) = f(g \cdot x)(f \in A \otimes \mathbf{C};$ $g, x \in G)$. The transpose of the canonical projection $G \to H \backslash G$ identifies then $B \otimes \mathbf{C}$ with the ring of invariants of H.

Let $b_1, \cdots, b_s$ be a finite system of generators of B, P_i a finite dimensional vector subspace of A such that $P_i \otimes \mathbf{C}$ is invariant under G and $b_i \in P_i$ [25, Theorem 12], P the direct sum of the P_i, $v = (b_1, \cdots, b_s)$, and π the natural representation of G in $V = P \otimes \mathbf{C}$. From (i) it is clear that $G_v = H$. Let $X = v \cdot G$, and μ be the map of $H \backslash G$ into the Zariski closure $\bar{X}$ of X defined by $g \to v \cdot g$. It is rational, defined over k, everywhere regular. Let B' be the ring of regular functions on $\bar{X}$. Since X is Zariski dense in $\bar{X}$, the elements of B' are determined by their restriction to X, and μ induces an injective homomorphism ${}^t\mu$ of B' into $B \otimes \mathbf{C}$. Let us prove now that ${}^t\mu$ is surjective. For this, it is enough to show that $b_i \in {}^t\mu(B') \, (i = 1, \cdots, s)$. In P_i, we may always find a basis $f_1, \cdots, f_t$ such that $f_j(e) = 0(j \geq 2)$. We have $b_i \cdot g = \sum_j a_j(g) \cdot f_j$, where the a_j's are regular functions on G, therefore

---

[2] Propositions 2.4, 2.5 and their proofs are also valid if $\mathbf{C}$ is replaced by a universal field of arbitrary characteristic.

[3] In our case, this also follows directly from the fact that G_k is Zariski dense in G [26].

$$b_i(g) = (b_i \cdot g)(e) = \sum_j a_j(g) f_j(e) = a_1(g) \cdot f_1(e) \ .$$

This implies that $b_i = {}^t\mu(x_1 \cdot f_1(e))$, where x_1 is the first coordinate function with respect to the basis (f_j). Thus $b_i \in {}^t\mu(B')$.

Thus, v is a regular map of $H \backslash G$ into $\bar{X}$, defined over k, whose transpose is an isomorphism of the rings of regular functions. Since $H \backslash G$ and $\bar{X}$ are affine, this implies that v is a biregular, bijective, hence also that $X = \bar{X}$.   q.e.d.

The following proposition is a slight strengthening of Proposition 5 in [8, Exp. 10] and is proved in the same way. It will be used only in 3.7.

2.5. PROPOSITION. *Let G be a connected algebraic group, H an algebraic subgroup, and k a field of definition for G and H. Then there exists a rational representation $\pi\colon G \to \mathbf{GL}(V)$, defined over k, a point $v \in V_k$, ($v \neq 0$), such that H is the set of elements of G which leave the 1-dimensional subspace $[v]$ spanned by v invariant.*[2]

Again let A be as in 2.4, and I the ideal of H in $A \otimes C$. It is invariant under H, and has a finite system of generators belonging to A. By [25, Theorem 12] there exists a finite dimensional subspace P of A such that $P \otimes C$ is invariant under G, and that $P \cap I$ contains a system of generators of I. Let $d = \dim P \cap I$ and π the natural representation of G in $V = \bigwedge^d (P \otimes C)$. Then $v = P \cap I$ fulfills our condition.

3. Reductive algebraic groups. Affine homogeneous spaces

3.1. An *algebraic torus* (a torus, in the terminology of [1]) is an algebraic group which is birationally isomorphic to a direct product of groups C^* (multiplicative group of non-zero complex numbers), or equivalently [1], a connected group which is diagonalizable. A connected subgroup of $\mathbf{GL}(n, C)$ is an algebraic torus if and only if its Lie algebra is algebraic, commutative, and consists of semi-simple elements (or is reductive in $\mathfrak{gl}(n, C)$). An algebraic group G will be called *reductive* if $G^0 = T \cdot G'$ where T is an algebraic torus, central in G^0, and G' is semi-simple. For an algebraic group $G \subset \mathbf{GL}(n, C)$ the following conditions are equivalent:

(i) G is reductive;

(ii) G is fully reducible;

(iii) $\mathfrak{g}$ is reductive in $\mathfrak{gl}(n, C)$.

In fact, G is reductive, or fully reducible, if and only if G^0 is, and this equivalence follows from the previous remarks and the characterization of fully reducible subalgebras of $\mathfrak{gl}(n, C)$ [5, § 6, No. 5]. Let π be a rational representation of G. If G is a torus, then so is $\pi(G)$, [1; 9.4], therefore, $\pi(G)$ is reductive if G is. The reductive algebraic groups are the algebraic groups, all of whose rational representations are completely reducible.

3.2. Proposition. *Let G be an algebraic group defined over* **R.** *Then the following conditions are equivalent:*

(i) *G is reductive;*

(ii) *$G_{\mathbf{R}}$ is reductive;*

(iii) $\mathfrak{g}$ *is the complexification of the (real) Lie algebra of a maximal compact subgroup of G^0.*

Let $\mathfrak{g}$ and $\mathfrak{g}_{\mathbf{R}}$ be the Lie algebras of G and $G_{\mathbf{R}}$, over **C** and **R** respectively. Then $\mathfrak{g} = \mathfrak{g}_{\mathbf{R}} \otimes \mathbf{C}$. Thus $\mathfrak{g}_{\mathbf{R}}$ is fully reducible if and only if $\mathfrak{g}$ is, and the equivalence of (i) and (ii) follows from 3.1 and 1.2. Let now $G^0 = T \cdot G'$ be reductive, A and B be maximal compact subgroups of T and G' respectively. Then $A \times B$ is a maximal compact subgroup of $T \times G'$, which moreover contains the (finite) kernel of the natural homomorphism of $T \times G'$ onto G^0, therefore $A \cdot B$ is a maximal compact subgroup of G^0. Since the Lie algebras $\mathfrak{a}$ and $\mathfrak{b}$ of A and B are real forms of the Lie algebras of T and G', this shows that (i) $\Rightarrow$ (iii). If now $\mathfrak{g}$ is the complexification of the Lie algebra of a compact group, it is certainly fully reducible, hence (iii) $\Rightarrow$ (i).

3.3. Proposition. *Let G be a reductive connected algebraic group, $\pi\colon G \to \mathbf{GL}(V)$ a rational representation. Then the invariant polynomials separate the closed orbits of G.*

Let X, Y be two distinct closed orbits. These are two affine subvarieties with empty intersection. There exists therefore a polynomial P vanishing on X, and a polynomial Q vanishing on Y such that $P + Q = 1$. We have then $P \cdot g + Q \cdot g = 1$ for any $g \in G$. Let K be a maximal compact subgroup of G, dk the Haar measure on K with total measure 1, and

$$P^* = \int_K P \cdot k \, dk \qquad Q^* = \int_K Q \cdot k \, dk .$$

We still have $P^* + Q^* = 1$, and $P^*(x) = Q^*(y) = 0 (x \in X, y \in Y)$, hence $P^* = 1$ on Y. Moreover P^* and Q^* are invariant under K. Since the natural representation of G on the space of polynomials on V of a given degree is rational, and G is the smallest algebraic subgroup of G containing K (3.2), P^* and Q^* are invariant under G, whence our contention.

3.4. Remark. The property (iii) also characterizes reductive groups among connected complex Lie groups. In fact, let G be complex, connected, K a maximal compact subgroup, and assume that $\mathfrak{g} = \mathfrak{k} \otimes \mathbf{C}$. Of course $\mathfrak{k} = \mathfrak{c} \times \mathscr{D}\mathfrak{k}$ with $\mathfrak{c}$ the Lie algebra of a torus (usual sense) and $\mathscr{D}\mathfrak{k}$ a compact semi-simple Lie algebra, therefore $\mathfrak{g} = (\mathfrak{c} \otimes \mathbf{C}) \times (\mathscr{D}\mathfrak{k} \otimes \mathbf{C})$, $G = T \cdot G'$, with G' complex semi-simple, T isomorphic to a product of $\mathbf{C}^*$'s, and $T \cap G'$ central, hence finite. $T \cap G'$ is also the intersection of any two

maximal compact subgroups of T and G', therefore is uniquely characterized by the maximal compact subgroup K. This implies that if G and G^*
are complex, connected, reductive, and have isomorphic maximal compact
subgroups, then they are isomorphic (as complex Lie groups at first).
Moreover, it is known that a reductive group has only one structure of
algebraic group compatible with the given complex analytic structure.
(For a more general theorem, see G. Hochschild and G. D. Mostow, Amer.
J. Math. 88 (1961), p.p. 111–136, § 8.) Therefore a complex analytic isomorphism between two reductive groups is necessarily birational, biregular.

3.5. THEOREM. *Let G be a connected reductive algebraic group, and
H an algebraic subgroup. Then G/H is an affine algebraic variety if and
only if H is reductive.*

This result is not new. Since a non-singular affine variety is a Stein
manifold, it is a consequence of the two following assertions, where G and
H are as in the theorem:

(a) If G/H is a Stein manifold, then H is reductive.

(b) If H is reductive, G/H is affine.

The assertion (a) is due to Matsushima [18]. Assertion (b) is stated in
[18] without proof, and attributed to Iwahori-Sugiura. For the sake of
completeness, we insert here a proof of (b), and also one for (a), which
is shorter than that of [18], although based on a similar idea. A consequence of (a) and (b) is that a quotient G/H, with G connected, reductive,
is Stein if and only if it is an affine variety. We also remark that the
statement 3.5 makes sense in arbitrary characteristic, but we do not know
whether it holds true.

3.6. PROOF OF (a). G/H° is a finite Galois covering of G/H, hence is a
Stein manifold [6], and we may assume H to be connected. We denote by
$H_i(X)$ the ith singular homology group of the space X, with complex
coefficients.

Let $n = \dim_\mathbb{C} G$, $m = \dim_\mathbb{C} H$. We claim first that H reductive is equivalent to $H_m(H) \neq 0$. As a manifold H is the topological product of a
euclidean space by a maximal compact subgroup, hence H reductive implies
$H_m(H) \neq 0$. A complex analytic group being the quotient by a finite group
of the semi-direct product of a semi-simple Lie by a solvable group, it is
enough to prove the converse for H solvable. A maximal compact subgroup K of H is then commutative, of real dimension $\geq m$. Since H is a
complex linear group, $\mathfrak{k}$ and $i \cdot \mathfrak{k}$ are linearly independent over $\mathbf{R}$, hence
$\mathfrak{h} = \mathfrak{k} + i\mathfrak{k}$ (direct), and $\mathfrak{h} \cong \mathfrak{k} \otimes \mathbf{C}$.

G is a locally trivial fibering, with typical fibre H, base G/H, and
structural group H, therefore the Betti numbers of G are majorized by

those of $(G/H) \times H$ (as follows from the existence of a spectral sequence). We have $H_n(G) \neq 0$; since G/H is Stein, $H_i(G/H) = 0$ $(i > n - m = \dim_C G/H)$ [6], hence by the Künneth rule, $H_i(H) \neq 0$ for some $i \geq m$. But G is Stein (it is algebraic), hence so is H (as a closed submanifold of G), and $H_i(H) = 0$ for $i > m$; we must then have $H_m(H) \neq 0$. As was already proved, this implies that H is reductive.

3.7. PROOF OF (b). Let $K \supset L$ be maximal compact subgroups of G and H respectively and ρ a faithful representation of K in $\mathbf{GL}(n, \mathbf{R})$. It follows from 3.4 that ρ extends to a biregular birational isomorphism of G onto the smallest complex algebraic group containing $\rho(K)$. We identify G with the latter. Thus G is defined over $\mathbf{R}$, and H, the smallest complex algebraic group containing L, is also defined over $\mathbf{R}$. The group $G_\mathbf{R}$, being real algebraic, with identity component K, is compact and consequently $G_\mathbf{R} = K$. By 2.5, there exists a rational representation $\pi: G \to \mathbf{GL}(V)$ defined over $\mathbf{R}$, and a non-zero element $v \in V_\mathbf{R}$ such that H is the subgroup of G leaving $\mathbf{C} \cdot v$ invariant. This yields a real 1-dimensional representation of L. Since L is compact, its image has at most 2 elements. By 1,10, applied to H, viewed as real reductive algebraic group, we have $H/H^0 \cong L/L^0$, the subgroup H' of H leaving v fixed has index ≤ 2, and G/H' is either equal to G/H or a two-fold covering of G/H. But the quotient of an affine variety by a finite group is an affine variety (see Serre, Symposium de Topologia Algebrica, Mexico, 1958, 24–53, § 13), therefore it is enough to prove that G/H' is affine. Changing slightly the notation, we may assume H to be the isotropy group of v. Let now $X = v \cdot G$, and $\bar{X}$ its closure. We want to prove that $X = \bar{X}$. The set $Y = \bar{X} - X$ is algebraic, defined over $\mathbf{R}$. Let $S_1, \cdots, S_q$ be polynomials with real coefficients which generate the ideal of Y and $S = S_1^2 + \cdots + S_q^2$. On $V_\mathbf{R}$, this polynomial vanishes on $Y_\mathbf{R}$ only. Therefore, for $x \in X_\mathbf{R}$, S has a constant sign on $x \cdot K$. We let G operate in the usual fashion on the ring A of polynomials over V, and consider as in 3.3 the average S^* of S over K:

$$S^* = \int_K S \cdot k \, dk \, ,$$

where dk is a Haar measure on K. It is invariant under K, hence also under G; consequently, S^* is constant on X, and therefore on $\bar{X}$. Since S does not vanish on the connected set $x \cdot K (x \in X_\mathbf{R})$, S^* is not zero on $\bar{X}$; on the other hand, Y is invariant under K, and S vanishes on Y, therefore S^* vanishes on Y. This is a contradiction, unless $Y = \varnothing$, $X = \bar{X}$. Thus X is an affine algebraic set. The map $g \to v \cdot g$ induces an isomorphism of $H \backslash G$ onto X and allows one to identify $H \backslash G$ with X (2.2).

For future reference, we state here a consequence of 2.4, 2.5, 3.4 to

3.7, the only one to be used in the sequel.

3.8. THEOREM. *Let G be a connected reductive algebraic group, defined over $\mathbf{Q}$, and H an algebraic subgroup. Then the two following conditions are equivalent:*

(i) *H is reductive, defined over $\mathbf{Q}$;*

(ii) *there exists a rational representation $\pi\colon G \to \mathrm{GL}(V)$ of G, defined over $\mathbf{Q}$, and an element $v \in V_{\mathbf{Q}}$, such that $v \cdot G$ is closed and $G_v = H$.*

In fact if (i) is true, then $H\backslash G$ is affine by 3.5, and (ii) follows from 2.4. If (ii) is true, then $H\backslash G$ may be identified with $v \cdot G$ (see end of the preceding proof) hence is affine, and H is reductive by 3.6. Finally, since $H = G_v$, $(v \in V_{\mathbf{Q}})$, H is defined over $\mathbf{Q}$.

4. Siegel domains

4.1. Let G be an open subgroup of a real algebraic, reductive group, $G = K \cdot A \cdot N$ an Iwasawa decomposition of G (1.11), and Σ the set of simple restricted roots, in the ordering for which the roots of $\mathrm{ad}_{\mathfrak{g}}\mathfrak{a}$ in $\mathfrak{n}$ are positive. Let $A_t = \{a \in A, \lambda(\log a) \leqq t; \lambda \in \Sigma\}$. A *Siegel domain* in G is a subset of the form $K \cdot A_t \cdot \omega$ where ω is compact in N. It will usually be denoted by $\mathfrak{S}_{t,\omega}$, or simply by $\mathfrak{S}$ if t, ω need not be specified. Clearly, $\mathfrak{S}_{t,\omega} \subset \mathfrak{S}_{t',\omega'}$ if $t \leqq t'$, $\omega \subset \omega'$.

It is understood once and for all that the choice of a Siegel domain presupposes an Iwasawa decomposition which, unless otherwise stated, will be written $K \cdot A \cdot N$ or equivalently, pre-supposes a Cartan involution θ, a maximal subalgebra on which $\theta = -\mathrm{Id}$, and an ordering of the restricted roots.

The union of the Siegel domains, with respect to a fixed Iwasawa decomposition, is G, and any finite union of such Siegel domains is contained in a Siegel domain.

Let us say that two families A, B of subsets of G are equivalent if every element of A (resp. B) is contained in an element of B (resp. A). It is clear that we get a family equivalent to the set of $\mathfrak{S}_{t,\omega}$ if we replace above A_t by $a \cdot A_0 (a \in A)$, or if we allow λ to run through all the positive roots in the definition of A_t.

4.2. PROPOSITION. *Let G be an open subgroup of a real algebraic, reductive group, $G = K \cdot A \cdot N$ an Iwasawa decomposition of G, and ω a compact set in N. Then the set of elements $a \cdot n \cdot a^{-1} (a \in A_t,\ n \in \omega)$ is relatively compact in N.*

Let $(x_i)(1 \leqq i \leqq m = \dim \mathfrak{n})$ be a basis of $\mathfrak{n}$, such that $[h, x_i] = \lambda_i(h)x_i\ (h \in \mathfrak{a})$, where λ_i is a linear form on $\mathfrak{a}$ which is >0 for the given ordering $(1 \leqq i \leqq m)$, and that $\lambda_i \leqq \lambda_j$ if $i \leqq j$. For each j, the elements $x_i (i \geq j)$ span then an ideal of $\mathfrak{n}$ and it follows from standard properties of simply connected nilpotent groups that $(t_1, \cdots, t_m) \rightarrow \exp(t_1 x_1) \cdots \exp(t_m x_m)$ is an analytic homeomorphism of $\mathbf{R}^m$ onto N. Further, if $n = \prod_i \exp(t_i x_i)$ and $a \in A$, then

$$a \cdot n \cdot a^{-1} = \prod_i \exp t_i \lambda_i (\log a) x_i \ .$$

The λ_i's are positive roots, therefore linear combinations with positive coefficients of the simple roots. Hence if $a \in A_t$ there exists a constant t' such that $\lambda_i(\log a) \leq t'$ for all i. If further $n \in \omega$, then the t_i's are also bounded, and the proposition is proved.

4.3. PROPOSITION. *Let G be an open subgroup of a real algebraic, semisimple Lie group. Then any Siegel domain has finite Haar measure.*

Let $G = K \cdot A \cdot N$ be the underlying Iwasawa decomposition, and dk, da, dn Haar measures on K, A and N. Then, in the notation of 4.2,

$$dg = \exp[\sigma(\log a)] \, dk \cdot da \cdot dn \qquad (\sigma = \lambda_1 + \cdots + \lambda_m)$$

is a Haar measure on G [9, Lemma 35], hence

$$\int_{\mathfrak{S}_{t,\omega}} dg = c \cdot \int_{\lambda(\log a) \leq t, \lambda \in \Sigma} \exp[\sigma(\log a)] da \ .$$

Let $\lambda_1, \cdots, \lambda_r$ be the simple roots. Then $\sigma = m_1 \lambda_1 + \cdots + m_r \lambda_r$ with $m_i > 0 (1 \leq i \leq r)$. Since G is semi-simple, $r = \dim \mathfrak{a}$, the λ_i form a coordinate system on $\mathfrak{a}$, and the exponential $\mathfrak{a} \rightarrow A$ carries the euclidean measure on the Haar measure, we have, up to a constant factor

$$\int_{\lambda(\log a) \leq t, \lambda \in \Sigma} \exp[\sigma(\log a)] da = \prod_i \int_{-\infty}^{t} \exp(m_i \lambda_i) d\lambda_i < \infty \ .$$

4.4. *Siegel domains in* $\mathbf{GL}(n, \mathbf{R})$. In $\mathbf{GL}(n, \mathbf{R})$ we consider the usual Iwasawa decomposition where $K = \mathbf{O}(n)$, A is the subgroup of diagonal matrices with positive coefficients, and N the group of upper triangular matrices with diagonal coefficients equal to 1. If we denote the diagonal coefficients $a_1 \cdots a_n$ of an element $a \in A$ by $\exp \mu_1, \cdots, \exp \mu_n$, the simple positive roots are the linear forms $\mu_i - \mu_{i+1} (i = 1, \cdots, n-1)$. Therefore the subset $\mathfrak{S}_{t,u} = \{k \cdot a \cdot \nu\}$ where $k \in \mathbf{O}(n)$, $a = \mathrm{diag}(a_1, \cdots, a_n)$, $a_i \leq t \cdot a_{i+1} (i = 1, \cdots, n-1)$, $\nu = (n_{ij}) \in N$, $|n_{ij}| \leq u(i < j)$ is a Siegel domain in the sense of 4.1. Clearly $\mathfrak{S}_{t,u} \subset \mathfrak{S}_{t',u'}$ if $t \leq t'$, $u \leq u'$ and $\mathbf{GL}(n, \mathbf{R})$ is the union of the $\mathfrak{S}_{t,u} (t, u > 0)$. If t and u are big enough (in fact $t > 2 \cdot 3^{-1/2}$, $u > 1/2$ will do), $\mathfrak{S}_{t,u}$, or its intersection with $\mathbf{SL}(n, \mathbf{R})$, or with the group of matrices of determinant ± 1, will be called a *standard Siegel domain* of the group in question. These domains will be important for us, because of the following result; for which references are given below.

4.5. PROPOSITION. *Let $\mathfrak{S}$ be a standard Siegel domain of* $\mathbf{GL}(n, \mathbf{R})$. *Then*

(a) $\mathbf{GL}(n, \mathbf{R}) = \mathfrak{S} \cdot \mathbf{SL}(n, \mathbf{Z})$.

(b) *For* $x, y \in \mathbf{GL}(n, \mathbf{Q})$, *the intersection* $\mathfrak{S}^{-1} \cdot \mathfrak{S} \cap x \cdot \mathbf{SL}(n, \mathbf{Z}) \cdot y$ *is finite.*

Let P_n be the space of real symmetric positive non-degenerate $n \times n$ matrices, and $\pi: X \to {}^t X \cdot X$ the usual projection of $\mathbf{GL}(n, \mathbf{R})$ onto P_n. We let as usual $\mathbf{GL}(n, \mathbf{R})$ act on the right on P_n by $F \to F[X] = {}^t X \cdot F \cdot X$ ($F \in P_n$, $X \in \mathbf{GL}(n, \mathbf{R})$). For $t, u > 0$, the *Siegel domain* $\mathfrak{S}'_{t,u}$ in P_n is the set of matrices $D[T]$ where

$$D = \text{diag}\,(d_1, \cdots, d_n) \qquad (d_i \leqq t \cdot d_{i+1}; i = 1, \cdots, n-1)$$
$$T = (t_{ij}) \in N \qquad (|\,t_{ij}\,| \leqq u, i < j)\,.$$

Clearly, $\mathfrak{S}_{t,u} = \pi^{-1}(\mathfrak{S}'_{t^2,u})$; the Siegel domains of $\mathbf{GL}(n, \mathbf{R})$ defined above are the inverse images of the Siegel domains in P_n. From the definitions, it is clear that $q \cdot \mathfrak{S}_{t,u} = \mathfrak{S}_{t,u}$, and $\mathfrak{S}'_{t,u}[q \cdot I] = \mathfrak{S}'_{t,u}$ for $q > 0$. Since $g \in \mathbf{GL}(n, \mathbf{Q})$ may be written as $g = q \cdot g'$, where g' has integral coefficients, and $\pi(X \cdot Y) = \pi(X)[Y]$, Proposition 4.5 follows from

(a') *For* $t \geq 4/3, u \geq 1/2, P_n = \bigcup_{S \in \mathbf{SL}(n, \mathbf{Z})} \mathfrak{S}'_{t,u}[S].$

(b') *For each positive number* q, *the set of integral matrices S with determinant smaller than q in absolute value for which* $\mathfrak{S}'_{t,u}[S] \cap \mathfrak{S}'_{t,u} \neq \varnothing$, *is finite.*

The statement (a'), which is more or less implicit in Hermite's work [14], is proved in [16]; in fact S is allowed in [16] to have determinant -1, but it is obvious from the proof that it may be taken in $\mathbf{SL}(n, \mathbf{Z})$. The second assertion is a well known theorem of Siegel [28].

Clearly, the interior of a standard Siegel domain contains a standard Siegel domain, a finite union of standard Siegel domains is contained in a standard Siegel domain, and every element of $\mathbf{GL}(n, \mathbf{R})$ belongs to a standard Siegel domain.

5. A finiteness lemma

5.1. Let G be a locally compact group, acting continuously on the right on a locally compact space M, and G_x the isotropy group of a point $x \in M$. Then $\mu: g \to x \cdot g$ induces a continuous bijective map of the space $G_x \backslash G$ of cosets $G_x \cdot g$ onto the orbit $x \cdot G$ of x. It follows from a theorem of Arens [19, p. 65] that if G is countable at infinity, and $x \cdot G$ is *closed*, then μ is a homeomorphism of $G_x \backslash G$, endowed with the quotient topology, onto $x \cdot G$, endowed with the induced topology. Since a compact set of $G_x \backslash G$ is always the image of a compact set of G, this can also be expressed by saying that, under the assumptions made, the set of $g \in G$ for which $x \cdot g$ belongs to a fixed compact set of M is of the form $G_x \cdot \Omega$, with Ω compact in G. Of this we shall need the following special case.

5.2. Proposition. *Let G be a Lie group with finitely many connected components, π a continuous complex or real linear representation of G in a finite dimensional vector space V, $v \in V$ a point whose orbit $v \cdot \pi(G)$ is closed, and Q a compact subset of V. Then $\{g \in G, v \cdot \pi(g) \in Q\} = G_x \cdot \Omega$, with Ω compact.*

By a *lattice* Γ in a finite dimensional vector space over $\mathbf{Q}$ or $\mathbf{R}$ we mean as usual a discrete additive subgroup which is generated by a vector space basis of V. A subspace W of V is said to be rational with respect to Γ if $W \cap \Gamma$ is a lattice in W, or, equivalently, if W is defined by linear equations with integral coefficients in a coordinate system in which Γ is the lattice of integral points.

5.3. Lemma. *Let G be a subgroup of finite index in a semi-simple real algebraic group, θ a Cartan involution of G, $G = K \cdot A \cdot N$ an Iwasawa decomposition of G compatible with θ, and $\mathfrak{S}$ a Siegel domain of G (with respect to the decomposition $K \cdot A \cdot N$). Let π be a real representation of G in a finite dimensional vector space V, and Γ a lattice in V with respect to which the maximal eigenspaces V_i of $\pi(A)$ are rational. Let $v \in V$ be a point whose orbit is closed and whose isotropy group G_v is stable under θ. Then $v \cdot \pi(\mathfrak{S}) \cap \Gamma$ is finite.*

We denote by $|v|$ the norm of $v \in V$ for a euclidean norm with respect to which the elements of $\pi(A)$ are self-adjoint (1.1). The space V is the direct sum of the V_i's, which are mutually orthogonal. We denote by E_i the orthogonal projection of V onto V_i, and by μ_i the weight of $\mathfrak{a}$ in V_i. Since V_i is rational with respect to Γ, the subgroup of Γ spanned by the intersections $\Gamma \cap V_i$ has finite index, and $E_i(\Gamma)$ is also a *lattice* in V_i. There exists therefore a constant $c > 0$ such that $w \in \Gamma$, $E_i(w) \neq 0$ implies $|E_i(w)| \geqq c$.

Let now $x = k_x \cdot a_x \cdot n_x \in \mathfrak{S}$, and $y_x = x \cdot a_x^{-1}$, $z_x = y_x \cdot a_x^{-1}$. In the sequel, we shall drop π, and write $w \cdot g$ instead of $w \cdot \pi(g)(w \in V, g \in G)$.

Let $w = v \cdot x$, and $w_i = E_i(w)$. Then

$$(1) \quad E_i(v \cdot y_x) = \exp\left[-\mu_i(\log a_x)\right] \cdot w_i , \quad E_i(v \cdot z_x) = \exp\left[-2\mu_i(\log a_x)\right]w_i .$$

It follows from 4.2 that

$$\{y_x\}_{x \in \mathfrak{S}} = K \cdot \{a \cdot n \cdot a^{-1}\}_{a \cdot n \in \mathfrak{S}}$$

is relatively compact. Consequently, $\{v \cdot y_x)_{x \in \mathfrak{S}}$ is relatively compact and there exists a constant c' such that $|w \cdot a_x^{-1}| = |v \cdot y_x| \leqq c'$. Hence

$$(2) \qquad\qquad |E_i(v \cdot y_x)| = |w_i| \exp\left[-\mu_i(\log a_x)\right] \leqq c' .$$

Assume now $w \in \Gamma$, and fix an i. There are two possibilities:

   (i) $w_i = 0$; then $E_i(v \cdot y_x) = E_i(v \cdot z_x) = 0$;

   (ii) $w_i \neq 0$; then, by the above, $|w_i| \geqq c$, hence, using (1),

$$(3) \qquad |E_i(v \cdot z_x)| = |E_i(v \cdot y_x)|^2 / |w_i| \leqq c'^2/c \ .$$

In both cases, $|E_i(v \cdot z_x)|$ has a bound which is independent of i and of $x \in \mathfrak{S}$. This proves therefore that if $v \cdot x \in \Gamma$, $x \in \mathfrak{S}$, then $v \cdot z_x$ belongs to some compact set of V. Since $v \cdot G$ is closed by assumption, 5.2 shows the existence of a compact subset Q of G such that

$$(4) \qquad v \cdot x \in \Gamma, \ x \in \mathfrak{S} \Rightarrow z_x \in G_v \cdot Q \ .$$

We now fix such an x, and drop the index x. We have

$$z = k \cdot a \cdot n \cdot a^{-2} = k \cdot a^{-1} \cdot a^2 \cdot n \cdot a^{-2} \in G_v \cdot Q \ .$$

But $\theta(k) = k$, $\theta(a) = a^{-1}$, and, by assumption, $\theta(G_v) = G_v$, hence

$$(5) \qquad \begin{aligned} \theta(z) &= k \cdot a \cdot \theta(a^2 \cdot n \cdot a^{-2}) = x \cdot n^{-1} \cdot \theta(a^2 \cdot n \cdot a^{-2}) \in G_v \cdot \theta(Q) \\ x &\in G_v \cdot \theta(Q) \cdot \theta(a^2 \cdot n \cdot a^{-2})^{-1} \cdot n \ . \end{aligned}$$

By definition of $\mathfrak{S}$, the element n lies in a fixed compact set. Further if we have $\mathfrak{S} = \mathfrak{S}_{t,\omega}$ in the notation of 4.1, then $a^2 \in A_{2t}$ and therefore, by 4.2, $a^2 \cdot n \cdot a^{-2}$ lies in a fixed compact set. Then so does $\theta(a^2 \cdot n \cdot a^{-2})^{-1}$, and (5) shows the existence of a compact set $Q' \subset G$ such that

$$v \cdot x \in \Gamma, \ x \in \mathfrak{S} \Rightarrow x \in G_v \cdot Q' \ .$$

But then $v \cdot x$ belongs to $\Gamma \cap v \cdot Q'$, which is finite.

REMARK. The lemma and its proof are valid if G is reductive, provided $\pi(A)$ is real diagonalizable, as will always happen if π is a rational representation. In fact we shall use the following form of the lemma (only for $G = \mathrm{GL}(n, \mathbf{C})$, and we could also limit ourselves to the case $G = \mathrm{SL}(n, \mathbf{C})$.

5.4. LEMMA. *Let G be a reductive complex algebraic group defined over $\mathbf{Q}$, θ a Cartan involution of $G_{\mathbf{R}}$, $G_{\mathbf{R}} = K \cdot A \cdot N$ an Iwasawa decomposition of $G_{\mathbf{R}}$ compatible with θ, where A is contained in an algebraic torus T of G which is defined over $\mathbf{Q}$ and is isomorphic over $\mathbf{Q}$ to a product of groups $\mathbf{C}^*$, and $\mathfrak{S}$ a Siegel domain. Let $\pi: G \to \mathrm{GL}(V)$ be a rational representation of G which is defined over $\mathbf{Q}$, and Γ a lattice of $V_{\mathbf{Q}}$. Let v be a point of $V_{\mathbf{R}}$ whose orbit under G is closed and whose isotropy group in $G_{\mathbf{R}}$ is stable under θ. Then $v \cdot \pi(\mathfrak{S}) \cap \Gamma$ is finite.*

By 2.3, $v \cdot \pi(G_{\mathbf{R}})$ is closed. Moreover $\pi(T)$ is diagonalizable *over* $\mathbf{Q}$ [26, Proposition 5], which means that a maximal eigenspace of $\pi(A)$ has a basis in $V_{\mathbf{Q}}$, hence that it is "rational with respect to Γ". Thus, taking the above remark into account, all conditions under which the proof of 5.3 is valid, are fulfilled.

5.5. EXAMPLE. Let $G = \mathbf{SL}(n, \mathbf{R})$, θ be the involution $g \to {}^t g^{-1}$, and $\mathfrak{S}$ a standard Siegel domain (4.4). Let V be the space of symmetric real $n \times n$ matrices, and Γ the lattice of integral symmetric matrices. For π, we take the usual representation $F \to {}^t X \cdot F \cdot X = F[X]$. Here A is the group of positive diagonal matrices with det 1, and $\pi(A)$ is diagonal with respect to the standard basis of Γ; the maximal eigenspaces of $\pi(A)$ (which are in fact 1-dimensional) are then rational with respect to Γ. Let c be a strictly positive number, p, q two positive integers whose sum is n, and v the diagonal matrix with p entries equal to c, and q entries equal to $-c$. Then G_v is the proper orthogonal group of the quadratic form v, and is clearly stable under θ. Moreover, the orbit $v \cdot G$ of v consists of all quadratic forms with determinant $(-1)^q \cdot c^n$ and signature (p, q), hence is closed. The lemma applies, and shows that $v[\mathfrak{S}] \cap \Gamma$ is finite. But the matrices $v[g](g \in \mathfrak{S})$ are those of the forms of determinant $(-1)^q \cdot c^n$ and signature (p, q), which are reduced in the sense of Hermite (if we agree to call reduced those positive quadratic forms whose matrix is of the type ${}^t X \cdot X$ (with $X \in \mathfrak{S}$). This is a somewhat bigger set than the one considered by Hermite). The lemma in this case means therefore that the integral reduced forms with given non-zero determinant and signature are finite in number, a well known statement of Hermite [14, p. 127]. Since $\mathbf{SL}(n, \mathbf{R}) = \mathfrak{S} \cdot \mathbf{SL}(n, \mathbf{Z})$, it implies that the number of proper classes of integral quadratic forms with a given non-zero determinant is finite.

If we consider, instead of π, the natural representation of $\mathbf{SL}(n, \mathbf{R})$ in the space of homogeneous forms of degree $m \geq 2$, then the lemma also applies, and shows that the number of classes of integral forms belonging to some closed orbit is finite. Such generalizations of Hermite's result have been given by C. Jordan (Journal Ec. Polytechnique, 48 (1880), pp. 151–168), and H. Poincaré (*ibidem*, 51 (1882), pp. 45–91).

6. A fundamental set for arithmetic subgroups

6.1. Let G be an algebraic group defined over $\mathbf{Q}$. Then $G_{\mathbf{Z}}$ is a discrete subgroup of $G_{\mathbf{R}}$. It is known [29], and will follow from 6.3, that if we change the imbedding, that is, if we replace G by its image G' under a rational injective homomorphism ρ, defined over $\mathbf{Q}$, then $G'_{\mathbf{Z}}$ is commensurable with $\rho(G_{\mathbf{Z}})$. (We recall that two subgroups of a group are commensurable if their intersection has finite index in both.) Therefore the commensurable class of $G_{\mathbf{Z}}$ in $G_{\mathbf{R}}$ has an intrinsic meaning.

6.2. PROPOSITION. *Let G be a connected algebraic group defined over*

$\mathbf{Q}$, $\pi\colon G \to \mathbf{GL}(V)$ *a rational representation defined over* $\mathbf{Q}$. *Then the subgroup of* $G_{\mathbf{Z}}$ *leaving a lattice* Γ *in* $V_{\mathbf{Q}}$ *invariant has finite index in* $G_{\mathbf{Z}}$.

Let H be the identity component of $G \cap \mathbf{SL}(n, \mathbf{C})$. Then $H_{\mathbf{Z}}$ has finite index in $G_{\mathbf{Z}}$, we may therefore assume $G \subset \mathbf{SL}(n, \mathbf{C})$. In this case, the coefficients $\pi(g)_{\mu\nu}$ of $\pi(g)$, with respect to a basis of Γ, are polynomials in those of g, with rational coefficients (2.2). Let $g'_{ij} = g_{ij} - \delta_{ij}$. Then

$$y_{\mu\nu}(g) = \pi(g)_{\mu\nu} - \delta_{\mu\nu} = P_{\mu\nu}(g'_{11}, \cdots, g'_{nn}) , \qquad (1 \leqq \mu, \nu \leqq \dim V) ,$$

where the $P_{\mu\nu}$ are polynomials with rational coefficients and no constant term. Let m be a common multiple of the denominators of the $P_{\mu\nu}$, and M be the congruence subgroup of the elements in $G_{\mathbf{Z}}$ which are $\equiv \mathrm{Id} \bmod \mathrm{m}$. Then M has finite index in $G_{\mathbf{Z}}$, and $\pi(g)_{\mu\nu} \in \mathbf{Z}$ for $g \in M$.

6.3. COROLLARY. *There exists a lattice in* $V_{\mathbf{Q}}$ *containing* Γ *which is invariant under* $G_{\mathbf{Z}}$. *If* π *is faithful,* $G_{\mathbf{Z}}$ *is commensurable with the subgroup of* G *leaving* Γ *invariant.*

We keep the notation of the previous proof. The subgroup M is invariant in $G_{\mathbf{Z}}$, therefore, for any $g \in G_{\mathbf{Z}}$, the lattice $\Gamma \cdot \pi(g)$ is invariant under M. Then the sum of the lattices $\Gamma \cdot \pi(g_i)$, where g_i runs through a system of representatives of $G_{\mathbf{Z}}/M$, is a lattice invariant under $G_{\mathbf{Z}}$. The second assertion follows from 6.2 applied to π and to π^{-1}.

6.4. COROLLARY. *Let* $G = H \cdot N$ *be the semi-direct product of a subgroup* H *and of an invariant subgroup* N, *both defined over* $\mathbf{Q}$. *Then* $H_{\mathbf{Z}} \cdot N_{\mathbf{Z}}$ *has finite index in* $G_{\mathbf{Z}}$.

We may assume G to be connected. The map $g = h \cdot n \to h (h \in H, n \in N)$ is a rational homomorphism defined over $\mathbf{Q}$, hence (6.2) $G_{\mathbf{Z}}$ has a subgroup M of finite index whose image is in $H_{\mathbf{Z}}$. Then $M \subset H_{\mathbf{Z}} \cdot N_{\mathbf{Z}}$.

6.5. THEOREM. *Let* $G \subset \mathbf{GL}(n, \mathbf{C})$ *be a reductive algebraic group defined over* $\mathbf{Q}$, $a \in \mathbf{SL}(n, \mathbf{R})$ *such that* $a \cdot G_{\mathbf{R}} \cdot a^{-1}$ *is self-adjoint (see 1.9), and* $\mathfrak{S}$ *a standard Siegel domain of* $\mathbf{GL}(n, \mathbf{R})$ *(see 4.4). Then there exist finitely many elements* $b_1, \cdots, b_m \in \mathbf{SL}(n, \mathbf{Z})$ *such that the interior* U *of* $\bar{U} = \bigcup_{i=1}^{i=m} (a^{-1} \cdot \mathfrak{S} \cdot b_i) \cap G_{\mathbf{R}}$ *has the following properties:*

(i) $G_{\mathbf{R}} = U \cdot G_{\mathbf{Z}}$;

(ii) $K \cdot U = U$ *for a suitable maximal compact subgroup* K *of* $G_{\mathbf{R}}$;

(iii) $U^{-1} \cdot U \cap x \cdot G_{\mathbf{Z}} \cdot y$ *is finite for any* $x, y \in G_{\mathbf{Q}}$.

The group $G_{\mathbf{Z}}$ *is finitely generated.*

By 3.8, we may find a rational representation $\pi\colon \mathbf{GL}(n, \mathbf{C}) \to \mathbf{GL}(V)$, defined over $\mathbf{Q}$, for which there exists $v \in V_{\mathbf{Q}}$ whose isotropy group is G and whose orbit is closed. Using 6.3, we take in $V_{\mathbf{Q}}$ a lattice Γ invariant under $GL(n, \mathbf{Z})$; replacing v by a multiple if necessary, we many assume $v \in \Gamma$. The group $G_{\mathbf{R}}$ is real algebraic, reductive (3.2); by 1.9, there exists $a \in SL(n, \mathbf{R})$ such that $a \cdot G_{\mathbf{R}} \cdot a^{-1} = G'_{\mathbf{R}}$ is self-adjoint. Let $v' = v \cdot \pi(a^{-1})$. Then $v' \cdot \pi(GL(n, \mathbf{C})) = v \cdot \pi(GL(n, \mathbf{C}))$ is closed, hence (2.3) so is $v' \cdot \pi(GL(n, \mathbf{R}))$. The isotropy group of v' in $GL(n, \mathbf{R})$ is $G'_{\mathbf{R}}$, and is by construction invariant under the Cartan involution $\theta: g \to {}^t g^{-1}$, which underlies the definition of the standard Siegel domain $\mathfrak{S}$. Moreover, in this case, A is the subgroup of diagonal matrices with positive real eigenvalues, and it belongs to the algebraic torus $D(n)$ of all diagonal matrices of $GL(n, \mathbf{C})$, which is defined over $\mathbf{Q}$. Consequently, all conditions of 5.4 are fulfilled, and we may assert that $v' \cdot \pi(\mathfrak{S}) \cap \Gamma$ is finite. *A fortiori*, $v' \cdot \pi(\mathfrak{S}) \cap v \cdot \pi(SL(n, \mathbf{Z}))$ is finite. Let then $b_1, \cdots, b_m \in SL(n, \mathbf{Z})$ be such that

$$(1) \qquad v \cdot \pi(\mathfrak{S}) \cap v \cdot \pi(SL(n, \mathbf{Z})) \subset \{v \cdot \pi(b_1^{-1}), \cdots, v \cdot \pi(b_m^{-1})\} \; .$$

Let now

$$(2) \qquad H = \{g \in GL(n, \mathbf{R}) \mid v' \cdot \pi(g) = v\} \; .$$

Then

$$(3) \qquad H = a \cdot G_{\mathbf{R}} = G'_{\mathbf{R}} \cdot a \; .$$

Let $h \in H$. By the classical reduction theory (4.5),

$$h = s \cdot b \qquad\qquad (s \in \mathfrak{S}, \, b \in SL(n, \mathbf{Z})) \; .$$

The equality $v' \cdot \pi(h) = v$ yields $v' \cdot \pi(s) = v \cdot \pi(b^{-1})$, hence, by (1), there exists an index i, $(1 \leq i \leq m)$ such that

$$v' \cdot \pi(s) = v \cdot \pi(b^{-1}) = v \cdot \pi(b_i^{-1}) \; .$$

We have then $b_i^{-1} \cdot b \in G \cap SL(n, \mathbf{Z}) \subset G_{\mathbf{Z}}$, hence $b \in b_i G_{\mathbf{Z}}$, and

$$
(4) \qquad
\begin{aligned}
& H \subset \bigcup_i \mathfrak{S} \cdot b_i \cdot G_{\mathbf{Z}} \; , \\
& G_{\mathbf{R}} = a^{-1} \cdot H \subset \bigcup_i a^{-1} \cdot \mathfrak{S} \cdot b_i \cdot G_{\mathbf{Z}} \; , \\
& G_{\mathbf{R}} = \bar{U} \cdot G_{\mathbf{Z}}, \qquad\qquad \left(\bar{U} = \bigcup_i (a^{-1} \cdot \mathfrak{S} \cdot b_i) \cap G_{\mathbf{R}}\right) \; .
\end{aligned}
$$

The equality (4) is *a fortiori* true if we replace $\mathfrak{S}$ by a standard Siegel domain containing $\mathfrak{S}$ in its interior, hence (i) is proved.

The group $K' = G'_{\mathbf{R}} \cap O(n)$ is maximal compact in $G'_{\mathbf{R}}$ (1.10), and $K = a^{-1} \cdot K' \cdot a$ is maximal compact in $G_{\mathbf{R}}$. Clearly, $K' \cdot (\mathfrak{S} b_i \cap G'_{\mathbf{R}}) \subset \mathfrak{S} b_i \cap G'_{\mathbf{R}}$, and therefore $K \cdot \bar{U} = \bar{U}$, $K \cdot U = U$.

Let now $x, y \in G_{\mathbf{Q}}$, and $u \in U^{-1} \cdot U \cap x \cdot G_{\mathbf{Z}} \cdot y$. There exist, then, two indices $i, j (1 \leq i, j \leq m)$ such that

$$u \in (a^{-1} \cdot \mathfrak{S} \cdot b_i)^{-1} \cdot (a^{-1} \cdot \mathfrak{S} \cdot b_j) = b_i^{-1} \cdot \mathfrak{S}^{-1} \cdot \mathfrak{S} \cdot b_j \; ,$$

whence

$$b_i \cdot u \cdot b_j^{-1} \in \mathfrak{S}^{-1} \cdot \mathfrak{S} \,,$$

and the finiteness of the number of the possible u's follows from Siegel's theorem (4.5). For $x = y = e$, this implies that $U^{-1} \cdot U \cap G_{\mathbf{Z}}$ is finite, in other words that U meets only a finite number of its right translates under $G_{\mathbf{Z}}$. Since $G_{\mathbf{Z}} \cap (G_{\mathbf{R}})^0$ is of finite index in $G_{\mathbf{Z}}$, the finite generation of $G_{\mathbf{Z}}$ follows from the following well known elementary lemma:

6.6. LEMMA. *Let H be a group which operates on a connected topological space M, and U an open set such that $U \cdot H = M$. Then $J = \{h \in H \mid U \cdot h \cap U \neq \varnothing\}$ is a set of generators for H.*

In fact let H' be the subgroup generated by J. Then $U \cdot H'$ is open. If $U \cdot h \cap U \cdot h' \neq \varnothing \, (h \in H, h' \in H')$, then $U \cdot h \cdot h'^{-1} \cap U \neq \varnothing$, hence $h \in J \cdot h' \subset H'$, from which it follows first that $U \cdot H' = M$, and then that $H' = H$.

6.7. REMARKS.

(1) Let us call fundamental set an open subset of $G_{\mathbf{R}}$ satisfying properties (i), (ii), (iii). In view of the inclusion properties of standard Siegel domains listed at the end of 4.5, the class C of fundamental sets constructed in 6.5 has the following properties: it covers $G_{\mathbf{R}}$, any finite union of such sets is contained in a set of C, any such set contains the closure of an element of C.

The main part of 6.5 will be extended to algebraic groups in 6.12. It will be shown later that, at any rate for a suitable a, the set U has finite Haar measure when G is semi-simple.

(2) Let K be a maximal compact subgroup such that $K \cdot U = U$, and π the natural projection of $G_{\mathbf{R}}$ onto $P = K \backslash G_{\mathbf{R}}$. Clearly, $U' = \pi(U)$ has the following properties: (i') $P = U' \cdot G_{\mathbf{Z}}$; (iii') for $x, y \in G_{\mathbf{Q}}$, the set of $g \in G_{\mathbf{Z}}$ for which $U' \cap U' \cdot x \cdot g \cdot y \neq \varnothing$ is finite. Conversely, the inverse image of a set U' in P having the properties (i'), (iii') has the properties (i), (ii), (iii), so that the construction of fundamental sets in $G_{\mathbf{R}}$ or in $K \backslash G_{\mathbf{R}}$ are essentially equivalent questions. In the special case where G is the orthogonal group of an integral non-degenerate indefinite quadratic form F, $K \backslash G_{\mathbf{R}}$ is the space of majorizing positive forms of F in the sense of Hermite. If we take, furthermore, the natural representation of $\mathbf{GL}(n, \mathbf{C})$ in the space of symmetric matrices, then our construction reduces to that of Hermite (always with the minor difference that instead of using an arbitrary Siegel domain $\mathfrak{S}$ of $\mathbf{GL}(n, \mathbf{R})$, Hermite uses the inverse image of the space of reduced positive forms in his sense, that is, the domain given by 4.5a'). In this case, the two properties of U' above, and the resulting finite generation of $G_{\mathbf{Z}}$, are also proved by Hermite [14, pp. 201–

233, § VIII].

6.8. LEMMA. *Let $M \subset G$ be reductive subgroups of $\mathbf{GL}(n, \mathbf{C}) = G'$, both defined over $\mathbf{Q}$. Let $\pi: G' \to \mathbf{GL}(W)$ be a rational representation, defined over $\mathbf{Q}$, Γ a lattice in $W_{\mathbf{Q}}$ invariant under $G'_{\mathbf{Z}}$, $w' \in \Gamma$ a point whose orbit under G' is closed and whose isotropy group in G' is M. Then $w' \cdot G_{\mathbf{R}} \cap \Gamma$ consists of a finite number of orbits of $G_{\mathbf{Z}}$.*

By 1.9, there exists $a \in \mathbf{SL}(n, \mathbf{R})$ such that $a \cdot G_{\mathbf{R}} \cdot a^{-1}$ and $a \cdot M_{\mathbf{R}} \cdot a^{-1}$ are self-adjoint. By 6.5, there exists a finite number of elements $b_i \in G'_{\mathbf{Z}}$ such that $G_{\mathbf{R}} = \bigcup_i (G_{\mathbf{R}} \cap a^{-1} \cdot \mathfrak{S} \cdot b_i) \cdot G_{\mathbf{Z}}$ where $\mathfrak{S}$ is a standard Siegel domain of $\mathbf{GL}(n, \mathbf{R})$. It suffices therefore to show that

$$w \cdot (G_{\mathbf{R}} \cap a^{-1} \cdot \mathfrak{S} \cdot b_i) \cap \Gamma$$

is finite, hence, *a fortiori*, that $w \cdot a^{-1} \cdot \mathfrak{S} \cap \Gamma$ is finite. Let $w' = w \cdot a^{-1}$. Then $G'_{w'} = a \cdot M \cdot a^{-1}$, hence $(G'_{w'})_{\mathbf{R}}$ is self-adjoint. Further, the group A of diagonal matrices which underlies the definition of $\mathfrak{S}$ belongs to an algebraic torus which is diagonal over $\mathbf{Q}$ hence 5.4 applies and yields the finiteness of $w \cdot a^{-1} \cdot \mathfrak{S} \cap \Gamma = w' \cdot \mathfrak{S} \cap \Gamma$.

6.9. THEOREM. *Let G be a reductive algebraic group defined over $\mathbf{Q}$, $\pi: G \to \mathbf{GL}(V)$ a rational representation defined over $\mathbf{Q}$, Γ a lattice in $V_{\mathbf{Q}}$ invariant under $G_{\mathbf{Z}}$, and X a closed orbit of G. Then $X \cap \Gamma$ consists of a finite number of orbits of $G_{\mathbf{Z}}$.*

Since G^0 has finite index in G, we may assume G to be connected. We assume $X \cap \Gamma \neq \varnothing$ (otherwise there is nothing to prove), take $v \in \Gamma \cap X$, and put $H = G_v$. The group H and the orbit $X = v \cdot G$ are defined over $\mathbf{Q}$. Since X is closed, H is reductive (3.8). We prove first:

(*) There exists a rational representation $\pi': G \to \mathbf{GL}(W)$ defined over $\mathbf{Q}$, a point $w \in W_{\mathbf{Q}}$ such that $G_w = H$, $w \cdot \pi'(G) = X'$ is closed, and that $X' \cap \Gamma'$ consists of a finite number of orbits of $G_{\mathbf{Z}}$ for any lattice $\Gamma' \subset W_{\mathbf{Q}}$ invariant under $G_{\mathbf{Z}}$.

Let $G' = \mathbf{GL}(n, \mathbf{C})$. The group H being reductive, defined over $\mathbf{Q}$, there exists by 3.8 a rational representation $\rho: G' \to \mathbf{GL}(W)$ defined over $\mathbf{Q}$, and a point $w \in W_{\mathbf{Q}}$ such that $w \cdot \rho(G')$ is closed and $H = G'_w$. We claim that the restriction π' of ρ to G fulfills our conditions. The orbit $X' = w \cdot \pi'(G)$ is closed because if we identify $w \cdot \rho(G')$ with $H \backslash G'$, the orbit X' is the inverse image of a point in the projection $H \backslash G' \to G \backslash G'$. Let now $\Gamma' \subset W_{\mathbf{Q}}$ be a lattice invariant under $G_{\mathbf{Z}}$. It is contained in a lattice invariant under $G'_{\mathbf{Z}}$ (6.3), therefore it is enough to consider $X' \cap \Gamma'$ for lattices invariant under $G'_{\mathbf{Z}}$. Of course, $X' \cap \Gamma' \subset X' \cap W_{\mathbf{R}}$, which consists of a finite number of closed orbits of $G_{\mathbf{R}}$ (2.3); therefore, it suffices to show that for $w' \in X' \cap \Gamma'$, the intersection $w' \cdot \pi'(G_{\mathbf{R}}) \cap \Gamma'$ consists of a finite number of orbits of G_Z. There exists $g \in G$ such that $w' = w \cdot \pi'(g)$, hence $G'_{w'} = g^{-1} \cdot H \cdot g \subset G$. The group $G'_{w'}$ is reductive, since H is, and defined over $\mathbf{Q}$, since $w' \in W_{\mathbf{Q}}$. Moreover $w' \cdot \rho(G') = w \cdot \rho(G')$ is closed. The fact that $w' \cdot \pi'(G_R) \cap \Gamma'$ is the union of a finite number of orbits of G_Z is now a consequence of 6.8 (with $M = G'_{w'}$).

The maps $g \to v \cdot \pi(g)$ and $g \to w \cdot \pi'(g)$ induce isomorphisms of $H \backslash G$ with X and X', defined over $\mathbf{Q}$, whence an equivariant isomorphism $\varphi \colon X \to X'$, defined over $\mathbf{Q}$. Let $x_1, \cdots, x_r$ and $y_1, \cdots, y_s$ be coordinates in V and W with respect to bases of Γ and Γ' respectively. The function $y_i(\varphi(x))(x \in X)$ is a regular function, defined over $\mathbf{Q}$. Since X is an affine algebraic set, $y_i(\varphi(x))$ may be written as a polynomial $P_i(x_1, \cdots, x_r)$ in the x_i's, with rational coefficients $(1 \leq i \leq s)$. If q is a common multiple of the denominators of those coefficients, then $q \cdot y_i(\varphi(x))$ is integral whenever $x_1, \cdots, x_r$ are, therefore

$$\varphi(X \cap \Gamma) \subset X' \cap \frac{1}{q} \Gamma' \,.$$

The lattice $(1/q)\Gamma'$ is of course also invariant under G_Z, therefore $(1/q)\Gamma' \cap X'$ consists of a finite number of orbits of G_Z by (*). Since φ is G-equivariant, its restriction to $X \cap \Gamma$ is G_Z-equivariant, hence $X \cap \Gamma$ is also a finite number of orbits of G_Z.

6.10. In order to go from the reductive to the general case in the two following sections, we shall use the following facts: a connected algebraic group G defined over a field k (of characteristic 0) is the semi-direct product of a reductive group H and of an invariant unipotent group N, both defined over k.[4] If G is unipotent, defined over $\mathbf{Q}$, then G_R/G_Z is compact. The latter fact is completely elementary; in fact, G being nilpotent, defined over $\mathbf{Q}$, its Lie algebra has a basis (x_i) $(i = 1, \cdots, \dim G)$ consisting of matrices with integral coefficients, such that for each $j \geq 1$, the elements $x_i(i \geq j)$ span an ideal. Let, further, m be the degree of the ambient linear group. Then $x_i^m = 0$ $(i = 1, \cdots, \dim G)$, and $g_i = \exp(m! \cdot x_i) \in G_Z$. Using induction on $\dim G$, it is immediately seen that the quotient of G_R by the subgroup generated by the g_i's is compact.

6.11. COROLLARY. *Let G, G' be algebraic groups defined over $\mathbf{Q}$ and $\mu \colon G \to G'$ a surjective rational homomorphism defined over $\mathbf{Q}$, with finite kernel (an isogeny). Then $\mu(G_Z)$ and G'_Z are commensurable.*

Here again, it is enough to prove this when G and G' are connected. It

---

[4] The corresponding statement for Lie algebras is proved in [7b, Chap. V, §4, Prop. 5]. For the global version, see G. D. Mostow, Amer. J. Math. 78 (1956), 200–221, Theorem 6.1.

follows from 6.3 that we may assume $\mu(G_Z)\subset G_Z'$. By adding one coordinate (see 2.1), we may have G' closed in $\mathbf{M}(n', \mathbf{C})$.

Let now G be reductive. We let G act on $V = \mathbf{M}(n', \mathbf{C})$ by right translations $x \to x \cdot \mu(g)$, and get in this way a rational representation $G \to \mathbf{GL}(V)$ defined over $\mathbf{Q}$. The subgroup G_Z' is the intersection of the closed orbit G' of G with the lattice $\mathbf{M}(n', \mathbf{Z})$ which is invariant under G_Z, and belongs to $V_\mathbf{Q}$. By 6.9, G_Z' consists of a finite number of cosets of $\mu(G_Z)$, hence $[G_Z': \mu(G_Z)] < \infty$.

If G is unipotent, $G_\mathbf{R}/G_Z$ is compact; then so is $G_\mathbf{R}'/\mu(G_Z)$; this space is a covering of $G_\mathbf{R}'/G_Z'$ with a discrete fibre which has $[G_Z': \mu(G_Z)]$ elements. This number must be finite.

In the general case, $G = H\cdot N$, $G' = H'\cdot N'$ with H, $H'=\mu(H)$ reductive, defined over $\mathbf{Q}$, N, $N'=\mu(N)$ unipotent, invariant, defined over $\mathbf{Q}$ (6.10). By the above $\mu(H_Z\cdot N_Z)$ has finite index in $H_Z'\cdot N_Z'$, and 6.11 follows from 6.4.

6.12. THEOREM. *Let G be an algebraic group defined over $\mathbf{Q}$. There exists an open set U in $G_\mathbf{R}$ such that $U\cdot G_Z = G_\mathbf{R}$, and that for any $x, y \in G_\mathbf{Q}$, the intersection $U^{-1}\cdot U\cap x\cdot G_Z\cdot y$ is finite. The group G_Z is finitely generated.*

It is enough to prove this for G connected. If G is reductive, see 6.5. If G is unipotent, then (6.10) there exists a relatively compact open subset U such that $G_\mathbf{R} = U\cdot G_Z$. Then U clearly satisfies our second condition.

In the general case, we use the decomposition $G = H\cdot N$ of 6.10. Let A and B be open subsets in H and N satisfying our two conditions in H and N, with B, moreover, relatively compact. We assert that $U = A\cdot B$ verifies our contention. In fact

$$G = H\cdot N = A\cdot H_Z\cdot N \subset A\cdot N\cdot H_Z \subset A\cdot B\cdot N_Z\cdot H_Z \subset A\cdot B\cdot G_Z \,,$$

which proves the first condition. The group $H_Z\cdot N_Z$ has finite index in G_Z (6.4). Therefore, our second assertion is equivalent to the finiteness of $U^{-1}\cdot U\cap(x\cdot H_Z\cdot N_Z\cdot y)$ for arbitrary $x, y \in G_\mathbf{Q}$. The group G being isomorphic to $H\cdot N$ over $\mathbf{Q}$, we also have $G_\mathbf{Q} = H_\mathbf{Q}\cdot N_\mathbf{Q}$, and may write $x = a\cdot b$, $y = c\cdot d(a, c \in H_\mathbf{Q}, b, d \in N_\mathbf{Q})$. Let $h \in H_Z, n \in N_Z$. We have

$$A\cdot B\cdot x\cdot h\cdot n\cdot y = A\cdot a\cdot h\cdot c\cdot B'\cdot n'$$
$$(B' = (a\cdot h\cdot c)^{-1}\cdot B\cdot a\cdot b\cdot h\cdot c \;; \quad n' = c^{-1}\cdot n\cdot c\cdot d) \,,$$

hence $A\cdot B\cap A\cdot B\cdot x\cdot h\cdot n\cdot y \neq \varnothing$ is equivalent to

$$(1) \qquad A\cdot a\cdot h\cdot c\cap A \neq \varnothing \,, \qquad B'\cdot n'\cap B \neq \varnothing \,.$$

By our assumption on A, the possible h's are finite in number. Since B is relatively compact, $B'^{-1} \cdot B$ is also relatively compact, hence its intersection with the discrete set $c^{-1} \cdot N_Z \cdot c \cdot d$ is finite. Thus there is a finite number of possibilities for h, and for each of them, a finite number of possible n's.

The above implies in particular that $U^{-1} \cdot U \cap G_Z$ is finite. The finite generation of G_Z then follows from 6.6, as in 6.5.

6.13. REMARK. As was mentioned in the introduction, the construction of U in 6.5 is a generalization of Hermite's procedure in the case of indefinite quadratic forms [14]. As is well known, the latter had been adapted, notably by Siegel, to many other cases, the most inclusive one being that of the automorphism group of a rational involutorial semi-simple algebra [24, 29]. This case represents essentially all classical groups with center reduced to the identity. However, U is constructed there in the symmetric space $K \backslash G_R$, rather than in G_R, but this is a minor difference (see 6.7). This implies, of course, the finite generation of G_Z. The finiteness of the volume of G_R/G_Z, or, equivalently, of $(K \backslash G_R)/G_Z$, in that case is also proved in [24], generalizing earlier results of Siegel.

In order to construct U, we take a rational representation of the ambient linear group $\mathrm{GL}(n, \mathbf{C})$, such that the representation space has a rational point with closed orbit and isotropy group G, whose existence follows from 2.5 and 3.5b. In this respect, we point out that 3.5a is used only in proving 6.9, and there only to ascertain that H is reductive. This last fact is obvious in 6.11, where $H = (e)$ and follows from [4] in 11.6, where H runs through the centralizers of semi-simple elements; as these are the only two applications of 6.9 made in this work, we see that, except for 6.9 in full generality, this paper can be made independent of 3.5a.

7. The finiteness of the volume for semi-simple groups

In this paragraph, we often write b^a for $a \cdot b \cdot a^{-1}$, where a, b are elements of a group.

For the sake of reference, we first sketch the proof of an elementary lemma on nilpotent Lie groups (see also [13, Lemma 1]):

7.0. LEMMA. *Let N be a connected, simply connected, real or complex nilpotent Lie group, $\mathfrak{n}^{(i)}$ a strictly decreasing sequence of ideals of $\mathfrak{n}$ such that $[\mathfrak{n}, \mathfrak{n}^{(i)}] \subset \mathfrak{n}^{(i+1)}$, $\mathfrak{n}_1$ and $\mathfrak{n}_2$ two mutually complementary subspaces such that $\mathfrak{n}^{(i)} = \mathfrak{n}_1 \cap \mathfrak{n}^{(i)} + \mathfrak{n}_2 \cap \mathfrak{n}^{(i)}$ $(i = 0, \cdots; \mathfrak{n}^{(0)} = \mathfrak{n})$. Then $(x, y) \to \exp x \cdot \exp y (x \in \mathfrak{n}_2, y \in \mathfrak{n}_1)$ is an analytic homeomorphism of $\mathfrak{n}_2 \times \mathfrak{n}_1$ onto N.*

Let s be the biggest index such that $\mathfrak{n}^{(s)} \neq 0$. Then $\mathfrak{n}^{(s)}$ is central, generates a closed simply connected central subgroup $N^{(s)}$ of N, and

$N/N^{(s)}$ is simply connected. Let $\mathfrak{n}'_1$ and $\mathfrak{n}'_2$ be supplementary subspaces to $\mathfrak{n}^{(s)} \cap \mathfrak{n}_1$ and $\mathfrak{n}^{(s)} \cap \mathfrak{n}_2$ in $\mathfrak{n}_1$ and $\mathfrak{n}_2$ respectively. Proceeding by induction on s, we may assume the lemma to be true for $\mathfrak{n}_1/(\mathfrak{n}_1 \cap \mathfrak{n}^{(s)})$ and $\mathfrak{n}_2/(\mathfrak{n}_2 \cap \mathfrak{n}^{(s)})$ in $\mathfrak{n}/\mathfrak{n}^{(s)}$. This implies immediately that $(x, y) \to \exp x \cdot \exp y$ is an analytic homeomorphism of $\mathfrak{n}'_2 \times \mathfrak{n}'_1$ onto $M = \exp \mathfrak{n}'_2 \cdot \exp \mathfrak{n}'_1$, and that $(m, z) \to m \cdot z$ is a homeomorphism of $M \times N^{(s)}$ onto N. The lemma then follows readily from the following facts; the exponential is an analytic homeomorphism of $\mathfrak{n}$ onto N, $\exp (a + b) = \exp a \cdot \exp b$ if $[a, b] = 0$ $(a, b \in \mathfrak{n})$, and $\mathfrak{n}^{(s)}$ is central.

7.1. Let G be a real algebraic semi-simple Lie group, θ a Cartan involution of G, and $G = K \cdot A \cdot N$ an Iwasawa decomposition of G compatible with θ (1.11). We use the notation of 4.1, except for the fact that Σ will denote the set of all positive roots for the ordering defined by N. Moreover, given $x = k \cdot a \cdot n (k \in K, a \in A, n \in N)$, we put $H(x) = \log a$ and $\nu(x) = n$.

Let M be the centralizer and M^* the normalizer of A in K. Then $M^*/M = W$ is a finite group, the "restricted" Weyl group of G. It acts by inner automorphisms on A or $\mathfrak{a}$, and this representation is faithful.

7.2. LEMMA. *We keep the above notation. Then $M^*/(M^* \cap G^0) = G/G^0$ and each coset of M^* modulo $M^* \cap G^0$ contains an element which normalizes N. The group G is the disjoint union of the subsets $Nm_w M_w AN$, where m_w runs through a system of representatives of the elements w of W.*

By Bruhat's lemma, [10], G^0 is the disjoint union of the subsets $Nm_w (M \cap G^0)A \cdot N$, where w runs through $M(M^* \cap G^0)/M$. Our second assertion follows from this and the first assertion. By a theorem of E. Cartan (see e.g. [12, Lemma 33]), given $k \in K$, there exists $k' \in K^0$ such that $kAk^{-1} = k'Ak'^{-1}$. Therefore M^* meets each connected component of G. Moreover, $M(M^* \cap G^0)/M$ is transitive on the Weyl chambers of $\mathfrak{a}$, therefore each coset of M^* modulo $M^* \cap G^0$ contains an element which leaves the positive Weyl chamber invariant, hence normalizes N. This proves the first assertion.

REMARK. The above is also valid in a semi-simple Lie group with finitely many connected components, whose identity component has a finite center.

7.3. Let $w \in W, \Sigma(w)$ be the set of positive roots whose transforms under w are negative, and $\Phi(w)$ the set of roots which are linear combinations of elements of $\Sigma(w) \cup (-\Sigma(w))$. Then

$$\mathfrak{g}_w = \sum_{\alpha \in \Phi(w)} \left(\mathfrak{g}_\alpha + [\mathfrak{g}_\alpha, \mathfrak{g}_{-\alpha}] \right)$$

is clearly a subalgebra. Let $\mathfrak{n}_w = \mathfrak{g}_w \cap \mathfrak{n}$, $\mathfrak{a}_w = \mathfrak{g}_w \cap \mathfrak{a}$, $\mathfrak{m}_w = \mathfrak{g}_w \cap \mathfrak{m}$, $\mathfrak{k}_w = \mathfrak{g}_w \cap \mathfrak{k}$. As usual, for $\alpha \in \mathfrak{a}^*$, h_α is the element of $\mathfrak{a}$ defined by $(h_\alpha, h) = \alpha(h)$, where $(\ ,\ )$ is the scalar product defined by the Killing form. Then $\mathfrak{g}_w$ is semi-simple,

$$(1) \qquad\qquad \mathfrak{g}_w = \mathfrak{k}_w + \mathfrak{a}_w + \mathfrak{n}_w$$

is an Iwasawa decomposition of $\mathfrak{g}_w$, and

$$(2) \qquad\qquad \mathfrak{a}_w = \sum_{\alpha \in \Sigma(w)} \mathbf{R} \cdot h_\alpha = \sum_{\alpha \in \Phi(w)} \mathbf{R} \cdot h_\alpha \ .$$

In fact, it is elementary and known that if $x \in \mathfrak{g}_\alpha$, $y \in \mathfrak{g}_{-\alpha}$, then

$$(3) \qquad \begin{aligned} &B(x, \theta(x)) \neq 0 &&(x \neq 0)\ , \\ &[x, \theta(x)] = h_\alpha B(x, \theta(x)), \quad [x, y] - h_\alpha B(x, y) \in \mathfrak{m}\ , \end{aligned}$$

where B is the Killing form, from which we deduce (2) and

$$(4) \qquad \mathfrak{m}_w + \mathfrak{a}_w = \sum_{\alpha \in \Phi(w)} [\mathfrak{g}_\alpha, \mathfrak{g}_{-\alpha}], \quad \mathfrak{g}_w = \mathfrak{m}_w + \mathfrak{a}_w + \sum_{\alpha \in \mathfrak{P}(w)} \mathfrak{g}_\alpha\ ,$$

so that (1) follows from

$$(5) \qquad \mathfrak{g}_\alpha + \mathfrak{g}_{-\alpha} = \mathfrak{g}_\alpha + \theta(\mathfrak{g}_\alpha) = \mathfrak{g}_\alpha + (\mathfrak{g}_\alpha + \mathfrak{g}_{-\alpha}) \cap \mathfrak{k}\ .$$

The algebra $\mathfrak{g}_w$ is stable under θ, hence reductive (1.4). By (4), $\mathfrak{m}_w + \mathfrak{a}_w$ belongs to the derived algebra $\mathscr{D}\mathfrak{g}_w$ of $\mathfrak{g}_w$. By (2), for each $\alpha \in \Phi(w)$, there exists $h \in \mathfrak{a}_w$ which does not annihilate α, hence $\mathfrak{g}_\alpha = [h, \mathfrak{g}_\alpha] \subset \mathscr{D}\mathfrak{g}$. Thus $\mathfrak{g}_w = \mathscr{D}\mathfrak{g}_w$, and $\mathfrak{g}_w$ is semi-simple.

Up to 7.7, G is a real algebraic semi-simple Lie group, θ a Cartan involution of G, $G = K \cdot A \cdot N$ an Iwasawa decomposition compatible with θ, Σ the set of the positive roots in the ordering defined by N. The Siegel domains of G are always defined with respect to the given Iwasawa decomposition.

7.4. **LEMMA.** *Let $\mathfrak{S}$ be a Siegel domain of G, $x \in G$ and t a positive real number. Then $\mathfrak{S}x \cap K \cdot A_t \cdot N$ is contained in a Siegel domain of G.*

For any Siegel domain $\mathfrak{S}'$ and elements $a \in A$, $n \in N$, the sets $\mathfrak{S}' \cdot a$ and $\mathfrak{S}' \cdot n$ belong to Siegel domains. In view of Bruhat's lemma (7.2) we may assume $x = m^{-1}$, $m \in M^*$. Let w be the element of W defined by m. We use the notation of 7.3. Let further $\mathfrak{n}'_w$ be the sum of the $\mathfrak{g}_\alpha$ where $\alpha \in \Sigma$, $\alpha \notin \Phi(w)$, $\mathfrak{a}'_w$ be the subspace of $\mathfrak{a}$ on which the elements of $\Phi(w)$ are all zero. Let A_w, A'_w, N_w, N'_w be the exponentials of $\mathfrak{a}_w$, $\mathfrak{a}'_w$, $\mathfrak{n}_w$, $\mathfrak{n}'_w$. Then $A = A_w \cdot A'_w$, and the group A'_w centralizes the analytic group $G(w)$ with Lie algebra $\mathfrak{g}_w$. Let further $\alpha_1, \cdots, \alpha_s$ be the positive roots arranged in increasing order. Then

$$\mathfrak{n}^{(i)} = \sum_{j \geq i} \mathfrak{g}_{\alpha_i}\ , \qquad \mathfrak{n}_1 = \mathfrak{n}_w, \mathfrak{n}_2 = \mathfrak{n}'_w\ ,$$

satisfy the assumptions of 7.0, hence $(n, n') \to \exp n \cdot \exp n'$ is an analytic homeomorphism of $\mathfrak{n}_w \times \mathfrak{n}'_w$ onto $N = N_w \cdot N'_w$. Let $y \in \mathfrak{S}$ and $z = y \cdot m^{-1}$. We may write

$$
(1) \quad
\begin{aligned}
y &= k \cdot a \cdot n & (k \in K,\, a \in A,\, n \in N)\,, \\
a &= a_1 \cdot a_2, \quad n = n_1 \cdot n_2 & (a_1 \in A_w,\, a_2 \in A'_w,\, n_1 \in N_w,\, n_2 \in N'_w)\,.
\end{aligned}
$$

In the sequel, we say that an element, which is a function of y, *is bounded* if it stays within some compact set when y varies subject to our conditions. We prove first that a_1 is bounded. We have

$$
z = k \cdot a \cdot n \cdot m^{-1} = k \cdot n^a \cdot a \cdot m^{-1} = k \cdot n^a \cdot m^{-1} \cdot m \cdot a \cdot m^{-1}\,,
$$

$$
z = k \cdot m^{-1} \cdot n^{ma} \cdot w(a)\,,
$$

hence

$$
(2) \qquad H(z) = H(n^{ma}) + \log w(a)\,.
$$

By 4.2, n^a is bounded, hence so is n^{ma}. There exists therefore $t_1 \geqq t$ such that

$$
(3) \qquad \alpha(\log w(a)) \leqq t_1 \qquad (\alpha \in \Sigma)\,.
$$

Let now $\alpha \in \Sigma(w)$. Then ${}^t w(\alpha) < 0$, and, by (3)

$$
(4) \qquad -\alpha(\log a) = -{}^t w(\alpha)(\log w(a)) \leqq t_1\,.
$$

By definition of a Siegel domain, there exists t_0 such that $\alpha(\log a) \leqq t_0$, for $\alpha > 0$. For t'' big enough, we have therefore

$$
(5) \qquad |\alpha(\log a)| \leqq t'' \qquad (\alpha \in \Sigma(w))\,.
$$

But $\alpha(\log a) = \alpha(\log a_1)$ for $\alpha \in \Sigma(w)$, and the roots $\alpha \in \Sigma(w)$ span the dual of $\mathfrak{a}_w$, therefore (5) implies that $\log a_1$ varies in a compact set. Thus a_1 is bounded, as was contended. But then, $w(a_1)$ is also bounded, and (3) shows the existence of t_3 such that

$$
(6) \qquad \alpha(\log w(a_2)) \leqq t_3 \qquad (\alpha \in \Sigma)\,.
$$

Since a_2 commutes with n_1, we may write

$$
(7) \qquad z = k \cdot m^{-1} \cdot (m \cdot a_1 \cdot n_1 \cdot m^{-1}) \cdot w(a_2) \cdot n'_2 \qquad (n'_2 = m \cdot n_2 \cdot m^{-1})\,.
$$

The element n is bounded (by definition of a Siegel domain), hence so are n_1, n_2 and n'_2. The product $a_1 \cdot n_1$ belongs to $G(w)$ and ${}^t w(\Sigma(w)) = -\Sigma(w^{-1})$; hence

$$
m \cdot a_1 \cdot n_1 \cdot m^{-1} \in m \cdot G(w) \cdot m^{-1} = G(w^{-1})\,,
$$

$$
m \cdot a_1 \cdot n_1 \cdot m^{-1} = k_3 \cdot a_3 \cdot n_3\,,
$$

$$
(8) \qquad (k_3 \in K \cap G(w^{-1}),\, a_3 \in A \cap G(w^{-1}),\, n_3 \in N \cap G(w^{-1}))\,,
$$

where a_3 and n_3 are bounded, since the left hand side is. The element a_2

commutes with $G(w)$, hence $w(a_2)$ commutes with $G(w^{-1})$, and we have by (7)

$$z = k \cdot m^{-1} \cdot k_3 \cdot a_3 \cdot w(a_2) \cdot n_3 \cdot n_2' \; ;$$

therefore,

$$H(z) = \log a_3 + \log w(a_2) \; .$$

We have already seen that a_3, n_3 and n_2' are bounded. Taking (6), into account, this shows the existence of $s > 0$ and of a compact set ω' in N such that

$$z \in K \cdot A_s \cdot \omega' \; ,$$

which proves the lemma.

7.5. **LEMMA.** *Let G_1 be an algebraic semi-simple subgroup of G, and assume it satisfies the following condition:* (a) *$\theta(G_1) = G_1$, $G_1 = K_1 \cdot A_1 \cdot N_1 (K_1 = K \cap G_1, A_1 = A \cap G_1, N_1 = N \cap G_1)$ is an Iwasawa decomposition of G_1, and the restrictions to $\mathfrak{a}_1$ of the positive roots on $\mathfrak{a}$ are $\geqq 0$ for the ordering associated to $\mathfrak{n}_1$. Let $\mathfrak{S}$ be a Siegel domain of G, and $x \in G$. Then there exists a Siegel domain $\mathfrak{S}_1$ of G_1 (for the Iwasawa decomposition $K_1 \cdot A_1 \cdot N_1$) and a finite number of elements $x_1, \cdots, x_q \in G_1$ such that $\mathfrak{S}x \cap G_1 \subset \bigcup_i \mathfrak{S}_1 \cdot x_i$.*

In this proof, the Siegel domains of G_1 are defined with respect to the decomposition $K_1 \cdot A_1 \cdot N_1$. Let

$$\mathfrak{n}_0 = \sum\nolimits_{\alpha \in \Sigma; \alpha(\mathfrak{a}_1) = 0} \mathfrak{g}_\alpha \; , \qquad \mathfrak{n}' = \sum\nolimits_{\alpha \in \Sigma; \alpha(\mathfrak{a}_1) \neq 0} \mathfrak{g}_\alpha \; .$$

Then, clearly, $\mathfrak{n}_0$ is a subalgebra and $\mathfrak{n}'$ an ideal of $\mathfrak{n}$. By 7.0, $(n, n') \to \exp n \cdot \exp n'$ is an analytic homeomorphism of $\mathfrak{n}_0 \times \mathfrak{n}'$ onto N. For any $\alpha \in \Sigma$ whose restriction to $\mathfrak{a}_1$ is not zero, let $\mathfrak{g}_\alpha'$ be a supplementary subspace to $\mathfrak{g}_1 \cap \mathfrak{g}_\alpha$ in $\mathfrak{g}_\alpha$, and let $\mathfrak{n}_2$ be the sum of all the $\mathfrak{g}_\alpha'$. Then $\mathfrak{n}' = \mathfrak{n}_1 + \mathfrak{n}_2$, where $\mathfrak{n}_1$ is a subalgebra, and it follows from 7.0 that $(n_2, n_1) \to \exp n_2 \cdot \exp n_1$ is an analytic homeomorphism of $\mathfrak{n}_2 + \mathfrak{n}_1$ onto N'. Thus $(n_0, n_2, n_1) \to \exp n_0 \cdot \exp n_2 \cdot \exp n_1$ is an analytic homeomorphism of $\mathfrak{n}_0 + \mathfrak{n}_2 + \mathfrak{n}_1$ onto $N = N_0 \cdot N_2 \cdot N_1$.

By 7.2, we have $x = a \cdot u \cdot m^{-1} \cdot v (a \in A; u, v \in N; m \in M^*)$. Since $\mathfrak{S} \cdot a \cdot u$ is contained in a Siegel domain, we may assume $x = m^{-1} \cdot v$. Let us write $v = v' \cdot v'' (v' \in N_0 \cdot N_2; v'' \in N_1)$. Then $\mathfrak{S}x \cap G_1 = (\mathfrak{S} \cdot m^{-1} \cdot v' \cap G_1) \cdot v''$. We may therefore assume $v = v' \in N_0 \cdot N_2$. In this case, 7.5 will follow from the more precise lemma:

7.6. **LEMMA.** *We keep the above notation and assume*

$$x = m^{-1} \cdot v \qquad\qquad (m \in M^*, v \in N_0 \cdot N_2) \; .$$

Let M_1^* be the normalizer and M_1 the centralizer of A_1 in K_1, and

$x_i (1 \leq i \leq \text{ord } M_1^*/M_1)$ *a set of representatives of the cosets of* M_1^* *modulo* M_1. *Then there exists a Siegel domain* $\mathfrak{S}_1$ *of* G_1 *such that* $\mathfrak{S} \cdot x \cap G_1 \subset \bigcup_i \mathfrak{S}_1 \cdot x_i$.

Let $y = k \cdot a \cdot n (k \in K, a \in A, n \in N)$ be an element of $\mathfrak{S}$ such that $z = y \cdot x \in G_1$. Then

$$z = k \cdot a \cdot n \cdot m^{-1} \cdot v = k_1 \cdot a_1 \cdot n_1 \qquad (k_1 \in K_1, a_1 \in A_1, n_1 \in N_1) .$$

We first show that $a_1 \cdot n_1 \cdot a_1^{-1}$ is bounded. We have

$$z = k \cdot n^a \cdot a \cdot m^{-1} \cdot v = k \cdot m^{-1} \cdot m \cdot n^a \cdot m^{-1} \cdot m \cdot a \cdot m^{-1} v$$
$$= k \cdot m^{-1} \cdot n^{ma} \cdot a^m \cdot v ,$$
$$z \in K \cdot n^{ma} \cdot a^m \cdot v = K \cdot e^{H(u)} \cdot \nu(u) \cdot a^m \cdot v \qquad\qquad (u = n^{ma}) ,$$

where $\nu(u)$ is the component in N of u (see 7.1), whence

(1) $$a_1 = e^{H(u)} \cdot a^m , \qquad n_1 \cdot v^{-1} = (a^m)^{-1} \cdot \nu(u) \cdot a^m .$$

By 4.2, n^a is bounded, hence so are $u = n^{ma}$, $\nu(u)$, and therefore by (1), also $a_1 \cdot (a^m)^{-1}$, and

(2) $$a_1 \cdot n_1 \cdot v^{-1} \cdot a_1^{-1} = (a_1 \cdot n_1 \cdot a_1^{-1}) \cdot (a_1 \cdot v^{-1} \cdot a_1^{-1}) .$$

It is clear from the definitions that a_1 normalizes N_1 and $N_0 \cdot N_2$, therefore the two factors on the right hand side belong to N_1 and $N_0 \cdot N_2$ respectively. Since the map $(n', n'') \to n' \cdot n''$ is a homeomorphism of $N_1 \times N_0 \cdot N_2$ onto N, we see that both factors on the right hand side of (2) are bounded.

Let, for $c > 0$,

$$V_{i,c} = \{z \in \mathfrak{S}x \cap G_1 \,|\, z \cdot x_i^{-1} \in K \cdot A_c \cdot N\} .$$

We want to prove the existence of $t > 0$ such that

(3) $$\mathfrak{S}x \cap G_1 \subset \bigcup_i V_{i,t} .$$

The restricted Weyl group W_1 of G_1 is transitive on the Weyl chambers, therefore, given z, there exists an i such that $x_i \cdot a_1 \cdot x_i^{-1} \in A_1^-$, where $A_1^- = A_{1,0}$ denotes the exponential of the negative Weyl chamber. We have

$$z \cdot x_i^{-1} = k_1 \cdot a_1 \cdot n_1 \cdot x_i^{-1} = k_1 \cdot n_1^{a_1} \cdot a_1 \cdot x_i^{-1} \in K_1 \cdot n_1^{x_i \cdot a_1} \cdot a_1^{x_i} ,$$
$$z \cdot x_i^{-1} \in K_1 \cdot e^{H(u)} \cdot a_1^{x_i} \cdot \left(a_1^{x_i^{-1}} \cdot \nu(u) \cdot a_1^{x_i}\right) , \qquad\qquad (u = n_1^{x_i \cdot a_1}) ,$$

which implies

$$H(z \cdot x_i^{-1}) = H(u) + H(x_i \cdot a_1 \cdot x_i^{-1}) .$$

But $x_i \cdot a_i \cdot x_i^{-1} \in A_1^-$ by assumption, hence, taking 7.5(a) into account, $\alpha(x_i \cdot a_1 \cdot x_i^{-1}) \leq 0$ for $\alpha \in \Sigma$. Moreover, $n_1^{a_1}$ is bounded, as was proved above, hence u is bounded. There exists, therefore, $t_i > 0$ such that $H(z \cdot x_i^{-1}) \subset \log A_{1,t_i}$ for all $z \in \mathfrak{S}x \cap G_1$ for which

$$x_i \cdot a_1 \cdot x_i^{-1} = x_i \cdot e^{H(z)} \cdot x_i^{-1} \in A_1^- \ .$$

This proves (3). In view of the condition (a) of 7.5, $A_{1,t} \subset A_t$, and it follows from 7.4 that we may find a Siegel domain $\mathfrak{S}_i'$ of G such that $z \cdot x_i^{-1} \in \mathfrak{S}_i'$ when $z \in V_{i,t}$; we have then $V_{i,t} \subset \mathfrak{S}_i' \cdot x_i$ or equivalently

$$(4) \qquad\qquad V_{i,t} \subset (\mathfrak{S}_i' \cap G_1) \cdot x_i \ .$$

But in view of 7.5(a), $\mathfrak{S}_i' \cap G_1$ is a Siegel domain of G_1. Since a finite union of Siegel domains (with respect to a fixed Iwasawa decomposition) is contained in a Siegel domain, the lemma follows from (3), (4).

7.7. LEMMA. *Let G' be a real algebraic semi-simple subgroup of G. Then there exists $a \in G$ such that $G_1 = a \cdot G' \cdot a^{-1}$ verifies condition* (a) *of 7.5.*

Let $\mathfrak{g} = \mathfrak{k} + \mathfrak{p}$ be the Cartan decomposition associated to θ. By 1.1 and 1.8, there exists $b \in G$ such that $G_2 = b \cdot G' \cdot b^{-1}$ is stable under θ. Then $\mathfrak{g}_2 = (\mathfrak{k} \cap \mathfrak{g}_2) + (\mathfrak{p} \cap \mathfrak{g}_2)$ is a Cartan decomposition of the Lie algebra $\mathfrak{g}_2$ of G_2. Let $\mathfrak{a}_2$ be a maximal subalgebra of $\mathfrak{p} \cap \mathfrak{g}_2$ and $\mathfrak{a}'$ a maximal subalgebra of $\mathfrak{p}$ containing $\mathfrak{a}_2$. By E. Cartan's conjugacy theorem (see e.g. [12, Lemma 33]), there exists $k \in K$ such that $k \cdot \mathfrak{a}' \cdot k^{-1} = \mathfrak{a}$. The group $G_3 = k \cdot G_2 \cdot k^{-1}$ is then still invariant under θ, and $\mathfrak{g}_3 \cap \mathfrak{a} = \mathfrak{a}_3$ is a maximal subalgebra of $\mathfrak{p}_3 = \mathfrak{g}_3 \cap \mathfrak{p}$. Let us now choose orderings on $\mathfrak{a}^*$ and $\mathfrak{a}_3^*$ such that the restriction to $\mathfrak{a}_3$ of a positive element of $\mathfrak{a}^*$ is $\geqq 0$, for instance take the lexicographic orderings with respect to a basis whose first elements span $\mathfrak{a}_3$. Let then $m \in M^*$ be such that $\mathrm{Ad}\, m$ transforms the positive Weyl chamber of $\mathfrak{a}$ for this ordering into the positive Weyl chamber for the ordering defined by N. Then $a = m \cdot k \cdot b$ fulfills our conditions.

7.8. THEOREM. *Let G be a semi-simple algebraic Lie group, defined over $\mathbf{Q}$. Then $G_{\mathbf{R}}/G_{\mathbf{Z}}$ has finite Haar measure.*

We may assume G to be contained in $\mathrm{SL}(n, \mathbf{C})$ (see 2.1), and also to be connected, since $(G^0)_{\mathbf{R}}$ has finite index in $G_{\mathbf{R}}$. We now apply 7.7, with G' and G replaced respectively by $G_{\mathbf{R}}$ and $\mathrm{SL}(n, \mathbf{R})$, θ and $K \cdot A \cdot N$ by the standard Cartan involution and Iwasawa decomposition of $\mathrm{SL}(n, \mathbf{R})$, and take $a \in \mathrm{SL}(n, \mathbf{R})$ such that $G' = a \cdot G \cdot a^{-1}$ is self-adjoint and satisfies condition (a) of 7.5. Let $\mathfrak{S}$ be a standard Siegel domain of $\mathrm{SL}(n, \mathbf{R})$. By 6.5, there exist finitely many elements $b_i \in \mathrm{SL}(n, \mathbf{Z})$ such that $G_{\mathbf{R}} = \bar{U} \cdot G_{\mathbf{Z}}$ with

$$\bar{U} = \mathbf{U}_i\, (a^{-1} \cdot \mathfrak{S} \cdot b_i) \cap G_{\mathbf{R}} \ .$$

We want to prove that $\bar{U}$ has finite Haar measure on $G_{\mathbf{R}}$. It is enough to show that $(a^{-1} \cdot \mathfrak{S} \cdot b_i) \cap G_{\mathbf{R}}$ has finite measure, hence also that $\mathfrak{S} \cdot b_i \cdot a^{-1} \cap G_{\mathbf{R}}'$ has finite Haar measure on $G_{\mathbf{R}}'$. By 7.5 (with G and G_1 replaced by

$\mathbf{SL}(n, \mathbf{R})$ and $G_{\mathbf{R}}'$), $\mathfrak{S} \cdot b_i \cdot a \cap G_{\mathbf{R}}'$ is contained in a finite union of translates of a Siegel domain of $G_{\mathbf{R}}'$. Since a Siegel domain of a semi-simple Lie group has finite volume (4.3), our contention is proved.

8. Remarks on characters of algebraic groups

8.1. Let G be an algebraic group. The group of *rational characters* of G, that is of rational homomorphisms of G into $\mathbf{C}^*$, is denoted by $X(G)$. It is a finitely generated commutative group, free when G is connected [23; 26]. If K is a field of definition for G, then $X_K(G)$ will be the group of rational characters defined over K.

A rational homomorphism $f: G \to G'$ obviously induces a homomorphism $f^0: X(G') \to X(G)$, which maps $X_K(G')$ into $X_K(G)$ if K is a field of definition for G, G' and f.

Let G be reductive, connected, defined over K, $S = Z(G)^0$ the identity component of its center, and G' the derived group of G. Then the restriction map $X(G) \to X(S)$ is injective, and identifies $X(G)$ and $X_K(G)$ with subgroups of finite index of $X(S)$ and $X_K(S)$ respectively. In fact, we have $X(G') = 1$, since G' is semi-simple, hence the restriction is injective. Conversely, given χ in $X(S)$ or in $X_K(S)$, χ^m ($m =$ order of $G' \cap S$) is trivial on $S \cap G'$, and therefore extends to a character of G.

8.2. Let now $G = T$ be an algebraic torus, K a field of definition for T. Let $\Gamma(T)$ be the group of rational homomorphisms of $\mathbf{C}^*$ into T, and $\Gamma_K(T)$ the group of rational homomorphisms of $\mathbf{C}^*$ into T which are defined over K. Both $X(T)$ and $\Gamma(T)$ are free, of rank $n = \dim T$. Given $a \in \Gamma(T), b \in X(T)$, the homomorphism $b \circ a: \mathbf{C}^* \to \mathbf{C}^*$ has the form $x \to x^m (m \in \mathbf{Z})$. The map $(a, b) \to \langle a, b \rangle = m$ is an integral valued non-degenerate bilinear form on $\Gamma(T) \times X(T)$, which puts these two groups in duality [8; Exp. 9, No. 5]. A rational homomorphism $f: T \to T'$ induces a homomorphism $f_0: \Gamma(T) \to \Gamma(T')$, and we have $\langle f_0 a, b \rangle = \langle a, f^0 b \rangle$ $(a \in \Gamma(T), b \in X(T'))$.

8.3. An algebraic torus T is said to *split over* K if it is defined over K and isomorphic over K to a product of groups $\mathbf{C}^*$. If $T \subset \mathbf{GL}(n, \mathbf{C})$, this is equivalent to the existence of $x \subset \mathbf{GL}(n, K)$ such that $x \cdot T \cdot x^{-1}$ is diagonal [26, Prop. 5]. If T splits over K, then $X_K(T) = X(T)$, $\Gamma_K(T) = \Gamma(T)$, the subtori and the homomorphic images (over K) of T split over K. Therefore, if $a \in \Gamma_K(T), a \neq 0$, then $a(\mathbf{C}^*)$ is a one-dimensional subtorus of T which splits over K, and if $\Gamma_K(T) = \Gamma(T)$, then T splits over K. That the two latter conditions are also equivalent to $X_K(T) = X(T)$ follows from the following lemma:

8.4. LEMMA. *Let T be an algebraic torus, K a field of definition for T. Then*

(a) *Given any subtorus S of T, defined over K, there exists a subtorus S', defined over K, such that $S \cdot S' = T$, and $S \cap S'$ is finite.*

(b) *$X_K(T)$ and $\Gamma_K(T)$ have equal ranks. In particular, $X_K(T) \neq \{0\}$ if and only if T contains a subtorus $S \neq (e)$ which splits over K.*

The torus T certainly splits over $\bar{K}$ [1, Chap. II], therefore there exists a finite Galois extension K' of K over which T splits. Let A be the Galois group of K' over K. It operates in a natural fashion on $\Gamma(T)$, $X(T)$, and we have

$$\langle \sigma(a), \sigma(b) \rangle = \langle a, b \rangle \qquad (a \in \Gamma(T), b \in X(T), \sigma \in A) .$$

The corresponding linear representations ρ, ρ' of A in $\Gamma(T) \otimes \mathbf{Q}$ and $X(T) \otimes \mathbf{Q}$ are therefore contragredient to each other. The fixed points of A in $\Gamma(T)$ and $X(T)$ are $\Gamma_K(T)$ and $X_K(T)$ respectively. Consequently, the ranks of $\Gamma_K(T)$ and $X_K(T)$ are equal to the dimension of the fixed point sets of ρ, ρ'. Since ρ and ρ' are contragredient to each other, these dimensions are equal, whence the first part of (b). The second one follows then from 8.3. Let now S be a subtorus of T, defined over K. Then $\Gamma(S)$ may be identified with a submodule of $\Gamma(T)$, which is invariant under A. Since A is finite, there is a subspace V of $\Gamma(T) \otimes \mathbf{Q}$, supplementary to $\Gamma(S) \otimes \mathbf{Q}$, and invariant under A. The images $\gamma(\mathbf{C}^*)$, $(\gamma \in \Gamma(T) \cap V)$, are subtori defined over K' which are permuted by A. They generate a subtorus S' of T, invariant under A, hence defined over K, and such that $\Gamma(S') = V \cap \Gamma(T)$. We have then $\Gamma(S) \cap \Gamma(S') = (0)$, and $\Gamma(S) \mid \Gamma(S')$ has finite index in $\Gamma(T)$, whence (a).

9. The finiteness of the volume

9.1. Let G be a Lie group, H a discrete subgroup. Then a Haar measure on G induces on G/H a measure μ, such that $g(\mu) = \chi(g) \cdot \mu$, where $\chi(g) = |\det \mathrm{Ad}\, g|$. If $\mu(G/H) < \infty$, then $\chi(g) = 1 (g \in G)$, μ is invariant and G is unimodular (any left invariant Haar measure is right invariant).

9.2. **LEMMA.** *Let G^*, G be connected algebraic groups defined over $\mathbf{Q}$, $\pi: G^* \to G$ an isogeny over $\mathbf{Q}$. Then $G_\mathbf{R}^*$ is unimodular and $G_\mathbf{R}^*/G_\mathbf{Z}^*$ has finite invariant measure, if and only if $G_\mathbf{R}$ is unimodular, and $G_\mathbf{R}/G_\mathbf{Z}$ has finite invariant measure.*

It is clear that $G_\mathbf{R}$ is unimodular if and only if $G_\mathbf{R}^*$ is so. Let N be the kernel of π. By 6.11, $G_\mathbf{Z}^*$ has a subgroup M of finite index, whose image $\pi(M)$ is a subgroup of finite index of $G_\mathbf{Z}$. We have $G_\mathbf{R}^*/M \cdot N \cong G_\mathbf{R}/\pi(M)$, and

G_R^*/G_Z^* (resp. G_R/G_Z) has finite measure if and only if $G_R^*/M \cdot N$ (resp. $G_R/\pi(M)$) has, whence the lemma.

9.3. Let T be an algebraic torus, defined over $\mathbf{Q}$. We shall use here and in § 10 the fact that if $X_\mathbf{Q}(T) = 1$, then T_R/T_Z is compact, proved by Ono [22]. (It is not stated explicitly there, but is an immediate consequence of the compactness of the quotient $J(T)/T_\mathbf{Q}$ of the idele group of T by the principal ideles.)

On the other hand, if T is diagonal, then T_Z is obviously finite. By 8.3, it follows that T_Z is finite whenever T splits over $\mathbf{Q}$. In that case, $\mu(T_R/T_Z)$ is of course infinite ($T \neq (e)$).

9.4. **Theorem.** *Let G be an algebraic group, defined over $\mathbf{Q}$. Then G_R is unimodular, and G_R/G_Z of finite invariant measure if and only if $X_\mathbf{Q}(G^0) = 1$.*

It is clearly enough to prove this when G is connected. Assume first that $G = T$ is a torus. If $X_\mathbf{Q}(G) = 1$, then T_R/T_Z is compact by Ono's result (9.3). If $X_\mathbf{Q}(G) \neq 1$, then, by 8.4, we have $T = S \cdot S'$, where S, S' are subtori of strictly positive dimension, and S splits over $\mathbf{Q}$. We have a natural isogeny $S \times S' \to T$, and it follows from 9.2, 9.3 that $\mu(T_R/T_Z)$ is infinite.

Let now G be reductive, $G = S \cdot G'$ its standard decomposition, where $S = Z(G)^0$, and G' is the derived group of G. Therefore G is the quotient of $S \times G'$ by a finite group. By 7.8, G'_R/G'_Z has finite invariant measure. By 8.1, $X_\mathbf{Q}(G)$ and $X_\mathbf{Q}(S)$ have equal ranks. Our assertion in this case follows therefore from the above and 9.2.

In the general case, $G = H \cdot N$ is the semi-direct product of a reductive group H and of an invariant unipotent group N, both defined over $\mathbf{Q}$, and N_R/N_Z is compact (6.10). Of course, $X(N) = 1$, hence $X(G) = X(H)$, $X_\mathbf{Q}(G) = X_\mathbf{Q}(H)$.

Let $X_\mathbf{Q}(G) = 1$. Then $X_\mathbf{Q}(H) = 1$, and H_R/H_Z has finite invariant measure by the above. Moreover, the determinant of $\operatorname{Ad} h \,|\, \mathfrak{n}\,(h \in H)$ is one, since $h \to \det(\operatorname{Ad} h \,|\, \mathfrak{n})$ is an element of $X_\mathbf{Q}(H)$. Therefore, G_R is unimodular, and a Haar measure on G_R is the product of Haar measures on H_R and N_R. We have $H_R = A \cdot H_Z$, $N_R = B \cdot N_Z$ with A, B open, of finite Haar measure, whence $G_R = A \cdot N \cdot H_Z = A \cdot B \cdot G_Z$ with $A \cdot B$ of finite measure.

Assume now G_R to be unimodular, and G_R/G_Z to have finite invariant measure. Since, G_R, H_R, N_R are unimodular, we must have $\det(\operatorname{Ad} h \,|\, \mathfrak{n}) = 1 (h \in H)$, and the Haar measure on G_R is the product of Haar measures on H_R and N_R. The subgroup $H_Z \cdot N_Z$ has finite index in G_Z (6.4), and $G_R/H_Z \cdot N_Z$ has finite measure. The projection $G_R/N_Z \to G_R/N_R \cong H_R$ is a fiber map with compact fibre N_R/N_Z, which commutes with the action of

H_Z defined by right translations. Therefore H_R/H_Z must also have finite invariant measure; by the above, this implies $X_Q(H) = 1$, hence $X_Q(G) = 1$.

19. Closed conjugacy classes

10.1. PROPOSITION. *Let G be a real or complex algebraic group, $x \in G$, $y \in \mathfrak{g}$, and $C(x) = \{g \cdot x \cdot g^{-1}, g \in G\}$, $C(y) = \mathrm{Ad}\, G(y)$ the conjugacy classes of x and y. Then*

(a) *If x (resp. $\mathrm{ad}\, y$) is semi-simple, $C(x)$ (resp. $C(y)$) is closed.*

(b) *If G is reductive, and x (resp. $\mathrm{ad}\, y$) is not semi-simple, $G(x)$ (resp. $C(y)$) is not closed.*

PROOF OF (a) Going over to the complexification, if necessary, it is enough, by 2.3, to consider the case where G is complex algebraic. For an endomorphism A of a vector space, we denote by $C(A, \lambda)$ its characteristic polynomial $\det(A - \lambda \cdot \mathrm{Id.})$ and by $M_A(\lambda)$ its minimal polynomial, where λ is an indeterminate. Let

$$P_y = \{z \in \mathfrak{g}, C(\mathrm{ad}\, z, \lambda) = C(\mathrm{ad}\, y, \lambda), M_{\mathrm{ad}y}(\mathrm{ad}\, z) = 0\} .$$

This is an algebraic subset of $\mathfrak{g}$, clearly invariant under G. The minimal polynomial of $\mathrm{ad}\, z(z \in P_y)$ divides the minimal polynomial of $\mathrm{ad}\, y$, hence has only simple factors, and $\mathrm{ad}\, z$ is also semi-simple. In particular, the dimension of the centralizer $Z(z)$ of z in G is equal to the multiplicity of the eigen-value zero of $\mathrm{ad}\, z$. But $\mathrm{ad}\, z$ has the same eigenvalues as $\mathrm{ad}\, y$, hence $\dim Z(z) = \dim Z(y)$. The orbit $\mathrm{Ad}\, G(z)$ of z, whose dimension is equal to $\dim G - \dim Z(z)$, has therefore the same dimension as $\mathrm{Ad}\, G(y)$, and P_y is a disjoint union of orbits of the same dimension. Since the boundary of an orbit is a union of orbits of strictly smaller dimension [1, § 15], it follows that $\mathrm{Ad}\, G(z)$ is closed for every $z \in P_y$.

The proof for x is entirely analogous. We introduce the set

$$P_x = \{z \in G, C(\mathrm{Ad}\, z, \lambda) = C(\mathrm{Ad}\, x, \lambda), M_{\mathrm{Ad}x}(\mathrm{Ad}\, z) = 0\} .$$

This is an algebraic set, invariant under G. The dimension of $Z(z)$ will be equal to the multiplicity of the eigenvalue one of $\mathrm{Ad}\, z$, hence equal to $\dim Z(x)$, and P_x consists again of orbits of the same dimension.

PROOF OF (b) Let first $G = \mathbf{SL}(2, \mathbf{R})$ or $\mathbf{SL}(2, \mathbf{C})$ and $x \neq 1$ be unipotent (resp. $y \neq 0$ be nilpotent). After a suitable inner automorphism, we may assume

$$x = \begin{pmatrix} 1 & 1 \\ 0 & 1 \end{pmatrix} \quad \left(\text{resp. } y = \begin{pmatrix} 0 & 1 \\ 0 & 0 \end{pmatrix} \right) .$$

Let $g_t = \mathrm{diag}\,(t, t^{-1})$. Then

$$g_t \cdot x \cdot g_t^{-1} = \begin{pmatrix} 1 & t^2 \\ 0 & 1 \end{pmatrix} \quad \left(\text{resp. Ad } g_t(y) = \begin{pmatrix} 0 & t^2 \\ 0 & 0 \end{pmatrix}\right),$$

hence the identity (resp. the origin) belongs to $\bar{C}(x)$ (resp. $\bar{C}(y)$).

Let now G be reductive, and $\text{ad}\, y$ be not semi-simple. We may write $y = z + y'$, with z central, $y' \in \mathscr{D}\mathfrak{g}$, hence $\text{ad}\, y' = \text{ad}\, y$ not semi-simple. It is clearly enough to show that the conjugacy class of y' in the semi-simple part of $\mathfrak{g}$ is not closed; we may therefore assume G to be semi-simple. We have then $y = s + n$, with $[s, n] = 0$, $\text{ad}\, s$ semi-simple, $\text{ad}\, n$ nilpotent and not zero [5, §6, No. 3]. The centralizer $\mathfrak{z}(s)$ of s in $\mathfrak{g}$ is reductive, its center consists of semi-simple elements [4, Prop. 4.1], $\mathscr{D}\mathfrak{z}(s)$ is semi-simple, and contains n. By the Jacobson-Morosow theorem [15], there exists a three-dimensional subalgebra $\mathfrak{m}$, isomorphic to $\mathfrak{sl}(2, \mathbf{R})$, or $\mathfrak{sl}(2, \mathbf{C})$, containing n. The analytic subgroup M generated by $\mathfrak{m}$ in $\text{Ad}\, \mathfrak{g}$ is a homomorphic, locally isomorphic, image of $\text{SL}(2, \mathbf{R})$ or $\text{SL}(2, \mathbf{C})$. By the above, the conjugacy class of n in $\mathfrak{m}$ has zero in its closure; since M centralizes s, it follows that the semi-simple element s belongs to the closure of $\bar{C}(y)$.

For x, the proof is similar. We write $x = z \cdot x'$, with z central, semi-simple, and x' in the semi-simple part of G, and not semi-simple. It is enough to show that $C(x')$ is not closed, hence we may assume G to be semi-simple. We have $x = x_s \cdot x_u$ with $x_s \cdot x_u = x_u \cdot x_s$, x_s semi-simple, $x_u \neq e$ unipotent [1, §8]. Applying the Jacobson-Morosow theorem to $\log x_u$, in the derived algebra of $\mathfrak{z}(x_s)$, we see that x_u belongs to a group M, which is a homomorphic, locally isomorphic image of $\text{SL}(2, \mathbf{R})$ or $\text{SL}(2, \mathbf{C})$, and centralizes x_s. By the above, the conjugacy class of x_u in M contains e, hence $\bar{C}(x)$ contains x_s.

10.2. REMARK. The proof of (b) shows more precisely that the semi-simple part of x (resp. y) belongs to $\bar{C}(x)$ (resp. $\bar{C}(y)$), and, in the real case, that it belongs to the closure of the conjugacy class of x (resp. y) with respect to the identity component G^0 of G, hence with respect to any open subgroup. Now, if G is a real (resp. complex) semi-simple Lie group, then $\text{Ad}\, \mathfrak{g}$ is of finite index in a real algebraic group (resp. is an algebraic group). If G is moreover linear, it is of finite index in a real (resp. complex) algebraic group. Consequently, 10.1 yields the following:

10.3. COROLLARY. *Let G be a real or complex semi-simple Lie group, and $y \in \mathfrak{g}$. Then $\text{Ad}\, G(y)$ is closed if and only if $\text{ad}\, y$ is semi-simple. If G is linear, the conjugacy class in G of an element $x \in G$ is closed if and only if x is semi-simple.*

11. Groups of units with compact fundamental sets

11.1. LEMMA. *Let G be a locally compact separable group, M a locally compact separable space on which G operates continuously on the right, $m \in M$, and H a closed subgroup. If $m \cdot H$ is closed, and $H \backslash G$ is compact, then $m \cdot G$ is closed.*

It follows from our assumption that $G = H \cdot K$ with K compact. If $m \cdot h_j \cdot k_j \to p(h_j \in H, k_j \in K)$, then, assuming, as we may, $k_j \to k \in K$, we have $m \cdot h_j \to p \cdot k^{-1}$. Since $m \cdot H$ is closed, there exists $h \in H$ such that $p \cdot k^{-1} = m \cdot h$, whence $p = m \cdot h \cdot k \in m \cdot G$.

11.2. Let G act on itself by inner automorphisms. If H is discrete, the conjugacy class in H of $h \in H$ is of course a closed subset of G; therefore, if $H \backslash G$ is compact, then the conjugacy class in G of any element of H is closed. Lemma 11.1 was suggested by this remark, which is due to Selberg. Together with 10.3, it shows that if G is a linear semi-simple Lie group, H a discrete subgroup such that $H \backslash G$ is compact, then any element of H is semi-simple.

11.3. PROPOSITION. *Let G be a connected algebraic group defined over $\mathbf{Q}$, and $\pi\colon G \to \mathrm{GL}(V)$ a rational representation of G defined over $\mathbf{Q}$. If $G_{\mathbf{R}}/G_{\mathbf{Z}}$ is compact, then the orbit under $G_{\mathbf{R}}$ of an element $v \in V_{\mathbf{Q}}$ is closed.*

By 6.3, there exists a lattice $\Gamma \subset V_{\mathbf{Q}}$ which contains v and is invariant under $G_{\mathbf{Z}}$. Therefore, $v \cdot \pi(G_{\mathbf{Z}})$ is a discrete set, and 11.3 follows from 11.1.

11.4. LEMMA. *Let G be a connected reductive algebraic group, k a field of definition for G. Then the following conditions are equivalent:*

(a) $X_k(G) = 1$, *and* $\mathfrak{g}_k$ *consists of semi-simple elements;*

(b) $X_k(G) = 1$, *and* G_k *consists of semi-simple elements:*

(c) $X_k(S) = 1$ *for every algebraic subtorus S of G which is defined over k.*

If k is a number field, and J its ring of integers, these conditions are equivalent to:

(d) $X_k(G) = 1$, *and G_J consists of semi-simple elements.*

(a) $\Rightarrow$ (b). The group G, being algebraic, contains the unipotent and semi-simple parts of its elements [1, Chap. II], therefore, if G_k contains a non-semi-simple element, it also contains a unipotent element $g \neq e$. But then $\log g$ is a non-zero nilpotent element of $\mathfrak{g}_k$.

(b) $\Rightarrow$ (c). Assume (c) to be false. By 8.4, there exists then a one-dimensional subtorus S of G which splits over k. It is not central, since otherwise 8.4 and 8.1 would imply $X_k(G) \neq 1$. The image of S in $\mathrm{Aut}\,\mathfrak{g}$

then also splits, which means that it can be diagonalized over k. The weights of the Lie algebra $\mathfrak{s}_k$ of S_k in $\mathfrak{g}_k$ are then elements of $(\mathfrak{s}_k)^*$, and $\mathfrak{g}_k$ is a direct sum of subspaces $\mathfrak{g}_{k,\alpha}(\alpha \in (\mathfrak{s}_k{}^*))$, where as usual, $\mathfrak{g}_{k,\alpha} = \{x \in \mathfrak{g}_k \mid [s, x] = \alpha(s)x,\ s \in \mathfrak{s}_k\}$. Since S is not central, $\mathfrak{g}_{k,\alpha} \neq 0$ for at least one $\alpha \neq 0$. But then it is well known that an element $x \in \mathfrak{g}_{k,\alpha}$ is nilpotent. If $x \neq 0$, then e^x is a unipotent element in G_k, different from the identity, which contradicts (b).

(c) $\Rightarrow$ (a). By 8.1, (c) implies that $X_k(G) = 1$. Assume now that $\mathfrak{g}_k$ does not consist of semi-simple elements. Being algebraic, it contains then a nilpotent element $x \neq 0$ [7a, p. 165], which necessarily belongs to $\mathscr{D}\mathfrak{g}$. Applying the Jacobson-Morosow theorem to $\mathscr{D}\mathfrak{g}_k$, we get a three dimensional subalgebra $\mathfrak{m} \subset \mathscr{D}\mathfrak{g}_k$, containing x, isomorphic over k to the Lie algebra of $SL(2, k)$. There exists therefore in G an algebraic subgroup M, with Lie algebra $\mathfrak{m} \otimes C$, defined over k, and which is the image of $SL(2, C)$ under a rational homomorphism with finite kernel, defined over k. The image of the group of diagonal matrices in $SL(2, C)$ is then a one dimensional subtorus of G, which splits over k. Let now k be a number field, J the ring of integers of k. Clearly, $(a) \Rightarrow (d)$. Assume now (a) to be false. As remarked at the beginning of the proof, there exists then $x \in \mathfrak{g}_k$ which is nilpotent and not zero. Let s be a positive rational integer such that $x^s = 0$, t an element of J such that $t \cdot x$ has integral coefficients, and $m = s! \cdot t$. Then e^{mx} is a unipotent element, different from the identity contained in G_J, in contradiction with (d).

11.5. COROLLARY. *Let G be a connected reductive algebraic group. If G satisfies the conditions of 11.4, then every connected subgroup of G which is defined over k is reductive, and satisfies those conditions.*

Let M be a connected algebraic subgroup of G, defined over k. As recalled in 6.10, $M = H \cdot N$ is the semi-direct product of a reductive group H and of a unipotent invariant subgroup N, both defined over k. In our case $N_k = (e)$ by the condition (b), hence $N = (e)$, and $M = H$ is reductive. Then it clearly satisfies (c).

11.6. THEOREM. *Let G be a reductive algebraic group, defined over $\mathbf{Q}$. Then $G_\mathbf{R}/G_\mathbf{Z}$ is compact if and only if $X_\mathbf{Q}(G^0) = 1$ and $G_\mathbf{Q}$ consists of semi-simple elements.*

The group G^0 has finite index in G, therefore $G_\mathbf{R}/G_\mathbf{Z}$ is compact if and only if $(G^0)_\mathbf{R}/(G^0)_\mathbf{Z}$ is. In the sequel, we assume G to be connected. Our assertion is then that the conditions (a) to (d) of 11.4 are equivalent to the compactness of $G_\mathbf{R}/G_\mathbf{Z}$.

Let first $G_\mathbf{R}/G_\mathbf{Z}$ be compact. If $X_\mathbf{Q}(G) \neq 1$, there exists a non-trivial one dimensional rational representation $\pi: G \to C^*$, defined over $\mathbf{Q}$. But then $\pi(G) = \mathbf{C}^*$, and $\pi(G_\mathbf{R})$ contains the multiplicative group $\mathbf{R}^+$ of strictly positive real numbers. The orbit under $G_\mathbf{R}$ of any element in $\mathbf{Q}^*$ (in fact in $\mathbf{C}^*$) has the origin in its closure, and is not closed, in contradiction with 11.3. Thus $X_\mathbf{Q}(G) = 1$. The condition on $G_\mathbf{Q}$ follows from 10.1 and 11.3.

From now on, G is assumed to verify the conditions of 11.4, and we prove the compactness of $G_\mathbf{R}/G_\mathbf{Z}$ by induction on dim G. There is nothing to prove in dimension zero, therefore we may assume 11.6 to be true for all groups (connected or not) of dimension strictly smaller than dim G. In particular, in view of 11.5, $H_\mathbf{R}/H_\mathbf{Z}$ is compact for every proper algebraic subgroup H of G which is defined over $\mathbf{Q}$.

We have $G = S \cdot G'$ with $S = Z(G)^0$, G' semi-simple invariant, S and G' defined over $\mathbf{Q}$. If $0 < \dim S < \dim G$, then, by the above $S_\mathbf{R} = A \cdot S_\mathbf{Z}$, $G'_\mathbf{R} = B \cdot G'_\mathbf{Z}$ with A and B compact, hence $S_\mathbf{R} \cdot G'_\mathbf{R} = A \cdot B \cdot S_\mathbf{Z} \cdot G'_\mathbf{Z}$. Since $S_\mathbf{R} \cdot G'_\mathbf{R}$ has finite index in $G_\mathbf{R}$, and $S_\mathbf{Z} \cdot G'_\mathbf{Z} \subset G_\mathbf{Z}$, it follows that $G_\mathbf{R}/G_\mathbf{Z}$ is compact. If $G = S$, see 9.3.

Let now $G = G'$ be semi-simple. Every element of $\mathfrak{g}_\mathbf{Q}$ is semi-simple (11.4), and therefore its conjugacy class under G is closed (10.1). The adjoint representation is consequently a locally faithful rational representation of G, defined over $\mathbf{Q}$, in which all rational points have closed orbits. Since, as pointed out above, $H_\mathbf{R}/H_\mathbf{Z}$ is compact for every proper algebraic subgroup of G which is defined over $\mathbf{Q}$, it will be enough, in order to conclude the proof of 11.6, to prove the following lemma.

11.7. LEMMA. *Let G be a connected semi-simple algebraic group defined over $\mathbf{Q}$. Assume that $H_\mathbf{R}/H_\mathbf{Z}$ is compact for every proper algebraic subgroup of G defined over $\mathbf{Q}$, and that there exists a locally faithful rational representation $\pi\colon G \to \mathrm{GL}(V)$ defined over $\mathbf{Q}$ in which all points of $V_\mathbf{Q}$ have closed orbits. Then $G_\mathbf{R}/G_\mathbf{Z}$ is compact.*

The representation π being fully reducible, we may take out the trivial representations, and assume

$$(1) \qquad\qquad G_x \neq G \qquad\qquad (x \in V, x \neq 0).$$

We fix a lattice Γ in $V_\mathbf{Q}$ which is invariant under $G_\mathbf{Z}$ (see 6.3), and take coordinates in V with respect to a basis of Γ. Let P be the set of polynomials on V which are invariant under G. Since π is defined over $\mathbf{Q}$, we have $P = P' \otimes \mathbf{C}$, where P' is the set of invariant polynomials with rational coefficients. By the theorem of invariants, applied to $\pi\colon \mathfrak{g}_\mathbf{Q} \to \mathfrak{gl}(V_\mathbf{Q})$,, [5, § 6, No. 9], P' is a finitely generated algebra over $\mathbf{Q}$. It is therefore generated by 1, and by finitely many homogeneous polynomials $P_1, \cdots, P_s$, which we may assume to have integral coefficients and degrees ≥ 1.

Let $\sigma: V \to \mathbf{C}^s$ be the map $v \to (P_1(v), \cdots, P_s(v))$. It is continuous, maps Γ, $V_{\mathbf{Q}}$, $V_{\mathbf{R}}$ in $\mathbf{Z}^s$, $\mathbf{Q}^s$, $\mathbf{R}^s$ respectively. By assumption, any $v \in V_{\mathbf{Q}}$ has a closed orbit. Since the P_i's, together with 1, generate P over C, and the invariant polynomials separate the closed orbits (3.3), we see that

$$(2) \qquad v, v' \in V_{\mathbf{Q}}, \ \sigma(v) = \sigma(v') \Rightarrow v' \in v \cdot G \ .$$

The origin is a closed orbit, hence is the only closed orbit on which all P_i's are zero, and

$$(3) \qquad\qquad \sigma(v) \neq 0 \qquad\qquad (v \in V_{\mathbf{Q}} - 0) \ .$$

The group G being semi-simple, connected, $\pi(G)$ consists of transformations of determinant one, and $\pi(G_{\mathbf{R}})$ leaves invariant the euclidean measure on V, identified with $\mathbf{R}^n$ by means of the basis chosen above. By Minkowski's classical idea, there exists a compact set C in $V_{\mathbf{R}}$ containing zero such that $C \cdot \pi(g) \cap \Gamma \neq \{0\}$ for any $g \in G_{\mathbf{R}}$. (For the sake of completeness, we recall the proof: let C_0 be a compact set in $V_{\mathbf{R}}$ with measure strictly greater than 1. Then the projection $V_{\mathbf{R}} \to V_{\mathbf{R}}/\Gamma$ is not injective on $C_0 \cdot \pi(g)$ whence the existence of $z \in \Gamma - 0$ such that $C_0 \cdot \pi(g) \cap (C_0 \cdot \pi(g) + z)$ $\neq \varnothing$. Therefore $C = \{x - y; x, y \in C_0\}$ fulfills our condition.) The image $\sigma(C)$ of C is compact, $\sigma(C) \cap \mathbf{Z}^s$ is finite, and we may find finitely many elements $w_j \in \Gamma - 0 (1 \leq j \leq t)$ such that

$$\sigma(C) \cap \sigma(\Gamma - 0) = \{\sigma(w_1), \cdots, \sigma(w_t)\} \ .$$

By (2), the intersection $\sigma^{-1}(\sigma(v)) \cap \Gamma (v \in V_{\mathbf{Q}})$, belongs to the orbit of v, which is closed by assumption, hence (6.9) consists of a *finite* number of orbits of $G_{\mathbf{Z}}$. There exists therefore a finite number of elements $v_1, \cdots, v_m \in \Gamma - 0$ such that

$$(4) \qquad \bigcup_i v_i \cdot \pi(G_{\mathbf{Z}}) = \sigma^{-1}(\sigma(C) \cap \sigma(\Gamma - 0)) \cap \Gamma \ .$$

Let now $g \in G_{\mathbf{R}}$. There exist $c \in C$ and $v \in \Gamma - 0$ such that $c \cdot \pi(g^{-1}) = v$. The polynomials P_i being invariant under G, we have then

$$\sigma(v) = \sigma(c \cdot \pi(g^{-1})) = \sigma(c) \in \sigma(C) \cap \sigma(\Gamma - 0) \ ,$$

hence $v \in \sigma^{-1}(\sigma(C) \cap \sigma(\Gamma - 0))$; by (4), there exists an index $i (1 \leq i \leq m)$ such that

$$c \cdot \pi(g^{-1}) = v \in v_i \cdot \pi(G_{\mathbf{Z}}) \ .$$

Given $g \in G_{\mathbf{R}}$, we have thus found an index i, and $b \in G_{\mathbf{Z}}$, such that $v_i \cdot \pi(b \cdot g) \in C$. This shows that

$$G_{\mathbf{R}} = \bigcup_i G_{\mathbf{Z}} \cdot X_i \qquad (X_i = \{g \in G_{\mathbf{R}} \,|\, v_i \cdot \pi(g) \in C\}) \ .$$

Let G_i be the isotropy group of $v_i (1 \leq i \leq m)$. The orbit $v_i \cdot \pi(G)$ is closed, hence so is $v_i \cdot \pi(G_{\mathbf{R}})$ (see 2.3); there exists a compact set $B_i \subset G_{\mathbf{R}}$ such that

$X_i = G_{iR} \cdot B_i$ (5.2). We have then

$$G_R = \bigcup_i G_Z \cdot G_{iR} \cdot B_i \qquad\qquad (B_i \text{ compact}) .$$

But G_i is defined over $\mathbf{Q}$, since $v_i \in V_Q$, and is a proper subgroup since $v_i \neq 0$ (see (1)). Therefore $G_{iR} = G_{iZ} \cdot C_i$, with C_i compact, and finally $G_R = G_Z \cdot A$ with $A = \bigcup_i C_i \cdot B_i$ compact.

REMARK. The preceding proof is an adaptation of A. Weil's argument, pertaining to groups of automorphisms of algebras with involution "which do not represent zero" [30, Theorem 4.1.1], and was in part suggested by it.

11.8. THEOREM. *Let G be an algebraic group defined over $\mathbf{Q}$. Then G_R/G_Q is compact if and only if $X_Q(G^0) = 1$, and every unipotent element of G_Z, or, equivalently, of G_Q, belongs to the radical of G.*[5]

As in 11.6, we may restrict ourselves to the case of a connected group G. Then $G = H \cdot N$ is the semi-direct product of a reductive group H and of an invariant unipotent group N, both defined over $\mathbf{Q}$, and N_R/N_Z is compact (see 6.10 for references). In view of 11.6, it will be enough to show:

(i) G_R/G_Z is compact if and only if H_R/H_Z is compact.

(ii) H verifies the conditions of 11.4 if and only if $X_Q(G) = 1$, and every unipotent element of G_Q (resp. of G_Z) belongs to N.

PROOF OF (i). Let G_R/G_Z be compact. Since $H_Z \cdot N_Z$ is of finite index in G_Z (6.4), we have $G_R = K \cdot H_Z \cdot N_Z$, with K compact, whence $H_R = \pi(K) \cdot H_Z$, where π is the natural projection of G onto H. Let now $H_R = A \cdot H_Z$ with A compact. Then

$$G_R = A \cdot H_Z \cdot N_R = A \cdot N_R \cdot H_Z = A \cdot B \cdot N_Z \cdot H_Z ,$$

with B compact, since N_R/N_Z is compact, whence $G_R = K \cdot G_Z$, with $K = A \cdot B$ compact.

PROOF OF (ii). A unipotent group has only the trivial rational character, hence $X(G)$ and $X_Q(G)$ are naturally isomorphic to $X(H)$ and $X_Q(H)$. Assume H to verify the conditions of 11.4. Then $X_Q(G) = 1$. Moreover, since a rational representation maps unipotent elements into unipotent elements [1, Chap. II], the unipotent elements of G_Q belong to the kernel of π, that is to N.

Assume now that $X_Q(G) = 1$, and that every unipotent element of G_Z belongs to N. We have then $X_Q(H) = 1$, by the initial remark of the proof of (ii). If $x \in G_Q$ is unipotent, then a suitable power x^m of x is a

---

[5] Another proof of Theorem 11.8 has been given by G. D. Mostow and T. Tamagawa, to appear in Ann. of Math.

unipotent element of G_Z (see the proof of (d) $\Rightarrow$ (a) in 11.4). It belongs to N by assumption, hence so does the one parameter group generated by $\log x^m = m \cdot \log x$, and therefore $x \in N$. Thus every unipotent element of G_Q is in N. Since H_Q contains the semi-simple and unipotent parts of its elements [1, Chap. II], we see that H_Q has only semi-simple elements.

12. Groups over number fields

12.1. *Notation.* K is a number field, J the ring of algebraic integers of K, Φ the set of distinct isomorphisms of K into $\mathbf{C}$, and $\bar{\sigma}$ the composition of $\sigma \in \Phi$ with the complex conjugation. Φ' will be a subset of Φ which contains exactly one representative of each pair $(\sigma, \bar{\sigma})$. As usual, x^σ is the image of $x \in K$ under σ, and K^σ the image of K. The completion of K^σ with respect to the absolute value in $\mathbf{C}$ is denoted by L_σ. Thus $L_\sigma = \mathbf{C}$ if $\sigma \neq \bar{\sigma}$, and $L_\sigma = \mathbf{R}$ otherwise. This notation is used throughout this paragraph.

12.2. Let $G \subset \mathbf{GL}(n, \mathbf{C})$ be a connected algebraic group defined over K, and I the ideal of polynomials on $\mathbf{M}(n, \mathbf{C})$, with coefficients in K, which vanish on G. The algebraic group defined by the ideal I^σ is denoted by G^σ. It has K^σ as a field of definition. For a subset ψ of Φ, we put

$$(1) \qquad G_\psi = \prod_{\sigma \in \psi} G^\sigma, \qquad G_{\psi,r} = \prod_{\sigma \in \psi} G^\sigma_{L_\sigma}.$$

In G_ψ, or in $G_{\psi,r}$ we identify G_J (resp. G_K) with the set of elements $(x^\sigma)_{\sigma \in \psi} (x \in G_J,$ resp. $x \in G_K)$.

There exists an algebraic group $G' = R_{K/Q}G$, the *group obtained from G by restriction of the ground field from K to $\mathbf{Q}$*, which is defined over $\mathbf{Q}$, and is isomorphic over $\bar{K}$ to G_Φ. It is essentially unique up to isomorphism over $\mathbf{Q}$ [30, Chap. I], and is isomorphic to G_Φ by an isomorphism μ' of the form $(\mu^\sigma)_{\sigma \in \Phi}$, where $\mu: G' \to G$ is a rational homomorphism defined over K, which verifies

$$(2) \qquad \mu'(G'_Z) = G_J, \qquad \mu'(G'_Q) = G_K \qquad (G' = R_{K/Q}G).$$

The homomorphism μ also induces in a natural way an isomorphism of $X_K(G)$ onto $X_Q(G')$. Let $\sigma \neq \bar{\sigma}$. The standard embedding of $\mathbf{GL}(n, \mathbf{C})$ into $\mathbf{GL}(2n, \mathbf{R})$ induces an isomorphism of $G^\sigma_{L_\sigma}$ onto the set of real points of an algebraic group $R_{C/R}G^\sigma \subset \mathbf{GL}(2n, \mathbf{C})$; from this, one deduces the existence of an isomorphism of real algebraic groups $\beta: G'_R \to G_{\Phi',r}$, which also maps G'_Z and G'_Q onto G_J and G_K respectively.

In general, G_J is not discrete in $G^\sigma_{L_\sigma}$, however, it is clear from the above, that G_J is discrete in G_Φ or in $G_{\Phi'}$. *More generally, if ψ contains all $\sigma \in \Phi'$ for which $G^\sigma_{L_\sigma}$ is not compact, then G_B is discrete in $G_{\psi,r}$.*

To see this, it is enough to show that given $\delta > 0$, there are only finitely many $b \in J$ which occur as coefficients of matrices in G_J, and which verify $|b^\sigma| < \delta (b \in \psi)$. If $\bar{\sigma} \in \psi$, then $|b^\sigma| < \delta$. If $\sigma, \bar{\sigma} \notin \psi$, then $G_{L_\sigma}^\tau$ is compact, hence $|b^\sigma| < \varepsilon$, where ε depends only on $G_{L_\sigma}^\tau$. Our assertion follows then from the familiar fact that J has only finitely many elements b, all of whose conjugates $b^\sigma (b \in \Phi)$ are in absolute value under a given bound.

When this condition is fulfilled, G_J is called an arithmetically defined subgroup, or a group of units of $G_{\psi,r}$. In view of the possibility of restricting the groundfield, it is clear that there is no essential loss in generality in limiting oneself to the case $K = \mathbf{Q}$, $J = \mathbf{Z}$, and that the main results of the preceding paragraphs extend automatically to the groups of units considered here. We state this formally for some of them for the convenience of reference, and leave the reformulation of the others to the reader.

12.3. THEOREM. *We keep the notation of 11.1. Let G be a connected algebraic group defined over K, and ψ a subset of Φ' containing all σ for which $G_{L_\sigma}^\tau$ is not compact. Then*

(a) *G_J is finitely generated.*

(b) *$G_{\psi,r}$ is the union of open subsets U having the following properties:*
  (i) *$G_{\psi,r} = U \cdot G_J$;*
  (ii) *$K \cdot U = U$ for a suitable maximal compact subgroup of $G_{\psi,r}$;*
  (iii) *$U^{-1} \cdot U \cap x \cdot G_J \cdot y$ is finite for $x, y \in G_K$.*

(c) *$G_{\psi,r}/G_J$ has finite Haar measure if and only if $X_K(G) = 1$; it is compact if and only if $X_K(G) = 1$, and every unipotent element of G_K or, equivalently, of G_J, belongs to the radical of G.*

The first assertion follows from 12.2 and 6.5. When $\psi = \Phi'$, (b) and (c) follow from 12.2, 6.5, 6.7, 9.4 and 11.8. Let now $\psi \neq \Phi'$ and θ be the complement of ψ in Φ'. Then $G_{\theta,r}$ is compact, and $G_{\Phi',r} = G_{\psi,r} \times G_{\theta,r}$. Let U be an open set verifying the properties (i) to (iii) for $\psi = \Phi'$. Since $G_{\theta,r}$ is compact, and invariant, it belongs to all maximal compact subgroups of $G_{\Phi',r}$, and the maximal compact subgroups of $G_{\Phi',r}$ are the products of $G_{\theta,r}$ with the maximal compact subgroups of $G_{\psi,r}$. By (ii) we have $G_{\theta,r} \cdot U = U$, hence $U = G_{\theta,r} \times U'$ with U' open in $G_{\psi,r}$; it is then obvious that U' has the properties (i) to (iii).

Let us write A and B for the images of G_J in $G_{\psi,r}$ and $G_{\Phi',r}$ under the canonical imbeddings. We have then clearly $A \cdot G_{\theta,r} = B \cdot G_{\theta,r}$, therefore $G_{\psi,r}/A = G_{\Phi',r}/A \cdot G_{\theta,r} = G_{\Phi',r}/B \cdot G_{\theta,r}$ is the base space of a fibration of $G_{\Phi',r}/B$ with fibre $G_{\theta,r}$. Since $G_{\theta,r}$ is compact, $G_{\psi,r}/A$ is compact (or of finite measure) if and only if $G_{\Phi',r}/B$ is, and (c) follows from the above.

12.4. COROLLARY. *Let G be a connected algebraic group defined over*

K.　We keep the assumptions of 11.3, *and assume moreover that there is at least one* $\sigma \in \Phi$ *for which* $G^{\tau}_{L_\sigma}$ *is compact.　Then* $G_{\psi,r}/G_J$ *is compact if and only if* $X_K(G) = 1$.

If $G^{\tau}_{L_\sigma}$ is compact, then it is reductive, and all its elements are semi-simple.　Of course, $x \in G_K$ is semi-simple if and only if x^σ is semi-simple $(\sigma \in \Phi)$, in our case, G_K consists therefore of semi-simple elements, and 12.4 follows from 12.3(c).

13. Appendix: Remarks on algebraic groups

This appendix contains some remarks about algebraic groups which are actually not needed in the paper, but bring natural complements to some auxiliary results proved in §§ 1, 8.　In A, unlike in §2, the universal field underlying the definition of an algebraic group may have arbitrary characteristic.

A.　Algebraic tori

13.1.　Let T be an algebraic torus, K a field of definition for T.　Then there always exists a separable finite Galois extension K' of K over which T splits (see Ono, Ann. of Math. 74 (1961), 61–139, Prop. 1.2.1).　This being taken into account, it is clear that 8.2, 8.3, and 8.4 go over without change to the general case.　It follows from 8.4 that if S is a subtorus of T, defined over K, and if $\chi \in X_K(S)$, then there exists an integer m such that χ^m extends to a rational character of T.　In fact we take S' as in 8.4a, and m such that χ^m is trivial on the finite group $S \cap S'$.　Since the character groups are finitely generated, this can also be expressed by saying that the injection $i : S \to T$ induces a homomorphism i^0 of $X(T)$ (resp. $X_K(T)$) onto a subgroup of finite index of $X(S)$ (resp. $X_K(S)$).

13.2.　PROPOSITION.　*Let* T *be an algebraic torus defined over a field* K.　*Then* T *contains two subtori* T_c, T_a *defined over* K *such that* $X_K(T_c) = 1$, T_a *splits over* K, $T_c \cap T_a$ *is finite and* $T = T_c \cdot T_a$.　*If* S *is an algebraic torus and* $f : S \to T$ *a rational homomorphism, both defined over* K, *then* $f(S_c) \subset T_c$ *and* $f(S_a) \subset T_a$.

Let K' be a Galois extension of K over which T splits, and A the Galois group of K' over K.　Let T_c be the identity component of the intersection of the kernels of the characters defined over K, and T_a be the subtorus generated by the images of the elements $\gamma \in \Gamma_K(T)$.　They are defined over K', and invariant under A.　Since K' is separable over K, T_c and T_a are defined over K.　By 8.2, T_a splits over K.　By 13.1, $X_K(T_c) = 1$.　Thus T_c has no non-trivial subtorus which splits over K (8.4), and $T_c \cap T_a$ is finite.　Lemma 8.4 also implies that dim T_c + dim T_a = dim T, whence

$T = T_a \cdot T_c$. Let now $f: S \to T$ be a rational homomorphism defined over K. We have $f_0(\Gamma_K(S)) \subset \Gamma_K(T)$, hence $f(S_a) \subset T_a$. If $\chi \in X_K(T)$, then $\chi \circ f \in X_K(S)$, hence $S_c \in \mathrm{Ker}\,(\chi \circ f)$, and $f(S_c) \in \mathrm{Ker}\,\chi$, which implies $f(S_a) \subset T_a$.

B. Real algebraic reductive groups

13.3. PROPOSITION. *Let $T \subset \mathbf{GL}(n, \mathbf{C})$ be an algebraic torus defined over $\mathbf{R}$, and $t_{\mathbf{R}}$ the Lie algebra of $T_{\mathbf{R}}$. Then $T_{c,\mathbf{R}}$ is a torus, $T_{a,\mathbf{R}}$ is real diagonalizable, the Lie algebra of $T_{c,\mathbf{R}}$ (resp. $T_{a,\mathbf{R}}$) is the set of elements of $t_{\mathbf{R}}$ with purely imaginary (resp. real) eigenvalues.*

Let t' (resp. t'') be the set of elements of $t_{\mathbf{R}}$ with real (resp. purely imaginary) eigenvalues. Then $t_{\mathbf{R}} = t' + t''$, and t', t'' are algebraic (see 1.4). The irreducible real algebraic group T' with Lie algebra t' is then a real algebraic torus which splits over $\mathbf{R}$, and is contained in $T_{a,\mathbf{R}}$. The analytic subgroup of $\mathbf{GL}(n, \mathbf{R})$ generated by t'' is closed, and belongs to a compact group hence is a torus in the usual sense (compact connected commutative Lie group). Since a compact linear group is algebraic [7b, p. 230], T'' is also the irreducible real algebraic subgroup of $T_{\mathbf{R}}$ with Lie algebra t''. Clearly, every element of $X_{\mathbf{R}}(T)$ is trivial on T'', therefore $T' = T_{a,\mathbf{R}}$, $T'' = T_{c,\mathbf{R}}$.

13.4. The preceding proposition shows that the decomposition $m = m_t + m_p$ of a fully reducible commutative algebraic Lie algebra corresponds to the global decomposition of 13.3. In particular (13.3) it is compatible with rational representations, and has therefore an intrinsic meaning, independent of the imbedding. From this and 1.10, 1.11, it follows that the notion of Cartan involution of a real reductive algebraic group is independent from the imbedding in $\mathbf{GL}(n, \mathbf{R})$, up to birational isomorphism. Since a real representation of a semi-simple Lie algebra is always rational, the following proposition generalizes a fact mentioned in 1.1:

13.5. PROPOSITION. *Let G be a real algebraic, reductive group, θ a Cartan involution of G, and $\rho: G \to \mathbf{GL}(m, \mathbf{R})$ a rational representation. Then there exists a Cartan involution θ' of $\mathbf{GL}(m, \mathbf{R})$ such that $\rho(\theta(g)) = \theta'(\rho(g))$.*

Let c be the center of g. The image g' of g is the direct product of $\rho(c)$ and of $\rho(\mathcal{D}g) = \mathcal{D}g'$. By 13.4, $\rho(c)$ is completely reducible, and therefore g' is reductive in $\mathfrak{gl}(m, \mathbf{R})$. By a general theorem [7a, p. 140], g' is algebraic. Let G' be the real algebraic subgroup of $\mathbf{GL}(m, \mathbf{R})$ with Lie algebra g', and M be the centralizer of $\rho(c)$. The group M is algebraic, and its Lie algebra m is reductive [4]. Thus (1.2) G' and M are real algebraic, reductive. As was recalled in 1.1, the image of a Cartan decomposition of $\mathcal{D}g$

is a Cartan decomposition of $\mathscr{D}\mathfrak{g}'$. It follows therefore from 1.6 and the conjugacy of Cartan decompositions of $\mathscr{D}\mathfrak{g}'$ that we may find a Cartan involution θ'' of $\mathbf{GL}(m, \mathbf{R})$ which leaves M, G' invariant and induces the given Cartan involution of $\mathscr{D}\mathfrak{g}'$. By 13.4 and 1.4, the $\mathfrak{k}$- and $\mathfrak{p}$-parts of $\rho(\mathfrak{c})$ with respect to θ'' are necessarily the images of the $\mathfrak{k}$- and $\mathfrak{p}$-parts of θ. The Cartan involution θ'' verifies therefore $\rho \circ \theta = \theta'' \circ \rho$ on $\mathfrak{g}$, hence also on G^0, and leaves G' invariant. It is defined by a positive non-degenerate quadratic form F which is invariant under the identity component of $\rho(K)$, where K is the fixed point set of θ in G. Let F' be the average of F, over $\rho(K)$. The corresponding Cartan involution θ' will then fulfill our conditions.

INSTITUTE FOR ADVANCED STUDY
COLUMBIA UNIVERSITY

BIBLIOGRAPHY

1. A. BOREL, *Groupes linéaires algébriques*, Ann. of Math., 64 (1956), 20–80.
2. ———, *Some properties of the adele groups attached to algebraic groups*, Bull. Amer. Math. Soc., 67 (1961), 583–585.
3. A. BOREL and HARISH-CHANDRA, *Arithmetic subgroups of algebraic groups*, Bull. Amer. Math. Soc., 67 (1961), 579–583.
4. A. BOREL and G. D. MOSTOW, *On semi-simple automorphisms of Lie algebras*, Ann. of Math., 61 (1955), 389–405.
5. N. BOURBAKI, Groupes et algèbres de Lie, Chap. I: Algèbres de Lie, Actualités Sci. Ind. 1285, Paris, 1960.
6. H. CARTAN, Séminaire E.N.S., 1951–52, Exp. XX (by J-P. Serre).
7. C. CHEVALLEY, Théorie des groupes de Lie, (a) T. II, Groupes algébriques, Actualités Sci. Ind. 1152, Paris, 1951; (b) T. III, Théorèmes généraux sur les algèbres de Lie, Actualités Sci. Ind. 1155, Paris, 1955.
8. ———, Séminaire sur la classification des groupes de Lie algébriques, Paris, 1958.
9. HARISH-CHANDRA, *Representations of a semi-simple Lie group on a Banach space* I, Trans. Amer. Math. Soc., 75 (1953), 185–243.
10. ———, *On a lemma of Bruhat*, J. math. pures appl. Série g, T. 35 (1956), 203–210.
11. ———, *The characters of semi-simple Lie groups*, Trans. Amer. Math. Soc., 83 (1956), 98–163.
12. ———, *Representations of semi-simple Lie groups* VI, Amer. J. Math., 78 (1956), 564–628.
13. ———, *A formula for semi-simple Lie groups*, Amer. J. Math., 79 (1957), 733–760.
14. C. HERMITE, Oeuvres complètes, Volume 1, Gauthier-Villars, Paris, 1905.
15. N. JACOBSON, *Completely reducible Lie algebras of linear transformations*, Proc. Amer. Math. Soc., 2 (1951), 105–113.
16. A. KORKINE and G. ZOLOTAREFF, *Sur les formes quadratiques*, Math. Ann., 6 (1873), 366–389.
17. S. LANG, Introduction to algebraic geometry, Interscience Tracts in Pure and Applied Math. 5, New York, 1958.
18. Y. MATSUSHIMA, *Espacse homogènes de Stein des groupes de Lie complexes*, Nagoya Math. J., 16 (1960), 205–218.

19. D. Montgomery and L. Zippin, *Topological transformation groups*, Interscience Tracts in Pure and Applied Math. 1, New York, 1955.

20. G. D. Mostow, *Some new decomposition theorems for semi-simple groups*, Mem. Amer. Math. Soc., 14 (1955), 31–54.

21. ———, *Self-adjoint group*, Ann. of Math., 62 (1955), 44–55.

22. T. Ono, *Sur une propriété arithmétique des groupes commutatifs*, Bull. Soc. Math. France 85 (1957), 307–323.

23. ———, *On some arithmetic properties of linear groups*, Ann. of Math., 70 (1959), 266–290.

24. K. G. Ramanathan, *Quadratic forms over involutorial division algebras* II, Math. Ann., 143 (1961), 293–332.

25. M. Rosenlicht, *Some basic theorems on algebraic groups*, Amer. J. Math., 78 (1956), 401–443.

26. ———, *Some rationality questions on algebraic groups*, Annali di Matematica Ser. IV, T. 43, 1957, 25–50.

27. Séminaire S. Lie, Groupes et algèbres de Lie, Paris, 1955.

28. C. L. Siegel, *Einheiten quadratischer Formen*, Abh. Math. Sem. Hamburg, 13 (1939), 209–239.

29. A. Weil, Discontinuous subgroups of classical groups, Notes, Chicago University, 1958.

30. ———, Adeles and algebraic groups, Notes, The Institute for Advanced Study, Princeton, 1961.

31. H. Whitney, *Elementary structure of real algebraic varieties*, Ann. of Math., 66 (1957), 545–556.

INVARIANT EIGENDISTRIBUTIONS ON SEMISIMPLE LIE GROUPS

BY HARISH-CHANDRA[1]

Communicated by Raoul Bott, September 13, 1962.

1. Let M be an oriented separable differentiable manifold of dimension n. (We do not assume that M is connected.) Let $C_c^\infty(M)$ denote the space of all complex-valued C^∞ functions on M with compact support. A distribution T on M is a linear mapping $T: C_c^\infty(M) \to \mathbf{C}$ which is continuous in the topology of Schwartz. More explicitly, this means the following. Let U be any open and relatively compact set in M. Then we can select differential operators[2] $D_1, \cdots, D_r$ on M such that

$$|T(f)| \leq \sum_i \sup |D_i f| \qquad (f \in C_c^\infty(U)).$$

Let G be a group acting on M. We denote by x^g the transform of $x \in M$ by $g \in G$. We assume that, for a fixed g, the mapping $x \to x^g$ of M is of class C^∞. Then for any $f \in C_c^\infty(M)$, the function $f^g: x \to f(x^{g^{-1}})$ is again in $C_c^\infty(M)$ and if T is a distribution, the mapping $T^g: f \to T(f^{g^{-1}})$ $(f \in C_c^\infty(M))$ is also a distribution. We say T is invariant (under G) if $T^g = T$ for all $g \in G$.

Now G operates in a natural way on the spaces[2] of differential operators and differential forms on M. For example if D is a differential operator, $D^g f = (Df^{g^{-1}})^g$ $(f \in C_c^\infty(M), g \in G)$. Fix a (real) differential form ω on M of degree n which is invariant under G and which is everywhere positive (with respect to the given orientation of M). Then for every differential operator D on M, we define its adjoint D^* to be the (unique) differential operator satisfying the relation

$$\int_M Df \cdot \phi \omega = \int_M f D^* \phi \cdot \omega$$

for all $f, \phi \in C_c^\infty(M)$. If T is a distribution, the mapping $f \to T(D^*f)$ $(f \in C_c^\infty(M))$ is also a distribution which we denote by DT. Now ω defines a positive Borel measure μ on M. For example if U is an open set in M,

---

[1] This work was supported by a grant from the Sloan foundation and a contract with the U. S. Army.

[2] All differential operators and differential forms are meant to be C^∞ unless explicitly mentioned otherwise.

117

"

$$\mu(U) = \int_U \omega.$$

Let F be a function on M which is locally summable (with respect to μ). Then corresponding to F, we get a distribution

$$T_F: f \to \int fF d\mu = \int_M fF \cdot \omega \qquad (f \in C_c^\infty(M)).$$

If T is a distribution, we say $T = F$ if $T = T_F$.

2. Let G be a connected semisimple Lie group. Take $M = G$, $x^g = gxg^{-1}$ $(x, g \in G)$ and ω the invariant differential form corresponding to the Haar measure dx on G. Let $\mathfrak{Z}$ be the algebra of all differential operators on G which are invariant under both left and right translations of G. Then $\mathfrak{Z}$ is abelian. Let $l = \mathrm{rank}\ G$. t being an indeterminate, we denote by $D(x)$ the coefficient of t^l in $\det(t + 1 - \mathrm{Ad}(x))$ $(x \in G)$. Then D is an analytic function on G and an element $x \in G$ is called regular if $D(x) \neq 0$. Let G' be the set of all regular elements in G. Then G' is an open and dense subset of G whose complement is of measure zero.

Let Θ be a distribution on G. We say that it is invariant if $\Theta^x = \Theta$ $(x \in G)$ and that it is an eigendistribution of $\mathfrak{Z}$ if $z\Theta = \chi(z)\Theta$ $(z \in \mathfrak{Z})$ for some homomorphism χ of $\mathfrak{Z}$ into $\mathbf{C}$.

THEOREM 1. *Let Θ be an invariant eigendistribution of $\mathfrak{Z}$ on G. Then Θ is a locally summable function which is analytic on G'.*

This answers, in particular, a question raised in [3, p. 396].

3. Now assume that the center of G is finite. Fix a maximal compact subgroup K of G and let $\mathcal{E}_K$ denote the set of all equivalence classes of irreducible finite-dimensional representations of K. For any $\mathfrak{b} \in \mathcal{E}_K$, let $\xi_\mathfrak{b}$ be the character of $\mathfrak{b}$ and $\mathfrak{b}^*$ the class contragradient to $\mathfrak{b}$ so that[3] $\xi_{\mathfrak{b}^*}(k) = \mathrm{conj}\ \xi_\mathfrak{b}(k)$ $(k \in K)$. For any $f \in C_c^\infty(G)$, define

$$f_\mathfrak{b}(x) = d(\mathfrak{b}) \int_K \xi_\mathfrak{b}(k)f(kx)dk \qquad (x \in G),$$

where $d(\mathfrak{b})$ is the degree of any representation in the class $\mathfrak{b}$ and dk is the normalized Haar measure of K. Then $f_\mathfrak{b} \in C_c^\infty(G)$ and the series $\sum_{\mathfrak{b} \in \mathcal{E}_K} f_\mathfrak{b}$ converges in $C_c^\infty(G)$ to f. If T is any distribution on G, the mapping $f \to T(f_{\mathfrak{b}^*})$ $(f \in C_c^\infty(G))$ is also a distribution, which we denote by $T_\mathfrak{b}$. Since

---

[3] conj c stands for the complex conjugate for $c \in \mathbf{C}$.

$$T(f) = \sum_{\mathfrak{b} \in \mathcal{E}_K} T_{\mathfrak{b}}(f) \qquad\qquad (f \in C_c^{\infty}(G)),$$

it is clear that $T_{\mathfrak{b}} \neq 0$ for some $\mathfrak{b} \in \mathcal{E}_K$, if $T \neq 0$.

Now suppose T is an eigendistribution of $\mathfrak{Z}$ on G. Then the same holds for $T_{\mathfrak{b}}$ ($\mathfrak{b} \in \mathcal{E}_K$). But since $T_{\mathfrak{b}}$ transforms, under left translations by elements of K, according to $\mathfrak{b}$, it follows easily that it satisfies an elliptic differential equation on G with analytic coefficients. Therefore $T_{\mathfrak{b}}$ is an analytic function.

4. Let $\mathfrak{g}$ be the Lie algebra of G and $\mathfrak{g}_c$ its complexification. Let G_c be the simply connected complex-analytic group corresponding to $\mathfrak{g}_c$. Assume that G is the real analytic subgroup of G_c corresponding to $\mathfrak{g}$ and rank $G = $ rank K. Fix a maximal connected abelian subgroup A of K and let $\mathfrak{a}$ denote its Lie algebra. Then A is a Cartan subgroup of G and $A' = A \cap G'$ is open and dense in A. Let $\mathfrak{a}_c$ denote the complexification of $\mathfrak{a}$, P the set of all positive roots (under some fixed order) and W the Weyl group of $(\mathfrak{g}_c, \mathfrak{a}_c)$. Then there exists an analytic function Δ on A such that

$$\Delta(\exp H) = \prod_{\alpha \in P} (e^{\alpha(H)/2} - e^{-\alpha(H)/2}) \qquad\qquad (H \in \mathfrak{a}).$$

Let $\hat{A}$ denote the character group of A. For any $\hat{a} \in \hat{A}$, define

$$\sigma(\hat{a}) = \prod_{\alpha \in P} \langle \alpha, \lambda \rangle$$

where λ is the linear function on $\mathfrak{a}_c$ such that $\hat{a}(\exp H) = e^{\lambda(H)} (H \in \mathfrak{a})$ and $\langle \alpha, \lambda \rangle$ denotes the usual scalar product defined under the Killing form of $\mathfrak{g}_c$. W operates on $\hat{A}$ in a natural way by duality. An element $\hat{a} \in \hat{A}$ is called regular if its transforms $\hat{a}^s$ ($s \in W$) are all distinct. Then $\hat{a}$ is singular or regular according as $\sigma(\hat{a}) = 0$ or not. Moreover $\sigma(\hat{a}^s) = \epsilon(s)\sigma(\hat{a})$ ($s \in W$, $\hat{a} \in \hat{A}$), where $\epsilon(s) = 1$ or -1 and is independent of $\hat{a}$.

If Θ is an invariant eigendistribution of $\mathfrak{Z}$ on G, one can, in view of Theorem 1, speak of the value $\Theta(x)$ of Θ at any point $x \in G'$. Define the function D as in §2.

THEOREM 2. *Fix a regular element $\hat{a} \in \hat{A}$. Then there exists exactly one invariant eigendistribution $\Theta_{\hat{a}}$ of $\mathfrak{Z}$ on G such that:*

(1) *The function $|D|^{1/2}\Theta_{\hat{a}}$ remains bounded on G';*

(2) $\Theta_{\hat{a}} = (-1)^q \sigma(\hat{a}) \Delta^{-1} \sum_{s \in W} \epsilon(s) \hat{a}^s$ *pointwise on A'.*

Here $q = \frac{1}{2}(\dim G - \dim K)$.

For $f, g \in C_c^{\infty}(G)$, let $f * g$ denote their convolution product so that

$$(f * g)(x) = \int_G f(y)g(y^{-1}x)dy \qquad\qquad (x \in G).$$

Also let $\bar{f}(x) = \mathrm{conj}(f(x^{-1}))$.

THEOREM 3. *Put* $\Theta = \Theta_{\hat{a}}$ *for a fixed regular element* $\hat{a}$ *in* $\hat{A}$. *Then* $\Theta(\bar{f} * f) \geq 0$ *for every* $f \in C_c^\infty(G)$. *Moreover the analytic functions* Θ_b ($b \in \mathcal{E}_K$) *all lie in* $L_2(G)$.

It is obvious from its definition that $\Theta \neq 0$. Hence we can choose $b \in \mathcal{E}_K$ such that $\Theta_b \neq 0$. Let V be the smallest closed subspace of $L_2(G)$ containing Θ_b, which is invariant under the left-regular representation λ of G. Then $V \neq \{0\}$ and it is easy to show that V is the orthogonal sum of a finite number of subspaces which are all invariant and irreducible under λ. This shows that each of the corresponding irreducible representations belongs to the discrete series.

Define $\Theta_{\hat{a}} = 0$ if $\hat{a}$ is a singular element of $\hat{A}$ and let $\mathfrak{H}$ be the smallest closed subspace of $L_2(G)$ which contains every C^∞ eigenfunction of $\mathfrak{Z}$ lying in $L_2(G)$. For any $f \in C_c^\infty(G)$ and $x \in G$, let f_x denote the function $y \to f(yx)$ ($y \in G$).

THEOREM 4. *The series*

$$\sum_{\hat{a} \in \hat{A}} \Theta_{\hat{a}}(f) \qquad\qquad (f \in C_c^\infty(G))$$

converges absolutely and the function

$$f^\sharp : x \to \sum_{\hat{a} \in \hat{A}} \Theta_{\hat{a}}(f_x) \qquad\qquad (x \in G)$$

lies in $\mathfrak{H}$. *Moreover the Haar measure of* G *can be so normalized that* $f - f^\sharp$ *is orthogonal to* $\mathfrak{H}$ *for every* $f \in C_c^\infty(G)$.

Theorem 4 shows that our method gives the entire discrete series.

5. The proofs of these results are rather long. We shall only give a brief outline of the main steps in the proofs of Theorems 1 and 2. As before, let $\mathfrak{g}_c$ be the complexification of the Lie algebra $\mathfrak{g}$ of G and $S(\mathfrak{g}_c)$ the symmetric algebra over $\mathfrak{g}_c$. G operates on $\mathfrak{g}_c$ by means of the adjoint representation. Let $I(\mathfrak{g}_c)$ be the subalgebra of all invariants of G in $S(\mathfrak{g}_c)$. Now we take (in the set up of §1) $M = \mathfrak{g}$ and ω the differential form corresponding to the Euclidean measure dX on $\mathfrak{g}$. For $p \in S(\mathfrak{g}_c)$, define the differential operator $\partial(p)$ on $\mathfrak{g}$ as in [4, §2] and identify $\mathfrak{g}_c$ with its dual under the Killing form Ω given by $\Omega(X) = \mathrm{tr}(\mathrm{ad}\, X)^2 (X \in \mathfrak{g}_c)$. Let $\mathfrak{g}'$ be the set of all regular elements of $\mathfrak{g}$. Then $\mathfrak{g}'$ is open and dense in $\mathfrak{g}$ and its complement is of measure zero.

A subset U of $\mathfrak{g}$ is called completely invariant, if it satisfies the following condition. C being any compact subset of U, $\mathrm{Cl}(C^G) \subset U$. Here $C^G = \bigcup_{x \in G} C^x$ and Cl denotes closure. If U is an open and completely invariant subset of $\mathfrak{g}$, we can take $M = U$ in §1.

LEMMA 1. *Let T be a distribution on a completely invariant open subset U of $\mathfrak{g}$ such that:*
   (1) $T^x = T (x \in G)$,
   (2) *There exists an ideal $\mathfrak{U}$ in $I(\mathfrak{g}_c)$ such that* $\dim I(\mathfrak{g}_c)/\mathfrak{U} < \infty$ *and* $\partial(u)T = 0$ *for* $u \in \mathfrak{U}$.
   Then T is a locally summable function on U, which is analytic on $U' = U \cap \mathfrak{g}'$.

This is proved by induction on $\dim \mathfrak{g}$. Let $\mathfrak{N}$ be the set of all $X \in \mathfrak{g}$ such that $\mathrm{ad}\, X$ is nilpotent. The most important step in the proof of Lemma 1 is the following result.

LEMMA 2. *Let T be an invariant distribution on $\mathfrak{g}$ such that[4] $\mathrm{Supp}\, T \subset \mathfrak{N}$ and $\partial(\Omega)T = 0$. Then $T = 0$.*

The proof of this makes use of a result of Kostant [6, Corollary 3.7 and Lemma 5.1] from which it follows (see [2, 2.3]) that $\mathfrak{N}$ is the union of a finite number of G-orbits.
   In order to obtain Theorem 1, we have now to lift the result of Lemma 1 to the group. For this one needs the following fact.

LEMMA 3. *Let D be a polynomial differential operator [4, §2] on $\mathfrak{g}$ such that $D^x = D\ (x \in G)$ and $Dp = 0$ for $p \in I(\mathfrak{g}_c)$. Then $DT = 0$ for every invariant distribution T on $\mathfrak{g}$.*

The proof again proceeds by induction on $\dim \mathfrak{g}$. The crucial part is the following lemma.

LEMMA 4. *Let T be a distribution and D a polynomial differential operator on $\mathfrak{g}$. We assume that:*
   (1) $T^x = T\ (x \in G)$,
   (2) $D^x = D$ *and* $Dp = 0\ (x \in G,\ p \in I(\mathfrak{g}_c))$,
   (3) $\mathrm{Supp}\, DT \subset \mathfrak{N}$.
Then $DT = 0$.

First one shows that it is sufficient to consider the case when T is tempered. (This requires a result of Borel, according to which, we can always find a discrete subgroup Γ of G such that G/Γ is compact. See Remark (2) at the bottom of p. 582 of [1].) Now we use the

---

[4] Supp T denotes the support of T.

theory of Fourier transforms. Put $B(X, Y) = \operatorname{tr}(\operatorname{ad} X \operatorname{ad} Y)(X, Y \in \mathfrak{g})$ and define

$$\hat{f}(Y) = \int e^{iB(Y,X)} f(X) dX \qquad (f \in C_c^\infty(\mathfrak{g}),\ Y \in \mathfrak{g}).$$

Then for any tempered distribution τ, its Fourier transform $\hat\tau$ is defined by $\hat\tau(f) = \tau(\hat f)$ $(f \in C_c^\infty(\mathfrak{g}))$. Let J be the ideal of $I(\mathfrak{g}_c)$ spanned by all homogeneous elements of degree ≥ 1. Then $\mathfrak{N}$ is exactly the set of zeros of J in $\mathfrak{g}$. Let $p_1, \cdots, p_r$ be an ideal basis for J. Then for every j $(1 \leq j \leq r)$, we can choose an integer $m_j \geq 0$ such that $p_j^{m_j} DT = 0$ around the origin. Since Supp $DT \subset \mathfrak{N}$ and DT is invariant, it follows that $p_j^{m_j} DT = 0$. Let $\mathfrak{U}$ be the ideal in $I(\mathfrak{g}_c)$ generated by $p_j^{m_j}$ $(1 \leq j \leq r)$. Then $\dim I(\mathfrak{g}_c)/\mathfrak{U} < \infty$ and $uDT = 0$ for $u \in \mathfrak{U}$. Hence we conclude from Lemma 1 that $(DT)^\frown$ is a locally summable function. Now define $\hat D$ as in [4, p. 91]. Then $(DT)^\frown = \hat D \hat T$ and it is easy to see that $\hat D$ also verifies condition (2) of Lemma 4. From this it follows without difficulty that $\hat D \sigma = 0$ on $\mathfrak{g}'$ for any invariant distribution σ on $\mathfrak{g}$. Hence $\hat D \hat T = 0$ on $\mathfrak{g}'$. But since $\hat D \hat T$ is a locally summable function, this implies that $\hat D \hat T = 0$ and therefore $DT = 0$.

6. Now we come to Theorem 2. So assume that rank $\mathfrak{g} = \operatorname{rank} \mathfrak{k}$ where $\mathfrak{k}$ is the Lie algebra of K. Put $\mathfrak{a}' = \mathfrak{a} \cap \mathfrak{g}'$ and $\pi = \prod_{\alpha \in P} \alpha$. Then π is a polynomial function on $\mathfrak{a}_c$.

LEMMA 5. *Fix $H_0 \in \mathfrak{a}'$ and let T be a tempered and invariant distribution on $\mathfrak{g}$ such that*

$$\partial(p)T = p(iH_0)T \qquad\qquad (p \in I(\mathfrak{g}_c)).$$

Then if[5] $T(H) = 0$ for $H \in \mathfrak{a}'$, we can conclude that $T = 0$.

LEMMA 6. *Fix $H_0 \in \mathfrak{a}'$. Then there exists exactly one tempered and invariant distribution T on $\mathfrak{g}$ such that:*
(1) $\partial(p)T = p(iH_0)T$ $(p \in I(\mathfrak{g}_c))$,
(2) $T(H) = \pi(H)^{-1} \sum_{s \in W} \epsilon(s) e^{iB(H_0, sH)}$ $(H \in \mathfrak{a}')$.

The uniqueness of T follows from Lemma 5. The existence is proved as follows. Put

$$\tau(f) = \pi(H_0) \sum_{s \in W} \int_G \hat f((sH_0)^x) dx \qquad (f \in C_c^\infty(\mathfrak{g})).$$

Then τ is a tempered and invariant distribution and $\partial(p)\tau = p(iH_0)\tau$

---

[5] In view of Lemma 1, we can speak of the value $T(X)$ of T at any point X in $\mathfrak{g}'$.

for $p \in I(\mathfrak{g}_c)$ (see [**5**, pp. 225–226]). Moreover it can be shown that τ satisfies condition (2) of Lemma 6 up to a nonzero constant factor.

Theorem 2 is obtained by lifting the result of Lemma 6 to the group.

REFERENCES

1. A. Borel and Harish-Chandra, *Arithmetic subgroups of algebraic groups*, Bull. Amer. Math. Soc. **67** (1961), 579–583.

2. ———, *Arithmetic subgroups of algebraic groups*, Ann. of Math. (2) **75** (1962), 485–535.

3. Harish-Chandra, *On the characters of a semisimple Lie group*, Bull. Amer. Math. Soc. **61** (1955), 389–396.

4. ———, *Differential operators on a semisimple Lie algebra*, Amer. J. Math. **79** (1957), 87–120.

5. ———, *Fourier transforms on a semisimple Lie algebra*. I, Amer. J. Math. **79** (1957), 193–257.

6. B. Kostant, *The principal three-dimensional subgroup and the Betti numbers of a complex simple Lie group*, Amer. J. Math. **81** (1959), 973–1032.

COLUMBIA UNIVERSITY

Reprinted from
Bull. Amer. Math. Soc.
69 (1963), 117–123

INVARIANT DISTRIBUTIONS ON LIE ALGEBRAS.*

By Harish-Chandra.

1. Introduction. The object of this paper is to start laying the ground-work for the proof of Theorem 1 of [4(e)], which says, roughly, that the irreducible characters of a semisimple Lie group, which had so far only been known to be distributions [4(b)], are actually functions. Our method consists in splitting the problem into two parts. The first one deals with an analogous question on the Lie algebra (see Lemma 1 of [4(e)]) and the second with the task of lifting this result from the algebra to the group itself. In this paper we shall be mainly concerned with one of the principal steps [4(e), Lemma 2] needed in the first part.

Our entire approach to the problem is based on a simple result (Theorem 1) about a C^∞ mapping π of a manifold M *onto* another manifold N, which has the maximum rank everywhere.[1] This allows one to transfer a distribution from N to M. Consider the following example. Let G be the orthogonal group of degree n operating in the usual way on $\boldsymbol{R}^n$. Identify $\boldsymbol{R}$ with a subspace of $\boldsymbol{R}^n$ under the mapping $t \to (t, 0, 0, \cdots, 0)$ $(t \in \boldsymbol{R})$. Let $\boldsymbol{R}_+$ be the set of all positive real numbers. Now let $M = G \times \boldsymbol{R}_+$, N the complement of the origin in $\boldsymbol{R}^n$ and π the mapping $(x, r) \to xr$ $(x \in G, \ r \in \boldsymbol{R}_+)$ of M onto N. Let f be a C^∞ function on N which is invariant under G. Then $f(xr)$ is independent of x. So in this way we get a function σ_f on $\boldsymbol{R}_+$. D being a differential operator on N which is invariant under G, one can show that there exists a differential operator $\Delta(D)$ on $\boldsymbol{R}_+$ such that $\sigma_{Df} = \Delta(D)\sigma_f$ for any invariant C^∞ function f on N. Theorem 1 enables us to prove similar results about distributions (see Theorems 2 and 3).

The most significant result of this paper is contained in Theorem 4. This will be used in subsequent papers to prove the theorem about characters, mentioned above.

The contents of this paper are as follows. In §2 we collect some well-known facts about manifolds, differential operators and distributions. §3 is devoted to the proof of Theorem 1, which, as we have just indicated, is rather basic for our purposes. In §5 we introduce invariant differential operators and distributions and consider some of their elementary properties. The

---

* Received October 25, 1963.

[1] I understand from Prof. L. Schwartz that he had independently obtained a similar result some years ago.

271

results of §7 are important for applications. They enable us to construct the above-mentioned mapping $D \to \Delta(D)$. The main result concerning $\Delta(D)$ and the invariant functions, is proved in Lemma 10. In Theorems 2 and 3 (of §9), this is extended to invariant distributions. This general theory is then applied to the case of a semisimple Lie algebra to prove Theorem 4. Our argument depends on a detailed consideration of some differential operators associated with the Killing form (see Lemmas 30 and 31).

Lemma 16 was pointed out to me by Borel who also gave me the proof of Lemma 17. These lemmas play an essential role in the proof of Theorem 4.

This work was supported, in part, by a grant from the Sloan Foundation and a contract with the U. S. Army.

2. Differential operators and distributions. By a manifold M we mean a Hausdorff and locally Euclidean space satisfying the second axiom of countability, all of whose connected components have the same dimension m and are endowed with a C^∞ structure. It is oriented if each of its connected components is oriented. Let $C^\infty(M)$ denote the space of all C^∞ functions from M to $\boldsymbol{C}$ and $\mathcal{E}$ the space of all $\boldsymbol{C}$-linear mappings of $C^\infty(M)$ into itself. Then $\mathcal{E}$ is an associative algebra. If X is a (C^∞) vector-field on M and $g \in C^\infty(M)$, the mappings $E_X \colon f \to Xf$ and $E_g \colon f \to gf$ $(f \in C^\infty(M))$ are both in $\mathcal{E}$. Let $\mathfrak{D} = \mathfrak{D}(M)$ denote the subalgebra of $\mathcal{E}$ generated by E_X and E_g for all X and g. Then $\mathfrak{D}$ is called the algebra of differential operators on M. It would be convenient to denote E_X and E_g by X and g respectively. However, if it is necessary to avoid confusion, we shall write the product of two differential operators D_1, D_2 as $D_1 \circ D_2$. Thus $E_X E_g = X \circ g$ while $E_{Xg} = Xg$. If $D \in \mathfrak{D}$ and $f \in C^\infty(M)$, we shall often denote the value of Df at a point $p \in M$ by $f(p; D)$.

For any $D \in \mathfrak{D}$, $\operatorname{Supp} D$ is the closed subset of M defined as follows. Let $p \in M$. Then $p \notin \operatorname{Supp} D$ if and only if $p \notin \operatorname{Supp} Df$ for every $f \in C^\infty(M)$. An indexed family $\{F_\alpha\}_{\alpha \in A}$ of sets in M is called scattered if, for every compact subset K of M, $K \cap F_\alpha = \varnothing$ for all except a finite number of α in A. A subset Φ of $\mathfrak{D}$ is scattered if the family $\{\operatorname{Supp} D\}_{D \in \Phi}$ is scattered. Let $C_c^\infty(M)$ denote the subspace of all functions in $C^\infty(M)$ whose support is compact. Put $\nu_D(f) = \sup_M |Df|$ $(D \in \mathfrak{D})$ and for any scattered subset Φ of $\mathfrak{D}$, define $\nu_\Phi(f) = \sum_{D \in \Phi} \nu_D(f)$ $(f \in C_c^\infty(M))$. Let S be the class of all scattered subsets of $\mathfrak{D}$. Then the seminorms $\{\nu_\Phi ; \Phi \in S\}$ define the structure of a locally convex space on $C_c^\infty(M)$. The corresponding topology is called the Schwartz topology. K being any compact subset of M, let $C_K^\infty(M)$ denote the subspace of all functions in $C_c^\infty(M)$ whose support is contained in K. It is easy to see that the collections $\{\nu_D; D \in \mathfrak{D}\}$ and $\{\nu_\Phi; \Phi \in S\}$ define the same topology on $C_K^\infty(M)$. U being an open subset of M, any function in $C_c^\infty(U)$ can be extended to a function on M by setting it zero outside U. In this way $C_c^\infty(U)$ becomes a subspace of $C_c^\infty(M)$ and it is easy to see that the injection $C_c^\infty(U) \to C_c^\infty(M)$ is continuous.

A distribution T on M is a continuous linear mapping of $C_c^\infty(M)$ into $\boldsymbol{C}$.

LEMMA 1. *Let V be a complex locally convex space and T a linear mapping of $C_c^\infty(M)$ into V. Assume that T is continuous on $C_K^\infty(M)$ for every compact subset K of M. Then T is continuous.*

This is an immediate consequence of Theorem II (p. 69) of [9].

LEMMA 2. *Let T be a linear mapping of $C_c^\infty(M)$ into $\boldsymbol{C}$. Then it is a distribution if and only if it satisfies the following condition. Given any open and relatively compact subset U of M, we can choose a finite number of differential operators $D_1, \cdots, D_r$ on M such that*

$$|T(f)| \leqq \sum_{1 \leqq i \leqq r} \sup |D_i f| \qquad (f \in C_c^\infty(U)).$$

This is well known and is an easy consequence of Lemma 1.

Let M_1, M_2 be two manifolds and $M_1 \times M_2$ their Cartesian product. For $f_i \in C_c^\infty(M_i)$ $(i = 1, 2)$ let $f_1 \times f_2$ denote the function $(p_1, p_2) \to f_1(p_1) f_2(p_2)$ $(p_i \in M_i; i = 1, 2)$ on $M_1 \times M_2$. In this way we get a linear mapping of $C_c^\infty(M_1) \otimes C_c^\infty(M_2)$ into $C_c^\infty(M_1 \times M_2)$.

LEMMA 3. *The image of $C_c^\infty(M_1) \otimes C_c^\infty(M_2)$ under this mapping is dense in $C_c^\infty(M_1 \times M_2)$.*

This is well known (see [9, p. 107, Theorem III]).

It follows from Lemma 3 that given $D_i \in \mathfrak{D}(M_i)$ $(i = 1, 2)$, there exists exactly one differential operator D on $M_1 \times M_2$ such that $D(f_1 \times f_2) = D_1 f_1 \times D_2 f_2$ $(f_i \in C_c^\infty(M_i), i = 1, 2)$. We denote D by $D_1 \times D_2$.

3. Proof of Theorem 1. Let ω be a (C^∞) differential form on M of degree m. Fix a point $p \in M$ and let $(x_1, \cdots, x_m)$ be a coordinate system on some open and connected neighborhood U of p in M. Then there exists a unique function $\alpha \in C^\infty(U)$ such that $\omega = \alpha\, dx_1 \wedge dx_2 \wedge \cdots \wedge dx_m$ on U. We say $\omega \neq 0$ at p if $\alpha(p) \neq 0$. Moreover if M is oriented and ω is real, we say $\omega > 0$ at p if $\epsilon\alpha(p) > 0$. Here $\epsilon = 1$ or -1 according as the coordinate system $(x_1, \cdots, x_m)$ is positive or negative with respect to the given orientation of M. We write $\omega > 0$ if ω is positive everywhere.

Now suppose $\omega > 0$. Then it defines a (positive) Borel measure μ on[2] M. For example if f is a continuous function on M with compact support, we have

$$\int f d\mu = \int_M f\omega.$$

Let M and N be two oriented manifolds of dimensions m and n respectively and $\pi: M \to N$ a surjective C^∞ mapping such that rank $d\pi = n$ everywhere.

THEOREM 1.[1] *Let ω_M and ω_N be two differential forms on M and N respectively of degree m and n. Assume that ω_N is nowhere zero. Then for every $\alpha \in C_c^\infty(M)$, there exists a unique function $f_\alpha \in C_c^\infty(N)$ such that*

$$\int_M (F \circ \pi) \alpha \omega_M = \int_N F f_\alpha \omega_N$$

for all $F \in C_c^\infty(N)$. Moreover $\operatorname{Supp} f_\alpha \subset \pi(\operatorname{Supp} \alpha)$ and $\alpha \to f_\alpha$ is a continuous mapping of $C_c^\infty(M)$ into $C_c^\infty(N)$. Finally if ω_M is nowhere zero, this mapping is surjective.

COROLLARY 1. *Let F be a continuous function on N. Then*

$$\int_M (F \circ \pi) \alpha \omega_M = \int_N F f_\alpha \omega_N \qquad (\alpha \in C_c^\infty(M)).$$

COROLLARY 2. *Assume $\omega_M > 0$, $\omega_N > 0$ and let μ_M and μ_N denote the corresponding Borel measures on M and N respectively. Suppose F is a Borel measurable function on N. Then F is locally summable (with respect to μ_N) if and only if $F \circ \pi$ is locally summable (with respect to μ_M) and, in case this is so, we have*

$$\int_M (F \circ \pi) \alpha \, d\mu_M = \int_N F f_\alpha \, d\mu_N \qquad (\alpha \in C_c^\infty(M)).$$

Since ω_N is nowhere zero, the uniqueness of f_α in Theorem 1 follows from its definition. Hence we are mainly concerned with its existence. We shall first prove Theorem 1 and its corollaries in a special case.

For any $a > 0$, let I_a denote the open interval $(-a, a)$ in $\boldsymbol{R}$ and J_a the closed interval $[-a, a]$. Let π denote the projection $(x_1, \cdots, x_m) \to (x_1, \cdots, x_n)$ of $\boldsymbol{R}^m$ on $\boldsymbol{R}^n$.

---

[2] It is convenient to express this relationship by writing $\omega \sim \mu$. Similarly in other cases.

LEMMA 4.   *For any $\alpha \in C_c^\infty(I_a^m)$ define $g_\alpha \in C_c^\infty(I_a^n)$ by*

$$g_\alpha(x_1, \cdots, x_n) = \int \alpha(x_1, \cdots, x_m) dx_{n+1} \cdots dx_m.$$

Then $\alpha \to g_\alpha$ is a continuous mapping of $C_c^\infty(I_a^m)$ onto $C_c^\infty(I_a^n)$ and $\operatorname{Supp} g_\alpha \subset \pi(\operatorname{Supp} \alpha)$.

It is obvious that $g_\alpha \in C_c^\infty(I_a^n)$ and $\operatorname{Supp} g_\alpha \subset \pi(\operatorname{Supp} \alpha)$. For any b $(0 < b < a)$, put $C_b^\infty(I_a^m) = C_K^\infty(I_a^m)$ where $K = J_b^m$. Then it is easy to verify that the mapping is continuous on $C_b^\infty(I_a^m)$. Therefore we conclude from Lemma 1 that it is also continuous on $C_c^\infty(I_a^m)$. So it only remains to show that it is surjective. For this we may assume that $m > n$. Fix $\gamma \in C_c^\infty(I_a^{m-n})$ such that

$$\int \gamma(x_{n+1}, \cdots, x_m) dx_{n+1} \cdots dx_m = 1.$$

Then for a given $\beta \in C_c^\infty(I_a^n)$, put $\alpha = \beta \times \gamma$. Then $\alpha \in C_c^\infty(I_a^m)$ and $\beta = g_\alpha$.

Now in Theorem 1 first consider the case when $M = I_a^m$, $N = I_a^n$ and π is the above projection. Let

$$dx = dx_1 \wedge dx_2 \wedge \cdots \wedge dx_m \text{ and } dy = dx_1 \wedge dx_2 \wedge \cdots \wedge dx_n.$$

Then $\omega_M = \gamma_M \, dx$ and $\omega_N = \gamma_N \, dy$ where γ_M and γ_N are C^∞ functions on M and N respectively and γ_N is nowhere zero. For any $\alpha \in C_c^\infty(M)$, define f_α on N as follows:

$$f_\alpha = \epsilon_M \epsilon_N \gamma_N^{-1} g_{\gamma_M \alpha}.$$

Here ϵ_M, ϵ_N are constants equal to ± 1 chosen in such a way that $\epsilon_M \, dx > 0$ on M and $\epsilon_N \, dy > 0$ on N. Then f_α obviously fulfills the condition of Theorem 1. Moreover it follows from Lemma 4 that $\alpha \to f_\alpha$ is a continuous mapping of $C_c^\infty(M)$ into $C_c^\infty(N)$ and $\operatorname{Supp} f_\alpha \subset \pi(\operatorname{Supp} \alpha)$. Now assume ω_M is nowhere zero. Then the endomorphisms $\alpha \to \gamma_M \alpha$ and $g \to \gamma_N^{-1} g$ of $C_c^\infty(M)$ and $C_c^\infty(N)$ respectively are surjective. Hence we conclude from Lemma 4 that the mapping $\alpha \to f_\alpha$ is surjective. Finally Corollaries 1 and 2 are, in this case, immediate consequences of Fubini's Theorem.

We come now to the general case. Fix a point $p \in M$ and choose a small open connected neighborhood U_0 of p in M. Let $V_0 = \pi(U_0)$. Since rank $d\pi = n$ everywhere, V_0 is open in N and if U_0 is sufficiently small, we can choose a coordinate system $(x_1, \cdots, x_m)$ on U_0 such that $x_i(p) = 0$ $(1 \leq i \leq m)$ and $x_j = y_j \circ \pi$ $(1 \leq j \leq n)$ where $(y_1, \cdots, y_n)$ is a coordinate system on V_0. Fix a small number $a > 0$ and let U be the set of all $q \in U_0$ where $|x_i(q)| < a$ $(1 \leq i \leq m)$. If a is sufficiently small, $q \to (x_1(q), \cdots, x_m(q))$

is a diffeomorphism of U with $I_a{}^m$. Put $V = \pi(U)$. Then it is clear from our discussion above that Theorem 1 and its corollaries hold for (U, V, π). This shows that for every $p \in M$, we can choose an open neighborhood U_p of p in M such that Theorem 1 and its corollaries hold for $(U_p, \pi(U_p), \pi)$.

Fix a compact subset K of M. Then $\{U_p\}_{p \in K}$ is an open covering of K in M. We can therefore select a finite subcovering $\{U_i\}_{1 \leq i \leq r}$ from it. Choose $\beta_i \in C_o{}^\infty(U_i)$ such that $\beta_1 + \cdots + \beta_r = 1$ on K. Put $V_i = \pi(U_i)$ and for any $\alpha \in C_K{}^\infty(M)$, define $\alpha_i = \beta_i \alpha$. Then $\alpha = \alpha_1 + \cdots + \alpha_r$ and $\alpha_i \in C_o{}^\infty(U_i)$. Let $f_{\alpha_i} \in C_o{}^\infty(V_i)$ be the function which corresponds to α_i under Theorem 1 for (U_i, V_i, π). Put $f_\alpha = \sum_{1 \leq i \leq r} f_{\alpha_i}$. Then if F is a continuous function on N,

$$\int (F \circ \pi) \alpha \omega_M = \sum_{1 \leq i \leq r} \int_M (F \circ \pi) \alpha_i \omega_M.$$

But

$$\int_M (F \circ \pi) \alpha_i \omega_M = \int_{U_i} (F \circ \pi) \alpha_i \omega_M = \int_{V_i} F f_{\alpha_i} \omega_N = \int_N F f_{\alpha_i} \omega_N$$

from Corollary 1 applied to (U_i, V_i, π). Hence

$$\int_M (F \circ \pi) \alpha \omega_M = \sum_i \int_N F f_{\alpha_i} \omega_N = \int F f_\alpha \omega_N.$$

Moreover $\operatorname{Supp} f_\alpha \subset \bigcup_i \operatorname{Supp} f_{\alpha_i}$. But $\operatorname{Supp} f_{\alpha_i} \subset \pi(\operatorname{Supp} \alpha_i) \subset \pi(\operatorname{Supp} \alpha)$ since Theorem 1 holds for (U_i, V_i, π). Hence $\operatorname{Supp} f_\alpha \subset \pi(\operatorname{Supp} \alpha)$ and the mapping $\alpha \to f_\alpha$ is continuous on $C_K{}^\infty(M)$. Now assume ω_M is nowhere zero and fix $g \in C_o{}^\infty(N)$. Then we can select the compact set K above in such a way that $\operatorname{Supp} g \subset \pi(K)$. Now $\{V_i\}_{1 \leq i \leq r}$ is an open covering for $\pi(K)$ in N. Choose $\rho_i \in C_o{}^\infty(V_i)$ such that $\rho_1 + \rho_2 + \cdots + \rho_r = 1$ on $\pi(K)$. Put $g_i = \rho_i g$. Then $g = g_1 + \cdots + g_r$. Since Theorem 1 is applicable to (U_i, V_i, π), we can choose $\alpha_i \in C_o{}^\infty(U_i)$ such that $g_i = f_{\alpha_i}$. Then if $\alpha = \alpha_1 + \cdots + \alpha_r$, it is obvious that $g = f_\alpha$. This proves Theorem 1 and Corollary 1 above.

Now we come to Corollary 2. Fix a point $q \in N$ and a point $p \in M$ such that $\pi(p) = q$. Select U_p as above and define $V = \pi(U_p)$. Then Corollary 2 holds for (U_p, V, π) and therefore $F \circ \pi$ is locally summable on V if and only if F is locally summable on U_p. Since q was any point of N, the first statement of Corollary 2 follows. Now assume that F is locally summable. Then if $\alpha \in C_o{}^\infty(M)$,

$$\int_M (F \circ \pi) \alpha \, d\mu_M = \sum_i \int_{U_i} (F \circ \pi) \alpha_i \, d\mu_M$$

$$= \sum_i \int_{V_i} F f_{\alpha_i} \, d\mu_N = \int_N F f_\alpha \, d\mu_N$$

in the above notation.

We keep to the notation of Theorem 1.

LEMMA 5. *Let T be a distribution on N. Then the mapping $\tau: \alpha \to T(f_\alpha)$ $(\alpha \in C_c^\infty(M))$ is a distribution on M. Moreover if ω_M is nowhere zero, T is completely determined by τ.*

The first assertion follows from the continuity of the mapping $\alpha \to f_\alpha$ and the second from its surjectivity (when ω_M is nowhere zero).

4. Adjoint of a differential operator. Assume that M is oriented and fix, once for all, a differential form $\omega > 0$. Then for every $D \in \mathfrak{D}(M)$, there exists a unique differential operator D^*, called the adjoint of D (with respect to ω), such that

$$\int_M Df \cdot \phi\omega = \int_M f \cdot D^*\phi \cdot \omega$$

whenever f and ϕ are two C^∞ functions on M, at least one of which has compact support. The mapping $D \to D^*$ is an anti-automorphism of $\mathfrak{D}(M)$ of order 2.

Let $\mathfrak{X}$ be the space of all distributions on M. Then we turn $\mathfrak{X}$ into a $\mathfrak{D}(M)$-module as follows. If $T \in \mathfrak{X}$ and $D \in \mathfrak{D}(M)$, the mapping $f \to T(D^*f)$ $(f \in C_c^\infty(M))$ is a distribution which, by definition, is DT. Let μ be the Borel measure on M corresponding to ω (see §3). If F is a function on M which is locally summable (with respect to μ), the mapping

$$T_F: f \to \int fF\, d\mu \qquad\qquad (f \in C_c^\infty(M))$$

is a distribution. If we identify two locally summable functions whose difference is almost everywhere zero, then the mapping $F \to T_F$ is injective. Let T be a distribution on M. Then we say that T is a function if $T = T_F$ for some locally summable function F. In that case we also write $T = F$.

If each connected component of M is endowed with an analytic structure, we say that M is an analytic manifold. Suppose M is analytic and D is a differential operator on M. Then D is analytic if for any analytic function f on an open subset U of M, the function Df is also analytic on U. Assume that the differential form ω above is analytic. Then if D is analytic, so is also D^*.

5. Invariant differential operators and distributions. Let G be a group and M a manifold. Assume that G operates on M. This means that for every $g \in G$, there is given a diffeomorphism $x \to x^g$ $(x \in M)$ of M such that $x^1 = x$ and $(x^{g_2})^{g_1} = x^{g_1 g_2}$ $(g_1, g_2 \in G; x \in M)$. Let Ω_r denote the space of all (C^∞) differential forms on M of degree r. Then G operates linearly on $C^\infty(M)$, $\mathfrak{D}(M)$ and Ω_r as follows. Fix $g \in G$. Then

$$f^g(x) = f(x^{g^{-1}}) \qquad\qquad (f \in C^\infty(M), x \in M),$$
$$D^g f = (Df^{g^{-1}})^g \qquad\qquad (D \in \mathfrak{D}(M), f \in C^\infty(M)),$$
$$\omega^g(X_1, \cdots, X_r) = (\omega(X_1^{g^{-1}}, \cdots, X_r^{g^{-1}}))^g \qquad (\omega \in \Omega_r)$$

where $X_1, \cdots, X_r$ are any r vector-fields on M. It is obvious that $f \to f^g$ $(f \in C_c^\infty(M))$ is a continuous endomorphism of $C_c^\infty(M)$. Therefore if T is a distribution on M, the mapping $T^g \colon f \to T(f^{g^{-1}})$ is also a distribution. In particular if μ is a measure on M, the same holds for μ^g.

Now assume M is oriented and we have fixed ω as in §4. Let μ be the measure on M corresponding to ω (see §3). Assume that $\mu^g = \mu$ for all $g \in G$ and fix $D \in \mathfrak{D}(M)$. Then it is easy to verify that $(D^g)^* = (D^*)^g$ and $(DT)^g = D^g T^g$ $(g \in G)$ for any distribution T on M. Let F be a locally summable function on M. Then if $T = F$, it follows that $T^g = F^g$ where F^g is the function $x \to F(x^{g^{-1}})$ $(x \in M)$.

A distribution T is called invariant, if $T^g = T$ for all $g \in G$. Similarly a differential operator D is called invariant if $D^g = D$ $(g \in G)$.

Actually, we shall mostly be concerned with the case when G is a Lie group and the mapping $(g, x) \to x^g$ of $G \times M$ into M is C^∞. Let $\mathfrak{g}$ be the Lie algebra of G. Then for any $X \in \mathfrak{g}$, we define a vector-field $\tau(X)$ on M as follows:

$$f(x; \tau(X)) = \{df^{\exp tX}(x)/dt\}_{t=0} \qquad\qquad (f \in C^\infty(M), x \in M).$$

It is easy to verify that $[\tau(X), \tau(Y)] = \tau([X, Y])$ $(X, Y \in \mathfrak{g})$. A distribution T on an open subset U of M is called locally invariant if $\tau(X)T = 0$ for all $X \in \mathfrak{g}$. Similarly a differential operator D on U is said to be locally invariant if $\tau(X) \circ D = D \circ \tau(X)$ for $X \in \mathfrak{g}$.

LEMMA 6. $\tau(X)^* = -\tau(X)$ $(X \in \mathfrak{g})$ *and every invariant distribution (or differential operator) on M is locally invariant.*

Fix $f, \phi \in C_c^\infty(M)$. Then

$$\int_M f^g \cdot \phi \cdot \omega = \int_M f \cdot \phi^{g^{-1}} \cdot \omega \qquad\qquad (g \in G).$$

Hence if $X \in \mathfrak{g}$,

$$\int_M \tau(X)f \cdot \phi\omega = \{(d/dt) \int_M f^{\exp tX} \cdot \phi \cdot \omega\}_{t=0}$$

$$= \{(d/dt) \int_M f \phi^{\exp(-tX)} \cdot \omega\}_{t=0} = -\int_M f \cdot \tau(X)\phi \cdot \omega.$$

This proves that $\tau(X)^* = -\tau(X)$. Moreover it is easy to verify that

$$t^{-1}(f^{\exp tX} - f) \to \tau(X)f$$

in $C_c^\infty(M)$ as $t \to 0$. From this the assertion about an invariant distribution follows immediately. Let D be an invariant differential operator. Then if $\exp tX = x(t)$, we have

$$(Df)^{x(t)} = Df^{x(t)}.$$

Since D defines a continuous endomorphism of $C_c^\infty(M)$, it is clear that

$$\tau(X)(Df) = \lim_{t \to 0} t^{-1}\{(Df)^{x(t)} - Df\}$$
$$= D(\tau(X)f).$$

This proves that D is locally invariant.

We now consider an example. Take $M = G$ and define $x^g = gx$ $(g, x \in G)$ and let dx denote the left-invariant Haar measure on G and $\omega > 0$ the corresponding left-invariant differential form. We write $X' = \tau(X)$ $(X \in \mathfrak{g})$ in this case so that

$$f(x;X') = -\{df(\exp tX \cdot x)/dt\}_{t=0} \qquad (f \in C^\infty(G), x \in G).$$

LEMMA 7. *Let T be a distribution on an open connected subset U of G. Suppose $X'T = 0$ for all $X \in \mathfrak{g}$. Then T is a constant.*

Select a base $X_1, \cdots, X_n$ for $\mathfrak{g}$ over $\boldsymbol{R}$ and put $D = X'^2_1 + \cdots + X'^2_n$. Then obviously D is an elliptic differential operator on G. Therefore since $DT = 0$, we can conclude that there exists a C^∞ function F on U such that $T = F$. But then $X'F = 0$ for all X in $\mathfrak{g}$. Since U is connected, this implies that F is a constant.

6. Some notations and elementary facts. Let E be a vector space of finite dimension over an infinite field k and $P(E)$ the algebra of all polynomial functions from E to k. Let E' be the space of all linear functions on E. Define $S(E) = P(E')$. We call $S(E)$ and $P(E)$ respectively the symmetric and polynomial algebras over E. Since E is the dual of E', $E \subset S(E)$. Let F be another finite-dimensional vector space over k and $i: F \to E$ a linear mapping. Then i can be extended uniquely to a homomorphism $S(F) \to S(E)$ which we still denote by i. If i is injective on F then it is also injective on $S(F)$. In particular if $F \subset E$ and i is the inclusion mapping, we can identify $S(F)$ under i with the subalgebra of $S(E)$ generated by 1 and F.

3

We denote by $P_d(E)$ the space of all homogeneous polynomials on E of degree d. Similarly $S_d(E) = P_d(E')$. If $F \subset E$, it is clear that $S_d(F) = S(F) \cap S_d(E)$.

Now suppose E is the direct sum of two subspaces E_1, E_2. Let α_i denote the projection of E on E_i $(i = 1, 2)$ corresponding to the sum $E = E_1 + E_2$. Then given any polynomial function p on E_i, we can extend it to a polynomial function on E by setting $p(X) = p(\alpha_i X)$ $(X \in E)$. This defines an isomorphism of $P(E_i)$ into $P(E)$ and we can identify $P(E_i)$ with its image by means of it. In this way $P(E_1)$ and $P(E_2)$ become subalgebras of E and it is obvious that the mapping $(p_1, p_2) \to p_1 p_2$ $(p_i \in P(E_i))$ defines an isomorphism of $P(E_1) \otimes P(E_2)$ onto $P(E)$.

Let A be an automorphism of E. Then A can be extended uniquely to an automorphism $q \to q^A$ of $S(E)$. Moreover, for any $p \in P(E)$, define p^A to be the function $X \to p(A^{-1}X)$ $(X \in E)$. Then $p \to p^A$ is an automorphism of $P(E)$.

7. The radial component.[3] Assume k has characteristic zero and let $\mathfrak{g}$ be a Lie algebra over k. For any $X \in \mathfrak{g}$ let L_X denote the endomorphism of $S(\mathfrak{g})$ corresponding to multiplication by X in $S(\mathfrak{g})$. Also let d_X denote the unique derivation of $S(\mathfrak{g})$ which coincides on $\mathfrak{g}$ with $\mathrm{ad}\,X$. It is immediate that $d_X L_Y - L_Y d_X = L_{[X, Y]}$. For $Y \in \mathfrak{g}$, put [4]

$$\sigma_Y(X) = L_{[X, Y]} + d_X \qquad\qquad (X \in \mathfrak{g}).$$

One checks without difficulty that σ_Y is a representation of $\mathfrak{g}$ on $S(\mathfrak{g})$. Let $\mathfrak{G}$ be the universal enveloping algebra of $\mathfrak{g}$. We can extend σ_Y (uniquely) to a representation of $\mathfrak{G}$ which we again denote by σ_Y. Define a linear mapping Γ_Y of $\mathfrak{G} \otimes S(\mathfrak{g})$ into $S(\mathfrak{g})$, given by

$$\Gamma_Y(g \otimes p) = \sigma_Y(g) p \qquad\qquad (g \in \mathfrak{G}, p \in S(\mathfrak{g})).$$

Let λ denote the canonical mapping of $S(\mathfrak{g})$ onto $\mathfrak{G}$ (see [4(a), p. 192]). For any linear subspace U of $\mathfrak{g}$, define $\mathfrak{S}_d(U) = \lambda(S_d(U), \mathfrak{S}(U)) = \lambda(S(U))$ and $\mathfrak{S}_+(U) = \sum_{d \geq 1} \mathfrak{S}_d(U)$. It is obvious that for a fixed $Y \in \mathfrak{g}$, Γ_Y defines a linear mapping of $\mathfrak{S}_{d_1}(\mathfrak{g}) \otimes S_{d_2}(\mathfrak{g})$ into $\sum_{0 \leq d \leq d_1 + d_2} S_d(\mathfrak{g})$.

Fix an element $X_0 \in \mathfrak{g}$ and let $\mathfrak{z}_{X_0}$ denote the centralizer of X_0 in $\mathfrak{g}$. Put $\mathfrak{g}_{X_0} = [X_0, \mathfrak{g}]$ and choose linear subspaces U and V in $\mathfrak{g}$ such that

$$\mathfrak{g} = U + \mathfrak{g}_{X_0} = V + \mathfrak{z}_{X_0}$$

---

[3] Cf. [4(d), §5], [4(c), §3] and [1, §2].

[4] Cf. Lemma 3 of [4(c)].

where both sums are direct. For any $u \in U$, consider the linear mapping

$$\phi_u : (w, v) \to w + [X_0 + u, v] \qquad (w \in U, v \in V)$$

of $U \times V$ into $\mathfrak{g}$. It follows from the definition of U and V that ϕ_0 is bijective. Let $\mathcal{E}(\mathfrak{g})$ be the space of all endomorphisms of $\mathfrak{g}$ and I the identity mapping of $\mathfrak{g}$. Since

$$\phi_u(w, v) = \phi_0(w, v) + [u, v] \qquad (u, w \in U; v \in V),$$

it is clear that

$$u \to \phi_u \phi_0^{-1} - I$$

is a linear mapping of U into $\mathcal{E}(\mathfrak{g})$. Put

$$q(u) = \det(\phi_u \phi_0^{-1}) \qquad (u \in U).$$

Then $q \in P(U)$ and $q(0) = 1$.

LEMMA 8.[5] *Let u be an element in U such that $q(u) \neq 0$. Then Γ_{X_0+u} defines a bijective linear mapping of $\mathfrak{S}(V) \otimes S(U)$ onto $S(\mathfrak{g})$.*

We prove by induction on $d \geq 0$ that

$$\sum_{d_1+d_2 \leq d} \Gamma_{X_0+u}(\mathfrak{S}_{d_1}(V) \otimes S_{d_2}(U)) = \sum_{r \leq d} S_r(\mathfrak{g}).$$

If $d = 0$, this is obvious. So assume $d \geq 1$. Obviously the left side is contained in the right side. Hence in view of our induction hypothesis, it is enough to prove that

$$\sum_{d_1+d_2 \leq d} \Gamma_{X_0+u}(\mathfrak{S}_{d_1}(V) \otimes S_{d_2}(U)) + \mathscr{S}_{d-1} \supset S_d(\mathfrak{g})$$

where $\mathscr{S}_r = \sum_{s \leq r} S_s(\mathfrak{g})$. Now $q(u) \neq 0$ and therefore

$$\mathfrak{g} = \phi_u \phi_0^{-1} \mathfrak{g} = U + [X_0 + u, V].$$

Hence $S_d(\mathfrak{g})$ is spanned by elements α of the form

$$\alpha = \prod_{1 \leq i \leq d_1} w_i \prod_{1 \leq j \leq d_2} [X_0 + u, v_j]$$

where $w_i \in U$, $v_j \in V$ and $d_1 + d_2 = d$. Put $g = (-1)^{d_1} \lambda(v_1 v_2 \cdots v_{d_2}) \in \mathfrak{S}_{d_2}(V)$ and $p = w_1 w_2 \cdots w_{d_1} \in S_{d_1}(U)$. Then

$$\Gamma_{X_0+u}(g \otimes p) = \sigma_{X_0+u}(g) p \equiv \alpha \bmod \mathscr{S}_{d-1}.$$

This proves our assertion.

---

[5] Cf. Lemma 4 of [4(c)].

Now since ϕ_0 is bijective, $\dim \mathfrak{g} = \dim U + \dim V$. Therefore the spaces $\mathscr{S}_d$ and $\sum\limits_{d_1+d_2 \leqq d} \mathfrak{S}_{d_1}(V) \otimes S_{d_2}(U)$ have the same dimension. This proves that Γ_{X_0+u} is injective.

Let A and B be two vector spaces over k and suppose $\dim A < \infty$. Then $p: A \to B$ is called a polynomial mapping if the subspace B_1 of B spanned by $p(A)$, has finite dimension and p, viewed as a mapping of A into B_1, is a polynomial mapping in the usual sense.

LEMMA 9.[6] *Fix an element $p \in S(\mathfrak{g})$. Then there exists an integer $r \geqq 0$ and a polynomial mapping $\gamma_p: U \to \mathfrak{S}(V) \otimes S(U)$ such that*

$$\Gamma_{X_0+u}(\gamma_p(u)) = q(u)^r p$$

for $u \in U$.

Fix an integer $d \geqq 0$ and let

$$W_d = \sum_{d_1+d_2 \leqq d} \mathfrak{S}_{d_1}(V) \otimes S_{d_2}(U)$$

and let $\Gamma(u)$ denote the restriction Γ_{X_0+u} on W_d. We have seen that $\dim W_d = \dim \mathscr{S}_d$. Fix a bijective linear mapping C of $\mathscr{S}_d$ onto W_d and put $A(u) = C\Gamma(u)$. Then it is obvious that $u \to A(u)$ is a polynomial mapping of U into the space $\mathcal{E}(W_d)$ of all endomorphisms of W_d. Let $Q(u) = \det A(u)$. Then it follows from the proof of Lemma 8 that $Q(u) = 0$ implies $q(u) = 0$ $(u \in U)$. Let $\bar{k}$ be the algebraic closure of k, $\bar{\mathfrak{g}} = \mathfrak{g} \otimes_k \bar{k}$ and $\bar{U}$, $\bar{V}$ the subspaces of $\bar{\mathfrak{g}}$ spanned by U and V respectively. We can regard Q and q as polynomial functions on $\bar{U}$. By applying the above reasoning to $\bar{\mathfrak{g}}$, $\bar{U}$, $\bar{V}$ instead of $\mathfrak{g}$, U, V, we deduce that $Q(\bar{u}) \neq 0$ unless $q(\bar{u}) = 0$ $(\bar{u} \in \bar{U})$. Since $\bar{k}$ is algebraically closed, this implies that Q divides q^r in $P(\bar{U})$ (and therefore also in $P(U)$) for some integer $r \geqq 0$.

Now let $N = \dim W_d$ and t an indeterminate. Then

$$\det(t - A(u)) = (-1)^N Q(u) + t \sum_{1 \leqq j \leqq N} Q_j(u) t^{j-1} \qquad (u \in U)$$

where $Q_j \in P(U)$. Put

$$B(u) = (-1)^{N+1} \sum_{1 \leqq j \leqq N} Q_j(u) A(u)^{j-1}.$$

Then B is a polynomial mapping of U into $\mathcal{E}(W_d)$ and

$$A(u)B(u) = B(u)A(u) = Q(u)I$$

---

[6] Cf. Lemma 5 of [4(c)].

where I is the identity mapping of W_d.

We can assume that d is so large that $p \in \mathscr{S}_d$. Then

$$\Gamma_{X_0+u}(B(u)Cp) = C^{-1}A(u)B(u)Cp = Q(u)p.$$

But $q^r = Qq_1$ $(q_1 \in P(U))$, as we saw above. Put

$$\gamma_p(u) = q_1(u)B(u)Cp.$$

Then

$$\Gamma_{X_0+u}(\gamma_p(u)) = q(u)^r p \qquad\qquad (u \in U).$$

Let U' be the set of all points $u \in U$ where $q(u) \neq 0$. Then $0 \in U'$ since $q(0) = 1$. For any $p \in S(\mathfrak{g})$, let $\beta_p(u)$ denote the unique element in $\mathfrak{S}(V) \otimes S(U)$ such that $p = \Gamma_{X_0+u}(\beta_p(u))$ $(u \in U')$. Moreover let $\alpha_u(p)$ denote the unique element in $S(U)$ such that

$$\beta_p(u) - (1 \otimes \alpha_u(p)) \in \mathfrak{S}_+(V) \otimes S(U) \qquad\qquad (u \in U').$$

We shall call $\alpha_u(p)$ the **radial component**[3] of p at u (with respect to X_0, U, V).

Remark. It is obvious from the proof of Lemma 8 that if $p \in S_d(\mathfrak{g})$ then $\alpha_u(p) \in \sum_{0 \leqq r \leqq d} S_r(U)$ for $u \in U'$.

8. The mapping Δ . Let $\mathfrak{g}$ be a Lie algebra over $\boldsymbol{R}$ and fix an element X_0 in $\mathfrak{g}$. Select U and V as in §7 so that $\mathfrak{g} = U + \mathfrak{g}x_0 = V + \mathfrak{z}x_0$. Let $\mathfrak{g}_c$ denote the complexification of $\mathfrak{g}$ and U_c, V_c the subspaces of $\mathfrak{g}_c$ spanned by U, V respectively. Now we take $k = \boldsymbol{C}$ in §7 and denote by $\mathfrak{G}$ the universal enveloping algebra of $\mathfrak{g}_c$. As before, U' is the set of all points $u \in U$ where $q(u) \neq 0$ (see Lemma 8).

$\mathfrak{g}$, being a real Euclidean space, is a manifold. For any $p \in S(\mathfrak{g}_c)$, we define, as usual (see [4(c), §2] the differential operator $\partial(p)$ on $\mathfrak{g}$. Let D be a differential operator on an open subset Ω of $\mathfrak{g}$. Then we define a differential operator $\Delta(D)$ on the open subset $U' \cap (\Omega - X_0)$ of the Euclidean space U as follows. As usual let D_Y denote the local expression of D at $Y \in \Omega$ (see [4(c), p. 90]). Fix $u \in U' \cap (\Omega - X_0)$ and choose $p_u \in S(\mathfrak{g}_c)$ such that $D_{X_0+u} = \partial(p_u)$. Then

$$\Delta_u(D) = \partial(\alpha_u(p_u)).$$

Here $\Delta_u(D)$ is the local expression of $\Delta(D)$ at u and $\alpha_u(p_u)$ is the radial component of p_u at u (see §7). Lemma 9 ensures that this does, indeed, define a differential operator on $U' \cap (\Omega - X_0)$.

For any $X \in \mathfrak{g}$, define a vector field $\tau(X)$ (cf. §5) on $\mathfrak{g}$ by

$$f(Y;\tau(X)) = -\{df(\exp(\operatorname{tad} X)Y)/dt\}_{t=0}$$
$$= -f(Y;\partial([X,Y])) \quad (f \in C^\infty(\mathfrak{g}), Y \in \mathfrak{g}).$$

It is easy to verify that $[\tau(X),\tau(Y)] = \tau([X,Y])$ for $X, Y \in \mathfrak{g}$. Hence τ can be extended uniquely to a homomorphism of $\mathfrak{G}$ into the algebra of all differential operators on $\mathfrak{g}$.

A C^∞ function f on an open subset Ω of $\mathfrak{g}$ is called locally invariant if $\tau(X)f = 0$ for all $X \in \mathfrak{g}$. The significance of $\Delta(D)$ is given by the following result (cf. [4(c), Lemma 6]).

LEMMA 10. *Fix a point $u_0 \in U'$ and let f be a locally invariant C^∞ function on an open neighborhood Ω of $X_0 + u_0$ in $\mathfrak{g}$. Then if D is any differential operator on Ω,*

$$f(X_0 + u_0;D) = f(X_0 + u_0;\Delta(D)).$$

Choose $p \in S(\mathfrak{g}_c)$ such that $D_{X_0+u_0} = \partial(p)$. Then $\Delta_{u_0}(D) = \partial(\alpha_{u_0}(p))$. Hence it is enough to prove that

$$f(X_0 + u_0;\partial(p)) = f(X_0 + u_0;\partial(\alpha_{u_0}(p)))$$

for $p \in S(\mathfrak{g}_c)$.

Let G be a Lie group with Lie algebra $\mathfrak{g}$. Then G operates on $\mathfrak{g}$ as follows: $X^x = \operatorname{Ad}(x)X$ $(x \in G, X \in \mathfrak{g})$. We sometimes also write xX for X^x. As usual (see [4(d), §4]), we regard elements of $\mathfrak{G}$ as left-invariant differential operators on G. Select open neighborhoods G_0 and Ω_0 of 1 and $X_0 + u_0$ in G and Ω respectively, such that $\Omega_0{}^{G_0} \subset \Omega$. Put $F(x:X) = F(xX)$ $(x \in G_0, X \in \Omega_0)$ for any $F \in C^\infty(\Omega)$. Then we know [4(c), Lemma 3] that

$$F(x;g:X;\partial(p)) = F(x:X;\partial(\sigma_X(g)p)) = F(x:X;\partial(\Gamma_X(g \otimes p)))$$

for $g \in \mathfrak{G}$, $p \in S(\mathfrak{g}_c)$.

The mapping $X \to -X$ of $\mathfrak{g}_c$ can be extended uniquely to an anti-automorphism of $\mathfrak{G}$ which we denote by $*$.

LEMMA 11. *Let $F \in C^\infty(\Omega)$. Then*

$$F(x;g:X) = F(x:X;\tau(g^*))$$

for $x \in G_0$, $X \in \Omega_0$ and $y \in \mathfrak{G}$.

Let $Y_1, \cdots, Y_r \in \mathfrak{g}$. Then

$$\{\partial^r F(x \exp t_1 Y_1 \cdots \exp t_r Y_r : X)/\partial t_1 \cdots \partial t_r\}_{t_1 = \cdots = t_r = 0}$$
$$= \{\partial^r F(x:(\exp t_1 Y_1 \cdots \exp t_r Y_r)X)/\partial t_1 \cdots \partial t_r\}_{t_1 = \cdots = t_r = 0}$$

for $x \in G_0$ and $X \in \Omega_0$. Lemma 11 follows immediately from this fact.

Now we return to Lemma 10. Fix $p \in S(\mathfrak{g}_c)$. Then, from Lemma 9, we can choose $g_i \in \mathfrak{S}_+(V_c)$, $p_j \in S(U_c)$, $q_{ij} \in P(U_c)$ $(1 \leq i, j \leq N)$ and an integer $r \geq 0$ such that

$$p = \alpha_u(p) + q(u)^{-r} \sum_{1 \leq i, j \leq N} q_{ij}(u) \Gamma_{X_0+u}(g_i \otimes p_j)$$

for $u \in U'$. Therefore

$$f(X_0 + u; \partial(p)) = f(X_0 + u; \partial(\alpha_u(p)))$$
$$+ q(u)^{-r} \sum_{i,j} q_{ij}(u) f(1; g_i : X_0 + u; \partial(p_j))$$

for $u \in U' \cap (\Omega - X_0)$. But g_i and therefore also $g_i{}^*$, are in $\mathfrak{S}_+(V_c)$. Hence

$$f(1; g_i : X_0 + u) = f(X_0 + u; \tau(g_i{}^*)) = 0$$

as f is locally invariant. Therefore

$$f(1; g_i : X_0 + u; \partial(p_j)) = 0 \qquad\qquad (1 \leq i, j \leq N)$$

and so

$$f(X_0 + u; \partial(p)) = f(X_0 + u; \partial(\alpha_u(p)))$$

for $u \in U' \cap (\Omega - X_0)$. This proves Lemma 10.

9. Application to invariant distributions. From now on we shall assume that G is unimodular. Then $\mathfrak{g}$ is also unimodular, i.e. $\operatorname{tr} \operatorname{ad} X = 0$ for $X \in \mathfrak{g}$. Let dX denote the Euclidean measure on $\mathfrak{g}$ and dx the Haar measure on G (both normalized in some fixed but arbitrary way). Then dx is both left- and right-invariant and dX is invariant under G. Fix an orientation of $\mathfrak{g}$ and let $\omega_{\mathfrak{g}}$ be the positive differential form on $\mathfrak{g}$ of degree $n = \dim \mathfrak{g}$, such that [2] $\omega_{\mathfrak{g}} \sim dX$. Similarly let $\omega_G \neq 0$ be a real and left-invariant differential form on G of degree n. We can orient G in such a way that $\omega_G > 0$. Moreover we may assume that $\omega_G \sim dx$.

Now take $M = \mathfrak{g}$ and $\omega = \omega_{\mathfrak{g}}$ in the set up of §5. Then the definitions of $\tau(X)$ $(X \in \mathfrak{g})$ of Sections 5 and 8 actually coincide and we conclude from Lemma 6 that $\tau(g^*) = \tau(g)^*$ $(g \in \mathfrak{S})$. Moreover g^* is the adjoint of the differential operator g on G, with respect to ω_G.

LEMMA 12. *Let π denote the mapping $(x, u) \to x(X_0 + u)$ of $G \times U$ into $\mathfrak{g}$. Then $\operatorname{rank} d\pi = n$ everywhere on $G \times U'$.*

Fix a point (x, u) in $G \times U'$. Since the tangent space at any point of a Euclidean space E can be canonically identified with E, the tangent space of $G \times U$ at (x, u) is $\mathfrak{g} \times U$. Similarly the tangent space of $\mathfrak{g}$ at $Y = x(X_0 + u)$ is $\mathfrak{g}$. Let $f \in C^\infty(\mathfrak{g})$. Then

$$\{(d/dt)f(\pi(x \, \exp tX, u))\}_{t=0} = f(Y; \partial([X, X_0 + u]^x))$$

for $X \in \mathfrak{g}$ and

$$\{(d/dt)f(\pi(x, u + tw))\}_{t=0} = f(Y; \partial(w^x))$$

for $w \in U$. Hence

$$(d\pi)_{x,u}(X, w) = ([X, X_0 + u] + w)^x$$

and therefore if $v \in V$, $w \in U$, it is clear that

$$(d\pi)_{x,u}(v, w) = (\phi_u(w, -v))^x$$

in the notation of §7. But since $u \in U'$, ϕ_u is a bijective mapping of $U \times V$ onto $\mathfrak{g}$. This proves that $(d\pi)_{x,u}$ maps $V \times U$ onto $\mathfrak{g}^x = \mathfrak{g}$ and therefore rank $(d\pi)_{x,u} = n$.

Fix an orientation of U and let $\omega_U > 0$ denote a differential form on U such that $\omega_U \sim du$, where du is a Euclidean measure on U. Now take $M = G \times U'$, $N = \pi(G \times U')$, $\omega_M \sim dx \, du$, $\omega_N \sim dX$. Then the following result is an immediate consequence of Theorem 1 and Lemma 12.

LEMMA 13. *Put $\Omega = \pi(G \times U') = (X_0 + U')^G$. Then Ω is open in $\mathfrak{g}$ and for every $\alpha \in C_c^\infty(G \times U')$, there exists a unique element $f_\alpha \in C_c^\infty(\Omega)$ such that*

$$\int_{G \times U'} F(x(X_0 + u))\alpha(x : u) \, dx \, du = \int_\Omega F(X)f_\alpha(X) \, dX$$

for every $F \in C_c^\infty(\Omega)$. The mapping $\alpha \to f_\alpha$ of $C_c^\infty(G \times U')$ into $C_c^\infty(\Omega)$ is continuous and surjective and $\operatorname{Supp} f_\alpha \subset \pi(\operatorname{Supp} \alpha)$.

COROLLARY 1. *Let F be a locally summable function on Ω. Then the function $(x, u) \to F(x(X_0 + u))$ is locally summable on $G \times U'$ and*

$$\int_{G \times U'} F(x(X_0 + u))\alpha(x : u) \, dx \, du = \int_\Omega F(X)f_\alpha(X) \, dX$$

for every $\alpha \in C_c^\infty(G \times U')$.

This follows from Corollary 2 of Theorem 1.

COROLLARY 2. *Let Ω_0 be an open subset of Ω and T a distribution on Ω_0. Select open subsets G_0, U_0 of G and U' respectively such that $\pi(G_0 \times U_0) \subset \Omega_0$. Then the mapping*

$$\sigma_T': \alpha \to T(f_\alpha) \qquad\qquad (\alpha \in C_o^\infty(G_0 \times U_0))$$

is a distribution on $G_0 \times U_0$. *Moreover* $\sigma_T' = 0$ *implies that* $T = 0$ *on* $\pi(G_0 \times U_0)$.

This follows by taking $M = G_0 \times U_0$ and $N = \pi(G_0 \times U_0)$ in Lemma 5.

For any $X \in \mathfrak{g}$, define the vector field X' on G as in Lemma 7. We denote the differential operator X' on G as in Lemma 7. We denote the differential operator $X' \times 1$ on $G \times U$ also by X'.

LEMMA 14. $f_{X'\alpha} = \tau(X)f_\alpha$ *for* $X \in \mathfrak{g}$ *and* $\alpha \in C_o^\infty(G \times U')$.

Fix α. Then for any $F \in C_o^\infty(\Omega)$,

$$\int F(x(X_0 + u))\alpha(yx : u)\,dx\,du$$
$$= \int F(y^{-1}x(X_0 + u))\alpha(x : u)\,dx\,du$$
$$= \int F^y(x(X_0 + u))\alpha(x : u)\,dx\,du$$
$$= \int F^y(X)f_\alpha(X)\,dX$$
$$= \int F(X)f_\alpha{}^{y^{-1}}(X)\,dX$$

if y is sufficiently near 1 in G. Now for a given $Y \in \mathfrak{g}$, put $y_t = \exp tY$ ($t \in \mathbf{R}$). Then it is clear that

$$-\int F(x(X_0 + u))\alpha(x ; Y' : u)\,dx\,du$$
$$= \{(d/dt)\int F(x(X_0 + u))\alpha(y_t x : u)\,dx\,du\}_{t=0}$$
$$= \{(d/dt)\int F(X)f_\alpha{}^{y_t^{-1}}(X)\,dX\}_{t=0}$$
$$= -\int F(X)f_\alpha(X ; \tau(Y))\,dX.$$

This proves that $f_{Y'\alpha} = \tau(Y)f_\alpha$.

THEOREM 2. *Let* G_0 *and* U_0 *be open subsets of* G *and* U' *respectively. Assume that* G_0 *is connected and not empty. Then if* T *is a locally invariant distribution on* $\Omega_0 = \pi(G_0 \times U_0)$, *there exists a unique distribution* σ_T *on* U_0 *such that*

$$T(f_\alpha) = \sigma_T(\beta_\alpha) \qquad\qquad (\alpha \in C_o^\infty(G_0 \times U_0))$$

where β_α *is the function in* $C_o^\infty(U_0)$ *given by*

$$\beta_\alpha(u) = \int \alpha(x : u)\,dx \qquad\qquad (u \in U_0).$$

Moreover $\sigma_T = 0$ *implies that* $T = 0$.

Define the distribution σ_T' on $G_0 \times U_0$ as in Corollary 2 of Lemma 13. Fix $X \in \mathfrak{g}$. Then, as we have seen above, X' is a vector-field on $G \times U$.

We claim $X'\sigma_T' = 0$. Since dx is left-invariant, $(X')^* = -X'$ and therefore we have to prove that $\sigma_T'(X'\alpha) = 0$ for $\alpha \in C_c^\infty (G_0 \times U_0)$. Fix α. Then

$$\sigma_T'(X'\alpha) = T(f_{X'\alpha}) = T(\tau(X)f_\alpha) = 0$$

from Lemma 6, since T is locally invariant.

Fix an element $\beta \in C_c^\infty (U_0)$ and consider the mapping

$$S_\beta : \gamma \to \sigma_T'(\gamma \times \beta) \qquad\qquad (\gamma \in C_c^\infty (G_0)).$$

Then S_β is a distribution on G_0 and it is obvious from the above result that $X'S_\beta = 0$ $(X \in \mathfrak{g})$. Therefore from Lemma 7, there exists a complex number $C(\beta)$ such that

$$S_\beta(\gamma) = C(\beta) \int \gamma \, dx \qquad\qquad (\gamma \in C_c^\infty (G)).$$

Now fix $\gamma_0 \in C_c^\infty (G_0)$ such that $\int \gamma_0 \, dx = 1$. Then

$$C(\beta) = S_\beta(\gamma_0) = \sigma_T'(\gamma_0 \times \beta).$$

The mapping $\beta \to \sigma_T'(\gamma_0 \times \beta)$ $(\beta \in C_c^\infty (U_0))$ is obviously a distribution on U_0, which we call σ_T. Then

$$\sigma_T'(\gamma \times \beta) = \sigma_T(\beta) \int \gamma \, dx \qquad\qquad (\gamma \in C_c^\infty (G_0), \beta \in C_c^\infty (U_0)).$$

Now for any $\alpha \in C_c^\infty (G_0 \times U_0)$ define β_α as in the statement of Theorem 2. Then the mapping $\sigma : \alpha \to \sigma_T(\beta_\alpha)$ is obviously a distribution on $G_0 \times U_0$ and $\sigma_T'(\gamma \times \beta) = \sigma(\gamma \times \beta)$ for $\beta \in C_c^\infty (U_0)$, $\gamma \in C_c^\infty (G_0)$. Hence we conclude from Lemma 3 that $\sigma_T' = \sigma$. This proves that

$$T(f_\alpha) = \sigma_T'(\alpha) = \sigma_T(\beta_\alpha) \qquad\qquad (\alpha \in C_c^\infty (G_0 \times U_0)).$$

Moreover for a given $\beta \in C_c^\infty (U_0)$, set $\alpha = \gamma_0 \times \beta$. Then $\beta = \beta_\alpha$. This shows that the mapping $\alpha \to \beta_\alpha$ of $C_c^\infty (G_0 \times U_0)$ into $C_c^\infty (U_0)$ is surjective. Hence the uniqueness of σ_T in Theorem 2 is obvious. Finally suppose $\sigma_T = 0$. Then $\sigma_T'(\alpha) = \sigma_T(\beta_\alpha) = 0$ $(\alpha \in C_c^\infty, G_0 \times U_0))$ and therefore $\sigma_T' = 0$. But, as we have seen (Corollary 2 of Lemma 13), this implies that $T = 0$.

We keep to the notation of Theorem 2.

THEOREM 3.[7]   *Assume further that* $1 \in G_0$. *Then if* D *is a differential operator and* T *a distribution on* Ω_0 *and they are both locally invariant, we have* $\sigma_{DT} = \Delta(D)\sigma_T$.

We observe that DT is also locally invariant and therefore σ_{DT} is defined. Moreover $U_0 \subset \Omega_0 - X_0$ and therefore $\Delta(D)$ is defined on U_0. Theorems 2

---

[7] This can be considered as a generalization of Lemma 10 to distributions.

and, 3 provide the main justification for the consideration of the differential operator $\Delta(D)$.

COROLLARY. *Let D_1, D_2 be two differential operators and T a distribution on Ω_0. Suppose all three of them are locally invariant. Then*

$$\Delta(D_1 D_2)\sigma_T = \Delta(D_1)\Delta(D_2)\sigma_T.$$

This an immediate consequence of Theorem 3.

The proof of Theorem 3 requires some preparation.

LEMMA 15. *Let D be a locally invariant differential operator on an open set Ω in $\mathfrak{g}$. Fix an element $X \in \Omega$ and an open connected neighborhood G_0 of 1 in G such that $yX \in \Omega$ for all $y \in G_0$. Then*

$$D_{yX} = (D_X)^y \qquad\qquad (y \in G_0).$$

For any $x \in G$, let $p \to p^x$ $(p \in S(\mathfrak{g}_c))$ denote the unique automorphism of $S(\mathfrak{g}_c)$ which coincides with $\mathrm{Ad}(x)$ on $\mathfrak{g}_c$. In this way G operates on $S(\mathfrak{g}_c)$ and it is obvious that for any $d \geq 0$, the space $S_d(\mathfrak{g}_c)$ is stable under G. Hence for a given p in $S(\mathfrak{g}_c)$, the elements p^x $(x \in G)$ span a finite-dimensional space V. The corresponding representation on V being obviously continuous, it follows, from the theory of Lie groups, that it is analytic. It is easy to verify that $\{dp^{\exp tY}/dt\}_{t=0} = d_Y p$ $(Y \in \mathfrak{g})$. Moreover it is clear that $\partial(Y^x) = (\partial(Y))^x$ $(Y \in \mathfrak{g}, x \in G)$ and therefore $\partial(p^x) = (\partial(p))^x$ $(p \in S(\mathfrak{g}_c))$.

We can choose $p_i \in S(\mathfrak{g}_c)$ and $a_i \in C^\infty(\Omega)$ $(1 \leq i \leq r)$ such that

$$D = \sum_{1 \leq i \leq r} a_i \partial(p_i).$$

Moreover we may obviously assume that $p_1, \cdots, p_r$ are linearly independent over C and the subspace of $S(\mathfrak{g}_c)$ spanned by them is stable under G. Let $p_i^x = \sum_j p_j c_{ji}(x)$ $(1 \leq i \leq r, x \in G)$ where c_{ji} are analytic functions on G. Let $C(x)$ denote the matrix $(c_{ij}(x))_{1 \leq i,j \leq r}$. Then $x \to C(x)$ is a matrix representation of G. Let $Y \to C(Y)$ $(Y \in \mathfrak{g})$ denote the corresponding representation of $\mathfrak{g}$ and $c_{ij}(Y)$ $(1 \leq i, j \leq r)$ the coefficients of the matrix $C(Y)$.

Put $g_i(y) = \sum_j c_{ij}(y^{-1}) a_j(yX)$ $(y \in G_0)$. Then if $Y \in \mathfrak{g}$,

$$-g_i(y; Y') = \{dg_i(\exp tY \cdot y)/dt\}_{t=0}$$
$$= -\sum_j c_{ij}(y^{-1})\{a_j(yX; \tau(Y)) + \sum_k c_{jk}(Y) a_k(yX)\}.$$

We claim that if $a \in C^\infty(\Omega)$ and $p \in S(\mathfrak{g}_c)$,

$$\tau(Y) \circ a - a \circ \tau(Y) = \tau(Y)a,$$
$$\tau(Y) \circ \partial(p) - \partial(p) \circ \tau(Y) = \partial(d_Y p).$$

The first assertion is obvious since $\tau(Y)$ is a vector field. The second is seen as follows. Let $f \in C^\infty(\mathfrak{g})$. Then

$$\tau(Y)(\partial(p)f) = \{d(\partial(p)f)^{y_t}/dt\}_{t=0}$$

where $y_t = \exp tY$. But $(\partial(p)f)^{y_t} = \partial(p^{y_t})f^{y_t}$ and therefore

$$\tau(Y)(\partial(p)f) = \partial(d_Y p)f + \partial(p)(\tau(Y)f).$$

Now since D is locally invariant, we have

$$0 = \tau(Y) \circ D - D \circ \tau(Y) = \sum_i \{(\tau(Y)a_i)\partial(p_i) + a_i\partial(d_Y p_i)\}.$$

But $d_Y p_i = \sum_j p_j c_{ji}(Y)$. Hence

$$\sum_i \{\tau(Y)a_i + \sum_j c_{ij}(Y)a_j\}\partial(p_i) = 0.$$

From this it is clear that

$$\tau(Y)a_i + \sum_j c_{ij}(Y)a_j = 0$$

But this obviously implies that $g_i(y; Y') = 0$. Since G_0 is connected, this shows that g_i is a constant. Hence

$$g_i(y) = g_i(1) = a_i(X).$$

Therefore

$$a_i(X) = \sum_j c_{ij}(y^{-1})a_j(yX)$$

or

$$a_i(yX) = \sum_j c_{ij}(y)a_j(X) \qquad\qquad (y \in G).$$

But then

$$(D_X)^y = \sum_i a_i(X)\partial(p_i{}^y) = \sum_{j,i} c_{ji}(y)a_i(X)\partial(p_j)$$

$$= \sum_j a_j(yX)\partial(p_j) = D_{yX}.$$

This proves Lemma 15.

Now we come to the proof of Theorem 3. Fix $\alpha \in C_c^\infty(G_0 \times U_0)$ and $F \in C_c^\infty(\Omega_0)$. Then

$$\int FD^*f_\alpha\, dX = \int DF \cdot f_\alpha\, dX = \int F(x(X_0 + u); D)\alpha(x: u)\, dx\, du$$

from the definition of f_α. But

$$F(x(X_0 + u); D) = F(x(X_0 + u); D_{x(X_0 + u)})$$
$$= F(x(X_0 + u); (D_{X_0 + u})^x) \qquad (x \in G_0, u \in U_0)$$

from Lemma 15. However

$$F(x(X_0 + u) ; (D_{X_0+u})^x) = F(x : X_0 + u ; D_{X_0+u})$$

in the notation of Lemma 11. Moreover we know (see the proof of Lemma 10) that there exist elements $g_i \in \mathfrak{S}_+(V_c)$, $p_i \in S(U_c)$ and $a_i \in C^\infty(U_0)$ $(1 \leqq i \leqq r)$ such that

$$D_{X_0+u} = \Delta_u(D) + \sum_{1 \leqq i \leqq r} a_i(u) \partial(\Gamma_{X_0+u}(g_i \otimes p_i)) \qquad (u \in U_0).$$

Hence

$$F(x : X_0 + u ; D_{X_0+u})$$
$$= F(x : X_0 + u ; \Delta(D)) + \sum_i a_i(u) F(x ; g_i : X_0 + u ; \partial(p_i))$$

for $x \in G_0$ and $u \in U_0$. Put

$$\alpha_i(x : u) = \alpha(x ; g_i^* : u ; \partial(p_i)^* \circ a_i), \qquad (1 \leqq i \leqq r)$$
$$\alpha_0(x : u) = \alpha(x : u ; \Delta(D)^*),$$

where, as usual, * denotes adjoint. Then

$$\int F D^* f_\alpha \, dX = \sum_{0 \leqq i \leqq r} \int F(x(X_0 + u)) \alpha_i(x : u) \, dx \, du$$
$$= \sum_{0 \leqq i \leqq r} \int F f_{\alpha_i} \, dX.$$

This shows that

$$D^* f_\alpha = \sum_{0 \leqq i \leqq r} f_{\alpha_i}$$

and therefore

$$T(D^* f_\alpha) = \sum_{0 \leqq i \leqq r} \sigma_T(\beta_{\alpha_i})$$

from Theorem 2. However g_i and therefore also g_i^*, are in $\mathfrak{S}_+(V_c)$. Hence

$$\int \alpha(x ; g_i^* : u) \, dx = 0 \qquad (u \in U_0)$$

from the right-invariance of the measure dx. This shows that $\beta_{\alpha_i} = 0$ for $i \geqq 1$. Moreover it is obvious that $\beta_{\alpha_0} = \Delta(D)^* \beta_\alpha$. Therefore

$$T(D^* f_\alpha) = \sigma_T(\Delta(D)^* \beta_\alpha)$$

and so we can conclude from Theorem 2 that $\sigma_{DT} = \Delta(D) \sigma_T$. This completes the proof of Theorem 3.

10. Statement of Theorem 4. Now we intend to apply the results of §9 to prove a theorem about distributions on a semisimple Lie algebra. So let us assume that $\mathfrak{g}$ is semisimple and G is connected. Identify $\mathfrak{g}_c$ with its dual, by means of the bilinear form $B(X, Y) = \mathrm{tr}(\mathrm{ad}\, X \, \mathrm{ad}\, Y) (X, Y \in \mathfrak{g}_c)$. In view of the invariance of B under G, this identification is compatible with the action of G. Thus elements of $S(\mathfrak{g}_c)$ become polynomial functions on $\mathfrak{g}_c$. Let $I(\mathfrak{g}_c)$ denote the subalgebra of all invariants of G in $S(\mathfrak{g}_c)$. It is easy to see that $I(\mathfrak{g}_c)$ is the set of all $p \in S(\mathfrak{g}_c)$ such that $d_X p = 0$ for all $X \in \mathfrak{g}$. By the Killing form (of $\mathfrak{g}$), we mean the polynomial function $\omega : X \to \operatorname{tr}(\operatorname{ad} X)^2 (X \in \mathfrak{g}_c)$. If θ is any automorphism of $\mathfrak{g}_c$, it is clear that $\operatorname{ad}(\theta X) = \theta(\operatorname{ad} X)\theta^{-1}$ and therefore $\omega(\theta X) = \omega(X) (X \in \mathfrak{g}_c)$. Hence, in particular, $\omega \in I(\mathfrak{g}_c)$.

An element $X \in \mathfrak{g}_c$ is called nilpotent if $\operatorname{ad} X$ is nilpotent. Let $\mathfrak{N}$ denote the set of all nilpotent elements of $\mathfrak{g}$.

THEOREM 4. *Let T be a distribution on an open subset Ω of $\mathfrak{g}$. Assume that*

1) *T is locally invariant,*

2) *$\operatorname{Supp} T \subset \mathfrak{N} \cap \Omega$,*

3) *$\partial(\omega) T = 0$, where ω is the Killing form of $\mathfrak{g}$.*
Then $T = 0$.

The proof of this theorem is rather long and requires considerable preparation. It is obvious that $\mathfrak{N}^x = \mathfrak{N} (x \in G)$. Hence G operates on $\mathfrak{N}$.

LEMMA 16. *$\mathfrak{N}$ is the union of a finite number of G-orbits.*

This follows from a result of Kostant (Corollary 3.7 and Lemma 5.1 of [6]) and Proposition 2.3 of [2].

Let $\mathfrak{N}_q$ denote the union of all G-orbits in $\mathfrak{N}$ of topological dimension $\leqq q$. Also put $\mathfrak{N}_{-1} = \emptyset$ and let $^c\mathfrak{N}_q$ denote the complement of $\mathfrak{N}_q$ in $\mathfrak{N}$.

LEMMA 17. *$\mathfrak{N}_q$ is closed in $\mathfrak{N}$ and if $X_0 \in \mathfrak{N}_q \cap {}^c\mathfrak{N}_{q-1}$, then X_0^G is open in $\mathfrak{N}_q (q \geqq 0)$.*

This is an easy consequence of the theory of algebraic groups. We shall give a proof (which is due to Borel) in § 18.

COROLLARY. *For any $X_0 \in \mathfrak{N}$, the orbit X_0^G is locally compact.*

Let q be the topological dimension of X_0^G. Then X_0^G is open in the closed subset $\mathfrak{N}_q$ of $\mathfrak{N}$. Since $\mathfrak{N}$ is obviously closed in $\mathfrak{g}$, $\mathfrak{N}_q$ is locally compact and our assertion follows.

Fix $X_0 \in \mathfrak{N}$ and let $Z(X_0)$ denote the centralizer of X_0 in G. Then $Z(X_0)$ is a closed subgroup of G. Let $x \to x^*$ denote the natural mapping of G on $G^* = G/Z(X_0)$ and define $x^* X_0 = X_0^{x^*} = xX_0$.

LEMMA 18. *$x^* \to x^* X_0$ is a homeomorphism of G^* on X_0^G.*

This follows from a theorem of Arens (see [7, p. 65]), if we take the corollary of Lemma 17 into account.

Let $\mathfrak{z}_{X_0}$ denote, as before, the centralizer of X_0 in $\mathfrak{g}$.

COROLLARY. *Let q be the topological dimension of $X_0{}^G$. Then*

$$q = \dim G - \dim Z(X_0) = \dim \mathfrak{g} - \dim \mathfrak{z}_{X_0}.$$

This is obvious from Lemma 18.

We observe that in order to prove Theorem 4 it would be sufficient to obtain the following result.

LEMMA 19. *Fix an integer $q \geq 0$ and let T be a locally invariant distribution on an open subset Ω of $\mathfrak{g}$ such that $\partial(\omega)T = 0$ and $\operatorname{Supp} T \subset \mathfrak{N}_q \cap \Omega$. Then $\operatorname{Supp} T \subset \mathfrak{N}_{q-1} \cap \Omega$.*

11. Some elementary facts about distributions. Let E be a finite-dimensional vector space over $\boldsymbol{R}$ and E_c its complexification. Fix a point $X_0 \in E$ and let δ_{X_0} denote the distribution $\delta_{X_0}: f \to f(X_0)$ $(f \in C_c{}^\infty(E))$ on E.

LEMMA 20. *Let σ be a distribution on an open neighborhood U of X_0 in E and suppose $\operatorname{Supp} \sigma \subset \{X_0\}$. Then there exists a unique $p \in S(E_c)$ such that*[8] *$\sigma = \partial(p)\delta_{X_0}$.*

This is a well-known and elementary result (see [9, Theorem 35, p. 99]).

LEMMA 21. *Let σ be a distribution on an open set U in E and X_0 a point in U. Let η be a function in $C^\infty(U)$ such that $\eta(X) = 0$ for every $X \in \operatorname{Supp} \sigma$. Then there exists an integer $k \geq 0$ such that $X_0 \notin \operatorname{Supp}(\eta^k \sigma)$.*

Fix an open neighborhood V of X_0 in U such that V is relatively compact in U. Let $X_1, \cdots, X_n$ be a base for E over $\boldsymbol{R}$ and let F_m denote the set of all monomials $X_1{}^{m_1} \cdots X_n{}^{m_n} \in S(E_c)$ with $m_1 + \cdots + m_n \leq m$. If D is any differential operator on U, we can choose $m \geq 0$ and $a_p \in C^\infty(U)$ $(p \in F_m)$ such that

$$D = \sum_{p \in F_m} a_p \partial(p).$$

Hence if $f \in C_c{}^\infty(V)$, it is clear that

$$|Df| \leq c \max_{p \in F_m} \sup |\partial(p)f|$$

where $c = \sum_{p \in F_m} \sup_V |a_p|$.

---

[8] Here the adjoint is defined with respect to the Euclidean measure on E.

Now, from Lemma 2, we can choose differential operators $D_1, \cdots, D_N$ on U such that

$$| \sigma(f) | \leq \sum_{1 \leq j \leq N} \sup | D_j f | \qquad (f \in C_c^\infty(V)).$$

Hence it is obvious that we can choose an integer $m \geq 0$ and a number $b \geq 0$ such that

$$| \sigma(f) | \leq b \max_{p \in F_m} \sup | \partial(p) f | \qquad (f \in C_c^\infty(V)).$$

First assume that η is a real-valued function. Fix $f \in C_c^\infty(V)$ and for any $\alpha \in C_c^\infty(\boldsymbol{R})$, define

$$\tau_f(\alpha) = \sigma(\alpha(\eta)f)$$

where $\alpha(\eta)$ denotes the function $X \to \alpha(\eta(X))$ on U. Then τ_f is obviously a distribution on $\boldsymbol{R}$. We claim $\operatorname{Supp} \tau_f \subset \{0\}$. For suppose $0 \notin \operatorname{Supp} \alpha$. Then if $X \in \operatorname{Supp}(\alpha(\eta)f)$, it is clear that $\eta(X) \in \operatorname{Supp} \alpha$ and $X \in V$. Hence $\eta(X) \neq 0$ and therefore, in view of our hypothesis, $X \notin \operatorname{Supp} \sigma$. This shows that $\operatorname{Supp}(\alpha(\eta)f) \cap \operatorname{Supp} \sigma = \emptyset$ and therefore $\tau_f(\alpha) = \sigma(\alpha(\eta)f) = 0$. Hence $\operatorname{Supp} \tau_f \subset \{0\}$.

On the other hand

$$| \tau_f(\alpha) | = | \sigma(\alpha(\eta)f) | \leq b \max_{p \in F_m} \sup | \partial(p)(\alpha(\eta)f) |$$
$$\leq c \max_{0 \leq r \leq m} \sup | d^r\alpha/dt^r |$$

where c is a positive number independent of α and t is the coordinate variable on $\boldsymbol{R}$. Therefore (see [9, pp. 99-100]) we can choose complex numbers $c_r(f)$ $(0 \leq r \leq m)$ such that

$$\tau_f(\alpha) = \sum_{0 \leq r \leq m} c_r(f)(d^r\alpha/dt^r)_0 \qquad (\alpha \in C_c^\infty(\boldsymbol{R})),$$

where $(d^r\alpha/dt^r)_0$ is the value of $d^r\alpha/dt^r$ at $t = 0$. Put $\alpha = t^{m+1}\beta$ for any $\beta \in C_c^\infty(\boldsymbol{R})$. Then $(d^r\alpha/dt^r)_0 = 0$ $(0 \leq r \leq m)$ and therefore

$$0 = \sigma(\alpha(\eta)f) = \sigma(\eta^{m+1}\beta(\eta)f).$$

Since V is relatively compact in U, we can choose a compact set J in $\boldsymbol{R}$ such that $\eta(V) \subset J$. Choose $\beta \in C_c^\infty(\boldsymbol{R})$ such that $\beta = 1$ on J. Then if $X \in V$, $\beta(\eta(X)) = 1$ and therefore $\beta(\eta)f = f$. Hence

$$\sigma(\eta^{m+1}f) = \sigma(\eta^{m+1}\beta(\eta)f) = 0.$$

This proves that $\eta^{m+1}\sigma = 0$ on V.

Now we come to the general case. Let $\eta = \eta_R + \sqrt{-1}\,\eta_I$ where η_R and

η_I are real-valued functions on U. Then by the above result, we can choose an integer $l \geqq 0$ such that $X_0 \notin \operatorname{Supp} \eta_R{}^l \sigma$, $X_0 \notin \operatorname{Supp} \eta_I{}^l \sigma$. Therefore

$$\eta^{2l} \sigma = (\eta_R + \sqrt{-1}\, \eta_I)^{2l} \sigma = 0$$

around X_0. This completes the proof.

12. First step in the proof of Theorem 4. We now return to the proof of Lemma 19. First assume $q = 0$. Then if $X_0 \in \mathcal{N}_0$, it follows from the corollary of Lemma 18 that $\mathfrak{g} = \mathfrak{z}_{X_0}$. Since $\mathfrak{g}$ is semisimple, this implies that $X_0 = 0$. Therefore $\mathcal{N}_0 = \{0\}$ and so if $0 \notin \Omega$, it is clear that $\operatorname{Supp} T = \emptyset$ and hence $T = 0$. So let us assume that $0 \in \Omega$. Then $T = \partial(p)\delta_0$ for some $p \in S(\mathfrak{g}_c)$, by Lemma 20. But $\partial(\omega)T = 0$ and therefore $\partial(\omega p)\delta_0 = 0$. This, however, implies, again by Lemma 20, that $\omega p = 0$. Since $S(\mathfrak{g}_c)$ is obviously an integral domain, we get $p = 0$ and therefore $T = \partial(p)\delta_0 = 0$.

So we can now assume that $q \geqq 1$. Fix an element $X_0 \in \mathcal{N}_q \cap {}^\circ\mathcal{N}_{q-1}$. Then we have to prove that $X_0 \notin \operatorname{Supp} T$. Choose U and V as in §8 and use the notation of §9. Then $0 \in U'$.

LEMMA 22. *We can choose an open neighborhood U_0 of zero in U' with the following property. Let x be an element in G. Then $xX_0 - X_0 \notin U_0$ unless $xX_0 = X_0$.*

I give here an improved version of my original proof, which was suggested by Kneser. It follows from Lemma 18 that $x^* \to x^* X_0$ is a univalent and regular analytic mapping of G^* into $\mathfrak{g}$. Hence $X_0{}^G$ can be regarded as an analytic submanifold of $\mathfrak{g}$ (see [3(a), p. 85]). Its tangent space at X_0 is obviously $\mathfrak{g}_{X_0}$. On the other hand $X_0 + U$ is an affine variety in $\mathfrak{g}$, whose tangent space at X_0 is U. Since $\mathfrak{g}_{X_0} \cap U = \{0\}$, $X_0{}^G$ and $X_0 + U$ are transversal at X_0. Therefore we can choose an open neighborhood G_0^* of 1^* in G^* such that $G_0^* X_0 \cap (X_0 + U) = \{X_0\}$. On the other hand, by Lemma 18, there exists an open neighborhood $\mathfrak{g}_0$ of X_0 in $\mathfrak{g}$ such that $G_0^* X_0 = \mathfrak{g}_0 \cap X_0{}^G$. Now select an open neighborhood U_0 of zero in U' such that $X_0 + U_0 \subset \mathfrak{g}_0$. Then

$$X_0{}^G \cap (X_0 + U_0) \subset G_0^* X_0 \cap (X_0 + U) = \{X_0\}.$$

This proves the lemma.

In order to prove that $X_0 \notin \operatorname{Supp} T$, we may obviously assume that $X_0 \in \Omega$. By Lemma 17, $X_0{}^G$ is open in $\mathcal{N}_q$. Hence we can choose an open neighborhood Ω_0 of X_0 in Ω such that $\Omega_0 \cap \mathcal{N}_q \subset X_0{}^G$. Let G_0 be an open connected neighborhood of 1 in G. We can assume that G_0 and U_0 (as defined in

4

Lemma 22) are so small that $\pi(G_0 \times U_0) \subset \Omega_0$. (Here π has the same meaning as in §9.) Define the distribution σ_T on U_0 as in Theorem 2.

LEMMA 23. $\operatorname{Supp} \sigma_T \subset \{0\}$.

Fix $\beta \in C_o^\infty(U_0)$ such that $0 \notin \operatorname{Supp} \beta$. Then we have to show that $\sigma_T(\beta) = 0$. Choose $\gamma \in C_o^\infty(G_0)$ such that $\int \gamma \, dx = 1$ and put $\alpha = \gamma \times \beta$. Then $\sigma_T(\beta) = T(f_\alpha)$. But we know that $\operatorname{Supp} f_\alpha \subset \pi(\operatorname{Supp} \alpha)$. Hence if $X \in \operatorname{Supp} f_\alpha$, we have $X = x(X_0 + u)$ for some $x \in \operatorname{Supp} \gamma$ and $u \in \operatorname{Supp} \beta$. We claim $X \notin X_0^G$. For otherwise $X = yX_0$ for some $y \in G$. Therefore $X_0 + u = zX_0$ where $z = x^{-1}y$. Hence $u = zX_0 - X_0 \in U_0$. But $u \neq 0$ since $u \in \operatorname{Supp} \beta$. So we get a contradiction with the definition of U_0 (see Lemma 22). This shows that $\operatorname{Supp} f_\alpha \cap X_0^G = \emptyset$. However

$$\operatorname{Supp} T \cap \operatorname{Supp} f_\alpha \subset \mathfrak{N}_q \cap \Omega_0 \subset X_0^G.$$

Therefore $\operatorname{Supp} T \cap \operatorname{Supp} f_\alpha = \emptyset$ and so $T(f_\alpha) = 0$. This proves the lemma.

COROLLARY. *There exists a unique element $z \in S(U_c)$ such that*

$$\sigma_T(\beta) = \beta(0; \partial(z))$$

for all $\beta \in C_o^\infty(U_0)$.

This is an immediate consequence of Lemma 20.

Remark. In order to prove that $X \notin \operatorname{Supp} T$, it would, in view of Theorem 2, be enough to show that $\sigma_T = 0$. Hence it would be sufficient to establish that $z = 0$ in the above corollary.

13. Theorems of Jacobson-Morosow and Mostow. In this section we recall some known facts about semisimple Lie algebras.

LEMMA 24 (Jacobson-Morosow). *Let $\mathfrak{g}$ be a semisimple Lie algebra over a field k of characteristic zero. If $X \neq 0$ is any nilpotent element in $\mathfrak{g}$, we can choose two elements $H, Y \in \mathfrak{g}$ such that*

$$[H, X] = 2X, \quad [H, Y] = -2Y, \quad [X, Y] = H.$$

For a proof see [5, p. 108, Theorem 3].

LEMMA 25. *Let $\mathfrak{l}$ be a three dimensional Lie algebra over k spanned by three elements H, X, Y with the relations*

$$[H, X] = 2X, \quad [H, Y] = -2Y, \quad [X, Y] = H.$$

Then $\mathfrak{l}$ is semisimple. Let ρ be a representation of $\mathfrak{l}$ on a finite-dimensional *vector space V. Then 1) ρ is fully reducible, 2) $\rho(X)$, $\rho(Y)$ are nilpotent, 3) $\rho(H)$ is semisimple and all its eigenvalues are rational integers. Now suppose V is irreducible and $\dim V = m + 1$ $(m \geq 0)$. Then the eigenvalues of $\rho(H)$ are $m - 2r$ $(0 \leq r \leq m)$ and each of them has multiplicity 1. Let V_m be the subspace of V corresponding to the eigenvalue m. Then V_m is also the space of all $v \in V$ such that $\rho(X)v = 0$. Finally ρ is absolutely irreducible and, apart from equivalence, there is exactly one irreducible representation of $\mathfrak{l}$ of degree d for any integer $d \geq 1$.*

All this is quite elementary and well known.

Let $\mathfrak{g}$ be a semisimple Lie algebra over $\mathbf{R}$. A Cartan involution θ of $\mathfrak{g}$ is an isomorphism of $\mathfrak{g}$ such that $\theta^2 = 1$ and the quadratic form $B(X, \theta(X)) = \mathrm{tr}(\mathrm{ad}\, X\, \mathrm{ad}\, \theta X)$ $(X \in \mathfrak{g})$ is negative-definite on $\mathfrak{g}$. The following result is due to Mostow [8, Theorem 6, p. 53].

LEMMA 26 (Mostow). *Let $\mathfrak{g}$ be a semisimple Lie algebra over $\mathbf{R}$ and $\mathfrak{g}_1$ a semisimple subalgebra of $\mathfrak{g}$. Let θ_1 be a Cartan involution of $\mathfrak{g}_1$. Then θ_1 can be extended to a Cartan involution θ of $\mathfrak{g}$.*

14. Some computations. We now return to the notation of § 12. Since $q = \dim \mathfrak{g} - \dim \mathfrak{z}_{X_0} \geq 1$, $X_0 \neq 0$. Therefore by Lemma 24, we can choose $H_0, Y_0 \in \mathfrak{g}$ such that $[H_0, X_0] = 2X_0$, $[H_0, Y_0] = -2Y_0$, $[X_0, Y_0] = H_0$. Let $\mathfrak{l} = \mathbf{R}H_0 + \mathbf{R}X_0 + \mathbf{R}Y_0$. It is easy to see that $\mathfrak{l}$ is a three-dimensional subalgebra of $\mathfrak{g}$ and the linear mapping

$$\theta_0 : (X_0, Y_0, H_0) \to (-Y_0, -X_0, -H_0)$$

defines a Cartan involution of $\mathfrak{l}$. Therefore, from Lemma 26, we can extend it to a Cartan involution θ of $\mathfrak{g}$. Introduce the structure of a real Hilbert space on $\mathfrak{g}$ by means of the positive-definite quadratic form

$$\| Z \|^2 = -B(Z, \theta(Z)) \qquad (Z \in \mathfrak{g}).$$

LEMMA 27. *Let $X \in \mathfrak{g}$. Then the Hilbert-space adjoint of $\mathrm{ad}\, X$ is $-\mathrm{ad}\, \theta(X)$.*

Let $\langle Y, Z \rangle = -B(Y, \theta(Z)) = -B(\theta(Y), Z)$ $(Y, Z \in \mathfrak{g})$ denote the scalar product in the Hilbert space $\mathfrak{g}$. Then

$$\begin{aligned}
\langle [X, Y], Z \rangle &= -B([\theta(X), \theta(Y)], Z) \\
&= B(\theta(Y), [\theta(X), Z]) \\
&= -\langle Y, [\theta(X), Z] \rangle
\end{aligned}$$

and this is equivalent to our assertion.

We can take (in § 12) U and V to be the orthogonal complements of $\mathfrak{g}_{x_0}$ and $\mathfrak{z}_{x_0}$ respectively in $\mathfrak{g}$. Then we have the following result.

LEMMA 28. $U = \theta(\mathfrak{z}_{x_0})$ and $V = \theta(\mathfrak{g}_{x_0})$.

Let $Z \in \mathfrak{z}_{x_0}$ and $Y \in \mathfrak{g}$. Then

$$\langle \theta(Z), [X_0, Y] \rangle = - \langle [\theta(X_0), \theta(Z)], Y \rangle$$

from Lemma 27. But $[\theta(X_0), \theta(Z)] = \theta([X_0, Z]) = 0$. This proves that $\theta(\mathfrak{z}_{x_0})$ is orthogonal to $\mathfrak{g}_{x_0}$. But

$$\dim \theta(\mathfrak{z}_{x_0}) = \dim \mathfrak{z}_{x_0} = \dim \mathfrak{g} - \dim \mathfrak{g}_{x_0} = \dim U.$$

Hence $\theta(\mathfrak{z}_{x_0}) = U$.

Since B is invariant under θ, it is clear that θ is unitary. Therefore the orthogonal complement of $\theta(\mathfrak{g}_{x_0})$ is $\theta^2(\mathfrak{z}_{x_0}) = \mathfrak{z}_{x_0}$. This proves that $V = \theta(\mathfrak{g}_{x_0})$.

Let ρ denote the representation $Z \to \operatorname{ad} Z$ $(Z \in \mathfrak{l})$ of $\mathfrak{l}$ on $\mathfrak{g}$. Since $\theta(\mathfrak{l}) = \mathfrak{l}$, it follows from Lemma 27 that $\rho(\mathfrak{l})$ is a self-adjoint family of linear transformations on $\mathfrak{g}$. Therefore we can write $\mathfrak{g} = \sum\limits_{1 \leq i \leq r} \mathfrak{g}_i$ where $\mathfrak{g}_i \neq \{0\}$ is a subspace of $\mathfrak{g}$ which is invariant and irreducible under ρ and the sum is orthogonal. Furthermore we may assume that $\mathfrak{g}_1 = \mathfrak{l}$.

Note that $\theta(X_0) = - Y_0$. Hence $U = \theta(\mathfrak{z}_{x_0}) = \mathfrak{z}_{Y_0}$ where $\mathfrak{z}_{Y_0}$ is the centralizer of Y_0 in $\mathfrak{g}$. Similarly $V = \theta(\mathfrak{g}_{x_0}) = [Y_0, \mathfrak{g}]$. Therefore $Y_0 \in U$ and $H_0 = - [Y_0, X_0] \in V$. Since $\mathfrak{z}_{x_0}$ is the kernel of the endomorphism $\rho(X_0)$ of $\mathfrak{g}$, it is clear that

$$\mathfrak{z}_{x_0} = \sum\limits_{1 \leq i \leq r} \mathfrak{z}_{x_0} \cap \mathfrak{g}_i.$$

Let $\lambda_i + 1 = \dim \mathfrak{g}_i$ so that $\lambda_i \geq 0$ $(1 \leq i \leq r)$. We know from Lemma 25 that $\dim(\mathfrak{z}_{x_0} \cap \mathfrak{g}_i) = 1$. Let w'_i be a unit vector in $\mathfrak{z}_{x_0} \cap \mathfrak{g}_i$. Then $\rho(H_0) w'_i = \lambda_i w'_i$ from Lemma 25. Put $w_i = \theta(w'_i)$. Then

$$[H_0, w_i] = - \theta([H_0, w'_i]) = - \lambda_i \theta(w'_i) = - \lambda_i w_i$$

and $w_1, \cdots, w_r$ is an orthonormal basis for $U = \theta(\mathfrak{z}_{x_0})$.

LEMMA 29. Let $n = \dim \mathfrak{g}$. Then

$$n = \sum\limits_{1 \leq i \leq r} (\lambda_i + 1), \qquad r = \dim \mathfrak{z}_{x_0}.$$

This is obvious since $\mathfrak{g} = \sum\limits_i \mathfrak{g}_i$.

Let $(u_1, \cdots, u_r)$ denote the Cartesian coordinates in U with respect to $(w_1, \cdots, w_r)$ so that $u = \sum\limits_i u_i w_i$ $(u \in U)$. Then $\partial(w_i) = \partial/\partial u_i$.

LEMMA 30. *Let D denote the vector field on $\mathfrak{g}$ defined by $D_X = \partial(X)$ $(X \in \mathfrak{g})$. Then D is invariant and*

$$\Delta(D) = \sum_{1 \leq i \leq r} (1 + \tfrac{1}{2}\lambda_i) d_i$$

where $d_i = u_i \partial/\partial u_i$ $(1 \leq i \leq r)$.

Since $(D_X)^x = (\partial(X))^x = \partial(xX) = D_{xX}$ $(x \in G, X \in \mathfrak{g})$, it follows that D is invariant.

Now fix $u \in U'$. Then

$$\Gamma_{X_0+u}(H_0 \otimes 1) = [H_0, X_0 + u] = 2X_0 + [H_0, u].$$

Therefore

$$\Gamma_{X_0+u}(\tfrac{1}{2}H_0 \otimes 1 + 1 \otimes (u - \tfrac{1}{2}[H_0, u])) = X_0 + u.$$

Now $[H_0, U] \subset U$ since $U = \mathfrak{z}_{Y_0}$. Therefore

$$\Delta_u(D) = \partial(u) - \tfrac{1}{2}\partial([H_0, u]).$$

But $u = \sum_i u_i w_i$ and so

$$u - \tfrac{1}{2}[H_0, u] = \sum_i (1 + \tfrac{1}{2}\lambda_i) u_i w_i.$$

Therefore

$$\Delta_u(D) = \sum_i (1 + \tfrac{1}{2}\lambda_i) u_i \partial(w_i) \qquad (u \in U').$$

LEMMA 31. *Fix integers $m_i \geq 0$, $(1 \leq i \leq r)$, put*

$$d = \partial(w_1^{m_1} w_2^{m_2} \cdots w_r^{m_r})$$

and define D and d_i $(1 \leq i \leq r)$ as above. Then [9] $\{d, d_i\} = m_i d$ *and*

$$(d \circ \Delta(D)^*)_0 = - \sum_{1 \leq i \leq r} (1 + \tfrac{1}{2}\lambda_i)(m_i + 1) d.$$

It is clear that

$$\{\partial(w_j), d_i\} = \delta_{ji} \partial(w_i).$$

Hence for a fixed i,

$$\{d, d_i\} = \{\partial(w_i^{m_i}), d_i\} d'$$

where $d' = \prod_{j \neq i} \partial(w_j^{m_j})$. But it is obvious from the formula above that

$$\{\partial(w_i^{m_i}), d_i\} = m_i \partial(w_i^{m_i})$$

and therefore $\{d, d_i\} = m_i d$. Moreover $d_i^* = -(d_i + 1)$. Hence

$$\Delta(D)^* = -\sum_i (1 + \tfrac{1}{2}\lambda_i)(d_i + 1)$$

from Lemma 30. Also

---

[9] If D_1, D_2 are two differential operators, we denote $D_1 \circ D_2 - D_2 \circ D_1$ by $\{D_1, D_2\}$.

$$dd_i = \{d, d_i\} + d_i d = (d_i + m_i)d.$$

Therefore since $(d_i d)_0 = 0$, we have $(dd_i)_0 = m_i d$. Hence

$$(d \circ \Delta(D)^*)_0 = -\sum_i (1 + \tfrac{1}{2}\lambda_i)(m_i + 1)d.$$

Let $\mathfrak{z}_I$ be the centralizer of I in $\mathfrak{g}$.

LEMMA 32. *The following four conditions are mutually equivalent.*

(1) ω *is identically zero on* $U = \mathfrak{z}_{Y_0}$;

(2) ω *is identically zero on* $\theta(U) = \mathfrak{z}_{X_0}$;

(3) $\lambda_i > 0 \; (1 \leqq i \leqq r)$;

(4) $\mathfrak{z}_I = \{0\}$.

We know that ω is invariant under θ. Hence (1) and (2) are equivalent. Since X_0, Y_0 generate I as a Lie algebra, it is obvious that $\mathfrak{z}_I = \mathfrak{z}_{X_0} \cap \mathfrak{z}_{Y_0}$. Let ρ_i denote the representation of I on $\mathfrak{g}_i$ defined under $\rho \; (1 \leqq i \leqq r)$. Then ρ_i is the zero representation if and only if $\lambda_i = 0$. On the other hand $\dim \mathfrak{z}_I$ is the number of times the zero representation of I occurs in the complete reduction of ρ. Hence

$$\dim \mathfrak{z}_I = \sum_{\lambda_i = 0} (\lambda_i + 1).$$

This proves that (3) and (4) are equivalent.

Now assume (2) holds. Then $B(X, Y) = 0$ for $X, Y \in \mathfrak{z}_{X_0}$ and therefore $\mathfrak{z}_{X_0}$ and $\theta(\mathfrak{z}_{X_0}) = \mathfrak{z}_{Y_0}$ are mutually orthogonal. But then $\mathfrak{z}_I = \mathfrak{z}_{X_0} \cap \mathfrak{z}_{Y_0} = \{0\}$. Therefore (2) implies (4).

Conversely suppose (3) holds. Then $\mathfrak{z}_{Y_0} = U = \sum_i \mathbf{R} w_i$ and

$$B(w_i, w_j) = - \langle \theta(w_i), w_j \rangle \qquad\qquad (1 \leqq i, j \leqq r).$$

We have seen that $w'_i = \theta(w_i)$ and w_j are eigenvectors of $\rho(H_0)$ belonging to the eigenvalues λ_i and $-\lambda_j$ respectively. It follows from (3) that these two eigenvalues are distinct. Since $\rho(H_0)$ is self-adjoint (by Lemma 27), the vectors w'_i and w_j are orthogonal. Hence $B(w_i, w_j) = 0 \; (1 \leqq i, j \leqq r)$. This shows that (3) implies (1) and so the lemma is proved.

Now consider the differential operator $\Delta(\partial(\omega))$ on U'. It is clear (see the remark at the end of §7) that its degree is $\leqq 2$. As usual $\Delta_0(\partial(\omega))$ denotes its local expression at zero.

LEMMA 33. *The homogeneous part of degree 2 of* $\Delta_0(\partial(\omega))$ *is zero if and only if* $\mathfrak{z}_I = \{0\}$.

Choose a base v_i $(1 \leqq i \leqq s)$ for V. Then $[X_0, v_i]$, w_j $(1 \leqq i \leqq s, 1 \leqq j \leqq r)$ is a base for $\mathfrak{g}$. Hence

$$\omega = \sum_{1 \leqq i, j \leqq s} a^{ij}[X_0, v_i][X_0, v_j] + \sum_{1 \leqq i, j \leqq r} b^{ij} w_i w_j$$
$$+ \sum_{1 \leqq i \leqq s} \sum_{1 \leqq j \leqq r} c^{ij}[X_0, v_i] w_j$$

where $a^{ij} = a^{ji}$, $b^{ij} = b^{ji}$ and c^{ij} are in $\boldsymbol{R}$. Let $\cdot$ denote product in $\mathfrak{G}$. Then

$$\Gamma_{X_0}(v_i \cdot v_j \otimes 1) = [X_0, v_i][X_0, v_j] - [v_i, [X_0, v_j]],$$

and

$$\Gamma_{X_0}(v_i \otimes w_j) = - [X_0, v_i] w_j + [v_i, w_j].$$

Therefore if λ denotes, as usual, the canonical mapping of $S(\mathfrak{g}_c)$ onto $\mathfrak{G}$, we get

$$\sum_{1 \leqq i, j \leqq s} a^{ij} \Gamma_{X_0}(\lambda(v_i v_j) \otimes 1) - \sum_{1 \leqq i \leqq s} \sum_{1 \leqq j \leqq r} c^{ij} \Gamma_{X_0}(v_i \otimes w_j)$$
$$= \sum_{i,j} a^{ij}[X_0, v_i][X_0, v_j] - \sum_{i,j} a^{ij}[v_i, [X_0, v_j]]$$
$$+ \sum_{i,j} c^{ij}[X_0, v_i] w_j - \sum_{i,j} c^{ij}[v_i, w_j]$$
$$= \omega - \sum_{1 \leqq i, j \leqq r} b^{ij} w_i w_j + \mu$$

where μ is an element of $S(\mathfrak{g}_c)$ of degree $\leqq 1$. This shows that the homogeneous part of degree 2 of $\Delta_0(\partial(\omega))$ is $\sum_{i,j} b^{ij} \partial(w_i w_j) = \sum_{i,j} b^{ij} \partial^2/\partial u_i \partial u_j$.

So it remains to show that $b^{ij} = 0$ $(1 \leqq i, j \leqq r)$ if and only if $\mathfrak{z}_{\mathrm{I}} = \{0\}$. Now $\mathfrak{g}_{X_0}$ is orthogonal to $U = \theta(\mathfrak{z}_{X_0})$. Therefore $B(X, Z) = 0$ for $X \in \mathfrak{g}_{X_0}$ and $Z \in \mathfrak{z}_{X_0}$. Hence if $v \in V$, $[X_0, v]$, regarded as a linear function on $\mathfrak{g}$, vanishes identically on $\mathfrak{z}_{X_0}$. Therefore if $X \in \mathfrak{z}_{X_0}$,

$$\omega(X) = \sum_{1 \leqq i, j \leqq r} b^{ij} B(w_i, X) B(w_j, X).$$

Now $w_1, \cdots, w_r$ is an orthonormal base for U. Hence $\theta(w_i)$ $(1 \leqq i \leqq r)$ is a base for $\theta(U) = \mathfrak{z}_{X_0}$. Let $X = \sum_{1 \leqq i \leqq r} t_i \theta(w_i)$ $(t \in \boldsymbol{R})$. Then it follows from the above orthonormality that $B(w_i, X) = - t_i$. Hence

$$\omega(X) = \sum_{1 \leqq i, j \leqq r} b^{ij} t_i t_j.$$

The required result is now obvious from Lemma 32.

COROLLARY 1. *Suppose $\mathfrak{z}_{\mathrm{I}} \neq \{0\}$. Then if $z \in S(U_c)$ and*

$$(\partial(z) \circ \Delta(\partial(\omega))^*)_0 = 0,$$

we can conclude that $z = 0$.

For otherwise suppose $z \neq 0$. Let m be the degree of z and z_0 the homogeneous part of degree m of z. Put $\Delta = \Delta(\partial(\omega))$. Then

$$\Delta = \sum_{1 \leq i, j \leq r} a_{ij} \partial^2/\partial u_i \partial u_j + \sum_{1 \leq i \leq r} a_i \partial/\partial u_i + a$$

where $a_{ij} = a_{ji}$, a_i and a are C^∞ functions on U'. It follows from Lemma 33 that

$$\sum_{i,j} a_{ij}(0) \partial^2/\partial u_i \partial u_j \neq 0.$$

Moreover it is clear that

$$\Delta^* = \sum_{i,j} a_{ij} \partial^2/\partial u_i \partial u_j + \sum_i b_i \partial/\partial u_i + b$$

where b_i and b are also C^∞ functions on U'. Since

$$\partial(z) \circ \Delta^* = \Delta^* \circ \partial(z) + \{\partial(z), \Delta^*\},$$

it is clear that the homogeneous part of degree $m + 2$ of $(\partial(z) \circ \Delta^*)_0$ is

$$\sum_{i,j} a_{ij}(0) \partial^2/\partial u_i \partial u_j \circ \partial(z_0) \neq 0.$$

Therefore $(\partial(z) \circ \Delta^*)_0 \neq 0$ and so we get a contradiction.

COROLLARY 2. *If* $\mathfrak{z}_I \neq \{0\}$, *then* $z = 0$ *in the corollary of Lemma 23 and therefore* $\sigma_T = 0$.

For since $\partial(\omega) T = 0$, we conclude from Theorem 3 that $\Delta \sigma_T = 0$ where $\Delta = \Delta(\partial(\omega))$. But then it follows from the corollary of Lemma 23 that $\beta(0; \partial(z) \circ \Delta^*) = 0$ for every $\beta \in C_o^\infty(U_0)$. This however implies that $(\partial(z) \circ \Delta^*)_0 = 0$. Our assertion now follows from Corollary 1 above.

Thus in order to complete the proof of Lemma 17, we may, from now on, assume that $\mathfrak{z}_I = \{0\}$. Then, by Lemma 32, ω vanishes identically on $U = \theta(\mathfrak{z}_{x_0})$ and $\mathfrak{z}_{x_0}$. Therefore

$$\omega(X_0 + u) = \omega(X_0) + \omega(u) + 2B(X_0, u) = 2B(X_0, u)$$

for $u \in U$. Now w_1 could be chosen to be any unit vector such that $\theta(w_1) \in \mathfrak{g}_1 \cap \mathfrak{z}_{x_0} = \boldsymbol{R} X_0$. (We recall that $\mathfrak{g}_1 = I$.) So we may take

$$w_1 = \| Y_0 \|^{-1} Y_0.$$

Then

$$u_1 = \langle w_1, u \rangle = -B(\theta(w_1), u) = \| Y_0 \|^{-1} B(X_0, u).$$

Hence $\omega(X_0 + u) = 2cu_1$ $(u \in U)$ where $c = \| X_0 \| = \| Y_0 \|$. We also note that $\dim \mathfrak{g}_1 = \dim I = 3$ and therefore $\lambda_1 = 2$.

Now ω vanishes identically on $\mathcal{N}$. Therefore, by Lemma 21, there exists an integer $k \geqq 0$ such that $\omega^k T = 0$ around X_0. On the other hand Lemma 19 would be proved if we can show that $T = 0$ around X_0. Therefore it would clearly be sufficient to obtain the following result.

LEMMA 34. *Fix a point $X_0 \in \mathcal{N}$ and an integer $k \geqq 0$ and let q be the topological dimension of the orbit $X_0{}^G$. Let T be a locally invariant distribution on some open neighborhood Ω of X_0 in $\mathfrak{g}$ such that*

    1) $\omega^k T = 0, \ \partial(\omega) T = 0$

    2) $\operatorname{Supp} T \subset \mathcal{N}_q \cap \Omega.$

Then $T = 0$ around X_0.

The case $k = 0$ being trivial, we assume $k \geqq 1$ and use induction. Also, in view of our results above, we may suppose that $q \geqq 1$ and $\mathfrak{z}_\mathfrak{l} = \{0\}$. Finally, we may assume that $X_0 \in \operatorname{Supp}(\omega^{k-1} T)$ for otherwise the assertion of the lemma would follow immediately from the induction hypothesis.

15. Some algebraic lemmas. We now study some simple algebraic properties of the differential operator D of Lemma 30.

LEMMA 35. *Let $p \in S_m(\mathfrak{g}_c)$. Then $D \circ p = p(D + m)$ and $\partial(p) \circ D = (D + m) \circ \partial(p)$.*

For any $f \in C^\infty(\mathfrak{g})$ and $t \in \mathbf{R}$, let f_t denote the function $X \to f(tX)$ $(X \in \mathfrak{g})$. Then

$$f(X; D) = f(X; \partial(X)) = \{df(tX)/dt\}_{t=1} \qquad (X \in \mathfrak{g})$$

and therefore $Df = (df_t/dt)_{t=1}$. Hence

$$D(pf) = \{d(pf)_t/dt\}_{t=1}.$$

But it is obvious that $p_t = t^m p$ and therefore $(pf)_t = p_t f_t = t^m p f_t$. Therefore

$$D(pf) = \{d(t^m p f_t)/dt\}_{t=1} = mpf + pDf$$

and this proves the first statement.

Now we come to the second part. Let $X_1, \cdots, X_m \in \mathfrak{g}$. Then
$$f_t(X; \partial(X_1 X_2 \cdots X_m)) = \{\partial^m f_t(X + t_1 X_1 + \cdots + t_m X_m)/\partial t_1 \cdots \partial t_m\}_{t_1 = \cdots = t_m = 0}$$
$$= \{t^m \partial^m f(tX + t_1 X_1 + \cdots + t_m X_m)/\partial t_1 \cdots \partial t_m\}_{t_1 = \cdots = t_m = 0}$$
$$= t^m f(tX; \partial(X_1 \cdots X_m)).$$

This shows that $\partial(p) f_t = t^m (\partial(p) f)_t$ for $p \in S_m(\mathfrak{g}_c)$. Therefore

$$\partial(p)(Df) = (d(\partial(p)f_t)/dt)_{t=1} = m\partial(p)f + (d(\partial(p)f)_t/dt)_{t=1}$$
$$= m\partial(p)f + D(\partial(p)f).$$

This completes the proof of Lemma 35.

LEMMA 36. *Put* $n = \dim \mathfrak{g}$. *Then*

$$\{\partial(\omega), \omega\} = 4(D + n/2).$$

Let X_i $(1 \leq i \leq n)$ be a base for $\mathfrak{g}$ and X^i $(1 \leq i \leq n)$ the dual base so that $B(X^i, X_j) = \delta_j{}^i$. Let $X^i = \sum_j g^{ij}X_j$ and $X_i = \sum_j g_{ij}X^j$ $(g_{ij}, g^{ij} \in \mathbf{R})$. Then $g^{ij} = B(X^i, X^j)$, $g_{ij} = B(X_i, X_j)$ and the matrices (g^{ij}) and (g_{ij}) are inverse to each other. Moreover

$$\partial(X_i)X_j = B(X_i, X_j) = g_{ij}.$$

Now let $X = \sum_i t^i X_i$ $(t^i \in \mathbf{R})$. Then

$$\omega(X) = \sum_{i,j} t^i t^j B(X_i, X_j) = \sum_{i,j} g_{ij} t^i t^j.$$

But $t^i = B(X^i, X)$. Hence

$$\omega = \sum_{i,j} g_{ij} X^i X^j = \sum_{i,j} g^{ij} X_i X_j$$

and therefore

$$\{\partial(X_k), \omega\} = \sum_{i,j} g^{ij}(g_{ki}X_j + X_i g_{kj})$$

$$= 2X_k.$$

Hence

$$\{\partial(\omega), \omega\} = \sum_k (\{\partial(X^k), \omega\} \circ \partial(X_k) + \partial(X^k) \circ \{\partial(X_k), \omega\})$$

$$= 2\sum_k (X^k \partial(X_k) + \partial(X_k) \circ X^k)$$

$$= 2n + 4\sum_k X^k \partial(X_k).$$

So it remains to verify that $\sum_k X^k \partial(X_k) = D$. This is seen as follows. If $X \in \mathfrak{g}$, then

$$\left(\sum_k X^k \partial(X_k)\right)_X = \sum_k B(X^k, X)\partial(X_k) = \partial(X) = D_X$$

since $B(X^k, X) = t^k$ if $X = \sum_k t^k X_k$ $(t^k \in \mathbf{R})$.

COROLLARY. *Let* l *be any integer* ≥ 1. *Then*

$$\{\partial(\omega), \omega^l\} = 4l\omega^{l-1}(D + n/2 + l - 1).$$

If $l = 1$, this follows from Lemma 36. So assume $l \geq 2$ and use induction. Then $\omega^l = \omega\omega^{l-1}$ and therefore

$$\{\partial(\omega),\omega^l\} = \{\partial(\omega),\omega\}\circ\omega^{l-1} + \omega\{\partial(\omega),\omega^{l-1}\}$$
$$= 4(D+n/2)\circ\omega^{l-1} + 4(l-1)\omega^{l-1}(D+n/2+l-2)$$

by induction hypothesis. But $(D+n/2)\circ\omega^{l-1} = \omega^{l-1}(D+n/2+2l-2)$ from Lemma 35. From this the required result follows immediately.

16. Completion of the proof of Theorem 4. Now we return to the proof of Lemma 34. Put $T_0 = (D+n/2+k-1)T$. Since D is invariant, T_0 is a locally invariant distribution on Ω and $\operatorname{Supp} T_0 \subset \operatorname{Supp} T$. Moreover

$$0 = \partial(\omega)(\omega^k T) - \omega^k(\partial(\omega)T)$$
$$= 4k\omega^{k-1}T_0$$

from the corollary of Lemma 36. This shows that $\omega^{k-1}T_0 = 0$. Moreover

$$\partial(\omega)T_0 = \partial(\omega)(D+n/2+k-1)T$$
$$= (D+n/2+k+1)\partial(\omega)T = 0$$

from Lemma 35. Thus T_0 satisfies all the conditions of Lemma 34 with k replaced by $k-1$. Hence by induction hypothesis $T_0 = 0$ around X_0. So by replacing Ω, if necessary, by a smaller open neighborhood of X_0, we may assume that $T_0 = 0$ on Ω. Therefore $(D+n/2+k-1)T = 0$.

Let δ_0 denote the distribution $\beta \to \beta(0)$ $(\beta \in C_c^\infty(U_0))$ on U_0. Then, from Lemmas 20 and 23, we can choose $z \in S(U_c)$ such that $\sigma_T = \partial(z)\delta_0$. If $p \in I(\mathfrak{g}_c)$, it is obvious that $\Delta(p) = q$ where q is the polynomial function $u \to p(X_0+u)$ $(u \in U')$. Since $\omega(X_0+u) = 2cu_1$ $(u \in U)$, it follows from Theorem 3 that $\sigma_{\omega^m T} = (2cu_1)^m \partial(z)\delta_0$ for any integer $m \geq 0$. Therefore since $\omega^k T = 0$, $X_0 \in \operatorname{Supp}(\omega^{k-1}T)$ and $(D+n/2+k-1)T = 0$, we obtain the following result by taking into account Theorems 2 and 3.

LEMMA 37. *$u_1^k\partial(z)\delta_0 = 0$ but $u_1^{k-1}\partial(z)\delta_0 \neq 0$. Moreover*

$$(\Delta(D)+n/2+k-1)\partial(z)\delta_0 = 0.$$

The following lemma is proved by a simple calculation.

LEMMA 38. *$u_1^p\partial(w_1^{m_1}w_2^{m_2}\cdots w_r^{m_r})\delta_0 = 0$ if $p > m_1$ and*

$$= (-1)^p \frac{m_1!}{(m_1-p)!}\,\partial(w_1^{m_1-p}w_2^{m_2}\cdots w_r^{m_r})\delta_0$$

if $p \leq m_1$.

Put $d(m) = \partial(w_1^{m_1}\cdots w_r^{m_r})$ for $m = (m_1,\cdots,m_r)$ and suppose

$$\partial(z) = \sum_m c(m)d(m)$$

where $c(m)$ are complex numbers which are zero for all m except a finite number. Then

$$0 = u_1{}^k \partial(z)\delta_0$$

$$= \sum_{m_1 \geq k} \sum_{m_2, \cdots, m_r \geq 0} c(m)(-1)^k (m_1!/(m_1-k)!)\partial(w_1{}^{m_1-k}w_2{}^{m_2}\cdots w_r{}^{m_r})\delta_0$$

and therefore it follows from Lemma 20 that $c(m) = 0$ if $m_1 \geq k$. Similarly

$$u_1{}^{k-1}\partial(z)\delta_0$$

$$= \sum_{m_1 \geq k-1} \sum_{m_2, \cdots, m_r \geq 0} c(m)(-1)^{k-1}(m_1!/(m_1-k+1)!)\partial(w_1{}^{m_1-k+1}w_2{}^{m_2}\cdots w_r{}^{m_r})\delta_0.$$

Therefore it is obvious from the above result and Lemma 37 that $c(m) \neq 0$ for some m with $m_1 = k-1$.

On the other hand

$$\sum_m c(m)(\Delta(D) + n/2 + k - 1)d(m)\delta_0 = 0$$

from Lemma 37. It follows from Lemma 31 that

$$\Delta(D)d(m)\delta_0 = -\sum_{1 \leq i \leq r}(1 + \tfrac{1}{2}\lambda_i)(m_i + 1)d(m)\delta_0.$$

Therefore

$$\sum_m c(m)\{n/2 + k - 1 - \sum_i (1 + \tfrac{1}{2}\lambda_i)(m_i + 1)\}d(m)\delta_0 = 0$$

and this implies that $c(m) = 0$ unless

$$\sum_i (1 + \tfrac{1}{2}\lambda_i)(m_i + 1) = n/2 + k - 1.$$

But $\sum_i (1 + \tfrac{1}{2}\lambda_i) = \tfrac{1}{2}(n+r)$ from Lemma 29. Therefore if $c(m) \neq 0$, we must have

$$\sum_i (1 + \tfrac{1}{2}\lambda_i)m_i + r/2 = k - 1.$$

Since $r \geq 1$ and $\lambda_1 = 2$, this implies that $2m_1 < k - 1$. However we have seen above that $c(m) \neq 0$ for some m with $m_1 = k - 1$. This contradiction completes the proof of Lemma 34 and therefore also of Theorem 4.

17. Proof of Theorem 5. It is now easy to generalize Theorem 4 as follows.

THEOREM 5. *Let T be a distribution on an open subset Ω of $\mathfrak{g}$. Assume that*

1) *T is locally invariant,*
2) *$\operatorname{Supp} T \subset \mathfrak{n} \cap \Omega$,*

3) *T satisfies an equation of the form* $\sum\limits_{0 \leq k \leq r} c_k \partial(\omega^k) T = 0$ *where c_k are complex numbers not all zero.*

Then $T = 0$.

Let $\mathfrak{X}$ be the space of distributions on Ω spanned over $\boldsymbol{C}$ by $\partial(\omega^k) T$ ($k \geq 0$). Then, in view of condition (3), $\dim \mathfrak{X} < \infty$. Since $T \in \mathfrak{X}$, it would be enough to prove that $\mathfrak{X} = \{0\}$. So let us assume that $\mathfrak{X} \neq \{0\}$. Since $\partial(\omega)$ defines a linear transformation in $\mathfrak{X}$, we can choose $T_0 \neq 0$ in $\mathfrak{X}$ such that $\partial(\omega) T_0 = c T_0$ for some $c \in \boldsymbol{C}$. It is clear that every distribution in $\mathfrak{X}$ satisfies conditions (1) and (2). Fix a point X_0 in the support of T_0. Since ω vanishes identically on $\mathfrak{N}$, it follows from Lemma 21 that we can choose an open neighborhood Ω_0 of X_0 in Ω and an integer $k \geq 0$ such that $\omega^k T_0 = 0$ on Ω_0.

LEMMA 39. *For any differential operator D on Ω_0, define*

$$d_\omega D = \{\partial(\omega), D\}.$$

Then if $DT_0 = 0$ on Ω_0, the same holds for $(d_\omega D) T_0$.

Let $D' = d_\omega D$. Then

$$D'T_0 = \partial(\omega)(DT_0) - D(\partial(\omega) T_0)$$
$$= (\partial(\omega) - c)(DT_0) = 0$$

on Ω_0, since $\partial(\omega) T_0 = c T_0$.

Therefore we conclude by induction on r that $(d_\omega{}^r \omega^k) T_0 = 0$ on Ω_0. But $d_\omega{}^{2k} \omega^k = 2k! 2^{2k} \partial(\omega^k)$ (see [4(c), p. 99]). Hence $\partial(\omega^k) T = 0$ on Ω_0. However $\partial(\omega^k) T_0 = c^k T_0$. Therefore since $X_0 \in \operatorname{Supp} T_0$, we must have $c = 0$. But then $T_0 = 0$ from Theorem 4 and so we get a contradiction.

18. Appendix. The results of this section were pointed out to me by A. Borel and they are reproduced here with his permission.

We use the terminology of [2, §§ 1-2].

LEMMA 40. *Let $G \subset \boldsymbol{GL}(n, \boldsymbol{R})$ be a real algebraic Lie group. G' an open subgroup of G, $f : G \to \boldsymbol{GL}(m, \boldsymbol{R})$ a rational representation of G and $X = f(G') \cdot x$ ($x \in \boldsymbol{R}^m$) an orbit of G'. Let q be the topological dimension of X. Then X is open in [10] $\operatorname{Cl} X$ and $\operatorname{Cl} X \cap {}^c X$ is the union of orbits of G' of dimension $< q$.*

---

[10] For any set U, $\operatorname{Cl} U$ stands for the topological closure of U and ${}^c U$ for the complement of U.

Let $G_c \subset \boldsymbol{GL}(n, \boldsymbol{C})$ be the smallest complex algebraic group containing G. Then f extends uniquely to a rational representation $f_c \colon G_c \to \boldsymbol{GL}(m, \boldsymbol{C})$ [3(b), Prop. 4, p. 109]. Consider the orbit $Y = f_c(G_c) \cdot x$ in $\boldsymbol{C}^m$. On each irreducible component of G_c, the map $g \to f_c(g) \cdot x$ is rational and regular and therefore Y is a complex manifold of complex dimension q. Since the image of an irreducible variety under a rational regular map contains a non-empty Zariski-open subset of its Zariski closure, it follows that Y contains an open everywhere dense subset of $\mathrm{Cl}\, Y$ and that $V = \mathrm{Cl}\, Y$ is an algebraic set of algebraic dimension q. By homogeneity, Y is open in V and consists of simple points and $V \cap {}^\circ Y$ is an algebraic set of algebraic dimension $q' < q$. Also since G is Zariski-dense in G_c, by definition of G_c, it is clear that V is the smallest algebraic set containing $f(G) \cdot x$.

Let $a \in Y_{\boldsymbol{R}}$. Then a is simple on V and there exists a neighborhood U of a in $\boldsymbol{R}^m$ such that $U \cap V_{\boldsymbol{R}}$ is a connected manifold of dimension q (see Whitney [10, §11]). On the other hand the isotropy group of a in G' is contained in the set of real points of an algebraic group of complex dimension $r - q$ where $r = \dim G$. Hence its dimension is $\leq r - q$. Therefore $f(G') \cdot a$ has dimension $\geq q$ and so it contains $U \cap V_{\boldsymbol{R}}$ if U is taken sufficiently small. This shows that a G'-orbit in $Y_{\boldsymbol{R}}$ is open in $V_{\boldsymbol{R}}$. Therefore ${}^\circ X \cap Y_{\boldsymbol{R}}$, being a union of such orbits, is open in $V_{\boldsymbol{R}}$. Hence X is closed in $Y_{\boldsymbol{R}}$. This proves that $\mathrm{Cl}\, V \cap Y_{\boldsymbol{R}} = X$ and therefore $\mathrm{Cl}\, X \subset X \cup (V_{\boldsymbol{R}} \cap {}^\circ Y_{\boldsymbol{R}})$. Since $V \cap {}^\circ Y$ is an algebraic set of algebraic dimension $q' < q$, the topological dimension of $V_{\boldsymbol{R}} \cap {}^\circ Y_{\boldsymbol{R}} = (V \cap {}^\circ Y)_{\boldsymbol{R}}$ is $\leq q'$. (This follows by induction on dimension from [10, Theorem 1 and Lemma 9].) Since $(V \cap {}^\circ Y)_{\boldsymbol{R}}$ is stable under G, our assertion follows immediately.

COROLLARY. *Let A be a real algebraic subset of $\boldsymbol{R}^m$ which is the union of a finite number of G'-orbits. Let A_q denote the union of all G'-orbits in A of topological dimension $\leq q$. Then A_q is closed and each G'-orbit of dimension q is open in A_q.*

It is clear from the above lemma that A_q is closed. Let $X = f(G') \cdot x$ ($x \in A$) be an orbit of dimension q. Then by Lemma 40, $A_q \cap {}^\circ X$ is closed and hence X is open in A_q.

It is now easy to give a proof of Lemma 17. Take G to be the group of all automorphisms of $\mathfrak{g}$ and G' the connected component of 1 in G. Then G' is an open subgroup of the real algebraic Lie group G. Let $n = \dim \mathfrak{g}$. Define polynomial functions p_i ($0 \leq i < n$) on $\mathfrak{g}$ by the relation

$$\det(t - \operatorname{ad} X) = \sum_{0 \leq i < n} p_i(X) t^i + t^n \qquad (X \in \mathfrak{g}),$$

where t is an indeterminate. Then obviously $\mathfrak{n}$ is the set of all common zeros of p_i $(0 \leq i < n)$ in $\mathfrak{g}$. Hence $\mathfrak{n}$ is a real algebraic subset of $\mathfrak{g}$. Lemma 17 is an immediate consequence of the above corollary if we take $A = \mathfrak{n}$.

INSTITUTE FOR ADVANCED STUDY.

REFERENCES.

[1] F. A. Berezin, "Laplace operators on semisimple Lie groups," *American Mathematical Society Translations*, vol. 21 (1962), pp. 239-339.

[2] A. Borel and Harish-Chandra, "Arithmetic subgroups of algebraic groups," *Annals of Mathematics*, vol. 75 (1962), pp. 485-535.

[3] C. Chevalley, (a) *Theory of Lie groups*, Princeton University Press, 1946.
　　　　(b) *Groupes Algébriques*, Actualités Scientifiques et Industrielles 1152, Paris, 1951.

[4] Harish-Chandra, (a) "Representations of a semisimple Lie group on a Banach space," *Transactions of the American Mathematical Society*, vol. 75 (1953), pp. 185-243.
　　　　(b) "Representations of semisimple Lie groups, III," *Transactions of the American Mathematical Society*, vol. 76 (1954), pp. 234-253.
　　　　(c) "Differential operators on a semisimple Lie algebra," *American Journal of Mathematics*, vol. 79 (1957), pp. 87-120.
　　　　(d) "The characters of semisimple Lie groups," *Transactions of the American Mathematical Society*, vol. 83 (1956), pp. 98-163.
　　　　(e) "Invariant eigendistributions on semisimple Lie groups," *Bulletin of the American Mathematical Society*, vol. 69 (1963), pp. 117-123.

[5] N. Jacobson, "Completely reducible Lie algebras of linear transformations," *Proceedings of the American Mathematical Society*, vol. 2 (1951), pp. 105-113.

[6] B. Kostant "The principal three-dimensional subgroup and the Betti numbers of a complex simple Lie group," *American Journal of Mathematics*, vol. 81 1959), pp. 973-1032.

[7] D. Montgomery and L. Zippin, *Topological transformation groups*, Interscience Tracts in Pure and Applied Mathematics, I, New York, 1955.

[8] G. D. Mostow, *Some new decomposition theorems for semisimple groups*, Memoirs of the American Mathematical Society, vol. 14 (1955), pp. 31-54.

[9] L. Schwartz, *Théorie des distributions* I, Paris, Hermann, 1950.

[10] H. Whitney, "Elementary structure of real algebraic varieties," *Annals of Mathematics*, vol. 66 (1957), pp. 545-556.

Reprinted from
Amer. J. of Math.
86 (1964), 271–309

INVARIANT DIFFERENTIAL OPERATORS AND DISTRIBUTIONS ON A SEMISIMPLE LIE ALGEBRA.

By Harish-Chandra.

1. Introduction. This paper deals with the second main step in the proof of Lemma 1 of [2(d)]. As indicated in [2(d), p. 121] this proof proceeds by induction on $\dim \mathfrak{g}$. Let H be a semisimple element of $\mathfrak{g}$ and $\mathfrak{z}$ the centralizer of H in $\mathfrak{g}$. Then $\dim \mathfrak{z} < \dim \mathfrak{g}$ if H is not in the center of $\mathfrak{g}$. In this paper we give a method of reducing the problem from $\mathfrak{g}$ to $\mathfrak{z}$, locally around H. Since the induction hypothesis is applicable to $\mathfrak{z}$, this enables us to resolve the question around H. Even when $\mathfrak{g}$ is semisimple, $\mathfrak{z}$ is only reductive and therefore, in order to permit induction, we are forced to start from the assumption that $\mathfrak{g}$ is reductive rather than semisimple.

The basic tools for effecting the above-mentioned reduction are contained in Lemma 17, its corollaries and Theorem 2. The proof of Theorem 3 provides a good illustration of how our method actually works in practice.

This work was supported, in part, by a grant from the Sloan Foundation and a contract with the U. S. Army.

2. Some algebraic results. Let k be a field of characteristic zero and $\mathfrak{g}$ a Lie algebra over k. We use the notation of [2(e), §§ 6-7]. If $n = \dim \mathfrak{g}$, we can select $p_i \in P(\mathfrak{g})$ $(0 \leq i \leq n)$ such that

$$\det(t - \operatorname{ad} X) = \sum_{0 \leq i \leq n} p_i(X) t^i \qquad (X \in \mathfrak{g})$$

for an indeterminate t. Note that $p_n = 1$. Let l be the least integer such that $p_l \neq 0$. Then l is called the rank of $\mathfrak{g}$ and an element $X \in \mathfrak{g}$ is called regular (in $\mathfrak{g}$) if $p_l(X) \neq 0$. Similarly X is singular if $p_l(X) = 0$.

Now assume that $\mathfrak{g}$ is reductive and $\mathfrak{z}$ is a subalgebra of $\mathfrak{g}$ such that 1) $\mathfrak{z}$ is reductive in $\mathfrak{g}$ and 2) $\operatorname{rank} \mathfrak{z} = \operatorname{rank} \mathfrak{g}$. Put

$$\zeta(Z) = \det(\operatorname{ad} Z)_{\mathfrak{g}/\mathfrak{z}} \qquad (Z \in \mathfrak{z}),$$

where $(\operatorname{ad} Z)_{\mathfrak{g}/\mathfrak{z}}$ denotes the endomorphism of the factor space $\mathfrak{g}/\mathfrak{z}$ defined by $\operatorname{ad} Z$. Let $\mathfrak{z}'$ be the set of all $Z \in \mathfrak{z}$ where $\zeta(Z) \neq 0$. Since $\mathfrak{z}$ is reductive in $\mathfrak{g}$, we can select a linear subspace $\mathfrak{q}$ of $\mathfrak{g}$ such that 1) $\mathfrak{g} = \mathfrak{z} + \mathfrak{q}$, the sum being direct and 2) $[\mathfrak{z}, \mathfrak{q}] \subset \mathfrak{q}$.

Received November 14, 1963.

534

Let c be the center and $\mathfrak{g}_1$ the derived algebra of $\mathfrak{g}$. Then $\mathfrak{g}_1$ is semi-simple and $\mathfrak{g}$ is the direct sum of $\mathfrak{g}_1$ and c.

LEMMA 1. $\mathfrak{q}$ *is the set of all* $Y \in \mathfrak{g}_1$ *such that* $\mathrm{tr}(\mathrm{ad}\, Z\, \mathrm{ad}\, Y) = 0$ *for every* $Z \in \mathfrak{z}$. *Moreover* $\zeta \neq 0$.

For the proof of this lemma, it is clearly permissible to extend the ground field k to its algebraic closure. Hence we may assume that k is algebraically closed. Fix a Cartan subalgebra $\mathfrak{h}$ of $\mathfrak{z}$. Since $\mathrm{rank}\,\mathfrak{z} = \mathrm{rank}\,\mathfrak{g}$, $\mathfrak{h}$ is also a Cartan subalgebra of $\mathfrak{g}$. Let Q be the set of all roots of $(\mathfrak{g}, \mathfrak{h})$. For any $\alpha \in Q$, select an element $X_\alpha \neq 0$ in $\mathfrak{g}$ such that $[H, X_\alpha] = \alpha(H) X_\alpha$ for all $H \in \mathfrak{g}$. Then since $[\mathfrak{h}, \mathfrak{z}] \subset \mathfrak{z}$ and $[\mathfrak{h}, \mathfrak{q}] \subset \mathfrak{q}$, it is obvious that

$$\mathfrak{z} = \mathfrak{h} + \sum_{\alpha \in Q_1} k X_\alpha, \qquad \mathfrak{q} = \sum_{\alpha \in Q_2} k X_\alpha$$

where Q_1 is the set of those $\alpha \in Q$ for which $X_\alpha \in \mathfrak{z}$ and Q_2 the complement of Q_1 in Q. Since $\mathfrak{z}$ is reductive, it is obvious that Q_1, and therefore also Q_2, are stable under the mapping $\alpha \to -\alpha$ of roots. Since $X_\alpha \in \mathfrak{g}_1$ $(\alpha \in Q)$ and since X_α is orthogonal to

$$\mathfrak{h} + \sum_{\beta \in Q, \beta \neq -\alpha} k X_\beta$$

under the Killing form, while $\mathrm{tr}(\mathrm{ad}\, X_\alpha\, \mathrm{ad}\, X_{-\alpha}) \neq 0$, the first assertion of the lemma is now obvious. Furthermore

$$\zeta(H) = \prod_{\alpha \in Q_2} \alpha(H) \neq 0$$

if H is a regular element of $\mathfrak{h}$. This proves that $\zeta \neq 0$.

Put $\mathfrak{Q} = \mathfrak{S}(\mathfrak{q})$, $\mathfrak{Q}_d = \mathfrak{S}_d(\mathfrak{q})$ $(d \geq 0)$ and $\mathfrak{Q}_+ = \mathfrak{S}_+(\mathfrak{q})$.

LEMMA 2. *For any* $Z \in \mathfrak{z}'$, Γ_Z *defines a bijective mapping of* $\mathfrak{Q} \otimes S(\mathfrak{z})$ *onto* $S(\mathfrak{g})$.

COROLLARY 1. *Fix* $p \in S(\mathfrak{g})$. *Then for any* $Z \in \mathfrak{z}'$, *there exist unique elements* $\alpha_Z(p) \in S(\mathfrak{z})$ *and* $\beta_Z(p) \in \mathfrak{Q}_+ \otimes S(\mathfrak{z})$ *such that*

$$p - \alpha_Z(p) = \Gamma_Z(\beta_Z(p)).$$

Moreover [1] $d^\circ \alpha_Z(p) \leq d^\circ p$.

COROLLARY 2. *For any* $p \in S(\mathfrak{g})$ *we can choose an integer* $r \geq 0$ *such that the mappings* $Z \to \zeta(Z)^r \alpha_Z(p)$ *and* $Z \to \zeta(Z)^r \beta_Z(p)$ $(Z \in \mathfrak{z}')$ *can be extended to polynomial mappings of* $\mathfrak{z}$ *into* $S(\mathfrak{z})$ *and* $\mathfrak{Q}_+ \otimes S(\mathfrak{z})$ *respectively.*

---

[1] $d^\circ q$ denotes the degree of q.

The proof of all this is substantially the same as in $[2(a), \S 7]$.

Remark. Assume $\mathfrak{z}$ is abelian. Then it is a Cartan subalgeba of $\mathfrak{g}$ and so we denote it by $\mathfrak{h}$ in this case. Since $\mathfrak{g} = \mathfrak{h} + \mathfrak{q}$, it is clear that $\mathfrak{G} = \mathfrak{S}(\mathfrak{q})\mathfrak{S}(\mathfrak{h})$. Moreover it is clear from its definition (see $[2(e), \S 7]$) that

$$\sigma_H(H_0)p = d_H p = 0$$

for $H, H_0 \in \mathfrak{h}$ and $p \in S(\mathfrak{h})$. Hence $\Gamma_H(\mathfrak{Q}\mathfrak{S}_+(\mathfrak{h}) \otimes S(\mathfrak{h})) = \{0\}$ for $H \in \mathfrak{h}$. Therefore since

$$\mathfrak{G}_+ = \mathfrak{S}_+(\mathfrak{g}) = \mathfrak{Q}\mathfrak{S}_+(\mathfrak{h}) + \mathfrak{Q}_+,$$

we conclude that

$$\Gamma_H(\mathfrak{G}_+ \otimes S(\mathfrak{h})) = \Gamma_H(\mathfrak{Q}_+ \otimes S(\mathfrak{h})) \qquad\qquad (H \in \mathfrak{h}).$$

LEMMA 3. *Let τ be an automorphism of $\mathfrak{g}$ such that $\mathfrak{z}^\tau = \mathfrak{z}$. Then $\mathfrak{z}'$ is invariant under τ and*

$$\alpha_{Z^\tau}(p^\tau) = (\alpha_Z(p))^\tau, \qquad\qquad \beta_{Z^\tau}(p^\tau) = (\beta_Z(p))^\tau$$

for $Z \in \mathfrak{z}'$ and $p \in S(\mathfrak{g})$.

Since τ is an automorphism of $\mathfrak{g}$ which leaves $\mathfrak{z}$ invariant, it is clear that $\zeta(Z^\tau) = \zeta(Z)$ $(Z \in \mathfrak{z})$ and τ maps $\mathfrak{q}$ into itself. Hence $\mathfrak{z}'$ is invariant under τ. We can extend τ uniquely to an automorphism of $\mathfrak{G}$ and then to a bijective linear mapping of $\mathfrak{G} \otimes S(\mathfrak{g})$ by setting $(g \otimes q)^\tau = g^\tau \otimes q^\tau$ $(g \in \mathfrak{G}, q \in S(\mathfrak{g}))$. Then it is obvious that

$$(\Gamma_X(u))^\tau = \Gamma_{X^\tau}(u^\tau)$$

for $X \in \mathfrak{g}$ and $u \in \mathfrak{G} \otimes S(\mathfrak{g})$. From this the second statement of the lemma follows immediately.

3. Some elementary facts about differential operators. Let E be a vector space over $\boldsymbol{C}$ of finite dimension and U a nonempty open subset of E. Let $\mathfrak{H}(U)$ denote the ring of all holomorphic functions on U. For any $X \in E$, let $\partial(X)$ denote, as usual, the derivation of $\mathfrak{H}(U)$ over $\boldsymbol{C}$ defined as follows. If $f \in \mathfrak{H}(U)$, $\partial(X)f$ is the function

$$Y \rightarrow \{df(Y + tX)/dt\}_{t=0} \qquad\qquad (Y \in U, t \in \boldsymbol{C}).$$

For any $g \in \mathfrak{H}(U)$, we denote the endomorphism $f \rightarrow gf$ of $\mathfrak{H}(U)$ again by g. Let $\mathfrak{D}_h(U)$ denote the algebra of all endomorphisms of $\mathfrak{H}(U)$ over $\boldsymbol{C}$ generated by $\partial(X)$ and g $(X \in E, g \in \mathfrak{H}(U))$. We call $\mathfrak{D}_h(U)$ the algebra of holomorphic differential operators on U. For $f \in \mathfrak{H}(U)$, $D \in \mathfrak{D}_h(U)$ and $X \in U$, we denote, as usual, by $f(X;D)$ the value of Df and X.

Now assume U is connected. Then $\mathfrak{H}(U)$ is an integral domain. Let $\mathfrak{M}(U)$ be the quotient field of $\mathfrak{H}(U)$. For any $X \in E$, $\partial(X)$ can be extended uniquely to a derivation of $\mathfrak{M}(U)$ over $\boldsymbol{C}$. Moreover corresponding to any $g \in \mathfrak{M}(U)$, we have the endomorphism $f \to gf$ $(f \in \mathfrak{M}(U))$ of $\mathfrak{M}(U)$ over $\boldsymbol{C}$, which we again denote by g. Let $\mathfrak{D}_m(U)$ be the algebra of all endomorphisms of $\mathfrak{M}(U)$ generated by $\partial(X)$ and $g(X \in E, g \in \mathfrak{M}(U))$. Then $\mathfrak{D}_m(U)$ is called the algebra of meromorphic differential operators on U. Fix an element $D \in \mathfrak{D}_m(U)$. It is easy to see that if $Df = 0$ for all $f \in \mathfrak{H}(U)$, then $D = 0$. From this it follows without difficulty that there exists an injective homomorphism $D \to D_m$ of $\mathfrak{D}_h(U)$ into $\mathfrak{D}_m(U)$ such that $D_m f = Df$ for $D \in \mathfrak{D}_h(U)$ and $f \in \mathfrak{H}(U)$. We may therefore identify $\mathfrak{D}_h(U)$ with a subalgebra of $\mathfrak{D}_m(U)$ under this homomorphism. In particular, we have the algebras $\mathfrak{D}_h(E) \subset \mathfrak{D}_m(E)$.

Let E' be the space dual to E. Then E' and $\partial(E)$ are both contained in $\mathfrak{D}_h(E)$. We denote by $\mathfrak{D}(E)$ the subalgebra of $\mathfrak{D}_h(E)$ generated by 1 and $E' + \partial(E)$. $\mathfrak{D}(E)$ is called the algebra of polynomial differential operators on E. The linear mapping $X \to \partial(X)$ $(X \in E)$ can be extended uniquely to a homomorphism $\partial: S(E) \to \mathfrak{D}(E)$.

For any $p \in P(E)$ and $q \in S(E)$, let $\langle p, q \rangle$ denote the value of $\partial(q)p$ at the origin so that $\langle p, q \rangle = p(0; \partial(q))$.

LEMMA 4. *The bilinear form $(p, q) \to \langle p, q \rangle$ on $P(E) \times S(E)$ is nondegenerate.*

Select a base $(X_1, \cdots, X_n)$ for E and let $(\lambda_1, \cdots, \lambda_n)$ be the dual base for E'. Then it is easy to verify that

$$\langle \lambda_1^{m_1} \lambda_2^{m_2} \cdots \lambda_n^{m_n}, X_1^{m_1'} \cdots X_n^{m_n'} \rangle = 0$$

unless $m_i = m_i'$ $(1 \leq i \leq n)$ and

$$\langle \lambda_1^{m_1} \cdots \lambda_n^{m_n}, X_1^{m_1} \cdots X_n^{m_n} \rangle = m_1! m_2! \cdots m_n!.$$

From this the assertion of the lemma is obvious.

COROLLARY 1. *The homomorphism $\partial: S(E) \to \mathfrak{D}(E)$ is injective.*

This is an immediate consequence of the lemma.

COROLLARY 2. *Fix $X_0 \in E$ and let q be an element of $S(E)$ such that $p(X_0; \partial(q)) = 0$ for all $p \in P(E)$. Then $q = 0$.*

Define $p_{X_0}(X) = p(X_0 + X)$ for $X \in E$ and $p \in P(E)$. Then $p \to p_{X_0}$ is an automorphism of $P(E)$ and $\langle p_{X_0}, q \rangle = p(X_0; \partial(q)) = 0$ $(p \in P(E))$. Hence $q = 0$ from Lemma 4.

It is clear that

$$\partial(X) \circ \lambda - \lambda \circ \partial(X) = \lambda(X) \qquad\qquad (X \in E, \lambda \in E').$$

Therefore it follows that $\mathfrak{D}(E)$ is spanned by elements of the form $p\partial(q)$ $(p \in P(E), q \in S(E))$. Fix $D \in \mathfrak{D}(E)$ and $X_0 \in E$. Then it follows from Corollary 2 above, that there exists exactly one element $D_{X_0} \in \partial(S(E))$ such that $p(X_0; D) = p(X_0; D_{X_0})$ for all $p \in P(E)$. As usual, we call D_{X_0} the *local expression* of D at X_0.

COROLLARY 3. *Suppose D is an element in $\mathfrak{D}(E)$ such that $Dp = 0$ for every $p \in P(E)$. Then $D = 0$.*

Fix $X \in E$. Then $p(X; D_X) = p(X; D) = 0$ for every $p \in S(E)$. Hence $D_X = 0$. Now we can write D in the form $D = \sum_{1 \le i \le r} p_i \partial(q_i)$ where $p_i \in P(E)$, $q_i \in S(E)$ and $q_1, \cdots, q_r$ are linearly independent over $\boldsymbol{C}$. Then

$$D_X = \sum_i p_i(X) \partial(q_i) = 0$$

and so it follows from Corollary 1 above that $p_i(X) = 0$ $(1 \le i \le r)$. This being true for every $X \in E$, we conclude that $p_i = 0$ $(1 \le i \le r)$ and therefore $D = 0$.

COROLLARY 4. *Let γ denote the linear mapping of $P(E) \otimes S(E)$ into $\mathfrak{D}(E)$ such that $\gamma(p \otimes q) = p\partial(q)$ $(p \in P(E), q \in S(E))$. Then γ is bijective.*

It is clear from what we have seen above that γ is surjective. On the other hand suppose $D = \sum_{1 \le i \le r} \gamma(p_i \otimes q_i)$ $(p_i \in P(E), q_i \in S(E))$, where $q_1, \cdots, q_r$ are linearly independent over $\boldsymbol{C}$. Then if $D = 0$, we have seen above (during the proof of Corollary 3) that $p_1 = p_2 = \cdots = p_r = 0$. This proves that γ is injective.

We know that E' and $\partial(E)$ are both contained in $\mathfrak{D}(E)$. It follows from Corollary 4 of Lemma 4 that $E' \cap \partial(E) = \{0\}$. Let $\mathfrak{T}(E)$ denote the tensor algebra over $E' + \partial(E)$ and ϕ_E the homomorphisms of $\mathfrak{T}(E)$ into $\mathfrak{D}(E)$ which corresponds to the inclusion mapping of $E' + \partial(E)$ into $\mathfrak{D}(E)$. Let $\mathfrak{U}(E)$ be the ideal in $\mathfrak{T}(E)$ generated by elements of one of the following three types.

$$(1) \quad \lambda \otimes \partial(X) - \partial(X) \otimes \lambda + \lambda(X) \qquad (\lambda \in E', X \in E),$$

$$(2) \quad \partial(X) \otimes \partial(Y) - \partial(Y) \otimes \partial(X) \qquad (X, Y \in E),$$

$$(3) \quad \lambda \otimes \mu - \mu \otimes \lambda \qquad (\lambda, \mu \in E').$$

Then clearly $\mathfrak{U}(E)$ lies in the kernel of ϕ_E.

LEMMA 5. $\mathfrak{U}(E)$ *is exactly the kernel of* ϕ_E *and therefore*

$$\mathfrak{D}(E) \cong \mathfrak{T}(E)/\mathfrak{U}(E).$$

Select a base $(X_1, \cdots, X_n)$ for E and the dual base $(\lambda_1, \cdots, \lambda_n)$ for E'. Define two injective linear mappings α and β of $P(E)$ and $\partial(S(E))$ respectively into $\mathfrak{T}(E)$ as follows.

$$\alpha(1) = 1, \alpha(\lambda_i{}^m) = \alpha(\lambda_i{}^{m-1}) \otimes \lambda_i \qquad (1 \leq i \leq n, m \geq 1)$$

$$\alpha(\lambda_1{}^{m_1}\lambda_2{}^{m_2} \cdots \lambda_n{}^{m_n}) = \alpha(\lambda_1{}^{m_1}) \otimes \alpha(\lambda_2{}^{m_2}) \otimes \cdots \otimes \alpha(\lambda_n{}^{m_n})$$

$$\beta(1) = 1, \beta(\partial(X_i{}^m)) = \beta(\partial(X_i{}^{m-1})) \otimes \partial(X_i) \qquad (1 \leq i \leq n, m \geq 1)$$

$$\beta(\partial(X_1{}^{m_1}X_2{}^{m_2} \cdots X_n{}^{m_n})) = \beta(\partial(X_1{}^{m_1})) \otimes \beta(\partial(X_2{}^{m_2})) \otimes \cdots \otimes \beta(\partial(X_n{}^{m_n})).$$

It is obvious that $\phi_E(\alpha(p)) = p$, $\phi_E(\beta(\partial(q))) = \partial(q)$ for $p \in P(E)$ and $q \in S(E)$. Now fix $\xi \in \mathfrak{T}(E)$. It is clear from the definition of $\mathfrak{U}(E)$ that we can choose elements $p_i \in P(E)$, $q_i \in S(E)$ $(1 \leq 1 \leq r)$ such that $q_1, \cdots, q_r$ are linearly independent over $\boldsymbol{C}$ and

$$\xi \equiv \sum_i \alpha(p_i) \otimes \beta(\partial(q_i)) \bmod \mathfrak{U}(E).$$

Then

$$\phi_E(\xi) = \sum_i p_i \partial(q_i)$$

and therefore if ξ lies in the kernel of ϕ_E, we can conclude, from Corollary 4 of Lemma 4, that $p_1 = \cdots = p_r = 0$. Hence $\xi \in \mathfrak{U}(E)$ and this proves our assertion.

Now suppose E is the direct sum of two subspaces E_1 and E_2. Let π_i denote the projection of E on E_i $(i = 1, 2)$ corresponding to this sum. Then π_i can be extended uniquely to a homomorphism of $S(E)$ onto $S(E_i)$. We call π_i the projection of $S(E)$ on $S(E_i)$. Note that $S(E_i) \subset S(E)$ (see $[2(e), \S 6]$ and $\pi_i p = p$ for $p \in S(E_i)$.

Consider an element $\xi \in \mathfrak{D}(E_1)$. Then

$$\xi = \sum_{1 \leq i \leq r} p_i \partial(q_i)$$

where $p_i \in P(E_1)$, $q_i \in S(E_1)$. Since $P(E_1) \subset P(E)$, $S(E_1) \subset S(E)$ $[2(e), \S 6]$, the right side can also be interpreted as an element of $\mathfrak{D}(E)$. We claim that in this way, we get an isomorphism of $\mathfrak{D}(E_1)$ into $\mathfrak{D}(E)$. Let $n_1 = \dim E_1$ and assume that the above base $(X_1, \cdots, X_n)$ of E is so chosen that $(X_1, \cdots, X_{n_1})$ is a base for E_1 while the remaining elements form a base for E_2. Define $\mathfrak{X}(E_1)$ and $\mathfrak{U}(E_1)$ corresponding to E_1 as above. Then since $E_1' + \partial(E_1)$ is a subspace of $E' + \partial(E)$, we can identify $\mathfrak{X}(E_1)$ with the subalgebra of $\mathfrak{X}(E)$ generated by 1 and $E_1' + \partial(E_1)$. In order to prove our contention, it would be enough to verify that $\mathfrak{U}(E_1) = \mathfrak{X}(E_1) \cap \mathfrak{U}(E)$. It is obvious that $\mathfrak{U}(E) \supset \mathfrak{U}(E_1)$. So now fix an element $\xi \in \mathfrak{X}(E_1) \cap \mathfrak{U}(E)$. Then we can choose $p_i \in P(E_1)$, $q_i \in S(E_1)$ $(1 \leq i \leq r)$ such that $q_1, \cdots, q_r$ are linearly independent over $\boldsymbol{C}$ and

$$\xi \equiv \sum_i \alpha(p_i) \otimes \beta(\partial(q_i)) \bmod \mathfrak{U}(E_1).$$

Then

$$0 = \phi_E(\xi) = \sum_i p_i \partial(q_i)$$

and therefore $p_1 = \cdots = p_r = 0$ as before. This proves that $\xi \in \mathfrak{U}(E_1)$.

Thus $\mathfrak{D}(E_i)$ can be identified with the subalgebra of $\mathfrak{D}(E)$ generated by $(1, E_i' + \partial(E_i))$ under the homomorphism which preserves every element of $E_i' + \partial(E_i)$ $(i = 1, 2)$.

Since $\partial(X_i)\lambda_j = \delta_{ij}$, it is obvious that any element of $\mathfrak{D}(E_1)$ commutes with every element of $\mathfrak{D}(E_2)$. Hence we can define a homomorphism ψ of $\mathfrak{D}(E_1) \otimes \mathfrak{D}(E_2)$ into $\mathfrak{D}(E)$ such that $\psi(D_1 \otimes D_2) = D_1 D_2$ $(D_i \in \mathfrak{D}(E_i),$ $i = 1, 2)$. We shall now prove that ψ is bijective. Since

$$E' + \partial(E) = \sum_{i=1,2} (E_i' + \partial(E_i)),$$

it is obvious that ψ is surjective. Let $\mathfrak{B}$ be the ideal in $\mathfrak{X}(E)$ generated by elements of the form $a \otimes b - b \otimes a$ $(a \in \mathfrak{X}(E_1), b \in \mathfrak{X}(E_2))$. Since $\phi_E(\mathfrak{X}(E_i)) = \mathfrak{D}(E_i)$ $(i = 1, 2)$, it follows from Lemma 5 that $\mathfrak{B} \subset \mathfrak{U}(E)$. Consider the subspace $\mathfrak{X}_{12} = \mathfrak{X}(E_1) \otimes \mathfrak{X}(E_2)$ of $\mathfrak{X}(E)$. Then it is obvious that

$$\mathfrak{X}(E) = \mathfrak{X}_{12} + \mathfrak{B}.$$

We claim that

$$\mathfrak{U}(E) \cap \mathfrak{X}_{12} = \mathfrak{U}_{12}$$

where $\mathfrak{U}_{12} = \mathfrak{X}(E_1) \otimes \mathfrak{U}(E_2) + \mathfrak{U}(E_1) \otimes \mathfrak{X}(E_2)$. It is clear that $\mathfrak{U}_{12}$ is contained in the left side. Now fix $\xi \in \mathfrak{X}_{12}$. Then, in view of what we have seen above, we can choose $p_{ij} \in P(E_1)$, $q_j \in S(E_1)$, $p_{ik}' \in P(E_2)$, $q_k' \in S(E_2)$ $(1 \leq i \leq r, 1 \leq j \leq s, 1 \leq k \leq t)$ such that $(q_1, \cdots, q_s)$ and $(q_1', \cdots, q_s')$ are separately linearly independent over $\boldsymbol{C}$ and

$$\xi \equiv \sum_{i,j,k} \alpha(p_{ij}) \otimes \beta(\partial(q_j)) \otimes \alpha(p_{ik}') \otimes \beta(\partial(q_k')) \bmod \mathfrak{U}_{12}.$$

Therefore

$$\phi_E(\xi) = \sum_{i,j,k} p_{ij}\partial(q_j) \circ p_{ik}'\partial(q_k') = \sum_{i,j,k} p_{ij}p_{ik}'\partial(q_jq_k')$$

since p_{ik}' and $\partial(q_j)$ commute. But the homomorphism of $S(E_1) \otimes S(E_2)$ into $S(E)$ which maps $q \otimes q'$ onto qq' $(q \in S(E_1),\ q' \in S(E_2))$ is obviously bijective. Hence q_jq_k' $(1 \leq j \leq s, 1 \leq k \leq t)$ are linearly independent in $S(E)$. Therefore if $\phi_E(\xi) = 0$, it follows that

$$\sum_i p_{ij}p_{ik}' = 0$$

for every j, k. But this implies that

$$\sum_i p_{ij} \otimes p_{ik}' = 0$$

in $P(E_1) \otimes P(E_2)$ (see [2(e), §6]. Therefore

$$\sum_i \alpha(p_{ij}) \otimes b \otimes \alpha(p_{ik}') \otimes c = 0$$

for any $b, c \in \mathfrak{X}(E)$. Therefore $\xi \equiv 0 \bmod \mathfrak{U}_{12}$ and this proves our assertion.

Now suppose $D \in \mathfrak{D}(E_1) \otimes \mathfrak{D}(E_2)$ and $\psi(D) = 0$. We have to show that $D = 0$. Choose $\xi_i \in \mathfrak{D}(E_1), \eta_i \in \mathfrak{D}(E_2)$ $(1 \leq i \leq r)$ such that $D = \sum_i \xi_i \otimes \eta_i$. Select $a_i \in \mathfrak{X}(E_1)$, $b_i \in \mathfrak{X}(E_2)$ such that $\phi_E(a_i) = \xi_i$, $\phi_E(b_i) = \eta_i$ $(1 \leq i \leq r)$. Then

$$v = \sum_i a_i \otimes b_i \in \mathfrak{X}_{12}$$

and

$$\phi_E(v) = \sum_i \xi_i\eta_i = \psi(D) = 0.$$

Therefore $v \in \mathfrak{U}_{12}$. Put $\phi_{12} = \phi_{E_1} \otimes \phi_{E_2}$. Then ϕ_{12} is a homomorphism of $\mathfrak{X}_{12}$ onto $\mathfrak{D}(E_1) \otimes \mathfrak{D}(E_2)$ and it is obvious that $\mathfrak{U}_{12}$ is contained in the kernel of ϕ_{12}. Hence

$$D = \sum_i \xi_i \otimes \eta_i = \phi_{12}(v) = 0.$$

Let $U \supset V$ be two open, nonempty and connected subsets of E. It is easy to see that there exists an injective homomorphism $D \to D_V$ of $\mathfrak{D}_m(U)$ into $\mathfrak{D}_m(V)$ such that g_V is the restriction of g on V $(g \in \mathfrak{M}(U))$ and $(\partial(X))_V$ is just the derivation $\partial(X)$ of $\mathfrak{M}(V)$ $(X \in E)$. We may therefore identify $\mathfrak{D}_m(U)$ with a subalgebra of $\mathfrak{D}_m(V)$ under this homomorphism. Then it is obvious that $\mathfrak{D}_h(U) \subset \mathfrak{D}_h(V)$. Hence, in particular, $\mathfrak{D}(E) \subset \mathfrak{D}_h(E) \subset \mathfrak{D}_h(U)$.

Fix an element $D \in \mathfrak{D}_h(U)$ and a point $X_0 \in U$. Then $D = \sum_{1 \leq i \leq r} f_i \partial(q_i)$ where $f_i \in \mathfrak{H}(U)$ and $q_i \in S(E)$. Put $q = \sum_i f_i(X_0) q_i$. Then it is obvious that $f(X_0; D) = f(X_0; \partial(q))$ for every $f \in \mathfrak{H}(U)$. Taking into account Corollary 2 of Lemma 4, we obtain the following result. There exists a unique element $D_{X_0} \in \partial(S(E))$ such that $f(X_0; D) = f(X_0; D_{X_0})$ for all $f \in \mathfrak{H}(U)$. D_{X_0} is called the local expressions of D at X_0.

Now fix $D \in \mathfrak{D}_m(U)$ and $X_0 \in U$. We say that D is *defined* at X_0, if we can choose an open connected neighborhood V of X_0 in U such that $D \in \mathfrak{D}_h(V)$. If this is so, we again denote by D_{X_0} the local expression of D at X_0. (The value of D_{X_0} is independent of the choice of V.)

Let $Q(E)$ denote the quotient field of $P(E)$. Then $Q(E) \subset \mathfrak{M}(E)$. Let $\mathfrak{D}_r(E)$ be the subalgebra of $\mathfrak{D}_m(E)$ generated by $Q(E)$ and $\mathfrak{D}(E)$. We call $\mathfrak{D}_r(E)$ the algebra of rational differential operators on E.

LEMMA 6. *Let D be an element in $\mathfrak{D}_r(E)$ and X_0 a point in E. Suppose D is defined at X_0. Then we can choose $p \in P(E)$ such that $p(X_0) \neq 0$ and $pD \in \mathfrak{D}(E)$. Moreover*

$$D_{X_0} = p(X_0)^{-1}(pD)_{X_0}$$

for any such p.

Choose an open connected neighborhood U of X_0 in E such that $D \in \mathfrak{D}_h(U)$. Since $\mathfrak{D}_r(E) = Q(E)\mathfrak{D}(E)$, we can choose $p_0 \in P(E)$ $(p_0 \neq 0)$ such that $p_0 D \in \mathfrak{D}(E)$. Therefore we can select $q_i \in S(E)$, $f_i \in \mathfrak{H}(U)$ and $p_i \in P(E)$ $(1 \leq i \leq r)$ such that $q_1, \cdots, q_r$ are linearly independent over $\boldsymbol{C}$ and

$$p_0 D = \sum_{1 \leq i \leq r} p_i \partial(q_i),$$

$$D = \sum_{1 \leq i \leq r} f_i \partial(q_i),$$

where the second equation is meant to hold in $\mathfrak{D}_h(U)$. Then it is clear that

$$(p_0 D)_X = \sum_i p_0(X) f_i(X) \partial(q_i) = \sum_i p_i(X) \partial(q_i) \qquad (X \in U).$$

Therefore $p_i = p_0 f_i$ $(1 \leq i \leq r)$ on U. This implies that X_0 does not lie on the polar divisor of any one of the rational functions p_i/p_0 $(1 \leq i \leq r)$. Let q denote the h. c. f. of $(p_0, p_1, \cdots, p_r)$ in $P(E)$ aid put $p = p_0/q$. Then $p \in P(E)$ and it is clear that $pD \in \mathfrak{D}(E)$ and X_0 does not lie in the divisor of zeros of p. Therefore $p(X_0) \neq 0$. The second statement of the lemma is obvious.

The group $\boldsymbol{GL}(E)$ operates on E. If $f \in \mathfrak{H}(E)$ and $x \in \boldsymbol{GL}(E)$ we denote by f^x the function $X \to f(x^{-1}X)$ $(X \in E)$. Then $f \to f^x$ is an automorphism of $\mathfrak{H}(E)$ which can be extended uniquely to an automorphism of $\mathfrak{M}(E)$ over $\mathbf{C}$. For any $D \in \mathfrak{D}_m(E)$, define D^x by

$$D^x g = (Dg^{x^{-1}})^x \qquad\qquad (g \in \mathfrak{M}(E)).$$

Then $D \to D^x$ is an automorphism of $\mathfrak{D}_m(E)$ which leaves $P(E)$ and $Q(E)$ invariant. Also $(\partial(X))^x = \partial(xX)$ $(X \in E)$. Hence x leaves $\partial(S(E))$, $\mathfrak{D}(E)$ and $\mathfrak{D}_r(E)$ invariant.

LEMMA 7. *Let $D \in \mathfrak{D}_r(E)$, $x \in \mathbf{GL}(E)$ and $X \in E$. Suppose D is defined at X. Then D^x is defined at xX and $(D^x)_{xX} = (D_X)^x$.*

Choose $p \in P(E)$ such that $p(X) \neq 0$ and $\xi = pD \in \mathfrak{D}(E)$. Then $D = p^{-1}\xi$ and therefore $D^x = (p^x)^{-1}\xi^x$. Now $p^x D^x = \xi^x \in \mathfrak{D}(E)$ and $p^x(xX) = p(X) \neq 0$. Hence D^x is defined at xX. Moreover

$$(D^x)_{xX} = (p^x(xX))^{-1}(\xi^x)_{xX} = p(X)^{-1}(\xi^x)_{xX}.$$

Now fix $q \in P(E)$. Then $\xi^x q = (\xi q^{x^{-1}})^x$ and $(\xi_X)^x q = (\xi_X q^{x^{-1}})^x$. Hence

$$q(Y;\xi^x) = q^{x^{-1}}(x^{-1}Y;\xi), \qquad q(Y;(\xi_X)^x) = q^{x^{-1}}(x^{-1}Y;\xi_X)$$

for $Y \in E$. Putting $Y = xX$, we get

$$q(xX;(\xi^x)_{xX}) = q^{x^{-1}}(X;\xi) = q(xX;(\xi_X)^x).$$

This being true for every $q \in P(E)$, we conclude from Corollary 2 of Lemma 4 that $(\xi^x)_{xX} = (\xi_X)^x$. Hence

$$(D^x)_{xX} = (p(X)^{-1}\xi_X)^x = (D_X)^x.$$

COROLLARY. *Let U be a subset of E with the following two properties.*

1) *If $p \in P(E)$ and p vanishes identically on U then $p = 0$.*

2) *D is defined at every point of $U \cup x^{-1}U$.*

Then if $D_{xX} = (D_X)^x$ for $X \in U$, we can conclude that $D = D^x$.

It follows from the above lemma that both D and D^x are defined at every point $Y \in U$ and $(D - D^x)_Y = 0$. Choose $p \in P(E)$ $(p \neq 0)$ such that $\Delta = p(D - D^x) \in \mathfrak{D}(E)$. Then if $q \in P(E)$,

$$q(Y;\Delta) = q(Y;\Delta_Y) = 0$$

for $Y \in U$. Hence it follows from condition 1) that $\Delta q = 0$. But then we can conclude from Corollary 3 of Lemma 4 that $\Delta = 0$ and therefore $D - D^x = 0$.

LEMMA 8. *Fix an element $\zeta \neq 0$ in $Q(E)$. Then there exists a unique automorphism μ_ζ of $\mathfrak{D}_r(E)$ such that $\mu_\zeta(q) = q$ for $q \in Q(E)$ and*

$$\mu_\zeta(\partial(X)) = \partial(X) + (1/2)\zeta^{-1}(\partial(X)\zeta) \qquad (X \in E).$$

Moreover $(\mu_\zeta(D))^a = \mu_{\zeta x}(D^x)$ $(x \in \boldsymbol{GL}(E), D \in \mathfrak{D}_r(E))$ *and*

$$\mu_{\zeta_1 \zeta_2} = \mu_{\zeta_1} \mu_{\zeta_2} \qquad (\zeta_i \in Q(E), \zeta_i \neq 0, i = 1, 2).$$

Let U be a non-empty open convex set in E such that U does not meet the divisor of ζ. Fix a point $X_0 \in U$ and define

$$\log \zeta(X) = c + \int_{X_0}^{X} \zeta^{-1} \, d\zeta \qquad (X \in U)$$

where the integral is taken along the straight line-segment from X_0 to X and c is a constant such that $\zeta(X_0) = e^c$. Put

$$\eta = \exp\big((1/2)\log\zeta\big).$$

Then η is a holomorphic function on U and $\zeta = \eta^2$ on U. Let ν_η denote the automorphism of $\mathfrak{D}_m(U)$ given by

$$\nu_\eta(D) = \eta^{-1}D \circ \eta \qquad (D \in \mathfrak{D}_m(U)).$$

It is clear that

$$\begin{aligned}
\nu_\eta(\partial(X)) &= \eta^{-1}\partial(X) \circ \eta \\
&= \partial(X) + \eta^{-1}(\partial(X)\eta) \\
&= \partial(X) + (1/2)\zeta^{-1}(\partial(X)\zeta)
\end{aligned}$$

for $X \in E$. Moreover $\nu_\eta(q) = q$ for $q \in Q(E)$. Hence ν_η obviously defines an automorphism of $\mathfrak{D}_r(E)$. Since the uniqueness of μ_ζ is clear, this proves the first statement of the lemma.

Let α_1, α_2 be two automorphisms of $\mathfrak{D}_r(E)$. If $\alpha_1(\partial(X)) = \alpha_2(\partial(X))$ $(X \in E)$ and $\alpha_1(q) = \alpha_2(q)$ $(q \in Q(E))$, then it is obvious that $\alpha_1 = \alpha_2$. The remaining statements of the lemma follow immediately from this fact.

Remark. We shall usually write $\zeta^{-\frac{1}{2}}D \circ \zeta^{\frac{1}{2}}$ instead of $\mu_\zeta(D)$ $(D \in \mathfrak{D}_r(E))$. It is clear from the above proof that if D, ζ and ζ^{-1} are all defined at a point $X_0 \in E$, then $\zeta^{\frac{1}{2}}D \circ \zeta^{\frac{1}{2}}$ is also defined at X_0.

4. The main theorem on invariant differential operators. Let $\mathfrak{g}$ be a reductive Lie algebra over $\boldsymbol{C}$ and G its connected adjoint group. Then $G \subset \boldsymbol{GL}(\mathfrak{g})$. Let $\mathfrak{F}_r(\mathfrak{g})$ be the subalgebra of those elements $D \in \mathfrak{D}_r(\mathfrak{g})$ for which $D^x = D$ $(x \in G)$.

LEMMA 9. *For any $X \in \mathfrak{g}$, let $\tau(X)$ denote the element of $\mathfrak{D}(\mathfrak{g})$ given by*

$$(\tau(X))_Y = \partial([Y, X]) \qquad (Y \in \mathfrak{g}).$$

Then if $D \in \mathfrak{D}_r(\mathfrak{g})$, the following two conditions on D are equivalent.

1) $D \in \mathfrak{F}_r(\mathfrak{g})$.

2) $\tau(X) \circ D - D \circ \tau(X) = 0$ *for all* $X \in \mathfrak{g}$.

Fix $p \in P(\mathfrak{g})$ $(p \neq 0)$ such that $\xi = pD$ lies in $\mathfrak{D}(\mathfrak{g})$. Let U_0 be a non-empty open set in $\mathfrak{g}$ such that p is nowhere zero on U_0. Fix $Y_0 \in U_0$ and choose open connected neighborhoods U and V of Y_0 and 1 in U_0 and G respectively such that $V^{-1} = V$ and $xY \in U_0$ for $x \in V$ and $Y \in U$. Then if $q \in P(\mathfrak{g})$, the functions $(x, Y) \to (Dq^x)(Y)$ and $(x, Y) \to (Dq)^x(Y)$ are both holomorphic on $V \times U$.

Now suppose $D \in \mathfrak{F}_r(\mathfrak{g})$. Fix $q \in P(\mathfrak{g})$ and $X \in \mathfrak{g}$. Then

$$(Dq)^{\exp tX}(Y) = (Dq^{\exp tX})(Y) \qquad (Y \in U)$$

if t is sufficiently near zero in $\mathbf{C}$. From this it follows that $D_1 q = 0$ on U. Here $D_1 = \tau(X) \circ D - D \circ \tau(X)$. But $D_1 \in \mathfrak{D}_r(\mathfrak{g})$ and therefore $D_1 q \in Q(\mathfrak{g})$. Hence $D_1 q = 0$. But then we can conclude from Corollary 3 of Lemma 4 that $D_1 = 0$.

Conversely suppose 2) holds. Then the proof of Lemma 15 of [2(e)] is applicable and we conclude that

$$D_{xX} = (D_X)^x \qquad (x \in V, X \in U).$$

But then $D = D^x$ $(x \in V)$ from the corollary of Lemma 7. Since G is connected, V generates G and therefore it follows that $D \in \mathfrak{F}_r(\mathfrak{g})$.

Now we shall use the notation of §2 (with $k = \mathbf{C}$). In particular $\mathfrak{z}$ and ζ have the same meaning as before. Let D be an element of $\mathfrak{D}_r(\mathfrak{g})$ which is defined at some point of $\mathfrak{z}$. Then we shall associate to it an element $\Delta \in \mathfrak{D}_r(\mathfrak{g})$ as follows. Choose $q \in P(\mathfrak{g})$ such that 1) q does not vanish identically on $\mathfrak{z}$ and 2) $D' = qD$ is in $\mathfrak{D}(\mathfrak{g})$. This is possible from Lemma 6. For any $Z \in \mathfrak{z}$, define $p_Z \in S(\mathfrak{g})$ by the condition $(D')_Z = \partial(p_Z)$. Then it follows from Corollary 2 of Lemma 2, that there exists an element $\Delta' \in \mathfrak{D}_r(\mathfrak{z})$ which is defined everywhere on $\mathfrak{z}'$ and

$$(\Delta')_Z = \partial(\alpha_Z(p_Z)) \qquad (Z \in \mathfrak{z}').$$

We define $\Delta = q_{\mathfrak{z}}^{-1}\Delta'$ where $q_{\mathfrak{z}}$ denotes the restriction of q on $\mathfrak{z}$. It is easy to verify that Δ does not depend on the choice of q. It is also clear from Lemma

6 that Δ is defined at all points of $\mathfrak{z}'$ at which D is defined. We shall denote Δ by $\delta_{\mathfrak{g}/\mathfrak{z}}'(D)$.

Let $\mathfrak{J}(\mathfrak{g}) = \mathfrak{D}(\mathfrak{g}) \cap \mathfrak{J}_r(\mathfrak{g})$.

LEMMA 10. *If $D \in \mathfrak{J}(\mathfrak{g})$, then $\delta_{\mathfrak{g}/\mathfrak{z}}'(D)$ is invariant under the normalizer of $\mathfrak{z}$ in G.*

Let $\Delta = \delta_{\mathfrak{g}/\mathfrak{z}}'(D)$. Then Δ is defined everywhere on $\mathfrak{z}'$. Fix $x \in G$ such that $x\mathfrak{z} = \mathfrak{z}$. Then by Lemma 3, $x\mathfrak{z}' = \mathfrak{z}'$. So, in view of the corollary of Lemma 7, it would be enough to prove that

$$(\Delta_Z)^x = \Delta_{xZ} \qquad\qquad (Z \in \mathfrak{z}').$$

For $X \in \mathfrak{g}$, define $p_X \in S(\mathfrak{g})$ by the condition $D_X = \partial(p_X)$. Then $p_{xX} = (p_X)^x$ since $D \in \mathfrak{J}(\mathfrak{g})$ (Lemma 7). Hence

$$\Delta_{xZ} = \partial(\alpha_{xZ}(p_{xZ})) = \partial((\alpha_Z(p_Z))^x)$$
$$= (\Delta_Z)^x \qquad\qquad (Z \in \mathfrak{z}')$$

from Lemma 3.

Let $\mathfrak{h}$ be a Cartan subalgebra of $\mathfrak{z}$ and $\mathfrak{h}'$ the set of those elements in $\mathfrak{h}$ which are regular in $\mathfrak{g}$. It is obvious that $\mathfrak{h}' \subset \mathfrak{z}'$. Now fix $D \in \mathfrak{D}(\mathfrak{g})$. Then $\Delta = \delta_{\mathfrak{g}/\mathfrak{z}}'(D)$ is defined everywhere on $\mathfrak{z}'$ and it is easy to see that the element $\delta_{\mathfrak{g}/\mathfrak{h}}'(\Delta)$ of $\mathfrak{D}_r(\mathfrak{h})$ is defined everywhere on $\mathfrak{h}'$.

LEMMA 11. $\delta_{\mathfrak{g}/\mathfrak{h}}'(D) = \delta_{\mathfrak{z}/\mathfrak{h}}'(\delta_{\mathfrak{g}/\mathfrak{z}}'(D))$ *for $D \in \mathfrak{D}(\mathfrak{g})$.*

Fix a point $H \in \mathfrak{h}'$ and let $\Delta = \delta_{\mathfrak{g}/\mathfrak{z}}'(D)$ and $\square = \delta_{\mathfrak{z}/\mathfrak{h}}'(\Delta)$. Choose

$$p \in S(\mathfrak{g}) \text{ such that } D_H = \partial(p). \text{ Then } \Delta_H = \partial(\alpha_H(p)). \text{ Now}$$

$$p - \alpha_H(p) \in \Gamma_H(\mathfrak{Q}_+ \otimes S(\mathfrak{z}))$$

from the definition of $\alpha_H(p)$. Moreover if $\square_H = \partial(q)$ $(q \in S(\mathfrak{h}))$, we have

$$\alpha_H(p) - q \in \Gamma_H(\mathfrak{G}_+ \otimes S(\mathfrak{h})).$$

On the other hand if we apply Lemma 2 to $(\mathfrak{z}, \mathfrak{h})$ instead of $(\mathfrak{g}, \mathfrak{h})$, we get

$$S(\mathfrak{z}) = \Gamma_H(\mathfrak{S}(\mathfrak{z}) \otimes S(\mathfrak{h})).$$

Therefore

$$p - \alpha_H(p) \in \Gamma_H(\mathfrak{Q}_+ \mathfrak{S}(\mathfrak{z}) \otimes S(\mathfrak{h})) \subset \Gamma_H(\mathfrak{G}_+ \otimes S(\mathfrak{h})).$$

This proves that

$$p - q = (p - \alpha_H(p)) + (\alpha_H(p) - q) \in \Gamma_H(\mathfrak{G}_+ \otimes S(\mathfrak{h})).$$

Hence if we take into account the remark after Lemma 2, it becomes clear that $(\delta_{\mathfrak{g}/\mathfrak{h}}'(D))_H = \vartheta(q) = \Box_H$. This proves that $\delta_{\mathfrak{g}/\mathfrak{h}}'(D) = \Box$.

Fix a Cartan subalgebra $\mathfrak{h}$ of $\mathfrak{g}$ and let W denote the Weyl group of $(\mathfrak{g}, \mathfrak{h})$. Let $\mathfrak{F}_r(\mathfrak{h})$ denote the subalgebra of those elements of $\mathfrak{D}_r(\mathfrak{h})$ which are invariant under W. Put $\mathfrak{F}(\mathfrak{h}) = \mathfrak{D}(\mathfrak{h}) \cap \mathfrak{F}_r(\mathfrak{h})$, $J(\mathfrak{g}) = P(\mathfrak{g}) \cap \mathfrak{F}(\mathfrak{g})$ and $J(\mathfrak{h}) = P(\mathfrak{h}) \cap \mathfrak{F}(\mathfrak{h})$. Also let $\pi = \prod_{\alpha > 0} \alpha$ where α runs over all roots of $(\mathfrak{g}, \mathfrak{h})$ which are positive (under some fixed order). Then $\pi \in P(\mathfrak{h})$.

THEOREM 1.[2] *Put*

$$\delta_{\mathfrak{g}/\mathfrak{h}}(D) = \pi \delta_{\mathfrak{g}/\mathfrak{h}}'(D) \circ \pi^{-1}$$

for $D \in \mathfrak{F}(\mathfrak{g})$. *Then* $\delta_{\mathfrak{g}/\mathfrak{h}}(D) \in \mathfrak{F}(\mathfrak{h})$ *and* $\delta_{\mathfrak{g}/\mathfrak{h}}$ *is a homomorphism of* $\mathfrak{F}(\mathfrak{g})$ *into* $\mathfrak{F}(\mathfrak{h})$. *Moreover an element* D *of* $\mathfrak{F}(\mathfrak{g})$ *lies in the kernel of this homomorphism if and only if* $Dp = 0$ *for all* $p \in J(\mathfrak{g})$.

We shall write δ and δ' for $\delta_{\mathfrak{g}/\mathfrak{h}}$ and $\delta_{\mathfrak{g}/\mathfrak{h}}'$ respectively. It is obvious that $\delta(D)$ lies in $\mathfrak{D}_r(\mathfrak{h})$ and its definition is independent of the order of roots.

Let $\mathfrak{c}$ be the center and $\mathfrak{g}_1$ the derived algebra of $\mathfrak{g}$ so that $\mathfrak{g} = \mathfrak{c} + \mathfrak{g}_1$, where the sum is direct. Then $\mathfrak{c} \subset \mathfrak{h}$ and $\mathfrak{h} = \mathfrak{c} + \mathfrak{g}_1 \cap \mathfrak{h}$. We can choose a compact real form $\mathfrak{g}_{10}$ of $\mathfrak{g}_1$ such that $\mathfrak{g}_{10} \cap \mathfrak{h}$ is a Cartan subalgebra of $\mathfrak{g}_1$ (see, for example [3, p. 155]). Choose a real subspace $\mathfrak{c}_0$ of $\mathfrak{c}$ such that $\dim_R \mathfrak{c}_0 = \dim_C \mathfrak{c}$ and $\mathfrak{c}_0$ spans $\mathfrak{c}$ over C. Then $\mathfrak{g}_0 = \mathfrak{c}_0 + \mathfrak{g}_{10}$ is a real form of $\mathfrak{g}$ and $\mathfrak{h}_0 = \mathfrak{h} \cap \mathfrak{g}_0$ is a Cartan subalgebra of $\mathfrak{g}_0$. Let $\mathfrak{h}'$ be the set of all points H in $\mathfrak{h}$ where $\pi(H) \neq 0$. Then $\delta(D)$ is defined everywhere on $\mathfrak{h}'$. It is clear that if $q \in P(\mathfrak{h})$ and $q = 0$ on $\mathfrak{h}_0' = \mathfrak{h}_0 \cap \mathfrak{h}'$, then $q = 0$. Hence in order to prove that δ is a homomorphism of $\mathfrak{F}(\mathfrak{g})$ into $\mathfrak{D}_r(\mathfrak{h})$, it is enough to verify that

$$(\delta(D_1 D_2) - \delta(D_1) \circ \delta(D_2))_H = 0$$

for $H \in \mathfrak{h}_0'$ and $D_1, D_2 \in \mathfrak{F}(\mathfrak{g})$. This is proved in the same way as Lemma 7 of [2(a)].

For a given $s \in W$, we can choose x in G such that $\mathrm{Ad}(x) = s$ on $\mathfrak{h}$. Then if $D \in \mathfrak{F}(\mathfrak{g})$, it follows from Lemma 10 that

$$(\delta'(D))^s = (\delta'(D))^x = \delta'(D).$$

Moreover $\pi^s = \epsilon(s)\pi$ where $\epsilon(s) = 1$ or -1. Therefore it is clear that $\delta(D) \in \mathfrak{F}_r(\mathfrak{h})$. Hence in order to prove that $\delta(D) \in \mathfrak{F}(\mathfrak{h})$, it would be enough to verify that $\delta(D) \in \mathfrak{D}(\mathfrak{h})$.

---

[2] This theorem is a generalization of Theorem 1 of [2(a)].

6

We know (see §3) that $\mathfrak{D}(\mathfrak{g}) = \mathfrak{D}(\mathfrak{c})\mathfrak{D}(\mathfrak{g}_1)$. Hence $D = \sum_{1 \leq i \leq r} \xi_i \eta_i$ where $\xi_i \in \mathfrak{D}(\mathfrak{c})$, $\eta_i \in \mathfrak{D}(\mathfrak{g}_1)$ and $\xi_1, \cdots, \xi_r$ are linearly independent over C. Then since $D \in \mathfrak{J}(\mathfrak{g})$, it is obvious that $\eta_i \in \mathfrak{J}(\mathfrak{g}_1)$. Moreover $\delta(\xi) = \xi$ for $\xi \in \mathfrak{D}(\mathfrak{c})$. Hence it follows from what we have proved above, that

$$\delta(D) = \sum_i \xi_i \delta(\eta_i).$$

So it would be sufficient to prove that $\delta(\eta_i) \in \mathfrak{D}(\mathfrak{h}_1)$ where $\mathfrak{h}_1 = \mathfrak{h} \cap \mathfrak{g}_1$. Thus in order to complete the proof of the first part of Theorem 1 we can confine ourselves to the case when $\mathfrak{g}$ is semisimple. This will be done in §6.

We now come to the second assertion of the theorem. Fix $p \in J(\mathfrak{g})$ and $D \in \mathfrak{J}(\mathfrak{g})$. Then it follows from the defintion of $\delta'(D)$ that $p(H;D) = p(H;\delta'(D))$ for $H \in \mathfrak{h}'$ (see [2(a), Lemma 6]). Therefore if $\delta(D) = 0$, it is obvious that $Dp = 0$ on $\mathfrak{h}'$. Let $n = \dim \mathfrak{g}$. Then it is easy to verify that the holomorphic mapping $(x, H) \to xH$ is of rank n on $G \times \mathfrak{h}'$. Therefore $(\mathfrak{h}')^G$ is open in $\mathfrak{g}$ and hence $Dp = 0$. Conversely suppose $\delta(D) \neq 0$. It follows from Corollary 2 of Lemma 2 that we can choose an integer $m \geq 0$ such that $\nabla = \pi^{2m}\delta'(D) \in \mathfrak{J}(\mathfrak{h})$. It is obvious that $\nabla \neq 0$ and so we can select a point $H_0 \in \mathfrak{h}_0'$ such that $\nabla_H \neq 0$. Since $\mathfrak{g}_{10}$ is a *compact* real form of $\mathfrak{g}_1$, $\mathfrak{h}_0$ is stable under W. Moreover since H_0 is regular in $\mathfrak{g}$, $sH_0 \neq H_0$ for $s \neq 1$ in W. Fix an open neighborhood U of H_0 in $\mathfrak{h}_0'$ such that $sU \cap U = \emptyset$ for $s \neq 1$ in W. Then we can choose $f_1 \in C_c^\infty(U)$ such that $f_1(H_0; \nabla) = f_1(H_0; \nabla_{H_0}) \neq 0$. Put $f(H) = \sum_{s \in W} f_1(sH)$ $(H \in \mathfrak{h}_0)$. Then f is invariant under W and $f(H_0; \nabla) = f_1(H_0; \nabla) \neq 0$. On the other hand we can select $q_1 \in P(\mathfrak{h})$ such that

$$|f(sH_0; \nabla) - q_1(sH_0; \nabla)| \leq (1/2)|f(H_0; \nabla)|$$

for $s \in W$. Put

$$q = [W]^{-1} \sum_{s \in W} q_1{}^s$$

where $[W]$ denotes the order of W. Since f and ∇ are both invariant under W, it is clear that

$$f(H_0; \nabla) - q(H_0; \nabla) = [W]^{-1} \sum_{s \in W} \{f(sH_0; \nabla) - q_1(sH_0; \nabla)\}$$

and therefore

$$|f(H_0; \nabla) - q(H_0; \nabla)| \leq (1/2)|f(H_0; \nabla)|.$$

This shows that $q(H_0; \nabla) \neq 0$. But $q \in J(\mathfrak{h})$ and therefore by Chevalley's Theorem (see [2(a), Lemma 9]), there exists an element $p \in J(\mathfrak{g})$ whose restriction on $\mathfrak{h}$ is q. Therefore

$$p(H_0; D) = q(H_0; \delta'(D)) = \pi(H_0)^{-2m} q(H_0; \nabla) \neq 0$$

and this proves that $Dp \neq 0$.

5. A simple lemma. Let E be a vector space over $\boldsymbol{C}$ of finite dimension. Assume that we are given a nondegenerate symmetric bilinear form B on E. We identify E with its dual under B. Let $(-1)^{\frac{1}{2}}$ denote a fixed square-root of -1.

LEMMA 12. *There exists a unique automorphism* $D \to \hat{D}$ *of* $\mathfrak{D}(E)$ *such that*

$$\hat{X} = -(-1)^{\frac{1}{2}}\partial(X), \quad (\partial(X))^{\smallfrown} = -(-1)^{\frac{1}{2}}X \qquad (X \in E).$$

Similarly there exists a unique anti-automorphism $D \to D^*$ *of* $\mathfrak{D}(E)$ *such that*

$$X^* = X, \quad \partial(X)^* = -\partial(X) \qquad (X \in E).$$

Let x *be an element of* $\boldsymbol{GL}(E)$ *which leaves* B *invariant. Then*

$$(D^x)^* = (D^*)^x, \quad (D^x)^{\smallfrown} = (\hat{D})^x$$

for $D \in \mathfrak{D}(E)$.

We use the notation of §3. Since $E = E'$, $(E + \partial(E), 1)$ generates $\mathfrak{D}(E)$ and therefore the uniqueness of both $*$ and $\smallfrown$ is obvious. We shall now verify their existence. $\mathfrak{T}(E)$ is now the tensor algebra over $E + \partial(E)$. Define two linear mappings $*$ and $\smallfrown$ of $E + \partial(E)$ onto itself according to the statement of the lemma. Then obviously $\smallfrown$ and $*$ can be extended uniquely to an automorphism and an anti-automorphism respectively of $\mathfrak{T}(E)$. It is easy to verify that the set of generators of $\mathfrak{U}(E)$ (described in §3) is mapped onto itself by each of these operations. Therefore we get the corresponding operations in $\mathfrak{D}(E) = \mathfrak{T}(E)/\mathfrak{U}(E)$. The last assertion of the lemma follows from the obvious fact that if two automorphisms (or anti-automorphisms) of $\mathfrak{D}(E)$ coincide on $E + \partial(E)$, then they must be identical.

6. Completion of the proof of Theorem 1. Let $\mathfrak{g}$ be a semisimple Lie algebra over $\boldsymbol{C}$ and $\mathfrak{h}$ a Cartan subalgebra of $\mathfrak{g}$. Put $B(X, Y) = \operatorname{tr}(\operatorname{ad} X \operatorname{ad} Y)$ $(X, Y \in \mathfrak{g})$ and let $B_{\mathfrak{h}}$ denote the restrictions of B on $\mathfrak{h}$. Then both B and $B_{\mathfrak{h}}$ are symmetric and non-degenerate and so we can identify $\mathfrak{g}$ and $\mathfrak{h}$ with their dual spaces respectively under these bilinear forms. Put $\delta = \delta_{\mathfrak{g}/\mathfrak{h}}$ (in the notation of Theorem 1) and define $\tau(X)$ $(X \in \mathfrak{g})$ as in Lemma 9.

LEMMA 13. *Let* D *be an element of* $\mathfrak{J}(\mathfrak{g})$. *Then* $\hat{D}$ *and* D^* *are also in* $\mathfrak{J}(\mathfrak{g})$. *Moreover* $\delta(D) \in \mathfrak{J}(\mathfrak{h})$ *and*

$$\delta(\hat{D}) = (\delta(D))^{\hat{}}, \qquad \delta(D^*) = \delta(D)^*.$$

Finally $(\tau(Y))^{\hat{}} = \tau(Y)$ *and* $(\tau(Y))^* = -\tau(Y)$ *for* $Y \in \mathfrak{g}.$

We can choose, as before, a compact real form $\mathfrak{g}_0$ of $\mathfrak{g}$ such that $\mathfrak{h}_0 = \mathfrak{h} \cap \mathfrak{g}_0$ is a Cartan subalgebra of $\mathfrak{g}_0$. Let G_0 be the real analytic subgroup of G corresponding to $\mathfrak{g}_0$. Then G_0 is compact. Define the space $\mathscr{B}(\mathfrak{g}_0)$ as in [2(a), §§ 2 and 5], and for any $f \in \mathscr{B}(\mathfrak{g}_0)$, put

$$\phi_f(H) = \pi(H) \int_{G_0} f(xH)\,dx \qquad\qquad (H \in \mathfrak{h}_0)$$

where the Haar measure dx on G_0 is so normalized as to make the total measure of G_0 equal to 1. Then we know [2(a), Theorem 3] that $\phi_f \in \mathscr{B}(\mathfrak{h}_0)$. As before, let $\mathfrak{h}_0'$ be the set of all regular elements in $\mathfrak{h}_0$. Then if $D \in \mathfrak{D}(\mathfrak{g})$, we know from § 4 that $\delta(D) = \pi \delta_{\mathfrak{g}/\mathfrak{h}}'(D) \circ \pi^{-1}$ is defined everywhere on $\mathfrak{h}_0'$ and it follows from the proof of Lemma 15 of [2(a)] that

$$\phi_{Df} = \delta(D)\phi_f$$

on $\mathfrak{h}_0'$. Let dX and dH denote Euclidean measures on $\mathfrak{g}_0$ and $\mathfrak{h}_0$ respectively. We normalize them in such a way that

$$\int_{\mathfrak{h}_0} f(X)\,dX = \int_{\mathfrak{g}_0} |\pi(H)|^2\,dH \int_{G_0} f(xH)\,dx$$

for $f \in C_c(\mathfrak{g}_0)$ (see [2(a), p. 105]). Now for $f \in \mathscr{B}(\mathfrak{g}_0)$ and $g \in \mathscr{B}(\mathfrak{h}_0)$, define their Fourier transforms $\hat{f}$ and $\hat{g}$ by

$$\hat{f}(Y) = \int f(X)\exp((-1)^{\frac{1}{2}}B(X,Y))\,dX \qquad\qquad (Y \in \mathfrak{g}_0)$$

and

$$\hat{g}(H') = \int g(H)\exp((-1)^{\frac{1}{2}}B(H,H'))\,dH \qquad\qquad (H' \in \mathfrak{h}_0)$$

Then there exists [2(a), Theorem 3] a number $c \neq 0$ such that

$$\phi_{\hat{f}} = c(\phi_f)^{\hat{}}$$

for all $f \in \mathscr{B}(\mathfrak{g}_0)$.

We know [2(a), Lemma 1] that $(Df)^{\hat{}} = \hat{D}\hat{f}$ for $D \in \mathfrak{D}(\mathfrak{g})$ and $f \in \mathscr{B}(\mathfrak{g}_0)$. Similarly D^* is the adjoint of D and corresponding statements hold for elements of $\mathfrak{D}(\mathfrak{h})$.

Since B is invariant under G, it follows immediately that

$$(f^y)^{\hat{}} = (\hat{f})^y$$

for $y \in G_0$ and $f \in \mathscr{B}(\mathfrak{g}_0)$. Moreover it follows from Lemma 12 that the operations of G on $\mathfrak{D}(\mathfrak{g})$ commute with both $\hat{}$ and $*$. Hence $\mathfrak{J}(\mathfrak{g})$ is mapped into itself by $\hat{}$ and $*$.

Fix $Y \in \mathfrak{g}_0$ and put $y_t = \exp tY$ $(t \in \boldsymbol{R})$. Then

$$f(X; \tau(Y)) = \{df^{y_t}(X)/dt\}_{t=0} = \{df(y_t^{-1}X)/dt\}_{t=0} \qquad (X \in \mathfrak{g}_0)$$

Hence if $f \in C_c^\infty(\mathfrak{g}_0)$, it is obvious that $(\tau(Y)f)\hat{}$ is the function

$$X \to \{d\hat{f}(y_t^{-1}X)/dt\}_{t=0} = \hat{f}(X; \tau(Y)).$$

This proves that $(\tau(Y))\hat{}\hat{f} = \tau(Y)\hat{f}$. Since $C_c^\infty(\mathfrak{g}_0)$ is dense in $\mathscr{E}(\mathfrak{g}_0)$, this shows that $(\tau(Y))\hat{} = \tau(Y)$ for $Y \in \mathfrak{g}_0$ and therefore also for $Y \in \mathfrak{g}$. Moreover we know from Lemma 6 of $[2(e)]$ that $\tau(X)^* = -\tau(X)$ $(X \in \mathfrak{g})$.

Now fix $D \in \mathfrak{J}(\mathfrak{g})$ and first let us assume that $\delta(D) \in \mathfrak{D}(\mathfrak{h})$. Then we have seen in §4 that $\delta(D) \in \mathfrak{J}(\mathfrak{h})$. But $(Df)\hat{} = \hat{D}\hat{f}$ and therefore

$$\phi_{\hat{D}\hat{f}} = \phi_{(Df)}\hat{} = c(\phi_{Df})\hat{} = c(\delta(D)\phi_f)\hat{}$$
$$= c(\delta(D))\hat{}(\phi_f)\hat{}$$

for $f \in \mathscr{E}(\mathfrak{g}_0)$. Also

$$\phi_{\hat{D}\hat{f}} = \delta(\hat{D})\phi_{\hat{f}} = c\delta(\hat{D})(\phi_f)\hat{}$$

on $\mathfrak{h}_0{}'$. Let $\mathscr{E}_W(\mathfrak{h}_0)$ denote the set of all functions on $\mathscr{E}(\mathfrak{h}_0)$ which are invariant under W. Then it follows from $[2(a)$, Theorem 4$]$ and the above result that

$$\delta(D)\hat{}(\pi g) = \delta(\hat{D})(\pi g)$$

on $\mathfrak{h}_0{}'$ for every $g \in \mathscr{E}_W(\mathfrak{h}_0)$. Fix an element $H_0 \in \mathfrak{h}_0{}'$ and an open neighborhood U of H_0 in $\mathfrak{h}_0{}'$ such that $sU \cap U = \emptyset$ for $s \neq 1$ in W. For any $\beta \in C_c^\infty(U)$, put

$$g_\beta = \pi^{-1} \sum_{s \in W} \epsilon(s)\beta^s.$$

Then $g_\beta \in C_c^\infty(\mathfrak{h}_0{}')$ and it is invariant under W. Let ξ be an element of $\mathfrak{J}_r(\mathfrak{h})$ which is defined everywhere on $\mathfrak{h}_0{}'$. Then it is clear that $\xi\beta^s = (\xi\beta)^s = 0$ on U if $s \neq 1$. Hence if we take $g = g_\beta$, in our result above, we get

$$\delta(D)\hat{}\beta = \delta(\hat{D})\beta$$

on U. This proves that $(\delta(D)\hat{} - \delta(\hat{D}))_{H_0} = 0$. But since $\delta(D)\hat{} - \delta(\hat{D})$ lies in $\mathfrak{D}_r(\mathfrak{h})$ and $\mathfrak{h}_0{}'$ is open in $\mathfrak{h}_0$, it follows that $\delta(D)\hat{} = \delta(\hat{D})$.

Now let D be an arbitrary element of $\mathfrak{J}(\mathfrak{g})$. Since π^2 is invariant under W, by Chevalley's Theorem $[2(a)$, Lemma 9$]$ there exists an element $\eta \in J(\mathfrak{g})$ such that $\eta(H) = \pi(H)^2$ $(H \in \mathfrak{h})$. Moreover in view of Corollary 2 of Lemma

2, we can choose an integer $m \geqq 0$ such that $\delta(D) \circ \pi^{2m}$ lies in $\mathfrak{D}(\mathfrak{h})$. Put $D_1 = D \circ \eta^m$. Then

$$\delta(D_1) = \delta(D) \circ \delta(\eta^m) = \delta(D) \circ \pi^{2m} \in \mathfrak{D}(\mathfrak{h})$$

and therefore $\delta(\hat{D}_1) = \delta(D_1)\hat{\;}$ by the proof above. Let q be the degree of π. Then η is homogeneous of degree $2q$. Therefore it is obvious from Lemma 12 that $\hat{\eta} = (-1)^q \partial(\eta)$ in $\mathfrak{D}(\mathfrak{g})$. Hence $\hat{D}_1 = (-1)^{qm} \hat{D} \circ \partial(\eta)^m$ and therefore

$$\delta(\hat{D}_1) = (-1)^{qm} \delta(\hat{D}) \circ \delta(\partial(\eta^m))$$
$$= (-1)^{qm} \delta(\hat{D}) \circ \partial(\pi^{2m})$$

from [2(a), Theorem 1]. Now suppose

$$\delta(\hat{D}) = \sum_{1 \leqq i \leqq r} q_i \partial(p_i)$$

where $q_i \in Q(\mathfrak{h})$, $p_i \in S(\mathfrak{h})$ and $p_1, \cdots, p_r$ are linearly independent over $\boldsymbol{C}$. Then

$$\sum_i q_i \partial(p_i \pi^{2m}) = (-1)^{qm} \delta(D_1)\hat{\;} \in \mathfrak{D}(\mathfrak{h}),$$

and we conclude easily from Corollary 4 of Lemma 4 that $q_i \in P(\mathfrak{h})$. Hence $\delta(\hat{D}) \in \mathfrak{D}(\mathfrak{h})$. But D was an arbitrary element of $\mathfrak{F}(\mathfrak{g})$ and $D \to \hat{D}$ is an automorphism of $\mathfrak{F}(\mathfrak{g})$. Therefore $\delta(D) \in \mathfrak{F}(\mathfrak{h})$ for every $D \in \mathfrak{F}(\mathfrak{g})$ and it follows from our earlier proof that $\delta(\hat{D}) = \delta(D)\hat{\;}$.

It remains to show that $\delta(D^*) = \delta(D)^*$ for $D \in \mathfrak{F}(\mathfrak{g})$. Fix H_0 and U and for any $\beta \in C_c^\infty(U)$ define g_β as above. Let $\mathscr{E}_{G_0}(\mathfrak{g}_0)$ be the set of all functions in $\mathscr{E}(\mathfrak{g}_0)$ which are invariant under G_0. Then by Theorem 4 of $[2(a)]$ there exists a unique element $f_\beta \in \mathscr{E}_{G_0}(\mathfrak{g}_0)$ such that $f_\beta = g_\beta$ on $\mathfrak{h}_0$. Then $\phi_{f_\beta} = \pi g_\beta = \sum_{s \in W} \epsilon(s)\beta^s$. But if $F_1, F_2 \in \mathscr{E}_{G_0}(\mathfrak{g}_0)$, it is clear that

$$\int F_1 F_2 \, dX = \int |\pi(H)|^2 F_1(H) F_2(H) \, dH$$
$$= (-1)^q \int \phi_{F_1} \phi_{F_2} \, dH$$

where q is the degree of π. Hence if $\alpha, \beta \in C_c^\infty(U)$,

$$\int Df_\alpha \cdot f_\beta \, dX = (-1)^q [W] \int \phi_{Df_\alpha} \cdot \beta \, dH$$

since $(\phi_{Df_\alpha})^s = \epsilon(s)\phi_{Df_\alpha}$ $(s \in W)$. But $\phi_{Df_\alpha} = \delta(D)\phi_{f_\alpha}$ and $\phi_{f_\alpha} = \alpha$ on U. Hence

$$\int Df_\alpha \cdot f_\beta \, dX = (-1)^q [W] \int \delta(D)\dot{\alpha} \cdot \beta \, dH.$$

On the other hand D^* is also in $\mathfrak{F}(\mathfrak{g})$ and therefore

$$\int Df_\alpha \cdot f_\beta \, dX = \int f_\alpha \cdot D^* f_\beta \, dX = (-1)^q [W] \int \alpha \cdot \delta(D^*)\beta \, dH.$$

This proves that $(\delta(D)^* - \delta(D^*))_H = 0$ for all $H \in U$. But since U is open in $\mathfrak{h}_0$, this implies that $\delta(D)^* = \delta(D^*)$. The proof of Lemma 13 and therefore also of Theorem 1, is now quite complete.

7. Application to invariant distributions. We now return to the notation of §2 and take $k = \boldsymbol{R}$. Let G be a connected Lie group with Lie algebra $\mathfrak{g}$. Fix a point $Z_0 \in \mathfrak{z}'$ and an open neighborhood Ω of Z_0 in $\mathfrak{g}$.

LEMMA 14. *Let f be a locally invariant C^∞ function on Ω. Then*[3]

$$f(Z_0; D) = f(Z_0; \delta_{\mathfrak{g}/\mathfrak{z}}{}'(D))$$

for $D \in \mathfrak{D}(\mathfrak{g}_c)$.

This is proved in the same way as Lemma 10 of [2(e)].

LEMMA 15. *Let π denote the mapping $(x, Z) \to xZ$ of $G \times \mathfrak{z}$ into $\mathfrak{g}$. Then* $\operatorname{rank} d\pi = n$ *everywhere on $G \times \mathfrak{z}'$. Here $n = \dim \mathfrak{g}$.*

Since $\zeta(Z) \neq 0$, it is clear that $[Z, \mathfrak{q}] = \mathfrak{q}$ for $Z \in \mathfrak{z}'$. Our assertion can now be proved in the same way as Lemma 12 of [2(e)].

Let dx denote the Haar measure on G and dX and dZ the Euclidean measures on $\mathfrak{g}$ and $\mathfrak{z}$ respectively.

LEMMA 16. *Put $\Omega = \pi(G \times \mathfrak{z}') = G\mathfrak{z}'$. Then Ω is open in $\mathfrak{g}$ and for every $\alpha \in C_c^\infty(G \times \mathfrak{z}')$, there exists a unique element $f_\alpha \in C_c^\infty(\Omega)$ such that*

$$\int_{G \times \mathfrak{z}'} F(xZ)\alpha(x:Z)\,dx\,dZ = \int_\Omega F(X)f_\alpha(X)\,dX$$

for every $F \in C_c^\infty(\Omega)$. The mapping $\alpha \to f_\alpha$ of $C_c^\infty(G \times \mathfrak{z}')$ into $C_c^\infty(\Omega)$ is continuous and surjective and $\operatorname{Supp} f_\alpha \subset \pi(\operatorname{Supp} \alpha)$.

The proof of this is similar to that of Lemma 13 of [2(e)].

COROLLARY 1. *Let F be a locally summable function on Ω. Then the function $(x, Z) \to F(xZ)$ is locally summable on $G \times \mathfrak{z}'$ and*

$$\int_{G \times \mathfrak{z}'} F(xZ)\alpha(x:Z)\,dx\,dZ = \int_\Omega F(X)f_\alpha(X)\,dX$$

for every $\alpha \in C_c^\infty(G \times \mathfrak{z}')$.

This follows from Corollary 2 of Theorem 1 of [2(e)].

---

[3] $\mathfrak{g}_c$ denotes the complexification of $\mathfrak{g}$. Similarly in other cases. For convenience we write $\delta_{\mathfrak{g}/\mathfrak{z}}{}'$ instead of $\delta_{\mathfrak{g}_c/\mathfrak{z}_c}{}'$.

LEMMA 17. *Let G_0 and $\mathfrak{z}_0$ be open subsets of G and $\mathfrak{z}'$ respectively. We assume that G_0 is connected and not empty. Let T be a locally invariant distribution on $\Omega_0 = G_0 \mathfrak{z}_0$. Then there exists a unique distribution σ_T on $\mathfrak{z}_0$ such that*

$$T(f_\alpha) = \sigma_T(\beta_\alpha) \qquad\qquad (\alpha \in C_c^\infty(G_0 \times \mathfrak{z}_0))$$

where $\qquad\qquad \beta_\alpha(Z) = \int \alpha(x:Z)\,dx \qquad\qquad (Z \in \mathfrak{z}_0).$

Moreover σ_T is locally invariant (with respect to $\mathfrak{z}$) and $\sigma_T = 0$ implies that $T = 0$.

Everything except the local invariance of σ_T is proved in the same way as Theorem 2 of [2(e)]. Now fix $\beta \in C_c^\infty(\mathfrak{z}_0)$ and $Y \in \mathfrak{z}$. We have to prove that $\sigma_T(\tau(Y)\beta) = 0$. Select $\gamma \in C_c^\infty(G_0)$ such that $\int \gamma(x)\,dx = 1$ and put $\alpha = \gamma \times \beta$ and $\alpha_1 = \gamma \times \tau(Y)\beta$. Then if $F \in C_c^\infty(\Omega_0)$, we have

$$\int f_{\alpha_1}(X) F(X)\,dX = \int \gamma(x)\beta(Z;\tau(Y)) F(xZ)\,dx dZ.$$

But it is clear that

$$\int \beta(Z;\tau(Y)) F(xZ)\,dx dZ = \int \beta(Z) F(x;Y:Z)\,dx dZ$$

where $F(x:Z) = F(xZ)$ $(x \in G_0, Z \in \mathfrak{z}_0)$. Therefore

$$\int f_{\alpha_1} F\,dX = -\int \gamma(x;Y)\beta(Z) F(x:Z)\,dx dZ = -\int f_{\alpha_2} F\,dX$$

where $\alpha_2 = Y\gamma \times \beta$. This proves that $f_{\alpha_1} = -f_{\alpha_2}$ and therefore

$$\sigma_T(\tau(Y)\beta) = T(f_{\alpha_1}) = -T(f_{\alpha_2}) = -\sigma_T(\beta_{\alpha_2}).$$

But since

$$\int \gamma(x;Y)\,dx = 0,$$

it is clear that $\beta_{\alpha_2} = 0$ and therefore $\sigma_T(\tau(Y)\beta) = 0$.

COROLLARY 1. *Assume that $G_0 = G$ and T is invariant under G. Let Ξ be the subgroup of all $\xi \in G$ such that $\xi\mathfrak{z}_0 = \mathfrak{z}_0$. Then σ_T is invariant under Ξ.*

Let Ξ_1 be the normalizer of $\mathfrak{z}$ in G and Ξ_0 the analytic subgroup of G corresponding to $\mathfrak{z}$. Then Ξ_0 is the connected component of 1 in Ξ_1 and if Z_G is the center of G, $\Xi_1/Z_G\Xi_0$ is finite (see [2(b), Lemma 15]). Since $\mathfrak{z}$ is reductive, the measure dZ is invariant under Ξ_0 and therefore also under Ξ_1. Now we may obviously assume that $\mathfrak{z}_0$ is not empty. Since it is open in $\mathfrak{z}$, it spans $\mathfrak{z}$ over $\boldsymbol{R}$. Hence $\Xi \subset \Xi_1$ and this proves that dZ is invariant under Ξ.

Now fix γ as above and for given $\beta \in C_c^\infty(\mathfrak{z}_0)$ and $\xi \in \Xi$, define $\beta_1(Z)$

$= \beta(\xi Z)$ $(Z \in \mathfrak{z}_0)$. Then $\sigma_T(\beta_1) = T(f_{\alpha_1})$ where $\alpha_1 = \gamma \times \beta_1$. On the other hand

$$\int f_{\alpha_1} F \, dX = \int \gamma(x) \beta(\xi Z) F(xZ) \, dx dZ$$
$$= \int \gamma(x\xi) \beta(Z) F(xZ) \, dx dZ$$

for any $F \in C_c^\infty(\Omega_0)$. Put $\alpha(x:Z) = \gamma(x\xi)\beta(Z)$ $(x \in G, Z \in \mathfrak{z}_0)$. Then the above equation implies that $f_{\alpha_1} = f_\alpha$ and therefore

$$\sigma_T(\beta_1) = T(f_{\alpha_1}) = T(f_\alpha) = \sigma_T(\beta),$$

since

$$\int \alpha(x:Z) \, dx = \beta(Z) \qquad\qquad (Z \in \mathfrak{z}_0).$$

This proves the corollary.

COROLLARY 2.[4] *For any* $D \in \mathfrak{F}(\mathfrak{g}_c)$, *write* $\Delta(D) = \delta_{\mathfrak{g}/\mathfrak{z}}'(D)$. *Then* $\sigma_{DT} = \Delta(D)\sigma_T$ *under the conditions of Lemma 17.*

Put $T' = DT$. Then T' is also locally invariant and $T'(f_\alpha) = T(D^*f_\alpha)$ for $\alpha \in C_c^\infty(G_0 \times \mathfrak{z}_0)$. Fix $F \in C_c^\infty(\Omega_0)$ and put $F_1 = DF$. Then

$$\int f_\alpha(X;D^*) F(X) \, dX = \int f_\alpha F_1 \, dX$$
$$= \int \alpha(x:Z) F_1(xZ) \, dx dZ.$$

As usual, define $F(x:Z) = F(xZ)$ $(x \in G_0, Z_0 \in \mathfrak{z}_0)$. Then since $D \in \mathfrak{F}(\mathfrak{g}_c)$, it is clear that

$$F_1(xZ) = F(xZ;D) = F(xZ;D^x).$$

Moreover from Corollary 2 of Lemma 2, we can choose $g_i \in \mathfrak{S}_+(\mathfrak{q}_c)$, $p_i \in S(\mathfrak{z}_c)$ and $a_i \in C^\infty(\mathfrak{z}_0)$ $(1 \leq i \leq r)$ such that

$$D_Z = (\Delta(D))_Z + \sum_i a_i(Z) \partial(\Gamma_Z(g_i \otimes p_i))$$

for $Z \in \mathfrak{z}_0$. Since

$$(D^x)_{xZ} = (D_Z)^x$$

it follows that

$$F(xZ;D^x) = F(x:Z;\Delta(D)) + \sum_i F(x;g_i:Z;\xi_i)$$

for $x \in G_0$ and $Z \in \mathfrak{z}_0$. Here ξ_i is the differential operator $a_i \partial(p_i)$ on $\mathfrak{z}_0$. Put $\alpha_0(x:Z) = \alpha(x:Z;\Delta(D)^*)$ and $\alpha_i(x:Z) = \alpha(x;g_i^*:Z;\xi_i^*)$ $(1 \leq i \leq r)$, where the star denotes the adjoint, as usual. Then

$$\int D^*f_\alpha \cdot F \, dX = \sum_{0 \leq i \leq r} \int \alpha_i(x:Z) F(xZ) \, dx dZ$$

---

[4] This can be regarded as a generalization of Lemma 14 to distributions.

$$= \sum_{0 \leq i \leq r} \int f_{\alpha_i} F \, dX.$$

This proves that

$$D^* f_\alpha = \sum_{0 \leq i \leq r} f_{\alpha_i}$$

and therefore

$$T'(f_\alpha) = \sum_{0 \leq i \leq r} \sigma_T(\beta_{\alpha_i}).$$

But since $g_i \in \mathfrak{S}_+(\mathfrak{q}_c)$ $(1 \leq i \leq r)$, it is clear that $g_i h = 0$ where h is the constant function 1 on G. Hence

$$\int g_i {}^* f \, dx = \int f \cdot g_i h \, dx = 0$$

for any $f \in C_c^\infty(G)$. This shows that $\beta_{\alpha_i} = 0$ for $i \geq 1$ and therefore

$$T'(f_\alpha) = \sigma_T(\beta_{\alpha_0}) = \sigma_T(\Delta(D)^* \beta_\alpha).$$

Since the mapping $\alpha \rightarrow \beta_\alpha$ of $C_c^\infty(G_0 \times \mathfrak{z}_0)$ into $C_c^\infty(\mathfrak{z}_0)$ is obviously surjective, this proves that $\sigma_{T'} = \Delta(D)\sigma_T$.

COROLLARY 3. *Let* $D_1, D_2 \in \mathfrak{F}(\mathfrak{g}_c)$. *Then*

$$\Delta(D_1 D_2)\sigma_T = \Delta(D_1)\Delta(D_2)\sigma_T$$

in the notation of Corollary 2 above.

This follows from the fact that $(D_1 D_2) T = D_1(D_2 T)$.

8. An explicit formula. Let $p \rightarrow p_{\mathfrak{z}}$ $(p \in S(\mathfrak{g}_c)$ denote the natural projection of $S(\mathfrak{g}_c)$ onto $S(\mathfrak{z}_c)$ with respect to the decomposition $\mathfrak{g}_c = \mathfrak{z}_c + \mathfrak{q}_c$ (see § 3). Define $\zeta \in P(\mathfrak{z}_c)$ as in § 2 and put [5]

$$\zeta^{-\frac{1}{2}} D \circ \zeta^{\frac{1}{2}} = \mu_\zeta(D) \qquad\qquad (D \in \mathfrak{D}(\mathfrak{z}_c)$$

in the notation of Lemma 8.

Let $I(\mathfrak{g}_c)$ be the algebra of all invariants of G in $S(\mathfrak{g}_c)$. We keep to the notation of § 7.

THEOREM 2. *Let* T *be a locally invariant distribution on* Ω_0. *Then*

$$\Delta(\partial(p))\sigma_T = (\zeta^{-\frac{1}{2}}\partial(p_{\mathfrak{z}}) \circ \zeta^{\frac{1}{2}})\sigma_T$$

for $p \in I(\mathfrak{g}_c)$.

Let $\mathfrak{c}$ be the center and $\mathfrak{g}_1$ the derived algebra of $\mathfrak{g}$. We can obviously choose a real, symmetric and nondegenerate bilinear form B on $\mathfrak{g}$ such that

---

[5] Actually it is not difficult to show that there exists an invariant polynomial function η on $\mathfrak{z}$ such that $\eta^2 = \zeta$.

1) $B(X, Y) = \operatorname{tr}(\operatorname{ad} X \operatorname{ad} Y)$ for $X, Y \in \mathfrak{g}_1$ and 2) $B(X, Y) = 0$ for $X \in \mathfrak{c}$ and $Y \in \mathfrak{g}_1$. Then B is invariant under G. Extend this form on $\mathfrak{g}_c$ by linearity over C and identify $\mathfrak{g}_c$ with its dual under B. Then $S(\mathfrak{g}_c) = P(\mathfrak{g}_c)$ and it follows from Lemma 1 that $B_{\mathfrak{z}}$, the restriction of B on $\mathfrak{z}_c$, is also non-degenerate. We identify $\mathfrak{z}_c$ with its dual under $B_{\mathfrak{z}}$. Then it is easy to verify that $p_{\mathfrak{z}}$ is simply the restriction of the polynomial function p on $\mathfrak{z}_c$ ($p \in S(\mathfrak{g}_c)$).

Let ω denote the polynomial function $X \to B(X, X)$ on $\mathfrak{g}_c$. We shall first prove the following result.

LEMMA 18. $\Delta(\partial(\omega)) = \zeta^{-\frac{1}{2}}\partial(\omega_{\mathfrak{z}}) \circ \zeta^{\frac{1}{2}}$.

Fix a Cartan subalgebra $\mathfrak{h}$ of $\mathfrak{z}$ and let $\mathfrak{h}'$ be the set of all elements of $\mathfrak{h}$ which are regular in $\mathfrak{g}$. Put

$$D = \Delta(\partial(\omega)) - \zeta^{-\frac{1}{2}}\partial(\omega_{\mathfrak{z}}) \circ \zeta^{\frac{1}{2}}.$$

Let Ξ be the analytic subgroup of G corresponding to $\mathfrak{z}$. It follows from Lemma 10 that D is invariant under Ξ. Moreover $\mathfrak{z_0}' = \bigcup_{\xi \in \Xi} \xi \mathfrak{h}'$ is an open and non-empty subset of $\mathfrak{z}'$ (Lemma 15). Therefore since $D \in \mathfrak{D}_r(\mathfrak{z}_c)$, it would be sufficient to verify that $D_{\xi H} = 0$ for $\xi \in \Xi$ and $H \in \mathfrak{h}'$. But $D_{\xi H} = (D_H)^\xi$ because D is invariant under Ξ. Hence we have only to prove that $D_H = 0$ for $H \in \mathfrak{h}'$.

Let P be the set of all positive roots of $(\mathfrak{g}_c, \mathfrak{h}_c)$ (under some fixed order). For any root α define X_α as in the proof of Lemma 1. We normalize X_α and $X_{-\alpha}$ in such a way that $B(X_\alpha, X_{-\alpha}) = 1$. The restriction $B_{\mathfrak{h}}$ of B on $\mathfrak{h}_c$ is also nondegenerate. We identify $\mathfrak{h}_c$ with it dual under $B_{\mathfrak{h}}$.

Define subsets $P_{\mathfrak{z}}$ and $P_{\mathfrak{q}}$ of P as follows. A root $\alpha \in P$ belongs to $P_{\mathfrak{z}}$ or $P_{\mathfrak{q}}$ according as X_α lies in $\mathfrak{z}_c$ or $\mathfrak{q}_c$. Then (see [2(a)], p. 98])

$$\omega = \omega_{\mathfrak{z}} + 2 \sum_{\alpha \in P_{\mathfrak{q}}} X_\alpha X_{-\alpha},$$

$$\omega_{\mathfrak{z}} = \omega_{\mathfrak{h}} + 2 \sum_{\alpha \in P_{\mathfrak{z}}} X_\alpha X_{-\alpha},$$

where $\omega_{\mathfrak{h}}$ is the restriction of ω on $\mathfrak{h}_c$. A simple computation shows that

$$\Gamma_H(\lambda(X_\alpha X_{-\alpha}) \otimes 1) = -\alpha(H)^2 X_\alpha X_{-\alpha} + \alpha(H)H_\alpha \qquad (\alpha \in P, H \in \mathfrak{h})$$

where λ is the canonical mapping of $S(\mathfrak{g}_c)$ onto $\mathfrak{G}$. Here H_α is the element in $\mathfrak{h}_c$ which corresponds to α under $B_{\mathfrak{h}}$. Therefore

$$\omega = \omega_{\mathfrak{z}} + 2 \sum_{\alpha \in P_{\mathfrak{q}}} \alpha(H)^{-1}H_\alpha - 2 \sum_{\alpha \in P_{\mathfrak{q}}} \alpha(H)^{-2}\Gamma_H(\lambda(X_\alpha X_{-\alpha}) \otimes 1)$$

for $H \in \mathfrak{h}'$. This implies that

$$(\Delta(\partial(\omega)))_H = \partial(\omega_{\mathfrak{z}}) + 2 \sum_{\alpha \in P_q} \alpha(H)^{-1}\partial(H_\alpha) \qquad (H \in \mathfrak{h}').$$

We find in the same way that

$$\omega_{\mathfrak{z}} = \omega_{\mathfrak{h}} + 2 \sum_{\alpha \in P_{\mathfrak{z}}} \alpha(H)^{-1}H_\alpha - 2 \sum_{\alpha \in P_{\mathfrak{z}}} \alpha(H)^{-2}\Gamma_H(\lambda(X_\alpha X_{-\alpha}) \otimes 1)$$

$$(H \in \mathfrak{h}').$$

Now fix a point $H_0 \in \mathfrak{h}'$. Choose an open convex neighborhood U_c of H_0 in $\mathfrak{z}_c$ such that 1) every point of U_c is regular in $\mathfrak{g}_c$ and 2) there exists a holomorphic function η on U_c such that $\zeta = \eta^2$ on U_c. Put $U = U_c \cap \mathfrak{z}$. Then if ϕ is a C^∞ function on U, it follows from Lemma 3 of $[2(\text{a})]$ that

$$\phi(H;\partial(\omega_{\mathfrak{z}})) = \phi(H;\partial(\omega_{\mathfrak{h}})) + 2 \sum_{\alpha \in P_{\mathfrak{z}}} \alpha(H)^{-1}\phi(H;\partial(H_\alpha))$$

$$- 2 \sum_{\alpha \in P_{\mathfrak{z}}} \alpha(H)^{-2}\phi(1;\lambda(X_\alpha X_{-\alpha}):H) \qquad (H \in U \cap \mathfrak{h}).$$

Now fix $f \in C^\infty(U)$. It is enough to show that $f(H_0;D) = 0$. Put $g = \eta f$. Then

$$f(H;D) = f(H;\Delta(\partial(\omega))) - \eta(H)^{-1}g(H;\partial(\omega_{\mathfrak{z}})) \qquad (H \in U \cap \mathfrak{h}).$$

Put

$$\pi = \prod_{\alpha \in P} \alpha, \quad \pi_q = \prod_{\alpha \in P_q} \alpha \text{ and } \pi_{\mathfrak{z}} = \prod_{\alpha \in P_{\mathfrak{z}}} \alpha.$$

It is obvious that $\zeta(H) = (-1)^r \pi_q(H)^2$ for $H \in \mathfrak{h}_c$. (Here r is the number of roots in P_q.) Since the set $\mathfrak{h}_c \cap U_c$ is convex and therefore connected, it is clear that $\eta(H) = \epsilon \pi_q(H)$ $(H \in \mathfrak{h}_c \cap U_c)$ where ϵ is a constant such that $\epsilon^4 = 1$.

Now $g(1;\lambda(X_\alpha X_{-\alpha}):H) = g(H;\tau(\lambda(X_\alpha X_{-\alpha})))$ for $\alpha \in P_{\mathfrak{z}}$ (see Lemma 11 of $[2(\text{e})]$). Since ζ and therefore also η are locally invariant (with respect to $\mathfrak{z}$), it is clear that

$$\tau(\lambda(X_\alpha X_{-\alpha}))g = \eta\tau(\lambda(X_\alpha X_{-\alpha}))f.$$

Hence we conclude from our formula above that

$$g(H;\delta(\omega_{\mathfrak{z}})) = \epsilon f(H;\partial(\omega_{\mathfrak{h}}) \circ \pi_q) + 2 \sum_{\alpha \in P_{\mathfrak{z}}} \alpha(H)^{-1}\epsilon f(H;\partial(H_\alpha) \circ \pi_q)$$

$$- 2\epsilon \sum_{\alpha \in P_{\mathfrak{z}}} \alpha(H)^{-2}\pi_q(H)f(1;\lambda(X_\alpha X_{-\alpha}):H)$$

for $H \in U \cap \mathfrak{h}$. On the other hand, it follows from our formula for $(\Delta(\partial(\omega)))_H$ that

$$f(H;\Delta(\partial(\omega))) = f(H;\partial(\omega_{\mathfrak{z}})) + 2 \sum_{\alpha \in P_q} \alpha(H)^{-1}f(H;\partial(H_\alpha))$$

$$= f(H;\partial(\omega_{\mathfrak{h}})) + 2 \sum_{\alpha \in P} \alpha(H)^{-1} f(H;\partial(H_\alpha))$$

$$- 2 \sum_{\alpha \in P_{\mathfrak{z}}} \alpha(H)^{-2} f(1;\lambda(X_\alpha X_{-\alpha}):H)$$

This proves that

$$\pi_{\mathfrak{q}}(H) f(H;D) = -f(H;\square) \qquad (H \in U \cap \mathfrak{h})$$

where $\square$ is the element of $\mathfrak{D}_r(\mathfrak{h}_c)$ given by [6]

$$\square = \{\partial(\omega_{\mathfrak{h}}), \pi_{\mathfrak{q}}\} + 2 \sum_{\alpha \in P_{\mathfrak{z}}} \alpha^{-1}(\partial(H_\alpha)\pi_{\mathfrak{q}}) - 2\pi_{\mathfrak{q}} \sum_{\alpha \in P_{\mathfrak{q}}} \alpha^{-1}\partial(H_\alpha).$$

We shall now verify that $\square = 0$ and this would prove the required result.

Let $\alpha_1, \alpha_2, \cdots, \alpha_r$ be all the distinct roots in $P_{\mathfrak{q}}$. Then $\pi_{\mathfrak{q}} = \alpha_1 \alpha_2 \cdots \alpha_r$ and $\{\partial(\omega_{\mathfrak{h}}), \alpha_i\} = 2\partial(\alpha_i)$ (see [2(a), p. 98]). Hence

$$\{\partial(\omega_{\mathfrak{h}}), \pi_{\mathfrak{q}}\} = 2 \sum_{1 \leq i \leq r} \alpha_1 \cdots \alpha_{i-1}\partial(\alpha_i) \circ (\alpha_{i+1} \cdots \alpha_r)$$

$$= 2\pi_{\mathfrak{q}} \sum_{1 \leq i \leq r} \alpha_i^{-1}\partial(\alpha_i) + q$$

where $q \in S(\mathfrak{h}_c)$. Hence

$$\square = q + 2 \sum_{\alpha \in P_{\mathfrak{z}}} \alpha^{-1}(\partial(H_\alpha)\pi_{\mathfrak{q}}) \in Q(\mathfrak{h}_c).$$

Therefore applying the differential operator $\square$ to the constant function 1, we get

$$\square = \square 1 = \partial(\omega_{\mathfrak{h}})\pi_{\mathfrak{q}} + 2 \sum_{\alpha \in P_{\mathfrak{z}}} \alpha^{-1}(\partial(H_\alpha)\pi_{\mathfrak{q}}).$$

On the other hand if f_0 denotes the constant function 1 on U, we have

$$\pi_{\mathfrak{q}}(H) f_0(H;D) = -f_0(H;\square) = -\square(H) \qquad (H \in U \cap \mathfrak{h}).$$

But it follows from our formula for $(\Delta(\partial(\omega)))_H$ that

$$f_0(H;D) = -\eta(H)^{-1}\eta(H;\partial(\omega_{\mathfrak{z}})).$$

Therefore

$$\epsilon\square(H) = \eta(H;\partial(\omega_{\mathfrak{z}})) \qquad (H \in U \cap \mathfrak{h}).$$

But since η is locally invariant (with respect to $\mathfrak{z}$), we conclude from Lemma 14 that

$$\eta(H;\partial(\omega_{\mathfrak{z}})) = \eta(H;\delta_{\mathfrak{z}/\mathfrak{h}}'(\partial(\omega_{\mathfrak{z}}))).$$

---

[6] If D_1, D_2 are two differential operators, we denote $D_1 \circ D_2 - D_2 \circ D_1$, as usual, by $\{D_1, D_2\}$.

But $\delta_{\mathfrak{z}/\mathfrak{h}'}\,(\partial(\omega_{\mathfrak{z}})) = \pi_{\mathfrak{z}}^{-1}\partial(\omega_{\mathfrak{h}})\circ\pi_{\mathfrak{z}}$ from [2(a), Lemma 8]. Hence

$$\epsilon\square(H) = \eta(H;\pi_{\mathfrak{z}}^{-1}\,\partial(\omega_{\mathfrak{h}})\circ\pi_{\mathfrak{z}}) = \epsilon\pi_{\mathfrak{z}}(H)^{-1}\pi(H;\partial(\omega_{\mathfrak{h}})).$$

However we know that $\partial(\omega_{\mathfrak{h}})\pi = 0$ [2(a), p. 101]. Therefore $\square = 0$ and so the lemma is proved.

Now we come to the proof of Theorem 2. It is obviously enough to consider the case when p is homogeneous. Let Λ denote the derivation of $\mathfrak{J}(\mathfrak{g}_c)$ given by $\Lambda D = (1/2)\{\partial(\omega),D\}$ $(D \in \mathfrak{J}(\mathfrak{g}_c))$. Similarly let ν_0 denote the derivation of $\mathfrak{D}_r(\mathfrak{z}_c)$ defined by $\nu_0\xi = (1/2)\{\Delta(\partial(\omega)),\xi\}$ $(\xi \in \mathfrak{D}_r(\mathfrak{z}_c))$. Then it follows from Corollary 3 of Lemma 17 that

$$\Delta(\Lambda D)\sigma_T = (\nu_0\Delta(D))\sigma_T \qquad\qquad (D \in \mathfrak{J}(\mathfrak{g}_c))$$

for a locally invariant distribution T on Ω_0. Hence we conclude by induction on m, that

$$\Delta(\Lambda^m D)\sigma_T = (\nu_0{}^m\Delta(D))\sigma_T \qquad\qquad (m \geqq 0).$$

Let p be a homogeneous element in $I(\mathfrak{g}_c)$ of degree k. It is clear from its definition that $\Delta(p) = p_{\mathfrak{z}}$. Moreover $\Lambda^k p = k!\partial(p)$ (see [2(a), p. 99]). Therefore

$$k!\Delta(\partial(p))\sigma_T = (\nu_0{}^k p_{\mathfrak{z}})\sigma_T.$$

But $\Delta(\partial(\omega)) = \mu_\zeta(\partial(\omega_{\mathfrak{z}}))$ from Lemma 18 and $p_{\mathfrak{z}} = \mu_\zeta(p_{\mathfrak{z}})$. Therefore since μ_ζ is an automorphism of $\mathfrak{D}_r(\mathfrak{z}_c)$, it is clear that

$$\nu_0{}^k p_{\mathfrak{z}} = \mu_\zeta(\nu^k p_{\mathfrak{z}})$$

where ν is the derivation of $\mathfrak{D}_r(\mathfrak{z}_c)$ given by

$$\nu\xi = (1/2)\{\partial(\omega_{\mathfrak{z}}),\xi\} \qquad\qquad (\xi \in \mathfrak{D}_r(\mathfrak{z}_c)).$$

Moreover $p_{\mathfrak{z}}$ is also a homogeneous element of $S(\mathfrak{z}_c)$ of degree k and $\nu^k p_{\mathfrak{z}} = k!\partial(p_{\mathfrak{z}})$ [2(a), p. 99]. Therefore

$$\mu_\zeta(\nu^k p_{\mathfrak{z}}) = k!\mu_\zeta(\partial(p_{\mathfrak{z}}))$$

and this proves that

$$\Delta(\partial(p))\sigma_T = \mu_\zeta(\partial(p_{\mathfrak{z}}))\sigma_T.$$

9. An elementary result. For $X \in \mathfrak{g}_c$, define the derivation d_X of $S(\mathfrak{g}_c)$ as in [2(e), §7]. Then it is obvious that $I(\mathfrak{g}_c)$ consists of all $p \in S(\mathfrak{g}_c)$ such that $d_X p = 0$ for every $X \in \mathfrak{g}_c$. Since $[\mathfrak{z},\mathfrak{q}] \subset \mathfrak{q}$, it follows that if $p \in I(\mathfrak{g}_c)$ then $d_Z p_{\mathfrak{z}} = 0$ $(Z \in \mathfrak{z}_c)$ and therefore $p_{\mathfrak{z}} \in I(\mathfrak{z}_c)$. Fix a Cartan subalgebra $\mathfrak{h}$ of $\mathfrak{z}$ and let $W(\mathfrak{g})$ and $W(\mathfrak{z})$ denote the Weyl groups of $(\mathfrak{g}_c, \mathfrak{h}_c)$ and $(\mathfrak{z}_c, \mathfrak{h}_c)$ respectively. Then $W(\mathfrak{z})$ is generated by the Weyl reflexions s_α corresponding to the roots α in $P_\mathfrak{z}$ (in the notation of the proof of Lemma 18). Hence we can regard $W(\mathfrak{z})$ as a subgroup of $W(\mathfrak{g})$. Let $w(\mathfrak{g}/\mathfrak{z})$ denote the index $[W(\mathfrak{g}) : W(\mathfrak{z})]$.

LEMMA 19. *Let J denote the image of $I(\mathfrak{g}_c)$ in $I(\mathfrak{z}_c)$ under the mapping $p \to p_\mathfrak{z}$ $(p \in I(\mathfrak{g}_c))$. Then $I(\mathfrak{z}_c)$ is a free abelian module over J of rank $w(\mathfrak{g}/\mathfrak{z})$.*

Let $\mathfrak{p}$ be the subspace of $\mathfrak{z}$ complementary to $\mathfrak{h}$ such that $[\mathfrak{h}, \mathfrak{p}] \subset \mathfrak{p}$ (see Lemma 1). Then we have the decompositions $\mathfrak{z} = \mathfrak{h} + \mathfrak{p}$ and $\mathfrak{g} = \mathfrak{h} + (\mathfrak{p} + \mathfrak{q})$. Let $E_\mathfrak{z}$ and $E_\mathfrak{g}$ denote the corresponding projections of $S(\mathfrak{z}_c)$ and $S(\mathfrak{g}_c)$ respectively on $S(\mathfrak{h}_c)$ (see §3). Define $p_\mathfrak{h} = E_\mathfrak{g} p$ $(p \in I(\mathfrak{g}_c))$, $q_\mathfrak{h} = E_\mathfrak{z} q$ $(q \in I(\mathfrak{z}_c))$, $J_\mathfrak{g} = E_\mathfrak{g} I(\mathfrak{g}_c)$, $J_\mathfrak{z} = E_\mathfrak{z} I(\mathfrak{z}_c)$. Then by Chevalley's Theorem [2(a), Lemma 9], $J_\mathfrak{z}$ is the algebra of all invariants of $W(\mathfrak{z})$ in $S(\mathfrak{h}_c)$ and $E_\mathfrak{z}$ defines an isomorphism of $I(\mathfrak{z}_c)$ onto $J_\mathfrak{z}$. Similarly $E_\mathfrak{g}$ defines an isomorphism of $I(\mathfrak{g}_c)$ onto $J_\mathfrak{g}$ which is, in fact, the algebra of all invariants of $W(\mathfrak{g})$ in $S(\mathfrak{h}_c)$. Since $(p_\mathfrak{z})_\mathfrak{h} = p_\mathfrak{h}$ for $p \in I(\mathfrak{g}_c)$, it is clear that under the above isomorphism of $I(\mathfrak{z}_c)$ onto $J_\mathfrak{z}$, J is mapped onto $J_\mathfrak{g}$. Therefore it would be sufficient to show that $J_\mathfrak{z}$ is a free abelian module over $J_\mathfrak{g}$ of rank $w(\mathfrak{g}/\mathfrak{z})$. The proof of this is substantially the same as that of Lemma 8 of [2(c)].

10. An example. As an application of the above theory, we shall conclude this paper with a proof of the following theorem.

THEOREM 3. *Let $\mathfrak{g}$ be a reductive Lie algebra over $\boldsymbol{R}$ and S the set of all singular elements of $\mathfrak{g}$. Let T be a distribution on an open subset Ω of $\mathfrak{g}$. Assume that:*

1) *T is locally invariant;*

2) *There exists an ideal $\mathfrak{U}$ in $I(\mathfrak{g}_c)$ such that $\dim I(\mathfrak{g}_c)/\mathfrak{U} < \infty$ and $\partial(u)T = 0$ for $u \in \mathfrak{U}$;*

3) *$\operatorname{Supp} T \subset \Omega \cap S$.*

Then $T = 0$.

We use induction on $\dim \mathfrak{g}$. Let us assume that contrary to our assertion, $T \neq 0$. Consider the space $\mathfrak{X}$ of all distributions on Ω of the form $\partial(p)T$ $(p \in I(\mathfrak{g}_c))$. Then it is obvious from condition 2) that $\dim \mathfrak{X} < \infty$. It is also obvious that every distribution in $\mathfrak{X}$ satisfies all the conditions of the theorem and $\mathfrak{X} \neq \{0\}$ since $T \in \mathfrak{X}$.

For any p in $I(\mathfrak{g}_c)$, let $\rho(p)$ denote the linear transformation $T' \to \partial(p)T'$ $(T' \in \mathfrak{X})$ in $\mathfrak{X}$. Then ρ is a representation of the abelian algebra $I(\mathfrak{g}_c)$ on the finite-dimensional space $\mathfrak{X} \neq \{0\}$. Hence we can choose an element $T_0 \neq 0$ in $\mathfrak{X}$ and a homomorphism χ of $I(\mathfrak{g}_c)$ into C such that $\partial(p)T_0 = \chi(p)T_0$ for every $p \in I(\mathfrak{g}_c)$. We intend to derive a contradiction by proving that $T_0 = 0$.

Let us change notation and write T instead of T_0. Fix an element $Z_0 \in \Omega$. We have to show that $T = 0$ around Z_0. Let $\mathfrak{c}$ be the center and $\mathfrak{g}_1$ the derived algebra of $\mathfrak{g}$ so that $\mathfrak{g}$ is the direct sum of $\mathfrak{c}$ and $\mathfrak{g}_1$. If $\mathfrak{g} = \mathfrak{c}$, it is clear that S is empty and therefore $T = 0$. Hence we may suppose that $\mathfrak{g}_1 \neq \{0\}$. First assume that $\dim \mathfrak{g}_1 < \dim \mathfrak{g}$. Choose open connected sets $\mathfrak{c}_0$ and Ω_1 in $\mathfrak{c}$ and $\mathfrak{g}_1$ respectively such that $Z_0 \in \mathfrak{c}_0 + \Omega_1 \subset \Omega$. Then it would be enough (see $[2(e)$, Lemma 3$]$) to show that $T(\alpha \times \beta) = 0$ for $\alpha \in C_c^\infty(\mathfrak{c}_0)$ and $\beta \in C_c^\infty(\Omega_1)$. Since $\mathfrak{c} \subset I(\mathfrak{g}_c)$, $\partial(C)T = \chi(C)T$ for $C \in \mathfrak{c}$. Now fix β and put $T_\beta(\alpha) = T(\alpha \times \beta)$ for $\alpha \in C_c^\infty(\mathfrak{c}_0)$. Then T_β is a distribution on $\mathfrak{c}_0$. Let $C_1, \cdots, C_r$ be a base for $\mathfrak{c}$ over $\mathbf{R}$. Then $\partial(C_1^2 + C_2^2 + \cdots + C_r^2)$ is an elliptic differential operator on $\mathfrak{c}$. Since

$$\partial(C_1^2 + \cdots + C_r^2)T_\beta = \chi(C_1^2 + \cdots + C_r^2)T_\beta,$$

it follows that T_β is an analytic function on $\mathfrak{c}_0$. Let dC denote the Euclidean measure over $\mathfrak{c}$. Then it follows from the differential equations $\partial(C)T_\beta = \chi(C)T_\beta$ $(C \in \mathfrak{c})$, that

$$T_\beta(\alpha) = \nu(\beta) \int \alpha(C) e^{\chi(C)} dC \qquad (\alpha \in C_c^\infty(\mathfrak{c}_0))$$

where $\nu(\beta)$ is a complex number independent of α. Choose $\alpha_0 \in C_c^\infty(\mathfrak{c}_0)$, such that

$$\int \alpha_0(C) e^{\chi(C)} dC = 1.$$

Then

$$\nu(\beta) = T_\beta(\alpha_0) = T(\alpha_0 \times \beta)$$

for $\beta \in C_c^\infty(\Omega_1)$. Hence ν is a distribution on Ω_1. It is obvious that $\tau(X)\nu = 0$ for $X \in \mathfrak{g}_1$ and $\partial(p)\nu = \chi(p)\nu$ for $p \in I(\mathfrak{g}_{1c})$. Moreover if $X \in \mathfrak{g}_1$ and $C \in \mathfrak{c}$, then $X + C$ is singular in $\mathfrak{g}$ if and only if X is singular in $\mathfrak{g}_1$. Hence $\mathrm{Supp}\,\nu \subset \Omega_1 \cap S_1$ where $S_1 = \mathfrak{g}_1 \cap S$. Since $\dim \mathfrak{g}_1 < \dim \mathfrak{g}$, our induction hypothesis is applicable to $\mathfrak{g}_1$ and therefore $\nu = 0$. But then

$$T(\alpha \times \beta) = \nu(\beta) \int \alpha(C) e^{\chi(C)} dC = 0$$

for $\alpha \in C_c^\infty(\mathfrak{c}_0)$ and $\beta \in C_c^\infty(\Omega_1)$. This proves that $T = 0$ around Z_0.

So we can now assume that $\mathfrak{g}_1 = \mathfrak{g}$ and therefore $\mathfrak{g}$ is semisimple. Then $Z_0 = H_0 + X_0$ where H_0 is a semisimple and X_0 a nilpotent element of $\mathfrak{g}$ and $[H_0, X_0] = 0$ (see [1, p. 79]). First assume $H_0 \neq 0$. Let $\mathfrak{z}$ be the centralizer of H_0 in $\mathfrak{g}$. Then since H_0 is contained in some Cartan subalgebra of $\mathfrak{g}$, it is clear that $\mathfrak{z}$ is reductive in $\mathfrak{g}$ and $\operatorname{rank} \mathfrak{z} = \operatorname{rank} \mathfrak{g}$. Define $\mathfrak{q}$ as in Lemma 1. Since $X_0 \in \mathfrak{z}$, $\operatorname{ad} X_0$ maps $\mathfrak{q}$ into itself. Also the eigenvalues of $\operatorname{ad} Z_0 = \operatorname{ad}(H_0 + X_0)$ in $\mathfrak{q}$ are the same as those of $\operatorname{ad} H_0$. Hence none of them can be zero. Therefore $\zeta(Z_0) \neq 0$ and so $Z_0 \in \mathfrak{z}'$ in the notation of § 2.

Choose open connected neighborhoods G_0 and $\mathfrak{z}_0$ of 1 and Z_0 in G and $\mathfrak{z}'$ respectively such that $\Omega_0 = G_0 \mathfrak{z}_0 \subset \Omega$. We now use the notation of §§ 7-9. Then in view of Lemma 17, it would be enough to prove that $\sigma_T = 0$. Fix $\beta \in C_c^\infty(\mathfrak{z}_0)$ and assume that every element of $\operatorname{Supp} \beta$ is regular in $\mathfrak{z}$. Choose $\gamma \in C_c^\infty(G_0)$ such that $\int \gamma \, dx = 1$ and put $\alpha = \gamma \times \beta$. Then $\sigma_T(\beta) = T(f_\alpha)$. Let $X \in \operatorname{Supp} f_\alpha$. Then by Lemma 16, $X = xZ$ for some $x \in G_0$ and $Z \in \operatorname{Supp} \beta$. Therefore Z is regular in $\mathfrak{z}$ and $\zeta(Z) \neq 0$. This obviously means that Z, and hence also X, is regular in $\mathfrak{g}$. Therefore $\operatorname{Supp} f_\alpha$ does not meet S and hence $T(f_\alpha) = 0$. This proves that every point of $\operatorname{Supp} \sigma_T$ is singular in $\mathfrak{z}$. Since ζ takes only real and non-zero values on $\mathfrak{z}_0$, it is clear $|\zeta|^2 = \zeta^2$ on $\mathfrak{z}_0$ and $|\zeta|^{\frac{1}{2}}$, $|\zeta|^{-\frac{1}{2}}$ are analytic functions on $\mathfrak{z}_0$. Therefore it is clear from Lemma 8 that $\mu_\zeta(D) = |\zeta|^{-\frac{1}{2}} D \circ |\zeta|^{\frac{1}{2}}$ $(D \in \mathfrak{D}(\mathfrak{z}_c))$ as differential operators on $\mathfrak{z}$. Put $\sigma = |\zeta|^{\frac{1}{2}} \sigma_T$. Then σ is a locally invariant distribution on $\mathfrak{z}_0$ and $\partial(u_\mathfrak{z})\sigma = 0$ for $u \in \mathfrak{U}$ (Lemma 17 and Theorem 2). Put $\mathfrak{W} = I(\mathfrak{z}_c)\mathfrak{U}_\mathfrak{z}$ where $\mathfrak{U}_\mathfrak{z}$ is the space of all elements in $I(\mathfrak{z}_c)$ of the form $u_\mathfrak{z}$ $(u \in \mathfrak{U})$. Then it follows from Lemma 19 and condition 2) of Theorem 3, that $\dim I(\mathfrak{z}_c)/\mathfrak{W} < \infty$. Moreover it is clear that $\partial(v)\sigma = 0$ $(v \in \mathfrak{W})$. Since $H_0 \neq 0$ and $\mathfrak{g}$ is semisimple, we know that $\dim \mathfrak{z} < \dim \mathfrak{g}$. Therefore $\sigma = 0$ by induction hypothesis. This proves that $\sigma_T = |\zeta|^{-\frac{1}{2}} \sigma = 0$ and therefore $T = 0$ on Ω_0.

Thus we have now shown that no element of $\mathfrak{g}$, which is not nilpotent, can lie in $\operatorname{Supp} T$. Therefore $\operatorname{Supp} T \subset \mathfrak{N} \cap \Omega$ in the notation of Theorem 4 of [2(e)]. Since the Killing form ω lies in $I(\mathfrak{g}_c)$, we have $\partial(\omega)T = \chi(\omega)T$. Therefore $T = 0$ from Theorem 5 of [2(e)]. This completes the proof of Theorem 3.

The Institute for Advanced Study.

7

REFERENCES.

[1] N. Bourbaki, *Groupes et Algèbres de Lie, Chapitre I, Algèbres de Lie*, 1960, Hermann, Paris.

[2] Harish-Chandra, (a) "Differential operators on a semisimple Lie algebra," *American Journal of Mathematics*, vol. 79 (1957), pp. 87-120.

(b) "Fourier transforms on a semisimple Lie algebra II," *American Journal of Mathematics*, vol. 79 (1957), pp. 653-686.

(c) "Spherical functions on a semisimple Lie group I," *American Journal of Mathematics*, vol. 80 (1958), pp. 241-310.

(d) "Invariant eigendistributions on semisimple Lie groups," *Bulletin of the American Mathematical Society*, vol. 69 (1963), pp. 117-123.

(e) "Invariant distributions on Lie algebras," *American Journal of Mathematics*, vol. 86 (1964), pp. 271-309.

[3] S. Helgason, *Differential Geometry and Symmetric Spaces*, 1962, Academic Press, New York.

Reprinted from
Amer. J. of Math.
86 (1964), 534–564

Some Results on an Invariant Integral on a Semi-Simple Lie Algebra

By HARISH-CHANDRA

1. Introduction

Let $\mathfrak{h}$ be a Cartan subalgebra of a real semi-simple Lie algebra $\mathfrak{g}$, and $\mathfrak{h}'$ the set of all regular elements of $\mathfrak{h}$. Then, corresponding to any $f \in C_c^\infty(\mathfrak{g})$, we have defined in [1 (c), Theorem 3, p. 225] a C^∞ function ψ_f on $\mathfrak{h}'$. This paper is concerned with the behaviour of ψ_f around the singular points. Since ψ_f is of class C^∞ on the closure of each connected component of $\mathfrak{h}'$, it is enough to observe its jumps around a semi-regular point H_0 in $\mathfrak{h}$. However the centralizer of H_0 is reductive and its derived algebra $\mathfrak{l}$ is of dimension 3. So this can be done by means of a rather elementary computation[1] on $\mathfrak{l}$. Our main result, in this direction, is contained in Theorem 1. Let π be the product of all the positive roots. Then it follows, in particular, that $\partial(\pi)\psi_f$ is a continuous function on $\mathfrak{h}$ (see § 11). Let $(\partial(\pi)\psi_f)_0$ denote its value at the origin. Then, as shown in [1 (d)], there exists a real number c such that

$$\left(\partial(\pi)\psi_f\right)_0 = cf(0)$$

for all $f \in C_c^\infty(\mathfrak{g})$. Theorem 3, which had been stated without proof in [1 (e), p. 760], gives the sign of c. As we shall see in another paper, all these results can be lifted to a connected Lie group G corresponding to $\mathfrak{g}$.

Let J be an invariant distribution on $\mathfrak{g}$ whose support is contained in the singular set of $\mathfrak{g}$. The rest of this paper is devoted to the proofs of two theorems (Theorems 4 and 5) which, roughly speaking, assert the following. If J has a certain special form and if its support does not contain any semi-regular element, then $J = 0$. Theorem 4 constitutes the third main step in the proof of Lemma 1 of [1 (f)] while Theorem 5 will be needed later for the construction of the discrete series of G (see [1 (f), Theorems 2 and 3]). The proofs of these theorems depend very much on a detailed consideration of the singularities of ψ_f.

This work has been supported at various stages by grants from the Sloan and the National Science Foundations and a contract with the U. S. Army.

2. Some elementary calculations

For any $f \in C_c^\infty(\mathbf{R}^2)$, put

---

[1] This method of utilizing the semi-regular elements has already been introduced in [1 (d)].

$$g_f(\theta) = \theta \int_0^\infty f(\theta e^t, \theta e^{-t})(e^t - e^{-t})dt \qquad\qquad (\theta \in \mathbf{R}')$$

where $\mathbf{R}'$ denotes the complement of zero in $\mathbf{R}$. It is obvious that g_f is a C^∞ function on $\mathbf{R}'$.

LEMMA 1. $\mathrm{Lim}_{\theta\to 0}\, dg_f(\theta)/d\theta = -2f(0, 0)$.

Fix f and put $g = g_f$ and $g' = dg/d\theta$. Then we find by differentiation that

$$g'(\theta) = \int_0^\infty \{f + \theta e^t f_x + \theta e^{-t} f_y\}(e^t - e^{-t})dt$$

where (x, y) are the coordinates in $\mathbf{R}^2$, $f_x = \partial f/\partial x$, $f_y = \partial f/\partial y$ and we have to make the substitution $x = \theta e^t$, $y = \theta e^{-t}$ inside the integral. For any $F \in C_c(\mathbf{R}^2)$, define $M(F) = \sup_{x,y} |F(x, y)|$ and choose $T > 0$ such that $|f(x, y)| = 0$ unless $|x| + |y| \leq T$. Then

$$g'(\theta) = \int_0^\infty (e^t f + \theta e^{2t} f_x)dt - \int_0^\infty e^{-t} f dt + \int_0^\infty (\theta f_y - \theta f_x - \theta e^{-2t} f_y)dt \ .$$

Since $|f(x, y)| \leq M(f)$ and $\int_0^\infty e^{-t} dt = 1$, it follows from Lebesgue's theorem that

$$\mathrm{Lim}_{\theta\to 0} \int_0^\infty e^{-t} f dt = f(0, 0) \ .$$

Similarly

$$\mathrm{Lim}_{\theta\to 0}\, \theta \int_0^\infty e^{-2t} f_y dt = 0 \ .$$

Moreover $f_y(\theta e^t, \theta e^{-t}) = 0$ unless $|\theta| e^t \leq T$. Therefore if $|\theta| \leq T$, it is obvious that

$$\int_0^\infty |f_y| \, dt \leq M(f_y) \log (T/|\theta|) \ ,$$

and so we conclude that

$$\mathrm{Lim}_{\theta\to 0}\, \theta \int_0^\infty f_y dt = 0 \ .$$

Similarly

$$\mathrm{Lim}_{\theta\to 0}\, \theta \int_0^\infty f_x dt = 0 \ .$$

Hence it would be sufficient to prove that

$$\mathrm{Lim}_{\theta\to 0} \int_0^\infty (e^t f + \theta e^{2t} f_x)dt = -f(0, 0) \ .$$

Now if $x = \theta e^t$, $y = \theta e^{-t}$, we have

$$d(xf)/dt = (f + xf_x)\theta e^t - \theta^2 f_y \ ,$$

and therefore

$$e^t f + \theta e^{2t} f_x = (f + x f_x) e^t = \theta^{-1} (d(xf)/dt) + \theta f_y .$$

Moreover

$$\theta^{-1} \int_0^\infty (d(xf)/dt) dt = -\theta^{-1}(xf)_{t=0} = -f(\theta, \theta) .$$

Hence the required result follows immediately.

COROLLARY. $dg_f/d\theta$ can be extended to a continuous function on **R**.

This is obvious.

LEMMA 2.[2] $\mathrm{Lim}_{\theta \to +0}\, g_f(\theta) = \int_0^\infty f(x, 0) dx, \quad \mathrm{Lim}_{\theta \to -0}\, g_f(\theta) = -\int_{-\infty}^0 f(x, 0) dx.$

We have seen above that

$$\theta \int_0^\infty f e^{-t} dt \to 0$$

as $\theta \to 0$. Hence it is enough to consider the integral $\theta \int_0^\infty f e^t dt$. First assume that $0 < \theta \leq T$. Then this is clearly equal to

$$\int_\theta^\infty f(x, \theta^2 x^{-1}) dx = \int_\theta^T f(x, \theta^2 x^{-1}) dx ,$$

and it follows from Lebesgue's theorem that

$$\mathrm{Lim}_{\theta \to +0} \int_\theta^\infty f(x, \theta^2 x^{-1}) dx = \int_0^T f(x, 0) dx = \int_0^\infty f(x, 0) dx .$$

The other case, when $\theta < 0$, is handled in a similar way.

The above lemma shows that g_f cannot, in general, be extended to a continuous function on **R**.

Let I be the simple Lie algebra of dimension 3 spanned over **R** by the elements H, X, Y which satisfy the relations $[H, X] = 2X$, $[H, Y] = -2Y$, $[X, Y] = H$. Let L be the (connected) adjoint group of I, and K the analytic subgroup of L corresponding to the subalgebra $\mathbf{R}(X - Y)$ of I. Then K is compact. For any $f \in C_c(\mathrm{I})$, put

$$\bar{f}(Z) = \int_K f(kZ) dk \qquad\qquad (Z \in \mathrm{I})$$

where dk is the normalized Haar measure of K. Define (see [1 (d), p. 681])

$$\Psi_f(\theta) = \theta \int_0^\infty \bar{f}(\theta(e^t X - e^{-t} Y))(e^t - e^{-t}) dt \qquad\qquad (\theta \in \mathbf{R}') .$$

LEMMA 3. For any $f \in C_c^\infty(\mathrm{I})$, we have

---

[2] $\theta \to +0$ means that $\theta > 0$ and $\theta \to 0$. Moreover $\theta \to -0$ signifies that $-\theta \to +0$.

$$\mathrm{Lim}_{\theta \to 0}\, d\Psi_f(\theta)/d\theta = -2f(0) \ .$$

This follows immediately from Lemma 1.

COROLLARY. *Suppose* $f \in C_c^\infty(\mathfrak{l})$ *and* k *is any non-negative integer.* *Then* $d^{2k+1}\Psi_f(\theta)/d\theta^{2k+1}$ *can be extended to a continuous function on* **R**.

Put $f_k = \partial(\Omega^k)f$ in the notation of [1 (d), § 6]. Then we know from [1 (d), Lemma 17] that

$$d^{2k+1}\Psi_f(\theta)/d\theta^{2k+1} = (-1)^k 2^{3k} d\Psi_{f_k}(\theta)/d\theta \ .$$

Therefore our assertion follows from the above lemma.

3. The semi-regular elements

Let $\mathfrak{g}$ be a reductive Lie algebra over **R** and l the rank of $\mathfrak{g}$. An element $X \in \mathfrak{g}$ is called semi-regular if

(1) $\operatorname{ad} X$ is semi-simple, and

(2) the centralizer $\mathfrak{z}$ of X in $\mathfrak{g}$ is of dimension $l + 2$.

Suppose X is semi-regular. Then it follows from (1) that $\mathfrak{z}$ is reductive in $\mathfrak{g}$. Therefore $\mathfrak{l} = [\mathfrak{z}, \mathfrak{z}]$ is semi-simple and $\mathfrak{z} = \sigma + \mathfrak{l}$ where σ is the center of $\mathfrak{z}$ and the sum is direct. We can choose a Cartan subalgebra $\mathfrak{h}$ of $\mathfrak{g}$ containing X. Then $\sigma \subset \mathfrak{h} \subset \mathfrak{z}$ and therefore $\mathfrak{h} = \sigma + \mathfrak{l} \cap \mathfrak{h}$ and $\mathfrak{l} \cap \mathfrak{h}$ is a Cartan subalgebra of $\mathfrak{l}$. But $\dim \mathfrak{z} = l + 2 = \dim \mathfrak{h} + 2$ and therefore

$$\dim \mathfrak{l} = 2 + \dim (\mathfrak{l} \cap \mathfrak{h}) = 2 + \operatorname{rank} \mathfrak{l} \ .$$

Since $\mathfrak{l}$ is semi-simple, this implies that $\dim \mathfrak{l} = 3$. Now, up to isomorphism, there are only two real semi-simple Lie algebras of dimension 3. One of them is compact while the other is not. We say that X is a semi-regular element of compact or non-compact type according as $\mathfrak{l}$ is compact or not.

4. Real and imaginary roots

Let $\mathfrak{h}$ be a Cartan subalgebra of $\mathfrak{g}$. For any root[3] α of $(\mathfrak{g}, \mathfrak{h})$, select an element $X_\alpha \neq 0$ in $\mathfrak{g}_c$ such that $[H, X_\alpha] = \alpha(H)X_\alpha$ for $H \in \mathfrak{h}_c$. We normalize X_α and $X_{-\alpha}$ in such a way that $\operatorname{tr}(\operatorname{ad} X_\alpha \operatorname{ad} X_{-\alpha}) = 1$. Put $H_\alpha = [X_\alpha, X_{-\alpha}] \in \mathfrak{h}_c$. Then $\alpha(H) = \operatorname{tr}(\operatorname{ad} H \operatorname{ad} H_\alpha)$ for $H \in \mathfrak{h}_c$. We call α real (imaginary) if $\alpha(H)$ is real (pure imaginary) for every $H \in \mathfrak{h}$. A root which is neither real nor imaginary will be called complex.

Now fix α and consider the subalgebra $\mathfrak{l}_c = CH_\alpha + CX_\alpha + CX_{-\alpha}$ of $\mathfrak{g}_c$. It is clear that $\mathfrak{l}_c \subset [\mathfrak{g}_c, \mathfrak{g}_c]$. Put $\mathfrak{l} = \mathfrak{g} \cap \mathfrak{l}_c$ and let η denote the conjugation of $\mathfrak{g}_c$ with respect to $\mathfrak{g}$.

LEMMA 4. *The following three conditions are mutually equivalent.*

---

[3] By a root of $(\mathfrak{g}, \mathfrak{h})$, we mean a root of $(\mathfrak{g}_c, \mathfrak{h}_c)$. Similarly for the Weyl group.

(1) $\mathfrak{l}_c$ *is stable under* η.

(2) $\dim_R \mathfrak{l} = 3$.

(3) α *is either real or imaginary.*

Since $\dim_C \mathfrak{l}_c = 3$, it is obvious that (1) and (2) are equivalent. For any linear function λ on $\mathfrak{h}_c$, define another linear function $\eta\lambda$ on $\mathfrak{h}_c$ by[4] $(\eta\lambda)(H) = \operatorname{conj} \lambda(H) \, (H \in \mathfrak{h})$. Since η leaves $\mathfrak{h}$ fixed, it is obvious that if β is a root, the same holds for $\eta\beta$. Moreover $\eta(H_\beta) = H_{\eta\beta}$ and $\eta(X_\beta) = c_\beta X_{\eta\beta}$ where c_β is a non-zero complex number. Therefore if (1) holds, it is clear that $\eta\alpha = \alpha$ or $-\alpha$. If $\eta\alpha = \alpha$, then $\eta(H_\alpha) = H_\alpha$ and therefore $H_\alpha \in \mathfrak{h}$. This obviously means that α is real. Similarly α is imaginary if $\eta\alpha = -\alpha$. Conversely suppose (3) holds. Then $\eta\alpha = \pm\alpha$, and therefore it is obvious that (1) holds.

We say that α is singular if $\dim_R \mathfrak{l} = 3$ and the algebra $\mathfrak{l}$, which is then obviously semi-simple, is not compact. On the other hand we say α is compact if $\dim_R \mathfrak{l} = 3$ and $\mathfrak{l}$ is compact.

LEMMA 5. *Let* α *be a real root. Then it is singular and we can choose* X_α *and* $X_{-\alpha}$ *in such a way that*

$$\mathfrak{l} = \mathbf{R} H_\alpha + \mathbf{R} X_\alpha + \mathbf{R} X_{-\alpha} \, .$$

Clearly $H_\alpha \in \mathfrak{l}$ and, since all eigenvalues of $\operatorname{ad} H_\alpha$ are real, it is obvious that $\mathfrak{l}$ cannot be compact. Hence α is singular. The rest follows from [1 (c), Lemma 46].

COROLLARY. *Every compact root is imaginary.*

This follows immediately from Lemmas 4 and 5.

Let W denote the Weyl group[3] of $(\mathfrak{g}, \mathfrak{h})$ and, for any root α, s_α the Weyl reflexion $H \to H - 2(\alpha(H)/\alpha(H_\alpha))H_\alpha$ on $\mathfrak{h}_c$. Let G be a connected Lie group with Lie algebra $\mathfrak{g}$. We denote by W_G the subgroup of those elements $s \in W$ for which we can select an element $x \in G$ such that[5] $xH = sH$ for all $H \in \mathfrak{h}$. (It is clear that W_G depends only on $\mathfrak{g}$ and not on the choice of G.)

LEMMA 6. *Let* α *be a root which is either real or compact. Then* $s_\alpha \in W_G$.

This follows from Corollary 2 of [1 (c), Lemma 46].

LEMMA 7. *Fix an element* $H_0 \in \mathfrak{h}$, *and let* Ξ *denote the centralizer of* H_0 *in* G, *and* $x \to x^*$ *the natural projection of* G *on* $G^* = G/\Xi$. *Then we can select a neighborhood* U *of* H_0 *in* $\mathfrak{h}$ *with the following property. Given any compact subset* ω *of* $\mathfrak{g}$, *there exists a compact subset* Ω^* *in* G^* *such that* $x^* \in \Omega^*$ *whenever* $xH \in \omega \, (x \in G)$ *for some* $H \in U$.

---

[4] As usual, $\operatorname{conj} c$ denotes the complex conjugate of a number $c \in \mathbf{C}$.

[5] As usual, we write $xX = X^x = \operatorname{Ad}(x)X \, (x \in G, X \in \mathfrak{g}_c)$.

For the purpose of this lemma, we can obviously assume that $\mathfrak{g}$ is semi-simple and G is its connected adjoint group. Define V as in [1 (e), Lemma 11] and fix a real number $t \neq 0$ such that $tH_0 \in V$. Then it is clear that Ξ is also the centralizer of $a_0 = \exp tH_0 \in G$. Therefore we can choose a neighborhood U of H_0 in $\mathfrak{h}$ such that $tU \subset V$ and $B = \exp(tU)$ fulfills the condition of Theorem 1 of [1 (e)]. Given ω, put $V_0 = V \cap (t\omega)$. Then since ω is compact, $\exp V_0$ is contained in a compact subset C of G. Therefore by [1 (e), Theorem 1] we can select a compact set Ω^* in G^* such that $x^* \in \Omega^*$ whenever $xax^{-1} \in C$ for some $x \in G$ and $a \in B$. Now suppose $xH \in \omega$ for some $x \in G$ and $H \in U$. Then $t(xH) \in V_0$ since $V^x = V$, and therefore $\exp(txH) = xax^{-1} \in C$ where $a = \exp tH \in B$. Hence $x^* \in \Omega^*$.

Fix an order in the space spanned over $\mathbf{R}$ by all the roots of $(\mathfrak{g}, \mathfrak{h})$ and define $\pi_Q = \prod_{\alpha \in Q} \alpha$ for any set Q of positive roots. ($\pi_Q = 1$ if Q is empty.) Then π_Q is a polynomial function on $\mathfrak{h}_c$. Let $\mathfrak{h}'(Q)$ denote the set of all points $H \in \mathfrak{h}$ where $\pi_Q(H) \neq 0$. Consider the set S of all positive singular roots. Then $S = S_R \cup S_I$ where S_R is the set of all positive real roots and S_I the set of all imaginary roots in S. We write $\pi = \pi_Q$ and $\mathfrak{h}' = \mathfrak{h}'(Q)$ if Q is the set of all positive roots. Also define $\pi_R = \pi_{S_R}$, $\pi_I = \pi_{S_I}$, $\mathfrak{h}'(R) = \mathfrak{h}'(S_R)$, $\mathfrak{h}'(I) = \mathfrak{h}'(S_I)$ and[6] $\varepsilon_R(H) = \operatorname{sign} \pi_R(H)$ $(H \in \mathfrak{h})$. Finally for $s \in W$, $\varepsilon(s)$ is defined to be ± 1 by the relation $\pi^s = \varepsilon(s)\pi$.

5. Statement of Theorem 1

Let G be a connected Lie group with Lie algebra $\mathfrak{g}$. By a Cartan subgroup A of G, we mean the centralizer (in G) of some Cartan subalgebra $\mathfrak{h}$ of $\mathfrak{g}$. It is clear that A is closed in G and $\mathfrak{h}$ is the Lie algebra of A. We call A the Cartan subgroup of G corresponding to $\mathfrak{h}$.

LEMMA 8. *Let A be a Cartan subgroup of G, A_0 the connected component of 1 in A and Z the center of G. Then ZA_0 is an open subgroup of A lying in the center of A and A/ZA_0 is finite. Also A/Z is abelian and, in case G has a finite-dimensional faithful representation, A itself is abelian. Let H be a regular element of $\mathfrak{g}$ contained in the Lie algebra of A. Then A is the centralizer of H in G.*

The above lemma is merely the corrected version of [1 (a), Lemma 3], which, as pointed out to me by Langlands, is inaccurate[7]. The proof of [1 (a), p. 104] just gives Lemma 8.

As before let $\mathfrak{h}$ be a Cartan subalgebra of $\mathfrak{g}$, and A the corresponding

---

[6] Let $t \in \mathbf{R}$. Then $\operatorname{sign} t = 1, -1$ or 0 according as $t > 0, t < 0$ or $t = 0$.

[7] The mistake lies in the statement "$\cdots$ it is obviously permissible for our proof to replace G by G/Z." on p. 104 of [1 (a)].

Cartan subgroup of G. Let $x \to x^*$ denote the natural projection of G on $G^* = G/A$ and define $x^*H = xH$ $(x \in G, H \in \mathfrak{h}_c)$. dx^* being the invariant measure on G^*, we put

$$\psi_f(H) = \varepsilon_R(H)\pi(H)\int_{G^*} f(x^*H)dx^* \qquad (H \in \mathfrak{h}')$$

for $f \in C_c^\infty(G)$. It follows from Lemmas 7 and 8 that ψ_f is defined and of class C^∞ on $\mathfrak{h}'$. As usual let $\mathfrak{D}(\mathfrak{h}_c)$ denote the algebra of polynomial differential operators on $\mathfrak{h}$. Then W operates on $\mathfrak{D}(\mathfrak{h}_c)$ (see [1 (h), §3]).

THEOREM 1. *Fix a point $H_0 \in \mathfrak{h}$ and an element $D \in \mathfrak{D}(\mathfrak{h}_c)$. Let $S_I(H_0)$ denote the set of all $\alpha \in S_I$ such that $\alpha(H_0) = 0$. Assume that $D^{s_\alpha} = -D$ for every $\alpha \in S_I(H_0)$. Then for any $f \in C_c^\infty(\mathfrak{g})$, $D\psi_f$ can be extended to a continuous function around[8] H_0.*

The proof requires some preparation which will be undertaken in the next few sections.

6. Some elementary properties of ψ_f

We keep to the notation of §§ 4–5.

LEMMA 9. *Let π_0 be the product of all positive imaginary roots. Then*

$$\varepsilon_R(H)\pi(H) = \varepsilon\pi_0(H)\,|\,\pi(H)\,|\,|\,\pi_0(H)\,|^{-1} \qquad (H \in \mathfrak{h}')$$

where ε is a constant and $\varepsilon^2 = 1$.

It is clear that $\pi = \pi_0\pi_c\pi_R$ where π_c is the product of all positive complex roots. If α is a complex root then so is also $\eta\alpha$ and $\eta\alpha \neq \pm\alpha$. Moreover $\eta\alpha(H) = \operatorname{conj}\alpha(H)$ for $H \in \mathfrak{h}$. Hence we can choose a set C of positive complex roots with the following properties. If α and β are in C, then $\beta \neq \pm\eta\alpha$ and every complex positive root, which is not in C, is of the form $\pm\eta\alpha$ for some $\alpha \in C$. For each $\alpha \in C$ choose $\varepsilon_\alpha = \pm 1$ in such a way that $\varepsilon_\alpha\eta\alpha > 0$. Then it is clear that $\pi_c = \varepsilon \prod_{\alpha \in C} (\alpha \cdot \eta\alpha)$ where $\varepsilon = \prod_{\alpha \in C} \varepsilon_\alpha$. This proves that $\pi_c(H) = \varepsilon\,|\,\pi_c(H)\,|$ for $H \in \mathfrak{h}$. Therefore

$$\varepsilon_R(H)\pi(H) = \varepsilon_R(H)\pi_0(H)\pi_c(H)\pi_R(H)$$
$$= \varepsilon\pi_0(H)\,|\,\pi_c(H)\pi_R(H)\,| \qquad (H \in \mathfrak{h})\,,$$

and this proves the lemma.

COROLLARY. *Let s be an element of W which maps $\mathfrak{h}$ into itself. Then $\pi_0{}^s = \varepsilon_0(s)\pi_0$ where $\varepsilon_0(s) = \pm 1$. Moreover*

$$(\varepsilon_R\pi)^s = \varepsilon_0(s)\varepsilon_R\pi$$

---

[8] This means that there exists a neighborhood V of H_0 in $\mathfrak{h}$ and a continuous function F on V such that $F = D\psi_f$ on $V \cap \mathfrak{h}'$.

on $\mathfrak{h}$.

If α is an imaginary root, the same holds for $s\alpha$. Therefore the corollary follows from Lemma 9.

LEMMA 10. *Suppose α and β are two roots and α is real while β is imaginary. Then s_α and s_β map $\mathfrak{h}$ into itself and $\alpha(H_\beta) = \beta(H_\alpha) = 0$.*

For any two roots γ, δ we know that $\gamma(H_\delta) = \delta(H_\gamma)$ is real. Moreover

$$s_\gamma H = H - 2\big(\gamma(H)/\gamma(H_\gamma)\big)H_\gamma \qquad\qquad (H \in \mathfrak{h}_c) \ .$$

If γ is real, then $H_\gamma \in \mathfrak{h}$ and $\gamma(H)$ is real for $H \in \mathfrak{h}$. Similarly if γ is imaginary, then $(-1)^{1/2}H_\gamma \in \mathfrak{h}$ and $(-1)^{1/2}\gamma(H)$ is real for $H \in \mathfrak{h}$. Hence in either case $s_\gamma H \in \mathfrak{h}$ for $H \in \mathfrak{h}$.

Now since α is real, H_α lies in $\mathfrak{h}$ and therefore $\beta(H_\alpha)$ is imaginary. But we know that $\alpha(H_\beta) = \beta(H_\alpha)$ is real. Hence $\alpha(H_\beta) = \beta(H_\alpha) = 0$.

COROLLARY. $\varepsilon_0(s_\alpha) = 1$ *if α is real. Similarly $\varepsilon_0(s_\beta) = -1$ if β is imaginary.*

It is obvious from the above lemma that $\pi_0{}^{s_\alpha} = \pi_0$ and $\pi_R{}^{s_\beta} = \pi_R$. Hence $\varepsilon_0(s_\alpha) = 1$ and $\varepsilon_0(s_\beta) = \varepsilon(s_\beta) = -1$ from Lemma 9.

We now use the notation of Theorem 1 of [1 (h)].

LEMMA 11. *For any $f \in C_c^\infty(\mathfrak{g})$, ψ_f can be extended uniquely to a C^∞ function on $\mathfrak{h}'(I)$. Moreover $\psi_{f^s} = \varepsilon_0(s)\psi_f$ for $s \in W_G$ and*

$$\psi_{Df} = \delta_{\mathfrak{g}/\mathfrak{h}}(D)\psi_f$$

on $\mathfrak{h}'(I)$ for $D \in \mathfrak{F}(\mathfrak{g}_c)$.

The first statement follows from Theorem 3 of [1(c)] and the second is proved in the same way as Lemma 34 of [1 (c)]. Now we come to the last part. Fix H_0 in $\mathfrak{h}'$. Then A is the centralizer of H_0 in G (Lemma 8). Select an open neighborhood U of H_0 in $\mathfrak{h}'$ satisfying the condition of Lemma 7. Then there exists a compact subset Ω^* of G^* with the property that $x^*H \notin \operatorname{Supp} f$ $(x^* \in G^*,$ $H \in U)$ unless $x^* \in \Omega^*$. Let dx and da denote the Haar measures of G and A normalized in such a way that $dx = dx^*da$. Since Ω^* is compact, we can choose an open set V in G and a function $\gamma \in C_c^\infty(G)$ such that $V^* \supset \Omega^*$ and

$$\int_A \gamma(xa)da = 1$$

for $x \in V$. Then if $g \in C_c^\infty(G)$ and $\operatorname{Supp} g \subset \operatorname{Supp} f$, it is clear that

$$\int_{G^*} g(x^*H)dx^* = \int_{\Omega^*} g(x^*H)dx^* = \int_G \gamma(x)g(xH)dx$$

for $H \in U$. Fix $D \in \mathfrak{F}(\mathfrak{g}_c)$ and put $g = Df$. Then

$$\int_{G^*} f(x^*H; D)dx^* = \int_{G} \gamma(x)f(xH; D)dx \qquad (H \in U).$$

Now put $\mathfrak{q} = [\mathfrak{h}, \mathfrak{g}]$, $\delta(D) = \delta_{\mathfrak{g}/\mathfrak{h}}(D)$ and $\delta'(D) = \delta_{\mathfrak{g}/\mathfrak{h}}{}'(D)$. Then if $f(x:X) = f(xX)$ $(x \in G, X \in \mathfrak{g})$, it follows from [1 (h), § 2] that

$$f(xH; D) = f(xH; D^x) = f(x:H; D)$$
$$= f\big(x:H; \delta'(D)\big) + \sum_{1 \le i, j \le N} a_{ij}(H)f\big(x; g_i:H; \partial(p_j)\big)$$

for $x \in G$ and $H \in \mathfrak{h}'$. Here a_{ij} are C^∞ functions on $\mathfrak{h}'$, $g_i \in \mathfrak{S}_+(\mathfrak{q}_c)$, $p_j \in S(\mathfrak{h}_c)$ and $(g_1, \cdots, g_N)$, $(p_1, \cdots, p_N)$ are (separately) linearly independent over C. Moreover it follows from Lemma 3 of [1 (h)] that $g_i{}^a = g_i$ for $a \in A$. Therefore if $H \in U$, we have

$$\int_{G^*} f(x^*H; D)dx^* = \int_{G} \gamma(x)f\big(x:H; \delta'(D)\big)dx$$
$$+ \sum_{1 \le i, j \le N} a_{ij}(H) \int_{G} \gamma(x; g_i{}^*)f\big(x:H; \partial(p_j)\big)dx$$

where $g_i{}^*$ denotes, as usual, the adjoint of g_i. It is obvious that $f(x:H; \partial(p_j)) = f(xa:H; \partial(p_j))$ for $a \in A$. On the other hand if we put $\gamma(x:a) = \gamma(xa)$ $(x \in G, a \in A)$, it is clear that

$$\int \gamma(x:a)da = 1$$

for $x \in V$. Therefore

$$0 = \int \gamma(x; g_i{}^*:a)da = \int \gamma(xa; g_i{}^*)da \qquad (x \in V),$$

since $g_i{}^*$ is also invariant under A. This proves that

$$\int_{G^*} f(x^*H; D)dx^* = \int_{G} \gamma(x)f\big(x:H; \delta'(D)\big)dx \qquad (H \in U),$$

and therefore $\psi_{Df} = \delta(D)\psi_f$ on U. Since $\mathfrak{h}'$ is dense in $\mathfrak{h}'(I)$, and $\delta(D)$ is a polynomial differential operator on $\mathfrak{h}$ (see [1 (h), Theorem 1]), it is now obvious that $\psi_{Df} = \delta(D)\psi_f$ on $\mathfrak{h}'(I)$.

COROLLALY. $\psi_{f^{s_\alpha}} = \psi_f$ if α is a real root.

For $s_\alpha \in W_G$ (Lemma 6), and $\varepsilon_0(s_\alpha) = 1$ by the corollary of Lemma 10.

7. The Cartan subalgebras $\mathfrak{a}$ and $\mathfrak{b}$

Fix a semi-regular element H_0 in $\mathfrak{g}$ of non-compact type, and let $\mathfrak{z}$ denote the centralizer of H_0 in $\mathfrak{g}$. Then $\mathfrak{z} = \sigma + \mathfrak{l}$ in the notation of § 3. Moreover since $\mathfrak{l}$ is non-compact, we can choose a base (H', X', Y') for $\mathfrak{l}$ over R such that

$$[H', X'] = 2X', \qquad [H', Y'] = -2Y', \qquad [X', Y'] = H'.$$

Then $\mathbf{R}H'$ and $\mathbf{R}(X' - Y')$ are two Cartan subalgebras of $\mathfrak{l}$. Hence $\mathfrak{a} = \mathbf{R}H' + \sigma$

and $\mathfrak{b} = \mathbf{R}(X' - Y') + \sigma$ are two Cartan subalgebras of $\mathfrak{z}$ and therefore also of $\mathfrak{g}$.Clearly $H_0 \in \sigma = \mathfrak{a} \cap \mathfrak{b}$. Let L be the analytic subgroup of G corresponding to $\mathfrak{l}$.

LEMMA 12. *Any Cartan subalgebra of* $\mathfrak{g}$ *containing* H_0 *is conjugate to* $\mathfrak{a}$ *or* $\mathfrak{b}$ *under* L. *Moreover* $\mathfrak{a}$ *and* $\mathfrak{b}$ *are not conjugate under any automorphism of* $\mathfrak{g}$.

Let $\mathfrak{h}$ be a Cartan subalgebra of $\mathfrak{g}$ containing H_0. Then $\mathfrak{h} \subset \mathfrak{z}$ and therefore $\mathfrak{h} = \sigma + \mathfrak{l} \cap \mathfrak{h}$. Hence $\mathfrak{l} \cap \mathfrak{h}$, being a Cartan subalgebra of $\mathfrak{l}$, is conjugate under L to $\mathbf{R}H'$ or $\mathbf{R}(X' - Y')$ (see [1 (d), p. 681]). Since L leaves σ pointwise fixed, the first assertion of the lemma is now obvious.

For the proof of the second part, we need some simple facts about quadratic forms. Let Q be a non-degenerate quadratic form on a real vector space V of finite dimension. Let U be a subspace of V such that the restriction of Q on U is again non-degenerate. Then we can choose subspaces U_+ and U_- of U such that

(1) U_+ and U_- are orthogonal under Q,

(2) $U = U_+ + U_-$, and

(3) Q is positive-definite on U_+ and negative-definite on U_-.

It is well known that the integer $j_U(Q) = \dim U_+ - \dim U_-$ is independent of the choice of U_+ and U_-.

Now it is obvious that, for the proof of the second statement of Lemma 12, we may assume that $\mathfrak{g}$ is semi-simple. Let ω denote the Killing form of $\mathfrak{g}$. Then $\mathfrak{l}$ and σ are orthogonal under ω. In order to see this, put $B(X, Y) = \mathrm{tr}(\mathrm{ad}\, X\, \mathrm{ad}\, Y)$ $(X, Y \in \mathfrak{g})$. Then if $H \in \sigma$ and $X, Y \in \mathfrak{l}$, we have

$$B(H, [X, Y]) = -B([X, H], Y) = 0$$

since $[X, H] = 0$. Therefore our assertion follows from the fact that $[\mathfrak{l}, \mathfrak{l}] = \mathfrak{l}$.

The mapping $(H', X', Y') \to (-H', -Y', -X')$ defines a Cartan involution θ of $\mathfrak{l}$ and $\theta(H') = -H', \theta(X' - Y') = X' - Y'$. Therefore $\omega(H') = \mathrm{tr}\,(\mathrm{ad}\,H')^2 > 0$ while $\omega(X' - Y') = \mathrm{tr}\,(\mathrm{ad}\,(X' - Y'))^2 < 0$. The restriction of ω on the Cartan subalgebra $\mathfrak{a}$ is, of course, non-degenerate. Therefore since $\mathfrak{a} = \mathbf{R}H' + \sigma$, it is clear that

$$j_\mathfrak{a}(\omega) = 1 + j_\sigma(\omega) \ .$$

Similarly since $\mathfrak{b} = \mathbf{R}(X' - Y') + \sigma$, we get $j_\mathfrak{b}(\omega) = -1 + j_\sigma(\omega)$. Hence $j_\mathfrak{a}(\omega) \neq j_\mathfrak{b}(\omega)$. On the other hand since ω is invariant under any automorphism τ of $\mathfrak{g}$, it is clear that $j_\mathfrak{a}(\omega) = j_{\tau\mathfrak{a}}(\omega)$. Hence $\mathfrak{b} \neq \tau\mathfrak{a}$.

Now consider the automorphism[9] $\nu = \exp\{-(-1)^{1/2}(\pi/4)\,\mathrm{ad}\,(X' + Y')\}$ of

---

[9] Here, of course, π denotes the least positive number such that $\sin(\pi/2) = 1$.

$\mathfrak{g}_c$. Clearly it maps $\mathfrak{l}_c$ into itself and leaves σ_c pointwise fixed. Also $\nu(H') = (-1)^{1/2}(X' - Y')$. Hence $\nu(\mathfrak{a}_c) = \mathfrak{b}_c$. We can extend this to an isomorphism $q \to q^\nu$ $(q \in S(\mathfrak{a}_c))$ of $S(\mathfrak{a}_c)$ onto $S(\mathfrak{b}_c)$. Similarly for $p \in P(\mathfrak{a}_c)$, let p^ν denote the polynomial function $H \to p(\nu^{-1}(H))$ $(H \in \mathfrak{b}_c)$ on $\mathfrak{b}_c$. Then we can define (see [1 (h), Lemma 5]) an isomorphism $D \to D^\nu$ of $\mathfrak{D}(\mathfrak{a}_c)$ onto $\mathfrak{D}(\mathfrak{b}_c)$ such that $(p\partial(q))^\nu = p^\nu\partial(q^\nu)$ $\big(p \in P(\mathfrak{a}_c),\ q \in S(\mathfrak{a}_c)\big)$. It is clear that if λ is a root of $(\mathfrak{g}, \mathfrak{a})$, then λ^ν is a root of $(\mathfrak{g}, \mathfrak{b})$.

LEMMA 13. *There exists exactly one root α of $\mathfrak{a}$ such that $\alpha(H') = 2$ and $\alpha = 0$ on σ. α is real, while $\beta = \alpha^\nu$ is a singular imaginary root of $\mathfrak{b}$. $\beta(X' - Y') = -2(-1)^{1/2}$ and $\beta = 0$ on σ. Let σ_α be the hyperplane in $\mathfrak{a}$ given by the equation $\alpha = 0$. Similarly let σ_β be the hyperplane in $\mathfrak{b}$ defined by $\beta = 0$. Then $\sigma_\alpha = \sigma_\beta = \sigma = \mathfrak{a} \cap \mathfrak{b}$.*

Let $H = H_\sigma + tH'$ $(H_\sigma \in \sigma,\ t \in \mathbf{R})$ be an element of $\mathfrak{a}$. Then $[H, X'] = t[H', X'] = 2tX'$. Since $\mathfrak{a} = \mathbf{R}H' + \sigma$, the existence and uniqueness of α is clear. $\beta = \alpha = 0$ on σ, since ν leaves σ pointwise fixed. Moreover $\beta(X' - Y') = \alpha(\nu^{-1}(X' - Y')) = -(-1)^{1/2}\alpha(H') = -2(-1)^{1/2}$. Therefore since $\mathfrak{b} = \sigma + \mathbf{R}(X' - Y')$, β is imaginary. Since α is real and $\mathfrak{l}_c$ is invariant under ν, it follows from Lemma 5 that β is singular. Finally the last statement of the lemma is obvious since $\mathfrak{a} = \mathbf{R}H' + \sigma$ and $\mathfrak{b} = \mathbf{R}(X' - Y') + \sigma$.

8. Proof of Theorem 2

THEOREM 2. *Let σ' be the set of all semi-regular elements in σ. Then there exists a locally constant[10] function c on σ' with the following property. Let D be an element in $\mathfrak{D}(\mathfrak{a}_c)$. Then[11]*

$$\psi_f^{\mathfrak{a}}(H; D) = c(H)\,\{\lim_{\varphi \to +0} \psi_f^{\mathfrak{b}}(H_\varphi; D^\nu) - \lim_{\varphi \to -0} \psi_f^{\mathfrak{b}}(H_\varphi; D^\nu)\}$$

for[2] $f \in C_c^\infty(\mathfrak{g})$ and $H \in \sigma'$. Here

$$H_\varphi = H + \varphi(X' - Y') \qquad\qquad (\varphi \in \mathbf{R},\ H \in \sigma)\ .$$

Since α is a real root, we know from Lemma 11 that $\psi_f^{\mathfrak{a}}$ is of class C^∞ around every point of σ'. Corresponding to the decomposition $\mathfrak{a} = \sigma + \mathbf{R}H'$, we have $\mathfrak{D}(\mathfrak{a}_c) = \mathfrak{D}(\sigma_c)\mathfrak{D}(CH')$ [1 (h), §3]. Similarly $\mathfrak{D}(\mathfrak{b}_c) = \mathfrak{D}(\sigma_c)\mathfrak{D}(C(X' - Y'))$ corresponding to the decomposition $\mathfrak{b} = \sigma + \mathbf{R}(X' - Y')$.

It is obvious that $\mathfrak{z} = \sigma + \mathfrak{l}$ is the centralizer of σ in $\mathfrak{g}$. Moreover $\mathfrak{z}$ is also the centralizer of any element in σ'. Now we change our notation slightly and fix an arbitrary element $H_0 \in \sigma'$. Let $A_\mathfrak{h}$ denote the Cartan subgroup of G

---

[10] This means that c is constant on some neighborhood of every point in σ'. Therefore it is constant on each connected component of σ'.

[11] We denote by $\psi_f^{\mathfrak{a}}$ the integral ψ_f of §5 corresponding to $\mathfrak{h} = \mathfrak{a}$. Similarly for $\psi_f^{\mathfrak{b}}$. The symbols $\varepsilon_R^{\mathfrak{a}}, \pi^{\mathfrak{b}}$ etc. have a corresponding meaning.

corresponding to $\mathfrak{h}$ ($\mathfrak{h} = \mathfrak{a}$ or $\mathfrak{b}$), Ξ the centralizer of H_0 in G and Ξ_0 the connected component of 1 in Ξ. Then $\Xi_1(\mathfrak{h}) = \Xi_0 A_\mathfrak{h}$ is an open subgroup of Ξ, and $[\Xi : \Xi_1(\mathfrak{h})] = r(\mathfrak{h}) \leq 2$ [1 (d), Lemma 7]. Let U be an open and convex neighborhood of H_0 in $\mathfrak{z}$. We assume that U is so small that

(1) $U \cap \mathfrak{h}$ satisfies the condition of Lemma 7, and

(2) if γ is a root of $\mathfrak{h}$ which vanishes at H_0, then no root $\delta \neq \pm\gamma$ of $\mathfrak{h}$ ever takes the value zero on $U \cap \mathfrak{h}$ ($\mathfrak{h} = \mathfrak{a}$ or $\mathfrak{b}$).

Since H_0 is semi-regular, this is evidently possible. Let $x \to \bar{x}$ denote the natural mapping of G on $\bar{G} = G/\Xi$. Let Z be the center of G. Since $\mathfrak{z}$ is reductive and $\Xi/\Xi_0 Z$ is finite (see [1 (d), Lemma 15]), Ξ is unimodular. Therefore there exists an invariant measure $d\bar{x}$ on $\bar{G}$. Now fix $\mathfrak{h}$ and drop it from the notation. Normalize the invariant measures dx^* and $d\xi^*$ on $G^* = G/A$ and $\Xi^* = \Xi/A$ respectively, in such a way that $dx^* = d\bar{x}d\xi^*$. Put $B_0 = \exp(-H_0)\exp(U \cap \mathfrak{h})$. Since $U \cap \mathfrak{h}$ is convex, B_0 is a connected neighborhood of 1 in A and therefore $B_0 \subset \Xi_0$. Let Ξ_0^* denote the image of Ξ_0 in Ξ^*, and select elements $y_1 = 1, \cdots, y_r$ $(r = r(\mathfrak{h}))$ in Ξ such that $\Xi = \bigcup_i y_i \Xi_1$. G operates on the left on $G^* = G/A$ in the usual way and Ξ^* is the disjoint union of $y_i \Xi_0^*$ ($1 \leq i \leq r$). Moreover $\Xi_0^* \cong \Xi_0/\Xi_0 \cap A$. If $\mathfrak{h} = \mathfrak{b}$ and $r = r(\mathfrak{b}) = 2$, we may further assume that $y_2 H = s_\beta H$ for $H \in \mathfrak{b}$ (see [1 (d), Lemma 7]). Then y_2 normalizes both $\mathfrak{l}$ and $\mathfrak{b}$.

Now fix a bounded open set ω in $\mathfrak{g}$. Then (Lemma 7) we can select a compact set $\bar{\Omega}$ in $\bar{G}$ such that $\bar{x} \in \bar{\Omega}$ whenever $xH \in \omega$ for some $x \in G$ and $H \in U \cap \mathfrak{h}$. Let dx and $d\xi$ denote the Haar measures on G and Ξ normalized in such a way that $dx = d\bar{x}d\xi$. Select $\gamma \in C_c^\infty(G)$ such that

$$\int_\Xi \gamma(x\xi)d\xi = 1$$

if $\bar{x} \in \bar{\Omega}$ ($x \in G$), and for any $f \in C_c^\infty(\omega)$, put

$$g_f(Z) = \sum_{1 \leq i \leq r} \int_G \gamma(x) f(xy_i Z)dx \qquad (Z \in \mathfrak{z}) .$$

If $H \in \mathfrak{h}'$, it is clear that

$$\int_{G^*} f(x^*H)dx^* = \int_{\bar{G}} f(H : \bar{x})d\bar{x}$$

where

$$f(H : \bar{x}) = \int_{\Xi^*} f(x(\xi^*H))d\xi^* = \sum_i \int_{\Xi_0^*} f(xy_i(\xi^*H))d\xi^* \qquad (x \in G) .$$

Hence if $H \in U \cap \mathfrak{h}'$, we have

$$\int_{G^*} f(x^*H)dx^* = \int_{\bar{G}} f(H : \bar{x})d\bar{x} = \int_{\Xi_0^*} g_f(\xi^*H)d\xi^*$$

for $f \in C_c^\infty(\omega)$. Hence

$$\psi_f(H) = \varepsilon_R(H)\pi(H) \int_{\Xi_0^*} g_f(\xi^* H)d\xi^* \qquad (H \in U \cap \mathfrak{h}')$$

for $f \in C_c^\infty(\omega)$. Since σ is the center of $\mathfrak{z}$, $A_\sigma = \exp \sigma$ is a closed analytic subgroup of Ξ_0 contained in $\Xi_0 \cap A$. Moreover $\Xi_0 = LA_\sigma$ where L is the analytic subgroup of G corresponding to $\mathfrak{l}$. Hence $\Xi_0/A_\sigma = L/L \cap A_\sigma$. Therefore it is clear that

$$\Xi_0/\Xi_0 \cap A \cong L/L \cap A .$$

Since $L \cap A$ is the centralizer of $\mathfrak{h}$ in L, it is obvious that it is also the Cartan subgroup of L corresponding to $\mathfrak{l} \cap \mathfrak{h}$. Hence $d\xi^*$ is just the invariant measure on $L/L \cap A$.

Now put $L_\mathfrak{a}^* = L/L \cap A_\mathfrak{a}$, $L_\mathfrak{b}^* = L/L \cap A_\mathfrak{b}$ and define

$$\bar{g}(Z) = \int_{K_1} g(Z^k)dk \qquad (Z \in \mathfrak{z}, g \in C_c^\infty(\mathfrak{z})) .$$

Here K_1 is the compact one-parameter subgroup of $\mathrm{Ad}(G)$ corresponding to $X' - Y'$ and dk is the normalized Haar measure of K_1. Since $\mathfrak{b} = \sigma + \mathbf{R}(X' - Y')$, any element H of $\mathfrak{b}$ can be written in the form $H = H_\sigma + \varphi(X' - Y')$ where H_σ is the component of H in σ and $\varphi = (-1)^{1/2}\beta(H)/2 \in \mathbf{R}$.

LEMMA 14. *For $g \in C_c^\infty(\mathfrak{z})$, put*

$$J_\mathfrak{a}(g:H) = \int_{-\infty}^{\infty} \bar{g}(H + tX')dt \qquad (H \in \mathfrak{a})$$

and

$$J_\mathfrak{b}(g:H) = \varphi \int_0^{\infty} \bar{g}(H_\sigma + \varphi(e^t X' - e^{-t} Y'))(e^t - e^{-t})dt$$

if $H \in \mathfrak{b}$ and $\varphi = (-1)^{1/2}\beta(H)/2 \neq 0$. Then there exist positive numbers $c_\mathfrak{a}'$ and $c_\mathfrak{b}'$ such that for any $g \in C_c^\infty(\mathfrak{z})$,

$$|\alpha(H)| \int_{L_\mathfrak{a}^*} g(\xi^* H)d\xi^* = c_\mathfrak{a}' J_\mathfrak{a}(g : H) \qquad (H \in \mathfrak{a}')$$

and

$$\beta(H) \int_{L_\mathfrak{b}^*} g(\xi^* H)d\xi^* = -(-1)^{1/2}c_\mathfrak{b}' J_\mathfrak{b}(g : H) \qquad (H \in \mathfrak{b}') .$$

For proof see [1 (d), pp. 682–683].

Now first take $\mathfrak{h} = \mathfrak{a}$ and put $\pi_\mathfrak{a}^\mathfrak{a} = \alpha^{-1}\pi^\mathfrak{a}$, $\pi_{R,\mathfrak{a}} = \alpha^{-1}\pi_R^\mathfrak{a}$. Since α is a real root, these are both polynomial functions on $\mathfrak{a}_c$ which are nowhere zero on the convex set $U \cap \mathfrak{a}$. Hence

$$\varepsilon_R^\mathfrak{a}(H) = \mathrm{sign}\, \pi_{R,\mathfrak{a}}(H_0)\cdot\mathrm{sign}\, \alpha(H) \qquad (H \in U \cap \mathfrak{a}) .$$

On the other hand $r(\mathfrak{a}) = 1$, so that

$$g_f{}^{\mathfrak{a}}(Z) = \int_G \gamma(x) f(xZ) dx \qquad (Z \in \mathfrak{z}) .$$

Therefore

$$\psi_f{}^{\mathfrak{a}}(H) = (\operatorname{sign} \pi_{R,\alpha}(H_0)) \pi_\alpha{}^{\mathfrak{a}}(H) \, |\, \alpha(H)\,| \int_{L_{\mathfrak{a}^*}} d\xi^* \int_G \gamma(x) f\big(x(\xi^* H)\big) dx$$

for $H \in U \cap \mathfrak{a}'$ and $f \in C_c^\infty(\omega)$. Put $c_\alpha = \operatorname{sign} \pi_{R,\alpha}(H_0) \cdot c_\alpha'$. Then it follows from Lemma 14 that

$$\psi_f{}^{\mathfrak{a}}(H) = c_\alpha J_\alpha(f_\gamma : H) \qquad (H \in U \cap \mathfrak{a}', f \in C_c^\infty(\omega))$$

where

$$f_\gamma(Z) = \int_G \gamma(x) f(xZ) dx \qquad (Z \in \mathfrak{z}) .$$

Now take $\mathfrak{h} = \mathfrak{b}$. Then since β is an imaginary root of $\mathfrak{b}$, $\varepsilon_R{}^{\mathfrak{b}}(H) = \varepsilon_R{}^{\mathfrak{b}}(H_0)$ for $H \in U \cap \mathfrak{b}$ and therefore

$$\psi_f{}^{\mathfrak{b}}(H) = \varepsilon_R{}^{\mathfrak{b}}(H_0) \pi^{\mathfrak{b}}(H) \int_{L_{\mathfrak{b}^*}} d\xi^* \sum_{1 \le i \le r(\mathfrak{b})} \int_G \gamma(x) f\big(xy_i(\xi^* H)\big) dx$$

for $H \in U \cap \mathfrak{b}'$ and $f \in C_c^\infty(\omega)$. We have seen that y_i normalizes both I and $\mathfrak{b}$, and therefore also L and $A_{\mathfrak{b}}$. Hence it is clear that

$$\int_{L_{\mathfrak{b}^*}} f\big(xy_i(\xi^* H)\big) d\xi^* = \int_{L_{\mathfrak{b}^*}} f\big(x\xi^*(y_i H)\big) d\xi^* .$$

But $y_2 H = s_\beta H$ if $r(\mathfrak{b}) = 2$. Put $\pi_\beta{}^{\mathfrak{b}} = \beta^{-1} \pi^{\mathfrak{b}}$. Then it follows from Lemma 14 that

$$\psi_f{}^{\mathfrak{b}}(H) = \begin{cases} (-1)^{1/2} c_{\mathfrak{b}} \pi_\beta{}^{\mathfrak{b}}(H) J_{\mathfrak{b}}(f_\gamma : H) & \text{if } r(\mathfrak{b}) = 1 , \\ (-1)^{1/2} c_{\mathfrak{b}} \pi_\beta{}^{\mathfrak{b}}(H) \{J_{\mathfrak{b}}(f_\gamma : H) - J_{\mathfrak{b}}(f_\gamma : s_\beta H)\} & \text{if } r(\mathfrak{b}) = 2 . \end{cases}$$

Here $c_{\mathfrak{b}} = -\varepsilon_R(H_0) c_{\mathfrak{b}}'$. Thus we have obtained the following result.

LEMMA 15. *There exist non-zero real numbers c_α and $c_{\mathfrak{b}}$ (independent of γ and ω) such that*

$$\psi_f{}^{\mathfrak{a}}(H) = c_\alpha \pi_\alpha{}^{\mathfrak{a}}(H) J_\alpha(f_\gamma : H) \qquad (H \in U \cap \mathfrak{a}')$$

and

$$\psi_f{}^{\mathfrak{b}}(H) = \begin{cases} (-1)^{1/2} c_{\mathfrak{b}} \pi_\beta{}^{\mathfrak{b}}(H) J_{\mathfrak{b}}(f_\gamma : H) & \text{if } r(\mathfrak{b}) = 1 , \\ (-1)^{1/2} c_{\mathfrak{b}} \pi_\beta{}^{\mathfrak{b}}(H) \{J_{\mathfrak{b}}(f_\gamma : H) - J_{\mathfrak{b}}(f_\gamma : s_\beta H)\} & \text{if } r(\mathfrak{b}) = 2 , \end{cases}$$

for $H \in U \cap \mathfrak{b}'$ and $f \in C_c^\infty(\omega)$.

COROLLARY. *Let D be an element in $\mathfrak{D}(\mathfrak{b}_c)$ such that $D'^\beta = -D$. Then for any $f \in C_c^\infty(\mathfrak{g})$, $D\psi_f{}^{\mathfrak{b}}$ can be extended to a continuous function around any point $H_0 \in \sigma'$.*

Put $D_1 = D \circ \pi_\beta{}^{\mathfrak{b}}$. Since $\pi_\beta{}^{\mathfrak{b}}$ is invariant under s_β, it is clear that $D_1'^\beta =$

$-D_1$. Now fix $f \in C_c^\infty(\mathfrak{g})$ and choose ω so large that Supp $f \subset \omega$. Then, in view of Lemma 15, it would be sufficient to show that $J_{\mathfrak{b}}(f_\gamma : H; D_1)$ can be extended to a continuous function around H_0. But $\mathfrak{D}(\mathfrak{b}_c) = \mathfrak{D}(\sigma_c)\mathfrak{D}(\mathbf{C}(X' - Y'))$ and, since s_β leaves $\mathfrak{D}(\sigma_c)$ fixed, D_1 can be written as a finite sum of elements of the form $D' = \Delta \circ \beta^i \partial(X' - Y')^j$ where $\Delta \in \mathfrak{D}(\sigma_c)$ and $i + j$ is odd. Hence it would be enough to show that the above extension is possible for

$$J_{\mathfrak{b}}(f_\gamma : H; D') = \beta(H)^i J_{\mathfrak{b}}(\Delta f_\gamma : H; \partial(X' - Y')^j)$$

on $U \cap \mathfrak{b}$. Since β vanishes on σ, this can be done, in case $i \geq 1$, by defining the extended function to be zero on[12] $U \cap \sigma$. So let us now assume that $i = 0$ and j is odd. Then our assertion follows immediately[12] from the corollary of Lemma 3.

We now come to the proof of Theorem 2. Put $D_+ = 2^{-1}(D + D^{s\alpha})$ and $D_- = 2^{-1}(D - D^{s\alpha})$. Then $D = D_+ + D_-$, $(D_-)^{s\alpha} = -D_-$ and

$$(D_-\psi_f{}^\alpha)^{s\alpha} = -D_-\psi_f{}^\alpha$$

from the corollary of Lemma 11. Therefore since $D_-\psi_f{}^\alpha$ is of class C^∞ on $\mathfrak{a}'(I)$ (Lemma 11), it is clear that $D_-\psi_f{}^\alpha = 0$ on σ'. On the other hand $(D_-{}^\nu)^{s\beta} = -D_-{}^\nu$ and so it follows from the corollary of Lemma 15 that

$$\mathrm{Lim}_{\varphi \to +0} \psi_f{}^{\mathfrak{b}}(H_\varphi; D_-{}^\nu) - \mathrm{Lim}_{\varphi \to -0} \psi_f{}^{\mathfrak{b}}(H_\varphi; D_-{}^\nu) = 0$$

for $H \in \sigma'$. Therefore it would be enough to consider the case when $D = D_+$ or $D^{s\alpha} = D$.

Fix $f \in C_c^\infty(\mathfrak{g})$, $H_0 \in \sigma'$ and choose ω so large that Supp $f \subset \omega$. Then since $U \cap \mathfrak{a} \subset \mathfrak{a}'(I)$, it follows from Lemmas 15 and 11 that

$$\psi_f{}^\alpha(H; D) = c_\mathfrak{a} J_\mathfrak{a}(f_\gamma : H; D_1)$$

for $H \in U \cap \mathfrak{a}$ where $D_1 = D \circ \pi_\mathfrak{a}{}^\alpha$. Now $D_1{}^{s\alpha} = D_1$. Hence D_1 is a finite linear combination of elements of the form $D' = \Delta \circ \alpha^i \partial(H')^j$ where $\Delta \in \mathfrak{D}(\sigma_c)$ and $i + j$ is even. Clearly

$$(D')^\nu = (-1)^{j/2} \Delta \circ \beta^i \partial(X' - Y')^j \; .$$

If $i \geq 1$, it is obvious that

$$J_\mathfrak{a}(f_\gamma : H; D') = \alpha(H)^i J_\mathfrak{a}(\Delta f_\gamma : H; \partial(H')^j) = 0$$

on $U \cap \sigma$. Similarly

$$\mathrm{Lim}_{\varphi \to 0} J_{\mathfrak{b}}(f_\gamma : H_\varphi; D'^\nu) = 0$$

for $H \in U \cap \sigma$. Now suppose $i = 0$ so that j is even. Then it follows from

---

[12] Here we have to bear in mind Lemma 3 of [1 (d)].

Lemmas 17 and 18 of [1 (d)] that[13]

$$\begin{aligned}
J_{\mathfrak{a}}(f_\gamma : H; D') &= J_{\mathfrak{a}}(\Delta f_\gamma : H; \partial(H')^j) \\
&= (-1)^{j/2}\{\mathrm{Lim}_{\varphi \to +0}\, J_{\mathfrak{b}}(\Delta f_\gamma : H_\varphi; \partial(X' - Y')^j) \\
&\quad - \mathrm{Lim}_{\varphi \to -0}\, J_{\mathfrak{b}}(\Delta f_\gamma : H_\varphi; \partial(X' - Y')^j)\} \\
&= \mathrm{Lim}_{\varphi \to +0}\, J_{\mathfrak{b}}(f_\gamma : H_\varphi; D'^\nu) - \mathrm{Lim}_{\varphi \to -0}\, J_{\mathfrak{b}}(f_\gamma : H_\varphi; D'^\nu)
\end{aligned}$$

for $H \in U \cap \sigma$. Now put $c = -(-1)^{1/2}c_{\mathfrak{a}}/r(\mathfrak{b})c_{\mathfrak{b}}$ in the notation of Lemma 15. Then it follows from this lemma that

$$\psi_f{}^{\mathfrak{a}}(H; D') = c\{\mathrm{Lim}_{\varphi \to +0}\, \psi_f{}^{\mathfrak{b}}(H_\varphi; D'^\nu) - \mathrm{Lim}_{\varphi \to -0}\, \psi_f{}^{\mathfrak{b}}(H_\varphi; D'^\nu)\}$$

for $H \in U \cap \sigma$. The statement of Theorem 2 is now obvious.

9. Some elementary facts about ψ_f

Let $\mathfrak{h}_i\ (i = 1, 2)$ be two Cartan subalgebras of $\mathfrak{g}$, and assume that they are conjugate under G. Fix $y \in G$ such that $\mathfrak{h}_2 = \mathfrak{h}_1{}^y$. Let A_i be the Cartan subgroup of G corresponding to $\mathfrak{h}_i$ and $x \to x_i{}^*$ the natural projection of G on $G_i{}^* = G/A_i\ (i = 1, 2)$. Then $A_2 = yA_1y^{-1}$ and therefore

$$xA_1y^{-1} = xy^{-1}A_2 \qquad\qquad (x \in G)\,.$$

Thus we get a homeomorphism $x_1{}^* \to (x_1{}^*)^\nu = (xy^{-1})_2{}^*\ (x \in G)$ of $G_1{}^*$ onto $G_2{}^*$ which commutes with the operations of G on the two homogeneous spaces. Hence if $dx_1{}^*$ is an invariant measure on $G_1{}^*$, it goes over into an invariant measure $d(x_1{}^*)^\nu$ on $G_2{}^*$ under this mapping. Moreover it is clear that y defines an isomorphism $D \to D^\nu$ of $\mathfrak{D}(\mathfrak{h}_{1,c})$ onto $\mathfrak{D}(\mathfrak{h}_{2,c})$ and $(\mathfrak{h}_1{}'(I))^\nu = \mathfrak{h}_2{}'(I)$. For any $f \in C_c{}^\infty(G)$, put $\psi_{f,i} = \psi_f{}^{\mathfrak{h}_i}$, $\pi_i = \pi^{\mathfrak{h}_i}$, $\varepsilon_{R,i} = \varepsilon_R{}^{\mathfrak{h}_i}\ (i = 1, 2)$.

LEMMA 16. *There exists a real number $a \ne 0$ such that*

$$\psi_{f,2}(yH) = a\psi_{f,1}(H)$$

for all $f \in C_c{}^\infty(\mathfrak{g})$ and $H \in \mathfrak{h}_1{}'(I)$.

For the proof of this lemma we can obviously assume that a root α of $\mathfrak{h}_1$ is positive if and only if the root α^ν of $\mathfrak{h}_2$ is positive. Then

$$\varepsilon_{R,2}(yH)\pi_2(yH) = \varepsilon_{R,1}(H)\pi_1(H) \qquad\qquad (H \in \mathfrak{h}_1)\,.$$

Moreover we can assume that

$$\psi_{f,i}(H) = \varepsilon_{R,i}(H)\pi_i(H) \int_{G_i{}^*} f(x_i{}^*H)dx_i{}^* \qquad\qquad (H \in \mathfrak{h}_i{}',\, i = 1, 2)$$

where $dx_2{}^*$ is the invariant measure $d(x_1{}^*)^\nu$ on $G_2{}^*$. Then it is clear that $\psi_{f,2}(yH) = \psi_{f,1}(H)$ for $H \in \mathfrak{h}_1{}'$ and therefore also for $H \in \mathfrak{h}_1{}'(I)$.

---

[13] Here we have to take into account the difference between our definition of $J_{\mathfrak{b}}(g : H)$ in Lemma 14 and that of $\Psi_f(\theta)$ in [1 (d), Lemma 16].

Corollary. $\psi_{f,2}(yH; D^v) = a\psi_{f,1}(H; D)$ for $H \in \mathfrak{h}'(I)$, $f \in C_c^\infty(\mathfrak{g})$ and $D \in \mathfrak{D}(\mathfrak{h}_{1,c})$.

This is an immediate consequence of Lemma 16.

10. Proof of Theorem 1

Now we come to the proof of Theorem 1. If $S_I(H_0)$ is empty, it is obvious that $H_0 \in \mathfrak{h}'(I)$ and, therefore, the required result follows from Lemma 11. So we may assume that $S_I(H_0)$ is not empty. If H_0 is semi-regular, our assertion again follows from the corollaries of Lemmas 15 and 16.

Now let H_0 be any element in $\mathfrak{h}$ and U an open and convex neighborhood of H_0 in $\mathfrak{h}$. We assume that U is so small that if $\alpha \in S_I$ and $\alpha(H_0) \neq 0$, then α never takes the value zero on U. Let U' be the set of all points in U which are either regular or semi-regular. Then it follows from what we have seen above that $g = D\psi_f$ can be extended to a continuous function on U'.

Fix a euclidean norm on $\mathfrak{h}$ and put

$$\nu(g) = \sup \left| g\big(H_1; \partial(H_2)\big) \right|$$

where H_1 and H_2 vary in $\mathfrak{h}'$ and $\mathfrak{h}$ under the sole restriction that $\| H_2 \| \leq 1$. Then we know [1 (c), Theorem 3] that $\nu(g) < \infty$. Let H_1, H_2 be two points in $U \cap \mathfrak{h}'$. We claim that

$$| g(H_2) - g(H_1) | \leq \nu(g) \| H_2 - H_1 \| .$$

Let J denote the interval $1 \leq t \leq 2$ in $\mathbf{R}$ and put $H_t = (2 - t)H_1 + (t - 1)H_2$ $(t \in J)$. First assume that the straight line-segment joining H_1 and H_2 lies entirely in U'. Then $F(t) = g(H_t)$ is a continuous function of $t \in J$. Moreover there are only a finite number of values of t in J such that $\pi(H_t) = 0$. We call such values singular. Then $dF(t)/dt = g(H_t; \partial(H_2 - H_1))$ for all non-singular t and therefore

$$| g(H_2) - g(H_1) | \leq \int_J | g\big(H_t; \partial(H_2 - H_1)\big) | \, dt \leq \nu(g) \| H_2 - H_1 \| .$$

On the other hand H_1, H_2 being any two points in $U \cap \mathfrak{h}'$, we can choose $H_i' \in U \cap \mathfrak{h}'$ arbitrarily near H_i $(i = 1, 2)$ such that the straight line-segment from H_1' to H_2' lies entirely in U' (see Lemma 19 of §14 below). Then by the above proof,

$$| g(H_2') - g(H_1') | \leq \nu(g) \| H_2' - H_1' \|$$

and therefore, by continuity, we obtain the same result for H_1, H_2. Since $U \cap \mathfrak{h}'$ is dense in U, it is now obvious that g can be extended to a continuous function on U. This completes the proof of Theorem 1.

11. Statement of Theorem 3

An element D in $\mathfrak{D}(\mathfrak{h}_c)$ (or in $S(\mathfrak{h}_c)$) is called skew if $D'^\alpha = -D$ for every root α. It follows immediately from Theorem 1 that, if D is skew, $D\psi_f$ can be extended to a continuous function on $\mathfrak{h}$ for any $f \in C_c^\infty(\mathfrak{g})$. We denote by $\psi_f(H; D)$ the value of this extended function at any point H in $\mathfrak{h}$. Put $\varpi = \prod_{\alpha>0} H_\alpha$ where α runs over all positive roots. Then ϖ is skew.

We recall that a Cartan subalgebra $\mathfrak{h}$ of $\mathfrak{g}$ is called fundamental if it has no real roots (see [1 (c), Lemma 33]).

THEOREM 3.[14] *Let* c *denote the center of* $\mathfrak{g}$. *Then if* $\mathfrak{h}$ *is not fundamental*

$$\psi_f(H; D) = 0 \qquad\qquad (f \in C_c^\infty(\mathfrak{g}),\ H \in c)$$

for any skew element $D \in \mathfrak{D}(\mathfrak{h}_c)$. *Now assume* $\mathfrak{h}$ *is fundamental and put*

$$q = s_I + \frac{1}{2} r_c$$

where s_I *is the number of singular imaginary positive roots and* r_c *the number of complex positive roots. Then* r_c *is even and there exists a positive number* c *such that*

$$\psi_f(H; \partial(\varpi)) = c(-1)^q f(H)$$

for all $f \in C_c^\infty(\mathfrak{g})$ *and* $H \in c$.

We use induction on $\dim \mathfrak{g}$. Assuming that $\mathfrak{h}$ is not fundamental, let α be a real root of $\mathfrak{h}$. Then it follows from the corollary of Lemma 11 that $\psi_f'^\alpha = \psi_f$ and therefore $(D\psi_f)'^\alpha = -D\psi_f$ on $\mathfrak{h}'(I)$. But since $D\psi_f$ is continuous on $\mathfrak{h}$ and s_α leaves c pointwise fixed, it is now obvious that $\psi_f(H; D) = 0$ for $H \in c$.

So let us now suppose that $\mathfrak{h}$ is fundamental. We have seen during the proof of Lemma 9 that r_c is even. Let $\mathfrak{g}_1$ be the derived algebra of $\mathfrak{g}$ so that $\mathfrak{h} = c + \mathfrak{h}_1$ where $\mathfrak{h}_1 = \mathfrak{h} \cap \mathfrak{g}_1$. Fix $H \in c$ and consider the function $f_H : X \to f(H + X)$ $(X \in \mathfrak{g}_1)$ on $\mathfrak{g}_1$. Since $\varpi \in S(\mathfrak{h}_{1,c})$, it is clear that, in order to prove the last statement of the theorem, we can replace $\mathfrak{g}$, $\mathfrak{h}$, f, H by $\mathfrak{g}_1$, $\mathfrak{h}_1$, f_H, 0 respectively. Therefore if $\dim \mathfrak{g}_1 < \dim \mathfrak{g}$, our assertion follows by the induction hypothesis. Hence we may now assume that $\mathfrak{g}$ is semi-simple.

Put $B(X, Y) = \operatorname{tr}(\operatorname{ad} X \operatorname{ad} Y)$ $(X, Y \in \mathfrak{g}_c)$, and let $B_{\mathfrak{h}}$ denote the restriction of B on $\mathfrak{h}_c$. Then B and $B_{\mathfrak{h}}$ are both non-degenerate and we can identify $\mathfrak{g}_c$ and $\mathfrak{h}_c$ with their respective duals by means of these bilinear forms. It is then obvious that $\varpi = \pi$.

---

[14] Cf. [1 (e), p. 760].

12. Recapitulation of some work of de Rham[15]

Let p, q be two non-negative integers such that $p + q = n \geq 2$. Let $(x_1, \cdots, x_n)$ denote the usual coordinates in $\mathbf{R}^n$ and put

$$u = \sum_{1 \leq i \leq p} x_i^2 - \sum_{p < i \leq n} x_i^2 , \qquad \square = \sum_{1 \leq i \leq p} (\partial/\partial x_i)^2 - \sum_{p < i \leq n} (\partial/\partial x_i)^2 .$$

Define a function ξ on $\mathbf{R}$ as follows.

(1) $\xi(t) = (-1)^{(p-1)/2} a_n^{-1} Y(t) |t|^{-1/2}$ if $n \equiv p \equiv 1 \bmod 2$,

(2) $\xi(t) = (-1)^{p/2} a_n^{-1} Y(-t) |t|^{-1/2}$ if $n \equiv q \equiv 1 \bmod 2$,

(3) $\xi(t) = -(-1)^{p/2} b_n^{-1} \log |t|$ if $n \equiv p \equiv 0 \bmod 2$,

(4) $\xi(t) = (-1)^{(p-1)/2} c_n^{-1} Y(t)$ if $p \equiv q \equiv 1 \bmod 2$.

Here $t \in \mathbf{R}$, $Y(t) = 0$ or 1 according as $t \leq 0$ or $t > 0$ and[9]

$$a_n = 2^{n-1} \pi^{(n-1)/2} \Gamma\big((n-1)/2\big) ,$$
$$b_n = \pi c_n = 2^n \pi^{n/2} \Gamma(n/2) .$$

Then $\Xi(X) = \xi(u(X))$ $(X \in \mathbf{R}^n)$ is a locally summable function on $\mathbf{R}^n$ with respect to the measure $dX = dx_1 \cdots dx_n$ and

$$f(0) = \int \Xi \square^{[n/2]} f \, dX$$

for every $f \in C_c^\infty(\mathbf{R}^n)$ [3, pp. 365–366].[16] (Here $[t]$ stands for the greatest integer not exceeding t.) Define the space $\mathcal{C}(\mathbf{R}^n)$ as usual [1 (b), p. 91]. Then the above formula remains true for $f \in \mathcal{C}(\mathbf{R}^n)$ [1 (c), p. 249].

13. Proof of Theorem 3

Define $\mathcal{C}(\mathfrak{g})$ as in [1 (b), p. 91] and put

$$\psi_f(H) = \varepsilon_R(H) \pi(H) \int_{G^*} f(x^* H) \, dx^* \qquad (H \in \mathfrak{h}')$$

for $f \in \mathcal{C}(\mathfrak{g})$. Since $C_c^\infty(\mathfrak{g})$ is dense in $\mathcal{C}(\mathfrak{g})$, it follows from [1 (c), Theorem 3] and Theorem 1 that $\partial(\pi)\psi_f$ can be extended to a continuous function on $\mathfrak{h}$. Let $\psi_f(0; \partial(\pi))$ denote, as before, the value of this function at zero.

LEMMA 17. *There exists a complex number c_0 such that*

$$\psi_f(0; \partial(\pi)) = c_0 f(0)$$

for all $f \in \mathcal{C}(\mathfrak{g})$.

This is an immediate consequence[17] of Theorem 2 of [1 (d)]. Thus in order

---

[15] In this section we correct the values of some numerical constants appearing in [1(c), § 10] (see also [3, p. 365]).

[16] De Rham considers only the case when $p > 0$, $q > 0$ and $n \geq 3$. However it is a simple matter to verify that our formula remains true under the more general conditions $p \geq 0$, $q \geq 0$, $n \geq 2$.

[17] Another independent proof of this result will be given later in another paper. The reader should note that Lemma 17 is used in this paper only for the proof of Theorem 3.

to prove Theorem 3, it is enough to verify that $(-1)^q c_0$ is real and positive.

Fix a Cartan involution θ of $\mathfrak{g}$ such that $\mathfrak{h} = \theta(\mathfrak{h})$ (see [1 (a), p. 100]) and let $\mathfrak{g} = \mathfrak{k} + \mathfrak{p}$ be the corresponding Cartan decomposition of $\mathfrak{g}$. Let K be the analytic subgroup of G corresponding to $\mathfrak{k}$. First we prove the following result.

LEMMA 18. *Assume $\mathfrak{h}$ is fundamental and define q as in Theorem 3. Then*

$$q = 2^{-1} \left(\dim (G/K) - \operatorname{rank} G + \operatorname{rank} K \right) .$$

Select compatible orders (see [1 (c), p. 195]) on the spaces of real-valued linear functions on $\mathfrak{h} \cap \mathfrak{p}$ and $\mathfrak{h} \cap \mathfrak{p} + (-1)^{1/2}(\mathfrak{h} \cap \mathfrak{k})$, and let P denote the set of all positive roots in this order. Define P_+, P_0 and P_- as in [1 (c), pp. 215–216]. Then P_+ is the set of all complex positive roots and P_0 the set of all positive singular imaginary roots (see [1 (c), p. 256, Corollary 1]). Therefore $r_c = |P_+|$ and $s_I = |P_0|$, where $|Q|$ denotes the number of elements in any subset Q of P. We regard $\mathfrak{g}$ as a real Hilbert space under the positive definite quadratic form $\|X\|^2 = -B(X, \theta(X))$ $(X \in \mathfrak{g})$, and let $\mathfrak{m}$ denote the orthogonal complement of $\mathfrak{h} \cap \mathfrak{p}$ in the centralizer of $\mathfrak{h} \cap \mathfrak{p}$ in $\mathfrak{g}$. Then $\dim (\mathfrak{m} \cap \mathfrak{p}) = 2|P_0|$. Put $\mathfrak{n} = \mathfrak{g} \cap \left(\sum_{\alpha \in P_+} C X_\alpha \right)$ in the notation of § 4. η being the conjugation of $\mathfrak{g}_c$, with respect to $\mathfrak{g}$, we may assume that $\eta(X_\alpha) = X_{\eta\alpha}$ for $\alpha \in P_+$. Hence it is clear that $\dim \mathfrak{n} = |P_+|$. Now for $X \in \mathfrak{p}$ and $Y \in \mathfrak{k}$, $\operatorname{ad} X$ and $\operatorname{ad} Y$ are self-adjoint and skew-adjoint respectively [1 (g), Lemma 27]. Therefore it follows by consideration of the eigenvalues of $\operatorname{ad} H$ $(H \in \mathfrak{h} \cap \mathfrak{p})$ that

$$\mathfrak{g} = (\mathfrak{h} \cap \mathfrak{p}) + \mathfrak{m} + \mathfrak{n} + \theta(\mathfrak{n})$$

where the sum is orthogonal. Since $\mathfrak{n} + \theta(\mathfrak{n})$ is stable under θ, it is clear that

$$\mathfrak{p} = \mathfrak{h} \cap \mathfrak{p} + \mathfrak{m} \cap \mathfrak{p} + \mathfrak{p} \cap (\mathfrak{n} + \theta(\mathfrak{n})) .$$

Now $\operatorname{ad} X$ is nilpotent for $X \in \mathfrak{n}$. Hence $\mathfrak{k} \cap \mathfrak{n} = \mathfrak{p} \cap \mathfrak{n} = \{0\}$. Therefore the mappings $X \to X + \theta(X)$ and $X \to X - \theta(X)$ of $\mathfrak{n}$ into $\mathfrak{k}$ and $\mathfrak{p}$ respectively are both injective. This means that $\dim(\mathfrak{k} \cap (\mathfrak{n} + \theta(\mathfrak{n})))$ and $\dim(\mathfrak{p} \cap (\mathfrak{n} + \theta(\mathfrak{n})))$ are both $\geq \dim \mathfrak{n}$. However

$$\mathfrak{n} + \theta(\mathfrak{n}) = \mathfrak{k} \cap (\mathfrak{n} + \theta(\mathfrak{n})) + \mathfrak{p} \cap (\mathfrak{n} + \theta(\mathfrak{n}))$$

and $\dim (\mathfrak{n} + \theta(\mathfrak{n})) = 2 \dim \mathfrak{n}$. Therefore

$$\dim (\mathfrak{k} \cap (\mathfrak{n} + \theta(\mathfrak{n}))) = \dim (\mathfrak{p} \cap (\mathfrak{n} + \theta(\mathfrak{n}))) = \dim \mathfrak{n} = |P_+| .$$

This proves that

$$\dim \mathfrak{p} = \dim (\mathfrak{h} \cap \mathfrak{p}) + \dim (\mathfrak{m} \cap \mathfrak{p}) + \dim \mathfrak{p} \cap (\mathfrak{n} + \theta(\mathfrak{n}))$$
$$= \dim (\mathfrak{h} \cap \mathfrak{p}) + 2|P_0| + |P_+| ,$$

and therefore

$$2q = 2|P_0| + |P_+| = \dim \mathfrak{p} - \dim (\mathfrak{h} \cap \mathfrak{p})$$
$$= \dim (\mathfrak{g}/\mathfrak{k}) - \dim \mathfrak{h} + \dim (\mathfrak{h} \cap \mathfrak{k}) \; .$$

But since $\mathfrak{h}$ is fundamental, $\dim (\mathfrak{h} \cap \mathfrak{k}) = \operatorname{rank} \mathfrak{k}$ [1 (c), § 8] and so the lemma is proved.

The constant c_0 of Lemma 17 is obviously independent of the order of roots. Hence we may use the order introduced during the proof of Lemma 18. Moreover we can obviously assume that G is the connected adjoint group of $\mathfrak{g}$.

First assume that $\operatorname{rank} \mathfrak{g} = \operatorname{rank} \mathfrak{k}$ so that $\mathfrak{h} \subset \mathfrak{k}$. Then $q = (\dim \mathfrak{p})/2$ from Lemma 18. If $q = 0$, G is compact and c_0 is positive (see [1 (b), Lemma 16 and the corollary on p. 110]). Hence we may assume that $\mathfrak{p} \neq \{0\}$.

Let ω be the Killing form of $\mathfrak{g}$. Then $\mathfrak{k}$ and $\mathfrak{p}$ are mutually orthogonal under ω and ω is positive-definite on $\mathfrak{p}$ and negative-definite on $\mathfrak{k}$. These facts should be kept in mind during the following discussion.

Let $\mathfrak{h}_i (1 \leq i \leq N, \mathfrak{h}_1 = \mathfrak{h})$ be a maximal set of Cartan subalgebras of $\mathfrak{g}$ such that $\theta(\mathfrak{h}_i) = \mathfrak{h}_i$ and no two are conjugate under G [1(a), p. 101]. Fix euclidean measures dX and $d_i H$ on $\mathfrak{g}$ and $\mathfrak{h}_i$ respectively. Define $\psi_{f,i} = \psi_f{}^{\mathfrak{h}_i} (f \in \mathcal{C}(\mathfrak{g}))$, $\pi_i = \pi^{\mathfrak{h}_i}$ and $\varepsilon_{R,i} = \varepsilon_R{}^{\mathfrak{h}_i}$ and let ω_i denote the restriction of ω on $\mathfrak{h}_i$. Put $\Xi(X) = \xi(\omega(X)) (X \in \mathfrak{g})$ where ξ is the function on $\mathbf{R}$ defined in § 12 corresponding to $n = \dim \mathfrak{g}$ and $p = \dim \mathfrak{p}$. Then if $m = [n/2]$ and dX is suitably normalized, we have

$$f(0) = \int_{\mathfrak{g}} \Xi \, \partial(\omega^m) f \, dX$$

for all $f \in \mathcal{C}(\mathfrak{g})$. Moreover if $d_i H (1 \leq i \leq N)$ are suitably normalized, we have[4]

$$\int_{\mathfrak{g}} \Xi \, \partial(\omega^m) f \, dX = \sum_{1 \leq i \leq N} \int_{\mathfrak{h}_i'} \varepsilon_{R,i}(H) \operatorname{conj} \pi_i(H) \cdot \Xi(H) \psi_{f,i}(H; \partial(\omega_i{}^m)) \, d_i H$$

for $f \in \mathcal{C}(\mathfrak{g})$ (see [1 (c), p. 250]). Let W denote the Weyl group of $(\mathfrak{g}, \mathfrak{h})$. Since every root of $\mathfrak{h}$ is imaginary, $\mathfrak{h}$ is invariant under W (Lemma 10). Moreover we have seen in [1 (c), § 11] that f can be so chosen in $\mathcal{C}(\mathfrak{g})$ that $f(0) = 1$, $\psi_{f,i} = 0$ for $i \neq 1$ and the function $H \to \sum_{s \in W} \varepsilon(s) \psi_f(sH) (H \in \mathfrak{h}')$ has an extension φ_f on $\mathfrak{h}$ which lies in $\mathcal{C}(\mathfrak{h})$. Then since ω_1 is invariant under W, it is clear that

$$1 = f(0) = \int_{\mathfrak{h}'} \operatorname{conj} \pi(H) \cdot \Xi(H) \psi_f(H; \partial(\omega_1{}^m)) \, d_1 H$$

$$= w^{-1}(-1)^{|P|} \int_{\mathfrak{h}} \pi(H) \Xi(H) \varphi_f(H; \partial(\omega_1{}^m)) \, d_1 H$$

where w is the order of W. Let $l = \dim \mathfrak{h}$. Then $n = l + 2|P|$ and therefore $n \equiv l \bmod 2$. Similarly since $l = \operatorname{rank} \mathfrak{k}$, we have $l \equiv \dim \mathfrak{k} \bmod 2$. Hence $n \equiv \dim \mathfrak{k} \equiv l \bmod 2$. Also ω_1 is negative definite on $\mathfrak{h}$. First suppose $l \geq 2$. Then it is clear that case (2) or case (3) of § 12 is applicable to both $\partial(\omega)$ and $\partial(\omega_1)$

(on $\mathfrak{g}$ and $\mathfrak{h}$ respectively) according as l is odd or even. Therefore it follows from the formulas of § 12 that there exists a positive number a_0 such that

$$\int \Xi(H)g(H;\,\partial(\omega_1{}^k))d_1H = a_0(-1)^q\,g(0)$$

for any $g \in \mathcal{C}(\mathfrak{h})$. (Here $k = [l/2]$.)

Let R and L respectively denote the mappings $v \to v \circ \partial(\omega_1)$ and $v \to \partial(\omega_1) \circ v$ ($v \in \mathfrak{D}(\mathfrak{h}_c)$) of $\mathfrak{D}(\mathfrak{h}_c)$ into itself. Put $D = L - R$. Then L, D and R commute and if p is a polynomial in $S(\mathfrak{h}_c)$ of degree s, then $D^{s+1}p = 0$ (see [1 (b), p. 99]). Now

$$\pi \circ \partial(\omega_1{}^m) = R^m\pi = (L - D)^m\pi = \sum_{0 \leq s \leq m} C_s{}^m(-1)^s\,L^{m-s}D^s\pi$$

where $C_s{}^m = m!\,(s!\,(m - s)!)^{-1}$. Put $r = |P|$. Since $n = l + 2r$, it is clear that $m = k + r$. On the other hand π is of degree r and therefore $D^s\pi = 0$ for $s > r$. Hence

$$\pi \circ \partial(\omega_1{}^m) = \sum_{0 \leq s \leq r} C_s{}^{k+r}(-1)^s\,L^{k+r-s}D^s\pi = \partial(\omega_1{}^k) \circ \eta$$

where

$$\eta = \sum_{0 \leq s \leq r} C_s{}^{k+r}(-1)^s\,L^{r-s}D^s\pi = \sum_{0 \leq s \leq r} C_s{}^{k+r}(-1)^s\,(R + D)^{r-s}D^s\pi$$
$$= \sum_{0 \leq t \leq r} \left\{\sum_{0 \leq s \leq t} C_s{}^{k+r}C_{r-t}{}^{r-s}(-1)^s\right\} R^{r-t}D^t\pi\,.$$

As usual let d_0 denote the local expression at zero (see [1 (b), p. 90]) of an element d in $\mathfrak{D}(\mathfrak{h}_c)$. Then $(R^{r-t}D^t\pi)_0 = (D^t\pi)_0 \circ \partial(\omega_1{}^{r-t})$ for $0 \leq t \leq r$. We claim $(D^t\pi)_0 = 0$ if $t < r$. For $(D^t\pi)_0 = \partial(p)$ where p is an element of degree $\leq t$ in $S(\mathfrak{h}_c)$ (see [1 (b), p. 99]) such that $p^s = \varepsilon(s)p\,(s \in W)$. But since $t < r$, it follows from Lemma 10 of [1 (b)] that $p = 0$. Hence

$$\eta_0 = \sum_{0 \leq s \leq r} C_s{}^{k+r}(-1)^s(D^r\pi)_0\,.$$

But

$$\sum_{0 \leq s \leq r} C_s{}^{k+r}(-1)^s = (-1)^r C_r{}^{k+r-1}$$

as is seen by comparing the coefficients of x^r in the identity

$$(1 - x)^{k+r-1} = (1 - x)^{k+r}(1 + x + x^2 + \cdots)\,.$$

Moreover $D^r\pi = 2^r \cdot r!\,\partial(\pi)$ [1 (b), p. 99]. Hence

$$\eta_0 = (-1)^r k(k + 1) \cdots (k + r - 1)2^r\,\partial(\pi)\,.$$

But then since $\eta\varphi_f \in \mathcal{C}(\mathfrak{h})$, we have

$$\int_{\mathfrak{h}} \pi(H)\Xi(H)\varphi_f(H;\,\partial(\omega_1{}^m))\,dH = \int \Xi(H)\varphi_f(H;\,\partial(\omega_1{}^k) \circ \eta)\,dH$$
$$= a_0(-1)^q\varphi_f(0;\,\eta) = (-1)^{r+q}a\varphi_f(0;\,\partial(\pi))$$

where $a = k(k + 1) \cdots (k + r - 1)2^r a_0$ and $dH = d_1 H$. This proves that

$$1 = f(0) = (-1)^q w^{-1} a \varphi_f(0; \partial(\pi)) .$$

But

$$\psi_f(sH; \partial(\pi^s)) = \varepsilon(s)\psi_f(sH; \partial(\pi)) \qquad (H \in \mathfrak{h}') ,$$

and so it is clear that $\varphi_f(0; \partial(\pi)) = w\,\psi_f(0; \partial(\pi)) = wc_0$ from Lemma 17. This proves that $1 = (-1)^q ac_0$. Since a is positive, the required result is now obvious.

On the other hand if $l = 1$, $\mathfrak{g}$ is isomorphic to the Lie algebra $\mathfrak{l}$ of § 2 and so our assertion is a consequence of Lemma 3.

Hence we may now suppose that rank $\mathfrak{g} >$ rank $\mathfrak{k}$. Then $\mathfrak{h} \cap \mathfrak{p} \neq \{0\}$. Let $\mathfrak{z}$ be the centralizer of $\mathfrak{h} \cap \mathfrak{p}$ in $\mathfrak{g}$. Since $\theta(\mathfrak{z}) = \mathfrak{z}$, ad Z $(Z \in \mathfrak{z})$ is a self-adjoint family of linear transformations in $\mathfrak{g}$ [1 (g), Lemma 27]. Hence $\mathfrak{z}$ is reductive in $\mathfrak{g}$. Also dim $\mathfrak{z} <$ dim $\mathfrak{g}$ since $\mathfrak{g}$ is semi-simple. Therefore Theorem 3 holds for $\mathfrak{z}$ by induction hypothesis. Put $H_\alpha' = 2H_\alpha/\alpha(H_\alpha)$ for $\alpha \in P$. Then $\alpha(H_\alpha)$ is positive and H_α' is the unique element $H \in \mathfrak{h}_c$ such that

(1) $s_\alpha H = -H$, and

(2) $\alpha(H) = 2$.

Put $P_\mathfrak{z} = P_0 \cup P_-$. Then $P_\mathfrak{z}$ is the set of all positive roots of $(\mathfrak{z}, \mathfrak{h})$ and it is clear from the above remark that whether we regard $\alpha \in P_\mathfrak{z}$ as a root of $\mathfrak{g}$ or a root of $\mathfrak{z}$, we get the same element H_α'. On the other hand we know that $c_0 = a(-1)^{|P+|/2}c_1'$ (see [1 (d), p. 679]) where a is a positive number and c_1' is defined as in the proof of Theorem 3 of [1(d)]. Applying Theorem 3 to $\mathfrak{z}$, we conclude easily that c_1' has the same sign as $(-1)^{|P_0|}$. Since $q = |P_0| + |P_+|/2$, it follows that $(-1)^q c_0 > 0$. This completes the proof of Theorem 3.

14. Some simple lemmas

Let E be a real euclidean space of finite dimension, U an open subset, and F a closed subset of E. Assume that $U \cap F$ is open in E and dense in F. Let f be a C^∞ function on U. We say that f is of class C^∞ on F if for every $p \in S(E_c)$, there exists a continuous function f_p on F which coincides with $\partial(p)f$ on $F \cap U$. It is obvious that f_p, if it exists, is unique. We shall often write $f(X; \partial(p))$ for $f_p(X)$ $(X \in F)$ and sometimes even denote f_p by $\partial(p)f$.

LEMMA 19. *Let U be a non-empty, open and convex subset of E, and U_1 the complement in U of the union of a finite number of affine subspaces of dimension $\leq r < n = \dim E$. Let U' be an open and dense subset of U_1. Then given any two points H_1, H_2 in U_1, we can choose H_i' in U' arbitrarily near $H_i (i = 1, 2)$ with the following property. Let J denote the straight line-segment joining H_1' to H_2'. Then $J \subset U_1$ if $r < n - 1$ and the complement of $J \cap U_1$ in J is a finite set if $r = n - 1$.*

The proof of this is substantially the same as that given in [1 (c), p. 253].

COROLLARY. *U_1 is connected if $r \leq n - 2$.*

This is obvious from the above lemma.

LEMMA 20. *Suppose U is as above and f is a C^∞ function on U such that $\sup_U |\partial(p)f| < \infty$ for every $p \in S(E_c)$. Then f is of class C^∞ on[18] $\mathrm{Cl}(U)$.*

Fix $p \in S(E_c)$. We have to show that $g = \partial(p)f$ can be extended to a continuous function on $\mathrm{Cl}(U)$. Since g also satisfies the conditions of the lemma, it would be enough to prove that f can be so extended. Fix a euclidean norm on E and a point $X_0 \in \mathrm{Cl}(U)$. Given $\varepsilon > 0$, let V_ε denote the set of all $X \in E$ with $|X - X_0| < \varepsilon$. Then $V_\varepsilon \cap U$ is a non-empty convex set. If X_1, X_2 are two points in $V_\varepsilon \cap U$, it is clear that

$$f(X_2) - f(X_1) = \int_0^1 \left(df(tX_2 + (1 - t)X_1)/dt\right) dt .$$

Hence

$$|f(X_2) - f(X_1)| \leq \mathrm{Sup}_U |\partial(X_2 - X_1)f| .$$

Now let $Y_1, \cdots, Y_n$ be an orthonormal base for E. Then if $X = \sum_i c_i Y_i$ ($c_i \in \mathbf{R}$), it is clear that $|X|^2 = \sum_i c_i^2$ and

$$\partial(X)f = \sum_i c_i \partial(Y_i)f .$$

Therefore

$$\mathrm{Sup}_U |\partial(X)f| \leq |X| \max_i \mathrm{sup}_U |\partial(Y_i)f| .$$

This proves that

$$|f(X_2) - f(X_1)| \leq \mathrm{Sup}_U |\partial(X_2 - X_1)f| \leq 2\varepsilon \max_i \mathrm{sup}_U |\partial(Y_i)f| .$$

It is now obvious that f can be extended to a continuous function on $\mathrm{Cl}(U)$.

Assume that we are given a euclidean norm on E. Then we can define a scalar product in E and identify E with its dual by means of it. Let dE denote the euclidean measure on E corresponding to this norm. For example if $(Y_1, \cdots, Y_n)$ is an orthonormal base for E, and Q the set of all points $X = \sum_i c_i Y_i$ ($c_i \in \mathbf{R}$) with $|c_i| < 1/2$, then

$$\int_Q dE = 1 .$$

(This holds in particular if $n = 0$, in which case $E = Q = \{0\}$.) Let F be an affine subspace of E. Then $F = X + F_0$ where F_0 is a linear subspace of E and X a point in F. The induced norm on F_0 defines the euclidean measure dF_0 on F_0. We denote by dF the measure on F obtained from dF_0 under the mapping

---

[18] As usual $\mathrm{Cl}(U)$ denotes the closure of U.

$Y \to X + Y$ $(Y \in F_0)$. Since dF_0 is invariant under translations by elements of F_0, it is clear that dF is independent of the choice of X.

Let E' be an open and dense subset of E and f a continuous function on E'. Then by the support of f in E we mean the closure (in E) of the support of f in E'.

LEMMA 21. *Let α_i be non-zero real linear functions on E and c_i real numbers $(1 \leq i \leq r)$. We denote by σ_i the hyperplane $\alpha_i = c_i$ in E and assume that $\sigma_1, \cdots, \sigma_r$ are all distinct. Let E' be the complement of $\bigcup_i \sigma_i$ in E and σ_i' the complement of $\bigcup_{j \neq i} \sigma_j$ in σ_i. Let f be a function on E' which is of class C^∞ on the closure of each connected component of E' and whose support (in E) is compact. For $X \in \sigma_i'$, put*

$$f^+(X) = \mathrm{Lim}_{t \to +0} f(X + t\alpha_i) , \quad f^-(X) = \mathrm{Lim}_{t \to +0} f(X - t\alpha_i) .$$

Then for any $Y \in E$,

$$\int_E \partial(Y) f \, dE = -\sum_{1 \leq i \leq r} \int_{\sigma_i'} \{f^+ - f^-\} \, \alpha_i(Y) \, |\alpha_i|^{-1} \, d\sigma_i .$$

It is clear from our assumptions that $\partial(Y)f$ is bounded on E' and f^+, f^- are both bounded on σ_i'. Since $\bigcup_j \sigma_j$ and $\sigma_i \cap \bigcup_{j \neq i} \sigma_j$ are both of measure zero in E and σ_i respectively, our integrals are well defined.

For $\varepsilon = \pm 1$, let $E_i(\varepsilon)$ denote the set of all points in E where $\varepsilon(\alpha_i - c_i) > 0$. Put

$$E(\varepsilon_1, \cdots, \varepsilon_r) = \bigcap_i E_i(\varepsilon_i)$$

where $\varepsilon_i = \pm 1$ $(1 \leq i \leq r)$. Then

$$\int_E \partial(Y) f \, dE = \sum_{\varepsilon_1, \cdots, \varepsilon_r} \int_{E(\varepsilon_1, \cdots, \varepsilon_r)} \partial(Y) f \, dE .$$

But if $\lambda_i = |\alpha_i|^{-1} \alpha_i$,

$$\int_{E(\varepsilon_1, \cdots, \varepsilon_r)} \partial(Y) f \, dE = -\sum_{1 \leq i \leq r} \int_{\sigma_i \cap \bigcap_{j \neq i} E_j(\varepsilon_j)} \varepsilon_i \lambda_i(Y) f_{\varepsilon_i} \, d\sigma_i ,$$

where

$$f_{\varepsilon_i}(X) = \mathrm{Lim}_{t \to +0} f(X + \varepsilon_i t \lambda_i) \qquad\qquad (X \in \sigma_i') .$$

However σ_i' is the disjoint union of the sets $\sigma_i \cap \bigcap_{j \neq i} E_j(\varepsilon_j)$ for all possible values of $\varepsilon_j (j \neq i)$. Therefore

$$\int_E \partial(Y) f \, dE = -\sum_{1 \leq i \leq r} \int_{\sigma_i'} \{f^+ - f^-\} \lambda_i(Y) \, d\sigma_i ,$$

and this proves the lemma.

15. Distributions involving ψ_f

We now go back to the notation of §§ 10 and 11. Let us recall that G operates on $\mathfrak{g}$ so that $X^x = xX = \mathrm{Ad}(x)X$ ($x \in G$, $X \in \mathfrak{g}$).

LEMMA 22. *Let $f \in C_c^\infty(\mathfrak{g})$. Then the function ψ_f is of class C^∞ on the closure of each connected component of $\mathfrak{h}'(I)$. Its support in $\mathfrak{h}$ is compact and contained in $\mathrm{Cl}(\mathfrak{h} \cap (\mathrm{Supp}\, f)^g)$. Let g be a function on $\mathfrak{h}$ which is locally summable with respect to the euclidean measure dH on $\mathfrak{h}$. Then for any $D \in \mathfrak{D}(\mathfrak{h}_c)$, the mapping*

$$f \longrightarrow \int_{\mathfrak{h}} g D\psi_f \, dH \qquad\qquad (f \in C_c^\infty(\mathfrak{g}))$$

is an invariant distribution on $\mathfrak{g}$.

Let $\mathrm{Supp}\,\psi_f$ denote the support of ψ_f in $\mathfrak{h}$. Fix $H_0 \in \mathrm{Supp}\,\psi_f$, and let U be a neighborhood of H_0 in $\mathfrak{h}$. Then it is clear that $xH \in \mathrm{Supp}\, f$ for some $x \in G$ and $H \in U \cap \mathfrak{h}'$. Hence U meets $(\mathrm{Supp}\, f)^g$. This proves that $\mathrm{Supp}\,\psi_f \subset \mathrm{Cl}(\mathfrak{h} \cap (\mathrm{Supp}\, f)^g)$. On the other hand we have the following simple lemma.

LEMMA 23. *Let B be a bounded subset of $\mathfrak{g}$. Then $\mathfrak{h} \cap B^g$ is bounded in $\mathfrak{h}$.*

Choose bounded sets C and B_1 in $\mathfrak{c}$ and $\mathfrak{g}_1$ respectively such that $B \subset C + B_1$. Then $B^g \subset C + B_1^g$ and therefore

$$\mathfrak{h} \cap B^g \subset C + \mathfrak{h}_1 \cap B_1^g$$

in the notation of §11. So it would be enough to prove that $\mathfrak{h}_1 \cap B_1^g$ is bounded. Hence we may assume that $\mathfrak{g}$ is semi-simple.

Let $l = \dim \mathfrak{h}$, $n = \dim \mathfrak{g}$, and r the number of positive roots so that $n = l + 2r$. Let p_j ($1 \leq j \leq 2r$) be the polynomial functions on $\mathfrak{g}_c$ given by

$$\det (t - \mathrm{ad}X) = t^l \{t^{2r} + \sum_{1 \leq j \leq 2r} t^{2r-j} p_j(X)\} \qquad\qquad (X \in \mathfrak{g}_c)$$

where t is an indeterminate. It is obvious that p_j is invariant under G and therefore it remains bounded on B^g. On the other hand if $H \in \mathfrak{h}$, we have

$$\det (t - \mathrm{ad}\, H) = t^l \prod_{\alpha > 0} (t^2 - \alpha(H)^2)$$

where α runs over all positive roots. This shows that the elementary symmetric functions in $\alpha(H)^2$ ($\alpha > 0$) remain bounded as H varies in $\mathfrak{h} \cap B^g$. Hence the numbers $\alpha(H)$ themselves remain bounded. But since $\mathfrak{g}$ is semi-simple, this implies that $\mathfrak{h} \cap B^g$ is bounded.

It follows from Lemma 23 that $\mathrm{Supp}\,\psi_f$ is compact. Moreover we conclude from Theorem 3 of [1 (c)] that $\mathrm{Sup}_{\mathfrak{h}'} |D\psi_f| < \infty$ for every $D \in \mathfrak{D}(\mathfrak{h}_c)$. The first assertion of Lemma 22 now follows from Lemmas 11 and 20. The last statement is an easy consequence of [1 (c), Theorem 3].

16. Reduction of the proofs of Theorems 4 and 5
to that of Lemma 28

Let $\mathfrak{g}$ be a semi-simple Lie algebra over $\mathbf{R}$ and $\mathfrak{h}$ a Cartan subalgebra of $\mathfrak{g}$. We identify $\mathfrak{g}_c$ and $\mathfrak{h}_c$ with their respective duals as usual (see [1 (b), p. 93]). Let $I(\mathfrak{g}_c)$ be the subalgebra of all invariants in $S(\mathfrak{g}_c)$. For any $p \in I(\mathfrak{g}_c)$, we denote by $p_{\mathfrak{h}}$ the restriction of the polynomial function p on $\mathfrak{h}_c$.

Let $\mathfrak{h}_i$ $(1 \le i \le r)$ be a maximal set of Cartan subalgebras of $\mathfrak{g}$, no two of which are conjugate under G. Put

$$\psi_{f,i} = \psi_f{}^{\mathfrak{h}i}, \qquad \pi_i = \pi^{\mathfrak{h}i}, \qquad p_i = p_{\mathfrak{h}_i} \qquad (1 \le i \le r)$$

for $f \in C_c^\infty(\mathfrak{g})$ and $p \in I(\mathfrak{g}_c)$. Fix a euclidean measure $d_i H$ on $\mathfrak{h}_i$ and an open subset Ω of $\mathfrak{g}$ which is invariant under G. Let ω be the Killing form of $\mathfrak{g}$.

THEOREM 4. *Suppose we are given, for each i $(1 \le i \le r)$, a function g_i on $\mathfrak{h}_i'(S)$ (see § 4) which is of class C^∞ on the closure of each connected component of $\mathfrak{h}_i'(S)$. Define*

$$J(f) = \sum_{1 \le i \le r} \int_{\mathfrak{h}_i} \{\partial(\omega_i)\psi_{f,i} \cdot g_i - \psi_{f,i}\partial(\omega_i)g_i\}d_i H$$

for $f \in C_c^\infty(\mathfrak{g})$. Then J is an invariant distribution on $\mathfrak{g}$. Moreover if $\operatorname{Supp} J \cap \operatorname{Cl}(\Omega)$ contains no semi-regular element of $\mathfrak{g}$ of non-compact type, then $J = 0$ on Ω.

We keep to the above notation.

THEOREM 5. *Fix $p \in I(\mathfrak{g}_c)$ and define*

$$J_p(f) = \sum_{1 \le i \le r} \int_{\mathfrak{h}_i} \{\partial(\omega_i p_i)(\pi_i \psi_{f,i}) \cdot g_i - \partial(p_i)(\pi_i \psi_{f,i}) \cdot \partial(\omega_i)g_i\}d_i H$$

for $f \in C_c^\infty(\mathfrak{g})$. Then J_p is an invariant distribution on $\mathfrak{g}$. Now consider the following two conditions.

(1) *$(\operatorname{Supp} J_1) \cap \operatorname{Cl}(\Omega)$ contains no semi-regular element of $\mathfrak{g}$ of non-compact type.*

(2) *g_i is of class C^∞ on the closure of each connected component of $\mathfrak{h}_i'(R)$ $(1 \le i \le r)$.*

Then under (1) we can assert that $J_1 = 0$ on Ω. Moreover if both (1) and (2) hold, we can conclude that $J_p = 0$ on Ω for every $p \in I(\mathfrak{g}_c)$.

It is convenient to prove these two theorems together. The fact that J and J_p are invariant distributions follows from Lemma 22. Now fix a Cartan involution θ of $\mathfrak{g}$ and let $\mathfrak{g} = \mathfrak{k} + \mathfrak{p}$ be the corresponding Cartan decomposition of $\mathfrak{g}$. Put

$$\langle X, Y \rangle = \operatorname{tr}(\operatorname{ad} X \operatorname{ad} Y), \qquad \langle X, Y \rangle_+ = -\langle X, \theta(Y) \rangle \qquad (X, Y \in \mathfrak{g}).$$

Then $\langle X, Y\rangle_+$ is a positive-definite scalar product on $\mathfrak{g}$, which corresponds to the euclidean norm $\|X\|^2 = \langle X, X\rangle_+$. Fix a Cartan subalgebra $\mathfrak{h}$ of $\mathfrak{g}$ such that $\theta(\mathfrak{h}) = \mathfrak{h}$ and put

$$[F_1, F_2] = \partial(\omega_{\mathfrak{h}})F_1 \cdot F_2 - F_1 \partial(\omega_{\mathfrak{h}})F_2$$

for two C^∞ functions F_1, F_2 on $\mathfrak{h}'$. Now define

$$J_{\mathfrak{h}}(f) = \int_{\mathfrak{h}} [\psi_f, g]dH, \qquad J_{p,\mathfrak{h}}(f) = \int_{\mathfrak{h}} [\partial(p_{\mathfrak{h}})(\pi\psi_f), g]dH$$

for $f \in C_c^\infty(\mathfrak{g})$ and $p \in I(\mathfrak{g}_c)$. Here g is a given function on $\mathfrak{h}'(S)$ which is of class C^∞ on the closure of each connected component of $\mathfrak{h}'(S)$ and the euclidean measure dH on $\mathfrak{h}$ is normalized as in § 14. For any linear function λ on $\mathfrak{h}_c$, let H_λ denote the corresponding element in $\mathfrak{h}_c$ so that $\lambda(H) = \mathrm{tr}(\mathrm{ad}\, H_\lambda\, \mathrm{ad}\, H)$ for $H \in \mathfrak{h}_c$. Then $\alpha(H_\alpha) > 0$ for every root α. Put $|\alpha| = (\alpha(H_\alpha))^{1/2} > 0$. For any singular root α, let σ_α denote the hyperplane $\alpha = 0$ in $\mathfrak{h}$. Define $\varepsilon_\alpha = 1$ or $(-1)^{1/2}$ according as α is real or imaginary, and put $\alpha^* = \varepsilon_\alpha\alpha$. Then $H_{\alpha^*} \in \mathfrak{h}$. Let σ_α' denote the set of all semi-regular elements in σ_α. Then σ_α' is the set of those points of σ_α where the polynomial function $\pi_\alpha = \alpha^{-1}\pi$ does not vanish. We define the euclidean measure $d\sigma_\alpha$ on σ_α as in § 14.

LEMMA 24. *Let F be a function on $\mathfrak{h}'(S)$ which is of class C^∞ on the closure of each connected component of $\mathfrak{h}'(S)$ and whose support in $\mathfrak{h}$ is compact. Then for $H' \in \mathfrak{h}_c$, we have*

$$\int_{\mathfrak{h}} \partial(H')F\,dH = -\sum_{\alpha \in S} \varepsilon_\alpha^{-1}|\alpha|^{-1}\int_{\sigma_\alpha'} (F^+ - F^-)\alpha(H')\,d\sigma_\alpha$$

where

$$F^+(H) = \mathrm{Lim}_{t \to +0} F(H + tH_{\alpha^*}),$$
$$F^-(H) = \mathrm{Lim}_{t \to -0} F(H + tH_{\alpha^*})$$

for $H \in \sigma_\alpha'$.

Fix $\alpha \in S$. Then α^* is a real-valued linear function on $\mathfrak{h}$ and σ_α is defined by the equation $\alpha^* = 0$. Also

$$\langle H_{\alpha^*}, H\rangle_+ = -\langle \theta(H_{\alpha^*}), H\rangle = \varepsilon_\alpha^2\langle H_{\alpha^*}, H\rangle = \varepsilon_\alpha^2\alpha^*(H) \qquad (H \in \mathfrak{h}).$$

(Here we use the fact that H_α lies in $\mathfrak{h}_c \cap \mathfrak{p}_c$ or $\mathfrak{h}_c \cap \mathfrak{k}_c$ according as α is real or imaginary.) Therefore we conclude from Lemma 21 that

$$\int_{\mathfrak{h}} \partial(H')F\,dH = -\sum_{\alpha \in S} |\alpha|^{-1}\varepsilon_\alpha^2\int_{\sigma_\alpha'} (F^+ - F^-)\alpha^*(H')\,d\sigma_\alpha.$$

But $\varepsilon_\alpha^2\alpha^* = \varepsilon_\alpha^{-1}\alpha$ and so our assertion is now obvious.

Put $F_-^+ = F^+ - F^-$.

COROLLARY. *Let F_1, F_2 be two functions on $\mathfrak{h}'(S)$ which are both of class C^∞ on the closure of each connected component of $\mathfrak{h}'(S)$. Suppose the support of F_1 in $\mathfrak{h}$ is compact. Then*

$$\int_{\mathfrak{h}} [F_1, F_2] dH = -\sum_{\alpha \in S} \varepsilon_\alpha^{-1} |\alpha|^{-1} \int_{\sigma_\alpha'} (\partial(\alpha) F_1 \cdot F_2 - F_1 \cdot \partial(\alpha) F_2)_-^{+} d\sigma_\alpha .$$

Let P be the set of all positive roots of $(\mathfrak{g}, \mathfrak{h})$. Then

$$\operatorname{tr}(\operatorname{ad} H)^2 = 2 \sum_{\alpha \in P} \alpha(H)^2 \qquad\qquad (H \in \mathfrak{h}_c) .$$

This shows that $\omega_{\mathfrak{h}} = 2 \sum_{\alpha \in P} \alpha^2$. Moreover if λ is any linear function on $\mathfrak{h}_c$, then

$$\lambda(H) = \operatorname{tr}(\operatorname{ad} H \operatorname{ad} H_\lambda) = 2 \sum_{\alpha \in P} \alpha(H)\alpha(H_\lambda) = 2 \sum_{\alpha \in P} \lambda(H_\alpha)\alpha(H)$$

for $H \in \mathfrak{h}_c$. Hence

$$\lambda = 2 \sum_{\alpha \in P} \lambda(H_\alpha)\alpha .$$

Now

$$\begin{aligned}
[F_1, F_2] &= 2 \sum_{\alpha \in P} (\partial(\alpha)^2 F_1 \cdot F_2 - F_1 \cdot \partial(\alpha)^2 F_2) \\
&= 2 \sum_{\alpha \in P} \partial(\alpha)(\partial(\alpha) F_1 \cdot F_2 - F_1 \cdot \partial(\alpha) F_2) .
\end{aligned}$$

Therefore it follows from Lemma 24 that

$$\int_{\mathfrak{h}} [F_1, F_2] dH$$

$$= -2 \sum_{\alpha \in P} \sum_{\beta \in S} \varepsilon_\beta^{-1} |\beta|^{-1} \int_{\sigma_\beta'} (\partial(\alpha) F_1 \cdot F_2 - F_1 \partial(\alpha) F_2)_-^{+} \beta(H_\alpha) d\sigma_\alpha .$$

The required result is obtained by observing that

$$2 \sum_{\alpha \in P} \beta(H_\alpha)\partial(\alpha) = \partial(\beta) .$$

LEMMA 25. *Put $\psi_{p,f} = \partial(p_{\mathfrak{h}})(\pi \psi_f)$ for $p \in I(\mathfrak{g}_c)$ and $f \in C_c^\infty(\mathfrak{g})$. Then*

$$\begin{aligned}
J_{\mathfrak{h}}(f) = \sum_{\alpha \in S_R} |\alpha|^{-1} \int_{\sigma_\alpha'} \psi_f (\partial(\alpha) g)_-^{+} d\sigma_\alpha \\
+ (-1)^{1/2} \sum_{\beta \in S_I} |\beta|^{-1} \int_{\sigma_\beta'} (\partial(\beta)\psi_f \cdot g - \psi_f \partial(\beta) g)_-^{+} d\sigma_\beta
\end{aligned}$$

and

$$\begin{aligned}
J_{p,\mathfrak{h}}(f) = -\sum_{\alpha \in S_R} |\alpha|^{-1} \int_{\sigma_\alpha'} \partial(\alpha)\psi_{p,f} \cdot g_-^{+} d\sigma_\alpha \\
+ (-1)^{1/2} \sum_{\beta \in S_I} |\beta|^{-1} \int_{\sigma_\beta'} (\partial(\beta)\psi_{p,f} \cdot g - \psi_{p,f} \partial(\beta) g)_-^{+} d\sigma_\beta .
\end{aligned}$$

Take $F_1 = \psi_f$ and $F_2 = g$ in the corollary of Lemma 24. Then if $\alpha \in S_R$, ψ_f is of class C^∞ around every point of σ_α' and $\partial(\alpha)\psi_f = 0$ on σ_α' (see Lemma

11 and its corollary). Hence

$$(\partial(\alpha)\psi_f \cdot g - \psi_f \cdot \partial(\alpha)g)_-{}^+ = - \psi_f(\partial(\alpha)g)_-{}^+$$

on σ_α', and from this the above expression for $J_{\mathfrak{h}}(f)$ follows immediately. In the same way $\psi_{p,f}$ is of class C^∞ on $\mathfrak{h}'(I)$ and $(\psi_{p,f})'{}^\alpha = - \psi_{p,f}$ for $\alpha \in S_R$. Therefore $\psi_{p,f} = 0$ on σ_α'. The above formula for $J_{p,\mathfrak{h}}(f)$ is now obvious from these facts.

For $f \in C_c^\infty(\mathfrak{g})$, $D \in \mathfrak{D}(\mathfrak{h}_c)$ and any singular imaginary root β, let $(D\psi_f)^\beta$ denote the function

$$H \longrightarrow \mathrm{Lim}_{t\to+0}\, \psi_f\!\left(H - (-1)^{1/2}tH_\beta;\, D\right) \qquad\qquad (H \in \sigma_\beta')$$

on σ_β'. Then it is clear that for $\beta \in S_I$, we have

$$(\psi_f\partial(\beta)g)_-{}^+ = - \psi_f{}^\beta(\partial(\beta)g)^- + \psi_f{}^{-\beta}(\partial(\beta)g)^+$$

on σ_β'. Moreover it follows from the corollary of Lemma 15 that $\partial(\beta)\psi_f$ and $\psi_{p,f}$ can be extended to continuous functions around any point of σ_β'. These facts should be kept in mind in the following discussion.

Now fix $\alpha \in S_R$. Then we can choose the elements X_α, $X_{-\alpha}$ of §4 in such a way that

(1) they lie in $\mathfrak{g}$, and

(2) $\theta(X_\alpha) = - X_{-\alpha}$

(see [1 (c), Lemma 46, p. 255]). Put $X_{\pm\alpha}' = \sqrt{2}\,|\alpha|^{-1}X_{\pm\alpha}$, and let ν denote the automorphism of $\mathfrak{g}_c$ given by[9] (cf. §7)

$$\nu = \exp\{-(-1)^{1/2}(\pi/4)\,\mathrm{ad}\,(X_\alpha' + X_{-\alpha}')\}\,.$$

Put $\mathfrak{a} = \mathfrak{h}$ and $\mathfrak{b} = \mathfrak{g} \cap \nu(\mathfrak{a}_c)$. Then $\mathfrak{a} = \sigma_\alpha + RH_\alpha'$ and $\mathfrak{b} = \sigma_\alpha + R(X_\alpha' - X_{-\alpha}')$ where $H_\alpha' = 2\,|\alpha|^{-2}H_\alpha$. Hence $\theta(\mathfrak{b}) = \mathfrak{b}$. Put $\beta = \alpha^\nu$ in the notation of §7. Then we know from Lemma 13 that β is a singular imaginary root of $\mathfrak{b}$ and $\sigma_\alpha = \sigma_\beta = \mathfrak{a} \cap \mathfrak{b}$. Moreover, from Theorem 2, there exists a locally constant function c on σ_β' such that

$$D\psi_f{}^\alpha = c\{(D^\nu\psi_f{}^\mathfrak{b})^\beta - (D^\nu\psi_f{}^\mathfrak{b})^{-\beta}\}$$

on σ_β' for any $D \in \mathfrak{D}(\mathfrak{a}_c)$ and $f \in C_c^\infty(\mathfrak{g})$. Therefore

$$\psi_f{}^\alpha = c(\psi_f{}^\beta - \psi_f{}^{-\beta})$$

and

$$\partial(\alpha)\psi_{p,f}{}^\alpha = c\{(\partial(\beta)\psi_{p,f}{}^\mathfrak{b})^\beta - (\partial(\beta)\psi_{p,f}{}^\mathfrak{b})^{-\beta}\}$$

on σ_β' for all $p \in I(\mathfrak{g}_c)$ and $f \in C_c^\infty(\mathfrak{g})$.

Now we come to the proofs of Theorems 4 and 5. In view of Lemma 16, its corollary and [1 (a), p. 100], we may assume that $\theta(\mathfrak{h}_i) = \mathfrak{h}_i$ $(1 \le i \le r)$. Moreover we can obviously suppose that the euclidean measure d_iH is normalized as in § 14. Let $S_{i,R}$ and $S_{i,I}$ denote the sets of all positive roots of $\mathfrak{h}_i$ which are real and singular imaginary respectively. Fix i and $\alpha \in S_{i,R}$ and define $\mathfrak{a} = \mathfrak{h}_i$ and $\mathfrak{b}$ as above. Then $\mathfrak{b}$ is conjugate to $\mathfrak{h}_j$ under G for some j. Therefore if we take Lemma 25 and the above discussion into account and make use of the corollary of Lemma 16, we get the following result.

LEMMA 26. *Let Q_i denote the set of all singular imaginary roots of $\mathfrak{h}_i$ ($1 \leq i \leq r$). Then for each $\beta \in Q_i$, we can choose functions g_β, $g_\beta{}'$, h_β and $h_\beta{}'$ on $\sigma_\beta{}'$ with the following properties. They are of class C^∞ on the closure of each connected component of $\sigma_\beta{}'$ and*

$$J(f) = \sum_{1 \leq i \leq r} \sum_{\beta \in Q_i} \int_{\sigma_\beta{}'} \{\partial(\beta)\psi_{f,i} \cdot h_\beta + \psi_{f,i}{}^\beta g_\beta\} d\sigma_\beta \,,$$

$$J_p(f) = \sum_{1 \leq i \leq r} \sum_{\beta \in Q_i} \int_{\sigma_\beta{}'} \{\psi_{p,f,i} \cdot h_\beta{}' + (\partial(\beta)\psi_{p,f,i})^\beta g_\beta{}'\} d\sigma_\beta$$

for all $f \in C_c^\infty(\mathfrak{g})$ and $p \in I(\mathfrak{g}_c)$. Moreover under condition (2) of Theorem 5, we can assume that $h_\beta{}' = 0$ ($\beta \in Q_i$, $1 \leq i \leq r$).

Only the last statement remains to be proved. We go back to the notation of Lemma 25. If g is of class C^∞ on the closure of each connected component of $\mathfrak{h}'(R)$, then

$$(\psi_{p,f}\partial(\beta)g)_-^+ = \psi_{p,f}(\partial(\beta)g)_-^+ = 0$$

on $\sigma_\beta{}'$ for $\beta \in S_I$. From this the required assertion follows immediately.

Define $W_{G,i}$ corresponding to $\mathfrak{h}_i$ (see § 4). It is obvious that if $\beta \in Q_i$ and $s \in W_{G,i}$ then $s\beta$ is also in Q_i and $\sigma_{s\beta} = s\sigma_\beta$. For any function h on $\sigma_\beta{}'$, let h^s denote the function $H \to h(s^{-1}H)$ on $\sigma_{s\beta}{}'$. $\varepsilon(s)$ and $\varepsilon_0(s)$ are defined as in §§ 4 and 6 respectively.

LEMMA 27. *We can assume that the functions g_β, $g_\beta{}'$, h_β and $h_\beta{}'$ ($\beta \in Q_i$, $1 \leq i \leq r$) of Lemma 26 satisfy, in addition, the relations*

    (1) $h_{-\beta} = -h_\beta$, $(h_\beta)^s = \varepsilon_0(s)h_{s\beta}$, $(g_\beta)^s = \varepsilon_0(s)g_{s\beta}$,

    (2) $h_{-\beta}{}' = h_\beta{}'$, $(h_\beta{}')^s = \varepsilon_0(s)\varepsilon(s)h_{s\beta}{}'$, $(g_\beta{}')^s = \varepsilon_0(s)\varepsilon(s)g_{s\beta}{}'$

for $\beta \in Q_i$ and $s \in W_{G,i}$ ($1 \leq i \leq r$).

We can obviously replace h_β by $(h_\beta - h_{-\beta})/2$ ($\beta \in Q_i$) without disturbing the conditions of Lemma 26. Hence we may assume that $h_{-\beta} = -h_\beta$ and similarly $h_{-\beta}{}' = h_\beta{}'$ for $\beta \in Q_i$ ($1 \leq i \leq r$). Now any $s \in W_{G,i}$ leaves both $\mathfrak{h}_i$ and $\mathfrak{h}_i \cap \mathfrak{p} + (-1)^{1/2}(\mathfrak{h}_i \cap \mathfrak{k})$ invariant. Hence $s\theta(H) = \theta(sH)$ and therefore $\|sH\| = \|H\|$ for $H \in \mathfrak{h}_i$. This shows that under the mapping $H \to sH$ ($H \in \sigma_\beta$), the measure $d\sigma_\beta$ goes into $d\sigma_{s\beta}$. Therefore

$$\int_{\sigma_\beta{}'} \partial(\beta)\psi_{f,i} \cdot h_\beta d\sigma_\beta = \int_{\sigma_{s\beta}{}'} (\partial(\beta)\psi_{f,i})^s (h_\beta)^s d\sigma_{s\beta}$$

for $f \in C_c^\infty(\mathfrak{g})$. But from Lemma 11

$$(\partial(\beta)\psi_{f,i})^s = \partial(s\beta)(\psi_{f,i})^s = \varepsilon_0(s)\partial(s\beta)\psi_{f,i}$$

on $\sigma_{s\beta}$. Similarly

$$(\psi_{f,i}{}^\beta)^s = \varepsilon_0(s)\psi_{f,i}{}^{s\beta}$$

on $\sigma_{s\beta}$. Put

$$\bar{h}_\beta = w_i^{-1}\sum_{s \in W_{G,i}} \varepsilon_0(s)(h_{s^{-1}\beta})^s ,$$

$$\bar{g}_\beta = w_i^{-1}\sum_{s \in W_{G,i}} \varepsilon_0(s)(g_{s^{-1}\beta})^s \qquad\qquad (\beta \in Q_i) ,$$

where w_i is the order of $W_{G,i}$. Then it is clear that we can replace h_β, g_β by $\bar{h}_\beta, \bar{g}_\beta$ respectively ($\beta \in Q_i$, $1 \le i \le r$) in Lemma 26. The new functions $\bar{h}_\beta$ and $\bar{g}_\beta$ obviously satisfy condition (1) of Lemma 27. The second part is proved in a similar fashion. We have only to observe that $p_i{}^s = p_i$ $(p \in I(\mathfrak{g}_c),\ s \in W_{G,i})$ and therefore

$$(\psi_{p,f,i})^s = \varepsilon(s)\varepsilon_0(s)\psi_{p,f,i}$$

on $\mathfrak{h}_i'(R)$. Similarly

$$\big((\partial(\beta)\psi_{p,f,i})^\beta\big)^s = \varepsilon(s)\varepsilon_0(s)\big(\partial(s\beta)\psi_{p,f,i}\big)^{s\beta}$$

for $\beta \in Q_i$.

Now assume that for every $\beta \in Q_i$ ($1 \le i \le r$), we are given two functions h_β and g_β on σ_β' such that

(1) $h_{-\beta} = -h_\beta$, $(h_\beta)^s = \varepsilon_0(s)h_{s\beta}$, $(g_\beta)^s = \varepsilon_0(s)g_{s\beta}$ $(s \in W_{G,i})$,

(2) h_β and g_β are locally summable on σ_β with respect to the euclidean measure $d\sigma_\beta$.

Put

$$T(f) = \sum_{1 \le i \le r} \sum_{\beta \in Q_i} \int_{\sigma_\beta'} (\partial(\beta)\psi_{f,i} \cdot h_\beta + \psi_{f,i}{}^\beta g_\beta)d\sigma_\beta .$$

One proves easily (see § 15) that T is an invariant distribution on $\mathfrak{g}$. Let $S_{I,i}$ denote, as before, the positive singular imaginary roots of $\mathfrak{h}_i$.

LEMMA 28.   *Fix i, $\lambda \in S_{I,i}$ and an element $H_0 \in \sigma_\lambda'$. Suppose $T = 0$ around H_0. Then $h_\lambda = g_\lambda = g_{-\lambda} = 0$ almost everywhere around H_0 on σ_λ' with respect to the measure $d\sigma_\lambda$.*

Assuming this lemma, we shall first complete the proofs of Theorems 4 and 5. Fix i ($1 \le i \le r$), $\lambda \in S_{I,i}$ and an element $H_0 \in \sigma_\lambda' \cap \mathrm{Cl}(\Omega)$. Then H_0 is a semi-regular element of non-compact type in $\mathrm{Cl}(\Omega)$ and therefore $J = J_1 = 0$ around H_0 by hypothesis. Now we may suppose that the functions $g_\beta, g_\beta', h_\beta, h_\beta'$ of Lemma 26 satisfy the additional conditions of Lemma 27. Then it follows from Lemmas 26 and 28 that $h_\lambda = g_\lambda = g_{-\lambda} = 0$ around H_0 on σ_λ'. This being true for any $H_0 \in \sigma_\lambda' \cap \mathrm{Cl}(\Omega)$, we conclude that $h_\lambda = g_\lambda = g_{-\lambda} = 0$ on $\sigma_\lambda' \cap \mathrm{Cl}(\Omega)$.

Now fix $f \in C_c^\infty(\Omega)$. Then $\mathrm{Supp}\,\psi_{f,i} \subset \mathfrak{h}_i \cap \mathrm{Cl}(\Omega)$ (see §15). Hence it follows from Lemma 26 that $J(f) = 0$. This proves that $J = 0$ on Ω.

In order to prove Theorem 5, we observe that $\psi_{1,f,i} = \pi_i \psi_{f,i} = 0$ on σ_β' for $\beta \in Q_i$. On the other hand $\pi_i = \beta \pi_\beta$ where $\pi_\beta = \beta^{-1}\pi_i$. Then since $\partial(\beta)\beta = |\beta|^2$, it is clear that

$$\left(\partial(\beta)\psi_{1,f,i}\right)^\beta = |\beta|^2 \pi_\beta(\psi_{f,i})^\beta$$

on σ_β'. Hence

$$J_1(f) = \sum_i \sum_{\beta \in Q_i} \int_{\sigma_\beta'} (\psi_{f,i})^\beta g_\beta'' d\sigma_\beta$$

where $g_\beta'' = |\beta|^2 \pi_\beta g_\beta'$ on σ_β'. Now $(\pi_\beta)^s = (s\beta)^{-1}\pi_i^s = \varepsilon(s)\pi_{s\beta}$ for $s \in W_{G,i}$. Hence Lemma 28 is applicable with (h_β, g_β) replaced by $(0, g_\beta'')$. Therefore $g_\beta'' = 0$ on $\sigma_\beta' \cap \mathrm{Cl}(\Omega)$ for every $\beta \in Q_i$ $(1 \le i \le r)$. So we conclude from the above expression for $J_1(f)$ that $J_1 = 0$ on Ω. On the other hand under condition (2) of Theorem 5, we may assume that $h_\beta' = 0$ (see Lemma 26). Therefore

$$J_p(f) = \sum_i \sum_{\beta \in Q_i} \int_{\sigma_\beta'} (\partial(\beta)\psi_{p,f,i})^\beta g_\beta' d\sigma_\beta$$

for $p \in I(\mathfrak{g}_c)$. However $g_\beta'' = |\beta|^2 \pi_\beta g_\beta' = 0$ on $\sigma_\beta' \cap \mathrm{Cl}(\Omega)$. Since π_β is nowhere zero on σ_β', it follows that $g_\beta' = 0$ on $\sigma_\beta' \cap \mathrm{Cl}(\Omega)$. Hence it is obvious that $J_p = 0$ on Ω.

17. Some questions of local summability

In order to prove Lemma 28 we need some preparation which will also be useful later in another paper. We go back to the notation of §5. So $\mathfrak{g}$ is reductive and $\mathfrak{h}$ is a Cartan subalgebra of $\mathfrak{g}$. Put $\mathfrak{g}_\mathfrak{h} = \bigcup_{x \in G} x\mathfrak{h}'$ and let $\mathfrak{g}'$ denote the set of all regular elements of $\mathfrak{g}$.

LEMMA 29. *Let g be a complex-valued function on $\mathfrak{h}'$. Then there exists a unique function F on $\mathfrak{g}'$ such that*

(1) $F(xX) = F(X)$ *for $x \in G$ and $X \in \mathfrak{g}'$;*

(2) $F = [W_G]^{-1}\pi^{-1}\sum_{s \in W_G} \varepsilon(s)g^s$ *on $\mathfrak{h}'$;*

(3) $F = 0$ *on $\mathfrak{g}'$ outside $\mathfrak{g}_\mathfrak{h}$.*

Here $[W_G]$ denotes the order of W_G and g^s is the function $H \to g(s^{-1}H)$ on $\mathfrak{h}'$ $(s \in W_G)$. The proof is obvious from the fact that two points of $\mathfrak{h}'$ are conjugate under G if and only if they are conjugate under W_G.

Let us denote by F_g the function F of the above lemma. Since the mapping $(x^*, H) \to x^*H$ of $G^* \times \mathfrak{h}'$ into $\mathfrak{g}$ is locally a homeomorphism (see [1 (h), Lemma

15]), it follows that F_g is measurable if g is measurable. Now fix euclidean measures dX and dH on $\mathfrak{g}$ and $\mathfrak{h}$ respectively. Since the set of singular elements of $\mathfrak{g}$ is of measure zero, F_g is defined almost everywhere on $\mathfrak{g}$. Similarly g is defined almost everywhere on $\mathfrak{h}$.

LEMMA 30. *Assume that g is locally summable on $\mathfrak{h}$. Then F_g is locally summable on $\mathfrak{g}$.*

Given a compact set V in $\mathfrak{g}$, we can choose $f \in C_c^\infty(\mathfrak{g})$ such that $f \geq 0$ everywhere and $f \geq 1$ on V. Then

$$\int_V |F_g|\, dX = \int_\mathfrak{g} |F_g|\, f dX = \int_{\mathfrak{g}_\mathfrak{h}} |F_g|\, f dX .$$

Now we can normalize the measure $d\dot{x}^*$ on G^* in such a way that

$$\int_{\mathfrak{g}_\mathfrak{h}} \alpha\, dX = \int_{G^* \times \mathfrak{h}} |\pi(H)|^2 \alpha(x^* H) dx^* dH$$

for any measurable function $\alpha \geq 0$ on $\mathfrak{g}$ (see [1 (c), p. 241]). But then, F_g being an invariant function, it follows that

$$\int_{\mathfrak{g}_\mathfrak{h}} |F_g|\, f dX \leq [W_G]^{-1} \sum_{s \in W_G} \int |\psi_f(H)|\, |g(sH)|\, dH = \int |\psi_f|\, |g|\, dH$$

from Lemma 11. However the right side is finite (see § 15). This proves the lemma.

COROLLARY 1. *There exists a number $c \neq 0$ such that*

$$\int f F_g dX = c \int \psi_f(H) \varepsilon_R(H) g(H) dH$$

for any locally summable function g on $\mathfrak{h}$ and $f \in C_c^\infty(\mathfrak{g})$. Let $2q$ be the number of imaginary roots of $(\mathfrak{g}, \mathfrak{h})$. Then $(-1)^q c > 0$ and

$$\int \alpha\, dX = c \int \pi(H)^2 \alpha(x^* H) dx^* dH$$

for $\alpha \in C_c(\mathfrak{g}_\mathfrak{h})$.

It follows from the above proof that

$$\int f F_g dX = \int |\pi(H)|^2 f(x^* H) F_g(H) dx^* dH$$

$$= \int |\pi(H)|^2 f(x^* H) g(H) \pi(H)^{-1} dx^* dH ,$$

since the integral $\int f(x^* H) dx^*$ $(H \in \mathfrak{h}')$ remains unchanged if we replace H by sH $(s \in W_G)$ [1 (c), p. 238]. Now

$$\operatorname{conj} \pi(H) = (-1)^q \pi(H) \qquad\qquad (H \in \mathfrak{h})$$

by Lemma 9. Hence

$$\int f F_g dX = (-1)^q \int \pi(H) f(x^*H) g(H) dx^* dH$$

$$= (-1)^q \int \psi_f(H) \varepsilon_R(H) g(H) dH \ .$$

Similarly

$$\int \alpha dX = \int |\pi(H)|^2 \alpha(x^*H) dx^* dH$$

$$= (-1)^q \int \pi(H)^2 \alpha(x^*H) dx^* dH$$

for $\alpha \in C_c(\mathfrak{g}_\mathfrak{h})$.

Since π^2 is invariant under W, there exists, by Chevalley's theorem [1 (b), p. 100], a unique invariant polynomial function η on $\mathfrak{g}_c$ whose restriction on $\mathfrak{h}_c$ is π^2.

COROLLARY 2. *The function $|\eta|^{-1/2}$ is locally summable on $\mathfrak{g}$.*

Let $\mathfrak{h}_i$ $(1 \leq i \leq r)$ be a maximal set of Cartan subalgebras of $\mathfrak{g}$, no two of which are conjugate under G. Put $\mathfrak{g}_i = \mathfrak{g}_{\mathfrak{h}_i}$. Then $\mathfrak{g}'$ is the disjoint union of $\mathfrak{g}_1, \cdots, \mathfrak{g}_r$. Hence if χ_i is the characteristic function of the open set $\mathfrak{g}_i$, it is clear that $\sum_i \chi_i = 1$ almost everywhere on $\mathfrak{g}$. So it would be enough to verify that $\chi_i |\eta|^{-1/2}$ is locally summable on $\mathfrak{g}$ for every i. Now fix i and put $\mathfrak{h} = \mathfrak{h}_i$ and $\chi = \chi_i$. Take $g(H) = \pi(H) |\pi(H)|^{-1}$ $(H \in \mathfrak{h}')$ in Lemma 30. Then it is clear that $F_g = \chi |\eta|^{-1/2}$ and therefore $\chi |\eta|^{-1/2}$ is locally summable on $\mathfrak{g}$.

Now fix an element $H_0 \in \mathfrak{h}$ and let $\mathfrak{z}$ denote the centralizer of H_0 in $\mathfrak{z}$. Then $\mathfrak{z}$ is reductive in $\mathfrak{g}$ and $\mathfrak{h} \subset \mathfrak{z}$ so that rank $\mathfrak{z}$ = rank $\mathfrak{g}$. Define $\mathfrak{z}'$ as in [1 (h), § 2] and fix an open neighborhood $\mathfrak{z}_0$ of H_0 in $\mathfrak{z}'$. We use the notation of [1 (h), Lemma 16].

LEMMA 31. *Let g be a locally summable function on $\mathfrak{h}$. Then the function $Z \rightarrow F_g(Z)$ $(Z \in \mathfrak{z}_0)$ is locally summable on $\mathfrak{z}_0$ and*

$$\int f_\alpha F_g dX = \int_{\mathfrak{z}_0} \beta_\alpha(Z) F_g(Z) dZ$$

for $\alpha \in C_c^\infty(G \times \mathfrak{z}_0)$. Here

$$\beta_\alpha(Z) = \int_G \alpha(x : Z) dx \qquad\qquad (Z \in \mathfrak{z}_0) \ .$$

F_g is locally summable on $\mathfrak{g}$ by Lemma 30. Moreover it is invariant under G. Hence our assertions follow immediately from Corollary 2 of Theorem 1 of [1 (g)].

18. The proof of Lemma 28

Now we come to the proof of Lemma 28. Let us use the notation of § 7. The mapping $(H', X', Y') \to (-H', -Y', -X')$ defines a Cartan involution of $\mathfrak{l}$ which, by Mostow's theorem (see [1 (g), § 13]), can be extended to a Cartan involution θ' of $\mathfrak{g}$. On the other hand it is clear (see § 9) that the statement of Lemma 28 is really independent of the choice of the Cartan involution θ, since any two such involutions are conjugate under G. Hence without loss of generality we may assume that $\theta = \theta'$. It is obvious that $\sigma = \sigma_\lambda$ and since λ is singular imaginary, it follows from Lemmas 12 and 13 that $\mathfrak{b}$ is conjugate to $\mathfrak{h}_i$ under L. Hence in view of the results of § 9, we can assume that $\mathfrak{b} = \mathfrak{h}_i$ and $\mathfrak{a} = \mathfrak{h}_j$ for some $j \neq i$.

Put $\zeta(Z) = \det (\operatorname{ad} Z)_{\mathfrak{g}/\mathfrak{z}}$ $(Z \in \mathfrak{z})$ as in [1 (h), § 2] and let $\mathfrak{z}'$ be the set of all $Z \in \mathfrak{z}$ where $\zeta(Z) \neq 0$. Select an open neighborhood $\mathfrak{z}_0$ of H_0 in $\mathfrak{z}'$ and put $\Omega_0 = (\mathfrak{z}_0)^\sigma$. Then (see [1 (h), Lemma 15]) Ω_0 is an open neighborhood of H_0 in $\mathfrak{g}$. Let $\mathfrak{h}$ be a Cartan subalgebra of $\mathfrak{g}$ and suppose $\psi_f{}^{\mathfrak{h}} \neq 0$ for some $f \in C_c^\infty(\Omega_0)$. Then (see § 15) $\mathfrak{h}'$ must meet Ω_0. Let $\mathfrak{g}'$ denote, as usual, the set of all regular elements in $\mathfrak{g}$ and $\mathfrak{l}'$ the set of those elements of $\mathfrak{l}$ which are regular in $\mathfrak{l}$ (or, equivalently, in $\mathfrak{z}$). Then $\mathfrak{z} \cap \mathfrak{g}' \subset \sigma + \mathfrak{l}'$ and every element of $\mathfrak{l}'$ is conjugate under L to some element in $\mathbf{R}H'$ or $\mathbf{R}(X' - Y')$ (see [1 (d), p. 681]). This means that $x\mathfrak{h}'$ meets $\mathfrak{a}' \cup \mathfrak{b}'$ for some $x \in G$. But two distinct Cartan subalgebras cannot have a regular element in common. Therefore $x\mathfrak{h}$ is either $\mathfrak{a}$ or $\mathfrak{b}$. This proves that

$$T(f) = T_\mathfrak{a}(f) + T_\mathfrak{b}(f) \qquad\qquad (f \in C_c^\infty(\Omega_0)) \,,$$

where

$$T_\mathfrak{h}(f) = \sum_{\beta \in S_I{}^\mathfrak{h}} \int_{\sigma_\beta{}'} (2\partial(\beta)\psi_f{}^{\mathfrak{h}} \cdot h_\beta + \psi_f{}^\beta g_\beta + \psi_f{}^{-\beta} g_{-\beta}) d\sigma_\beta$$

for $\mathfrak{h} = \mathfrak{a}$ or $\mathfrak{b}$.

Let τ denote the linear function on $\mathfrak{a}$ defined as follows. $\tau = 0$ on σ and $\tau(H') = 2$. Then by Lemma 13, τ is a root and we may assume that τ is positive. Obviously τ is real and $H' = 2 |\tau|^{-2} H_\tau$.

LEMMA 32. *Let $\mathfrak{h}$ be either $\mathfrak{a}$ or $\mathfrak{b}$ and let $A_\mathfrak{h}{}^*$ denote the normalizer of $\mathfrak{h}$ in G. Then any element x of G such that $x\mathfrak{h}'$ meets $\mathfrak{z}$, must lie in $LA_\mathfrak{h}{}^*$.*

Suppose $xH \in \mathfrak{z}$ for some $H \in \mathfrak{h}'$. Then $xH \in \mathfrak{g}'$ and since rank $\mathfrak{z} =$ rank $\mathfrak{g}$, the centralizer of xH in $\mathfrak{g}$ is actually contained in $\mathfrak{z}$. Hence $x\mathfrak{h} \subset \mathfrak{z}$. But any Cartan subalgebra of $\mathfrak{z}$ is conjugate under L to either $\mathfrak{a}$ or $\mathfrak{b}$. Since $\mathfrak{a}$ and $\mathfrak{b}$ are not conjugate under G (Lemma 12), it follows that $x\mathfrak{h} = y\mathfrak{h}$ for some $y \in L$. Hence $y^{-1} x \in A_\mathfrak{h}{}^*$.

Corollary. *Suppose $H \in \mathrm{Supp}\ \psi_f{}^{\mathfrak{h}}$ for some $f \in C_c^{\infty}(\Omega_0)$. Then $sH \in \mathrm{Cl}(\mathfrak{z}_0{}^L)$ for some $s \in W_G{}^{\mathfrak{h}}$.*

Let V be any neighborhood of H in $\mathfrak{h}$. Then $\psi_f{}^{\mathfrak{h}}$ cannot vanish identically on $V \cap \mathfrak{h}'$. Hence $x(V \cap \mathfrak{h}')$ meets $\mathfrak{z}_0$ for some $x \in G$. Then by above lemma, we can choose $y \in L$ and $s \in W_G{}^{\mathfrak{h}}$ such that $y^{-1}xH_1 = sH_1$ for all $H_1 \in \mathfrak{h}$. Hence $sV = y^{-1}x V$ meets $y^{-1}\mathfrak{z}_0 \subset \mathfrak{z}_0{}^L$. This proves that $sH \in \mathrm{Cl}(\mathfrak{z}_0{}^L)$.

Let $\zeta_{\mathfrak{b}}$ denote the restriction of ζ on $\mathfrak{b}_c$. Since $\mathfrak{b} = \sigma + \mathbf{R}(-1)^{1/2}H_\lambda$ and since $\zeta_{\mathfrak{b}}$ is invariant under the Weyl reflexion s_λ, it is clear that $\zeta_{\mathfrak{b}}$ can be written uniquely in the form

$$\zeta_{\mathfrak{b}} = \sum_{k \geq 0} (-1)^k 2^{3k} \lambda^{2k} q_k$$

where $q_k \in S(\sigma_c)$. Define $\omega_{\mathfrak{l}}(Z) = \mathrm{tr}\ \rho(Z)^2\ (Z \in \mathfrak{l}_c)$ where ρ is the adjoint representation of $\mathfrak{l}_c$. Then $\omega_{\mathfrak{l}}$ is a quadratic form on $\mathfrak{l}$ which is invariant under L and $\omega_{\mathfrak{l}} = -8\lambda^2$ on $\mathfrak{b} \cap \mathfrak{l}$. Hence it is obvious that[19]

$$\zeta = \sum_{k \geq 0} q_k \omega_{\mathfrak{l}}{}^k \ .$$

For any number $\varepsilon > 0$, let $\mathfrak{l}_\varepsilon$ denote the set of all $Z \in \mathfrak{l}$ with $|\omega_{\mathfrak{l}}(Z)| < 8\varepsilon^2$. Then $\mathfrak{l}_\varepsilon$ is an open neighborhood of zero in $\mathfrak{l}$ which is invariant under L and $\mathrm{Cl}(\mathfrak{l}_\varepsilon) \subset \mathfrak{l}_\delta$ if $\varepsilon < \delta$. Now

$$\zeta(H + Z) = \sum_{k \geq 0} q_k(H) \omega_{\mathfrak{l}}(Z)^k$$

for $H \in \sigma$ and $Z \in \mathfrak{l}$. Therefore since $\zeta(H_0) \neq 0$, we can choose an open neighborhood U_σ of H_0 in $\sigma' = \sigma_\lambda'$ and a number $\delta > 0$ such that

$$|\zeta(H + Z) - \zeta(H_0)| \leq |q_0(H) - q_0(H_0)| + \sum_{k \geq 1} |q_k(H)|\ 2^{3k}\delta^{2k}$$
$$\leq |\zeta(H_0)|/2$$

for $H \in \mathrm{Cl}(U_\sigma)$ and $Z \in \mathfrak{l}_\delta$. This shows that $U_\sigma + \mathfrak{l}_\delta \subset \mathfrak{z}'$.

Choose an open and convex neighborhood $U_{\mathfrak{h}}$ of H_0 in $\mathfrak{h}$ ($\mathfrak{h} = \mathfrak{a}$ or $\mathfrak{b}$) such that

    (1) $U_{\mathfrak{h}} = sU_{\mathfrak{h}}$ if $sH_0 = H_0$ and $U_{\mathfrak{h}} \cap sU_{\mathfrak{h}} = \varnothing$ if $sH_0 \neq H_0\ (s \in W_G{}^{\mathfrak{h}})$;

    (2)   only one positive root of $\mathfrak{h}$ can ever take the value zero on $U_{\mathfrak{h}}$.

This is clearly possible. Since $\omega_{\mathfrak{l}}(tH') = 8t^2\ (t \in \mathbf{R})$, it is obvious that by choosing U_σ and δ sufficiently small, we can arrange that

$$\mathrm{Cl}(U_\sigma) + \mathfrak{h} \cap \mathfrak{l}_\delta \subset U_{\mathfrak{h}} \ .$$

Now choose a positive number $\varepsilon < \delta$ and take $\mathfrak{z}_0 = U_\sigma + \mathfrak{l}_\varepsilon$. Then $\mathfrak{z}_0{}^L = \mathfrak{z}_0$ and

$$\mathrm{Cl}(\mathfrak{z}_0) \cap \mathfrak{h} \subset (\mathrm{Cl}(U_\sigma) + \mathfrak{l}_\delta) \cap \mathfrak{h} \subset U_{\mathfrak{h}} \ .$$

Therefore we get the following result from the corollary of Lemma 32.

---

[19] Here we make use of the decomposition $\mathfrak{z} = \sigma + \mathfrak{l}$ and the results of [1 (g), §6].

LEMMA 33. *Let $f \in C_c^\infty(\Omega_0)$ where $\Omega_0 = \mathfrak{z}_0^a$. Then*

$$\operatorname{Supp} \psi_f^{\mathfrak{h}} \subset \bigcup_{s \in W_G \mathfrak{h}} s U_\mathfrak{h}$$

for $\mathfrak{h} = \mathfrak{a}$ or $\mathfrak{b}$.

We now claim that $T_\mathfrak{a} = 0$ on Ω_0. Fix $f \in C_c^\infty(\Omega_0)$. It is obvious that $\pm s\tau$ are the only roots of $\mathfrak{a}$ which vanish at some point of $s U_\mathfrak{a}$ ($s \in W_G^\mathfrak{a}$). Since τ is real, the same holds for $s\tau$. Hence it follows from Lemma 33 that $\sigma_\beta \cap \operatorname{Supp} \psi_f^\mathfrak{a} = \varnothing$ for $\beta \in S_I^\mathfrak{a}$ and therefore $T_\mathfrak{a}(f) = 0$.

On the other hand let q be the number of positive roots of $\mathfrak{b}$ which are conjugate to $\pm\lambda$ under $W_G = W_G^\mathfrak{b}$. It is clear that they are all in $S_I^\mathfrak{b}$. Let W_0 be the stabilizer of H_0 in W_G and W_1 the subgroup of those elements of W_G which map $\sigma = \sigma_\lambda$ into itself. It is clear that $W_1 \supset W_0$ and $s\lambda = \pm\lambda (s \in W_1)$ and $s U \cap \sigma_\lambda = \varnothing$ ($s \in W_G$) unless $s \in W_1$. Hence it follows from Lemmas 11 and 33 and our assumptions on h_β, $h_{-\beta}$, g_β, $g_{-\beta}$ ($\beta \in S_I^\mathfrak{b}$) that

$$T_\mathfrak{b}(f) = q\int_{\sigma_\lambda'} \{2\partial(\lambda)\psi_f^\mathfrak{b} \cdot h_\lambda + \psi_f^\lambda g_\lambda + \psi_f^{-\lambda} g_{-\lambda}\} d\sigma_\lambda$$

$$= q[W_1 : W_0]\int_{U_\mathfrak{b} \cap \sigma_\lambda'} \{2\partial(\lambda)\psi_f^\mathfrak{b} \cdot h_\lambda + \psi_f^\lambda g_\lambda + \psi_f^{-\lambda} g_{-\lambda}\} d\sigma_\lambda$$

for $f \in C_c^\infty(\Omega_0)$. Moreover we can, by hypothesis, choose an open neighborhood Ω of H_0 in Ω_0 such that $T = 0$ on Ω. Therefore $T(f) = T_\mathfrak{b}(f) = 0$ and hence

$$\int_{U_\mathfrak{b} \cap \sigma} \{2\partial(\lambda)\psi_f^\mathfrak{b} \cdot h_\lambda + \psi_f^\lambda g_\lambda + \psi_f^{-\lambda} g_{-\lambda}\} d\sigma = 0$$

for $f \in C_c^\infty(\Omega)$. (Here $d\sigma = d\sigma_\lambda$.)

We now apply Lemma 31 and Corollary 1 of Lemma 30 to $\mathfrak{h} = \mathfrak{b}$. Then if $\alpha \in C_c^\infty(G \times \mathfrak{z}_0)$ and u is a locally summable function on $\mathfrak{b}$, we have

$$c\int_\mathfrak{b} \psi_{f_\alpha}^\mathfrak{b}(H)\varepsilon_R^\mathfrak{b}(H)u(H)dH = \int_\mathfrak{g} f_\alpha F_u dX = \int_{\mathfrak{z}_0} \beta_\alpha(Z)F_u(Z)dZ .$$

Let $\mathfrak{z}_\mathfrak{b}$ denote the set of all elements in $\sigma + I'$ which are conjugate under L to some element of $\mathfrak{b}$. Then $\mathfrak{z}_\mathfrak{b}$ is open in $\mathfrak{z}$. Let $d\xi^*$ denote the invariant measure on $L^* = L/L \cap A_\mathfrak{b}$. ($A_\mathfrak{b}$ denotes, as usual, the Cartan subgroup of G corresponding to $\mathfrak{b}$.) Then by Corollary 1 of Lemma 30, there exists a number $c_1 \neq 0$ such that

$$\int \gamma(Z)dZ = c_1\int_{L^* \times \mathfrak{b}} \lambda(H)^2\gamma(\xi^*H)d\xi^*dH$$

for $\gamma \in C_c(\mathfrak{z}_\mathfrak{b})$. Hence, from Lemma 14, there exists a number $c_2 \neq 0$ (independent of u) such that

$$\int \gamma(Z)F_u(Z)dZ = c_2\int \lambda(H)J_\mathfrak{b}(\gamma : H)F_u(H)dH$$

for $\gamma \in C_c^\infty(\mathfrak{z})$. This shows that

$$\int_{\mathfrak{b}} \psi_{f_\alpha}{}^{\mathfrak{b}}(H)\varepsilon_R{}^{\mathfrak{b}}(H)u(H)dH = c_3 \int_{\mathfrak{b}} u(H)\sum_{s \in W_G} \varepsilon(s)\pi_\lambda(sH)^{-1}J_{\mathfrak{b}}(\beta_\alpha : sH)dH$$

where $c_3 = c^{-1}c_2[W_G]^{-1}$ and $\pi_\lambda = \lambda^{-1}\pi^{\mathfrak{b}}$. This being true for any u, we conclude that

$$\psi_{f_\alpha}{}^{\mathfrak{b}}(H) = c_3\varepsilon_R{}^{\mathfrak{b}}(H)\sum_{s \in W_G} \varepsilon(s)\pi_\lambda(sH)^{-1}J_{\mathfrak{b}}(\beta_\alpha : sH)$$

for $H \in \mathfrak{b}'$. On the other hand $\mathrm{Supp}\,\beta_\alpha \subset \mathfrak{z}_0 = U_\sigma + I_\varepsilon$. Therefore $J_{\mathfrak{b}}(\beta_\alpha : H) = 0$ unless $H \in \mathfrak{b} \cap \mathrm{Cl}(\mathfrak{z}_0) \subset U_{\mathfrak{b}}$. This implies that

$$\psi_{f_\alpha}{}^{\mathfrak{b}}(H) = c_4\sum_{s \in W_0} \varepsilon(s)\pi_\lambda(sH)^{-1}J_{\mathfrak{b}}(\beta_\alpha : sH)$$

for $H \in U_{\mathfrak{b}}' = U_{\mathfrak{b}} \cap \mathfrak{b}'$. Here $c_4 = c_3\varepsilon_R{}^{\mathfrak{b}}(H_0)$.

Since λ is the only positive root of $\mathfrak{b}$ which vanishes at H_0, the stabilizer of H_0 in W is $\{1, s_\lambda\}$ (see [2, Theorem 2.15, p. 249]). Hence $W_0 = \{1, s_\lambda\}$ or $\{1\}$ according as s_λ lies in W_G or not. Therefore since $(\pi_\lambda)^{s_\lambda} = \pi_\lambda$, we get

$$\psi_{f_\alpha}{}^{\mathfrak{b}}(H) = c_4\pi_\lambda(H)^{-1}\sum_{s \in W_0} \varepsilon(s)J_{\mathfrak{b}}(\beta_\alpha : sH) \qquad (H \in U_{\mathfrak{b}}') \, .$$

Now put $H_\varphi = H + \varphi(X' - Y')$ $(\varphi \in \mathbf{R})$ for $H \in \sigma$ as in §8. Since $(X' - Y') = -(-1)^{1/2}2|\lambda|^{-2}H_\lambda$, it is clear that

$$J_{\mathfrak{b}}(\gamma : H_\varphi; \partial(\lambda)) = (-1)^{1/2}2^{-1}|\lambda|^2 dJ_{\mathfrak{b}}(\gamma : H_\varphi)/d\varphi$$

for $\gamma \in C_c^\infty(\mathfrak{z})$. Therefore it follows from Lemma 1 that

$$J_{\mathfrak{b}}(\gamma : H; \partial(\lambda)) = -(-1)^{1/2}|\lambda|^2\gamma(H) \qquad (H \in \sigma) \, .$$

Moreover since π_λ is invariant under s_λ, it is clear that $\partial(\lambda)(\pi_\lambda^{-1}) = 0$ on σ'. Therefore it follows that

$$\psi_{f_\alpha}{}^{\mathfrak{b}}(H; \partial(\lambda)) = c_5\pi_\lambda(H)^{-1}\beta_\alpha(H) \qquad (H \in U_{\mathfrak{b}} \cap \sigma)$$

where $c_5 = -(-1)^{1/2}|\lambda|^2 c_4[W_0]$. Similarly[2]

$$\psi_{f_\alpha}{}^{\pm\lambda}(H) = c_4\pi_\lambda(H)^{-1}\sum_{s \in W_0} \varepsilon(s)\,\mathrm{Lim}_{\varphi \to \pm 0}\,J_{\mathfrak{b}}(\beta_\alpha : sH_\varphi)$$

for $H \in U_{\mathfrak{b}} \cap \sigma$. Put

$$j_\gamma{}^\pm(H) = \mathrm{Lim}_{\varphi \to \pm 0}\,J_{\mathfrak{b}}(\gamma : H_\varphi) \qquad (H \in \sigma)$$

for $\gamma \in C_c^\infty(\mathfrak{z})$. Then if $W_0 = \{1\}$, we get

$$\psi_{f_\alpha}{}^{\pm\lambda} = c_4\pi_\lambda^{-1}j_{\beta_\alpha}{}^\pm$$

on $U_{\mathfrak{b}} \cap \sigma$. On the other hand if $W_0 = \{1, s_\lambda\}$, we have

$$\psi_{f_\alpha}{}^\lambda = -\psi_{f_\alpha}{}^{-\lambda} = c_4\pi_\lambda^{-1}\{j_{\beta_\alpha}{}^+ - j_{\beta_\alpha}{}^-\}$$

on $U_{\mathfrak{b}} \cap \sigma$. But in this case $s_\lambda \in W_G$. Therefore since s_λ leaves σ pointwise fixed and $\varepsilon_0(s_\lambda) = -1$ (see the corollary of Lemma 10), it follows from our assumptions that $g_{-\lambda} = -g_\lambda$. Hence in either case

$$\int_{\sigma_{\mathfrak{b}} \cap \sigma} \{2\partial(\lambda)\psi_{f_\alpha}{}^{\mathfrak{b}} \cdot h_\lambda + \psi_{f_\alpha}{}^{\lambda} \cdot g_\lambda + \psi_{f_\alpha}{}^{-\lambda} \cdot g_{-\lambda}\}d\sigma$$

$$= \int_{\sigma_{\mathfrak{b}} \cap \sigma} \pi_\lambda^{-1}\{2c_5\beta_\alpha h_\lambda + c_6 j_{\beta_\alpha}{}^+ g_\lambda + c_6 j_{\beta_\alpha}{}^- g_{-\lambda}\}d\sigma$$

where $c_6 = c_4[W_0]$.

Fix open neighborhoods G_0 and $\mathfrak{z}_1$ of 1 and H_0 in G and $\mathfrak{z}_0$ respectively, so small that $(\mathfrak{z}_1)^{G_0} \subset \Omega$. Select an element $v \in C_e^\infty(G_0)$ such that $\int v(x)dx = 1$. Then if we take $\alpha = v \times \gamma$ $(\gamma \in C_e^\infty(\mathfrak{z}_1))$, it is clear that $\beta_\alpha = \gamma$ and $\mathrm{Supp}\, f_\alpha \subset \Omega$. Therefore

$$\int_{\sigma_{\mathfrak{b}} \cap \sigma} \pi_\lambda^{-1}\{2c_5\gamma h_\lambda + c_6 j_\gamma{}^+ g_\lambda + c_6 j_\gamma{}^- g_{-\lambda}\}d\sigma = 0$$

for every $\gamma \in C_e^\infty(\mathfrak{z}_1)$. Choose open neighborhoods σ_0 and I_0 of H_0 and zero in U_σ and I respectively, such that $\sigma_0 + I_0 \subset \mathfrak{z}_1$. Put $\gamma = \alpha \times \beta$ where $\alpha \in C_e^\infty(\sigma_0)$ and $\beta \in C_e^\infty(I_0)$. Then it follows from Lemma 2 that

$$j_\gamma{}^\pm = c^\pm(\beta)\alpha$$

on σ. Here

$$c^+(\beta) = \int_0^\infty \bar{\beta}(tX')dt, \quad c^-(\beta) = -\int_{-\infty}^0 \bar{\beta}(tX')dt$$

and

$$\bar{\beta}(Z) = \int_{K_1} \beta(Z^k)dk \qquad (Z \in I)$$

in the notation of §8. Therefore the above result implies that

$$\int_{\sigma \cap \sigma_{\mathfrak{b}}} \alpha \pi_\lambda^{-1}\{2c_5\beta(0)h_\lambda + c_6 c^+(\beta)g_\lambda + c_6 c^-(\beta)g_{-\lambda}\}d\sigma = 0 \,.$$

Now $\sigma_0 \subset U_\sigma \subset U_{\mathfrak{b}}$. Therefore since this holds for every $\alpha \in C_e^\infty(\sigma_0)$, we conclude that

$$2c_5\beta(0)h_\lambda + c_6 c^+(\beta)g_\lambda + c_6 c^-(\beta)g_{-\lambda} = 0$$

almost everywhere on σ_0. But $c_5 \neq 0$, $c_6 \neq 0$ and β is an arbitrary element in $C_e^\infty(I_0)$. Therefore (see the proof of Lemma 34 below) $h_\lambda = g_\lambda = g_{-\lambda} = 0$ almost everywhere on σ_0. This completes the proof of Lemma 28.

19. Some results on I

We conclude this paper with a few simple results on the three-dimensional semi-simple Lie algebra I, which will be needed later on. We use the notation of §2 and put $\mathfrak{a} = \mathbf{R}H$, $\mathfrak{b} = \mathbf{R}(X - Y)$. For any $f \in C_e^\infty(I)$, define

$$\psi_f{}^a(t) = \int_0^\infty \bar{f}(tH + xX)dx \qquad\qquad (t \in \mathbf{R}) \,,$$

$$\psi_f{}^b(\theta) = \theta \int_0^\infty \bar{f}(\theta(e^t X - e^{-t} Y))(e^t - e^{-t})dt \qquad\qquad (\theta \in \mathbf{R}')$$

and

$$c^\pm(f) = \mathrm{Lim}_{\theta \to \pm 0}\, \psi_f{}^b(\theta) \,.$$

Then by Lemma 2,

$$c^+(f) = \int_0^\infty \bar{f}(xX)dx, \quad c^-(f) = -\int_{-\infty}^0 \bar{f}(xX)dx$$

and

$$\psi_f{}^a(0) = \int_{-\infty}^\infty \bar{f}(xX)dx = c^+(f) - c^-(f) \,.$$

LEMMA 34. *Let* $\mathfrak{l}_0$ *be an open neighborhood of zero in* $\mathfrak{l}$ *and* a, a^+, a^- *three complex numbers. Suppose*

$$af(0) + a^+c^+(f) + a^-c^-(f) = 0$$

for all $f \in C_c^\infty(\mathfrak{l}_0)$*. Then* $a = a^+ = a^- = 0$.

We can choose $f \in C_c^\infty(\mathfrak{l}_0)$ such that $f(0) = c^-(f) = 0$ while $c^+(f) \neq 0$ $\big($see [1 (d), p. 685]$\big)$. This shows that $a^+ = 0$. We prove similarly that $a^- = 0$. But then it is obvious that $a = 0$

LEMMA 35. *Let* g, g^+ *be two* C^∞ *functions on the closed interval* $[0, \infty)$ *and* g^- *a* C^∞ *function on the closed interval* $(-\infty, 0]$*. Put*

$$T(f) = \int_0^\infty (D_t{}^2\psi_f{}^a \cdot g - \psi_f{}^a D_t{}^2 g)dt - \int_0^\infty (D_\theta{}^2\psi_f{}^b \cdot g^+ - \psi_f{}^b D_\theta{}^2 g^+)d\theta$$

$$- \int_{-\infty}^0 (D_\theta{}^2\psi_f{}^b \cdot g^- - \psi_f{}^b D_\theta{}^2 g^-)d\theta$$

for $f \in C_c^\infty(\mathfrak{l})$ *where* $D_t = d/dt$ *and* $D_\theta = d/d\theta$*. Then* T *is an invariant distribution on* $\mathfrak{l}$ *and the following three conditions are mutually equivalent.*

(1) $T = 0$.

(2) $0 \notin \mathrm{Supp}\, T$.

(3) $g^+(0) = g^-(0)$ *and* $(D_t g)_0 = (D_\theta g^+)_0 = (D_\theta g^-)_0$.

Here the subscript 0 *denotes the value at zero.*

That T is an invariant distribution follows from Lemma 22. Fix $f \in C_c^\infty(\mathfrak{l})$. Then it is clear by integration that

$$T(f) = -(D_t\psi_f{}^a \cdot g - \psi_f{}^a D_t g)_0 + \mathrm{Lim}_{\theta \to +0}\, (D_\theta\psi_f{}^b \cdot g^+ - \psi_f{}^b \cdot D_\theta g^+)$$

$$- \mathrm{Lim}_{\theta \to -0}\, (D_\theta\psi_f{}^b \cdot g^- - \psi_f{}^b \cdot D_\theta g^-) \,.$$

Since $(D_t \psi_f{}^a)_0 = 0$ (Lemma 11) and $(D_\theta \psi_f{}^b)_0 = -2f(0)$ (Lemma 1), we get

$$T(f) = c^+(f)\{(D_t g)_0 - (D_\theta g^+)_0\} - c^-(f)\{(D_t g)_0 - (D_\theta g^-)_0\}$$
$$- 2f(0)\{g^+(0) - g^-(0)\} .$$

Our assertion now follows immediately from Lemma 34.

Let Ω denote the Killing form of $\mathfrak{l}$ and identify $\mathfrak{l}$ with its dual by means of Ω in the usual way. Then $\Omega = 8^{-1}H^2 + 2^{-1}XY \in S(\mathfrak{l}_c)$.

COROLLARY 1. *Put* $g_r = 2^{-3r}D_t^{2r}g$, $g_r^\pm = (-1)^r 2^{-3r}D_\theta^{2r}g^\pm$, $f_r = \partial(\Omega^r)f$ *and*

$$T_r(f) = \int_0^\infty (\psi_{f_r}{}^a \cdot g - \psi_f{}^a g_r)dt + \int_0^\infty (\psi_{f_r}{}^b g^+ - \psi_f{}^b g_r^+)d\theta$$
$$+ \int_{-\infty}^0 (\psi_{f_r}{}^b g^- - \psi_f{}^b g_r^-)d\theta$$

for $f \in C_c^\infty(\mathfrak{l})$ *and* $r \geq 0$. *Then the following three conditions are equivalent.*

(1) $T_r = 0$ *for all* $r \geq 0$.

(2) $0 \notin \operatorname{Supp} T_r$ *for every* $r \geq 0$.

(3) $(D_\theta^{2r}g^+)_0 = (D_\theta^{2r}g^-)_0$ *and*

$$(-1)^r(D_t^{2r+1}g)_0 = (D_\theta^{2r+1}g^+)_0 = (D_\theta^{2r+1}g^-)_0 \qquad (r \geq 0) .$$

Define

$$J_r(f) = \int_0^\infty (\psi_{f_1}{}^a \cdot g_r - \psi_f{}^a \cdot g_{r+1})dt + \int_0^\infty (\psi_{f_1}{}^b \cdot g_r^+ - \psi_f{}^b \cdot g_{r+1}^+)d\theta$$
$$+ \int_{-\infty}^0 (\psi_{f_1}{}^b \cdot g_r^- - \psi_f{}^b \cdot g_{r+1}^-)d\theta$$

for $f \in C_c^\infty(\mathfrak{l})$ and $r \geq 0$. Since

$$\psi_{f_r}{}^a \cdot g - \psi_f{}^a \cdot g_r = \sum_{0 \leq k < r} (\psi_{f_{k+1}}{}^a \cdot g_{r-k-1} - \psi_{f_k}{}^a \cdot g_{r-k}) ,$$

it is obvious that

$$T_r(f) = \sum_{0 \leq k < r} J_{r-k-1}(f_k)$$

and therefore

$$T_r = \sum_{0 \leq k < r} \partial(\Omega^{r-k-1})J_k \qquad (r \geq 0) .$$

On the other hand we know from [1 (d), Lemma 17] that

$$\psi_{f_1}{}^a = 2^{-3}D_t^2 \psi_f{}^a, \quad \psi_{f_1}{}^b = -2^{-3}D_\theta^2 \psi_f{}^b .$$

Therefore if we fix r and apply Lemma 35 to (g_r, g_r^+, g_r^-) in place of (g, g^+, g^-), we find that the following three conditions are equivalent.

(a) $J_r = 0$.

(b) $J_r = 0$ around zero.

(c) $g_r^+(0) = g_r^-(0)$ and $(D_t g_r)_0 = (D_\theta g_r^+)_0 = (D_\theta g_r^-)_0 .$

In view of the above relation between T_r and $J_k (0 \leq k < r)$, this shows that (3) implies (1). Moreover (1) obviously implies (2). So let us now assume that (2) holds. Then we have to prove (3). In consequence of the above formula for T_r and the equivalence of (a), (b), (c), it would be enough to show that $0 \notin \mathrm{Supp}\, J_r$ for every $r \geq 0$. We shall do this by induction on r. Fix $r \geq 0$. Then by induction hypothesis and the equivalence of (a) and (b), it follows that $J_k = 0$ $(0 \leq k < r)$. Therefore

$$T_{r+1} = \sum_{0 \leq k \leq r} \partial(\Omega^{r-k}) J_k = J_r .$$

But then by (2), $J_r = 0$ around zero. This proves the desired result and so the corollary is established.

COROLLARY 2. *Assume that g^+, g^- are defined and analytic on $\mathbf{R}$ and $0 \notin \mathrm{Supp}\, T_r$ for every $r \geq 0$. Then $g^+ = g^-$.*

This follows from the fact that $(D_\theta^r g^+)_0 = (D_\theta^r g^-)_0$ for all $r \geq 0$.

INSTITUTE FOR ADVANCED STUDY

REFERENCES

1. HARISH-CHANDRA, (a) *The characters of semisimple Lie groups*, Trans. Amer. Math. Soc., **83** (1956), 98-163.
   (b) *Differential operators on a semisimple Lie algebra*, Amer. Jour. Math., **79** (1957), 87-120.
   (c) *Fourier transforms on a semisimple Lie algebra*, I, Amer. Jour. Math., **79** (1957), 193-257.
   (d) *Fourier transforms on a semisimple Lie algebra*, II, Amer. Jour. Math., **79** (1957), 653-686.
   (e) *A formula for semisimple Lie groups*, Amer. Jour. Math., **79** (1957), 733-760.
   (f) *Invariant eigendistributions on semisimple Lie groups*, Bull. Amer. Math. Soc., **69** (1963), 117-123.
   (g) *Invariant distributions on Lie algebras*, Amer. Jour. Math., **86** (1964), 271-309.
   (h) *Invariant differential operators and distributions on a semisimple Lie algebra*, to appear in Amer. Jour. Math.
2. S. HELGASON, Differential Geometry and Symmetric Spaces, Academic Press, New York, 1962.
3. G. de RHAM, *Solution élémentaire d'opérateurs différentiels du second ordre*, Ann. de l'Institut Fourier, **8** (1958), 337-366.

(Received January 31, 1964)

Reprinted from
Ann. of Math.
80 (1964), 551-593

INVARIANT EIGENDISTRIBUTIONS
ON A SEMISIMPLE LIE ALGEBRA
by HARISH-CHANDRA

§ 1. INTRODUCTION

Let $\mathfrak{g}$ be a semisimple Lie algebra over $\mathbf{R}$ and T an invariant distribution on $\mathfrak{g}$ which is an eigendistribution of all invariant differential operators on $\mathfrak{g}$ with constant coefficients. Then the first result of this paper (Theorem 1) asserts that T is a locally summable function F which is analytic on the regular set $\mathfrak{g}'$ of $\mathfrak{g}$ (cf. Lemma 1 of $[3(g)]$). The second result (Theorem 5) can be stated as follows. Let D be an invariant analytic differential operator on $\mathfrak{g}$ such that $Df=0$ for every invariant C^∞ function f on $\mathfrak{g}$. Then $DS=0$ for any invariant distribution S on $\mathfrak{g}$ (cf. $[3(g)$, Lemma 3]). This will be needed in the next paper of this series, in order to lift the first result, from $\mathfrak{g}$ to the corresponding group G (see $[3(g)$, Theorem 1]), by means of the exponential mapping.

Proof of Theorem 1 proceeds by induction on $\dim \mathfrak{g}$. In § 2 we show that there exists an analytic function F on $\mathfrak{g}'$ such that $T=F$ on $\mathfrak{g}'$. Moreover we verify that F is locally summable on $\mathfrak{g}$ and therefore it defines a distribution T_F on $\mathfrak{g}$. Thus it remains to prove that $\theta=T-T_F$ is actually zero. The results of § 3 enable us to reduce this to the verification of the fact that no semisimple element H of $\mathfrak{g}$ lies in Supp θ. If $H \neq 0$, this follows easily from $[3\,(i)$, Theorem 2] and the induction hypothesis. Hence we conclude (see Corollary 1 of Lemma 8) that Supp $\theta \subset \mathcal{N}$ where $\mathcal{N}$ is the set of all nilpotent elements of $\mathfrak{g}$. Let ω be the Killing form of $\mathfrak{g}$. Then $\partial(\omega)T-cT$ $(c \in \mathbf{C})$. Since $T=\theta+T_F$, we get

$$(\partial(\omega)-c)\theta=J$$

where $J=-(\partial(\omega)-c)T_F$. By making use of $[3\,(j)$, Theorem 4], one proves that $J=0$ and therefore it follows from $[3\,(h)$, Theorem 5] that $\theta=0$.

In § 8 we study the function F in greater detail. Let $\mathfrak{a}$ be a Cartan subalgebra of $\mathfrak{g}$ and $\pi^{\mathfrak{a}}$ the product of all the positive roots of $(\mathfrak{g}, \mathfrak{a})$. Define $g_{\mathfrak{a}}(H)=\pi^{\mathfrak{a}}(H)F(H)$ for $H \in \mathfrak{a}'=\mathfrak{a} \cap \mathfrak{g}'$. Then we show that $\partial(\pi^{\mathfrak{a}})g_{\mathfrak{a}}$ can be extended to a continuous function $h_{\mathfrak{a}}$ on $\mathfrak{a}$ and if $\mathfrak{b}$ is another Cartan subalgebra of $\mathfrak{g}$, then $h_{\mathfrak{a}}=h_{\mathfrak{b}}$ on $\mathfrak{a} \cap \mathfrak{b}$ (Theorem 3). These results will be used in subsequent papers for a detailed study of the irreducible characters of a semisimple Lie group. In § 10 we apply Theorem 3 to give a new and simpler proof of the main result of $[3\,(e)]$.

609

The rest of this paper is devoted to the proof of the second result mentioned at the beginning. It depends, in an essential way, on Theorem 1 and the theory of Fourier transforms for distributions. However, since the given distribution S is not assumed to be tempered, one has to construct a method of reducing the problem to the tempered case. This is done by means of Lemma 29. The last seven sections (§§ 15-21) are devoted to the proof of this lemma.

This work was partially supported by grants from the Sloan and the National Science Foundations.

§ 2. BEHAVIOUR OF T ON THE REGULAR SET

We use the terminology of [3 (h)] and [3 (i)]. Let $\mathfrak{g}$ be a reductive Lie algebra over $\mathbf{R}$ and $\mathfrak{g}'$ the set of all regular elements of $\mathfrak{g}$. Let $I(\mathfrak{g}_c)$ denote the subalgebra of all invariants in $S(\mathfrak{g}_c)$ (see [3 (i), § 9]). Fix a Euclidean measure $d\mathrm{X}$ on $\mathfrak{g}$.

Lemma 1. — Let T *be a distribution on an open subset* Ω *of* $\mathfrak{g}$. *Assume that:*

1) T *is locally invariant;*

2) *There exists an ideal* $\mathfrak{U}$ *in* $I(\mathfrak{g}_c)$ *such that* $\dim(I(\mathfrak{g}_c)/\mathfrak{U})<\infty$ *and* $\partial(u)\mathrm{T}=0$ *for* $u\in\mathfrak{U}$. *Then there exists an analytic function* F *on* $\Omega'=\Omega\cap\mathfrak{g}'$ *such that*

$$\mathrm{T}(f)=\int f\mathrm{F}d\mathrm{X}$$

for all $f\in C_c^\infty(\Omega')$.

Fix a point $\mathrm{H}_0\in\Omega'$. It is obviously enough to show that T coincides with an analytic function around H_0. Fix a connected Lie group G with Lie algebra $\mathfrak{g}$ and let $\mathfrak{h}$ and A be the centralizers of H_0 in $\mathfrak{g}$ and G respectively. Then $\mathfrak{h}$ is a Cartan subalgebra of $\mathfrak{g}$ and A is the corresponding Cartan subgroup of G [3(j), Lemma 8]. Let $x\to x^\bullet$ denote the natural projection of G on $G^\bullet=G/A$. As usual we define $x^\bullet\mathrm{H}=x\mathrm{H}$ ($x\in G$, $\mathrm{H}\in\mathfrak{h}$). Then if $n=\dim\mathfrak{g}$ and $\mathfrak{h}'=\mathfrak{h}\cap\mathfrak{g}'$, the mapping $\varphi:(x^\bullet,\mathrm{H})\to x^\bullet\mathrm{H}$ has rank n everywhere on $G^\bullet\times\mathfrak{h}'$ [3(i), Lemma 15]. Therefore we can select open connected neighborhoods G_0 and $\mathfrak{h}_0$ of 1 and H_0 in G and $\mathfrak{h}'$ respectively such that $\Omega_0=G_0\mathfrak{h}_0\subset\Omega$ and φ is univalent on $G_0^\bullet\times\mathfrak{h}_0$. Then Ω_0 is open in $\mathfrak{g}$ and φ defines an analytic diffeomorphism φ_0 of $G_0^\bullet\times\mathfrak{h}_0$ onto Ω_0.

Fix a Euclidean measure $d\mathrm{H}$ on $\mathfrak{h}$ and let σ_T denote the distribution on $\mathfrak{h}_0$ which corresponds to T under Lemma 17 of [3(i)]. As usual let π denote the product of all the positive roots of $(\mathfrak{g}, \mathfrak{h})$. Then if $\sigma=\pi\sigma_\mathrm{T}$, we conclude from Theorem 2 of [3(i)] that $\partial(u_\mathfrak{h})\sigma=0$ for $u\in\mathfrak{U}$. Let $\mathfrak{U}_\mathfrak{h}$ denote the image of $\mathfrak{U}$ under the homomorphism $p\to p_\mathfrak{h}$ of $I(\mathfrak{g}_c)$ into $S(\mathfrak{h}_c)$. Put $\mathfrak{B}=S(\mathfrak{h}_c)\mathfrak{U}_\mathfrak{h}$. Since $\dim(I(\mathfrak{g}_c)/\mathfrak{U})<\infty$, it follows from Lemma 19 of [3(i)] that $\dim(S(\mathfrak{h}_c)/\mathfrak{B})<\infty$. Moreover it is obvious that $\partial(v)\sigma=0$ for $v\in\mathfrak{B}$. Therefore from the corollary of [3(b), Lemma 27], we get the following result.

Lemma 2. — We can choose linear functions λ_i *and polynomial functions* p_i *on* $\mathfrak{h}_c$ ($1\le i\le r$) *such that*

$$\sigma(\beta)=\int\beta g d\mathrm{H} \qquad\qquad (\beta\in C_c^\infty(\mathfrak{h}_0))$$

where

$$g(H) = \sum_{1 \leq i \leq r} p_i(H) e^{\lambda_i(H)} \qquad (H \in \mathfrak{h}_c).$$

This shows that

$$T(f_\alpha) = \int \beta_\alpha \pi^{-1} g \, dH \qquad (\alpha \in C_c^\infty(G_0 \times \mathfrak{h}_0))$$

in the notation of [3(*i*), Lemma 17].

Since φ_0 is an analytic diffeomorphism, we can now define an analytic function F on Ω_0 as follows:

$$F(x^* H) = g(H) \pi(H)^{-1} \qquad (x^* \in G_0^*, \, H \in \mathfrak{h}_0).$$

Then if $\alpha \in C_c^\infty(G_0 \times \mathfrak{h}_0)$, we have

$$\int f_\alpha F \, dX = \int \alpha(x : H) F(xH) \, dx \, dH = \int \beta_\alpha \pi^{-1} g \, dH = T(f_\alpha).$$

Since the mapping $\alpha \to f_\alpha$ of $C_c^\infty(G_0 \times \mathfrak{h}_0)$ into $C_c^\infty(\Omega_0)$ is surjective [3(*h*), Theorem 1], this implies that $T = F$ on Ω_0 and so Lemma 1 is proved.

Lemma 3. — The function F of Lemma 1 is locally summable on Ω.

Let $l = \mathrm{rank}\ \mathfrak{g}$ and t an indeterminate. We denote by $\eta(X)$ $(X \in \mathfrak{g}_c)$ the coefficient of t^l in $\det(t - \mathrm{ad}\, X)$. Then we know (see [3(*j*), Corollary 2 of Lemma 30]) that $|\eta|^{-1/2}$ is locally summable on $\mathfrak{g}$. Since the singular set of $\mathfrak{g}$ is of measure zero, it would be enough to show that there exists a neighborhood V (in Ω) of any given point $X_0 \in \Omega$, such that $|\eta|^{1/2}|F|$ is bounded on $V \cap \Omega'$.

Fix X_0 in Ω and a positive-definite quadratic form Q on $\mathfrak{g}$. For $\varepsilon > 0$, let Ω_ε be the set of all $X \in \mathfrak{g}$ such that $Q(X - X_0) < \varepsilon^2$. Then $\Omega_\varepsilon \subset \Omega$ if ε is sufficiently small. Put

$$p(X) = (Q(X - X_0) - \varepsilon^2) \eta(X) \qquad (X \in \mathfrak{g}).$$

Then p is a polynomial function on $\mathfrak{g}$. Let $\mathfrak{g}''$ be the set of all points $X \in \mathfrak{g}$ where $p(X) \neq 0$. By a theorem of Whitney [4, Theorem 4, p. 547] $\mathfrak{g}''$ has only a finite number of connected components. It is obvious that any connected component of $\Omega_\varepsilon' = \Omega_\varepsilon \cap \mathfrak{g}'$ is also a connected component of $\mathfrak{g}''$. Hence Ω_ε' has only a finite number of connected components ([1]). So it would be enough to show that $|\eta|^{1/2}|F|$ remains bounded on a connected component Ω^0 of Ω_ε'.

We now fix an element $H_0 \in \Omega^0$ and use the notation of the proof of Lemma 1. In particular φ is the mapping $(x^*, H) \to x^* H$ of $G^* \times \mathfrak{h}'$ into $\mathfrak{g}'$. Let U denote the connected component of $(1^*, H_0)$ in $\varphi^{-1}(\Omega^0)$. We claim that $\varphi(U) = \Omega^0$. Since φ is everywhere regular, $\varphi(U)$ is open in Ω^0. Therefore since Ω^0 is connected, it would be enough to show that $\varphi(U)$ is closed in Ω^0. So let (x_k^*, H_k) $(k \geq 1)$ be a sequence in U such that $X_k = x_k^* H_k$ converges to some point $X \in \Omega^0$. Then $\eta(H_k) = \eta(x_k^* H_k) \to \eta(X) \neq 0$.

([1]) This proof was pointed out to me by A. Borel.

611

Moreover $X_k \in \Omega^0 \subset \Omega_\epsilon$. Since Ω_ϵ is a bounded set in g, we can conclude from Lemma 23 of $[3(j)]$ that H_k remains bounded. Hence by selecting a subsequence, we can arrange that H_k converges to some $H' \in \mathfrak{h}$. But then $\eta(H') = \eta(X) \neq 0$ and therefore $H' \in \mathfrak{h}'$. Hence $[3(j), \text{Lemma } 8]$ A is the centralizer of H' in G and therefore $[3(j), \text{Lemma } 7]$ x_k^* remains within a compact subset of G^*. So again by selecting a subsequence we can assume that $x_k^* \to x^*$ for some $x^* \in G^*$. Then $(x_k^*, H_k) \to (x^*, H')$ in $G^* \times \mathfrak{h}'$. Since $X_k \to X$, it follows that $x^* H' = X \in \Omega^0$ and therefore $(x^*, H') \in \varphi^{-1}(\Omega^0)$. But U, being a connected component of $\varphi^{-1}(\Omega^0)$, is closed in $\varphi^{-1}(\Omega^0)$. Hence $(x^*, H') \in U$ and $X = x^* H' \in \varphi(U)$. This proves that $\varphi(U)$ is closed in Ω^0 and therefore $\varphi(U) = \Omega^0$.

Now choose G_0, $\mathfrak{h}_0$ as in the proof of Lemma 1. We may assume that $G_0^* \times \mathfrak{h}_0 \subset U$. Moreover we recall (see Lemma 2) that g is defined and analytic on $\mathfrak{h}$. Consider the function $v : (x^*, H) \to F(x^* H) - \pi(H)^{-1} g(H)$ on U. It is obviously analytic and it vanishes identically on $G_0^* \times \mathfrak{h}_0$. Therefore, since U is connected, $v = 0$. This shows that

$$|\eta(x^* H)|^{1/2} |F(x^* H)| = |g(H)|$$

for $(x^*, H) \in U$. However $\varphi(U) = \Omega^0$ is contained in the bounded set Ω_ϵ. Therefore if V is the projection of U on $\mathfrak{h}$, it follows from $[3(j), \text{Lemma } 23]$ that V is bounded. Hence g is bounded on V and therefore $|\eta|^{1/2}|F|$ is bounded on $\varphi(U) = \Omega^0$. This proves Lemma 3.

Corollary. — *Let* $p \in I(g_c)$. *Then* $\partial(p)F$ *is also locally summable on* Ω.

Since F is analytic and $T = F$ on Ω', it is clear that $\partial(p)T = \partial(p)F$ on Ω'. However the distribution $\partial(p)T$ obviously also satisfies all the conditions of Lemma 1. Therefore our assertion follows by applying Lemma 3 to $(\partial(p)T, \partial(p)F)$ in place of (T, F).

Φ being a locally summable function on Ω, define the distribution T_Φ on Ω by

$$T_\Phi(f) = \int f \Phi dX \qquad\qquad (f \in C_c^\infty(\Omega)).$$

We intend to show (under some mild extra conditions) that $T = T_F$.

Let Ω_a be the set of all points $X \in \Omega$ such that T coincides around X with an analytic function. Clearly Ω_a is open and there exists an analytic function F_a on Ω_a such that $T = F_a$ on Ω_a. Moreover $\Omega_a \supset \Omega'$ from Lemma 1 and therefore $F_a = F$ on Ω'. But then, since the singular set of g has measure zero, it is obvious that $T_{F_a} = T_F$. Hence we shall write F instead of F_a.

We say that an element $H \in g$ is of compact type if 1) ad H is semisimple and 2) the derived algebra of the centralizer $\mathfrak{z}$ of H in g is compact. (It follows from 1) that $\mathfrak{z}$ is reductive in g and therefore $[\mathfrak{z}, \mathfrak{z}]$ is semisimple.)

Lemma 4. — *Every element of* Ω *of compact type lies in* Ω_a.

Fix an element H_0 in Ω of compact type and let $\mathfrak{z}$ denote the centralizer of H_0 in g. Then it is clear that $\mathfrak{z}$ satisfies the conditions of $[3(i), \S 2]$. Define ζ and $\mathfrak{z}'$ as in $[3(i), \S 2]$. Let Ξ be the analytic subgroup of G corresponding to $\mathfrak{z}$ and $x \to x^*$ the natural mapping of G on $G^* = G/\Xi$. Since $\mathfrak{z}$ is reductive, Ξ is unimodular and therefore there exists an invariant measure dx^* on G^*. Select open neighborhoods G_0 and $\mathfrak{z}_0$

of 1 and H_0 in G and $\mathfrak{z}'$ respectively such that $\mathfrak{z}_0^{G_0} \subset \Omega$ and G_0 is connected. Let G_0^* denote the image of G_0 in G^*. Then if G_0 and $\mathfrak{z}_0$ are sufficiently small, the following conditions hold (see [3(e), pp. 654-655]).

1) There exists an analytic mapping ψ of G_0^* into G such that $(\psi(x^*))^* = x^* \; (x^* \in G_0^*)$ and ψ is regular on G_0^*.

2) The mapping $\varphi : (x^*, Z) \to \psi(x^*)Z$ of $G_0^* \times \mathfrak{z}_0$ into Ω is univalent.
Put $\Omega_0 = \varphi(G_0^* \times \mathfrak{z}_0)$. Then Ω_0 is open in Ω and φ is an analytic diffeomorphism of $G_0^* \times \mathfrak{z}_0$ onto Ω_0. Moreover since $\mathfrak{z}_1 = [\mathfrak{z}, \mathfrak{z}]$ is compact, $\mathfrak{z}_{00} = \bigcap_{\xi \in \Xi} \mathfrak{z}_0^\xi$ is open. Hence by replacing $\mathfrak{z}_0$ by $\mathfrak{z}_{00}$, we can assume that $\mathfrak{z}_0^\Xi = \mathfrak{z}_0$.

Let σ_T be the distribution on $\mathfrak{z}_0$ which corresponds to T under Lemma 17 of [3(i)]. Since $|\zeta|^{1/2}$ is an analytic function on $\mathfrak{z}_0$, $\sigma = |\zeta|^{1/2}\sigma_T$ is also a distribution on $\mathfrak{z}_0$. Moreover since $\zeta^2 > 0$ on $\mathfrak{z}_0$ it follows from Theorem 2 of [3(i)] that $\partial(u_\mathfrak{z})\sigma = 0$ for $u \in \mathfrak{U}$. Let $\mathfrak{U}_\mathfrak{z}$ denote the image of $\mathfrak{U}$ in $I(\mathfrak{z}_c)$ under the mapping $p \to p_\mathfrak{z}$ of $I(\mathfrak{g}_c)$ into $I(\mathfrak{z}_c)$. Put $\mathfrak{B} = I(\mathfrak{z}_c)\mathfrak{U}_\mathfrak{z}$. Then $\partial(v)\sigma = 0$ for $v \in \mathfrak{B}$ and it follows from Lemma 19 of [3(i)] that $\dim(I(\mathfrak{z}_c)/\mathfrak{B}) < \infty$.

Let $\mathfrak{c}_\mathfrak{z}$ be the center and $\mathfrak{z}_1$ the derived algebra of $\mathfrak{z}$. We identify $\mathfrak{z}_1$ with its dual under the Killing form ω_1 of $\mathfrak{z}_1$. Select a base $H_1, \ldots, H_r$ for $\mathfrak{c}_\mathfrak{z}$ over $\mathbf{R}$ and put

$$\omega = H_1^2 + \ldots + H_r^2 - \omega_1.$$

Then $\omega \in I(\mathfrak{z}_c)$ and since $\mathfrak{z}_1$ is compact, $\square = \partial(\omega)$ is an elliptic differential operator on $\mathfrak{z}$. Let $N = \dim(I(\mathfrak{z}_c)/\mathfrak{B})$. Then we can choose complex numbers $c_1, \ldots, c_N$ such that

$$\omega^N + c_1\omega^{N-1} + \ldots + c_N \in \mathfrak{B}.$$

Hence

$$(\square^N + c_1\square^{N-1} + \ldots + c_N)\sigma = 0.$$

This shows that σ satisfies an elliptic differential equation with constant coefficients. Therefore there exists an analytic function g on $\mathfrak{z}_0$ such that

$$\sigma(\beta) = \int \beta g \, dZ \qquad\qquad (\beta \in C_c^\infty(\mathfrak{z}_0)).$$

Since ζ is invariant under Ξ, it follows from [3(i), Lemma 17] that g is locally invariant (with respect to $\mathfrak{z}$). Therefore since Ξ is connected and $\mathfrak{z}_0^\Xi = \mathfrak{z}_0$, it follows that g is invariant under Ξ.

Now consider the analytic function F_0 on Ω_0 defined by

$$F_0(\varphi(x^*, Z)) = |\zeta(Z)|^{-1/2}g(Z) \qquad\qquad (x^* \in G_0^*, \, Z \in \mathfrak{z}_0)$$

Then if $\alpha \in C_c^\infty(G_0 \times \mathfrak{z}_0)$, we have (see [3($i$), § 7])

$$\int f_\alpha F_0 \, dX = \int \alpha(x : Z)F_0(xZ) \, dx \, dZ.$$

However if $x \in G_0$, it is clear that $x = \psi(x^*)\xi$ where $\xi \in \Xi$. Therefore

$$F_0(xZ) = F_0(\varphi(x^*, \xi Z)) = |\zeta(Z)|^{-1/2}g(Z) \qquad\qquad (Z \in \mathfrak{z}_0)$$

613

since ζ and g are invariant under Ξ. This shows that

$$\int f_\alpha F_0 dX = \int \alpha(x : Z)|\zeta(Z)|^{-1/2} g(Z) dx dZ$$
$$= \int \beta_\alpha |\zeta|^{-1/2} g dZ = \sigma_T(\beta_\alpha) = T(f_\alpha)$$

from Lemma 17 of [3(i)]. Hence $T = F_0$ on Ω_0 and this proves that $H_0 \in \Omega_a$.

§ 3. SOME PROPERTIES OF COMPLETELY INVARIANT SETS

We keep to the above notation. An element $H \in \mathfrak{g}$ is called semisimple if ad H is semisimple. Moreover $X \in \mathfrak{g}$ is called nilpotent if $X \in \mathfrak{g}_1 = [\mathfrak{g}, \mathfrak{g}]$ and ad X is nilpotent. It is obvious that if X is both semisimple and nilpotent then $X = 0$.

Lemma 5. — Any element $Y \in \mathfrak{g}$ can be written uniquely in the form $Y = H + X$ where H is a semisimple and X a nilpotent element of $\mathfrak{g}$ and $[H, X] = 0$.

Let $\mathfrak{c}$ be the center of $\mathfrak{g}$. Since $\mathfrak{g} = \mathfrak{c} + \mathfrak{g}_1$, the lemma follows from well-known facts about semisimple Lie algebras (see Bourbaki [2, p. 79]). H and X respectively are called the semisimple and the nilpotent components of Y.

Lemma 6. — Let $\mathfrak{z}$ be a subalgebra of $\mathfrak{g}$ which is reductive in $\mathfrak{g}$. An element Z of $\mathfrak{z}$ is semisimple (or nilpotent) in $\mathfrak{z}$, if and only if the same holds in $\mathfrak{g}$.

Let $\mathfrak{c}_\mathfrak{z}$ be the center of $\mathfrak{z}$. Since $\mathfrak{z}$ is reductive in $\mathfrak{g}$, every element of $\mathfrak{c}_\mathfrak{z}$ is semisimple in $\mathfrak{g}$. The lemma follows easily from this (see [2, p. 79]).

Corollary. — Let $Z \in \mathfrak{z}$. Then the semisimple component of Z in $\mathfrak{z}$ is the same as in $\mathfrak{g}$. Similarly for the nilpotent component.

This is obvious from Lemma 5.

Lemma 7. — Let U_1 be a neighborhood of zero in $\mathfrak{g}_1$ and X a nilpotent element of $\mathfrak{g}$. Then we can choose $x \in G$ such that $xX \in U_1$.

We may assume that $X \neq 0$. Then by the Jacobson-Morosow theorem [3(h), Lemma 24], we can choose $H \in \mathfrak{g}_1$ sucht hat $[H, X] = 2X$. Put $a_t = \exp(-tH) \in G$ $(t \in \mathbf{R})$. Then $a_t X = e^{-2t} X$ and therefore $a_t X \in U_1$ if t is positive and sufficiently large.

Corollary. — Let H denote the semisimple component of an element $Z \in \mathfrak{g}$. Then [1] $H \in Cl(Z^G)$.

Let X be the nilpotent component of Z so that $Z = H + X$. Consider the centralizer $\mathfrak{z}$ of H in $\mathfrak{g}$. Then $\mathfrak{z}$ is reductive in $\mathfrak{g}$ and $X \in \mathfrak{z}$. Hence X is nilpotent in $\mathfrak{z}$ (Lemma 6). Let Ξ be the analytic subgroup of G corresponding to $\mathfrak{z}$. Then by Lemma 7, applied to $\mathfrak{z}$, we have

$$H \in H + Cl(X^\Xi) = Cl(Z^\Xi) \subset Cl(Z^G).$$

Let Ω be a subset of $\mathfrak{g}$. We say that Ω is completely invariant if it has the following property: C being any compact subset of Ω, $Cl(C^G) \subset \Omega$.

---

[1] ClS denotes the closure of S.

Lemma 8. — Let Ω be a completely invariant subset of $\mathfrak{g}$ and Z an element in Ω. Then if H is the semisimple component of Z, $H \in \Omega$.

This is obvious from the corollary of Lemma 7.

Let $\mathcal{N}$ be the set of all nilpotent elements of $\mathfrak{g}$.

Corollary 1. — Let S be the set of all semisimple elements of Ω and Φ an invariant subset of Ω which is closed in Ω. Then $\Phi \cap S = \varnothing$ implies that $\Phi = \varnothing$. Similarly $\Phi \cap S \subset \{o\}$ implies that $\Phi \subset \Omega \cap \mathcal{N}$.

For suppose $Z \in \Phi$. Then if H is the semisimple component of Z, $H \in \mathrm{Cl}(Z^G) \subset \Omega$. Since Φ is invariant and closed in Ω, it follows that $H \in \Phi \cap S$. The two statements of the corollary are now obvious.

Corollary 2. — Let Ω_0 be an open and invariant subset of Ω. Assume that $S \subset \Omega_0$. Then $\Omega_0 = \Omega$.

This follows from Corollary 1 by taking Φ to be the complement of Ω_0 in Ω.

Let $\mathfrak{c}$ be the center of $\mathfrak{g}$. Fix an open and completely invariant subset Ω of $\mathfrak{g}$ and a point $X_0 = C_0 + Z_0$ $(C_0 \in \mathfrak{c}, Z_0 \in \mathfrak{g}_1)$ in Ω. Select a relatively compact and open neighborhood $\mathfrak{c}_0$ of C_0 in $\mathfrak{c}$ such that $\mathrm{Cl}(\mathfrak{c}_0) + Z_0 \subset \Omega$.

Lemma 9. — Let Ω_1 be the set of all $Z \in \mathfrak{g}_1$ such that $Z + \mathrm{Cl}\,\mathfrak{c}_0 \subset \Omega$. Then Ω_1 is an open and completely invariant neighborhood of Z_0 in $\mathfrak{g}_1$.

It is obvious that Ω_1 is an open neighborhood of Z_0 in $\mathfrak{g}_1$. Fix a compact set Q in Ω_1. Then $\mathrm{Cl}\,\mathfrak{c}_0 + Q$ is a compact subset of Ω and therefore

$$\mathrm{Cl}(\mathrm{Cl}\,\mathfrak{c}_0 + Q)^G = \mathrm{Cl}\,\mathfrak{c}_0 + \mathrm{Cl}(Q^G) \subset \Omega,$$

since Ω is completely invariant. This shows that $\mathrm{Cl}(Q^G) \subset \Omega_1$ and therefore Ω_1 is also completely invariant.

Lemma 10. — The following three conditions on Ω are equivalent:

1) $\Omega \cap \mathcal{N} \neq \varnothing$;

2) $o \in \Omega$;

3) $\mathcal{N} \subset \Omega$.

Let $X \in \Omega \cap \mathcal{N}$. By Lemma 8, $o \in \Omega$. Hence 1) implies 2). Now assume $o \in \Omega$. Then if $X \in \mathcal{N}$, it follows from Lemma 7 that $X^x \in \Omega$ for some $x \in G$. Since Ω is invariant, this means that $X \in \Omega$. Therefore 2) implies 3). It is obvious that 3) implies 1).

§ 4. THE MAIN PART OF THE PROOF OF THEOREM 1

We shall now begin the proof of the following theorem (cf. [3(g), Lemma 1]).

Theorem 1. — Let $\mathfrak{g}$ be a reductive Lie algebra over $\mathbf{R}$, Ω an open and completely invariant subset of $\mathfrak{g}$ and T a distribution on Ω. Assume that:

1) T *is invariant;*

2) *There exists an ideal $\mathfrak{U}$ in $I(\mathfrak{g}_c)$ such that $\dim(I(\mathfrak{g}_c)/\mathfrak{U}) < \infty$ and $\partial(u)T = 0$ for $u \in \mathfrak{U}$.*

Then T is a locally summable function on Ω which is analytic on $\Omega' = \Omega \cap \mathfrak{g}'$.

We use induction on dim $\mathfrak{g}$. Let F be the analytic function on Ω' corresponding to Lemma 1. Then by Lemma 3, F is locally summable on Ω and we have to show that $T = T_F$.

Let $\mathfrak{c}$ be the center and $\mathfrak{g}_1$ the derived algebra of $\mathfrak{g}$. First assume that $\mathfrak{c} \neq \{0\}$. Fix a point $X_0 = C_0 + Z_0$ ($C_0 \in \mathfrak{c}$, $Z_0 \in \mathfrak{g}_1$) in Ω. We have to prove that $T = T_F$ around X_0. Select an open and relatively compact neighborhood $\mathfrak{c}_0$ of C_0 in $\mathfrak{c}$ such that $(\text{Cl}\,\mathfrak{c}_0) + Z_0 \subset \Omega$. Let Ω_1 be the set of all elements $Z \in \mathfrak{g}_1$ such that $\text{Cl}\,\mathfrak{c}_0 + Z \subset \Omega$. Then by Lemma 9, Ω_1 is also completely invariant.

Fix Euclidean measures dC and dZ on $\mathfrak{c}$ and $\mathfrak{g}_1$ respectively such that $dX = dC\,dZ$ for $X = C + Z$ ($C \in \mathfrak{c}$, $Z \in \mathfrak{g}_1$) and, for any $\alpha \in C_c^\infty(\mathfrak{c}_0)$, consider the distribution θ_α on Ω_1 given by

$$\theta_\alpha(\beta) = T(\alpha \times \beta) \qquad\qquad (\beta \in C_c^\infty(\Omega_1)).$$

Then if G_1 is the analytic subgroup of G corresponding to $\mathfrak{g}_1$, it is clear that θ_α is invariant under G_1. Moreover $I(\mathfrak{g}_c) = S(\mathfrak{c}_c)I(\mathfrak{g}_{1c})$ since $\mathfrak{g} = \mathfrak{c} + \mathfrak{g}_1$. Put $\mathfrak{U}_1 = \mathfrak{U} \cap I(\mathfrak{g}_{1c})$. Then it is obvious that

$$\dim(I(\mathfrak{g}_{1c})/\mathfrak{U}_1) \leq \dim(I(\mathfrak{g}_c)/\mathfrak{U}) < \infty$$

and $\partial(u)\theta_\alpha = 0$ for $u \in \mathfrak{U}_1$. Therefore, since $\dim \mathfrak{g}_1 < \dim \mathfrak{g}$, it follows by the induction hypothesis that θ_α coincides on Ω_1 with a locally summable function g_α. Put $\Omega_1' = \Omega_1 \cap \mathfrak{g}_1'$ where $\mathfrak{g}_1'$ is the set of those elements of $\mathfrak{g}_1$ which are regular in $\mathfrak{g}_1$. Since $\mathfrak{g}' = \mathfrak{c} + \mathfrak{g}_1'$, it is clear that $\mathfrak{c}_0 + \Omega_1' \subset \Omega'$. Moreover since $T = T_F$ on Ω', it follows that

$$\theta_\alpha(\beta) = T(\alpha \times \beta) = T_F(\alpha \times \beta) = \int \alpha(C)\beta(Z)F(C + Z)\,dC\,dZ$$

for $\beta \in C_c^\infty(\Omega_1')$. Since g_α is analytic on Ω_1' (by the induction hypothesis), it is clear from the above relation that

$$g_\alpha(Z) = \int \alpha(C)F(C + Z)\,dC \qquad\qquad (Z \in \Omega_1').$$

But since g_α and F are locally summable on Ω_1 and Ω respectively, we can now conclude that

$$T(\alpha \times \beta) = \theta_\alpha(\beta) = \int \beta(Z)\alpha(C)F(C + Z)\,dC\,dZ = T_F(\alpha \times \beta)$$

for $\beta \in C_c^\infty(\Omega_1)$. This proves (see [3($h$), Lemma 3]) that $T = T_F$ on $\mathfrak{c}_0 + \Omega_1$.

So now we can assume that $\mathfrak{c} = \{0\}$ and therefore $\mathfrak{g}$ is semisimple. Fix a semisimple element $H_0 \neq 0$ in Ω. We shall first prove that $T = T_F$ around H_0. Let $\mathfrak{z}$ be the centralizer of H_0 in $\mathfrak{g}$ and Ξ the analytic subgroup of G corresponding to $\mathfrak{z}$. Define ζ and $\mathfrak{z}'$ as in [3(i), § 2]. Then $\zeta(H_0) \neq 0$. Let $\Omega_\mathfrak{z}$ be the set of all $Z \in \mathfrak{z} \cap \Omega$ such that $|\zeta(Z)| > |\zeta(H_0)|/2$. Then $\Omega_\mathfrak{z}$ is an open neighborhood of H_0 in $\mathfrak{z}'$. Moreover since ζ is invariant under Ξ, it follows easily that $\Omega_\mathfrak{z}$ is completely invariant in $\mathfrak{z}$. Let σ_T be the distribution on $\Omega_\mathfrak{z}$ corresponding to T under [3(i), Lemma 17] with $G_0 = G$ and $\mathfrak{z}_0 = \Omega_\mathfrak{z}$. Then by Corollary 1 of [3(i), Lemma 17], σ_T is invariant under Ξ. Now $\zeta^2 > 0$ on $\Omega_\mathfrak{z}$. Hence $\sigma = |\zeta|^{1/2}\sigma_T$ is also an invariant distribution on $\Omega_\mathfrak{z}$ and it follows from Theorem 2 of $[3(i)]$ that $\partial(u_\mathfrak{z})\sigma=0$ for $u\in\mathfrak{U}$. Let $\mathfrak{U}_\mathfrak{z}$ denote the image of $\mathfrak{U}$ under the homomorphism $p\to p_\mathfrak{z}$ of $I(\mathfrak{g}_c)$ into $I(\mathfrak{z}_c)$. Then if $\mathfrak{B}=I(\mathfrak{z}_c)\mathfrak{U}_\mathfrak{z}$, it is clear from Lemma 19 of $[3(i)]$ that $\dim(I(\mathfrak{z}_c)/\mathfrak{B})<\infty$. On the other hand $\dim\,\mathfrak{z}<\dim\mathfrak{g}$ since $\mathfrak{g}$ is semisimple and $H_0\neq0$. Therefore the induction hypothesis is applicable to $(\sigma,\Omega_\mathfrak{z},\mathfrak{B})$ in place of $(T,\Omega,\mathfrak{U})$. Let $\Omega'_\mathfrak{z}$ be the set of all points in $\Omega_\mathfrak{z}$ which are regular in $\mathfrak{z}$. Then σ coincides with a locally summable function g on $\Omega_\mathfrak{z}$ which is analytic on $\Omega'_\mathfrak{z}$. This shows that

$$T(f_\alpha)=\sigma_T(\beta_\alpha)=\int\beta_\alpha|\zeta|^{-1/2}gdZ \qquad (\alpha\in C_c^\infty(G\times\Omega_\mathfrak{z}))$$

in the notation of $[3(i),\text{ Lemma }17]$. On the other hand since $\Omega_\mathfrak{z}\subset\mathfrak{z}'$, it is clear that $\Omega'_\mathfrak{z}\subset\Omega'$. Moreover $T=T_F$ on Ω'. Therefore

$$T(f_\alpha)=T_F(f_\alpha)=\int\alpha(x:Z)F(xZ)dxdZ$$

for $\alpha\in C_c^\infty(G\times\Omega'_\mathfrak{z})$. However T is invariant and therefore the same holds for F. Hence

$$T(f_\alpha)=\int\beta_\alpha(Z)F(Z)dZ.$$

This proves that $g(Z)=|\zeta(Z)|^{1/2}F(Z)$ for $Z\in\Omega'_\mathfrak{z}$. Now fix $\alpha\in C_c^\infty(G\times\Omega_\mathfrak{z})$. Then

$$T(f_\alpha)=\int\beta_\alpha|\zeta|^{-1/2}gdZ=\int_{\Omega'_\mathfrak{z}}\beta_\alpha|\zeta|^{-1/2}gdZ$$
$$=\int_{G\times\Omega'_\mathfrak{z}}\alpha(x:Z)F(xZ)dxdZ=T_F(f_\alpha)$$

from Corollary 2 of $[3(h),\text{ Theorem }1]$. This proves that $T=T_F$ around H_0.

Put $\theta=T-T_F$. Then θ is an invariant distribution on Ω.

Lemma 11. — Let $\mathcal{N}$ be the set of all nilpotent elements of $\mathfrak{g}$. Then

$$\text{Supp }\theta\subset\mathcal{N}\cap\Omega.$$

It follows from the above proof that no semisimple element of Ω, other than zero, can lie in Supp θ. Therefore our assertion follows immediately by taking $\Phi=\text{Supp }\theta$ in Corollary 1 of Lemma 8.

As usual we identify $\mathfrak{g}_c$ with its dual under the Killing form ω of $\mathfrak{g}$.

Lemma 12. — Assume that there exists a complex number c and an integer $r\geq0$ such that $(\partial(\omega)-c)^rT=0$. Then $T=T_F$.

We shall prove this by induction on r. If $r=0$ then $T=0$ and our statement is true. So assume that $r\geq1$. Put $T_0=(\partial(\omega)-c)T$. Then T_0 satisfies all the conditions of Theorem 1 and $(\partial(\omega)-c)^{r-1}T_0=0$. Moreover since $T=F$ on Ω' and F is analytic on Ω', it is obvious that $T_0=(\partial(\omega)-c)F$ on Ω'. Therefore it follows by the induction hypothesis that $T_0=T_{F_0}$ where $F_0=(\partial(\omega)-c)F$ (see also the corollary of Lemma 3). Hence

$$(\partial(\omega)-c)(\theta+T_F)=T_{F_0}$$

and therefore

$$(\partial(\omega)-c)\theta=T_{\partial(\omega)F}-\partial(\omega)T_F.$$

Lemma 13. — $T_{\partial(\omega)F}\partial - (\omega)T_F = 0$.

Assuming this for a moment, we shall complete the proof of Lemma 12. For then we have $(\partial(\omega) - c)\theta = 0$ and therefore we conclude from $[3(h),$ Theorem 5] that $\theta = 0$. Hence $T = T_F$.

The proof of Lemma 13 is based on Theorem 4 of $[3(j)]$ and requires some preparation. Select a system of generators $(^1)$ $(p_1, \ldots, p_m)$ for the algebra $I(g_c)$ over $\mathbf{C}$.

Lemma 14. — *Fix* $X_0 \in g$ *and for any* $\varepsilon > 0$, *let* $U_{X_0}(\varepsilon)$ *denote the set of all* $X \in g$ *such that* $|p_i(X) - p_i(X_0)| < \varepsilon$ $(1 \le i \le m)$. *Then* $U_{X_0}(\varepsilon)$ *is open and completely invariant.*

$U_{X_0}(\varepsilon)$ is obviously open. Let C be a compact subset of $U_{X_0}(\varepsilon)$. Then it is clear that we can choose a $(0 < a < \varepsilon)$ such that

$$\sup_{X \in C} |p_i(X) - p_i(X_0)| \le a \qquad (1 \le i \le m).$$

Since p_i is invariant, it is obvious that $|p_i(Y) - p_i(X_0)| \le a$ for any $Y \in \mathrm{Cl}(C^G)$ and therefore $U_{X_0}(\varepsilon)$ is completely invariant.

Now put $J_0 = T_{\partial(\omega)F} - \partial(\omega)T_F$ and fix $X_0 \in \Omega$. We have to prove that $J_0 = 0$ around X_0. Define $\Omega(\varepsilon) = \Omega \cap U_{X_0}(\varepsilon)$ for $\varepsilon > 0$. Then $\Omega(\varepsilon)$ is an open and completely invariant neighborhood of X_0. We shall now use the notation of $[3(j),$ Theorem 4]. Put $\Phi_i(\varepsilon) = \mathfrak{h}_i \cap \Omega(\varepsilon)$ and $\Phi_i = \bigcap_{\varepsilon > 0} \Phi_i(\varepsilon)$ $(1 \le i \le r)$. If $H \in \Phi_i$, it is clear that $p(H) = p(X_0)$ for $p \in I(g_c)$. Hence it follows from Chevalley's theorem $[3(c),$ Lemma 9] that Φ_i is a finite set. For each $H \in \Phi_i$, choose two open convex neighborhoods U_H, V_H of H in $\mathfrak{h}_i$ such that $\mathrm{Cl}\,U_H \subset V_H \subset \Phi_i(1)$ and $V_H \cap V_{H'} = \emptyset$ for $H \ne H'$ $(H, H' \in \Phi_i)$. Then $\mathrm{Cl}\,V_H$ is compact (see the proof of Lemma 23 of $[3(j)]$). Put

$$U_i = \bigcup_{H \in \Phi_i} U_H, \qquad V_i = \bigcup_{H \in \Phi_i} V_H$$

and select $\alpha_H \in C_c^\infty(V_H)$ such that $\alpha_H = 1$ on U_H $(H \in \Phi_i)$. Define

$$\alpha_i = \sum_{H \in \Phi_i} \alpha_H.$$

Let F_i denote the restriction of F on $\Omega' \cap \mathfrak{h}_i = \Omega \cap \mathfrak{h}_i'$. Fix i and let P_c be the set of all complex positive roots of $\mathfrak{h}_i$. Let Q be a connected component of $\mathfrak{h}_i'(S)$ and Q_1 the set consisting of all regular and semiregular points of Q. If β is a root of $(g, \mathfrak{h}_i)$ which vanishes at some point H_0 in Q_1, then it is clear that β is compact and therefore H_0 is of compact type in g. Obviously Q_1 is open in $\mathfrak{h}_i$. Therefore by Lemma 4, F_i can be extended to an analytic function on $Q_1 \cap \Omega$ which we again denote by F_i.

Now fix $H \in \Phi_i$ and consider $Q_1 \cap V_H$. Then $Q_1 \cap V_H$ is connected (see the corollary of $[3(j),$ Lemma 19]). Also $V_H \subset \Omega$ and therefore F_i is analytic on the connected set $Q_1 \cap V_H$. Hence by Lemma 2, there exists an analytic function h_H on $\mathfrak{h}_i$ such that $\pi_i F_i = h_H$ on $Q_1 \cap V_H$. Then $\alpha_H \pi_i F_i = \alpha_H h_H$ on $Q_1 \cap \Omega$ and therefore

$$\alpha_i \pi_i F_i = \sum_{H \in \Phi_i} \alpha_H h_H$$

$(^1)$ Since g is semisimple, it follows from the theory of invariants that $I(g_c)$ is finitely generated.

618

on $Q_1 \cap \Omega$. Put $g_i' = \alpha_i \pi_i F_i$. Then the above result shows that g_i' is of class C^∞ on $\mathrm{Cl}(Q_1) = \mathrm{Cl}(Q)$.

Choose $\varepsilon > 0$ so small that $\Phi_i(\varepsilon) \subset U_i$ $(1 \le i \le r)$. Then from Corollary 1 of $[3(j),$ Lemma 30] we can choose numbers c_i $(1 \le i \le r)$ such that

$$\int f u \, d\mathrm{X} = \sum_{1 \le i \le r} c_i \int_{\mathfrak{h}_i} \psi_{l,i} \, \varepsilon_{\mathrm{R},i} \pi_i u_i d_i \mathrm{H} \qquad (f \in C_c^\infty(\Omega))$$

for any invariant and locally summable function u on Ω. (Here u_i is the restriction of u on $\Omega \cap \mathfrak{h}_i$.) Now suppose $f \in C_c^\infty(\Omega(\varepsilon))$. Since $\Omega(\varepsilon)$ is completely invariant, it follows from $[3(j),$ Lemma 22] that

$$\mathrm{Supp} \, \psi_{l,i} \subset \Omega(\varepsilon) \cap \mathfrak{h}_i = \Phi_i(\varepsilon) \subset U_i.$$

Hence

$$\int f u \, d\mathrm{X} = \sum_{1 \le i \le r} c_i \int \psi_{l,i} \, \varepsilon_{\mathrm{R},i} \alpha_i \pi_i u_i d_i \mathrm{H}.$$

Now take $u = \mathrm{F}$. Then $c_i \varepsilon_{\mathrm{R},i} \alpha_i \pi_i u_i = c_i \varepsilon_{\mathrm{R},i} g_i' = g_i$ (say). On the other hand, by the corollary of Lemma 3, we can also take $u = \partial(\omega)\mathrm{F}$. Then it follows from $[3(c),$ Lemma 3] that $u_i = \pi_i^{-1} \partial(\omega_i)(\pi_i \mathrm{F}_i)$ on $\mathfrak{h}_i' \Omega$ and therefore

$$c_i \varepsilon_{\mathrm{R},i} \alpha_i \pi_i u_i = \partial(\omega_i) g_i$$

on $U_i \cap \mathfrak{h}_i'$. Therefore

$$\mathrm{J}_0(f) = \int (\mathrm{F} \partial(\omega) f - \partial(\omega)\mathrm{F} . f) d\mathrm{X}$$

$$= \sum_{1 \le i \le r} \int_{\mathfrak{h}_i} (\partial(\omega_i)\psi_{l,i} . g_i - \psi_{l,i} . \partial(\omega_i) g_i) d_i \mathrm{H}$$

from $[3(d),$ Theorem 3]. Now define J as in $[3(j),$ Theorem 4], corresponding to the above functions g_i $(1 \le i \le r)$. Then the above result shows that $\mathrm{J} = \mathrm{J}_0$ on $\Omega(\varepsilon)$. Since g_i is obviously of class C^∞ on the closure of each connected component of $\mathfrak{h}_i'(\mathrm{S})$, Theorem 4 of $[3(j)]$ is applicable. Fix an open and relatively compact neighborhood V of X_0 in $\Omega(\varepsilon)$. Then since $\Omega(\varepsilon)$ is completely invariant, $\mathrm{Cl}(\mathrm{V}^{\mathrm{G}}) \subset \Omega(\varepsilon)$. Let $\mathscr{S}$ denote the set of all semiregular elements of $\Omega(\varepsilon)$ of noncompact type. Then in order to prove that $\mathrm{J}_0 = 0$ on V, it is enough, from $[3(j),$ Theorem 4], to verify that $\mathrm{Supp} \, \mathrm{J}_0 \cap \mathscr{S} = \varnothing$. However $\mathrm{J}_0 = (\partial(\omega) - c)\theta$ and so it follows from Lemma 11 that

$$\mathrm{Supp} \, \mathrm{J}_0 \subset \mathrm{Supp} \, \theta \subset \mathscr{N} \cap \Omega.$$

Since zero is the only semisimple element in $\mathscr{N}$, it is clear that $\mathscr{N} \cap \mathscr{S} \subset \{0\}$. Therefore we may assume that $\mathscr{S}$ contains zero. But then it follows from $[3(j),$ § 4] that $\mathfrak{g}$ is isomorphic to the three dimensional noncompact semisimple algebra I of $[3(j),$ § 2]. We shall consider this case in detail in the next section.

§ 5. SOME COMPUTATIONS ON I

So we now assume that $g=1$ and $o \in \Omega$. Then we have to show that $J_0 = o$ around zero. Hence we take $X_0 = o$ (see § 4). Then it follows from Lemma 10 that $\mathcal{N} \subset \Omega(\varepsilon)$. Now $\mathcal{N}$ is also the singular set of g in the present case. Therefore $\mathrm{Supp}\, J \subset \mathcal{N}$. However $J = J_0$ on $\Omega(\varepsilon)$ and so it is obvious that $J = J_0$ on Ω.

Lemma 15. — *We can choose complex numbers a, a^+ and a^- such that*

$$J(f) = af(o) + a^+ c^+(f) + a^- c^-(f) \qquad (f \in C_c^\infty(g))$$

in the notation of $[3(j),\ \mathrm{Lemma}\ 34]$.

This is obvious from $[3(j),\ \mathrm{Lemmas}\ 2,\ 3\ \mathrm{and}\ 26]$.

Corollary. — $\omega J = o$.

Since $\omega = o$ on $\mathcal{N}$, this is an immediate consequence of Lemma 15.

We have seen in § 4 that $\mathrm{Supp}\, J_0 \subset \mathcal{N}$. Fix an element $X \ne o$ in $\mathcal{N}$. We shall first prove that $J_0 = o$ around X. By the Jacobson-Morosow theorem $[3(h),\ \mathrm{Lemma}\ 24]$, we can select H, Y in g such that

$$[H,\ X] = 2X, \qquad [H,\ Y] = -2Y, \qquad [X,\ Y] = H.$$

Then $\mathfrak{z}_X = \mathbf{R}X$ is the centralizer of X in g and $g_X = [X,\ g] = \mathbf{R}H + \mathbf{R}X$. Take $U = \mathbf{R}Y$ and $V = \mathbf{R}H + \mathbf{R}Y$ so that $g = U + g_X = \mathfrak{z}_X + V$. We now use the notation of $[3(h),\ \S\ 7]$. Then $4\omega = 2^{-1}H^2 + 2XY$, $\omega(X + tY) = 8t$ and

$$\Gamma_{X+tY}(Y^2 \otimes 1) = H^2 - 2Y, \qquad \Gamma_{X+tY}(H \otimes Y) = 2(XY - Y - tY^2)$$

for $t \in \mathbf{R}$. Hence

$$\Gamma_{X+tY}\left(\frac{1}{2}Y^2 \otimes 1 + H \otimes Y + 1 \otimes (3Y + 2tY^2)\right) = 4\omega.$$

This means that

$$4\Delta(\partial(\omega)) = 3D + 2tD^2$$

on U' in the notation of $[3(h),\ \S\ 8]$. (Here $D = d/dt$.) On the other hand $(\partial(\omega) - c)\theta = J_0$ and θ and J_0 are both invariant distributions. Hence

$$(\Delta - c)\sigma_\theta = \sigma_{J_0}$$

in the notation of $[3(h),\ \mathrm{Theorem}\ 3]$ where $\Delta = \Delta(\partial(\omega))$.

Since $\mathrm{Supp}\,\theta \subset \mathcal{N}$, we can regard σ_θ as a distribution on an open neighborhood U_0 of the origin in $\mathbf{R}$ and assume that $\mathrm{Supp}\,\sigma_\theta \subset \{o\}$ (see $[3(h),\ \mathrm{Lemma}\ 23]$). If $\sigma_\theta = o$, it follows from $[3(h),\ \mathrm{Theorem}\ 2]$ that $0 = o$ around X and therefore the same holds for J_0. Hence we may assume that $o \in \mathrm{Supp}\,\sigma_\theta$. Then (see $[3(h),\ \mathrm{Lemma}\ 20]$)

$$\sigma_\theta = \sum_{0 \le k \le m} a_k D^k \delta$$

where δ denotes the Dirac distribution $\beta \to \beta(0)$ $(\beta \in C_c^\infty(U_0))$ and a_k are complex numbers $(a_m \neq 0)$. Now $\omega J = \omega J_0 = 0$ on Ω. Since $\omega(X + tY) = 8t$, it follows that $t\sigma_{J_0} = 0$ on U_0. Hence

$$\sum_{0 \leq k \leq m} a_k t(3D + 2tD^2 - 4c)D^k\delta = 0.$$

But it is easy to verify that

$$tD^k\delta = -kD^{k-1}\delta,$$

$$t^2 D^k\delta = k(k-1)D^{k-2}\delta \qquad\qquad (k \geq 0)$$

where $D^\nu\delta$ should be interpreted to mean zero if $\nu < 0$. Therefore

$$\sum_{0 \leq k \leq m} a_k\{(k+1)(2k+1)D^k\delta + 4ckD^{k-1}\delta\} = 0.$$

But since the distributions $D^k\delta$ $(k \geq 0)$ on U_0 are linearly independent, we conclude that $(m+1)(2m+1)a_m = 0$. However this is impossible since $m \geq 0$ and $a_m \neq 0$. This contradiction shows that $\sigma_\theta = 0$ and therefore $\theta = J_0 = 0$ around X. This proves that $\operatorname{Supp} \theta \subset \{0\}$ and $\operatorname{Supp} J_0 \subset \{0\}$.

Now $J = J_0$ on Ω. Hence it follows (see [3(e), p. 685]) that $a^+ = a^- = 0$ in Lemma 15 and therefore $J_0 = J = a\delta_0$ on Ω. Here δ_0 is the Dirac distribution $f \to f(0)$ $(f \in C_c^\infty(\mathfrak{g}))$ on $\mathfrak{g}$. But since $\operatorname{Supp} \theta \subset \{0\}$, we conclude from [3(h), Lemma 20] that $\theta = \partial(p)\delta_0$ where $p \in S(\mathfrak{g}_c)$. On the other hand $(\partial(\omega) - c)\theta = J_0 = a\delta_0$ on Ω. Therefore $(\omega - c)p = a$ again from [3(h), Lemma 20]. Since ω is homogeneous of degree 2, this is possible only if $p = a = 0$. Therefore $\theta = J_0 = 0$ and so Lemma 13 is now proved.

§ 6. COMPLETION OF THE PROOF OF THEOREM 1

It remains to complete the proof of Theorem 1 in case $\mathfrak{g}$ is semisimple. Let $\mathfrak{X}$ be the vector space of all distributions on Ω of the form $\partial(p)T$ $(p \in I(\mathfrak{g}_c))$. Then it is clear that

$$\dim \mathfrak{X} \leq \dim(I(\mathfrak{g}_c)/\mathfrak{U}) < \infty$$

and every element of $\mathfrak{X}$ satisfies all the conditions of Theorem 1. The mapping $S \to \partial(\omega)S$ $(S \in \mathfrak{X})$ is obviously an endomorphism of $\mathfrak{X}$. Hence we can choose a base T_j $(1 \leq j \leq N)$ for $\mathfrak{X}$ over $\mathbf{C}$ with the following property. There exist complex numbers c_j and integers $r_j \geq 0$ such that

$$(\partial(\omega) - c_j)^{r_j}T_j = 0 \qquad\qquad (1 \leq j \leq N).$$

Then Lemma 12 is applicable to T_j. Let F_j be the analytic function on Ω' such that $T_j = F_j$ on Ω' (Lemma 1). Then F_j is locally summable on Ω (Lemma 3) and $T_j = T_{F_j}$ (Lemma 12). Since (T_j) $(1 \leq j \leq N)$ is a base for $\mathfrak{X}$, $T = \sum_j a_j T_j$ for some $a_j \in \mathbf{C}$. Then if $F = \sum_j a_j F_j$, it is obvious that $T = T_F$. This proves Theorem 1.

§ 7. SOME CONSEQUENCES OF THEOREM 1

We shall now derive some consequences of Theorem 1. Define $\mathfrak{I}(\mathfrak{g}_c)$ as in $[3(i),$ § 4]. We keep to the notation of Theorem 1.

Lemma 16. — Fix $D \in \mathfrak{I}(\mathfrak{g}_c)$. Then the distribution DT also satisfies the conditions of Theorem 1. Hence DF is locally summable on Ω and $DT = T_{DF}$.

Corollary. — Let D^* denote, as usual, the adjoint of D. Then

$$\int f DF dX = \int D^* f . F dX \qquad\qquad (f \in C_c^\infty(\Omega)).$$

This is merely a restatement of the relation $DT = T_{DF}$.

Since the distribution DT is obviously invariant, it is enough to verify that the dimension of the space of all distributions of the form $\partial(p)(DT)$ $(p \in I(\mathfrak{g}_c))$ is finite. This requires some preparation.

Let us now use the notation of $[3(i),$ § 3]. For any $p \in S(E)$, let r_p and d_p denote the endomorphisms $D \to D \circ \partial(p)$ and (^{1}) $D \to \{\partial(p), D\}$ $(D \in \mathfrak{D}(E))$ respectively of $\mathfrak{D}(E)$.

Lemma 17. — Fix $p \in S(E)$. Then for every $D \in \mathfrak{D}(E)$ we can choose an integer $N \geq 0$ such that $d_p^N D = 0$.

Let A be the set of all $p \in S(E)$ for which the lemma holds. We claim that A is a subalgebra of $S(E)$. Observe that d_p, r_p, d_q, r_q $(p, q \in S(E))$ all commute with each other and

$$d_{pq} = d_p d_q + r_p d_q + d_p r_q.$$

Now fix p, q in A and $D \in \mathfrak{D}(E)$ and choose an integer $N \geq 0$ such that $d_p^N D = d_q^N D = 0$. Then it is obvious that $(d_p + d_q)^{2N} D = 0$ and

$$d_{pq}^{3N} D = (d_p d_q + r_p d_q + d_p r_q)^{3N} D = 0.$$

This shows that $p + q$ and pq are both in A and therefore A is a subalgebra. On the other hand if $p \in P(E)$, $q \in S(E)$ and $X \in E$, it is obvious that

$$d_X^N(p \partial(q)) = (d_X^N p) \partial(q) = 0$$

if $N > d^0 p$. This shows that $E \subset A$ and therefore $A = S(E)$.

We now return to the proof of Lemma 16. Let $\mathfrak{X}$ denote the space of all distributions of the form $\partial(p)T$ $(p \in I(\mathfrak{g}_c))$. Then $\dim \mathfrak{X} < \infty$. Since the algebra $I(\mathfrak{g}_c)$ is abelian, we can choose a base $T_1, \ldots, T_m$ for $\mathfrak{X}$ over $\mathbf{C}$ and homomorphisms $\chi_1, \ldots, \chi_m$ of $I(\mathfrak{g}_c)$ into $\mathbf{C}$ such that

$$(\partial(p) - \chi_i(p))^m T_i = 0 \qquad\qquad (1 \leq i \leq m).$$

Since T is a linear combination of T_i, it would be enough to prove Lemma 16 under the additional assumption that

$$(\partial(p) - \chi(p))^m T = 0 \qquad\qquad (p \in I(\mathfrak{g}_c))$$

(^{1}) As usual $\{D_1, D_2\} = D_1 \circ D_2 - D_2 \circ D_1$ for two differential operators D_1, D_2.

for some integer $m \geq 0$ and some homomorphism χ of $I(\mathfrak{g}_c)$ into $\mathbf{C}$. Now fix $p \in I(\mathfrak{g}_c)$ and choose $N \geq 0$ so large that $d_p^N D = 0$ (in the notation of Lemma 17 with $E = \mathfrak{g}_c$). Then

$$(\partial(p) - \chi(p))^{N+m} \circ D = (d_p + r_p - \chi(p))^{N+m} D$$

$$= \sum_{0 \leq k \leq N+m} C_k^{N+m} (r_p - \chi(p))^{N+m-k} d_p^k D,$$

where C_k^{N+m} stands for the usual binomial coefficient. Now consider

$$((r_p - \chi(p))^{N+m-k} d_p^k D) T = (d_p^k D)((\partial(p) - \chi(p))^{N+m-k} T).$$

If $k \geq N$, $d_p^k D = 0$ and if $k \leq N$, $(\partial(p) - \chi(p))^{N+m-k} T = 0$. Hence

$$(\partial(p) - \chi(p))^{N+m} (DT) = 0.$$

Choose $p_1, \ldots, p_l$ in $I(\mathfrak{g}_c)$ such that $I(\mathfrak{g}_c) = \mathbf{C}[p_1, \ldots, p_l]$. Then we can choose an integer $M \geq 0$ such that

$$(\partial(p_i) - \chi(p_i))^M DT = 0 \qquad\qquad (1 \leq i \leq l).$$

But this implies that the space of all distributions of the form $\partial(p) DT$ $(p \in I(\mathfrak{g}_c))$ has dimension at most M^l. This proves that Theorem 1 is applicable to DT.

It is obvious that $DT = DF$ on Ω'. Hence by applying Theorem 1 to DT we conclude that DF is locally summable on Ω and $DT = T_{DF}$.

§ 8. FURTHER PROPERTIES OF F

Fix a Cartan subalgebra $\mathfrak{h}$ of $\mathfrak{g}$ and let us use the notation of $[3(j), \S 4]$. Define the analytic function g on $\mathfrak{h}' \cap \Omega$ by

$$g(H) = \pi(H) F(H) \qquad\qquad (H \in \mathfrak{h}' \cap \Omega).$$

Theorem 2. — g can be extended to an analytic function on $\mathfrak{h}'(R) \cap \Omega$.

Fix an element $H_0 \in \mathfrak{h}'(R) \cap \Omega$. It is enough to show that there exists an analytic function g_1 on an open neighborhood U of H_0 in $\mathfrak{h}'(R) \cap \Omega$ such that $g_1 = g$ on $U \cap \mathfrak{h}'$. First assume that H_0 is semiregular. Let β be the unique positive root of $\mathfrak{h}$ which vanishes at H_0. Then clearly β is imaginary. If β is compact, the required result follows immediately from Lemma 4. Hence we may assume that β is singular. Define $\mathfrak{a}$ and $\mathfrak{b}$ as in $[3(j), \S 7]$ corresponding to H_0. Then it follows from $[3(j), \text{Lemmas } 12 \text{ and } 13]$ that we may assume that $\mathfrak{h} = \mathfrak{b}$.

Let $\mathfrak{z}$ be the centralizer of H_0 in $\mathfrak{g}$. Define ζ and $\mathfrak{z}'$ as usual $[3(i), \S\S 2, 7]$. We now use the notation of $[3(j), \S 7]$. Let $\mathfrak{z}_0$ be the set of all $Z \in \mathfrak{z} \cap \Omega$ such that $|\zeta(Z)| > |\zeta(H_0)|/2$. Then $\mathfrak{z}_0$ is an open neighborhood of H_0 in $\mathfrak{z}'$ which is completely invariant (with respect to $\mathfrak{z}$). Now σ is the center of $\mathfrak{z}$. Fix an open and convex neighborhood σ_0 of H_0 in σ such that $\mathrm{Cl}\,\sigma_0$ is compact and contained in $\mathfrak{z}_0$. Let Ω_I denote the set of all $Z \in I$ such that $\mathrm{Cl}\,\sigma_0 + Z \subset \mathfrak{z}_0$. Then by Lemma 9, Ω_I is a completely invariant and open neighborhood of zero in I.

Now apply Lemma 17 of $[3(i)]$ with $G_0 = G$ and put $T_\delta = |\zeta|^{1/2}\sigma_T$. Then it follows from Theorem 2 of $[3(i)]$ that $\partial(u_\delta)T_\delta = 0$ $(u \in \mathfrak{U})$. Put $\mathfrak{B} = I(\mathfrak{z}_c)\mathfrak{U}_\delta$ where $\mathfrak{U}_\delta$ is the image of $\mathfrak{U}$ in $I(\mathfrak{z}_c)$ under the mapping $p \to p_\delta$ $(p \in I(\mathfrak{g}_c))$. Then by $[3(i), \text{Lemma } 19]$, $\dim (I(\mathfrak{z}_c)/\mathfrak{B}) < \infty$. Let $\omega_\mathfrak{l}$ denote the Killing form of $\mathfrak{l}$. Then, if we identify $\mathfrak{l}$ with its dual under $\omega_\mathfrak{l}$, we have $\omega_\mathfrak{l} \in I(\mathfrak{l}_c) \subset I(\mathfrak{z}_c)$. Hence we can choose complex numbers $c_1, \ldots, c_r$ such that

$$\sum_{0 \leq k \leq r} c_k \omega^{r-k} \in \mathfrak{B}$$

where $c_0 = 1$. This proves that

$$\sum_{0 \leq k \leq r} c_k \partial(\omega_\mathfrak{l})^{r-k} T_\delta = 0.$$

Now fix $\gamma \in C_c^\infty(\sigma_0)$ and let τ_γ denote the distribution

$$\tau_\gamma : f \to T_\delta(\gamma \times f) \qquad\qquad (f \in C_c^\infty(\Omega_\mathfrak{l}))$$

on $\Omega_\mathfrak{l}$. Obviously τ_γ is invariant under L and it is clear from the above relation that

$$\sum_{0 \leq k \leq r} c_k \partial(\omega_\mathfrak{l})^{r-k} \tau_\gamma = 0.$$

Since $I(\mathfrak{l}_c) = \mathbf{C}[\omega_\mathfrak{l}]$, Theorem 1 and Lemma 16 are both applicable to $(\mathfrak{l}, \Omega_\mathfrak{l}, \tau_\gamma)$ in place of $(\mathfrak{g}, \Omega, T)$. Let $\Omega_\mathfrak{l}'$ be the set of those points of $\Omega_\mathfrak{l}$ which are regular in $\mathfrak{l}$. Fix a Euclidean measure $d\mathfrak{l}$ on $\mathfrak{l}$. Then we can choose an analytic function φ_γ on $\Omega_\mathfrak{l}'$ which is locally summable on $\Omega_\mathfrak{l}$ and such that

$$\tau_\gamma(f) = \int f\varphi_\gamma d\mathfrak{l} \qquad\qquad (f \in C_c^\infty(\Omega_\mathfrak{l})).$$

Hence it follows from Lemma 16 that

$$\int \{ \partial(\omega_\mathfrak{l})^k f \cdot \varphi_\gamma - f \cdot \partial(\omega_\mathfrak{l})^k \varphi_\gamma \} d\mathfrak{l} = 0$$

for $k \geq 0$ and $f \in C_c^\infty(\Omega_\mathfrak{l})$. For any $\varepsilon > 0$, let $\Omega_\mathfrak{l}(\varepsilon)$ denote the set of all $Z \in \mathfrak{l}$ with $|\omega_\mathfrak{l}(Z)| < 8\varepsilon^2$. If ε is sufficiently small, it is obvious that tH' and $t(X'-Y')$ both lie in $\Omega_\mathfrak{l}$ whenever $|t| \leq \varepsilon$ $(t \in \mathbf{R})$. Since $\Omega_\mathfrak{l}$ is completely invariant under L, we can conclude (see $[3(e), \text{p. } 681]$) that $\mathrm{Cl}\,\Omega_\mathfrak{l}(\varepsilon) \subset \Omega_\mathfrak{l}$. It follows from Lemma 2 that there exist three analytic functions g_γ, g_γ^+, g_γ^- on $\mathbf{R}$ such that

$$g_\gamma(t) = t\varphi_\gamma(tH') \qquad\qquad (0 < t \leq \varepsilon)$$

$$g_\gamma^+(\theta) = \theta\varphi_\gamma(\theta(X'-Y')) \qquad\qquad (0 < \theta \leq \varepsilon)$$

$$g_\gamma^-(\theta) = \theta\varphi_\gamma(\theta(X'-Y')) \qquad\qquad (-\varepsilon \leq \theta < 0).$$

Now define the distributions T_k $(k \geq 0)$ on $\mathfrak{l}$ as in Corollary 1 of $[3(j), \text{Lemma } 35]$ with (g, g^+, g^-) replaced by $(g_\gamma, g_\gamma^+, g_\gamma^-)$. Then it follows from $[3(e), \text{Lemma } 16]$ and $[3(c), \text{Theorem } 1]$ that

$$T_k(f) = c \int \{ \partial(\omega_\mathfrak{l})^k f \cdot \varphi_\gamma - f\partial(\omega_\mathfrak{l})^k \varphi_\gamma \} d\mathfrak{l} = 0$$

for $f \in C_c^\infty(\Omega_1(\varepsilon/2))$. (Here c is a positive constant.) Therefore we conclude from the corollaries of [3(j), Lemma 35] that $g_\gamma^+ = g_\gamma^-$ and

$$(-1)^k (d^{2k+1} g_\gamma^- / dt^{2k+1})_0 = (d^{2k+1} g_\gamma^+ / d\theta^{2k+1})_0 \qquad (k \geq 0)$$

where the subscript o denotes the value at zero.

On the other hand let F_3 denote the restriction of F to $\mathfrak{z}_0$. Then by Corollary 2 of [3(h), Theorem 1], F_3 is locally summable on $\mathfrak{z}_0$ and since F is obviously invariant under G, we have

$$T(f_\alpha) = \int f_\alpha F \, dX = \int \beta_\alpha F_3 \, dZ \qquad (\alpha \in C_c^\infty(\mathfrak{z}_0))$$

in the notation of [3(i), Lemma 17]. This proves that $\sigma_T = F_3$ and therefore $T_3 = |\zeta|^{1/2} F_3$. Now $\mathfrak{a} = \sigma + \mathbf{R}H'$ and $\mathfrak{b} = \mathfrak{h} = \sigma + \mathbf{R}(X' - Y')$. Let τ and λ be the unique positive roots of $\mathfrak{a}$ and $\mathfrak{b}$ respectively which vanish at H_0. We may assume that $\tau(H') = 2$, $\lambda(X' - Y') = -2(-1)^{1/2}$ and the positive roots of $\mathfrak{a}$ go into positive roots of $\mathfrak{b}$ under the automorphism ν of [3(j), § 7]. Put $\pi_\tau^{\mathfrak{a}} = \tau^{-1} \pi^{\mathfrak{a}}$, $\pi_\lambda^{\mathfrak{b}} = \lambda^{-1} \pi^{\mathfrak{b}}$. Then it is clear that

$$|\zeta(H)|^{1/2} = |\pi_\tau^{\mathfrak{a}}(H)| \qquad (H \in \mathfrak{a}),$$

$$|\zeta(H)|^{1/2} = |\pi_\lambda^{\mathfrak{b}}(H)| \qquad (H \in \mathfrak{b}).$$

Let I denote the open interval $(-\varepsilon, \varepsilon)$ in $\mathbf{R}$. Put $\mathfrak{a}(\varepsilon) = \sigma_0 + IH'$ and $\mathfrak{b}(\varepsilon) = \sigma_0 + I(X' - Y')$. Then $\mathfrak{a}(\varepsilon)$ and $\mathfrak{b}(\varepsilon)$ are both connected sets. Since $\mathfrak{a}(\varepsilon) \subset \mathfrak{z}_0$, it is obvious that no positive root of $(\mathfrak{g}, \mathfrak{a})$ other than τ can vanish anywhere on $\mathfrak{a}(\varepsilon)$. Hence $|\pi_\tau^{\mathfrak{a}}(H)| / \pi_\tau^{\mathfrak{a}}(H)$ is a continuous function on $\mathfrak{a}(\varepsilon)$. But since its fourth power is 1 (see [3(j), Lemma 9]), it must be a constant. Put $c = |\pi_\tau^{\mathfrak{a}}(H_0)| / \pi_\tau^{\mathfrak{a}}(H_0)$. Since $\pi_\lambda^{\mathfrak{b}} = (\pi_\tau^{\mathfrak{a}})^\nu$ and H_0 remains fixed under ν, it is clear that

$$c = |\pi_\lambda^{\mathfrak{b}}(H_0)| / \pi_\lambda^{\mathfrak{b}}(H_0).$$

Hence we conclude by a similar argument that

$$|\pi_\lambda^{\mathfrak{b}}(H)| = c \pi_\lambda^{\mathfrak{b}}(H) \qquad (H \in \mathfrak{b}(\varepsilon)).$$

This shows that

$$t |\zeta(H + tH')|^{1/2} = 2^{-1} c \pi^{\mathfrak{a}}(H + tH') \qquad (|t| < \varepsilon)$$

$$\theta |\zeta(H + \theta(X' - Y'))|^{1/2} = 2^{-1}(-1)^{1/2} c \pi^{\mathfrak{b}}(H + \theta(X' - Y')) \qquad (|\theta| < \varepsilon)$$

for $H \in \sigma_0$. Now put

$$g^{\mathfrak{a}}(H) = \pi^{\mathfrak{a}}(H) F(H) \qquad (H \in \mathfrak{a}' \cap \Omega),$$

$$g^{\mathfrak{b}}(H) = \pi^{\mathfrak{b}}(H) F(H) \qquad (H \in \mathfrak{b}' \cap \Omega).$$

and fix a Euclidean measure $d\sigma$ on σ such that $d\sigma dI$ is equal to the Euclidean measure dZ on $\mathfrak{z}$ used above. Since $T_3 = |\zeta|^{1/2} F_3$, it is obvious that

$$\varphi_\gamma(Y) = \int \gamma(H) |\zeta(H + Y)|^{1/2} F(H + Y) \, d\sigma \qquad (Y \in \Omega_1').$$

625

Hence

$$g_\gamma(t) = 2^{-1}c \int g^\mathfrak{a}(H + tH')\gamma(H)d\sigma \qquad (0 < t < \varepsilon),$$

$$g_\gamma^+(\theta) = 2^{-1}(-1)^{1/2}c \int g^\mathfrak{b}(H + \theta(X'-Y'))\gamma(H)d\sigma \qquad (0 < \theta < \varepsilon),$$

$$g_\gamma^-(\theta) = 2^{-1}(-1)^{1/2}c \int g^\mathfrak{b}(H + \theta(X'-Y'))\gamma(H)d\sigma \qquad (-\varepsilon < \theta < 0).$$

On the other hand if J is the open interval $(0, \varepsilon)$ in $\mathbf{R}$, it is clear that $\sigma_0 \pm JH'$ are connected sets contained in $\mathfrak{a}' \cap \Omega$. Let $\mathfrak{a}^\pm$ denote the connected component of $\mathfrak{a}' \cap \Omega$ containing $\sigma_0 \pm JH'$. Similarly let $\mathfrak{b}^\pm$ be the connected component of $\mathfrak{b}' \cap \Omega$ containing $\sigma_0 \pm J(X'-Y')$. Then by Lemmas 1 and 2, there exist analytic functions $g_\pm^\mathfrak{a}$ and $g_\pm^\mathfrak{b}$ on $\mathfrak{a}$ and $\mathfrak{b}$ respectively such that $g^\mathfrak{a} = g_+^\mathfrak{a}$ on $\mathfrak{a}^+$, $g^\mathfrak{a} = g_-^\mathfrak{a}$ on $\mathfrak{a}^-$, $g^\mathfrak{b} = g_+^\mathfrak{b}$ on $\mathfrak{b}^+$ and $g^\mathfrak{b} = g_-^\mathfrak{b}$ on $\mathfrak{b}^-$. It is then obvious that

$$g_\gamma(t) = 2^{-1}c \int g_+^\mathfrak{a}(H + tH')\gamma(H)d\sigma \qquad (t \in \mathbf{R}),$$

$$g_\gamma^\pm(\theta) = 2^{-1}(-1)^{1/2}c \int g_\pm^\mathfrak{b}(H + \theta(X'-Y'))\gamma(H)d\sigma \qquad (\theta \in \mathbf{R}).$$

On the other hand we have seen above that $g_\gamma^+ = g_\gamma^-$ for every $\gamma \in C_c^\infty(\sigma_0)$. Therefore it is clear that $g_+^\mathfrak{b} = g_-^\mathfrak{b}$. This shows that $g = g^\mathfrak{b} = g_+^\mathfrak{b}$ on $\mathfrak{b}(\varepsilon) \cap \mathfrak{b}'$. Since $\mathfrak{b}(\varepsilon)$ is a neighborhood of H_0 in $\mathfrak{b}$, our assertion is proved in this case.

Moreover since

$$(d^{2k+1}g_\gamma/dt^{2k+1})_0 = (-1)^k(d^{2k+1}g_\gamma^+/d\theta^{2k+1})_0 \qquad (k \geq 0)$$

and $\nu(H') = (-1)^{1/2}(X'-Y')$, we find in the same way that

$$g_+^\mathfrak{a}(H; \partial(H')^{2k+1}) = g_+^\mathfrak{b}(H; \partial(\nu(H'))^{2k+1})$$

for $H \in \sigma$.

We now use the notation of $[3(j), \S\ 8]$.

Lemma 18. — *Let s_τ be the Weyl reflexion in $\mathfrak{a}$ corresponding to τ. Then $(g^\mathfrak{a})^{s_\tau} = -g^\mathfrak{a}$. If D is an element in $\mathfrak{D}(\mathfrak{a}_c)$ such that $D^{s_\tau} = -D$, then $Dg^\mathfrak{a}$ can be extended to a continuous function on $\mathfrak{a}(\varepsilon)$ and* $(^1)$

$$g^\mathfrak{a}(H; D) = g^\mathfrak{b}(H; D^\nu)$$

for $H \in \sigma_0$.

Since τ is real we know from $[3(j),$ Lemma 6] that $s_\tau \in W_G^\mathfrak{a}$. Therefore since F is invariant under G, it is obvious that $(g^\mathfrak{a})^{s_\tau} = -g^\mathfrak{a}$ and hence $(Dg^\mathfrak{a})^{s_\tau} = Dg^\mathfrak{a}$. This implies that $(Dg_+^\mathfrak{a})^{s_\tau} = Dg_-^\mathfrak{a}$ and therefore $Dg_+^\mathfrak{a} = Dg_-^\mathfrak{a}$ on σ. It is now clear that $Dg^\mathfrak{a}$ can be extended to a continuous function on $\mathfrak{a}(\varepsilon)$. So it remains to show that

$$g_+^\mathfrak{a}(H; D) = g^\mathfrak{b}(H; D^\nu)$$

for $H \in \sigma_0$. Since $\mathfrak{D}(\mathfrak{a}_c) = \mathfrak{D}(\sigma_c)\mathfrak{D}(\mathbf{C}H')$ and since s_τ leaves σ pointwise fixed, it is sufficient to consider the case when $D = \Delta \circ \tau^i \partial(H')^j$. Here $\Delta \in \mathfrak{D}(\sigma_c)$ and $i+j$ is odd.

$(^1)$ $g^\mathfrak{a}(H; D)$ denotes, as usual, the value of the continuous function $Dg^\mathfrak{a}$ at H. Similarly in other cases.

Now Δ and τ commute. Therefore, if $i \geq 1$, our assertion is obvious from the fact that τ and λ are both zero on σ. So we may assume that $i = 0$ so that j is odd. It is enough to verify that

$$g_+^a(H; \partial(H')^j) = g^b(H; \partial(\nu(H'))^j) \qquad (H \in \sigma_0)$$

since the required relation would then follow by applying the differential operator Δ to this equation. However $g^b = g_+^b$ on $b(\varepsilon)$ and so this follows from the result proved above.

Now we return to the proof of Theorem 2. Fix a point $H_0 \in \mathfrak{h}'(R) \cap \Omega$ and an open convex neighborhood U of H_0 in $\mathfrak{h}'(R) \cap \Omega$. Let U_1 be the set consisting of all regular and semiregular elements of U. Then U_1 is open, and if β is a root of $(\mathfrak{g}, \mathfrak{h})$ which vanishes at some point of U_1, it is clear that β is imaginary. Hence it follows from the above proof that there exists an analytic function g_1 on U_1 such that $g_1 = g$ on $U_1 \cap \mathfrak{h}'$. Now fix a connected component U_2 of $U_1 \cap \mathfrak{h}' = U \cap \mathfrak{h}'$. Then by Lemma 2 there exists an analytic function g_2 on $\mathfrak{h}$ such that $g = g_2$ on U_2. Since U_1 is connected (see the corollary of [3(j), Lemma 19]), we conclude that $g_1 = g_2$ and therefore $g = g_2$ on $U \cap \mathfrak{h}'$. Since g_2 is analytic on $\mathfrak{h}$, we have shown that g can be extended to an analytic function on U. Thus Theorem 2 is proved.

We denote the extended analytic function on $\mathfrak{h}'(R) \cap \Omega$ again by g.

Lemma 19 ([1]). — *Let H_0 be a point in $\mathfrak{h} \cap \Omega$ and D an element in $\mathfrak{D}(\mathfrak{h}_c)$ such that $D^{s_\alpha} = -D$ for every real root α of $(\mathfrak{g}, \mathfrak{h})$ which vanishes at H_0. Then Dg can be extended to a continuous function around H_0.*

Fix an open, convex and relatively compact neighborhood U of H_0 in $\Omega \cap \mathfrak{h}$. By taking it sufficiently small we can arrange that no real root α of $(\mathfrak{g}, \mathfrak{h})$ vanishes anywhere on U unless $\alpha(H_0) = 0$. Let U_0 be the set consisting of all regular and semiregular elements of U. Then, as before, U_0 is open and connected and it follows from Theorem 2 and Lemma 18 that there exists a continuous function g_0 on U_0 such that $Dg = g_0$ on $U_0 \cap \mathfrak{h}'(R)$. The set $U \cap \mathfrak{h}'$ has only a finite number of connected components, say $U_1, \ldots, U_r$. By Lemma 2 we can choose an analytic function g_i on $\mathfrak{h}$ such that $g = g_i$ on $U_i (1 \leq i \leq r)$. This shows that Dg is of class C^∞ on $\mathrm{Cl}\,U_i$ (see [3(j), § 14]). But $\mathrm{Cl}\,U_i = \mathrm{Cl}(U_i \cap U_0)$ and $Dg = g_0$ on $U_i \cap U_0$. Therefore g_0 is also of class C^∞ on $\mathrm{Cl}\,U_i (1 \leq i \leq r)$. Fix a Euclidean norm on $\mathfrak{h}$ and put

$$\nu(g_0) = \sup |g_0(H_1; \partial(H_2))|$$

where H_1, H_2 vary in $U \cap \mathfrak{h}'$ and $\mathfrak{h}$ respectively under the sole restriction that $\|H_2\| \leq 1$. Then it is obvious from what we have said above that $\nu(g_0) < \infty$. Moreover (see [3(j), § 10])

$$|g_0(H_1) - g_0(H_2)| \leq \nu(g_0) \|H_1 - H_2\|$$

for any two points H_1, H_2 in $U \cap \mathfrak{h}'$. Obviously this means that Dg can be extended to a continuous function on U.

([1]) Cf. [3(j), Theorem 1].

Corollary. — *Let* D *be an element of* $\mathfrak{D}(\mathfrak{h}_c)$ *such that* $D^{s_\alpha} = -D$ *for every real root* α *of* $\mathfrak{h}$. *Then* Dg *can be extended to a continuous function on* $\mathfrak{h} \cap \Omega$.

This is obvious from the above lemma. We denote the extended function again by Dg. Moreover $g(H; D)$ $(H \in \mathfrak{h} \cap \Omega)$ will stand for the value of Dg at H.

Put $\varpi = \prod_{\alpha > 0} H_\alpha$ where α runs over all positive roots of $(\mathfrak{g}, \mathfrak{h})$. Then $\varpi \in S(\mathfrak{h}_c)$ and $\varpi^{s_\alpha} = -\varpi$ for every root α. Hence $\partial(\varpi)g$ is a continuous function on $\mathfrak{h} \cap \Omega$. Since the differential operator $\partial(\varpi) \circ \pi$ is obviously independent of the choice of positive roots of $\mathfrak{h}$, it is clear that the function $\partial(\varpi)g$ also does not depend on this choice. Corresponding to any Cartan subalgebra $\mathfrak{a}$ of $\mathfrak{g}$, we define $\varpi^\mathfrak{a}$, $g^\mathfrak{a}$ and $\partial(\varpi^\mathfrak{a})g^\mathfrak{a}$ in an analogous way.

Theorem 3. — *Let* $\mathfrak{a}$ *and* $\mathfrak{b}$ *be two Cartan subalgebras of* $\mathfrak{g}$. *Then*

$$\partial(\varpi^\mathfrak{a})g^\mathfrak{a} = \partial(\varpi^\mathfrak{b})g^\mathfrak{b}$$

on $\mathfrak{a} \cap \mathfrak{b} \cap \Omega$.

Before giving the proof we derive a consequence of this theorem. Let g_T denote the function g of Theorem 2 corresponding to the distribution T. For any $D \in \mathfrak{Z}(\mathfrak{g}_c)$, DT also fulfills the conditions of Theorem 1 (Lemma 16). Hence we can consider the corresponding function g_{DT}. It follows from [3(*i*), Theorem 1] that $g_{DT} = \delta_{\mathfrak{g}/\mathfrak{h}}(D)g_T$. Therefore

$$\partial(\varpi)g_{DT} = (\partial(\varpi) \circ \delta_{\mathfrak{g}/\mathfrak{h}}(D))g$$

can also be extended to a continuous function on $\mathfrak{h} \cap \Omega$.

Corollary. — $(\partial(\varpi^\mathfrak{a}) \circ \delta_{\mathfrak{g}/\mathfrak{a}}(D))g^\mathfrak{a} = (\partial(\varpi^\mathfrak{b}) \circ \delta_{\mathfrak{g}/\mathfrak{b}}(D))g^\mathfrak{b}$ *on* $\mathfrak{a} \cap \mathfrak{b} \cap \Omega$ *for any* $D \in \mathfrak{Z}(\mathfrak{g}_c)$.

This follows by applying Theorem 3 to DT instead of T.

We shall prove Theorem 3 by induction on $\dim \mathfrak{g}$. Fix a point $H_0 \in \mathfrak{a} \cap \mathfrak{b} \cap \Omega$. We have to show that $g^\mathfrak{a}(H_0; \partial(\varpi^\mathfrak{a})) = g^\mathfrak{b}(H_0; \partial(\varpi^\mathfrak{b}))$. Let $\mathfrak{c}$ be the center and $\mathfrak{g}_1$ the derived algebra of $\mathfrak{g}$ and first suppose that $\mathfrak{c} \neq \{0\}$. Let $H_0 = C_0 + H_1$ where $C_0 \in \mathfrak{c}$ and $H_1 \in \mathfrak{g}_1$. Then it is clear that $H_1 \in \mathfrak{a}_1 \cap \mathfrak{b}_1$ where $\mathfrak{h}_1 = \mathfrak{h} \cap \mathfrak{g}_1$ ($\mathfrak{h} = \mathfrak{a}$ or $\mathfrak{b}$). Choose an open and relatively compact neighborhood $\mathfrak{c}_0$ of C_0 in $\mathfrak{c}$ and let Ω_1 be the set of all $Z \in \mathfrak{g}_1$ such that $\mathrm{Cl}\,\mathfrak{c}_0 + Z \subset \Omega$. Then (Lemma 9) Ω_1 is an open and completely invariant neighborhood of H_1 in $\mathfrak{g}_1$, if $\mathfrak{c}_0$ is sufficiently small. Fix $\alpha \in C_c^\infty(\mathfrak{c}_0)$ and consider the distribution

$$\tau_\alpha : f \to T(\alpha \times f) \qquad\qquad (f \in C_c^\infty(\Omega_1))$$

on Ω_1. Put $\mathfrak{U}_1 = \mathfrak{U} \cap I(\mathfrak{g}_{1c})$. Then it is clear that

$$\dim\,(I(\mathfrak{g}_{1c})/\mathfrak{U}_1) \leq \dim\,(I(\mathfrak{g}_c)/\mathfrak{U}) < \infty$$

and $\partial(u_1)\tau_\alpha = 0$ for $u_1 \in \mathfrak{U}_1$. Hence Theorem 1 also holds if we replace $(\mathfrak{g}, \Omega, T)$ by $(\mathfrak{g}_1, \Omega_1, \tau_\alpha)$. Since $\dim \mathfrak{g}_1 < \dim \mathfrak{g}$, Theorem 3 applies to τ_α by the induction hypothesis. Put

$$g_\alpha^\mathfrak{b}(H) = \int \alpha(C)g^\mathfrak{b}(C + H)\,dC \qquad\qquad (H \in \mathfrak{h}' \cap \Omega_1)$$

where dC is a Euclidean measure on $\mathfrak{c}$. Then we conclude that

$$g_\alpha^a(H;\ \partial(\varpi^a))=g_\alpha^b(H;\ \partial(\varpi^b))$$

for $H\in\mathfrak{a}\cap\mathfrak{b}\cap\Omega_1$. Since this is true for every $\alpha\in C_c^\infty(\mathfrak{c}_0)$, it is clear that $\partial(\varpi^a)g^a$ and $\partial(\varpi^b)g^b$ coincide around H_0 on $\mathfrak{a}\cap\mathfrak{b}\cap\Omega$.

So now we can assume that $\mathfrak{c}=\{o\}$ and therefore $\mathfrak{g}$ is semisimple. Then we identify $\mathfrak{g}$ and $\mathfrak{h}$ with their respective duals by means of the Killing form (see $[3(i),\ \S\ 6]$) so that $\varpi^b=\pi^b$ $(\mathfrak{h}=\mathfrak{a}$ or $\mathfrak{b})$. First assume that $H_0\neq o$ and let $\mathfrak{z}$ be the centralizer of H_0 in $\mathfrak{g}$. Then $\dim\mathfrak{z}<\dim\mathfrak{g}$ and we can identify $\mathfrak{z}$ with its dual by means of the restriction (to $\mathfrak{z}$) of the Killing form of $\mathfrak{g}$. Define ζ and $\mathfrak{z}'$ as in $[3(i),\ \S\ 2]$ and put $\Omega_\mathfrak{z}=\mathfrak{z}'\cap\Omega$. Then $\Omega_\mathfrak{z}$ is an open neighborhood of H_0 in $\mathfrak{z}$ which is completely invariant (with respect to $\mathfrak{z}$). Take $G_0=G$ and $\mathfrak{z}_0=\Omega_\mathfrak{z}$ in Lemma 17 of $[3(i)]$ and let σ_T denote the corresponding distribution on $\Omega_\mathfrak{z}$. Then

$$\sigma_T(\beta_\alpha)=T(f_\alpha)=\int f_\alpha F\,dX \qquad (\alpha\in C_c^\infty(G\times\Omega_\mathfrak{z})).$$

But since F is invariant under G, we conclude from Corollary 2 of $[3(h),\ \text{Theorem}\ 1]$ that the function $F_\mathfrak{z}:Z\to F(Z)\,(Z\in\Omega_\mathfrak{z})$ is locally summable on $\Omega_\mathfrak{z}$ and

$$\int f_\alpha F\,dX=\int\beta_\alpha F_\mathfrak{z}\,dZ.$$

This shows that $\sigma_T=F_\mathfrak{z}$.

Let $\mathfrak{h}=\mathfrak{a}$ or $\mathfrak{b}$. Then $\mathfrak{h}\subset\mathfrak{z}$. Define $\mathfrak{q}$ as in $[3(i),\ \S\ 2]$. P^b being the set of all positive roots of $(\mathfrak{g},\ \mathfrak{h})$, let $P_\mathfrak{z}^b$ and $P_\mathfrak{q}^b$ denote the subsets of those $\alpha\in P^b$ for which X_α lies in $\mathfrak{z}_c$ and $\mathfrak{q}_c$ respectively. Let $\pi_\mathfrak{z}^b$ and $\pi_\mathfrak{q}^b$ be the products of all roots in $P_\mathfrak{z}^b$ and $P_\mathfrak{q}^b$ respectively. Then $\pi^b=\pi_\mathfrak{z}^b\pi_\mathfrak{q}^b$ and it is clear that $(\pi_\mathfrak{q}^b)^{s_\alpha}=\pi_\mathfrak{q}^b$ for all $\alpha\in P_\mathfrak{z}^b$. Hence, by Chevalley's theorem $[3(c),\ \text{Lemma}\ 9]$, there exists an invariant polynomial function p on $\mathfrak{z}$ such that $p(H)=\pi_\mathfrak{q}^a(H)$ for $H\in\mathfrak{a}$. But

$$\zeta(H)=(-1)^q(\pi_\mathfrak{q}^a(H))^2 \qquad (H\in\mathfrak{a})$$

where $q=2^{-1}\dim\mathfrak{q}$ is the number of roots in $P_\mathfrak{q}^a$. Therefore $\zeta=(-1)^q p^2$ again by Chevalley's theorem. Let $p_\mathfrak{h}$ denote the restriction of p to $\mathfrak{h}$. Then since ζ coincides with $(-1)^q(\pi_\mathfrak{q}^b)^2$ on $\mathfrak{b}$, it is clear that $p_\mathfrak{b}=\varepsilon\pi_\mathfrak{q}^b$ where $\varepsilon=\pm1$.

Now put $T_\mathfrak{z}=p\sigma_T$. Then it follows from Theorem 2 of $[3(i)]$ (see also $\S\ 4$) that Theorem 1 still holds if we replace $(\mathfrak{g},\ \Omega,\ T)$ by $(\mathfrak{z},\ \Omega_\mathfrak{z},\ T_\mathfrak{z})$. Put

$$g_\mathfrak{z}^b(H)=\pi_\mathfrak{z}^b(H)p(H)F(H) \qquad (H\in\mathfrak{h}'\cap\Omega_\mathfrak{z}).$$

Since $\dim\mathfrak{z}<\dim\mathfrak{g}$, both Theorem 3 and its corollary are applicable to $T_\mathfrak{z}$. Moreover $\delta_{\mathfrak{z}/\mathfrak{h}}(\partial(p))=\partial(p_\mathfrak{h})$ $[3(c),\ \text{Theorem}\ 1]$ and so we conclude that

$$\partial(\pi_\mathfrak{z}{}^a p_a)g_\mathfrak{z}^a=\partial(\pi_\mathfrak{z}{}^b p_b)g_\mathfrak{z}^b$$

on $\Omega_\mathfrak{z}\cap\mathfrak{a}\cap\mathfrak{b}$. However $\pi_\mathfrak{z}{}^a p_a=\pi^a$ and $\pi_\mathfrak{z}{}^b p_b=\varepsilon\pi^b$. Therefore $g_\mathfrak{z}^a=g^a$ and $g_\mathfrak{z}^b=\varepsilon g^b$ on $\Omega_\mathfrak{z}\cap\mathfrak{a}'$ and $\Omega_\mathfrak{z}\cap\mathfrak{b}'$ respectively. So it follows that $\partial(\pi^a)g^a=\partial(\pi^b)g^b$ on $\Omega_\mathfrak{z}\cap\mathfrak{a}\cap\mathfrak{b}$. Since $H_0\in\Omega_\mathfrak{z}\cap\mathfrak{a}\cap\mathfrak{b}$, our assertion is proved in this case.

629

4

321

So it remains to consider the case when $H_0 = 0$. Hence we may assume that $0 \in \Omega$. For any Cartan subalgebra $\mathfrak{h}$ of $\mathfrak{g}$, put $c(\mathfrak{h}) = g^{\mathfrak{h}}(0; \partial(\pi^{\mathfrak{h}}))$.

Lemma 20. — Let $\mathfrak{a}$ and $\mathfrak{b}$ be two Cartan subalgebras of $\mathfrak{g}$. Then, if $\mathfrak{a} \cap \mathfrak{b} \neq \{0\}$, $c(\mathfrak{a}) = c(\mathfrak{b})$.

Since Ω is an open neighborhood of zero in $\mathfrak{g}$, we can choose $H \neq 0$ in $\mathfrak{a} \cap \mathfrak{b}$ such that $tH \in \mathfrak{a} \cap \mathfrak{b} \cap \Omega$ for $0 \leq t \leq 1$. Then in view of what we have proved above, it is clear that

$$g^{\mathfrak{a}}(tH; \partial(\pi^{\mathfrak{a}})) = g^{\mathfrak{b}}(tH; \partial(\pi^{\mathfrak{b}})) \qquad\qquad (0 < t \leq 1).$$

Making t tend to zero we get $c(\mathfrak{a}) = c(\mathfrak{b})$.

Lemma 21. — Let $\mathfrak{h}$ be a Cartan subalgebra of $\mathfrak{g}$. $\cdot$ Then $c(\mathfrak{h}) = c(\mathfrak{h}^x)$ for any x in G.

Let $\mathfrak{a} = \mathfrak{h}^x$. Without loss of generality we may assume that $\pi^{\mathfrak{a}} = (\pi^{\mathfrak{h}})^x$. Then it is clear that

$$g^{\mathfrak{a}}(H^x) = g^{\mathfrak{b}}(H) \qquad\qquad (H \in \Omega \cap \mathfrak{h}')$$

and therefore

$$g^{\mathfrak{a}}(H^x; \partial(\pi^{\mathfrak{a}})) = g^{\mathfrak{b}}(H; \partial(\pi^{\mathfrak{b}}))$$

for $H \in \Omega \cap \mathfrak{h}'$. We obtain the required result by making H tend to zero.

Fix a Cartan involution θ of $\mathfrak{g}$ and let $\mathfrak{g} = \mathfrak{k} + \mathfrak{p}$ be the corresponding Cartan decomposition. If $\mathfrak{h}$ is a Cartan subalgebra of $\mathfrak{g}$ which is stable under θ, we put

$$l_+(\mathfrak{h}) = \dim(\mathfrak{h} \cap \mathfrak{p}), \qquad l_-(\mathfrak{h}) = \dim(\mathfrak{h} \cap \mathfrak{k}).$$

Then $l_+(\mathfrak{h}) + l_-(\mathfrak{h}) = \dim \mathfrak{h} = l$ where $l = \operatorname{rank} \mathfrak{g}$. Let $l_+ = \sup_{\mathfrak{h}} l_+(\mathfrak{h})$, $l_- = \sup_{\mathfrak{h}} l_-(\mathfrak{h})$ where $\mathfrak{h}$ runs over all Cartan subalgebras stable under θ. Fix two Cartan subalgebras $\mathfrak{h}_+$, $\mathfrak{h}_-$, both stable under θ, such that $l_+ = l_+(\mathfrak{h}_+)$, $l_- = l_-(\mathfrak{h}_-)$.

Lemma 22. — Let $\mathfrak{h}$ be a Cartan subalgebra of $\mathfrak{g}$ which is stable under θ. Then $c(\mathfrak{h}) = c(\mathfrak{h}_+)$ if $l_+(\mathfrak{h}) > 0$ and $c(\mathfrak{h}) = c(\mathfrak{h}_-)$ if $l_-(\mathfrak{h}) > 0$.

Let K be the analytic subgroup of G corresponding to $\mathfrak{k}$. Then $\mathfrak{h}_+ \cap \mathfrak{p}$ and $\mathfrak{h}_- \cap \mathfrak{k}$ are maximal abelian subspaces of $\mathfrak{p}$ and $\mathfrak{k}$ respectively. Since any two maximal abelian subspaces of $\mathfrak{p}$ (or $\mathfrak{k}$) are conjugate under K, it is clear that we can choose $k_1, k_2 \in K$ such that

$$(\mathfrak{h} \cap \mathfrak{p})^{k_1} \subset \mathfrak{h}_+ \cap \mathfrak{p}, \qquad (\mathfrak{h} \cap \mathfrak{k})^{k_2} \subset \mathfrak{h}_- \cap \mathfrak{k}.$$

Then

$$\dim(\mathfrak{h}^{k_1} \cap \mathfrak{h}_+) \geq \dim(\mathfrak{h} \cap \mathfrak{p})^{k_1} = l_+(\mathfrak{h})$$

and similarly

$$\dim(\mathfrak{h}^{k_2} \cap \mathfrak{h}_-) \geq l_-(\mathfrak{h}).$$

Hence our assertion follows from Lemmas 20 and 21.

Lemma 23. — $c(\mathfrak{h}_+) = c(\mathfrak{h}_-)$.

We may obviously assume that $l \geq 1$. If $l_-(\mathfrak{h}_+) + l_+(\mathfrak{h}_-) \geq 1$, our statement follows from Lemma 22. Hence we may assume that $\mathfrak{h}_+ \subset \mathfrak{p}$ and $\mathfrak{h}_- \subset \mathfrak{k}$. Then $\mathfrak{h}_+$ is not fundamental in $\mathfrak{g}$ and so there exists a positive real root α of $(\mathfrak{g}, \mathfrak{h}_+)$ (see [3(d),

630

Lemma 33]). We assume, as we may (see [3(d), Lemma 46]), that $\theta(X_\alpha)=-X_{-\alpha}$ and X_α, $X_{-\alpha}$ are in $\mathfrak{g}$. Take $H'=a^{-2}H_\alpha$, $X'=a^{-1}X_\alpha$, $Y'=a^{-1}X_{-\alpha}$ where $a=(\alpha(H_\alpha)/2)^{1/2}$. Define the automorphism ν of $\mathfrak{g}_c$ as in [3(j), § 7]. Then $\theta(X')=-Y'$ and $\mathfrak{b}=\nu((\mathfrak{h}_+)_c)\cap\mathfrak{g}=\sigma_\alpha+\mathbf{R}(X'-Y')$ is a Cartan subalgebra of $\mathfrak{g}$ which is stable under θ. Here σ_α is the hyperplane consisting of all points $H\in\mathfrak{h}_+$ where $\alpha(H)=0$. Since $\mathfrak{b}\cap\mathfrak{p}=\sigma_\alpha$ and $\mathfrak{b}\cap\mathfrak{k}=\mathbf{R}(X'-Y')$, it is obvious that $l_+(\mathfrak{b})=l-1$ and $l_-(\mathfrak{b})=1$. Hence, if $l\geq2$, it follows from Lemma 22 that $c(\mathfrak{h}_+)=c(\mathfrak{b})=c(\mathfrak{h}_-)$. On the other hand if $l=1$, zero is a semiregular element of $\mathfrak{g}$ and our assertion follows immediately from Lemmas 18 and 21.

We shall now finish the proof of Theorem 3. Choose x,y in G such that $\mathfrak{a}^x$ and $\mathfrak{b}^y$ are stable under θ (see [3(b), p. 100]). Then it is clear from Lemmas 21, 22 and 23 that $c(\mathfrak{a})=c(\mathfrak{b})$. The proof of Theorem 3 is now complete.

§ 9. THE DIFFERENTIAL OPERATOR $\nabla_\mathfrak{g}$ AND THE FUNCTION $\nabla_\mathfrak{g}F$

Lemma 24. — There exists a unique differential operator $\nabla_\mathfrak{g}$ on $\mathfrak{g}'$ with the following two properties:

1) $\nabla_\mathfrak{g}$ *is invariant under* G.

2) *Let $\mathfrak{h}$ be a Cartan subalgebra of $\mathfrak{g}$.* *Then*
$$f(H;\nabla_\mathfrak{g})=f(H;\partial(\varpi^\mathfrak{h})\circ\pi^\mathfrak{h})$$
for $f\in C^\infty(\mathfrak{g})$ and $H\in\mathfrak{h}'$.
Moreover $\nabla_\mathfrak{g}$ is analytic.

Since two distinct Cartan subalgebras cannot have a regular element in common, the uniqueness is obvious. The existence is proved as follows. Fix a Cartan subalgebra $\mathfrak{a}$ of $\mathfrak{g}$ and define $\mathfrak{g}_\mathfrak{a}=(\mathfrak{a}')^G$. Then $\mathfrak{g}_\mathfrak{a}$ is an open subset of $\mathfrak{g}$. Let A be the Cartan subgroup of G corresponding to $\mathfrak{a}$ and $x\to x^*$ the natural projection of G on $G^*=G/A$. Then the mapping $\varphi:(x^*,H)\to x^*H$ of $G^*\times\mathfrak{a}'$ onto $\mathfrak{g}_\mathfrak{a}$ (in the notation of § 2) is everywhere regular. Define $W_G=\widetilde{A}/A$ where $\widetilde{A}$ is the normalizer of $\mathfrak{a}$ in G. Then W_G operates on G^* and $\mathfrak{a}$ as follows. Fix $s\in W_G$ and choose $y\in\widetilde{A}$ lying in the coset s. Then
$$sH=H^y, \qquad x^*s=(xy)^*$$
for $H\in\mathfrak{a}$ and $x\in G$. It is clear that the complete inverse image under φ of a point $x^*H\in\mathfrak{g}_\mathfrak{a}\,(x^*\in G^*, H\in\mathfrak{a}')$ consists of the elements $(x^*s, s^{-1}H)\,(s\in W_G)$, which are all distinct. Since φ is locally an analytic diffeomorphism and since $\partial(\varpi^\mathfrak{a})\circ\pi^\mathfrak{a}$ is obviously invariant under W_G, it is clear that there exists an analytic differential operator ∇ on $\mathfrak{g}_\mathfrak{a}$ such that
$$f(x^*H;\nabla)=f(x^*:H;\partial(\varpi^\mathfrak{a})\circ\pi^\mathfrak{a}) \qquad (x^*\in G^*, H\in\mathfrak{a}')$$
for $f\in C^\infty(\mathfrak{g}_\mathfrak{a})$. Here $f(x^*:H)=f(x^*H)$ as usual. It is easy to verify that ∇ satisfies the two conditions of the lemma on $\mathfrak{g}_\mathfrak{a}$.

631

Now select a maximal set $\mathfrak{h}_1, \ldots, \mathfrak{h}_r$ of Cartan subalgebras of $\mathfrak{g}$, no two of which are conjugate under G. Put $\mathfrak{g}_i = (\mathfrak{h}_i')^G$ and define a differential operator ∇_i on $\mathfrak{g}_i$ as above corresponding to $\mathfrak{a} = \mathfrak{h}_i$. Since $\mathfrak{g}'$ is the disjoint union of the open sets $\mathfrak{g}_1, \ldots, \mathfrak{g}_r$, we can define $\nabla_\mathfrak{g}$ by setting $\nabla_\mathfrak{g} = \nabla_i$ on $\mathfrak{g}_i$ $(1 \leq i \leq r)$.

Lemma 25. — For any $D \in \mathfrak{J}(\mathfrak{g}_c)$, $(\nabla_\mathfrak{g} \circ D)F$ *can be extended to a continuous function on* Ω.

We shall use induction on dim $\mathfrak{g}$. In view of Lemma 16, it is enough to consider the case when $D = 1$. Define $\mathfrak{c}$ and $\mathfrak{g}_1$ as in § 4 and first assume that $\mathfrak{c} \neq \{0\}$. Fix a point $X_0 = C_0 + Z_0$ $(C_0 \in \mathfrak{c}, Z_0 \in \mathfrak{g}_1)$ in Ω. We have to show that $\nabla_\mathfrak{g} F$ can be extended to a continuous function around X_0. Select an open, connected and relatively compact neighborhood $\mathfrak{c}_0$ of C_0 in $\mathfrak{c}$ such that $(\mathrm{Cl}\,\mathfrak{c}_0) + Z_0 \subset \Omega$. Define Ω_1 to be the set of all $Z \in \mathfrak{g}_1$ such that $\mathrm{Cl}\,\mathfrak{c}_0 + Z \subset \Omega$. Then, by Lemma 9, Ω_1 is also open and completely invariant in $\mathfrak{g}_1$. Since $S(\mathfrak{c}_c) \subset I(\mathfrak{g}_c)$, it is clear that

$$\dim(S(\mathfrak{c}_c)/\mathfrak{U} \cap S(\mathfrak{c}_c)) \leq \dim(I(\mathfrak{g}_c)/\mathfrak{U}) < \infty.$$

Let E be the space of all analytic functions χ on $\mathfrak{c}_0$ such that $\partial(u)\chi = 0$ for $u \in \mathfrak{U} \cap S(\mathfrak{c}_c)$. Then (see the proof of Lemma 13 of $[3(c)]$) dim $E < \infty$. Let χ_j $(1 \leq j \leq N)$ be a base for E over $\mathbf{C}$. Fix $Z \in \Omega_1' = \Omega_1 \cap \mathfrak{g}'$. Then it is obvious that $F(Z + C; \partial(u)) = 0$ for $u \in \mathfrak{U} \cap S(\mathfrak{c}_c)$ and $C \in \mathfrak{c}_0$. Therefore

$$F(C + Z) = \sum_{1 \leq j \leq N} \chi_j(C) F_j(Z) \qquad (C \in \mathfrak{c}_0)$$

where $F_j(Z) \in \mathbf{C}$. Since F is analytic on Ω', it is obvious that F_j $(1 \leq j \leq N)$ are analytic functions on Ω_1'.

Fix Euclidean measures dC and dZ on $\mathfrak{c}$ and $\mathfrak{g}_1$ respectively such that $dX = dC\,dZ$ for $X = C + Z$ $(C \in \mathfrak{c}, Z \in \mathfrak{g}_1)$ and for any $\alpha \in C_c^\infty(\mathfrak{c}_0)$ define the distribution θ_α on Ω_1 by

$$\theta_\alpha(\beta) = T(\alpha \times \beta) \qquad (\beta \in C_c^\infty(\Omega_1)).$$

Then, as we have seen in § 4, the induction hypothesis is applicable to $(\mathfrak{g}_1, \theta_\alpha, \Omega_1)$ in place of $(\mathfrak{g}, T, \Omega)$. Put

$$F_\alpha(Z) = \sum_{1 \leq j \leq N} F_j(Z) \int \chi_j(C)\alpha(C)\,dC \qquad (Z \in \Omega_1').$$

Then $\nabla_{\mathfrak{g}_1} F_\alpha$ can be extended to a continuous function on Ω_1. Since this is true for every $\alpha \in C_c^\infty(\mathfrak{c}_0)$, the same holds for $\nabla_{\mathfrak{g}_1} F_j$, $1 \leq j \leq N$ (see $[3(e)]$, Lemma 20]). But it is obvious that

$$F(C + Z; \nabla_\mathfrak{g}) = \sum_{1 \leq j \leq N} \chi_j(C) F_j(Z; \nabla_{\mathfrak{g}_1})$$

for $C \in \mathfrak{c}_0$ and $Z \in \Omega_1'$. Hence $\nabla_\mathfrak{g} F$ extends to a continuous function on $\mathfrak{c}_0 + \Omega_1$, which proves our assertion.

So now we may assume that $\mathfrak{g}$ is semisimple. Let Ω^0 be the set of all points $X_0 \in \Omega$ such that $\nabla_\mathfrak{g} F$ can be extended to a continuous function around X_0. Clearly Ω^0 is an open and invariant subset of Ω. Therefore, in view of Corollary 2 of Lemma 8, it would be enough to show that every semisimple element of Ω lies in Ω^0.

632

Fix a semisimple element $H_0 \in \Omega$. First assume that $H_0 \neq 0$. Let $\mathfrak{z}$ be the centralizer of H_0 in $\mathfrak{g}$. Define q and ζ as in [3(i), § 2]. Then as we have seen during the proof of Theorem 3, there exists an invariant polynomial function p on $\mathfrak{z}$ such that $\zeta = (-1)^q p^2$ where $q = (\dim q)/2$. Let $\mathfrak{h}$ be a Cartan subalgebra of $\mathfrak{z}$. We identify $\mathfrak{g}$, $\mathfrak{z}$, $\mathfrak{h}$ with their respective duals by means of the Killing form of $\mathfrak{g}$. Define $\pi_{\mathfrak{z}}$ and π_q as in the proof of Theorem 3. Since $\zeta = (-1)^q \pi_q{}^2$ on $\mathfrak{h}$, it follows from [3(c), Theorem 1] that

$$\delta'_{\mathfrak{z}/\mathfrak{h}}(\partial(p) \circ p) = \pi_{\mathfrak{z}}^{-1} \partial(\pi_q) \circ \pi.$$

Hence if $H \in \mathfrak{h} \cap \Omega'$, we get

$$F(H; \nabla_{\mathfrak{g}}) = F(H; \partial(\pi) \circ \pi)$$
$$= F(H; \partial(\pi_{\mathfrak{z}}) \circ \pi_{\mathfrak{z}} \circ \delta'_{\mathfrak{z}/\mathfrak{h}}(\partial(p) \circ p)).$$

On the other hand define $\Omega_{\mathfrak{z}}$, σ_T and $F_{\mathfrak{z}}$ as before (see the proof of Theorem 3) and put $T_{\mathfrak{z}} = p\sigma_T$. Then by [3($i$), Theorem 2 and Lemma 19], the induction hypothesis is applicable to $(\mathfrak{z}, \Omega_{\mathfrak{z}}, T_{\mathfrak{z}})$ in place of $(\mathfrak{g}, \Omega, T)$. On the other hand we have seen during the proof of Theorem 3 that $\sigma_T = F_{\mathfrak{z}}$. Therefore $T_{\mathfrak{z}} = pF_{\mathfrak{z}}$ and so by the induction hypothesis $(\nabla_{\mathfrak{z}} \circ \partial(p))(pF_{\mathfrak{z}})$ extends to a continuous function $g_{\mathfrak{z}}$ on $\Omega_{\mathfrak{z}}$.

Let Ξ denote the analytic subgroup of G corresponding to $\mathfrak{z}$ and $x \to x^*$ the natural projection of G on $G^* = G/\Xi$. Select open connected neighborhoods G_0 and $\mathfrak{z}_0$ of 1 and H_0 in G and $\Omega_{\mathfrak{z}}$ respectively and let G_0^* denote the image of G_0 in G^*. Then if G_0 and $\mathfrak{z}_0$ are sufficiently small, we can define ψ, φ and Ω_0 as in the proof of Lemma 4. Define a function g on Ω_0 as follows:

$$g(\varphi(x^*, Z)) = g_{\mathfrak{z}}(Z) \qquad\qquad (x^* \in G_0^*,\ Z \in \mathfrak{z}_0).$$

Since φ is an analytic diffeomorphism of $G_0^* \times \mathfrak{z}_0$ with Ω_0, g is obviously continuous. Fix $X \in \Omega_0 \cap \mathfrak{g}'$. We claim that $g(X) = F(X; \nabla_{\mathfrak{g}})$. Let $X = \varphi(x^*, H)$ $(x^* \in G_0^*, H \in \mathfrak{z}_0)$. Then it is clear that $g(X) = g(H)$. Similarly, since $\nabla_{\mathfrak{g}} F$ is invariant under G, it follows that $F(X; \nabla_{\mathfrak{g}}) = F(H; \nabla_{\mathfrak{g}})$. Hence it would be enough to show that $g(H) = F(H; \nabla_{\mathfrak{g}})$. Obviously H is regular in both $\mathfrak{g}$ and $\mathfrak{z}$. Let $\mathfrak{h}$ be the centralizer of H in $\mathfrak{z}$. Then $\mathfrak{h}$ is a Cartan subalgebra of $\mathfrak{z}$ and $H \in \mathfrak{h} \cap \Omega'$. Therefore, as we have seen above,

$$F(H; \nabla_{\mathfrak{g}}) = F(H; \partial(\pi_{\mathfrak{z}}) \circ \pi_{\mathfrak{z}} \circ \delta'_{\mathfrak{z}/\mathfrak{h}}(\partial(p) \circ p)).$$

Put $F'_{\mathfrak{z}} = (\partial(p) \circ p)F_{\mathfrak{z}}$. Since $F_{\mathfrak{z}}$ is invariant under Ξ and $\partial(p) \circ p \in \mathfrak{I}(\mathfrak{z}_c)$, it follows from [3($i$), Lemma 14] that

$$F'_{\mathfrak{z}}(H') = F_{\mathfrak{z}}(H'; \delta'_{\mathfrak{z}/\mathfrak{h}}(\partial(p) \circ p)) \qquad\qquad (H' \in \mathfrak{h}' \cap \Omega_{\mathfrak{z}}),$$

and therefore

$$g_{\mathfrak{z}}(H) = F'_{\mathfrak{z}}(H; \nabla_{\mathfrak{z}}) = F_{\mathfrak{z}}(H; \partial(\pi_{\mathfrak{z}}) \circ \pi_{\mathfrak{z}} \circ \delta'_{\mathfrak{z}/\mathfrak{h}}(\partial(p) \circ p))$$
$$= F(H; \nabla_{\mathfrak{g}})$$

from the definition of $\nabla_{\mathfrak{z}}$. This proves that $\nabla_{\mathfrak{g}} F = g$ on $\Omega_0 \cap \mathfrak{g}'$ and therefore $H_0 \in \Omega^0$.

So in order to complete the proof of Lemma 25, we may assume that $0 \in \Omega$. Then, by Lemma 10, $\mathcal{N} \subset \Omega$ and it follows from Corollary 1 of Lemma 8 that $\nabla_{\mathfrak{g}} F$ can be extended to a continuous function g on ([1]) $\Omega \cap {}^c \mathscr{N}$. Hence it would be sufficient to prove the following result.

Lemma 26. — There exists a number c with following property. If (X_k) $(k \geq 1)$ is a sequence in Ω' which converges to some element X in $\mathscr{N}$, then $g(X_k) \to c$.

Define $\mathfrak{h}_i$ and $\mathfrak{g}_i$ $(1 \leq i \leq r)$ as in the proof of Lemma 24 and $c(\mathfrak{h}_i)$ as in § 8. Then $c(\mathfrak{h}_1) = \ldots = c(\mathfrak{h}_r) = c$ (say) from the results of § 8. Since $\mathfrak{g}'$ is the union of $\mathfrak{g}_1, \ldots, \mathfrak{g}_r$ we can select, for each k, an index i_k and elements $x_k \in G$, $H_k \in \mathfrak{h}'_{i_k}$ such that $X_k = x_k H_k$. Since $X_k \to X$, it is clear (see the proof of [3(j), Lemma 23]) that $H_k \to 0$. Hence it follows from the definition of $\nabla_\mathfrak{g}$ and c that

$$g(H_k) = F(H_k; \nabla_\mathfrak{g}) \to c.$$

But since $g = \nabla_\mathfrak{g} F$ is invariant under G, $g(X_k) = g(H_k)$ and therefore $g(X_k) \to c$.

§ 10. A DIGRESSION

We shall now apply Theorem 3 to give a new proof of the main result of [3(e)]. We keep to the notation of Theorem 1.

Lemma 27. — Assume that F is locally constant on Ω'. Then T is locally constant on Ω.

Given any point $H_0 \in \Omega$, we have to show that T coincides with a constant around H_0. In view of Corollary 2 of Lemma 8, it would be sufficient to consider the case when H_0 is semisimple. However we first prove the following lemma.

Lemma 28. — There exists a number $a > 0$ such that

$$\partial(\varpi^\mathfrak{h})\pi^\mathfrak{h} = a$$

for every Cartan subalgebra $\mathfrak{h}$ of $\mathfrak{g}$.

Take $T = 1$ and $\Omega = \mathfrak{g}$ in Theorem 1. Then $\partial(\varpi^\mathfrak{h})g^\mathfrak{h} = \partial(\varpi^\mathfrak{h})\pi^\mathfrak{h}$ in the notation of Theorem 3. But $\partial(\varpi^\mathfrak{h})\pi^\mathfrak{h}$ is obviously a constant which we denote by $a(\mathfrak{h})$. Since zero belongs to every Cartan subalgebra $\mathfrak{h}$, it follows from Theorem 3 that $a(\mathfrak{h})$ is actually independent of $\mathfrak{h}$. Hence we may denote it by a. On the other hand we know (see [3(c)], p. 110]) that $a(\mathfrak{h}) > 0$. This proves the lemma ([2]).

Let us now return to the proof of Lemma 27. Since F is locally constant on Ω', it follows from Lemma 28 that $\nabla_\mathfrak{g} F = aF$. Therefore we conclude from Lemma 25 that F can be extended to a continuous function on Ω. Since $T = T_F$, this proves that T is locally constant on Ω.

Now we know that the distribution T' of [3(d), Lemma 30] is locally constant on $\mathfrak{g}'$ (see [3(d), p. 235]). Hence from Lemma 27, it is a constant. This gives a new proof of Lemma 17 of [3(j)].

([1]) ^{c}S denotes the complement of any set S.
([2]) It is obviously possible to give a direct proof of Lemma 28.

§ 11. **PROOF OF THEOREM 4**

In order to prove that the irreducible unitary characters of G are actually functions [3(g), Theorem 1], we have to develop a method of lifting our results from $\mathfrak{g}$ to G. The remainder of this paper is devoted to this task.

We use the notation of [3(i), Theorem 1].

Theorem 4. — Let Ω be a completely invariant open set in $\mathfrak{g}$ and T an invariant distribution on Ω. Let D be a differential operator in $\mathfrak{J}(\mathfrak{g}_c)$ such that $Dp = 0$ for all $p \in J(\mathfrak{g}_c)$. Then $DT = 0$.

We proceed by induction on dim $\mathfrak{g}$. Let $\mathfrak{c}$ be the center and $\mathfrak{g}_1$ the derived algebra of $\mathfrak{g}$ and first assume that $\mathfrak{c} \neq \{0\}$. Then $\mathfrak{J}(\mathfrak{g}_c) = \mathfrak{D}(\mathfrak{c}_c)\mathfrak{J}(\mathfrak{g}_{1c})$ (see [3(i), § 3]). Hence $D = \sum_{1 \leq i \leq r} \xi_i D_i$ where $\xi_i \in \mathfrak{D}(\mathfrak{c}_c)$, $D_i \in \mathfrak{J}(\mathfrak{g}_{1c})$ and $\xi_1, \ldots, \xi_r$ are linearly independent over **C**.

Fix a point $X_0 \in \Omega$ and let $X_0 = C_0 + Z_0$ ($C_0 \in \mathfrak{c}$, $Z_0 \in \mathfrak{g}_1$). Define $\mathfrak{c}_0$ and Ω_1 as in the proof of Lemma 25. Since $J(\mathfrak{g}_c) = P(\mathfrak{c}_c)J(\mathfrak{g}_{1c})$, we conclude that

$$\sum_i (\xi_i q)(D_i p_1) = 0$$

for all $q \in P(\mathfrak{c}_c)$ and $p_1 \in J(\mathfrak{g}_{1c})$. Fix $p_1 \in J(\mathfrak{g}_{1c})$. Then it follows from the above result that

$$\sum_i (D_i p_1)\xi_i = 0$$

in $\mathfrak{D}(\mathfrak{g}_c)$. Therefore we can conclude (see [3(i), § 3]) that $D_i p_1 = 0 \ (1 \leq i \leq r)$.

Now fix $\alpha \in C_c^\infty(\mathfrak{c}_0)$. Then if $\beta \in C_c^\infty(\Omega_1)$, we have

$$(DT)(\alpha \times \beta) = \sum_i T(\xi_i^* \alpha \times D_i^* \beta) = \sum_i (D_i T_i)(\beta)$$

where $T_i(\beta) = T(\xi_i^* \alpha \times \beta)$. Since dim $\mathfrak{g}_1 <$ dim $\mathfrak{g}$, Theorem 4 holds for (Ω_1, T_i, D_i) in place of (Ω, T, D) by the induction hypothesis. Hence $D_i T_i = 0$. In view of [3(h), Lemma 3] this shows that $DT = 0$ on $\mathfrak{c}_0 + \Omega_1$ and therefore $X_0 \notin \mathrm{Supp}\ DT$. Since X_0 was an arbitrary point in Ω, this proves that $DT = 0$.

Hence we may now assume that $\mathfrak{c} = \{0\}$ and therefore $\mathfrak{g}$ is semisimple. Let $H_0 \neq 0$ be a semisimple element in Ω. We intend to show that $H_0 \notin \mathrm{Supp}\ DT$. Let $\mathfrak{z}$ be the centralizer of H_0 in $\mathfrak{g}$. Define ζ and $\mathfrak{z}'$ as usual (see [3(i), § 2]) and let $\Omega_\mathfrak{z}$ be the set of all $Z \in \Omega \cap \mathfrak{z}$ such that $|\zeta(Z)| > |\zeta(H_0)|/2$. Then $\Omega_\mathfrak{z}$ is open and completely invariant in $\mathfrak{z}$. Take $G_0 = G$ and $\mathfrak{z}_0 = \Omega_\mathfrak{z}$ in [3(i), Lemma 17] and let σ_T be the corresponding distribution on $\Omega_\mathfrak{z}$. Let Ξ be the analytic subgroup of G corresponding to $\mathfrak{z}$. Then σ_T is invariant under Ξ (see Corollary 1 of [3(i), Lemma 17]). Now it follows from [3(i), Lemma 10 and Corollary 2 of Lemma 2] that $D_1 = \zeta^m \delta'_{\mathfrak{g}/\mathfrak{z}}(D) \in \mathfrak{J}(\mathfrak{z}_c)$, if m is a sufficiently large positive integer. Moreover

$$\sigma_{DT} = \delta'_{\mathfrak{g}/\mathfrak{z}}(D)\sigma_T$$

by Corollary 2 of [3(i), Lemma 17].   Fix a Cartan subalgebra $\mathfrak{h}$ of $\mathfrak{z}$.   Then

$$\delta'_{\mathfrak{z}/\mathfrak{h}}(D_1) = \zeta_{\mathfrak{h}}^m \delta'_{\mathfrak{z}/\mathfrak{h}}(D)$$

from [3(i), Lemma 11] where $\zeta_{\mathfrak{h}}$ is the restriction of ζ on $\mathfrak{h}$.   Moreover we know from [3(i), Theorem 1] that $\delta'_{\mathfrak{z}/\mathfrak{h}}(D)=0$.   Therefore, by applying [3(i), Theorem 1] to $\mathfrak{z}_c$ instead of $\mathfrak{g}_c$, we conclude that $D_1 p_1 = 0$ for all $p_1 \in J(\mathfrak{z}_c)$.   Therefore, since dim $\mathfrak{z} <$ dim $\mathfrak{g}$, we conclude from the induction hypothesis that $D_1 \sigma_T = \zeta^m \sigma_{DT} = 0$.   Since ζ is nowhere zero on $\Omega_{\mathfrak{z}}$, this implies that $\sigma_{DT} = 0$ and therefore $DT = 0$ around H_0.

In view of the above result, it follows from Corollary 1 of Lemma 8 that Supp $DT \subset \Omega \cap \mathcal{N}$.   Hence, in order to complete the proof of Theorem 4, we may assume that $\Omega \cap \mathcal{N} \neq \varnothing$.   But then $\mathcal{N} \subset \Omega$ from Lemma 10.

Lemma 29. — *We can select a function $f \in C^\infty(\mathfrak{g})$ such that:*

1) *f is invariant under* G;
2) *Supp $f \subset \Omega$;*
3) *$f = 1$ on some neighborhood of zero in* $\mathfrak{g}$;
4) *the distribution fT on $\mathfrak{g}$ is tempered.*

Notice that since Supp $f \subset \Omega$, the distribution $fT : g \to T(fg)$ $(g \in C_c^\infty(\mathfrak{g}))$ is well defined on $\mathfrak{g}$.   The proof of this lemma is rather long and therefore, in order not to interrupt our main line of argument, we shall postpone it until later (see § 19).

We have seen above that Supp $DT \subset \mathcal{N}$.   Choose an open neighborhood V of zero in Ω such that $f = 1$ on V.   Fix a point $X \in \mathcal{N}$.   Then, by Lemma 7, we can choose $y \in G$ such that $y^{-1}X \in V$.   Now T and fT are both invariant distributions which obviously coincide on V.   Hence they also coincide on V^y.   Therefore, in order to show that $DT = 0$ around X, it would be sufficient to verify that $D(fT) = 0$ around X.   This means that in order to complete the proof of Theorem 4, it is enough to prove that $D(fT) = 0$.   Therefore, replacing T by fT, we may now assume that T is an invariant and tempered distribution on $\mathfrak{g}$.   Moreover we know from the above proof that Supp $DT \subset \mathcal{N}$.

Define the space $\mathscr{C}(\mathfrak{g})$ as in [3(c), p. 91] and for any $f \in \mathscr{C}(\mathfrak{g})$ define its Fourier transform $\hat{f}$ by

$$\hat{f}(Y) = \int_{\mathfrak{g}} \exp((-1)^{1/2} B(Y, X)) f(X) dX \qquad (Y \in \mathfrak{g})$$

where dX is a fixed Euclidean measure on $\mathfrak{g}$ and $B(Y, X) = \mathrm{tr}(\mathrm{ad}\, Y \,\mathrm{ad}\, X)$ $(X, Y \in \mathfrak{g}_c)$ as usual.   If σ is any tempered distribution on $\mathfrak{g}$, its Fourier transform $\hat{\sigma}$ is also a tempered distribution on $\mathfrak{g}$ given by $\hat{\sigma}(f) = \sigma(\hat{f})$ $(f \in \mathscr{C}(\mathfrak{g}))$.   Since $f \to \hat{f}$ is a topological mapping of $\mathscr{C}(\mathfrak{g})$ onto itself, $\hat{\sigma} = 0$ implies that $\sigma = 0$.

As usual we identify $\mathfrak{g}_c$ with its dual under B and use the notation of [3(i), Lemma 12].   Then the mapping $\alpha : \Delta \to (\hat{\Delta})^*$ $(\Delta \in \mathfrak{D}(\mathfrak{g}_c))$ is an anti-automorphism of $\mathfrak{D}(\mathfrak{g}_c)$ and therefore α^2 is an automorphism.   However it is easy to check that α^2 leaves $\mathfrak{g}_c + \partial(\mathfrak{g}_c)$ fixed pointwise and therefore it must be the identity.   The relation

$\alpha^2\Delta = \Delta\ (\Delta\in\mathfrak{D}(\mathfrak{g}_c))$ implies that $\Delta^* = (\alpha\Delta)\hat{\ }$. On the other hand $(\alpha\Delta)^* = \hat{\Delta}$ from the definition of α. Therefore

$$(\Delta\sigma)\hat{\ }(f) = \sigma(\Delta^*\hat{f}) = \sigma(((\alpha\Delta)f)\hat{\ })$$
$$= \hat{\sigma}((\alpha\Delta)f) = (\hat{\Delta}\hat{\sigma})(f) \qquad (f\in\mathscr{C}(\mathfrak{g})),$$

for any tempered distribution σ. This proves that $(\Delta\sigma)\hat{\ } = \hat{\Delta}\hat{\sigma}$. Similarly, since B is invariant under G, $(f^x)\hat{\ } = (\hat{f})^x$ for $f\in\mathscr{C}(\mathfrak{g})$ and $x\in G$. Therefore $(\sigma^x)\hat{\ } = (\hat{\sigma})^x$.

Thus $\hat{T}$ is an invariant distribution on $\mathfrak{g}$ and $(DT)\hat{\ } = \hat{D}\hat{T}$. Fix a Cartan subalgebra $\mathfrak{h}$ of $\mathfrak{g}$. Then it follows from $[3(i),$ Theorem $1]$ and our hypothesis on D, that $\delta'_{\mathfrak{g}/\mathfrak{h}}(D) = 0$. Therefore we conclude from $[3(i),$ Lemma $13]$ that $\hat{D}\in\mathfrak{I}(\mathfrak{g}_c)$ and $\hat{D}p = 0$ for every $p\in J(\mathfrak{g}_c) = I(\mathfrak{g}_c)$. So the above proof is also applicable to $(\hat{D}, \hat{T})$ instead of (D, T). Hence $\text{Supp}\ \hat{D}\hat{T}\subset\mathscr{N}$.

Now put $\sigma = DT$ and fix an element $p\in I(\mathfrak{g}_c)$ such that p vanishes at zero. Then it follows from Lemma 7 that $p = 0$ on $\mathscr{N}$. Hence (see $[3(h),$ Lemma $21]$) we can choose an integer $m\geq 0$ such that $p^m\sigma\hat{\ } = 0$ around zero. Then, if we take $\Omega = \mathfrak{g}$ and $\Phi = \text{Supp}(p^m\sigma)$ in Corollary 1 of Lemma 8, we can conclude that $p^m\sigma = 0$. Choose a finite number of homogeneous elements $p_1, \ldots, p_l$ of positive degrees in $I(\mathfrak{g}_c)$ such that $I(\mathfrak{g}_c) = \mathbf{C}[p_1, \ldots, p_l]$. Fix an integer $m\geq 0$ such that $p_i^m\sigma = 0\ (1\leq i\leq l)$. Then, if $\mathfrak{B}$ is the ideal in $I(\mathfrak{g}_c)$ generated by $p_1^m, \ldots, p_l^m$, it is obvious that $\dim(I(\mathfrak{g}_c)/\mathfrak{B})\leq m^l$ and $v\sigma = 0$ for $v\in\mathfrak{B}$. On the other hand, by $[3(i),$ Lemmas 12 and $13]$, $\Delta\to\hat{\Delta}\ (\Delta\in\mathfrak{D}(\mathfrak{g}_c))$ is an automorphism of $\mathfrak{D}(\mathfrak{g}_c)$ of order 4 which maps $I(\mathfrak{g}_c)$ onto $\partial(I(\mathfrak{g}_c))$. Therefore $\hat{\mathfrak{B}}$ is an ideal in $\partial(I(\mathfrak{g}_c))$ and $\hat{v}\hat{\sigma} = (v\sigma)\hat{\ } = 0$ for $v\in\mathfrak{B}$. Hence we conclude from Theorem 1, applied to $\hat{\sigma}$ instead of T, that $\hat{\sigma}$ is a locally summable function on $\mathfrak{g}$. But $\hat{\sigma} = \hat{D}\hat{T}$ and therefore, as we have seen above, $\text{Supp}\ \hat{\sigma}\subset\mathscr{N}$. Since $\mathscr{N}$ is of measure zero in $\mathfrak{g}$, it follows that $\hat{\sigma} = 0$ and therefore $DT = \sigma = 0$. This proves Theorem 4.

§ 12. ANALYTIC DIFFERENTIAL OPERATORS

For applications we have to generalize Theorem 4 to the case when D is an analytic differential operator on Ω. For this we need some preparation.

Let E be a vector space over $\mathbf{R}$ of finite dimension, Ω a non-empty open subset of E and $\mathfrak{D}_\infty(\Omega : E)$ the algebra of all differential operators on Ω. Then any such operator D can be written in the form

$$D = \sum_{1\leq i\leq r} f_i\partial(p_i)$$

where $f_i\in C^\infty(\Omega)$ and $p_i\in S(E)$. For any $X\in\Omega$, D_X denotes, as usual the local expression of D at X (see $[3(c),$ p. $90]$) so that

$$D_X = \sum_i f_i(X)\partial(p_i).$$

Let $\mathscr{A}(\Omega)$ be the algebra of all analytic functions on Ω. Then $\mathscr{A}(\Omega)$ is a subalgebra of $C^{\infty}(\Omega)$. We denote by $\mathfrak{D}_a(\Omega : E)$ the subalgebra of $\mathfrak{D}_{\infty}(\Omega : E)$ generated by $\mathscr{A}(\Omega) \cup \partial(S(E))$. If Ω is empty, we define $\mathfrak{D}_{\infty}(\Omega : E) = \mathfrak{D}_a(\Omega : E) = \{o\}$.

Let Ω_1 be an open subset of Ω. Then we get a homomorphism

$$j : \mathfrak{D}_{\infty}(\Omega : E) \rightarrow \mathfrak{D}_{\infty}(\Omega_1 : E)$$

as follows. If $D \in \mathfrak{D}_{\infty}(\Omega : E)$, then $j(D)$ is the restriction of D on Ω_1. It is clear that j maps $\mathfrak{D}_a(\Omega : E)$ into $\mathfrak{D}_a(\Omega_1 : E)$. We say that an element $D \in \mathfrak{D}_{\infty}(\Omega : E)$ is analytic on Ω_1 if $j(D) \in \mathfrak{D}_a(\Omega : E)$. In particular D is analytic if $D \in \mathfrak{D}_a(\Omega : E)$.

§ 13. EXTENSION OF SOME RESULTS TO ANALYTIC DIFFERENTIAL OPERATORS

Now let $\mathfrak{g}$ be a reductive Lie algebra over **R** and Ω a non-empty open set in $\mathfrak{g}$. If Ω is invariant, G operates on $\mathfrak{D}_{\infty}(\Omega : \mathfrak{g})$ (see [3(h), § 5]). We denote by $\mathfrak{I}_{\infty}(\Omega : \mathfrak{g})$ the subalgebra consisting of all invariant elements and put $\mathfrak{I}_a(\Omega : \mathfrak{g}) = \mathfrak{I}_{\infty}(\Omega : \mathfrak{g}) \cap \mathfrak{D}_a(\Omega : \mathfrak{g})$.

Fix $\mathfrak{z}$ and define ζ and $\mathfrak{z}'$ as in [3(i), §§ 2, 7]. Put $\Omega_{\mathfrak{z}} = \Omega \cap \mathfrak{z}'$ and for any $D \in \mathfrak{D}_{\infty}(\Omega : \mathfrak{g})$ define an element $\Delta(D) \in \mathfrak{D}_{\infty}(\Omega_{\mathfrak{z}} : \mathfrak{z})$ as follows. Fix $Z \in \Omega_{\mathfrak{z}}$ and choose $p_Z \in S(\mathfrak{g}_c)$ such that $D_Z = \partial(p_Z)$. Then, corresponding to Corollary 1 of [3(i), Lemma 2], $\alpha_Z(p_Z) \in S(\mathfrak{z}_c)$. It follows from Corollary 2 of [3(i), Lemma 2] that there exists a unique element $\nabla \in \mathfrak{D}_{\infty}(\Omega_{\mathfrak{z}} : \mathfrak{z})$ such that $\nabla_Z = \partial(\alpha_Z(p_Z))$ for $Z \in \Omega_{\mathfrak{z}}$. We define $\Delta(D) = \nabla$. (In case $\Omega_{\mathfrak{z}}$ is empty, $\Delta(D) = o$ by definition.)

Let $\delta'_{\mathfrak{g}/\mathfrak{z}}$ denote the mapping $D \rightarrow \Delta(D)$ of $\mathfrak{D}_{\infty}(\Omega : \mathfrak{g})$ into $\mathfrak{D}_{\infty}(\Omega_{\mathfrak{z}} : \mathfrak{z})$ (cf. [3(i), § 4]).

Lemma 30. — $\delta'_{\mathfrak{g}/\mathfrak{z}}$ *maps* $\mathfrak{D}_a(\Omega : \mathfrak{g})$ *into* $\mathfrak{D}_a(\Omega_{\mathfrak{z}} : \mathfrak{z})$. *Moreover, if* Ω *is invariant,* $\delta'_{\mathfrak{g}/\mathfrak{z}}$ *maps* $\mathfrak{I}_{\infty}(\Omega : \mathfrak{g})$ *into* $\mathfrak{I}_{\infty}(\Omega_{\mathfrak{z}} : \mathfrak{z})$.

The first statement is obvious from Corollary 2 of [3(i), Lemma 2]. Moreover, if Ω is invariant, then $\Omega_{\mathfrak{z}}$ is invariant in $\mathfrak{z}$ and the second assertion follows from [3(i), Lemma 3].

Let $\mathfrak{h}$ be a Cartan subalgebra of $\mathfrak{z}$.

Lemma 31. — $\delta'_{\mathfrak{g}/\mathfrak{h}}(D) = \delta'_{\mathfrak{z}/\mathfrak{h}}(\delta'_{\mathfrak{g}/\mathfrak{z}}(D))$ *for* $D \in \mathfrak{D}_{\infty}(\Omega : \mathfrak{g})$.

The proof of this is the same as that of [3(i), Lemma 11].

Lemma 32. — *Let f be a locally invariant* C^{∞} *function on an open subset* Ω_0 *of* Ω. *Then*

$$f(Z; D) = f(Z; \delta'_{\mathfrak{g}/\mathfrak{z}}(D))$$

for $Z \in \Omega_0 \cap \mathfrak{z}'$ *and* $D \in \mathfrak{D}_{\infty}(\Omega : \mathfrak{g})$.

This is proved in the same way as Lemma 14 of [3(i)].

Lemma 33. — *Let D be a differential operator on an open subset* Ω *of* $\mathfrak{g}$. *Then the following two conditions on D are equivalent.*

1) *For every Cartan subalgebra* $\mathfrak{h}$ *of* $\mathfrak{g}$, $\delta'_{\mathfrak{g}/\mathfrak{h}}(D) = o$.

2) *If* Ω_0 *is an open subset of* Ω *and f a locally invariant* C^{∞} *function on* Ω_0, *then* $Df = o$.

Assume 1) holds and let f be a locally invariant C^∞ function on Ω_0. Since $\Omega_0' = \Omega_0 \cap \mathfrak{g}'$ is dense in Ω_0, it is enough to verify that $Df = 0$ on Ω_0'. Fix $H_0 \in \Omega_0'$ and let $\mathfrak{h}$ be the centralizer of H_0 in $\mathfrak{g}$. Then $\mathfrak{h}$ is a Cartan subalgebra of $\mathfrak{g}$ and since f is locally invariant, it follows from Lemma 32 that $f(H_0; D) = f(H_0; \delta_{\mathfrak{g}/\mathfrak{h}}'(D)) = 0$. This proves that $Df = 0$ on Ω_0'.

Conversely assume that 2) holds. Fix a Cartan subalgebra $\mathfrak{h}$ of $\mathfrak{g}$ and a point $H_0 \in \Omega \cap \mathfrak{h}'$. Let A be the Cartan subgroup of G corresponding to $\mathfrak{h}$. We now use the notation of the proof of Lemma 1. Then φ defines an analytic diffeormorphism of $G_0^* \times \mathfrak{h}_0$ with Ω_0. Fix $\beta \in C^\infty(\mathfrak{h}_0)$ and define $f \in C^\infty(\Omega_0)$ by the relation $f(x^* H) = \beta(H)$ $(x^* \in G_0^*, H \in \mathfrak{h}_0)$. Then it is obvious that f is locally invariant and therefore

$$0 = f(H; D) = f(H; \delta_{\mathfrak{g}/\mathfrak{h}}'(D)) = \beta(H; \delta_{\mathfrak{g}/\mathfrak{h}}'(D)) \qquad (H \in \mathfrak{h}_0)$$

Since β was an arbitrary element of $C^\infty(\mathfrak{h}_0)$, this implies that $\delta_{\mathfrak{g}/\mathfrak{h}}'(D) = 0$ on $\mathfrak{h}_0$. Hence in particular $(\delta_{\mathfrak{g}/\mathfrak{h}}'(D))_{H_0} = 0$. This shows that 2) implies 1).

Corollary. — *Assume Ω is invariant. Then either one of the two conditions above is equivalent to the following.*

3) *For every invariant function f in $C^\infty(\Omega)$, $Df = 0$.*

Obviously 2) implies 3). Now assume 3) holds. Fix a Cartan subalgebra $\mathfrak{h}$ of $\mathfrak{g}$ and a point $H_0 \in \Omega \cap \mathfrak{h}'$. Let $\mathfrak{h}_0$ be an open neighborhood of H_0 in $\mathfrak{h}' \cap \Omega$. We assume that $\mathfrak{h}_0$ is relatively compact in $\mathfrak{h}'$ and $s\mathfrak{h}_0 \cap \mathfrak{h}_0 = \varnothing$ for $s \neq 1$ in W_G (see § 9 for the definition of W_G). Fix $\beta_0 \in C_c^\infty(\mathfrak{h}_0)$ and put

$$\beta(H) = \sum_{s \in W_G} \beta_0(sH) \qquad (H \in \mathfrak{h}).$$

Then $\beta^s = \beta$ $(s \in W_G)$ and the mapping $\varphi : G^* \times \mathfrak{h}' \to \mathfrak{g}$ is everywhere regular. Put $\mathfrak{g}_\mathfrak{h} = \varphi(G^* \times \mathfrak{h}') = (\mathfrak{h}')^G$. The group W_G operates on $G^* \times \mathfrak{h}'$ on the right as follows:

$$(x^*, H)s = (x^*s, s^{-1}H) \qquad (s \in W_G)$$

in the notation of the proof of Lemma 24. Since no point of $G^* \times \mathfrak{h}'$ is left fixed by s if $s \neq 1$, it follows that φ defines an analytic diffeomorphism of the quotient manifold $(G^* \times \mathfrak{h}')/W_G$ with $\mathfrak{g}_\mathfrak{h}$. Now define a function F on $G^* \times \mathfrak{h}'$ by

$$F(x^* : H) = \beta(H) \qquad (x^* \in G^*, H \in \mathfrak{h}').$$

Then $F(x^*s : s^{-1}H) = \beta(s^{-1}H) = \beta(H) = F(x^* : H)$ and therefore F defines a C^∞ function f on $\mathfrak{g}_\mathfrak{h}$. Since $\mathfrak{h}_0$ is relatively compact in $\mathfrak{h}'$ it follows from [3(j), Lemma 7] that $\mathrm{Cl}(\mathfrak{h}_0^G) \subset \mathfrak{g}_\mathfrak{h}$ and therefore we can extend f to a C^∞ function on $\mathfrak{g}$ by defining it to be zero outside $\mathfrak{g}_\mathfrak{h}$. Then it is clear that f is invariant and therefore $Df = 0$ on Ω by 3). But

$$f(H; D) = f(H; \delta_{\mathfrak{g}/\mathfrak{h}}'(D)) = \beta_0(H; \delta_{\mathfrak{g}/\mathfrak{h}}'(D)) \qquad (H \in \mathfrak{h}_0)$$

because $f = \beta = \beta_0$ on $\mathfrak{h}_0$. Since β_0 was arbitrary in $C_c^\infty(\mathfrak{h}_0)$, this shows that $(\delta_{\mathfrak{g}/\mathfrak{h}}'(D))_{H_0} = 0$. Therefore 3) implies 1) and the corollary is proved.

§ 14. PROOF OF THEOREM 5

We shall now prove the following generalization of Theorem 4.

Theorem 5. — Let Ω be a completely invariant open set in $\mathfrak{g}$ and T an invariant distribution on Ω. Let D be an analytic and invariant differential operator on Ω such that $Df = 0$ for every invariant C^∞ function f on Ω. Then $DT = 0$.

We again use induction on $\dim \mathfrak{g}$. Define $\mathfrak{c}$ and $\mathfrak{g}_1$ as in § 4 and fix a semisimple element $H_0 \in \Omega$ such that $H_0 \notin \mathfrak{c}$. We shall prove that $DT = 0$ around H_0. Let $\mathfrak{z}$ denote the centralizer of H_0 in $\mathfrak{g}$ and define ζ and $\mathfrak{z}'$ as in [3(i), § 2]. Then $\Omega_{\mathfrak{z}} = \Omega \cap \mathfrak{z}'$ is an open and completely invariant set in $\mathfrak{z}$. Let σ_T and σ_{DT} be the distributions on $\Omega_{\mathfrak{z}}$ corresponding to T and DT respectively under [3(i), Lemma 17] with $G_0 = G$ and $\mathfrak{z}_0 = \Omega_{\mathfrak{z}}$. Then it would be enough to show that $\sigma_{DT} = 0$. However it is easy to prove (cf. Corollary 2 of [3(i), Lemma 17]) that $\sigma_{DT} = \Delta \sigma_T$ where $\Delta = \delta'_{\mathfrak{g}/\mathfrak{z}}(D)$. Now σ_T is an invariant distribution on $\Omega_{\mathfrak{z}}$ (see Corollary 1 of [3(i), Lemma 17]) and $\Delta \in \mathfrak{I}_a(\Omega_{\mathfrak{z}} : \mathfrak{z})$ by Lemma 30. Therefore since $\dim \mathfrak{z} < \dim \mathfrak{g}$, it follows by induction hypothesis that $\Delta \sigma_T = 0$ (see Lemma 31 and the corollary of Lemma 33).

Now fix $C_0 \in \mathfrak{c} \cap \Omega$. We claim that $T = 0$ around C_0. Applying the translation by $-C_0$ to the whole problem, we are reduced to the case when $C_0 = 0$. Let

$$D = \sum_{1 \le i \le r} a_i \, \partial(p_i)$$

where $p_1, \ldots, p_r$ are linearly independent homogeneous elements in $S(\mathfrak{g}_c)$ and $a_1, \ldots, a_r$ are analytic functions on Ω. Let V be the subspace of $S(\mathfrak{g}_c)$ spanned by $p_i^x \, (1 \le i \le r, x \in G)$. Then obviously $\dim V < \infty$ and we may, without loss of generality, assume that $(p_1, \ldots, p_r)$ is a base for V. Then

$$p_i^x = \sum_j c_{ji}(x) p_j \qquad\qquad (x \in G)$$

where c_{ji} are analytic functions on G. Since $D = D^x$, it follows that $D_{xX} = (D_X)^x$ and therefore

$$\sum_i a_i(xX) \, \partial(p_i) = \sum_i a_i(X) \, \partial(p_i^x) \qquad\qquad (X \in \Omega).$$

This shows that

$$a_i(xX) = \sum_j c_{ij}(x) a_j(X) \qquad\qquad (x \in G, X \in \Omega).$$

For any integer m, let $\mathfrak{D}_m$ denote the subspace of $\mathfrak{D}(\mathfrak{g}_c)$ spanned by elements of the form $p \partial(q)$ where p and q are homogeneous elements in $P(\mathfrak{g}_c)$ and $S(\mathfrak{g}_c)$ respectively and $\deg p - \deg q = m$. Then if $\Delta \in \mathfrak{D}_m$ and Q is a homogeneous polynomial function on $\mathfrak{g}$, it is clear that ΔQ is homogeneous and

$$\deg(\Delta Q) = \deg Q + m.$$

Choose an open and convex neighborhood Ω_0 of zero in Ω such that each a_i ($1 \leq i \leq r$) can be expanded in a power series around zero, which converges absolutely on Ω_0. Then

$$a_i(X) = \sum_{v \geq 0} q_{vi}(X) \qquad (X \in \Omega_0)$$

where q_{vi} is a homogeneous polynomial function on g of degree v. It is obvious from our result above that

$$q_{vi}(xX) = \sum_{1 \leq j \leq r} c_{ij}(x) q_{vj}(X) \qquad (x \in G,\ X \in \Omega_0)$$

and therefore

$$_vD = \sum_i q_{vi} \partial(p_i)$$

lies in $\mathfrak{I}(g_c)$. On the other hand it is clear that $\mathfrak{D}(g_c)$ is the direct sum of $\mathfrak{D}_m$ for all m $(-\infty < m < \infty)$ and each $\mathfrak{D}_m$ is stable under G. Let $_vD_m$ denote the component of $_vD$ in $\mathfrak{D}_m$ is this direct sum. Then it is clear that $_vD_m \in \mathfrak{I}(g_c)$. Moreover $_vD_m \neq 0$ implies that $v = m + \deg p_i$ for some i. Hence if $m_0 = \sup_i \deg p_i$, it follows that $_vD_m = 0$ for $v > m + m_0$. Put

$$D_m = \sum_{v \geq 0} {_vD_m}.$$

Then $D_m \in \mathfrak{I}(g_c) \cap \mathfrak{D}_m$. Moreover if p is a homogeneous element in $J(g_c)$, then by hypothesis

$$0 = Dp = \sum_{m \geq -m_0} D_m p$$

on Ω_0. Since $D_m p$ is homogeneous of degree $m + \deg p$, it is clear that $D_m p = 0$. Therefore $D_m T = 0$ by Theorem 4.

Now fix $f \in C_c^\infty(\Omega_0)$. It is clear that for any $p \in S(g_c)$, the series

$$\sum_{v \geq 0} |\partial(p) q_{vi}| \qquad (1 \leq i \leq r)$$

converge uniformly on any compact subset of Ω_0. Hence it follows without difficulty that the series

$$\sum_{m \geq -m_0} D_m^* f$$

converges in $C_c^\infty(\Omega_0)$ to $D^* f$. (Here the star denotes adjoint, as usual.) This implies that the series

$$\sum_{m \geq -m_0} T(D_m^* f)$$

converges to $T(D^* f)$. But $T(D_m^* f) = 0$ since $D_m T = 0$. Therefore $T(D^* f) = 0$. This means that $DT = 0$ on Ω_0.

The above proof shows that Supp DT contains no semisimple element of Ω.

Hence it follows from Corollary 1 of Lemma 8 that $DT = 0$. This completes the proof of Theorem 5.

Remark. — I do not know whether Theorem 5 continues to hold when the condition of analyticity of D is dropped.

§ 15. SOME PREPARATION FOR THE PROOF OF LEMMA 29

Let G be a connected semisimple Lie group with a faithful finite-dimensional representation, $\mathfrak{g}$ its Lie algebra over **R**, θ a Cartan involution of $\mathfrak{g}$ and $\mathfrak{g} = \mathfrak{k} + \mathfrak{p}$ the corresponding Cartan decomposition. Let $\mathfrak{a}$ be a maximal abelian subspace of $\mathfrak{p}$. We introduce an order in the space of all (real) linear functions on $\mathfrak{a}$ and denote by Σ the set of all positive roots of $(\mathfrak{g}, \mathfrak{a})$ (see [3(f), p. 244]). Let $\mathfrak{a}^+$ be the set of all points $H \in \mathfrak{a}$ where $\alpha(H) \geq 0$ for every $\alpha \in \Sigma$. Put $A = \exp \mathfrak{a}$ and $A^+ = \exp(\mathfrak{a}^+)$ in G. The exponential mapping from $\mathfrak{a}$ to A is bijective. We denote its inverse by log. Introduce a partial order in A as follows. Given two elements h_1, h_2 in A, we write $h_1 \succ h_2$ if $h_1 h_2^{-1} \in A^+$. Let $l = \dim \mathfrak{a}$. Then we can choose a simple system of roots $\alpha_1, \ldots, \alpha_l$ in Σ (see [3(d), Lemma 1]).

Lemma 34. — *Fix some norm* ν *on the finite-dimensional space* $\mathfrak{g}$. *Then for any number* $a \geq 0$, *we can choose two numbers* b, c $(b \geq a, c \geq 1)$ *with the following property. Suppose* $X \in \mathfrak{g}$, $h \in A^+$ *and* $\nu(X) \leq a$. *Then there exist elements* $X_0 \in \mathfrak{g}$ *and* $h_0 \in A^+$ *such that*

1) $X^h = X_0^{h_0}$, $\nu(X_0) \leq b$, $1 \prec h_0 \prec h$;

2) $\max_{1 \leq i \leq l} \exp \alpha_i(\log h_0) \leq c(1 + \nu(X^h))^l$.

(In case $l = 0$, $\max_i \exp \alpha_i(\log h_0)$ should be taken to mean 1.) We shall give a proof of this lemma in § 20.

As usual put $B(X, Y) = \operatorname{tr}(\operatorname{ad} X \operatorname{ad} Y)$ $(X, Y \in \mathfrak{g})$. Then the quadratic form

$$||X||^2 = -B(X, \theta(X)) \qquad\qquad (X \in \mathfrak{g})$$

is positive-definite and defines the structure of a real Hilbert space on $\mathfrak{g}$. For any $a > 0$, let ω_a denote the set of all $X \in \mathfrak{g}$ with $||X|| < a$ and put $\Omega_a = (\omega_a)^G$.

Lemma 35. — *Suppose* $a > b > 0$. *Then* $\operatorname{Cl}\Omega_b \subset \Omega_a$.

We shall give a proof of this in § 21.

Corollary. — Ω_a *is an open and completely invariant subset of* $\mathfrak{g}$.

It is obvious that Ω_a is open and invariant. Fix $X_0 \in \Omega_a$. Then $X_0 = Y_0^{x_0}$ where $||Y_0|| < a$ and $x_0 \in G$. Choose b such that $||Y_0|| < b < a$. Then $X_0 \in \Omega_b$ and $\operatorname{Cl}(\Omega_b) \subset \Omega_a$ by Lemma 35. This shows that every point of Ω_a has an open invariant neighborhood whose closure (in $\mathfrak{g}$) is contained in Ω_a. From this it is clear that Ω_a is completely invariant.

For any linear transformation T in $\mathfrak{g}$, let T^* denote its adjoint (in the sense of Hilbert-space theory). Put

$$||x||^2 = \operatorname{tr}(\operatorname{Ad}(x)^* \operatorname{Ad}(x)) \qquad\qquad (x \in G).$$

Choose a base $(H_1, \ldots, H_l)$ for $\mathfrak{a}$ over $\mathbf{R}$ dual to $(\alpha_1, \ldots, \alpha_l)$ so that

$$\alpha_i(H_j) = \delta_{ij} \qquad (1 \leq i, j \leq l)$$

and define $m(\alpha) = \sum_i \alpha(H_i)$ for $\alpha \in \Sigma$. Since $H_i \in \mathfrak{a}^+$, it is clear that $m(\alpha)$ is a positive integer.

Lemma 36. — Given $a > 0$, we can choose numbers b, c $(b \geq a, c \geq 1)$ such that the following condition holds. For any $X \in \Omega_a$, we can select $x \in G$ such that

$$1) \quad \|X^{x^{-1}}\| \leq b, \qquad 2) \quad \|x\| \leq c(1 + \|X\|)^m$$

where $m = l \max\limits_{\alpha \in \Sigma} m(\alpha)$.

(If $l = 0$ then $m = 0$ by definition.) Choose b_0, c_0 such that Lemma 34 holds for (b_0, c_0) instead of (b, c) with the norm $\nu(Z) = \|Z\|$ $(Z \in \mathfrak{g})$. Let K be the analytic subgroup of G corresponding to $\mathfrak{k}$. Then K is compact, $G = KA^+K$ and $\mathrm{Ad}(k)$ is unitary for $k \in K$. Hence if $x = k_1 h k_2$ $(k_1, k_2 \in K, h \in A^+)$, it follows that $\|x\| = \|h\|$. However $\mathrm{Ad}(h)$ is self-adjoint [1] and its eigenvalues are 1 and $e^{\pm \alpha(\log h)}$ $(\alpha \in \Sigma)$. Since $\alpha(\log h) \geq 0$, it is clear that

$$\|h\| \leq n^{1/2} \max_{\alpha \in \Sigma} e^{\alpha(\log h)}$$

where $n = \dim \mathfrak{g}$. But $\alpha = \sum\limits_{1 \leq i \leq} m_i \alpha_i$ where $m_i = \alpha(H_i)$ are integers ≥ 0. Hence

$$\alpha(\log h) \leq m(\alpha) \max_i \alpha_i(\log h) \leq m_0 \max_i \alpha_i(\log h)$$

where $m_0 = \max\limits_{\alpha \in \Sigma} m(\alpha)$. Therefore

$$\|h\| \leq n^{1/2} \left(\max_i e^{\alpha_i(\log h)} \right)^{m_0}$$

(This holds also if $l = 0$. We define $m_0 = 0$ in that case.)

Now fix $X \in \Omega_a$ and choose $Y \in \mathfrak{g}$, $y \in G$ such that $\|Y\| < a$ and $X = Y^y$. Let $y = k_1 h k_2$ $(k_1, k_2 \in K, h \in A^+)$. Replacing (Y, y) by $(Y^{k_2}, k_1 h)$, we can assume that $y = k_1 h$. Select $Y_0 \in \mathfrak{g}$ and $h_0 \in A^+$ such that $Y^h = Y_0^{h_0}$, $\|Y_0\| \leq b_0$, $1 \prec h_0 \prec h$ and

$$\max_i e^{\alpha_i(\log h_0)} \leq c_0(1 + \|Y^h\|)^l.$$

This is possible from the definition of b_0, c_0. Then

$$\|h_0\| \leq n^{1/2} c_0^{m_0}(1 + \|Y^h\|)^m.$$

Now put $x = k_1 h_0$. Then $X = Y^y = Y_0^x$ and therefore $\|X^{x^{-1}}\| \leq b_0$. Moreover

$$\|x\| = \|h_0\| \leq n^{1/2} c_0^{m_0}(1 + \|X\|)^m.$$

Therefore we can take $b = b_0$ and $c = n^{1/2} c_0^{m_0}$ in the lemma.

---

[1] The facts stated here are all well known. They can be found in [3(a)].

§ 16. SOME INEQUALITIES

For $t \geq 1$, let $G(t)$ denote the set of all $x \in G$ with $||x|| \leq t$. Then $G(t)$ is obviously compact.

Lemma 37. — Let μ denote the Haar measure of G. Then there exists a number $c > 0$ and an integer $M \geq 0$ such that

$$\mu(G(t)) \leq ct^M$$

for $t \geq 1$.

The statement is trivial if G is compact. Hence we may assume that $l \geq 1$. Put $A^+(t) = A^+ \cap G(t)$. Then it is clear that $G(t) = KA^+(t)K$ and therefore (see [3(a), Lemma 22])

$$\mu(G(t)) = \int_{A^+(t)} D(h)dh$$

where dh is the (suitably normalized) Haar measure on A,

$$D(h) = \prod_{\alpha \in \Sigma} (e^{\alpha(\log h)} - e^{-\alpha(\log h)})^{m_\alpha} \qquad\qquad (h \in A)$$

and m_α is the multiplicity of α (m_α is the dimension of the space $\mathfrak{g}_\alpha$ consisting of all $X \in \mathfrak{g}$ such that $[H, X] = \alpha(H)X$ for all H in $\mathfrak{a}$.) Put $2\rho = \sum_{\alpha \in \Sigma} m_\alpha \alpha$. Then it is obvious that

$$D(h) \leq e^{2\rho(\log h)} \qquad\qquad (h \in A^+).$$

Now $2\rho = \sum_{1 \leq i \leq l} m_i \alpha_i$ where m_i are positive integers. Put $\tau_i = \alpha_i(\log h)$. Then $dh = c_1 d\tau_1 \ldots d\tau_l$ where c_1 is a positive constant and

$$e^{2\rho(\log h)} = \exp(m_1 \tau_1 + \ldots + m_l \tau_l).$$

Now if $h \in A^+(t)$, we have

$$1 \leq e^{\alpha_i(\log h)} \leq || h || \leq t$$

and therefore $0 \leq \tau_i \leq \log t$. Hence

$$\mu(G(t)) \leq \int_{A^+(t)} e^{2\rho(\log h)} dh \leq ct^M$$

where $c = c_1/(m_1 m_2 \ldots m_l)$ and $M = m_1 + \ldots + m_l$.

Lemma 38. — There exists a compact neighborhood U of 1 in G and two constants $a_1, c_1 > 0$ with the following property. For any $t \geq 1$, we can choose a finite set of points x_i ($1 \leq i \leq N(t)$) in G such that

1) $G(t) \subset \bigcup_i x_i U$;
2) $||x_i|| \leq a_1 t$;
3) $N(t) \leq c_1 t^M$.

By a theorem of Borel [1, Theorem C], there exists a discrete subgroup Γ of G such that $\Gamma \backslash G$ is compact. Choose a compact neighborhood U of 1 in G such that $U = U^{-1}$ and $G = \Gamma U$. Put

$$\Gamma(t) = \Gamma \cap (G(t)U).$$

Select a compact neighborhood $V = V^{-1}$ of 1 in U such that $V^2 \cap \Gamma = \{1\}$ $(V^2 = VV)$. Then the union

$$\underset{\gamma \in \Gamma(t)}{U} \gamma V = \Gamma(t) V$$

is disjoint and

$$\Gamma(t) V \subset G(t) U V \subset G(t) U^2.$$

Choose $a_1 \geq 1$ so large that $U^2 \subset G(a_1)$. Then

$$\Gamma(t) V \subset G(t) G(a_1) \subset G(ta_1)$$

since $\|xy\| \leq \|x\| . \|y\|$ $(x, y \in G)$. Hence

$$\mu(\Gamma(t) V) \leq \mu(G(ta_1)) \leq ca_1^M t^M$$

from Lemma 37. But since the above union was disjoint,

$$\mu(\Gamma(t) V) = N(t) \mu(V)$$

where $N(t)$ is the number of elements in $\Gamma(t)$. Hence $N(t) \leq c_1 t^M$ where $c_1 = ca_1^M / \mu(V)$. Let $x_i (1 \leq i \leq N(t))$ be all the elements of $\Gamma(t)$. Since $\Gamma(t) \subset G(t) U \subset G(ta_1)$, it follows that $\|x_i\| \leq a_1 t$. Finally since $G = \Gamma U$, it is obvious that

$$G(t) \subset \Gamma(t) U = \underset{i}{U} x_i U.$$

§ 17. PROOF OF LEMMA 39

Fix a number $a > 0$ and let $\Omega = \Omega_a$ in the notation of Lemma 35. For $0 \leq s < t$, let $\Omega(s, t)$ denote the set of all $X \in \Omega$ with $s \leq \|X\| < t$. Also put $\Omega(t) = \Omega(0, t)$.

Lemma 39. — *Let* T *be an invariant distribution on* $\mathfrak{g}$. *Then there exist elements* $p_1, \ldots, p_r$ *in* $S(\mathfrak{g}_c)$ *and an integer* $\nu \geq 0$ *such that*

$$|T(f)| \leq (1+t)^\nu \underset{1 \leq i \leq r}{\Sigma} \sup |\partial(p_i) f|$$

for all $f \in C_c^\infty(\Omega(t))$ *and* $t > 0$.

This requires some preparation. As before let $\omega_t (t > 0)$ be the set of all points $X \in \mathfrak{g}$ with $\|X\| < t$.

Lemma 40. — *Define* b, c *and* m *as in Lemma 36 and for any* $t \geq 0$ *let* G_t *denote the set of all* $x \in G$ *with* $\|x\| \leq c(1+t)^m$. *Then* $\omega_b^{G_t} \supset \Omega(t)$.

This is obvious from Lemma 36.

Define U and M as in Lemma 38.

Lemma 41. — *There exist two numbers* $c_1, c_2 > 0$ *with the following property. For any* $t > 0$, *we can choose a finite set* F_t *of points in* G *such that:*

    1) $G_t \subset F_t U$;

    2) $\|x\| \leq c_1(1+t)^m$ *for* $x \in F_t$;

    3) $[F_t] \leq c_2(1+t)^{mM}$.

Here $[F_t]$ *denotes the number of elements in* F_t.

This follows immediately from Lemma 38 if we note that $G_t = G(t')$ where $t' = c(1+t)^m$.

Now choose $b_1 > b$ such that $\mathrm{Cl}(\omega_b)^U \subset \omega_{b_1}$ and fix $\alpha \in C_c^\infty(\omega_{b_1})$ such that $0 \leq \alpha \leq 1$ and $\alpha = 1$ on ω_b^U. For any $t > 0$, put

$$\varphi_t = \sum_{x \in F_t} \alpha^x.$$

Since $\Omega(t) \subset \omega_b^{G_t} \subset (\omega_b^U)^{F_t}$ and $\alpha = 1$ on ω_b^U, it is clear that $\varphi_t \geq 1$ on $\Omega(t)$. Put $\alpha_x = \alpha^x / \varphi_t$ on $\Omega(t)$ $(x \in F_t)$.

Lemma 42. — Given $p \in S(\mathfrak{g}_c)$, we can choose a number $c(p) \geq 0$ and an integer $m(p) \geq 0$ such that

$$\sup |\partial(p)\alpha^x| \leq c(p)(1+t)^{m(p)}$$

for $x \in F_t$ and $t > 0$.

Let V be the subspace of $S(\mathfrak{g}_c)$ spanned by p^y $(y \in G)$ and let $p_1, \ldots, p_r$ be a base for V. Then

$$p^y = \sum_{1 \leq i \leq r} a_i(y) p_i$$

where a_i are analytic functions on G. We can choose $c' \geq 0$ and an integer $\nu \geq 0$ such that (see [1] [3(d), p. 203])

$$|a_i(y)| \leq c' \|y\|^\nu \qquad\qquad (y \in G,\ 1 \leq i \leq r).$$

Then

$$\partial(p)\alpha^x = (\partial(p^{x^{-1}})\alpha)^x = \sum_i a_i(x^{-1})(\partial(p_i)\alpha)^x.$$

Hence

$$\sup |\partial(p)\alpha^x| \leq c_0 \|x^{-1}\|^\nu$$

where $c_0 = c' \sum_i \sup |\partial(p_i)\alpha|$. If $x = k_1 h k_2$ $(k_1, k_2 \in K, h \in A)$, it is obvious that $\|x\| = \|h\|$. Moreover θ is a unitary transformation of $\mathfrak{g}$ and therefore since $\theta \mathrm{Ad}(h)\theta^{-1} = \mathrm{Ad}(h^{-1})$, it is clear that $\|h\| = \|h^{-1}\| = \|x^{-1}\|$. This shows that $\|x^{-1}\| = \|x\|$ and therefore

$$\sup |\partial(p)\alpha^x| \leq c_0 \|x\|^\nu \leq c_0 c_1^\nu (1+t)^{m\nu}$$

since $\|x\| \leq c_1(1+t)^m$ for $x \in F_t$. So we can take $c(p) = c_0 c_1^\nu$ and $m(p) = m\nu$.

Corollary 1. — $\sup |\partial(p)\varphi_t| \leq c(p) c_2(1+t)^{m(p)+mM}$ $(t > 0)$.

This is obvious since $[F_t] \leq c_2(1+t)^{mM}$.

Corollary 2. — Given $p \in S(\mathfrak{g}_c)$, we can choose $c'(p) \geq 0$ and an integer $\mu(p) \geq 0$ such that

$$\sup_{\Omega(t)} |\partial(p)\alpha_x| \leq c'(p)(1+t)^{\mu(p)}$$

for $x \in F_t$ and $t > 0$.

Since $\alpha_x = \alpha^x / \varphi_t$ and $\varphi_t \geq 1$ on $\Omega(t)$, this is an immediate consequence of Lemma 42 and Corollary 1 above.

---

(1) The proof of Lemma 6 of [3(d)] is clearly independent of the assumption that rank $\mathfrak{g}$ = rank $\mathfrak{k}$ which was made at the beginning of § 3 of [3(d)].

Now we come to the proof of Lemma 39.   Put $f_x = \alpha_x f$ $(x \in F_t)$.   Since $\sum\limits_{x \in F_t} \alpha_x = 1$ on $\Omega(t)$, it is obvious that

$$f = \sum_{x \in F_t} f_x$$

and therefore

$$T(f) = \sum_{x \in F_t} T(f_x).$$

But $T(f_x) = T((f_x)^{x^{-1}})$, since T is invariant.   Moreover

$$\operatorname{Supp} f_x \subset \operatorname{Supp} f \cap \operatorname{Supp} \alpha_x \subset \operatorname{Supp} \alpha^x.$$

Hence

$$\operatorname{Supp}(f_x)^{x^{-1}} \subset \operatorname{Supp} \alpha \subset \omega_{b_1}.$$

Since ω_{b_1} is relatively compact in $\mathfrak{g}$, we can select $p_1, \ldots, p_r \in S(\mathfrak{g}_c)$ such that

$$|T(g)| \leq \sum_{1 \leq i \leq r} \sup |\partial(p_i) g| \qquad (g \in C_c^\infty(\omega_{b_1})).$$

Therefore

$$|T(f_x)| = |T((f_x)^{x^{-1}})| \leq \sum_i \sup |\partial(p_i) f_x^{x^{-1}}|.$$

But

$$\sup |\partial(p_i) f_x^{x^{-1}}| = \sup |\partial(p_i^x) f_x|.$$

Let V be the subspace of $S(\mathfrak{g}_c)$ spanned by p_i^y $(y \in G, 1 \leq i \leq r)$ and let q_j $(1 \leq j \leq s)$ be a base for V.   Then

$$p_i^y = \sum_{1 \leq j \leq s} a_{ij}(y) q_j \qquad (y \in G)$$

where a_{ij} are analytic functions on G.   Moreover we can choose $c_0 \geq 0$ and an integer $\nu \geq 0$ such that $|a_{ij}(y)| \leq c_0 \|y\|^\nu$ for $y \in G$ (see [3(d), p. 203]).   Then

$$\sum_{1 \leq i \leq r} \sup |\partial(p_i) f_x^{x^{-1}}| \leq c_0 \|x\|^\nu r \sum_{1 \leq j \leq s} \sup |\partial(q_j) f_x|.$$

We can obviously select q_{kj}, q'_{kj} in $S(\mathfrak{g}_c)$ $(1 \leq k \leq u, 1 \leq j \leq s)$ such that

$$\partial(q_j)(\beta\gamma) = \sum_{1 \leq k \leq u} \partial(q_{kj}) \beta \cdot \partial(q_{kj}') \gamma \qquad (1 \leq j \leq s)$$

for any two C^∞ functions β and γ on $\mathfrak{g}$.   Then since $f_x = \alpha_x f$, we get

$$\sum_j \sup |\partial(q_j) f_x| \leq \sum_{k,j} \sup_{\Omega(t)} |\partial(q_{kj}) \alpha_x| \, |\partial(q_{kj}') f|.$$

Therefore

$$|T(f_x)| \leq c_0 r \|x\|^\nu \sum_{1 \leq k \leq u} \sum_{1 \leq j \leq s} \sup_{\Omega(t)} |\partial(q_{kj}) \alpha_x| \cdot \sup |\partial(q_{kj}') f|.$$

Now $\|x\| \leq c_1 (1+t)^m$ for $x \in F_t$ (Lemma 41).   Therefore we get the following result from Corollary 2 of Lemma 42.   There exists a number $c_3 \geq 0$ and an integer $m_3 \geq 0$ such that

$$|T(f_x)| \leq c_3 (1+t)^{m_3} \sum_{k,j} \sup |\partial(q_{kj}') f|$$

647

339

for $x\in F_t$, $f\in C_c^\infty(\Omega(t))$ and $t>0$. Since $f=\sum_{x\in F_t} f_x$ and $[F_t]\leq c_2(1+t)^{mM}$ (Lemma 41), we conclude that

$$|T(f)|\leq c_4(1+t)^{m_4}\sum_{k,j}\sup|\partial(q_k')f|$$

where $c_4=c_2c_3$ and $m_4=m_3+mM$. Obviously this implies the statement of Lemma 39.

§ 18. PROOF OF LEMMA 43

For any $a>0$ define Ω_a as in Lemma 35.

Lemma 43. — *Let* T *be an invariant distribution on* $\mathfrak{g}$ *and fix a number* $a>0$. *Then we can choose* $p_1,\ldots,p_r$ *in* $S(\mathfrak{g}_c)$ *and an integer* $d\geq 0$ *such that*

$$|T(f)|\leq\sum_{1\leq i\leq r}\sup(1+||X||)^d|f(X;\partial(p_i))|$$

for all $f\in C_c^\infty(\Omega_a)$.

We need some preliminary work. Fix a function $\alpha\in C_c^\infty(\mathbf{R})$ such that 1) $\alpha(-t)=\alpha(t)$, 2) $0\leq\alpha\leq 1$, 3) $\alpha(t)=1$ if $|t|\leq 1/2$ and $\alpha(t)=0$ if $|t|\geq 3/4$ $(t\in\mathbf{R})$. Put $\alpha_k(t)=\alpha(t-k)$ for any integer k and let

$$\beta=\sum_{-\infty<k<\infty}\alpha_k.$$

Fix $t_0\in\mathbf{R}$ and select an integer k_0 such that $|t_0-k_0|\leq 1/2$. Then $\alpha_{k_0}(t_0)=1$ and therefore $\beta(t_0)\geq 1$. Moreover $t_0\notin\mathrm{Supp}\,\alpha_k$ unless $|t_0-k|\leq 3/4$. Since the closed interval of length $3/2$ with t_0 at its center, can contain at most two integral points, it is clear that

$$|(d^m\beta/dt^m)_{t=t_0}|\leq 2\sup_t|(d^m\alpha/dt^m)|$$

for any integer $m\geq 0$. Therefore $1\leq\beta\leq 2$ everywhere and

$$\sup_t|(d^m\beta/dt^m)|\leq 2\sup_t|(d^m\alpha/dt^m)|<\infty.$$

Put $\gamma_k=\alpha_k/\beta$. Then it is clear that

$$\sup_t|(d^m\gamma_k/dt^m)|$$

is finite and independent of k. We denote it by c_m.

Since $0\notin\mathrm{Supp}\,\alpha_k$ if $k\neq 0$, it is clear that $\beta=1$ around the origin. Hence $B(s)=\beta(|s|^{1/2})$ $(s\in\mathbf{R})$ is a C^∞ function on $\mathbf{R}$ and

$$\sup_s|(d^mB/ds^m)|=\sup_t|(d/2tdt)^m\beta|.$$

Since $\beta=1$ around zero, it is clear that

$$\sup_t|t^{-p}(d^q\beta/dt^q)|<\infty$$

for two integers p and q $(p \geq 0, q \geq 1)$. Hence it follows that

$$\sup_{s} |(d^m B/ds^m)| < \infty.$$

Similarly if $A_k(s) = \alpha_k(|s|^{1/2})$ $(s \in \mathbf{R})$, one sees that A_k is a function of class C^∞. Moreover if $k \geq 0$, it follows in the same way that

$$\sup_{s} |(d^m A_k/ds^m)| = \sup_{t \geq 0} |(d/2tdt)^m \alpha_k|$$

$$\leq \sup_{t \geq 0} |(d/2(t+k)dt)^m \alpha| \leq c'_m$$

where c'_m is a positive number independent of k.

Now put $g(X) = \beta(\|X\|) = B(\|X\|^2)$, $h_k(X) = \alpha_k(\|X\|) = A_k(\|X\|^2)$ for $X \in \mathfrak{g}$ and $k \geq 0$. Since $Q : X \to \|X\|^2$ is a quadratic form on $\mathfrak{g}$, it is obvious that g and h_k are C^∞ functions on $\mathfrak{g}$.

Lemma 44. — Let p be an element in $S(\mathfrak{g}_c)$ of degree $\leq d$. Then we can choose a number $c_p \geq 0$ such that

$$|g(X; \partial(p))| \leq c_p(1 + \|X\|)^d, \qquad |h_k(X; \partial(p))| \leq c_p(1 + \|X\|)^d$$

for $X \in \mathfrak{g}$ and $k \geq 0$.

One proves by an easy induction on d that there exist polynomial functions q_j $(0 \leq j \leq d)$ on $\mathfrak{g}$ of degrees $\leq d$ such that

$$g(X; \partial(p)) = \sum_{0 \leq j \leq d} q_j(X)(d^j B/ds^j)_{s = \|X\|^2},$$

$$h_k(X; \partial(p)) = \sum_{0 \leq j \leq d} q_j(X)(d^j A_k/ds^j)_{s = \|X\|^2}$$

for $X \in \mathfrak{g}$ and $k \geq 0$. Our assertion now follows immediately from the facts proved above.

Put $g_k = h_k/g$ $(k \geq 0)$. Since $g \geq 1$, g_k is also of class C^∞.

Corollary. — We can choose $c'_p \geq 0$ such that

$$|g_k(X; \partial(p))| \leq c'_p(1 + \|X\|)^d$$

for $X \in \mathfrak{g}$ and $k \geq 0$.

This is obvious from Lemma 44 if we take into account the fact that $g \geq 1$.

We now come to the proof of Lemma 43. Since $\alpha_k(t) = 0$ for $k < 0$ and $t \geq 0$, it follows that $\sum_{k \geq 0} g_k = 1$. Fix $f \in C_c^\infty(\Omega_a)$ and put $f_k = g_k f$. Then $\sum_{k \geq 0} f_k = f$. It is clear that if $X \in \mathrm{Supp}\, g_k$, then $|\|X\| - k| \leq 3/4$. Therefore $f_k = 0$ if k is large. Hence

$$T(f) = \sum_{k \geq 0} T(f_k).$$

Define $\Omega(s, t)$ and $\Omega(t)$ $(0 \leq s < t)$ for $\Omega = \Omega_a$ as in the beginning of § 17. Then $\operatorname{Supp} f_k \subset \Omega(k+1)$. Therefore, by Lemma 39, we can choose $p_1, \ldots, p_r$ in $S(\mathfrak{g}_c)$ and an integer $\nu \geq 0$ such that

$$|T(f_k)| \leq (2+k)^\nu \sum_{1 \leq i \leq r} \sup |\partial(p_i) f_k|$$

for all $f \in C_c^\infty(\Omega_a)$ and all $k \geq 0$. Moreover since $\|X\| \geq k-1$ if $X \in \operatorname{Supp} f_k$, it follows that

$$(2+k)^{\nu+2} \sup |\partial(p_i) f_k| \leq \sup(3 + \|X\|)^{\nu+2} |f_k(X; \partial(p_i))|.$$

Choose q_{ij}, q'_{ij} in $S(\mathfrak{g}_c)(1 \leq i \leq r, 1 \leq j \leq s)$ such that

$$\partial(p_i)(F_1 F_2) = \sum_j \partial(q_{ij}) F_1 \cdot \partial(q'_{ij}) F_2 \qquad (1 \leq i \leq r)$$

for any two C^∞ functions F_1, F_2 on $\mathfrak{g}$. Then

$$|f_k(X; \partial(p_i))| \leq \sum_j |g_k(X; \partial(q_{ij}))| \, |f(X; \partial(q'_{ij}))|$$

since $f_k = g_k f$. Therefore there exist, from the corollary of Lemma 44, an integer $d_0 \geq 0$ and a number $c \geq 0$ such that

$$|f_k(X; \partial(p_i))| \leq c(1 + \|X\|)^{d_0} \sum_j |f(X; \partial(q'_{ij}))|$$

for all $f \in C_c^\infty(\Omega_a)$, $k \geq 0$, $X \in \mathfrak{g}$ and $1 \leq i \leq r$. Hence

$$(2+k)^{\nu+2} \sup |\partial(p_i) f_k| \leq 3^{\nu+2} c \sum_j \sup(1 + \|X\|)^d |\partial(q'_{ij}) f|$$

where $d = d_0 + \nu + 2$. Put

$$c_0 = 3^{\nu+2} c \sum_{k \geq 0} (k+2)^{-2} < \infty.$$

Then it follows that

$$|T(f)| \leq \sum_{k \geq 0} |T(f_k)| \leq c_0 \sum_{i,j} \sup(1 + \|X\|)^d |\partial(q'_{ij}) f|$$

for $f \in C_c^\infty(\Omega_a)$. This completes the proof of Lemma 43.

§ 19. COMPLETION OF THE PROOF OF LEMMA 29

As usual we identify $\mathfrak{g}_c$ with its dual under the Killing form. Call an element $p \in S(\mathfrak{g}_c)$ real if $p(X)$ is real for $X \in \mathfrak{g}$. Then we can select $p_1, \ldots, p_r$ in $I(\mathfrak{g}_c)$ such that 1) p_i is real and homogeneous of degree ≥ 1 and 2) $I(\mathfrak{g}_c) = \mathbf{C}[p_1, \ldots, p_r]$. Put

$$q(X) = \sum_{1 \leq i \leq r} p_i(X)^2 \qquad (X \in \mathfrak{g}).$$

Lemma 45. — We can choose a number $\delta > 0$ *such that* $q(X) < \delta$ $(X \in \mathfrak{g})$ *implies that* $X \in \Omega_a$.

Suppose this is false. Then we can choose a sequence $X_k \in \mathfrak{g}$ $(k \geq 1)$ such that $q(X_k) \to 0$ and $X_k \notin \Omega_a$. Let Y_k and Z_k respectively be the semisimple and nilpotent components of X_k (see § 3). Then $Y_k \in Cl(X_k^G)$ from the corollary of Lemma 7. Therefore $q(Y_k) = q(X_k)$. Since Ω_a is open and invariant and $X_k \notin \Omega_a$, it is clear that $Y_k \notin \Omega_a$. Therefore $q(Y_k) = q(X_k) \to 0$ and $Y_k \notin \Omega_a$.

Let $\mathfrak{h}_1, \ldots, \mathfrak{h}_m$ be a maximal set of Cartan subalgebras of $\mathfrak{g}$, no two of which are conjugate under G. Y_k, being semisimple, lies in some Cartan subalgebra of $\mathfrak{g}$ which must be conjugate to $\mathfrak{h}_j$ for some j. Hence we can choose $x_k \in G$ and an index j_k such that $Y_k^{x_k} \in \mathfrak{h}_{j_k}$. By choosing a subsequence we may assume that $H_k = Y_k^{x_k} \in \mathfrak{h}$ $(k \geq 1)$ where $\mathfrak{h}$ is a fixed Cartan subalgebra of $\mathfrak{g}$. Then $q(H_k) = q(Y_k) \to 0$ and therefore it is obvious that $p(H_k) \to 0$ for any $p \in I(\mathfrak{g}_c)$ which is homogeneous of degree ≥ 1. Now define q_j $(1 \leq j \leq n)$ in $I(\mathfrak{g}_c)$ by

$$\det(t - \mathrm{ad}\ X) = t^n + \sum_{1 \leq j \leq n} q_j(X) t^{n-j} \qquad (X \in \mathfrak{g}),$$

where t is an indeterminate. Then q_j is homogeneous of positive degree and therefore $q_j(H_k) \to 0$. However

$$\det(t - \mathrm{ad}\ H) = t^l \prod_{\alpha > 0} (t - \alpha(H)^2) \qquad (H \in \mathfrak{h})$$

where $l = \dim \mathfrak{h}$ and α runs over all positive roots of $(\mathfrak{g}, \mathfrak{h})$. Therefore $\alpha(H_k) \to 0$ for every root α and hence $H_k \to 0$. But then $\|H_k\| < a$ if k is large and therefore $Y_k = x_k^{-1} H_k \in \Omega_a$, giving a contradiction with our earlier result. This proves Lemma 45.

Corollary 1. — *There exists a C^∞ function g on $\mathfrak{g}$ such that:*

1) *g is invariant and* $\mathrm{Supp}\ g \subset \Omega_a$;

2) *$g = 1$ around zero;*

3) *for any $p \in S(\mathfrak{g}_c)$, we can choose $c_p, m_p \geq 0$ such that*

$$|g(X; \partial(p))| \leq c_p(1 + \|X\|)^{m_p} \qquad (X \in \mathfrak{g}).$$

Select a C^∞ function F on **R** such that 1) $F(t) = F(-t)$, 2) $F(t) = 1$ if $|t| \leq \delta/3$ and $F(t) = 0$ if $|t| \geq \delta/2$ $(t \in \mathbf{R})$. Put

$$g(X) = F(q(X)) \qquad (X \in \mathfrak{g}).$$

If $X \in \mathrm{Supp}\ g$, it is clear that $q(X) \leq \delta/2$ and therefore $X \in \Omega_a$. Moreover $g(X) = 1$ if $q(X) \leq \delta/3$. Fix $p \neq 0$ in $S(\mathfrak{g}_c)$ and let $d = d^\circ p$. Then it is clear that

$$g(X; \partial(p)) = \sum_{0 \leq j \leq d} (d^j F / dt^j)_{t = q(X)} p_j(X) \qquad (X \in \mathfrak{g})$$

where p_j $(0 \leq j \leq d)$ are suitable elements in $S(\mathfrak{g}_c)$. Hence g obviously satisfies condition 3).

Corollary 2. — *Let T be an invariant distribution on Ω_a. Then $g^2 T$ is a tempered distribution on $\mathfrak{g}$.*

Put $T_k = g^k T$ $(k = 1, 2)$. Then T_k is an invariant distribution on $\mathfrak{g}$. We now apply Lemma 43 to T_1. So we can choose an integer $d \geq 0$ and elements $p_i \in S(\mathfrak{g}_c)$ $(1 \leq i \leq r)$ such that

$$|T_1(f)| \leq \sum_{1 \leq i \leq r} \sup (1 + \|X\|)^d |f(X; \partial(p_i))|$$

for $f \in C_c^\infty(\Omega_a)$. Therefore if $f \in C_c^\infty(\mathfrak{g})$, we have

$$|T_2(f)| = |T_1(f_1)| \leq \sum_i \sup(1 + ||X||)^d |f_1(X; \partial(p_i))|$$

where $f_1 = gf$. Now select $p_{ij}, q_{ij} \in S(\mathfrak{g}_c)$ $(1 \leq i \leq r, 1 \leq j \leq s)$ in such a way that

$$\partial(p_i)(\varphi_1 \varphi_2) = \sum_j \partial(p_{ij})\varphi_1 \cdot \partial(q_{ij})\varphi_2 \qquad (1 \leq i \leq r)$$

for any two C^∞ functions φ_1, φ_2 on $\mathfrak{g}$. Then

$$\partial(p_i)f_1 = \sum_j \partial(p_{ij})g \cdot \partial(q_{ij})f.$$

Therefore, by condition 3) of Corollary 1 above, it is obvious that there exist $c, m \geq 0$ such that

$$|T_2(f)| \leq c \sum_{i,j} \sup(1 + ||X||)^{d+m} |f(X; \partial(q_{ij}))|$$

for $f \in C_c^\infty(\mathfrak{g})$. This proves that T_2 is tempered.

We can now complete the proof of Lemma 29. Since Ω is an open neighborhood of zero, we can choose $a > 0$ such that $X \in \Omega$ whenever $||X|| < a$ $(X \in \mathfrak{g})$. Therefore $\Omega_a \subset \Omega$. Now take $f = g^2$ where g is defined as in Corollary 1 of Lemma 45. Then it follows from Corollary 2 above that fT is a tempered distribution on $\mathfrak{g}$. This proves Lemma 29.

§ 20. PROOF OF LEMMA 34

We shall now begin the proof of Lemma 34. Since any two norms on $\mathfrak{g}$ are equivalent, it is enough to consider the case when $v(X) = ||X||$ $(X \in \mathfrak{g})$. The case $l = 0$ being trivial, we assume $l \geq 1$ and use induction. For any (real-valued) linear function λ on $\mathfrak{a}$, let $\mathfrak{g}_\lambda$ denote the space of all $X \in \mathfrak{g}$ such that $[H, X] = \lambda(H)X$ for all $H \in \mathfrak{a}$. We denote by E_λ the orthogonal projection of $\mathfrak{g}$ on $\mathfrak{g}_\lambda$. Then $\mathfrak{g}_\lambda = \{0\}$ unless $\lambda = 0$ or $\pm \alpha$ for some $\alpha \in \Sigma$. Since ad H is self-adjoint for $H \in \mathfrak{a}$ (see [3(h), Lemma 27]), the spaces $\mathfrak{g}_\lambda$ and $\mathfrak{g}_\mu$ $(\lambda \neq \mu)$ are mutually orthogonal. Therefore if

$$E_+ = \sum_{\alpha \in \Sigma} E_\alpha, \qquad E_- = \sum_{\alpha \in \Sigma} E_{-\alpha},$$

it is clear that $E_+ + E_0 + E_- = 1$.

Let S denote the set $\{1, 2, \ldots, l\}$ and for any subset Q of S, let Σ_Q denote the set of all $\alpha \in \Sigma$ which are linear combinations of α_i $(i \in Q)$. Define $\mathfrak{n}_Q = \sum_{\alpha \in \Sigma_Q} \mathfrak{g}_\alpha$ and let $\mathfrak{g}_Q$ be the subalgebra of $\mathfrak{g}$ generated by $\mathfrak{n}_Q + \theta(\mathfrak{n}_Q)$. Then $\theta(\mathfrak{g}_Q) = \mathfrak{g}_Q$ and therefore $\mathfrak{g}_Q = \mathfrak{k}_Q + \mathfrak{p}_Q$ where $\mathfrak{k}_Q = \mathfrak{k} \cap \mathfrak{g}_Q$, $\mathfrak{p}_Q = \mathfrak{p} \cap \mathfrak{g}_Q$.

Lemma 46. — $\mathfrak{g}_Q$ is semisimple.

Let $<X, Y> = -B(X, \theta(Y))$ $(X, Y \in \mathfrak{g})$ denote the scalar product in the Hilbert space $\mathfrak{g}$ and, for any linear function λ on $\mathfrak{a}$, let H_λ denote the element in $\mathfrak{a}$ such that

$<H, H_\lambda> = \lambda(H)$ for all $H \in \mathfrak{a}$. We know (see [3(d), Lemma 3]) that if $X \in \mathfrak{g}_\lambda$ and $\|X\| = 1$, then $[\theta(X), X] = H_\lambda$.

First we claim that $\mathfrak{g}_Q$ is reductive in $\mathfrak{g}$. Let U be any subspace of $\mathfrak{g}$ such that $[\mathfrak{g}_Q, U] \subset U$. Since $\mathfrak{g}_Q = \theta(\mathfrak{g}_Q)$, ad $\mathfrak{g}_Q$ is a self-adjoint family of transformations in $\mathfrak{g}$ [3(h), Lemma 27]. Hence if V is the orthogonal complement of U in $\mathfrak{g}$, V is stable under ad $\mathfrak{g}_Q$. This proves our assertion. Therefore $\mathfrak{g}'_Q = [\mathfrak{g}_Q, \mathfrak{g}_Q]$ is semisimple. Now fix $\alpha \in \Sigma_Q$ and $X \in \mathfrak{g}_\alpha$ with $\|X\| = 1$. Then $[\theta(X), X] = H_\alpha \in \mathfrak{g}_Q$ and therefore $[H_\alpha, X] = \alpha(H_\alpha)X \in \mathfrak{g}'_Q$. Since $\alpha(H_\alpha) = \|H_\alpha\|^2 > 0$, this proves that $\mathfrak{g}_\alpha \subset \mathfrak{g}'_Q$. However $\mathfrak{g}'_Q$ is obviously stable under θ and so we conclude that $\mathfrak{n}_Q + \theta(\mathfrak{n}_Q) \subset \mathfrak{g}'_Q$. But, in view of the definition of $\mathfrak{g}_Q$, this implies that $\mathfrak{g}'_Q = \mathfrak{g}_Q$. This proves that $\mathfrak{g}_Q$ is semisimple.

Let F_Q denote the orthogonal projection of $\mathfrak{g}$ on $\mathfrak{g}_Q$. We have seen above that $H_\alpha \in \mathfrak{a} \cap \mathfrak{g}_Q$ for $\alpha \in \Sigma_Q$. Put $\mathfrak{a}_Q = \sum_{i \in Q} RH_{\alpha_i}$ and let $\mathfrak{b}_Q$ denote the orthogonal complement of $\mathfrak{a}_Q$ in $\mathfrak{a}$. Then $\mathfrak{a}_Q = \sum_{\alpha \in \Sigma_Q} RH_\alpha \subset \mathfrak{g}_Q$.

Lemma 47. — $\mathfrak{a}_Q = \mathfrak{a} \cap \mathfrak{g}_Q$. *Moreover F_Q commutes with θ and E_λ for any linear function λ on $\mathfrak{a}$.*

Let $H \in \mathfrak{b}_Q$. Then $\alpha_i(H) = <H_{\alpha_i}, H> = 0$ $(i \in Q)$ and therefore $\alpha(H) = 0$ for $\alpha \in \Sigma_Q$. Hence H commutes with $\mathfrak{n}_Q + \theta(\mathfrak{n}_Q)$ and therefore also with $\mathfrak{g}_Q$. Since $\mathfrak{g}_Q$ is semisimple, it follows that $\mathfrak{g}_Q \cap \mathfrak{b}_Q = \{0\}$. Therefore since $\mathfrak{a}_Q \subset \mathfrak{g}_Q$, it is obvious that $\mathfrak{a} \cap \mathfrak{g}_Q = \mathfrak{a}_Q$.

Let $\mathfrak{m}_Q$ be the set of all $X \in \mathfrak{g}_Q$ such that $[H, X] \in \mathfrak{g}_Q$ for all $H \in \mathfrak{a}$. Then $\mathfrak{m}_Q$ is a subalgebra of $\mathfrak{g}_Q$ which contains $\mathfrak{n}_Q + \theta(\mathfrak{n}_Q)$. Hence $\mathfrak{m}_Q = \mathfrak{g}_Q$. Therefore $\mathfrak{g}_Q$ is stable under ad H $(H \in \mathfrak{a})$ and this implies that $E_\lambda \mathfrak{g}_Q \subset \mathfrak{g}_Q$ for any linear function λ on $\mathfrak{a}$. This shows that F_Q commutes with E_λ. Similarly since $\mathfrak{g}_Q$ is stable under θ, F_Q commutes with θ.

Corollary. — $\mathfrak{a}_Q$ *is maximal abelian in* $\mathfrak{p}_Q$ *and* $\mathfrak{a}_Q = F_Q \mathfrak{a}$.

Since $\mathfrak{g}_0 + \mathfrak{n}_Q + \theta(\mathfrak{n}_Q)$ is a subalgebra of $\mathfrak{g}$, it must contain $\mathfrak{g}_Q$. Therefore

$$X = E_0 X + \sum_{\alpha \in \Sigma_Q} E_\alpha X + \sum_{\alpha \in \Sigma_Q} E_{-\alpha} X \qquad (X \in \mathfrak{g}_Q).$$

Now suppose $X \in \mathfrak{p}_Q$ and it commutes with $\mathfrak{a}_Q$. Then

$$0 = [H, X] = \sum_{\alpha \in \Sigma_Q} \alpha(H) E_\alpha X - \sum_{\alpha \in \Sigma_Q} \alpha(H) E_{-\alpha} X \qquad (H \in \mathfrak{a}_Q)$$

and therefore $\alpha(H)E_{\pm\alpha}X = 0$ for $H \in \mathfrak{a}_Q$ and $\alpha \in \Sigma_Q$. But $H_\alpha \in \mathfrak{a}_Q$ for $\alpha \in \Sigma_Q$ and $\alpha(H_\alpha) = \|H_\alpha\|^2 > 0$. Hence $E_{\pm\alpha}X = 0$ $(\alpha \in \Sigma_Q)$ and therefore $X = E_0 X \in \mathfrak{g}_0$. This means that $X \in \mathfrak{g}_0 \cap \mathfrak{p} = \mathfrak{a}$ since $\mathfrak{a}$ is maximal abelian in $\mathfrak{p}$. But then $X \in \mathfrak{a} \cap \mathfrak{g}_Q = \mathfrak{a}_Q$. This proves that $\mathfrak{a}_Q$ is maximal abelian in $\mathfrak{p}_Q$.

Since F_Q commutes with θ and E_0 and $\mathfrak{a} \subset \mathfrak{p}$, it is clear that $F_Q \mathfrak{a} \subset \mathfrak{p}_Q \cap \mathfrak{g}_0$. But since $\mathfrak{a}_Q$ is maximal abelian in $\mathfrak{p}_Q$, $\mathfrak{p}_Q \cap \mathfrak{g}_0 = \mathfrak{a}_Q$. This proves that $F_Q \mathfrak{a} = \mathfrak{a}_Q$.

Let l_Q denote the number of elements in Q. Then dim $\mathfrak{a}_Q = l_Q$. Let G_Q and A_Q be the analytic subgroups of G corresponding to $\mathfrak{g}_Q$ and $\mathfrak{a}_Q$ respectively. If $Q \neq S$, Lemma 34 holds for $(\mathfrak{g}_Q, \mathfrak{a}_Q)$ instead of $(\mathfrak{g}, \mathfrak{a})$ by the induction hypothesis. Let A_Q^+ be the set of all $h \in A_Q$ such that $\alpha_i(\log h) \geq 0$ $(i \in Q)$. Then we obviously have the following result.

Lemma 48. — Assume that $Q \neq S$. Then there exist numbers $b_Q, c_Q \geq 1$ with the following properties. Suppose $X \in \mathfrak{g}_Q$, $\|X\| \leq 1$ and $h \in A_Q^+$. Then we can choose $X_0 \in \mathfrak{g}_Q$, $h_0 \in A_Q^+$ such that:

    1) $X^h = X_0^{h_0}$, $\|X_0\| \leq b_Q$, $0 \leq \alpha_i(\log h_0) \leq \alpha_i(\log h)$ $(i \in Q)$,

    2) $\max\limits_{i \in Q} \exp(\alpha_i(\log h_0)) \leq c_Q (1 + \|X_0^{h_0}\|)^{l_Q}$.

Let $A^+(Q)$ be the set of all $h \in A^+$ such that $\alpha_j(\log h) = 0$ $(j \notin Q)$. For any $h \in A$, define

$$h_Q = \exp(\sum_{i \in Q} \alpha_i(\log h) H_i).$$

Then $\alpha_i(\log h) = \alpha_i(\log h_Q)$ $(i \in Q)$ and therefore $\log h - \log h_Q$ commutes with $\mathfrak{g}_Q$ so that $X^h = X^{h_Q}$ $(X \in \mathfrak{g}_Q)$. Moreover if $h \in A^+$, it is clear that $1 \prec h_Q \prec h$ and $h_Q \in A^+(Q)$.

Corollary. — Suppose $X \in \mathfrak{g}_Q$, $\|X\| \leq 1$ and $h \in A^+(Q)$. Then we can choose $X_0 \in \mathfrak{g}_Q$ and $h_0 \in A^+(Q)$ such that

    1) $X^h = X_0^{h_0}$, $\|X_0\| \leq b_Q$, $1 \prec h_0 \prec h$,

    2) $\max\limits_{1 \leq i \leq l} \exp \alpha_i(\log h_0) \leq c_Q (1 + \|X_0^{h_0}\|)^{l_Q}$.

Put $h' = \exp(\sum_{i \in Q} \alpha_i(\log h) F_Q H_i)$. Then $h' \in A_Q^+$ and $(h')_Q = h$ from the corollary of Lemma 47. Hence $X^{h'} = X^h$. Choose $h_0' \in A_Q^+$ and $X_0 \in \mathfrak{g}_Q$ such that the conditions of Lemma 48 hold for (X, h', X_0, h_0') in place of (X, h, X_0, h_0). Then if we put $h_0 = (h_0')_Q$ all the conditions of the corollary are fulfilled.

For any $i \in S$ and $Z \in \mathfrak{g}$, define

$$\mu(i : Z) = \max_{\substack{\alpha \in \Sigma \\ \alpha(H_i) \neq 0}} \|E_\alpha Z\|$$

and let $Q(Z)$ be the set of all $i \in S$ for which $\mu(i : Z) \geq 1$. Moreover for any subset Q of S, let Σ_Q' denote the complement of Σ_Q in Σ.

Lemma 49. — Let Z be an element of $\mathfrak{g}$. Then $\|E_\alpha Z\| < 1$ for every $\alpha \in (\Sigma_{Q(Z)})'$.

Suppose $\|E_\alpha Z\| \geq 1$ for some $\alpha \in \Sigma$. We have to show that $\alpha \in \Sigma_{Q(Z)}$. Fix $i \in S$ such that $\alpha(H_i) \neq 0$. Then

$$\mu(i : Z) \geq \|E_\alpha Z\| \geq 1$$

and therefore $i \in Q(Z)$. Since this holds for every i for which $\alpha(H_i) \neq 0$, it is clear that $\alpha \in \Sigma_{Q(Z)}$.

Put $F_Q' = 1 - F_Q$ for any subset Q of S. Fix $X \in \mathfrak{g}$ and $h \in A^+$ and assume that $\|X\| \leq 1$. Put $Q_0 = Q(X^h)$ and let s denote the number of elements in Σ.

Lemma 50. — $\|F_{Q_0}' X^a\| \leq 1 + s^{1/2}$ for any $a \in A$ such that $1 \prec a \prec h$.

Let λ be a linear function on $\mathfrak{a}$ such that $\mathfrak{g}_\lambda \neq \{0\}$. Then

$$E_\lambda X^a = e^{\lambda(\log a)} E_\lambda X.$$

Now $a \succ 1$ and therefore $\lambda(\log a) \leq 0$ if $\lambda \leq 0$. Therefore since F'_{Q_0} commutes with E_0 and E_-, it is obvious that

$$||(E_0 + E_-)F'_{Q_0}X^a|| \leq ||(E_0 + E_-)X^a|| \leqslant ||X|| \leq 1.$$

On the other hand $\alpha(\log a) \leq \alpha(\log h)$ $(\alpha \in \Sigma)$ since $a \prec h$. Therefore

$$||E_+ F'_{Q_0}X^a||^2 = \sum_{\alpha \in \Sigma'_{Q_0}} ||E_\alpha X^a||^2 \leq \sum_{\alpha \in \Sigma'_{Q_0}} ||E_\alpha X^h||^2 \leq s$$

from Lemma 49. Since

$$F'_{Q_0}X^a = (E_+ + E_0 + E_-)F'_{Q_0}X^a,$$

our assertion is now obvious.

Lemma 51. — *For any* $(^1)$ $Q < S$, *select* b_Q *and* c_Q *corresponding to Lemma 48 and define*

$$b_0 = 1 + s^{1/2} + \max_{Q < S} b_Q, \qquad c_0 = \max_{Q < S} c_Q.$$

Let $X \in \mathfrak{g}, h \in A^+$ *and suppose that* $||X|| \leq 1$ *and* $Q(X^h) \neq S$. *Then we can choose* $X_0 \in \mathfrak{g}$ *and* $h_0 \in A^+$ *such that*

1) $X^h = X_0^{h_0}, \ 1 \prec h_0 \prec h, \ ||X_0|| \leq b_0$;
2) $\max\limits_{1 \leq i \leq l} \exp \alpha_i(\log h_0) \leq c_0(1 + ||X_0^{h_0}||)^{l-1}$;

Put $Q = Q(X^h)$. Then $X^h = F_Q X^h + F'_Q X^h$. But since F_Q commutes with $\mathrm{Ad}(h)$ (Lemma 47), we have

$$F_Q X^h = (F_Q X)^h = X_Q^{h_Q}$$

where $X_Q = F_Q X$. Since $Q < S$, we can apply the corollary of Lemma 48 to (X_Q, h_Q). Hence we can choose $X_1 \in \mathfrak{g}_Q$ and $h_0 \in A^+(Q)$ such that:

1) $X_Q^{h_Q} = X_1^{h_0}, \ ||X_1|| \leq b_Q, \ 1 \prec h_0 \prec h_Q$;
2) $\max\limits_{1 \leq i \leq l} \exp \alpha_i(\log h_0) \leq c_Q(1 + ||X_1^{h_0}||)^{l_Q}$.

Then

$$X^h = X_1^{h_0} + F'_Q X^h = (X_1 + F'_Q X^{h_2})^{h_0}$$

where $h_2 = hh_0^{-1}$. Since $1 \prec h_0 \prec h_Q \prec h$, it follows that $1 \prec h_2 \prec h$. Put

$$X_0 = X_1 + F'_Q X^{h_2}.$$

Then

$$||X_0|| \leq ||X_1|| + ||F'_Q X^{h_2}|| \leq b_Q + 1 + s^{1/2} \leq b_0$$

from Lemma 50. Moreover

$$X_1^{h_0} = X_Q^{h_Q} = F_Q X^h.$$

Therefore

$$||X_1^{h_0}|| \leq ||X^h|| = ||X_0^{h_0}||.$$

Hence

$$\max_{1 \leq i \leq l} \exp \alpha_i(\log h_0) \leq c_Q(1 + ||X_1^{h_0}||)^{l_Q} \leq c_0(1 + ||X_0^{h_0}||)^{l-1}$$

and so the lemma is proved.

$(^1)$ $Q < S$ means that Q is a subset of S and $Q \neq S$.

Put $c = 2^l c_0$ and $b = b_0$. Then in order to prove Lemma 34, it is obviously enough to verify the following result.

Lemma 52. — *Let* $X \in \mathfrak{g}$ *and* $h \in A^+$ *and suppose* $\|X\| \leq 1$. *Then we can choose* $X_0 \in \mathfrak{g}$ *and* $h_0 \in A^+$ *such that:*

1) $X^h = X_0^{h_0}$, $\|X_0\| \leq b$, $1 \prec h_0 \prec h$;

2) $\max\limits_{1 \leq i \leq l} \exp \alpha_i(\log h_0) \leq c(1 + \|X_0^{h_0}\|)^l$.

If $Q(X^h) < S$, our statement follows immediately from Lemma 51. So we may assume that

$$\mu(i : X^h) \geq 1 \qquad\qquad (1 \leq i \leq l).$$

Let Ω be the set of all $a \in A^+$ such that 1) $1 \prec a \prec h$ and 2) $\mu(i : X^a) \geq 1/2$ ($1 \leq i \leq l$). Obviously Ω is a compact set containing h. Put

$$f(a) = \sum_{1 \leq i \leq l} \mu(i : X^a) \qquad\qquad (a \in \Omega).$$

Then f is a continuous function on Ω which must take its minimum at some point $a_0 \in \Omega$. First suppose $a_0 = 1$. Then $1 \in \Omega$ and therefore

$$\mu(i : X) \geq 1/2 \qquad\qquad (1 \leq i \leq l).$$

Now fix $i \in S$ and choose $\alpha \in \Sigma$ such that $\alpha(H_i) \neq 0$ and $\|E_\alpha X\| \geq 1/2$. Then

$$\|E_+ X^h\| \geq e^{\alpha(\log h)} \|E_\alpha X\| \geq 2^{-1} e^{\alpha_i(\log h)}.$$

Therefore

$$\max_i e^{\alpha_i(\log h)} \leq 2 \|E_+ X^h\| \leq 2 \|X^h\|.$$

Since $b \geq 1$ and $c = 2^l c_0 \geq 2$, we can take $X_0 = X$ and $h_0 = h$ in this case.

So now assume that $a_0 \neq 1$. Then we claim that $\mu(i : X^{a_0}) = 1/2$ for some i. For otherwise suppose $\mu(i : X^{a_0}) > 1/2$ for every i. Choose j such that $\alpha_j(\log a_0) \neq 0$. Put $a_\varepsilon = a_0(\exp(-\varepsilon H_j))$ where ε is a small positive number. If ε is sufficiently small, it is clear that $a_\varepsilon \in \Omega$. Hence $f(a_\varepsilon) \geq f(a_0)$. On the other hand since

$$\|E_\alpha X^{a_\varepsilon}\| = e^{-\varepsilon \alpha(H_j)} \|E_\alpha X^{a_0}\| \qquad\qquad (\alpha \in \Sigma),$$

it is clear that

$$\mu(i : X^{a_\varepsilon}) \leq \mu(i : X^{a_0})$$

for every i. Moreover $e^{-\varepsilon \alpha(H_j)} < 1$ if $\alpha(H_j) \neq 0 (\alpha \in \Sigma)$ and therefore since $\mu(j : X^{a_0}) \geq 1/2$, it is obvious that

$$\mu(j : X^{a_\varepsilon}) < \mu(j : X^{a_0}).$$

But this implies that $f(a_\varepsilon) < f(a_0)$ and so we get a contradiction. Hence $\mu(i : X^{a_0}) = 1/2$ for some i and therefore $Q(X^{a_0}) < S$. But then by Lemma 51 we can choose $X_0 \in \mathfrak{g}$ and $a_1 \in A^+$ such that $X^{a_0} = X_0^{a_1}$, $\|X_0\| \leq b_0$, $1 \prec a_1 \prec a_0$ and

$$\max_{1 \leq i \leq l} \exp \alpha_i(\log a_1) \leq c_0(1 + \|X_0^{a_1}\|)^{l-1}.$$

Now put $h_0 = ha_0^{-1}a_1$. Then

$$X^h = (X^{a_0})^{ha_0^{-1}} = (X_0^{a_1})^{ha_0^{-1}} = X_0^{h_0}$$

and therefore

$$\|X^h\| \geq \|E_\alpha X^{a_0}\| \exp \alpha(\log(ha_0^{-1})) \qquad (\alpha \in \Sigma).$$

Fix $i \in S$. Then since $\mu(i : X^{a_0}) \geq 1/2$, we can select $\alpha \in \Sigma$ such that $\alpha(H_i) \neq 0$ and $\|E_\alpha X^{a_0}\| \geq 1/2$. Therefore since $1 \prec a_0 \prec h$, we have

$$\|X^h\| \geq 2^{-1} \exp \alpha_i(\log(ha_0^{-1})).$$

On the other hand

$$e^{\alpha_i(\log a_1)} \leq c_0(1 + \|X_0^{a_1}\|)^{l-1} = c_0(1 + \|X^{a_0}\|)^{l-1}.$$

Therefore since $h_0 = ha_0^{-1}a_1$, we get

$$e^{\alpha_i(\log h_0)} \leq 2c_0 \|X^h\| (1 + \|X^{a_0}\|)^{l-1}.$$

But since $1 \prec a_0 \prec h$, we have (see the proof of Lemma 50)

$$\|E_+ X^{a_0}\| \leq \|E_+ X^h\| \leq \|X^h\|$$

and

$$\|(E_0 + E_-)X^{a_0}\| \leq \|X\| \leq 1.$$

Therefore

$$\|X^{a_0}\| \leq 1 + \|X^h\|$$

and hence

$$e^{\alpha_i(\log h_0)} \leq 2c_0 \|X^h\|(2 + \|X^h\|)^{l-1} \leq c(1 + \|X^h\|)^l.$$

Since $\|X_0\| \leq b_0 = b$, Lemma 51 (and therefore also Lemma 34) is proved.

§ 21. **PROOF OF LEMMA** 35

We have still to prove Lemma 35. Fix $a > b > 0$ and let x_i and X_i $(i \geq 1)$ be two sequences in G and $\mathfrak{g}$ respectively such that $\|X_i\| \leq b$ and $x_i X_i$ converges to some $Y \in \mathfrak{g}$. We have to prove that $Y \in \Omega_a$. Let $x_i = k_i h_i k_i'$ $(k_i, k_i' \in K; h_i \in A^+)$. Replacing (x_i, X_i) by $(k_i h_i, k_i' X_i)$ we may assume that $x_i = k_i h_i$. Moreover by selecting a subsequence we can arrange that $k_i \to k$ and $X_i \to X$ $(k \in K, X \in \mathfrak{g})$. Then by replacing (x_i, X_i, Y) by $(k^{-1}x_i, X_i, k^{-1}Y)$, we are reduced to the case when $k = 1$. Now

$$X_i^{x_i} - X_i^{h_i} = (1 - \mathrm{Ad}(k_i^{-1}))X_i^{x_i}.$$

Since $X_i^{x_i} \to Y$ and $k_i \to 1$, it is clear that $\|X_i^{x_i} - X_i^{h_i}\| \to 0$. Hence $X_i^{h_i} \to Y$.

By selecting a subsequence we can obviously arrange that the following condition holds. There exists a subset Q of S such that $\alpha_j(\log h_i) \to t_j$ $(t_j \in \mathbf{R})$ for $j \in Q$ and $\alpha_j(\log h_i) \to +\infty$ for $j \notin Q$ $(1 \leq j \leq l)$ as $i \to \infty$. Then it is clear that

$$E_{-\alpha}X_i^{h_i} = e^{-\alpha(\log h_i)}E_{-\alpha}X_i \to 0$$

for $\alpha \in \Sigma_Q'$. Put

$$E = E_0 + \sum_{\alpha \in \Sigma_Q}(E_\alpha + E_{-\alpha})$$

and

$$h = \exp\left(\sum_{j \in Q} t_j H_j\right).$$

Then it is clear that

$$EX_i^{h_i} \to EX^h.$$

On the other hand if

$$E'_+ = \sum_{\alpha \in \Sigma'_Q} E_\alpha,$$

we have

$$\mathbf{1} = E + E'_+ + \sum_{\alpha \in \Sigma'_Q} E_{-\alpha}.$$

Therefore since $E_{-\alpha} X_i^{h_i} \to 0 \; (\alpha \in \Sigma'_Q)$, we conclude that

$$(E + E'_+) X_i^{h_i} \to Y.$$

Therefore $Y = EY + E'_+ Y$ and $EY = EX^h$. Now select $H \in \mathfrak{a}$ such that $\alpha_j(H) = 0$ for $j \in Q$ and $\alpha_j(H) > 0$ for $j \notin Q$ $(1 \leq j \leq l)$. Then $\alpha(H) > 0$ for $\alpha \in \Sigma'_Q$ and therefore

$$\mathrm{Ad}(\exp(-tH)) E'_+ Y \to 0$$

as $t \to +\infty$. Put $y_t = (h \exp tH)^{-1}$. Then

$$Y^{y_t} = EX + (E'_+ Y)^{y_t} \to EX$$

as $t \to +\infty$. Since $\|EX\| \leq \|X\| \leq b$, it follows that $\|Y^{y_t}\| < a$ if t is sufficiently large and positive. Therefore $Y \in \Omega_a$ and this proves Lemma 35.

The Institute for Advanced Study,
      Princeton, New Jersey.

REFERENCES

[1] A. Borel, Compact Clifford-Klein forms of symmetric spaces, *Topology*, **2** (1963), 111-122.

[2] N. Bourbaki, *Groupes et algèbres de Lie*, chapitre Ier : « Algèbres de Lie », 1960, Hermann, Paris.

[3] Harish-Chandra :

    a) Representations of semisimple Lie groups VI, *Amer. Jour. Math.*, **78** (1956), 564-628.

    b) The characters of semisimple Lie groups, *Trans. Amer. Math. Soc.*, **83** (1956), 98-163.

    c) Differential operators on a semisimple Lie algebra, *Amer. Jour. Math.*, **79** (1957), 87-120.

    d) Fourier transforms on a semisimple Lie algebra, I, *Amer. Jour. Math.*, **79** (1957), 193-257.

    e) Fourier transforms on a semisimple Lie algebra, II, *Amer. Jour. Math.*, **79** (1957), 653-686.

    f) Spherical functions on semisimple Lie groups, I, *Amer. Jour. Math.*, **80** (1958), 241-310.

    g) Invariant eigendistributions on semisimple Lie groups, *Bull. Amer. Math. Soc.*, **69** (1963), 117-123.

    h) Invariant distributions on Lie algebras, *Amer. Jour. Math.*, **86** (1964), 271-309.

    i) Invariant differential operators and distributions on a semisimple Lie algebra, *Amer. Jour. Math.*, **86** (1964), 534-564.

    j) Some results on an invariant integral on a semisimple Lie algebra, *Ann. of Math.*, **80** (1964), 551-593.

[4] H. Whitney, Elementary structure of real algebraic varieties, *Ann. of Math.*, **66** (1957), 545-556.

Manuscrit reçu le 8 avril 1964.

Reprinted from
Publ. Math. IHES
No. **27** (1965), 5–54

INVARIANT EIGENDISTRIBUTIONS
ON A SEMISIMPLE LIE GROUP

BY

HARISH-CHANDRA

1. **Introduction.** Let G be a connected semisimple Lie group and $\mathfrak{Z}$ the algebra of all differential operators on G which commute with both left and right translations of G. One of the main objects of this paper is to show that every invariant eigendistribution T of $\mathfrak{Z}$ on G, is actually a locally summable function F which is analytic on the regular set G' of G (Theorem 2). In particular, this implies that the character of an irreducible unitary representation of G is a function.

In the second part we examine the behavior of F around the singular points of G (see §§19, 20). This is done by applying the results of [4(n), §§8, 9]. The third part is devoted to the detailed study of an invariant integral on G, which had been introduced in [4(h), Theorem 2]. Here we have to make use of [4(m), Theorem 1]. The full significance of Theorem 3 for harmonic analysis on G will appear only in later papers. Roughly speaking, it is the group analogue of [4(g), Theorem 3].

Our methods are substantially the same as those introduced in [4(l)] and [4(n)], although they have now to be applied to the group G instead of its Lie algebra $\mathfrak{g}$. Here Theorem 2 of [4(l)] gets replaced by Lemma 22, which is based on Theorem 1 and this, in its turn, depends on Theorem 5 of [4(n)]. The results of §3 enable us to limit ourselves to the semisimple points of G and the reduction procedure, outlined above, can be applied to any such point a provided it does not lie in the center Z of G. However, if $a \in Z$, the translation by a^{-1} reduces the problem to the case $a = 1$. Then we use the results of §14 to transform it, by means of the exponential mapping, into an analogous question on $\mathfrak{g}$ around zero, which has already been discussed in [4(n)]. This general pattern of proof applies to most results of this paper (e.g., Theorems 1 and 2). However, sometimes it is more convenient to reduce the problem around a directly to the corresponding question around zero on the centralizer $\mathfrak{z}$ of a in $\mathfrak{g}$ (see, for example, the proofs of Lemmas 31, 35, 37 and 40).

Theorem 3 is proved by making use of the elementary solution of a certain elliptic "Laplacian" and imitating the argument of [4(g), pp. 208–211]. The Appendix contains a few simple lemmas which are needed in the proof of this theorem.

Received by the editors June 2, 1964.

457

This work was partially supported by grants from the Sloan and the National Science Foundations.

2. The mapping Γ_x. Let G be a Lie group, $\mathfrak{g}$ its Lie algebra over R and $\mathfrak{G}$ the universal enveloping algebra of $\mathfrak{g}_c$. For any $X \in \mathfrak{g}_c$, let L_X and R_X, respectively, denote the endomorphisms $g \to Xg$ and $g \to gX$ $(g \in \mathfrak{G})$ of $\mathfrak{G}$. Fix $x \in G$ and define[1] (cf. [4(e), p. 114])

$$\sigma_x(X) = L_{x^{-1}X} - R_X \qquad (X \in \mathfrak{g}_c).$$

Note that L_X and R_Y commute and[2]

$$[L_X, L_Y] = L_{[X,Y]}, \quad [R_X, R_Y] = - R_{[X,Y]}$$

for $X, Y \in \mathfrak{g}_c$. Hence it follows immediately that σ_x is a representation of $\mathfrak{g}_c$ on $\mathfrak{G}$. It may, therefore, be extended (uniquely) to a representation of $\mathfrak{G}$ which we shall again denote by σ_x. Let Γ_x denote the linear mapping of $\mathfrak{G} \otimes \mathfrak{G}$ into $\mathfrak{G}$ such that

$$\Gamma_x(g_1 \otimes g_2) = \sigma_x(g_1)g_2 \qquad (g_1, g_2 \in \mathfrak{G}).$$

Let λ denote the canonical mapping (see [4(b), p. 192]) of $S(\mathfrak{g}_c)$ onto $\mathfrak{G}$. We say that an element $g \in \mathfrak{G}$ is homogeneous of degree d if $g \in \mathfrak{G}_d = \lambda(S_d(\mathfrak{g}_c))$ in the notation of [4(k), §6]. Put

$$_d\mathfrak{G} = \sum_{0 \leq m \leq d} \mathfrak{G}_m.$$

Then it is obvious that Γ_x defines a linear mapping of $\mathfrak{G}_{d_1} \otimes \mathfrak{G}_{d_2}$ into $_{d_1+d_2}\mathfrak{G}$.

Let $x \to x^a$ $(x \in G)$ be an automorphism of G. Then it defines an automorphism of $\mathfrak{g}$ which can be extended uniquely to an automorphism $g \to g^a$ $(g \in \mathfrak{G})$ of g.

LEMMA 1. *For any $x \in G$ and $g_1, g_2 \in \mathfrak{G}$,*

$$\Gamma_{x^a}(g_1^a \otimes g_2^a) = (\Gamma_x(g_1 \otimes g_2))^a.$$

Let A denote the automorphism $g \to g^a$ of $\mathfrak{G}$. Then one verifies from the definitions that

$$\sigma_{x^a}(X^a) = A\sigma_x(X)A^{-1}$$

for $X \in \mathfrak{g}$. Our assertion is an immediate consequence of this fact.

LEMMA 2. *Suppose X_i and Y_j $(1 \leq i \leq r, 1 \leq j \leq s)$ are elements in $\mathfrak{g}_c$. Fix $x \in G$ and put $X_i' = x^{-1}X_i - X_i$ $(1 \leq i \leq r)$. Then*

$$\Gamma_x(\lambda(X_1X_2\cdots X_r) \otimes \lambda(Y_1Y_2\cdots Y_s)) \equiv \lambda(X_1'X_2'\cdots X_r' \, Y_1Y_2\cdots Y_s) \mod {}_{(r+s-1)}\mathfrak{G}.$$

It follows by an easy induction on r that

---

(1) As usual $xX = X^x = \mathrm{Ad}(x)X$ for $x \in G$ and $X \in \mathfrak{g}_c$.

(2) $[A, B] = AB - BA$ for two endomorphisms A and B of a vector space.

$$\Gamma_x(X_1 X_2 \cdots X_r \otimes Y_1 Y_2 \cdots Y_s) \equiv X_1{'} X_2{'} \cdots X_r{'} Y_1 Y_2 \cdots Y_s \bmod_{(r+s-1)} \mathfrak{G},$$

where all the products are in $\mathfrak{G}$. Hence our assertion is an immediate consequence of the following well-known fact (see [4(a), p. 902]).

LEMMA 3. *Let $Z_1, \cdots, Z_d$ be elements of $\mathfrak{g}_c$ and $(i_1, i_2, \cdots, i_d)$ a permutation of $(1, 2, \cdots, d)$. Then*

$$Z_1 Z_2 \cdots Z_d - Z_{i_1} Z_{i_2} \cdots Z_{i_d} \in {}_{(d-1)}\mathfrak{G}.$$

As usual we regard elements of $\mathfrak{G}$ as left-invariant differential operators on G. Moreover, for every $X \in \mathfrak{g}$, let $\rho(X)$ denote the right-invariant vector-field on G given by[3]

$$f(x; \rho(X)) = (df(\exp tX \cdot x)/dt)_{t=0} \qquad (x \in G, f \in C^\infty(G)).$$

A simple argument shows that $[\rho(X), \rho(Y)] = -\rho([X, Y])$ and therefore ρ can be extended uniquely to an anti-homomorphism of $\mathfrak{G}$ into the algebra of all differential operators on G. We define

$$f(g; x) = f(x; \rho(g))$$

for $x \in G$, $g \in \mathfrak{G}$ and $f \in C^\infty(G)$. If $X_1, \cdots, X_r \in \mathfrak{g}$, then[4]

$$f(X_1 X_2 \cdots X_r; x) = f(x; \rho(X_1 X_2 \cdots X_r)) = f(x; \rho(X_r) \cdots \rho(X_2) \rho(X_1))$$

$$= \{\partial^r f(\exp t_1 X_1 \cdots \exp t_r X_r \cdot x)/\partial t_1 \cdots \partial t_r\}_{t_1 = \cdots = t_r = 0}$$

$$= f(x; (X_1 X_2 \cdots X_r)^{x^{-1}})$$

since $\exp t_1 X_1 \cdots \exp t_r X_r \cdot x = x(\exp t_1 X_1 \cdots \exp t_r X_r)^{x^{-1}}$. Therefore

$$f(g; x) = f(x; g^{x^{-1}}) \qquad (g \in \mathfrak{G}).$$

It is obvious that X and $\rho(Y)$ $(X, Y \in \mathfrak{g})$ commute (in the algebra of differential operators on G) and therefore g_1 and $\rho(g_2)$ $(g_1, g_2 \in \mathfrak{G})$ also commute.

Now G operates on itself by means of inner automorphisms so that $y^x = xyx^{-1}$ $(x, y \in G$, see [4(k), §5]). Let Ω_0 and G_0 be two open sets in G and f a C^∞-function on $\Omega = \Omega_0{}^{G_0}$. Put $f(x : y) = f(y^x)$ $(x \in G_0, y \in \Omega_0)$. The significance of the mapping Γ_y arises from the following lemma.

LEMMA 4. *Let $g_1, g_2 \in \mathfrak{G}$. Then*

$$f(x; g_1 : y; g_2) = f(x : y; \Gamma_y(g_1 \otimes g_2))$$

for $x \in G_0$ and $y \in \Omega_0$.

---

[3] We use here the notation of [4(k), §2].

[4] For any $x \in G$, we extend $\mathrm{Ad}(x)$ to an automorphism $g \to g^x$ of $\mathfrak{G}$ and define $y^x = xyx^{-1}$ $(y \in G)$.

If $X \in \mathfrak{g}$, it is clear that

$$y^{\exp tX} = y \exp(tX^{y^{-1}}) \exp(-tX) \qquad (y \in G, t \in R).$$

On the other hand if $x \in G_0$ and $y \in \Omega_0$, it is clear that

$$f(x \exp tX : y) = f(x : y^{\exp tX})$$

provided $|t|$ is small. Therefore

$$f(x; X : y) = f(x : y; X^{y^{-1}} - X) = f(x : y; \rho(X) - X).$$

Since $\rho(X)$ commutes with g_2, it follows by differentiation with respect to y that

$$f(x; X : y; g_2) = f(x : y; \rho(X) \circ g_2 - g_2 X)$$
$$= f(x : y; X^{y^{-1}} g_2 - g_2 X) = f(x : y; \sigma_y(X) g_2).$$

Hence if $X_1, \cdots, X_r \in \mathfrak{g}$ and $g_1 = X_1 X_2 \cdots X_r$, it follows by induction on r that

$$f(x; g_1 : y; g_2) = f(x : y; \sigma_y(g_1) g_2).$$

The statement of the lemma is now obvious.

3. **Completely invariant sets.** Assume that G is connected. Consider the polynomial

$$\det(t + 1 - \mathrm{Ad}(x)) = \sum_{0 \le j \le n} t^j D_j(x) \qquad (x \in G),$$

where t is an indeterminate and $n = \dim G$. Then D_j are analytic functions on G and $D_n = 1$. Let l be the least integer such that $D_l \ne 0$. Then $l = \mathrm{rank}\, G = \mathrm{rank}\, \mathfrak{g}$ and an element $x \in G$ is called singular or regular according as $D_l(x) = 0$ or not. Let G' be the set of all regular elements. Then it is obvious that G' is open and dense in G and the set of singular elements is of measure zero with respect to the left-invariant Haar measure of G.

We say that G is reductive if $\mathfrak{g}$ is reductive. An element $x \in G$ is called semi-simple if the endomorphism $\mathrm{Ad}(x)$ of $\mathfrak{g}$ is semisimple. Let $\mathfrak{z}_x$ denote the centralizer of x in $\mathfrak{g}$. We assume, from now on, that G is reductive.

LEMMA 5. *Let x be an element of G. Then $x \in G'$ if and only if $\mathfrak{z}_x$ is a Cartan subalgebra of $\mathfrak{g}$. Moreover, if x is semisimple then $\mathfrak{z}_x$ is reductive in $\mathfrak{g}$ and* rank $\mathfrak{z}_x$ = rank $\mathfrak{g}$. *Finally, a regular element is always semisimple.*

It is clearly enough to consider the case when $\mathfrak{g}$ is semisimple. The first and last statements follow from [4(e), Lemma 5]. Put $B(X, Y) = \mathrm{tr}\,(\mathrm{ad}\, X\, \mathrm{ad}\, Y)\, (X, Y \in \mathfrak{g})$ and let B_x denote the restriction of the bilinear form B on $\mathfrak{z}_x$. Now assume that x is semisimple. An elementary argument (see [2, p. 391]) shows that B_x is non-degenerate. Hence it follows from [2, Proposition 3.4] and [3] that $\mathfrak{z}_x$ is reductive in $\mathfrak{g}$. Finally rank $\mathfrak{z}_x = \operatorname{rank}\mathfrak{g}$ from [2, Proposition 4.6].

COROLLARY. *If x is semisimple, it is contained in a Cartan subgroup of G.*

Let $\mathfrak{h}$ be a Cartan subalgebra of $\mathfrak{z}_x$ and A the centralizer of $\mathfrak{h}$ in G. Since rank $\mathfrak{z}_x = \operatorname{rank}\mathfrak{g}$, A is a Cartan subgroup of G and $x \in A$.

Let $\mathscr{N}$ denote the set of all nilpotent elements (see [4(n), §3]) of $\mathfrak{g}$. Put $\mathscr{N}_G = \exp \mathscr{N} \subset G$. The mapping $X \to \exp \operatorname{ad} X$ ($X \in \mathscr{N}$) is known (see [4(h), §3]) to be univalent on $\mathscr{N}$.

LEMMA 6. *Every $x \in G$ can be written uniquely in the form $x = hn$, where h is a semisimple element of G, $n \in \mathscr{N}_G$ and h, n commute with each other. Let Z_x denote the centralizer of x in G. Then h and n lie in the center of Z_x.*

It is obviously enough to consider the case when $\mathfrak{g}$ is semisimple and G is the connected component of 1 in the adjoint group G_0 of $\mathfrak{g}$. Then G_0 is the set of all real points of a linear algebraic group defined over R. Therefore the lemma follows from well-known results on algebraic groups (see, for example, [1, §8]). h and n, respectively, are called the semisimple and unipotent components of x.

COROLLARY(5). $h \in \operatorname{Cl}(x^G)$.

Choose $X \in \mathscr{N}$ such that $n = \exp X$ and let $\mathfrak{z}$ denote the centralizer of h in $\mathfrak{g}$. Then $X \in \mathfrak{z}$. We may obviously assume that $X \neq 0$. Then by the Jacobson-Morosow theorem, we can choose $H \in \mathfrak{z}$ such that $[H, X] = 2X$ (see [4(n), §3]). Put $y_t = \exp(-tH)$. Then

$$x^{y_t} = (h \exp X)^{y_t} = h \exp(e^{-2t} X) \to h$$

as $t \to +\infty$. This proves the corollary.

Let U be a subset of G. We say that U is completely invariant (cf. [4(n), §3]) if it has the following property. If C is any compact subset of U, then $\operatorname{Cl}(C^G) \subset U$.

LEMMA 7. *Let U be a completely invariant subset of G and V an invariant subset of U which is closed in U. Then if V contains no semisimple element of U, V is empty.*

Suppose $x \in V$. Then it follows from the corollary of Lemma 6 that the semisimple component of x also lies in V. Hence the lemma.

Now assume that $\mathfrak{g}$ is semisimple. For any $c > 0$, let $\mathfrak{g}(c)$ denote the set of all $X \in \mathfrak{g}$ such that(6) $|\operatorname{Im}\lambda| < c$ for every eigenvalue λ of $\operatorname{ad} X$. Clearly $\mathfrak{g}(c)$ is an open and completely invariant neighborhood of zero in $\mathfrak{g}$ and $\mathscr{N} \subset \mathfrak{g}(c)$. Moreover, if $X \in \mathfrak{g}(c)$ then $tX \in \mathfrak{g}(c)$ for $0 \leq t \leq 1$. Hence $\mathfrak{g}(c)$ is connected.

LEMMA 8. *Assume that $c \leq \pi$. Then the exponential mapping from $\mathfrak{g}$ into G is everywhere regular and univalent on $\mathfrak{g}(c)$. Moreover, $\exp \mathfrak{g}(c)$ is completely invariant in G.*

The proof of the first part is the same as that of [4(h), Lemma 11]. In order to obtain the second part we use the notation of [4(h), Lemma 12] and first prove the following lemma.

LEMMA 9. *Assume $c \leq \pi$ and let X_r $(r \geq 1)$ be a sequence in $\mathfrak{g}(c)$. Then if $\| \exp X_r \|$ remains bounded, the same holds for $\| X_r \|$.*

We keep to the notation of the proof of [4(h), Lemma 12]. Then $X_r = \mathrm{Ad}(u_r)Y_r$, $Y_r = H_r + Z_r$ and therefore $|\mathrm{Im}\,\alpha(H_r)| < c \leq \pi$ for any $\alpha \in P$. On the other hand $\| \exp X_r \| = \| \exp Y_r \|$ and $\mathrm{ad}\,Y_r$ has the same eigenvalues as $\mathrm{ad}\,H_r$. This shows that $|e^{\alpha(H_r)}|$ remains bounded for every $\alpha \in P$. In view of the above result this implies that $\| H_r \|$ itself remains bounded. The rest of the proof now goes through in the same way as for Lemma 12 of [4(h)].

Now fix a compact set C in $V = \exp \mathfrak{g}(c)$. We have to show that $\mathrm{Cl}(C^G) \subset V$. Let X_k and x_k $(k \geq 1)$ be sequences in $\mathfrak{g}(c)$ and G respectively such that $\exp X_k \in C$ and $(\exp X_k)^{x_k}$ converges to some point y in G. We have to verify that $y \in V$. Let $\log$ denote the inverse mapping from V to $\mathfrak{g}(c)$. Then $\log C$ is compact. Hence, in view of Lemma 9, we can, by choosing suitable subsequences, arrange that $X_k \to X$ and $x_k X_k \to Y$, where $X \in \log C$ and $Y \in \mathfrak{g}$. But it is obvious that $\mathrm{ad}\,X$ and $\mathrm{ad}\,Y$ have the same eigenvalues. Hence $Y \in \mathfrak{g}(c)$ and therefore $y = \exp Y \in V$. This completes the proof of Lemma 8.

4. Some algebraic results([7]). We return again to the case when $\mathfrak{g}$ is reductive. Fix a semisimple element $a \in G$ and let $\mathfrak{z} = \mathfrak{z}(a)$ denote the centralizer of a in $\mathfrak{g}$ and $\Xi = \Xi(a)$ the analytic subgroup of G corresponding to $\mathfrak{z}$. Put

$$v_a(y) = \det(\mathrm{Ad}(ay)^{-1} - 1)_{\mathfrak{g}/\mathfrak{z}} \quad (y \in \Xi).$$

Then v_a is an analytic function on Ξ and $v_a(1) \neq 0$. Let $\Xi' = \Xi'(a)$ be the set of all points $y \in \Xi$, where $v_a(y) \neq 0$. Then Ξ' is an open neighborhood of 1 in Ξ.

In view of Lemma 5, $\mathfrak{z}$ satisfies the conditions of [4(l), §2]. Define $\mathfrak{q}$ as in [4(l), §2] and put $\mathfrak{Q} = \mathfrak{S}(\mathfrak{q}_c)$ and $\mathfrak{Q}_+ = \mathfrak{S}_+(\mathfrak{q}_c)$ in the notation of [4(k), §7].

LEMMA 10. *Fix $y \in \Xi'$. Then Γ_{ay} defines a bijective mapping of $\mathfrak{Q} \otimes \mathfrak{S}(\mathfrak{z}_c)$ onto $\mathfrak{S}$. Moreover,*

$$\sum_{d_1 + d_2 \leq d} \Gamma_{ay}(\mathfrak{S}_{d_1}(\mathfrak{q}_c) \otimes \mathfrak{S}_{d_2}(\mathfrak{z}_c)) = {}_d\mathfrak{S} \qquad (d \geq 0).$$

Put $W_d = \sum_{d_1 + d_2 \leq d} \mathfrak{S}_{d_1}(\mathfrak{q}_c) \otimes \mathfrak{S}_{d_2}(\mathfrak{z}_c)$. Since $\mathfrak{g}$ is the direct sum of $\mathfrak{q}$ and $\mathfrak{z}$, it is clear that $\dim W_d = \dim {}_d\mathfrak{S}$. Hence it would be sufficient to prove that

---

(7) The results of this section are similar to those of [4(l), §2]. See also [4(e)].

$\Gamma_{ay}(W_d) = {}_d\mathfrak{G}$. We do this by induction on d. It is obvious that $\Gamma_{ay}(W_d) \subset {}_d\mathfrak{G}$. Hence it would be sufficent to show that

$$_d\mathfrak{G} \subset \Gamma_{ay}(W_d) + {}_{(d-1)}\mathfrak{G}.$$

Fix two integers $d_1, d_2 \geq 0$ such that $d_1 + d_2 = d$ and suppose $Y_i \in \mathfrak{q}$ $(1 \leq i \leq d_1)$ and $Z_j \in \mathfrak{z}$ $(1 \leq j \leq d_2)$. Let $q = Y_1 Y_2 \cdots Y_{d_1} \in S(\mathfrak{q}_c)$ and $z = Z_1 Z_2 \cdots Z_{d_2} \in S(\mathfrak{z}_c)$. (Here we have to take $q = 1$ if $d_1 = 0$ and $z = 1$ if $d_2 = 0$.) Define λ as in §2. Then it would be enough to verify that

$$\lambda(qz) \in \Gamma_{ay}(W_d) + {}_{(d-1)}\mathfrak{G}.$$

If $d_1 = 0$, this is obvious since $\Gamma_{ay}(1 \otimes \lambda(z)) = \lambda(z)$. Hence we may assume $d_1 > 0$. Now ay commutes with a and therefore $\mathfrak{z}^{ay} = \mathfrak{z}$ and $\mathfrak{q}^{ay} = \mathfrak{q}$ (see [4(1), §2]). Therefore since $v_a(y) \neq 0$, we can choose $Y_i' \in \mathfrak{q}$ such that $(\mathrm{Ad}(ay)^{-1} - 1)Y_i' = Y_i$ $(1 \leq i \leq d_1)$. Put $q' = Y_1' \cdots Y_r' \in S(\mathfrak{q}_c)$. Then

$$\Gamma_{ay}(\lambda(q') \otimes \lambda(z)) \equiv \lambda(qz) \quad \mathrm{mod} \ {}_{(d-1)}\mathfrak{G}$$

from Lemma 2. This proves the lemma.

COROLLARY 1. *Fix* $g \in \mathfrak{G}$. *Then for any* $y \in \Xi'$, *there exist unique elements* $\alpha_y(g) \in \mathfrak{S}(\mathfrak{z}_c)$ *and* $\beta_y(g) \in \mathfrak{Q}_+ \otimes \mathfrak{S}(\mathfrak{z}_c)$ *such that*

$$g = \alpha_y(g) + \Gamma_{ay}(\beta_y(g)).$$

Moreover, if $g \in {}_d\mathfrak{G}$, *then* $d^0 \alpha_y(g) \leq d$ *and*

$$\beta_y(g) \in \sum_{d_2 \geq 0} \sum_{1 \leq d_1 \leq d - d_2} \mathfrak{S}_{d_1}(\mathfrak{q}_c) \otimes \mathfrak{S}_{d_2}(\mathfrak{z}_c).$$

This is obvious from Lemma 10.

Let M be an analytic manifold and f a mapping of M into a complex vector space V. We say that f is analytic if the subspace U of V spanned by the image $f(M)$ is of finite dimension and f, viewed as a mapping of M into U, is analytic in the usual sense.

COROLLARY 2. *Given* $g \in \mathfrak{G}$, *we can choose an integer* $r \geq 0$ *such that the mappings* $y \to v_a(y)^r \alpha_y(g)$ *and* $y \to v_a(y)^r \beta_y(g)$ $(y \in \Xi')$ *can be extended to analytic mappings of* Ξ *into* $\mathfrak{S}(\mathfrak{z}_c)$ *and* $\mathfrak{Q}_+ \otimes \mathfrak{S}(\mathfrak{z}_c)$, *respectively.*

Let $d = d^0 g$. If $d = 0$ our statement is obvious. So we assume that $d \geq 1$ and use induction. We may obviously assume that $g = \lambda(qz)$ in the notation of the proof of Lemma 10. Let $A(y)$ denote the restriction of $(\mathrm{Ad}(ay)^{-1} - 1)$ on $\mathfrak{q}$ $(y \in \Xi)$. Then if t is an indeterminate,

$$\det(t - A(y)) = \sum_{0 \leq k \leq m} D_k(y)t^k.$$

Here $m = \dim \mathfrak{q}$, D_k $(0 \leq k \leq m)$ are analytic functions on Ξ, $D_m = 1$ and

$$D_0(y) = (-1)^m \det A(y) = (-1)^m v_a(y).$$

Therefore

$$v_a(y) = B(y) A(y) = A(y) B(y),$$

where

$$B(y) = (-1)^{m+1} \sum_{0 \le k < m} D_{k+1}(y) A(y)^k.$$

Put $Y_i(y) = B(y) Y_i$ $(1 \le i \le d_1)$ and $q(y) = \prod_{1 \le i \le d_1} Y_i(y) \in S(\mathfrak{q}_c)$ $(y \in \Xi)$. Then it follows from Lemma 2 that

$$v(y) = \Gamma_{ay}(\lambda(q(y)) \otimes z) - v_a(y)^{d_1} \lambda(qz) \in {}_{(d-1)}\mathfrak{G}.$$

Therefore

$$v(y) = \sum_{1 \le l \le r} a_i(y) v_l$$

where a_i are analytic functions on Ξ and $d^0 v_i < d$. The required result now follows immediately by applying the induction hypothesis to v_i $(1 \le i \le r)$.

COROLLARY 3. *Let Z_a denote the centralizer of a in G. Then if $x \in Z_a$, $y \in \Xi'$ and $g \in \mathfrak{G}$, we have*

$$\alpha_{xyx^{-1}}(g^x) = (\alpha_y(g))^x.$$

This follows immediately from Lemma 1.

5. **The mapping $\delta_{a,G/\Xi}$.** We keep to the notation of §4. Let U_G be an open neighborhood of a in G. Put $U_\Xi = \Xi' \cap (a^{-1} U_G)$. Then U_Ξ is an open neighborhood of 1 in Ξ. For any differential operator D on U_G, we define a differential operator $\Delta(D)$ on U_Ξ as follows:

$$(\Delta(D))_y = \alpha_y(D_{ay}) \qquad (y \in U_\Xi).$$

Here D_{ay} and $(\Delta(D))_y$ denote, as usual, the local expressions (see [4(e), p. 112]) of D at ay and $\Delta(D)$ at y, respectively. Corollary 2 of Lemma 10 insures that there does exist a differential operator $\Delta(D)$ on U_Ξ satisfying the above relation and it is analytic if D is analytic. Finally, if we assume that U_G and D are invariant under G (see [4(j), §2]), it follows from Corollary 3 of Lemma 10 that U_Ξ and $\Delta(D)$ are invariant under Z_a. We shall denote the mapping $D \to \Delta(D)$ by δ_a or, if necessary, by $\delta_{a,G/\Xi}$.

Let b be an element of U_Ξ which is regular in Ξ and let $\mathfrak{h}$ denote the centralizer of b in $\mathfrak{z}$. Then $\mathfrak{h}$ is a Cartan subalgebra of $\mathfrak{z}$ and therefore also of $\mathfrak{g}$ (see Lemma 5). Let $A_\mathfrak{h}$ denote the Cartan subgroup of G corresponding to $\mathfrak{h}$ (see [4(m), §5]). Then a, b are in $A_\mathfrak{h}$. Let A be the connected component of 1 in $A_\mathfrak{h}$.

LEMMA 11. *The following two conditions on an element c of A are mutually equivalent.*

(1) $c \in b^{-1} U_\Xi$ *and* $\det(\mathrm{Ad}(bc)^{-1} - 1)_{\mathfrak{z}/\mathfrak{h}} \ne 0$.

(2) $c \in (ab)^{-1} U_G$ *and* $\det(\mathrm{Ad}(abc)^{-1} - 1)_{\mathfrak{g}/\mathfrak{h}} \ne 0$.

Since $\mathrm{Ad}(a) = 1$ on $\mathfrak{z}$, it is clear that

$$\det(\mathrm{Ad}(abc)^{-1} - 1)_{\mathfrak{g}/\mathfrak{h}} = \det(\mathrm{Ad}(abc)^{-1} - 1)_{\mathfrak{g}/\mathfrak{z}}\det(\mathrm{Ad}(bc)^{-1} - 1)_{\mathfrak{z}/\mathfrak{h}}$$

for $c \in A$. Now suppose (1) holds. Then $bc \in U_{\Xi}$ and therefore $abc \in U_G$ and $\det(\mathrm{Ad}(abc)^{-1} - 1)_{\mathfrak{g}/\mathfrak{z}} \neq 0$. Therefore (1) implies (2). Conversely, assume that (2) holds. Then $bc \in a^{-1}U_G$ and

$$\nu_a(bc)\det(\mathrm{Ad}(bc)^{-1} - 1)_{\mathfrak{z}/\mathfrak{h}} \neq 0.$$

Hence $bc \in U_{\Xi}$ and (1) holds.

COROLLARY. *ab is regular in G and $\mathfrak{h}$ is the centralizer of ab in* g.

Take $c = 1$ in Lemma 11. Then condition (1) obviously holds and therefore $\det(\mathrm{Ad}(ab)^{-1} - 1)_{\mathfrak{g}/\mathfrak{h}} \neq 0$ by (2). Since $\mathrm{Ad}(ab) = 1$ on $\mathfrak{h}$, it follows that $\mathfrak{h}$ is the centralizer of ab in g. Therefore ab is regular in G by Lemma 5.

Let U_A be the set of all $c \in A$ satisfying the conditions of Lemma 11.

LEMMA 12. *Let D be a differential operator on U_G. Then*[8]

$$\delta_{ab,G/A}(D) = \delta_{b,\Xi/A}(\delta_{a,G/\Xi}(D)).$$

It follows from Lemma 11 that both sides are differential operators on U_A. Let $\Delta_1 = \delta_{a,G/\Xi}(D)$, $\Delta_2 = \delta_{b,\Xi/A}(\Delta_1)$ and $\Delta = \delta_{ab,G/A}(D)$. We have to prove that $\Delta_2 = \Delta$. Let $\mathfrak{m} = [\mathfrak{h},\mathfrak{z}]$ and $\mathfrak{p} = \mathfrak{q} + \mathfrak{m}$. Then $\mathfrak{g} = \mathfrak{h} + \mathfrak{p}$ and $\mathfrak{z} = \mathfrak{h} + \mathfrak{m}$ where both sums are direct.

Fix $h \in U_A$. Then

$$(\Delta_2)_h - (\Delta_1)_{bh} \in \Gamma_{bh}(\mathfrak{S}_+(\mathfrak{m}_c) \otimes \mathfrak{S}(\mathfrak{h}_c)).$$

On the other hand $bh \in U_{\Xi}$ and therefore

$$(\Delta_1)_{bh} - D_{abh} \in \Gamma_{abh}(\mathfrak{S}_+(\mathfrak{q}_c) \otimes \mathfrak{S}(\mathfrak{z}_c)).$$

Since $\mathrm{Ad}(a) = 1$ on $\mathfrak{z}$, it is clear that

$$\sigma_{abh}(z) = \sigma_{bh}(z) \quad (z \in \mathfrak{S}(\mathfrak{z}_c))$$

and therefore

$$(\Delta_2)_h - D_{abh} \in \Gamma_{abh}(\mathfrak{S}_+ \otimes \mathfrak{S}(\mathfrak{z}_c)),$$

where $\mathfrak{S}_+ = \mathfrak{S}_+(\mathfrak{g}_c)$. But

$$\Gamma_{abh}(\mathfrak{S}(\mathfrak{m}_c) \otimes \mathfrak{S}(\mathfrak{h}_c)) = \Gamma_{bh}(\mathfrak{S}(\mathfrak{m}_c) \otimes \mathfrak{S}(\mathfrak{h}_c)) = \mathfrak{S}(\mathfrak{z}_c)$$

from Lemma 10 (applied to (Ξ, b) instead of (G, a)), since $\det(\mathrm{Ad}(bh)^{-1} - 1)_{\mathfrak{z}/\mathfrak{h}} \neq 0$. Hence

$$\begin{aligned}
\Gamma_{abh}(\mathfrak{S}_+ \otimes \mathfrak{S}(\mathfrak{z}_c)) &= \sigma_{abh}(\mathfrak{S}_+)\mathfrak{S}(\mathfrak{z}_c) \\
&= \sigma_{abh}(\mathfrak{S}_+)\sigma_{abh}(\mathfrak{S}(\mathfrak{m}_c))\mathfrak{S}(\mathfrak{h}_c) \\
&= \sigma_{abh}(\mathfrak{S}_+)\mathfrak{S}(\mathfrak{h}_c).
\end{aligned}$$

---

[8] Cf. [4(I), Lemma 11].

But $\mathfrak{G} = \mathfrak{S}(\mathfrak{p}_c)\mathfrak{S}(\mathfrak{h}_c)$ and since $\mathfrak{h}$ is abelian, it is clear that

$$\sigma_{abh}(\mathfrak{S}_+(\mathfrak{h}_c))\mathfrak{S}(\mathfrak{h}_c) = \{0\}.$$

Therefore since

$$\mathfrak{G}_+ = \mathfrak{S}_+(\mathfrak{p}_c) + \mathfrak{S}(\mathfrak{p}_c)\mathfrak{S}_+(\mathfrak{h}_c),$$

we conclude that

$$\Gamma_{abh}(\mathfrak{G}_+ \otimes \mathfrak{S}(\mathfrak{z}_c)) = \Gamma_{abh}(\mathfrak{S}_+(\mathfrak{p}_c) \otimes \mathfrak{S}(\mathfrak{h}_c)).$$

This shows that

$$(\Delta_2)_h - D_{abh} \in \Gamma_{abh}(\mathfrak{S}_+(\mathfrak{p}_c) \otimes \mathfrak{S}(\mathfrak{h}_c))$$

and therefore $(\Delta_2)_h = \Delta_h$ from the definition of Δ.

6. The case when a is regular. Let $\mathfrak{Z}$ be the algebra of all differential operators on G which are invariant under both left and right translations of G. It is obvious that $\mathfrak{Z}$ consists of the center of $\mathfrak{G}$ and therefore $\mathfrak{Z}$ is abelian.

Let G' be the set of all regular elements of G. Fix $a \in G'$ and let $\mathfrak{h}$ denote the centralizer of a in $\mathfrak{g}$ and A the analytic subgroup of G corresponding to $\mathfrak{h}$. Then $\mathfrak{h}$ is a Cartan subalgebra of $\mathfrak{g}$ and

$$v_a(h) = \det(\mathrm{Ad}(ah)^{-1} - 1)_{\mathfrak{g}/\mathfrak{h}} \qquad (h \in A).$$

Hence $A' = A \cap (a^{-1}G')$ is the set of all points $h \in A$ where $v_a(h) \neq 0$. Let W be the Weyl group of $(\mathfrak{g}_c, \mathfrak{h}_c)$. Then W operates on $\mathfrak{S}(\mathfrak{h}_c)$. Let $I(\mathfrak{h}_c)$ be the algebra of all invariants of W in $\mathfrak{S}(\mathfrak{h}_c)$. We have a canonical isomorphism γ of $\mathfrak{Z}$ onto $I(\mathfrak{h}_c)$ (see [4(e), Lemma 19]). Thus for every $z \in \mathfrak{Z}$, $\gamma(z)$ is a differential operator on A which is invariant under the translations of A.

LEMMA 13. $\delta_{a,G/A}(z) = |v_a|^{-1/2}\gamma(z) \circ |v_a|^{1/2}$ on A' for any $z \in \mathfrak{Z}$.

This is substantially the same as the first statement of [4(e), Theorem 2, p. 125].

7. Application to invariant distributions(9). Fix a semisimple element $a \in G$ and define Ξ and Ξ' as in §4.

LEMMA 14. Consider the mapping $\phi: (x,y) \to (ay)^x$ of $G \times \Xi$ into G. Then if $n = \dim G$, ϕ is everywhere of rank n on $G \times \Xi'$.

We identify the tangent space of $G \times \Xi$ at a point (x,y) with $\mathfrak{g} \times \mathfrak{z}$ in the usual way. Then a simple calculation shows that

$$(d\phi)_{x,y}(X,Z) = (Z + (\mathrm{Ad}(ay)^{-1} - 1)X)^x$$

for $X \in \mathfrak{g}$ and $Z \in \mathfrak{z}$. But

$$\mathfrak{z} + (\mathrm{Ad}(ay)^{-1} - 1)\mathfrak{q} = \mathfrak{z} + \mathfrak{q} = \mathfrak{g}$$

if $y \in \Xi'$ and therefore our assertion is obvious.

---

(9) The results of this section are similar to those of [4(l), §7].

Let dx denote the Haar measure on G. We orient G and fix a left-invariant differential form $\omega_G > 0$ of degree n on G, corresponding to the measure dx. Then the set up of $[4(k), \S 5]$ is applicable to $M = G$, if we define $y^x = xyx^{-1}$ $(x, y \in G)$ as above.

Let U be an open neighborhood of 1 in Ξ' which is invariant under Ξ (i.e., $U^y = U$ for $y \in \Xi$). Put $\Omega = \phi(G \times U) = (aU)^G$. Then by Lemma 14, Ω is open in G. Let dx, dy denote the Haar measures on G and Ξ, respectively. Now take $M = G \times U$, $N = \Omega$, $\pi = \phi$ in Theorem 1 of $[4(k)]$ and let ω_M and ω_N be the differential forms corresponding to the measures $dx\,dy$ and dx, respectively. Let $\alpha \to f_\alpha$ denote the corresponding mapping of $C_c^\infty(G \times U)$ onto $C_c^\infty(\Omega)$.

LEMMA 15. *Let T be an invariant distribution on Ω. Then there exists a unique distribution σ_T on U such that $T(f_\alpha) = \sigma_T(\beta_\alpha)$ $(\alpha \in C_c^\infty(G \times U))$, where*

$$\beta_\alpha(y) = \int \alpha(x : y)\,dx \qquad (y \in U).$$

Moreover, σ_T is invariant under Ξ and $\sigma_T = 0$ implies that $T = 0$.

Define $T'(\alpha) = T(f_\alpha)$ $(\alpha \in C_c^\infty(G \times U))$. Then (see $[4(k),$ Lemma 5$]$) T' is a distribution on $G \times U$. Fix $x_0 \in G$ and let α_{x_0} denote the function $(x, y) \to \alpha(x_0 x : y)$ on $G \times U$. We claim that $T'(\alpha) = T'(\alpha_{x_0})$. For if $F \in C_c^\infty(\Omega)$, we have

$$\int f_{\alpha_{x_0}} F\,dx \;=\; \int \alpha_{x_0}(x : y)\,F((ay)^x)\,dx\,dy \;=\; \int \alpha(x : y)\,F^{x_0}((ay)^x)\,dx\,dy$$

$$= \int f_\alpha F^{x_0}\,dx = \int f_\alpha^{x_0^{-1}} \quad F\,dx.$$

Hence $f_{\alpha_{x_0}} = f_\alpha^{x_0^{-1}}$ and therefore $T'(\alpha_{x_0}) = T(f_{\alpha_{x_0}}) = T(f_\alpha) = T'(\alpha)$. Now fix $\beta \in C_c^\infty(U)$ and put $T'_\beta(\gamma) = T'(\gamma \times \beta)$ $(\gamma \in C_c^\infty(G))$. Then T'_β is a distribution on G which is invariant under the left translations of G. Hence $T'_\beta = c(\beta)$, where $c(\beta)$ is a constant (see $[4(k),$ Lemmas 6 and 7$]$). Now select $\gamma_0 \in C_c^\infty(G)$ such that $\int \gamma_0\,dx = 1$. Then

$$c(\beta) = T'_\beta\,(\gamma_0) = T'(\gamma_0 \times \beta) \qquad (\beta \in C_c^\infty(U)).$$

This shows that the mapping $\beta \to c(\beta)$ is a distribution on U which we denote by σ_T. Then

$$T'(\gamma \times \beta) = \sigma_T(\beta) \int \gamma\,dx \qquad (\gamma \in C_c^\infty(G), \beta \in C_c^\infty(U))$$

and therefore we conclude from $[4(k),$ Lemma 3$]$ that $T'(\alpha) - \sigma_T(\beta_\alpha) = 0$ for $\alpha \in C_c^\infty(G)$. Since $\beta_\alpha = \beta$ for $\alpha = \gamma_0 \times \beta$, the mapping $\alpha \to \beta_\alpha$ of $C_c^\infty(G \times U)$ into $C_c^\infty(U)$ is surjective. Finally the mapping $\alpha \to f_\alpha$ of $C_c^\infty(G \times U)$ into $C_c^\infty(\Omega)$ is also surjective (see $[4(k),$ Theorem 1$]$) and so all the statements of the lemma, except the invariance of σ_T under Ξ, are now obvious.

Fix $\zeta \in \Xi$ and define $\alpha^\zeta(x:y) = \alpha(x:y^{\zeta^{-1}})$. Then we claim that $T'(\alpha) = T'(\alpha^\zeta)$ for $\alpha \in C_c^\infty(G \times U)$. This is seen as follows.

$$\int f_{\alpha^\zeta} F\, dx = \int \alpha(x:y^{\zeta^{-1}}) F((ay)^x)\, dx\, dy = \int \alpha(x:y) F((ay)^{x\zeta})\, dx\, dy$$

$$= \int \alpha(x\zeta^{-1}:y) F((ay)^x)\, dx\, dy$$

for any $F \in C_c^\infty(\Omega)$. Hence if $\alpha'(x:y) = \alpha(x\zeta^{-1}:y)$, it is clear that $f_{\alpha^\zeta} = f_{\alpha'}$. Therefore

$$T'(\alpha^\zeta) = T'(\alpha') = \sigma_T(\beta_{\alpha'}).$$

But

$$\beta_{\alpha'}(y) = \int \alpha(x\zeta^{-1}:y)\, dx = \beta_\alpha(y) \qquad (y \in U)$$

by the right-invariance of dx. Hence $T'(\alpha^\zeta) = \sigma_T(\beta_\alpha) = T'(\alpha)$. On the other hand

$$\beta_{\alpha^\zeta}(y) = \int \alpha(x:y^{\zeta^{-1}})\, dx = \beta_\alpha(y^{\zeta^{-1}}) \qquad (y \in U).$$

Therefore $\beta_{\alpha^\zeta} = (\beta_\alpha)^\zeta$. Now for a given $\beta \in C_c^\infty(U)$, we can choose $\alpha \in C_c^\infty(G \times U)$ such that $\beta = \beta_\alpha$. Then

$$\sigma_T(\beta) = T'(\alpha) = T'(\alpha^\zeta) = \sigma_T(\beta_{\alpha^\zeta}) = \sigma_T(\beta^\zeta).$$

This shows that σ_T is invariant under Ξ.

COROLLARY. *Let D be an invariant differential operator on Ω. Then $\sigma_{DT} = \Delta\sigma_T$, where $\Delta = \delta_a(D)$.*

It follows from Corollary 2 of Lemma 10 and the definition of Δ, that we can select $q_i \in \mathfrak{S}_+(q_c)$, $v_i \in \mathfrak{S}(\mathfrak{z}_c)$ and $a_i \in C^\infty(U)$ $(1 \leq i \leq r)$ such that

$$D_{ay} = \Delta_y + \sum_{1 \leq i \leq r} a_i(y)\Gamma_{ay}(q_i \otimes v_i) \qquad (y \in U).$$

Moreover, $\sigma_{DT}(\beta_\alpha) = T(D^* f_\alpha)$ for $\alpha \in C_c^\infty(G \times U)$, where the star denotes the adjoint as usual. Fix $F \in C_c^\infty(\Omega)$. Then

$$\int D^* f_\alpha \cdot F\, dx = \int f_\alpha DF\, dx = \int \alpha(x:y) F((ay)^x; D)\, dx\, dy.$$

Put $F(x:u) = F(u^x)$ for any pair $(x,u) \in G \times G$ such that $u^x \in \Omega$. Then it is clear from Lemma 4 that

$$F((ay)^x; D) = F((ay)^x; D^x) = F(x:ay; D)$$

$$= F(x:ay; \Delta_y) + \sum_{1 \leq i \leq r} a_i(y) F(x; q_i: ay; v_i)$$

for $x \in G$ and $y \in U$. Put

$$\alpha_0(x:y) \;=\; \alpha(x:y;\Delta^*),$$

$$\alpha_i(x:y) \;=\; \alpha(x;q_i^*:y;(a_iv_i)^*) \qquad (1 \leq i \leq r).$$

Then

$$\int D^*f_\alpha \cdot F\, dx \;=\; \sum_{0 \leq i \leq r} \int \alpha_i(x:y) F((ay)^x)\, dx\, dy$$

$$=\; \sum_{0 \leq i \leq r} \int f_{\alpha_i} F\, dx.$$

This proves that

$$D^*f_\alpha \;=\; \sum_{0 \leq i \leq r} f_{\alpha_i}$$

and therefore

$$T(D^*f_\alpha) \;=\; \sum_{0 \leq i \leq r} \sigma_T(\beta_{\alpha_i}).$$

Now $\beta_{\alpha_0} = \Delta^*\beta_\alpha$ and if j denotes the distribution on G corresponding to the constant function 1, it is obvious that $q_i j = 0$ since $q_i \in \mathfrak{G}_+$. Hence it follows that $\beta_{\alpha_i} = 0$ $(1 \leq i \leq r)$ and therefore

$$T(D^*f_\alpha) \;=\; \sigma_T(\Delta^*\beta_\alpha).$$

This proves that $\sigma_{DT} = \Delta\sigma_T$.

For any $X \in \mathfrak{g}$, let $\tau_G(X)$ denote the vector-field on G defined by

$$\tau_G(X)f \;=\; (df^{\exp tX}/dt)_{t=0} \quad (f \in C^\infty(G)).$$

Let V be an open subset of G. Then the local invariance of a differential operator, a distribution or a C^∞-function on V is defined as in [4(k), §5]. Since $[\tau_G(X), \tau_G(Y)] = \tau_G([X, Y])$ $(X, Y \in \mathfrak{g})$, τ_G can be extended (uniquely) to a homomorphism of $\mathfrak{G}$ into the algebra of all differential operators on G.

Let G_0 and U_0 be open neighborhoods of 1 in G and Ξ', respectively, and put $\Omega_0 = (aU_0)^{G_0}$. Define the mapping $\alpha \to f_\alpha$ of $C_c^\infty(G_0 \times U_0)$ onto $C_c^\infty(\Omega_0)$ as above. Then the following result is proved in the same way as [4(l), Lemma 17] and [4(k), Theorem 3].

LEMMA 16. *Assume that G_0 is connected and T is a locally invariant distribution on Ω_0. Then there exists a unique distribution σ_T on U_0 such that $T(f_\alpha) = \sigma_T(\beta_\alpha)$ $(\alpha \in C_c^\infty(G_0 \times U_0))$, where*

$$\beta_\alpha(y) \;=\; \int \alpha(x:y)\, dx \quad (y \in U_0).$$

Moreover, σ_T is locally invariant (with respect to Ξ) and $\sigma_T = 0$ implies that

$T = 0$. Finally, $\sigma_{DT} = \delta_a(D)\sigma_T$ for any locally invariant differential operator D on Ω_0.

8. **Some preparation for Theorem 1.** Let G_0, Ω_0 and Ω be three open subsets of G such that $\Omega_0{}^{G_0} \subset \Omega$. For any $f \in C^\infty(\Omega)$, we write $f(x:y) = f(y^x)\,(x \in G_0, y \in \Omega_0)$ as in §2.

LEMMA 17. *Let $f \in C^\infty(\Omega)$. Then*

$$f(x;g:y) = f(x:y;\tau_G(g^*))$$

for $x \in G_0$, $y \in \Omega_0$ and $g \in \mathfrak{G}$.

The proof is the same as that of [4(k), Lemma 11].

LEMMA 18. *Let D be a differential operator and f a locally invariant C^∞-function on an open subset Ω of G. Fix a semisimple element $a \in \Omega$ and define $\Omega_\Xi = a^{-1}\Omega \cap \Xi'$ in the notation of §4. Then*

$$f(ay;D) = f(ay;\delta_a(D)) \qquad (y \in \Omega_\Xi).$$

Fix $y_0 \in \Omega_\Xi$ and choose open neighborhoods G_0 and Ω_0 of 1 and ay_0, respectively, in G such that $\Omega_0{}^{G_0} \subset \Omega$. Put $\Delta = \delta_a(D)$. Then it follows from the definition of Δ that

$$D_{ay_0} - \Delta_{y_0} = \sum_{1 \le i \le r} \Gamma_{ay_0}(g_i \otimes v_i),$$

where $g_i \in \mathfrak{G}_+$ and $v_i \in \mathfrak{S}(\mathfrak{z}_c)$. Therefore we conclude from Lemma 4 that

$$f(ay_0 ; D_{ay_0} - \Delta_{y_0}) = \sum_{1 \le i \le r} f(1;g_i:ay_0;v_i).$$

But

$$f(1;g_i:x) = f(x;\tau_G(g_i^*)) = 0 \qquad (x \in \Omega_0)$$

from Lemma 17 since f is locally invariant and $g_i^* \in \mathfrak{G}_+$. Therefore $f(1;g_i:x;v_i) = 0$ for $x \in \Omega_0$ and hence

$$f(ay_0;D_{ay_0} - \Delta_{y_0}) = 0.$$

This proves the lemma.

LEMMA 19. *Let D and Ω be as above. Then the following two conditions on D are equivalent.*

(1) $\delta_a(D) = 0$ *for every regular element a in Ω.*

(2) *For any open subset Ω_0 of Ω and a locally invariant C^∞-function f on Ω_0, $Df = 0$.*

Suppose (1) holds and Ω_0 and f are given as in condition (2). Fix a regular element $a \in \Omega_0$. Then it follows from Lemma 18 that

$$f(a;D) = f(a;\delta_a(D)) = 0.$$

Therefore $Df = 0$ on $\Omega_0' = \Omega_0 \cap G'$. Since Ω_0' is obviously dense in Ω_0, we conclude that $Df = 0$.

Conversely, assume that (2) holds and fix $a \in \Omega \cap G'$. Let $\mathfrak{h}$ be the centralizer of a in $\mathfrak{g}$ and A the analytic subgroup of G corresponding to $\mathfrak{h}$. Put $\Omega_A = a^{-1}\Omega \cap A'$ where A' is the set of all $h \in A$ where

$$v_a(h) = \det(\mathrm{Ad}(ah)^{-1} - 1)_{\mathfrak{g}/\mathfrak{h}} \neq 0.$$

Then $\delta_a(D)$ is a differential operator on Ω_A. Let $x \to x^*$ denote the natural projection of G on $G^* = G/A$. Since A is abelian, $(ah)^x$ ($x \in G$, $h \in A$) depends only on x^* and so we may denote it by $(ah)^{x^*}$. It follows from Lemma 14 that the mapping $\psi : (x^*, h) \to (ah)^{x^*}$ of $G^* \times \Omega_A$ into G is everywhere regular. Fix a point $h_0 \in \Omega_A$. Then we can choose open neighborhoods G_0^* and U of 1^* and h_0 in G^* and Ω_A, respectively, such that $\Omega_0 = \psi(G_0^* \times U) \subset \Omega$ and ψ defines an analytic diffeomorphism of $G_0^* \times U$ onto the open neighborhood Ω of ah_0 in Ω. Fix $\beta \in C_\infty(U)$ and define $f \in C^\infty(\Omega_0)$ by $f(\psi(x^*, h)) = \beta(h)$ ($x^* \in G_0^*$, $h \in U$). Then it is obvious that f is locally invariant and therefore $Df = 0$ by (2). On the other hand we know from Lemma 18 that

$$f(ah_0; u) = f(ah_0; D) = 0,$$

where u is the local expression of $\delta_a(D)$ at h_0. Since $u \in \mathfrak{S}(\mathfrak{h}_c)$ and $f(ah) = \beta(h)$ ($h \in U$), it is obvious that $\beta(h_0; u) = 0$. This being true for every $\beta \in C^\infty(U)$, we conclude that $u = 0$. Since h_0 was an arbitrary point of Ω_A, this proves that $\delta_a(D) = 0$. Therefore (2) implies (1).

9. First part of the proof of Theorem 1. We shall now begin the proof of the following theorem (cf. [4(n), Theorem 5]).

THEOREM 1. *Let Ω be a completely invariant open set in G and D an analytic differential operator on Ω. Assume that:*
(1) D is invariant under G,
(2) $\delta_a(D) = 0$ for every regular element $a \in \Omega$.
Then $DT = 0$ for every invariant distribution T on Ω.

We use induction on $\dim G$. By replacing (U, V) in Lemma 7 with $(\Omega, \mathrm{Supp}\, DT)$, it becomes obvious that it would be enough to verify that no semisimple element of Ω lies in $\mathrm{Supp}\, DT$. Let Z denote the center of G. Fix a semisimple element a in Ω and first assume that $a \notin Z$. Put $\Omega_\Xi = a^{-1}\Omega \cap \Xi'$ in the notation of §4. Then it is obvious that Ω_Ξ is a completely invariant open neighborhood of 1 in Ξ. Corresponding to Lemma 15, we get an invariant distribution σ_T on Ω_Ξ. Moreover, $\sigma_{DT} = \Delta\sigma_T$, where $\Delta = \delta_a(D)$ (see the corollary of Lemma 15). Fix an element $b \in \Omega_\Xi$ which is regular in Ξ. Then $ab \in \Omega' = \Omega \cap G'$ (see the corollary of Lemma 11) and therefore

$$\delta_{b,\Xi/A}(\Delta) = \delta_{ab}(D) = 0$$

in the notation of Lemma 12. Moreover, as we have seen in §5, Δ is analytic and invariant under Ξ. Now $\dim \Xi < \dim G$, since $a \notin Z$. Therefore we conclude from the induction hypothesis that $\Delta \sigma_T = 0$. But then $a \notin \operatorname{Supp} DT$ by Lemma 15.

So now we may assume that $a \in \Omega \cap Z$. It follows from its definition (see §2) that the mapping Γ_x depends only on $\operatorname{Ad}(x)$ $(x \in G)$. Therefore if we apply the translation by a^{-1} to the whole problem, we are reduced to the case $a = 1$. So we may assume that $1 \in \Omega$ and it remains to show that $1 \notin \operatorname{Supp} DT$.

Let c be the center and g_1 the derived algebra of g. Choose an open and relatively compact neighborhood c_0 of zero in c such that the exponential mapping is univalent on c_0. Moreover, select a number c $(0 < c \leq \pi)$ and define $g_1(c)$ as in Lemma 8. Put $g_0 = c_0 + g_1(c)$. Then g_0 is an open and completely invariant neighborhood of zero in g and the exponential mapping is everywhere regular on g_0. Now suppose $\exp(C_1 + X_1) = \exp(C_2 + X_2)$, where $C_i \in c_0$ and $X_i \in g_1(c)$ $(i = 1, 2)$. Then $\exp(\operatorname{ad} X_1) = \exp(\operatorname{ad} X_2)$ and so it follows from Lemma 8 that $X_1 = X_2$. Hence $\exp C_1 = \exp C_2$ and therefore $C_1 = C_2$ from the definition of c_0. This proves that the exponential mapping defines an analytic diffeomorphism of g_0 onto the open set $\exp g_0$ in G. Let $\log$ denote its inverse and put $U = \log \Omega_0$, where $\Omega_0 = \exp g_0 \cap \Omega$. Let V be a compact subset of U. Since g_0 is completely invariant, $\operatorname{Cl}(V^G) \subset g_0$. Moreover,

$$\exp\left(\operatorname{Cl}(V^G)\right) \subset \operatorname{Cl}(\exp V^G) = \operatorname{Cl}((\exp V)^G) \subset \Omega$$

since Ω is completely invariant. Hence it follows that $\operatorname{Cl}(V^G) \subset U$ and this shows that U is completely invariant.

Now, in order to complete the proof, we need some preparation which will be undertaken in the next section.

10. **Reduction to** g. Put

$$\xi(X) = \left| \det\{(e^{\operatorname{ad} X/2} - e^{-\operatorname{ad} X/2})/\operatorname{ad} X\} \right|^{1/2} \qquad (X \in g).$$

Then ξ is analytic around every point $X_0 \in g$, where $\xi(X_0) \neq 0$. Moreover, the exponential mapping of g into G is regular at X_0 if and only if $\xi(X_0) \neq 0$ (see, for example, [5, p. 95]).

Let U be an open subset of g such that the exponential mapping is regular and univalent on U and put $U_G = \exp U$. Then U_G is open in G and the exponential mapping defines an analytic diffeomorphism of U onto U_G. For any function ϕ on U, let f_ϕ denote the function on U_G given by

$$f_\phi(\exp X) = \xi(X)^{-1}\phi(X) \qquad (X \in U).$$

Then f_ϕ is C^∞ or analytic if and only if the same holds for ϕ. In particular, $f \to f_\phi$ defines a linear topological mapping of $C_c^\infty(U)$ onto $C_c^\infty(U_G)$. Moreover, it is obvious that, for any differential operator D on U_G, there exists a unique differential operator $\Delta(D)$ on U such that $Df_\phi = f_{\Delta(D)\phi}$ for $\phi \in C^\infty(U)$. Finally, D is analytic if and only if $\Delta(D)$ is analytic.

As usual let dX denote the Euclidean measure on $\mathfrak{g}$ and dx the Haar measure on G. Then if dX is suitably normalized, we have the relation (see [5, p. 95])

$$dx = \xi(X)^2 dX \qquad (x = \exp X, \ X \in U).$$

Hence it follows that

$$\int \phi_1 \phi_2 \, dX = \int f_{\phi_1} f_{\phi_2} \, dx$$

for $\phi_1 \in C^\infty(U)$ and $\phi_2 \in C_c^\infty(U)$.

LEMMA 20. $\Delta(D^*) = \Delta(D)^*$ *for any differential operator D on U_G.*

Fix D and write Δ for $\Delta(D)$. Then if $\phi_1, \phi_2 \in C_c^\infty(U)$, we have

$$\int D^* f_{\phi_1} \cdot f_{\phi_2} \, dx = \int f_{\phi_1} \cdot Df_{\phi_2} \, dx = \int f_{\phi_1} f_{\Delta\phi_2} \, dx$$

$$= \int \phi_1 \cdot \Delta\phi_2 \, dX = \int \Delta^* \phi_1 \cdot \phi_2 \, dX$$

$$= \int f_{\Delta^*\phi_1} \cdot f_{\phi_2} \, dx.$$

This shows that $D^* f_{\phi_1} = f_{\Delta^*\phi_1}$ and from this our assertion follows immediately.

For any distribution T on U_G, let τ_T denote the distribution on U given by $\tau_T(\phi) = T(f_\phi)$ ($\phi \in C_c^\infty(U)$). Then it follows from Lemma 20 that $\tau_{DT} = \Delta(D)\tau_T$.

Now assume that U is invariant under G. Since $\exp(X^x) = (\exp X)^x$ ($x \in G, X \in \mathfrak{g}$), U_G is also invariant. Moreover, since ξ is obviously invariant under G, it is clear that $(f_\phi)^x = f_{\phi^x}$ and $\Delta(D^x) = (\Delta(D))^x$ ($x \in G$) for $\phi \in C^\infty(U)$ and any differential operator D on U. Similarly $\tau_{T^x} = (\tau_T)^x$.

11. Completion of the proof of Theorem 1. We are now ready to finish the proof of Theorem 1. Define U as in §9. Then $U_G = \exp U = \Omega_0$ and, corresponding to T, we get an invariant distribution τ_T on U. Since D is an invariant and analytic differential operator on U_G, $\Delta = \Delta(D)$ is also invariant and analytic on U. Let ϕ be any invariant C^∞-function on U. Then

$$f_{\Delta\phi} = Df_\phi = 0$$

from Lemma 19. Hence $\Delta\phi = 0$. However, since U is completely invariant (see §9), we conclude from [4(n), Theorem 5] that $\tau_{DT} = \Delta\tau_T = 0$. Obviously this implies that $DT = 0$ on $U_G = \Omega_0$ and therefore $1 \notin \operatorname{Supp} DT$. This completes the proof of Theorem 1.

12. Two isomorphisms. Let $\mathfrak{m}$ be a subalgebra of $\mathfrak{g}$ such that (1) $\mathfrak{m}$ is reductive in $\mathfrak{g}$ and (2) rank $\mathfrak{m}$ = rank $\mathfrak{g}$. As before, let $\mathfrak{Z} = \mathfrak{Z}(\mathfrak{g})$ be the center of $\mathfrak{G} = \mathfrak{G}(\mathfrak{g}_c)$

and $\mathfrak{Z}(\mathfrak{m})$ the center of $\mathfrak{S}(\mathfrak{m}_c)$. We shall now define a homomorphism $\mu = \mu_{\mathfrak{g}/\mathfrak{m}}$ of $\mathfrak{Z}$ into $\mathfrak{Z}(\mathfrak{m})$.

Fix a Cartan subalgebra $\mathfrak{h}$ of $\mathfrak{m}$. Then $\mathfrak{h}$ is also a Cartan subalgebra of $\mathfrak{g}$. Let W and $W(\mathfrak{m})$ denote the Weyl groups of $(\mathfrak{g},\mathfrak{h})$ and $(\mathfrak{m},\mathfrak{h})$, respectively. Then $W(\mathfrak{m})$ is a subgroup of W. Let $I(\mathfrak{h}_c)$ and $I_{\mathfrak{m}}(\mathfrak{h}_c)$ denote the algebras of invariants of W and $W(\mathfrak{m})$, respectively, in $\mathfrak{S}(\mathfrak{h}_c)$. Then $I_{\mathfrak{m}}(\mathfrak{h}_c) \supset I(\mathfrak{h}_c)$. Let $\gamma: \mathfrak{Z} \to I(\mathfrak{h}_c)$ and $\gamma_{\mathfrak{m}}: \mathfrak{Z}(\mathfrak{m}) \to I_{\mathfrak{m}}(\mathfrak{h}_c)$ denote the canonical isomorphisms (see [4(e), Lemma 19]). We define $\mu(z) = \gamma_{\mathfrak{m}}^{-1}(\gamma(z))$ $(z \in \mathfrak{Z})$. Since any two Cartan subalgebras of $\mathfrak{m}_c$ are conjugate under the connected complex adjoint group of $\mathfrak{m}_c$, it follows easily from [4(e), §6] that μ is independent of the choice of $\mathfrak{h}$.

LEMMA 21. $\mathfrak{Z}(\mathfrak{m})$ *is a free abelian module over* $\mu_{\mathfrak{g}/\mathfrak{m}}(\mathfrak{Z})$ *of rank* $[W:W(\mathfrak{m})]$.

It is enough to show that $I_{\mathfrak{m}}(\mathfrak{h}_c)$ is a free abelian module over $I(\mathfrak{h}_c)$ of rank $[W:W(\mathfrak{m})]$. The proof of this is substantially the same as that of Lemma 8 of [4(i)].

If $\mathfrak{h}$ is a Cartan subalgebra of $\mathfrak{g}$, we can take $\mathfrak{m} = \mathfrak{h}$. Then it is clear that $\mu_{\mathfrak{g}/\mathfrak{h}} = \gamma$. As usual let $I(\mathfrak{g}_c)$ denote the algebra of all invariants of G in $S(\mathfrak{g}_c)$. Then we have the Chevalley isomorphism $j: p \to p_{\mathfrak{h}}$ of $I(\mathfrak{g}_c)$ onto [10] $I(\mathfrak{h}_c)$ (see [4(l), §9]). For any $z \in \mathfrak{Z}$, let p_z denote the element $j^{-1}(\gamma(z)) \in I(\mathfrak{g}_c)$. Then $z \to p_z$ is an isomorphism of $\mathfrak{Z}$ onto $I(\mathfrak{g}_c)$. It follows again from the results of [4(e), §6] that this isomorphism is independent of the choice of $\mathfrak{h}$. We shall call it the canonical isomorphism of $\mathfrak{Z}$ onto $I(\mathfrak{g}_c)$.

13. **A consequence of Theorem 1.** We use the notation of §5.

LEMMA 22. *Let* U_G *be a completely invariant open set in* G. *Fix a semisimple element* $a \in U_G$ *and define* $U_\Xi = \Xi' \cap (a^{-1} U_G)$ *as in* §5. *Then* U_Ξ *is completely invariant under* Ξ. *Let* σ *be an invariant distribution on* U_Ξ. *Then* [11]

$$\delta_a(z)\sigma = |v_a|^{-1/2} \mu_{\mathfrak{g}/\mathfrak{z}}(z)(|v_a|^{1/2}\sigma)$$

for $z \in \mathfrak{Z}$.

It is obvious that U_Ξ is an open and completely invariant subset of Ξ. Therefore, in view of Theorem 1 and Lemma 19, it is enough to prove the following result.

LEMMA 23. *Let* V *be an open subset of* U_Ξ *and* f *a* C^∞ *-function on* V *which is locally invariant under* Ξ. *Then*

$$\delta_a(z)f = |v_a|^{-1/2} \mu_{\mathfrak{g}/\mathfrak{z}}(z)(|v_a|^{1/2}f) \qquad (z \in \mathfrak{Z}).$$

Let V' be the set of those elements of V which are regular in Ξ. Since V' is

---

[10] Since $\mathfrak{h}$ is abelian, we may identify $S(\mathfrak{h}_c)$ with $\mathfrak{S}(\mathfrak{h}_c)$ under the canonical mapping λ of $S(\mathfrak{g}_c)$ onto $\mathfrak{S}$.

[11] Cf. [4(l), Theorem 2].

dense in V, it is enough to verify that the above equation holds on V'. Fix $b \in V'$ and $z \in \mathfrak{Z}$. Then we have to show that

$$f(b; \delta_a(z)) = f(b; |v_a|^{-1/2} \mu(z) \circ |v_a|^{1/2}),$$

where $\mu = \mu_{\mathfrak{g}/\mathfrak{z}}$. Let $\mathfrak{h}$ be the centralizer of b in $\mathfrak{z}$ and A the analytic subgroup of G corresponding to $\mathfrak{h}$. Let A' denote the set of all points $h \in A$ where

$$\det(\mathrm{Ad}(bh)^{-1} - 1)_{\mathfrak{g}/\mathfrak{h}} \neq 0$$

and put $U_A = A' \cap b^{-1} U_\Xi$ and $V_A = (b^{-1} V) \cap U_A$. Then V_A is an open neighborhood of 1 in A'. Moreover,

$$\begin{aligned}
f(bh; \delta_a(z)) &= f(bh; \delta_{b, \Xi/A}(\delta_a(z))) \\
&= f(bh; \delta_{ab}(z)) \qquad (h \in V_A)
\end{aligned}$$

from Lemmas 18 and 12. But since ab is regular in G (see the corollary of Lemma 11), it follows from Lemma 13 that

$$\delta_{ab}(z) = |v_{ab}|^{-1/2} \gamma(z) \circ |v_{ab}|^{1/2}$$

on V_A. Therefore

$$f(bh; \delta_a(z)) = |v_{ab}(h)|^{-1/2} F(h; \gamma(z)),$$

where

$$F(h) = |v_{ab}(h)|^{1/2} f(bh) \qquad (h \in V_A).$$

Now put $f_1(y) = |v_a(y)|^{1/2} f(y)$ for $y \in V$. Then

$$\begin{aligned}
f(bh; |v_a|^{-1/2} \mu(z) \circ |v_a|^{1/2}) &= |v_a(bh)|^{-1/2} f_1(bh; \mu(z)) \\
&= |v_a(bh)|^{-1/2} f_1(bh; \delta_{b, \Xi/A}(\mu(z))) \qquad (h \in V_A)
\end{aligned}$$

from Lemma 18. On the other hand it follows from Lemma 13 (applied to Ξ) and the definition of $\mu(z)$ that

$$\delta_{b, \Xi/A}(\mu(z)) = |v_{b, \Xi}|^{-1/2} \gamma(z) \circ |v_{b, \Xi}|^{1/2}$$

on U_A, where

$$v_{b, \Xi}(h) = \det(\mathrm{Ad}(bh)^{-1} - 1)_{\mathfrak{z}/\mathfrak{h}} \qquad (h \in A).$$

Therefore

$$f_1(bh; \delta_{b, \Xi/A}(\mu(z))) = |v_{b, \Xi}(h)|^{-1/2} F_1(h; \gamma(z)),$$

where

$$F_1(h) = |v_{b, \Xi}(h)|^{1/2} f_1(bh) = |v_{b, \Xi}(h) v_a(bh)|^{1/2} f(bh) \qquad (h \in V_A).$$

But since

$$v_{b, \Xi}(h) v_a(bh) = v_{ab}(h),$$

we have $F = F_1$. This shows that

$$f(bh; |v_a|^{-1/2}\mu(z) \circ |v_a|^{1/2}) = |v_a(bh)v_{b,\mathbf{z}}(h)|^{-1/2} F_1(h; \gamma(z))$$
$$= |v_{ab}(h)|^{-1/2} F(h; \gamma(z)) = f(bh; \delta_a(z))$$

for $h \in V_A$. Putting $h = 1$, we get the required result.

14. The relation between $\mathfrak{Z}$ and $\partial(I(\mathfrak{g}_c))$. We now use the notation of §10. For any open subset V of U and a function ϕ on V, we define, as before, the function f_ϕ on $V_G = \exp V$ by $f_\phi(\exp X) = \xi(X)^{-1}\phi(X)$ $(X \in V)$. Let $z \to p_z$ denote the canonical isomorphism of $\mathfrak{Z}$ onto $I(\mathfrak{g}_c)$ (see §12).

LEMMA 24.   *Let ϕ be a locally invariant C^∞-function on an open subset V of U. Then*

$$z f_\phi = f_{\partial(p_z)\phi}$$

for $z \in \mathfrak{Z}$.

Fix $z \in \mathfrak{Z}$ and let V' be the set of all regular points in V. Consider the differential operator $\Delta(z)$ on U corresponding to z (see §12). Then it is enough to prove that $\Delta(z)\phi = \partial(p_z)\phi$ on V'. Fix a point $H_0 \in V'$ and let $\mathfrak{h}$ denote the centralizer of H_0 in $\mathfrak{g}$. Then $\mathfrak{h}$ is a Cartan subalgebra of $\mathfrak{g}$. Let $\mathfrak{h}_0$ be an open and connected neighborhood of H_0 in $\mathfrak{h} \cap V'$. Then it would be sufficient to show that

$$\phi(H; \Delta(z)) = \phi(H; \partial(p_z)) \qquad (H \in \mathfrak{h}_0).$$

Since ϕ is locally invariant, it follows from [4(l), Lemma 14] and [4(f), Theorem 1] that

$$\phi(H; \partial(p_z)) = \phi(H; \delta_{\mathfrak{g}/\mathfrak{h}}'(\partial(p_z))) = \pi(H)^{-1}\phi(H; \partial(q) \circ \pi) \qquad (H \in \mathfrak{h}_0).$$

Here π denotes, as usual, the product of all the positive roots of $(\mathfrak{g}, \mathfrak{h})$ and $q = (p_z)_{\mathfrak{h}}$ in the notation of [4(l), §8]. Let A be the analytic subgroup of G corresponding to $\mathfrak{h}$ and put $A' = A \cap G'$ and

$$v(h) = \det(\mathrm{Ad}(h)^{-1} - 1)_{\mathfrak{g}/\mathfrak{h}} \qquad (h \in A).$$

Since $\mathfrak{h}_0 \subset U$, it follows that $\xi(H) \neq 0$ and therefore $\exp H \in A'$ for $H \in \mathfrak{h}_0$. Moreover, it is clear that f_ϕ is locally invariant with respect to G. Therefore we conclude from Lemma 17 and [4(e), Theorem 2] that

$$f_\phi(\exp H; z) = |v(\exp H)|^{-1/2} f_\phi(\exp H; \gamma(z) \circ |v|^{1/2})$$

for $H \in \mathfrak{h}_0$. But it is obvious that

$$|v(\exp H)|^{1/2} = \xi(H)|\pi(H)|$$

and therefore

$$|v(\exp H)|^{1/2} f_\phi(\exp H) = |\pi(H)|\phi(H) \qquad (H \in \mathfrak{h}_0).$$

If r is the number of positive roots of $(\mathfrak{g}, \mathfrak{h})$, we know that $\det(\mathrm{ad} H) = (-1)^r \pi(H)^2$ $(H \in \mathfrak{h})$. This shows that $\pi(H)^2$ is real. Therefore since $\mathfrak{h}_0$ is connected and π is nowhere zero on $\mathfrak{h}_0$, it follows that $|\pi(H)| = \varepsilon\pi(H)(H \in \mathfrak{h}_0)$, where $\varepsilon = |\pi(H_0)|/\pi(H_0)$ Moreover, $j(p_z) = \gamma(z)$ in the notation of §12. Hence it is clear that

$$f_\phi(\exp H; z) = \pi(H)^{-1}\xi(H)^{-1}\phi(H; \partial(q) \circ \pi)$$

$$= \xi(H)^{-1}\phi(H; \partial(p_z)) \qquad (H \in \phi_0).$$

On the other hand

$$\phi(X; \Delta(z)) = \xi(X)f_\phi(\exp X; z) \qquad (X \in V)$$

from the definition of $\Delta(z)$. Therefore

$$\phi(H; \Delta(z)) = \phi(H; \partial(p_z)) \qquad (H \in \mathfrak{h}_0)$$

and this proves the lemma.

COROLLARY. *Assume that U is completely invariant. Then if T is an invariant distribution on U_G,*

$$\tau_{zT} = \partial(p_z)\tau_T \qquad (z \in \mathfrak{Z}).$$

We know (see §10) that $\tau_{zT} = \Delta(z)\tau_T$ and it follows from Lemma 24 and [4(n), Theorem 5] that $\Delta(z)\tau_T = \partial(p_z)\tau_T$. Hence the corollary.

15. Proof of Theorem 2. We now come to one of the main results of this paper (cf. [4(j), Theorem 1]).

THEOREM 2. *Let Ω be a completely invariant open set in G and T a distribution on Ω. We assume that:*
(1) T is invariant;
(2) there exists an ideal $\mathfrak{U}$ in $\mathfrak{Z}$ such that $\dim \mathfrak{Z}/\mathfrak{U} < \infty$ and $uT = 0$ for $u \in \mathfrak{U}$.
Then T is a locally summable function which is analytic on $\Omega' = \Omega \cap G'$.

We shall use induction on $\dim G$. Let Ω_0 be the set of all points $a \in \Omega$ with the following property. There exists an open neighborhood U of a in Ω and a locally summable function F on U such that F is analytic on $U \cap G'$ and $T = F$ on U. Clearly Ω_0 is an open and invariant subset of Ω. It would be enough to prove that $\Omega_0 = \Omega$. But then, in view of Lemma 7, we have only to verify that Ω_0 contains all semisimple points of Ω.

LEMMA 25. $\Omega' \subset \Omega_0$.

Fix $a \in \Omega'$ and let $\mathfrak{h}$ denote the centralizer of a in $\mathfrak{g}$. Then $\mathfrak{h}$ is a Cartan subalgebra of $\mathfrak{g}$. Consider the analytic subgroup A of G corresponding to $\mathfrak{h}$ and put $\Omega_A = a^{-1}\Omega \cap A'$, where A' is the set of all $h \in A$ such that

$$v_a(h) = \det(\mathrm{Ad}(ah)^{-1} - 1)_{\mathfrak{g}/\mathfrak{h}} \neq 0.$$

Let σ_T denote the distribution on Ω_A corresponding to T under Lemma 15. Put $\sigma = |v_a|^{1/2} \sigma_T$. Then we conclude from Lemma 13 and the corollary of Lemma 15 that $\gamma(u)\sigma = 0$ for $u \in \mathfrak{U}$. Since $\mathfrak{S}(\mathfrak{h}_c)$ is a finite module over $\gamma(\mathfrak{Z}) = I(\mathfrak{h}_c)$ (Lemma 21), it follows that $\mathfrak{B} = \mathfrak{S}(\mathfrak{h}_c)\gamma(\mathfrak{U})$ has finite codimension in $\mathfrak{S}(\mathfrak{h}_c)$. Fix a base $H_1, \cdots, H_l$ for $\mathfrak{h}$ over R and put

$$\square = H_1^2 + \cdots + H_l^2.$$

Then if $N = \dim \mathfrak{S}(\mathfrak{h}_c)/\mathfrak{B}$, it is obvious that we can choose $c_i \in C$ $(1 \leq i \leq N)$ such that

$$v = \square^N + \sum_{1 \leq k \leq N} c_k \square^{N-k} \in \mathfrak{B}.$$

Now v is an analytic differential operator on A which is obviously elliptic. Therefore since $v\sigma = 0$, we conclude that σ coincides with an analytic function g on Ω_A. Put $G^* = G/A$ and define the mapping $\psi: G^* \times \Omega_A \to \Omega$ as in the proof of Lemma 19. Then we can choose open neighborhoods G_0 and V of 1 in G and Ω_A, respectively, such that ψ defines an analytic diffeomorphism of $G_0^* \times V$ onto the open subset $U = \psi(G_0^* \times V)$ of Ω. Define the analytic function F on U by

$$F((ah)^{x^*}) = \left| v_a(h) \right|^{-1/2} g(h) \qquad (x^* \in G_0^*, \ h \in V).$$

Then by Lemma 15, we get

$$T(f_\alpha) = \sigma_T(\beta_\alpha) = \int \beta_\alpha \left| v_a \right|^{-1/2} g \, dh \qquad (\alpha \in C_c^\infty(G_0 \times V)),$$

where dh is the Haar measure on A. On the other hand, it follows from the definition of f_α that

$$\int f_\alpha F \, dx = \int \alpha(x : h) F((ah)^x) dx \, dh$$

$$= \int \beta_\alpha \left| v_a \right|^{-1/2} g \, dh.$$

This shows that $T = F$ on U and therefore $a \in \Omega_0$.

It is clear from the above lemma that there exists an analytic function F on Ω' such that $T = F$ on Ω'. Now fix a semisimple element $a \in \Omega$ and let us use the notation of §4. Z being the center of G, first assume that $a \notin Z$ so that $\dim \mathfrak{z} < \dim \mathfrak{g}$. Put $\Omega_\Xi = a^{-1}\Omega \cap \Xi'$. Then Ω_Ξ is an open and completely invariant neighborhood of 1 in Ξ. Let σ_T denote the distribution on Ω_Ξ which corresponds to T under Lemma 15. Then σ_T is invariant under Ξ and it follows from the corollary of Lemma 15 that $\delta_a(u)\sigma_T = 0$ for $u \in \mathfrak{U}$. But then we conclude from Lemma 22 that

$$\mu(u)\sigma = 0 \qquad (u \in \mathfrak{U}).$$

Here $\sigma = \left| v_a \right|^{1/2}\sigma_T$ and $\mu = \mu_{\mathfrak{g}/\mathfrak{z}}$. Since $\mathfrak{Z}(\mathfrak{z})$ is a finite module over $\mu(\mathfrak{Z})$ (Lemma 21), it is clear that $\mathfrak{B} = \mathfrak{Z}(\mathfrak{z})\mu(\mathfrak{U})$ has finite codimension in $\mathfrak{Z}(\mathfrak{z})$. Let Ω_Ξ' be the set of those elements in Ω_Ξ which are regular in Ξ. Then it follows by the induction hypothesis that $\sigma = g$, where g is a locally summable function on Ω_Ξ which is analytic on Ω_Ξ'.

Let ϕ denote the mapping $(x, y) \to (ay)^x$ of $G \times \Omega_{\Xi}$ into Ω. Then $U = \phi(G \times \Omega_{\Xi})$ is an open neighborhood of a in Ω (Lemma 14). Moreover, it is easy to verify that $\phi(G \times \Omega_{\Xi}') = U'$, where $U' = U \cap G' = U \cap \Omega'$. Since $T = F$ on Ω', we have

$$T(f_a) = \int f_a F \, dx = \int \alpha(x : y) F((ay)^x) \, dx dy$$

for $\alpha \in C_c^{\infty}(G \times \Omega_{\Xi}')$ in the notation of Lemma 15. However,

$$T(f_a) = \sigma_T(\beta_a) = \int \alpha(x : y) |v_a(y)|^{-1/2} g(y) \, dx dy.$$

This shows that the analytic function

$$(x, y) \to F((ay)^x) - |v_a(y)|^{-1/2} g(y)$$

is zero on $G \times \Omega_{\Xi}'$ and therefore $F \circ \phi$ is locally summable on $G \times \Omega_{\Xi}$. Hence F is locally summable on U (see [4(k), Corollary 2 of Theorem 1]) and

$$\int f_a F \, dx = \int \alpha(x : y) F((ay)^x) \, dx dy$$

$$= \int \alpha(x : y) |v_a(y)|^{-1/2} g(y) \, dx dy$$

$$= \sigma_T(\beta_a) = T(f_a)$$

for $\alpha \in C_c^{\infty}(G \times \Omega_{\Xi})$. This proves that $T = F$ on U and therefore $a \in \Omega_0$.

It remains to consider the case when $a \in Z$. Then by a translation by a^{-1}, we are reduced to the case $a = 1$. Then, as we have seen in §9, there exists an open and completely invariant neighborhood U of zero in $\mathfrak{g}$ such that the exponential mapping of $\mathfrak{g}$ into G is univalent and regular on U and $U_G = \exp U \subset \Omega$. Let τ_T be the distribution on U corresponding to T (see §10). Then we know from the corollary of Lemma 24 that $\partial(p_u)\tau_T = 0$ for $u \in \mathfrak{U}$. Let $\mathfrak{B}$ denote the image of $\mathfrak{U}$ in $I(\mathfrak{g}_c)$ under the canonical isomorphism $z \to p_z$ of $\mathfrak{Z}$ onto $I(\mathfrak{g}_c)$. Then

$$\dim I(\mathfrak{g}_c)/\mathfrak{B} = \dim \mathfrak{Z}/\mathfrak{U} < \infty$$

and so we conclude from [4(n), Theorem 1] that $\tau_T = \Phi$, where Φ is a locally summable function on U. Define the function f_{Φ} on U_G as in §10. Then it is obvious that f_{Φ} is locally summable on U_G and $T = f_{\Phi}$ on U_G. But since $T = F$ on $U_G \cap \Omega'$, it follows that $f_{\Phi} = F$ almost everywhere on U_G. Hence F is locally summable on U_G and $T = F$ on U_G. This shows that $1 \in \Omega_0$ and so the proof of Theorem 2 is now complete.

The above theorem shows that F is locally summable on Ω and $T = F$ on Ω. Fix $z \in \mathfrak{Z}$. Then the distribution zT also satisfies all the conditions of Theorem 2 and it is obvious that $zT = zF$ on Ω'. Hence zF is also locally summable on Ω and $zT = zF$ on Ω. Thus we obtain the following corollary (cf. [4(n), Lemma 16]).

COROLLARY. *For any $z \in \mathfrak{Z}$, the function zF on Ω' is locally summable on Ω and $zT = zF$. Hence*

$$\int f \cdot zF \, dx = \int z^* f \cdot F \, dx$$

for $f \in C_c^\infty(\Omega)$.

16. **Some elementary facts about reductive groups.** As before let $\mathfrak{c}$ be the center and $\mathfrak{g}_1$ the derived algebra of $\mathfrak{g}$. Fix a Cartan subalgebra $\mathfrak{h}$ of $\mathfrak{g}$. Then $\mathfrak{h}_1 = \mathfrak{h} \cap \mathfrak{g}_1$ is a Cartan subalgebra of the semisimple Lie algebra $\mathfrak{g}_1$. We can choose a Cartan involution θ of $\mathfrak{g}_1$ such that $\theta(\mathfrak{h}_1) = \mathfrak{h}_1$ [4(e), p. 100]. We extend θ to an automorphism of $\mathfrak{g}$ by defining $\theta(C) = C$ for $C \in \mathfrak{c}$. Let $\mathfrak{k}$ and $\mathfrak{p}$ be the subspaces of $\mathfrak{g}$ corresponding to the eigenvalues 1 and -1 of θ. Then $\mathfrak{c} \subset \mathfrak{k}$ and $\mathfrak{p} \subset \mathfrak{g}_1$. Moreover, since $\theta(\mathfrak{h}) = \mathfrak{h}$, it is clear that $\mathfrak{h} = \mathfrak{h} \cap \mathfrak{k} + \mathfrak{h} \cap \mathfrak{p}$.

Let K be the analytic subgroup of G corresponding to $\mathfrak{k}$ and Z the center of G.

LEMMA 26. *The mapping $\phi : (k, X) \to k \exp X$ $(k \in K, X \in \mathfrak{p})$ is an analytic diffeomorphism of $K \times \mathfrak{p}$ onto G. Moreover, $Z \subset K$ and K/Z is compact.*

It is easy to verify (see [4(d), p. 614]) that ϕ is everywhere regular. Let C, G_1 and K_1 be the analytic subgroups of G corresponding to $\mathfrak{c}$, $\mathfrak{g}_1$ and $\mathfrak{k}_1 = \mathfrak{k} \cap \mathfrak{g}_1$, respectively. Then $G = CG_1$ and $G_1 = K_1 \exp \mathfrak{p}$ (see e.g. [5, pp. 214–215]). Therefore since $CK_1 = K$, it follows that ϕ is surjective. Now suppose

$$k_1 \exp X_1 = k_2 \exp X_2 \quad (k_i \in K, X_i \in \mathfrak{p}, \ i = 1, 2).$$

Put $k = k_2^{-1} k_1$. Then $k \exp X_1 = \exp X_2$ and therefore

$$\mathrm{Ad}(k) \exp(\mathrm{ad}\, X_1) = \exp(\mathrm{ad}\, X_2).$$

Since $\mathrm{Ad}(G)$ is semisimple, we conclude [5, pp. 214–215] that $X_1 = X_2$. Hence $k_1 = k_2$. This proves that ϕ is univalent and so it is an analytic diffeomorphism.

Let Z_1 be the center of G_1. Then we know that $Z_1 \subset K_1$ and K_1/Z_1 is compact [5, p.214]. Since $K = CK_1$ and $Z = CZ_1$, it follows that $Z \subset K$ and K/Z is compact.

COROLLARY. 1. *θ can be extended to an automorphism of G such that*

$$\theta(k \exp X) = k \exp(-X) \quad (k \in K, \ X \in \mathfrak{p}).$$

First assume that G is simply connected. Then our statement is obvious. Moreover, θ leaves Z pointwise fixed since $Z \subset K$. Therefore if Z_0 is any closed subgroup of Z, it defines an automorphism of G/Z_0. From this our assertion follows immediately in the general case.

COROLLARY 2 ([12]). *Let $Y' = \mathrm{Ad}(k \exp X)Y$, where $Y, Y' \in \mathfrak{g}, k \in K$ and $X \in \mathfrak{p}$. Then if Y and Y' are both eigenvectors of θ, $[X, Y] = 0$.*

([12]) This result was pointed out to me by A. Borel.

Since

$$\theta(Y') = \mathrm{Ad}(k \exp(-X))\theta(Y),$$

it is clear that

$$e^{2\,\mathrm{ad}X} Y = \varepsilon Y,$$

where $\varepsilon = \pm 1$. Moreover, it follows from [4(k), Lemma 27] that ad X is semisimple and all its eigenvalues are real. Therefore it is obvious that $\varepsilon = 1$ and $[X, Y] = 0$.

COROLLARY 3. *Let $\mathfrak{a}$ be a subset of $\mathfrak{h}$ such that $\mathfrak{a} = \theta(\mathfrak{a})$ and let Ξ and $\mathfrak{z}$ be the centralizers of $\mathfrak{a}$ in G and $\mathfrak{g}$, respectively. Then they are both invariant under θ, $\mathfrak{z}$ is reductive in $\mathfrak{g}$ and*

$$\Xi = \Xi_K \exp(\mathfrak{z} \cap \mathfrak{p}),$$

where $\Xi_K = \Xi \cap K$.

For the proof we can obviously replace $\mathfrak{a}$ by the linear subspace of $\mathfrak{g}$ spanned by it. Then $\mathfrak{a} = \mathfrak{a} \cap \mathfrak{k} + \mathfrak{a} \cap \mathfrak{p}$. The invariance of Ξ and $\mathfrak{z}$ under θ is obvious and therefore (see [4(g), Lemma 10]) $\mathfrak{z}$ is reductive in $\mathfrak{g}$. The last statement follows from Corollary 2.

Let A be the Cartan subgroup of G corresponding to $\mathfrak{h}$.

COROLLARY 4. $A = A_K A_{\mathfrak{p}}$, *where $A_K = A \cap K$ and $A_{\mathfrak{p}} = \exp(\mathfrak{h} \cap \mathfrak{p})$. Moreover, $Z \subset A_K$ and A_K/Z is compact.*

The first statement follows from Corollary 3 if we take $\mathfrak{a} = \mathfrak{h}$. It is obvious from Lemma 26 that $Z \subset A_K$. Moreover, since K/Z is compact and A_K is closed in K, it follows that A_K/Z is compact.

COROLLARY 5. *Suppose every root of $(\mathfrak{g},\mathfrak{h})$ is imaginary (see [4(m), §4]). Then A is connected and contained in K.*

For then it is obvious that $\mathfrak{h} \cap \mathfrak{p} = \{0\}$ and therefore $A = A_K$. Since $\mathfrak{k}$ is reductive and its derived algebra is compact, the connected component of 1 in A is maximal abelian in K. This shows that A is connected.

17. **Complex semisimple groups.** Let $\mathfrak{g}_c$ be a complex semisimple Lie algebra and G_c a complex analytic group corresponding to it. Fix a Cartan subalgebra $\mathfrak{h}_c$ of $\mathfrak{g}_c$. Then we can choose a compact real form $\mathfrak{u}$ of $\mathfrak{g}_c$ such that $\mathfrak{h} = \mathfrak{h}_c \cap \mathfrak{u}$ is a Cartan subalgebra of $\mathfrak{u}$ (see [5, p. 155]). Let η denote the conjugation of $\mathfrak{g}_c$ with respect to $\mathfrak{u}$. Then if we regard $\mathfrak{g}_c$ as a Lie algebra over R, η is a Cartan involution of $\mathfrak{g}_c$ and $\mathfrak{g}_c = \mathfrak{u} + (-1)^{1/2}\mathfrak{u}$ is the corresponding Cartan decomposition. Let U be the real analytic subgroup of G_c corresponding to $\mathfrak{u}$. Then U is compact and by Lemma 26, the mapping

$$(u, X) \to u \exp(-1)^{1/2}X \qquad (u \in U, \ X \in \mathfrak{u})$$

is an analytic diffeomorphism of $U \times \mathfrak{u}$ onto G_c.

LEMMA 27. *Let $\mathfrak{a}$ be a subset of $\mathfrak{h}_c$ such that $\eta(\mathfrak{a}) = \mathfrak{a}$ and let $\mathfrak{z}_c$ and Ξ_c denote the centralizers of $\mathfrak{a}$ in $\mathfrak{g}_c$ and G_c, respectively. Then $\mathfrak{z}_c$ is reductive in $\mathfrak{g}_c$ and Ξ_c is connected.*

We may obviously replace $\mathfrak{a}$ by the subspace $\mathfrak{a}_c$ spanned by it over C. It follows from Corollary 3 of Lemma 26 that $\mathfrak{z}_c$ is reductive in $\mathfrak{g}_c$ and

$$\Xi_c = \Xi \exp((-1)^{1/2}\mathfrak{z}),$$

where $\Xi = \Xi_c \cap U$ and $\mathfrak{z} = \mathfrak{z}_c \cap \mathfrak{u}$. It is clear that Ξ is the centralizer of $\mathfrak{a}_c \cap \mathfrak{u}$ in U and therefore it is connected (see [5, p. 247]). This proves that Ξ_c is connected.

COROLLARY. *Let A_c be the Cartan subgroup of G_c corresponding to $\mathfrak{h}_c$. Then A_c is connected.*

By definition A_c is the centralizer of $\mathfrak{h}_c$ in G_c. Hence the corollary follows by taking $\mathfrak{a} = \mathfrak{h}_c$.

The following lemma, together with its proof, was pointed out to me by Borel.

LEMMA 28. *$\mathfrak{z}_c$ being as above, put $\mathfrak{z}_{1c} = [\mathfrak{z}_c, \mathfrak{z}_c]$ and let Ξ_{1c} be the complex analytic subgroup of G_c corresponding to $\mathfrak{z}_{1c}$. Then if G_c is simply connected, the same holds for Ξ_{1c}.*

Put $\mathfrak{h}_R = (-1)^{1/2}(\mathfrak{h}_c \cap \mathfrak{u})$ and $\mathfrak{a}_R = (-1)^{1/2}(\mathfrak{a}_c \cap \mathfrak{u})$. Introduce compatible orders (see [4(g), p. 195]) in the spaces of linear functions on $\mathfrak{h}_R$ and $\mathfrak{a}_R$. Let P be the set of all positive roots of $(\mathfrak{g}_c, \mathfrak{h}_c)$ under this order. Let $\bar{\alpha}$ denote the restriction of α on $\mathfrak{a}_c$ for any root α and let P_0 denote the set of those $\alpha \in P$ for which $\bar{\alpha} = 0$. Consider the set $(\alpha_1, \alpha_2, \cdots, \alpha_l)$ of simple roots in P and assume that $\alpha_i \in P_0$ $(1 \leq i \leq m)$ and $\alpha_i \notin P_0$ $(m < i \leq l)$. We claim that $(\alpha_1, \cdots, \alpha_m)$ is a set of fundamental roots for $(\mathfrak{z}_c, \mathfrak{h}_c)$. Fix $\alpha \in P_0$. Then $\alpha = \sum_{1 \leq i \leq l} r_i \alpha_i$, where r_i are integers ≥ 0. Hence

$$\sum_{1 \leq i \leq l} r_i \bar{\alpha}_i = \bar{\alpha} = 0.$$

Now $\bar{\alpha}_i = 0$ $(1 \leq i \leq m)$ and $\bar{\alpha}_i > 0$ $(m < i \leq l)$ by the compatibility of our orders. So it is obvious that $r_i = 0$ for $m < i \leq l$. Since $(\alpha_1, \cdots, \alpha_m)$ are linearly independent, this proves our assertion.

For any root α, let H_α denote, as usual, the element in $\mathfrak{h}_R$ such that $\mathrm{tr}(\mathrm{ad}\, H \, \mathrm{ad}\, H_\alpha) = \alpha(H)$ for $H \in \mathfrak{h}_c$. Put

$$H_i = 2\alpha_i(H_{\alpha_i})^{-1} H_{\alpha_i} \qquad (1 \leq i \leq l).$$

Then it is clear that H_i $(1 \leq i \leq m)$ form a base for $\mathfrak{h}_c \cap \mathfrak{z}_{1c}$ over C. Now suppose t_i $(1 \leq i \leq l)$ are complex numbers such that

$$\exp\left(2\pi(-1)^{1/2} \sum_{1 \leq i \leq m} t_i H_i\right) = 1$$

in Ξ_{1c}. Then since G_c is simply connected, we can conclude (see Weyl [8]) that t_i are rational integers. This implies that Ξ_{1c} is simply connected.

LEMMA 29. *Assume that G_c is simply connected and let λ be a linear function on $\mathfrak{h}_c$. Then there exists a character ξ_λ of A_c such that*

$$\xi_\lambda(\exp H) = e^{\lambda(H)} \qquad (H \in \mathfrak{h}_c)$$

if and only if $2\lambda(H_\alpha)/\alpha(H_\alpha)$ is a rational integer for every root α. Put

$$\rho = \frac{1}{2} \sum_{\alpha \in P} \alpha,$$

where P is the set of positive roots under some order. Then the above condition is fulfilled for $\lambda = \rho$.

This is well known (see Weyl [8]).

18. **Acceptable groups.** Let G be a connected Lie group with the Lie algebra $\mathfrak{g}$ over R which we assume, as before, to be reductive. Let j be the inclusion mapping of $\mathfrak{g}$ into $\mathfrak{g}_c$ and G_c a complex analytic group corresponding to $\mathfrak{g}_c$. We say that G_c is a complexification of G if j can be extended to a homomorphism of G into G_c.

Define $\mathfrak{c}$ and $\mathfrak{g}_1$ as in §9 and let C and G_1, respectively, be the corresponding analytic subgroups of G. We call G_1 the semisimple part of G. Similarly let C_c and G_{1c} denote the complex analytic subgroups of G_c corresponding to $\mathfrak{c}_c$ and $\mathfrak{g}_{1c}$, respectively. We say that G_c is quasisimply connected (q. s. c.) if $C_c \cap G_{1c} = \{1\}$ and G_{1c} is simply connected. Moreover, G itself is called q. s. c. if it has a q. s. c. complexification. Assume that $G_1 \cap C$ is finite. Then G always has a complexification. Moreover, since the center of a complex semisimple group is finite, it is clear that there exists a q. s. c. covering group $\tilde{G}$ which covers G only finitely many times.

Let A be the Cartan subgroup of G corresponding to $\mathfrak{h}$. Consider a complexification G_c of G and let A_c denote its Cartan subgroup corresponding to $\mathfrak{h}_c$. Then A_c is connected (corollary of Lemma 27) and it is obvious that $j(A) \subset A_c$. Let λ be a linear function on $\mathfrak{h}_c$. Then there exists at most one complex-analytic homomorphism ξ_λ of A_c into C such that

$$\xi_\lambda(\exp H) = e^{\lambda(H)} \qquad (H \in \mathfrak{h}_c).$$

Then $\xi_\lambda \circ j$ is a homomorphism of A into C, which is easily seen to be independent of the particular choice of G_c (so long as it can be defined by means of G_c at all). We shall write ξ_λ instead of $\xi_\lambda \circ j$. If α is a root of $(\mathfrak{g}, \mathfrak{h})$, it is obvious that ξ_α always exists.

Let P be the set of all positive roots of $(\mathfrak{g}, \mathfrak{h})$ in some order and put

$$\rho = \frac{1}{2} \sum_{\alpha \in P} \alpha.$$

If G is q. s. c., we can take G_c to be a q. s. c. complexification of G. Then it follows from Lemma 29 that ξ_ρ exists.

Let W denote the Weyl group of $(\mathfrak{g}, \mathfrak{h})$. Then it is well known (see [8]) that $s\rho - \rho$ $(s \in W)$ is an integral linear combination of the roots. Therefore the condition that ξ_ρ should be defined is independent of the order of roots. Moreover, since any two Cartan subalgebras of $\mathfrak{g}_c$ are conjugate under the (connected) adjoint group of $\mathfrak{g}_c$, it follows that the above condition also does not depend on the choice of $\mathfrak{h}$. We shall say that G is *acceptable* if this condition is satisfied. Similarly a complexification G_c of G is called acceptable if ξ_ρ can be defined on A_c.

Let $\mathfrak{m}$ be the centralizer of $\mathfrak{h} \cap \mathfrak{p}$ in $\mathfrak{g}$ and M the analytic subgroup of G corresponding to $\mathfrak{m}$. Introduce compatible orders (see [4(g), p. 195]) on the spaces of real linear functions on $\mathfrak{h} \cap \mathfrak{p}$ and $\mathfrak{h} \cap \mathfrak{p} + (-1)^{1/2} \mathfrak{h} \cap \mathfrak{k}$ respectively. We assume that P is the set of positive roots under this order. Let P_M denote the set of those $\alpha \in P$ which vanish identically on $\mathfrak{h} \cap \mathfrak{p}$. Put

$$\rho_M = \frac{1}{2} \sum_{\alpha \in P_M} \alpha.$$

LEMMA 30. *Suppose G is acceptable. Then the same holds for M and in fact*

$$\xi_{\rho_M}(h) = \xi_\rho(h_1) \qquad (h \in A \cap M),$$

where $h = h_1 h_2$ $(h_1 \in A_K, h_2 \in A_\mathfrak{p})$ in the notation of §16.

Let P_+ be the complement of P_M in P. Then it is easy to verify that if $\alpha \in P_+$, the same holds for $-\theta\alpha$. This shows that $\rho - \rho_M = 0$ on $\mathfrak{h} \cap \mathfrak{k}$. Let $\mathfrak{m}_1$ be the set of all $X \in \mathfrak{m}$ such that $\operatorname{tr}(\operatorname{ad} H \operatorname{ad} X) = 0$ for $H \in \mathfrak{h} \cap \mathfrak{p}$. Then $\theta(\mathfrak{m}_1) = \mathfrak{m}_1$ and $(\mathfrak{h} \cap \mathfrak{p}) \cap \mathfrak{m}_1 = \{0\}$. Hence if M_1 is the analytic subgroup of G corresponding to $\mathfrak{m}_1$, it is clear that $M = A_\mathfrak{p} M_1$ and $A_\mathfrak{p} \cap M_1 = \{1\}$. Now $\mathfrak{m}$ is reductive (Corollary 3 of Lemma 26) and $\mathfrak{h} \cap \mathfrak{p}$ lies in the center of $\mathfrak{m}$. Therefore since $\rho = \rho_M$ on $\mathfrak{h} \cap \mathfrak{k}$ and $A_\mathfrak{p}$ is simply connected, the statement of the lemma follows immediately by considering an acceptable complexification of G.

19. **Behavior of F around singular points.** From now on we assume that G is acceptable. Put

$$\Delta_A(h) = \xi_\rho(h) \prod_{\alpha \in P} (1 - \xi_\alpha(h)^{-1}) \qquad (h \in A).$$

(We shall often drop the subscript A if there is no risk of confusion.) Then $A' = A \cap G'$ is the set of all points $h \in A$, where $\Delta(h) \neq 0$. Put

$$\Delta_R'(h) = \prod_{\alpha \in P_R} (1 - \xi_\alpha(h)^{-1}) \qquad (h \in A),$$

where P_R is the set of all real roots (see [4(m), §4]) in P. Let $A'(R)$ be the set of those $h \in A$ where $\Delta_R'(h) \neq 0$. We now use the notation of §15.

LEMMA 31([13]). *Put $\Phi_A(h) = \Delta_A(h)F(h)$ $(h \in A' \cap \Omega)$. Then Φ_A can be extended to an analytic function on $A'(R) \cap \Omega$.*

([13]) Cf. [4(n), Theorem 2] and [4(e), Theorem 8].

Fix a point $a \in A \cap \Omega$. Then a is semisimple. We now use the notation of §4 and define $\Omega_\Xi = a^{-1}\Omega \cap \Xi'$. Put $\sigma = |v_a|^{1/2}\sigma_T$ as in §15, and let Ω_Ξ' be the set of all elements in Ω_Ξ which are regular in Ξ. We denote by g, as before, the analytic function on Ω_Ξ' such that g is locally summable on Ω_Ξ and $\sigma = g$ on Ω_Ξ. Since T is invariant, the same holds for F and, as we have seen during the proof of Theorem 2,

$$F(ay) = |v_a(y)|^{-1/2}g(y) \qquad (y \in \Omega_\Xi').$$

Now Ω_Ξ is an open and completely invariant neighborhood of 1 in Ξ. Hence (see §9) we can choose an open and completely invariant neighborhood U of zero in $\mathfrak{z}$ such that the exponential mapping defines an analytic diffeomorphism of U onto an open subset U_Ξ of Ω_Ξ. Consider the function $\xi_\mathfrak{z}$ on $\mathfrak{z}$ (see §10) and, for any $\phi \in C_c^\infty(U)$, define $f_\phi \in C_c^\infty(U_\Xi)$ by

$$f_\phi(\exp Z) = \xi_\mathfrak{z}(Z)^{-1}\phi(Z) \qquad (Z \in U).$$

Let τ be the distribution on U given by $\tau(\phi) = \sigma(f_\phi)$ $(\phi \in C_c^\infty(U))$. Define $\mathfrak{B} = \mathfrak{Z}(\mathfrak{z}) \cdot \mu(\mathfrak{U})$, where $\mu = \mu_{\mathfrak{g}/\mathfrak{z}}$ (in the notation of §12). Then we know from Lemmas 21 and 22 that $\mathfrak{Z}(\mathfrak{z})$ is a finite module over $\mu(\mathfrak{Z})$ and $v\sigma = 0$ $(v \in \mathfrak{B})$. Let $z \to p_z$ denote the canonical isomorphism of $\mathfrak{Z}(\mathfrak{z})$ onto $I(\mathfrak{z}_c)$ (see §12). Then $\partial(p_v)\tau = 0$ $(v \in \mathfrak{B})$ from the corollary of Lemma 24. Hence Theorem 1 of $[4(n)]$ is applicable to $(\mathfrak{z}, U, \tau)$. Let $\mathfrak{z}'$ denote the set of all elements of $\mathfrak{z}$ which are regular in $\mathfrak{z}$ and let ψ be the analytic function on $U' = U \cap \mathfrak{z}'$ such that $\tau = \psi$.

LEMMA 32. $\psi(Z) = \xi_\mathfrak{z}(Z)|v_a(\exp Z)|^{1/2}F(a\exp Z)$ $(Z \in U')$.

Fix $\phi \in C_c^\infty(U)$. Then

$$\int \phi\psi \, dZ = \tau(\phi) = \sigma(f_\phi) = \int f_\phi g \, dy$$

$$= \int \phi(Z)\xi_\mathfrak{z}(Z)g(\exp Z)\,dZ.$$

Here dy is the Haar measure of Ξ and dZ the Euclidean measure on $\mathfrak{z}$ and they are related (see §10) by the equation

$$dy = \xi_\mathfrak{z}(Z)^2 dZ \qquad (y = \exp Z, \, Z \in U).$$

Since $\exp(U') \subset \Omega_\Xi'$, we have

$$g(\exp Z) = |v_a(\exp Z)|^{1/2}F(a\exp Z) \qquad (Z \in U')$$

and so our assertion is now obvious.

Let $P_\mathfrak{z}$ be the set of all roots $\alpha \in P$ such that $\xi_\alpha(a) = 1$. Put $P_{\mathfrak{z},R} = P_\mathfrak{z} \cap P_R$ and let $\mathfrak{h}'(R)$ be the set of all points $H \in \mathfrak{h}$, where $\prod_{\alpha \in P_{\mathfrak{z},R}} \alpha(H) \neq 0$. Then we know from $[4(n), \text{Theorem 2}]$ that there exists an analytic function u on $\mathfrak{h}'(R) \cap U$ such that

$$u(H) = \pi_{\mathfrak{z}}(H)\psi(H) \qquad (H \in \mathfrak{h} \cap U'),$$

where $\pi_{\mathfrak{z}} = \prod_{\alpha \in P_{\mathfrak{z}}} \alpha$.

LEMMA 33. *Let $\mathfrak{h}_0$ be an open and connected neighborhood of zero in $U \cap \mathfrak{h}$. Then*

$$\pi_{\mathfrak{z}}(H)\xi_{\mathfrak{z}}(H)\left| v_a(\exp H)\right|^{1/2} = c\Delta(a \exp H) \qquad (H \in \mathfrak{h}_0),$$

where c is a constant. Let P' be the complement of $P_{\mathfrak{z}}$ in P and p the number of roots in P'. Then $p = (\dim \mathfrak{g} - \dim \mathfrak{z})/2$ and

$$c^2 = (-1)^p \operatorname{sign} v_a(1).$$

Finally

$$c = \left| v_a(1)\right|^{1/2} \left\{ \xi_\rho(a) \prod_{\alpha \in P'} (1 - \xi_\alpha(a)^{-1}) \right\}^{-1}.$$

Put $\rho' = (1/2) \sum_{\alpha \in P'} \alpha$. Then it is clear that

$$v_a(\exp H) = \prod_{\alpha \in P'} \{(\xi_\alpha(a \exp H)^{-1} - 1)(\xi_\alpha(a \exp H) - 1)\}$$

$$= (-1)^p \xi_{2\rho'}(a) \left\{ e^{\rho'(H)} \prod_{\alpha \in P'} (1 - \xi_\alpha(a \exp H)^{-1}) \right\}^2$$

for $H \in \mathfrak{h}$. Since $v_a(\exp H)$ is real and $\neq 0$ for $H \in \mathfrak{h}_0$, it is clear that

$$\left| v_a(\exp H)\right|^{-1/2} e^{\rho'(H)} \prod_{\alpha \in P'} (1 - \xi_\alpha(a \exp H)^{-1})$$

is an analytic function on $\mathfrak{h}_0$ whose fourth power is a constant. Therefore since $\mathfrak{h}_0$ is connected, we conclude that

$$\left| v_a(\exp H)\right|^{1/2} = c_1 e^{\rho'(H)} \prod_{\alpha \in P'} (1 - \xi_\alpha(a \exp H)^{-1}) \qquad (H \in \mathfrak{h}_0),$$

where

$$c_1 = \left| v_a(1)\right|^{1/2} \prod_{\alpha \in P'} (1 - \xi_\alpha(a)^{-1})^{-1}.$$

A similar argument shows that

$$\pi_{\mathfrak{z}}(H)\xi_{\mathfrak{z}}(H) = \prod_{\alpha \in P_{\mathfrak{z}}} (e^{\alpha(H)/2} - e^{-\alpha(H)/2}) \qquad (H \in \mathfrak{h}_0).$$

Hence

$$\pi_{\mathfrak{z}}(H)\xi_{\mathfrak{z}}(H)\left| v_a(\exp H)\right|^{1/2} = c\Delta(a \exp H) \qquad (H \in \mathfrak{h}_0),$$

where $c = c_1 \xi_\rho(a)^{-1}$.

It is obvious that $\dim \mathfrak{g} - \dim \mathfrak{z} = 2p$. Since $\xi_\alpha(a) = 1$ for $\alpha \in P_{\mathfrak{z}}$, it is clear that $\xi_{2\rho'}(a) = \xi_{2\rho}(a)$. Now

$$c_1^2 \prod_{\alpha \in P'} (1 - \xi_\alpha(a^{-1}))^2 = \left| v_a(1)\right| = v_a(1) \operatorname{sign} v_a(1)$$

$$= (-1)^p (\operatorname{sign} v_a(1)) \xi_{2\rho'}(a) \prod_{\alpha \in P'} (1 - \xi_\alpha(a^{-1}))^2.$$

This shows that $c^2 = (-1)^p \operatorname{sign} v_a(1)$.

It follows from Lemmas 32 and 33 that

$$u(H) = c\Phi_A(a \exp H) \qquad (H \in \mathfrak{h}_0 \cap U').$$

Now put $V = a \exp \mathfrak{h}_0$ and $v(a \exp H) = c^{-1}u(H)$ $(H \in \mathfrak{h}_0 \cap \mathfrak{h}'(R))$. Then V is an open neighborhood of a in $A \cap \Omega$, v is an analytic function on $V \cap A'(R)$ and $v = \Phi_A$ on V. This proves Lemma 31.

For any root α, let s_α denote the Weyl reflexion corresponding to α. The Weyl group W of $(\mathfrak{g}, \mathfrak{h})$ operates on $\mathfrak{h}_c$ and therefore also on $\mathfrak{S}(\mathfrak{h}_c)$.

LEMMA 34. *Fix a point $a \in A \cap \Omega$ and suppose v is an element in $\mathfrak{S}(\mathfrak{h}_c)$ such that $v^{s_\alpha} = -v$ for every real root α for which $\xi_\alpha(a) = 1$. Then $v\Phi_A$ can be extended to a continuous function around a.*

We keep to the above notation. Then by Lemma 19 of [4(n)] $\partial(v)u$ can[10] be extended to a continuous function around zero. Since $\Phi_A(a \exp H) = c^{-1}u(H)$ $(H \in \mathfrak{h}_0 \cap U')$, our assertion is now obvious.

For any root α, define H_α as in [4(m), §4] and put $\varpi = \prod_{\alpha \in P} H_\alpha \in \mathfrak{S}(\mathfrak{h}_c)$. Then, by Lemma 34, $\varpi\Phi_A$ can be extended to a continuous function Ψ_A on A.

LEMMA 35. *Let A and B be two Cartan subgroups of G. Then $\Psi_A = \Psi_B$ on $A \cap B \cap \Omega$.*

Fix $a \in A \cap B \cap \Omega$ and let $\mathfrak{a}$ and $\mathfrak{b}$ be the Cartan subalgebras corresponding to $\mathfrak{a}$ and $\mathfrak{b}$, respectively. Define $\mathfrak{z}$, U and ψ as in Lemma 32. Then $\mathfrak{a}$, $\mathfrak{b}$ are Cartan subalgebras of $\mathfrak{z}$. Put $\mathfrak{h} = \mathfrak{a}$ or $\mathfrak{b}$ and define[14]

$$\varpi_{\mathfrak{z}}{}^{\mathfrak{h}} = \varpi_{\mathfrak{z}} = \prod_{\alpha \in P_{\mathfrak{z}}} H_\alpha, \qquad \varpi_{\mathfrak{g}/\mathfrak{z}}{}^{\mathfrak{h}} = \varpi_{\mathfrak{g}/\mathfrak{z}} = \prod_{\alpha \in P'} H_\alpha$$

in the notation introduced above. Then $\varpi = \varpi_{\mathfrak{z}} \cdot \varpi_{\mathfrak{g}/\mathfrak{z}}$. Since $\varpi^{s_\alpha} = -\varpi$, $\varpi_{\mathfrak{z}}{}^{s_\alpha} = -\varpi_{\mathfrak{z}}$ for any $\alpha \in P_{\mathfrak{z}}$, it is clear that $\varpi_{\mathfrak{g}/\mathfrak{z}}$ is invariant under the Weyl group of $(\mathfrak{z}, \mathfrak{h})$. Therefore, by Chevalley's theorem [4(f), Lemma 9], there exists an element $\eta \in I(\mathfrak{z}_c)$ such that the projection $\eta_\mathfrak{a}$ of η in $\mathfrak{S}(\mathfrak{a}_c) = S(\mathfrak{a}_c)$ (see [4(l), §8]) is $\varpi_{\mathfrak{g}/\mathfrak{z}}{}^\mathfrak{a}$.

Let G_c be an acceptable complexification of G and Ξ_c the analytic subgroup of G_c corresponding to $\mathfrak{z}_c$. Then we can choose $y \in \Xi_c$ such that $(\mathfrak{a}_c)^y = \mathfrak{b}_c$. Thus we have an isomorphism $D \to D^y$ of $\mathfrak{D}(\mathfrak{a}_c)$ onto $\mathfrak{D}(\mathfrak{b}_c)$ (see [4(l), §3]). Since the definition of Ψ_B is obviously independent of the order of roots, we may assume that the positive roots of $\mathfrak{a}$ are mapped into positive roots of $\mathfrak{b}$ under this isomorphism. Define j as in §18. Then it is obvious that $yj(a)y^{-1} = j(a)$. Therefore it follows from Lemma 33 that $c_A = c_B$ and $\eta_\mathfrak{b} = \varpi_{\mathfrak{g}/\mathfrak{z}}{}^\mathfrak{b}$. (Here c_A and c_B are the constants which correspond to c of Lemma 33 for the cases $\mathfrak{h} = \mathfrak{a}$ and $\mathfrak{h} = \mathfrak{b}$, respectively.)

---

[14] We use a similar notation in other cases. For example $\pi_{\mathfrak{z}}{}^{\mathfrak{h}} = \pi_{\mathfrak{z}}$ and $\varpi^{\mathfrak{h}} = \varpi$.

Now put

$$u^{\mathfrak{b}}(H) = \pi_3^{\mathfrak{b}}(H)\psi(H) \qquad (H \in U' \cap \mathfrak{h}).$$

Then it follows from the corollary of Theorem 3 of [4(n)] and [4(f), Theorem 1] (both applied to $\mathfrak{z}$) that

$$\partial(\varpi_3{}^\mathfrak{a} \cdot \eta_\mathfrak{a})u^\mathfrak{a} = \partial(\varpi_3{}^\mathfrak{b} \cdot \eta_\mathfrak{b})u^\mathfrak{b}$$

on $\mathfrak{a} \cap \mathfrak{b} \cap U$. This proves that $\partial(\varpi^\mathfrak{a})u^\mathfrak{a} = \partial(\varpi^\mathfrak{b})u^\mathfrak{b}$ on $\mathfrak{a} \cap \mathfrak{b} \cap \Omega$. But if U_0 is an open convex neighborhood of zero in U, we know that

$$u^\mathfrak{a}(H) = c_A\Phi_A(a\exp H) \qquad (H \in U_0' \cap \mathfrak{a}),$$

$$u^\mathfrak{b}(H) = c_B\Phi_B(a\exp H) \qquad (H \in U_0' \cap \mathfrak{b}),$$

where $U_0' = U_0 \cap U'$. Therefore since $c_A = c_B \neq 0$, we conclude that $\Psi_A(a) = \Psi_B(a)$.

20. The function $\nabla_G F$. We write $\varpi = \varpi_A$ for a given Cartan subgroup A.

LEMMA 36. *There exists a unique differential operator ∇_G on G' with the following properties.*

(1) *∇_G is invariant under G.*

(2) *Let A be a Cartan subgroup of G. Then*

$$f(h; \nabla_G) = f(h; \varpi_A \circ \Delta_A)$$

for $f \in C^\infty(G)$ and $h \in A \cap G'$.

Moreover, ∇_G is analytic.

The proof is similar to that of [4(n), Lemma 24]. Since two distinct Cartan subgroups cannot have a regular element in common, the uniqueness is obvious. The existence is proved as follows. Fix a Cartan subgroup A of G and define $G_A = \bigcup_{x \in G} xA'x^{-1}$, where $A' = A \cap G'$. Let $\mathfrak{h}$ be Cartan subalgebra of A, $\tilde{A}$ the normalizer of $\mathfrak{h}$ in G and $\tilde{A}_K = \tilde{A} \cap K$ in the notation of §16. Then by Corollary 2 of Lemma 26, $\tilde{A} = \tilde{A}_K A_\mathfrak{p}$ and if A_0 is the center of A, it follows (see §16) that

$$W_A = \tilde{A}/A_0 \simeq \tilde{A}_K/A_0 \cap K$$

is both compact and discrete and therefore it is finite. Let $x \to x^*$ denote the natural projection of G on $G^* = G/A_0$. Define $h^{x^*} = h^x$ $(h \in A, x \in G)$. Then the mapping $\phi: (x^*, h) \to h^{x^*}$ of $G^* \times A$ into G is everywhere regular on $G^* \times A'$. Hence $G_A = \phi(G^* \times A')$ is open in G. Now W_A operates on G^* and A as follows. Let y be an element in A whose image in W_A is s. Then

$$x^*s = (xy)^* \ (x \in G), \quad h^s = yhy^{-1}.$$

Define

$$(x^*, h)s = (x^*s, h^{s^{-1}}) \qquad (x^* \in G^*, \ h \in A').$$

In this way W_A operates on the right on $G^* \times A'$ without fixed points and the quotient space $(G^* \times A')/W_A$ may be identified with G_A by means of ϕ. By making use of the homomorphism $j : G \to G_c$ (see §18) one proves without difficulty that the differential operator $\varpi_A \circ \Delta_A$ on A is invariant under W_A. The rest of the proof now goes through exactly as in [4(n), §9].

LEMMA 37. *For any $z \in \mathfrak{Z}$, $(\nabla_G \circ z)F$ can be extended to a continuous function on Ω.*

Since the distribution zT also satisfies all the conditions of Theorem 2, it is enough to consider the case $z = 1$. Let Ω_0 be the set of all points $x_0 \in \Omega_0$ for which there exists an open neighborhood V of x_0 in Ω and a continuous function v on V such that $v = \nabla_G F$ on $V \cap G'$. Obviously Ω_0 is an open and invariant subset of Ω. Hence, in view of Lemma 7, it would be enough to prove that every semisimple element of Ω is contained in Ω_0.

Fix a semisimple element $a \in \Omega$ and let us use the notation of Lemma 32. Let $\mathfrak{z}'$ be the set of those elements of $\mathfrak{z}$ which are regular in $\mathfrak{z}$. Define the differential operator $\nabla_{\mathfrak{z}}$ on $\mathfrak{z}'$ as in [4(n), §9] and fix an open and convex neighborhood U_0 of zero in U and put $U_0' = U_0 \cap U'$. Let $\mathfrak{a}$ be a Cartan subalgebra of $\mathfrak{z}$. Then, as we have seen in §19, there exists a unique element $\eta \in I(\mathfrak{z}_c)$ such that $\eta_\mathfrak{a} = \varpi_{\mathfrak{g}/\mathfrak{z}}{}^\mathfrak{a}$. Let c denote the constant of Lemma 33 corresponding to $\mathfrak{h} = \mathfrak{a}$.

LEMMA 38. $F(a \exp Z; \nabla_G) = c\psi(Z; \nabla_{\mathfrak{z}} \circ \partial(\eta)) \ (Z \in U_0')$.

Fix $H_0 \in U_0'$ and let $\mathfrak{h}$ be the centralizer of H_0 in $\mathfrak{z}$. Then $\mathfrak{h}$ is a Cartan subalgebra of $\mathfrak{z}$ and therefore also of $\mathfrak{g}$. Moreover, $a \exp H_0 \in \Omega \cap G'$. Let A be the Cartan subgroup of G corresponding to $\mathfrak{h}$. Then

$$F(a \exp H_0; \nabla_G) = F(a \exp H_0 ; \varpi_A \circ \Delta_A) ,$$

from the definition of ∇_G. Put $\mathfrak{h}_0 = \mathfrak{h} \cap U_0$. Then we have seen in §19 that

$$\Delta_A(a \exp H)F(a \exp H) = c_A \pi_{\mathfrak{z}}(H)\psi(H) \qquad (H \in \mathfrak{h}_0 \cap U'),$$

where c_A is a constant. Moreover, by a suitable choice of positive roots of $(\mathfrak{g},\mathfrak{h})$ we can arrange (see the proof of Lemma 35) that $c_A = c$ and $\eta_\mathfrak{h} = \varpi_{\mathfrak{g}/\mathfrak{z}}{}^\mathfrak{h}$. Then it follows from [4(f), Theorem 1] and the definition of $\nabla_{\mathfrak{z}}$ that

$$\psi(H;\nabla_{\mathfrak{z}} \circ \partial(\eta)) = \psi(H;\partial(\varpi) \circ \pi_{\mathfrak{z}}) \qquad (H \in \mathfrak{h}_0 \cap U'),$$

where $\varpi = \varpi_A$. Therefore it is clear that

$$F(a \exp H_0; \nabla_G) = c\psi(H_0; \nabla_{\mathfrak{z}} \circ \partial(\eta))$$

and this proves our assertion.

It follows from Lemma 38 and [4(n), Lemma 25] that there exists an open neighborhood V_Ξ of 1 in Ω_Ξ and a continuous function g_0 on V_Ξ such that

$$F(ay; \nabla_G) = g_0(y) \qquad (y \in V_\Xi' = V_\Xi \cap \Omega_\Xi')$$

in the notation of §19. Let $x \to \bar{x}$ denote the natural mapping of G on $\bar{G} = G/\Xi$. Select open neighborhoods $\bar{G}_0$ and V_0 of $\bar{1}$ and 1 in $\bar{G}$ and V_Ξ, respectively. If they are sufficiently small the following conditions hold. There exists an analytic mapping ϕ of $\bar{G}_0$ into G such that: (1) $\overline{\phi(\bar{x})} = \bar{x}$ for $\bar{x} \in \bar{G}_0$ and (2) the mapping $\alpha:(\bar{x}, y) \to (ay)^{\phi(\bar{x})}$ of $\bar{G}_0 \times V_0$ into G is univalent and regular and $V = \alpha(\bar{G}_0 \times V_0) \subset V_\Xi$. Then V is an open neighborhood of a in G and α defines an analytic diffeomorphism of $\bar{G}_0 \times V_0$ on V. Define a function F_0 on V by

$$F_0(\alpha(\bar{x}, y)) = g_0(y) \qquad (\bar{x} \in \bar{G}_0, y \in V_0).$$

Then F_0 is continuous and since $\nabla_G F$ is invariant under G, it is obvious that $F_0 = \nabla_G F$ on $V \cap G'$. This shows that $a \in \Omega_0$ and therefore Lemma 37 is proved.

21. An elementary result. Let $\mathfrak{h}$ be a Cartan subalgebra of $\mathfrak{g}$ and W the Weyl group of $(\mathfrak{g}, \mathfrak{h})$.

LEMMA 39. *Let λ be a linear function on $\mathfrak{h}_c$. Then there exists an invariant analytic function f_λ on $\mathfrak{g}$ such that*

$$\pi(H)f_\lambda(H) = \sum_{s \in W} \varepsilon(s)e^{\lambda(sH)} \quad (H \in \mathfrak{h}).$$

Moreover, f_λ is unique.

Let $\mathfrak{h}'$ be the set of all elements $H \in \mathfrak{h}$, where $\pi(H) \neq 0$. Since $(\mathfrak{h}')^G$ is an open subset of $\mathfrak{g}$, the uniqueness of f_λ is obvious. Therefore it remains to prove its existence. For this we may obviously assume that $\mathfrak{g}$ is semisimple and G is the connected adjoint group of $\mathfrak{g}$. Now we use the notation of §16. Let G_c be the (connected) complex adjoint group of $\mathfrak{g}_c$ and U the real analytic subgroup of G_c corresponding to the compact real form $\mathfrak{u} = \mathfrak{k} + (-1)^{1/2}\mathfrak{p}$ of $\mathfrak{g}_c$. Then U is compact. Put $B(X, Y) = \mathrm{tr}(\mathrm{ad}\, X\, \mathrm{ad}\, Y)\,(X, Y \in \mathfrak{g}_c)$ as usual and consider

$$f(X:Y) = \int_U \exp B(uX, Y)\, du \quad (X, Y \in \mathfrak{g}_c),$$

where du is the normalized Haar measure on U. Then f is obviously a holomorphic function on $\mathfrak{g}_c \times \mathfrak{g}_c$ and it is clear that

$$f(X;\tau(Z):Y) = 0 \quad (Z \in \mathfrak{u})$$

in the notation of [4(I), §4]. Since f is holomorphic in X, this implies that $f(xX:Y) = f(X:Y)$ for $x \in G_c$.

Let H_λ denote the element in $\mathfrak{h}_c$ such that $B(H, H_\lambda) = \lambda(H)$ for all $H \in \mathfrak{h}_c$. Then we know from [4(f), Theorem 2] that

$$\pi(H_\lambda)\pi(H)f(H:H_\lambda) = c \sum_{s \in W} \varepsilon(s)e^{\lambda(sH)} \quad (H \in \mathfrak{h}_c),$$

where c is a number $\neq 0$ independent of H and λ. Therefore we can take

$$f_\lambda(X) = c^{-1}\pi(H_\lambda)f(X:H_\lambda) \quad (X \in \mathfrak{g}).$$

22. The invariant integral on G. We now return to the notation of §19. Let A_0 denote the center of A and $x \to x^*$ the natural projection of G on $G^* = G/A_0$. Put $h^{x^*} = h^x$ $(h \in A, x \in G)$ and let dx^* denote the invariant measure on G^*. For any $f \in C_c^\infty(G)$, put

$$F_f(h) = \varepsilon_R(h)\Delta(h) \int_{G^*} f(h^{x^*}) \, dx^* \qquad (h \in A'),$$

where $A' = A \cap G'$ and $\varepsilon_R(h) = \operatorname{sign} \Delta_R{}'(h)$. Then F_f is a C^∞-function on A' and if γ is the canonical isomorphism of $\mathfrak{Z}$ onto $I(\mathfrak{h}_c)$ (see §6), we have [4(h), Theorem 3]

$$F_{zf} = \gamma(z)F_f \qquad (z \in \mathfrak{Z}, \ f \in C_c^\infty(G)).$$

Let S_I denote the set of all positive singular imaginary roots of $(\mathfrak{g}, \mathfrak{h})$ (see [4(m), §4]). Define

$$\Delta_I{}'(h) = \prod_{\alpha \in S_I} (1 - \xi_\alpha(h)^{-1}) \qquad (h \in A)$$

and let $A'(I)$ be the set of those points $h \in A$ where $\Delta_I{}'(h) \neq 0$.

LEMMA 40. *Fix* $f \in C_c^\infty(G)$. *Then* F_f *can be extended to a* C^∞-*function on* $A'(I)$. *Let a be a point in* A *and* v *an element in* $\mathfrak{S}(\mathfrak{h}_c)$ *such that* $v^{s_\alpha} = -v$ *for every singular imaginary root* α *for which* $\xi_\alpha(a) = 1$. *Then* vF_f *can be extended to a continuous function around* a.

Let $\mathfrak{z}$ and Ξ_1 denote the centralizers of a in $\mathfrak{g}$ and G, respectively, and Ξ the connected component of 1 in Ξ_1. Then if Z is the center of G, $\Xi_1/Z\Xi$ is finite (see [4(g₂), Lemma 15]). Choose an open neighborhood B of 1 in A with the following property (see [4(h), Theorem 1]). If $h \in B$ and $x \in G$ vary in such a way that $(ah)^x$ stays inside some compact subset of G, then the coset $\bar{x} = x\Xi_1$ remains within a compact subset of $\bar{G} = G/\Xi_1$. Let $x \to \bar{x}$ denote the natural projection of G on $\bar{G}$. Since $\mathfrak{z}$ is reductive and $\Xi_1/Z\Xi$ is finite, it follows that the group Ξ_1 is unimodular. Hence we have an invariant measure $d\bar{x}$ on $\bar{G}$. Let dy^* denote the invariant measure on $\Xi_1^* = \Xi_1/A_0$. Then if $d\bar{x}$ and dy^* are suitably normalized, we have

$$F_f(ah) = \varepsilon_R(ah)\Delta(ah) \int_{\bar{G}} d\bar{x} \int_{\Xi_1^*} f(x(ah)^{y^*} x^{-1}) \, dy^* \qquad (h \in B')$$

for $f \in C_c^\infty(G)$. Here $B' = B \cap a^{-1}A'$. Now fix an open and relatively compact subset G_0 of G and choose a compact set $\bar{\Omega}$ in $\bar{G}$ such that $(aB)^x \cap G_0 = \varnothing$ $(x \in G)$ unless $\bar{x} \in \bar{\Omega}$. Let dy denote the Haar measure of Ξ_1 and choose $\gamma \in C_c^\infty(G)$ such that

$$\int_{\Xi Z} \gamma(xy) \, dy = 1$$

if $\bar{x} \in \bar{\Omega}$ $(x \in G)$. Then if dy is suitably normalized, we have

$$F_f(ah) = \varepsilon_R(ah)\Delta(ah) \int_G \gamma(x) \, dx \int_{\Xi^*} f(x(ah^{y^*})x^{-1}) \, dy^* \qquad (h \in B')$$

for $f \in C_c^\infty(G_0)$. Here $\Xi^* = \Xi/\Xi \cap A_0$. Now fix $f \in C_c^\infty(G_0)$ and put

$$g_0(y) = \int_G \gamma(x) f(x(ay)x^{-1}) \, dx \qquad (y \in \Xi).$$

Then $g_0 \in C_c^\infty(\Xi)$.

We now use the notation of §19. Select an open and connected neighborhood $\mathfrak{h}_0$ of zero in $\mathfrak{h}$ such that $\exp \mathfrak{h}_0 \subset B$, $\xi_\alpha(a \exp H) \neq 1$ ($\alpha \in P'$) and

$$(1 - e^{-\alpha(H)})/\alpha(H) \neq 0 \qquad (\alpha \in P_{\mathfrak{i}})$$

for $H \in \mathfrak{h}_0$. Then if $\mathfrak{h}_0'$ is the set of all points $H \in \mathfrak{h}_0$, where $\pi_{\mathfrak{i}}(H) \neq 0$, it is clear that $\exp \mathfrak{h}_0' \subset B'$ and

$$F_f(a \exp H) = \varepsilon_R(a \exp H)\Delta(a \exp H) \int_{\Xi^*} g_0((\exp H)^{y^*}) dy^* \qquad (H \in \mathfrak{h}_0').$$

A simple argument shows (see §21) that there exists an analytic function D_a on $\mathfrak{z}$ such that: (1) D_a is invariant under Ξ and (2) $\Delta(a \exp H) = \pi_{\mathfrak{i}}(H)D_a(H)$ for $H \in \mathfrak{h}$. Fix an open and completely invariant neighborhood $\mathfrak{z}_0$ of zero in $\mathfrak{z}$ such that the exponential mapping (from $\mathfrak{z}$ to Ξ) is regular and univalent on $\mathfrak{z}_0$ and select a C^∞-function u on $\mathfrak{z}$ such that: (1) u is invariant under Ξ, (2) $\mathrm{Supp}\, u \subset \mathfrak{z}_0$, and (3) $u = 1$ around zero. This is possible (see §9 and [4(n), Corollary 1 of Lemma 45]). Now put

$$g(Z) = u(Z)D_a(Z)g_0(\exp Z) \qquad (Z \in \mathfrak{z}).$$

Then $g \in C_c^\infty(\mathfrak{z}_0)$. Since $\mathfrak{h}_0$ is connected and $\xi_\alpha(a \exp H) \neq 1$ for $\alpha \in P'$ and $H \in \mathfrak{h}_0$, it is clear that

$$\varepsilon_R(a \exp H) = \varepsilon_{\mathfrak{z}, R}(H)\varepsilon_a \qquad (H \in \mathfrak{h}_0),$$

where

$$\varepsilon_{\mathfrak{z}, R}(H) = \mathrm{sign} \prod_{\alpha \in Pz \cap P_R} \alpha(H)$$

and ε_a is a constant. Therefore

$$F_f(a \exp H) = \varepsilon_a \varepsilon_{\mathfrak{z}, R}(H)\pi_{\mathfrak{i}}(H) \int_{\Xi^*} g(y^*H) \, dy^* \qquad (H \in U \cap \mathfrak{h}_0'),$$

where U is an open neighborhood of zero in $\mathfrak{z}$ such that $u = 1$ on U. The second assertion of the lemma now follows by applying Theorem 1 of [4(m)] to $(\mathfrak{z}, \mathfrak{h})$ and g. Moreover, this obviously implies the first assertion.

COROLLARY. ϖF_f can be extended to a continuous function on A.

Since $\varpi^{s_\alpha} = - \varpi$ for every root α, this is an immediate consequence of Lemma 40.

23. **Statement of Theorem 3.** Define G_A as in §20. Since A_0 is abelian and A/A_0 is finite (see the proof of Lemma 36), the Haar measure dh of A is bi-invariant. We keep to the notation of §22.

LEMMA 41. *There exists a number $c > 0$ such that*

$$\int_G f(x)\,dx = c \int_A |\Delta(h)|^2\,dh \int_{G^*} f(h^{x^*})\,dx^*$$

for $f \in C_c(G_A)$.

We observe that

$$\det(\mathrm{Ad}(h)^{-1} - 1)_{\mathfrak{g}/\mathfrak{h}} = (-1)^r \Delta(h)^2 \qquad (h \in A),$$

where r is the number of positive roots of $(\mathfrak{g},\mathfrak{h})$. From this our assertion follows in the usual way (see the proof of Lemma 36 and [4(c), p. 508]).

COROLLARY. *Let $f \in C_c^\infty(G)$. Then*

$$\int_A |\Delta(h)F_f(h)|\,dh \leq c^{-1} \int_{G_A} |f(x)|\,dx.$$

This is obvious from the above lemma.

We now use the notation of §16. Let $\mathfrak{m}$ be the centralizer of $\mathfrak{h} \cap \mathfrak{p}$ in $\mathfrak{g}$ and M the analytic subgroup of G corresponding to $\mathfrak{m}$. Let P_M be the set of all positive roots of $(\mathfrak{g},\mathfrak{h})$ which vanish identically on $\mathfrak{h} \cap \mathfrak{p}$. Then P_M is also the set of all positive imaginary roots of $(\mathfrak{g},\mathfrak{h})$ or, equivalently, the set of positive roots of $(\mathfrak{m},\mathfrak{h})$. Put

$$\Delta_M(h) = \xi_\rho(h_1) \prod_{\alpha \in P_M} (1 - \xi_\alpha(h^{-1})) \qquad (h \in A),$$

where $h = h_1 h_2$ ($h_1 \in A_K$, $h_2 \in A_\mathfrak{p}$). It follows from [4(h), Theorem 2] that

$$\int_A |\Delta_M(h)F_f(h)|\,dh < \infty \qquad (f \in C_c^\infty(G)).$$

THEOREM 3. *Let v be a seminorm*[15] *on the complex vector space $C_c^\infty(G)$ and $\mathfrak{Z}_0$ a subalgebra of $\mathfrak{Z}$ containing 1. Assume that $\mathfrak{Z}$ is a finite module over $\mathfrak{Z}_0$ and*

$$\int_A |\Delta_M(h)F_f(h)|\,dh \leq v(f) \qquad (f \in C_c^\infty(G)).$$

Then for any $u \in \mathfrak{S}(\mathfrak{h}_c)$, we can choose a finite set of elements $z_1, \cdots, z_N \in \mathfrak{Z}_0$ such that

$$\sup_{h \in A'} |F_f(h;u)| \leq \sum_{1 \leq i \leq N} v(z_i f) \qquad (f \in C_c^\infty(G)).$$

REMARK. The above form of this theorem suggested itself to me after a conversation with R. P. Langlands. My original version was less comprehensive.

Let Z be the center of G and V a subset of A such that $VZ = A$. Put $V' = V \cap A'$. We claim that it would be sufficient to prove the following lemma for a conveniently chosen V.

LEMMA 42. *For any $u \in \mathfrak{S}(\mathfrak{h}_c)$, we can choose $z_1, \cdots, z_N$ in $\mathfrak{Z}_0$ such that*

---

[15] Here we ignore completely the topology of $C_c^\infty(G)$.

$$\sup_{h \in V'} \left| F_f(h;u) \right| \leq \sum_{1 \leq i \leq N} v(z_i f) \qquad (f \in C_c^\infty(G)).$$

Put

$$v_0(f) = \int_A \left| \Delta_M(h) F_f(h) \right| dh \qquad (f \in C_c^\infty(G)).$$

Then v_0 is a seminorm on $C_c^\infty(G)$ which satisfies the condition of Theorem 3. For any $y \in Z$, let f_y denote the function $x \to f(xy)$ on G. Then it is clear that $\xi_\rho(y) F_{f_y}(h) = F_f(hy)$ $(h \in A')$ and $\Delta_M(hy) = \xi_\rho(y)\Delta_M(h)$. Since $\left| \xi_\rho(y) \right| = 1$, it follows that $v_0(f_y) = v_0(f)$. Therefore if Lemma 42 holds for v_0, we can conclude that

$$\sup_{h \in V'} \left| F_f(hy;u) \right| = \sup_{h \in V'} \left| F_{f_y}(h;u) \right|$$

$$\leq \sum_{1 \leq i \leq N} v_0(z_i f_y) = \sum_i v_0(z_i f) \qquad (y \in Z)$$

since $z_i f_y = (z_i f)_y$. Now fix a seminorm v as in Theorem 3. Then $v_0(g) \leq v(g)$ for $g \in C_c^\infty(G)$. Hence

$$\sup_{h \in V'} \left| F_f(hy;u) \right| \leq \sum_i v(z_i f) \qquad (y \in Z,\ f \in C_c^\infty(G)).$$

But since $V'Z = A'$, the assertion of Theorem 3 now follows immediately.

24. Reduction to $\mathfrak{h}$ in a special case. So now it remains to prove Lemma 42. First *assume that every root of* $(\mathfrak{g},\mathfrak{h})$ *is imaginary*. Then $M = G$ and it follows from Corollary 5 of Lemma 26 that A/Z is compact. So we can take V to be compact. Let $\mathscr{S}$ be the set of all seminorms σ on([15]) $C_c^\infty(G)$ with the following property. We can choose a finite number of elements $z_1, \cdots, z_N$ in $\mathfrak{Z}_0$ such that

$$\sigma(f) \leq \sum_{1 \leq i \leq N} v(z_i f) \qquad (f \in C_c^\infty(G)).$$

Then since V is compact, it would obviously be enough to prove the following result.

LEMMA 43. *Given $h_0 \in A$, we can choose an open neighborhood U of h_0 in A with the following property. For any $u \in \mathfrak{S}(\mathfrak{h}_c)$, there exists an element $\sigma_u \in \mathscr{S}$ such that*

$$\sup_{h \in A' \cap U} \left| F_f(h;u) \right| \leq \sigma_u(f) \quad (f \in C_c^\infty(G)).$$

Let $\mathfrak{c}$ be the center and $\mathfrak{g}_1$ the derived algebra of $\mathfrak{g}$. Then $\mathfrak{h} = \mathfrak{c} + \mathfrak{h}_1$, where $\mathfrak{h}_1 = \mathfrak{h} \cap \mathfrak{g}_1$. Since every root of $(\mathfrak{g},\mathfrak{h})$ is imaginary, $-\operatorname{tr}(\operatorname{ad} H)^2$ $(H \in \mathfrak{h}_1)$ is a positive-definite quadratic form on $\mathfrak{h}_1$. We extend it to a positive-definite quadratic form Q on $\mathfrak{h}$ in such a way that $\mathfrak{c}$ and $\mathfrak{h}_1$ are orthogonal under Q and, moreover, regard $\mathfrak{h}$ as a real Hilbert space under the norm $\| H \|^2 = Q(H)$ $(H \in \mathfrak{h})$.

Let us now introduce the notation of §19 corresponding to $a = h_0$. Fix a number

c $(0 < c \leqq 1)$ and let V be the set of all $H \in \mathfrak{h}$ with $\| H \| < c$. We assume that c is so small that:

(1) $\left| (e^{\alpha(H)/2} - e^{-\alpha(H)/2} \xi_\alpha(h_0^{-1})) \right| \geqq (1/2) \left| 1 - \xi_\alpha(h_0^{-1}) \right|$ for every root α of $(\mathfrak{g}, \mathfrak{h})$ and $H \in V$.

(2) The exponential mapping of V into A is univalent.

(3) $\left| \{ (e^{\alpha(H)/2} - e^{-\alpha(H)/2}) / \alpha(H) \} \right| \geqq 1/2$ for $\alpha \in P_{\mathfrak{z}}$ and $H \in V$.

Let $\mathfrak{h}'$ be the set of all $H \in \mathfrak{h}$ where $\pi_{\mathfrak{z}}(H) \neq 0$. Then $V' = V \cap \mathfrak{h}'$ consists of a finite number of connected components, say $V_1, \cdots, V_q$. Put

$$U = h_0 \exp(aV),$$

where a is a positive number $(0 < a \leq 1)$. Then it is clear that

$$U \cap A' = \bigcup_{1 \leqq i \leqq q} U_i$$

where $U_i = h_0 \exp(aV_i)$. Then it would be sufficient to prove the following lemma.

LEMMA 44. *Fix i $(1 \leqq i \leqq q)$. Then we can select a number a $(0 < a \leqq 1)$ with the following property. For any $u \in \mathfrak{S}(\mathfrak{h}_c)$ we can choose $\sigma \in \mathscr{S}$ such that*

$$\sup_{H \in aV_i} \left| F_f(h_0 \exp H; u) \right| \leqq \sigma(f) \qquad (f \in C_c^\infty(G)).$$

Put

$$\phi_f(H) = F_f(h_0 \exp H) \qquad (f \in C_c^\infty(G), \ H \in V').$$

Then it follows from [4(h), Theorem 3] that[10] $\phi_{zf} = \partial(\gamma(z))\phi_f$ for $z \in \mathfrak{z}$. Moreover, it is obvious that Lemma 44 is equivalent to the following.

LEMMA 45. *Fix i $(1 \leq i \leq q)$. Then we can select a $(0 < a \leqq 1)$ with the following property. For any $u \in S(\mathfrak{h}_c)$, we can choose $\sigma \in \mathscr{S}$ such that*

$$\sup_{H \in aV_i} \left| \phi_f(H; \partial(u)) \right| \leqq \sigma(f)$$

for all $f \in C_c^\infty(G)$.

We may assume that $i = 1$. Let L be the rank of $\mathfrak{z}_1 = [\mathfrak{z}, \mathfrak{z}]$. Then we can choose L roots $\alpha_1, \cdots, \alpha_L$ of $(\mathfrak{z}, \mathfrak{h})$ with the following property. If α is a root of $(\mathfrak{z}, \mathfrak{h})$ such that $(-1)^{1/2}\alpha(H) > 0$ for $H \in V_1$, then $\alpha = \sum_{1 \leqq i \leqq L} m_i \alpha_i$, where m_i are rational integers $\geqq 0$. Put $t_i(H) = (-1)^{1/2}\alpha_i(H)$ $(1 \leq i \leq L, \ H \in \mathfrak{h})$ and choose a base H_j $(1 \leqq j \leqq l_1)$ for $\mathfrak{h}_1 = \mathfrak{h} \cap \mathfrak{g}_1$ such that $t_i(H_j) = \delta_{ij}$ $(1 \leqq i \leqq L, \ 1 \leqq j \leqq l_1)$. Let H_j $(l_1 < j \leqq l)$ be an orthonormal base for $\mathfrak{c}$. Extend $(t_1, \cdots, t_L)$ to a Cartesian coordinate system $(t_1, \cdots, t_l)$ on $\mathfrak{h}$ by defining $t_i(H_j) = \delta_{ij}$ $(1 \leq i, j \leq l)$. Then a point $H \in V$ lies in V_1 if and only if $t_i(H) > 0$ $(1 \leq i \leq L)$. Define

$$\tau(H) = \begin{cases} \min_{1 \leqq l \leqq L} \left| t_l(H) \right| & \text{if } L > 0, \\ c & \text{if } L = 0 \end{cases} \qquad (H \in \mathfrak{h}).$$

Clearly $\|H\|^2 \geq 2|\alpha(H)|^2$ for any root α of $(\mathfrak{g}, \mathfrak{h})$. Hence if $H_0 \in V_1/2$ and $\|H - H_0\| \leq \tau(H_0)/2$ $(H \in \mathfrak{h})$, it is clear that $\|H\| \leq \|H_0\| + \tau(H_0)/2 < c$. Therefore $H \in V$. Moreover,

$$\left|t_i(H - H_0)\right| = \left|\alpha_i(H - H_0)\right| \leq \|H - H_0\| \leq \frac{1}{2}\tau(H_0) \leq \frac{1}{2}t_i(H_0) \qquad (1 \leq i \leq L).$$

Therefore $t_i(H) \geq \tau(H_0)/2$ $(1 \leq i \leq L)$ and so $H \in V_1$. This also shows that $\tau(H) \geq \tau(H_0)/2$. Thus we have obtained the following result.

LEMMA 46. *Fix $H_0 \in V_1/2$ and let H be an element in $\mathfrak{h}$ such that*

$$\|H - H_0\| \leq \tau(H_0)/2.$$

Then $H \in V_1$ and $\tau(H) \geq \tau(H_0)/2$.

Fix a function ψ on R of class C^∞ such that $\psi = 1$ on the interval $(-\infty, 0]$, $\psi = 0$ on the interval $[1, +\infty)$ and $0 \leq \psi \leq 1$ everywhere.

LEMMA 47. *For any real number ε $(0 < \varepsilon \leq 1/2)$ define*

$$\Psi_\varepsilon(H) = \psi(\varepsilon^{-1}\|H\| - 2) \qquad (H \in \mathfrak{h}).$$

Then for any element $u \in S(\mathfrak{h}_c)$ of degree $\leq d$, we can choose a number $b > 0$ such that

$$\left|\Psi_\varepsilon(H; \partial(u))\right| \leq b\varepsilon^{-d}$$

for all $H \in \mathfrak{h}$ and $0 < \varepsilon \leq 1/2$.

This is an immediate consequence of Lemma 55 of the Appendix (see also [6, p. 281]). Observe that $\Psi_\varepsilon(H) = 0$ unless $\|H\| \leq 3\varepsilon$.

25. Proof of a weaker result. Now first we prove the following weaker form[16] of Lemma 45.

LEMMA 48. *Given $u \in S(\mathfrak{h}_c)$, we can choose an integer $q \geq 0$ and $\sigma \in \mathscr{S}$ such that*

$$\sup_{H \in V_1/2} \left\{\prod_{1 \leq i \leq L} t_i(H)\right\}^q \left|\phi_f(H; \partial(u))\right| \leq \sigma(f)$$

for all $f \in C_c^\infty(G)$.

Let $\omega_1 \in \mathfrak{Z}$ be the Casimir operator (see [4(e), p. 140]) corresponding to $\mathfrak{g}_1$. Put

$$\omega = \omega_1 - \sum_{l_1 < j \leq l} H_j^2 \in \mathfrak{Z}.$$

Then it is easy to verify (see [4(e), p. 144]) that[17] $\gamma(\omega) + <\rho, \rho>$ is homo-

---

[16] Cf. [4(g), p. 206].

[17] $\langle\rho, \rho\rangle = \rho(H_\rho)$, where H_ρ is the unique element in $\mathfrak{h}_{1_c}$ such that $\text{tr}(\text{ad}H\,\text{ad}H_\rho) = \rho(H)$ for all $H \in \mathfrak{h}$.

geneous of degree 2 and $D = \partial(\gamma(\omega)) + <\rho, \rho>$ is an elliptic differential operator on $\mathfrak{h}$.

Now $\mathfrak{Z}$ being a finite module over $\mathfrak{Z}_0$, we can choose $v_1 = 1, v_2, \cdots, v_r$ in $\mathfrak{Z}$ such that

$$\mathfrak{Z} = \sum_{1 \leq i \leq r} \mathfrak{Z}_0 v_i.$$

Fix an integer $m \geq 1$. Then we have an equation of the form

$$\omega_0^{mr} + \sum_{1 \leq j \leq r} z_j \omega_0^{m(r-j)} = 0,$$

where $\omega_0 = \omega + <\rho, \rho>$ and $z_j \in \mathfrak{Z}_0$.

For the proof of Lemma 48, we may obviously assume that $u \neq 0$. Let d be the degree of u. Fix a Euclidean measure dH on $\mathfrak{h}$ such that dH corresponds, locally, to the Haar measure dh on A, under the exponential mapping. Then if m is sufficiently large, there exists a function E_0 on $\mathfrak{h}$ of class $C^{2m(r-1)+d}$ such that

$$D^{mr} E_0 = \delta$$

in the sense of the theory of distributions on the Euclidean space $\mathfrak{h}$ (with respect to the measure dH). Here δ is the Dirac measure on $\mathfrak{h}$ concentrated at zero and E_0 is of class C^∞ everywhere except at the origin (see Lemma 57, §29). Put $E = \partial(u)^* E_0$, where the star denotes adjoint. It follows by applying the homomorphism γ to the relation above that

$$D^{mr} + \sum_{1 \leq j \leq r} \partial(\gamma(z_j)) D^{m(r-j)} = 0.$$

Since $D^* = D$, we find, by taking adjoints, that

$$D^{mr} + \sum_{1 \leq j \leq r} \partial(\gamma(z_j))^* D^{m(r-j)} = 0.$$

Put $E_j = - D^{m(r-j)} E$ $(1 \leq j \leq r)$. Then E_j is a function of class $C^{2m(j-1)}$ and

$$\partial(u)^* \delta = \sum_{1 \leq j \leq r} \partial(\gamma(z_j))^* E_j.$$

Clearly E_j is of class C^∞ everywhere except at zero. Put $E_{j,\varepsilon} = \Psi_\varepsilon E_j$ for any ε $(0 < \varepsilon \leq 1/3)$ in the notation of Lemma 47. Then it is clear that

$$\sum_{1 \leq j \leq r} \partial(\gamma(z_j))^* E_{j,\varepsilon} = \partial(u)^* \delta + \beta_\varepsilon,$$

where $\beta_\varepsilon \in C_c^\infty(\mathfrak{h})$ and $\operatorname{Supp} \beta_\varepsilon \subset \operatorname{Supp} \Psi_\varepsilon$. Now $\Psi_\varepsilon(H) = 1$ if $\|H\| \leq 2\varepsilon$. Hence $\beta_\varepsilon(H) = 0$ unless $2\varepsilon \leq \|H\| \leq 3\varepsilon$.
Therefore

$$\sup_H |\beta_\varepsilon(H)| = \sup_{2\varepsilon \leq \|H\| \leq 3\varepsilon} |\beta_\varepsilon(H)|.$$

Making use of Lemma 47 and the explicit formula for E_0 (see §29), we find that

$$\sup_{H} \left| \beta_\varepsilon(H) \right| \le b_1 \varepsilon^{-p+1} \left| \log \varepsilon \right| \le b_2 \varepsilon^{-p}$$

where b_1, b_2 are positive numbers and p an integer ≥ 0, all independent of ε $(0 < \varepsilon \le 1/3)$.

Now fix $H_0 \in V_1/2$ and put $\varepsilon_0 = \tau(H_0)/6$, $E_{j,H_0} = E_{j,\varepsilon_0}$ and $\beta_{H_0} = \beta_{\varepsilon_0}$ $(1 \le j \le r)$. Then

$$\sum_{1 \le i \le r} \partial(\gamma(z_i))^* E_{i,H_0} = \partial(u)^* \delta + \beta_{H_0},$$

$$\sup \left| \beta_{H_0} \right| \le b_3 \tau(H_0)^{-p},$$

where $b_3 = 6^p b_2$. Now $\operatorname{Supp} E_{i,H_0}$ and $\operatorname{Supp} \beta_{H_0}$ are both contained in $\operatorname{Supp} \Psi_{\varepsilon_0}$. Moreover, $\left\| H \right\| \le \tau(H_0)/2$ if $H \in \operatorname{Supp} \Psi_{\varepsilon_0}$. Hence if $H - H_0 \in \operatorname{Supp} \Psi_{\varepsilon_0}$, it follows from Lemma 46 that $H \in V_1$ and $\tau(H) \ge \tau(H_0)/2$. Let $V(H_0)$ be the set of all $H \in V$ such that $\tau(H) \ge \tau(H_0)/2$. Then it is clear that $V(H_0) \subset V_1$ and

$$\phi_f(H_0; \partial(u)) = \sum_{1 \le i \le r} \int_{V(H_0)} \phi_f(H; \partial(\gamma(z_i))) E_{i,H_0}(H - H_0)\, dH$$

$$- \int_{V(H_0)} \phi_f(H) \beta_{H_0}(H - H_0)\, dH.$$

On the other hand it follows from the definition of V (see §24) that we can choose a number $c_1 > 0$ such that

$$\left| \Delta(H_0 \exp H) \right| \ge c_1 \left| \pi_{\mathfrak{z}}(H) \right| \qquad (H \in V).$$

Let q_1 be the number of roots in $P_{\mathfrak{z}}$. Then it follows from our definition of $t_1, \cdots, t_L$ that

$$\left| \pi_{\mathfrak{z}}(H) \right| \ge \tau(H)^{q_1} \qquad (H \in V_1).$$

Therefore

$$\left| \Delta(h_0 \exp H) \right| \ge c_1 \tau(H)^{q_1} \qquad (H \in V_1).$$

Hence

$$\tau(H_0)^{q_1} \left| \int_{V(H_0)} \phi_f(H; \partial(\gamma(z_i))) E_{i,H_0}(H - H_0)\, dH \right|$$

$$\le c_2 \int_{V(H_0)} \left| \phi_{z_i f}(H) \Delta(h_0 \exp H) \right| \left| E_{i,H_0}(H - H_0) \right| dH,$$

where $c_2 = 2^{q_1} c_1^{-1}$. Moreover, since $\left| \Psi_\varepsilon \right| \le 1$ and E_i are continuous functions on $\mathfrak{h}$, it is clear that

$$\sup_{H \in V(H_0)} \left| E_{i,H_0}(H - H_0) \right| \le \sup_{\|H\| \le 2} \left| E_i(H) \right| \le c_3,$$

where c_3 is a positive number independent of H_0 or i. Hence

$$\tau(H_0)^{q_1}\left|\int_{V(H_0)}\phi_f(H;\partial(\gamma(z_i)))E_{i,H_0}(H-H_0)\,dH\right|\le c_2 c_3 \nu(z_i f) \qquad (1\le i\le r).$$

Similarly since $\sup|\beta_{H_0}|\le b_3\tau(H_0)^{-p}$, we get

$$\left|\tau(H_0)\right|^{p+q_1}\left|\int_{V(H_0)}\phi_f(H)\beta_{H_0}(H-H_0)\,dH\right|\le c_2 b_3 \nu(f).$$

Moreover, $\tau(H_0)\le c\le 1$. Hence

$$\tau(H_0)^{p+q_1}\left|\phi_f(H_0;\ \partial(u))\right|\le c_2 b_3 \nu(f)+c_2 c_3 \sum_{1\le i\le r}\nu(z_i f)$$

for $H_0\in V_1/2$ and $f\in C_c^\infty(G)$. Now

$$\prod_{1\le i\le L}t_i(H)\le c^{L-1}\tau(H) \qquad (H\in V_1).$$

This is obvious if $L\ge 1$ and is also true if $L=0$. Therefore the statement of Lemma 48 follows immediately if we take $q=p+q_1$.

26. **Proof of Lemma 45**[18]. Now we come to the proof of Lemma 45. If $L=0$, it is an immediate consequence of Lemma 48. So we may assume that $L\ge 1$.

By a monomial T we mean a function on $\mathfrak{h}$ of the form $t_1^{q_1}t_2^{q_2}\cdots t_L^{q_L}$, where $q_1,\cdots,q_L$ are integers ≥ 0. The degree of T is the integer $q_1+q_2+\cdots+q_L$ and we denote it by d^0T. Since $S(\mathfrak{h}_c)$ is a finite module over $I(\mathfrak{h}_c)=\gamma(\mathfrak{Z})$ (see [4(f), Lemma 11]), it is also a finite module over $\gamma(\mathfrak{Z}_0)$. Hence we can choose u_j $(1\le j\le r)$ in $S(\mathfrak{h}_c)$ such that $u_1=1$ and

$$S(\mathfrak{h}_c)=\sum_{1\le j\le r}\gamma(\mathfrak{Z}_0)u_j.$$

We say that a monomial T has property (P) if there exists a number $a=a(T)$ $(0<a\le 1)$ and $\sigma\in\mathscr{S}$ such that

$$\text{(P)}\qquad \sup_{H\in aV_1}T(H)\left|\phi_f(H;\partial(u_j))\right|\le\sigma(f)\qquad (1\le j\le r)$$

for all $f\in C_c^\infty(G)$. Now suppose T has property (P) and put

$$\sigma_T(f)=\max_{1\le j\le r}\ \sup_{H\in aV_1}\ T(H)\left|\phi_f(H;\partial(u_j))\right|\qquad (f\in C_c^\infty(G)),$$

where $a=a(T)$. Then it is obvious that $\sigma_T\in\mathscr{S}$ and, for a given $u\in S(\mathfrak{h}_c)$, we can select $z_i\in\mathfrak{Z}_0$ $(1\le i\le r)$ such that $u=\sum_{1\le i\le r}\gamma(z_i)u_i$. Hence

$$\phi_f(H;\partial(u))=\sum_i\phi_{z_i f}(H;\partial(u_i))\qquad (H\in V',\ f\in C_c^\infty(G)).$$

---

[18] Cf. [4(g), pp. 208–211].

So it is clear that

$$\sigma_{T,u}(f) = \sup_{H \in aV_1} T(H)\big|\phi_f(H;\partial(u))\big| \leq \sum_{1 \leq j \leq r} \sigma_T(z_j f)$$

for all $f \in C_c^\infty(G)$ and therefore $\sigma_{T,u} \in \mathscr{S}$.

Hence, in order to prove Lemma 45, it is obviously enough to obtain the following result.

LEMMA 49. *The monomial 1 has property* (P).

It is clear from Lemma 48 that monomials with property (P) actually do exist. Let T be a monomial with property (P) of the lowest possible degree. We claim that $T = 1$. For otherwise suppose $d^0 T > 0$. Then, without loss of generality, we may assume that $T = t_1^{q_1} t_2^{q_2} \cdots t_L^{q_L}$ and $q_1 \geq 1$. Put $T_2 = t_2^{q_2} \cdots t_L^{q_L}$ so that $T = t_1^{q_1} T_2$ and $d^0 T_2 < d^0 T$. Let $a = a(T)$ and, for any $f \in C_c^\infty(G)$, put

$$\psi_{f,i}(H) = T_2(H)\phi_f(H;\partial(u_i)) \qquad (H \in aV_1, \ 1 \leq i \leq r).$$

We recall that $H_1, \cdots, H_l$ is a base for $\mathfrak{h}$ over R such that $t_i(H_j) = \delta_{ij}$ $(1 \leq i, j \leq l)$. Now choose $z_{ij} \in \mathfrak{Z}_0$ $(1 \leq i, j \leq r)$ such that

$$H_1 u_i = \sum_{1 \leq j \leq r} \gamma(z_{ij})u_j \qquad (1 \leq i \leq r).$$

Then

$$\partial\psi_{f,i}/\partial t_1 = \sum_j \psi_{z_{ij}f,j}$$

on aV_1 and therefore

$$\big|t_1^{q_1}(\partial\psi_{f,i}/\partial t_1)\big| \leq \sum_j \sigma_T(z_{ij}f) \qquad (f \in C_c^\infty(G))$$

on aV_1. Here

$$\sigma_T(f) = \max_j \ \sup_{H \in aV_1} T(H)\big|\phi_f(H;\partial(u_j))\big| \qquad (f \in C_c^\infty(G))$$

and it is clear that $\sigma_T \in \mathscr{S}$ since T has property (P). Put

$$\sigma(g) = \sum_{1 \leq i,j \leq r} \sigma_T(z_{ij}g) \qquad (g \in C_c^\infty(G)).$$

Then σ also lies in $\mathscr{S}$.

For any $b > 0$, let W_b denote the set of all $H \in \mathfrak{h}$ such that $\big|t_i(H)\big| \leq b$ $(1 \leq i \leq l)$. Choose a_1 $(0 < a_1 \leq 1)$ so small that $W_{a_1} \subset aV$ and a_2 $(0 < a_2 \leq a_1)$ such that $a_2 V \subset W_{a_1}$. Suppose $H \in a_2 V_1$. Then $H' = H + (a_1 - t_1(H))H_1 \in W_{a_1} \subset aV$. Since $t_i(H) > 0$ $(1 \leq i \leq L)$ and $t_1(H) \leq \|H\| < a_2 c \leq a_2 \leq a_1$, it follows that $t_i(H') > 0$ $(1 \leq i \leq L)$ and therefore $H' \in aV_1$. But aV_1 being convex, the whole line segment joining H to H' lies in aV_1. On the other hand, we have seen above that

$$\big|(\partial\psi_{f,i}/\partial t_1)\big| \leq t_1^{-q_1}\sigma(f) \qquad (f \in C_c^\infty(G))$$

on aV_1. Hence, by integrating on this line segment, we get

$$\left|\psi_{f,i}(H') - \psi_{f,i}(H)\right| \leq \sigma(f)\int_{t_1(H)}^{a_1} s^{-q_1}ds$$

since $t_1(H') = a_1$. Moreover,

$$\left|t_1(H')^{q_1}\psi_{f,i}(H')\right| = T(H')\left|\phi_f(H';\partial(u_i))\right| \leq \sigma_T(f).$$

Therefore

$$\left|\psi_{f,i}(H')\right| \leq a_1^{-q_1}\sigma_T(f).$$

This shows that

$$\left|\psi_{f,i}(H)\right| \leq a_1^{-q_1}\sigma_T(f) + \sigma(f)\int_{t_1(H)}^{a_1} s^{-q_1}ds$$

for $H \in a_2V_1$ and $f \in C_c^{\infty}(G)$.

Now first suppose that $q_1 \geq 2$. Then

$$\int_{t_1(H)}^{a_1} s^{-q_1}ds = (q_1 - 1)^{-1}(t_1(H)^{1-q_1} - a_1^{1-q_1}).$$

Hence if $T_1 = t_1^{q_1-1}T_2$, it is clear that

$$\left|T_1(H)\phi_f(H;\partial(u_i))\right| \leq a_1^{-q_1}\sigma_T(f) + \sigma(f)$$

for $H \in a_2V_1$, $f \in C_c^{\infty}(G)$ and $1 \leq i \leq r$. This shows that T_1 has property (P). But since $d^0T_1 = d^0T - 1 < d^0T$, this gives a contradiction. So the case $q_1 \geq 2$ is impossible.

Hence $q_1 = 1$. Then

$$\int_{t_1(H)}^{a_1} s^{-1}ds = \log(a_1/t_1(H))$$

and therefore

$$\left|\psi_{f,i}(H)\right| \leq a_1^{-1}\sigma_T(f) + \sigma(f)\log(a_1/t_1(H)) \qquad (1 \leq i \leq r)$$

for $H \in a_2V_1$ and $f \in C_c^{\infty}(G)$. Put

$$\sigma'(g) = \sigma(g) + a_1^{-1}\sigma_T(g) \quad (g \in C_c^{\infty}(G)).$$

Then $\sigma' \in \mathscr{S}$ and

$$\left|\psi_{f,i}\right| \leq \sigma'(f)\{1 + \log(a_1/t_1)\}$$

on a_2V_1. Put

$$\sigma_2(g) = \sum_{1 \leq i, \, \leq r} \sigma'(z_{ij}g) \quad (g \in C_c^{\infty}(G)).$$

Then $\sigma_2 \in \mathscr{S}$ and since

$$\partial \psi_{f,i}/\partial t_1 = \sum_j \psi_{z_{ij}f,j}$$

we conclude that

$$\left| (\partial \psi_{f,i}/\partial t_1) \right| \leq \sigma_2(f)\{1 + \log(a_1 t_1^{-1})\}$$

on $a_2 V_1$. Now choose numbers a_3, a_4 $(0 < a_4 \leq a_3 \leq a_2)$ such that $W_{a_3} \subset a_2 V$ and $a_4 V \subset W_{a_3}$. For any $H \in a_4 V_1$, define

$$H'' = H + (a_3 - t_1(H_1))H_1.$$

Then $H'' \in a_2 V_1$ and so again by integrating along the line segment joining H and H'', we conclude that

$$\left| \psi_{f,i}(H'') - \psi_{f,i}(H) \right| \leq \sigma_2(f)\int_{t_1(H)}^{a_3} (1 + \log(a_1 s^{-1}))\, ds$$

$$\leq b\sigma_2(f)$$

for $H \in a_4 V_1$, $f \in C_c^\infty(G)$ and $1 \leq i \leq r$. Here

$$b = \int_0^{a_3} (1 + \log(a_1 s^{-1}))\, ds < \infty.$$

Now $t_1(H'') = a_3$ and

$$t_1(H'')\left| \psi_{f,i}(H'') \right| \leq \sigma_T(f)$$

since $T = t_1 T_2$. Therefore

$$\left| \psi_{f,i}(H) \right| \leq a_3^{-1}\sigma_T(f) + b\sigma_2(f) \qquad (1 \leq i \leq r)$$

for $H \in a_4 V_1$ and $f \in C_c^\infty(G)$. This shows that T_2 has property (P) and therefore again, since $d^\circ T_2 = d^\circ T - 1 < d^\circ T$, we get a contradiction. This proves Lemma 49 and hence also Lemma 45.

27. **Proof of Theorem 3 in the general case.** Now we come to the general case and use of the notation of §16. Let $_0M$ be the centralizer of $\mathfrak{h} \cap \mathfrak{p}$ in G and M the connected component of 1 in $_0M$. Then $A \subset {_0M}$ and, by Lemma 30, M is acceptable.

Let G_c be a complexification of G and define j as in §18. Put

$$\Phi_0 = j(K) \cap \exp((-1)^{1/2}(\mathfrak{h} \cap \mathfrak{p})).$$

Lemma 50. Φ_0 *is a finite group. Let* Φ *be a finite subset of* G *such that* $j(\Phi) = \Phi_0$. *Then* $A = \Phi A^0 Z$, *where* A^0 *is the connected component of* 1 *in* A.

Since $\mathfrak{h} \cap \mathfrak{p} \subset \mathfrak{g}_1$, we may obviously assume, for the proof of this lemma, that $\mathfrak{g}$ is semisimple and G_c is simply connected. Then $j(K)$ is compact and

$\Phi_0 \subset j(A) \cap j(K) = j(A_K)$. Extend θ to a complex-analytic automorphism of G_c. Then since $\theta = -1$ on $\mathfrak{p}$, it is clear that $a^2 = 1$ for every $a \in \Phi_0$. Therefore since $j(A_K)$ is a compact abelian group, it follows that Φ_0 is finite.

Now $A = A_K A_{\mathfrak{p}}$ and $A_{\mathfrak{p}} \subset A^0$ (see Corollary 4 of Lemma 26). Hence, in order to prove the second statement, it would be enough to verify that $j(A_K) = \Phi_0 j(A_K{}^0)$, where $A_K{}^0$ is the analytic subgroup of G corresponding to $\mathfrak{h} \cap \mathfrak{k}$.

Let A_c be the Cartan subgroup of G_c corresponding to $\mathfrak{h}_c$. Put $\mathfrak{u} = \mathfrak{k} + (-1)^{1/2}\mathfrak{p}$ and and define U and η as in §17. Then if $a \in j(A_K)$, it is clear that $a \in U \cap A_c = \exp(\mathfrak{h}_c \cap \mathfrak{u})$. But $\mathfrak{h}_c \cap \mathfrak{u} = \mathfrak{h} \cap \mathfrak{k} + (-1)^{1/2}(\mathfrak{h} \cap \mathfrak{p})$ and therefore $a = a_1 a_2$, where $a_1 \in j(A_K{}^0)$ and $a_2 \in \Phi_0$. This proves that $j(A_K) = \Phi_0 j(A_K{}^0)$.

LEMMA 51. *Let $a \in \Phi$ and $m \in M$. Then a and m commute.*

Since M is connected, this follows from the fact that its Lie algebra $\mathfrak{m}$ commutes with $\mathfrak{h} \cap \mathfrak{p}$.

Fix an order in the space of (real-valued) linear functions λ on $\mathfrak{h} \cap \mathfrak{p}$ and, for any such λ, let $\mathfrak{g}_\lambda$ denote the space of all $X \in \mathfrak{g}$ such that $[H, X] = \lambda(H)X$ for all $H \in \mathfrak{h} \cap \mathfrak{p}$. Put $\mathfrak{n} = \sum_{\lambda > 0} \mathfrak{g}_\lambda$. Then $\mathfrak{n}$ is a nilpotent subalgebra of $\mathfrak{g}$. Let N be the analytic subgroup of G corresponding to $\mathfrak{n}$. It is clear that $_0M$ normalizes $\mathfrak{n}$. Put

$$d(m) = \left| \det(\mathrm{Ad}(m))_{\mathfrak{n}} \right|^{1/2} \qquad (m \in {}_0M),$$

where the subscript $\mathfrak{n}$ denotes restriction on $\mathfrak{n}$. Put $G_0 = \mathrm{Ad}(G)$ and let K_0 denote the image of K in G_0 under the homomorphism $x \to \mathrm{Ad}(x)$. Then K_0 is compact. For any $x \in G$ and $y_0 \in G_0$, define $x^{y_0} = yxy^{-1}$ where y is any element of G such that $y_0 = \mathrm{Ad}(y)$. Put

$$\tilde{f}(x) = \int_{K_0} f(x^{k_0})\,dk_0, \quad g_f(m) = d(m) \int_N \tilde{f}(mn)\,dn$$

for $f \in C_c^\infty(G)$, $x \in G$ and $m \in {}_0M$. Here dk_0 and dn are the Haar measures on K_0 and N, respectively, and $\int_{K_0} dk_0 = 1$.

Introduce an order on the space of real-valued linear functions on $(-1)^{1/2}(\mathfrak{h} \cap \mathfrak{k}) + \mathfrak{h} \cap \mathfrak{p}$ which is compatible (see [4(g), p. 195]) with the one already chosen above. We may assume, without loss of generality, that the set P of positive roots of $(\mathfrak{g}, \mathfrak{h})$ is defined with respect to this order. Since every root of $(\mathfrak{m}, \mathfrak{h})$ is imaginary, it follows from Corollary 5 of Lemma 26 that $A \cap M = A^0$. Let $m \to m^*$ denote the natural projection of M on $M^* = M/A^0$ and define

$$F_g{}^M(h) = \Delta_M(h) \int_{M^*} g(h^{m^*})\,dm^* \qquad (h \in A^0 \cap M'),$$

$$v_M(g) = \int_{A^0} \left| \Delta_M(h) F_g{}^M(h) \right| dh \qquad (g \in C_c^\infty(M)),$$

where dm^* is the invariant measure on M^* and M' is the set of those elements of M which are regular in M.

Let $\mathfrak{M}$ be the subalgebra of $\mathfrak{G}$ generated by $(1, \mathfrak{m}_c)$ and $\mathfrak{Z}_M$ the center of $\mathfrak{M}$. Then we have the isomorphism $\mu = \mu_{\mathfrak{g}/\mathfrak{m}}$ of $\mathfrak{Z}$ into $\mathfrak{Z}_M$ (see §12). Moreover, $G = KMN$ from [4(g), Lemma 11].

LEMMA 52. *For any $a \in \Phi$, put*

$$g_{f,a}(m) = g_f(am) \qquad (m \in M, f \in C_c^\infty(G)).$$

Then $g_{f,a} \in C_c^\infty(M)$ and

$$g_{zf,a} = \mu(z)g_{f,a} \qquad (z \in \mathfrak{Z}).$$

Moreover, if dm^* and dn are suitably normalized, we have the relation

$$F_f(ah) = \xi_\rho(a)F_{g_f,}{}^M(h)$$

for $f \in C_c^\infty(G)$, $h \in A^0 \cap (a^{-1}G')$ and $a \in \Phi$.

Although the proof of this lemma is not difficult, it is rather long. Hence we postpone it to another paper.

We can now complete the proof of Theorem 3. It is clear that

$$\Delta_M(ah) = \xi_\rho(a)\Delta_M(h) \qquad (a \in \Phi, \ h \in A).$$

Hence we conclude from Lemma 52 that

$$v_M(g_{f,a}) = \int_{A^0} |\Delta_M(h)F_f(ah)| \, dh \le v(f).$$

We have seen (Lemma 30) that M is acceptable and every root of $(\mathfrak{m}, \mathfrak{h})$ is imaginary. Moreover, $\mathfrak{Z}_M$ is a finite module over $\mu(\mathfrak{Z}_0)$ by Lemma 21. Hence Theorem 3 holds for $(M, A^0, \mu(\mathfrak{Z}_0), v_M)$ in place of $(G, A, \mathfrak{Z}_0, v)$. Therefore for any $u \in \mathfrak{S}(\mathfrak{h}_c)$, we can, in view of Lemma 52, choose a finite set of elements $z_1, \cdots, z_r \in \mathfrak{Z}_0$ such that

$$\sup_{h \in A'} |F_f(h;u)| \le \max_{a \in \Phi} \sum_{1 \le i \le r} v_M(\mu(z_i)g_{f,a}) \le \sum_{1 \le i \le r} v(z_i f)$$

for $f \in C_c^\infty(G)$. This proves Theorem 3.

28. **The local summability of $|D|^{-1/2}$.** Let $l = \operatorname{rank} G$ and put $D = D_l$ in the notation of §3. Then D is an analytic function on G and

$$D(h) = \det(1 - \operatorname{Ad}(h))_{\mathfrak{g}/\mathfrak{h}} = (-1)^p \Delta(h)^2 \qquad (h \in A),$$

where p is the number of positive roots of $(\mathfrak{g}, \mathfrak{h})$.

LEMMA 53. *$|D|^{-1/2}$ is locally summable on G.*

Let C be a compact subset of G. Then we can choose $f \in C_c^\infty(G)$ such that $f \geq 0$ everywhere and $f \geq 1$ on C. Then it is clear that

$$\int_C |D|^{-1/2}\, dx \leq \int_G |D|^{-1/2}\, f\, dx.$$

Let $A_i\ (1 \leq i \leq r)$ be a maximal set of Cartan subgroups of G no two of which are conjugate under G. Put

$$G_i = \bigcup_{x \in G} x A_i' x^{-1},$$

where $A_i' = A_i \cap G'$. Then G' is the disjoint union of $G_1, \cdots, G_r$ (see [4(e), Lemma 5]). Hence it would be enough to verify that

$$\int_{G_i} |D|^{-1/2}\, f\, dx < \infty \qquad (1 \leq i \leq r).$$

So fix i and put $A = A_i$. Then $G_i = G_A$ in the notation of §23 and it follows from Lemma 41 and [4(h), Theorem 2] that

$$\int_{G_A} |D|^{-1/2} f\, dx = c \int_A |F_f(h)|\, dh < \infty.$$

This proves the lemma.

29. **Appendix.** Put $\rho(x) = (x_1{}^2 + x_2{}^2 + \cdots + x_n{}^2)^{1/2} \geq 0$ for $x \in R^n$.

LEMMA 54. *Let α be a real number and $D = \partial^k/\partial x_{i_1} \partial x_{i_2} \cdots \partial x_{i_k}$. Then*

$$D\rho^\alpha = \sum_{0 \leq j \leq k} p_j \rho^{\alpha - j - k},$$

$$D(\rho^\alpha \log \rho) = \sum_{0 \leq j \leq k} P_j \rho^{\alpha - j - k} + (\log \rho) \sum_{0 \leq j \leq k} Q_j \rho^{\alpha - j - k},$$

where p_j, P_j and Q_j are homogeneous polynomials in $(x_1, \cdots, x_n)$ of degree j.

This follows by an easy induction on k.

COROLLARY 1. *If $\alpha > k$, then ρ^α and $\rho^\alpha \log \rho$ are functions of class C^k on R^n.*

This is obvious from the lemma.

COROLLARY 2. $\rho^{k-1} D\rho$ *remains bounded on* R^n.

We know that

$$\rho^{k-1} D\rho = \sum_{0 \leq j \leq k} p_j \rho^{-j},$$

where p_j is a homogeneous polynomial of degree j in $(x_1, \cdots, x_n)$. Our assertion therefore follows from the obvious fact that $|p_j| \rho^{-j}$ is bounded on R^n.

The following lemma is implicit in the paper of Morrey and Nirenberg [6, p. 281].

LEMMA 55. *Let h be a function in $C^\infty(\mathbf{R})$ which is constant on the intervals $(-\infty, 0]$ and $[1, +\infty)$. Choose two numbers r, δ such that $0 < \delta \leqq r \leqq 1$ and put*

$$H_{r,\delta}(x) = h(\delta^{-1}(\rho(x) - r)) \qquad (x \in \mathbf{R}^n).$$

Then for each integer $k \geqq 0$, there exists a number $c_k \geqq 0$, independent of r and δ, such that

$$\left| DH_{r,\delta} \right| \leqq c_k \delta^{-k}$$

for $D = \partial^k/\partial x_{i_1} \partial x_{i_2} \cdots \partial x_{i_k}$ $(1 \leqq i_1, \cdots, i_k \leqq n)$.

We use induction on k. If $k = 0$, we can take $c_0 = \sup |h|$. So let us assume that $k \geqq 1$ and put $h'(t) = dh/dt$ $(t \in \mathbf{R})$. Then h' also satisfies the conditions of the lemma and

$$\partial H_{r,\delta}/\partial x_i = \delta^{-1} H_{r,\delta}' \cdot \partial \rho/\partial x_i \qquad (1 \leqq i \leqq n),$$

where

$$H_{r,\delta}'(x) = h'(\delta^{-1}(\rho(x) - r)) \qquad (x \in \mathbf{R}^n).$$

Now $H_{r,\delta}'(x) = 0$ unless $r \leqq \rho(x) \leqq r + \delta$ and therefore

$$\sup_x \left| DH_{r,\delta} \right| \leqq \delta^{-1} \sup_{\rho \geqq \delta} \left| D'(H_{r,\delta}' \cdot \partial \rho/\partial x_{i_k}) \right|,$$

where $D' = \partial^{k-1}/\partial x_{i_1} \cdots \partial x_{i_{k-1}}$. Hence if we expand

$$D'(H_{r,\delta}' \cdot \partial \rho/\partial x_{i_k})$$

by means of the Leibniz formula, make use of Corollary 2 of Lemma 54 and apply the induction hypothesis to $H_{r,\delta}'$, we get the required assertion.

Put $\Delta = \sum_{1 \leqq i \leqq n} (\partial/\partial x_i)^2$ and let δ denote the Dirac measure concentrated at the origin.

LEMMA 56. *If n is odd*

$$\Delta^{l+(n-1)/2} \, \rho^{2l-1} = c_l \delta \qquad (l \geqq 1)$$

and if n is even

$$\Delta^{l+n/2}(\rho^{2l}\log \rho) = c_l' \delta \qquad (l \geqq 0).$$

Here c_l and c_l' are nonzero numbers and the above relations are meant in the sense of the theory of distributions.

This is well known (see [7, p. 47]).

LEMMA 57. *Fix integers $d \geqq 0$ and $r \geqq 1$. Then we can choose an integer $m \geqq 1$ and a function e on $\mathbf{R}^n$ of class $C^{2m(r-1)+d}$ such that*

$$\Delta^{mr} e = \delta.$$

Choose m so large that $2m > d + n$. First suppose n is odd. Then $l = mr - (n-1)/2$ is an integer and

$$2l = 2mr - n + 1 > d + 1.$$

Hence $l \geqq 1$. Put $e = c_l^{-1} \rho^{2l-1}$. Then

$$\Delta^{mr} e = \Delta^{l+(n-1)/2} e = \delta$$

and $2l - 1 = 2mr - n = 2m(r-1) + 2m - n > 2m(r-1) + d$. Hence e is of class $C^{2m(r-1)+d}$ by Corollary 1 of Lemma 54.

On the other hand if n is even put $l = mr - n/2$. Then $2l = 2mr - n > d$ and therefore l is positive. Now put

$$e = (c_l')^{-1} \rho^{2l} \log \rho.$$

Then $2l = 2mr - n > 2m(r-1) + d$ and therefore again e is of class $C^{2m(r-1)+d}$.

References

1. A. Borel, *Groupes linéaires algébriques*, Ann. of Math. (2) **64** (1956), 20–82.

2. A. Borel and G. D. Mostow, *On semi-simple automorphisms of Lie algebras*, Ann. of Math. (2) **61** (1955), 389–405.

3. N. Bourbaki, *Groupes et algèbres de Lie*, Chapitre I, *Algèbres de Lie*, Hermann, Paris, 1960.

4. (a) Harish-Chandra, *On representation of Lie algebras*, Ann. of Math. (2) **50** (1949), 900–915.

    (b) ———, *Representations of a semisimple Lie group on a Banach space*. I, Trans. Amer. Math. Soc. **75** (1953), 185–243.

    (c) ———, *The Plancherel formula for complex semisimple Lie groups*, Trans. Amer. Math. Soc. **76** (1954), 485–528.

    (d) ———, *Representations of semisimple Lie groups*. VI, Amer. J. Math. **78** (1956), 564–628.

    (e) ———, *The characters of semisimple Lie groups*, Trans. Amer. Math. Soc. **83** (1956), 98–163.

    (f) ———, *Differential operators on a semisimple Lie algebra*, Amer. J. Math. **79** (1957), 87–120.

    (g) ———, *Fourier transforms on a semisimple Lie algebra*. I, Amer. J. Math. **79** (1957), 193–257.

    (g₂) ———, *Fourier transforms on a semisimple Lie algebra*. II, Amer. J. Math. **79** (1957), 653–686.

    (h) ———, *A formula for semisimple Lie groups*, Amer. J. Math. **79** (1957), 733–760.

    (i) ———, *Spherical functions on a semisimple Lie group*. I, Amer. J. Math. **80** (1958) 241–310.

    (j) ———, *Invariant eigendistributions on semisimple Lie groups*, Bull. Amer. Math. Soc. **69** (1963), 117–123.

(k) ———, *Invariant distributions on Lie algebras*, Amer. J. Math. **86** (1964), 271–309.

(l) ———, *Invariant differential operators and distributions on a semisimple Lie algebra*, Amer. J. Math. **86** (1964), 534–564.

(m) ———, *Some results on an invariant integral on a semi-simple Lie algebra*, Ann. of Math. (2) **80** (1964), 551–593.

(n) ———, *Invariant eigendistributions on a semisimple Lie algebra*, Inst. Hautes Études Sci. Publ. Math. No. 27.

5. S. Helgason, *Differential geometry and symmetric spaces*, Academic Press, New York, 1962.

6. C. B. Morrey and L. Nirenberg, *On the analyticity of solutions of linear elliptic systems of partial differential equations*, Comm. Pure Appl. Math. **10** (1957), 271–290.

7. L. Schwartz, *Théorie des distributions*. I, Hermann, Paris, 1950.

8. H. Weyl, *The structure and representation of continuous groups*, The Institute for Advanced Study, Princeton, N. J., 1935.

THE INSTITUTE FOR ADVANCED STUDY,
     PRINCETON, NEW JERSEY

Reprinted from
Trans. Amer. Math. Soc.
119 (1965), 457–508

DISCRETE SERIES FOR SEMISIMPLE LIE GROUPS I

CONSTRUCTION OF INVARIANT EIGENDISTRIBUTIONS

BY

HARISH-CHANDRA

The Institute for Advanced Study, Princeton, N.J., U.S.A.

Table of contents

§ 1. Introduction

Let G be a connected semisimple Lie group with a compact Cartan subgroup B, and B^* the character group of B. Let $\mathfrak{g}$ and $\mathfrak{b}$ denote the Lie algebras of G and B respectively. Then every $b^* \in B^*$ defines a linear function $\lambda = \log b^*$ on $\mathfrak{b}_c$ by the relation

$$\langle b^*, \exp H \rangle = e^{\lambda(H)} \quad (H \in \mathfrak{b}).$$

Let W be the Weyl group of $(\mathfrak{g}, \mathfrak{b})$. We say that b^* is regular if $s\lambda \neq \lambda$ for every $s \neq 1$ in W. Let $B^{*\prime}$ denote the set of all regular elements of B^* and define $\mathfrak{B}$ as in [2 (m), § 1]. Then corresponding to every $b^* \in B^{*\prime}$, we construct in Theorem 3 an invariant eigendistribution Θ_{b^*} of $\mathfrak{B}$ on G (cf. [2 (h), Theorem 2]). We shall see later in another paper that those irreducible characters of G which correspond to the discrete series (see [2 (a), § 5]) are actually finite linear combinations of these distributions (cf. [2 (h), Theorems 3 and 4]).

The second main result of this paper is contained in Theorem 4 which gives an alternative formula for the distribution Θ_{b^*}. This will be needed for the determination of the contribution of the discrete series to the Plancherel formula of G.

Our method consists in first proving analogous results on $\mathfrak{g}$ and then lifting them to G, roughly speaking, by means of the exponential mapping. Theorem 1 is the $\mathfrak{g}$-analogue of Theorem 4 and its proof depends very much on Theorem 5 of [2 (k)]. Then in § 8 we introduce the notion of a tempered distribution on an open subset of a Euclidean space (see also [2 (c), p. 90]) and prove some elementary results which are then applied in § 14 to certain tempered and invariant eigendistributions on a reductive subalgebra $\mathfrak{z}$ of $\mathfrak{g}$ containing $\mathfrak{b}$. Lemma 28 asserts the uniqueness of such distributions and the existence is proved in Theorem 2 and Lemma 37. Lemma 41 contains the key result required for the reduction of the proof of Theorem 4 from the group to the Lie algebra.

The rest of this paper is devoted to the proofs of Theorems 3 and 4. The uniqueness part of Theorem 3 is relatively easy and follows from Lemma 28. However the problem of existence is more delicate. Lemma 50 contains the main step required in its solution. Lemma 59 gives a rather explicit formula for Θ_{b^*} which will be useful in later work. The main burden of the proof of Theorem 4 rests on Lemma 66.

Let L' be the set of all linear functions λ on $\mathfrak{b}$ of the form $\lambda = \log b^*$ $(b^* \in B^{*\prime})$ and write $\Theta_\lambda = \Theta_{b^*}$. Define $\varpi \in S(\mathfrak{b}_c)$ as in [2 (k), § 11]. Then we show in § 29 that for any $f \in C_c^\infty(G)$, the series

$$\sum_{\lambda \in L'} \varpi(\lambda)\, \Theta_\lambda(f)$$

converges absolutely and its sum represents a distribution T on G. We shall see later that, apart from a constant factor, T is just the contribution of the discrete series to the Plancherel formula of G (cf. [2 (h), Theorem 4]).

This work was partially supported by a grant from the National Science Foundation.

Part I. Theory on the Lie algebra

§ 2. Reduction of Theorem 1 to the semisimple case

We use the notation and terminology of [2 (l)]. Let $\mathfrak{g}$ be a reductive Lie algebra over $\mathbf{R}$, Ω a completely invariant open subset of $\mathfrak{g}$, T a distribution on Ω satisfying the conditions of [2 (l), Theorem 1] and F the corresponding analytic function on $\Omega' = \Omega \cap \mathfrak{g}'$. Then we have seen in [2 (l), § 9] that $\Phi = \nabla_{\mathfrak{g}} F$ extends to a continuous function on Ω.

Let $\mathfrak{h}$ be a Cartan subalgebra of $\mathfrak{g}$. For any function ϕ on Ω' let $\phi_{\mathfrak{h}}$ denote its restriction on $\mathfrak{h} \cap \Omega'$.

LEMMA 1. *Let* $D \in \mathfrak{D}(\mathfrak{h}_c)$. *Then the function* $D\Phi_{\mathfrak{h}}$ *is locally bounded* (1) *on* $\mathfrak{h} \cap \Omega$.

Fix a point $H_0 \in \mathfrak{h} \cap \Omega$ and select a positive-definite quadratic form Q on $\mathfrak{h}$. For any $\varepsilon > 0$, consider the set $\mathfrak{h}(\varepsilon)$ of all $H \in \mathfrak{h}$ such that $Q(H - H_0) < \varepsilon^2$. Then if ε is sufficiently small, $\mathfrak{h}(\varepsilon) \subset \Omega$. Moreover the set $\mathfrak{h}'(\varepsilon) = \mathfrak{h}(\varepsilon) \cap \Omega'$ has only a finite number of connected components. It follows from [2 (l), Lemma 2] that $D\Phi_{\mathfrak{h}}$ remains bounded on each connected component of $\mathfrak{h}'(\varepsilon)$ and therefore also on $\mathfrak{h}'(\varepsilon)$. Obviously this implies the statement of the lemma.

---

(1) This means that $D\Phi_{\mathfrak{h}}$ remains bounded on $C \cap \Omega'$ for any compact subset C of $\mathfrak{h} \cap \Omega$.

COROLLARY. *For any $D \in \mathfrak{J}(\mathfrak{g}_c)$, $D\Phi$ is locally summable on Ω.*

Fix $\mathfrak{h}$ as above. Then by [2 (j), Lemma 14],

$$(D\Phi)_{\mathfrak{h}} = \delta_{\mathfrak{g}/\mathfrak{h}}{}'(D)\,\Phi_{\mathfrak{h}} = \pi^{-1}(\delta_{\mathfrak{g}/\mathfrak{h}}(D) \circ \pi)\,\Phi_{\mathfrak{h}}.$$

But $\delta_{\mathfrak{g}/\mathfrak{h}}(D) \circ \pi \in \mathfrak{D}(\mathfrak{h}_c)$ by [2 (j), Theorem 1] and therefore we conclude from the above lemma that $\pi(D\Phi)_{\mathfrak{h}}$ is locally bounded on $\mathfrak{h} \cap \Omega$.

Let $m = (n-l)/2$ where $n = \dim \mathfrak{g}$, $l = \operatorname{rank} \mathfrak{g}$. Then $m = d^0\,\pi$. Let t be an indeterminate and $\eta(X)$ the coefficient of t^l in $\det(t - adX)$ $(X \in \mathfrak{g}_c)$. Then η is an invariant polynomial function on $\mathfrak{g}_c$ and $\eta(H) = (-1)^m \pi(H)^2$ $(H \in \mathfrak{h}_c)$. Moreover it follows from the above result (see the proof of Lemma 3 of [2 (l)]) that $|\eta|^{\frac{1}{2}}|D\Phi|$ is locally bounded on Ω. Therefore since $|\eta|^{-\frac{1}{2}}$ is locally summable on $\mathfrak{g}$ [2 (k), Corollary 2 of Lemma 30], our assertion is now obvious.

Let $\nabla_{\mathfrak{g}}{}^*$ denote the adjoint of $\nabla_{\mathfrak{g}}$. Then $\nabla_{\mathfrak{g}}{}^*$ is also an invariant and analytic differential operator on $\mathfrak{g}'$.

LEMMA 2. *Put $f(x:H) = f(H^x)$ $(x \in G,\ H \in \mathfrak{h})$ for $f \in C^\infty(\mathfrak{g})$. Then*

$$f(H^x;\ \nabla_{\mathfrak{g}}{}^*) = (-1)^m f(x:H;\ \pi^{-1}\partial(\varpi) \circ \pi^2) \quad (x \in G,\ H \in \mathfrak{h}')$$

where $m = \tfrac{1}{2}(\dim \mathfrak{g} - \operatorname{rank} \mathfrak{g})$.

Put $\mathfrak{g}_{\mathfrak{h}} = (\mathfrak{h}')^G$. Then $\mathfrak{g}_{\mathfrak{h}}$ is an open subset of $\mathfrak{g}'$. Fix $g \in C_c^\infty(\mathfrak{g}_{\mathfrak{h}})$. Then

$$\int \nabla_{\mathfrak{g}}{}^* f \cdot g\, dX = \int f \cdot \nabla_{\mathfrak{g}} g\, dX$$

and therefore we conclude from Corollary 1 of Lemma 30 of [2 (k)] that

$$\int \pi(H)^2 f(x^* H;\ \nabla_{\mathfrak{g}}{}^*)\, g(x^* H)\, dx^*\, dH = \int \pi(H)^2 f(x^* H)\, g(x^* H;\ \nabla_{\mathfrak{g}})\, dx^*\, dH.$$

Now define $\phi(x^* : H) = \phi(x^* H)$ $(x^* \in G^*,\ H \in \mathfrak{h})$ for $\phi = f$ or g. Then it follows from the definition of $\nabla_{\mathfrak{g}}$ [2 (l), Lemma 24] that

$$g(x^* H;\ \nabla_{\mathfrak{g}}) = g(x^* : H;\ \partial(\varpi) \circ \pi).$$

Therefore

$$\int \pi(H)^2 f(x^* H)\, g(x^* H;\ \nabla_{\mathfrak{g}})\, dx^*\, dH = (-1)^m \int \pi(H)^2 f(x^* : H;\ \pi^{-1}\partial(\varpi) \circ \pi^2)\, g(x^* H)\, dx^*\, dH$$

since ϖ is homogeneous of degree m. The differential operator $\pi^{-1}\partial(\varpi) \circ \pi^2$ being invariant under the Weyl group of $(\mathfrak{g}, \mathfrak{h})$, there exists (see the proof of Lemma 24 of [2 (1)]) a unique invariant differential operator D on $\mathfrak{g}_\mathfrak{h}$ such that

$$f(x^* H; D) = (-1)^m f(x^* : H; \pi^{-1}\partial(\varpi)\circ\pi^2)$$

for $x^* \in G^*$, $H \in \mathfrak{h}'$ and $f \in C^\infty(\mathfrak{g})$. Hence it is clear that

$$\int \nabla_\mathfrak{g}^* f \cdot g \, dX = \int Df \cdot g \, dX.$$

This being true for every $g \in C_c^\infty(\mathfrak{g}_\mathfrak{h})$, we conclude that $\Delta_\mathfrak{g}^* = D$ on $\mathfrak{g}_\mathfrak{h}$ and therefore

$$f(H^{x^*}; \nabla_\mathfrak{g}^*) = (-1)^m f(x^* : H; \pi^{-1}\partial(\varpi)\circ\pi^2)$$

for $x^* \in G^*$, $H \in \mathfrak{h}'$. This is equivalent to the statement of the lemma.

COROLLARY. $f(H^x; \nabla_\mathfrak{g}^* \circ \eta^{-1}) = f(x : H; \pi^{-1}\partial(\varpi))$ $(x \in G, H \in \mathfrak{h}')$.

Since $\eta(H) = (-1)^m \pi(H)^2$, this is obvious from Lemma 2.

By Chevalley's theorem [2 (c), Lemma 9], there exists a unique element $p \in I(\mathfrak{g}_c)$ such that $p_\mathfrak{h} = (\varpi^\mathfrak{h})^2$ for every Cartan subalgebra $\mathfrak{h}$ of $\mathfrak{g}$. (Here we have used the notation of [2 (i), § 8] and [2 (1), Theorem 3].) Put $\square = \partial(p)$.

LEMMA 3. *Let f be a locally invariant C^∞ function on an open subset U of $\mathfrak{g}'$. Then*

$$(\nabla_\mathfrak{g}^* \circ \eta^{-1} \circ \nabla_\mathfrak{g})f = \square f.$$

Fix a point $H_0 \in U$ and let $\mathfrak{h}$ be the centralizer of H_0 in $\mathfrak{g}$. Then $\mathfrak{h}$ is a Cartan subalgebra of $\mathfrak{g}$ and it follows from the corollary of Lemma 2 that

$$f(H; \nabla_\mathfrak{g}^* \circ \eta^{-1} \circ \nabla_\mathfrak{g}) = f_1(H; \pi^{-1}\partial(\varpi))   (H \in \mathfrak{h} \cap U)$$

where $f_1 = \nabla_\mathfrak{g} f$. However

$$f_1(H) = f(H; \partial(\varpi)\circ\pi)   (H \in \mathfrak{h} \cap U)$$

from the definition of $\nabla_\mathfrak{g}$. Therefore

$$f(H; \nabla_\mathfrak{g}^* \circ \eta^{-1} \circ \nabla_\mathfrak{g}) = f(H; \pi^{-1}\partial(\varpi^2)\circ\pi).$$

On the other hand since f is locally invariant, we have

$$f(H; \square) = f(H; \delta_{\mathfrak{g}/\mathfrak{h}}'(\square)) = f(H; \pi^{-1}\partial(\varpi^2)\circ\pi)   (H \in \mathfrak{h} \cap U)$$

from [2 (c), Theorem 1] and the definition of $\square$. This shows that

$$f(H_0; \nabla_{\mathfrak{g}}^{*} \circ \eta^{-1} \circ \nabla_{\mathfrak{g}}) = f(H_0; \square)$$

and so the lemma is proved.

COROLLARY. $\square F = (\nabla_{\mathfrak{g}}^{*} \circ \eta^{-1} \circ \nabla_{\mathfrak{g}}) F = \nabla_{\mathfrak{g}}^{*}(\eta^{-1} \Phi).$

This is obvious since F is invariant and $\nabla_{\mathfrak{g}} F = \Phi$.

For any $\varepsilon > 0$ let $\mathfrak{g}(\varepsilon)$ denote the set of all $X \in \mathfrak{g}$ where $|\eta(X)| > \varepsilon^2$. Let u be a measurable function on $\mathfrak{g}'$ which is integrable (with respect to the Euclidean measure dX) on $\mathfrak{g}(\varepsilon)$ for every $\varepsilon > 0$. Then we define [1]

$$\mathrm{p.v.} \int u \, dX = \lim_{\varepsilon \to 0} \int_{\mathfrak{g}(\varepsilon)} u \, dX$$

provided this limit exists and is finite.

THEOREM 1. *For any $f \in C_c^{\infty}(\Omega)$ we have*

$$\int f \square F \, dX = \mathrm{p.v.} \int \eta^{-1} \nabla_{\mathfrak{g}} f \cdot \Phi \, dX.$$

Since $\square \in \mathfrak{I}(\mathfrak{g}_c)$, it follows from [2 (l), Lemma 16] that $\square F$ is locally summable on Ω. Hence the left side of the above equation is well defined. Now consider the right side. Let V_δ $(0 < \delta \leqslant \delta_0)$ be a family of invariant measurable functions on $\mathfrak{g}$ with the following properties.

1) There exists a number a such that $|V_\delta(X)| \leqslant a$ for $X \in \mathfrak{g}$ and all δ.
2) $V_\delta(X) = 0$ if $|\eta(X)| < \delta^2$ $(X \in \mathfrak{g},\ 0 < \delta \leqslant \delta_0)$.
3) $\lim_{\delta \to 0} V_\delta(X) = 1$ for $X \in \mathfrak{g}'$.

Fix a Cartan subalgebra $\mathfrak{h}$ of $\mathfrak{g}$ and put $\mathfrak{g}_\mathfrak{h} = (\mathfrak{h}')^G$ as before. Then we can choose a real number $c = c(\mathfrak{h}) \neq 0$ such that

$$\int g \, dX = c \int \pi(H)^2 g(x^* H) \, dx^* \, dH$$

for $g \in C_c(\mathfrak{g}_\mathfrak{h})$ in the notation of Corollary 1 of [2 (k), Lemma 30]. Since $V_\delta \eta^{-1} \nabla_{\mathfrak{g}} f \cdot \Phi$ vanishes outside a compact subset of $\mathfrak{g}'$, it is obviously integrable on $\mathfrak{g}$. Therefore

$$\int_{\mathfrak{g}_\mathfrak{h}} V_\delta \eta^{-1} \nabla_{\mathfrak{g}} f \cdot \Phi \, dX = (-1)^m c \int_{\mathfrak{h}} V_\delta(H) \Phi(H) \, dH \int_{G^*} f(x^* H; \nabla_{\mathfrak{g}}) \, dx^*$$

---

[1] p.v. stands for "principal value".

if we recall that $\eta = (-1)^m \pi^2$ on $\mathfrak{h}$. On the other hand it follows from the definition of $\nabla_{\mathfrak{g}}$ that

$$\int_{G^*} f(x^* H; \nabla_{\mathfrak{g}})\, dx^* = \varepsilon_R(H)\, \psi_f(H : \partial(\varpi)) \quad (H \in \mathfrak{h}')$$

in the notation of [2 (k), § 5], Therefore since $\partial(\varpi)^* = (-1)^m \partial(\varpi)$, we get

$$\int_{\mathfrak{g}_{\mathfrak{h}}} V_\delta \eta^{-1} \nabla_{\mathfrak{g}} f \cdot \Phi\, dx = c \int_{\mathfrak{h}} V_{\delta,\mathfrak{h}}\, \varepsilon_R \Phi_{\mathfrak{h}} \partial(\varpi)^* \psi_f\, dH$$

where $V_{\delta,\mathfrak{h}}$ denotes the restriction of V_δ on $\mathfrak{h}$. Since Φ is continuous on Ω, it is clear (see [2 (k), § 15]) that

$$\int \left| \Phi_{\mathfrak{h}} \partial(\varpi)^* \psi_f \right| dH < \infty.$$

Therefore the following lemma is now obvious.

LEMMA 4. *Let* $f \in C_c^\infty(\Omega)$. *Then*

$$\lim_{\delta \to 0} \int_{\mathfrak{g}_{\mathfrak{h}}} V_\delta \eta^{-1} \nabla_{\mathfrak{g}} f \cdot \Phi\, dX = c \int_{\mathfrak{h}} \varepsilon_R \Phi_{\mathfrak{h}} \partial(\varpi)^* \psi_f\, dH.$$

Select a maximal set $\mathfrak{h}_i \ (1 \leqslant i \leqslant r)$ of Cartan subalgebras of $\mathfrak{g}$ no two of which are conjugate under G. Put $\mathfrak{g}_i = (\mathfrak{h}_i')^G$. Then $\mathfrak{g}'$ is the disjoint union of $\mathfrak{g}_1, \mathfrak{g}_2, \ldots, \mathfrak{g}_r$. Fix a Euclidean measure $d_i H$ on $\mathfrak{h}_i$ and put $c_i = c(\mathfrak{h}_i)$, $\Phi_i = \Phi_{\mathfrak{h}_i}$ and $\varpi_i = \varpi^{\mathfrak{h}_i}$. Then we have the following result in the notation of [1] [2 (k), § 16].

COROLLARY. *For any* $f \in C_c^\infty(\Omega)$,

$$\lim_{\delta \to 0} \int_{\mathfrak{g}} V_\delta \eta^{-1} \nabla_{\mathfrak{g}} f \cdot \Phi\, dX = \sum_{1 \leqslant i \leqslant r} c_i \int \varepsilon_{R,i}\, \Phi_i \partial(\varpi_i)^* \psi_{f,i}\, d_i H = \mathrm{p.v.} \int \eta^{-1} \nabla_{\mathfrak{g}} f \cdot \Phi\, dX.$$

The first equality is obvious from Lemma 4 and the second follows by taking V_δ to be the characteristic function of $\mathfrak{g}(\delta)$.

On the other hand (see the proof of Lemma 3),

$$F(H; \square) = F(H; \pi^{-1} \partial(\varpi)^2 \circ \pi) = \Phi(H; \pi^{-1} \partial(\varpi)) \quad (H \in \mathfrak{h}' \cap \Omega).$$

Therefore
$$\int_{\mathfrak{g}_{\mathfrak{h}}} f \cdot \square F\, dX = c \int \varepsilon_R \psi_f \partial(\varpi) \Phi_{\mathfrak{h}}\, dH$$

and so it is obvious that Theorem 1 is equivalent to the following lemma.

---

[1] $\varepsilon_{R,i}$ denotes ε_R for $\mathfrak{h} = \mathfrak{h}_i$.

LEMMA 5. *Let* $f \in C_c^\infty(\Omega)$. *Then*

$$\sum_{1 \leqslant i \leqslant r} c_i \int_{\mathfrak{h}_i} \varepsilon_{R,i} \left(\psi_{f,i} \, \partial(\varpi_i) \, \Phi_i - \partial(\varpi_i)^* \, \psi_{f,i} \cdot \Phi_i \right) d_i H = 0.$$

We shall now prove Theorem 1 by induction on dim $\mathfrak{g}$. Put

$$J(f) = \int f \square F \, dX - \text{p.v.} \int \eta^{-1} \nabla_\mathfrak{g}^* f \cdot \Phi \, dX$$

$$= \sum_{1 \leqslant i \leqslant r} c_i \int_{\mathfrak{h}_i} \varepsilon_{R,i} \left(\psi_{f,i} \, \partial(\varpi_i) \, \Phi_i - \partial(\varpi_i)^* \, \psi_{f,i} \cdot \Phi_i \right) d_i H$$

for $f \in C_c^\infty(\Omega)$. Then it follows from [2 (k), §15] that J is an invariant distribution on Ω. We have to prove that $J = 0$.

Let $\mathfrak{c}$ be the center and $\mathfrak{g}_1$ the derived algebra of $\mathfrak{g}$ and first assume that $\mathfrak{c} \neq \{0\}$. Fix a point $X_0 \in \Omega$. We have to show that $J = 0$ around X_0. Let $X_0 = C_0 + Z_0$ ($C_0 \in \mathfrak{c}$, $Z_0 \in \mathfrak{g}_1$). Select on open and relatively compact neighborhood $\mathfrak{c}_0$ of C_0 in $\mathfrak{c}$ such that $Z_0 + \text{Cl}(\mathfrak{c}_0) \subset \Omega$. Let Ω_1 be the set of all points $Z \in \mathfrak{g}_1$ such that $Z + \text{Cl}(\mathfrak{c}_0) \subset \Omega$. Then Ω_1 is an open and completely invariant neighborhood of Z_0 in $\mathfrak{g}_1$ (see [2 (l), Lemma 9]). It would be sufficient to prove (see [2 (i), Lemma 3]) that

$$J(\alpha \times g) = 0 \quad (\alpha \in C_c^\infty(\mathfrak{c}_0),\ g \in C_c^\infty(\Omega_1)).$$

Fix $\alpha \in C_c^\infty(\mathfrak{c}_0)$ and consider the distributions

$$T_\alpha(g) = T(\alpha \times g), \quad J_\alpha(g) = J(\alpha \times g) \quad (g \in C_c^\infty(\Omega_1))$$

on Ω_1. Then T_α and J_α are both invariant. Put $\mathfrak{U}_1 = \mathfrak{U} \cap I(\mathfrak{g}_{1c})$ where $\mathfrak{U}$ has the same meaning as in [2 (l), Theorem 1]. Then

$$\dim I(\mathfrak{g}_{1c})/\mathfrak{U}_1 \leqslant \dim I(\mathfrak{g}_c)/\mathfrak{U} < \infty$$

and $\partial(\mathfrak{U}_1) T_\alpha = \{0\}$. Hence Theorem 1 of [2 (l)] is also applicable to $(T_\alpha, \mathfrak{g}_1, \Omega_1)$ instead of $(T, \mathfrak{g}, \Omega)$. Put $\Omega_1' = \Omega_1 \cap \mathfrak{g}'$ and fix Euclidean measures dC and dZ on $\mathfrak{c}$ and $\mathfrak{g}_1$ respectively such that $dX = dC \, dZ$ ($X = C + Z$, $C \in \mathfrak{c}$, $Z \in \mathfrak{g}_1$). Let F_α be the analytic function on Ω_1' such that

$$T_\alpha(g) = \int F_\alpha g \, dZ \quad (g \in C_c^\infty(\Omega_1)).$$

Then it is clear that

$$F_\alpha(Z) = \int \alpha(C) \, F(C + Z) \, dC \quad (Z \in \Omega_1').$$

Put $\Phi_\alpha = \nabla_{\mathfrak{g}_1} F_\alpha$. If $\mathfrak{h}$ is any Cartan subalgebra of $\mathfrak{g}$, it is clear that $\mathfrak{h} = \mathfrak{c} + \mathfrak{h}_1$ where $\mathfrak{h}_1 = \mathfrak{h} \cap \mathfrak{g}_1$. Moreover π and $\partial(\varpi)$ are in $\mathfrak{D}(\mathfrak{h}_{1c})$ and $\square \in \partial(I(\mathfrak{g}_{1c}))$. Hence it follows without difficulty that

$$J_\alpha(g) = \int_{\mathfrak{g}_1} g \square F_\alpha \, dZ - \text{p.v.} \int_{\mathfrak{g}_1} \eta^{-1} \nabla_{\mathfrak{g}_1} g \cdot \Phi_\alpha \, dZ$$

for $g \in C_c^\infty(\mathfrak{g}_1)$. But since $\dim \mathfrak{g}_1 < \dim \mathfrak{g}$, we conclude from the induction hypothesis that $J_\alpha = 0$. This shows that $J(\alpha \times g) = 0$ for $\alpha \in C_c^\infty(\mathfrak{c}_0)$ and $g \in C_c^\infty(\Omega_1)$ and therefore $J = 0$ around X_0.

§ 3. Second reduction

Hence we may now assume that $\mathfrak{g}$ is semisimple and identify $\mathfrak{g}$ with its dual space by means of the Killing form ω of $\mathfrak{g}$. For any $p \in I(\mathfrak{g}_c)$, let p_i denote the restriction of p on $\mathfrak{h}_i$ and put $\pi_i = \pi^{\mathfrak{h}_i}$ $(1 \leqslant i \leqslant r)$. We also identify $\mathfrak{h}_i$ with its dual space by means of ω_i. Then $\varpi_i = \pi_i$. Put $\delta_i(D) = \delta_{\mathfrak{g}/\mathfrak{h}_i}(D)$ $(D \in \mathfrak{J}(\mathfrak{g}_c))$ in the notation of [2 (j), Theorem 1].

LEMMA 6. *Let* $D \in \mathfrak{J}(\mathfrak{g}_c)$, $p \in I(\mathfrak{g}_c)$ *and* $f \in C_c^\infty(\Omega)$. *Then*

$$\sum_{1 \leqslant i \leqslant r} c_i \int \varepsilon_{R,i} \, \partial(\omega_i p_i) \, (\pi_i \psi_{f,i}) \cdot \delta_i(D) \, \Phi_i d_i H$$

$$= \sum_{1 \leqslant i \leqslant r} c_i \int \varepsilon_{R,i} \, \partial(p_i) \, (\pi_i \psi_{f,i}) \cdot \delta_i(\partial(\omega) \circ D) \, \Phi_i d_i H$$

and

$$\sum_{1 \leqslant i \leqslant r} c_i \int \varepsilon_{R,i} \, \partial(\omega_i) \, \psi_{f,i} \cdot (\delta_i(D) \circ \pi_i \circ \partial(p_i)) \, \Phi_i d_i H$$

$$= \sum_{1 \leqslant i \leqslant r} c_i \int \varepsilon_{R,i} \, \psi_{f,i} (\delta_i(\partial(\omega) \circ D) \circ \pi_i \circ \partial(p_i)) \, \Phi_i d_i H.$$

We shall prove this in § 4.

COROLLARY 1. *For any* $k \geqslant 0$,

$$\sum_{1 \leqslant i \leqslant r} c_i \int \varepsilon_{R,i} \, \partial(\omega_i^k) \, \psi_{f,i} \cdot (\delta_i(D) \circ \pi_i \circ \partial(p_i)) \, \Phi_i d_i H$$

$$= \sum_{1 \leqslant i \leqslant r} c_i \int \varepsilon_{R,i} \, \psi_{f,i} (\partial(\omega_i^k) \circ \delta_i(D) \circ \pi_i \circ \partial(p_i)) \, \Phi_i d_i H.$$

Since $\psi_{\partial(\omega)f,i} = \partial(\omega_i) \psi_{f,i}$ and $\delta_i(\partial(\omega^k) \circ D) = \partial(\omega_i^k) \circ \delta_i(D)$, this follows immediately from the second statement of Lemma 6 by induction on k.

17 – 652923. *Acta mathematica.* 113. Imprimé le 11 mai 1965.

COROLLARY 2.

$$\sum_i c_i \int \varepsilon_{R,i}\, \partial(\omega_i^k)\,(\pi_i\,\psi_{f,i})\cdot \delta_i(D)\,\Phi_i\,d_iH = \sum_i c_i \int \varepsilon_{R,i}\,\pi_i\,\psi_{f,i}\cdot \delta_i(\partial(\omega^k)\circ D)\,\Phi_i\,d_iH$$

for $k \geqslant 0$.

This follows from the first statement of Lemma 6 by induction on k.

COROLLARY 3.

$$\sum_i c_i \int \varepsilon_{R,i}(\partial(\omega_i^j)\circ\pi_i\circ\partial(\omega_i^k))\,\psi_{f,i}\cdot \Phi_i\,d_iH = \sum_i \int \varepsilon_{R,i}\,\psi_{f,i}\,(\partial(\omega_i^k)\circ\pi_i\circ\partial(\omega_i^j))\,\Phi_i\,d_iH$$

for $j,\, k \geqslant 0$.

Apply Corollary 2 to $f_k = \partial(\omega)^k f$ with $D = 1$. Then since

$$\psi_{f_k,i} = \partial(\omega_i^k)\,\psi_{f,i},$$

we obtain
$$\sum_i c_i \int \varepsilon_{R,i}(\partial(\omega_i^j)\circ\pi_i\circ\partial(\omega_i^k))\,\psi_{f,i}\cdot \Phi_i\,d_iH$$

$$= \sum_i c_i \int \varepsilon_{R,i}\,\partial(\omega_i^k)\,\psi_{f,i}\cdot \pi_i\,\partial(\omega_i^j)\,\Phi_i\,d_iH.$$

Now apply Corollary 1 with $D = 1$ and $p = \omega^j$. This gives the required result.

We shall now complete the proof of Lemma 5 and therefore also of Theorem 1. Let Λ_i denote the derivation of $\mathfrak{D}(\mathfrak{h}_{ic})$ given by[1]

$$\Lambda_i\,\xi = \tfrac{1}{2}\,\{\partial(\omega_i),\,\xi\} \quad (\xi \in \mathfrak{D}(\mathfrak{h}_{ic})).$$

Then since π_i is homogeneous of degree m, it is clear that (see [2 (c), p. 99]) that

$$\Lambda_i^m\,\pi_i = m!\,\partial(\pi_i).$$

Therefore
$$\partial(\pi_i) = (m!\,2^m)^{-1} \sum_{0\leqslant k\leqslant m} C_k^m(-1)^{m-k}\,\partial(\omega_i^k)\circ\pi_i\circ\partial(\omega_i^{m-k})$$

where C_k^m denotes the usual binomial coefficient. Hence Lemma 5 follows immediately from Corollary 3 above.

§ 4. Third reduction

Fix $D \in \mathfrak{J}(\mathfrak{g}_c)$ and put

---

[1] As usual $\{D_1,\,D_2\} = D_1\circ D_2 - D_2\circ D_1$ for two differential operators $D_1,\, D_2$.

$$J(f) = \sum_i c_i \int \varepsilon_{R,i} \{\partial(\omega_i)\,(\pi_i\,\psi_{f,i})\cdot\delta_i(D)\,\Phi_i - \pi_i\,\psi_{f,i}\,\delta_i(\partial(\omega)\circ D)\,\Phi_i\}\,d_iH$$

and $$J'(f) = \sum_i c_i \int \varepsilon_{R,i} \{\partial(\omega_i)\,\psi_{f,i}\cdot\delta_i(D)\,(\pi_i\,\Phi_i) - \psi_{f,i}\,\delta_i(\partial(\omega)\circ D)\,(\pi_i\,\Phi_i)\}\,d_iH$$

for $f \in C_c^\infty(\Omega)$. Then J and J' are invariant distributions on Ω.

LEMMA 7. *No semiregular element of Ω of noncompact type lies in*

$$(\text{Supp } J) \cup (\text{Supp } J').$$

Assuming this result, we shall now prove Lemma 6. For $p \in I(\mathfrak{g}_c)$ and $f \in C_c^\infty(\Omega)$, define

$$J_p(f) = \sum_i c_i \int \varepsilon_{R,i} \{\partial(\omega_i\,p_i)\,(\pi_i\,\psi_{f,i})\cdot\delta_i(D)\,\Phi_i - \partial(p_i)\,(\pi_i\,\psi_{f,i})\cdot\delta_i(\partial(\omega)\circ D)\,\Phi_i\}\,d_iH.$$

Then J_p is an invariant distribution on Ω. We shall now show that $J_p = J' = 0$.

Fix a point $X_0 \in \Omega$ and, for any $\varepsilon > 0$, define $U_{X_0}(\varepsilon)$ as in [2 (l), Lemma 14] and put $\Omega(\varepsilon) = \Omega \cap U_{X_0}(\varepsilon)$. Then $\Omega(\varepsilon)$ is an open and completely invariant neighborhood of X_0 in $\mathfrak{g}$. Put

$$\mathfrak{h}_i(\varepsilon) = \mathfrak{h}_i \cap \Omega_i(\varepsilon), \quad \mathfrak{h}_i(0) = \bigcap_{\varepsilon>0} \mathfrak{h}_i(\varepsilon) \quad (1 \leqslant i \leqslant r).$$

Then we have seen during the proof of [2 (l), Lemma 13] that $\mathfrak{h}_i(0)$ is a finite set. For every $H \in \mathfrak{h}_i(0)$, select two open, convex neighborhoods U_H, V_H of H in $\mathfrak{h}_i$ such that $\text{Cl } U_H \subset V_H \subset \mathfrak{h}_i(1)$ and $V_H \cap V_{H'} = \varnothing$ for $H \neq H'$ $(H, H' \in \mathfrak{h}_i(0))$. Put

$$U_i = \bigcup_{H \in \mathfrak{h}_i(0)} U_H, \quad V_i = \bigcup_{H \in \mathfrak{h}_i(0)} V_H$$

and select $\alpha_H \in C_c^\infty(V_H)$ such that $\alpha_H = 1$ on U_H $(H \in \mathfrak{h}_i(0))$. Define

$$\alpha_i = \sum_{H \in \mathfrak{h}_i(0)} \alpha_H$$

and put $$g_i = c_i\,\varepsilon_{R,i}\,\alpha_i\,\delta_i(D)\,\Phi_i, \quad g_i' = c_i\,\varepsilon_{R,i}\,\alpha_i\,\delta_i(D)\,(\pi_i\,\Phi_i).$$

Then it follows from [2 (j), Theorem 1], [2 (l), Theorem 2] and [2 (l), § 4] that g_i and g_i' are functions of class C^∞ on the closure of each connected component of $\mathfrak{h}_i'(R)$.

Now choose $\varepsilon > 0$ so small that $\mathfrak{h}_i(\varepsilon) \subset U_i$ $(1 \leqslant i \leqslant r)$. Then if $f \in C_c^\infty(\Omega(\varepsilon))$, it is clear that $\text{Supp } \psi_{f,i} \subset U_i$. Since $\alpha_i = 1$ on U_i, it follows that

$$J_p(f) = \sum_i \int \{\partial(\omega_i\,p_i)\,\psi_{f,i}\cdot g_i - \partial(p_i)\,\psi_{f,i}\cdot\partial(\omega_i)\,g_i\}\,d_iH$$

and
$$J'(f) = \sum_i \int \{\partial(\omega_i)\,\psi_{f,\,i}\cdot g_i{}' - \psi_{f,\,i}\cdot\partial(\omega_i)\,g_i{}'\}\,d_i\,H$$

for $f\in C_c^\infty(\Omega(\varepsilon))$. Moreover $\Omega(\varepsilon)$ being completely invariant, we can choose an open neighborhood V of X_0 in $\mathfrak{g}$ such that $\mathrm{Cl}(V^G)\subset\Omega(\varepsilon)$. Now $J=J_1$ and therefore it follows from Lemma 7 and [2 (k), Theorem 5] that $J_p=0$ on V^G for $p\in I(\mathfrak{g}_c)$. Hence $X_0\notin\mathrm{Supp}\,J_p$. But X_0 was an arbitrary point of Ω. Therefore we conclude that $J_p=0$. This proves the first statement of Lemma 6.

Similarly by applying [2 (k), Theorem 4] we conclude that $J'=0$. This gives the second statement of Lemma 6 in the special case $p=1$. Now fix $p\in I(\mathfrak{g}_c)$ and consider the distribution $T_0=\partial(p)\,T$. Then T_0 also satisfies the conditions of [2 (l), Theorem 1] and therefore $T_0=F_0$ where $F_0=\partial(p)\,F$. Put $\Phi_0=\nabla_\mathfrak{g}F_0$ and let Φ_{0i} denote the restriction of Φ_0 on $\mathfrak{h}_i\cap\Omega'$ $(1\leqslant i\leqslant r)$.

LEMMA 8. $\qquad\qquad\qquad \Phi_{0i}=\partial(p_i)\,\Phi_i \quad (1\leqslant i\leqslant r)$.

Let F_i and F_{0i} respectively denote the restrictions of F and F_0 on $\mathfrak{h}_i\cap\Omega'$. Since F is an invariant function, we know [2 (c), Theorem 1] that

$$F_{0i}=\pi_i^{-1}\,\partial(p)\,(\pi_i\,F_i).$$

Therefore $\qquad\qquad \Phi_{0i}=\partial(\pi_i)\,(\pi_i\,F_{0i})=\partial(p_i\pi_i)\,(\pi_i\,F_i)=\partial(p_i)\,\Phi_i.$

Now if we apply the result $J'=0$ to the distribution T_0 (instead of T), we obtain

$$\sum_i c_i \int \varepsilon_{R,\,i}\,\{\partial(\omega_i)\,\psi_{f,\,i}\cdot\delta_i(D)\,(\pi_i\,\Phi_{0i}) - \psi_{f,\,i}\,\delta_i(\partial(\omega)\circ D)\,(\pi_i\,\Phi_{0i})\}\,d_i\,H = 0$$

for $f\in C_c^\infty(\Omega)$. In view of Lemma 8, this is equivalent to the second assertion of Lemma 6.

§ 5. New expressions for J and J'

Define η as in § 2. Then $\eta\in I(\mathfrak{g}_c)$ and $\eta_i=(-1)^m\pi_i^2$ $(1\leqslant i\leqslant r)$. Moreover $|\eta|^{\frac{1}{2}}$ and $|\eta|^{-\frac{1}{2}}$ are analytic functions on $\mathfrak{g}'$.

LEMMA 9. *Define J and J' as in § 4. Then*

$$J(f) = \mathrm{p.v.}\int \{\partial(\omega)\,(|\eta|^{\frac{1}{2}}f)\cdot D(|\eta|^{-\frac{1}{2}}\Phi) - |\eta|^{\frac{1}{2}}f(\partial(\omega)\circ D\circ|\eta|^{-\frac{1}{2}})\,\Phi\}\,dX$$

and
$$J'(f) = \int \{\partial(\omega)f \cdot D\Phi - f\partial(\omega)(D\Phi)\}\, dX$$

for $f \in C_c^\infty(\Omega)$.

Since η takes only real values on $\mathfrak{g}$, it is obvious that $|\eta_i|^{\frac{1}{2}} = \varepsilon_i \pi_i$ on $\mathfrak{h}_i'$ where ε_i is a locally constant function on $\mathfrak{h}_i'$ such that $\varepsilon_i^4 = 1$. Since Φ and $|\eta|^{-\frac{1}{2}}$ are invariant functions, it follows from [2 (j), Lemma 14] that

$$D'(|\eta|^{-\frac{1}{2}}\Phi) = \varepsilon_i^{-1}\pi_i^{-1}\delta_i(D')\Phi_i$$

on $\mathfrak{h}_i \cap \Omega'$ for any $D' \in \mathfrak{J}(\mathfrak{g}_c)$.

For any $f \in C_c^\infty(\Omega)$, let g_f denote the function on $\mathfrak{g}'$ given by

$$g_f = \partial(\omega)(|\eta|^{\frac{1}{2}}f) \cdot D(|\eta|^{-\frac{1}{2}}\Phi) - |\eta|^{\frac{1}{2}}f(\partial(\omega) \circ D \circ |\eta|^{-\frac{1}{2}})\Phi.$$

Fix a function $v \in C^\infty(\mathbf{R})$ such that $v(t) = 0$ if $|t| \leqslant \frac{1}{2}$ and $v(t) = 1$ if $|t| \geqslant 1$ $(t \in \mathbf{R})$. For any $\varepsilon > 0$, put $v_\varepsilon(t) = v(\varepsilon^{-2}t)$ and

$$V_\varepsilon(X) = v_\varepsilon(\eta(X)) \quad (X \in \mathfrak{g}).$$

Then V_ε is an invariant C^∞ function on $\mathfrak{g}$ and $V_\varepsilon = 1$ on $\mathfrak{g}(\varepsilon)$ (in the notation of § 2). Put $f_\varepsilon' = V_\varepsilon f$ and $f_\varepsilon = |\eta|^{\frac{1}{2}}f_\varepsilon'$. It is clear that f_ε and f_ε' are in $C_c^\infty(\Omega)$ and $f = f_\varepsilon'$ on $\mathfrak{g}(\varepsilon)$. Hence

$$\int_{\mathfrak{g}(\varepsilon)} g_f\, dX = \int_{\mathfrak{g}(\varepsilon)} g_{f_\varepsilon'}\, dX$$

$$= \sum_i c_i \int_{\mathfrak{h}_i(\varepsilon)} \varepsilon_{i,R}\varepsilon_i^{-1}\{\partial(\omega_i)\psi_{f_\varepsilon,i} \cdot \delta_i(D)\Phi_i - \psi_{f_\varepsilon,i}\delta_i(\partial(\omega) \circ D)\Phi_i\}\, d_i H$$

where $\mathfrak{h}_i(\varepsilon) = \mathfrak{h}_i \cap \mathfrak{g}(\varepsilon)$. However it is obvious that

$$\psi_{f_\varepsilon,i} = \varepsilon_i \pi_i \psi_{f,i}$$

on $\mathfrak{h}_i(\varepsilon)$. Therefore

$$\int_{\mathfrak{g}(\varepsilon)} g_f\, dX = \sum_i c_i \int_{\mathfrak{h}_i(\varepsilon)} \varepsilon_{i,R}\{\partial(\omega_i)(\pi_i\psi_{f,i}) \cdot \delta_i(D)\Phi_i - \pi_i\psi_{f,i}\delta_i(\partial(\omega) \circ D)\Phi_i\}\, d_i H.$$

Making $\varepsilon \to 0$ we get
$$\text{p.v.} \int g_f\, dX = J(f)$$

and this proves the first statement of the lemma.

We know from the corollary of Lemma 1 that the integral

$$\int \{\partial(\omega) f \cdot D\Phi - f\partial(\omega)(D\Phi)\}\, dX$$

is well defined. Moreover since Φ is an invariant function,

$$D'\Phi = \pi_i^{-1}\delta_i(D')(\pi_i\Phi_i) \quad (D' \in \mathfrak{I}(\mathfrak{g}_c))$$

on $\mathfrak{h}_i \cap \Omega'$ and therefore the above integral is equal to $J'(f)$. This proves the second statement of the lemma.

LEMMA 10. *For any $\varepsilon > 0$, define the function V_ε as above and put*

$$J_\varepsilon(f) = \int V_\varepsilon \{\partial(\omega)(|\eta|^{\frac{1}{2}}f) \cdot D(|\eta|^{-\frac{1}{2}}\Phi) - |\eta|^{\frac{1}{2}}f(\partial(\omega)\circ D\circ|\eta|^{-\frac{1}{2}})\Phi\}\, dX$$

and
$$J_\varepsilon'(f) = \int V_\varepsilon \{\partial(\omega) f \cdot D\Phi - f\partial(\omega)(D\Phi)\}\, dX$$

for $f \in C_c^\infty(\Omega)$. Then

$$J(f) = \lim_{\varepsilon\to 0} J_\varepsilon(f), \quad J'(f) = \lim_{\varepsilon\to 0} J_\varepsilon'(f).$$

Put $f_\varepsilon = |\eta|^{\frac{1}{2}} V_{\varepsilon/2} f$. Then $f_\varepsilon \in C_c^\infty(\Omega)$ and $J_\varepsilon(f) = J_\varepsilon(V_{\varepsilon/2}f)$. Hence it follows that

$$J_\varepsilon(f) = \sum_i c_i \int V_{\varepsilon,i}\, \varepsilon_{i,R}\, \varepsilon_i^{-1} \{\partial(\omega_i)\,\psi_{f_\varepsilon,i} \cdot \delta_i(D)\,\Phi_i - \psi_{f_\varepsilon,i}\,\delta_i(\partial(\omega)\circ D)\,\Phi_i\}\, d_i H$$

where $V_{\varepsilon,i}$ is the restriction of V_ε on $\mathfrak{h}_i$. On the other hand, it is clear that

$$\psi_{f_\varepsilon,i}(H) = \varepsilon_i \pi_i(H)\psi_{f,i}(H)$$

if $|\pi_i(H)| \geqslant \varepsilon/2 \ (H \in \mathfrak{h}_i)$. Hence

$$J_\varepsilon(f) = \sum_i c_i \int V_{\varepsilon,i}\, \varepsilon_{i,R} \{\partial(\omega_i)(\pi_i\psi_{f,i}) \cdot \delta_i(D)\,\Phi_i - \pi_i\psi_{f,i}\,\delta_i(\partial(\omega)\circ D)\,\Phi_i\}\, d_i H.$$

The two assertions of the lemma are now obvious.

LEMMA 11. *Put* [1]
$$\Psi_\varepsilon = (|\eta|^{\frac{1}{2}}\{\partial(\omega),V_\varepsilon\}\circ D\circ|\eta|^{-\frac{1}{2}})\,\Phi,$$
$$\Psi_\varepsilon' = (\{\partial(\omega),V_\varepsilon\}\circ D)\,\Phi$$
for $\varepsilon > 0$. Then

---

[1] See footnote 1, p. 250.

$$J_\varepsilon(f) = \int f \Psi_\varepsilon \, dX, \quad J_\varepsilon'(f) = \int f \Psi_\varepsilon' \, dX$$

for $f \in C_c^\infty(\Omega)$.

Since $\operatorname{Supp} V_\varepsilon \subset \mathfrak{g}'$, this follows from Lemma 10 if we observe that

$$(V_\varepsilon \partial(\omega) \circ |\eta|^{\frac{1}{2}})^* = |\eta|^{\frac{1}{2}} \partial(\omega) \circ V_\varepsilon, \quad (V_\varepsilon \partial(\omega))^* = \partial(\omega) \circ V_\varepsilon.$$

Fix a Cartan subalgebra $\mathfrak{h}$ of $\mathfrak{g}$ and let us use the notation introduced at the beginning of § 2. In particular $V_{\varepsilon,\mathfrak{h}}$ and $\omega_\mathfrak{h}$ denote the restrictions of V_ε and ω on $\mathfrak{h}$.

LEMMA 12. *For any $\varepsilon > 0$, we have*

$$(\Psi_\varepsilon)_\mathfrak{h} = (\{\partial(\omega_\mathfrak{h}), V_{\varepsilon,\mathfrak{h}}\} \circ \delta_{\mathfrak{g}/\mathfrak{h}}(D)) \, \Phi_\mathfrak{h}$$

and
$$(\Psi_\varepsilon')_\mathfrak{h} = (\pi^{-1} \{\partial(\omega_\mathfrak{h}), V_{\varepsilon,\mathfrak{h}}\} \circ \delta_{\mathfrak{g}/\mathfrak{h}}(D) \circ \pi) \, \Phi_\mathfrak{h}.$$

$|\eta|^{\frac{1}{2}}$, Φ and V_ε are invariant C^∞ functions on Ω'. Moreover there exists a locally constant function a on $\mathfrak{h}'$ such that $a^4 = 1$ and $|\eta|^{\frac{1}{2}} = a\pi$ on $\mathfrak{h}'$. The required relations now follow easily by a repeated use of [2 (j), Lemma 14] and [2 (c), Theorem 1].

§ 6. Proof of Lemma 7

We now come to the proof of Lemma 7. Fix a semiregular element $H_0 \in \Omega$ of noncompact type and let $\mathfrak{z}$ denote the centralizer of H_0 in $\mathfrak{g}$. Define ζ and $\mathfrak{z}'$ as in [2 (j), § 2] and put $\Omega_\mathfrak{z} = \Omega \cap \mathfrak{z}'$. Then $\Omega_\mathfrak{z}$ is an open and completely invariant neighborhood of H_0 in $\mathfrak{z}$. Fix a Euclidean measure dZ on $\mathfrak{z}$ and define

$$j = \sigma_J, j' = \sigma_{J'}, j_\varepsilon = \sigma_{J_\varepsilon} \text{ and } j_\varepsilon' = \sigma_{J_\varepsilon'} \quad (\varepsilon > 0)$$

in the notation of [2 (j), Lemma 17] corresponding to $G_0 = G$ and $\mathfrak{z}_0 = \Omega_\mathfrak{z}$. Since $J_\varepsilon = \Psi_\varepsilon$ and $J_\varepsilon' = \Psi_\varepsilon'$ (Lemma 11), it is obvious that

$$j_\varepsilon(\gamma) = \int \gamma(Z) \Psi_\varepsilon(Z) \, dZ, \quad j_\varepsilon'(\gamma) = \int \gamma(Z) \Psi_\varepsilon'(Z) \, dZ$$

for $\gamma \in C_c^\infty(\Omega_\mathfrak{z})$. Moreover

$$j(\gamma) = \lim_{\varepsilon \to 0} j_\varepsilon(\gamma), \quad j'(\gamma) = \lim_{\varepsilon \to 0} j_\varepsilon'(\gamma) \quad (\gamma \in C_c^\infty(\Omega_\mathfrak{z}))$$

from Lemma 10.

Now we use the notation of [2 (k), § 7]. In particular σ is the center of $\mathfrak{z}$ and $\mathfrak{a} = \mathbf{R}H' + \sigma$, $\mathfrak{b} = \mathbf{R}(X' - Y') + \sigma$ are two Cartan subalgebras of $\mathfrak{z}$. Fix Euclidean measures $d\sigma$, $d\mathfrak{a}$, $d\mathfrak{b}$ on σ, $\mathfrak{a}$, $\mathfrak{b}$ respectively such that

$$da = dt\, d\sigma, \quad db = d\phi\, d\sigma$$

where $t = \alpha/2$ and $\phi = (-1)^{\frac{1}{2}}\beta/2$ in the notation of [2 (k), Lemma 13]. Then $d\sigma$ can be so normalized that (see [2 (e), Lemma 3])

$$\int \gamma\, dZ = \frac{1}{2}\int_{a^+} \alpha J_\gamma{}^a\, da + (-1)^{\frac{1}{2}} \frac{1}{2}\int_b \beta J_\gamma{}^b\, db$$

for $\gamma \in C_c^\infty(\mathfrak{z})$. Here

$$J_\gamma{}^a(H) = J_a(\gamma : H) \quad (H \in a),$$
$$J_\gamma{}^b(H) = J_b(\gamma : H) \quad (H \in b'')$$

in the notation of [2 (k), Lemma 14], a^+ is the set of all points H in a where $\alpha(H) > 0$ and b'' is the set of those $H \in b$ where $\beta(H) \neq 0$. Therefore since Ψ_ε is an invariant C^∞ function on Ω, it is clear that

$$j_\varepsilon(\gamma) = \frac{1}{2}\int_{a^+} \alpha J_\gamma{}^a (\Psi_\varepsilon)_a\, da + (-1)^{\frac{1}{2}} \frac{1}{2}\int_b \beta J_\gamma{}^b (\Psi_\varepsilon)_b\, db$$

for $\gamma \in C_c^\infty(\Omega_\mathfrak{z})$. Now apply Lemma 12 and observe that $\mathrm{Supp}\, V_{\varepsilon,\mathfrak{h}} \subset \mathfrak{h} \cap \mathfrak{g}'$ and

$$(\partial(\omega_\mathfrak{h}) \circ V_{\varepsilon,\mathfrak{h}})^* = V_{\varepsilon,\mathfrak{h}} \partial(\omega_\mathfrak{h}) \quad (\mathfrak{h} = a \text{ or } b).$$

Then it follows that

$$j_\varepsilon(\gamma) = \frac{1}{2}\int_{a^+} V_{\varepsilon,a}\{\partial(\omega_a)(\alpha J_\gamma{}^a) \cdot \Phi_{0,a} - \alpha J_\gamma{}^a \cdot \partial(\omega_a)\Phi_{0,a}\}\, da$$
$$+ (-1)^{\frac{1}{2}} \frac{1}{2}\int_b V_{\varepsilon,b}\{\partial(\omega_b)(\beta J_\gamma{}^b) \cdot \Phi_{0,b} - \beta J_\gamma{}^b \cdot \partial(\omega_b)\Phi_{0,b}\}\, db$$

where $\Phi_{0,\mathfrak{h}} = \delta_{\mathfrak{g}/\mathfrak{h}}(D)\Phi_\mathfrak{h}$ ($\mathfrak{h} = a$ or b). Hence it is obvious that

$$j(\gamma) = \frac{1}{2}\int_{a^+} \{\partial(\omega_a)(\alpha J_\gamma{}^a) \cdot \Phi_{0,a} - \alpha J_\gamma{}^a \partial(\omega_a)\Phi_{0,a}\}\, da$$
$$+ (-1)^{\frac{1}{2}} \frac{1}{2}\int_b \{\partial(\omega_b)(\beta J_\gamma{}^b) \cdot \Phi_{0,b} - \beta J_\gamma{}^b \partial(\omega_b)\Phi_{0,b}\}\, db$$

for $\gamma \in C_c^\infty(\Omega_\mathfrak{z})$. Now $\omega_a = \omega_\sigma + |\alpha|^{-2}\alpha^2$ where ω_σ is the restriction of ω on σ. Similarly $\omega_b = \omega_\sigma + |\beta|^{-2}\beta^2$. Hence (see [2 (k), Lemma 21]) it follows that

$$j(\gamma) = \frac{1}{2|\alpha|^2}\int_{a^+} \partial(\alpha)\{\partial(\alpha)(\alpha J_\gamma{}^a) \cdot \Phi_{0,a} - \alpha J_\gamma{}^a \partial(\alpha)\Phi_{0,a}\}\, da$$
$$+ \frac{(-1)^{\frac{1}{2}}}{2|\beta|^2}\int_b \partial(\beta)\{\partial(\beta)(\beta J_\gamma{}^b) \cdot \Phi_{0,b} - \beta J_\gamma{}^b \partial(\beta)\Phi_{0,b}\}\, db.$$

Now $d\mathfrak{a} = d\sigma\, dt$, $d\mathfrak{b} = d\sigma\, d\phi$ and

$$\partial(\alpha) = \tfrac{1}{2}\,|\alpha|^2 \partial/\partial t, \quad \partial(\beta) = \tfrac{1}{2}(-1)^{\frac{1}{2}}\,|\beta|^2 \partial/\partial\phi$$

since

$$H' = 2\,|\alpha|^{-2}H_\alpha, \quad X' - Y' = -2\,(-1)^{\frac{1}{2}}\,|\beta|^{-2}H_\beta$$

in the notation of [2 (k), § 7]. Therefore

$$j(\gamma) = -\tfrac{1}{4}\int_\sigma \{\partial(\alpha)\,(\alpha J_\gamma^{\,\mathfrak{a}})\cdot\Phi_{0,\mathfrak{a}} - \alpha J_\gamma^{\,\mathfrak{a}}\cdot\partial(\alpha)\,\Phi_{0,\mathfrak{a}}\}^+\, d\sigma$$

$$+\tfrac{1}{4}\int_\sigma \{\partial(\beta)\,(\beta J_\gamma^{\,\mathfrak{b}})\cdot\Phi_{0,\mathfrak{b}} - \beta J_\gamma^{\,\mathfrak{b}}\,\partial(\beta)\,\Phi_{0,\mathfrak{b}}\}_-^{+}\, d\sigma.$$

Here

$$u_\mathfrak{a}^{+}(H) = \lim_{t\to+0} u_\mathfrak{a}(H + tH'), \quad u_\mathfrak{b}^{\pm}(H) = \lim_{\phi\to+0} u_\mathfrak{b}(H \pm \phi(X' - Y')) \quad (H\in\sigma;\, t,\, \phi\in\mathbf{R})$$

for two functions $u_\mathfrak{a}$ and $u_\mathfrak{b}$ on $\mathfrak{a}$ and $\mathfrak{b}$ respectively and $(u_\mathfrak{b})_-^{\,+} = u_\mathfrak{b}^{+} - u_\mathfrak{b}^{-}$. Since $\alpha = \beta = 0$ on σ and $|\alpha|^2 = |\beta|^2$ [2 (k), Lemma 13], it follows that

$$j(\gamma) = \tfrac{1}{4}\,|\alpha|^2 \int_\sigma \{(J_\gamma^{\,\mathfrak{b}}\Phi_{0,\mathfrak{b}})_-^{\,+} - (J_\gamma^{\,\mathfrak{a}}\Phi_{0,\mathfrak{a}})^{+}\}\, d\sigma.$$

However $\Phi_{0,\mathfrak{a}}$ and $\Phi_{0,\mathfrak{b}}$ are continuous functions on $\mathfrak{a}\cap\Omega_\delta$ and $\mathfrak{b}\cap\Omega_\delta$ respectively and $\Phi_{0,\mathfrak{a}} = \Phi_{0,\mathfrak{b}}$ on $\sigma\cap\Omega_\delta$ [2 (l), Lemma 18]. Therefore

$$j(\gamma) = \tfrac{1}{4}\,|\alpha|^2 \int_\sigma \{(J_\gamma^{\,\mathfrak{b}})_-^{\,+} - J_\gamma^{\,\mathfrak{a}}\}\,\Phi_{0,\mathfrak{a}}\, d\sigma \quad (\gamma\in C_c^\infty(\Omega_\delta)).$$

But $(J_\gamma^{\,\mathfrak{b}})_-^{\,+} = J_\gamma^{\,\mathfrak{a}}$ on σ [2 (k), § 10]. Hence $j = 0$ on Ω_δ.

Now put[1] $\pi_\alpha = \alpha^{-1}\pi^\mathfrak{a}$, $\pi_\beta = \beta^{-1}\pi^\mathfrak{b}$ and

$$\Phi_\mathfrak{h}{}' = \delta_{\mathfrak{a}/\mathfrak{h}}(D)\,(\pi^\mathfrak{h}\Phi_\mathfrak{h})$$

for $\mathfrak{h} = \mathfrak{a}$ or $\mathfrak{b}$. Then if $\gamma\in C_c^\infty(\Omega_\delta)$, we have

$$j_\varepsilon{}'(\gamma) = \frac{1}{2}\int_{\mathfrak{a}^+} \alpha J_\gamma^{\,\mathfrak{a}}(\Psi_\varepsilon{}')_\mathfrak{a}\,d\mathfrak{a} + (-1)^{\frac{1}{2}}\frac{1}{2}\int_\mathfrak{b} \beta J_\gamma^{\,\mathfrak{b}}(\Psi_\varepsilon{}')_\mathfrak{b}\,d\mathfrak{b}$$

$$= \frac{1}{2}\int_{\mathfrak{a}^+} V_{\varepsilon,\,\mathfrak{a}}\{\partial(\omega_\mathfrak{a})\,(\pi_\alpha^{-1}J_\gamma^{\,\mathfrak{a}})\cdot\Phi_\mathfrak{a}{}' - \pi_\alpha^{-1}J_\gamma^{\,\mathfrak{a}}\partial(\omega_\mathfrak{a})\,\Phi_\mathfrak{a}{}'\}\,d\mathfrak{a}$$

$$+\tfrac{1}{2}(-1)^{\frac{1}{2}}\int_\mathfrak{b} V_{\varepsilon,\,\mathfrak{b}}\{\partial(\omega_\mathfrak{b})\,(\pi_\beta^{-1}J_\gamma^{\,\mathfrak{b}})\cdot\Phi_\mathfrak{b}{}' - \pi_\beta^{-1}J_\gamma^{\,\mathfrak{b}}\partial(\omega_\mathfrak{b})\,\Phi_\mathfrak{b}{}'\}\,d\mathfrak{b}$$

---

[1] We assume, as we may, that $(\pi^\mathfrak{a})^\nu = \pi^\mathfrak{b}$ in the notation of [2 (k). § 7].

from Lemma 12. Hence

$$j'(\gamma) = \frac{1}{2|\alpha|^2} \int_{\mathfrak{a}^+} \partial(\alpha)\{\partial(\alpha)(\pi_\alpha^{-1}J_\gamma^\mathfrak{a})\cdot\Phi_\mathfrak{a}' - \pi_\alpha^{-1}J_\gamma^\mathfrak{a}\partial(\alpha)\Phi_\mathfrak{a}'\}\,d\mathfrak{a}$$

$$+ \frac{(-1)^{\frac{1}{2}}}{2|\beta|^2} \int_\mathfrak{b} \partial(\beta)\{\partial(\beta)(\pi_\beta^{-1}J_\gamma^\mathfrak{b})\cdot\Phi_\mathfrak{b}' - \pi_\beta^{-1}J_\gamma^\mathfrak{b}\partial(\beta)\Phi_\mathfrak{b}'\}\,d\mathfrak{b}$$

$$= -\tfrac{1}{4}\int_\sigma \{\partial(\alpha)(\pi_\alpha^{-1}J_\gamma^\mathfrak{a})\cdot\Phi_\mathfrak{a}' - \pi_\alpha^{-1}J_\gamma^\mathfrak{a}\partial(\alpha)\Phi_\mathfrak{a}'\}^+\,d\sigma$$

$$+ \tfrac{1}{4}\int_\sigma \{\partial(\beta)(\pi_\beta^{-1}J_\gamma^\mathfrak{b})\cdot\Phi_\mathfrak{b}' - \pi_\beta^{-1}J_\gamma^\mathfrak{b}\partial(\beta)\Phi_\mathfrak{b}'\}_-{}^+\,d\sigma.$$

Now $\pi_\alpha^{-1}J_\gamma^\mathfrak{a}$ is a C^∞ function on $\mathfrak{a}$ which is invariant under the Weyl reflexion s_α. Hence $\partial(\alpha)(\pi_\alpha^{-1}J_\gamma^\mathfrak{a})=0$ on σ. Moreover, $\partial(\beta)(\pi_\beta^{-1}J_\gamma^\mathfrak{b})$ is a continuous function on $\mathfrak{b}$ by [2 (k), Theorem 1] and $\partial(\alpha)\Phi_\mathfrak{a}'$, $\partial(\beta)\Phi_\mathfrak{b}'$ are continuous functions on $\mathfrak{a}\cap\Omega_\mathfrak{z}$, $\mathfrak{b}\cap\Omega_\mathfrak{z}$ respectively and they are equal on $\sigma\cap\Omega_\mathfrak{z}$ [2 (l), Lemma 18]. Finally $\Phi_\mathfrak{b}'$ is an analytic function on $\mathfrak{b}\cap\Omega_\mathfrak{z}$ [2 (l), Theorem 2]. Hence

$$j'(\gamma) = \tfrac{1}{4}\int_\sigma \pi_\alpha^{-1}J_\gamma^\mathfrak{a}\partial(\alpha)\Phi_\mathfrak{a}'\,d\sigma - \tfrac{1}{4}\int_\sigma (\pi_\beta^{-1}J_\gamma^\mathfrak{b})_-{}^+\partial(\beta)\Phi_\mathfrak{b}'\,d\sigma$$

$$= \tfrac{1}{4}\int_\sigma \{\pi_\alpha^{-1}J_\gamma^\mathfrak{a} - (\pi_\beta^{-1}J_\gamma^\mathfrak{b})_-{}^+\}\partial(\beta)\Phi_\mathfrak{b}'\,d\sigma.$$

But([1]) $\pi_\alpha = \pi_\beta$ and $J_\gamma^\mathfrak{a} = (J_\gamma^\mathfrak{b})_-{}^+$ on σ. Therefore $j' = 0$ on $\Omega_\mathfrak{z}$. In view of [2 (j), Lemma 17] this completes the proof of Lemma 7.

§ 7. A consequence of Theorem 1

We now return to the notation of § 2 so that $\mathfrak{g}$ is again reductive. For any $p \in I(\mathfrak{g}_c)$, let p_i denote the projection of p in $I(\mathfrak{h}_{ic})$ (see [2 (j), § 8]).

LEMMA 13. *Fix $p \in I(\mathfrak{g}_c)$. Then*

$$\sum_{1\leqslant i\leqslant r} c_i \int \varepsilon_{R,i}\{\partial(\varpi_i p_i)\psi_{f,i}\cdot\Phi_i - \psi_{f,i}\partial(\varpi_i p_i)^*\Phi_i\}\,d_i H = 0$$

for $f \in C_c^\infty(\Omega)$.

We note that $\partial(\varpi_i)^* = (-1)^m\partial(\varpi_i)$. Therefore applying Lemma 5 to $\partial(p)f$, instead of f, we get

---

$$\sum_i c_i \int \varepsilon_{i,R} \partial(\varpi_i p_i) \psi_{f,i} \cdot \Phi_i d_i H = \sum_i c_i \int \varepsilon_{i,R} \partial(p_i) \psi_{f,i} \partial(\varpi_i)^* \Phi_i d_i H = (-1)^m \int \partial(p) f \cdot \square F \, dX.$$

But it follows from the corollary of [2 (l), Lemma 16] that

$$\int \partial(p) f \cdot \square F \, dX = \int f \cdot \square (\partial(p)^* F) \, dX.$$

Hence we conclude from Lemma 8 and [2 (j), Lemma 13] that

$$(-1)^m \int f \square (\partial(p)^* F) \, dX = \sum_i c_i \int \varepsilon_{i,R} \psi_{f,i} \partial(p_i \varpi_i)^* \Phi_i d_i H.$$

The statement of Lemma 13 is now obvious.

§ 8. Some elementary facts about tempered distributions

Let E be a vector space over $\mathbf{R}$ of finite dimension. Define $S(E_c)$, $P(E_c)$ and $\mathfrak{D}(E_c)$ as usual (see [2 (j), § 3]). Let U be an open subset of E and T a distribution on U. We say that T is tempered if we can choose $D_i \in \mathfrak{D}(E_c)$ $(1 \leqslant i \leqslant r)$ such that

$$|T(f)| \leqslant \sum_i \sup |D_i f| \quad (f \in C_c^\infty(U)).$$

It is clear that if T is tempered, the same holds for DT for any $D \in \mathfrak{D}(E_c)$.

Fix a Euclidean measure dX on E and let g be a locally summable function on U. Then g will be said to be tempered (on U) if the distribution

$$f \to \int f g \, dX \quad (f \in C_c^\infty(U))$$

on U is tempered.

Introduce a Euclidean norm $\| \ \|$ on E.

LEMMA 14. *Let g be a measurable function on U such that*

$$\sup_{X \in U} |g(X)| \, (1 + \|X\|)^{-m} < \infty$$

for some $m \geqslant 0$. Then g is tempered.

We can choose $r \geqslant 0$ such that

$$c_1 = \int_E (1 + \|X\|)^{-r} \, dX < \infty.$$

Put $c_2 = \sup_{X \in U} |g(X)| (1 + \|X\|)^{-m}$. Then

$$\left| \int g f \, dX \right| \leqslant c_1 c_2 \nu_{m+r} (f) \quad (f \in C_c^\infty (U))$$

where

$$\nu_{m+r} (f) = \sup_{X \in U} |f(X)| (1 + \|X\|)^{m+r}.$$

Since $X \to \|X\|^2$ is a quadratic form on E, it is now clear that g is tempered.

A subset V of E is called full if $tX \in V$ whenever $X \in V$ and $t \geqslant 1$.

LEMMA 15. *Let V be a non-empty, open and full subset of E. Put*

$$g(X) = \sum_{1 \leqslant i \leqslant r} p_i(X) e^{\lambda_i(X)} \quad (X \in E)$$

where $\lambda_1, \ldots, \lambda_r$ are distinct linear functions on E_c and $p_i \in P(E_c)$ ($p_i \neq 0$). Then g is tempered on V if and only if[1]

$$\Re \lambda_i(X) \leqslant 0$$

for all $X \in V$ and $1 \leqslant i \leqslant r$.

We recall that $S(E_c)$ is the algebra of polynomial functions on the dual space E_c' of E_c. Fix $p \in S(E_c)$ and $\lambda \in E_c'$. Then

$$\partial(p) \circ e^\lambda = e^\lambda \partial(p_\lambda)$$

where p_λ is the polynomial function $\mu \to p(\lambda + \mu)$ ($\mu \in E_c'$). Therefore if $q \in P(E_c)$ and $\partial(p)(e^\lambda q) = 0$, we conclude that $\partial(p_\lambda) q = 0$. Now assume that $q \neq 0$ and let q_0 be the homogeneous component of q of the highest degree. Then it is clear that $p_\lambda(0) q_0 = 0$ and therefore $p(\lambda) = 0$. We shall need this fact presently.

Let us now turn to the proof of Lemma 15. If $\Re \lambda_i(X) \leqslant 0$ for $X \in V$ and $1 \leqslant i \leqslant r$, it follows from Lemma 14 that g is tempered on V. To prove the converse we use induction on r.

So let us assume that g is tempered on V. It would be enough to show that $\Re \lambda_1(X) \leqslant 0$ for $X \in V$. First suppose that $r \geqslant 2$. Then $\lambda_1 \neq \lambda_r$ and therefore we can choose $q \in S(E_c)$ such that $q(\lambda_r) = 0$ while $q(\lambda_1) \neq 0$. Put $p = q^d$ where $d > d^0 p_r$. Then

$$\partial(p) (e^{\lambda_i} p_i) = p_i' e^{\lambda_i} \quad (1 \leqslant i \leqslant r)$$

where $p_i' = \partial(p_{\lambda_i}) p_i$. Since $p(\lambda_1) = q(\lambda_1)^d \neq 0$, it follows from what we have seen above, that $p_1' \neq 0$. On the other hand $p_{\lambda_r} = (q_{\lambda_r})^d$ and $q_{\lambda_r}(0) = q(\lambda_r) = 0$. Therefore since $d > d^0 p_r$, it is obvious that $p_r' = 0$. Hence

---

[1] $\Re c$ denotes the real part of a complex number c.

$$\partial(p)g = \sum_{1 \leqslant i < r} p_i' \, e^{\lambda_i}.$$

Now $\partial(p)g$ is also tempered on V and $p_1' \neq 0$. Therefore we conclude from the induction hypothesis that $\Re\lambda_1(X) \leqslant 0$ for $X \in V$.

Thus it remains to consider the case $r = 1$. Fix $X_1 \in V$ and write λ and p instead of λ_1 and p_1 respectively. Then we have to prove that $\Re\lambda(X_1) \leqslant 0$. If $X_1 = 0$, this is obvious. So let us assume that $X_1 \neq 0$. Choose a linear subspace F of E complementary to $\mathbf{R}X_1$ and an open convex neighborhood U of zero in F such that $X_1 + U \subset V$. Then

$$tX_1 + U = t(X_1 + t^{-1}U) \subset t(X_1 + U) \subset V$$

for $t \geqslant 1$. Let J denote the open interval $(1, \infty)$ in $\mathbf{R}$. Fix $\alpha \in C_c^\infty(U)$ and for any $\beta \in C_c^\infty(J)$, consider the function $\gamma_\beta \in C_c^\infty(V)$ given by

$$\gamma_\beta(tX_1 + X_2) = \beta(t)\,\alpha(X_2) \quad (t \in \mathbf{R},\ X_2 \in F).$$

Put
$$\sigma(\beta) = \int \gamma_\beta\, g\, dX = \int \beta(t)\,\alpha(X_2)\,g(tX_1 + X_2)\,dt\,dX_2$$

where dX_2 is the Euclidean measure on F normalized in such a way that $dX = dt\,dX_2$ for $X = tX_1 + X_2$. Then

$$\sigma(\beta) = \int e^{ct} q(t)\,\beta(t)\,dt \quad (\beta \in C_c^\infty(J))$$

where $c = \lambda(X_1)$ and
$$q(t) = \int p(tX_1 + X_2)\,\alpha(X_2)\,e^{\lambda(X_2)}\,dX_2.$$

Since $p \neq 0$, α can obviously be so selected that $q \neq 0$. Moreover since g is tempered on V, it is easy to see that σ is a tempered distribution on J. Hence it would be sufficient to prove the following lemma.

LEMMA 16. *Fix $c \in \mathbf{C}$, $t_0 \in \mathbf{R}$ and let $q \neq 0$ be a (complex-valued) polynomial function on $\mathbf{R}$. Then if the function $q(t)\,e^{ct}$ $(t \in \mathbf{R})$ is tempered on the open interval $J = (t_0, \infty)$, we can conclude that $\Re c \leqslant 0$.*

Put
$$T(\beta) = \int \beta(t)\,q(t)\,e^{ct}\,dt \quad (\beta \in C_c^\infty(J))$$

and $D = d/dt$. Let
$$T_0 = (D - c)^d T$$

where $d = d^\circ q$. Then T_0 is also a tempered distribution on J. But

$$(D-c)^d (qe^{ct}) = e^{ct} D^d q = ae^{ct}$$

where a is a nonzero constant. Hence it would be enough to consider the case when $q=1$. Then T being tempered, we can choose a number $A \geqslant 0$ and an integer $r \geqslant 0$ such that

$$\left| \int \alpha(t) e^{ct} dt \right| \leqslant A \sum_{0 \leqslant m, n \leqslant r} \sup |t^m D^n \alpha|$$

for $\alpha \in C_c^\infty(J)$. Let $c = 2c_1 + (-1)^{\frac{1}{2}} c_2$ where $c_i \in \mathbf{R}$ $(i=1,2)$. We have to show that $c_1 \leqslant 0$. So let us assume that $c_1 > 0$. Put

$$\alpha(t) = \beta(t) e^{-c't}$$

where $c' = c_1 + (-1)^{\frac{1}{2}} c_2$ and $\beta \in C_c^\infty(J)$. Then

$$|D^n \alpha| = e^{-c_1 t} |(D - c')^n \beta|.$$

Therefore we can select a number $A_1 \geqslant 0$ such that

$$\left| \int \beta(t) e^{c_1 t} dt \right| \leqslant A_1 \sum_{0 \leqslant m, n \leqslant r} \sup e^{-c_1 t} |t^m D^n \beta|$$

for all $\beta \in C_c^\infty(J)$.

Now fix a function $f \in C^\infty(\mathbf{R})$ such that 1) $0 \leqslant f \leqslant 1$, 2) $f(t) = 0$ if $t \leqslant 0$ and 3) $f(t) = 1$ if $t \geqslant 1$. For any $M > t_0 + 2$, define

$$\beta_M(t) = f(t - t_0 - 1) f(M + 1 - t) \quad (t \in \mathbf{R}).$$

Then $\beta_M \in C_c^\infty(J)$ and
$$\int \beta_M(t) e^{c_1 t} dt \geqslant \int_{t_0+2}^M e^{c_1 t} dt.$$

On the other hand
$$\sup |e^{-c_1 t} t^m D^n \beta_M| \leqslant a_m b_n$$

where
$$a_m = \sup_{t \geqslant t_0} |t^m e^{-c_1 t}|, \quad b_n = 2^n \max_{0 \leqslant k \leqslant n} \sup |D^k f|^2.$$

Therefore
$$\left| \int \beta_M(t) e^{c_1 t} dt \right| \leqslant A' \sum_{0 \leqslant m \leqslant r} a_m \sum_{0 \leqslant n \leqslant r} b_n = B \quad \text{(say)}.$$

This proves that
$$B \geqslant \int_{t_0+2}^M e^{c_1 t} dt.$$

But as $M \to +\infty$, the right side tends to $+\infty$ giving a contradiction. This completes the proof.

Let U be an open subset of E and $\mathbf{C}(U)$ the space of all C^∞ functions f on U such that

$$\nu_D(f) = \sup |Df| < \infty$$

for all $D \in \mathfrak{D}(E_c)$. The seminorms ν_D $(D \in \mathfrak{D}(E_c))$ define the structure of a locally convex space on $C(U)$.

It is well known (see [3, p. 93]) that the inclusion mapping of $C_c^\infty(E)$ into $C(E)$ is continuous and the image is dense in $C(E)$. Hence tempered distributions on E are the same as continuous linear functions on $C(E)$.

§ 9. Proof of Lemma 17

We now return to the notation of §2. Let $\mathfrak{h}$ be a Cartan subalgebra of $\mathfrak{g}$ and define g as in [2 (l), Theorem 2].

LEMMA 17. *Suppose T is tempered on Ω. Then we can choose an integer $q \geqslant 0$ such that $\pi^q g$ is tempered on $\Omega \cap \mathfrak{h}'$.*

Let A be the Cartan subgroup of G corresponding to $\mathfrak{h}$ and $x \to x^*$ the natural projection of G on $G^* = G/A$. The group W_G operates on G^* on the right in the usual way (see [2 (1), § 9]). Fix an invariant measure dx^* on G^* and a function $\alpha_0 \in C_c^\infty(G^*)$ such that

$$\int \alpha_0(x^*) \, dx^* = 1.$$

Put
$$\alpha(x^*) = [W_G]^{-1} \sum_{s \in W_G} \alpha_0(x^* s).$$

Select a compact set C in G such that $\operatorname{Supp} \alpha \subset C^*$ and $C^* s = C^*$ for $s \in W_G$ and, for any $\beta \in C_c^\infty(\mathfrak{h}')$, define a function $f_\beta \in C_c^\infty(\mathfrak{g})$ as follows.

$$f_\beta(x^* H) = [W_G]^{-1} \alpha(x^*) \sum_{s \in W_G} \beta(s^{-1} H)$$

for $x^* \in C^*$ and $H \in \operatorname{Supp} \beta$ and $\operatorname{Supp} f_\beta \subset (\operatorname{Supp} \beta)^{C^*}$.

Now define $f(x:X) = f(X^x)$ as usual ($x \in G$, $X \in \mathfrak{g}$) for any $f \in C^\infty(\mathfrak{g})$. Fix $D \in \mathfrak{D}(\mathfrak{g}_c)$. Then

$$f(xH; D) = f(x : H; D^{x^{-1}}) \quad (H \in \mathfrak{h})$$

and it is clear that
$$D^{x^{-1}} = \sum_{1 \leqslant i \leqslant r} a_i(x) D_i \quad (x \in G)$$

where $a_1, \ldots, a_r$ are analytic functions on G and $D_1, \ldots, D_r$ are linearly independent elements in $\mathfrak{D}(\mathfrak{g}_c)$. Hence

$$f(xH; D) = \sum_i a_i(x) f(x : H; D_i).$$

On the other hand if $\mathfrak{q}=[\mathfrak{h},\mathfrak{g}]$, we can choose (see [2 (j), § 2]) an integer $m \geqslant 0$ and elements $q_{ij} \in \mathfrak{S}_+(\mathfrak{q}_c)$, $\xi_{ij} \in \mathfrak{D}(\mathfrak{h}_c)$ $(1 \leqslant j \leqslant N)$ such that

$$f(x:H;D_i) = \pi(H)^{-m} \sum_j f(x;q_{ij}:H;\xi_{ij}) \quad (1 \leqslant i \leqslant r)$$

for $f \in C^\infty(\mathfrak{g})$, $x \in G$ and $H \in \mathfrak{h}'$. Put $\alpha(x) = \alpha(x^*)$ $(x \in G)$. Then if $x \in C$ and $H \in \mathfrak{h}'$, we get

$$f_\beta(xH;D) = \pi(H)^{-m} \sum_{i,j} a_i(x)\,\alpha(x; q_{ij})\,\beta_0(H;\xi_{ij})$$

where
$$\beta_0(H) = [W_G]^{-1} \sum_{s \in W_G} \beta(s^{-1}H).$$

Since C is compact, it is obvious that

$$\sup |Df_\beta| \leqslant B \sum_{i,j} \sup |\pi^{-m} \xi_{ij} \beta_0| \quad (\beta \in C_c^\infty(\mathfrak{h}')),$$

where B is a constant which depends only on D. Thus we have obtained the following result.

LEMMA 18. *For any* $D \in \mathfrak{D}(\mathfrak{g}_c)$, *we can choose an integer* $m \geqslant 0$ *and a finite number of elements* $\xi_i \in \mathfrak{D}(\mathfrak{h}_c)$ $(1 \leqslant i \leqslant N)$ *such that*

$$\sup |Df_\beta| \leqslant \sum_{1 \leqslant i \leqslant N} \sup |\pi^{-m} \xi_i \beta|$$

for all $\beta \in C_c^\infty(\mathfrak{h}')$.

Now we come to the proof of Lemma 17. Since T is tempered, there exist $D_i \in \mathfrak{D}(\mathfrak{g}_c)$ $(1 \leqslant i \leqslant r)$ such that
$$|T(f)| \leqslant \sum_i \sup |D_i f|$$

for all $f \in C_c^\infty(\Omega)$. Therefore by Lemma 18, we can choose an integer $m_0 \geqslant 0$ and elements $\xi_j \in \mathfrak{D}(\mathfrak{h}_c)$ $(1 \leqslant j \leqslant N)$ such that

$$|T(f_\beta)| \leqslant \sum_{1 \leqslant i \leqslant r} \sup |D_i f_\beta| \leqslant \sum_{1 \leqslant j \leqslant N} \sup |\pi^{-m_0} \xi_j \beta|$$

for $\beta \in C_c^\infty(\Omega \cap \mathfrak{h}')$. On the other hand

$$T(f_\beta) = \int f_\beta F\,dx = c \int \varepsilon_R \psi_{f_\beta} g\,dH$$

where $c = c(\mathfrak{h})$ in the notation of § 2. Moreover

$$\psi_{f_\beta}(H) = \varepsilon_R(H)\,\pi(H) \int f_\beta(x^*H)\,dx^* = \varepsilon_R(H)\,\pi(H)\,\beta_0(H) \quad (H \in \mathfrak{h}').$$

Hence
$$T(f_\beta) = c \int \pi \beta_0 g \, dH = c \int \pi \beta g \, dH$$

if we take into account the fact that $g^s = \varepsilon(s) g$ $(s \in W_G)$. Put $\gamma = \pi^{m-1} \beta$ $(m \geqslant 1)$. Then

$$\left| \int \beta \pi^m g \, dH \right| = |c^{-1} T(f_\gamma)| \leqslant |c|^{-1} \sum_j \sup |\pi^{-m_0} \xi_j (\pi^{m-1} \beta)|$$

for $\beta \in C_c^\infty(\Omega \cap \mathfrak{h}')$. If m is sufficiently large, it is clear that $\pi^{-m_0} \xi_j \circ \pi^{m-1} \in \mathfrak{D}(\mathfrak{h}_c)$. This shows that $\pi^m g$ is tempered on $\Omega \cap \mathfrak{h}'$.

Fix a Euclidean norm $\|X\|$ $(X \in \mathfrak{g})$ on $\mathfrak{g}$ and for any Cartan subalgebra $\mathfrak{h}$ define $g^\mathfrak{h}$ as in [2 (l), Theorem 3].

LEMMA 19. *Suppose for every Cartan subalgebra $\mathfrak{h}$ of $\mathfrak{g}$ we can choose numbers $a \geqslant 0$ and $m \geqslant 0$ such that*
$$|g^\mathfrak{h}(H)| \leqslant a(1 + \|H\|)^m$$

for $H \in \Omega \cap \mathfrak{h}'(R)$. Then T is tempered.

We use the notation of Lemma 5 and put $g_i = g^{\mathfrak{h}_i}$ $(1 \leqslant i \leqslant r)$. Then

$$T(f) = \sum_i c_i \int \varepsilon_{R,i} g_i \psi_{f,i} d_i H \quad (f \in C_c^\infty(\Omega)).$$

Therefore we can choose $c \geqslant 0$ and an integer $M \geqslant 0$ such that

$$|T(f)| \leqslant c \sum_i \sup_{\mathfrak{h}_i'} (1 + \|H\|)^M |\psi_{f,i}(H)|$$

for $f \in C_c^\infty(\Omega)$. Our assertion now follows immediately from [2 (d), Theorem 3].

§ 10. An auxiliary result

Let $\mathfrak{g}$ be a reductive Lie algebra over $\mathbf{C}$, $\mathfrak{h}$ a Cartan subalgebra of $\mathfrak{g}$ and W the Weyl group of $(\mathfrak{g}, \mathfrak{h})$.

LEMMA 20. *Let λ be a linear function on $\mathfrak{h}$ and α a root of $(\mathfrak{g}, \mathfrak{h})$. Suppose $s\lambda = \lambda - c\alpha$ for some $s \in W$ and $c \neq 0$ in $\mathbf{C}$. Then*[1] *$s\lambda = s_\alpha \lambda$.*

For the proof we may obviously assume that $\mathfrak{g}$ is semisimple. Let $\mathfrak{F}$ be the real vector space consisting of all linear functions μ on $\mathfrak{h}$ such that[2] $\mu(H_\beta) \in \mathbf{R}$ for every

---

[1] As usual s_α denotes the Weyl reflexion corresponding to α.

[2] H_β has the same meaning as in [2 (k), § 4].

18 – 652923. Acta mathematica. 113. Imprimé le 11 mai 1965.

root β. Fix an order in $\mathfrak{F}$ and first assume that $\lambda \in \mathfrak{F}$. Then $\sigma\lambda \in \mathfrak{F}$ for every $\sigma \in W$. Select $\sigma_0 \in W$ such that $\sigma_0 \lambda \geqslant \sigma\lambda$ for all $\sigma \in W$. Then if we put $\lambda' = \sigma_0 \lambda$, $s' = \sigma_0 s \sigma_0^{-1}$ and $\alpha' = \sigma_0 \alpha$, we obviously get $s'\lambda' = \lambda' - c\alpha'$. Moreover the relation $s\lambda = s_\alpha \lambda$ is equivalent to $s'\lambda' = s_{\alpha'} \lambda'$. Hence without loss of generality, we may assume that $\lambda \geqslant \sigma\lambda$ for all $\sigma \in W$. Since λ and $s\lambda$ are both in $\mathfrak{F}$, it is clear that $c \in \mathbf{R}$. Replacing α by $-\alpha$, if necessary, we may assume that $\alpha > 0$. Then $c > 0$ since $\lambda \geqslant s\lambda$. Now consider

$$s_\alpha s\lambda = s_\alpha \lambda + c\alpha = \lambda - c'\alpha$$

where $c' = 2(\lambda(H_\alpha)/\alpha(H_\alpha)) - c$. We claim $c' = 0$. For otherwise $c' > 0$ since $s_\alpha s\lambda \leqslant \lambda$. Moreover

$$s_\alpha \lambda = \lambda - (c + c')\alpha = s\lambda - c'\alpha.$$

Therefore
$$s^{-1} s_\alpha \lambda = \lambda - c' s^{-1}\alpha, \quad s^{-1}\lambda = \lambda + c s^{-1}\alpha.$$

Since c and c' are both positive, it follows that at least one of the two elements $s^{-1} s_\alpha \lambda$, $s^{-1}\lambda$ is higher than λ in our order. But this contradicts the condition that $\lambda \geqslant \sigma\lambda$ for all $\sigma \in W$. Hence $c' = 0$. This shows that $s_\alpha s\lambda = \lambda$ and therefore $s\lambda = s_\alpha \lambda$.

Now consider the general case. Then $\lambda = \lambda_R + (-1)^{\frac{1}{2}} \lambda_I$ and $c = a + (-1)^{\frac{1}{2}} b$ where $\lambda_R, \lambda_I \in \mathfrak{F}$ and $a, b \in \mathbf{R}$. The relation $s\lambda = \lambda - c\alpha$ implies that

$$s\lambda_R = \lambda_R - a\alpha, \quad s\lambda_I = \lambda_I - b\alpha.$$

Hence if $ab \neq 0$, we get $s\lambda_R = s_\alpha \lambda_R$, $s\lambda_I = s_\alpha \lambda_I$ from the above proof. Therefore $s\lambda = s_\alpha \lambda$ in this case. Now suppose $a \neq 0$, $b = 0$. Then $s\lambda_R = s_\alpha \lambda_R$ and $s\lambda_I = \lambda_I$ again from the above proof. Let W_0 be the subgroup of all $\sigma \in W$ such that $\sigma\lambda_I = \lambda_I$. For $\mu_1, \mu_2 \in \mathfrak{F}$, let $\langle \mu_1, \mu_2 \rangle$ denote the usual scalar product defined by means of the Killing form of $\mathfrak{g}$ so that

$$\langle \mu_1, \mu_2 \rangle = \sum_\beta \mu_1(H_\beta) \mu_2(H_\beta)$$

where β runs over all roots of $(\mathfrak{g}, \mathfrak{h})$. Then $\langle \sigma\mu_1, \mu_2 \rangle = \langle \mu_1, \sigma^{-1}\mu_2 \rangle$ for $\sigma \in W$. Hence

$$\langle s\lambda_R, \lambda_I \rangle = \langle \lambda_R, s^{-1}\lambda_I \rangle = \langle \lambda_R, \lambda_I \rangle.$$

But $\lambda_R - s\lambda_R = a\alpha$ and $a \neq 0$. Therefore $\langle \alpha, \lambda_I \rangle = 0$ and this implies that $s_\alpha \lambda_I = \lambda_I$. Hence $s\lambda = s_\alpha \lambda$. The case $a = 0$, $b \neq 0$ can be reduced to the one above by replacing λ by $(-1)^{\frac{1}{2}} \lambda$.

We shall need the above result for the proof of Lemma 26.

§ 11. Proof of Lemma 21

We return to the notation of § 2. So $\mathfrak{g}$ is a reductive Lie algebra over $\mathbf{R}$ and $\mathfrak{g}_1 = [\mathfrak{g}, \mathfrak{g}]$. Let $\mathfrak{a}$ be a Cartan subalgebra of $\mathfrak{g}$ and $\mathfrak{a}_R$ the set of all points of $\mathfrak{a}_1 = \mathfrak{a} \cap \mathfrak{g}_1$ where every root of $(\mathfrak{g}, \mathfrak{a})$ takes a real value. Similarly let $\mathfrak{a}_I$ be the set of those points of $\mathfrak{a}$ where all roots of $(\mathfrak{g}, \mathfrak{a})$ take pure imaginary values. Define $\theta, \mathfrak{k}, \mathfrak{p}$ and K as in [2 (m), § 16] corresponding to $\mathfrak{a}$. Then it is clear that $\mathfrak{a}_R = \mathfrak{a} \cap \mathfrak{p}$, $\mathfrak{a}_I = \mathfrak{a} \cap \mathfrak{k}$ and therefore $\mathfrak{a} = \mathfrak{a}_R + \mathfrak{a}_I$, where the sum is direct.

Define $\mathfrak{a}'(R)$ as usual (see [2 (k), § 4]) and fix a connected component $\mathfrak{a}_R^+$ of $\mathfrak{a}'(R) \cap \mathfrak{a}_R$. Let P_R be the set of all real roots of $(\mathfrak{g}, \mathfrak{a})$ which take only positive values on $\mathfrak{a}_R^+$. We can introduce compatible orders (see [2 (d), p. 195]) in the spaces of real-valued linear functions on $\mathfrak{a}_R$ and $\mathfrak{a}_R + (-1)^{\frac{1}{2}} \mathfrak{a}_I$ in such a way that all roots in P_R are positive. Let P be the set of all positive roots of $(\mathfrak{g}, \mathfrak{a})$ under this order.

Let $\mathfrak{m}$ be the centralizer of $\mathfrak{a}_I$ in $\mathfrak{g}$. Then $\mathfrak{m}$ is reductive in $\mathfrak{g}$ (see [2 (m), Cor. 3 of Lemma 26]) and it is obvious that P_R is the set of all positive roots of $(\mathfrak{m}, \mathfrak{a})$.

LEMMA 21. *Suppose $\mathfrak{g}$ has a Cartan subalgebra $\mathfrak{h}$ such that every root of $(\mathfrak{g}, \mathfrak{h})$ is imaginary. Then $\mathfrak{a}_R$ is a Cartan subalgebra of $\mathfrak{m}_1 = [\mathfrak{m}, \mathfrak{m}]$ and $\mathfrak{a}_I$ is the center of $\mathfrak{m}$.*

We can choose $x \in G$ such that $\mathfrak{h}^x \subset \mathfrak{k}$ (see [2 (d), § 8]). Since $\mathfrak{h}^x$ is maximal abelian in $\mathfrak{k}$ and $\mathfrak{a}_I \subset \mathfrak{k}$, we can select $k \in K$ such that $\mathfrak{h}^{kx} \supset \mathfrak{a}_I$. Hence without loss of generality we may suppose that $\mathfrak{a}_I \subset \mathfrak{h} \subset \mathfrak{k}$.

Let Q be the set of all positive roots of $(\mathfrak{g}, \mathfrak{h})$ and Q_0 the subset consisting of those $\beta \in Q$ which vanish identically on $\mathfrak{a}_I$. Then it is clear that

$$\mathfrak{m}_c = \mathfrak{h}_c + \sum_{\beta \in Q_0} (\mathbf{C} X_\beta + \mathbf{C} X_{-\beta})$$

in the usual notation (see [(2 (k), § 4]). Since $\mathfrak{h} \subset \mathfrak{k}$, both $\mathfrak{k}$ and $\mathfrak{p}$ are stable under $\mathrm{ad}\,\mathfrak{h}$ and therefore, for any root γ, X_γ lies either in $\mathfrak{k}_c$ or in $\mathfrak{p}_c$. Hence it is obvious that

$$[\mathfrak{m}_c, \mathfrak{m}_c] \supset [\mathfrak{h}_c, \mathfrak{m}_c] = \sum_{\beta \in Q_0} (\mathbf{C} X_\beta + \mathbf{C} X_{-\beta}) \supset \mathfrak{m}_c \cap \mathfrak{p}_c.$$

This shows that $\qquad\qquad \mathfrak{m}_1 \supset \mathfrak{m} \cap \mathfrak{p} \supset \mathfrak{a}_R.$

On the other hand let $\mathfrak{c}_\mathfrak{m}$ denote the center of $\mathfrak{m}$ and put $l = \mathrm{rank}\,\mathfrak{g}$. Since $\mathfrak{a} \subset \mathfrak{m}$, it is clear that

$$l = \mathrm{rank}\,\mathfrak{m} = \dim \mathfrak{c}_\mathfrak{m} + \mathrm{rank}\,\mathfrak{m}_1.$$

But $\mathfrak{a}_I \subset \mathfrak{c}_\mathfrak{m}$, $\mathfrak{a}_R \subset \mathfrak{m}_1$ and $\dim \mathfrak{a}_I + \dim \mathfrak{a}_R = \dim \mathfrak{a} = l$. Therefore we conclude that $\mathfrak{c}_\mathfrak{m} = \mathfrak{a}_I$ and $\mathfrak{a}_R$ is a Cartan subalgebra of $\mathfrak{m}_1$.

Select a fundamental system $(\alpha_1, \ldots, \alpha_m)$ of positive roots of $(\mathfrak{m}, \mathfrak{a})$ and let W_R be the subgroup[1] of $W(\mathfrak{g}/\mathfrak{a})$ generated by[2] s_α for $\alpha \in P_R$. Then $s_{\alpha_1}, \ldots, s_{\alpha_m}$ generate W_R and $m = \dim \mathfrak{a}_R$ from Lemma 21.

LEMMA 22. *Let μ be a linear function on $\mathfrak{a}_c$ which takes only real values on $\mathfrak{a}_R + (-1)^{\frac{1}{2}} \mathfrak{a}_I$ and suppose $\mu \geqslant s_{\alpha_i} \mu$ $(1 \leqslant i \leqslant m)$. Then $\mu \geqslant s\mu$ for $s \in W_R$ and $\mu(H) \geqslant 0$ for $H \in \mathfrak{a}_R^+$.*

Define linear functions μ_j $(1 \leqslant j \leqslant m)$ on $\mathfrak{a}$ as follows.

$$s_{\alpha_i} \mu_j = \mu_j - \delta_{ij} \alpha_j \quad (1 \leqslant i \leqslant m)$$

and $\mu_j = 0$ on $\mathfrak{a}_I$. Then $\mu_j \geqslant s\mu_j$ $(s \in W_R)$ and $\mu_j(H) \geqslant 0$ for $H \in \mathfrak{a}_R^+$ (see [2 (g), p. 280]). Let

$$s_{\alpha_i} \mu = \mu - c_i \alpha_i \quad (1 \leqslant i \leqslant m)$$

where $c_i \in \mathbf{R}$. Then $c_i \geqslant 0$. Put $\mu_0 = \sum_j c_j \mu_j$. Then $\mu = \mu_0$ on $\mathfrak{a}_R$. Therefore it is clear that

$$\mu - s\mu = \mu_0 - s\mu_0 \geqslant 0 \quad (s \in W_R)$$

and $\mu(H) = \mu_0(H) \geqslant 0$ for $H \in \mathfrak{a}_R^+$.

§ 12. Recapitulation of some elementary facts

Fix a Cartan subalgebra $\mathfrak{h}$ of $\mathfrak{g}$ and let $j = j_{\mathfrak{h}}$ denote the Chevalley isomorphism $p \to p_{\mathfrak{h}}$ of $I(\mathfrak{g}_c)$ onto $I(\mathfrak{h}_c)$ [2 (j), § 9]. Let λ be a linear function on $\mathfrak{h}_c$. Since every element q of $S(\mathfrak{h}_c)$ is a polynomial function on the dual space of $\mathfrak{h}_c$, we can consider its value $q(\lambda)$ at λ. Let $\chi_\lambda = \chi_\lambda^{\mathfrak{h}}$ denote the homomorphism $p \to p_{\mathfrak{h}}(\lambda)$ $(p \in I(\mathfrak{g}_c))$ of $I(\mathfrak{g}_c)$ into $\mathbf{C}$.

Let $W = W(\mathfrak{g}/\mathfrak{h})$. Then W operates on $\mathfrak{D}(\mathfrak{h}_c)$. We say that λ is regular if $s\lambda = \lambda^s \neq \lambda$ for $s \neq 1$ in W. It is well known that λ is singular or regular according as $\varpi(\lambda) = 0$ or not. Moreover if λ' is another linear function on $\mathfrak{h}_c$, then $\chi_\lambda = \chi_{\lambda'}$ if and only if $\lambda' = s\lambda$ for some $s \in W$.

Conversely let $\chi \neq 0$ be a homomorphism of $I(\mathfrak{g}_c)$ into $\mathbf{C}$. Then $\xi : q \to \chi(j^{-1}(q))$ $(q \in I(\mathfrak{h}_c))$ is a homomorphism of $I(\mathfrak{h}_c)$ into $\mathbf{C}$. Since $S(\mathfrak{h}_c)$ is a finite module over $I(\mathfrak{h}_c)$ (see [2 (c), Lemma 11]), ξ can be extended to a homomorphism of $S(\mathfrak{h}_c)$. Hence there exists a linear function λ on $\mathfrak{h}_c$ such that $\xi(q) = q(\lambda)$ for all $q \in I(\mathfrak{h}_c)$. This shows

---

[1] $W(\mathfrak{g}/\mathfrak{a})$ denotes the Weyl group of $(\mathfrak{g}, \mathfrak{a})$.
[2] See footnote 1, p. 265.

that $\chi = \chi_\lambda$. Moreover, as we have seen above, λ is unique up to an operation of W. We say that χ is regular if λ is regular. Put $p_0 = j^{-1}(\varpi^2)$. Then

$$\chi(p_0) = \varpi(\lambda)^2.$$

Hence χ is regular if and only if $\chi(p_0) \neq 0$. We note that p_0 is actually independent of $\mathfrak{h}$ and therefore the concept of the regularity of χ does not depend on the choice of $\mathfrak{h}$.

Let $\mathfrak{a}, \mathfrak{b}$ be two Cartan subalgebras of $\mathfrak{g}$ and y an element of the connected complex adjoint group G_c of $\mathfrak{g}_c$ such that $\mathfrak{b}_c = (\mathfrak{a}_c)^y$. Then y defines an isomorphism $D \to D^y$ of $\mathfrak{D}(\mathfrak{a}_c)$ onto $\mathfrak{D}(\mathfrak{b}_c)$.

LEMMA 23. *Let λ be a linear function on $\mathfrak{a}_c$. Then*

$$\chi_\lambda{}^{\mathfrak{a}} = \chi_{\lambda y}{}^{\mathfrak{b}}.$$

This follows from the obvious fact that $j_\mathfrak{b}(p) = (j_\mathfrak{a}(p))^y$ for $p \in I(\mathfrak{g}_c)$.

LEMMA 24. *Let U be a non-empty open connected subset of $\mathfrak{h}$ and λ a regular linear function on $\mathfrak{h}_c$. Suppose g is an analytic function on U such that $\partial(q)g = q(\lambda)g$ for all $q \in I(\mathfrak{h}_c)$. Then there exist unique complex numbers $c_s\,(s \in W)$ such that*

$$g(H) = \sum_{s \in W} \varepsilon(s)\, c_s\, e^{\lambda(s^{-1}H)} \quad (H \in U).$$

For a proof see [2 (c), p. 102].

§ 13. Proof of Lemma 26

Let $\mathfrak{z}$ be a subalgebra of $\mathfrak{g}$ such that 1) $\mathfrak{z}$ is reductive in $\mathfrak{g}$ and 2) rank $\mathfrak{z}$ = rank $\mathfrak{g}$. Let $\Omega_\mathfrak{z}$ be an open and completely invariant subset of $\mathfrak{z}$ and χ a regular homomorphism of $I(\mathfrak{g}_c)$ into $\mathbb{C}$. Let Ξ denote the analytic subgroup of G corresponding to $\mathfrak{z}$ and define the isomorphism $p \to p_\mathfrak{z}$ of $I(\mathfrak{g}_c)$ into $I(\mathfrak{z}_c)$ as in [2 (j), § 9]. Consider a distribution $T_\mathfrak{z}$ on $\Omega_\mathfrak{z}$ such that

1) $T_\mathfrak{z}$ is invariant under Ξ,
2) $\partial(p_\mathfrak{z})T_\mathfrak{z} = \chi(p)T_\mathfrak{z}$ for all $p \in I(\mathfrak{g}_c)$.

Fix a Euclidean measure dZ on $\mathfrak{z}$ and let $\Omega_\mathfrak{z}'$ denote the set of those points of $\Omega_\mathfrak{z}$ which are regular in $\mathfrak{z}$. Then by [2 (j), Lemma 19] and [2 (l), Theorem 1], $T_\mathfrak{z}$ coincides with an analytic function on $\Omega_\mathfrak{z}'$. We denote by $T_\mathfrak{z}(Z)$ the value of this function at any point $Z \in \Omega_\mathfrak{z}'$.

Let $\mathfrak{h}$ be a Cartan subalgebra of $\mathfrak{z}$, P the set of all positive roots of $(\mathfrak{g}, \mathfrak{h})$ and $P_\mathfrak{z}$ the subset consisting of all positive roots of $(\mathfrak{z}, \mathfrak{h})$. Put

$$\pi_{\mathfrak{z}} = \prod_{\alpha \in P_{\mathfrak{z}}} \alpha, \quad g_{\mathfrak{z}}(H) = \pi_{\mathfrak{z}}(H) T_{\mathfrak{z}}(H) \quad (H \in \Omega_{\mathfrak{z}}' \cap \mathfrak{h}).$$

Let $\mathfrak{h}'(\mathfrak{z}:R)$ denote the set of those points of $\mathfrak{h}$ where no real root in $P_{\mathfrak{z}}$ takes the value zero. Then by [2 (l), Theorem 2], $g_{\mathfrak{z}}$ extends to an analytic function on $\Omega_{\mathfrak{z}} \cap \mathfrak{h}'(\mathfrak{z}:R)$. Put $W = W(\mathfrak{g}/\mathfrak{h})$ and select a linear function λ on $\mathfrak{h}_c$ such that $\chi = \chi_\lambda$ in the notation of § 12.

LEMMA 25. *There exist locally constant functions* c_s $(s \in W)$ *on* $\Omega_{\mathfrak{z}} \cap \mathfrak{h}'(\mathfrak{z}:R)$ *such that*

$$g_{\mathfrak{z}} = \sum_{s \in W} \varepsilon(s) c_s e^{s\lambda}$$

on $\Omega_{\mathfrak{z}} \cap \mathfrak{h}'(\mathfrak{z}:R)$.

Since $(\partial(p_{\mathfrak{z}}) - \chi_\lambda(p)) T = 0$ $(p \in I(\mathfrak{g}_c))$, it follows (see the proof of [2 (l), Lemma 1]) that

$$(\partial(q) - q(\lambda)) g_{\mathfrak{z}} = 0 \quad (q \in I(\mathfrak{h}_c)).$$

Hence our assertion is an immediate consequence of Lemma 24.

Put $\zeta(Z) = \det(\operatorname{ad} Z)_{\mathfrak{g}/\mathfrak{z}}$ $(Z \in \mathfrak{g})$ and fix an element $H_0 \in \mathfrak{z}$ such that $\zeta(H_0) \neq 0$. Then the centralizers of H_0 in $\mathfrak{z}$ and $\mathfrak{g}$ are the same. Hence H_0 is semiregular in $\mathfrak{z}$ if and only if it is so in $\mathfrak{g}$. Now assume $H_0 \in \Omega_{\mathfrak{z}}$, $\zeta(H_0) \neq 0$ and H_0 is semiregular of noncompact type. We shall now use the notation of [2 (k), § 7] without further comment. Then it is clear that $\mathfrak{a}$ and $\mathfrak{b}$ are Cartan subalgebras of $\mathfrak{z}$. Put $W = W(\mathfrak{g}/\mathfrak{a})$ and choose a linear function λ on $\mathfrak{a}_c$ such that $\chi = \chi_\lambda^{\mathfrak{a}}$. Define G_c as in § 12 and let Ξ_c denote its complex-analytic subgroup corresponding to $\operatorname{ad} \mathfrak{z}_c$. Then it is clear that the element ν of [2 (k), § 7] lies in Ξ_c. We assume that the orders of roots are so chosen that([1]) $(\varpi^{\mathfrak{a}})^\nu = \varpi^{\mathfrak{b}}$ and $(\pi_{\mathfrak{z}}^{\mathfrak{a}})^\nu = \pi_{\mathfrak{z}}^{\mathfrak{b}}$. Then it follows from Lemma 24 that

$$\partial(\varpi^{\mathfrak{a}}) g_{\mathfrak{z}}^{\mathfrak{a}} = \varpi^{\mathfrak{a}}(\lambda) \sum_{s \in W} c_s^{\mathfrak{a}} e^{s\lambda}$$

on $\Omega_{\mathfrak{z}} \cap \mathfrak{a}'(\mathfrak{z}:R)$ and $\qquad \partial(\varpi^{\mathfrak{b}}) g_{\mathfrak{z}}^{\mathfrak{b}} = \varpi^{\mathfrak{a}}(\lambda) \sum_{s \in W} c_s^{\mathfrak{b}} \exp((s\lambda)^\nu)$

on $\Omega_{\mathfrak{z}} \cap \mathfrak{b}'(\mathfrak{z}:R)$. Here $c_s^{\mathfrak{b}}$ are locally constant functions on $\Omega_{\mathfrak{z}} \cap \mathfrak{h}'(\mathfrak{z}:R)$ $(\mathfrak{h} = \mathfrak{a}$ or $\mathfrak{b})$. Put

$$c_s^{\pm\alpha}(H_0) = \lim_{t \to +0} c_s^{\mathfrak{a}}(H_0 \pm tH')$$

and note that $H_0 \in \Omega_{\mathfrak{z}} \cap \mathfrak{b}'(\mathfrak{z}:R)$.

LEMMA 26. *For any*([2]) $s \in W$,

([1]) Here the notation is obvious (cf. [2 (l), Theorem 3]).

([2]) See footnote 1, p. 265.

$$c_s^{\alpha}(H_0) + c_{s_\alpha s}^{\alpha}(H_0) = c_s^{-\alpha}(H_0) + c_{s_\alpha s}^{-\alpha}(H_0) = c_s^{b}(H_0) + c_{s_\alpha s}^{b}(H_0).$$

Put $\sigma = \mathfrak{a} \cap \mathfrak{b}$, $\pi_\alpha = \alpha^{-1}\pi^{\mathfrak{a}}$, $\pi_\beta = \beta^{-1}\pi^{\mathfrak{b}}$ and let U be an open and convex neighborhood of H_0 in $\Omega_{\mathfrak{z}}$. We assume that U is so small that π_α and π_β never take the value zero on $U \cap \mathfrak{a}$ and $U \cap \mathfrak{b}$ respectively. Since $c_s^{\mathfrak{a}}$ ($s \in W$) is locally constant on $\Omega_{\mathfrak{z}} \cap \mathfrak{a}'(\mathfrak{z}:R)$, it is clear that

$$c_s^{\mathfrak{a}}(H) = c_s^{\pm\alpha}(H_0) \quad (H \in U \cap \mathfrak{a}')$$

according as $\alpha(H)$ is positive or negative. Similarly $c_s^{\mathfrak{b}}(H) = c_s^{\mathfrak{b}}(H_0)$ for $H \in U \cap \mathfrak{b}$. Moreover $\varpi^{\mathfrak{a}}(\lambda) \neq 0$ since λ is regular. Therefore if we apply [2 (l), Lemma 18] with $D = \partial(\varpi^{\mathfrak{a}})$ and recall that ν leaves σ pointwise fixed, we get

$$\sum_{s \in W} c_s^{\alpha}(H_0) \exp(\lambda(s^{-1}H))$$

$$= \sum_{s \in W} c_s^{-\alpha}(H_0) \exp(\lambda(s^{-1}H)) = \sum_{s \in W} c_s^{\mathfrak{b}}(H_0) \exp(\lambda(s^{-1}H)) \quad (H \in U \cap \sigma).$$

Let μ_s denote the restriction of $s\lambda$ on σ.

LEMMA 27. *Suppose s_1, s_2 are two distinct elements in W. Then $\mu_{s_1} = \mu_{s_2}$ if and only if $s_2 = s_\alpha s_1$.*

Since s_α leaves σ pointwise fixed, it is clear that $\mu_{s_1} = \mu_{s_2}$ if $s_2 = s_\alpha s_1$. Conversely suppose $\mu_{s_1} = \mu_{s_2}$. Then it is obvious that $s_2\lambda - s_1\lambda = c\alpha$ for some $c \in \mathbb{C}$. Since λ is regular, $c \neq 0$. Therefore it follows from Lemma 20 that $s_1^{-1}s_2\lambda = s_\gamma\lambda$ where $\gamma = s_1^{-1}\alpha$. But then $s_\gamma = s_1^{-1}s_\alpha s_1$ and therefore $s_2\lambda = s_\alpha s_1\lambda$. Since λ is regular, this implies that $s_2 = s_\alpha s_1$.

Now if we take into account the elementary fact that the exponentials of distinct linear functions on σ are linearly independent, Lemma 26 follows immediately from the relations proved above.

§ 14. Tempered and invariant eigendistributions

Let $\mathfrak{c}$ be the center and $\mathfrak{g}_1$ the derived algebra of $\mathfrak{g}$. Fix a number $c > 0$ and put $\mathfrak{g}_0 = \mathfrak{c}_0 + \mathfrak{g}_1(c)$. Here $\mathfrak{c}_0$ is a nonempty, open, connected subset of $\mathfrak{c}$ and $\mathfrak{g}_1(c)$ is defined as in [2 (m), § 3]. Then $\mathfrak{g}_0$ is a completely invariant open set in $\mathfrak{g}$.

Now take $\Omega_{\mathfrak{z}} = \mathfrak{z} \cap \mathfrak{g}_0$ and assume that there exists a Cartan subalgebra $\mathfrak{h}$ of $\mathfrak{z}$ and a linear function λ on $\mathfrak{h}_c$ such that 1) every root of $(\mathfrak{g}, \mathfrak{h})$ is imaginary, 2) λ takes only pure imaginary values on $\mathfrak{h}$ and 3) $\chi = \chi_\lambda^{\mathfrak{h}}$ in the notation of § 12. Since

$(\mathfrak{g}, \mathfrak{h})$ has no real roots and $\mathfrak{g}_0 \cap \mathfrak{h}$ is connected, we conclude from Lemma 25 that

$$g_{\mathfrak{z}}^{\mathfrak{h}}(H) = \sum_{s \in W(\mathfrak{g}/\mathfrak{h})} \varepsilon(s)\, c_s \exp(\lambda(s^{-1}H)) \quad (H \in \mathfrak{g}_0 \cap \mathfrak{h})$$

where $c_s \in \mathbf{C}$. Let C denote the additive subgroup of $\mathbf{C}$ generated by c_s $(s \in W(\mathfrak{g}/\mathfrak{h}))$. Fix a Cartan subalgebra $\mathfrak{a}$ of $\mathfrak{z}$ and a connected component $\mathfrak{a}^+$ of $\mathfrak{g}_0 \cap \mathfrak{a}'(\mathfrak{z} : R)$. Select a linear function $\lambda_\mathfrak{a}$ on $\mathfrak{a}_c$ such that $\chi_{\lambda_\mathfrak{a}}{}^\mathfrak{a} = \chi$. Then by Lemma 25 there exist unique complex numbers $c_s(\mathfrak{a}^+)$ such that

$$g_{\mathfrak{z}}{}^\mathfrak{a} = \sum_{s \in W(\mathfrak{g}/\mathfrak{a})} \varepsilon(s)\, c_s(\mathfrak{a}^+) \exp(s\lambda_\mathfrak{a})$$

on $\mathfrak{a}^+$.

LEMMA 28. *Suppose $T_\mathfrak{z}$ is tempered on $\mathfrak{z}_0 = \mathfrak{z} \cap \mathfrak{g}_0$. Then for a given $s \in W(\mathfrak{g}/\mathfrak{a})$, $c_s(\mathfrak{a}^+) = 0$ unless* [1]

$$\Re \lambda_\mathfrak{a}(s^{-1}H) \leqslant 0$$

for all $H \in \mathfrak{a}^+$. Moreover $c_s(\mathfrak{a}^+) \in C$.

COROLLARY. *Under the above conditions $g_{\mathfrak{z}}{}^\mathfrak{h} = 0$ implies $T_\mathfrak{z} = 0$.*

This is obvious from the lemma since $C = \{0\}$ if $\mathfrak{g}_\mathfrak{z}{}^\mathfrak{h} = 0$.

Fix a real quadratic form Q on $\mathfrak{g}$ such that 1) $Q(X) = \operatorname{tr}(\operatorname{ad} X)^2$ for $X \in \mathfrak{g}_1$, 2) Q is negative-definite on $\mathfrak{c}$ and 3) $\mathfrak{g}_1$ and $\mathfrak{c}$ are orthogonal under Q. Let U be any subspace of $\mathfrak{g}$ such that the restriction of Q on U is nondegenerate. Then we denote by $i_U(Q)$ the index of Q on U (see the proof of Lemma 12 of [2 (k)]).

Since $\mathfrak{c} \subset \mathfrak{a}$, it is obvious that the restriction of Q on $\mathfrak{a}$ is nondegenerate. We shall prove Lemma 28 by induction on $i_\mathfrak{a}(Q)$. Let $l = \operatorname{rank} \mathfrak{g}$. It is obvious that $i_\mathfrak{a}(Q) \geqslant -l$. Now if $i_\mathfrak{a}(Q) = -l$, it follows that all roots of $(\mathfrak{g}, \mathfrak{a})$ (and therefore also of $(\mathfrak{z}, \mathfrak{a})$) are imaginary. Hence (see [2 (d), p. 237]) $\mathfrak{a}$ is conjugate to $\mathfrak{h}$ under Ξ and so our assertion is obvious in this case. Therefore we may assume that $i_\mathfrak{a}(Q) > -l$ so that $\mathfrak{a}_R \neq \{0\}$. Since $\mathfrak{c}_0$ is connected, it is clear that

$$\mathfrak{a}^+ = \mathfrak{a}_I \cap \mathfrak{g}_0 + \mathfrak{a}_R{}^+(\mathfrak{z})$$

where $\mathfrak{a}_R{}^+(\mathfrak{z})$ is a connected component of $\mathfrak{a}'(\mathfrak{z} : R) \cap \mathfrak{a}_R$.

LEMMA 29. *Let $\mathfrak{a}_R(\mathfrak{z})$ be the set of points in $\mathfrak{a} \cap [\mathfrak{z}, \mathfrak{z}]$ where every root of $(\mathfrak{z}, \mathfrak{a})$ takes real values. Similarly let $\mathfrak{a}_I(\mathfrak{z})$ be the set of those points of $\mathfrak{a}$ where all roots of $(\mathfrak{z}, \mathfrak{a})$ take pure imaginary values. Then $\mathfrak{a}_R(\mathfrak{z}) = \mathfrak{a}_R$ and $\mathfrak{a}_I(\mathfrak{z}) = \mathfrak{a}_I$.*

---

[1] See footnote 1, p. 260.

It is obvious that $\mathfrak{a}_I(\mathfrak{z}) \supset \mathfrak{a}_I$. Moreover we may assume without loss of generality that $\mathfrak{a}_I(\mathfrak{z}) \subset \mathfrak{h}$ (see the proof of Lemma 21). Fix $H \in \mathfrak{a}_I(\mathfrak{z})$. Since every root of $(\mathfrak{g}, \mathfrak{h})$ is imaginary, it is clear that every eigenvalue of $\mathrm{ad}\, H$ is pure imaginary. Hence $H \in \mathfrak{a}_I$. This proves that $\mathfrak{a}_I(\mathfrak{z}) = \mathfrak{a}_I$. Let $\mathfrak{m}$ be the centralizer of $\mathfrak{a}_I$ in $\mathfrak{g}$ and put $\mathfrak{m}_{\mathfrak{z}} = \mathfrak{m} \cap \mathfrak{z}$. Then it follows from Lemma 21 that

$$\mathfrak{a}_R(\mathfrak{z}) = \mathfrak{a} \cap [\mathfrak{m}_{\mathfrak{z}}, \mathfrak{m}_{\mathfrak{z}}] \subset \mathfrak{a} \cap [\mathfrak{m}, \mathfrak{m}] = \mathfrak{a}_R.$$

Since $\mathfrak{a} = \mathfrak{a}_R + \mathfrak{a}_I = \mathfrak{a}_R(\mathfrak{z}) + \mathfrak{a}_I(\mathfrak{z})$ and both sums are direct (see § 11), we conclude that $\mathfrak{a}_R(\mathfrak{z}) = \mathfrak{a}_R$.

Let $P_R(\mathfrak{z})$ be the set of all real roots of $(\mathfrak{z}, \mathfrak{a})$ which take only positive values on $\mathfrak{a}_R{}^+(\mathfrak{z})$. Then $P_R(\mathfrak{z})$ can be regarded as the set of all positive roots of $(\mathfrak{m}_{\mathfrak{z}}, \mathfrak{a})$ and if $m = \dim \mathfrak{a}_R$, we can choose a fundamental system $(\alpha_1, \ldots, \alpha_m)$ of roots in $P_R(\mathfrak{z})$ (see § 11). Let $W_R(\mathfrak{z}/\mathfrak{a})$ be the subgroup of $W(\mathfrak{g}/\mathfrak{a})$ generated by s_α $(\alpha \in P_R(\mathfrak{z}))$. Then $W_R(\mathfrak{z}/\mathfrak{a})$ is also generated by s_{α_i} $(1 \leqslant i \leqslant m)$ and $\mathfrak{a}_R{}^+(\mathfrak{z})$ is exactly the set of those $H \in \mathfrak{a}_R$ where $\alpha_i(H) > 0$ $(1 \leqslant i \leqslant m)$.

Now fix i and choose $H_R \in \mathfrak{a}_R$ such that $\alpha_i(H_R) = 0$, $\alpha_j(H_R) > 0$ $(j \neq i, 1 \leqslant j \leqslant m)$ and $\alpha(H_R) \neq 0$ for any real root $\alpha \neq \pm \alpha_i$ of $(\mathfrak{g}, \mathfrak{a})$. Then $H_R \in \mathrm{Cl}(\mathfrak{a}_R{}^+(\mathfrak{z}))$ and we can obviously choose a connected component $\mathfrak{a}_R{}^+$ of $\mathfrak{a}'(R) \cap \mathfrak{a}_R$ such that 1) $\mathfrak{a}_R{}^+ \subset \mathfrak{a}_R{}^+(\mathfrak{z})$ and 2) $H_R \in \mathrm{Cl}(\mathfrak{a}_R{}^+)$. Define P and P_R as in § 11 corresponding to $\mathfrak{a}_R{}^+$ and select $H_I \in \mathfrak{a}_I \cap \mathfrak{g}_0$ in such a way that $\alpha(H_I) \neq 0$ for $\alpha \in P$ unless $\alpha \in P_R$. This is obviously possible. Then it is clear that $H_0 = H_I + H_R \in \mathrm{Cl}(\mathfrak{a}^+)$ and the only root in P which vanishes at H_0 is α_i. Therefore H_0 is semiregular in $\mathfrak{z}$ and $\zeta(H_0) \neq 0$. Define ν and $\mathfrak{b}$ as in § 13. Then $\mathfrak{h}$ is a Cartan subalgebra of $\mathfrak{z}$ and, as we have seen during the proof of [2 (k), Lemma 12], $i_{\mathfrak{b}}(Q) = i_{\mathfrak{a}}(Q) - 2$. Therefore the induction hypothesis is applicable to $\mathfrak{b}$ and so it follows from Lemma 26 that

$$c_s(\mathfrak{a}^+) + c_{s_{\alpha_i} s}(\mathfrak{a}^+) \in C \quad (s \in W(\mathfrak{g}/\mathfrak{a})).$$

Now fix $s \in W(\mathfrak{g}/\mathfrak{a})$. Then it follows from Lemma 23 that we can choose $y \in G_c$ such that $\mathfrak{a}_c = (\mathfrak{h}_c)^y$ and $s\lambda_{\mathfrak{a}} = \lambda^y$. Let $(\beta_1, \ldots, \beta_r)$ be a maximal set of linearly independent roots of $(\mathfrak{g}, \mathfrak{h})$. Since λ takes only pure imaginary values on $\mathfrak{h}$, we can choose $a_j \in \mathbf{R}$ such that

$$\lambda - \sum_{1 \leqslant j \leqslant r} a_j \beta_j = 0$$

on $\mathfrak{h}_1 = \mathfrak{h} \cap \mathfrak{g}_1$. Hence

$$s\lambda_{\mathfrak{a}} = \lambda^y = \sum_j a_j \beta_j{}^y$$

on $\mathfrak{a}_1 = \mathfrak{a} \cap \mathfrak{g}_1$. Since $\beta_j{}^y$ is a root of $(\mathfrak{g}, \mathfrak{a})$, it follows that $s\lambda_{\mathfrak{a}}$ takes only real values on $\mathfrak{a}_R$. Moreover $\lambda = \lambda^\nu$ on $\mathfrak{c}$ and so $\lambda_\mathfrak{a}(s^{-1}H)$ is pure imaginary for $H \in \mathfrak{a}_I$. Fix a non-empty open subset U of $\mathfrak{a}_I$ such that 1) $U \subset \mathfrak{a}_I \cap \mathfrak{g}_0$ and 2) all the roots of $(\mathfrak{g}, \mathfrak{a})$ which take the value zero on U, are real. Also fix a connected component $\mathfrak{a}_R^+$ of $\mathfrak{a}_R^+(\mathfrak{z}) \cap \mathfrak{a}'(R)$. Then it is clear that

$$U + \mathfrak{a}_R^+ \subset \mathfrak{a}' \cap \mathfrak{z}_0.$$

Since $T_\mathfrak{z}$ is tempered on $\mathfrak{z}_0$, it follows from Lemma 17 that $(\pi_\mathfrak{z}^\mathfrak{a})^q g_\mathfrak{z}^\mathfrak{a}$ is tempered on $\mathfrak{a}' \cap \mathfrak{z}_0$ for some $q \geqslant 0$. Fix a function $\gamma \in C_c^\infty(U)$ and put

$$g_\gamma(H) = \int_U \gamma(H_I)\,(\pi_\mathfrak{z}^\mathfrak{a}(H + H_I))^q\, g_\mathfrak{z}^\mathfrak{a}(H + H_I)\, dH_I \quad (H \in \mathfrak{a}_R^+)$$

where dH_I is a Euclidean measure on $\mathfrak{a}_I$. Then it is obvious that g_γ is tempered on $\mathfrak{a}_R^+$. Let μ_s and ν_s respectively denote the restrictions of $s\lambda_\mathfrak{a}$ ($s \in W(\mathfrak{g}/\mathfrak{a})$) on $\mathfrak{a}_R$ and $\mathfrak{a}_I$. Then it is clear that

$$g_\gamma(H) = \sum_{s \in W(\mathfrak{g}/\mathfrak{a})} \varepsilon(s)\, c_s(\mathfrak{a}^+)\, e^{\mu_s(H)} \int \gamma(H_I)\,(\pi_\mathfrak{z}^\mathfrak{a}(H + H_I))^q\, e^{\nu_s(H_I)}\, dH_I$$

for $H \in \mathfrak{a}_R^+$. Fix $s_0 \in W(\mathfrak{g}/\mathfrak{a})$ and suppose $\mu_{s_0}(H) > 0$ for some $H \in \mathfrak{a}_R^+$. Let W_0 be the set of all $s \in W(\mathfrak{g}/\mathfrak{a})$ such that $\mu_s = \mu_{s_0}$. Then it follows from Lemma 15 that

$$\sum_{s \in W_0} \varepsilon(s)\, c_s(\mathfrak{a}^+) \int \gamma(H_I)\,(\pi_\mathfrak{z}^\mathfrak{a}(H + H_I))^q\, e^{\nu_s(H_I)}\, dH_I = 0.$$

This being true for every $\gamma \in C_c^\infty(U)$, we conclude that

$$\sum_{s \in W_0} \varepsilon(s)\, c_s(\mathfrak{a}^+)\, e^{\nu_s} = 0.$$

But since $\mu_s = \mu_{s_0}$ ($s \in W_0$), it follows that

$$\sum_{s \in W_0} \varepsilon(s)\, c_s(\mathfrak{a}^+)\, e^{s\lambda_\mathfrak{a}} = 0.$$

However $\lambda_\mathfrak{a}$ being regular, this implies that $c_s(\mathfrak{a}^+) = 0$ ($s \in W_0$). Therefore in particular $c_{s_0}(\mathfrak{a}^+) = 0$. Since $\mathfrak{a}_R^+$ was an arbitrary component of $\mathfrak{a}_R^+(\mathfrak{z}) \cap \mathfrak{a}'(R)$, the first assertion of Lemma 28 is now obvious.

It remains to show that $c_s(\mathfrak{a}^+) \in C$ for all $s \in W(\mathfrak{g}/\mathfrak{a})$. Suppose this is false. Let W_1 be the set of all $s \in W(\mathfrak{g}/\mathfrak{a})$ such that $c_s(\mathfrak{a}^+) \notin C$. We have seen above that

$$c_s(\mathfrak{a}^+) + c_{s_{\alpha_i} s}(\mathfrak{a}^+) \in C \quad (s \in W(\mathfrak{g}/\mathfrak{a}),\ 1 \leqslant i \leqslant m).$$

Therefore $s_{\alpha_i} s \in W_1$ whenever $s \in W_1$. This shows that W_1 is a union of cosets of the form $W_R(\mathfrak{z}/\mathfrak{a})\, s$.

Introduce compatible orders on the spaces of real-valued linear functions on $\mathfrak{a}_R$ and $\mathfrak{a}_R + (-1)^{\frac{1}{2}} \mathfrak{a}_I$ corresponding to some connected component $\mathfrak{a}_R^+$ of $\mathfrak{a}_R^+(\mathfrak{z}) \cap \mathfrak{a}'(R)$ (see § 11). We have seen that $s\lambda_\mathfrak{a}$ $(s \in W(\mathfrak{g}/\mathfrak{a}))$ takes only real values on $\mathfrak{a}_R + (-1)^{\frac{1}{2}} \mathfrak{a}_I$. Choose $\sigma \in W_1$ such that $\mu = \sigma\lambda_\mathfrak{a} \geqslant s\lambda_\mathfrak{a}$ for all $s \in W_1$. Then $\mu \geqslant s\mu$ for all $s \in W_R(\mathfrak{z}/\mathfrak{a})$ and therefore we conclude from Lemma 22 (applied to $(\mathfrak{z}, \mathfrak{a})$) that $\mu(H) \geqslant 0$ for $H \in \mathfrak{a}_R^+(\mathfrak{z})$. However $c_\sigma(\mathfrak{a}^+) \neq 0$ since $\sigma \in W_I$. Therefore it follows from the above proof that $\mu(H) \leqslant 0$ for $H \in \mathfrak{a}_R^+(\mathfrak{z})$. This shows that $\mu = 0$ on $\mathfrak{a}_R$ and therefore $s_{\alpha_i}\mu = \mu$ $(1 \leqslant i \leqslant m)$. But since $\lambda_\mathfrak{a}$ is regular and $m = \dim \mathfrak{a}_R \geqslant 1$, this is impossible. The proof of Lemma 28 is now complete.

§ 15. Proof of Lemma 30

We keep to the notation of § 14. Let $\mathfrak{z}_1, \mathfrak{z}_2$ be two subalgebras of $\mathfrak{g}$ and $\mathfrak{h}$ a Cartan subalgebra of $\mathfrak{g}$ such that:

1) $\mathfrak{z}_i$ is reductive in $\mathfrak{g}$ $(i = 1, 2)$ and $\mathfrak{z}_1 \supset \mathfrak{z}_2 \supset \mathfrak{h}$.
2) Every root of $(\mathfrak{g}, \mathfrak{h})$ is imaginary.
3) If $\mathfrak{a}$ is any Cartan subalgebra of $\mathfrak{z}_2$, then every real root of $(\mathfrak{z}_1, \mathfrak{a})$ is also a root of $(\mathfrak{z}_2, \mathfrak{a})$.

Define χ as in § 14.

Let T_i be a tempered invariant distribution on $\mathfrak{z}_i \cap \mathfrak{g}_0$ such that

$$\partial(p_{\mathfrak{z}_i}) T_i = \chi(p) T_i \quad (p \in I(\mathfrak{g}_c), \ i = 1, 2).$$

Consider the set P of positive roots of $(\mathfrak{g}, \mathfrak{h})$ and let P_i denote the subset of those $\beta \in P$ which are roots of $(\mathfrak{z}_i, \mathfrak{h})$ $(i = 1, 2)$. Then $P \supset P_1 \supset P_2$. Put $\pi_i - \prod_{\alpha \in P_i} \alpha$. Then it is clear that π_1/π_2 is a polynomial function on $\mathfrak{h}_c$ which is invariant under the Weyl reflexions s_α for $\alpha \in P_2$. Therefore by Chevalley's theorem [2(c), Lemma 9] there exists a unique invariant polynomial function η_0 on $\mathfrak{z}_2$ which coincides with π_1/π_2 on $\mathfrak{h}$.

Put $\mathfrak{g}_0' = \mathfrak{g}_0 \cap \mathfrak{g}'$ where $\mathfrak{g}'$ denotes, as before, the set of all regular elements of $\mathfrak{g}$.

LEMMA 30. *Suppose* $T_2 = \eta_0 T_1$ *pointwise on* $\mathfrak{h} \cap \mathfrak{g}_0'$. *Then* $T_2 = \eta_0 T_1$ *pointwise on* $\mathfrak{z}_2 \cap \mathfrak{g}_0'$.

Let $\mathfrak{a}$ be a Cartan subalgebra of $\mathfrak{z}_2$. It would be enough to show that $T_2 = \eta_0 T_1$ pointwise on $\mathfrak{a} \cap \mathfrak{g}_0'$. We shall do this by induction on $i_\mathfrak{a}(Q)$ as in § 14. Let Ξ_2 be the analytic subgroup of G corresponding to $\mathfrak{z}_2$. If $i_\mathfrak{a}(Q) = -l$, then $\mathfrak{a}$ is conjugate to $\mathfrak{h}$ under Ξ_2 and so our assertion is obvious. Hence we may assume that $i_\mathfrak{a}(Q) > -l$ so that $m = \dim \mathfrak{a}_R \geqslant 1$.

We use the notation of § 14 corresponding to $\mathfrak{z} = \mathfrak{z}_1, \mathfrak{z}_2$. In particular $\mathfrak{g}_{\mathfrak{z}_i}{}^{\mathfrak{a}}$ is defined corresponding to T_i and we put $g_i{}^{\mathfrak{a}} = g_{\mathfrak{z}_i}{}^{\mathfrak{a}}$, $\pi_i{}^{\mathfrak{a}} = \pi_{\mathfrak{z}_i}{}^{\mathfrak{a}}$ $(i = 1, 2)$. It follows from our assumptions on $\mathfrak{z}_1, \mathfrak{z}_2$ that

$$\mathfrak{a}'(\mathfrak{z}_1 : R) = \mathfrak{a}'(\mathfrak{z}_2 : R).$$

Fix a connected component $\mathfrak{a}_R{}^+(\mathfrak{z}_2)$ of $\mathfrak{a}'(\mathfrak{z}_2 : R) \cap \mathfrak{a}_R$ and let $P_R(\mathfrak{z}_2)$ be the set of all real roots of $(\mathfrak{z}_2, \mathfrak{a})$ which take only positive values on $\mathfrak{a}_R{}^+(\mathfrak{z}_2)$. Select the fundamental system $(\alpha_1, \ldots, \alpha_m)$ of roots in $P_R(\mathfrak{z}_2)$ as in § 14.

Choose a linear function $\lambda_\mathfrak{a}$ on $\mathfrak{a}_c$ such that $\chi = \chi_{\lambda_\mathfrak{a}}{}^{\mathfrak{a}}$. Then by Lemma 25 there exist complex numbers $c_s(i)$ $(s \in W(\mathfrak{g}/\mathfrak{a}))$ such that

$$g_i{}^{\mathfrak{a}} = \sum_{s \in W(\mathfrak{g}/\mathfrak{a})} \varepsilon(s) \, c_s(i) \, e^{s\lambda_\mathfrak{a}} \quad (i = 1, 2)$$

on $\mathfrak{a}^+ = \mathfrak{g}_0 \cap \mathfrak{a}_I + \mathfrak{a}_R{}^+(\mathfrak{z}_2)$. It is obvious that

$$\eta_0 = a \, \pi_1{}^{\mathfrak{a}} / \pi_2{}^{\mathfrak{a}}$$

on $\mathfrak{a}$ where a is a constant $(a = \pm 1)$. Therefore it would be sufficient to show that $g_2{}^{\mathfrak{a}} = a g_1{}^{\mathfrak{a}}$ on $\mathfrak{a}^+$.

Fix j $(1 \leqslant j \leqslant m)$. Then (see § 14) we can select an element $H_0 \in \mathrm{Cl}(\mathfrak{a}^+)$ such that 1) $\alpha_j(H_0) = 0$ and 2) $\alpha(H_0) \neq 0$ for any root $\alpha \neq \pm \alpha_j$ of $(\mathfrak{g}, \mathfrak{a})$. It is clear that H_0 is semiregular in each of the three algebras $\mathfrak{z}_1, \mathfrak{z}_2$ and $\mathfrak{g}$. Define ν and $\mathfrak{b}$ as in § 13. Then $\mathfrak{b} \subset \mathfrak{z}_2$ and $i_\mathfrak{b}(Q) = i_\mathfrak{a}(Q) - 2$. Hence our induction hypothesis is applicable to $\mathfrak{b}$ and so it follows from Lemma 26 that

$$c_s(2) + c_{s_{\alpha_j} s}(2) = a \left\{ c_s(1) + c_{s_{\alpha_j} s}(1) \right\}$$

for $s \in W(\mathfrak{g}/\mathfrak{a})$.

In order to complete the proof we have to show that $c_s(2) = a c_s(1)$ for all $s \in W(\mathfrak{g}/\mathfrak{a})$. Suppose this is false. Let W_1 be the set of all $s \in W(\mathfrak{g}/\mathfrak{a})$ such that $c_s(2) \neq a c_s(1)$. Then it follows from the above result that if $s \in W_1$, the same holds for $s_{\alpha_j} s$ $(1 \leqslant j \leqslant m)$. Define $W_R(\mathfrak{z}_2/\mathfrak{a})$ as in § 14. Then W_1 is a union of cosets of the form $W_R(\mathfrak{z}_2/\mathfrak{a}) s$. Fix a connected component $\mathfrak{a}_R{}^+$ of $\mathfrak{a}_R{}^+(\mathfrak{z}_2) \cap \mathfrak{a}'(R)$ and define an order in the space $\mathfrak{F}$ of real-valued linear functions on $\mathfrak{a}_R + (-1)^{\frac{1}{2}} \mathfrak{a}_I$ corresponding to $\mathfrak{a}_R{}^+$ as in § 11. We have seen in § 14 that $s\lambda_\mathfrak{a} \in \mathfrak{F}$ for all $s \in W(\mathfrak{g}/\mathfrak{a})$. Choose $\sigma \in W_1$ such that $\mu = \sigma \lambda_\mathfrak{a} \geqslant s \lambda_\mathfrak{a}$ for all $s \in W_1$. Then $\mu \geqslant s\mu$ for $s \in W_R(\mathfrak{z}_2/\mathfrak{a})$. Therefore by Lemma 22, $\mu(H) \geqslant 0$ for $H \in \mathfrak{a}_R{}^+$. On the other hand since $\sigma \in W_1$, it is clear that $c_\sigma(1)$ and $c_\sigma(2)$ cannot both be zero. Therefore it follows from Lemma 28 that $\mu(H) \leqslant 0$ for $H \in \mathfrak{a}_R{}^+$. But this implies that $\mu = 0$ on $\mathfrak{a}_R$ and therefore $s_{\alpha_j}\mu = \mu$ $(1 \leqslant j \leqslant m)$. Since $m \geqslant 1$ and $\lambda_\mathfrak{a}$ is regular, this is impossible and thus Lemma 30 is proved.

We continue our assumption that $\mathfrak{h}_I = \mathfrak{h}$ and define θ as in [2 (m), § 16] corresponding to $\mathfrak{h}$.

LEMMA 31. *Let $\mathfrak{z}_1$ be a subalgebra of $\mathfrak{g}$ such that $\theta(\mathfrak{z}_1) = \mathfrak{z}_1$ and $\mathfrak{z}_1 \supset \mathfrak{h}$. Fix an element $H_1 \in \mathfrak{h}$ and let $\mathfrak{z}_2$ be the centralizer of H_1 in $\mathfrak{z}_1$. Then $\mathfrak{z}_1, \mathfrak{z}_2$ satisfy all the conditions required above.*

Since $\theta = 1$ on $\mathfrak{h}$, it is clear that $\theta(\mathfrak{z}_i) = \mathfrak{z}_i \supset \mathfrak{h}$ and hence $\mathfrak{z}_i$ $(i = 1, 2)$ is reductive in $\mathfrak{g}$ (see [2 (d), Lemma 10]). Let $\mathfrak{a}$ be a Cartan subalgebra of $\mathfrak{z}_2$. Then we know from Lemma 29 that $\mathfrak{a}_R(\mathfrak{z}_i) = \mathfrak{a}_R$ and $\mathfrak{a}_I(\mathfrak{z}_i) = \mathfrak{a}_I$ $(i = 1, 2)$. Let $\mathfrak{m}$ be the centralizer of $\mathfrak{a}_I$ in $\mathfrak{z}_2$. Since $H_1 \in \mathfrak{a}_I$, $\mathfrak{m}$ is also the centralizer of $\mathfrak{a}_I$ in $\mathfrak{z}_1$. Therefore the real roots of $(\mathfrak{z}_i, \mathfrak{a})$ are the same as the roots of $(\mathfrak{m}, \mathfrak{a})$ (see § 11). This proves the lemma.

§ 16. The distribution T_λ

Let $\mathfrak{b}$ be a Cartan subalgebra of $\mathfrak{g}$ and assume that every root of $(\mathfrak{g}, \mathfrak{b})$ is imaginary. Consider the space $\mathfrak{F}$ of all linear functions on $\mathfrak{b}_c$ which take only pure imaginary values on $\mathfrak{b}$. Define π, ϖ and $W = W(\mathfrak{g}/\mathfrak{b})$ as usual (corresponding to $\mathfrak{b}$) and let $\mathfrak{F}'$ be the set of all $\lambda \in \mathfrak{F}$ where $\varpi(\lambda) \neq 0$. Consider the subgroup $W_k = W_k(\mathfrak{g}/\mathfrak{b})$ of W generated by s_β corresponding to the compact roots β of $(\mathfrak{g}, \mathfrak{b})$ (see [2 (k), § 4]). Then $W_k = W_G$ (see Cor. 2 of [2 (m), Lemma 6]) in the notation of [2 (k), § 4].

THEOREM 2. *For any $\lambda \in \mathfrak{F}'$, there exists a unique distribution T_λ on $\mathfrak{g}$ with the following properties:*

1) T_λ *is invariant and tempered.*
2) $\partial(p) T_\lambda = p_\mathfrak{b}(\lambda) T_\lambda$ $(p \in I(\mathfrak{g}_c))$.
3) $T_\lambda(H) = \pi(H)^{-1} \sum_{s \in W_k} \varepsilon(s) e^{\lambda(s^{-1}H)}$ $(H \in \mathfrak{b}')$.

The uniqueness is obvious from the corollary of Lemma 28. Hence only the existence requires proof.

First assume that $\mathfrak{g}$ is semisimple. We identify $\mathfrak{g}_c$ and $\mathfrak{b}_c$ with their respective duals by means of the Killing form of $\mathfrak{g}$ (see [2 (j), § 6]). Fix a Euclidean measure dX on $\mathfrak{g}$ and put

$$\hat{f}(Y) = \int f(X) \exp((-1)^{\frac{1}{2}} B(X, Y)) \, dX \quad (Y \in \mathfrak{g})$$

for $f \in C(\mathfrak{g})$. (As usual $B(X, Y) = \mathrm{tr}(\mathrm{ad}\, X\, \mathrm{ad}\, Y)$ for $X, Y \in \mathfrak{g}_c$.) Moreover for any $H_0 \in \mathfrak{b}'$ define

$$\tau_{H_0}(f) = \psi_{\hat{f}}(H_0) = \pi(H_0) \int_{G^*} f(x^* H_0)\, dx^* \quad (f \in C(\mathfrak{g}))$$

in the notation of [2 (k), § 5] (for $\mathfrak{h} = \mathfrak{b}$). Then we know from [2 (d), Theorem 3] that the integral is absolutely convergent and τ_{H_0} is an invariant and tempered distribution on $\mathfrak{g}$ which satisfies (see [2 (d), p. 226]) the differential equations

$$\partial(p)\, \tau_{H_0} = p((-1)^{\frac{1}{2}} H_0)\, \tau_{H_0} \quad (p \in I(\mathfrak{g}_c)).$$

Fix $H_0 \in \mathfrak{b}'$ and let $\mathfrak{T}_{H_0}$ denote the space of all invariant and tempered distributions T on $\mathfrak{g}$ such that
$$\partial(p)\, T = p((-1)^{\frac{1}{2}} H_0)\, T \quad (p \in I(\mathfrak{g}_c)).$$

For any $T \in \mathfrak{T}_{H_0}$, let g_T denote the analytic function on $\mathfrak{b}$ (see [2 (l), Theorem 2]) given by
$$g_T(H) = \pi(H)\, T(H) \quad (H \in \mathfrak{b}').$$
Then by Lemma 25,

$$g_T(H) = \sum_{s \in W} \varepsilon(s)\, c_s(T)\, \exp\, ((-1)^{\frac{1}{2}}\, B(sH_0, H)) \quad (H \in \mathfrak{b})$$

where $c_s(T)$ are uniquely determined complex numbers. It is clear that $g_T{}^t = \varepsilon(t) g_T$ and therefore $c_{ts}(T) = c_s(T)$ for $t \in W_G = W_k$ and $s \in W$. On the other hand the linear mapping $T \to g_T$ is injective from the corollary of Lemma 28. Hence it is obvious that

$$\dim \mathfrak{T}_{H_0} \leqslant [W : W_k].$$

On the other hand it is clear that $\tau_{sH_0} \in \mathfrak{T}_{H_0}\, (s \in W)$. Put $r = [W : W_k]$ and select $s_i \in W\ (1 \leqslant i \leqslant r)$ such that
$$W = \bigcup_{1 \leqslant i \leqslant r} W_k s_i.$$

Write $\tau_i = \tau_{s_i H_0}$. Then we claim that $\tau_1, \ldots, \tau_r$ are linearly independent over C. Put

$$\sigma_i(f) = \psi_f(s_i H_0) \quad (f \in C(\mathfrak{g})).$$

Since $f \to \hat{f}$ is a topological mapping of $C(\mathfrak{g})$ onto itself, it would be enough to verify that the tempered distributions $\sigma_1, \ldots, \sigma_r$ are linearly independent. Since $s_i H_0$ is semisimple, the orbit $(s_i H_0)^G$ is closed in $\mathfrak{g}$ (see [1, p. 523]). Therefore it follows from the definition of σ_i that
$$\mathrm{Supp}\, \sigma_i = (s_i H_0)^G.$$

Now we claim that $(s_i H_0)^G \cap (s_j H_0)^G = \varnothing$ if $i \neq j$. For otherwise $s_i H_0 = (s_j H_0)^x$ for some $x \in G$. Since H_0 is regular, this implies that $s_i = s s_j$ for some $s \in W_G = W_k$. But this is impossible from the definition of $(s_1, \ldots, s_r)$. This shows that the sets Supp σ_i are disjoint and non-empty and therefore the distributions σ_i $(1 \leqslant i \leqslant r)$ are linearly independent.

So it is now obvious that $\dim \mathfrak{T}_{H_0} = r$ and $\tau_1, \ldots, \tau_r$ is a base for $\mathfrak{T}_{H_0}$. Let a_s $(s \in W)$ be given complex numbers such that $a_{ts} = a_s$ $(t \in W_k)$. Then it follows from the above result that we can choose a unique element $T \in \mathfrak{T}_{H_0}$ such that $a_s = c_s(T)$. Hence, in particular, there exists a distribution T in $\mathfrak{T}_{H_0}$ such that

$$g_T(H) = \sum_{s \in W_k} \varepsilon(s) \exp ((-1)^k B(sH_0, H)).$$

This proves Theorem 2 when $\mathfrak{g}$ is semisimple.

Now we come to the general case. Define $\mathfrak{g}_1$ and $\mathfrak{c}$ as before (see § 2), put $\mathfrak{b}_1 = \mathfrak{b} \cap \mathfrak{g}_1$ and let λ_1 denote the restriction of λ on $\mathfrak{b}_{1c}$. Fix Euclidean measures dC and dZ on $\mathfrak{c}$ and $\mathfrak{g}_1$ respectively such that $dX = dC\,dZ$ for $X = C + Z$ $(C \in \mathfrak{c}, Z \in \mathfrak{g}_1)$. Since $\mathfrak{g}_1$ is semisimple, there exists, from the above proof, an invariant and tempered distribution T_1 on $\mathfrak{g}_1$ such that $\partial(p)T_1 = p_{\mathfrak{b}}(\lambda)T_1$ $(p \in I(\mathfrak{g}_{1c}))$ and

$$\pi(H)T_1(H) = \sum_{s \in W_k} \varepsilon(s) \exp (\lambda_1(s^{-1}H)) \quad (H \in \mathfrak{b}_1 \cap \mathfrak{g}').$$

Put
$$T_\lambda(f) = T_1(f_1) \quad (f \in C_c^\infty(\mathfrak{g}))$$

where
$$f_1(Z) = \int f(Z + C)\, e^{\lambda(C)}\, dC \quad (Z \in \mathfrak{g}_1).$$

Since λ takes only pure imaginary values on $\mathfrak{c}$, it is clear that T_λ satisfies all the conditions of Theorem 2.

Fix a Cartan subalgebra $\mathfrak{a}$ of $\mathfrak{g}$ and an element $y \in G_c$ such that $(\mathfrak{b}_c)^y = \mathfrak{a}_c$. For any $\lambda \in \mathfrak{F}'$, define the analytic function $g_\lambda{}^\mathfrak{a}$ on $\mathfrak{a}'(R)$ corresponding to T_λ as usual so that

$$g_\lambda{}^\mathfrak{a}(H) = \pi^\mathfrak{a}(H)T_\lambda(H) \quad (H \in \mathfrak{a}').$$

Fix a connected component $\mathfrak{a}^+$ of $\mathfrak{a}'(R)$. Then by Lemmas 25 and 28,

$$g_\lambda{}^\mathfrak{a} = \sum_{s \in W} \varepsilon(s)\, c(s : \lambda : \mathfrak{a}^+)\, e^{(s\lambda)^y}$$

on $\mathfrak{a}^+$ where $c(s : \lambda : \mathfrak{a}^+) \in \mathbf{Z}$.

LEMMA 32. *For fixed $s \in W$ and $\mathfrak{a}^+$, the integer $c(s:\lambda:\mathfrak{a}^+)$ $(\lambda \in \mathfrak{F}')$ depends only on the connected component of λ in $\mathfrak{F}'$.*

In view of the last part of the proof of Theorem 2, it is clear that it would be sufficient to consider the case when $\mathfrak{g}$ is semisimple. Define τ_{H_0} as above for $H_0 \in \mathfrak{b}'$. Then by [2 (l), Theorem 1] there exists an analytic function F_{H_0} on $\mathfrak{g}'$ such that

$$\tau_{H_0}(f) = \psi \hat{f}(H_0) = \int F_{H_0}(X) f(X)\, dX \quad (f \in C_c^\infty(\mathfrak{g})).$$

We know from Lemma 25 that

$$\pi^{\mathfrak{a}}(H) F_{H_0}(H) = \sum_{s \in W} \varepsilon(s) \, a_s(H_0) \exp((-1)^{\frac{1}{2}} B(sH_0, y^{-1}H))$$

for $H \in \mathfrak{a}^{+'} = \mathfrak{a}^+ \cap \mathfrak{g}'$. Here $a_s(H_0)$ are uniquely determined complex numbers. Moreover we know from [2 (d), pp. 229–231] that a_s, regarded as functions on $\mathfrak{b}'$, are locally constant. By considering, in particular, the case $\mathfrak{a} = \mathfrak{b}$, we get

$$\pi(H) F_{H_0}(H) = \sum_{s \in W} \varepsilon(s) \, b_s(H_0) \exp((-1)^{\frac{1}{2}} B(sH_0, H))$$

for $H, H_0 \in \mathfrak{b}'$. Here b_s are certain locally constant functions on $\mathfrak{b}'$.

Now define $s_1 = 1, s_2, \ldots, s_r$ as in the proof of Theorem 2 and put

$$b_{ij}(H_0) = b_{s_i s_j^{-1}}(s_j H_0) \quad (1 \leqslant i, j \leqslant r, H_0 \in \mathfrak{b}').$$

Fix $H_0 \in \mathfrak{b}'$. Since $b_{ts}(H_0) = b_s(H_0)$ $(t \in W_k)$ and $\tau_{s_i H_0}$ $(1 \leqslant i \leqslant r)$ are linearly independent, it follows from the proof of Theorem 2 that the matrix $(b_{ij}(H_0))_{1 \leqslant i, j \leqslant r}$ is non-singular. Let $(b^{ij}(H_0))_{1 \leqslant i, j \leqslant r}$ denote its inverse. Put $b^j = b^{j1}$ and

$$T_{H_0} = \sum_{1 \leqslant j \leqslant r} b^j(H_0) \tau_{s_j H_0} \quad (H_0 \in \mathfrak{b}').$$

Then it is obvious that $T_{H_0} \in \mathfrak{T}_{H_0}$ (in the notation of the proof of Theorem 2) and

$$\pi(H) T_{H_0}(H) = \sum_{s \in W_k} \varepsilon(s) \exp((-1)^{\frac{1}{2}} B(sH_0, H)) \quad (H \in \mathfrak{b}')$$

for $H_0 \in \mathfrak{b}'$. Hence it follows from Theorem 2 that $T_{H_0} = T_\lambda$ where λ is the element of $\mathfrak{F}'$ given by $\lambda(H) = (-1)^{\frac{1}{2}} B(H_0, H)$ $(H \in \mathfrak{b})$. Therefore

$$g_\lambda{}^{\mathfrak{a}}(H) = \sum_{1 \leqslant j \leqslant r} b^j(H_0) \, \pi^{\mathfrak{a}}(H) F_{s_j H_0}(H) \quad (H \in \mathfrak{a}')$$

and this shows that

$$c(s:\lambda:\mathfrak{a}^+) = \sum_{1 \leqslant j \leqslant r} \varepsilon(s_j) \, b^j(H_0) \, a_{ss_j^{-1}}(s_j H_0) \quad (s \in W).$$

Since b^j and a_s are locally constant on $\mathfrak{b}'$, the assertion of the lemma is now obvious.

$\mathfrak{F}^+$ being any connected component of $\mathfrak{F}'$, we denote by $c(s:\mathfrak{F}^+:\mathfrak{a}^+)$ the integer $c(s:\lambda:\mathfrak{a}^+)$ $(\lambda \in \mathfrak{F}^+)$. Put

$$\phi_\lambda = \varpi(\lambda)^{-1} \nabla_\mathfrak{g} F_\lambda \quad (\lambda \in \mathfrak{F}')$$

where F_λ is the analytic function on $\mathfrak{g}'$ corresponding to T_λ and $\nabla_\mathfrak{g}$ is defined as before (see § 2).

LEMMA 33.
$$\phi_\lambda = \sum_{s \in W} c(s:\mathfrak{F}^+:\mathfrak{a}^+)\, e^{(s\lambda)\nu}$$

on $\mathfrak{a}^+$ for $\lambda \in \mathfrak{F}^+$.

This is obvious from the definition of $\nabla_\mathfrak{g}$ and the above formula for $g_\lambda{}^\mathfrak{a}$.

For any $s \in W$ define an element $s^\nu \in W(\mathfrak{g}/\mathfrak{a})$ as follows:

$$(sH)^\nu = s^\nu H^\nu \quad (H \in \mathfrak{b}_c).$$

Then $s \to s^\nu$ is an isomorphism of $W(\mathfrak{g}/\mathfrak{a})$ whose inverse we denote by $t \to t^{\nu^{-1}}$ $(t \in W(\mathfrak{g}/\mathfrak{a}))$. Define the subgroup $W_G(\mathfrak{g}/\mathfrak{a})$ of $W(\mathfrak{g}/\mathfrak{a})$ as usual (see [2 (k), § 4]). We have seen above that $W_k = W_G = W_G(\mathfrak{g}/\mathfrak{b})$.

COROLLARY. *Fix* $s \in W$, $t \in W_G(\mathfrak{g}/\mathfrak{a})$ *and* $u \in W_k$. *Then*

$$c(t^{\nu^{-1}} su^{-1} : u\,\mathfrak{F}^+ : t\,\mathfrak{a}^+) = c(s:\mathfrak{F}^+:\mathfrak{a}^+).$$

Fix $\lambda \in \mathfrak{F}^+$. Then it is clear from Theorem 2 that $T_{u\lambda} = \varepsilon(u)\, T_\lambda$ and therefore $\phi_{u\lambda} = \phi_\lambda$. Moreover ϕ_λ is invariant under G and therefore its restriction on $\mathfrak{a}$ is invariant under $W_G(\mathfrak{g}/\mathfrak{a})$. Our assertion is an immediate consequence of these facts.

§ 17. Application of Theorem 1 to T_λ

Now we use the notation of § 2 and assume that $\mathfrak{h}_1 = \mathfrak{b}$. Let $m_i(R)$ denote the number of positive real roots of $(\mathfrak{g}, \mathfrak{h}_i)$ $(1 \leqslant i \leqslant r)$ and put $m = \tfrac{1}{2}$ (dim $\mathfrak{g}$ − rank $\mathfrak{g}$). For any $\lambda \in \mathfrak{F}'$, let $\phi_{\lambda,i}$ denote the restriction of ϕ_λ on $\mathfrak{h}_i$.

Define numbers $c_i > 0$ by the relation

$$\int_\mathfrak{g} f(X)\, dX = \sum_{1 \leqslant i \leqslant r} c_i(-1)^{m_i(I)} \int \varepsilon_{R,i}\, \pi_i\, \psi_{f,i}\, d_i H \quad (f \in C_c^\infty(\mathfrak{g}))$$

where $m_i(I)$ is the number of positive imaginary roots of $(\mathfrak{g}, \mathfrak{h}_i)$ (see [2 (k), Cor. 1 of Lemma 30]). Also put $dH = d_1 H$.

19 − 652923. *Acta mathematica.* 113. Imprimé le 11 mai 1965.

LEMMA 34. *For any $f \in C_c^\infty(\mathfrak{g})$ and $\lambda \in \mathfrak{F}'$,*

$$c_1 [W_k] \int_{\mathfrak{b}} \partial(\varpi) \, \psi_f \, e^\lambda \, dH = c_1 \int_{\mathfrak{b}} \partial(\varpi) \, \psi_f \sum_{s \in W_k} e^{s\lambda} \, dH$$

$$= \varpi(\lambda) \, T_\lambda(f) - \sum_{2 \leqslant i \leqslant r} (-1)^{m_i(R)} c_i \int_{\mathfrak{h}_i} \varepsilon_{R,i} \, \partial(\varpi_i) \, \psi_{f,i} \cdot \phi_{\lambda,i} \, d_i H.$$

Since the number of positive complex roots of $(\mathfrak{g}, \mathfrak{h}_i)$ is even (see the proof of Lemma 9 of [2 (k)]), it follows that

$$m_i(R) + m_i(I) \equiv m \bmod 2.$$

Hence $$(-1)^m \int_{\mathfrak{g}} f(X) \, dX = \sum_{1 \leqslant i \leqslant r} c_i (-1)^{m_i(R)} \int \varepsilon_{R,i} \, \pi_i \, \psi_{f,i} \, d_i H \quad (f \in C_c^\infty(\mathfrak{g})).$$

Moreover $\partial(\varpi) \, \psi_f$ is invariant under $W_k = W_G$ (see [2 (k), § 6]) and

$$\phi_{\lambda,1} = \sum_{s \in W_k} e^{s\lambda}.$$

Therefore our assertion follows from Theorem 1 and the corollary of Lemma 4, if we take into account the fact that $\Box F_\lambda = \varpi(\lambda)^2 F_\lambda$.

Fix a connected component $\mathfrak{F}^+$ of $\mathfrak{F}'$. Then for any $\mu \in \mathrm{Cl}(\mathfrak{F}^+)$, we define a distribution $T_{\mu, \mathfrak{F}^+} = T_\mu^+$ as follows:

$$T_\mu^+(f) = \lim_{\lambda \to \mu} T_\lambda(f) \quad (f \in C_c^\infty(\mathfrak{g}))$$

where $\lambda \in \mathfrak{F}^+$. Put $g_{\lambda,i} = g_\lambda^{\mathfrak{h}_i}$. Then

$$T_\lambda(f) = (-1)^m \sum_{1 \leqslant i \leqslant r} (-1)^{m_i(R)} c_i \int \varepsilon_{R,i} \, \psi_{f,i} \, g_{\lambda,i} \, d_i H$$

and so it is obvious that the above limit exists and

$$T_\mu^+(f) = (-1)^m \sum_{1 \leqslant i \leqslant r} (-1)^{m_i(R)} c_i \int \varepsilon_{R,i} \, \psi_{f,i} \, g_{\mu,i}^+ \, d_i H$$

where $g_{\mu,i}^+$ is defined as follows. Fix i and put $\mathfrak{a} = \mathfrak{h}_i$. Then

$$g_{\mu,i}^+ = \lim_{\lambda \to \mu} g_\lambda^{\mathfrak{a}} = \sum_{s \in W} \varepsilon(s) \, c \, (s : \mathfrak{F}^+ : \mathfrak{a}^+) \, e^{(s\mu)\nu}$$

on any connected component $\mathfrak{a}^+$ of $\mathfrak{a}'(R)$. We know from Lemma 28 that $c(s : \mathfrak{F}^+ : \mathfrak{a}^+) = 0$ $(s \in W)$ unless $\Re(s\lambda)^\nu(H) \leqslant 0$ for all $H \in \mathfrak{a}^+$ and $\lambda \in \mathfrak{F}^+$. Therefore it is clear from the above formulas and Lemma 19 that T_μ^+ is an invariant and tempered distribution on $\mathfrak{g}$. Since $\partial(p) T_\lambda = p_\mathfrak{b}(\lambda) T_\lambda$, it follows immediately by going over to the limit that

$$\partial(p) T_\mu^+ = p_\mathfrak{b}(\mu) T_\mu^+ \quad (p \in I(\mathfrak{g}_c)).$$

For any Cartan subalgebra $\mathfrak{a}$ of $\mathfrak{g}$ define the function $(\phi_\mu^+)_\mathfrak{a}$ on $\mathfrak{a}'(R)$ by

$$(\phi_\mu^+)_\mathfrak{a} = \sum_{s \in W} c(s : \mathfrak{F}^+ : \mathfrak{a}^+) \, e^{(s\mu)\nu}$$

on $\mathfrak{a}^+$.

LEMMA 35.
$$|(\phi_\mu^+)_\mathfrak{a}| \leqslant \sum_{s \in W} |c(s : \mathfrak{F}^+ : \mathfrak{a}^+)|$$

on $\mathfrak{a}^+$.

Fix $H \in \mathfrak{a}^+$. Then if $\lambda \in \mathfrak{F}^+$, it follows from Lemma 28 that

$$|\phi_\lambda(H)| \leqslant \sum_{s \in W} |c(s : \mathfrak{F}^+ : \mathfrak{a}^+)|.$$

Our assertion now follows by letting λ tend to μ.

For $\mathfrak{a} = \mathfrak{h}_i$ we denote the function $(\phi_\mu^+)_\mathfrak{a}$ by $\phi_{\mu,i}^+$.

LEMMA 36. *For any* $f \in C_c^\infty(\mathfrak{g})$,

$$c_1[W_k] \int_\mathfrak{b} \partial(\varpi) \, \psi_f \, e^\mu \, dH = \varpi(\mu) \, T_\mu^+(f) - \sum_{2 \leqslant i \leqslant r} (-1)^{m_i(R)} c_i \int \varepsilon_{R,i} \, \partial(\varpi_i) \, \psi_{f,i} \cdot \phi_{\mu,i}^+ \, d_i H.$$

Take a variable element $\lambda \in \mathfrak{F}^+$ which converges to μ. Then our assertion follows immediately from Lemma 34 by taking limits.

§ 18. Proof of Lemma 41

As in § 14, let $\mathfrak{z}$ be a subalgebra of $\mathfrak{g}$ such that 1) $\mathfrak{z} \supset \mathfrak{b}$ and 2) $\mathfrak{z}$ is reductive in $\mathfrak{g}$. Fix a Euclidean measure dZ on $\mathfrak{z}$ and let $W_k(\mathfrak{z}/\mathfrak{b})$ be the subgroup of $W(\mathfrak{z}/\mathfrak{b})$ generated by the Weyl reflexions corresponding to the compact roots of $(\mathfrak{z}, \mathfrak{b})$. Then $W(\mathfrak{z}/\mathfrak{b}) \subset W$ and $W_k(\mathfrak{z}/\mathfrak{b}) \subset W_k$. Define $\varpi_\mathfrak{z}$ and $\pi_\mathfrak{z}$ as in § 13 for $\mathfrak{h} = \mathfrak{b}$.

LEMMA 37. *Let* a_s $(s \in W_k)$ *be continuous functions* [1] *on* $\mathfrak{F}$ *such that* $a_{ts} = a_s$ *for* $t \in W_k \cap W(\mathfrak{z}/\mathfrak{b})$. *Then for any* $\lambda \in \mathfrak{F}'$, *there exists a unique distribution* $T_{\mathfrak{z},\lambda}$ *on* $\mathfrak{z}$ *such that:*

1) $T_{\mathfrak{z},\lambda}$ *is invariant and tempered.*
2) $\partial(p_\mathfrak{z}) T_{\mathfrak{z},\lambda} = p_\mathfrak{b}(\lambda) T_{\mathfrak{z},\lambda}$ $(p \in I(\mathfrak{g}_c))$.
3) $\pi_\mathfrak{z} T_{\mathfrak{z},\lambda} = \sum_{s \in W_k} \varepsilon(s) \, a_s(\lambda) \, e^{s\lambda}$ *pointwise on* $\mathfrak{b}'$.

---

[1] For most applications a_s will be constants.

The uniqueness is obvious from the corollary of Lemma 28. The existence is proved as follows. Applying Theorem 2 to $(\mathfrak{z}, \mathfrak{b})$ instead of $(\mathfrak{g}, \mathfrak{b})$, we conclude that there exists a unique invariant and tempered distribution τ_λ on $\mathfrak{z}$ such that $\partial(p)\tau_\lambda = p_\mathfrak{b}(\lambda)\tau_\lambda$ $(p \in I(\mathfrak{z}_c))$ and

$$\pi_\mathfrak{z}\tau_\lambda = \sum_{s \in W_k(\mathfrak{z}/\mathfrak{b})} \varepsilon(s)\, e^{s\lambda}$$

pointwise on $\mathfrak{b}'$. Put

$$T_{\mathfrak{z},\lambda} = [W_k(\mathfrak{z}/\mathfrak{b})]^{-1} \sum_{s \in W_k} \varepsilon(s)\, a_s(\lambda)\, \tau_{s\lambda} = \sum_{s \in W_k(\mathfrak{z}/\mathfrak{b})\setminus W_k} \varepsilon(s)\, a_s(\lambda)\, \tau_{s\lambda}$$

where the second sum is over a complete system of representatives. Then it is obvious that $T_{\mathfrak{z},\lambda}$ fulfills all the conditions of the lemma.

COROLLARY.

$$T_{\mathfrak{z},\lambda} = [W_k(\mathfrak{z}/\mathfrak{b})]^{-1} \sum_{s \in W_k} \varepsilon(s)\, a_s(\lambda)\, \tau_{s\lambda} = \sum_{s \in W_k(\mathfrak{z}/\mathfrak{b})\setminus W_k} \varepsilon(s)\, a_s(\lambda)\, \tau_{s\lambda}.$$

Fix a connected component $\mathfrak{F}^+$ of $\mathfrak{F}'$ and for any $\mu \in \mathrm{Cl}\,\mathfrak{F}^+$ define $T_{\mathfrak{z},\mu}{}^+$ and $\tau_\mu{}^+ = \tau_{\mu,\mathfrak{F}^+}$ by means of the limits

$$T_{\mathfrak{z},\mu}{}^+(f) = \lim_{\lambda \to \mu} T_{\mathfrak{z},\lambda}(f), \quad \tau_\mu{}^+(f) = \lim_{\lambda \to \mu} \tau_\lambda(f) \quad (f \in C_c^\infty(\mathfrak{z}))$$

where $\lambda \in \mathfrak{F}^+$. We have seen in §17 that $\tau_\mu{}^+$ is a tempered distribution and therefore it follows from the above corollary that the same holds for $T_{\mathfrak{z},\mu}{}^+$. In fact the following lemma is now obvious.

LEMMA 38. $$T_{\mathfrak{z},\mu}{}^+ = \sum_{s \in W_k(\mathfrak{z}/\mathfrak{b})\setminus W_k} \varepsilon(s)\, a_s(\mu)\, \tau_{s\mu,\,s\mathfrak{F}^+}.$$

Let P and $P_\mathfrak{z}$ respectively be the sets of all positive roots of $(\mathfrak{g}, \mathfrak{b})$ and $(\mathfrak{z}, \mathfrak{b})$ and let $P_{\mathfrak{g}/\mathfrak{z}}$ denote the complement of $P_\mathfrak{z}$ in P. Put

$$\pi_{\mathfrak{g}/\mathfrak{z}} = \prod_{\alpha \in P_{\mathfrak{g}/\mathfrak{z}}} \alpha, \quad \varpi_{\mathfrak{g}/\mathfrak{z}} = \prod_{\alpha \in P_{\mathfrak{g}/\mathfrak{z}}} H_\alpha.$$

It is clear that $\pi_\mathfrak{z}^2$, $\pi_{\mathfrak{g}/\mathfrak{z}}$ and $\varpi_{\mathfrak{g}/\mathfrak{z}}$ are all invariant under $W(\mathfrak{z}/\mathfrak{b})$. Hence by Chevalley's theorem [2 (c), Lemma 9], we can choose an invariant polynomial function $\eta_\mathfrak{z}$ on $\mathfrak{z}_c$ and an element $q_{\mathfrak{g}/\mathfrak{z}} = q \in I(\mathfrak{z}_c)$ such that $\eta_\mathfrak{z} = (-1)^r \pi_\mathfrak{z}^2$ on $\mathfrak{b}$ and $q_\mathfrak{b} = \varpi_{\mathfrak{g}/\mathfrak{z}}$. (Here r is the number of roots in $P_\mathfrak{z}$.) Let $\mathfrak{z}'$ be the set of all $Z \in \mathfrak{z}$ where $\eta_\mathfrak{z}(Z) \neq 0$ and define the invariant differential operator $\nabla_\mathfrak{z}$ on $\mathfrak{z}'$ as usual (see [2 (l), §9]). Fix $\lambda \in \mathfrak{F}'$. Then we know [2 (l), Lemma 25] that there exists a continuous function $S_{\mathfrak{z},\lambda}$ on $\mathfrak{z}$ such that

$$S_{\mathfrak{z},\lambda} = \varpi(\lambda)^{-1} \nabla_{\mathfrak{z}} (\partial(q_{\mathfrak{g}/\mathfrak{z}}) \, T_{\mathfrak{z},\lambda})$$

pointwise on $\mathfrak{z}'$.

LEMMA 39. *Fix $\lambda \in \mathfrak{F}'$. Then* [1]

$$\varpi(\lambda) \, T_{\mathfrak{z},\lambda}(f) = \mathrm{p.v.} \int \eta_{\mathfrak{z}}^{-1} S_{\mathfrak{z},\lambda} \nabla_{\mathfrak{z}} (\partial(q_{\mathfrak{g}/\mathfrak{z}})^* f) \, dZ \quad (f \in C_c^\infty(\mathfrak{z})),$$

in the notation of Theorem 1.

Put $\square_{\mathfrak{z}} = \partial(q_1)$ where q_1 is the unique element in $I(\mathfrak{z}_c)$ such that $(q_1)_\mathfrak{b} = \varpi_{\mathfrak{z}}^2$. Then $(q_1 q^2)_\mathfrak{b} = \varpi^2$. Hence if Q is the unique element in $I(\mathfrak{g}_c)$ such that $Q_\mathfrak{b} = \varpi^2$ and $Q_{\mathfrak{z}}$ is the projection (see [2 (j), § 8]) of Q on $\mathfrak{z}$, it is obvious that $Q_{\mathfrak{z}} = q_1 q^2$. Therefore

$$(\square_{\mathfrak{z}} \circ \partial(q^2)) \, T_{\mathfrak{z},\lambda} = \partial(Q_{\mathfrak{z}}) \, T_{\mathfrak{z},\lambda} = Q_\mathfrak{b}(\lambda) \, T_{\mathfrak{z},\lambda} = \varpi(\lambda)^2 \, T_{\mathfrak{z},\lambda}.$$

Hence if $T = (\square_{\mathfrak{z}} \circ \partial(q)) \, T_{\mathfrak{z},\lambda}$, it follows from Theorem 1 that

$$\varpi(\lambda)^2 \, T_{\mathfrak{z},\lambda}(f) = T(\partial(q)^* f) = \varpi(\lambda) \left\{ \mathrm{p.v.} \int \eta_{\mathfrak{z}}^{-1} S_{\mathfrak{z},\lambda} (\nabla_{\mathfrak{z}} \circ \partial(q)^*) f \, dZ \right\}$$

for $f \in C_c^\infty(\mathfrak{z})$. Since $\varpi(\lambda) \neq 0$, this implies the assertion of the lemma.

Let $\mathfrak{a}$ be a Cartan subalgebra of $\mathfrak{z}$ and $S_\lambda^{\mathfrak{a}}$ the restriction of $S_{\mathfrak{z},\lambda}$ on $\mathfrak{a}$. Then it follows from the definitions of $\nabla_{\mathfrak{z}}$ and q and [2 (c), Lemma 8] that [2]

$$S_\lambda^{\mathfrak{a}} = \varpi(\lambda)^{-1} \partial(\varpi^y) \, (\pi_{\mathfrak{z}}^{\mathfrak{a}} \, T_{\mathfrak{z},\lambda})$$

pointwise on $\mathfrak{a}'$.

On the other hand let $\mathfrak{F}_{\mathfrak{z}}'$ be the set of all $\lambda \in \mathfrak{F}$ where $\varpi_{\mathfrak{z}}(\lambda) \neq 0$ and $\mathfrak{F}_{\mathfrak{z}}^+$ a connected component of $\mathfrak{F}_{\mathfrak{z}}'$. Fix a connected component $\mathfrak{a}^+$ of $\mathfrak{a}'(\mathfrak{z}:R)$ (see § 13). Then corresponding to Lemma 32 and the corollary of Lemma 33 we have the following result for $\mathfrak{z}$.

LEMMA 40. *There exist integers $c_{\mathfrak{z}}(s : \mathfrak{F}_{\mathfrak{z}}^+ : \mathfrak{a}^+)$ $(s \in W(\mathfrak{z}/\mathfrak{a}))$ such that*

$$\pi_{\mathfrak{z}}^{\mathfrak{a}} \tau_\lambda = \sum_{s \in W(\mathfrak{z}/\mathfrak{a})} \varepsilon(s) \, c_{\mathfrak{z}}(s : \mathfrak{F}_{\mathfrak{z}}^+ : \mathfrak{a}^+) \, e^{s\lambda^y}$$

on $\mathfrak{a}^+ \cap \mathfrak{z}'$ for $\lambda \in \mathfrak{F}_{\mathfrak{z}}^+$. Moreover

$$c_{\mathfrak{z}}(st^y : t^{-1} \mathfrak{F}_{\mathfrak{z}}^+ : \mathfrak{a}^+) = c_{\mathfrak{z}}(s : \mathfrak{F}_{\mathfrak{z}}^+ : \mathfrak{a}^+)$$

for $t \in W_k(\mathfrak{z}/\mathfrak{b})$.

---

[1] As usual, the star denotes the adjoint here.

[2] Here y is an element in the complex analytic subgroup Ξ_c of G_c corresponding to $\mathrm{ad}\,\mathfrak{z}_c$ such that $(\mathfrak{b}_c)^y = \mathfrak{a}_c$. We also assume that $P_{\mathfrak{z}}^y$ is the set of positive roots of $(\mathfrak{z}, \mathfrak{a})$.

Now write $c_{\mathfrak{z}}(s:\mathfrak{F}^+:\mathfrak{a}^+)=c_{\mathfrak{z}}(s:\mathfrak{F}_{\mathfrak{z}}^+:\mathfrak{a}^+)$ for any connected component $\mathfrak{F}^+$ of $\mathfrak{F}'$ which is contained in $\mathfrak{F}_{\mathfrak{z}}^+$. Then it follows from the corollary of Lemma 37 that

$$\pi_{\mathfrak{z}}^{\mathfrak{a}}\,T_{\mathfrak{z},\lambda}=\sum_{t\in W_k(\mathfrak{z}/\mathfrak{b})\backslash\,W_k}\varepsilon(t)\,a_t(\lambda)\sum_{s\in W(\mathfrak{z}/\mathfrak{a})}\varepsilon(s)\,c_{\mathfrak{z}}(s:t\,\mathfrak{F}^+:\mathfrak{a}^+)\,e^{s(t\lambda)^{\nu}}$$

on $\mathfrak{a}^+\cap\mathfrak{z}'$ for any λ lying in a connected component $\mathfrak{F}^+$ of $\mathfrak{F}'$. Therefore it follows from the above formula for $S_{\lambda}^{\mathfrak{a}}$ that

$$S_{\lambda}^{\mathfrak{a}}=\sum_{t\in W_k(\mathfrak{z}/\mathfrak{b})\backslash\,W_k}a_t(\lambda)\sum_{s\in W(\mathfrak{z}/\mathfrak{a})}c_{\mathfrak{z}}(s:t\,\mathfrak{F}^+:\mathfrak{a}^+)\,e^{s(t\lambda)^{\nu}}$$

on $\mathfrak{a}^+\cap\mathfrak{z}'$.

Now fix $\mu\in\mathrm{Cl}(\mathfrak{F}^+)$. Then as λ tends to μ $(\lambda\in\mathfrak{F}^+)$, it is clear that the functions $S_{\lambda}^{\mathfrak{a}}$ converge uniformly on every compact subset of $\mathfrak{a}$. Hence we conclude (see Lemma 69 of § 30) that the functions $S_{\mathfrak{z},\lambda}$ converge uniformly on every compact subset of $\mathfrak{z}$. We denote the limit function by $S_{\mathfrak{z},\mu}^+$. It is obviously continuous and invariant.

Lemma 41. $$S_{\mathfrak{z},\mu}^+=\sum_{t\in W_k(\mathfrak{z}/\mathfrak{b})\backslash\,W_k}a_t(\mu)\sum_{s\in W(\mathfrak{z}/\mathfrak{a})}c_{\mathfrak{z}}(s:t\,\mathfrak{F}^+:\mathfrak{a}^+)\,e^{s(t\mu)^{\nu}}$$

on $\mathfrak{a}^+\cap\mathfrak{z}'$. *Moreover*

$$\varpi(\mu)\,T_{\mathfrak{z},\mu}^+(f)=\mathrm{p.v.}\int\eta_{\mathfrak{z}}^{-1}\,S_{\mathfrak{z},\mu}^+\nabla_{\mathfrak{z}}\,(\partial\,(q_{\mathfrak{g}/\mathfrak{z}})^*f)\,dZ$$

for $f\in C_c^{\infty}(\mathfrak{z})$.

The first statement is obvious from the above formula for $S_{\lambda}^{\mathfrak{a}}$ and the second follows from Lemma 39 if we take into account the corollary of Lemma 4.

Corollary. $S_{\mathfrak{z},\mu}^+=0$ *if* $a_t(\mu)=0$ $(t\in W_k)$.

Now suppose $\mathfrak{z}_1,\mathfrak{z}_2$ and η_0 are as in Lemma 30 (with $\mathfrak{h}=\mathfrak{b}$). Then since

$$\pi_{\mathfrak{z}_1}\,T_{\mathfrak{z}_1,\lambda}=\pi_{\mathfrak{z}_2}\,T_{\mathfrak{z}_2,\lambda}=\sum_{s\in W_k}\varepsilon(s)\,e^{s\lambda}$$

pointwise on $\mathfrak{b}'$ for $\lambda\in\mathfrak{F}'$, it follows from Lemma 30 that

$$T_{\mathfrak{z}_2,\lambda}=\eta_0\,T_{\mathfrak{z}_1,\lambda}$$

pointwise on $\mathfrak{g}'\cap\mathfrak{z}_2$. Fix a Cartan subalgebra $\mathfrak{a}$ of $\mathfrak{z}_2$ and an element y in the complex analytic subgroup Ξ_{2c} of G_c corresponding to $\mathrm{ad}\,\mathfrak{z}_{2c}$, such that $\mathfrak{b}_c^{\,y}=\mathfrak{a}_c$. We may assume that $P^{\,y}$ is the set of all positive roots of $(\mathfrak{g},\mathfrak{a})$. Then

$$S_{\delta_i,\lambda} = \varpi(\lambda)^{-1}\nabla_{\delta_i}(\partial(q_{\mathfrak{g}/\delta_i})\,T_{\delta_i,\lambda}) = \varpi(\lambda)^{-1}\partial(\varpi^{\nu})\,F_\lambda$$

pointwise on $\mathfrak{a}'$ where

$$F_\lambda(H) = \pi_{\delta_1}{}^{\mathfrak{a}}(H)\,T_{\delta_1,\lambda}(H) = \pi_{\delta_2}{}^{\mathfrak{a}}(H)\,T_{\delta_2,\lambda}(H) \quad (H\in\mathfrak{a}').$$

This shows that $S_{\delta_1,\lambda} = S_{\delta_2,\lambda}$ on $\mathfrak{z}_2$ and therefore we get the following result by taking limits.

LEMMA 42. *Fix* $\mu\in \mathrm{Cl}(\mathfrak{F}^+)$. *Then*

$$T_{\delta_1,\mu}{}^{+} = \eta_0\,T_{\delta_1,\mu}{}^{+}$$

pointwise on $\mathfrak{g}'\cap\mathfrak{z}_2$ *and* $$S_{\delta_1\mu}{}^{+} = S_{\delta_2\mu}{}^{+}$$

on $\mathfrak{z}_2$.

We now return to the notation of Lemma 41 and write $T_{\delta,\mu,\mathfrak{F}+} = T_{\delta,\mu}{}^{+}$ whenever it is convenient to do so. Let Ξ be the analytic subgroup of G corresponding to $\mathfrak{z}$ and $\mathfrak{h}_1,\mathfrak{h}_2,\ldots,\mathfrak{h}_r$ a maximal set of Cartan subalgebras of $\mathfrak{z}$ no two of which are conjugate under Ξ. Fix a Euclidean measure $d_i H$ on $\mathfrak{h}_i$ and define $\psi_{\mathfrak{z},f,i}$ $(f\in C_c^{\infty}(\mathfrak{z}))$ as in Lemma 5 for $(\mathfrak{z},\mathfrak{h}_i)$ instead of $(\mathfrak{g},\mathfrak{h}_i)$.

LEMMA 43. *Assume that the functions* a_t $(t\in W_\kappa)$ *remain bounded on* $\mathfrak{F}$. *Then there exists a number* $C\geqslant 0$ *with the following property. Let* $\mathfrak{F}^+$ *be a connected component of* $\mathfrak{F}'$ *and* μ *a point in* $\mathrm{Cl}(\mathfrak{F}^+)$. *Then*

$$\left|T_{\delta,\mu,\mathfrak{F}+}(f)\right| \leqslant C\sum_{1\leqslant i\leqslant r}\int_{\mathfrak{h}_i}\left|\psi_{\mathfrak{z},f,i}\right|d_i H.$$

Let $\mathfrak{a}$ be a Cartan subalgebra of $\mathfrak{z}$. It follows from Lemmas 28 and 40 that

$$\left|\pi_{\mathfrak{z}}{}^{\mathfrak{a}}(H)\,\tau_\lambda(H)\right| \leqslant \sum_{s\in W(\mathfrak{z}/\mathfrak{a})}\left|c_{\mathfrak{z}}(s:\mathfrak{F}^+:\mathfrak{a}^+)\right|$$

for $H\in\mathfrak{a}^+\cap\mathfrak{z}'$ and $\lambda\in\mathfrak{F}^+$. Put

$$g_\lambda{}^{\mathfrak{a}}(H) = \pi_{\mathfrak{z}}{}^{\mathfrak{a}}(H)\,T_{\delta,\lambda}(H) \quad (H\in\mathfrak{a}').$$

Then, in view of the corollary of Lemma 37, we can choose a number $a\geqslant 0$ such that

$$\left|g_\lambda{}^{\mathfrak{a}}(H)\right| \leqslant a$$

for $H\in\mathfrak{a}'$ and $\lambda\in\mathfrak{F}'$. Now put $g_{\lambda,i} = g_\lambda{}^{\mathfrak{h}_i}$ $(1\leqslant i\leqslant r)$. Then, as we have seen in §2, there exist real numbers $c_1,\ldots,c_r$ such that

$$T_{\mathfrak{z},\lambda}(f) = \sum_{1 \leqslant i \leqslant r} c_i \int \psi_{\mathfrak{z},f,i}\, g_{\lambda,i}\, \varepsilon_{\mathfrak{z},R,i}\, d_i H$$

for all $f \in C_c^\infty(\mathfrak{z})$ and $\lambda \in \mathfrak{F}'$. (Here $\varepsilon_{\mathfrak{z},R,i}$ is a locally constant function on $\mathfrak{h}_i'$ whose values are ± 1.) Therefore

$$|T_{\mathfrak{z},\lambda}(f)| \leqslant C \sum_{1 \leqslant i \leqslant r} \int |\psi_{\mathfrak{z},f,i}|\, d_i H$$

where $C = a\,\max_i |c_i|$. The statement of the lemma now follows by letting λ tend to μ $(\lambda \in \mathfrak{F}^+)$.

Part II. Theory on the group

§ 19. Statement of Theorem 3

We keep to the notation of § 16 and assume, moreover, that G is acceptable. Let B be the Cartan subgroup of G corresponding to $\mathfrak{b}$. Then B is connected and therefore abelian (see [2 (m), Cor. 5 of Lemma 26]). Let B^* denote the character group of B. For any $b^* \in B^*$, we denote by $\langle b^*, b \rangle$ the value of the character b^* at a point $b \in B$. It is obvious that there exists a unique element $\lambda \in \mathfrak{F}$ such that

$$\langle b^*, \exp H \rangle = e^{\lambda(H)} \quad (H \in \mathfrak{b}).$$

We shall denote λ by $\log b^*$. b^* is called singular or regular according as $\varpi(\lambda) = 0$ or not. We have seen that $W_k = W_G$. Now W_G operates on B as usual (see [2 (m), § 20]) and therefore, by duality, also on B^*. Then

$$\langle (b^*)^s, b \rangle = \langle b^*, b^{s^{-1}} \rangle \quad (b^* \in B^*,\ b \in B)$$

and $\log (b^*)^s = s\,(\log b^*)$ $(s \in W_G)$.

Define $\mathfrak{Z}$ as in [2 (m), § 6] and let $z \to p_z$ $(z \in \mathfrak{Z})$ denote the canonical isomorphism of $\mathfrak{Z}$ onto $I(\mathfrak{g}_c)$ (see [2 (m), § 12]). For $b^* \in B^*$, define

$$\chi_{b^*}(z) = \chi_\lambda^\mathfrak{b}(p_z) \quad (z \in \mathfrak{Z})$$

(in the notation of § 12) for $\lambda = \log b^*$. Then χ_{b^*} is a homomorphism of $\mathfrak{Z}$ into $\mathbb{C}$.

Let t be an indeterminate and l the rank of G. For any $x \in G$, we denote by $D(x)$ the coefficient of t^l in $\det(t + 1 - \mathrm{Ad}(x))$. Then D is an analytic function on G. As usual let G' denote the set of all regular elements in G (see [2 (m), § 3]). Fix a Haar measure dx on G and let Θ be a distribution on G. We say that Θ is an invariant eigendistribution of $\mathfrak{Z}$ if 1) $\Theta^x = \Theta$ $(x \in G)$ and 2) there exists a homomorphism

χ of $\mathfrak{Z}$ into $\mathbf{C}$ such that $z\Theta = \chi(z)\Theta$ for all $z \in \mathfrak{Z}$. In view of [2 (m), Theorem 2], we can speak of the value $\Theta(x)$ of such a distribution at a point $x \in G'$.

Let $B^{*\prime}$ denote the set of all regular elements in B^* and put $\Delta = \Delta_B$ in the notation of [2 (m), § 19].

THEOREM 3.   *Fix an element $b^* \in B^{*\prime}$. Then there exists exactly one invariant eigendistribution Θ on G such that:*

1) $$z\Theta = \chi_{b^*}(z)\Theta \quad (z \in \mathfrak{Z});$$

2) $$\sup_{x \in G'} |D(x)|^{\frac{1}{2}}|\Theta(x)| < \infty;$$

3) $$\Theta = \Delta^{-1} \sum_{s \in W_G} \varepsilon(s)(b^*)^s \text{ pointwise on } B' = B \cap G'.$$

§ 20. Proof of the uniqueness

In order to obtain the uniqueness in Theorem 3, it is sufficient to prove the following result.

LEMMA 44.   *Fix $b^* \in B^{*\prime}$ and let Θ be an invariant eigendistribution of $\mathfrak{Z}$ on G such that:*

1) $$z\Theta = \chi_{b^*}(z)\Theta \quad (z \in \mathfrak{Z});$$

2) $$\sup_{x \in G'} |D(x)|^{\frac{1}{2}}|\Theta(x)| < \infty;$$

3) $$\Theta = 0 \text{ pointwise on } B'.$$

Then $\Theta = 0$.

Fix a semisimple element $a \in G$. In view of [2 (m), Lemma 7], it would be sufficient to prove that $a \notin \operatorname{Supp} \Theta$. We now use the notation of [2 (m), § 4] and put $\sigma = |\nu_a|^{\frac{1}{2}}\sigma_\Theta$ in the notation of [2 (m), Lemma 15]. Since $z\Theta = \chi_{b^*}(z)\Theta$, we conclude from [2 (m), Lemma 22] that

$$\mu_{\mathfrak{g}/\mathfrak{z}}(z)\,\sigma = \chi_{b^*}(z)\,\sigma \quad (z \in \mathfrak{Z}).$$

Define $\mathfrak{g}_0 = \mathfrak{c}_0 + \mathfrak{g}_1(c)$ as in § 14 where $\mathfrak{c}_0$ is an open and convex neighborhood of zero in $\mathfrak{c}$. Then $\mathfrak{g}_0$ is an open and completely invariant neighborhood of zero in $\mathfrak{g}$ and if $\mathfrak{c}_0$ and c are sufficiently small, the exponential mapping of $\mathfrak{g}$ into G is univalent and regular on $\mathfrak{g}_0$ (see [2 (m), § 9]). Put $\mathfrak{z}_0 = \mathfrak{g}_0 \cap \mathfrak{z}$.

Now first assume that $a \in B$ and let Z_G denote the center of G. Then since

B/Z_G is compact [2 (m), § 16], every eigenvalue of $\mathrm{Ad}(a)$ has absolute value 1. Hence if c is sufficiently small, it is obvious that no eigenvalue of $(\mathrm{Ad}(a \exp Z))_{\mathfrak{g}/\mathfrak{z}}$ can be 1 for $Z \in \mathfrak{z}_0$. This shows that $\exp \mathfrak{z}_0 \subset \Xi'$. Let τ denote the distribution on $\mathfrak{z}_0$ obtained from σ by applying the procedure of [2 (m), § 10] to $\mathfrak{z}$ (in place of $\mathfrak{g}$). Since

$$\mu_{\mathfrak{g}/\mathfrak{z}}(z)\,\sigma = \chi_{b*}(z)\,\sigma \quad (z \in \mathfrak{Z}),$$

it follows from the corollary of [2 (m), Lemma 24] and the definition of $\mu_{\mathfrak{g}/\mathfrak{z}}$ [2 (m), § 12] that

$$\partial(p_{\mathfrak{z}})\,\tau = \chi(p)\,\tau \quad (p \in I(\mathfrak{g}_c)),$$

where $\chi = \chi_\lambda^b$ and $\lambda = \log b^*$. Now $\mathfrak{b} \subset \mathfrak{z}$ since $a \in B$. Therefore $T_{\mathfrak{z}} = \tau$ satisfies all the conditions of § 13. Let $\mathfrak{z}_0'$ be the set of those elements of $\mathfrak{z}_0$ which are regular in $\mathfrak{z}$. Then we know from [2 (m), Lemma 32] that

$$\tau(Z) = \xi_{\mathfrak{z}}(Z)\,|\nu_a\,(\exp Z)|^{\frac{1}{2}}\,\Theta\,(a \exp Z) \quad (Z \in \mathfrak{z}_0').$$

Let $\mathfrak{a}$ be a Cartan subalgebra of $\mathfrak{z}$ and A the corresponding Cartan subgroup of G. It is easy to verify that

$$|D(a \exp H)| = |\pi_{\mathfrak{z}}^{\mathfrak{a}}(H)\,\xi_{\mathfrak{z}}(H)|^2\,|\nu_a\,(\exp H)|$$

for $H \in \mathfrak{a}$ and therefore

$$|\pi_{\mathfrak{z}}^{\mathfrak{a}}(H)\,\tau(H)| = |D(a \exp H)|^{\frac{1}{2}}\,|\Theta\,(a \exp H)|$$

for $H \in \mathfrak{a}' \cap \mathfrak{z}_0$. Hence we conclude from Lemma 19 and condition 2) that τ is a tempered distribution on $\mathfrak{z}_0$. Moreover if we take $\mathfrak{a} = \mathfrak{b}$, it follows from condition 3) that $\tau = 0$ pointwise on $\mathfrak{z}_0 \cap \mathfrak{b}'$. Therefore (see the corollary of Lemma 29), $\tau = 0$ on $\mathfrak{z}_0$. This, in turn, implies that $\Theta = 0$ pointwise on $a \exp \mathfrak{z}_0' = G' \cap (a \exp \mathfrak{z}_0)$. But $V = (a \exp \mathfrak{z}_0)^G$ is open in G [2 (m), Lemma 14]. Hence $\Theta = 0$ on V.

Now we drop the assumption that $a \in B$. Define $\theta, \mathfrak{k}, \mathfrak{p}$ and K as in [2 (m), § 16] corresponding to $\mathfrak{h} = \mathfrak{b}$. Then $B \subset K$ [2 (m), Cor. 5 of Lemma 26]. Let $\mathfrak{a}$ be any Cartan subalgebra of $\mathfrak{z}$. We can choose $x \in G$ such that $\theta(\mathfrak{a}^x) = \mathfrak{a}^x$ and $\mathfrak{a}^x \cap \mathfrak{k} \subset \mathfrak{b}$ (see Lemma 45 below). Let A be the Cartan subgroup of G corresponding to $\mathfrak{h} = \mathfrak{a}^x$. Then $a^x \in A$. Let $a^x = a_0 \exp H$ where $a_0 \in A \cap K$ and $H \in \mathfrak{h} \cap \mathfrak{p}$ (see [2 (m), Cor. 4 of Lemma 26]). Since K is connected and K/Z_G is compact, we can choose $k \in K$ such that $b = a_0^k \in B$. Then

$$a^{kx} = b \exp Z_0$$

where $Z_0 = H^k \in \mathfrak{p} \subset [\mathfrak{g}, \mathfrak{g}]$. Let $\mathfrak{z}_b$ denote the centralizer of b in $\mathfrak{g}$. It is obvious that

$Z_0 \in \mathfrak{z}_b$. Moreover since $Z_0 \in \mathfrak{p}$, all the eigenvalues of $\mathrm{ad}\, Z_0$ are real [2 (i), Lemma 27]. Hence by applying the result obtained above to b, we conclude that

$$a^{kx} = b \exp Z_0 \notin \mathrm{Supp}\ \Theta.$$

Therefore since Θ is invariant, it follows that $a \notin \mathrm{Supp}\ \Theta$. This proves the lemma.

§ 21. Some elementary facts about Cartan subgroups

Let $\mathfrak{a}$ be a Cartan subalgebra of $\mathfrak{g}$ and A the corresponding Cartan subgroup of G. Define $\mathfrak{a}_R$ and $\mathfrak{a}_I$ as in § 11.

LEMMA 45. *Let A_I be the subgroup of all $a \in A$ such that all eigenvalues of* $\mathrm{Ad}(a)$ *have absolute values* 1. *Then* $(a, H) \to a \exp H$ $(a \in A_I, H \in \mathfrak{a}_R)$ *is a topological mapping of* $A_I \times \mathfrak{a}_R$ *onto* A. *Moreover for any* $a \in A_I$, *we can choose* $x \in G$ *such that* 1) $a^x \in B$, 2) $\theta(a^x) = a^x$ *and* 3) $(\mathfrak{a}_I)^x \subset \mathfrak{b}$. *Finally, x may be selected to lie in K if* $\theta(\mathfrak{a}) = \mathfrak{a}$.

It follows from [2 (b), p. 100] that we can choose $y \in G$ such that $\theta(\mathfrak{a}^y) = \mathfrak{a}^y$. Then $(\mathfrak{a}_I)^y$ is an abelian subspace of $\mathfrak{k}$. Since $\mathfrak{b}$ is maximal abelian in $\mathfrak{k}$ and K/Z_G is compact, we can choose $k \in K$ such that $(\mathfrak{a}_I)^{ky} \subset \mathfrak{b}$. Replacing $\mathfrak{a}$ by $\mathfrak{a}^{ky}$, we can now obviously assume that $\theta(\mathfrak{a}) = \mathfrak{a}$ and $\mathfrak{a}_I \subset \mathfrak{b}$. Then the first statement follows from the results of [2 (m), § 16]. Moreover it is clear that $A_I = A \cap K \subset K = B^K$. Fix $a \in A_I$ and choose $k \in K$ such that $b = a^k \in B$. Let $\mathfrak{z}$ be the centralizer of b in $\mathfrak{g}$ and Ξ the analytic subgroup of G corresponding to $\mathfrak{z}$. Then $\mathfrak{a}^k$ and $\mathfrak{b}$ are two Cartan subalgebras of $\mathfrak{z}$ and $\mathfrak{a}_I{}^k + \mathfrak{b} \subset \mathfrak{z} \cap \mathfrak{k}$. Since $\mathfrak{b}$ is maximal abelian in $\mathfrak{z} \cap \mathfrak{k}$, we can choose $\xi \in \Xi \cap K$ such that $(\mathfrak{a}_I)^{\xi k} \subset \mathfrak{b}$. Put $x = \xi k$. Then $a^x = b^\xi = b$ and $(\mathfrak{a}_I)^x \subset \mathfrak{b}$. Moreover since $x \in K$, it is clear that $\theta(a^x) = a^x$. The last statement follows from the fact that we can take $y = 1$ if $\theta(\mathfrak{a}) = \mathfrak{a}$.

COROLLARY. *An element a of G lies in B^G if and only if* 1) a *is semisimple and* 2) *all eigenvalues of* $\mathrm{Ad}(a)$ *have absolute value* 1.

Since $B \subset K$, it is obvious that any $a \in B^G$ fullfills these two conditions. Conversely suppose these conditions hold. Then by 1), a is contained in some Cartan subgroup A of G [2 (m), Cor. of Lemma 5]. Therefore by 2) $a \in A_I$. But then $a \in B^G$ by Lemma 45.

We write $A_R = \exp \mathfrak{a}_R$. By Lemma 45, every $h \in A$ can be written uniquely in the form $h = h_1 h_2$ $(h_1 \in A_I, h_2 \in A_R)$. We call h_1 and h_2 the components of h in A_I and A_R respectively.

§ 22. Proof of the existence

We now come to the proof of the existence of Θ in Theorem 3. In view of later applications, we shall consider a somewhat more general situation.

Fix a connected component $\mathfrak{F}^+$ of $\mathfrak{F}'$ and a point $b^* \in B^*$ such that

$$\lambda = \log b^* \in \mathrm{Cl}\,(\mathfrak{F}^+).$$

Select an open convex neighborhood c_0 of zero in c and define

$$\mathfrak{g}_0 = c_0 + \mathfrak{g}_1(c) \quad (0 < c \leqslant \pi = 3.14 \ldots)$$

as in § 14. We assume that c_0 is so small that the exponential mapping of $\mathfrak{g}$ into G is univalent and regular on $\mathfrak{g}_0$ (see [2 (m), § 9]).

Fix $b \in B$ and let $\mathfrak{z} = \mathfrak{z}_b$ denote the centralizer of b in $\mathfrak{g}$. Define $T_b{}^+ = T_{\mathfrak{z},\lambda}{}^+$ and $S_b{}^+ = S_{\mathfrak{z},\lambda}{}^+$ in the notation of § 18 corresponding to the constants $a_s = \langle (b^*)^s, b \rangle$ $(s \in W_G)$. (Here we have to observe that $b^t = b$ for $t \in W_k \cap W(\mathfrak{z}/\mathfrak{b})$ and therefore $a_{ts} = a_s$.)

Let $\Xi = \Xi(b)$ denote the analytic subgroup of G corresponding to $\mathfrak{z}$. Put $\mathfrak{z}_0 = \mathfrak{g}_0 \cap \mathfrak{z}$ and $\Xi_0(b) = \Xi_0 = \exp \mathfrak{z}_0$. Then Ξ_0 is an open and completely invariant subset of Ξ [2 (m), Lemma 8]. As usual define the function $\xi_{\mathfrak{z}}$ on $\mathfrak{z}$ by

$$\xi_{\mathfrak{z}}(Z) = \left| \det \left\{ (e^{\mathrm{ad}\, Z/2} - e^{-\mathrm{ad}\, Z/2})/\mathrm{ad}\, Z \right\} \right|^{\frac{1}{2}} \quad (Z \in \mathfrak{z}).$$

Then $\xi_{\mathfrak{z}}$ is analytic and nowhere zero on $\mathfrak{z}_0$. Put

$$\Phi_b{}^+ (\exp Z) = \xi_{\mathfrak{z}}(Z)^{-1} T_b{}^+ (Z) \quad (Z \in \mathfrak{g}_0 \cap \mathfrak{z}')$$

where $\mathfrak{z}'$ is the set of those elements of $\mathfrak{z}$ which are regular in $\mathfrak{z}$. Then $\Phi_b{}^+$ is a locally summable function on $\Xi_0(b)$.

Define the homomorphism $\mu_b = \mu_{\mathfrak{g}/\mathfrak{z}}$ as in [2 (m), § 12].

LEMMA 46. $\mu_b(z)\,\Phi_b{}^+ = \chi_{b*}(z)\,\Phi_b{}^+ \quad (z \in \mathfrak{Z})$

as a distribution on $\Xi_0(b)$.

This follows immediately from the corollary of [2 (m), Lemma 24] (applied to $\mathfrak{z}$) and the fact (see § 18) that $\partial(p_{\mathfrak{z}})\,T_b{}^+ = p_b(\lambda)\,T_b{}^+$ for $p \in I(\mathfrak{g}_c)$.

We have seen in [2 (m), § 22] that there exists an invariant analytic function D_b on $\mathfrak{z}$ such that

$$\Delta(b \exp H) = \pi_{\mathfrak{z}}(H)\,D_b(H) \quad (H \in \mathfrak{b}).$$

Put $\Xi_0{}''(b) = \Xi_0(b) \cap (b^{-1} G')$ and let $\mathfrak{z}''$ be the set of all points $Z \in \mathfrak{z}'$ where $D_b(Z) \neq 0$. Then it is clear that $\Xi_0{}''(b) = \exp(\mathfrak{g}_0 \cap \mathfrak{z}'')$. Put

$$\Theta_b{}^+ (\exp Z) = D_b(Z)^{-1} T_b{}^+(Z) \quad (Z \in \mathfrak{g}_0 \cap \mathfrak{z}'').$$

Then $\Theta_b{}^+$ is an analytic function on $\Xi_0''(b)$. Similarly define

$$\Psi_b^{\cdot\,+} (\exp Z) = S_b{}^+(Z) \quad (Z \in \mathfrak{z}_0).$$

Then $\Psi_b^{\cdot\,+}$ is a continuous function on $\Xi_0(b)$.

Define $$\nu_b(y) = \det(\mathrm{Ad}(by)^{-1} - 1)_{\mathfrak{g}/\mathfrak{z}} \quad (y \in \Xi)$$

as in [2 (m), § 14].

LEMMA 47. *Let* $\grave{\mathfrak{z}}$ *be the set of all points* $Z \in \mathfrak{z}$ *where* $\nu_b(\exp Z) \neq 0$. *Then there exists a locally constant function* ε_b *on* $\grave{\mathfrak{z}}$ *such that* $\varepsilon_b{}^4 = 1$ *and*

$$\xi_{\mathfrak{z}}(Z) |\nu_b(\exp Z)|^{\frac{1}{2}} = \varepsilon_b(Z) D_b(Z) \quad (Z \in \grave{\mathfrak{z}}).$$

It would be enough to verify that

$$\xi_{\mathfrak{z}}(Z)^4 \nu_b(\exp Z)^2 = D_b(Z)^4$$

for $Z \in \mathfrak{z}$. Since both sides are analytic functions on $\mathfrak{z}$ which are invariant under Ξ, it would be enough to do this when Z varies in some non-empty open subset of $\mathfrak{b}$. Hence our assertion follows from [2 (m), Lemma 33].

COROLLARY. $$|\nu_b(\exp Z)|^{\frac{1}{2}} \Theta_b{}^+ (\exp Z) = \varepsilon_b(Z) \Phi_b{}^+ (\exp Z)$$

for $Z \in \mathfrak{g}_0 \cap \mathfrak{z}''$.

This is obvious.

Put $\mathfrak{z}_0'' = \mathfrak{g}_0 \cap \mathfrak{z}''$ and let u be an element in G such that $\mathfrak{b}^u = \mathfrak{b}$.

LEMMA 48. *We have the relations*

$$\Theta_{b^u}{}^+ (\exp Z^u) = \Theta_b{}^+ (\exp Z), \quad \Psi_{b^u}^{\cdot\,+} (\exp Z^u) = \Psi_b^{\cdot\,+} (\exp Z)$$

for $Z \in \mathfrak{z}_0''$.

Since $\mathfrak{z}^u$ is the centralizer of $\mathfrak{b}^u$ in $\mathfrak{g}$, it is clear that $\Theta_{b^u}{}^+ (\exp Z^u)$ and $\Psi_{b^u}^{\cdot\,+} (\exp Z^u)$ are defined for $Z \in \mathfrak{z}_0''$. Let t be an element in W_G such that $H^u = tH$ for $H \in \mathfrak{b}$. It is obvious that $\pi_{\mathfrak{z}^u} = \gamma \pi_{\mathfrak{z}}^t$ where $\gamma = \pm 1$. Therefore since

$$\Delta(b^u \exp H^u) = \varepsilon(t) \Delta(b \exp H) \quad (H \in \mathfrak{b}),$$

it follows that $D_{b^u}(H^u) = \varepsilon(t) \gamma D_b(H)$. But the function

$$Z \to D_{b^u}(Z^u) - \varepsilon(t)\,\gamma\, D_b(Z) \quad (Z \in \mathfrak{z})$$

is obviously analytic and invariant under Ξ. Hence we can conclude that

$$D_{b^u}(Z^u) = \varepsilon(t)\,\gamma\, D_b(Z) \quad (Z \in \mathfrak{z}).$$

Now for any $\mu \in \mathfrak{F}^+$, let $T_{\mathfrak{z},\mu}$ be the distribution of Lemma 37 corresponding to the constants $a_s = \langle (b^*)^s, b \rangle$ $(s \in W_G)$. Similarly define $T_{\mathfrak{z}^u,\mu}$ on $\mathfrak{z}^u$ corresponding to the constants $a_s = \langle (b^*)^s, b^u \rangle$. Then

$$\pi_{\mathfrak{z}^u}(uH)\,T_{\mathfrak{z}^u,\mu}(uH) = \sum_{s \in W_G} \varepsilon(s)\,\langle (b^*)^s,\, b^u \rangle\, e^{s\mu(uH)}$$

$$= \varepsilon(t) \sum_{s \in W_G} \varepsilon(s)\,\langle (b^*)^s,\, b \rangle\, e^{s\mu(H)} = \varepsilon(t)\,\pi_{\mathfrak{z}}(H)\,T_{\mathfrak{z},\mu}(H) \quad (H \in \mathfrak{b}').$$

Hence
$$T_{\mathfrak{z}^u,\mu}(uH) = \varepsilon(t)\,\gamma\, T_{\mathfrak{z},\mu}(H) \quad (H \in \mathfrak{b}').$$

Now consider the distribution

$$T_\mu' : f \to \int f(Z)\, T_{\mathfrak{z}^u,\mu}(uZ)\, dZ \quad (f \in C_c^\infty(\mathfrak{z}))$$

on $\mathfrak{z}$. It is obviously invariant and tempered. Moreover it is clear that $p_{\mathfrak{z}^u} = (p_{\mathfrak{z}})^u$ for $p \in I(\mathfrak{g}_c)$. Let dZ' denote the Euclidean measure on $\mathfrak{z}^u$ which corresponds to dZ under the mapping $Z' = Z^u$ $(Z \in \mathfrak{z})$. Then

$$T_\mu'(\partial(p_{\mathfrak{z}})^* f) = \int f(u^{-1}Z';\, \partial(p_{\mathfrak{z}})^*)\, T_{\mathfrak{z}^u,\mu}(Z')\, dZ'$$

$$= \int f'(Z';\, \partial(p_{\mathfrak{z}^u})^*)\, T_{\mathfrak{z}^u,\mu}(Z')\, dZ'$$

$$= p_b(\mu) \int f'(Z')\, T_{\mathfrak{z}^u,\mu}(Z')\, dZ' = p_b(\mu)\, T_\mu'(f)$$

for $p \in I(\mathfrak{g}_c)$ and $f \in C_c^\infty(\mathfrak{z})$. Here f' denotes the function $Z' \to f(u^{-1}Z')$ $(Z' \in \mathfrak{z}^u)$ in $C_c^\infty(\mathfrak{z}^u)$. Hence it follows from the uniqueness assertion of Lemma 37 that

$$T_\mu' = \varepsilon(t)\,\gamma\, T_{\mathfrak{z},\mu}.$$

Therefore
$$T_{\mathfrak{z}^u,\mu}(f') = \varepsilon(t)\,\gamma\, T_{\mathfrak{z},\mu}(f)$$

and by making μ tend to λ, we conclude that

$$T_{b^u}^+(f') = \varepsilon(t)\,\gamma\, T_b^+(f) \quad (f \in C_c^\infty(\mathfrak{z})).$$

This proves that

$$T_{b^u}{}^+(Z^u) = \varepsilon(t)\,\gamma\,T_b{}^+(Z) \quad (Z \in \mathfrak{z}').$$

The first assertion of the lemma is now obvious.

Define $\nabla_{\mathfrak{z}}$, $\nabla_{\mathfrak{z}^u}$ and $\varpi_{\mathfrak{g}/\mathfrak{z}}$, $\varpi_{\mathfrak{g}/\mathfrak{z}^u}$ as in §18. It is clear that

$$\nabla_{\mathfrak{z}^u} f' = (\nabla_{\mathfrak{z}} f)'$$

for $f \in C_c^\infty(\mathfrak{z}')$. On the other hand $\varpi_{\mathfrak{g}/\mathfrak{z}^u} = \varepsilon(t)\,\gamma\,(\varpi_{\mathfrak{g}/\mathfrak{z}})^t$. Therefore it is clear that

$$q_{\mathfrak{g}/\mathfrak{z}^u} = \varepsilon(t)\,\gamma\,(q_{\mathfrak{g}/\mathfrak{z}})^u$$

in the notation of §18. Hence

$$\varpi(\mu)\,S_{\mathfrak{z}^u,\mu}(Z^u) = T_{\mathfrak{z}^u,\mu}(Z^u; \nabla_{\mathfrak{z}^u} \circ \partial(q_{\mathfrak{g}/\mathfrak{z}^u})) = T_{\mathfrak{z},\mu}(Z; \nabla_{\mathfrak{z}} \circ \partial(q_{\mathfrak{g}/\mathfrak{z}})) = \varpi(\mu)\,S_{\mathfrak{z},\mu}(Z)$$

for $Z \in \mathfrak{z}'$ and $\mu \in \mathfrak{F}^+$. This shows that

$$S_{\mathfrak{z}^u,\mu}(Z^u) = S_{\mathfrak{z},\mu}(Z)$$

and so by making μ tend to λ, we deduce that

$$S_{b^u}{}^+(Z) = S_b{}^+(Z) \quad (Z \in \mathfrak{z}).$$

Obviously this implies the second assertion of the lemma.

COROLLARY. *Let x be an element in G such that $b^x \in B$. Then*

$$\Theta_{b^x}{}^+(\exp Z^x) = \Theta_b{}^+(\exp Z),$$

$$\Psi'_{b^x}{}^+(\exp Z^x) = \Psi'_b{}^+(\exp Z) \quad (Z \in \mathfrak{z}_0'').$$

Since $b^x \in B$, it is clear that $B^{x^{-1}} \subset \Xi$. Hence $\mathfrak{b}$ and $\mathfrak{b}^{x^{-1}}$ are two fundamental Cartan subalgebras of $\mathfrak{z}$ and therefore we can choose $y \in \Xi$ such that $\mathfrak{b}^{yx^{-1}} = \mathfrak{b}$ (see [2 (d), p. 237]). Put $u = xy^{-1}$. Then $x = uy$ and $b^x = b^u$. Therefore

$$\Theta_{b^x}{}^+(\exp Z^x) = \Theta_{b^u}{}^+(\exp Z^{uy}) = \Theta_b{}^+(\exp Z^y)$$

by Lemma 48. Similarly

$$\Psi'_{b^x}{}^+(\exp Z^x) = \Psi'_b{}^+(\exp Z^y) \quad (Z \in \mathfrak{z}_0'').$$

Since $\Theta_b{}^+$ and $\Psi'_b{}^+$ are obviously invariant under Ξ, we get the required assertion.

Since $b^x = b^u$, we have obtained the following result during the above proof.

LEMMA 49. *If two elements of B are conjugate under G, then they are also conjugate under the normalizer of B in G.*

Now fix $a \in B^G$, define $\mathfrak{z}_a$ and $\Xi(a)$ as usual (see [2 (m), §4]) and put $\Xi_0(a) = \exp(\mathfrak{g}_0 \cap \mathfrak{z}_a)$, $\Xi_0''(a) = \Xi_0(a) \cap (a^{-1}G')$. Choose $x \in G$ such that $a^x \in B$ and define

$$\Theta_a^+(y) = \Theta_{a^x}^+(y^x) \quad (y \in \Xi_0''(a))$$

and
$$\Psi_a^+(y) = \Psi_{a^x}^+(y^x) \quad (y \in \Xi_0(a)).$$

It follows from the corollary of Lemma 48 that these definitions are independent of the choice of x.

We now define two functions Θ^+ and Ψ^+ on G' as follows. Fix $h \in G'$ and let $\mathfrak{a}$ be the centralizer of h in $\mathfrak{g}$ and A the corresponding Cartan subgroup of G. Define A_I and A_R as in §21 and let $h = h_1 h_2$ ($h_1 \in A_I$, $h_2 \in A_R$). Since every eigenvalue of ad H is real for $H \in \mathfrak{a}_R$ and since h is regular, it is clear that $h_2 \in \Xi_0''(h_1)$. We define

$$\Theta^+(h) = \Theta_{h_1}^+(h_2), \quad \Psi^+(h) = \Psi_{h_1}^+(h_2).$$

(Observe that $A_I \subset B^G$ from the corollary of Lemma 45.) If $x \in G$, it is obvious that

$$\Theta^+(h^x) = \Theta_{h_1^x}^+(h_2^x) = \Theta_{h_1}^+(h_2) = \Theta^+(h).$$

Similarly $\Psi^+(h^x) = \Psi^+(h)$. This shows that Θ^+ and Ψ^+ are invariant under G. We intend to prove that they are analytic on G'.

LEMMA 50. *Fix $b \in B$. Then there exists a number $c_b > 0$ with the following property. Let $\mathfrak{z}_b(c_b)$ be the set of all $Z \in \mathfrak{z}_b$ such that* [1] $|\operatorname{Im} \mu| < c_b$ *for every eigenvalue μ of* $(\operatorname{ad} Z)_{\mathfrak{g}/\mathfrak{z}_b}$. *Then*

$$\Theta_b^+(\exp Z) = \Theta^+(b \exp Z), \quad \Psi_b^+(\exp Z) = \Psi^+(b \exp Z)$$

for all $Z \in \mathfrak{g}_0 \cap \mathfrak{z}_b(c_b)$ such that $b \exp Z \in G'$.

It is obvious that if c_b is sufficiently small, $\nu_b(\exp Z) \neq 0$ for $Z \in \mathfrak{z}_b(c_b)$. Let $\mathfrak{z}_b'(c_b)$ be the set of those elements of $\mathfrak{z}_b(c_b)$ which are regular in $\mathfrak{z}_b$. Then for any $Z \in \mathfrak{g}_0 \cap \mathfrak{z}_b(c_b)$, the two conditions $b \exp Z \in G'$ and $Z \in \mathfrak{g}_0 \cap \mathfrak{z}_b'(c_b)$ are obviously equivalent. Hence, in particular,

$$\mathfrak{g}_0 \cap \mathfrak{z}_b'(c_b) \subset \mathfrak{g}_0 \cap \mathfrak{z}_b''.$$

Fix $Z_0 \in \mathfrak{g}_0 \cap \mathfrak{z}_b'(c_b)$ and let $\mathfrak{a}$ be the centralizer of Z_0 in $\mathfrak{z}_b$. Then $\mathfrak{a}$ is a Cartan subalgebra of $\mathfrak{g}$. Since $b = \theta(b)$, $\mathfrak{z}_b$ is stable under θ and therefore, by Lemmas 29 and 45, we can choose $y \in \Xi(b)$ such that $\mathfrak{a}^y$ is stable under θ and $(\mathfrak{a}_I)^y \subset \mathfrak{b}$. Put $H_0 = Z_0^y$.

---

[1] $\operatorname{Im} \mu$ denotes, as usual, the imaginary part of a complex number μ.

Since $\Theta_b{}^+$, Θ^+, $\Psi'_b{}^+$ and Ψ'^+ are all invariant under $\Xi(b)$, it would be enough to verify that

$$\Theta_b{}^+(\exp H_0) = \Theta^+(b \exp H_0), \quad \Psi'_b{}^+(\exp H_0) = \Psi'^+(b \exp H_0).$$

So we may assume that $Z_0 = H_0$, $y = 1$, $\theta(\mathfrak{a}) = \mathfrak{a}$ and $\mathfrak{a}_I = \mathfrak{a} \cap \mathfrak{k} \subset \mathfrak{b}$.

Let $H_0 = H_1 + H_2$ where $H_1 \in \mathfrak{a}_I$, $H_2 \in \mathfrak{a}_R$. Then $h = b \exp H_0 = h_1 h_2$ where $h_1 = b \exp H_1 \in A_I$ and $h_2 = \exp H_2$. (A is, as before, the Cartan subgroup of G corresponding to $\mathfrak{a}$.) It is clear that $H_1 \in \mathfrak{z}_b(c_b)$ and therefore $\nu_b(\exp H_1) \neq 0$. Hence $\mathfrak{z}_{h_1} \subset \mathfrak{z}_b$. Now put $\mathfrak{z}_1 = \mathfrak{z}_b$, $\mathfrak{z}_2 = \mathfrak{z}_{h_1}$. Then $\mathfrak{z}_2$ is the centralizer of H_1 in $\mathfrak{z}_1$ so that Lemma 31 is applicable.

For $\mu \in \mathfrak{F}'$, define the distributions $T_{i,\mu} = T_{\mathfrak{z}_i,\mu}$ and $S_{i,\mu} = S_{\mathfrak{z}_i,\mu}$ on $\mathfrak{z}_i$ ($i = 1, 2$) as in Lemma 37 corresponding to the constants $a_s = \langle (b^*)^s, b \rangle$ ($s \in W_G$). For any $f \in C_c^\infty(\mathfrak{z}_2)$, define $f_{H_1}(Z) = f(Z - H_1)$ ($Z \in \mathfrak{z}_2$) and put

$$T_{2,\mu}'(f) = T_{2,\mu}(f_{H_1}), \quad S_{2,\mu}'(f) = S_{2,\mu}(f_{H_1}).$$

Then
$$\pi_{\mathfrak{z}_2}(H) T_{2,\mu}'(H) = \sum_{s \in W_G} \varepsilon(s) \langle (b^*)^s, b \rangle e^{s\mu(H + H_1)} \quad (H \in \mathfrak{b}').$$

Moreover H_1 lies in the center of $\mathfrak{z}_2$ and

$$\langle (b^*)^s, h_1 \rangle = \langle (b^*)^s, b \rangle e^{s\lambda(H_1)} \quad (s \in W_G).$$

Now suppose μ tends to λ ($\mu \in \mathfrak{F}^+$). Then it follows from Lemma 38 that

$$\lim_{\mu \to \lambda} T_{2,\mu}'(f) = T_{h_1}{}^+(f)$$

and similarly (see the corollary of Lemma 41)

$$\lim_{\mu \to \lambda} S_{2,\mu}'(f) = S_{h_1}{}^+(f) \quad (f \in C_c^\infty(\mathfrak{z}_2)).$$

Define
$$T_i{}^+ = T_{\mathfrak{z}_i,\lambda}{}^+, \quad S_i{}^+ = S_{\mathfrak{z}_i,\lambda}{}^+ \quad (i = 1, 2)$$

in the notation of §18. Then it is clear from the above result that

$$T_{h_1}{}^+(f) = T_2{}^+(f_{H_1}), \quad S_{h_1}{}^+(f) = S_2{}^+(f_{H_1}) \quad (f \in C_c^\infty(\mathfrak{z}_2)).$$

Moreover $T_b{}^+ = T_1{}^+$, $S_b{}^+ = S_1{}^+$ by definition. Hence

$$\Theta^+(h) = \Theta_{h_1}{}^+(h_2) = D_{h_1}(H_2)^{-1} T_{h_1}{}^+(H_2) = D_{h_1}(H_2)^{-1} T_2{}^+(H_1 + H_2).$$

On the other hand $T_2{}^+ = \eta_0 T_1{}^+$ pointwise on $\mathfrak{g}' \cap \mathfrak{z}_2$ by Lemma 42 and

20 − 652923. Acta mathematica. 113. Imprimé le 12 mai 1965.

$$\Theta_b{}^+ (\exp H_0) = D_b (H_0)^{-1} T_1{}^+ (H_0).$$

Hence it would be enough to verify that $D_b(H)\eta_0(H) = D_{h_1}(H - H_1)$ for $H \in \mathfrak{a}$. Put

$$v(Z) = D_{h_1}(Z - H_1) - D_b(Z)\eta_0(Z) \quad (Z \in \mathfrak{z}_2).$$

Then v is an analytic function on $\mathfrak{z}_2$ which is invariant under $\Xi_2 = \Xi(h_1)$. So it would be enough to show that $v = 0$ on $\mathfrak{b}'$. But it follows from the definition of D_{h_1}, D_b and η_0 that

$$v(H) = \pi_{\delta_2}(H - H_1)^{-1} \Delta(h_1 \exp (H - H_1))$$

$$- \pi_{\delta_1}(H)^{-1}\Delta(b \exp H)\pi_{\delta_1}(H)\pi_{\delta_2}(H)^{-1} = 0 \quad (H \in \mathfrak{b}'),$$

since $\pi_{\delta_2}(H - H_1) = \pi_{\delta_2}(H)$. This proves the first statement of the lemma.

On the other hand,

$$\Psi^+(h) = \Psi_{h_1}{}^+(h_2) = S_{h_1}{}^+(H_2) = S_2{}^+(H_1 + H_2)$$

$$= S_1{}^+(H_1 + H_2) = S_b{}^+(H_0) = \Psi_b{}^+(\exp H_0)$$

since $S_1{}^+ = S_2{}^+$ on $\mathfrak{z}_2$ from Lemma 42. This proves the second statement.

COROLLARY. Θ^+ *and* Ψ^+ *are both analytic on* G'. *Moreover* Ψ^+ *can be extended to a continuous function on* G.

Let Ω be the set of all points $x_0 \in G$ with the following property. There exists an open neighborhood U of x_0 in G such that Θ^+ and Ψ^+ are both analytic on $U \cap G'$ and Ψ^+ extends to a continuous function on U. We have to verify that $\Omega = G$. Clearly Ω is an open and invariant subset of G. Therefore, in view of [2 (m), Lemma 7], it would be sufficient to verify that every semisimple element of G lies in Ω.

Fix a semisimple element $a \in G$. Then we can choose (see the corollary of [2 (m), Lemma 5]) a Cartan subgroup A of G containing a. Let $a = a_1 a_2$ where $a_1 \in A_I$, $a_2 \in A_R$. By Lemma 45 we can choose $x \in G$ such that $b = a_1{}^x \in B$. Since Ω is invariant, it would be enough to verify that $a^x \in \Omega$. Hence we may assume that $x = 1$ and $a = ba_2$ where $b = a_1 \in A_I \cap B$. Now put $V = \exp(\mathfrak{g}_0 \cap \mathfrak{z}_b(c_b)) \subset \Xi(b)$ in the notation of Lemma 50. Then V is an open neighborhood of 1 in $\Xi(b)$ and

$$\Theta^+(by) = \Theta_b{}^+(y), \quad \Psi^+(by) = \Psi_b{}^+(y)$$

for $y \in V' = V \cap (b^{-1}G')$. Moreover we note that $\Psi_b{}^+$ is continuous on V, $a_2 \in V$ and $v_b(a_2) \neq 0$.

Now let $x \to x^*$ denote the natural mapping of G on $G^* = G/\Xi(b)$ and fix open neighborhoods V_0 and G_0^* of a_2 and 1^* in V and G^* respectively. If V_0 and G_0^* are sufficiently small, we can choose an analytic mapping ϕ of G_0^* into G such that:

1) $(\phi(x^*))^* = x^*$ $(x^* \in G_0^*)$.

2) The mapping $\psi:(x^*, y) \to (by)^{\phi(x^*)}$ of $G_0^* \times V_0$ into G is univalent and regular.

This is evidently possible (see [2 (m), Lemma 14]). Put $U = \psi(G_0^* \times V_0)$. Then U is an open neighborhood of $a = ba_2$ in G and ψ defines an analytic diffeomorphism of $G_0^* \times V_0$ onto U. Put $V_0' = V_0 \cap V'$ and $U' = U \cap G'$. Then it is obvious that $\psi(G_0^* \times V_0') = U'$. Since Θ^+ and Ψ^+ are invariant functions, it is clear that

$$\Theta^+(\psi(x^*, y)) = \Theta^+(by) = \Theta_b^+(y), \qquad \Psi^+(\psi(x^*, y)) = \Psi^+(by) = \Psi_b^+(y)$$

for $x^* \in G_0^*$ and $y \in V_0'$. However Θ_b^+ and Ψ_b^+ are both analytic on V'. Therefore it follows that Θ^+ and Ψ^+ are analytic on U'. Similarly since Ψ_b^+ is continuous on V, we conclude that Ψ^+ can be extended to a continuous function on U. This proves the corollary.

Define the character ξ_ϱ of B as in [2 (m), § 18].

LEMMA 51. *Let Z_G be the center of G. Then*

$$\Theta^+(zx) = \xi_\varrho(z)^{-1} \langle b^*, z \rangle \Theta^+(x), \qquad \Psi^+(zx) = \langle b^*, z \rangle \Psi^+(x)$$

for $z \in Z_G$ and $x \in G'$.

Fix $h \in G'$ and let $\mathfrak{a}$ be the centralizer of h in $\mathfrak{g}$ and A the corresponding Cartan subgroup of G. Then $h = h_1 h_2$ $(h_1 \in A_I, h_2 \in A_R)$ and we can choose $y \in \bar{G}$ such that $h_1^y \in B$ (Lemma 45). The required result holds for $x = h$ if and only if it holds for $x = h^y$. Hence we may assume that $y = 1$ and therefore $h_1 \in B$. Then

$$\Theta^+(zh) = \Theta_{zh_1}^+(h_2) = D_{zh_1}(H_2)^{-1} T_{zh_1}^+(H_2),$$

$$\Psi^+(zh) = \Psi_{zh_1}^+(h_2) = S_{zh_1}^+(H_2) \quad (z \in Z_G)$$

where [1] $H_2 = \log h_2 \in \mathfrak{a}_R$. Now h_1 and zh_1 have the same centralizer $\mathfrak{z}$ in $\mathfrak{g}$ and so it is obvious from the definitions of the various distributions that

$$T_{zh_1}^+ = \langle b^*, z \rangle T_{h_1}^+, \qquad S_{zh_1}^+ = \langle b^*, z \rangle S_{h_1}^+.$$

On the other hand $\qquad\qquad \Delta(zb) = \xi_\varrho(z) \Delta(b) \quad (b \in B).$

---

[1] As usual log denotes the inverse of the exponential mapping of $\mathfrak{a}_R$ onto A_R.

Therefore it is clear that

$$D_{zh_1}(Z) = \xi_\varrho(z)\, D_{h_1}(Z) \quad (Z \in \mathfrak{z})$$

and now our assertions follow immediately.

LEMMA 52. *Let A be a Cartan subgroup of G and put $A' = A \cap G'$. Then*

$$\sup_{h \in A'} |\Delta_A(h)\, \Theta^+(h)| < \infty,$$

in the notation of [2 (m), § 19].

Since A_I/Z_G is compact, it would, in view of Lemma 51, be enough to prove the following result.

LEMMA 53. *For any $a \in A_I$, we can choose an open neighborhood U of 1 in A such that $U \supset A_R$ and*

$$\sup_{h \in U'} |\Delta_A(ah)\, \Theta^+(ah)| < \infty.$$

Here $U' = U \cap a^{-1}A'$.

By Lemma 45 we can select $x \in G$ such that $a^x \in B$. Hence, in view of the invariance of Θ^+, we may assume, without loss of generality, that $a \in B$. Then from Lemma 50,

$$\Theta^+(a \exp Z) = \Theta_a^+(\exp Z) = D_a(Z)^{-1} T_a^+(Z)$$

for all $Z \in \mathfrak{g}_0 \cap \mathfrak{z}_a(c_a)$ such that $a \exp Z \in G'$. Let $\mathfrak{a}$ be the Lie algebra of A. Then $\mathfrak{a} \subset \mathfrak{z}_a$. Put $\mathfrak{a}_0 = \mathfrak{a} \cap \mathfrak{g}_0 \cap \mathfrak{z}_a(c_a)$ and $U = \exp \mathfrak{a}_0$. Then $U \supset A_R$ and if $a \exp H \in G'$ ($H \in \mathfrak{a}_0$), it is clear that

$$|\Delta_A(a \exp H)\Theta^+(a \exp H)| = |\Delta_A(a \exp H)\Theta_a^+(\exp H)| = |\pi_{\mathfrak{z}_a}(H)\, T_a^+(H)|$$

from the corollary of Lemma 47 and [2 (m), Lemma 33]. Hence if we take into account Lemmas 28, 38 and 40, we get

$$\sup_{h \in U'} |\Delta_A(ah)\, \Theta^+(ah)| < \infty.$$

LEMMA 54. *Θ^+ is locally summable on G and*

$$\sup_{x \in G'} |D(x)|^{\frac{1}{2}} |\Theta^+(x)| < \infty.$$

Moreover

$$z\Theta^+ = \chi_{b*}(z)\, \Theta^+ \quad (z \in \mathfrak{z})$$

as a distribution on G.

Since there are only a finite number of non-conjugate Cartan subgroups of G, it follows from Lemma 52 that

$$\sup_{x \in G'} |D(x)|^{\frac{1}{2}} |\Theta^+(x)| < \infty.$$

Therefore Θ^+ is locally summable on G from [2 (m), Lemma 53].

Now fix $z \in \mathfrak{Z}$ and consider the distribution

$$T = z\,\Theta^+ - \chi_{b*}(z)\,\Theta^+$$

on G. We have to show that $T = 0$. In view of [2 (m), Lemma 7], it would be enough to verify that no semisimple element of G lies in Supp T.

Fix a semisimple element $h \in G$. Then h lies in some Cartan subgroup A of G [2 (m), Cor. of Lemma 5]. Let $h = h_1 h_2$ $(h_1 \in A_I, h_2 \in A_R)$. Then again by Lemma 45, there exists $x \in G$ such that $h_1^x \in B$. T being invariant, it would be sufficient to prove that $h^x \notin$ Supp T. Hence replacing (h, A) by (h^x, A^x), we may assume that $a = h_1 \in B$. Let σ_T and σ_{Θ^+} be the distributions on $\Xi'(a)$ corresponding to T and Θ^+ respectively under [2 (m), Lemma 15]. Then

$$\sigma_T = |\nu_a|^{-\frac{1}{2}} \mu_a(z) \left(|\nu_a|^{\frac{1}{2}} \sigma_{\Theta^+}\right) - \chi_{b*}(z)\,\sigma_{\Theta^+}$$

by [2 (m), Lemma 22] where $\mu_a = \mu_{\mathfrak{g}/\mathfrak{z}_a}$ as in Lemma 46. Let θ_a denote the function $y \to \Theta^+(ay)$ on $\Xi'(a)$. Then it follows from [2 (i), Cor. 2 of Theorem 1] that θ_a is locally summable and therefore $\sigma_{\Theta^+} = \theta_a$ from the definition of σ_{Θ^+}. Hence it follows from Lemma 50 and the corollary of Lemma 47 that the distribution $|\nu_a|^{\frac{1}{2}} \sigma_{\Theta^+}$ coincides on $V = \exp(\mathfrak{g}_0 \cap \mathfrak{z}_a(c_a))$ with the locally summable function $\varepsilon_a(0)\,\Phi_a^+$. Therefore we conclude from Lemma 46 that $\sigma_T = 0$ on V. Since V is an open subset of $\Xi'(a)$ containing h_2, we conclude [2 (m), Lemma 15] that $T = 0$ around $h = ah_2$. This proves Lemma 54.

LEMMA 55. $$\Theta^+(b) = \Delta(b)^{-1} \sum_{s \in W_G} \varepsilon(s) \langle (b^*)^s, b \rangle$$

for $b \in B'$.

Fix $b \in B'$. Then $\mathfrak{z}_b = \mathfrak{b}$ and therefore $D_b(H) = \Delta(b \exp H)$ and

$$T_b^+(H) = \sum_{s \in W_G} \varepsilon(s) \langle (b^*)^s, b \rangle\, e^{\lambda(s^{-1}H)} \quad (H \in \mathfrak{b}).$$

Hence $$\Theta^+(b) = \Theta_b^+(1) = D_b(0)^{-1} T_b^+(0) = \Delta(b)^{-1} \sum_{s \in W_G} \varepsilon(s) \langle (b^*)^s, b \rangle.$$

This shows that Θ^+ satisfies all the conditions of Theorem 3. Therefore in view of Lemma 44, the proof of Theorem 3 is now complete.

§ 23. Further properties of Θ

Let $\mathfrak{a}$ be a Cartan subalgebra of $\mathfrak{g}$ and A the corresponding Cartan subgroup of G. Put $\mathfrak{a}_R' = \mathfrak{a}_R \cap \mathfrak{a}'(R)$ and $A_R' = A_R \cap A'(R)$ in the notation of [2 (m), § 19]. Let A^+ be a connected component of $A'(R)$. Then it is obvious that $A^+ = A_I^+ A_R^+$ where A_I^+ is a connected component of A_I and $A_R^+ \subset A_R$.

Let us assume that $\theta(\mathfrak{a}) = \mathfrak{a}$. Then by Lemma 45 we can choose $k \in K$ such that $(A_I^+)^k \subset B$. Hence we may suppose that $A_I^+ \subset B$. Let $\mathfrak{z}$ denote the centralizer of A_I^+ in $\mathfrak{g}$. Then $\mathfrak{a}$ and $\mathfrak{b}$ are both Cartan subalgebras of $\mathfrak{z}$. Consider the complex-analytic subgroup Ξ_c of G_c corresponding to $\mathrm{ad}\,\mathfrak{z}_c$. We can choose $y \in \Xi_c$ such that $\mathfrak{b}_c^y = \mathfrak{a}_c$. Put $W(A^+) = W(\mathfrak{z}/\mathfrak{a})$. Since $\mathfrak{a}_I$ lies in the center of $\mathfrak{z}$, every root of $(\mathfrak{z}, \mathfrak{a})$ is real. Hence (see [2 (k), Lemma 6]) every element of $W(A^+)$ is induced on $\mathfrak{a}$ by some element of the analytic subgroup Ξ of G corresponding to $\mathfrak{z}$. Let W_Ξ be the subgroup of those elements of $W(\mathfrak{z}/\mathfrak{b})$ which can be induced on $\mathfrak{b}$ by some element of Ξ. Then $W_\Xi = W_k(\mathfrak{z}/\mathfrak{b})$ in the notation of § 18.

Put $\mathfrak{w}(A_I^+) = W_G \cap W(\mathfrak{z}/\mathfrak{b})$ and write $\mathfrak{w} = \mathfrak{w}(A_I^+)$ for simplicity.

LEMMA 56. *Suppose* t_1, t_2 *are two elements in* W_G *such that*

$$t_1^y \in W(A^+) t_2^y.$$

Then $t_1 \in \mathfrak{w}\, t_2$.

Put $t = t_1 t_2^{-1}$. Then $t \in (W(A^+))^{y^{-1}} \cap W_G = W(\mathfrak{z}/\mathfrak{b}) \cap W_G = \mathfrak{w}$.

COROLLARY. *Let* $r = [W_G : \mathfrak{w}]$ *and* $t_1, \ldots, t_r$ *a complete set of representatives in* W_G *for* $\mathfrak{w} \backslash W_G$. *Then the elements* st_i^y $(s \in W(A^+),\ 1 \leqslant i \leqslant r)$ *are all distinct.*

This is obvious from the above lemma.

LEMMA 57. *Fix an element* $b^* \in B^{*\prime}$ *and define* Θ *as in Theorem 3. Then there exist unique complex numbers* $c_{b^*}(s : t : A^+)$ $(s \in W(A^+), t \in W_G)$ *such that*

1) $c_{b^*}(su^y : u^{-1}t : A^+) = c_{b^*}(s : t : A^+)$ $(u \in \mathfrak{w})$,

2) $\Delta_A(h_1 h_2)\,\Theta(h_1 h_2) = \sum_{t \in \mathfrak{w}\backslash W_G} \varepsilon(t)\, \langle (b^*)', h_1 \rangle \sum_{s \in W(A^+)} \varepsilon(s)\, c_{b^*}(s : t : A^+)\, \exp\,(s\,(t\lambda)^y\,(H_2))$

for $h_1 \in A_I^+$, $h_2 \in A_R^+$. Here $\lambda = \log b^*$ and [1] $H_2 = \log h_2$.

Let $H_1 \in \mathfrak{a}_I$ and $H_2 \in \mathfrak{a}_R$. Then since s^{-1} and y^{-1} leave H_1 fixed, it is clear that

$$\langle (b^*)', \exp H_1 \rangle\, \exp\,(s\,(t\lambda)^y\,(H_2)) = \exp\,(s\,(t\lambda)^y\,(H_1 + H_2))$$

---

[1] See footnote 1, p. 299.

for $s \in W(A^+)$ and $t \in W_G$. Since λ is regular, the uniqueness is obvious from the corollary of Lemma 56. On the other hand the existence is seen as follows. We use the notation of § 18. Put $\mathfrak{a}^+ = \mathfrak{a}_I + \log A_R^+$. Then $\mathfrak{a}^+$ is a connected component of $\mathfrak{a}'(\mathfrak{z}:R)$ (see § 13).

LEMMA 58. *Put*

$$c(s:\mathfrak{F}^+:A^+) = \sum_{t \in \mathfrak{w}/W_\Xi} c_{\mathfrak{z}}(st^\mathfrak{y}:t^{-1}\mathfrak{F}^+:\mathfrak{a}^+)$$

for $s \in W(A^+)$ *and any connected component* $\mathfrak{F}^+$ *of* $\mathfrak{F}'$. *Then*

$$c_{b*}(s:t:A^+) = c(s:t\mathfrak{F}^+:A^+) \quad (s \in W(A^+), t \in W_G)$$

where $\mathfrak{F}^+$ *is the component of* $\log b^*$ *in* $\mathfrak{F}'$ $(b^* \in B^{*\prime})$.

In view of Lemma 40, the definition of $c(s:\mathfrak{F}^+:A^+)$ is legitimate and it is obvious that

$$c(su^\mathfrak{y}:u^{-1}\mathfrak{F}^+:A^+) = c(s:\mathfrak{F}^+:A^+) \quad (u \in \mathfrak{w}).$$

Therefore it would be sufficient to prove the following result.

LEMMA 59. *Fix* $b^* \in B^*$ *and a connected component* $\mathfrak{F}^+$ *of* $\mathfrak{F}'$ *such that* $\lambda = \log b^* \in \mathrm{Cl}\,\mathfrak{F}^+$ *and define* Θ^+, Ψ^+ *as in* § 22 *corresponding to* b^* *and* $\mathfrak{F}^+$. *Then*

$$\Delta_A(h_1 h_2)\,\Theta^+(h_1 h_2) = \sum_{t \in \mathfrak{w}\backslash W_G} \varepsilon(t)\,\langle (b^*)^t, h_1 \rangle \sum_{s \in W(A^+)} \varepsilon(s)\,c\,(s:t\mathfrak{F}^+:A^+)\,\exp\,(s\,(t\lambda)^\mathfrak{y}\,(H_2)),$$

and $$\Psi^+(h_1 h_2) = \sum_{t \in \mathfrak{w}\backslash W_G} \langle (b^*)^t, h_1 \rangle \sum_{s \in W(A^+)} c(s:t\mathfrak{F}^+:A^+)\,\exp\,(s\,(t\lambda)^\mathfrak{y}\,(H_2))$$

for $h_1 \in A_I^+$ *and* $h_2 \in A_R^+$. *Here* $H_2 = \log h_2$ *as before.*

Fix a point $b_0 \in A_I^+$. Then we can choose $H_0 \in \mathfrak{a}_I$ arbitrarily near zero such that 1) $\nu_{b_0}(\exp H_0) \neq 0$ and 2) every root of $(\mathfrak{z}_{b_0}, \mathfrak{a})$ which vanishes at H_0 is real. Put $b = b_0 \exp H_0$. Then $b \in A_I^+$ and it is obvious that $\mathfrak{z}_b = \mathfrak{z}$. This shows that the set V of those points $b \in A_I^+$ for which $\mathfrak{z}_b = \mathfrak{z}$, is dense in A_I^+. Fix a point $b \in V$. Then from Lemma 50,

$$\Theta^+(b\exp Z) = \Theta_b^+(\exp Z) = D_b(Z)^{-1} T_b^+(Z), \quad \Psi^+(b\exp Z) = \Psi_b^+(\exp Z) = S_b^+(Z)$$

for all $Z \in \mathfrak{g}_0 \cap \mathfrak{z}_b(c_b)$ such that $b \exp Z \in G'$. Put $U = \mathfrak{a}^+ \cap \mathfrak{g}_0 \cap \mathfrak{z}_b(c_b)$ and let U' be the set of all points $H \in U$ where $\Delta_A(b \exp H) \neq 0$. Recall that P is the set of all positive roots of $(\mathfrak{g}, \mathfrak{b})$. Then we may assume, without loss of generality, that $P^\mathfrak{y}$ is the set of all positive roots of $(\mathfrak{g}, \mathfrak{a})$. Then it is clear that

$$D_b(\exp H) = \Delta_A(b \exp H) \pi_\delta{}^\mathfrak{a}(H)^{-1}$$

and therefore

$$\Delta_A(b \exp H) \Theta^+(b \exp H) = \pi_\delta{}^\mathfrak{a}(H) T_b{}^+(H) \quad (H \in U').$$

On the other hand it follows from Lemmas 38 and 40 that

$$\pi_\delta{}^\mathfrak{a}(H) T_b{}^+(H) = \sum_{t \in W_\Xi \backslash W_G} \varepsilon(t) \langle (b^*)^t, b \rangle \sum_{s \in W(A^+)} \varepsilon(s) c_\delta(s:t\mathfrak{F}^+:\mathfrak{a}^+) \exp(s(t\lambda)^\nu(H))$$

for $H \in U'$. Now suppose $H = H_1 + H_2$ $(H_1 \in \mathfrak{a}_I, H_2 \in \mathfrak{a}_R)$. Since s^{-1} and y^{-1} leave $\mathfrak{a}_I$ pointwise fixed, it is clear that

$$\langle (b^*)^t, b \rangle \exp(s(t\lambda)^\nu(H)) = \langle (b^*)^t, h_1 \rangle \exp(s(t\lambda)^\nu(H_2))$$

for $s \in W(A^+)$ and $t \in W_G$. Here $h_1 = b \exp H_1$. Therefore since the function

$$h \to \Delta_A(h) \Theta^+(h) \quad (h \in A^+ \cap A')$$

extends to an analytic function on A^+ (see [2 (m), Lemma 31]), it is obvious that

$$\Delta_A(h_1 h_2) \Theta^+(h_1 h_2) = \sum_{t \in W_\Xi \backslash W_G} \varepsilon(t) \langle (b^*)^t, h_1 \rangle \sum_{s \in W(A^+)} \varepsilon(s) c_\delta(s:t\mathfrak{F}^+:\mathfrak{a}^+) \exp(s(t\lambda)^\nu(H_2))$$

for $h_1 \in A_I{}^+, h_2 \in A_R{}^+$. Our first assertion now follows immediately if we take into account Lemma 40.

Similarly we conclude from Lemmas 50 and 41 that

$$\Psi^+(b \exp H) = S_b{}^+(H) = \sum_{t \in W_\Xi \backslash W_G} \langle (b^*)^t, b \rangle \sum_{s \in W(A^+)} c_\delta(s:t\mathfrak{F}^+:\mathfrak{a}^+) \exp(s(t\lambda)^\nu(H))$$

for $H \in U'$. Since Ψ^+ extends to a continuous function on G (see the corollary of Lemma 50), this relation holds for all $H \in U$. Now $\log A_R{}^+ \subset U$ and V is dense in $A_I{}^+$. Therefore the second assertion of the lemma is now obvious.

LEMMA 60. $c(s:\mathfrak{F}^+:A^+) = 0$ *unless* $\Re \mu^\nu(s^{-1}H) \leqslant 0$ *for every* $\mu \in \mathfrak{F}^+$ *and* $H \in \mathfrak{a}^+$.

This is obvious from Lemma 58 and Lemma 28.

COROLLARY. *There exists a number* C *(independent of* b^* *and* $\mathfrak{F}^+$*) such that*

$$|D(x)|^{\frac{1}{2}}|\Theta^+(x)| \leqslant C \quad (x \in G')$$

and
$$|\Psi^+(x)| \leqslant C \quad (x \in G)$$

in the above notation.

Let $C(A^+)$ denote the maximum of $|c_3(s:\mathfrak{F}^+:\mathfrak{a}^+)|$ for all $s \in W(A^+)$ and all $\mathfrak{F}^+$. Then it follows from Lemmas 58, 59 and 60 that

$$|\Delta_A(h)\,\Theta^+(h)| \leqslant [\mathfrak{w}:W_\Xi]\,[W_G:\mathfrak{w}]\,[W(A^+)]\,C(A^+) \leqslant [W]^2\,C(A^+) \quad (h \in A' \cap A^+)$$

where $W = W(\mathfrak{g}/\mathfrak{b})$ as usual. Similarly

$$|\Psi^+(h)| \leqslant [W]^2\,C(A^+) \quad (h \in A^+).$$

It is clear that $C(zA^+) = C(A^+)$ for $z \in Z_G$. Therefore since A/Z_G and $\mathfrak{a}'(R)$ both have only a finite number of connected components,

$$C(A) = [W]^2 \operatorname*{Sup}_{A^+} C(A^+) < \infty.$$

Here A^+ runs over all connected components of $A'(R)$. This shows that

$$|\Delta_A(h)\,\Theta^+(h)| \leqslant C(A) \quad (h \in A')$$

and
$$|\Psi^+(h)| \leqslant C(A) \quad (h \in A).$$

But then since G has only a finite number of non-conjugate Cartan subgroups, our assertion is obvious.

§ 24. The distribution Θ_λ^*

Put $L = \log B^*$. Then L is a closed additive subgroup of $\mathfrak{F}$ which is, in fact, a lattice if B is compact. For any $\lambda \in L$, let ξ_λ denote the corresponding element of B^* so that $\xi_\lambda(\exp H) = e^{\lambda(H)}$ $(H \in \mathfrak{b})$. Fix $\lambda \in L$ and a connected component $\mathfrak{F}^+$ of $\mathfrak{F}'$ such that $\lambda \in \operatorname{Cl}\mathfrak{F}^+$. Then we denote by $\Theta_{\lambda,\mathfrak{F}^+}$ and $\Psi_{\lambda,\mathfrak{F}^+}$ respectively, the distributions Θ^+ and Ψ^+ of § 22 for $b^* = \xi_\lambda$. In particular if $\lambda \in L' = L \cap \mathfrak{F}'$, the component $\mathfrak{F}^+$ is uniquely determined and so in this case we denote them simply by Θ_λ and Ψ_λ.

Now fix $\lambda \in L'$ and suppose that $s\lambda \in L$ for every $s \in W = W(\mathfrak{g}/\mathfrak{b})$. Then we intend to study the distribution
$$\Theta_\lambda^* = \sum_{s \in W} \varepsilon(s)\,\Theta_{s\lambda}$$

more closely. Let us return to the notation of § 23 and define

$$\xi_{t,\lambda}(h_1 h_2) = \xi_{t\lambda}(h_1)\,\exp\left((t\lambda)^\nu(\log h_2)\right) \quad (t \in W)$$

for $h_1 \in A_I^+$ and $h_2 \in A_R$. Let $\mathfrak{m}$ be the centralizer of $\mathfrak{a}_R$ in $\mathfrak{g}$ and put

$$W_0 = W(\mathfrak{m}/\mathfrak{a})^{\nu^{-1}}.$$

Since $\mathfrak{a}_I$ lies in the center of $\mathfrak{z}$ and $\mathfrak{a}_R$ in the center of $\mathfrak{m}$, it is clear that $W(\mathfrak{z}/\mathfrak{a})$ and $W(\mathfrak{m}/\mathfrak{a})$ commute (as subgroups of $W(\mathfrak{g}/\mathfrak{a})$). Therefore $W(\mathfrak{z}/\mathfrak{b})$ and W_0 also commute in W.

LEMMA 61. *For any connected component* $\mathfrak{F}^+$ *of* $\mathfrak{F}'$, *define*

$$c^*(t:\mathfrak{F}^+:A^+) = [W_G:\mathfrak{w}] \sum_{s \in W(\mathfrak{z}/\mathfrak{b})} c(s^\nu:s^{-1}t\mathfrak{F}^+:A^+) \quad (t \in W).$$

Then
$$c^*(u^{-1}t:\mathfrak{F}^+:A^+) = c^*(t:\mathfrak{F}^+:A^+)$$

for $u \in W_0$ *and* $t \in W$. *Moreover,*

$$\Delta_A \Theta_\lambda^* = \sum_{t \in W} \varepsilon(t) c^*(t:\mathfrak{F}^+:A^+) \xi_{t,\lambda}$$

on A^+. *Here* $\mathfrak{F}^+$ *is the component of* $\mathfrak{F}'$ *containing* λ.

Fix $u \in W_0$ and $t \in W$. Since u and $W(\mathfrak{z}/\mathfrak{b})$ commute, it is clear that

$$c^*(u^{-1}t:\mathfrak{F}^+:A^+) = [W_G:\mathfrak{w}] \sum_{s \in W(\mathfrak{z}/\mathfrak{b})} c(s^\nu:u^{-1}s^{-1}t\mathfrak{F}^+:A^+)$$

$$= [W_G:W_\Xi] \sum_{s \in W(\mathfrak{z}/\mathfrak{b})} c_{\mathfrak{z}}(s^\nu:u^{-1}s^{-1}t\mathfrak{F}^+:\mathfrak{a}^+)$$

from Lemma 58. Define $\mathfrak{F}_{\mathfrak{z}}'$ as in §18 and for fixed $s \in W(\mathfrak{z}/\mathfrak{b})$ and $t \in W$, let $\mathfrak{F}_{\mathfrak{z}}^+$ be the unique connected component of $\mathfrak{F}_{\mathfrak{z}}'$ containing $s^{-1}t\mathfrak{F}^+$. Since u^{-1} leaves every root of $(\mathfrak{z}, \mathfrak{b})$ fixed, it is clear that $u^{-1}\mathfrak{F}_{\mathfrak{z}}^+ = \mathfrak{F}_{\mathfrak{z}}^+$. Hence

$$c_{\mathfrak{z}}(s^\nu:u^{-1}s^{-1}t\mathfrak{F}^+:\mathfrak{a}^+) = c_{\mathfrak{z}}(s^\nu:\mathfrak{F}_{\mathfrak{z}}^+:\mathfrak{a}^+) = c_{\mathfrak{z}}(s^\nu:s^{-1}t\mathfrak{F}^+:\mathfrak{a}^+)$$

in the notation of §18. This implies the first assertion of the lemma.

Now let $\mathfrak{F}^+$ be the component of $\mathfrak{F}'$ containing λ. Then it follows from Lemma 59 that

$$\Delta_A \Theta_\lambda^* = \sum_{u \in W} \varepsilon(u) \sum_{t \in \mathfrak{w} \backslash W_G} \varepsilon(t) \sum_{s \in W(\mathfrak{z}/\mathfrak{b})} \varepsilon(s) c(s^\nu:tu\mathfrak{F}^+:A^+) \xi_{stu,\lambda}$$

on A^+. From this the second assertion of the lemma follows immediately.

Now assume that G_c is an acceptable complexification (see [2 (m), §18]) of G and G is the real analytic subgroup of G_c corresponding to $\mathfrak{g}$. Let A_c and B_c be the Cartan subgroups of G_c corresponding to $\mathfrak{a}_c$ and $\mathfrak{b}_c$ respectively. Then W operates on

B_c and therefore also on B. Hence L is invariant under W. Similarly $W(\mathfrak{m}/\mathfrak{a})$ operates on A_c. Since it maps $\mathfrak{a}_I$ into itself and leaves $\mathfrak{a}_R$ pointwise fixed, it leaves every point in $A_I \cap \exp(-1)^{\frac{1}{2}}\mathfrak{a}_R$ fixed and maps $A_I^0 = \exp \mathfrak{a}_I$ into itself. Therefore (see [2 (m), Lemma 50]) $W(\mathfrak{m}/\mathfrak{a})$ operates on A and maps A_I^+ into itself. Now if $u \in W_0$ then $s = u^\nu \in W(\mathfrak{m}/\mathfrak{a})$ and

$$\xi_{ut,\lambda}(h_1 h_2) = \xi_{ut\lambda}(h_1)\, \exp((t\lambda)^\nu (\log\, h_2)) = \xi_{t,\lambda}((h_1 h_2)^{s^{-1}}) \quad (t \in W)$$

for $h_1 \in A_I^+$, $h_2 \in A_R$. Hence we obtain the following result from Lemma 61.

LEMMA 62. *Under the above conditions*

$$\Delta_A(h)\,\Theta_\lambda^*(h) = \sum_{t \in W_0 \backslash W} \varepsilon(t)\, c^*(t:\mathfrak{F}^+:A^+) \sum_{s \in W(\mathfrak{m}/\mathfrak{a})} \varepsilon(s)\, \xi_{t,\lambda}(h^s)$$

for $h \in A^+$.

Let P_+ be the set of all positive roots of $(\mathfrak{g}, \mathfrak{a})$ which do not vanish identically on $\mathfrak{a}_R$. Put $\sigma = \frac{1}{2} \sum_{\alpha \in P_+} \alpha$ and

$$\Delta_+(h) = e^{\sigma(\log h_2)} \prod_{\alpha \in P_+} (1 - \xi_\alpha(h^{-1})) \quad (h \in A)$$

in the notation of [2 (m), § 18]. Here h_2 is the component of h in A_R.

COROLLARY. $$\sup_{h \in A'} |\Delta_+(h)\,\Theta_\lambda^*(h)| < \infty.$$

In view of Lemma 51, it is enough to show that $\Delta_+(h)\,\Theta_\lambda^*(h)$ remains bounded for $h \in A^+ \cap A'$. In order to do this we can obviously assume that the set of positive roots of $(\mathfrak{g}, \mathfrak{a})$ is chosen as in [2 (m), § 27]. Define M, Δ_M and ξ_ϱ as in [2 (m), § 27]. Then

$$\Delta_A(h) = \Delta_M(h)\,\Delta_+(h) \quad (h \in A).$$

Moreover, it follows from Lemma 60, that

$$c(s^\nu : s^{-1} t\,\mathfrak{F}^+ : A^+) = 0 \quad (s \in W(\mathfrak{z}/\mathfrak{b}),\, t \in W)$$

unless $(t\lambda)^\nu(H_2) \leqslant 0$ for $H_2 \in \log A_R^+$. Therefore it is clear from Lemmas 61 and 62 that

$$|\Delta_+(h)\,\Theta_\lambda^*(h)| \leqslant \sum_{t \in W_0 \backslash W} |c^*(t:\mathfrak{F}^+:A^+)|\,\Big|\Delta_M(h_1)^{-1} \sum_{s \in W(\mathfrak{m}/\mathfrak{a})} \varepsilon(s)\, \xi_{t\lambda}(h_1^s)\Big|$$

for $h \in A' \cap A^+$ where h_1 is the component of h in A_I^+. Now choose

$$a \in A_I^+ \cap \exp(-1)^{\frac{1}{2}}\mathfrak{a}_R$$

such that $A_I^+ = aA_I^0$, Then (see [2 (m), § 23])

$$\Delta_M(ah) = \xi_\varrho(a)\,\Delta_M(h)$$

and
$$\sum_{s\,\in\,W(\mathfrak{m}/\mathfrak{a})} \varepsilon(s)\,\xi_{t\lambda}((ah)^s) = \xi_{t\lambda}(a)\sum_{s\,\in\,W(\mathfrak{m}/\mathfrak{a})} \varepsilon(s)\,\xi_{t\lambda}(h^s)$$

for $h \in A_I^0$ and $t \in W$. Therefore our assertion is obvious from [2 (b), Cor. 2, p. 139].

§ 25. Statement of Theorem 4

Fix a Haar measure dx on G and consider the distribution

$$\Theta^+(f) = \int f\,\Theta^+ dx \quad (f \in C_c^\infty(G))$$

as in Lemma 54. For any $\varepsilon > 0$, get $G(\varepsilon)$ denote the set of all $x \in G$ where $|D(x)| > \varepsilon^2$. Suppose u is a measurable function on G' which is integrable (with respect to dx) on $G(\varepsilon)$ for every $\varepsilon > 0$. Then we define [1]

$$\text{p.v.} \int u\,dx = \lim_{\varepsilon \to 0} \int_{G(\varepsilon)} u\,dx$$

provided this limit exists and is finite.

THEOREM 4. *Define* Θ^+ *and* Ψ^+ *as in* § 22 *and put* $\lambda = \log b^*$. *Then*

$$\varpi(\lambda)\,\Theta^+(f) = \text{p.v.} \int D^{-1}\Psi^+ \nabla_G f\, dx$$

where ∇_G *has the same meaning as in* [2 (m), § 20].

Before proceeding with the proof, we need some formulas on integrals (cf. § 2). Let $\mathfrak{a}_1 = \mathfrak{b}$, $\mathfrak{a}_2$, ..., $\mathfrak{a}_r$ be a maximal set of Cartan subalgebras of $\mathfrak{g}$ no two of which are conjugate under G. Let A_i be the Cartan subgroup of G corresponding to $\mathfrak{a}_i$. Put $G_i^* = G/A_{i0}$ where A_{i0} is the center of A_i and fix a Haar measure $d_i a$ on A_i and an invariant measure $d_i x^*$ on G_i^*. Also let

$$\Delta_i(a) = \Delta_{A_i}(a) \quad (a \in A_i)$$

in the usual notation (see [2 (m), § 19]).

LEMMA 63. *There exist numbers* $c_i > 0$ $(1 \leqslant i \leqslant r)$ *such that*

---

[1] See footnote 1, p. 246.

$$\int f(x)\,dx = \sum_{1 \leqslant i \leqslant r} c_i \int_{G_i^* \times A_i} |\Delta_i(a)|^2 f(a^{x^*})\,d_i x^*\,d_i a$$

for $f \in C_c(G)$ in the notation of $[2\,(m),\ \S\,22]$.

Put $G_i = A_i{}^G \cap G'$. Then G' is the disjoint union of $G_1, \ldots, G_r$ and our assertion is an immediate consequence of $[2\,(m),\ \text{Lemma } 41]$.

§ 26. A simple property of the function Δ

Let $\mathfrak{a}$ be a Cartan subalgebra of $\mathfrak{g}$ and A the corresponding Cartan subgroup of G. Suppose a is an element of A and α a root of $(\mathfrak{g}, \mathfrak{a})$. We say that a and α *commute* if $\xi_\alpha(a) = 1$ in the notation of $[2\,(m),\ \S\,19]$.

Put $m = \frac{1}{2}\,(\dim \mathfrak{g} - \operatorname{rank} \mathfrak{g})$ as in § 2. Then m is the number of positive roots of $(\mathfrak{g}, \mathfrak{a})$. For any $a \in A$, define the integer $m(R{:}a) \geqslant 0$ as follows. Let $a = a_1 a_2$ $(a_1 \in A_I,\ a_2 \in A_R)$. Then $m(R{:}a)$ is the number of positive real roots of $(\mathfrak{g}, \mathfrak{a})$ which commute with a_1. If α is a real root, $\alpha(H) = 0$ for $H \in \mathfrak{a}_I$. Hence it is clear that $m(R{:}a)$ depends only on the connected component of a_1 in A_I. Therefore the function $m(R){:}a \to m(R{:}a)$ is locally constant on A.

LEMMA 64. $\operatorname{conj} \Delta_A(a) = (-1)^{m + m(R{:}a)} \Delta_A(a)$ $(a \in A)$.

This result is obviously independent of the choice of positive roots. Hence we may select compatible orders on the spaces of real linear functions on $\mathfrak{a}_R$ and $\mathfrak{a}_R + (-1)^{\frac{1}{2}}\mathfrak{a}_I$ respectively and assume that P is the set of positive roots of $(\mathfrak{g}, \mathfrak{a})$ in this order. Let η denote the conjugation of $\mathfrak{g}_c$ with respect to $\mathfrak{g}$. Then it is clear that if α is a root, the same holds for $\eta\alpha$ and

$$\xi_{\eta\alpha}(a) = \operatorname{conj} \xi_\alpha(a) \quad (a \in A).$$

Let P_R, P_I and P_c respectively denote the sets of real, imaginary and complex roots in P (see $[2\,(k),\ \S\,4]$). We now use the notation of $[2\,(m),\ \S\,19]$. Then

$$\Delta(a) = \xi_\varrho(a)\,\Delta_I{}'(a)\,\Delta_+{}'(a)$$

where $\qquad \Delta_I{}'(a) = \prod_{\alpha \in P_I} (1 - \xi_\alpha(a)^{-1}), \quad \Delta_+{}'(a) = \prod_{\alpha \in P_+} (1 - \xi_\alpha(a)^{-1}) \quad (a \in A)$

and $P_+ = P_R \cup P_c$. Since P_+ is invariant under η, it is clear that $\Delta_+{}'(a)$ is real. On the other hand, $\eta\alpha = -\alpha$ for $\alpha \in P_I$. Therefore

$$\operatorname{conj} \Delta_I{}'(a) = (-1)^{m(I)} \xi_{2\varrho_I}(a)\,\Delta_I{}'(a)$$

where $m(I)$ is the number of roots in P_I and $\varrho_I = \frac{1}{2}\sum_{\alpha \in P_I}\alpha$. Now suppose $a = a_1 a_2$ $(a_1 \in A_I,\ a_2 \in A_R)$. Then conj $\xi_\varrho(a) = \xi_\varrho(a_1^{-1}a_2)$ and $\xi_{2\varrho_I}(a_2) = 1$. Hence

$$\text{conj } \Delta(a) = (-1)^{m(I)}\xi_\varrho(a_1^{-1}a_2)\,\xi_{2\varrho_I}(a_1)\,\Delta_I'(a)\,\Delta_+'(a)$$

$$= (-1)^{m(I)}\xi_{2\varrho}(a_1)^{-1}\xi_{2\varrho_I}(a_1)\Delta(a) = (-1)^{m(I)}\xi_{2\varrho_+}(a_1)^{-1}\Delta(a)$$

where $\varrho_+ = \frac{1}{2}\sum_{\alpha \in P_+}\alpha$. Now if $\alpha \in P_c$ then the same holds for $\eta\alpha$ and $\eta\alpha \neq \alpha$. Moreover,

$$\xi_\alpha(a_1)\,\xi_{\eta\alpha}(a_1) = |\xi_\alpha(a_1)|^2 = 1.$$

Hence $$\xi_{2\varrho_+}(a_1) = \xi_{2\varrho_R}(a_1)$$

where $\varrho_R = \frac{1}{2}\sum_{\alpha \in P_R}\alpha$. But for any $\alpha \in P_R$, $\xi_\alpha(a_1)$ is both real and unimodular. Therefore it is ± 1. Hence

$$\xi_{2\varrho_R}(a_1) = \prod_{\alpha \in P_R}\xi_\alpha(\alpha_1) = (-1)^q$$

where q is the number of roots $\alpha \in P_R$ such that $\xi_\alpha(a_1) = -1$. But then $q + m(R:a)$ is the total number of roots in P_R. We have seen above that the roots in P_c occur in pairs. Hence

$$q + m(R:a) + m(I) \equiv m \bmod 2.$$

This shows that $$q + m(I) \equiv m + m(R:a) \bmod 2$$

and therefore $$\text{conj } \Delta(a) = (-1)^{m(I)+q}\Delta(a) = (-1)^{m+m(R:a)}\Delta(a).$$

This proves the lemma.

§ 27. Reduction of Theorem 4 to Lemma 66

We now come to Theorem 4. Suppose V_ε $(0 < \varepsilon \leqslant \varepsilon_0)$ is a family of measurable functions on G such that (cf. § 2)

1) $$0 \leqslant V_\varepsilon \leqslant 1 \quad \text{and} \quad \lim_{\varepsilon \to 0} V_\varepsilon(x) = 1 \text{ for } x \in G',$$

2) $$V \text{ is invariant under } G.$$

3) $$V_\varepsilon(x) = 0 \text{ if } |D(x)| < \varepsilon^2 \quad (x \in G).$$

Fix $f \in C_c^\infty(G)$ and define $F_{f,i}$, $\varepsilon_{R,i}$ and ϖ_i on A_i $(1 \leqslant i \leqslant r)$ as in [2 (m), § 22] and let $m_i(R)$ be the locally constant function on A_i introduced in § 26. Since

$$D(a) = (-1)^m\Delta_i(a)^2 \quad (a \in A_i),$$

it is obvious from Lemmas 63 and 64 that

$$\int V_\varepsilon D^{-1} \Psi'^+ \nabla_G f \, dx = \sum_i c_i \int V_{\varepsilon, i} (-1)^{m_i(R)} \varepsilon_{R, i} \Psi'^+_i \varpi_i F_{f, i} d_i a$$

where $V_{\varepsilon, i}$ and Ψ'^+_i respectively denote the restrictions of V_ε and Ψ'^+ on A_i. Therefore the following lemma is now obvious (cf. Lemma 4) from [2 (f), Theorem 2].

LEMMA 65. *Fix* $f \in C_c^\infty(G)$. *Then*

$$\lim_{\varepsilon \to 0} \int V_\varepsilon D^{-1} \Psi'^+ \nabla_G f \, dx = \text{p.v.} \int D^{-1} \Psi'^+ \nabla_G f \, dx = \sum_{1 \leqslant i \leqslant r} c_i \int (-1)^{m_i(R)} \varepsilon_{R, i} \Psi'^+_i \varpi_i F_{f, i} d_i a.$$

Now put
$$T(f) = \varpi(\lambda) \Theta^+(f) - \text{p.v.} \int D^{-1} \Psi'^+ \nabla_G f \, dx$$

for $f \in C_c^\infty(G)$. It follows from [2 (f), Theorem 2] and the above lemma that T is an invariant distribution on G. We have to show that $T = 0$. Hence it is sufficient by [2 (m), Lemma 7] to verify that no semisimple element of G lies in Supp T.

Fix a function $v \in C^\infty(\mathbf{R})$ such that $0 \leqslant v \leqslant 1$, $v(t) = 0$ if $|t| \leqslant \frac{1}{2}$ and $v(t) = 1$ if $|t| \geqslant 1$ $(t \in \mathbf{R})$. For any $\varepsilon > 0$, put

$$V_\varepsilon(x) = v(2^{-1} \varepsilon^{-2} D(x)) \quad (x \in G).$$

Then it follows from Lemma 65 that

$$\lim_{\varepsilon \to 0} \int D^{-1} V_\varepsilon \Psi'^+ \nabla_G f \, dx = \text{p.v.} \int D^{-1} \Psi'^+ \nabla_G f \, dx.$$

Put
$$T_\varepsilon(f) = \varpi(\lambda) \Theta^+(f) - \int D^{-1} V_\varepsilon \Psi'^+ \nabla_G f \, dx \quad (f \in C_c^\infty(G))$$

for $\varepsilon > 0$. As usual let ∇_G^* denote the adjoint of ∇_G on G'. Since $D^{-1} V_\varepsilon \Psi'^+$ is a C^∞ function on G whose support is contained in G', if follows that the distribution T_ε is, in fact, a locally summable function given by the formula

$$T_\varepsilon = \varpi(\lambda) \Theta^+ - \nabla_G^*(D^{-1} V_\varepsilon \Psi'^+).$$

Moreover,
$$T(f) = \lim_{\varepsilon \to 0} T_\varepsilon(f) \quad (f \in C_c^\infty(G)).$$

Fix a semisimple element $a \in G$. Then a is contained in some Cartan subgroup A of G and $a = a_1 a_2$ where $a_1 \in A_I$, $a_2 \in A_R$. Let $\mathfrak{a}$ be the Lie algebra of A. By Lemma 45, we can choose $x \in G$ such that $\theta(\mathfrak{a}^x) = \mathfrak{a}^x$, $(\mathfrak{a}_I)^x \subset \mathfrak{b}$ and $a_1^x \in B$. Since T is invariant under G, it would be enough to verify that $a^x \notin \text{Supp } T$. Hence replacing

(a, A) by (a^z, A^z), we may assume that $\theta(\mathfrak{a}) = \mathfrak{a}$, $\mathfrak{a}_I \subset \mathfrak{b}$ and $a_1 \in B$. Then $a = b \exp H_0$ where $b = a_1 \in A_I \cap B$ and $H_0 = \log a_2 \in \mathfrak{a}_R$. Define $\mathfrak{z}_b(c_b)$ as in Lemma 50. Then $\mathfrak{z}_0 = \mathfrak{z}_b(c_b) \cap \mathfrak{g}_0$ is an open and completely invariant neighborhood of zero in $\mathfrak{z} = \mathfrak{z}_b$ and $H_0 \in \mathfrak{z}_0$. Put $\Xi = \Xi(b)$ and $\Xi_0 = \exp \mathfrak{z}_0$. Then Ξ_0 is an open and completely invariant neighborhood of 1 in Ξ (see [2 (m), Lemma 8]) and $\exp H_0 \in \Xi_0$. Let σ and σ_ε be the distributions on Ξ_0 corresponding to T and T_ε respectively under [2 (m), Lemma 15]. It would be sufficient to verify that $\sigma = 0$. It is obvious (see [2 (i), Cor. 2 of Theorem 1]) that σ_ε is the locally summable function

$$y \to T_\varepsilon(by) \quad (y \in \Xi_0)$$

on Ξ_0 and therefore

$$\sigma(g) = \lim_{\varepsilon \to 0} \sigma_\varepsilon(g) = \varpi(\lambda) \int g(y)\,\Theta^+(by)\,dy - \lim_{\varepsilon \to 0} \int g(y)\,\Psi^+(by;\ \nabla_G{}^* \circ D^{-1} V_\varepsilon)\,dy$$

for $g \in C_c^\infty(\Xi_0)$. (Here dy is the Haar measure on Ξ.) Let τ' be the distribution on $\mathfrak{z}_0$ which corresponds to σ under the process described in [2 (m), § 10]. Then by Lemma 50,

$$\tau'(f) = \varpi(\lambda) \int \xi_\delta(Z) f(Z)\,\Theta_b{}^+(\exp Z)\,dZ - \lim_{\varepsilon \to 0} \int \xi_\delta(Z)\,\Psi^+(b \exp Z;\ \nabla_G{}^* \circ D^{-1} V_\varepsilon)\,dZ$$

for $f \in C_c^\infty(\mathfrak{z}_0)$ and it would be sufficient to verify that $\tau' = 0$.

Put

$$S_\varepsilon{}^+(Z) = V_\varepsilon(b \exp Z)\,\Psi^+(b \exp Z) = V_\varepsilon(b \exp Z)\,S_b{}^+(Z) \quad (Z \in \mathfrak{z}_0)$$

in the notation of § 22.

> **LEMMA 66.** *We have* ([1])
>
> $$\Psi^+(b \exp Z;\ \nabla_G{}^* \circ D^{-1} V_\varepsilon) = D_b(Z)^{-1} S_\varepsilon{}^+(Z;\ \partial(q_{\mathfrak{g}/\mathfrak{z}}) \circ \nabla_\delta{}^* \circ \eta_\delta{}^{-1})$$
>
> *for* $Z \in \mathfrak{z}_0$ *in the notation of* § 22 *and Lemma 41.*

Assuming this for a moment, we shall first finish the proof of Theorem 4. Put $\tau = \xi_\delta^{-1} D_b \tau'$ and recall that

$$\Theta_b{}^+(\exp Z) = D_b(Z)^{-1} T_b{}^+(Z)$$

by definition (see § 22). Hence if we write $q = q_{\mathfrak{g}/\mathfrak{z}}$, we get

$$\tau(f) = \varpi(\lambda)\,T_b{}^+(f) - \lim_{\varepsilon \to 0} \int f(Z)\,S_\varepsilon{}^+(Z;\ \partial(q) \circ \nabla_\delta{}^* \circ \eta_\delta{}^{-1})\,dZ$$

([1]) See footnote 1, p. 285.

for $f \in C_c^\infty(\mathfrak{z}_0)$. But since S_ε^+ is a C^∞ function on $\mathfrak{z}_0$ and η_δ is nowhere zero on its support, it is clear that

$$\int f(Z)\, S_\varepsilon^+(Z; \partial(q) \circ \nabla_\delta^* \circ \eta_\delta^{-1})\, dZ = \int \eta_\delta^{-1}(\nabla_\delta \circ \partial(q)^*)\, f \cdot S_\varepsilon^+\, dZ.$$

Now as $\varepsilon \to 0$ the right side obviously tends (see Lemma 4) to the limit

$$\text{p.v.} \int \eta_\delta^{-1}(\nabla_\delta \circ \partial(q)^*)\, f \cdot S_b^+\, dZ.$$

Hence
$$\tau(f) = \varpi(\lambda)\, T_b^+(f) - \text{p.v.} \int \eta_\delta^{-1}(\nabla_\delta \circ \partial(q)^*)\, f \cdot S_b^+\, dZ = 0$$

by Lemma 41. This proves Theorem 4.

§ 28. Proof of Lemma 66

We have still to prove Lemma 66. This requires some preparation. Fix a Cartan subgroup A of G and define ϖ_A, Δ_A as in [2 (m), § 20]. Also put $A' = A \cap G'$ as usual.

LEMMA 67. *The differential operator ∇_G^* on G' is invariant under G and*

$$f(h; \nabla_G^*) = (-1)^m \Delta_A(h)^{-1} f(h; \varpi_A \circ \Delta_A^2) \quad (h \in A')$$

for $f \in C^\infty(G')$.

Since ∇_G is invariant, it is obvious that the same holds for ∇_G^*. Fix $h_0 \in A'$ and an open and relatively compact neighborhood U of h_0 in A'. Then $V = U^G$ is an open neighborhood of h_0 in G. Put $\Lambda = \Delta_A$ and let us use the notation of [2 (m), Lemma 41]. Then if $g \in C_c^\infty(V)$, it is clear that

$$\int g\, \nabla_G^* f\, dx = \int \nabla_G g \cdot f\, dx = c \int_A |\Delta(h)|^2\, dh \int_{G^*} g(h^{x^*}; \nabla_G)\, f(h^{x^*})\, dx^*$$

$$= c \int_{A \cap V} |\Delta(h)|^2\, dh \int_{G^*} g(x^*:h; \varpi_A \circ \Delta)\, f(x^*:h)\, dx^*$$

where $g(x^*:h) = g(h^{x^*})$ and $f(x^*:h) = f(h^{x^*})$ $(h \in A \cap V, x^* \in G^*)$. On the other hand

$$|\Delta|^2 = (-1)^{m+m(R)} \Delta^2$$

from Lemma 64 and it is obvious that

$$A \cap V = \bigcup_{s \in W_A} U^s$$

in the notation of [2 (m), § 20]. Hence $A \cap V$ is relatively compact in A'. Therefore (see [2 (f), Theorem 1]) there exists a compact set Ω^* in G^* such that $h^{x^*} \notin \operatorname{Supp} g$ for $h \in A \cap V$ and $x^* \in G^*$ unless $x^* \in \Omega^*$. Hence it is obvious that

$$\int g \, \nabla_G^* f \, dx = c \, (-1)^m \int_{A \cap V} |\Delta(h)|^2 \, dh \int_{G^*} g(h^{x^*}) f(x^* : h; \Delta^{-1} \varpi_A \circ \Delta^2) \, dx^*.$$

On the other hand, there exists (see [2 (m), § 20]) a unique differential operator ∇' on $G_A = (A')^G$ such that

$$\beta(h^x; \nabla') = \beta(x : h; \Delta^{-1} \varpi_A \circ \Delta^2)$$

for $x \in G$ and $h \in A'$. Here β is any C^∞ function on G_A and $\beta(x : h) = \beta(h^x)$. Therefore

$$\int g \nabla_G^* f \, dx = c \, (-1)^m \int_A |\Delta(h)|^2 \, dh \int_{G^*} g(h^{x^*}) f(h^{x^*}; \nabla') \, dx^* = (-1)^m \int g \, \nabla' f \, dx.$$

This shows that $\nabla_G^* = (-1)^m \nabla'$ on V and therefore

$$f(h_0; \nabla_G^*) = (-1)^m f(h_0; \Delta^{-1} \varpi_A \circ \Delta^2).$$

Thus the lemma is proved.

Now in Lemma 66, both sides are C^∞ functions on $\mathfrak{z}_0$ which are invariant under Ξ. Therefore it would be enough to show that they are equal on $\mathfrak{a}_0' = \mathfrak{a}' \cap \mathfrak{z}_0$ for any Cartan subalgebra $\mathfrak{a}$ of $\mathfrak{z}$. Fix $\mathfrak{a}$ and let A denote the corresponding Cartan subgroup of G. Since

$$V_\varepsilon(b \exp Z) \, \Psi^+(b \exp Z) = S_\varepsilon^+(Z) \quad (Z \in \mathfrak{z}_0)$$

and $D(a) = (-1)^m \Delta_A(a)^2 \; (a \in A)$, it follows from Lemma 67 that

$$\Psi^+(b \exp H; \nabla_G^* \circ D^{-1} V_\varepsilon) = \Delta_A(b \exp H)^{-1} S_\varepsilon^+(H; \partial(\varpi_A)) \quad (H \in \mathfrak{a}_0').$$

Let G_c denote, as before, the (connected) adjoint group of $\mathfrak{g}_c$ and Ξ_c the complex-analytic subgroup corresponding to $\operatorname{ad} \mathfrak{z}_c$. Select $y \in \Xi_c$ such that $\mathfrak{b}_c^y = \mathfrak{a}_c$. P being the set of positive roots of $(\mathfrak{g}, \mathfrak{b})$, we may assume that P^y is the set of all positive roots of $(\mathfrak{g}, \mathfrak{a})$. Then it is clear that

$$\Delta_A(b \exp H) = \pi_{\mathfrak{z}}^{\mathfrak{a}}(H) \, D_b(H) \quad (H \in \mathfrak{a}).$$

Hence $\qquad D_b(H) \, \Psi^+(b \exp H; \nabla_G^* \circ D^{-1} V_\varepsilon) = S_\varepsilon^+(H; (\pi_{\mathfrak{z}}^{\mathfrak{a}})^{-1} \partial(\varpi_A)) \quad (H \in \mathfrak{a}_0').$

Put $q = q_{\mathfrak{g}/\mathfrak{z}}$ and let $q_\mathfrak{a}$ denote the projection of q in $S(\mathfrak{a}_c)$ (see [2 (j), § 8]). Then

$$\varpi_A = \varpi^y = (\varpi_{\mathfrak{g}/\mathfrak{z}} \, \varpi_{\mathfrak{z}})^y = q_\mathfrak{a} \, \varpi_{\mathfrak{z}}^y$$

in the notation of § 18. Therefore since S_ε^+ is invariant under Ξ, it follows from the corollary of Lemma 2 and [2 (c), Theorem 1] that

$$S_\varepsilon^+(H; \partial(q) \circ \nabla_\delta^* \circ \eta_\delta^{-1}) = S_\varepsilon^+(H; (\pi_\delta{}^a)^{-1} \partial(\varpi_A)) \quad (H \in \mathfrak{a}_0').$$

This proves Lemma 66.

§ 29. Some convergence questions

We use the notation introduced at the beginning of § 24. Put

$$\chi_\lambda = \chi_{b^*} \quad \text{for} \quad \lambda = \log b^* \quad (b^* \in B^*).$$

LEMMA 68. *Let p be a (complex-valued) polynomial function on $\mathfrak{F}$. Then we can choose an element $z \in \mathfrak{Z}$ with the following property. If $\mathfrak{F}^+$ is a connected component of $\mathfrak{F}'$ and $\lambda \in L \cap \mathrm{Cl}\, \mathfrak{F}^+$, then*

$$|p(\lambda) \Theta_{\lambda, \mathfrak{F}^+}(f)| \leqslant \sum_{1 \leqslant i \leqslant r} c_i \int_{A_i} |F_{zf,i}|\, d_i a \quad (f \in C_c^\infty(G)).$$

Here the notation is the same as in Lemma 65.

Define $\mathfrak{c}$ and $\mathfrak{g}_1$ as in § 14 and let ω_1 be the Casimir operator corresponding to $\mathfrak{g}_1$ (see [2 (b), p. 140]). Then $\omega_1 \in \mathfrak{Z}$. Put $\omega_0 = \omega_1 - (H_1^2 + \dots + H_s^2)$ where $H_1, \dots, H_s$ is a base for $\mathfrak{c}$ over $\mathbf{R}$. Then a simple calculation shows that $\chi_\lambda(\omega_0) = \|\lambda\|^2 - c \;(\lambda \in L)$, where c is a real number (independent of λ) and $\mu \to \|\mu\| \;(\mu \in \mathfrak{F})$ is a Euclidean norm on $\mathfrak{F}$. Put $\omega = 1 + c + \omega_0$. Then $\chi_\lambda(\omega) = 1 + \|\lambda\|^2 \;(\lambda \in \mathfrak{F})$ and ω is a self-adjoint differential operator in $\mathfrak{Z}$. Now fix $\mathfrak{F}^+$ and write $\Theta_\lambda^+ = \Theta_{\lambda, \mathfrak{F}^+} \;(\lambda \in L^+ = L \cap \mathrm{Cl}\, \mathfrak{F}^+)$. Then

$$\Theta_\lambda^+(\omega^q f) = \chi_\lambda(\omega^q)\, \Theta_\lambda^+(f) = (1 + \|\lambda\|^2)^q\, \Theta_\lambda^+(f)$$

for any integer $q \geqslant 0 \;(\lambda \in L^+, f \in C_c^\infty(G))$. Define C as in the corollary of Lemma 60. Then it follows from Lemma 63 that

$$|\Theta_\lambda^+(f)| \leqslant \sum_i c_i \int_{A_i} |\Delta_i(a)\, \Theta_\lambda^+(a)\, F_{f,i}(a)|\, d_i a \leqslant C \sum_i c_i \int_{A_i} |F_{f,i}|\, d_i a \quad (f \in C_c^\infty(G)).$$

Replacing f by $\omega^q f$, we get

$$(1 + \|\lambda\|^2)^q\, |\Theta_\lambda^+(f)| \leqslant \sum_{1 \leqslant i \leqslant r} c_i \int |F_{zf,i}|\, d_i a$$

where $z = C \omega^q$. The assertion of the lemma is now obvious.

Now L is a closed additive subgroup of $\mathfrak{F}$. Let $d\lambda$ denote the Haar measure of L. It is clear from Lemmas 57 and 58 that for a fixed $f \in C_c^\infty(G)$, $\Theta_\lambda^+(f)$ $(\lambda \in L^+)$ is a measurable function of $\lambda \in L^+$

COROLLARY 1. *For any* $p \in S(\mathfrak{b}_c)$, *we can choose* $z \in \mathfrak{Z}$ *such that*

$$\int_{L^+} |p(\lambda)\,\Theta_\lambda^+(f)|\,d\lambda \leqslant \sum_{1 \leqslant i \leqslant r} c_i \int_{A_i} |F_{zf,i}|\,d_i a$$

for all $f \in C_c^\infty(G)$.

We can obviously choose an integer $q \geqslant 0$ such that

$$\alpha = \int_L (1 + \|\lambda\|)^{-q}\,d\lambda < \infty.$$

On the other hand, by the above lemma, we can select $z_0 \in \mathfrak{Z}$ such that

$$(1 + \|\lambda\|)^q\,|p(\lambda)\,\Theta_\lambda^+(f)| \leqslant \sum_i c_i \int |F_{z_0 f,i}|\,d_i a$$

for $\lambda \in L^+$ and $f \in C_c^\infty(G)$. Hence we can take $z = \alpha z_0$.

Define Θ_λ for $\lambda \in L'$ as in § 24 and let us agree to the convention that $\varpi(\lambda)\,\Theta_\lambda = 0$ if $\varpi(\lambda) = 0$ $(\lambda \in L)$.

COROLLARY 2. *Put*

$$T(f) = \int_L \varpi(\lambda)\,\Theta_\lambda(f)\,d\lambda \quad (f \in C_c^\infty(G)).$$

Then T *is an invariant distribution on* G *and, in fact, we can choose* $z \in \mathfrak{Z}$ *such that*

$$|T(f)| \leqslant \sum_{1 \leqslant i \leqslant r} c_i \int_{A_i} |F_{zf,i}|\,d_i a$$

for all $f \in C_c^\infty(G)$.

The second statement follows from Corollary 1 above and the rest is obvious from [2 (f), Theorem 2].

Now assume B is compact. Then L is discrete and therefore

$$T(f) = \sum_{\lambda \in L} \varpi(\lambda)\,\Theta_\lambda(f).$$

Put $q = \tfrac{1}{2}\dim(G/K)$. Then q is an integer (see [2 (k), Lemma 18]) and we shall see in another paper that there exists a number $c > 0$ such that $(-1)^q c\,T$ is precisely the contribution of the discrete series (see [2 (a), § 5]) to the Plancherel formula of G (see [2 (h), Theorem 4]). The proof of this fact depends on Theorem 4.

§ 30. Appendix

Let $\mathfrak{g}$ be a reductive Lie algebra over $\mathbf{R}$ and Ω a completely invariant open subset of $\mathfrak{g}$.

LEMMA 69. *Let F_k ($k \geqslant 1$) be a sequence of continuous and invariant functions on Ω. Then the following two conditions are equivalent.*

1) *For any Cartan subalgebra $\mathfrak{a}$ of $\mathfrak{g}$, F_k converges uniformly on every compact subset of $\mathfrak{a} \cap \Omega$.*

2) *F_k converges uniformly on every compact subset of Ω.*

Obviously 2) implies 1). So let us assume that 1) holds. Let Ω_0 be the set of all elements $X_0 \in \Omega$ with the following property. There exists an open neighborhood U of X_0 in Ω such that F_k converges uniformly on U. It would be sufficient to show that $\Omega_0 = \Omega$. Clearly Ω_0 is open and invariant. Therefore in view of [2 (1), Cor. 2 of Lemma 8], we have only to verify that every semisimple point of Ω lies in Ω_0.

Fix a semisimple element $H_0 \in \Omega$ and an open and relatively compact neighborhood U of H_0 in Ω. It would obviously be enough to show that F_k converges uniformly on $U' = U \cap \mathfrak{g}'$

Let $\mathfrak{a}_1, \ldots, \mathfrak{a}_r$ be a complete set of Cartan subalgebras of $\mathfrak{g}$ no two of which are conjugate under G. Then $V_i = \mathrm{Cl}(\mathfrak{a}_i \cap U^G)$ is a compact subset of $\mathfrak{a}_i \cap \Omega$ (see [2 (k), Lemma 23]). Now fix $X \in U'$. Then $X = H^x$ where $x \in G$ and $H \in V_i$ for some i. Hence

$$F_j(X) - F_k(X) = F_j(H) - F_k(H) \quad (j,\, k \geqslant 1).$$

However, the sequence F_k converges uniformly on $\bigcup_{1 \leqslant i \leqslant r} V_i$ by 1) and so the required result follows immediately.

References

[1]. BOREL, A., & HARISH-CHANDRA, Arithmetic subgroups of algebraic groups. *Ann. of Math.,* 75 (1962), 485–535.

[2]. HARISH-CHANDRA, (a) Representations of semisimple Lie groups, VI. *Amer. J. Math.,* 78 (1956), 564–628.

       (b) The characters of semisimple Lie groups. *Trans. Amer. Math. Soc.,* 83 (1956), 98–163.

       (c) Differential operators on a semisimple Lie algebra. *Amer. J. Math.,* 79 (1957), 87–120.

       (d) Fourier transforms on a semisimple Lie algebra, I. *Amer. J. Math.,* 79 (1957), 193–257.

(e) Fourier transforms on a semisimple Lie algebra, II. *Amer. J. Math.* 79 (1957), 653–686.

(f) A formula for semisimple Lie groups. *Amer. J. Math.*, 79 (1957), 733–760.

(g) Spherical functions on a semisimple Lie group, I. *Amer. J. Math.*, 80 (1958), 241–310.

(h) Invariant eigendistributions on semisimple Lie groups. *Bull. Amer. Math. Soc.*, 69 (1963), 117–123.

(i) Invariant distributions on Lie algebras. *Amer. J. Math.*, 86 (1964), 271–309.

(j) Invariant differential operators and distributions on a semisimple Lie algebra. *Amer. J. Math.*, 86 (1964), 534–564.

(k) Some results on an invariant integral on a semisimple Lie algebra. *Ann. of Math.*, 80 (1964), 551–593.

(l) Invariant eigendistributions on a semisimple Lie algebra. *Pub. Math. I.H.E.S.*, No. 27.

(m) Invariant eigendistributions on a semisimple Lie group. *Trans. Amer. Math. Soc.* To appear.

[3]. SCHWARTZ L., *Théorie des distributions*, II. Paris, Hermann, 1951.

Received October 27, 1964.

Reprinted from
Acta Math.
113 (1965), 241–318

Two theorems on semi-simple Lie groups

By Harish-Chandra

TABLE OF CONTENTS

1. Introduction

Let G be a connected semi-simple Lie group with a compact Cartan subgroup B and $\mathfrak{g}$, $\mathfrak{b}$ the corresponding Lie algebras. Then the character group of B may be identified with a lattice L in the space of all real-valued linear functions on $(-1)^{1/2}\mathfrak{b}$. Let L' be the set of all $\lambda \in L$ such that $\langle \lambda, \alpha \rangle \neq 0$ for every root α of $(\mathfrak{g}, \mathfrak{b})$. Then, for every $\lambda \in L'$, we have defined in [4(o)] an invariant eigendistribution Θ_λ of $\mathfrak{Z}$ on G. Let δ denote the Dirac measure on G concentrated at 1, and $A_1 = B, A_2, \cdots, A_r$ a complete set of Cartan subgroups, no two of which are conjugate under G. The first result of this paper (Theorem 3) expresses δ as a linear combination of $\Theta_\lambda(\lambda \in L')$ and certain invariant distributions T_i associated to $A_i(2 \leq i \leq r)$. This can be regarded as the first main step towards obtaining an explicit Plancherel formula for G.

Indeed when rank $G/K = 1$, it gives just the Plancherel formula (see § 24). The second result affirms a certain inequality (Theorems 4 and 5), which, as we shall see in subsequent papers, plays, together with [4(n), Th. 3], an important role in the harmonic analysis on G.

This paper is divided into two parts. Part I deals with the proof of Theorem 3. In § 8 we define the functions Θ_λ and Ψ_λ on G for all $\lambda \in L$, including the case when λ is singular. Then in § 11 we proceed to consider the series

$$\sum_{\lambda \in L} \int_{A^+} \Psi_\lambda \varpi F_f dh$$

for $f \in C_c^\infty(G)$. The results of Theorem 2, ensure its convergence and enable us, by means of a simple Poisson summation, to bring it into another form (see Lemma 22). This together with the main result (Theorem 4) of [4(h)] gives us Theorem 3. Moreover in § 10 we verify Lemma 52 of [4(n)] (which had been stated there without proof).

Part II is devoted to the proof of the inequality stated in Lemma 35. Here the essential point is contained in Lemma 39 and its proof depends on the maximum principle for elliptic differential operators of degree 2 (see [2, p. 326]). Theorem 6 is an immediate consequence of Theorem 3 and Lemma 39 and, as we shall see in another paper, it enables us to verify the second conjecture of [4(j), § 16] and thus complete the proof of the Plancherel formula for G/K.

Part I

2. An elementary lemma

Let M be a differentiable manifold (see [4(k), § 2]) and J the set consisting of all real numbers $t \geq 0$ together with ∞. A generalized semi-norm (g.s.n.) on M is a function ν from $C_c^\infty(M)$ to J such that

(1) $\nu(f_1 + f_2) \leq \nu(f_1) + \nu(f_2)$, and

(2) $\nu(cf) = |c| \nu(f)$ $(f, f_1, f_2 \in C_c^\infty(M), c \in C)$. (As usual, we define $0 \cdot \infty = 0$.)

Let U be an open subset of M. Then the restriction ν_U of ν on $C_c^\infty(U)$ is a g.s.n. on U. ν is said to be continuous if

(1) $\nu(f) < \infty$ for every $f \in C_c^\infty(M)$, and

(2) ν, regarded as a mapping of $C_c^\infty(M)$ into $\mathbf{R}$ is continuous.

Similarly we say that ν is locally continuous if every point $p \in M$ has an open neighborhood U (in M) such that ν_U is continuous.

LEMMA 1. *Let ν be a g.s.n. on M. Then ν is continuous if and only if it is locally continuous.*

Suppose ν is locally continuous. Fix a compact subset K of M and for each

$p \in K$, choose an open neighborhood U_p of p in M such that ν_{U_p} is continuous. Then $\{U_p\}_{p \in K}$ is an open covering for K from which we can select a finite subcovering $\{U_i\}_{1 \leq i \leq r}$. Choose $\alpha_i \in C_c^\infty(U_i)$ such that $\sum_{1 \leq i \leq r} \alpha_i = 1$ on K. Define $C_K^\infty(M)$ as usual (see [4(k), §2]) and, for any $f \in C_K^\infty(M)$, put $f_i = \alpha_i f$ ($1 \leq i \leq r$). Then $f = \sum_i f_i$ and therefore

$$\nu(f) \leq \sum_i \nu(f_i) < \infty .$$

Moreover $f \to f_i$ being a continuous mapping of $C_K^\infty(M)$, it is clear that the restriction of ν on $C_K^\infty(M)$ is continuous.

Since K was an arbitrary compact set in M, it is obvious that $\nu(f) < \infty$ for every compact $f \in C_c^\infty(M)$. Moreover we can conclude from [6, Th. II, p. 69] that ν is continuous.

3. Reduction of Theorem 1 to Lemma 2

Let $\mathfrak{g}$ be a reductive Lie algebra over $\mathbf{R}$ and $\mathfrak{h}$ a Cartan subalgebra of $\mathfrak{g}$. Define $\mathfrak{h}_I$ and $\mathfrak{h}_R$ as usual (see [4(o), §11]) and let $\mathfrak{S}(\mathfrak{g})$ denote the collection of all continuous semi-norms on $C_c^\infty(\mathfrak{g})$. Let P be the set of all positive roots of $(\mathfrak{g}, \mathfrak{h})$, and P_R and P_I respectively the subsets of all real and imaginary roots in P. Let W_I be the subgroup of $W = W(\mathfrak{g}/\mathfrak{h})$ generated by the Weyl reflexions s_β for $\beta \in P_I$. Suppose $\mathfrak{h}_i(i = 1, 2)$ are two linear subspaces of $\mathfrak{h}$ such that

(1) $\mathfrak{h} = \mathfrak{h}_1 + \mathfrak{h}_2$ where the sum is direct.

(2) $\mathfrak{h}_1^s = \mathfrak{h}_1$ for $s \in W_I$.

(3) Every root α of $(\mathfrak{g}, \mathfrak{h})$ takes only real values on $(-1)^{1/2}\mathfrak{h}_1 + \mathfrak{h}_2$.

Let Q be a subset of P_R and put $\pi_Q = \prod_Q \alpha$. Fix a euclidean measure dH_1 on $\mathfrak{h}_1$, and let $\mathfrak{F}_1$ denote the space of all linear functions on $\mathfrak{h}_{1c}$ which take only pure imaginary values on $\mathfrak{h}_1$. We now use the notation of [4(m), §5] and define ϖ as usual (see [4(m), §11]).

THEOREM 1. *Let L be a lattice[1] in $\mathfrak{F}_1$, and p an element in $S(\mathfrak{h}_c)$. Suppose L is stable under W_I and $p^{s_\alpha} = p$ for $\alpha \in Q \cup P_I$. Then we can choose $\nu \in \mathfrak{S}(\mathfrak{g})$ such that*

$$\sum_{\lambda \in L} \left| \int_{\mathfrak{h}_1} \psi_f(H_1 + H_2; \partial(p\varpi)) e^{\lambda(H_1)} dH_1 \right| \leq | \pi_Q(H_2) | \nu(f)$$

for all $f \in C_c^\infty(\mathfrak{g})$ and $H_2 \in \mathfrak{h}_2$.

Fix $H_2 \in \mathfrak{h}_2$. Then it is obvious that any root which vanishes identically on the set $H_2 + \mathfrak{h}_1$ must be real. Hence it follows from [4(m), Lem. 22] that the above integrals are well defined.

We shall prove the theorem by induction on dim $\mathfrak{g}$. Let $\mathfrak{c}$ be the center of

---

[1] Let V be a vector space over $\mathbf{R}$ of finite dimension. A lattice L in V is a discrete additive subgroup of V such that $V = \mathbf{R}L$.

$\mathfrak{g}$, and put $\mathfrak{g}_1 = \mathfrak{h}_1 + [\mathfrak{g}, \mathfrak{g}]$. First assume that $\mathfrak{h}_2 \subset \mathfrak{c}$ and $\mathfrak{h}_2 \neq \{0\}$. Then $P_R = Q = \varnothing$ and it would be sufficient to consider the case when $p = p_1 p_2 (p_i \in S(\mathfrak{h}_{i c}),$ $i = 1, 2)$ and $p_1^{s_\alpha} = p_1$ for $\alpha \in P_I = P$. Fix $H_2 \in \mathfrak{h}_2$ and define $f_{H_2}(X) = f(H_2 + X; \partial(p_2))$ $(X \in \mathfrak{g}_1)$. Then $f_{H_2} \in C_c^\infty(\mathfrak{g}_1)$ and

$$\psi_f(H_1 + H_2; \partial(p\varpi)) = \psi_{f_2}(H_1; \partial(p_1\varpi))$$

where $f_2 = f_{H_2}$. It is clear that $\mathfrak{h}_2 \not\subset \mathfrak{g}_1$, and so our induction hypothesis is applicable to $\mathfrak{g}_1$. Hence we can choose $\nu_1 \in \mathfrak{S}(\mathfrak{g}_1)$ such that

$$\sum_{\lambda \in L} \left| \int_{\mathfrak{h}_1} \psi_g(H_1; \partial(p_1\varpi)) e^{\lambda(H_1)} dH_1 \right| \leq \nu_1(g)$$

for all $g \in C_c^\infty(\mathfrak{g}_1)$. Now put

$$\nu(f) = \sup{}_{H_2 \in \mathfrak{h}_2} \nu_1(f_{H_2}) \qquad\qquad \left(f \in C_c^\infty(\mathfrak{g}) \right).$$

It follows easily from Lemma 1 that $\nu \in \mathfrak{S}(\mathfrak{g})$, and so it is clear that Theorem 1 holds with this ν.

Now suppose that $\mathfrak{h}_2 \not\subset \mathfrak{c}$. Let $\mathfrak{m}$ denote the centralizer of $\mathfrak{h}_2$ in $\mathfrak{g}$. Then $\dim \mathfrak{m} < \dim \mathfrak{g}$, and $\mathfrak{m}$ is also the centralizer of $\mathfrak{h}_R$. Choose a Cartan involution θ of $\mathfrak{g}$ such that $\theta(\mathfrak{h}) = \mathfrak{h}$ and $\theta = 1$ on $\mathfrak{c}$. Let us now use the notation of [4(n), § 27] and assume that K is compact. Then $\mathfrak{g} = \theta(\mathfrak{n}) + \mathfrak{m} + \mathfrak{n}$. For $f \in C_c^\infty(\mathfrak{g})$, put

$$\bar{f}(X) = \int_K f(X^k) dk \qquad\qquad (X \in \mathfrak{g})$$

and

$$g_f(Y) = \int_{\mathfrak{n}} f(Y + Z) dZ \qquad\qquad (Y \in \mathfrak{m}).$$

Here dk is the normalized Haar measure on K, and dZ a fixed euclidean measure on $\mathfrak{n}$.

Put $\varpi = \varpi_{\mathfrak{g}/\mathfrak{m}} \varpi_{\mathfrak{m}}$ as usual [4(o), § 18]. Since P_I is exactly the set of all positive roots of $(\mathfrak{m}, \mathfrak{h})$, there exists an element $u \in I(\mathfrak{m}_c)$ such that $u_{\mathfrak{h}} = p\varpi_{\mathfrak{g}/\mathfrak{m}}$ (see [4(l), § 9]). For $g \in C_c^\infty(\mathfrak{m})$, let φ_g denote the function on $\mathfrak{h}'(I)$ of [4(m), Lem. 11] corresponding to $(\mathfrak{m}, \mathfrak{h})$ instead of $(\mathfrak{g}, \mathfrak{h})$. Then

$$\psi_f(H) = \varphi_{g_f}(H) \qquad\qquad (H \in \mathfrak{h}'(I))$$

for all $f \in C_c^\infty(\mathfrak{g})$, provided dZ is suitably normalized (see [4(f), p. 224]). Hence [4(f), Th. 3]

$$\partial(p\varpi)\psi_f = \partial(\varpi_{\mathfrak{m}})\varphi_{\partial(u)g_f}.$$

Moreover since $p^{s_\alpha} = p$, it is obvious that $(p\varpi_{\mathfrak{g}/\mathfrak{m}})^{s_\alpha} = -p\varpi_{\mathfrak{g}/\mathfrak{m}}$ $(\alpha \in Q)$. Put $\mathfrak{m}_1 = \mathfrak{m} \cap \mathfrak{k} + [\mathfrak{m}, \mathfrak{m}]$. Then $\mathfrak{m}$ is the direct sum of $\mathfrak{h}_R$ and $\mathfrak{m}_1$. For any $\alpha \in P_R$, s_α can be extended to an automorphism of $\mathfrak{m}$ by defining it to be 1 on $\mathfrak{m}_1$. Then it is clear that $u^{s_a} = -u \, (\alpha \in Q)$. Therefore if we extend π_Q to a polynomial function on $\mathfrak{m}$ by setting $\pi_Q(H + Y) = \pi_Q(H) \, (H \in \mathfrak{h}_R, \; Y \in \mathfrak{m}_1)$, it follows (see Corollary 2 of Lemma 57 of the Appendix) that

$$g \longrightarrow \pi_Q^{-1}\partial(u)g$$

is a continuous mapping of $C_c^\infty(\mathfrak{m})$ into itself. Moreover $\partial(\varpi_\mathfrak{m})$ commutes with π_Q. Hence

$$\partial(p\varpi)\psi_f = \pi_Q\partial(\varpi_\mathfrak{m})\varphi_{h_f}$$

where $h_f = \pi_Q^{-1}\partial(u)g_f$. Now our induction hypothesis is applicable to $(\mathfrak{m}, \mathfrak{h}_1, \mathfrak{h}_2)$ in place of $(\mathfrak{g}, \mathfrak{h}_1, \mathfrak{h}_2)$. Therefore there exists a $\mu \in \mathcal{S}(\mathfrak{m})$ such that

$$\sum_{\lambda \in L} \left| \int_{\mathfrak{h}_1} \varphi_g(H_1 + H_2; \partial(\varpi_\mathfrak{m}))e^{\lambda(H_1)}dH_1 \right| \leq \mu(g)$$

for $H_2 \in \mathfrak{h}_2$ and $g \in C_c^\infty(\mathfrak{m})$. Put

$$\nu(f) = \mu(h_f) \qquad\qquad (f \in C_c^\infty(\mathfrak{g})) \; .$$

Then $\nu \in \mathcal{S}(\mathfrak{g})$ and the inequality of Theorem 1 holds for this ν.

So it remains to consider the case when $\mathfrak{h}_2 = \{0\}$. Then $P = P_I$ and therefore p is invariant under W. Hence by Chevalley's theorem, we can choose $u \in I(\mathfrak{g}_c)$ such that $u_\mathfrak{h} = p$ (see [4(l)], §9]). But then $\partial(p)\psi_f = \psi_{\partial(u)f}$ by [4(f), Th. 3]. Hence it would obviously be enough to consider the case $p = 1$.

Now we use the notation of [4(o), §17] with $\mathfrak{b} = \mathfrak{h}$. Fix a connected component $\mathfrak{F}^+$ of $\mathfrak{F}'$ as usual. Then if $\lambda \in \mathrm{Cl}(\mathfrak{F}^+)$, we have seen in [4(o), Lem. 36] that

$$c_1[W_k]\int_\mathfrak{b} \partial(\varpi)\psi_f \cdot e^\lambda dH$$

$$= \varpi(\lambda)T_\lambda^+(f) - \sum_{2 \leq i \leq r} (-1)^{m_i(R)}c_i \int \varepsilon_{R,i}\partial(\varpi_i)\psi_{f,i}\cdot\varphi_{\lambda,i}^+ d_iH$$

for $f \in C_c^\infty(\mathfrak{g})$.

LEMMA 2. *Fix $i \geq 2$, and put $L^+ = L \cap \mathrm{Cl}(\mathfrak{F}^+)$. Then there exists an element $\nu_i \in \mathcal{S}(\mathfrak{g})$ such that*

$$\sum_{\lambda \in L^+} \left| \int \varepsilon_{R,i}\partial(\varpi_i)\psi_{f,i}\cdot\varphi_{\lambda,i}^+ d_iH \right| \leq \nu_i(f)$$

for all $f \in C_c^\infty(\mathfrak{g})$.

Assuming this for a moment, let us finish the proof of Theorem 1. It would clearly be sufficient to verify the following lemma.

LEMMA 3. *There exists an element $\nu \in \mathcal{S}(\mathfrak{g})$ such that*

$$\sum_{\lambda \in L^+} | \varpi(\lambda)T_\lambda^+(f)| \leq \nu(f)$$

for all $f \in C_c^\infty(\mathfrak{g})$.

Let $\mathfrak{c}$ be the center and $\mathfrak{g}_1$ the derived algebra of $\mathfrak{g}$ and ω_1 the Killing form of $\mathfrak{g}_1$. Fix any negative-definite quadratic form q on $\mathfrak{c}$ and define

$$\omega(C + X) = q(C) + \omega_1(X) \qquad\qquad (C \in \mathfrak{c},\ X \in \mathfrak{g}_1)\,.$$

Then ω is a non-degenerate quadratic form on $\mathfrak{g}$, and we can identify $\mathfrak{g}$, and $\mathfrak{b}$ with their respective duals by means of ω. Then $\|\lambda\|^2 = \omega(\lambda)$ $(\lambda \in \mathfrak{F})$ is a positive-definite quadratic form on $\mathfrak{F}$. Since

$$\sum_{\lambda \in L}(1 + \|\lambda\|^2)^{-N} < \infty$$

for a sufficiently large positive integer N, it would obviously be enough to prove the following result.

LEMMA 4. *For any integer $N \geq 0$, we can choose $\nu \in \mathfrak{S}(\mathfrak{g})$ such that*

$$(1 + \|\lambda\|^2)^N\, |\, T_\lambda^+(f)\,| \leq \nu(f)$$

for all $\lambda \in \mathfrak{F}^+$ and $f \in C_c^\infty(\mathfrak{g})$.

Put $f_N = (1 + \partial(\omega))^N f$. Since $\omega \in I(\mathfrak{g}_c)$, it is clear that

$$T_\lambda^+(f_N) = (1 + \|\lambda\|^2)^N\, T_\lambda^+(f)\,.$$

Hence it is sufficient to consider the case $N = 0$. But then our assertion is an immediate consequence of [4(m), § 15] and [4(o), Lem. 43] with $\mathfrak{z} = \mathfrak{g}$.

4. Proof of Lemma 2

Fix $i \geq 2$, put $\mathfrak{h}_i = \mathfrak{a}$, and drop i from the notation so that $\varpi = \varpi_i$ and $\psi_f = \psi_{f,i}$. We may obviously assume (see [4(o), Lem. 45]) that $\mathfrak{a}_I \subset \mathfrak{b}$. Let $\mathfrak{z}$ be the centralizer of $\mathfrak{a}_I$ in $\mathfrak{g}$. Then

$$\varphi_\lambda^+ = \sum_{t \in W_k(\mathfrak{z}/\mathfrak{b})\backslash W_k} \sum_{s \in W(\mathfrak{z}/\mathfrak{a})} c_{\mathfrak{z}}(s : t\mathfrak{F}^+ : \mathfrak{a}^+) e^{s(t\lambda)^y}$$

on $\mathfrak{a}^+$ for $\lambda \in \mathfrak{F}^+$ in the notation of [4(o), § 18]. Since L is stable under $W = W(\mathfrak{g}/\mathfrak{b})$, it would be enough, in order to prove Lemma 2, to verify the following result.

LEMMA 5. *Fix $s \in W(\mathfrak{z}/\mathfrak{a})$. Then we can choose $\nu \in \mathfrak{S}(\mathfrak{g})$ such that*

$$\sum_{\lambda \in L^+} |\,c_{\mathfrak{z}}(s : \mathfrak{F}^+ : \mathfrak{a}^+)\,|\, \left|\int_{\mathfrak{a}^+} \partial(\varpi)\psi_f e^{s\lambda^y} dH\,\right| \leq \nu(f)$$

for all $f \in C_c^\infty(\mathfrak{g})$.

Put $\mathfrak{a}_1 = \mathfrak{a}_I$, $\mathfrak{a}_2 = \mathfrak{a}_R$, and let $\mathfrak{a}_2^+$ denote the projection of $\mathfrak{a}^+$ on $\mathfrak{a}_2$ with respect to the decomposition $\mathfrak{a} = \mathfrak{a}_1 + \mathfrak{a}_2$. Since $\mathfrak{c} \subset \mathfrak{b}$ and $\mathfrak{a}_2 \neq \{0\}$, it follows that $\mathfrak{a}_2 \not\subset \mathfrak{c}$. Hence, in view of the results of § 3, Theorem 1 holds for $(\mathfrak{a}, \mathfrak{a}_1, \mathfrak{a}_2)$ in place of $(\mathfrak{h}, \mathfrak{h}_1, \mathfrak{h}_2)$.

Put $\mathfrak{b}_1 = \mathfrak{a}_I$ and $\mathfrak{b}_2 = \mathfrak{b} \cap [\mathfrak{z}, \mathfrak{z}]$. Then y centralizes $\mathfrak{b}_1$, $\mathfrak{b}_2^y = (-1)^{1/2}\mathfrak{a}_2$ and $\mathfrak{b} = \mathfrak{b}_1 + \mathfrak{b}_2$, the sum being direct. Let $\mathfrak{F}_1$, $\mathfrak{F}_2$ be the subspaces of those $\lambda \in \mathfrak{F}$

which vanish identically ωn $\mathfrak{b}_2$ and $\mathfrak{b}_1$ respectively. We denote by p_i the projection of $\mathfrak{F}$ on $\mathfrak{F}_i$ corresponding to the direct sum $\mathfrak{F} = \mathfrak{F}_1 + \mathfrak{F}_2$.

Let $L_i = L \cap \mathfrak{F}_i$ and $L'_i = p_i L (i = 1, 2)$. Since $\mathfrak{b} \subset \mathfrak{k}$, there exists an element $k \in K$ such that $\mathrm{Ad}(k)X = \theta(X)$ for $X \in \mathfrak{g}$ (see [4(c), Lem. 15]). Hence we can choose $t_0 \in W(\mathfrak{g}/\mathfrak{a})$ such that $t_0 = \theta$ on $\mathfrak{a}$. Put $t = t_0^{\nu^{-1}} \in W$. Then $t(H_1 + H_2) = H_1 - H_2$ for $H_i \in \mathfrak{b}_i$ $(i = 1, 2)$. Since L is stable under W, it follows that

$$2p_1\lambda = \lambda + t\lambda \in L_1, \quad 2p_2\lambda = \lambda - t\lambda \in L_2 \qquad (\lambda \in L) ,$$

and therefore $2\lambda \in L_1 + L_2$. This shows that $L/(L_1 + L_2)$ and L'_i/L_i $(i = 1, 2)$ are finite. Hence L_i is a lattice in $\mathfrak{F}_i$ and the same holds for L'_i.

Fix a root β of $(\mathfrak{g}, \mathfrak{b})$. Then for any $\lambda \in L$,

$$\lambda - s_\beta\lambda = 2\{\lambda(H_\beta)/\beta(H_\beta)\}\beta$$

also lies in L. Therefore the additive group d_β of all real numbers of the form $2\lambda(H_\beta)/\beta(H_\beta)$ $(\lambda \in L)$ is discrete. Since L is a lattice in $\mathfrak{F}$, it is obvious that $d_\beta \neq \{0\}$. Let $c(\beta) > 0$ denote the generator of d_β.

We know that $\mathfrak{a}_2$ is a Cartan subalgebra of $[\mathfrak{z}, \mathfrak{z}]$ (see [4(o), Lem. 21]). Let Q be the set of all roots α of $(\mathfrak{z}, \mathfrak{a})$ which take only non-negative values on $\mathfrak{a}^+$. Put $\pi_Q = \prod_{\alpha \in Q} \alpha$ and select a fundamental system $(\alpha_1, \cdots, \alpha_m)$ in Q. Define $\beta_i = (s^{-1}\alpha_i)^{\nu^{-1}}$, $c_i = c(\beta_i)$ $(1 \leq i \leq m)$, and let d_i denote the additive subgroup of L_2 generated by $\beta'_i = c_i\beta_i$. Then if

$$L(s : \mathfrak{a}^+) = L_1 + d_1 + d_2 + \cdots + d_m ,$$

it is clear that

$$\mathrm{rank}\, L(s : \mathfrak{a}^+) = \mathrm{rank}\, L_1 + m = \mathrm{rank}\, L ,$$

since $m = \dim \mathfrak{a}_2 = \mathrm{rank}\, L_2$. Therefore $L/L(s : \mathfrak{a}^+)$ is finite.

Now we may obviously suppose, for the proof of our lemma, that $c_{\mathfrak{z}}(s : \mathfrak{F}^+ : \mathfrak{a}^+) \neq 0$. Then $s\lambda^\nu(H) \leq 0$ for $\lambda \in \mathrm{Cl}(\mathfrak{F}^+)$ and $H \in \mathfrak{a}_2^+$ [4(o), Lem. 28]. Since $\mathfrak{a}_2^+$ is precisely the set of those points $H \in \mathfrak{a}_2$ where $\alpha_i(H) > 0$ $(1 \leq i \leq m)$, we conclude that

$$p_2\lambda = -\sum_{1 \leq i \leq m} (a_i + k_i)\beta'_i \qquad (\lambda \in L^+)$$

where k_i are integers ≥ 0 and $0 \leq a_i < 1$. Let N be the order of the additive group $L/L(s : \mathfrak{a}^+)$. Then it is obvious that

$$\sum_{\lambda \in L^+} \left| \int_{\mathfrak{a}^+} \partial(\varpi)\psi_f e^{s\lambda^\nu} dH \right|$$

$$\leq N \sum_{\mu \in L'_1} \sum_{k_1, \cdots, k_m \geq 0} \int_{\mathfrak{a}_2^+} \exp\left(-\sum_i k_i\alpha'_i(H_2)\right) dH_2$$

$$\times \left| \int_{\mathfrak{a}_1} \psi_f(H_1 + H_2; \partial(\varpi)) e^{\mu(H_1)} dH_1 \right|$$

for $f \in C_c^\infty(\mathfrak{g})$. Here $\alpha_i' = c_i \alpha_i (1 \leq i \leq m)$, $H = H_1 + H_2 (H_i \in \mathfrak{a}_i, i = 1, 2)$ and dH_i is a euclidean measure on $\mathfrak{a}_i$ such that $dH = dH_1 dH_2$.

For any $\lambda \in \mathfrak{F}$, let $\bar{\lambda}$ denote its restriction on $\mathfrak{a}_1 = \mathfrak{b}_1$. Then the mapping $\lambda \to \bar{\lambda}$ is injective on $\mathfrak{F}_1$. Hence we may identify $\mathfrak{F}_1$ with the space of all real linear functions on $(-1)^{1/2}\mathfrak{a}_1$ under this mapping. Let α be an imaginary root of $(\mathfrak{g}, \mathfrak{a})$. Then $\beta = \alpha^{y^{-1}}$ is a root of $(\mathfrak{g}, \mathfrak{b})$ which vanishes identically on $\mathfrak{b}_2$, and therefore

$$s_\beta(p_1\lambda) = p_1(s_\beta\lambda) \qquad\qquad (\lambda \in L)\,.$$

This shows that L_1' is stable under s_β. But y leaves $\mathfrak{b}_1$ pointwise fixed and so $s_\alpha = s_\beta$ on $\mathfrak{b}_1$. Hence L_1' is stable under s_α. Moreover we have seen above that Theorem 1 is true for $(\mathfrak{a}, \mathfrak{a}_1, \mathfrak{a}_2)$. Therefore there exists a $\nu_0 \in \mathfrak{S}(\mathfrak{g})$ such that

$$\sum_{\mu \in L_1'} \left| \int_{\mathfrak{a}_1} \psi_f(H_1 + H_2; \partial(\varpi)) e^{\mu(H_1)} dH_1 \right| \leq |\pi_\varrho(H_2)| \nu_0(f)$$

for all $H_2 \in \mathfrak{a}_2$ and $f \in C_c^\infty(\mathfrak{g})$.

Now fix a bounded open subset Ω of $\mathfrak{g}$. Then we can choose a compact set ω in $\mathfrak{a}_2$ such that

$$\operatorname{Supp}\,\psi_f \subset \mathfrak{a}_1 + \omega$$

for all $f \in C_c^\infty(\Omega)$ (see [4(m), § 15]). Put $\omega^+ = \omega \cap \mathfrak{a}_2^+$. Since $t(1 - e^{-t})^{-1}\,(t \in \mathbf{R})$ is an analytic function, it is clear that we can choose a number $c_0 > 0$ such that

$$0 \leq \pi_\varrho(H_2) \sum_{k_1,\cdots,k_m \geq 0} \exp\left(-\sum_{1 \leq i \leq m} k_i \alpha_i'(H_2)\right) \leq c_0$$

for $H_2 \in \omega^+$. Therefore, if

$$c = Nc_0 \int_{\omega^+} dH_2,$$

we get

$$\sum_{\lambda \in L^+} \left| \int_{\mathfrak{a}^+} \partial(\varpi) \psi_f e^{s\lambda^y} dH \right| \leq c\nu_0(f)$$

for all $f \in C_c^\infty(\Omega)$. The assertion of Lemma 5 is now an immediate consequence of Lemma 1.

This completes the proof of Theorem 1.

5. Statement of Theorem 2

Let G be a connected and acceptable (see [4(n), § 18]) Lie group with Lie algebra $\mathfrak{g}$, and A the Cartan subgroup of G corresponding to $\mathfrak{h}$. Define A_I as in [4(o), § 21] and put $A_R = \exp \mathfrak{h}_R$. Then $(a, H) \to a \exp H\,(a \in A_I, H \in \mathfrak{h}_R)$ is a topological mapping of $A_I \times \mathfrak{h}_R$ onto A. As usual we denote by $\log$ the inverse of the exponential mapping from $\mathfrak{h}_R$ to A_R.

Assume that A_I is compact. Let $A_I^0 = \exp \mathfrak{h}_I$ be the connected component of 1 in A_I and $C(A_I^0)$ the character group of A_I^0. For any $a^* \in C(A_I^0)$, we denote by $\langle a^*, a \rangle$ the value of the character a^* at a point $a \in A_I^0$. Let $\log a^*$ denote the linear function λ on $(\mathfrak{h}_I)_c$ such that $< a^*, \exp H > = e^{\lambda(H)}$ $(H \in \mathfrak{h}_I)$. Then $\log a^* \in \mathfrak{F}_I$ where $\mathfrak{F}_I$ is the space of all real-valued linear functions on $(-1)^{1/2}\mathfrak{h}_I$.

Let α be a root of $(\mathfrak{g}, \mathfrak{h})$ and h a point in A. We say that α and h *commute* if $\xi_\alpha(h) = 1$ in the notation of [4(n), § 18]. Suppose α is real, and it commutes with some element $b \in A_I$. Then since $\alpha = 0$ on $\mathfrak{h}_I$, it follows that α commutes with every element of bA_I^0, i.e., it commutes with the connected component of b in A_I.

Define F_f as in [4(n), § 22] for $f \in C_c^\infty(G)$ and let da denote the Haar measure on A_I. Also let P, P_I, P_R, π_Q and ϖ have the same meaning as in § 3.

THEOREM 2. *Let $\mathfrak{S}(G)$ denote the set of all continuous semi-norms on $C_c^\infty(G)$. Fix $b \in A_I$, a subset Q of P_R and an element[2] $u \in S(\mathfrak{h}_c) = \mathfrak{S}(\mathfrak{h}_c) = \mathfrak{H}$. We assume that:*

(1) *Every root in Q commutes with b:*

(2) $u^{s_\alpha} = u$ *for $\alpha \in Q \cup P_I$.*

Then there exists an element $\nu \in \mathfrak{S}(G)$ such that

$$\sum_{a^* \in \sigma(A_I^0)} \left| \int_{A_I^0} F_f(ba \exp H; u\varpi)\langle a^*, a \rangle \, da \right| \leq | \pi_Q(H) | \, \nu(f)$$

for $H \in \mathfrak{h}_R$ and $f \in C_c^\infty(G)$.

It follows from [4(n), Lem. 40] that the above integrals are well defined. Moreover in view of Lemma 1, it would be enough to prove the following result.

LEMMA 6. *For any $x_0 \in G$, there exists an open neighborhood U of x_0 in G and a continuous semi-norm ν on U such that*

$$\sum_{a^* \in \sigma(A_I)} \left| \int_{A_I^0} F_f(ba \exp H; u\varpi)\langle a^*, a \rangle \, da \right| \leq | \pi_Q(H) | \, \nu(f)$$

for $H \in \mathfrak{h}_R$ and $f \in C_c^\infty(U)$.

Let Ω be the set of all points $x_0 \in G$ for which the above lemma holds. We have to show that $\Omega = G$. It is obvious that Ω is open and invariant. Therefore (see [4(n), Lemma 7]) it would be enough to verify that every semi-simple point of G lies in Ω.

6. An auxiliary result

LEMMA 7. *Let x_0 be an element of G. Then $x_0^G \cap A$ is a finite set. More-*

---

[2] Here the notation is the same as in [4(k), p. 280].

over if V is any subset of A, $(\mathrm{Cl}\,V)^{\sigma}$ is the set of all semi-simple elements in $\mathrm{Cl}(V^{\sigma})$.

Since every element of A is semi-simple, we may obviously assume that x_0 is semi-simple. Let $\mathfrak{z}$ denote the centralizer of x_0 in $\mathfrak{g}$ and Ξ the analytic subgroup of G corresponding to $\mathfrak{z}$. Let $\mathfrak{a}_i (1 \leq i \leq r)$ be a maximal set of Cartan subalgebras of $\mathfrak{z}$ such that

(1) no two of them are conjugate under Ξ, and

(2) for each i, there exists an element $x_i \in G$ such that $\mathfrak{a}_i = \mathfrak{h}^{x_i}$.

Fix $h \in x_0^G \cap A$. Then $h = x_0^y$ for some $y \in G$ and therefore $\mathfrak{h} \subset \mathfrak{z}^y$. This shows that $\mathfrak{h}^{y^{-1}} \subset \mathfrak{z}$, and therefore we can choose an index i and an element $\xi \in \Xi$ such that $\mathfrak{h}^{y^{-1}} = \mathfrak{a}_i^{\xi}$. Then $u = y\xi x_i$ normalizes $\mathfrak{h}$. Let A^* be the normalizer of A in G and A_0 the center of A. Then $\mathfrak{w} = A^*/A_0$ is a finite group which operates on A in the usual way (see [4(n), § 20]). Let s be the element of $\mathfrak{w}$ corresponding to u. Then

$$h^{s^{-1}} = x_0^{u^{-1}y} = x_i^{-1}x_0 x_i \; .$$

This proves that $x_i^{-1}x_0 x_i \in A$ and $h = (x_i^{-1}x_0 x_i)^s$. Hence $x_0^G \cap A$ is a finite set.

Now let x_0 be a semi-simple element of $\mathrm{Cl}(V^{\sigma})$ and put $\Phi = x_0^G \cap A$. In order to prove that $x_0 \in (\mathrm{Cl}\,V)^{\sigma}$, it would be sufficient to show that Φ meets $\mathrm{Cl}\,V$. Therefore if B is any neighborhood of 1 in A, we have to verify that ΦB meets V.

Let $\mathfrak{c}$ be the center and $\mathfrak{g}_1$ the derived algebra of $\mathfrak{g}$. Choose an open neighborhood $\mathfrak{c}_0$ of zero in $\mathfrak{c}$, and for any c $(0 < c \leq \pi = 3.14 \cdots)$ let $\mathfrak{g}_1(c)$ denote the set of all $X \in \mathfrak{g}_1$ such that $|\mu| < c$ for every eigenvalue μ of $\mathrm{ad}X$. Put $\mathfrak{g}_0 = \mathfrak{c}_0 + \mathfrak{g}_1(c)$. By choosing $\mathfrak{c}_0$ and c sufficiently small, we can assume that (see [4(n), § 9])

(1) the exponential mapping of $\mathfrak{g}$ into G is univalent and regular on $\mathfrak{g}_0$,

(2) $\exp(\mathfrak{g}_0 \cap \mathfrak{h}) \subset B$.

Define $\mathfrak{z}$ and Ξ as above, select an open neighborhood $\mathfrak{z}_0$ of zero in $\mathfrak{z}$, and put $\Xi_0 = \exp \mathfrak{z}_0$. We assume that $\mathfrak{z}_0 \subset \mathfrak{g}_0$ and

$$\nu_{x_0}(y) = \det \left(\mathrm{Ad}(x_0 y)^{-1} - 1\right)_{\mathfrak{g}/\mathfrak{z}} \neq 0$$

for $y \in \Xi_0$. Then by [4(n), Lem. 14] $(x_0 \Xi_0)^{\sigma}$ is an open neighborhood of x_0 in G. Since $x_0 \in \mathrm{Cl}(\dot{V}^{\sigma})$, it follows that $(x_0 \Xi_0)^{\sigma}$ meets V^{σ} or equivalently

$$(x_0 \Xi_0)^{\sigma} \cap V \neq \varnothing \; .$$

Select $h \in (x_0 \Xi_0)^{\sigma} \cap V$. Then $h = (x_0 y)^x$ for some $x \in G$ and $y \in \Xi_0$. Since $\nu_{x_0}(y) \neq 0$, the centralizer of $x_0 y$ in $\mathfrak{g}$ is actually contained in $\mathfrak{z}$. On the other hand the centralizer of h contains $\mathfrak{h}$. Therefore $\mathfrak{h} \subset \mathfrak{z}^x$. Hence we can choose an index i and an element $\xi \in \Xi$ such that $\mathfrak{h}^{x^{-1}} = \mathfrak{a}_i^{\xi}$. This means that

$u = x \xi x_i \in A^*$. Since $\mathfrak{h} \subset \mathfrak{z}^x$, it is obvious that $x_0^x \in A$. Hence

$$y^x = (x_0^x)^{-1} h \in A .$$

Fix $Z \in \mathfrak{z}_0$ such that $y = \exp Z$. Then $Z^x \in \mathfrak{g}_0$ and $\exp Z^x \in A$. Therefore it follows from the definition of $\mathfrak{g}_0$, that Z^x commutes with $\mathfrak{h}$, and hence $Z^x \in \mathfrak{h} \cap \mathfrak{g}_0$. Thus $y^x \in \exp(\mathfrak{h} \cap \mathfrak{g}_0) \subset B$. Now $x_0^x \in A \cap x_0^g = \Phi$. Therefore $h = (x_0 y)^x \in \Phi B$, and this proves that ΦB meets V.

Conversely it is obvious that every element of $(\mathrm{Cl} V)^g$ is semi-simple. Therefore the lemma is proved.

7. Proof of Theorem 2

Now we return to the proof of Lemma 6. As we have seen in § 5, we may assume that x_0 is semi-simple. First suppose that $x_0 \notin A_1^g$ where A_1 is the connected component of b in A. Then by Lemma 7, we can choose an open neighborhood U of x_0 in G such that U does not meet A_1^g. This means that $U^g \cap A_1 = \varnothing$. Now if $f \in C_o^\infty(U)$, it is clear that

$$\mathrm{Supp}\, F_f \subset \mathrm{Cl}(U^g \cap A)$$

and therefore, since A_1 is open in A, it follows that

$$\mathrm{Supp}\, F_f \cap A_1 = \varnothing .$$

Hence Lemma 6 holds for x_0 and U.

So we may assume that $x_0^y \in A_1$ for some $y \in G$. For the proof of Lemma 6, we can obviously replace x_0 by x_0^y, and therefore suppose that $x_0 \in A_1$.

LEMMA 8. *Let A_1 be a connected component of A and h_0 a point in A_1. Let $\mathfrak{z}$ denote the centralizer of h_0 in $\mathfrak{g}$, and Ω an open and invariant neighborhood of zero in $\mathfrak{g}$ such that:*

(1) *The exponential mapping of $\mathfrak{g}$ into G is univalent and regular on Ω.*

(2) $\det(\mathrm{Ad}(h_0 \exp Z) - 1)_{\mathfrak{g}/\mathfrak{z}} \neq 0$ *for $Z \in \mathfrak{z} \cap \Omega$.*

Then

$$\big(h_0 \exp(\mathfrak{z} \cap \Omega)\big)^g \cap A_1 \subset \Phi \exp(\mathfrak{h} \cap \Omega)$$

where $\Phi = h_0^g \cap A_1$.

As before, let Ξ be the analytic subgroup of G corresponding to $\mathfrak{z}$. Define $\mathfrak{a}_i = \mathfrak{h}^{x_i} (1 \leq i \leq r)$ as in § 6 (for $x_0 = h_0$), and suppose

$$a \in \big(h_0 \exp(\mathfrak{z} \cap \Omega)\big)^g \cap A_1 .$$

Then $a = (h_0 \exp Z)^x$ where $Z \in \mathfrak{z} \cap \Omega$ and $x \in G$. Let $\mathfrak{z}_a$ denote the centralizer of a in $\mathfrak{g}$. Then it follows from condition (2) that $\mathfrak{z}^x \supset \mathfrak{z}_a \supset \mathfrak{h}$. Therefore we can choose an index i and an element $y \in \Xi$ such that $\mathfrak{h}^{x^{-1}} = \mathfrak{a}_i^{y^{-1}}$. Put $u = yx^{-1}$. Then $\mathfrak{h}^u = \mathfrak{a}_i = \mathfrak{h}^{x_i}$. Hence

$$h_0 \exp Z^y = a^u \in A^u = A^{x_i} .$$

On the other hand $h_0 \in A^{x_i}$ since $\mathfrak{a}_i \subset \mathfrak{z}$. Hence $\exp Z^y \in A^{x_i}$. Since $Z^y \in \Omega$, it follows from condition (1) that Z^y commutes with $\mathfrak{a}_i$. Hence $Z^y \in \mathfrak{a}_i$ and therefore

$$a^u \in h_0 \exp(\mathfrak{a}_i \cap \Omega) .$$

This implies that

$$a \in h_0^{u^{-1}} \exp(\mathfrak{h} \cap \Omega) .$$

Therefore, since $a \in A_1$, the same holds for $h_0^{u^{-1}}$. Hence

$$a \in \Phi \exp(\mathfrak{h} \cap \Omega) .$$

This proves the lemma.

Coming back to the proof of Lemma 6, we may assume, as before, that $x_0 = h_0 \in A_1$ where A_1 is the connected component of b in A. Let $h_0, h_1, \cdots, h_r$ be all the distinct elements in $\Phi = h_0^g \cap A_1$ (see Lemma 7). Choose $x_i \in G$ such that $h_i = h_0^{x_i} (0 \leq i \leq r, x_0 = 1)$. Let $\mathfrak{z}$ and Ξ denote the centralizers of h_0 in $\mathfrak{g}$ and G respectively, and put $\mathfrak{z}_i = \mathfrak{z}^{x_i}$, $\Xi_i = \Xi^{x_i} (0 \leq i \leq r)$. Choose a compact and connected neighborhood ω_0 of 1 in A such that the following conditions hold:

(1) $h_i \omega_0 \cap h_j \omega_0 = \varnothing$ if $i \neq j$ $(0 \leq i, j \leq r)$.

(2) Given a compact set C in G and an index i, we can choose a compact set Ω_i in $\bar{G}_i = G/\Xi_i$ with the following property. If x is an element of G and

$$(h_i \omega_0)^x \cap C \neq \varnothing ,$$

then the coset $\bar{x} = x\Xi_i$ lies in Ω_i.

In view of Theorem 1 of [4(h)], this is possible.

Define $\mathfrak{g}_1(c)$ $(0 < c \leq \pi)$ as in § 6 and for any open neighborhood c' of zero in $\mathfrak{c}$, put $\mathfrak{g}(c', c) = c' + \mathfrak{g}_1(c)$. Fix $c_0 (0 < c_0 < \pi)$ and an open and relatively compact neighborhood $\mathfrak{c}_0$ of zero in $\mathfrak{c}$ with the following properties :

(1) $\Omega = \mathfrak{g}(\mathfrak{c}_0, c_0)$ satisfies the two conditions of Lemma 8.

(2) $\exp(\mathfrak{h} \cap \mathfrak{g}(\mathfrak{c}_0, c_0)) \subset \omega_0$.

This is clearly possible. Fix three open neighborhoods $\mathfrak{c}_i (1 \leq i \leq 3)$ of zero in $\mathfrak{c}$ such that

$$\mathrm{Cl}(\mathfrak{c}_i) \subset \mathfrak{c}_{i-1} \qquad\qquad (i = 1, 2, 3)$$

and three numbers $c_i (c_{i-1} > c_i > 0, i = 1, 2, 3)$. Put

$$\mathfrak{g}(i) = \mathfrak{g}(\mathfrak{c}_i, c_i), \quad \mathfrak{h}(i) = \mathfrak{h} \cap \mathfrak{g}(i) ,$$
$$\mathfrak{z}_j(i) = \mathfrak{z}_j \cap \mathfrak{g}(i) \qquad\qquad (0 \leq j \leq r, 0 \leq i \leq 3)$$

and select compact, connected neighborhoods $\mathfrak{h}_I^0$ and $\mathfrak{h}_R^0$ of zero in $\mathfrak{h}_I$ and $\mathfrak{h}_R$

respectively such that $\mathfrak{h}^0 = \mathfrak{h}^0_I + \mathfrak{h}^0_R \subset \mathfrak{h}(2)$. Then

$$\omega = \exp \mathfrak{h}^0 \subset \omega_0 \ .$$

Let $h_i = a_i \exp H_i (a_i \in A_I, H_i \in \mathfrak{h}_R, 0 \leq i \leq r)$. We can assume that $\mathfrak{h}^0_R$ is so small that

$$| \alpha(H_i + H) | \geq | \alpha(H_i) |/2 \qquad\qquad (0 \leq i \leq r)$$

for $\alpha \in P$ and $H \in \mathfrak{h}^0_R$. Moreover by choosing c_3 and c_3 sufficiently small, we may suppose that $\mathfrak{h}(3) \subset \mathfrak{h}^0$.

Put $\mathfrak{z}(i) = \mathfrak{z}_0(i)$ $(0 \leq i \leq 3)$. Then $(h_0 \exp \mathfrak{z}(3))^\sigma$ is an open neighborhood of h_0 in G by [4(n), Lem. 14]. Hence we can choose an open and relatively compact neighborhood U of h_0 in G such that $\mathrm{Cl}\, U \subset (h_0 \exp \mathfrak{z}(3))^\sigma$. Then for each $i (0 \leq i \leq r)$, we can select a compact set C_i in $\bar{G}_i = G/\Xi_i$ with the following property. If $x \in G$ and $(h_i \omega_0)^x$ meets U, then the coset $\bar{x} = x\Xi_i$ lies in C_i. Fix Haar measures dx and $d_i y$ on G and Ξ_i respectively. We normalize the invariant measure $d_i \bar{x}$ on $\bar{G}_i$ in such a way that $dx = d_i \bar{x} d_i y$. Since C_i is compact, we can choose $\gamma_i \in C_c^\infty(G)$ such that

$$\int \gamma_i(xy) d_i y = 1$$

for any point $x \in G$ whose coset $\bar{x} = x\Xi_i$ lies in C_i.

Fix a Haar measure dh on A and let A_0 denote the center of A, and $x \rightarrow x^*$ the natural projection of G on $G^* = G/A_0$. We normalize the invariant measures dx^* and $d_i y^*$ on G^* and $\Xi_i^* = \Xi_i/A_0$ respectively in such a way that $dx = dx^* dh$, $d_i y = d_i y^* dh$. Then it is obvious that $dx^* = d_i \bar{x} d_i y^*$.

It follows from Lemma 8 that $A_1 \cap U^\sigma \subset \Phi\omega$. Therefore, since ω is compact, we conclude that

$$A_1 \cap \mathrm{Supp}\, F_f \subset \Phi\omega \qquad\qquad (f \in C_c^\infty(U)) \ .$$

Now let us agree to use the notation of [4(n), § 22]. Then if $h \in h_i \omega_0 \cap A'$, it is clear that

$$\int f(h^{x^*}) dx^* = \int_{\bar{G}_i} d_i \bar{x} \int_{\Xi_i^*} f(xh^{\nu^*}x^{-1}) d_i y^* \qquad\qquad (f \in C_c^\infty(U))$$

But if $h^x \in U$ for some $x \in G$, then $(h_i \omega_0)^x$ meets U and therefore the coset $\bar{x} = x\Xi_i$ lies in C_i. Hence

$$\int f(h^{x^*}) dx^* = \int_G \gamma_i(x) dx \int_{\Xi_i^*} f(xh^{\nu^*}x^{-1}) d_i y^* \ .$$

Let Z_G denote the center of G, and $\Xi(h_i)$ the analytic subgroup of G corresponding to $\mathfrak{z}_i$. Then $\Xi(h_i)Z_G$ is of finite index in Ξ_i (see [4(g), Lem. 15]). Hence we can choose a finite subset F_i of Ξ_i such that Ξ_i^* is the disjoint union of $\xi\Xi(h_i)^*$ $(\xi \in F_i)$. Put

$$f_i(y) = \sum_{\xi \in F_i} \int \gamma_i(x) f(x h_i y^\xi x^{-1}) dx \qquad (y \in \Xi(h_i)) .$$

Since γ_i has compact support, it is obvious that $f \to f_i$ is a continuous mapping of $C_c^\infty(U)$ into $C_c^\infty(\Xi(h_i))$. Moreover

$$F_f(h_i h) = \Delta(h_i h) \varepsilon_R(h_i h) \int_{\Xi(h_i)^*} f_i(h^{y^*}) d_i y^* \qquad (h \in \omega_0 \cap h_i^{-1} A')$$

in the notation of [4(n), § 22].

Let P_i be the set of all positive roots of $(\mathfrak{z}_i, \mathfrak{h})$. Put $P_{R,i} = P_R \cap P_i$ and

$$\pi_{\mathfrak{z}i} = \prod_{\alpha \in P_i} \alpha, \quad \varepsilon_{R,i}(H) = \operatorname{sign} \prod_{\alpha \in P_{R,i}} \alpha(H) \qquad (H \in \mathfrak{h}) .$$

For $g \in C_c^\infty(\mathfrak{z}_i)$, define

$$\psi_{g,i}(H) = \varepsilon_{R,i}(H) \, \pi_{\mathfrak{z}i}(H) \int_{\Xi(h_i)^*} g(y^* H) d_i y^*$$

for any $H \in \mathfrak{h}$ such that $\pi_{\mathfrak{z}i}(H) \neq 0$. Moreover let P_R^0 denote the set of all roots in P_R which commute with b.

Fix an invariant C^∞ function β_i on $\mathfrak{z}_i$ such that

(1) $\beta_i = 1$ on $\mathfrak{z}_i(2)$, and

(2) $\operatorname{Supp} \beta_i \subset \mathfrak{z}_i(1)$.

It is not difficult to verify that this is possible (see [4(p), § 19]). Let D_{h_i} be the unique invariant analytic function on $\mathfrak{z}_i$ (see [4(n), § 21]) such that

$$\pi_{\mathfrak{z}i}(H) D_{h_i}(H) = \Delta(h_i \exp H) \qquad (H \in \mathfrak{h}) .$$

We define

$$g_{f,i}(Z) = \beta_i(Z) D_{h_i}(Z) f_i(\exp Z) \qquad (Z \in \mathfrak{z}_i)$$

and put $\varphi_{f,i} = \psi_{g,i}$ for $g = g_{f,i}$. Then

$$\operatorname{Supp} \varphi_{f,i} \subset \mathfrak{h} \cap \operatorname{Supp} \beta_i \subset \mathfrak{h}(1) .$$

Now let $H \in \operatorname{Supp} \varphi_{f,i}$. Then $H \in \mathfrak{h}(1)$, and we can choose a point H_0 in $\mathfrak{h}$ arbitrarily near H such that $f((h_i \exp H_0)^x) \neq 0$ for some $x \in G$. This implies that

$$h_i \exp H \in \operatorname{Cl}(A_1 \cap U^a) \subset \Phi\omega .$$

Since $\mathfrak{h}(1) \subset \mathfrak{h}(0)$ and $\exp \mathfrak{h}(0) \subset \omega_0$, it follows from the definitions of ω, ω_0 and $\mathfrak{g}(0)$ that $H \in \mathfrak{h}^0$. This proves that

$$\operatorname{Supp} \varphi_{f,i} \subset \mathfrak{h}^0 \qquad (f \in C_c^\infty(U), \ 0 \leq i \leq r) .$$

Now $h_i = a_i \exp H_i$ $(a_i \in A_I, H_i \in \mathfrak{h}_R)$ as before. Fix $H \in \mathfrak{h}$ such that $h_i \exp H \in A'$ and consider $\varepsilon_R(h_i \exp H)$. Then $h_i \exp H \in A_1$, and therefore it is clear that

$$\varepsilon_R(h_i \exp H) = \operatorname{sign} \prod_{\alpha \in P_R^0} \alpha(H_i + H).$$

Obviously $P_{R,i}$ is the set of those $\alpha \in P_R^0$ which vanish at H_i. Hence if $H \in \mathfrak{h}^0$,

$$\varepsilon_R(h_i \exp H) = \varepsilon_i \varepsilon_{R,i}(H)$$

where

$$\varepsilon_i = \operatorname{sign} \prod_{\alpha \in Q_i'} \alpha(H_i)$$

and Q_i' is the complement of $P_{R,i}$ in P_R^0.

We now claim that

$$F_f(h_i \exp H) = \varepsilon_i \varphi_{f,i}(H) \qquad\qquad \left(f \in C_c^\infty(U),\ H \in \mathfrak{h}(2)\right).$$

Since $\operatorname{Supp} F_f \subset \Phi \omega$ and $\operatorname{Supp} \varphi_{f,i} \subset \mathfrak{h}^0$, this is obvious if $H \notin \mathfrak{h}^0$. On the other hand if $H \in \mathfrak{h}^0$, it follows from our definitions that

$$F_f(h_i \exp H) = D_{h_i}(H)\varepsilon_i\varepsilon_{R,i}(H)\,\pi_{\delta_i}(H)\Big|_{\Xi(h_i)^*} f_i\big((\exp H)^{y^*}\big)d_i y^* = \varepsilon_i \varphi_{f,i}(H)\,,$$

since $\mathfrak{h}^0 \subset \mathfrak{h}(2)$. Hence

$$F_f(h_i \exp H;\, u\varpi) = \varepsilon_i \varphi_{f,i}(H;\, \partial(v_i \varpi_i)) \qquad\qquad (H \in \mathfrak{h}(2))$$

where $v_i = u\varpi_{\mathfrak{g}/\mathfrak{d}_i}$, and $\varpi_i = \varpi_{\delta_i}$.

Fix $H_R \in \mathfrak{h}_R$, and let χ_i denote the characteristic function of $h_i \omega$. Then it is clear that

$$\int_{A_I^0} F_f(ba \exp H_R;\, u\varpi)\langle a^*, a\rangle\, da$$

$$= \sum_{0 \le i \le r} \int_{A_I^0} \chi_i(ba \exp H_R)\, F_f(ba \exp H_R;\, u\varpi)\langle a^*, a\rangle\, da$$

$$= \sum_{0 \le i \le r} \langle a^*, a_i b^{-1}\rangle$$

$$\times \int_{A_I^0} \chi_i\big(h_i a \exp(H_R - H_i)\big)\, F_f\big(h_i a \exp(H_R - H_i);\, u\varpi\big)\langle a^*, a\rangle\, da$$

for $a^* \in C = C(A_I^0)$ since

$$ba \exp H_R = ba_i^{-1} h_i a \exp(H_R - H_i)\,.$$

Let J be the set of those indices i for which $H_R - H_i \in \mathfrak{h}_R^0$. Then if $\omega_I = \exp \mathfrak{h}_I^0$, we have

$$\int_{A_I^0} F_f(ba \exp H_R;\, u\varpi)\langle a^*, a\rangle\, da$$

$$= \sum_{i \in J} \langle a^*, a_i b^{-1}\rangle \int_{\omega_I} F_f\big(h_i a \exp(H_R - H_i);\, u\varpi\big)\langle a^*, a\rangle\, da$$

$$= \sum_{i \in J} \langle a^*, a_i b^{-1}\rangle\, \varepsilon_i \int_{\mathfrak{h}_I} \varphi_{f,i}\big(H_I + H_R - H_i;\, \partial(v_i \varpi_i)\big) e^{\lambda(H_I)} dH_I$$

where $\lambda = \log a^*$ and dH_I is the euclidean measure on $\mathfrak{h}_I$ which corresponds locally to da under the exponential mapping.

Let L be the set of all elements in $\mathfrak{F}_I$ of the form $\log a^*$ $(a^* \in C)$. Since A_I^0 is compact, L is a lattice in $\mathfrak{F}_I$. Define W_I as in Theorem 1. Then $\mathfrak{h}_I$ is stable under W_I and $L' = \sum_{s \in W_I} sL$ is also a lattice in $\mathfrak{F}_I$ (see Cor. 2 of Lem. 58 of the Appendix).

Let Q_i be the set of those roots in Q which commute with h_i. Then $Q_i = Q \cap P_i$. Since $\varpi_{\mathfrak{g}/\mathfrak{z}_i}$ is invariant under s_α for any root α of $(\mathfrak{z}_i, \mathfrak{h})$, it is clear that v_i is invariant under s_α for $\alpha \in Q_i \cup P_{I,i}$ where $P_{I,i} = P_I \cap P_i$. Hence by Theorem 1, there exists $\nu_i \in \mathfrak{S}(\mathfrak{z}_i)$ such that

$$\sum_{\mu \in L'} \left| \int_{\mathfrak{h}_I} \psi_{g,i}(H_I + H; \partial(v,\varpi_i)) e^{\mu(H_I)} dH_I \right| \leq \nu_i(g) \, | \pi_{Q_i}(H) |$$

for $g \in C_c^\infty(\mathfrak{z}_i)$ and $H \in \mathfrak{h}_R$. Therefore it is clear that

$$\sum_{a^* \in C} \left| \int_{A_I^0} F_f(ba \exp H_R; u\varpi) \langle a^*, a \rangle \, da \right| \leq \sum_{i \in J} \nu_i(g_{f,i}) \, | \pi_{Q_i}(H_R - H_i) |$$

for $f \in C_c^\infty(U)$. On the other hand, H_i obviously lies in the center of $\mathfrak{z}_i$ and therefore $\pi_{Q_i}(H_R - H_i) = \pi_{Q_i}(H_R)$. Moreover $H_R - H_i \in \mathfrak{h}_R^0$ if $i \in J$, and therefore

$$| \alpha(H_R) | \geq | \alpha(H_i) | / 2 \qquad\qquad (\alpha \in P)$$

from the definition of $\mathfrak{h}_R^0$. Put

$$c_i' = \prod_{\alpha \in Q_i'} 2^{-1} | \alpha(H_i) |$$

where Q_i' is the complement of Q_i in Q. Then $c_i' > 0$ and it is clear that

$$| \pi_{Q_i}(H_R) | \leq c \, | \pi_Q(H_R) | \qquad\qquad (i \in J)$$

where $c = \max_{0 \leq i \leq r} (c_i')^{-1}$. Hence

$$\sum_{a^* \in C} \left| \int_{A_I^0} F_f(ba \exp H_R; u\varpi) \langle a^*, a \rangle \, da \right| \leq c \sum_{0 \leq i \leq r} \nu_i(g_{f,i}) \, | \pi_Q(H_R) |$$

for all $H_R \in \mathfrak{h}_R$ and $f \in C_c^\infty(U)$. Since the mapping $f \to g_{f,i}$ of $C_c^\infty(U)$ into $C_c^\infty(\mathfrak{z}_i)$ is evidently continuous, the assertion of Lemma 6 is now obvious.

8. The functions Θ_λ and Ψ_λ

We now use the notation of [4(o), § 23]. Thus $\mathfrak{a}$, $\mathfrak{b}$ are two Cartan subalgebras of $\mathfrak{g}$, and A, B the corresponding Cartan subgroups of G. Let A_I^+ be a connected component of A_I, and assume that $A_I^+ \subset B = B_I$. Let $\mathfrak{z}$ denote the centralizer of A_I^+ in $\mathfrak{g}$, and $\mathfrak{F}_\mathfrak{z}'$ the set of all $\mu \in \mathfrak{F}$ such that $\varpi_\mathfrak{z}(\mu) \neq 0$. Let C and $C_\mathfrak{z}$ denote the sets of all connected components of $\mathfrak{F}'$ and $\mathfrak{F}_\mathfrak{z}'$ respectively. For any $\mu \in \mathfrak{F}$, let $C(\mu)$ denote the set of all $\mathfrak{F}^+ \in C$ such that $\mu \in \mathrm{Cl}(\mathfrak{F}^+)$. Similarly $C_\mathfrak{z}(\mu)$ is the set of all $\mathfrak{F}_\mathfrak{z}^+ \in C_\mathfrak{z}$ whose closure contains μ.

Fix $\mathfrak{F}^+ \in C$. Then $s\mathfrak{F}^+$ $(s \in W(\mathfrak{g}/\mathfrak{b}))$ are all the distinct elements in C.

Similarly if we fix $\mathfrak{F}_{\mathfrak{z}}^{+} \in C_{\mathfrak{z}}$, then $t\mathfrak{F}_{\mathfrak{z}}^{+}$ $(t \in W(\mathfrak{z}/\mathfrak{b}))$ are all the distinct elements in $C_{\mathfrak{z}}$. Let $i(\mathfrak{F}^{+})$ denote the unique element $\mathfrak{F}_{\mathfrak{z}}^{+} \in C_{\mathfrak{z}}$ which contains $\mathfrak{F}^{+}$. Clearly $i(t\mathfrak{F}^{+}) = ti(\mathfrak{F}^{+})$ $(t \in W(\mathfrak{z}/\mathfrak{b}))$. Hence $i^{-1}(t\mathfrak{F}_{\mathfrak{z}}^{+}) = ti^{-1}(\mathfrak{F}_{\mathfrak{z}}^{+})$ for $\mathfrak{F}_{\mathfrak{z}}^{+} \in C_{\mathfrak{z}}$ and $t \in W(\mathfrak{z}/\mathfrak{b})$. This shows that every component of $\mathfrak{F}_{\mathfrak{z}}'$ contains the same number of components of $\mathfrak{F}'$. Hence this number is[3]

$$N = [C]\,[C_{\mathfrak{z}}]^{-1} = [\,W(\mathfrak{g}/\mathfrak{b}) : W(\mathfrak{z}/\mathfrak{b})\,]\,.$$

Moreover it is obvious that $i(C(\mu)) = C_{\mathfrak{z}}(\mu)$ for $\mu \in \mathfrak{F}$.

Define $L = \log B^{*}$ as in [4(o), §24], and let A^{+} be a connected component of $A'(R)$ such that $A^{+} = A_{I}^{+} A_{R}^{+}$ where $A_{R}^{+} \subset A_{R}$. Put

$$c(s : \lambda : A^{+}) = [C(\lambda)]^{-1} \sum\nolimits_{\mathfrak{F}^{+} \in o(\lambda)} c(s : \mathfrak{F}^{+} : A^{+})$$

for $s \in W(\mathfrak{z}/\mathfrak{a}) = W(A^{+})$ and $\lambda \in L$ in the notation of [4(o), Lem. 58]. Similarly define

$$\Theta_{\lambda} = [C(\lambda)]^{-1} \sum\nolimits_{\mathfrak{F}^{+} \in o(\lambda)} \Theta_{\lambda,\mathfrak{F}^{+}}\,,$$
$$\Psi_{\lambda} = [C(\lambda)]^{-1} \sum\nolimits_{\mathfrak{F}^{+} \in o(\lambda)} \Psi_{\lambda,\mathfrak{F}^{+}}\,,$$

where $\Theta_{\lambda,\mathfrak{F}^{+}}$ and $\Psi_{\lambda,\mathfrak{F}^{+}}$ have the same meanings as in [4(o), §24]. Let $\xi_{s,\lambda}(s \in W(A^{+}))$ denote the function on A^{+} given by

$$\xi_{s,\lambda}(h_{1}h_{2}) = \langle b^{*}, h_{1}\rangle e^{s\lambda^{y}(\log h_{2})} \qquad\qquad (h_{1} \in A_{I}^{+}, h_{2} \in A_{R}^{+})\,.$$

Here b^{*} is the element of B^{*} corresponding to λ, and y is fixed as in [4(o), §23]. Then it follows from [4(o), Lem. 59] that

$$\Delta_{A}\Theta_{\lambda} = \sum\nolimits_{t \in \mathfrak{w} \backslash W_{G}} \varepsilon(t) \sum\nolimits_{s \in W(A^{+})} \varepsilon(s)\, c(s : t\lambda : A^{+})\, \xi_{s,t\lambda},$$
$$\Psi_{\lambda} = \sum\nolimits_{t \in \mathfrak{w} \backslash W_{G}} \sum\nolimits_{s \in W(A^{+})} c(s : t\lambda : A^{+})\xi_{s,t\lambda}$$

on A^{+}. Now define W_{Ξ} and $\mathfrak{a}^{+}$ as in [4(o), §23] and put

$$c_{\mathfrak{z}}(s : \lambda : \mathfrak{a}^{+}) = [C_{\mathfrak{z}}(\lambda)]^{-1} \sum\nolimits_{\mathfrak{F}_{\mathfrak{z}}^{+} \in C_{\mathfrak{z}}(\lambda)} c_{\mathfrak{z}}(s : \mathfrak{F}_{\mathfrak{z}}^{+} : \mathfrak{a}^{+})$$

for $s \in W(A^{+})$ and $\lambda \in L$ in the notation of [4(o), Lem. 40].

LEMMA 9. *We have the relations*

$$\Psi_{\lambda} = \sum\nolimits_{t \in W_{\Xi} \backslash W_{G}} \sum\nolimits_{s \in W(A^{+})} c_{\mathfrak{z}}(s : t\lambda : \mathfrak{a}^{+})\, \xi_{s,t\lambda},$$
$$\Delta_{A}\Theta_{\lambda} = \sum\nolimits_{t \in W_{\Xi} \backslash W_{G}} \varepsilon(t) \sum\nolimits_{s \in W(A^{+})} \varepsilon(s) c_{\mathfrak{z}}(s : t\lambda : \mathfrak{a}^{+})\, \xi_{s,t\lambda}$$

on A^{+} for $\lambda \in L$.

For we know from Lemmas 58 and 59 of [4(o)] that

$$\Psi_{\lambda,\mathfrak{F}^{+}} = \sum\nolimits_{t \in W_{\Xi} \backslash W_{G}} \sum\nolimits_{s \in W(A^{+})} c_{\mathfrak{z}}(s : t\mathfrak{F}^{+} : \mathfrak{a}^{+})\, \xi_{s,t\lambda}$$

if $\mathfrak{F}^{+} \in C(\lambda)$. (Here we have to make use of the obvious fact that $\xi_{suv,t\lambda} = \xi_{s,ut\lambda}$ for $s \in W(A^{+})$, $u \in \mathfrak{w}$ and $t \in W_{G}$.) Therefore

---

[3] We denote by $[S]$ the number of elements in a set S.

$$\Psi_\lambda = \sum_{t \in W_{\Xi} \backslash W_G} \sum_{s \in W(\varDelta^+)} \xi_{s,t\lambda} [C(t\lambda)]^{-1} \sum_{\mathfrak{F}^+ \in \sigma(t\lambda)} c_{\delta}(s : \mathfrak{F}^+ : \mathfrak{a}^+) \ .$$

On the other hand for every $\mu \in \mathfrak{F}$,

$$\begin{aligned}
\sum_{\mathfrak{F}^+ \in \sigma(\mu)} c_{\delta}(s : \mathfrak{F}^+ : \mathfrak{a}^+) &= \sum_{\mathfrak{F}^+ \in \sigma(\mu)} c_{\delta}(s : i(\mathfrak{F}^+) : \mathfrak{a}^+) \\
&= \sum_{\mathfrak{F}^+_{\delta} \in \sigma_{\delta}(\mu)} [C(\mu) \cap i^{-1}(\mathfrak{F}^+_{\delta})] c_{\delta}(s : \mathfrak{F}^+_{\delta} : \mathfrak{a}^+) \ .
\end{aligned}$$

Let $W_\mu(\mathfrak{g}/\mathfrak{b})$ be the subgroup of all $t \in W(\mathfrak{g}/\mathfrak{b})$ such that $t\mu = \mu$. Then $W_\mu(\mathfrak{g}/\mathfrak{b})$ is simply transitive on $C(\mu)$ (see Lemma 59 of the Appendix). Put $W_\mu(\mathfrak{z}/\mathfrak{b}) = W_\mu(\mathfrak{g}/\mathfrak{b}) \cap W(\mathfrak{z}/\mathfrak{b})$. Then $W_\mu(\mathfrak{z}/\mathfrak{b})$ is also simply transitive on $C_{\delta}(\mu)$ and

$$C(\mu) \cap i^{-1}(t\, \mathfrak{F}^+_{\delta}) = t(C(\mu) \cap i^{-1}(\mathfrak{F}^+_{\delta}))$$

for $t \in W_\mu(\mathfrak{z}/\mathfrak{b})$ and $\mathfrak{F}^+_{\delta} \in C_{\delta}(\mu)$. This shows that

$$[C(\mu) \cap i^{-1}(\mathfrak{F}^+_{\delta})] \qquad\qquad (\mathfrak{F}^+_{\delta} \in C_{\delta}(\mu))$$

is independent of $\mathfrak{F}^+_{\delta}$ and, therefore,

$$\begin{aligned}
[C(\mu) \cap i^{-1}(\mathfrak{F}^+_{\delta})] &= [C(\mu)] [C_{\delta}(\mu)]^{-1} \\
&= [W_\mu(\mathfrak{g}/\mathfrak{b}) : W_\mu(\mathfrak{z}/\mathfrak{b})] \ .
\end{aligned}$$

Hence

$$[C(\mu)]^{-1} \sum_{\mathfrak{F}^+ \in \sigma(\mu)} c_{\delta}(s : \mathfrak{F}^+ : \mathfrak{a}^+) = [C_{\delta}(\mu)]^{-1} \sum_{\mathfrak{F}^+_{\delta} \in C_{\delta}(\mu)} c_{\delta}(s : \mathfrak{F}^+_{\delta} : \mathfrak{a}^+)$$

and this implies the first statement of the lemma. The proof of the second is quite similar.

9. A simple lemma on nilpotent groups

Before proceeding further, we shall now give a proof of Lemma 52 of [4(n)]. For this we need some preparation which will also be useful later.

LEMMA 10[4]. *Let N be a connected, simply connected, real or complex nilpotent Lie group and $\mathfrak{n}$ its Lie algebra. Let $\mathfrak{n}_i (0 \leq i \leq r)$ be a sequence of ideals in $\mathfrak{n}$ such that*

(1) $\mathfrak{n}_0 = \mathfrak{n}$, $\mathfrak{n}_r = \{0\}$,

(2) $\mathfrak{n}_i \supset \mathfrak{n}_{i+1}$ and $[\mathfrak{n}, \mathfrak{n}_i] \subset \mathfrak{n}_{i+1}$ $(0 \leq i < r)$.

Let A_j $(1 \leq j \leq m)$ be endomorphisms of $\mathfrak{n}$ such that $A = \sum_{1 \leq j \leq m} A_j$ is nonsingular and $A_j \mathfrak{n}_i \subset \mathfrak{n}_i$ $(1 \leq j \leq m, 0 \leq i \leq r)$. Then

$$\varphi : X \longrightarrow \log (\exp A_1 X \cdot \exp A_2 X \cdots \exp A_m X)$$

is a bijective polynomial mapping of $\mathfrak{n}$ onto itself whose inverse is also a polynomial mapping.

It is obvious that φ is a polynomial mapping. If $r \leq 1$, $\mathfrak{n}$ is abelian and $\varphi(X) = AX$, and so our statement is obvious. So we may assume that $r \geq 2$ and use induction on r. Let $X \to X^*$ $(X \in \mathfrak{n})$ denote the projection of $\mathfrak{n}$ on

---

[4] See [4(h), Lem. 2] and [1, §7.0].

$\mathfrak{n}^* = \mathfrak{n}/\mathfrak{n}_{r-1}$ and A_j^* the linear transformation in $\mathfrak{n}^*$ corresponding to A_j. Then, since $\mathfrak{n}_{r-1}^* = \{0\}$, our induction hypothesis is applicable to $(\mathfrak{n}^*, \mathfrak{n}_i^*, A_j^*)$. Let V be a subspace of $\mathfrak{n}$ complementary to $\mathfrak{n}_{r-1}$, and E the projection of $\mathfrak{n}$ on V corresponding to the decomposition $\mathfrak{n} = V + \mathfrak{n}_{r-1}$. Then it is clear from our result on $\mathfrak{n}^*$, that we can choose a polynomial mapping ψ_0 of V into itself such that $EX = \psi_0(E\varphi(X))$ for $X \in \mathfrak{n}$.

Now $A\mathfrak{n}_{r-1} \subset \mathfrak{n}_{r-1}$. Therefore since A is non-singular, $A\mathfrak{n}_{r-1} = \mathfrak{n}_{r-1} = A^{-1}\mathfrak{n}_{r-1}$. Moreover, $\mathfrak{n}_{r-1}$ lies in the center of $\mathfrak{n}$ and hence

$$\varphi(Y + Z) = \varphi(Y) + AZ \qquad\qquad (Y \in \mathfrak{n}, \, Z \in \mathfrak{n}_{r-1}) .$$

Therefore if $X = Y + Z \, (Y \in V, \, Z \in \mathfrak{n}_{r-1})$, it follows that

$$AZ = \varphi(X) - \varphi(Y) .$$

On the other hand

$$Y = EX = \psi_0(E\varphi(X)) .$$

Therefore

$$\begin{aligned} X = Y + Z &= Y + A^{-1}(\varphi(X) - \varphi(Y)) \\ &= \psi_0(E\varphi(X)) + A^{-1}\varphi(X) - A^{-1}\varphi(\psi_0(E\varphi(X))) . \end{aligned}$$

Put

$$\psi(X) = \psi_0(EX) + A^{-1}X - A^{-1}\varphi(\psi_0(EX)) \qquad\qquad (X \in \mathfrak{n}) .$$

Then ψ is a polynomial mapping, and the above result can be expressed as

$$X = \psi(\varphi(X)) \qquad\qquad (X \in \mathfrak{n}) .$$

This proves that φ is univalent.

Now given $X \in \mathfrak{n}$, we can, by our induction hypothesis, choose $Y \in V$ such that $\varphi(Y) \equiv X \bmod \mathfrak{n}_{r-1}$. Put

$$Z = A^{-1}(X - \varphi(Y)) .$$

Then $Z \in \mathfrak{n}_{r-1}$ and $\varphi(Y + Z) = \varphi(Y) + AZ = X$. This shows that φ is surjective, and so the lemma is proved.

10. Proof of an earlier result

Let $\mathfrak{g}$ be a reductive Lie algebra over $\mathbf{R}$, and G a connected Lie group corresponding to it. Let A be a Cartan subgroup of G with Lie algebra $\mathfrak{h}$. Define $\mathfrak{h}_I, \mathfrak{h}_R$ and A_I, A_R as usual [4(o), § 21]. Fix an order in the space of (real-valued) linear functions on $\mathfrak{h}_R$ and define $\mathfrak{n}$ as in [4(n), § 27] corresponding to this order. Let $_0M$ be the centralizer of $\mathfrak{h}_R$ in G and N the analytic subgroup of G corresponding to $\mathfrak{n}$. It is clear that $_0M$ normalizes N. Let dn denote the Haar measure on N.

LEMMA 11. *Fix $m \in {}_0M$ and assume that $\det (\mathrm{Ad}\,(m^{-1}) - 1)_\mathfrak{n} \neq 0$. Then*

$$n \longrightarrow n' = n^{m^{-1}} \cdot n^{-1} \qquad\qquad (n \in N)$$

is an analytic diffeomorphism of N onto itself and

$$dn' = |\det (\mathrm{Ad}\,(m^{-1}) - 1)_\mathfrak{n}|\, dn\;.$$

Let φ denote this mapping. It is obviously analytic and $\varphi(n \exp tX) = n' \exp (t\,\mathrm{Ad}\,(nm^{-1})X) \exp (-t\,\mathrm{Ad}\,(n)X)$ for $n \in N$, $X \in \mathfrak{n}$ and $t \in \mathbf{R}$. Hence

$$(d\varphi)_n X = \mathrm{Ad}\,(n)(\mathrm{Ad}\,(m^{-1}) - 1)X$$

and therefore

$$\det (d\varphi)_n = \det (\mathrm{Ad}\,(m^{-1}) - 1)_\mathfrak{n} \neq 0\;.$$

which shows that φ is regular.

For any real linear function λ on $\mathfrak{h}_R$, let $\mathfrak{g}_\lambda$ denote the set of all $X \in \mathfrak{g}$ such that $[H, X] = \lambda(H)X$ for all $H \in \mathfrak{h}_R$. Then

$$\mathfrak{n} = \sum\nolimits_{\lambda > 0} \mathfrak{g}_\lambda$$

by definition. Let $\lambda_0 < \lambda_1 < \cdots < \lambda_r$ be all the linear functions $\lambda > 0$ such that $\mathfrak{g}_\lambda \neq \{0\}$. Put

$$\mathfrak{n}_i = \sum\nolimits_{i \leq j \leq r} \mathfrak{g}_{\lambda_j} \qquad\qquad (0 \leq i \leq r)$$

and define $\mathfrak{n}_{r+1} = \{0\}$. Let $A_1 = (\mathrm{Ad}\,(m^{-1}))_\mathfrak{n}$, $A_2 = -1$ and $A = A_1 + A_2$. Then all the conditions of Lemma 10 are fulfilled. Hence

$$\Phi(X) = \log \left(\exp \mathrm{Ad}\,(m^{-1})X \cdot \exp (-X)\right) \qquad\qquad (X \in \mathfrak{n})$$

is a bijective polynomial mapping of $\mathfrak{n}$ onto itself. The statements of Lemma 11 are now obvious.

COROLLARY. *Let $f \in C_c(G)$. Then*

$$\int_N f(m^n)dn = |\det (\mathrm{Ad}\,(m^{-1}) - 1)_\mathfrak{n}|^{-1} \int_N f(mn)dn\;.$$

Since $nmn^{-1} = mn^{m^{-1}} \cdot n^{-1}$, this is an immediate consequence of the above lemma.

Now we suppose that G is acceptable (see [4(n), § 18]) and use the notation of [4(n), § 27].

LEMMA 12. *Let $a \in A'$. Then*

$$\varepsilon_R(a)\Delta(a) = \Delta_M(a)\,|\det (\mathrm{Ad}\,(a^{-1}) - 1)_\mathfrak{n}|\, e^{\rho(\log a_2)}$$

where a_2 is the component of a in A_R.

Define P and P_M as in [4(n), § 27] and θ as in [4(n), § 16], and let P^+ denote the complement of P_M in P. Then $P^+ = P_R \cup P_c$ where P_R and P_c respectively are the sets of real and complex roots (see [4(m), § 4]) in P. It is clear that

$$\det\left(\mathrm{Ad}\,(a^{-1}) - 1\right)_{\mathfrak{n}} = \prod_{\alpha \in P^+}\left(\xi_\alpha(a^{-1}) - 1\right).$$

Now if $\alpha \in P^+$, the same holds for $-\theta\alpha$. Since

$$\xi_{-\theta\alpha}(a^{-1}) = \mathrm{conj}\,\xi_\alpha(a^{-1}),$$

it follows that

$$\det\left(\mathrm{Ad}\,(a^{-1}) - 1\right)_{\mathfrak{n}} = \prod_{\alpha \in P_R}\left(\xi_\alpha(a^{-1}) - 1\right)\prod_{\alpha \in P_c^+}|\,\xi_\alpha(a^{-1}) - 1\,|^2$$

where P_c^+ is the set of those roots $\alpha \in P_c$ for which $\alpha > -\theta\alpha$. This shows that

$$\left|\det\left(\mathrm{Ad}\,(a^{-1}) - 1\right)_{\mathfrak{n}}\right| = \varepsilon_R(a)\prod_{\alpha \in P^+}\left(1 - \xi_\alpha(a^{-1})\right)$$

and therefore

$$\Delta_M(a)\,\left|\det\left(\mathrm{Ad}\,(a^{-1}) - 1\right)_{\mathfrak{n}}\right|\,e^{\rho(\log a_2)}$$
$$= \varepsilon_R(a)\xi_\rho(a)\prod_{\alpha \in P}\left(1 - \xi_\alpha(a^{-1})\right) = \varepsilon_R(a)\Delta(a).$$

We keep to the notation of [4(n), § 27], and denote by dm^* the invariant measure on $M^* = M/M \cap A$.

COROLLARY 1. *Let $f \in C_c(G)$. Then*

$$\varepsilon_R(a)\Delta(a)\int_{M^*\times N} f(na^{m^*}n^{-1})dm^*dn = \Delta_M(a)e^{\rho(\log a_2)}\int_{M^*\times N} f(a^{m^*}n)dm^*dn$$

for $a \in A'$.

This is an immediate consequence of Lemma 12 and the corollary of Lemma 11.

Let Z_G denote the center of G, and dk the normalized Haar measure on the compact group K/Z_G.

COROLLARY 2. *It is possible to normalize dm^* in such a way that*

$$\varepsilon_R(a)\Delta(a)\int_{G^*} f(a^{x^*})dx^* = \Delta_M(a)e^{\rho(\log a_2)}\int_{M^*\times N}\bar{f}(a^{m^*}n)dm^*dn$$

for $f \in C_c(G)$ and $a \in A'$. Here

$$\bar{f} = \int_{K/Z_G} f^k dk$$

as in [4(n), § 27].

It follows from [4(f), p. 217] that

$$\int f(a^{x^*})dx^* = \int_{M^*\times N}\bar{f}(na^{m^*}n^{-1})dm^*dn \qquad\qquad (f \in C_c(G),\ a \in A')$$

if dm^* is suitably normalized. Our assertion now follows by applying Corollary 1 to $\bar{f}$.

Now define

$$g_f(m) = d(m) \int_N \bar{f}(mn)dn \qquad\qquad (m \in {}_0M,\, f \in C_c^\infty(G))$$

as in [4(n), § 27]. It follows from [4(f), Lem. 11] that $g_f \in C_c^\infty({}_0M)$ and $f \to g_f$ is a continuous mapping of $C_c^\infty(G)$ into $C_c^\infty({}_0M)$.

Let $\mathfrak{m}$ be the centralizer of $\mathfrak{h}_R$ in $\mathfrak{g}$. Then it is clear that

$$\mathfrak{g} = \theta(\mathfrak{n}) + \mathfrak{m} + \mathfrak{n}$$

where the sum is direct. Let $\mathfrak{G}$ be the universal enveloping algebra of $\mathfrak{g}_c$ and $\mathfrak{M}$ the subalgebra of $\mathfrak{G}$ generated by $(1, \mathfrak{m}_c)$.

LEMMA 13. *Let u be an element of $\mathfrak{G}$ such that $[H, u] = 0$ for $H \in \mathfrak{h}_R$. Then there exists a unique element $v \in \mathfrak{M}$ such that $u - v \in \mathfrak{G}\mathfrak{n}_c$.*

Let us use the notation of [4(k), § 7]. Then we know that the mapping $q \otimes m \otimes p \to qmp\,(q \in \mathfrak{S}(\theta(\mathfrak{n}_c)),\, m \in \mathfrak{M},\, p \in \mathfrak{S}(\mathfrak{n}_c))$ defines a linear bijection of $\mathfrak{S}(\theta(\mathfrak{n}_c)) \otimes \mathfrak{M} \otimes \mathfrak{S}(\mathfrak{n}_c)$ onto $\mathfrak{G}$. Hence $\mathfrak{M} \cap \mathfrak{G}\mathfrak{n}_c = \{0\}$ and so we have uniqueness. Moreover

$$\mathfrak{G} = \mathfrak{S}(\theta(\mathfrak{n}_c))\mathfrak{M} + \mathfrak{G}\mathfrak{n}_c \,,$$

and it is obvious that any element of $\mathfrak{S}(\theta(\mathfrak{n}_c))\mathfrak{M}$ which commutes with $\mathfrak{h}_R$ must lie in $\mathfrak{M}$. This implies the existence of v.

As usual let $\mathfrak{Z}$ denote the center of $\mathfrak{G}$ and $\mu = \mu_{\mathfrak{g}/\mathfrak{m}}$ the homomorphism of [4(n), Lem. 21].

COROLLARY. *Suppose $u \in \mathfrak{Z}$. Then $v^m = v$ for $m \in {}_0M$ and*

$$v = d^{-1}\mu(u) \circ d$$

as differential operators on ${}_0M$.

Fix $m \in {}_0M$. Then m normalizes both $\mathfrak{m}$ and $\mathfrak{n}$. Hence

$$u - v^m = (u - v)^m \in \mathfrak{G}\mathfrak{n}_c \,,$$

and therefore it follows from the above lemma that $v^m = v$. Define P, P_M and ρ as above and put $\mathfrak{H} = \mathfrak{S}(\mathfrak{h}_c)$ and $\rho_M = (1/2)\sum_{\alpha \in P_M} \alpha$. Let β and β_M denote the automorphisms of $\mathfrak{H}$ given by

$$\beta(H) = H + \rho(H), \quad \beta_M(H) = H + \rho_M(H) \qquad\qquad (H \in \mathfrak{h}) \,.$$

Put $\mathfrak{q}_c = \sum_{\alpha \in P_M} CX_\alpha$ where X_α has the usual meaning (see [4(m), § 4]). Then $\mathfrak{q}_c \subset \mathfrak{m}_c$. Now define the homomorphisms γ and $\gamma_\mathfrak{m}$ as in [4(n), § 12]. Then

$$v - \beta_M(\gamma_\mathfrak{m}(v)) \in \mathfrak{M}\mathfrak{q}_c$$

from the definition of $\gamma_\mathfrak{m}$. Therefore

$$u - \beta_M(\gamma_\mathfrak{m}(v)) \in \mathfrak{G}(\mathfrak{n}_c + \mathfrak{q}_c) \,,$$

and, in view of the definition of γ, this implies that

$$\beta_M(\gamma_{\mathfrak{m}}(v)) = \beta(\gamma(u)) \,.$$

Put $\beta_+ = \beta_M^{-1} \circ \beta$. Then since $\gamma(u) = \gamma_{\mathfrak{m}}(\mu(u))$ from the definition of μ, we get

$$\gamma_{\mathfrak{m}}(v) = \beta_+(\gamma_{\mathfrak{m}}(\mu(u))) \,.$$

Put $\mathfrak{m}_1 = \mathfrak{m} \cap \mathfrak{k} + [\mathfrak{m}, \mathfrak{m}]$. Then $\mathfrak{m} = \mathfrak{h}_R + \mathfrak{m}_1$. Moreover, since d is a homomorphism of $_0M$ into the multiplicative group of positive real numbers, $d(k) = 1$ for $k \in {}_0M \cap K$, and it follows easily that

$$d^{-1}X \circ d = X \,, \qquad d^{-1}H \circ d = H + \rho(H)$$

for $X \in \mathfrak{m}_1$ and $H \in \mathfrak{h}_R$. Let $\mathfrak{Z}_{\mathfrak{m}}$ denote the center of $\mathfrak{M}$. Then, since $\mathfrak{h}_R$ lies in the center of $\mathfrak{m}$, it follows that $d^{-1}\zeta \circ d \in \mathfrak{Z}_{\mathfrak{m}}$ for $\zeta \in \mathfrak{Z}_{\mathfrak{m}}$. Moreover since $\rho_M = 0$ on $\mathfrak{h}_R$, it is easy to see that

$$\gamma_{\mathfrak{m}}(d^{-1}\zeta \circ d) = \beta_+(\gamma_{\mathfrak{m}}(\zeta)) \qquad\qquad (\zeta \in \mathfrak{Z}_{\mathfrak{m}}) \,.$$

Hence

$$\gamma_{\mathfrak{m}}(v) = \gamma_{\mathfrak{m}}(d^{-1}\mu(u) \circ d) \,,$$

and this implies the assertion of the corollary.

LEMMA 14.　*Put $f(x: n) = f(xn)$ for $x \in G$ and $n \in N(f \in C_c^\infty(G))$. Then*

$$\int_N f(x; q: n)dn = 0$$

for $q \in \mathfrak{G}_{\mathfrak{n}_c}$.

This is obvious from the left-invariance of the Haar measure dn.

COROLLARY.　*Fix $z \in \mathfrak{Z}$ and $f \in C_c^\infty(G)$. Then*

$$g_{zf} = \mu(z)g_f \,.$$

Put $f_1 = zf$. Then it is clear that

$$g_{f_1}(m) = d(m) \int \bar{f}(mn: z)dn$$

$$= d(m) \int \bar{f}(m; z: n)dn$$

$$= d(m) \int \bar{f}(m; d^{-1}\mu(z) \circ d: n)dn \qquad (m \in {}_0M) \,,$$

from Lemma 14 and the corollary of Lemma 13. Our assertion now follows immediately.

It is obvious that Lemma 52 of [4(n)] is a direct consequence of Corollary 2 of Lemma 12 and the corollary of Lemma 14.

11. Convergence of a certain series

Now we return to the notation of §8 and assume that B is compact so that L is a lattice in $\mathfrak{F}$. For a fixed $f \in C_c^\infty(G)$, consider the series

$$\sum_{\lambda \in L} \int_{A^+} \Psi_\lambda \varpi F_f dh$$

where dh is the Haar measure on A. We shall prove that it converges absolutely. By Lemma 9 we have

$$\int_{A^+} \Psi_\lambda \varpi F_f dh = \sum_{t \in W_\Xi \backslash W_G} \sum_{s \in W(A^+)} c_{\tilde{\delta}}(s: t\lambda: \mathfrak{a}^+) \int_{A^+} \varpi F_f \cdot \xi_{s,t\lambda} dh \ .$$

Moreover L is stable under W_G. Hence it would be sufficient to verify that

$$\sum_{\lambda \in L} \sum_{s \in W(A^+)} |c_{\tilde{\delta}}(s: \lambda: \mathfrak{a}^+)| \left| \int_{A^+} \varpi F_f \cdot \xi_{s,\lambda} dh \right| < \infty \ .$$

Write $h = h_1 h_2 (h_1 \in A_I,\ h_2 \in A_R)$ for $h \in A$, and let dh_1, dh_2 denote the Haar measures on A_I and A_R respectively such that $dh = dh_1 dh_2$. Then it would be enough to prove the following result.

LEMMA 15. *Fix $s \in W(A^+)$ and $f \in C_c^\infty(G)$. Then*

$$\sum_{\lambda \in L} |c_{\tilde{\delta}}(s: \lambda: \mathfrak{a}^+)| \int_{A_R^+} dh_2 \left| \int_{A_I^+} \varpi F_f \cdot \xi_{s,\lambda} dh_1 \right| < \infty \ .$$

Let P_R^0 be the set of all roots in P_R which commute with A_I^+ (see §5) and define $\pi_0 = \pi_Q$ for $Q = P_R^0$ as in Theorem 2. Let L_1 be the set of all $\lambda \in L$ such that $\lambda^v = 0$ on $\mathfrak{a}_R$. Consider the set $Q(\mathfrak{a}^+)$ of those roots α of $(\mathfrak{z}, \mathfrak{a})$ which take only non-negative values on $\mathfrak{a}^+$. Let $(\alpha_1, \cdots, \alpha_m)$ be a fundamental system of roots in $Q(\mathfrak{a}^+)$. Then $\beta_i = (s^{-1}\alpha_i)^{v^{-1}} (1 \leq i \leq m)$ are roots of $(\mathfrak{g}, \mathfrak{b})$. Let $L_2(s, \mathfrak{a}^+)$ be the additive subgroup of L generated by $(\beta_1, \cdots, \beta_m)$ and put $L(s, \mathfrak{a}^+) = L_1 + L_2(s, \mathfrak{a}^+)$. Then $L/L(s, \mathfrak{a}^+)$ is finite (see Cor. 3 of Lem. 58 of the Appendix). Therefore, in order to prove Lemma 15, it would be enough to verify the following result.

LEMMA 16. *Fix $s \in W(A^+)$, $f \in C_c^\infty(G)$ and $\lambda_0 \in L$. Then the series*

$$\sum_{\lambda \in L(s,\mathfrak{a}^+)} |c_{\tilde{\delta}}(s: \lambda_0 + \lambda: \mathfrak{a}^+)| \int_{A_R^+} dh_2 \left| \int_{A_I^+} \varpi F_f \cdot \xi_{s,\lambda_0 + \lambda} dh_1 \right|$$

converges.

Fix $\mu \in L_2(s, \mathfrak{a}^+)$, put $\mu_0 = \lambda_0 + \mu$ and consider the series

$$\sum_{\lambda \in L_1} |c_{\tilde{\delta}}(s: \mu_0 + \lambda: \mathfrak{a}^+)| \int_{A_R^+} dh_2 \left| \int_{A_I^+} \varpi F_f \cdot \xi_{s,\mu_0 + \lambda} dh_1 \right| \ .$$

Since $\lambda \in L_1$, it is clear that $C_{\tilde{\delta}}(\mu_0 + \lambda) = C_{\tilde{\delta}}(\mu_0)$. Hence

$$c_{\tilde{\delta}}(s : \mu_0 + \lambda : \mathfrak{a}^+) = c_{\tilde{\delta}}(s : \mu_0 : \mathfrak{a}^+) \qquad\qquad (\lambda \in L_1) \ .$$

Let ξ_σ denote the character of B corresponding to any $\sigma \in L$. Then

$$\xi_{s,\mu_0+\lambda}(h_1 h_2) = \xi_{\mu_0+\lambda}(h_1) \exp\left(s\mu_0^y(\log h_2)\right).$$

Fix $b \in A_I^+$. Then $A_I^+ = bA_I^0$ where A_I^0 is, as before, the connected component of 1 in A_I. Hence

$$\int_{A_I^+} F_f(h_1 h_2; \varpi)\xi_{s,\mu_0+\lambda}(h_1 h_2)dh_1 = \xi_{s,\mu_0+\lambda}(bh_2)\int_{A_I^0} F_f(bah_2; \varpi)\xi_{\mu_0+\lambda}(a)da$$

for $h_2 \in A_R^+$. But, by Theorem 2, we can choose $\nu \in \mathcal{S}(G)$ such that

$$\sum_{a^*\in \mathcal{O}(A_I^0)}\left|\int_{A_I^0} F_g(bah_2; \varpi)\langle a^*, a\rangle da\right| \leq |\pi_0(\log h_2)|\,\nu(g)$$

for $g \in C_c^\infty(G)$ and $h_2 \in A_R$. Therefore

$$\sum_{\lambda\in L_1}\left|c_{\delta}(s : \mu_0 + \lambda : \mathfrak{a}^+)\int_{A_I^+} F_f(h_1 h_2; \varpi)\xi_{s,\mu_0+\lambda}(h_1 h_2)dh_1\right|$$
$$\leq |c_{\delta}(s : \mu_0 : \mathfrak{a}^+)|\,e^{s\mu_0^y(\log h_2)}\,|\pi_0(\log h_2)|\,\nu(f).$$

On the other hand, it follows from [4(o), Lem. 28] that $c(s : \lambda_0 + \mu : \mathfrak{a}^+) = 0$ unless we can choose real numbers $c_i \geq 0$ $(1 \leq i \leq m)$ such that

$$s(\lambda_0 + \mu)^y = -\sum_{1\leq i\leq m} c_i\alpha_i$$

on $\mathfrak{a}_R$. As μ varies in $L_2(s, \mathfrak{a}^+)$, the values of c_i remain fixed mod 1. Therefore since[5]

$$|\pi_0(H)|\prod_{1\leq i\leq m}(1 - e^{-\alpha_i(H)})^{-1} \qquad\qquad (H \in \mathfrak{a}_R^+)$$

remains bounded on any bounded subset of $\mathfrak{a}_R^+$, and

$$|c_{\delta}(s : \lambda_0 + \mu : \mathfrak{a}^+)| \leq c_0 = \sup_{\mathfrak{F}_{\delta}^+\in \mathcal{O}_{\delta}}|c_{\delta}(s : \mathfrak{F}_{\delta}^+ : \mathfrak{a}^+)|,$$

the following result is now obvious from Lemma 1.

LEMMA 17. *We can choose* $\nu_0 \in \mathcal{S}(G)$ *such that*

$$\sum_{\lambda\in L(s,\mathfrak{a}^+)}|c_{\delta}(s : \lambda_0 + \lambda : \mathfrak{a}^+)|\int_{A_R^+}dh_2\left|\int_{A_I^+}\varpi F_f\cdot\xi_{s,\mu_0+\lambda}dh_1\right| \leq \nu_0(f)$$

for all $f \in C_c^\infty(G)$.

Clearly this implies Lemma 16, and therefore also Lemma 15.

COROLLARY. *There exists an element* $\nu \in \mathcal{S}(G)$ *such that*

$$\sum_{s\in W(A^+)}\sum_{\lambda\in L}|c_{\delta}(s : \lambda : \mathfrak{a}^+)|\int_{A_R^+}dh_2\left|\int_{A_I^+}\varpi F_f\cdot\xi_{s,\lambda}dh_1\right| \leq \nu(f)$$

for all $f \in C_c^\infty(G)$.

Since $W(A^+)$ is a finite set, this follows from the above lemma by taking into account the fact that $L/L(s, \mathfrak{a}^+)$ is finite.

---

[5] $\mathfrak{a}_R{}^+ = \log A_R{}^+$ as usual.

LEMMA 18. *We have the relation*

$$\sum_{\lambda \in L} \int_{A^+} \Psi_\lambda \varpi F_f dh$$

$$= [W_G : W_\Xi] \sum_{s \in W(A^+)} \sum_{\lambda \in L} c_\delta(s : \lambda : \mathfrak{a}^+) \int_{A^+} \varpi F_f \cdot \xi_{s,\lambda} dh$$

for $f \in C_c^\infty(G)$.

This is an immediate consequence of Lemmas 15 and 9.

12. Transformation of the above series

Put $\mathfrak{b}_1 = \mathfrak{a}_I$ and $\mathfrak{b}_2 = \mathfrak{b} \cap [\mathfrak{z}_0, \mathfrak{z}_0]$ where $\mathfrak{z}_0$ is the centralizer of $\mathfrak{b}_1$ in $\mathfrak{g}$.

LEMMA 19. $\mathfrak{b}_2 = \mathfrak{b} \cap [\mathfrak{z}, \mathfrak{z}]$ *and $\mathfrak{b}$ is the direct sum of $\mathfrak{b}_1$ and $\mathfrak{b}_2$.*

We know from [4(o), Lem. 21] that $\mathfrak{a}_I$ is the center of $\mathfrak{z}_0$ and therefore it is clear that $\mathfrak{b}$ is the direct sum of $\mathfrak{b}_1$ and $\mathfrak{b}_2$. Fix $a \in A_I^+$, and let $\mathfrak{z}_a$ denote the centralizer of a in $\mathfrak{g}$. Then $\mathfrak{z}_a$ is reductive in $\mathfrak{g}$ and $\mathfrak{b} \subset \mathfrak{z}_a$. Hence we can also apply [4(o), Lem. 21] to $\mathfrak{z}_a$ in place of $\mathfrak{g}$, and conclude (see [4(o), Lem. 29]) that $\mathfrak{b}_1$ is the center of $\mathfrak{z}$. Clearly this implies that $\mathfrak{b} = \mathfrak{b}_1 + \mathfrak{b} \cap [\mathfrak{z}, \mathfrak{z}]$. Since $\mathfrak{z} \subset \mathfrak{z}_0$, this shows that $\mathfrak{b}_2 = \mathfrak{b} \cap [\mathfrak{z}, \mathfrak{z}]$.

Let B_i be the analytic subgroup of B corresponding to $\mathfrak{b}_i (i = 1, 2)$. Then $B_1 = A_I^0$. Let Ξ_0 be the analytic subgroup of G corresponding to $[\mathfrak{z}_0, \mathfrak{z}_0]$. Then Ξ_0 is closed in G (see Cor. 1 of Lemma 58 of the Appendix). Since $\mathfrak{b}_2$ is a Cartan subalgebra of $[\mathfrak{z}_0, \mathfrak{z}_0]$, we conclude that B_2 is closed in Ξ_0 and hence also in G. Hence B_1, B_2 are both compact, $B = B_1 B_2$ and $B_1 \cap B_2$ is a finite group.

Let G_c be a complexification of G and j the corresponding homomorphism of G into G_c (see [4(n), §18]). Let $Z(A)$ be the set of all $a \in A_I$ such that $j(a) \in \exp((-1)^{1/2} \mathfrak{a}_R)$ for every complexification G_c. Since A_I is compact, it is clear that $Z(A)$ is a finite subgroup of A_I.

LEMMA 20. *Put $Z(A_I^+) = A_I^+ \cap B_2$. Then $Z(A_I^+)$ and $B_1 \cap B_2$ are both contained in $Z(A)$ and $Z(A)A_I^0 = A_I$.*

Fix $a \in Z(A_I^+)$ and let G_c and $\mathfrak{z}$ be as above. Since $\mathfrak{z}$ centralizes A_I^+, it is clear that $j(a) = j(a)^y$. However $a = \exp H$ for some $H \in \mathfrak{b}_2$, and therefore

$$j(a)^y = \exp H^y \in \exp\left((-1)^{1/2} \mathfrak{a}_R\right)$$

since $\mathfrak{b}_2^y = (-1)^{1/2} \mathfrak{a}_R$. We prove in the same way that $B_1 \cap B_2 \subset Z(A)$.

Since $B = B_1 B_2$, it is obvious that $A_I^+ = Z(A_I^+)A_I^0$ and, in view of [4(o), Lem. 45], this implies the last assertion.

We keep to the notation of §11. Let λ be an element of L, and ξ_λ the corresponding element of B^*. Since $\mathfrak{b}_2^y = (-1)^{1/2} \mathfrak{a}_R$, it is clear that $\lambda \in L_1$ if and only if $\xi_\lambda = 1$ on B_2. Now fix $\lambda_0 \in L$ and $s \in W(A^+)$. Then, as we have seen in § 11,

$$c_{\hat{\delta}}(s : \lambda + \lambda_0 : \mathfrak{a}^+) = c_{\hat{\delta}}(s : \lambda_0 : \mathfrak{a}^+) \qquad\qquad (\lambda \in L_1)\,.$$

LEMMA 21. *Normalize the Haar measure dh_1 on A_I in such a way that the total measure of A_I^0 is 1. Then*[3,6]

$$\sum_{\lambda \in L_1} \int_{A_I^+} F_f(h_1 h_2;\, \varpi) \xi_{\lambda+\lambda_0}(h_1) dh_1$$
$$= [Z(A_I^+)]^{-1} \sum_{a \in Z(A_I^+)} F_f(ah_2;\, \varpi) \xi_{\lambda_0}(a)$$

for $f \in C_c^\infty(G)$ and $h_2 \in A_R$.

Assuming this for a moment, we get the following result from Lemma 18.

LEMMA 22. *Let $f \in C_c^\infty(G)$. Then*

$$\sum_{\lambda \in L} \int_{A^+} \Psi_\lambda \varpi F_f dh = [W_G : W_\Xi][Z(A_I^+)]^{-1} \sum_{s \in W(A^+)} \sum_{\lambda \in L/L_1} c_{\hat{\delta}}(s : \lambda : \mathfrak{a}^+)$$
$$\times \sum_{a \in Z(A_I^+)} \int_{A_R^+} F_f(ah_2;\, \varpi) \xi_\lambda(a) \exp\left(s\lambda^\nu(\log h_2)\right) dh_2\,.$$

The proof of Lemma 21 is based on Theorem 2. Put $F = B_1 \cap B_2$ and select $b \in Z(A_I^+)$ so that $A_I^+ = bA_I^0$. Then F is a finite group and $Z(A_I^+) = bF$. Fix $f \in C_c^\infty(G)$, $h_2 \in A_R$ and define

$$g(h_1) = \sum_{a \in Z(A_I^+)} F_f(ah_1 h_2;\, \varpi) \xi_{\lambda_0}(ah_1)$$
$$= \sum_{a \in F} F_f(bah_1 h_2;\, \varpi) \xi_{\lambda_0}(bah_1)$$

for $h_1 \in A_I^0 = B_1$. Then g is a continuous function on B_1. Define $C = C(A_I^0)$ as in Theorem 2 and let C_0 be the set of those characters $a^* \in C$ whose kernel contains F. Put

$$g^*(a^*) = \int_{A_I^0} g(h_1) \langle a^*, h_1 \rangle dh_1 \qquad\qquad (a^* \in C)\,.$$

Then it follows from Theorem 2 that $\sum_{a^* \in C} |g^*(a^*)| < \infty$, and so it is clear that

$$g(1) = \sum_{a^* \in C} g^*(a^*)\,.$$

On the other hand, it is obvious from the definition of g that $g^*(a^*) = 0$ unless $a^* \in C_0$. Therefore

$$g(1) = \sum_{a^* \in C_0} g^*(a^*)\,.$$

For any $\lambda \in L$, let a_λ^* denote the restriction of ξ_λ on A_I^0. Since B is compact and $B/B_2 \cong B_1/F$, it is clear that $\lambda \to a_\lambda^*$ $(\lambda \in L_1)$ is a bijection of L_1 onto C_0. Hence

---

[6] We recall that ϖF_f extends to a continuous function on A by [4(n), §22].

$$g(1) = \sum_{\lambda \in L_1} g^*(a_\lambda^*) ,$$

and our assertion now follows from the obvious fact that[7]

$$g^*(a_\lambda^*) = [Z(A_I^+)] \int_{A_I^+} F_f(h_1 h_2;\, \varpi) \xi_{\lambda + \lambda_0}(h_1) dh_1$$

for $\lambda \in L_1$.

Fix a norm on $\mathfrak{a}_R$, and put $\mathfrak{a}_R^+ = \log A_R^+$ as before.

LEMMA 23. *Fix* $s \in W(A^+)$, *and define* π_0 *as in* §11. *Then the series*

$$\sum_{\lambda \in L/L_1} |c_{\hat\delta}(s : \lambda : \mathfrak{a}^+)| \exp\big(s\lambda^\nu(H)\big) \qquad (H \in \mathfrak{a}_R^+)$$

converges uniformly on every compact subset of $\mathfrak{a}_R^+$. *Moreover we can choose* $c, N \geq 0$ *such that*

$$\sum_{\lambda \in L/L_1} |c_{\hat\delta}(s : \lambda : \mathfrak{a}^+)| \, |\pi_0(H)| \exp\big(s\lambda^\nu(H)\big) \leq c(1 + \|H\|)^N$$

for $H \in \mathfrak{a}_R^+$.

It follows from the results of §11 that we can select a finite number of real linear functions $\mu_i (1 \leq i \leq r)$ on $\mathfrak{a}_R$ and a number $c_0 \geq 0$ with the following properties. Let λ be any element in L such that $c_{\hat\delta}(s : \lambda : \mathfrak{a}^+) \neq 0$. Then $|c_{\hat\delta}(s : \lambda : \mathfrak{a}^+)| \leq c_0$ and we can choose integers $k_j \geq 0$ $(1 \leq j \leq m)$ and an index i such that

$$s\lambda^\nu = \mu_i - (k_1 \alpha_1 + \cdots + k_m \alpha_m)$$

on $\mathfrak{a}_R$. Moreover $\mu_i(H) \leq 0$ for $H \in \mathfrak{a}_R^+$ $(1 \leq i \leq r)$. Therefore the first assertion follows from the fact that the series

$$\sum_{k_1, \ldots, k_m \geq 0} \exp\big(-\sum_j k_j \alpha_j(H)\big)$$

converges uniformly on every compact subset of $\mathfrak{a}_R^+$ to

$$\prod_{1 \leq j \leq m} (1 - e^{-\alpha_j(H)})^{-1} .$$

Since

$$|\pi_0(H)| \sum_{1 \leq i \leq r} e^{\mu_i(H)} \prod_j (1 - e^{-\alpha_j(H)})^{-1} \leq r \, |\pi_0(H)| \prod_j (1 - e^{-\alpha_j(H)})^{-1}$$

on $\mathfrak{a}_R^+$, the second assertion is also obvious.

COROLLARY. *Put*

$$v_a(h_2) = [W_G : W_\Xi][Z(A_I^+)]^{-1}\pi_0(\log h_2) \sum_{s \in W(A^+)} \sum_{\lambda \in L/L_1} c_{\hat\delta}(s : \lambda : \mathfrak{a}^+)$$
$$\times \xi_\lambda(a) \exp\big(s\lambda^\nu(\log h_2)\big)$$

for $a \in Z(A_I^+)$ *and* $h_2 \in A_R^+$. *Then for a fixed* a, v_a *is an analytic function on* A_R^+, *and we can choose* $c, N \geq 0$ *such that*

$$|v_a(h_2)| \leq c(1 + \|\log h_2\|)^N$$

---

[7] We observe that $[Z(A_I^+)] = [F] = [Z(A_I^0)]$.

for $h_2 \in A_R^+$.

Since $s\lambda^\nu (s \in W(A^+), \lambda \in L)$ takes only real values on $\mathfrak{a}_R$, it is obvious that

$$|\exp (s\lambda^\nu (H_1 + (-1)^{1/2} H_2))| = \exp (s\lambda^\nu (H_1))$$

for $H_1, H_2 \in \mathfrak{a}_R$. Therefore the corollary follows immediately from the lemma.

LEMMA 24. *Let* $f \in C_c^\infty(G)$. *Then*

$$\sum_{\lambda \in L} \int_{A^+} \Psi_\lambda \varpi F_f \, dh = \sum_{a \in Z(A_I^+)} \int_{A_R^+} F_f(ah_2;\ \varpi) \pi_0 (\log h_2)^{-1} v_a(h_2) \, dh_2 \ .$$

This is an immediate consequence of Lemmas 15, 18 and 21.

13. Proof of Lemma 27

Let P be the set of all positive roots of $(\mathfrak{g}, \mathfrak{a})$ and P_R the subset consisting of all real roots in P. Put

$$\varpi_R = \prod_{\alpha \in P_R} H_\alpha, \qquad \varpi_0 = \prod_{\alpha \in P_0} H_\alpha$$

where P_0 is the complement of P_R in P. Then $\varpi = \varpi_R \varpi_0$ and, if[2] $u \in \mathfrak{S}((\mathfrak{a}_R)_c)$, it is obvious that

$$(u\varpi_0)^{s\beta} = -u\varpi_0$$

for any imaginary root β of $(\mathfrak{g}, \mathfrak{a})$. Hence, by [4(n), Lem. 40], $u\varpi_0 F_f$ is a continuous function on A. Therefore we have the following result.

LEMMA 25. *Fix* $a \in A_I$ *and* $f \in C_c^\infty(G)$. *Then the function*

$$h \longrightarrow F_f(ah;\ \varpi_0) \qquad\qquad (h \in A_R)$$

is a C^∞ *function on* A_R.

Put $W(G/A) = \tilde{A}/A_0$ where $\tilde{A}$ is the normalizer of A in G and A_0 the center of A. Then $W(G/A)$ operates on A (see [4(n), § 20]) and therefore also on $\mathfrak{a}$. Clearly $\mathfrak{a}_R$, A_I and $Z(A)$ are invariant under $W(G/A)$. For any $s \in W(G/A)$ and $a \in A_I$, define

$$\varepsilon_s(a) = \operatorname{sign} \prod_{\alpha \in P_R(s)} \xi_\alpha(a^s) \ ,$$

where $P_R(s)$ is the set of all $\alpha \in P_R$ such that $s^{-1}\alpha < 0$. Clearly ε_s is a homomorphism of A_I whose kernel contains A_I^0. There is an obvious homomorphism $s \to \sigma$ of $W(G/A)$ into $W(\mathfrak{g}/\mathfrak{a})$ such that $sH = \sigma H (H \in \mathfrak{a})$. Since $\xi_\alpha(a^s) = \xi_{\sigma^{-1}\alpha}(a)$, it is clear that $\varepsilon_s(a)$ depends only on σ and so we may also denote it by $\varepsilon_\sigma(a)$. Define ε_R on A' as in [4(n), § 22] and let h_1, h_2 denote the components in A_I and A_R respectively of $h \in A$. Moreover let

$$\varpi_R^s = \varepsilon_R(s)\varpi_R, \qquad \varpi_0^s = \varepsilon_0(s)\varpi \qquad\qquad (s \in W(G/A)) \ .$$

LEMMA 26. *Let* $h \in A'$ *and* $s \in W(G/A)$. *Then*

$$\varepsilon_R(h^s) = \varepsilon_R(h)\varepsilon_R(s)\varepsilon_s(h_1) .$$

Put $P_R^- = P_R(s)$ and let P_R^+ denote the complement of P_R^- in P_R. Then

$$\varepsilon_R(h^s) = \operatorname{sign} \Delta'_R(h^s)$$
$$= \operatorname{sign} \prod_{\alpha \in P_R^+}(1 - \xi_{s^{-1}\alpha}(h)^{-1}) \prod_{\alpha \in P_R^-}(1 - \xi_{-s^{-1}\alpha}(h))$$
$$= (-1)^r \operatorname{sign} \prod_{\alpha \in P_R^-} \xi_{-s^{-1}\alpha}(h) \cdot \operatorname{sign} \Delta'_R(h)$$

where r is the number of roots in P_R^- and Δ'_R has the same meaning as in [4(n), § 22]. But $(-1)^r = \varepsilon_R(s)$ and

$$\operatorname{sign} \xi_{-s^{-1}\alpha}(h) = \operatorname{sign} \xi_\alpha(h_1^s)$$

for $\alpha \in P_R$. Therefore our assertion is now obvious.

COROLLARY 1. $\varepsilon_{s_1 s_2}(a) = \varepsilon_{s_1}(a^{s_2})\varepsilon_{s_2}(a)$ $(s_1, s_2 \in W(G/A), a \in A_I)$.

It follows from the above lemma that

$$\varepsilon_s(h_1) = \varepsilon_R(s)\varepsilon_R(h^s)/\varepsilon_R(h) \qquad (s \in W(G/A), h \in A')$$

and therefore we conclude by direct computation that

$$\varepsilon_{s_1 s_2}(h_1) = \varepsilon_{s_1}(h_1^{s_2})\varepsilon_{s_2}(h_1) .$$

Moreover, it is clear that h can be so chosen that $h_1 \in aA_I^0$, and so our assertion follows immediately.

COROLLARY 2. $F_f(h^s) = \varepsilon_0(s)\varepsilon_s(h_1)F_f(h)$ for $f \in C_c^\infty(G), h \in A'$ and $s \in W(G/A)$.

Since $\varepsilon(s)\varepsilon_R(s) = \varepsilon_0(s)$, this is obvious from Lemma 26.

Let W_R be the subgroup of $W(\mathfrak{g}/\mathfrak{a})$ generated by s_α for $\alpha \in P_R$ and define

$$h^s = \exp sH \qquad (h \in A_R, s \in W_R)$$

where $H = \log h$.

COROLLARY 3. For $a \in Z(A)$, put $W_a = W(\mathfrak{z}/\mathfrak{a})$ where $\mathfrak{z}$ is the centralizer of aA_I^0 in $\mathfrak{g}$. Then $W_a \subset W_R$ and

$$F_f(ah^s; \varpi) = \varepsilon(s)F_f(ah; \varpi) \qquad (h \in A_R, s \in W_a)$$

for $f \in C_c^\infty(G)$. Moreover $\varepsilon_s(a) = \varepsilon_0(s) = 1$ $(s \in W_a)$.

Since every root of $(\mathfrak{z}, \mathfrak{a})$ is real, it is obvious that $W_a \subset W_R$. Let W^0 be the subgroup of those elements of $W(G/A)$ which leave aA_I^0 pointwise fixed. Then if $h \in A'$ and $h_1 \in aA_I^0$, it follows from Corollary 2 that

$$F_f(h^s; \varpi) = \varepsilon_R(s)\varepsilon_s(a)F_f(h; \varpi) \qquad (s \in W^0) .$$

On the other hand, we conclude from Corollary 1 that $s \to \varepsilon_s(a)$ is a homomorphism of W^0. Let P_a be the set of all roots in P_R which commute with a. Then P_a is also the set of all positive roots of $(\mathfrak{z}, \mathfrak{a})$. Let $(\alpha_1, \cdots, \alpha_m)$ be a fundamental system of roots in P_a. Fix $i(1 \le i \le m)$, and let $P_R(i)$ denote the set of all $\alpha \in P_R$ such that $s_{\alpha_i} \alpha < 0$. If such a root α lies in P_a, it is clear that $\alpha = \alpha_i$, otherwise $\xi_\alpha(a) = -1$. Hence

$$\operatorname{sign} \prod_{\alpha \in P_R(i)} \xi_\alpha(a) = (-1)^{r_i - 1},$$

where r_i is the number of elements in $P_R(i)$. By [4(m), Lem. 6], there exists an element $s_i \in W^0$ whose image in $W(\mathfrak{g}/\mathfrak{a})$ is s_{α_i}. Then $(-1)^{r_i} = \varepsilon_R(s_i)$, and therefore

$$\varepsilon_R(s_i)\varepsilon_{s_i}(a) = -1 = \varepsilon(s_{\alpha_i})$$

This shows that

$$F_f(ah^{s_i}; \varpi) = \varepsilon(s_{\alpha_i}) \, F_f(ah; \varpi) \qquad\qquad (h \in A_R, \, 1 \leqq i \leqq m)$$

and therefore, since W_a is generated by $s_{\alpha_1}, \cdots, s_{\alpha_m}$, our assertions follow if we take into account the results of [4(m), § 6].

Put $\pi_a = \prod_{\alpha \in P_a} \alpha \; (a \in Z(A))$ where P_a is defined as above.

COROLLARY 4. *Fix $a \in Z(A)$ and $f \in C_c^\infty(G)$. Then there exists a C^∞ function g on A_R such that*

$$\pi_a(\log h)g(h) = F_f(ah; \varpi) \qquad\qquad (h \in A_R) .$$

This follows immediately from Corollary 3 above and Corollary 2 of Lemma 57 of the Appendix.

Put $\varepsilon_A(h) = \varepsilon_R(h)\,(-1)^{m(R:h)} \; (h \in A')$ in the notation of [4(o), § 26], and define the Haar measures dh and dh_2 on A and A_R respectively as in Lemma 24.

LEMMA 27. *For each $a \in Z(A)$, we can choose a function v_a on A_R with the following properties:*

(1) *v_a is analytic around every point $h \in A_R$ where $\pi_a(\log h) \neq 0$.*

(2) *There exist numbers $c, N \geqq 0$ such that*
$$|v_a(h)| \leqq c(1 + \|\log h\|)^N \qquad\qquad (a \in Z(A), \, h \in A_R) .$$

(3) *$v_{a^s}(h^s) = \varepsilon_R(s)\varepsilon_s(a)v_a(h)$ for $a \in Z(A)$, $h \in A_R$ and $s \in W(G/A)$.*

(4) *For any $f \in C_c^\infty(G)$,*

$$\sum_{\lambda \in L} \int_A \varepsilon_A \Psi_\lambda \varpi F_f \, dh = \sum_{a \in Z(A)} \int_{A_R} F_f(ah_2; \varpi) \, |\pi_a(\log h_2)|^{-1} v_a(h_2) dh_2 .$$

This follows immediately from the results of § 12 if we take into account Corollaries 1 and 2 of Lemma 26 and the obvious fact that $|\pi_{a^s}(\log h_2^s)| = |\pi_a(\log h_2)| \; (a \in Z(A), \, h_2 \in A_R \text{ and } s \in W(G/A))$.

14. The distribution T_A

Put $\tilde{A}_R = Z(A)A_R$. Then $\tilde{A}_R$ is a closed subgroup of A and $(a, h) \to ah \, (a \in Z(A), \, h \in A_R)$ defines a topological isomorphism of $Z(A) \times A_R$ onto $\tilde{A}_R$.

Thus $\tilde{A}_R$ is a Lie group and $W(G/A)$ operates on $\tilde{A}_R$.

LEMMA 28.   *For any $g \in C_c^\infty(\tilde{A}_R)$, there exists a unique element $g_0 \in C_c^\infty(\tilde{A}_R)$ such that*

$$\pi_a(\log h)g_0(ah) = [W(G/A)]^{-1} \sum_{s \in W(G/A)} \varepsilon_R(s)\varepsilon_s(a)g(a^sh^s; \varpi_R)$$

for $a \in Z(A)$, $h \in A_R$. Moreover the mapping $g \to g_0$ of $C_c^\infty(\tilde{A}_R)$ into itself, is continuous.

Define

$$g_s(ah) = \varepsilon_s(a)g((ah)^s) \qquad\qquad (a \in Z(A),\, h \in A_R)$$

for $g \in C_c^\infty(G)$ and $s \in W(G/A)$. Then $g_{st} = (g_s)_t$ $(s,\, t \in W(G/A))$. Therefore, if $g' = [W(G/A)]^{-1} \sum_{s \in W(G/A)} g_s$, it is clear that $g' = (g')_s$, and hence

$$g'(a^sh^s; \varpi_R) = \varepsilon_R(s)\varepsilon_s(a)g'(ah; \varpi_R) \qquad\qquad (s \in W(G/A))$$

for $a \in Z(A)$ and $h \in A_R$. Our assertion is now an immediate consequence of Corollary 3 of Lemma 26 and Corollary 2 of Lemma 57 of the Appendix.

Define a function v on $\tilde{A}_R$ as follows:

$$v(ah) = v_a(h)\,\mathrm{sign}\,\pi_a(\log h) \qquad\qquad (a \in Z(A),\, h \in A_R)$$

and put

$$T_A(g) = \int_{\tilde{A}_R} g_0 v \, dh_2 \qquad\qquad (g \in C_c^\infty(\tilde{A}_R)),$$

v_a being defined, and the Haar measure dh_2 on $\tilde{A}_R$ being normalized, as in Lemma 27. Then it is obvious that T_A is a distribution on $\tilde{A}_R$ and $T_A(g_s) = T_A(g)$ $(s \in \overline{W(G/A)})$.

Lemma 29.   *For $f \in C_c^\infty(G)$, define the function φ_f on $\tilde{A}_R$ by*

$$\varphi_f(h) = F_f(h; \varpi_0) \qquad\qquad (h \in \tilde{A}_R).$$

Then $(\varphi_f)_s = \varphi_f$ $(s \in W(G/A))$ and $f \to \varphi_f$ is a continuous mapping of $C_c^\infty(G)$ into $C_c^\infty(\tilde{A}_R)$. Moreover

$$\sum_{\lambda \in L} \int_A \varepsilon_A \Psi_\lambda \varpi F_f \, dh = T_A(\varphi_f) \qquad\qquad (f \in C_c^\infty(G)).$$

It is clear from [4(h), Th. 2] that the mapping $f \to \varphi_f$ is continuous. The rest follows from Corollary 2 of Lemma 26 and condition (4) of Lemma 27.

15. Proof of Theorem 3

Fix a Haar measure dx on G. In view of the definition of Θ_λ and Ψ_λ (see §8), it follows from [4(o), Th. 4] that

$$\varpi(\lambda)\Theta_\lambda(f) = \mathrm{p.v.} \int D^{-1}\Psi_\lambda \nabla_G f \, dx$$

for $\lambda \in L$ and $f \in C_c^\infty(G)$. Let us now use the notation of [4(o), Lem. 65]. Then $A_1 = B$ and it follows from Lemma 29 and [4(o), Lem. 65] that

$$\sum_{\lambda \in L} \varpi(\lambda)\,\Theta_\lambda(f) = \sum_{1 \le i \le r} c_i T_{A_i}(\varphi_{f,i}) \qquad\qquad \left(f \in C_c^\infty(G)\right),$$

where $\varphi_{f,i}$ is the function φ_f of Lemma 29 corresponding to $A = A_i$. Now take $A = B$. Then $\tilde{A}_R = \{1\}$ and it is clear from Lemma 22 that in this case

$$\sum_{\lambda \in L} \int_B \Psi_\lambda \varpi F_f\, dh = [W_G] F_f(1;\, \varpi)\,,$$

provided the total measure of B is 1. Put $q = (1/2)(\dim \mathfrak{g} - \dim \mathfrak{k})$. Then by [4(m), Lemmas 17 and 18] and [4(h), p. 759] there exists a positive number c_0 such that $F_f(1;\, \varpi) = (-1)^q c_0 f(1)$ for all $f \in C_c^\infty(G)$. Put $c = [W_G] c_1 c_0$. Then $c > 0$ and it is clear that

$$\sum_{\lambda \in L} \varpi(\lambda)\,\Theta_\lambda(f) = (-1)^q c f(1) + \sum_{2 \le i \le r} c_i T_{A_i}(\varphi_{f,i})$$

for $f \in C_c^\infty(G)$. Put $T_i = (-1)^{q+1} c_i T_{A_i} (2 \le i \le r)$. Then we have obtained the following theorem which is one of the main steps in the proof of the Plancherel formula.

THEOREM 3. *Let $f \in C_c^\infty(G)$. Then*

$$cf(1) = (-1)^q \sum_{\lambda \in L} \varpi(\lambda)\Theta_\lambda(f) + \sum_{2 \le i \le r} T_i(\varphi_{f,i})\,.$$

PART II

16. The function Ξ

Let G be a connected Lie group corresponding to a reductive Lie algebra $\mathfrak{g}$ over $\mathbf{R}$. Define θ, $\mathfrak{k}$, $\mathfrak{p}$ and K as in [4(n), §16] and let Z_G denote the center of G. Then $Z_G \subset K$ and K/Z_G is compact. Let $I_c(G)$ denote the space of all continuous functions f on G such that

(1) $f(k_1 x k_2) = f(x)\ (k_1,\, k_2 \in K$ and $x \in G)$, and

(2) f, regarded as a function on G/Z_G, has compact support.

Fix a maximal abelian subspace $\mathfrak{a}$ of $\mathfrak{p}$ and let $\mathfrak{m}$ and M denote the centralizers of $\mathfrak{a}$ in $\mathfrak{g}$ and G respectively. Then $\theta(\mathfrak{m}) = \mathfrak{m}$ and therefore $\mathfrak{m}$ is reductive in $\mathfrak{g}$. Let M^0 denote the connected component of 1 in M. Then $M/M^0 Z_G$ is finite [4(g), Lem. 15]. Hence M is unimodular and therefore the homogeneous space $G^* = G/M$ has an invariant measure dx^*. Let A be the analytic subgroup of G corresponding to $\mathfrak{a}$ and $x \to x^*\ (x \in G)$ the projection of G on G^*. Put $a^{x^*} = xax^{-1}(x \in G)$,

$$\nu(a) = \left| \det\!\left(\mathrm{Ad}(a^{-1}) - 1\right)_{\mathfrak{g}/\mathfrak{m}} \right|^{1/2} \qquad\qquad (a \in A)$$

and fix a Haar measure da on A.

LEMMA 30. *There exists a unique continuous function $\Xi(x)$ on G with the following properties:*

(1) $\Xi(1) = 1$, $\Xi(k_1 x k_2) = \Xi(x)$ $(k_1, k_2 \in K; x \in G)$.

(2) *We can choose a number $c > 0$ such that*[8]

$$\int_{G/Z_G} f \Xi dx = c \int_{G^* \times A} \nu(a) f(a^{z^*}) dx^* da$$

for all $f \in I_c(G)$.

The uniqueness is obvious and therefore only the existence requires proof. Without loss of generality we may replace G by G/Z_G and thus assume that G is semi-simple and K is compact. Let $\mathfrak{g} = \mathfrak{k} + \mathfrak{a} + \mathfrak{n}$ be an Iwasawa decomposition of $\mathfrak{g}$ and $G = KAN$ the corresponding decomposition of G (see [5(a), p. 234]). Put

$$\sigma(H) = \frac{1}{2} \operatorname{tr}(\operatorname{ad} H)_\mathfrak{n} \qquad (H \in \mathfrak{a}) .$$

Let dk be the normalized Haar measure of K. Then if the Haar measures dx and dn of G and N respectively are suitably normalized, we have (see [4(a), Lem. 35])

$$dx = e^{2\sigma(\log a)} dk \, da \, dn$$

for $x = kan$ $(k \in K, a \in A, n \in N)$. Put $H(x) = \log a$.

LEMMA 31. *Under the above conditions*

$$\Xi(x) = \int_K e^{-\sigma(H(xk))} dk \qquad (x \in G) .$$

Let $\Xi_0(x)$ denote the value of the above integral. Then it is clear that Ξ_0 is an analytic function on G which satisfies condition (1) of Lemma 30. Now if $f \in I_c(G)$, it is obvious that

$$\int f \Xi_0 dx = \int f(x) e^{-\sigma(H(x))} dx = \int f(an) e^{\sigma(\log a)} da dn .$$

On the other hand by [4(i), p. 261], we can choose a number $c > 0$ such that

$$c\nu(a) \int f(a^{z^*}) dx^* = e^{\sigma(\log a)} \int f(an) dn \qquad (a \in A, \nu(a) \neq 0)$$

for all $f \in I_c(G)$. Therefore Ξ_0 also satisfies condition (2) of Lemma 30. In view of the uniqueness of Ξ, this proves Lemmas 30 and 31.

COROLLARY. *There exists a number $c > 0$ such that*

$$\int_G f \Xi dx = c \int_{A \times N} f(an) e^{\sigma(\log a)} da dn$$

---

[8] Here dx is the Haar measure of G/Z_G.

for all $f \in I_o(G)$.

We have already seen this during the above proof.

Since any two maximal abelian subspaces of $\mathfrak{p}$ are conjugate under K (see [5(a), p. 211]), it follows easily from Lemma 30 that Ξ is actually independent of the choice of $\mathfrak{a}$.

LEMMA 32. $\Xi(x) = \Xi(x^{-1})$ *for* $x \in G$. *Moreover, for any compact set* Ω *in* G, *we can choose a number* $c > 0$ *such that*

$$\Xi(y_1 x y_2) \leqq c\Xi(x) \qquad\qquad (x \in G)$$

for $y_1, y_2 \in \Omega$.

This is an immediate consequence of [4(i), Lemmas 47 and 48].

17. Fourier components of a distribution

Define $\mathfrak{G}$ and $\mathfrak{Z}$ as in [4(n), § 6], and let $\mathfrak{Z}_K$ denote the center of[2] $\mathfrak{G}(\mathfrak{k}_c)$. Put $\mathfrak{U} = \mathfrak{Z}_K\mathfrak{Z}$.

LEMMA 33. *Let* T *be a distribution on an open subset* Ω *of* G *and suppose that the dimension of the space of all distributions of the form* $uT\,(u \in \mathfrak{U})$ *is finite. Then* T *coincides with an analytic function on* Ω.

Let $\mathfrak{c}$ be the center and $\mathfrak{g}_1$ the derived algebra of $\mathfrak{g}$. We can choose a positive-definite quadratic form Q on $\mathfrak{g}$ such that

(1) $Q(X) = -\mathrm{tr}(\mathrm{ad}\,X\,\mathrm{ad}\,\theta(X))\ (X \in \mathfrak{g}_1)$, and

(2) $\mathfrak{c}$ and $\mathfrak{g}_1$ are orthogonal under Q.

We regard $\mathfrak{g}$ as a real Hilbert space under Q. It is clear that Q is invariant under K and $\mathfrak{p}$ is orthogonal to $\mathfrak{k}$. Let $\langle X, Y\rangle\,(X, Y \in \mathfrak{g})$ denote the scalar product in $\mathfrak{g}$ and put

$$B(X, Y) = -\langle X, \theta(Y)\rangle \qquad\qquad (X, Y \in \mathfrak{g})\,.$$

Then B is a symmetric bilinear form which is invariant under the adjoint group of $\mathfrak{g}$.

Select orthonormal bases $(X_1, \cdots, X_r)$ and $(Y_1, \cdots, Y_s)$ for $\mathfrak{k}$ and $\mathfrak{p}$ respectively and define

$$\omega_{\mathfrak{k}} = X_1^2 + \cdots + X_r^2, \qquad \omega_{\mathfrak{p}} = Y_1^2 + \cdots + Y_s^2$$

in $\mathfrak{G}$. Since Q is invariant under K and B is invariant under G, it is obvious that $\omega_{\mathfrak{k}} \in \mathfrak{Z}_K$ and

$$\omega = -\omega_{\mathfrak{k}} + \omega_{\mathfrak{p}} \in \mathfrak{Z}\,.$$

Hence $u = \omega + 2\omega_{\mathfrak{k}} \in \mathfrak{U}$, and therefore the distributions $u^k T(k \geqq 0)$ cannot all be linearly independent. This shows that we can choose an integer $m \geqq 1$ and complex numbers $c_1, \cdots, c_m$ such that $vT = 0$, where

$$v = u^m + \sum_{1 \le k \le m} c_k u^{m-k} .$$

It is obvious that u and therefore also v, are elliptic differential operators on G. Hence we conclude that T is an analytic function.

Now assume that K is compact and let $\mathfrak{S}_K$ denote the set of all equivalence classes of finite-dimensional irreducible representations of K. For $\mathfrak{d} \in \mathfrak{S}_K$, let $\xi_\mathfrak{d}$ denote the character of $\mathfrak{d}$ and $d(\mathfrak{d}) = \xi_\mathfrak{d}(1)$ the degree of $\mathfrak{d}$. Define

$$f_\mathfrak{d}(x) = d(\mathfrak{d}) \int_K f(xk) \xi_\mathfrak{d}(k^{-1}) dk \qquad\qquad (x \in G)$$

for any continuous function f on G. (Here dk is, as before, the normalized Haar measure of K.) Then if $f \in C_c^\infty(G)$, the series $\sum_{\mathfrak{d} \in \mathfrak{S}_K} f_\mathfrak{d}$ converges to f in $C_c^\infty(G)$ (see [4(b), § 4]) and therefore

$$T(f) = \sum_{\mathfrak{d} \in \mathfrak{S}_K} T(f_\mathfrak{d})$$

for any distribution T on G. Let $\mathfrak{d}^*$ denote the class in $\mathfrak{S}_K$ contragredient to $\mathfrak{d}$ so that $\xi_{\mathfrak{d}^*}(k) = \mathrm{conj}\, \xi_\mathfrak{d}(k)(k \in K)$. Define the distribution $T_\mathfrak{d}$ on G by

$$T_\mathfrak{d}(f) = T(f_{\mathfrak{d}^*}) \qquad\qquad (f \in C_c^\infty(G)) .$$

Then the above relation shows that

$$T(f) = \sum_{\mathfrak{d} \in \mathfrak{S}_K} T_\mathfrak{d}(f) .$$

$T_\mathfrak{d}$ will be called the $\mathfrak{d}^{\text{th}}$ Fourier component of T.

LEMMA 34. *Let T be a distribution on G such that the dimension of the space of all distributions of the form zT $(z \in \mathfrak{Z})$ is finite. Then $T_\mathfrak{d}$ is an analytic function for every $\mathfrak{d} \in \mathfrak{S}_K$.*

It is clear that if we regard elements of $\mathfrak{Z}_K$ as differential operators on K, then every irreducible character $\xi_\mathfrak{d}(\mathfrak{d} \in \mathfrak{S}_K)$ is an eigenfunction of all the operators in $\mathfrak{Z}_K$. Hence there exists a homomorphism $\alpha_\mathfrak{d}$ of $\mathfrak{Z}_K$ into C such that

$$\gamma \xi_\mathfrak{d} = \alpha_\mathfrak{d}(\gamma) \xi_\mathfrak{d} \qquad\qquad (\gamma \in \mathfrak{Z}_K) .$$

From this it follows without difficulty that $\gamma T_\mathfrak{d} = \alpha_\mathfrak{d}(\gamma) T_\mathfrak{d}$. Our assertion is therefore an immediate consequence of Lemma 33.

18. Reduction of Theorem 4 to a special case

We keep to the assumption that K is compact. Put $I_c^\infty(G) = C_c^\infty(G) \cap I_c(G)$. Fix a Cartan subalgebra $\mathfrak{h} = \theta(\mathfrak{h})$ of $\mathfrak{g}$ and let $A_\mathfrak{h}$ denote the corresponding Cartan subgroup of G. We now use the notation of [4(n), § 23].

THEOREM 4. *Assume that G is acceptable [4(n), § 18]. Then there exists a number $c > 0$ such that*

$$\int_{A_\mathfrak{h}} |\Delta_M(h) F_f(h)| \, dh \le c \int_G |f| \, \Xi dx$$

for all $f \in I_c^\infty(G)$.

Put $G_\mathfrak{h} = (A_\mathfrak{h}')^G$. One proves without difficulty (see [4(n), § 20]) that there exists a unique function $D_\mathfrak{h}$ on G with the following two properties:

(1) $D_\mathfrak{h}(h^x) = \det (1 - \mathrm{Ad}(h))_{\mathfrak{g}/\mathfrak{m}} (h \in A_\mathfrak{h}', x \in G)$,

(2) $D_\mathfrak{h} = 0$ outside $G_\mathfrak{h}$.

Moreover $D_\mathfrak{h}$ is analytic on $G_\mathfrak{h}$.

LEMMA 35.   *Theorem 4 is equivalent to the following statement. There exists a number $c > 0$ such that*

$$\int_K |D_\mathfrak{h}(xk)|^{-1/2}\, dk \leq c \Xi(x)$$

for almost all $x \in G$.

It follows from [4(n), Lem. 41] that we can normalize dx in such a way that

$$\int_G f\, |D_\mathfrak{h}|^{-1/2}\, dx = \int_{A_\mathfrak{h}} |\Delta_M F_f|\, dh$$

for $f \in C_c^\infty(G)(f \geq 0)$. Now suppose Theorem 4 holds. Then

$$\int_G f\, |D_\mathfrak{h}|^{-1/2}\, dx \leq c \int_G f \Xi\, dx$$

for $f \in I_c^\infty(G)(f \geq 0)$ and therefore it follows by standard arguments that

$$\int_G |f|\, |D_\mathfrak{h}|^{-1/2}\, dx \leq c \int_G |f|\, \Xi\, dx$$

for $f \in I_c(G)$. This implies the assertion of Lemma 35.

Put $M_\mathfrak{h} = (M' \cap A_\mathfrak{h})^M$ where M' is the set of those elements of M which are regular in M. Then the following lemma is an immediate consequence of [4(n), Lem. 52] (see also § 10).

LEMMA 36.   *Define $g_f(f \in C_c^\infty(G))$ as in [4(n), § 27]. Then*

$$g_f(k_1 m k_2) = g_f(m) \qquad\qquad (k_1, k_2 \in K \cap M;\, m \in M)$$

for $f \in I_c^\infty(G)$. Moreover the Haar measure dm on M can be so normalized that

$$\int_{A_\mathfrak{h}} |\Delta_M(h) F_f(h)|\, dh \leq \int_{M_\mathfrak{h}} |g_f(m)|\, dm$$

for all $f \in I_c^\infty(G)$.

We shall now prove Theorem 4 by induction on $\dim \mathfrak{g}$. Put $\mathfrak{a}_2 = \mathfrak{h}_R = \mathfrak{h} \cap \mathfrak{p}$ and extend $\mathfrak{a}_2$ to a maximal abelian subspace $\mathfrak{a}$ of $\mathfrak{p}$. Introduce compatible orders on the spaces of (real) linear functions on $\mathfrak{a}_2$ and $\mathfrak{a}$ respectively and, for any linear function α on $\mathfrak{a}$, let $\mathfrak{g}_\alpha$ denote the space of all $X \in \mathfrak{g}$ such that

$[H, X] = \alpha(H)X$ for all $H \in \mathfrak{a}$. Consider the set Σ of those $\alpha > 0$ for which $\mathfrak{g}_\alpha \neq \{0\}$. Put

$$\mathfrak{n}_1 = \sum_{\alpha \in \Sigma_1} \mathfrak{g}_\alpha, \quad \mathfrak{n}_2 = \sum_{\alpha \in \Sigma_2} \mathfrak{g}_\alpha, \quad \mathfrak{n} = \mathfrak{n}_1 + \mathfrak{n}_2 ,$$

where Σ_1 is the subset consisting of those $\alpha \in \Sigma$ which vanish identically on $\mathfrak{a}_2$, and Σ_2 is the complement of Σ_1 in Σ. Then $\mathfrak{n}_1, \mathfrak{n}_2, \mathfrak{n}$ are all nilpotent subalgebras of $\mathfrak{g}$, and $\mathfrak{n}_2$ is an ideal in $\mathfrak{n}$. Put

$$\mathfrak{m}_1 = \mathfrak{m} \cap \mathfrak{k} + [\mathfrak{m}, \mathfrak{m}] \cap \mathfrak{p}$$

where $\mathfrak{m}$ is, as before, the centralizer of $\mathfrak{h}_R$ in $\mathfrak{g}$. Then $\mathfrak{m} = \mathfrak{m}_1 + \mathfrak{a}_2$ where the sum is direct. Put $\mathfrak{a}_1 = \mathfrak{a} \cap \mathfrak{m}_1$ so that $\mathfrak{a} = \mathfrak{a}_1 + \mathfrak{a}_2$, and let $M_1, A_1, A_2, A,$ N_1, N_2, N be the analytic subgroups of G corresponding to $\mathfrak{m}_1, \mathfrak{a}_1, \mathfrak{a}_2, \mathfrak{a}, \mathfrak{n}_1, \mathfrak{n}_2, \mathfrak{n}$ respectively. Then $M = M_1 A_2$, $G = KAN$ and $M_1 = K_1 A_1 N_1$ where $K_1 = M_1 \cap K$. Moreover it is clear that

$$g_f(m_1 a_2) = e^{\sigma(\log a_2)} \int f(m_1 a_2 n_2) dn_2 \qquad (f \in I_c^\infty(G), \, m_1 \in M_1, \, a_2 \in A_2)$$

where dn_2 is the Haar measure on N_2.

Now assume that $\mathfrak{h}_R \neq \{0\}$. Then $\dim \mathfrak{m}_1 < \dim \mathfrak{g}$ and therefore Theorem 4 holds for M_1 (in place of G) by induction hypothesis. Put $M_{1\mathfrak{h}} = M_1 \cap M_\mathfrak{h}$, and let dm_1, da_1, dn_1 denote the Haar measures on M_1, A_1, N_1 respectively. Also let $I_c(M_1)$ denote the set of all functions $G \in C_c(M_1)$ which are invariant under both left and right translations by elements of K_1. Finally put

$$\sigma_1(H) = \operatorname{tr}(\operatorname{ad} H)_{\mathfrak{n}_1} \qquad (H \in \mathfrak{a}) .$$

Since $\mathfrak{h}_I = \mathfrak{h} \cap \mathfrak{k}$ is obviously a Cartan subalgebra of $\mathfrak{m}_1$, we get the following result, in view of the induction hypothesis, from Lemma 35 and the corollary of Lemma 31.

Lemma 37. *There exists a number $c_1 > 0$ such that*

$$\int_{M_{1\mathfrak{h}}} |g(m_1)| \, dm_1 \leq c_1 \int_{A_1 \times N_1} |g(a_1 n_1)| \, e^{\sigma_1(\log a_1)} da_1 dn_1$$

for all $g \in I_c(M_1)$.

We can now fix a Haar measure da_2 on A_2 such that $dm = dm_1 da_2 (m = m_1 a_2, \, m_1 \in M_1, \, a_2 \in A_2)$. Then it follows from Lemma 37 that

$$\int_{M_\mathfrak{h}} |g_f(m)| \, dm \leq c_1 \int |g_f(a_1 a_2 n_1)| \, e^{\sigma_1(\log a_1)} da_1 da_2 dn_1$$

$$\leq c_1 \int |f(a_1 a_2 n_1 n_2)| \, \exp \{\sigma_1(\log a_1) + \sigma(\log a_2)\} da_1 da_2 dn_1 dn_2$$

for $f \in I_c^\infty(G)$. Now it is clear that $[\mathfrak{m}, \mathfrak{n}_2] \subset \mathfrak{n}_2$ and $\mathfrak{a}_1 \subset [\mathfrak{m}, \mathfrak{m}]$. Hence $\operatorname{tr}(\operatorname{ad} H)_{\mathfrak{n}_2} = 0$ for $H \in \mathfrak{a}_1$, and therefore $\sigma = \sigma_1$ on $\mathfrak{a}_1$. Moreover since $\mathfrak{n}_2$ is an ideal in $\mathfrak{n}$, the mapping $(n_1, n_2) \to n_1 n_2$ defines an analytic diffeomorphism of $N_1 \times N_2$ onto N, and the Haar measure dn on N can be so normalized that $dn = dn_1 dn_2 (n = n_1 n_2)$. Similarly we can assume that $da = da_1 da_2$ where da is the Haar measure on A. Therefore

$$\int_{M_{\mathfrak{h}}} |g_f(m)|\, dm \leq c_1 \int |f(an)|\, e^{\sigma(\log a)} da\, dn \qquad\qquad \left(f \in I_c^{\infty}(G)\right).$$

The assertion of Theorem 4 is now obvious from Lemma 36 and the corollary of Lemma 31.

Thus it remains to consider the case when $\mathfrak{h}_R = \{0\}$. We write $\mathfrak{h} = \mathfrak{b}$ and $A_{\mathfrak{h}} = B$ in this case.

19. Second reduction

Let φ_B denote the characteristic function of the open set $G_B = (B')^G$ in G. In view of Lemma 35, it would be enough to prove the following result.

THEOREM 5. *There exists a number $c > 0$ such that*

$$\int_K \varphi_B(xk) dk \leq c\,\Xi(x)$$

for almost all $x \in G$.

Since $Z_G \subset B \subset K$, it is clear that the statement of Theorem 5 is not affected if G is replaced by G/Z_G. Therefore we may assume that G is semisimple. Let G_c be the simply connected complex-analytic group corresponding to $\mathfrak{g}_c$ and G_0 the real analytic subgroup of G_c with Lie algebra $\mathfrak{g}$. We may suppose, without loss of generality, that $G = G_0$ and G is not compact.

Let us use the notation of [4(o), § 24]. It is obvious that L is stable under $W = W(\mathfrak{g}/\mathfrak{b})$ in the present case.

LEMMA 38. *Fix $\lambda \in L'$ and define Θ_λ^* as in [4(o), § 24]. Then we can choose a number $c \geq 0$ such that*

$$\left| \Theta_\lambda^*(f) - \int_{G_B} f\Theta_\lambda^* dx \right| \leq c \int_G |f|\, \Xi\, dx$$

for all $f \in I_c^{\infty}(G)$.

Let $\mathfrak{a}_1 = \mathfrak{b}, \mathfrak{a}_2, \cdots, \mathfrak{a}_r$ be a complete set of Cartan subalgebras of $\mathfrak{g}$, no two of which are conjugate under G. We may assume that $\theta(\mathfrak{a}_i) = \mathfrak{a}_i (1 \leq i \leq r)$. Let A_i be the Cartan subgroup of G corresponding to $\mathfrak{a}_i$. Since $\mathfrak{a}_i (i \geq 2)$ is not conjugate to $\mathfrak{b}$, $\mathfrak{a}_i \cap \mathfrak{p} \neq \{0\}$ (see [4(f), § 8]). Therefore, by § 18, Theorem 4 is applicable to $\mathfrak{a}_i$. Put $G_i = (A_i')^G$. Then it follows from the corollary of Lemma 62 of [4(o)] and [4(n), Lem. 41] that

$$\left| \int_{G_i} f\Theta_\lambda^* dx \right| \leq c_i \int_G |f|\, \Xi\, dx \qquad\qquad (f \in I_c^{\infty}(G),\ 2 \leq i \leq r)$$

where c_i is a positive number independent of f. Since

$$\Theta_\lambda^*(f) = \sum_{1 \leq i \leq r} \int_{G_i} f\Theta_\lambda^* dx \;,$$

our assertion is now obvious.

It is clear from the corollary of Lemma 62 of [4(o)] that the function Θ_λ^* remains bounded on G_B. Hence we get the following corollary.

COROLLARY. *We can choose $c_0 \geq 0$ such that*

$$|\Theta_\lambda^*(f)| \leq c_0 \int_G |f|\, \varphi_B dx + c \int_G |f|\, \Xi dx$$

for all $f \in I_c^\infty(G)$.

LEMMA 39. $\Theta_\lambda^*(f) = 0$ *for* $\lambda \in L'$ *and* $f \in I_c^\infty(G)$.

Let P be the set of all positive roots of $(\mathfrak{g}, \mathfrak{b})$ and put

$$\rho = \frac{1}{2}\sum_{\alpha \in P} \alpha$$

as usual. Then $\rho \in L'$ and

$$\prod_{\alpha \in P}(e^{\alpha/2} - e^{-\alpha/2}) = \sum_{s \in W} \varepsilon(s) e^{s\rho}$$

on $\mathfrak{b}$. Hence it follows from the definition of Θ_ρ^* that

$$\Theta_\rho^* = [W_G]$$

on G_B.

Now assuming Lemma 39 in the special case $\lambda = \rho$, we shall deduce Theorem 5, and therefore also Theorem 4. Taking $\lambda = \rho$ in Lemmas 38 and 39, we get

$$\left|\int_{G_B} f dx\right| \leq c_1 \int_G |f|\, \Xi dx \qquad (f \in I_c^\infty(G))$$

where $c_1 = [W_G]^{-1}c$. From this we conclude by standard arguments that

$$\int |f|\, \varphi_B dx \leq c_1 \int |f|\, \Xi dx$$

for all $f \in I_c(G)$. The statement of Theorem 5 is now obvious.

20. Proof of Lemma 39

We now come to the proof of Lemma 39.

LEMMA 40. *For any $f \in C_c^\infty(G)$, define*

$$f^\natural(x) = \int_{K \times K} f(k_1 x k_2) dk_1 dk_2 \qquad (x \in G)\,.$$

Then there exists an analytic function Φ_λ on G such that

$$\int_G \Phi_\lambda f dx = \Theta_\lambda^*(f^\natural)$$

for all $f \in C_c^\infty(G)$.

This is an immediate consequence of Lemma 34. Moreover it is obvious that

$$z\Phi_\lambda = \chi_\lambda(z)\Phi_\lambda \qquad\qquad (z \in \mathfrak{Z})$$

where χ_λ is defined as in [4(o), § 29].

Put

$$\psi(x) = \int_K \varphi_B(xk)dk \qquad\qquad (x \in G) .$$

Then the corollary of Lemma 38 gives us the following result.

LEMMA 41. *$|\Phi_\lambda| \leq c_0\psi + c_1\Xi$ almost everywhere on G.*

On the other hand we also have the following lemma.

LEMMA 42. *Given $\varepsilon > 0$, we can choose a compact subset C of G such that $\psi \leq \varepsilon$ outside C.*

COROLLARY. $\lim_{x \to \infty} \Phi_\lambda(x) = 0$.

It follows from [4(i), Th. 3, p. 279] that $\lim_{x \to \infty} \Xi(x) = 0$. Hence the corollary follows immediately from Lemmas 41 and 42, if we take into account the fact that Φ_λ is continuous.

Assuming Lemma 42, we shall first show that $\Phi_\rho = 0$. Let ω denote the Casimir operator of $\mathfrak{g}$ (see [4(d), p. 140]). Then $\omega \in \mathfrak{Z}$ and, since $\chi_\rho(\omega) = 0$, it is clear that $\omega\Phi_\rho = 0$. Hence we conclude from the corollary of Lemma 42 and the maximum principle[8a] for elliptic differential operators of degree 2 (see [2, p. 326]) that $\Phi_\rho = 0$. The same fact can also be deduced from [2(a), Theorems 4.1 and 4.4]. This proves Lemma 39 in the case $\lambda = \rho$, and therefore establishes Theorems 4 and 5.

Now fix $\lambda \in L'$ as before.

LEMMA 43. *There exists a constant $c \geq 0$ such that*

$$|\Phi_\lambda| \leq c\Xi .$$

This is obvious from Lemma 41 and Theorem 5.

In order to complete the proof of Lemma 39, we have to undertake some preparation. Let Φ be an analytic function on G such that

(1) $\Phi(k_1xk_2) = \Phi(x)$ $(k_1, k_2 \in K, x \in G)$, and

(2)   the dimension of the space of all functions of the form $z\Phi(z \in \mathfrak{Z})$ is finite.

Let $\mathfrak{Q}$ be the centralizer of $\mathfrak{k}$ in $\mathfrak{G}$.

[8a] I am grateful to L. Hörmander for bringing this principle to my attention.

LEMMA 44. *Let V be the space of all functions of the form $q\Phi(q \in \mathfrak{Q})$. Then* $\dim V < \infty$.

Let $\mathfrak{Q}_0$ be the set of all $q \in \mathfrak{Q}$ such that $q\Phi = 0$. Then $\mathfrak{Q}_0$ is a left ideal in $\mathfrak{Q}$ and

$$\mathfrak{Q}_0 \supset \mathfrak{Q}_1 = \mathfrak{Q} \cap (\mathfrak{G}\mathfrak{k} + \mathfrak{G}\mathfrak{Z}_0)$$

where $\mathfrak{Z}_0 = \mathfrak{Z} \cap \mathfrak{Q}_0$. But we know (see [4(a), Th. 1, p. 195]) that $\mathfrak{Q}_1$ has finite codimension in $\mathfrak{Q}$. Hence $\dim \mathfrak{Q}/\mathfrak{Q}_0 < \infty$.

It follows from [4(i), Th. 1] that $\mathfrak{Q}/\mathfrak{Q} \cap \mathfrak{G}\mathfrak{k}$ is abelian, and therefore $\mathfrak{Q}_0$ is a two-sided ideal in $\mathfrak{Q}$. Let $q_i (1 \leq i \leq r)$ be a base for $\mathfrak{Q} \bmod \mathfrak{Q}_0$.

LEMMA 45. *Fix $\varphi \in V$. Then we can choose $f \in I_c^\infty(G)$ such that*

$$\varphi(x) = \int_G f(y)\Phi(yx)dy \qquad (x \in G) .$$

Put $\Phi_i = q_i\Phi (1 \leq i \leq r)$. Then $\Phi_1, \cdots, \Phi_r$ is a base for V over C. Therefore (see [4(g), Lem. 20]) we can choose $f \in C_c^\infty(G)$ such that

$$\int f\Phi_i dx = \varphi(1; q_i) \qquad (1 \leq i \leq r) .$$

Replacing f by $f^\natural$, we may assume that $f \in I_c^\infty(G)$. Now consider

$$\varphi_0(x) = \int_G f(y)\Phi(yx)dy - \varphi(x) \qquad (x \in G) .$$

It is obvious that

(1) φ_0 is analytic,

(2) it is invariant under both left and right translations by elements of K, and

(3) $q\varphi_0 = 0$ for $q \in \mathfrak{Q}_0$.

Moreover it follows from the definition of f that $\varphi_0(1; q_i) = 0 \ (1 \leq i \leq r)$. Hence $\varphi_0(1; q) = 0$ for all $q \in \mathfrak{Q}$.

Now fix $g \in \mathfrak{G}$. Then $\varphi_0(1; g) = \varphi_0(1; q) = 0$ where[9]

$$q = \int_K g^k dk \in \mathfrak{Q} .$$

But since φ_0 is analytic, this implies that $\varphi_0 = 0$. Hence the lemma.

Now take $\Phi = \Phi_\lambda$. Then in order to show that $\Phi_\lambda = 0$, it would be enough to verify that $V = \{0\}$. So assume that $V \neq \{0\}$. Since $\mathfrak{Q}/\mathfrak{Q} \cap \mathfrak{G}\mathfrak{k}$ is abelian, we can choose an element $\varphi \neq 0$ in V, and a homomorphism χ of $\mathfrak{Q}$ into C such that $q\varphi = \chi(q)\varphi$ for all $q \in \mathfrak{Q}$. Then by multiplying φ by a non-zero constant, we can assume (see [5(a), Th. 6.16, p. 432]) that

$$\varphi(x) = \int_K e^{(\Lambda - \sigma)(H(xk))} dk \qquad (x \in G)$$

____

[9] Here $g^x (g \in \mathfrak{G}, x \in G)$ has the same meaning as in [4(n), §2].

in the notation of Lemma 31, where Λ is a suitable linear function on $\mathfrak{a}_c$.

Extend $\mathfrak{a}$ to a Cartan subalgebra $\mathfrak{h}$ of $\mathfrak{g}$ so that $\mathfrak{h} = \mathfrak{a} + \mathfrak{h}_I$ where $\mathfrak{h}_I = \mathfrak{h} \cap \mathfrak{k}$. Define compatible orders on the space $\mathfrak{a}^*$ and $\mathfrak{h}^*$ of real linear functions on $\mathfrak{a}$ and $\mathfrak{a} + (-1)^{1/2}\mathfrak{h}_I$ respectively. We assume that the order in $\mathfrak{a}^*$ corresponds to the Iwasawa decomposition $\mathfrak{g} = \mathfrak{k} + \mathfrak{a} + \mathfrak{n}$. Let Q be the set of all positive roots of $(\mathfrak{g}, \mathfrak{h})$ under this order and Q_0 the subset of those $\alpha \in Q$ which vanish identically on $\mathfrak{a}$. Put

$$\sigma_0 = \frac{1}{2}\sum_{\alpha \in Q_0}\alpha\,, \qquad \sigma_+ = \frac{1}{2}\sum_{\alpha \in Q_+}\alpha$$

where Q_+ is the complement of Q_0 in Q. If $\alpha \in Q_+$, then $-\theta\alpha$ is also a root in Q_+. Hence $\theta\sigma_+ = -\sigma_+$ and $\theta\sigma_0 = \sigma_0$. Moreover $\sigma = \sigma_+$ on $\mathfrak{a}$.

Extend Λ to a linear function on $\mathfrak{h}$ by defining it to be zero on $\mathfrak{h}_I$ and put $\Lambda_0 = \Lambda + \sigma_0$. Moreover define

$$\chi_{\Lambda_0}^{\mathfrak{h}}(z) = \chi_{\Lambda_0}^{\mathfrak{h}}(p_z) \qquad\qquad (z \in \mathfrak{Z})$$

in the notations of [4(n), § 12] and [4(o), § 12].

LEMMA 46. $z\varphi = \chi_{\Lambda_0}^{\mathfrak{h}}(z)\varphi \ (z \in \mathfrak{Z})$.

This follows from the above integral formula for φ, if we take into account [4(i), Lemmas 1, 2, 4, 5 and 18].

COROLLARY. $\Lambda \in \mathfrak{a}^*$.

It is obvious that $z\varphi = \chi_\lambda(z)\varphi \ (z \in \mathfrak{Z})$. Choose y in G_c such that $\mathfrak{h}_c = \mathfrak{b}_c^y$. Since $\varphi(1) = 1$, we conclude from [4(o), Lem. 23] that $\lambda^y = \Lambda_0$, provided y is suitably chosen. However λ is a real linear combination of the roots of $(\mathfrak{g}, \mathfrak{b})$, and so it follows that $\Lambda_0 \in \mathfrak{h}^*$, and therefore $\Lambda \in \mathfrak{a}^*$.

Define Σ as in [4(i), p. 244] and let $\mathfrak{w}$ be the Weyl group of $(\mathfrak{g}, \mathfrak{a})$ (see [4(i), p. 249]). Put $A^+ = \exp\mathfrak{a}^+$ where $\mathfrak{a}^+$ is the set of all points $H \in \mathfrak{a}$ where $\alpha(H) \geq 0$ for all $\alpha \in \Sigma$.

LEMMA 47. *Let μ be a real linear function on $\mathfrak{a}$. Then there exists a number $c > 0$ such that*

$$\int_K e^{(\mu - \sigma)(H(hk))}dk \geq ce^{-\sigma(\log h)}\sum_{s \in \mathfrak{w}}e^{s\mu(\log h)}$$

for $h \in A^+$.

For the proof, we use the notation of [4(i), p. 288]. In view of [4(i), Cor. on p. 261] we can replace μ by $s\mu(s \in \mathfrak{w})$. Therefore we may assume that $\mu \geq s\mu$ for all $s \in \mathfrak{w}$. Then if $h \in A^+$, we have

$$e^{-(\mu-\sigma)(\log h)}\int_K e^{(\mu-\sigma)(H(hk))}dk = \int \exp\{(\mu-\sigma)(H(\bar{n}^k)) - (\mu+\sigma)(H(\bar{n}))\}d\bar{n}$$

$$\geq \int \exp\{-\sigma(H(\bar{n}^k)) - (\mu+\sigma)(H(\bar{n}))\}d\bar{n}$$

$$\geq \int \exp\{-(\mu+2\sigma)(H(\bar{n}))\}d\bar{n} = c_1 > 0$$

from [4(i), Lem. 43]. Since $s\mu(H) \leq \mu(H)$ $(s \in \mathfrak{w}, H \in \mathfrak{a}^+)$, our assertion follows if we take $c = [\mathfrak{w}]^{-1}c_1$.

Fix a norm on $\mathfrak{a}$.

COROLLARY. *Suppose there exist numbers $c_0, r \geq 0$ such that*

$$\int_K e^{(\mu-\sigma)(H(hk))}dk \leq c_0(1 + \|\log h\|)^r \Xi(h)$$

for $h \in A^+$. Then $\mu = 0$.

Again we may assume that $\mu \geq s\mu$ for $s \in \mathfrak{w}$. Then $\mu(H) \geq 0$ for $H \in \mathfrak{a}^+$, and therefore our assertion follows from [4(i), Th. 3] and the above lemma.

In view of Lemmas 32, 43 and 45 and the corollaries of Lemmas 46 and 47, we can now conclude that $\Lambda = 0$. However $\mathfrak{a} \neq \{0\}$ since G is not compact. Therefore $\mathfrak{h}$ is not fundamental, and $(\mathfrak{g}, \mathfrak{h})$ has a real root α [4(f), Lem. 33]. This means that $\Lambda_0(H_\alpha) = \Lambda(H_\alpha) = 0$. But since $\Lambda_0 = \lambda^\nu$ and $\lambda \in L'$, this is impossible. This contradiction completes the proof of Lemma 39.

21. Proof of Lemma 42

We now come to the proof of Lemma 42. Let us keep to the above notation and regard $\mathfrak{g}$ as a real Hilbert space with the norm

$$\|X\|^2 = -\mathrm{tr}\,(\mathrm{ad}\,X\,\mathrm{ad}\,\theta(X)) \qquad\qquad (X \in \mathfrak{g}) .$$

Let S^+ be the set of all $H \in \mathfrak{a}^+$ with $\|H\| = 1$. Then since $G = KA^+K$ and S^+ is compact, it would be sufficient to prove the following result.

LEMMA 48. *For any $H_0 \in S^+$, we can choose a neighborhood V of H_0 in S^+ with the following property. Given $\varepsilon > 0$, there exists a number $t_0 > 0$ such that*

$$\psi(\exp tH) \leq \varepsilon$$

for $t \geq t_0$ and $H \in V$.

Let us use notation of [4(i), p. 280]. Then $(\alpha_1, \cdots, \alpha_l)$ is a fundamental system of roots in Σ. Suppose $\alpha_i(H_0) = 0$ $(1 \leq i < m)$ and $\alpha_i(H_0) > 0$ $(m \leq i \leq l)$. Put

$$\Lambda = \sum_{m \leq i \leq l} \lambda_i .$$

Then $\langle \Lambda, \alpha_i \rangle = 0 \ (1 \leq i < m), \langle \Lambda, \alpha_i \rangle > 0 \ (m \leq i \leq l)$, and there exists an irreducible representation σ of G on a finite-dimensional complex Hilbert space U such that Λ is the highest weight of σ (with respect to $\mathfrak{a}$). We assume, as we may (see [4(i), p. 244]) that $\sigma(k)$ is unitary $(k \in K)$ and $\sigma(X)$ is self-adjoint $(X \in \mathfrak{p})$.

Let $\Lambda_0 = \Lambda > \Lambda_1 > \cdots > \Lambda_p$ be all the weights of σ (with respect to $\mathfrak{a}$), and $U_j \ (0 \leq j \leq p)$ the subspace of all $u \in U$ such that $\sigma(H)u = \Lambda_j(H)u$ for all $H \in \mathfrak{a}$. Let E_j denote the orthogonal projection of U on U_j, and put $g(k) = \operatorname{tr}(E_0\sigma(k)) \ (k \in K)$. Then g is an analytic function on K which is not identically zero since

$$g(1) = \operatorname{tr} E_0 \geq 1 \ .$$

For any $\delta > 0$, let K_δ denote the set of all $k \in K$ where $|g(k)| < \delta$. Then $K_0 = \bigcap_{\delta>0} K_\delta$ is the set of zeros of g in K. Let μ be the normalized Haar measure of K. Since g is analytic and not identically zero, it is clear that $\mu(K_0) = 0$. Hence for a given $\varepsilon > 0$, we can choose $\delta > 0$ such that $\mu(K_\delta) \leq \varepsilon$.

Now put $2\gamma = \min_{m \leq i \leq l} \alpha_i(H_0)$. Then $\gamma > 0$ and, moreover, $\Lambda(H_0) > 0$ (see the corollary of Lemma 49 below). Hence we can select a neighborhood V of H_0 in S^+ such that $\alpha_i(H) \geq \alpha_i(H_0)/2 \ (1 \leq i \leq l)$, $\Lambda(H) \geq \Lambda(H_0)/2$ for all $H \in V$. Put $\mu_j = \Lambda - \Lambda_j \ (1 \leq j \leq p)$. Then

$$\mu_j = \sum_{1 \leq i \leq l} c_{ji}\alpha_i$$

where c_{ji} are integers ≥ 0, and $c_{ji} > 0$ for some $i \geq m$ (see Lemma 49 below). Hence

$$\mu_j(H) \geq \gamma \qquad\qquad\qquad (1 \leq j \leq p)$$

for $H \in V$.

For any $h \in A$, let $K(h)$ denote the set of all $k \in K$ such that $hk \in G_B$. Now fix $H \in V$ and put $h_t = \exp tH \ (t \geq 0)$. Then if $k \in K(h_t)$, it is clear that

$$|\operatorname{tr} \sigma(h_t k)| \leq d$$

where $d = \dim U$. On the other hand

$$\begin{aligned}
\sigma(h_t k) &= \sum_{0 \leq j \leq p} E_j\sigma(h_t)\sigma(k) \\
&= e^{t\Lambda(H)}\{E_0\sigma(k) + \sum_{1 \leq j \leq p} e^{-t\mu_j(H)}E_j\sigma(k)\} \ .
\end{aligned}$$

Therefore

$$e^{-t\Lambda(H)} \operatorname{tr} \sigma(h_t k) = g(k) + \sum_{1 \leq j \leq p} e^{-t\mu_j(H)} \operatorname{tr}\left(E_j\sigma(k)\right) \ .$$

This shows that

$$|g(k)| \leq de^{-t\Lambda(H)} + de^{-\gamma t} \leq 2de^{-ct} \qquad\qquad (k \in K(h_t))$$

where $c = \min(\gamma, \Lambda(H_0)/2)$. Now choose $t_0 \geq 0$ so large that $2de^{-ct_0} < \delta$. Then the above result shows that $K(\exp tH) \subset K_\delta$ for $H \in V$ and $t \geq t_0$. On the other hand, it is obvious from the definition of ψ that

$$\psi(h) \leq \mu\big(K(h)\big) \qquad\qquad (h \in A) \,.$$

Therefore

$$\psi(\exp tH) \leq \mu(K_\delta) \leq \varepsilon$$

for $H \in V$ and $t \geq t_0$. This proves Lemma 48, and hence also Lemma 42.

22. Proofs of some elementary facts

We shall now verify the two statements used in the above proof. Let $\mathfrak{m}$ be the centralizer of H_0 in $\mathfrak{g}$ and Σ_+ the set of all $\alpha \in \Sigma$ such that $\alpha(H_0) > 0$. Then it is obvious that $\mathfrak{g} = \mathfrak{n}_+ + \mathfrak{m} + \mathfrak{n}_-$ where $\mathfrak{n}_- = \theta(\mathfrak{n}_+)$ and

$$\mathfrak{n}_+ = \sum\nolimits_{\alpha \in \Sigma_+} \mathfrak{g}_\alpha$$

(see § 18 for the definition of $\mathfrak{g}_\alpha$). Then $\mathfrak{n}_\pm$ are nilpotent subalgebras of $\mathfrak{g}$.

Define $\mathfrak{M} = \mathfrak{S}(\mathfrak{m}_c)$, $\mathfrak{N}_\pm = \mathfrak{S}((\mathfrak{n}_\pm)_c)$ in the notation of [4(k), § 7]. Then it follows from the above decomposition of $\mathfrak{g}$ that

$$\mathfrak{G} = \mathfrak{N}_- \mathfrak{M} \mathfrak{N}_+ \,.$$

We keep to the notation of § 21.

LEMMA 49. $\mu_j = \sum\nolimits_{\alpha \in \Sigma_+} c_j(\alpha)\alpha$ $(1 \leq j \leq p)$ where $c_j(\alpha)$ are integers ≥ 0.

Fix an element $u_0 \neq 0$ in $E_0 U = U_0$. Then since σ is irreducible, $U = \sigma(\mathfrak{G})u_0$. Moreover $\sigma(\mathfrak{n}_+)u_0 = \{0\}$, since Λ is the highest weight of σ. Therefore

$$U = \sigma(\mathfrak{G})u_0 = \sigma(\mathfrak{N}_- \mathfrak{M})u_0 \,.$$

On the other hand it is clear that

$$\mathfrak{m} = \theta(\mathfrak{n}_0) + \mathfrak{m}_0 + \mathfrak{n}_0 \,,$$

where $\mathfrak{m}_0$ is the centralizer of $\mathfrak{a}$ in $\mathfrak{g}$ and $\mathfrak{n}_0 = \sum\nolimits_{\alpha \in \Sigma_0} \mathfrak{g}_\alpha$. Here Σ_0 is the set of all $\alpha \in \Sigma$ such that $\alpha(H_0) = 0$.

Now fix $\alpha \in \Sigma_0$ and choose $X \in \mathfrak{g}_\alpha$ with $\|X\| = c$ where $c^2 = 2\langle \alpha, \alpha \rangle^{-1}$. Put $Y = -\theta(X)$ and $H = [X, Y]$. Then $\mathfrak{l} = \mathbf{R}H + \mathbf{R}X + \mathbf{R}Y$ is a three-dimensional subalgebra of $\mathfrak{g}$ and (see [4(f), Lem. 3])

$$[H, X] = 2X, \quad [H, Y] = -2Y, \quad [X, Y] = H \,.$$

Moreover, if $u \in U_0$, it is clear that $\sigma(X)u = \sigma(H)u = 0$ since Λ is the highest weight of σ and $\langle \Lambda, \alpha \rangle = 0$. Hence we conclude from the theory of finite-dimensional representations of $\mathfrak{l}$ (see [4(k), Lem. 25]) that $\sigma(Y)u = 0$. This shows that $\sigma(\mathfrak{n}_0)u = \sigma(\theta(\mathfrak{n}_0))u = \{0\}$ and therefore

$$\sigma(\mathfrak{m})u = \sigma(\mathfrak{m}_0)u \subset U_0$$

for $u \in U_0$. Hence we conclude that

$$U = \sigma(\mathfrak{N}_-)U_0$$

and the assertion of the lemma now follows immediately.

COROLLARY. $\Lambda(H_0) > 0$.

Since

$$\sigma(H) = \Lambda(H)E_0 + \sum_{1 \le j \le p} \big(\Lambda(H) - \mu_j(H)\big)E_j \qquad\qquad (H \in \mathfrak{a}),$$

it follows that

$$0 = \operatorname{tr} \sigma(H) = \Lambda(H)d - \sum_{1 \le j \le p} \mu_j(H)d_j$$

where $d_j = \operatorname{tr} E_j = \dim U_j$ $(1 \le j \le p)$. Hence, in particular,

$$\Lambda(H_0)d = \sum_{1 \le j \le p} \mu_j(H_0)d_j \ge 2\gamma \sum_{1 \le j \le p} d_j \,,$$

and therefore $\Lambda(H_0) > 0$ unless $U = U_0$. However $U = U_0$ implies that $d_j = 0$ $(1 \le j \le p)$ and therefore $\Lambda(H)d = 0$ for all $H \in \mathfrak{a}$. On the other hand $H_0 \ne 0$, and so it is obvious from its definition that $\Lambda \ne 0$. This proves the corollary.

23. Application to $I_c^\infty(G)$

We keep to the notation of Lemma 39. It follows from [4(o), § 29] that

$$\sum_{\lambda \in L} \varpi(\lambda)\Theta_\lambda(f) = [W]^{-1} \sum_{\lambda \in L'} \varpi(\lambda)\Theta_\lambda^*(f)$$

for $f \in C_c^\infty(G)$. Hence we get the following result from Theorem 3 and Lemma 39.

THEOREM 6. *Let $f \in I_c^\infty(G)$. Then*

$$cf(1) = \sum_{2 \le i \le r} T_i(\varphi_{f,i})$$

in the notation of Theorem 3, provided G is not compact.

We shall see in another paper how Theorem 6 permits us to complete the proof of the Plancherel formula for G/K. In [4(j), § 16], this proof had been reduced to the verification of two conjectures. The validity of the first of these follows from the work of Gindikin and Karpelevič [3] (see also Helgason [5(b), § 3]). Theorem 6 enables us to prove the second conjecture.

24. The case rank $G/K = 1$

Now we shall apply the above theory to the special case when rank $G/K=1$. We assume, as in § 19, that $B \subset K \subset G \subset G_c$, and G_c is simply connected. Fix a singular root β of $(\mathfrak{g}, \mathfrak{b})$ (see [4(m), § 4]) and a point $H_0 \in \mathfrak{b}$ such that $\pm\beta$ are the only roots which vanish at H_0. Then H_0 is semi-regular of non-compact type (see [4(m), § 3]). Let $\mathfrak{z}$ be its centralizer in $\mathfrak{g}$. Then $\mathfrak{z} = \sigma + \mathfrak{l}$ where σ is the center and $\mathfrak{l}$ the derived algebra of $\mathfrak{z}$. Moreover $\theta(\mathfrak{l}) = \mathfrak{l}$, and we may select a base H', X', Y' for $\mathfrak{l}$ over $\mathbf{R}$ such that (see [4(m), § 7])

$$[H', X'] = 2X', \quad [H', Y'] = -2Y', \quad [X', Y'] = H' ,$$

θ maps (H', X', Y') into $(-H', -Y', -X')$ and

$$H'_\beta = 2\beta(H_\beta)^{-1}H_\beta = (-1)^{1/2}(X' - Y') ,$$

where H_β has the usual meaning (see [4(m), § 4]). Then $\mathfrak{a} = \mathbf{R}H' + \sigma$ and $\mathfrak{b} = \mathbf{R}(X' - Y') + \sigma$ are both Cartan subalgebras of $\mathfrak{z}$. Since rank $\mathfrak{p} = 1$, it is clear that $(\mathfrak{a}, \mathfrak{b})$ is a complete set of non-conjugate Cartan subalgebras of $\mathfrak{g}$. Let A, B be the Cartan subgroups of G corresponding to $\mathfrak{a}$, $\mathfrak{b}$ respectively.

Put $y = \exp(-1)^{1/2}(\pi/4)(X' + Y') \in G_o$. Then $\mathfrak{b}^y_o = \mathfrak{a}_o$ and, if $\alpha = \beta^v$, $\alpha(H') = 2$ and $\alpha = 0$ on σ. Introduce an order in the space of real linear functions λ on $\mathbf{R}H' + (-1)^{1/2}\sigma$ such that $\lambda > 0$ whenever $\lambda(H') > 0$. Let Q be the set of all positive roots of $(\mathfrak{g}, \mathfrak{a})$ under this order. Then $\alpha \in Q$. We may assume that $P = Q^{v^{-1}}$ is the set of all positive roots of $(\mathfrak{g}, \mathfrak{b})$. Then $\beta \in P$ and $(H'_\beta)^v = H'$.

Now if $a \in Z(A) = K \cap \exp(-1)^{1/2}\mathbf{R}H'$, it is obvious that $a = \theta(a) = a^{-1}$ so that $a^2 = 1$. Since $SL(2, \mathbf{C})$ is simply connected, it follows immediately, from the theory of this group, that $a = \exp((-1)^{1/2}tH')$ where $t = 0$ or π. Put $\gamma = \exp((-1)^{1/2}\pi H')$. Then we conclude in the same way that

$$\gamma = \exp\left(\pi(X' - Y')\right) ,$$

and therefore $\gamma \in Z(A)$.

LEMMA 50. $\gamma \neq 1$ and $Z(A) = \{1, \gamma\}$.

We can choose a fundamental system of roots $(\alpha_1, \cdots, \alpha_l)$ for $(\mathfrak{g}, \mathfrak{a})$ (not necessarily in Q) such that $\alpha_1 = \alpha$. Fix a linear function λ on $\mathfrak{a}_o$ such that

$$\lambda(H') = 2\alpha(H_\alpha)^{-1}\lambda(H_\alpha) = 1, \quad \lambda(H_{\alpha_i}) = 0 \qquad (2 \leq i \leq l) .$$

Since G_o is simply connected, λ defines a character ξ_λ of the complex Cartan subgroup A_o of G_o corresponding to $\mathfrak{a}_o$, given by

$$\xi_\lambda(\exp H) = e^{\lambda(H)} \qquad (H \in \mathfrak{a}_o) .$$

Hence $\xi_\lambda(\gamma) = -1$. This proves the lemma.

Put $\mathfrak{b}_1 = \sigma = \mathfrak{a}_1$, $\mathfrak{b}_2 = \mathbf{R}(X' - Y')$, and let B_1, B_2 be the analytic subgroups of B corresponding to $\mathfrak{b}_1$, $\mathfrak{b}_2$ respectively. Then they are both compact and $B_1 \cap B_2 \subset Z(A)$ (see § 12).

LEMMA 51. A_I has one or two connected components according as γ lies in $B_1 \cap B_2$ or not.

Since $A_I = Z(A)B_1$, our assertion is obvious.

Clearly $A_I \subset B$ and since γ centralizes I, it is evident that $\mathfrak{z}$ is the centralizer in $\mathfrak{g}$ of each connected component of A_I.

We wish to compute the function Ψ_λ (see §8) for any $\lambda \in L$. In view of [4(o), Lemmas 58 and 59] the problem reduces to $\mathfrak{z}$ or, what is equivalent, to I. So let λ be a linear function on $C(X' - Y')$, and let

$$m = \lambda(H'_\beta) = (-1)^{1/2}\lambda(X' - Y')$$

be a real number $\neq 0$. Then $\partial(H'_\beta)e^\lambda = me^\lambda$ on $I \cap \mathfrak{b}$. Now define T_λ on I as in [4 (o), Th. 2]. Since $\lambda^\nu(H') = m$, we know from [4 (o), Lem. 32] that

$$tT_\lambda(tH') = c_+e^{mt} - c_-e^{-mt} \qquad (t > 0)$$

where c_+, c_- are constants depending only on the sign of m. Moreover T_λ is tempered and, therefore, we conclude from [4 (o), Lem. 28] that $c_+ = 0$ if $m > 0$ and $c_- = 0$ if $m < 0$. Define φ_λ on I as in [4 (o), Lem. 33]. Since $\partial(H') = (\partial/\partial t)$ and $H'^{\nu^{-1}} = H_\beta'$, it follows that

$$\varphi_\lambda(tH') = c_+e^{mt} + c_-e^{-mt} \qquad (t > 0)\ .$$

Moreover φ_λ is continuous on I so that $1 = \varphi_\lambda(0) = c_+ + c_-$ and therefore $\varphi_\lambda(tH') = e^{-|mt|}$. A similar argument shows that this is also true for $t < 0$.

LEMMA 52. *For any* $\lambda \in L$, *define* Ψ_λ *on* G *as in* §8. *Then*

$$\Psi_\lambda(a_1a_2) = \sum_{s \in W_G} \xi_{s\lambda}(a_1) \exp\left(-|(s\lambda)^\nu(\log a_2)|\right)$$

for $a_1 \in A_I$, $a_2 \in A_R$.

If $\lambda \in L'$, this follows immediately from the above analysis. The rest is proved by approximation arguments (see [4(o), §17]).

Let da_1 and db denote the normalized Haar measures on A_I and B respectively. Put $h_t = \exp tH'$ $(t \in \mathbf{R})$ and normalize the Haar measure da on A so that $da = da_1 dt$ if $a = a_1 h_t$ $(a_1 \in A_I, t \in \mathbf{R})$. Then we know from [4(o), §27] that

$$\varpi(\lambda)\Theta_\lambda(f) = c_B \int_B \Psi_\lambda \varpi^B F_f^B db + c_A \int_A (-1)\varepsilon_R \Psi_\lambda \varpi^A F_f^A da$$

for $\lambda \in L$ and $f \in C_c^\infty(G)$. Here c_A and c_B are positive numbers independent of λ and f. Now

$$\int_A \varepsilon_R \Psi_\lambda \varpi F_f da = 2 \int_0^\infty dt \int_{A_I} \Psi_\lambda(a_1h_t)F_f(a_1h_t; \varpi)da_1\ ,$$

where we have dropped the superscript A for convenience. Therefore it follows from Lemmas 15 and 52 that

$$\sum_{\lambda \in L} \int_A \varepsilon_R \Psi_\lambda \varpi F_f da$$

$$= 2[W_G] \sum_{\lambda \in L} \int_0^\infty e^{-t|\lambda(H'_\beta)|} dt \int_{A_I} \xi_\lambda(a_1) F_f(a_1 h_t;\ \varpi) da_1\ .$$

Let L_1 be the sublattice consisting of those $\lambda \in L$ which vanish at H'_β and L_0 the lattice generated by L_1 and β. As we have seen during the proof of Lemma 50, there exists an element $\lambda_0 \in L$ such that $\lambda_0(H'_\beta) = 1$.

LEMMA 53. *L/L_0 is an additive group of order 2 and $\lambda_0 \notin L_0$.*

Since $\beta(H'_\beta) = 2$, we can, for a given $\lambda \in L$, choose an integer k such that $(\lambda - k\beta)(H'_\beta) = 0$ or 1. This shows that either $\lambda \in L_0$ or $\lambda - \lambda_0 \in L_0$. Moreover it is obvious that $\lambda(H'_\beta) \equiv 0 \bmod 2$ for every $\lambda \in L_0$. Hence $\lambda_0 \notin L_0$.

LEMMA 54. *Fix $\mu \in L$. Then*

$$\sum_{\lambda \in L_1} \int_{A_I} \xi_{\mu+\lambda}(a_1) F_f(a_1 h_t;\ \varpi) da_1 = \frac{1}{2} \sum_{a \in Z(A)} \xi_\mu(a) F_f(a h_t;\ \varpi)$$

for $t \in \mathbf{R}$.

If $\gamma \in B_1$, then $A_I = B_1$ from Lemma 51, and our assertion follows from Lemma 21. On the other hand if $\gamma \notin B_1$, A_I has two connected components B_1 and γB_1, and therefore $\int_{B_1} da_1 = 1/2$. Moreover each of these components contains exactly one point of $Z(A)$. Therefore again our statement follows from Lemma 21.

COROLLARY.

$$\sum_{\lambda \in L} \int_A \varepsilon_R \Psi_\lambda \varpi F_f da$$

$$= [W_G] \sum_{a \in Z(A)} \int_0^\infty F_f(a h_t;\ \varpi)\{\coth t + \xi_{\lambda_0}(a)(\sinh t)^{-1}\} dt\ .$$

Since $\beta(H'_\beta) = 2$ and $1 + 2\sum_{k \geq 1} e^{-2kt} = \coth t$,

$$2\sum_{k \geq 0} e^{-(2k+1)t} = (\sinh t)^{-1} \qquad\qquad (t > 0)\ ,$$

our assertion follows immediately from Lemmas 53 and 54, if we take into account the results of § 12.

Define

$$\varphi_f^{\pm}(t) = \frac{1}{2}\{F_f(h_t;\ \varpi_0) \pm F_f(\gamma h_t;\ \varpi_0)\} \qquad\qquad (t \in \mathbf{R})$$

where ϖ_0 has the same meaning as in § 13. Then $\varphi_f^{\pm}$ are both even functions of t. Moreover, since $\xi_{\lambda_0}(\gamma) = -1$ and[10] $H_\alpha = (|\alpha|^2/2)H'$, it follows from the above corollary that

---

[10] As usual $|\alpha|^2 = \alpha(H_\alpha)$.

$$\sum_{\lambda \in L} \int_{\Delta} \varepsilon_R \Psi_\lambda \varpi F_f da = [W_G] |\alpha|^2 \int_0^\infty \{\coth t \cdot d\varphi_f^+/dt + (\sinh t)^{-1} d\varphi_f^-/dt\} dt .$$

Put

$$\hat{\varphi}_f^{\pm}(\lambda) = \int_{-\infty}^{\infty} \varphi_f^{\pm}(t) e^{(-1)^{1/2}\lambda t} dt \qquad\qquad (\lambda \in \mathbf{R}) .$$

Then it follows from the theory of Fourier transforms that

$$2 \int_0^\infty d\varphi_f^+/dt \cdot \coth t \, dt = \int_{-\infty}^{\infty} d\varphi_f^+/dt \cdot \coth t \, dt$$

$$= -\int_{-\infty}^{\infty} \frac{\lambda}{2} \coth \frac{\pi\lambda}{2} \cdot \hat{\varphi}_f^+(\lambda) d\lambda$$

$$= -\int_0^\infty \lambda \coth \frac{\pi\lambda}{2} \cdot \hat{\varphi}_f^+(\lambda) d\lambda ,$$

since φ_f^+ is an even function. Similarly

$$2 \int_0^\infty d\varphi_f^-/dt \cdot (\sinh t)^{-1} dt = -\int_0^\infty \lambda \tanh \frac{\pi\lambda}{2} \cdot \hat{\varphi}_f^-(\lambda) d\lambda .$$

Let C denote the character group of A_I and $C^{\pm}$ the subsets of those $a^* \in C$ for which $\langle a^*, \gamma \rangle = \pm 1$. Put

$$T_{a^*,\lambda}(f) = \int_{A_I \times \mathbf{R}} F_f(a_1 h_t) \langle a^*, a_1 \rangle e^{(-1)^{1/2}\lambda t} da_1 dt$$

$(a^* \in C, \lambda \in \mathbf{R})$ and $\varpi(a^*, \lambda) = \varpi(\Lambda)$ where Λ is the linear function on $\mathfrak{a}_c$ defined as follows. $\Lambda(H') = (-1)^{1/2}\lambda$ and

$$\langle a^*, \exp H \rangle = e^{\Lambda(H)} \qquad\qquad (H \in \mathfrak{b}_1 = \mathfrak{a}_I) .$$

Since every imaginary root of $(\mathfrak{g}, \mathfrak{a})$ is compact, F_f is a C^∞ function on A (see [4(n), Lem. 40]), and therefore

$$\int_{A_I \times \mathbf{R}} F_f(a_1 h_t; \varpi) \langle a^*, a_1 \rangle e^{(-1)^{1/2}\lambda t} da_1 dt = (-1)^m \varpi(a^*, \lambda) T_{a^*,\lambda}(f)$$

where $m = (1/2)(\dim \mathfrak{g} - \operatorname{rank} \mathfrak{g})$. Moreover since $H_\alpha = (|\alpha|^2/2)H'$, we conclude that

$$(-1)^m \varpi(a^*, \lambda) T_{a^*,\lambda}(f)$$
$$= -(-1)^{1/2} \frac{|\alpha|^2}{2} \lambda \int_{A_I \times \mathbf{R}} F_f(a_1 h_t; \varpi_0) \langle a^*, a_1 \rangle e^{(-1)^{1/2}\lambda t} da_1 dt .$$

But it is obvious that

$$\sum_{a^* \in 0^{\pm}} \int_{A_I \times \mathbf{R}} F_f(a_1 h_t; \varpi_0) \langle a^*, a_1 \rangle e^{(-1)^{1/2}\lambda t} da_1 dt = \hat{\varphi}_f^{\pm}(\lambda) .$$

Therefore we have obtained the following result.

LEMMA 55.

$$\sum_{\lambda \in L} \int_A (-1) \varepsilon_R \Psi_\lambda \varpi F_f \, da$$

$$= (-1)^m [W_G](-1)^{1/2} \Big\{ \sum_{a^* \in \sigma^+} \int_0^\infty \coth \frac{\pi \lambda}{2} \cdot \varpi(a^*, \lambda) T_{a^*, \lambda}(f) \, d\lambda$$

$$+ \sum_{a^* \in \sigma^-} \int_0^\infty \tanh \frac{\pi \lambda}{2} \cdot \varpi(a^*, \lambda) T_{a^*, \lambda}(f) \, d\lambda \Big\}$$

for $f \in C_o^\infty(G)$.

Put $q = (1/2) \dim G/K$ as before. Then

$$F_f^B(1; \varpi^B) = (-1)^q c f(1) \qquad\qquad \big(f \in C_o^\infty(G) \big)$$

where c is a positive number independent of f (see § 15). Moreover

$$\sum_{\lambda \in L} \int_B \Psi_\lambda \varpi^B F_f^B \, db = [W_G] F_f^B(1; \varpi) \, ,$$

as we saw in § 15. Therefore we get the following formula.

LEMMA 56. *For any* $f \in C_o^\infty(G)$,

$$c c_B f(1) = (-1)^q [W_G]^{-1} \sum_{\lambda \in L} \varpi(\lambda) \Theta_\lambda(f)$$

$$- c_A (-1)^{m+q} \Big\{ \sum_{a^* \in \sigma^+} (-1)^{1/2} \int_0^\infty \coth \frac{\pi \lambda}{2} \cdot \varpi(a^*, \lambda) T_{a^*, \lambda}(f) \, d\lambda$$

$$+ \sum_{a^* \in \sigma^-} (-1)^{1/2} \int_0^\infty \tanh \frac{\pi \lambda}{2} \cdot \varpi(a^*, \lambda) T_{a^*, \lambda}(f) \, d\lambda \Big\} \, .$$

This is substantially the Plancherel formula for G (cf. [7, p. 432]).

25. Appendix

Let E be a real vector space of finite dimension. Define the space $C(E)$ as usual [4(e), p. 91]. Let $\lambda \neq 0$ be a (real) linear function on E, c a real number, and H the hyperplane in E given by the equation $\lambda = c$. Let $C_H(E)$ be the subspace of all $f \in C(E)$ which vanish identically on H. Clearly $C_H(E)$ is closed in $C(E)$.

LEMMA 57. *For every* $f \in C_H(E)$ *there exists a unique element* $g \in C(E)$ *such that* $f = (\lambda - c)g$. *Let us denote* g *by* $(\lambda - c)^{-1} f$. *Then* $f \to (\lambda - c)^{-1} f$ *is a continuous mapping of* $C_H(E)$ *into* $C(E)$.

This is easy (see [5(b), p. 571]).

COROLLARY 1. *Let* λ_i *be non-zero linear functions on* E, *and* c_i *real numbers* $(1 \leq i \leq r)$. *We assume that the hyperplanes* $H_i: \lambda_i = c_i$ *are all distinct. Put*

$$q = \prod_{1 \leq i \leq r} (\lambda_i - c_i) \, ,$$

and let $C_0(E)$ *be the subspace of all* $f \in C(E)$ *which vanish identically on*

$\bigcup_{1 \leq i \leq r} H_i$. *Then for every* $f \in \mathcal{C}_0(E)$, *there exists a unique element* $g = q^{-1}f$ *in* $\mathcal{C}(E)$ *such that* $f = qg$. *Moreover* $f \to q^{-1}f$ *is a continuous mapping of* $\mathcal{C}_0(E)$ *into* $\mathcal{C}(E)$.

This follows immediately from the lemma by induction on r.

COROLLARY 2. *Consider* $V = C_e^\infty(E) \cap \mathcal{C}_0(E)$ *as a topological subspace of* $C_e^\infty(E)$. *Then* $q^{-1}f \in C_e^\infty(E)$ *for* $f \in V$ *and the mapping* $f \to q^{-1}f$ *of* V *into* $C_e^\infty(E)$ *is continuous.*

This is obvious from [4(k), Lem. 1].

LEMMA 58. *Let* G *be a real analytic subgroup of* $\mathrm{GL}(n, \mathbf{C})$. *Then if* G *is semi-simple, it is closed in* $\mathrm{GL}(n, \mathbf{C})$.

This is well known and follows from the fact that the center of G is finite and $\mathrm{Ad}\,(G)$ is the connected component of 1 in the group of all automorphisms of the Lie algebra $\mathfrak{g}$ of G.

We now use the notation of § 5. Let Z_G be the center of G. Then Z_G, being a closed subgroup of A_I, is compact.

COROLLARY 1. *Let* Ξ *be a semi-simple analytic subgroup of* G *whose Lie algebra is contained in* $[\mathfrak{g}, \mathfrak{g}]$. *Then it is closed in* G *and its center is finite.*

Let G_e be a complexification of G, and $j: G \to G_e$ the corresponding homomorphism. (It follows from the definition of acceptability [4(n), § 18] that such a complexification exists.) Put $\mathfrak{g}_1 = [\mathfrak{g}, \mathfrak{g}]$, and let G_{1e} be the complex analytic subgroup of G_e corresponding to $\mathfrak{g}_{1e}$. Then G_{1e} is closed in G_e and it has a faithful finite-dimensional representation σ. Hence it follows from Lemma 58 that $j(\Xi)$ is closed in G_{1e}. Moreover since Z_G is compact, the kernel of j is finite. Hence it is obvious that Ξ is closed in G. Finally $\sigma(j(\Xi))$, being a semi-simple linear group, has finite center. Therefore the same also holds for Ξ.

Put $L' = \sum_{s \in W_I} sL$ in the notation of § 7.

COROLLARY 2. L' *is a lattice in* $\mathfrak{F}_I$.

Let D be the kernel of the exponential mapping from $\mathfrak{h}_I$ to A_I^0. Since A_I^0 is compact, D is a lattice in $\mathfrak{h}_I$. Let $\mathfrak{c}$ be the center of $\mathfrak{g}$. Then $\exp \mathfrak{c}$ is a closed and hence compact subgroup of A_I^0. Hence $D \cap \mathfrak{c}$ is a lattice in $\mathfrak{c}$.

Let G_1 be the analytic subgroup of G corresponding to $\mathfrak{g}_1 = [\mathfrak{g}, \mathfrak{g}]$. It follows from Corollary 1 that $G_1 \cap Z_G$ is finite. Put

$$N = [G_1 \cap Z_G].$$

Let D_1 be the set of all $H \in \mathfrak{h}_I \cap \mathfrak{g}_1$ such that $e^{\mathrm{ad}\,H} = 1$. Then it is obvious that $sD_1 = D_1$ for $s \in W_I$ and $ND_1 \subset D \cap \mathfrak{g}_1$. Put

$$D_0 = ND_1 + D \cap \mathfrak{c} .$$

Then D_0 is a lattice in $\mathfrak{h}_I$ which is stable under W_I. Therefore since $D_0 \subset D' = \bigcap_{s \in W_I} sD \subset D$, D' is also a lattice in $\mathfrak{h}_I$. Let L'' be the set of all $\lambda \in \mathfrak{F}_I$ such that $e^{\lambda(H)} = 1$ for $H \in D'$. Then L'' is a lattice in $\mathfrak{F}_I$ and $L \subset L' \subset L''$. This shows that L' is a lattice in $\mathfrak{h}_I$.

Define L and $L(s, \mathfrak{a}^+)$ as in the proof of Lemma 15.

COROLLARY 3. $L/L(s, \mathfrak{a}^+)$ *is finite.*

We use the notation of § 11. Fix $\lambda \in L$. We claim that[10] $2N|\beta|^{-2}\lambda(H_\beta)$ is an integer for every root β of $(\mathfrak{g}, \mathfrak{h})$. This follows from the fact that $2\alpha(H_\beta)|\beta|^{-2}$ is an integer for every root α and therefore

$$\exp\left(2\pi(-1)^{1/2}2|\beta|^{-2}\operatorname{ad}H_\beta\right) = 1 .$$

Hence $N\lambda - \sum_{1 \leq i \leq m} 2N|\beta_i|^{-2}\lambda(H_{\beta_i})\beta_i \in L_1$ (see [4(o), Lem. 29]) and therefore $N\lambda \in L(s, \mathfrak{a}^+)$. This proves that $NL \subset L(s, \mathfrak{a}^+)$ and therefore $L/L(s, \mathfrak{a}^+)$ is finite.

We now use the notation of § 8.

LEMMA 59. $W_\mu(\mathfrak{g}/\mathfrak{h})$ *is simply transitive on* $C(\mu)$.

Let P be the set of all positive roots of $(\mathfrak{g}, \mathfrak{h})$, P_μ the subset of those $\alpha \in P$ for which $\mu(H_\alpha) = 0$ and P'_μ the complement of P_μ in P. Put $W_\mu = W_\mu(\mathfrak{g}/\mathfrak{h})$ and

$$\varpi_\mu = \prod_{\alpha \in P_\mu} \alpha , \qquad \varpi'_\mu = \prod_{\alpha \in P'_\mu} \alpha .$$

Let $\mathfrak{F}''$ be the set of all $\lambda \in \mathfrak{F}$ where $\varpi_\mu(\lambda) \neq 0$ and C_μ the set of all connected components of $\mathfrak{F}''$. Choose an element $H_0 \in \mathfrak{h}$ such that $\alpha(H_0) = (-1)^{1/2}\mu(H_\alpha)$ for all $\alpha \in P$, and let $\mathfrak{z}$ be the centralizer of H_0 in $\mathfrak{g}$. Then it is clear that $W_\mu = W(\mathfrak{z}/\mathfrak{h})$ (see [5(a), Th. 2.15, p. 249]) and therefore W_μ acts simply transitively on C_μ.

For any $\mathfrak{F}^+ \in C(\mu)$, let $\mathfrak{F}_\mu^+$ denote the unique element of C_μ which contains $\mathfrak{F}^+$. Clearly $(s\mathfrak{F}^+)_\mu = s\mathfrak{F}_\mu^+$ $(s \in W_\mu)$. Hence it would be enough to show that $\mathfrak{F}^+ \to \mathfrak{F}_\mu^+$ is an injection of $C(\mu)$ into C_μ.

Choose an open and convex neighborhood U of μ in $\mathfrak{F}$ such that $\varpi'_\mu(\lambda) \neq 0$ for $\lambda \in U$. Let $\mathfrak{F}_1, \cdots, \mathfrak{F}_r$ be all the distinct elements of $C(\mu)$. Since $\mu \in \mathrm{Cl}\mathfrak{F}_i$, it is obvious that $U \cap \mathfrak{F}_i$ $(1 \leq i \leq r)$ are all the distinct connected components of

$$U' = U \cap \mathfrak{F}' = U \cap \mathfrak{F}'' .$$

Hence $(\mathfrak{F}_i)_\mu \neq (\mathfrak{F}_j)_\mu$ $(1 \leq i, j \leq r)$ unless $i = j$.

INSTITUTE FOR ADVANCED STUDY

REFERENCES

1. A. BOREL and HARISH-CHANDRA, *Arithmetic subgroups of algebraic groups*, Ann. of Math., 75 (1962), 485–535.
2. R. COURANT and D. HILBERT, Methods of mathematical physics, vol. II (1962), Interscience Publishers, New York.
2 (a). H. FURSTENBERG, *A Poisson formula for semi-simple Lie groups*, Ann. of Math., 77 (1963), 335–386.
3. S. G. GINDIKIN and F. I. KARPELEVIČ, *Plancherel measure of Riemannian symmetric spaces of nonpositive curvature*, Soviet Math., 3 (1962), 962–965.
4. HARISH-CHANDRA, (a) *Representations of a semi-simple Lie group on a Banach space*: I, Trans. Amer. Math. Soc., 75 (1953), 185–243.

    (b) *Representations of semi-simple Lie groups*: III, Trans. Amer. Math. Soc., 76 (1954), 234–253.

    (c) *Representations of semi-simple Lie groups*: V, Amer. J. Math., 78 (1956), 1–41.

    (d) *The characters of semi-simple Lie groups*, Trans. Amer. Math. Soc., 83 (1956), 98–163.

    (e) *Differential operators on a semi-simple Lie algebra*, Amer. J. Math., 79 (1957), 87–120.

    (f) *Fourier transforms on a semi-simple Lie algebra*: I, Amer. J. Math., 79 (1957), 193–257.

    (g) *Fourier transforms on a semi-simple Lie algebra*: II, Amer. J. Math., 79 (1957), 653–686.

    (h) *A formula for semi-simple Lie groups*, Amer. J. Math., 79 (1957), 733–760.

    (i) *Spherical functions on a semi-simple Lie group*, I, Amer. J. Math., 80 (1958), 241–310.

    (j) *Spherical functions on a semi-simple Lie group*, II, Amer. J. Math., 80 (1958), 553–613.

    (k) *Invariant distributions on Lie algebras*, Amer. J. Math., 86 (1964), 271–309.

    (l) *Invariant differential operators and distributions on a semi-simple Lie algebra*, Amer. J. Math., 86 (1964), 534–564.

    (m) *Some results on an invariant integral on a semi-simple Lie algebra*, Ann. of Math., 80 (1964), 551–593.

    (n) *Invariant eigendistributions on a semi-simple Lie group*, Trans. Amer. Math. Soc., to appear.

    (o) *Discrete series for semi-simple Lie groups*: I, Acta Math., 113 (1965), 241–318.

    (p) *Invariant eigendistributions on a semi-simple Lie algebra*, Pub. Math. I.H.E.S., No. 27.
5. S. HELGASON, (a) Differential Geometry and Symmetric Spaces, Academic Press, New York, 1962.

    (b) *Fundamental solutions of invariant differential operators on symmetric spaces*, Amer. J. Math., 86 (1964), 565–601.
6. L. SCHWARTZ, Théorie des distributions: I, Hermann, Paris, 1950.
7. R. TAKAHASHI, *Sur les representations unitaires des groups de Lorentz généralisés*, Bull. Soc. Math. France, 91 (1963), 289–433.

(Received May 5, 1965)

Reprinted from
Ann. of Math.
83 (1966), 74–128

DISCRETE SERIES FOR SEMISIMPLE LIE GROUPS. II

EXPLICIT DETERMINATION OF THE CHARACTERS

BY

HARISH–CHANDRA

The Institute for Advanced Study, Princeton, N.J., U.S.A.

Table of Contents

1 – 662900. *Acta mathematica.* 116. Imprimé le 10 juin 1966.

§ 1. Introduction

Let G be a connected semisimple Lie group and K a maximal compact subgroup of G. We shall show in this paper that G has a discrete series (see [4 (d), § 5]) if and only if it has a compact Cartan subgroup B. Let $\mathcal{E}_d$ denote the set of all equivalence classes of irreducible unitary representations of G, which are square-integrable. For any $\omega \in \mathcal{E}_d$, let Θ_ω denote the character, χ_ω the infinitesimal character and $d(\omega)$ the formal degree (see [4 (d), § 3]) of ω. Then it is known [4 (d), § 5] that the distribution

$$T = \sum_{\omega \in \mathcal{E}_d} d(\omega)\, \Theta_\omega$$

represents the contribution of the discrete series to the Plancherel formula of G. We intend to obtain explicit formulas for $d(\omega)$, Θ_ω and T.

Let $\mathfrak{g}$ and $\mathfrak{b}$ be the Lie algebras of G and B respectively. Then the character group of B may be identified with a lattice L in the space of all real-valued linear functions $\mathfrak{F}$ on $(-1)^{\frac{1}{2}}\mathfrak{b}$. Put

$$\varpi(\lambda) = \prod_{\alpha > 0} \langle \lambda, \alpha \rangle \quad (\lambda \in \mathfrak{F}),$$

where α runs over all positive roots of $(\mathfrak{g}, \mathfrak{b})$ and the scalar product is defined, as usual, by means of the Killing form. Let L' denote the set of all $\lambda \in L$ where $\varpi(\lambda) \neq 0$. Then, for every $\lambda \in L'$, we have constructed in [4, (p)], an invariant eigendistribution Θ_λ of $\mathfrak{z}$ on G. Put $q = \frac{1}{2} \dim G/K$ and $\varepsilon(\lambda) = \operatorname{sign} \varpi(\lambda)$ $(\lambda \in L')$. Our main result (Theorem 16) states that the distributions $(-1)^q \varepsilon(\lambda)\Theta_\lambda$ $(\lambda \in L')$ are exactly the characters of the discrete series and

$$T = c^{-1}(-1)^q \sum_{\lambda \in L'} \varpi(\lambda)\, \Theta_\lambda$$

where c is a positive constant.

Let W be the Weyl group of $(\mathfrak{g}, \mathfrak{b})$ and $\tilde{B}$ the normalizer of B in G. Then $W_G = \tilde{B}/B$ may be regarded as a subgroup of W. Define χ_λ $(\lambda \in L')$ as usual (see [4 (p), § 29]) and let $\mathcal{E}_d(\lambda)$ denote the set of all $\omega \in \mathcal{E}_d$ such that $\chi_\omega = \chi_\lambda$. Then[1]

$$[\mathcal{E}_d(\lambda)] = [W(\lambda)][W_G]^{-1},$$

where $W(\lambda)$ is the set of all $s \in W$ such that $s\lambda \in L$.

Fix $\lambda \in L'$ and let $\Theta_{\lambda, \mathfrak{b}}$ $(\mathfrak{b} \in \mathcal{E}_K)$ denote the Fourier components of Θ_λ (see [4 (q), § 17]). Then $\Theta_{\lambda, \mathfrak{b}}$ are analytic functions and it is one of the principal steps of the proof to show that they lie in $L_2(G)$ (Theorem 12). This is done by means of Lemma 67, whose proof is based on two key results (Lemmas 42 and 43), which are derived from a study of certain differential equations.

This paper is divided into four parts. After recalling some known facts about representations on a locally convex space, we prove Theorem 1, which seems to play an important role in harmonic analysis. Then we introduce the space $C(G)$, which is the analogue of the Schwartz space over $\mathbf{R}^n$. Theorem 2 asserts that $C_c^\infty(G)$ is dense in $C(G)$ and Theorem 3 allows us to reduce certain problems from G to a proper subgroup. Theorem 4 contains a general result which implies the convergence of certain integrals (Theorems 5 and 6) and thus enables us to define the mapping $f \to F_f (f \in C(G))$ in § 18. A distribution on G is called tempered if it extends to a continuous linear function on $C(G)$. Theorem 7 gives a

---

[1] As usual, $[F]$ denotes the number of elements in a set F.

simple necessary and sufficient condition for an invariant eigendistribution to be tempered. This shows, in particular, that Θ_λ ($\lambda \in L'$) is tempered and Theorem 3 of [4 (q)] remains valid for $f \in C(G)$ (Theorem 8). This permits us to prove the second conjecture of [4 (k), § 16] and thus complete the proof of the Plancherel formula for G/K. Theorem 9 established a weak estimate for a Fourier component Θ_b of a tempered and invariant eigendistribution Θ. This will be required in Parts II and III.

The main problem in Part II is to determine the behaviour, at infinity, of a K-finite eigenfunction f of $\mathfrak{Z}$, which satisfies a weak inequality of the type mentioned above. The principal result (Lemma 43) is that f lies in $L_2(G)$ if and only if it lies in $C(G)$. This is proved by induction on dim G. By making use of the differential equations, one reduces the problem from G to a proper subgroup M (cf. [4 (k)]).

In Part III, we apply the above results to the task of determining the eigenfunctions of $\mathfrak{Z}$ in $C(G)$. Here Lemma 64 plays an important role. It enables us to show that such eigenfunctions do not exist unless rank G = rank K. As an application we obtain in § 33 a proof of a conjecture of Selberg.

These results are then utilized to determine all the characters of the discrete series. Here the fact that we work in $C(G)$, rather than $C_c^\infty(G)$, is decisive. First we show that Θ_ω ($\omega \in \mathcal{E}_d$) is tempered and there exists a $\lambda \in L'$ such that $\chi_\omega = \chi_\lambda$. Let $s_1, s_2, ..., s_r$ be a complete set of representatives for $W_G \backslash W(\lambda)$. Then we prove that

$$\Theta_\omega = \sum_{1 \leqslant i \leqslant r} c_i \Theta_{s_i \lambda},$$

where c_i are complex numbers. Moreover, by making use of the Schur orthogonality relations, it is possible to establish that

$$\sum_{1 \leqslant i \leqslant r} |c_i|^2 = 1.$$

On the other hand, one verifies that c_i are integers. This proves that

$$\Theta_\omega = (-1)^q \varepsilon(s_i \lambda) \Theta_{s_i \lambda}$$

for some i and $d(\omega) = c^{-1} [W_G] |\varpi(\lambda)|$. It should be noticed that the entire discussion of §§ 40, 41 is quite similar to Weyl's original treatment of the same problem in the compact case (see [10, § 3]). The main task here is to relate the Fourier analysis on G, so far as the discrete series is concerned, to that on B (see Theorem 14 and Lemma 81, together with its corollaries). This can be done only by operating in $C(G)$.

Part IV deals with certain inequalities which are needed for the proof of Lemma 21. They will also be useful later, when we come to the continuous series for G in another paper.

Some of the results of this paper have been announced in [4 (l)].

Part I. Analysis in the space $C(G)$

§ 2. Representations on a locally convex space

In this section we recall some elementary and well-known facts about representations on locally convex spaces (see [2, p. 109]).

Let V be a Hausdorff, locally convex (real or complex) vector space and G a locally compact topological group. A representation π of G on V is a mapping, which assigns, to every $x \in G$, a continuous endomorphism $\pi(x)$ of V such that the following two conditions hold:

1) $\pi(xy) = \pi(x)\pi(y)$ $(x, y \in G)$ and $\pi(1) = I$.

2) $(x, v) \to \pi(x)v$ is a continuous mapping of $G \times V$ into V.

It is easy to verify that the above two conditions imply the following [2, p. 110].

3) Let C be a compact set in G. Then for any neighborhood U_0 of zero in V, we can choose another neighborhood U of zero such that $\pi(C)U \subset U_0$.

Let S be the set of all continuous seminorms on V. Then the following immediate consequence of 3) will be frequently useful.

4) Given a compact set C in G and an element $\nu_0 \in S$, we can choose $\nu \in S$ such that

$$\nu_0(\pi(x)\,v) \leqslant \nu(v)$$

for all $x \in C$ and $v \in V$.

Conversely we have the following result.

LEMMA 1. *Suppose π satisfies, in addition to 1), the following two conditions.*

2') *The mapping $x \to \pi(x)v$ of G into V is continuous for every $v \in V$.*

3') *There exists a neighborhood U of 1 in G with the following property. Given $\nu_0 \in S$, we can choose $\nu \in S$ such that*

$$\nu_0(\pi(u)v) \leqslant \nu(v)$$

for all $u \in U$ and $v \in V$.

Then π is a representation of G on V.

Fix $x_0 \in G$ and $v_0 \in V$. Then

$$\pi(x_0 u)v - \pi(x_0)v_0 = \pi(x_0)\pi(u)(v - v_0) + \pi(x_0)(\pi(u)v_0 - v_0)$$

for $u \in U$ and $v \in V$. Hence it is clear that $\pi(x_0 u)v \to \pi(x_0)v_0$ as $u \to 1$ and $v \to v_0$. This proves our assertion.

Now assume that V is complex and complete, G is unimodular and dx is a Haar measure on G. π being a representation of G on V, we define, for any $f \in C_c(G)$, a linear transformation $\pi(f)$ in V by

$$\pi(f)v = \int f(x)\,\pi(x)\,v\,dx \quad (v \in V).$$

It follows from 3) that $\pi(f)$ is continuous. Define the convolution $f*g$ ($f, g \in C_c(G)$) as usual, so that

$$(f * g)(x) = \int f(y)\,g(y^{-1}x)\,dy \quad (x \in G).$$

Then $\pi(f*g) = \pi(f)\pi(g)$.

Let us now suppose that G is a Lie group. A vector $v \in V$ is said to be differentiable (under π) if the mapping $x \to \pi(x)v$ of G into V is of class C^∞. Let V^∞ denote the space of all differentiable vectors in V. $\mathfrak{g}$ being the Lie algebra of G, define

$$\pi(X)v = \lim_{t \to 0} t^{-1}(\pi(\exp tX)v - v) \quad (t \in \mathbf{R}, t \neq 0)$$

for $X \in \mathfrak{g}$ and $v \in V^\infty$. Then $\pi(X)$ is a linear transformation in V^∞ and $X \to \pi(X)$ is a representation of $\mathfrak{g}$ on V^∞. Let $\mathfrak{G}$ be the universal enveloping algebra of $\mathfrak{g}_c$. Then this extends uniquely to a representation of $\mathfrak{G}$ which we denote again by π.

As usual we regard elements of $\mathfrak{G}$ as left-invariant differential operators on G. Define the anti-isomorphism ϱ of $\mathfrak{G}$ onto the algebra of right-invariant differential operators on G as in [4 (o), § 2]. U being any open set in G, we write

$$f(g; x) = f(x; \varrho(g)), \quad f(g; x; g') = f(x; \varrho(g) \circ g') \quad (x \in U; g, g' \in \mathfrak{G})$$

for $f \in C^\infty(U)$.

LEMMA 2. *Let* $f \in C_c^\infty(G)$. *Then* $\pi(f)V \subset V^\infty$ *and*

$$\pi(X)\pi(f)v = -\pi(\varrho(X)f)v$$

for $X \in \mathfrak{g}$ *and* $v \in V$.

It is clear that

$$\pi(y)\pi(f)v = \pi(_yf)v,$$

where $_yf(x) = f(y^{-1}x)$. Fix $X \in \mathfrak{g}$ and put $y_t = \exp tX$ ($t \in \mathbf{R}$). Then

$$f(y_t^{-1}x) - f(x) = -t \int_0^1 f'(y_{st}^{-1}x)\,ds,$$

where $f' = \varrho(X)f$. Hence

$$t^{-1}(\pi(y_t) - 1)\pi(f)w = -\int_0^1 \pi(y_{st})v'\,ds,$$

where $v' = \pi(f')v$. The statements of the lemma are now obvious.

Let $g \to g^*$ be the anti-automorphism of $\mathfrak{G}$ such that $X^* = -X$ ($X \in \mathfrak{g}$).

CoROLLARY 1. $\pi(g)\pi(f)v = \pi(\varrho(g^*)f)v$ for $g \in \mathfrak{G}$.

This is obvious from the lemma.

CoROLLARY 2. Let $v \in V^\infty$ and $f \in C_c^\infty(G)$. Then

$$\pi(f)\pi(g)v = \pi(g^*f)v \quad (g \in \mathfrak{G}).$$

Define $f_y(x) = f(xy^{-1})$ $(x \in G)$ for a fixed $y \in G$. Then $\pi(f)\pi(y)v = \pi(f_y)v$ and from this we conclude, as in the proof of Lemma 2 that

$$\pi(f)\pi(X)v = -\pi(Xf)v \quad (X \in \mathfrak{g}).$$

Clearly this implies the required result.

Let f_j $(j \geqslant 1)$ be a sequence in $C_c^\infty(G)$. We say that it is a Dirac sequence if the following conditions hold. $f_j \geqslant 0$ and

$$\int f_j \, dx = 1$$

and, for any neighborhood U of 1, Supp $f_j \subset U$ for all j except a finite number. Let K be a compact subgroup of G. It is obvious that there always exists a Dirac sequence f_j $(j \geqslant 1)$ such that $f_j(kxk^{-1}) = f_j(x)$ $(k \in K, x \in G)$.

LEMMA 3. Let f_j $(j \geqslant 1)$ be a Dirac sequence in $C_c^\infty(G)$. Then

$$\lim_{j \to \infty} \pi(f_j)v = v$$

for every $v \in V$.

Fix $v \in V$, $\nu \in S$ and $\varepsilon > 0$. Then we can choose a neighborhood U of 1 in G such that $\nu(\pi(x)v - v) \leqslant \varepsilon$ for $x \in U$. Now

$$\pi(f_j)v - v = \int f_j(x)\,(\pi(x)v - v)\,dx,$$

and therefore $\nu(\pi(f_j)v - v) \leqslant \varepsilon$ if Supp $f_j \subset U$. This proves the lemma.

COROLLARY 1. V^∞ is dense in V.

This is obvious from Lemmas 2 and 3.

COROLLARY 2. Fix $v_0 \in V$ and let V_0 be the smallest closed subspace of V containing v_0 which is stable under $\pi(G)$. Then elements of the form $\pi(f)v_0$ $(f \in C_c^\infty(G))$ are dense in V_0.

Let W be the space consisting of all elements of the form $\pi(f)v_0$ $(f \in C_c^\infty(G))$. It is obvious that W is stable under $\pi(G)$. Therefore the same holds for $\mathrm{Cl}(W)$. But then since $v_0 \in \mathrm{Cl}(W)$ from Lemma 3, our assertion follows.

Let K be a compact subgroup of G and $\mathcal{E}_K$ the set of all equivalence classes of finite-dimensional irreducible representations of K. For any $\mathfrak{b} \in \mathcal{E}_K$, define a linear transformation $E_\mathfrak{b}$ in V by

$$E_\mathfrak{b} v = d(\mathfrak{b}) \int_K \mathrm{conj}\ \xi_\mathfrak{b}(k) \cdot \pi(k) v\, dk \quad (v \in V).$$

Here $d(\mathfrak{b})$ is the degree and $\xi_\mathfrak{b}$ the character of $\mathfrak{b}$ and dk is the normalized Haar measure of K. Then $E_\mathfrak{b}$ is a continuous projection. Put $V_\mathfrak{b} = E_\mathfrak{b} V$.

LEMMA 4. $\sum_{\mathfrak{b} \in \mathcal{E}_K} V_\mathfrak{b} \cap V^\infty$ *is dense in* V.

We shall give a proof of this lemma in § 6.

§ 3. Absolute convergence of the Fourier series

As before let S be the set of all continuous seminorms on a complex locally convex space V, which we assume to be complete. Let $\{v_j\}_{j \in J}$ be an indexed family of elements of V. We say that the series

$$\sum_{j \in J} v_j$$

converges, if the following condition holds. Define

$$s_F = \sum_{j \in F} v_j$$

for any finite subset F of J. Then for any neighborhood V_0 of zero in V, there should exist a finite subset F_0 of J such that $s_{F_1} - s_{F_2} \in V_0$ for any two finite subsets F_1, F_2 of J containing F_0. Since V is complete, the partial sums s_F then have a limit s in V. s is called the sum of the series and we write

$$s = \sum_{j \in J} v_j.$$

Moreover, the series is said to converge absolutely if

$$\sum_{j \in J} \nu(v_j) < \infty$$

for every $\nu \in S$. It is obvious that absolute convergence implies convergence.

Let π be a representation of a compact Lie group K on V and define $E_\mathfrak{b}(\mathfrak{b} \in \mathcal{E}_K)$ as in § 2. Then for any vector $v \in V$, we call $E_\mathfrak{b} v$ the $\mathfrak{b}$th Fourier component of v.

LEMMA 5. *Let v be a differentiable vector in V. Then the Fourier series*

$$\sum_{\mathfrak{b}\in\mathcal{E}_K} E_\mathfrak{b} v$$

converges absolutely to v.

Let $\mathfrak{k}$ be the Lie algebra of K and $\mathfrak{K}$ the universal enveloping algebra of $\mathfrak{k}_c$. Since K is compact, we can choose a positive-definite quadratic form Q on $\mathfrak{k}$ which is invariant under the adjoint representation of K. Let $X_1, \ldots, X_r$ be a base for $\mathfrak{k}$ over $\mathbf{R}$ orthonormal with respect to Q and put

$$\Omega = 1 - (X_1^2 + \ldots + X_r^2) \in \mathfrak{K}.$$

Also put $\alpha_\mathfrak{b} = d(\mathfrak{b}) \operatorname{conj} \xi_\mathfrak{b} \, (\mathfrak{b}\in\mathcal{E}_K)$ so that $E_\mathfrak{b} = \pi(\alpha_\mathfrak{b})$. It is obvious that Ω, regarded as a differential operator, commutes with both left and right translations of K. Fix a unitary representation σ in the class $\mathfrak{b}$. Then $\sigma(\Omega)$ commutes with $\sigma(k)$ $(k\in K)$ and therefore, by Schur's lemma, $\sigma(\Omega) = c(\mathfrak{b})\sigma(1)$ where $c(\mathfrak{b})\in\mathbf{C}$. However $\sigma(X_i)$ $(1\leqslant i\leqslant r)$ are obviously skew-adjoint operators. Hence $c(\mathfrak{b})$ is real and $\geqslant 1$. Therefore

$$\xi_\mathfrak{b}(k;\,\Omega) = \operatorname{tr}\,(\sigma(k)\sigma(\Omega)) = c(\mathfrak{b})\,\xi_\mathfrak{b}(k) \quad (k\in K)$$

and this shows that $\Omega\alpha_\mathfrak{b} = c(\mathfrak{b})\alpha_\mathfrak{b}$. Hence we conclude from Corollary 2 of Lemma 2 that

$$E_\mathfrak{b}\pi(\Omega)v = c(\mathfrak{b})\,E_\mathfrak{b}v \quad (\mathfrak{b}\in\mathcal{E}_K).$$

LEMMA 6. *Fix $v\in S$. Then we can select $v_0\in S$ such that*

$$\nu(E_\mathfrak{b}v) \leqslant c(\mathfrak{b})^{-m} d(\mathfrak{b})^2 \nu_0(\pi(\Omega^m)v)$$

for $\mathfrak{b}\in\mathcal{E}_K$, any integer $m\geqslant 0$ and any differentiable vector $v\in V$.

Since K is compact, we can choose $\nu_0\in S$ such that

$$\nu(\pi(k)u) \leqslant \nu_0(u)$$

for $k\in K$ and $u\in V$. Therefore

$$\nu(E_\mathfrak{b}u) = \nu(\pi(\alpha_\mathfrak{b})u) \leqslant d(\mathfrak{b})^2\nu_0(u)$$

since $\sup|\alpha_\mathfrak{b}| \leqslant d(\mathfrak{b})^2$. Now we have seen above that

$$E_\mathfrak{b}v = c(\mathfrak{b})^{-m}E_\mathfrak{b}\pi(\Omega^m)v \quad (m\geqslant 0),$$

if v is differentiable. Hence our assertion follows immediately.

LEMMA 7.
$$\sum_{\mathfrak{b}\in\mathcal{E}_K} d(\mathfrak{b})^2 c(\mathfrak{b})^{-m} < \infty$$
if m is sufficiently large.

Assuming this for a moment, we shall first finish the proof of Lemma 5. It is obvious from Lemmas 6 and 7 that the series

$$\sum_{\mathfrak{b}} E_{\mathfrak{b}} v$$

converges absolutely. Let v_0 denote its sum and put $u = v - v_0$. We have to show that $u = 0$. Fix $\mathfrak{b}_0 \in \mathcal{E}_K$. Since $E_{\mathfrak{b}_0}$ is continuous,

$$E_{\mathfrak{b}_0} v_0 = \sum_{\mathfrak{b}} E_{\mathfrak{b}_0} E_{\mathfrak{b}} v = E_{\mathfrak{b}_0} v$$

from the Schur orthogonality relations. This shows that $E_{\mathfrak{b}} u = 0$ for all $\mathfrak{b} \in \mathcal{E}_K$.

Now fix $\nu \in S$ and select $\nu_0 \in S$ as in the proof of Lemma 6. For a given $\varepsilon > 0$, choose a neighborhood K_0 of 1 in K such that $\nu(\pi(k) u - u) \leqslant \varepsilon$ for $k \in K_0$. Fix a function $f \geqslant 0$ in $C(K)$ such that $f = 0$ outiside K_0 and

$$\int f(k)\,dk = 1.$$

Then
$$\nu(\pi(f) u - u) \leqslant \int f(k)\,\nu(\pi(k) u - u)\,dk \leqslant \varepsilon.$$

Call a function $\beta \in C(K)$ K-finite, if the space spanned by the left and right translates of β under K has finite dimension. Then by the Peter–Weyl theorem, we can choose a K-finite function $\beta \in C(K)$ such that $\sup |\beta - f| \leqslant \varepsilon$. Then

$$\nu(\pi(f) u - \pi(\beta) u) \leqslant \varepsilon \sup_{k \in K} \nu(\pi(k) u) \leqslant \varepsilon \nu_0(u)$$

and therefore
$$\nu(\pi(\beta) u - u) \leqslant \varepsilon(\nu_0(u) + 1).$$

On the other hand, $\alpha_{\mathfrak{b}} * \beta = \beta * \alpha_{\mathfrak{b}}$ since $\alpha_{\mathfrak{b}}$ is a class function. Therefore

$$E_{\mathfrak{b}} \pi(\beta) u = \pi(\beta) E_{\mathfrak{b}} u = 0.$$

Moreover, since β is K-finite, we can choose a finite subset F of $\mathcal{E}_K$ such that

$$\beta = \sum_{\mathfrak{b} \in F} \alpha_{\mathfrak{b}} * \beta.$$

Therefore
$$\pi(\beta) u = \sum_{\mathfrak{b} \in F} E_{\mathfrak{b}} \pi(\beta) u = 0$$

and this shows that
$$\nu(u) \leqslant \varepsilon(\nu_0(u) + 1).$$

Making ε tend to zero, we get $\nu(u) = 0$. Since ν was an arbitrary element in S, this implies that $u = 0$.

§ 4. Proof of Lemma 7

It remains to verify Lemma 7. Let K_0 be the connected component of 1 in K. Since K is compact, the index $N = [K : K_0]$ is finite. Put $\mathcal{E} = \mathcal{E}_K$ and let $\mathcal{E}_0$ be the set of all equivalence classes of irreducible finite-dimensional representations of K_0. If $\mathfrak{b} \in \mathcal{E}$ and $\mathfrak{b}_0 \in \mathcal{E}_0$, we denote by $[\mathfrak{b} : \mathfrak{b}_0]$ the number of times $\mathfrak{b}_0$ occurs in the reduction of $\mathfrak{b}$ with respect to K_0. For a given $\mathfrak{b}_0 \in \mathcal{E}_0$, let $\mathcal{E}(\mathfrak{b}_0)$ denote the set of all $\mathfrak{b} \in \mathcal{E}$ such that $[\mathfrak{b} : \mathfrak{b}_0] \geqslant 1$. Then it is a simple consequence of the Frobenius reciprocity theorem that

$$\sum_{\mathfrak{b} \in \mathcal{E}(\mathfrak{b}_0)} [\mathfrak{b} : \mathfrak{b}_0]\, d(\mathfrak{b}) = N d(\mathfrak{b}_0),$$

where $d(\mathfrak{b}_0)$ is the degree of $\mathfrak{b}_0$. Let $\xi_{\mathfrak{b}_0}$ denote the character (on K_0) of a class $\mathfrak{b}_0 \in \mathcal{E}_0$. Then it is easy to see that
$$\Omega \xi_{\mathfrak{b}_0} = c(\mathfrak{b}_0)\, \xi_{\mathfrak{b}_0},$$

where $c(\mathfrak{b}_0) = c(\mathfrak{b})$ for any $\mathfrak{b} \in \mathcal{E}(\mathfrak{b}_0)$. Therefore

$$\sum_{\mathfrak{b} \in \mathcal{E}} c(\mathfrak{b})^{-m} d(\mathfrak{b})^2 \leqslant \sum_{\mathfrak{b}_0 \in \mathcal{E}_0} c(\mathfrak{b}_0)^{-m} \sum_{\mathfrak{b} \in \mathcal{E}(\mathfrak{b}_0)} d(\mathfrak{b}_0)^2 \leqslant N^2 \sum_{\mathfrak{b}_0 \in \mathcal{E}_0} c(\mathfrak{b}_0)^{-m} d(\mathfrak{b}_0)^2.$$

Hence it would be enough to consider the case when K is connected.

Fix a Cartan subgroup A of K with Lie algebra $\mathfrak{a}$. Then A is connected. Let L be the kernel of the exponential mapping of $\mathfrak{a}$ into A. Then L is a lattice in $\mathfrak{a}$. Consider the space $\mathfrak{F}$ of all real-valued linear functions on $(-1)^{\frac{1}{2}}\mathfrak{a}$ and the lattice L^* of all $\lambda \in \mathfrak{F}$ such that $e^{\lambda(H)} = 1$ for $H \in L$. Introduce an order in $\mathfrak{F}$ and put

$$\varrho = \tfrac{1}{2} \sum_{\alpha > 0} \alpha,$$

where α runs over all positive roots of $(\mathfrak{k}, \mathfrak{a})$. For any $\mathfrak{b} \in \mathcal{E}$, let $\lambda(\mathfrak{b})$ denote the highest weight of $\mathfrak{b}$ with respect to $\mathfrak{a}$. Then the following facts are well known. $\lambda(\mathfrak{b}) \in L^*$ and $\mathfrak{b}$ is completely determined by $\lambda(\mathfrak{b})$. Moreover there exists a polynomial function p on $\mathfrak{F}$ such that $d(\mathfrak{b}) = p(\lambda(\mathfrak{b}))$ for all $\mathfrak{b} \in \mathcal{E}$. Finally, there exists a positive-definite quadratic form q on $\mathfrak{F}$ such that
$$c(\mathfrak{b}) = 1 + q(\lambda(\mathfrak{b}) + \varrho) - q(\varrho) \quad (\mathfrak{b} \in \mathcal{E}).$$

We can obviously choose a compact set C in $\mathfrak{F}$ such that

$$q(\lambda + \varrho) - q(\varrho) \geqslant q(\lambda)/2$$

for $\lambda \in \mathfrak{F}$ outside C. Let F denote the set of all $\mathfrak{b} \in \mathcal{E}$ such that $\lambda(\mathfrak{b}) \in C$. Then F is finite since $L^* \cap C$ is finite. Let ${}^c F$ denote the complement of F in $\mathcal{E}$. Then

$$\sum_{\mathfrak{b}\in \mathscr{F}} c(\mathfrak{b})^{-m} d(\mathfrak{b})^2 \leqslant 2^m \sum_{\lambda\in L^*}(1+q(\lambda))^{-m}\,|p(\lambda)|^2 < \infty,$$

provided m is sufficiently large.

§ 5. Differentiable vectors and Fourier series in function spaces

If M is a differentiable manifold, the spaces $C_c^\infty(M)$ and $C^\infty(M)$, taken with their usual topologies, are locally convex and complete. Let us now return to the notation of § 2 and for any $f\in C^\infty(G)$ and $y\in G$, define $l(y)f$ to be the function $x\to f(y^{-1}x)$ $(x\in G)$. We claim l is a representation of G on $C^\infty(G)$. For any compact subset Ω of G and $g\in\mathfrak{G}$, put

$$\nu_{\Omega,\,g}(f)=\sup_\Omega |gf|\quad (f\in C^\infty(G)).$$

Then the seminorms $\nu_{\Omega,\,g}$, taken together for all Ω and g, define the topology of $C^\infty(G)$ and it is clear that

$$\nu_{\Omega,\,g}(l(y)f)=\nu_{y^{-1}\Omega,\,g}(f).$$

Moreover, g and $l(y)$ commute as linear transformations in $C^\infty(G)$. Therefore if $f_1=gf$,

$$\nu_{\Omega,\,g}(l(y)f-f)=\sup_\Omega |l(y)f_1-f_1|\to 0$$

as $y\to 1$. Hence we conclude from Lemma 1 that l is a representation.

LEMMA 8. *Fix $f\in C^\infty(G)$. Then f is a differentiable vector under l and $l(X)f=-\varrho(X)f$ for $X\in\mathfrak{g}$.*

Fix $X\in\mathfrak{g}$ and put $y_t=\exp tX$ $(t\in\mathbf{R})$. Then it would be enough to verify that

$$\lim_{t\to 0} t^{-1}(l(y_t)f-f)=-\varrho(X)f$$

in $C^\infty(G)$. Fix Ω and g as above and put $f_1=gf$. Then since g and $\varrho(X)$ commute, we have

$$\nu_{\Omega,\,g}(t^{-1}\{l(y_t)f-f\}+\varrho(X)f)=\sup_\Omega |t^{-1}\{l(y_t)f_1-f_1\}+f_2|,$$

where $f_2=\varrho(X)f_1$. But we have seen during the proof of Lemma 2 that

$$f_1(y_t^{-1}x)-f_1(x)=-t\int_0^1 f_2(y_{st}^{-1}x)\,ds\quad (x\in G).$$

Therefore $\qquad\displaystyle \sup_\Omega |t^{-1}\{l(y_t)f_1-f_1\}+f_2| \leqslant \int_0^1 \sup_\Omega |l(y_{st})f_2-f_2|\,ds\to 0$

as $t\to 0$.

For any $\alpha \in C(K)$ and $f \in C^{\infty}(G)$, define

$$\alpha * f = \int_K \alpha(k)\, l(k) f\, dk$$

and let $\alpha_b(b \in \mathcal{E}_K)$ have the same meaning as in § 3.

COROLLARY 1. *The series*

$$\sum_{b \in \mathcal{E}_K} \alpha_b * f \quad (f \in C^{\infty}(G))$$

converges absolutely to f in $C^{\infty}(G)$.

This is an immediate consequence of Lemmas 5 and 8.

COROLLARY 2. *Fix $f \in C_c^{\infty}(G)$. Then the series*

$$\sum_{b \in \mathcal{E}_K} \alpha_b * f$$

converges absolutely to f in $C_c^{\infty}(G)$.

Fix a compact set Ω in G such that $K\Omega = \Omega$ and $\operatorname{Supp} f \subset \Omega$. As usual, let $C_\Omega^{\infty}(G)$ denote the space of all functions $\phi \in C_c^{\infty}(G)$ whose support lies in Ω. Then $C_\Omega^{\infty}(G)$ is a closed subspace of $C_c^{\infty}(G)$ and the two topologies induced on it from $C^{\infty}(G)$ and $C_c^{\infty}(G)$ coincide. Therefore our assertion follows from Corollary 1.

Let $r(y)f$ $(y \in G,\ f \in C^{\infty}(G))$ denote the function $x \to f(xy)$ $(x \in G)$. Then one proves in the same way that r is a representation of G on $C^{\infty}(G)$ and every $f \in C^{\infty}(G)$ is differentiable under r. Moreover $l(g)f = \varrho(g^*)f$ and $r(g)f = gf$ $(g \in \mathfrak{G})$ in the notation of the corollaries of Lemma 2. Define

$$f * \alpha = \int_K \alpha(k^{-1})\, r(k) f\, dk \quad (\alpha \in C(K)).$$

Then the analogues of the two corollaries of Lemma 8 hold also for r.

Note that $l(x)$ and $r(y)$ $(x,\ y \in G)$ commute and hence

$$\alpha * (f * \beta) = (\alpha * f) * \beta \quad (\alpha,\ \beta \in C(K)).$$

We may therefore simply write $\alpha * f * \beta$.

Define a representation π of $G \times G$ on $C^{\infty}(G)$ as follows.

$$\pi((x,\ y))f = l(x) r(y) f \quad (x,\ y \in G,\ f \in C^{\infty}(G)).$$

Since
$$\pi((x,\ y))f - f = l(x)\,(r(y)f - f) + (l(x)f - f),$$

it follows from Lemma 1 that π is indeed a representation. It is obvious from Lemma 8 and its analogue for r, that every $f \in C^\infty(G)$ is differentiable under π.

LEMMA 9. *Let V be either one of the two spaces $C^\infty(G)$ or $C_c^\infty(G)$, taken with its usual topology. Then for any $f \in V$, the series*

$$\sum_{b_1, b_2 \in \mathcal{E}_K} \alpha_{b_1} * f * \alpha_{b_2}$$

converges absolutely to f in V.

If $V = C^\infty(G)$, this follows immediately from Lemma 5. The rest is proved in the same way as Corollary 2 of Lemma 8.

§ 6. Proof of Lemma 4

We now come to the proof of Lemma 4. Fix $v \in V$, $\nu \in S$ and $\varepsilon > 0$. Then, by Lemma 3, we can select $f \in C_c^\infty(G)$ such that

$$\nu(\pi(f)v - v) \leqslant \varepsilon.$$

Choose a compact set Ω in G such that $K\Omega = \Omega$ and $\mathrm{Supp}\, f \subset \Omega$. Put

$$\mu(g) = \int |g| \, dx \quad (g \in C_c^\infty(G))$$

and

$$\alpha_F = \sum_{b \in F} \alpha_b$$

for any finite subset F of $\mathcal{E}_K$. Then μ is a continuous seminorm on $C_c^\infty(G)$ and $\mathrm{Supp}\,(f - \alpha_F * f) \subset \Omega$. Therefore

$$\nu(\pi(f - \alpha_F * f)v) \leqslant c\mu(f - \alpha_F * f),$$

where

$$c = \sup_{x \in \Omega} \nu(\pi(x)v) < \infty.$$

Hence we conclude from Corollary 2 of Lemma 8 that

$$\nu(\pi(f - \alpha_F * f)v) \leqslant \varepsilon$$

if F is sufficiently large. Therefore

$$\nu(\pi(\alpha_F * f)v - v) \leqslant \nu(\pi(f - \alpha_F * f)v) + \nu(\pi(f)v - v) \leqslant 2\varepsilon.$$

Since

$$\pi(\alpha_F * f)v \in \sum_{b \in \mathcal{E}_K} V_b \cap V^\infty$$

from Lemma 2, the assertion of Lemma 4 is now obvious.

§ 7. Some elementary facts about σ and Ξ

Let $\mathfrak{g}$ be a reductive Lie algebra over $\mathbf{R}$. Then $\mathfrak{g} = \mathfrak{c} + \mathfrak{g}_1$ where $\mathfrak{c}$ is the center and $\mathfrak{g}_1$ the derived algebra of $\mathfrak{g}$. Let θ be an automorphism of $\mathfrak{g}$ such that $\theta^2 = 1$ and let $\mathfrak{k}$ and $\mathfrak{p}$ be the subspaces of $\mathfrak{g}$ corresponding to the eigenvalues 1 and -1 respectively of θ. We assume that the quadratic form

$$Q(X) = -\operatorname{tr}(\operatorname{ad} X \operatorname{ad} \theta(X)) \quad (X \in \mathfrak{g})$$

is positive-definite on $\mathfrak{g}_1$.

Let G be a connected Lie group with Lie algebra $\mathfrak{g}$ and K the analytic subgroup of G corresponding to $\mathfrak{k}$. We assume that:

1) K is compact,

2) The mapping $(k, X) \to k \exp X$ $(k \in K, X \in \mathfrak{p})$ defines an analytic diffeomorphism of $K \times \mathfrak{p}$ onto G.

Let $\log$ denote the inverse of the exponential mapping from $\mathfrak{p}$ to $\exp \mathfrak{p}$. Suppose we have a Euclidean norm on $\mathfrak{p}$ such that $\|X^k\| = \|X\|$ $(k \in K, X \in \mathfrak{p})$ and

$$\langle X, (\operatorname{ad} Z)^2 Y \rangle = \langle (\operatorname{ad} Z)^2 X, Y \rangle \quad (X, Y, Z \in \mathfrak{p})$$

for the corresponding scalar product. Then it is easy to see that $\mathfrak{c}_\mathfrak{p} = \mathfrak{c} \cap \mathfrak{p}$ and $\mathfrak{p}_1 = [\mathfrak{k}, \mathfrak{p}] = \mathfrak{p} \cap \mathfrak{g}_1$ are mutually orthogonal (under this norm). Put

$$\sigma(x) = \|X\|$$

for $x = k \exp X$ $(k \in K, X \in \mathfrak{p})$ and extend θ to an automorphism of G (see [4(o), § 16]). Then it is obvious that

$$\sigma(x) = \sigma(\theta(x)) = \sigma(x^{-1}) \quad (x \in G).$$

If $|C|$ $(C \in \mathfrak{c}_\mathfrak{p})$ is an arbitrary Euclidean norm on $\mathfrak{c}_\mathfrak{p}$ and we define

$$\|C + X\|^2 = |C|^2 - \operatorname{tr}(\operatorname{ad} X \operatorname{ad} \theta(X)) \quad (C \in \mathfrak{c}_\mathfrak{p}, X \in \mathfrak{p}_1)$$

then all the above conditions are fulfilled.

LEMMA 10. $\sigma(xy) \leqslant \sigma(x) + \sigma(y)$ for $x, y \in G$.

We may obviously assume that $x = \exp X$, $y = \exp Y$ $(X, Y \in \mathfrak{p})$. Then if $xy = k \exp Z$ $(k \in K, Z \in \mathfrak{p})$, it is clear that

$$\exp 2Z = \theta(xy)^{-1} xy = \exp Y \cdot \exp 2X \cdot \exp Y.$$

Now define $Z(t) \in \mathfrak{p}$ by

$$\exp 2Z(t) = \exp tY \cdot \exp 2X \cdot \exp tY \quad (t \in \mathbf{R}).$$

Then $Z(t)$ is an analytic function of t and it follows by differentiating with respect to t that (see [5(a), p. 95])

$$\{(1 - \exp(-2 \operatorname{ad} Z(t)))/2 \operatorname{ad} Z(t)\} Z'(t) = \tfrac{1}{2}(1 + \exp(-2 \operatorname{ad} Z(t))) Y,$$

where $Z'(t) = dZ(t)/dt$. Therefore

$$\langle Z(t), Z'(t) \rangle = \langle Z(t), Y \rangle.$$

Now suppose $Z(t) = 0$ for some $t \in \mathbf{R}$. Then

$$1 = \exp tY \cdot \exp 2X \cdot \exp tY$$

and therefore $X = -tY$. But then $Z = (1 - t) Y$ and so

$$\|Z\| \leqslant \|Y\| + \|tY\| = \|Y\| + \|X\|$$

which is equivalent to the assertion of the lemma. Hence we may assume that $Z(t) \neq 0$ for every $t \in \mathbf{R}$. Then $\|Z(t)\|$ is analytic in t and

$$\langle Z(t), Z'(t) \rangle = \|Z(t)\| d\|Z(t)\|/dt.$$

This shows that
$$\|Z(t)\| d\|Z(t)\|/dt = \langle Z(t), Y \rangle$$

and hence
$$|d\|Z(t)\|/dt| \leqslant \|Y\|$$

But then by integrating we get

$$\|Z(1)\| - \|Z(0)\| \leqslant \|Y\|.$$

However $Z(0) = X$ and $Z(1) = Z$. Therefore

$$\|Z\| \leqslant \|X\| + \|Y\|$$

and this proves the lemma.

Put $\mathfrak{k}_1 = \mathfrak{k} + \mathfrak{c} = \mathfrak{k} + \mathfrak{c}_\mathfrak{p}$ and let K_1 be the analytic subgroup of G corresponding to $\mathfrak{k}_1$. Define the function Ξ on G corresponding to K_1 as in [4(q), § 16]. Then by [4 (q), Lemma 31] Ξ is everywhere positive.

LEMMA 11. *There exists a number $r \geqslant 0$ such that*

$$\int_G \Xi(x)^2 (1 + \sigma(x))^{-r} dx < \infty.$$

Let G_0 and $C_\mathfrak{p}$ be the analytic subgoups of G corresponding to $\mathfrak{g}_0 = \mathfrak{k} + \mathfrak{p}_1$ and $\mathfrak{c}_\mathfrak{p}$ respectively. Then since $\Xi(yc) = \Xi(y)$ and

$$\sigma(yc) \geqslant \max\ (\sigma(y),\ \sigma(c))\quad (y \in G_0,\ c \in C_\mathfrak{p}),$$

it is clear that the above integral is majorized by

$$\int_{C_\mathfrak{p}} (1 + \sigma(c))^{-r/2}\, dc \int_{G_\bullet} \Xi(y)^2 (1 + \sigma(y))^{-r/2}\, dy,$$

where dc and dy are the Haar measures on $C_\mathfrak{p}$ and G_0 respectively. Now it is clear that

$$\int_{C_\mathfrak{p}} (1 + \sigma(c))^{-r/2}\, dc < \infty$$

if r is sufficiently large. Hence it would be enough to consider the case when $\mathfrak{c}_\mathfrak{p} = \{0\}$. Then we can obviously replace G by G/Z_G where Z_G is the center of G. So we may assume that G is semisimple. Define $\mathfrak{a}$, Σ, $\mathfrak{a}^+$, A and A^+ as in [4 (q), § 21] and let dh denote the Haar measure of A. Then it follows from [4 (d), Lemma 38] that

$$\int_G \Xi(x)^2 (1 + \sigma(x))^{-r}\, dx = c_1 \int_{A^+} D(h)\, \Xi(h)^2 (1 + \sigma(h))^{-r}\, dh,$$

where c_1 is a positive number and

$$D(\exp\ H) = \prod_{\alpha \in \Sigma} (e^{\alpha(H)} - e^{-\alpha(H)})^{m_\alpha}\quad (H \in \mathfrak{a}^+),$$

m_α being the multiplicity of α. Put $\varrho = \tfrac{1}{2} \sum_{\alpha \in \Sigma} m_\alpha$. Then we know from [4 (j), Theorem 3] that we can choose positive numbers c_2 and d such that

$$\Xi(h) \leqslant c_2\, e^{-\varrho(\log\ h)} (1 + \sigma(h))^d\quad (h \in A^+).$$

Therefore, since it is clear that

$$D(h) \leqslant e^{2\varrho(\log\ h)}\quad (h \in A^+),$$

we conclude that

$$\int_G \Xi^2 (1 + \sigma)^{-r}\, dx \leqslant c_1\, c_2^2 \int_{A^+} (1 + \sigma(h))^{2d-r}\, dh < \infty,$$

if r is sufficiently large. This proves Lemma 11.

Remark. Suppose $\mathfrak{c}_\mathfrak{p} = \{0\}$. Then one proves in the same way that $(1 + \sigma)^r \Xi \in L_p(G)$ for $p > 2$ and $r \in \mathbf{R}$.

§ 8. Proof of Theorem 1

We keep to the notation of § 7 and define $\mathfrak{G}$ as in § 2. Let $\mathfrak{Z}$ be the center of $\mathfrak{G}$ and $\mathfrak{K}$ the subalgebra of $\mathfrak{G}$ generated by $(1, \mathfrak{k}_c)$. The following theorem will play an important role in the harmonic analysis on G.

THEOREM 1.(¹) *Let V be a complex vector space of finite dimension and f a C^∞ function from G to V such that the functions zf $(z \in \mathfrak{Z}\mathfrak{K})$ span a finite-dimensional space. Fix a neighborhood U of 1 in G and let J be the space of all functions $\alpha \in C_c^\infty(G)$ such that $\mathrm{Supp}\ \alpha \subset U$ and $\alpha(kxk^{-1}) = \alpha(x)$ $(k \in K, x \in G)$. Then there exists an element $\alpha \in J$ such that $f * \alpha = f$.*

We regard f as an element of $C^\infty(G) \otimes V$ and extend the representation r of § 5 on this space by making G act trivially on V. Then, as we have seen in § 5, every element $\phi \in C^\infty(G) \otimes V$ is differentiable under r and $r(g)\phi = g\phi$ $(g \in \mathfrak{G})$. Let $\mathfrak{U}$ be the set of all $u \in \mathfrak{Z}\mathfrak{K}$ such that $uf = 0$. Then $\mathfrak{U}$ is a left ideal in $\mathfrak{Z}\mathfrak{K}$ of finite codimension. Let W be the smallest closed subspace of $C^\infty(G) \otimes V$ containing f, which is stable under $r(G)$. Then it is obvious that W contains $W_0 = r(\mathfrak{G})f$. We claim that $W = \mathrm{Cl}(W_0)$. For otherwise, by the Hahn-Banach theorem, we could choose a continuous linear function $\beta \neq 0$ on W such that $\beta = 0$ on W_0. Put

$$F(x) = \beta(r(x)f) \quad (x \in G).$$

Since f is differentiable under r, it is obious that $F \in C^\infty(G)$ and

$$F(x; g) = \beta(r(x)r(g)f) \quad (g \in \mathfrak{G}).$$

Therefore $uF = 0$ for $u \in \mathfrak{U}$. However $\mathfrak{U}$ contains elliptic differential operators (see the proof of Lemma 33 of [4 (q)]) and so we conclude that F is an analytic function. On the other hand,

$$F(1; g) = \beta(r(g)f) = 0 \quad (g \in \mathfrak{G})$$

since $\beta = 0$ on W_0. Hence $F = 0$ and this implies that $\beta = 0$ on W. This contradiction proves that $W = \mathrm{Cl}(W_0)$.

Put $W_1 = r(\mathfrak{K})f$. Then $\dim W_1 < \infty$ and therefore W_1 is closed in W. Moreover one proves in the same way as above that $r(K)f \subset W_1$ so that W_1 is stable under $r(K)$. Since $f \in W_1$, we can choose a finite subset F of $\mathcal{E}_K$ such that $f = f * \alpha_F$. (Here α_F has the same meaning as in § 6.) Put

---

(¹) In my original proof of this theorem, I had to impose a mild condition on f at infinity, in order to get a representation of G on a suitable Banach space containing f. It was noticed by H. Jacquet that the argument worked equally well for a representation on a locally convex space and therefore the extra condition could be dropped. The proof given here, which is simpler than the original version, although based on the same idea, was obtained during a discussion with A. Borel.

$$E_F = \sum_{b \in F} E_b$$

in the notation of § 2 (with $\pi = r$). We claim that $W_F = E_F W$ has finite dimension. Since W_1 is fully reducible under $r(K)$, it is obvious that the natural representation of $\Re$ on $\Re/\Re \cap \mathfrak{U}$ is semisimple. Moreover since dim $(\mathfrak{Z}\Re/\mathfrak{U}) < \infty$, it follows from [4 (a), Theorem 1, p. 195] that

$$W_0 = \sum_{b \in \mathcal{E}_K} E_b W_0$$

and dim $E_b W_0 < \infty$ for every $b \in \mathcal{E}_K$. On the other hand, W_0 is dense in W and therefore $E_b W_0$ is dense in $E_b W$. Hence $E_b W_0 = E_b W$ and this shows that dim $W_F < \infty$.

We have seen in § 2 that there exists a Dirac sequence α_j $(j \geqslant 1)$ with $\alpha_j \in J$. Then by Lemma 3, $f * \alpha_j \to f$ in W as $j \to \infty$. Let W_2 be the space of all elements in W of the form $f * \alpha$ $(\alpha \in J)$. Since $\alpha(kxk^{-1}) = \alpha(x)$ $(k \in K, \ x \in G)$ and $f \in W_F$, it is obvious that $W_2 \subset W_F$. Hence W_2 is a vector space of finite dimension and therefore it is closed in W. Therefore $f = \lim_{j \to \infty} f * \alpha_j \in W_2$ and this proves the theorem.

§ 9. The space $C(G)$

Fix an open set U in G and let $C^0(U)$ denote the space of all continuous functions f from U to C such that

$$\nu_r(f) = \sup_U \ (1 + \sigma)^r \Xi^{-1} |f| < \infty$$

for every $r \in \mathbf{R}$. Put

$$_{g_1}\nu_{r, \, g_2}(f) = \nu_r((\varrho(g_1) \circ g_2) f)$$

for $f \in C^\infty(U)$, g_1, $g_2 \in \mathfrak{G}$ and $r \in \mathbf{R}$. Let $C(U)$ be the subspace of those $f \in C^\infty(U)$ for which $_{g_1}\nu_{r, \, g_2}(f) < \infty$ for all r and (g_1, g_2). We topologize $C(U)$ by means of the seminorms $_{g_1}\nu_{r, \, g_2}$ $(g_1, g_2 \in \mathfrak{G}, \ r \in \mathbf{R})$. In this way $C(U)$ becomes a locally convex Hausdorff space which is easily seen to be complete.[1]

LEMMA 12. *Fix $a, b \in G$ and, for any function f on U, let f' denote the function on aUb given by*

$$f'(x) = f(a^{-1}xb^{-1}) \quad (x \in aUb).$$

Then $f \to f'$ defines a topological mapping of $C(U)$ onto $C(aUb)$.

This is an easy consequence of Lemma 10 and [4(q), Lemma 32].

Now let G' be a Lie group such that G is the connected component of 1 in G'. Moreover let U be an open subset of G' which meets only a finite number of connected components of G'. Then we can choose $a_i \in G'$ and open sets U_i in G such that U is the disjoint union

---

[1] $C(U) = \{0\}$ by convention, if U is empty.

of $a_i U_i (1 \leqslant i \leqslant r)$. For any $f \in C^\infty(U)$, let f_i denote the function on U_i given by $f_i(x) = f(a_i x)$ $(x \in U_i)$. Consider the space $C(U)$ of all $f \in C^\infty(U)$ such that $f_i \in C(U_i)$ $(1 \leqslant i \leqslant r)$ and let V denote the Cartesian product of $C(U_i)$ $(1 \leqslant i \leqslant r)$ with the natural topology. We topologize $C(U)$ in such a way that the mapping $f \to (f_1, ..., f_r)$ of $C(U)$ onto V becomes an isomorphism. It follows from Lemma 12 that the structure of $C(U)$, as a locally convex space, is independent of the choice of a_i and U_i. Moreover it is obvious that the injection of $C_c^\infty(U)$ into $C(U)$ is continuous.

By a tempered distributon T on U, we mean a continuous linear mapping of $C(U)$ into C.

Now assume that G'/G is finite. Then U can be any open subset of G.

THEOREM 2. *Suppose G'/G is finite. Then $C_c^\infty(G')$ is dense in $C(G')$.*

In view of this theorem, we can identify tempered distributions on G' with those distributions which are continuous in the relative topology of $C_c^\infty(G')$ as a subspace of $C(G')$. Moreover, it is obviously enough to prove this theorem in case $G' = G$. This requires some preparation which will be undertaken in the next few sections.

§ 10. The left- and right-regular representations on $C(G)$

Let S denote the set of all continuous seminorms on $C(G)$. For any $f \in C(G)$ and $y \in G$, define $l(y)f$ and $r(y)f$ as in § 5.

LEMMA 13. *$l(y)f$ and $r(y)f$ are in $C(G)$. Moreover for a given compact set Ω in G and $\nu \in S$, we can choose $\nu' \in S$ such that*

$$\nu(l(y)f) + \nu(r(y)f) \leqslant \nu'(f)$$

for $y \in \Omega$ and $f \in C(G)$.

Put $r(y)f = f_y$ and fix $g, g' \in \mathfrak{G}$. Then

$$f_y(g'; x; g) = f(g'; xy; g^{y^{-1}})$$

for $x, y \in G$. We can choose linearly independent elements g_i $(1 \leqslant i \leqslant p)$ in $\mathfrak{G}$ and analytic functions a_i on G such that

$$g^{y^{-1}} = \sum_{1 \leqslant i \leqslant p} a_i(y) g_i \quad (y \in G).$$

Then
$$f_y(g'; x; g) = \sum_i a_i(y) f(g'; xy; g_i).$$

If we apply a similar argument to $l(y)f$ and take into account Lemma 10 and [4 (q), Lemma 32], our assertions follow immediately.

LEMMA 14. *Fix $s \geqslant 0$ and put*

$$\nu_s(f) = \sup \, (1+\sigma)^s \, \Xi^{-1} |f| \quad (f \in C(G)).$$

Then for any $\varepsilon > 0$, we can choose a neighborhood U of 1 in G and an element $\nu' \in S$ such that

$$\nu_s(f_y - f) \leqslant \varepsilon \nu'(f)$$

for $y \in U$ and $f \in C(G)$. Moreover, ν' does not depend on ε.

Introduce a norm in $\mathfrak{g}$ and fix a base $X_1, ..., X_n$ for $\mathfrak{g}$ over $\mathbf{R}$. Then we can choose $c_0 > 0$ such that

$$\max_{1 \leqslant i \leqslant n} |c_i| \leqslant c_0 \Big| \sum_{1 \leqslant i \leqslant n} c_i X_i \Big|$$

for $c_i \in \mathbf{R}$ ($1 \leqslant i \leqslant n$). Now let $f \in C(G)$ and $X \in \mathfrak{g}$. Then it is clear that

$$f(x \exp X) - f(x) = \int_0^1 f(x \exp tX; X) \, dt \quad (x \in G)$$

and therefore

$$|f(x \exp X) - f(x)| \leqslant c_0 |X| \sum_i \int_0^1 |f(x \exp tX; X_i)| \, dt.$$

But then it is obvious that we can choose $c > 0$ such that

$$\nu_s(f_{\exp X} - f) \leqslant c |X| \sum_{1 \leqslant i \leqslant n} \nu_s(X_i f)$$

for $|X| \leqslant 1$ ($X \in \mathfrak{g}$) and $f \in C(G)$. Clearly this implies the assertion of the lemma.

COROLLARY 1. *Fix $\nu \in S$ and $\varepsilon > 0$. Then we can choose a neighborhood U of 1 in G and $\nu' \subset S$ such that*

$$\nu(l(y)f - f) + \nu(r(y)f - f) \leqslant \varepsilon \nu'(f)$$

for $y \in U$ and $f \in C(G)$. Moreover, ν' is independent of ε.

We use the notation of the proof of Lemma 13. Then

$$f_y(g'; x; g) - f(g'; x; g) = \sum_i (a_i(y) - 1) f(g'; xy; g_i) + \{f(g'; xy; g) - f(g'; x; g)\}.$$

Put $D = \varrho(g') \circ g$ and $D_i = \varrho(g') \circ g_i$. Then D and D_i define continuous endomorphisms of $C(G)$ and the above relation may be written as

$$D(r(y)f - f) = \sum_i (a_i(y) - 1) r(y) D_i f + (r(y) - 1) Df.$$

Therefore it is obvious from Lemmas 13 and 14 that, for a given $s \geqslant 0$, we can choose a neighborhood U of 1 in G and an element $\nu' \in S$ (independent of ε) such that

$$v_s(D(r(y)f - f)) \leqslant \varepsilon v'(f)$$

for $y \in U$ and $f \in \mathcal{C}(G)$. Since a similar argument holds for $l(y)$, our assertion follows.

COROLLARY 2. *Both l and r define representations of G on $\mathcal{C}(G)$.*

This is obvious from Lemmas 1 and 13 and Corollary 1 above.

LEMMA 15. *Every element $f \in \mathcal{C}(G)$ is a differentiable vector under both l and r. More-over, $r(X)f = Xf$ and $l(X)f = -\varrho(X)f$ for $X \in \mathfrak{g}$.*

Fix $f \in \mathcal{C}(G)$, $Y \in \mathfrak{g}$ and put

$$\phi_t = t^{-1}(r(y_t)f - f) - Yf \quad (t \in \mathbf{R}, \ t \neq 0),$$

where $y_t = \exp tY$. Then it follows from Lemma 8 that if D is any differential operator on G and Ω any compact subset of G, then

$$\sup_{\Omega} |D\phi_t| \to 0$$

as $t \to 0$. Fix g, $g' \in \mathfrak{G}$ and a number $m \geqslant 0$. Then it follows in particular that for any compact Ω,

$$\sup_{x \in \Omega} |\phi_t(g'; x; g)| \, \Xi(x)^{-1}(1 + \sigma(x))^m \to 0$$

as $t \to 0$.

On the other hand (see the proof of Lemma 2),

$$\phi_t(x) = \int_0^1 \{f(xy_{ts}; Y) - f(x; Y)\} \, ds.$$

Hence
$$\phi_t(g'; x; g) = \sum_i \int_0^1 a_i(y_{ts}) f' \, (xy_{ts}; g_i \, Y) \, ds - f'(x; gY),$$

where $f' = \varrho(g')f$ and $a_i, g_i \ (1 \leqslant i \leqslant p)$ have the same meaning as in the proof of Lemma 13. Fix a compact neighborhood $U = U^{-1}$ of 1 in G. Then we can choose a number c such that $1 + \sigma(y) \leqslant c$, $|a_i(y)| \leqslant c \ (1 \leqslant i \leqslant p)$ and $\Xi(xy) \leqslant c \Xi(x)$ for $y \in U$ and $x \in G$. Fix $\delta > 0$ such that $y_t \in U$ for $|t| \leqslant \delta$. Then

$$|\phi_t(g'; x; g)| \, \Xi(x)^{-1}(1 + \sigma(x))^m$$
$$\leqslant c^{m+2} \sum_i \sup_{u \in U} |f'(xu; g_i Y)| \, \Xi(xu)^{-1}(1 + \sigma(xu))^m + |f'(x; gY)| \, \Xi(x)^{-1}(1 + \sigma(x))^m$$

for $|t| \leqslant \delta$. Now fix $\varepsilon > 0$. Since $f' \in \mathcal{C}(G)$, we can choose a compact set Ω_0 in G such that

$$|f'(x; g_i Y)| \Xi(x)^{-1}(1+\sigma(x))^m \leqslant \varepsilon$$

outside Ω_0 for $0 \leqslant i \leqslant p$. (Here $g_0 = g$.) Put $\Omega = \Omega_0 U$. Then Ω is compact and it is clear that

$$|\phi_t(g'; x; g)| \Xi(x)^{-1}(1+\sigma(x))^m \leqslant (p+1)c^{m+2}\varepsilon$$

if $x \notin \Omega$ and $|t| \leqslant \delta$. Therefore, in view of our earlier result, we can now conclude that $\phi_t \to 0$ in $C(G)$ as $t \to 0$. This shows that f is differentiable under r and $r(Y)f = Yf$. The proof for l is similar.

Define $\alpha_{\mathfrak{b}}$ ($\mathfrak{b} \in \mathcal{E}_K$) as in § 5.

LEMMA 16. *For any $f \in C(G)$, the series*

$$\sum_{\mathfrak{b}_1, \mathfrak{b}_2 \in \mathcal{E}_K} \alpha_{\mathfrak{b}_1} * f * \alpha_{\mathfrak{b}_2}$$

converges absolutely to f in $C(G)$.

This is proved in the same way as Lemma 9.

§ 11. Spherical functions

Let $\mu = (\mu_1, \mu_2)$ be a (continuous) double representation[1] of K on a (complex) vector space V of finite dimension. Then by a μ-spherical function we mean a function ϕ from G to V such that

$$\phi(k_1 x k_2) = \mu_1(k_1)\phi(x)\mu_2(k_2) \quad (k_1, k_2 \in K; \ x \in G).$$

Fix a norm on V.

LEMMA 17. *For any two elements g, $g' \in \mathfrak{G}$, we can choose a finite number of $g_i \in \mathfrak{G}$ ($1 \leqslant i \leqslant p$) with the following property. If ϕ is any C^∞ μ-spherical function, then*

$$|\phi(g; x; g')| \leqslant \sum_{1 \leqslant j \leqslant p} |\phi(x; g_j)|$$

and

$$|\phi(g; x; g')| \leqslant \sum_{1 \leqslant j \leqslant p} |\phi(g_j; x)|$$

for $x \in G$.

Let $\mathfrak{a}$ be a maximal abelian subspace of $\mathfrak{p}$. Introduce an order in the space of real linear functions α on $\mathfrak{a}$ and, for any such α, let $\mathfrak{g}_\alpha$ denote the subspace of those $X \in \mathfrak{g}$ for which $[H, X] = \alpha(H)X$ for $H \in \mathfrak{a}$. Let Σ be the set of all positive roots of $(\mathfrak{g}, \mathfrak{a})$ and $(\alpha_1, ..., \alpha_l)$ the set of simple roots in Σ. Put

$$\mathfrak{n} = \sum_{\alpha > 0} \mathfrak{g}_\alpha.$$

Then $\mathfrak{g} = \mathfrak{k} + \mathfrak{a} + \mathfrak{n}$ and $\mathfrak{G} = \mathfrak{K}\mathfrak{A}\mathfrak{N}$ where[2] $\mathfrak{K} = \mathfrak{S}(\mathfrak{k}_c)$, $\mathfrak{A} = \mathfrak{S}(\mathfrak{a}_c)$ and $\mathfrak{N} = \mathfrak{S}(\mathfrak{n}_c)$.

---

[1] This means that V is a left K-module under μ_1 and a right K-module under μ_2. Moreover, the operations of K on the left, commute with those on the right.

[2] We use here the notation of [4 (m), p. 280].

Fix an integer $d \geqslant 0$ such that g, $g' \in {}_d\mathfrak{G}$ (see [4 (o), § 2] for the notation). Then we can choose a base B for ${}_d\mathfrak{G}$ such that every element $b \in B$ has the form $b = \varkappa u v$ where $\varkappa \in \mathfrak{K}$, $u \in \mathfrak{A}$, $v \in \mathfrak{N}$ and

$$v^a = \exp \left(\sum_{1 \leqslant i \leqslant l} m_i \, \alpha_i \, (\log a) \right) v \quad (a \in A = \exp \mathfrak{a}),$$

m_i being nonnegative integers. Then

$$g^k = \sum_{b \in B} a_b(k) \, b, \, g'^k = \sum_{b \in B} a_b'(k) \, b \quad (k \in K),$$

where a_b and a_b' are continuous functions on K.

Now since any two norms on V are equivalent, we may assume that $|\mu_1(k_1) v \mu_2(k_2)| = |v|$ for k_1, $k_2 \in K$ and $v \in V$. Put $A^+ = \exp \mathfrak{a}^+$ where $\mathfrak{a}^+$ is the set of those points $H \in \mathfrak{a}$ where $\alpha(H) \geqslant 0$ for $\alpha \in \Sigma$. Then $G = KA^+K$. Put

$$c = \sup_{k \in K} \max_{b \in B} \, (|a_b(k)|, \, |a_b'(k)|).$$

Then if $x = k_1 h k_2$ $(k_1, k_2 \in K; \, h \in A^+)$, it is clear that

$$|\phi(g \colon x; g')| \leqslant |\phi(g^{k_1^{-1}} \colon h; g'^{k_2})| \leqslant c^2 \sum_{b, \, b' \in B} |\phi(b \colon h; b')|.$$

Now $b = \varkappa_b u_b v_b$ $(b \in B)$ as above. Let us denote the representation of $\mathfrak{K}_c$ corresponding to μ_i again by μ_i $(i = 1, 2)$. For any endomorphism T of V, define

$$|T| = \sup_{|v| \leqslant 1} |Tv| \quad (v \in V)$$

as usual and put
$$c_1 = \sup_{b \in B} |\mu_1(\varkappa_b)|.$$

Then
$$|\phi(b \colon h; b')| = |\mu_1(\varkappa_b) \, \phi(h; u_b \, v_b^{h^{-1}} b')| \leqslant c_1 |\phi(h; u_b \, v_b \, b')|$$

since $\alpha_i(\log h) \geqslant 0$ $(1 \leqslant i \leqslant l)$. Hence

$$|\phi(g \colon x; g')| \leqslant c^2 \, c_1 \sum_{b, \, b' \in B} |\phi(h; u_b \, v_b \, b')|.$$

Now let g_j $(1 \leqslant j \leqslant p)$ be a base for the subspace of $\mathfrak{G}$ spanned by $(u_b \, v_b \, b')^k$ $(b, b' \in B, k \in K)$. Then it is clear that we can choose a number $c_2 \geqslant 0$ with the following property. If $b, b' \in B$, $k \in K$ and

$$(u_b \, v_b \, b')^k = \sum_{1 \leqslant j \leqslant p} \gamma_j g_j \quad (\gamma_j \in C),$$

then $|\gamma_j| \leqslant c_2$. This shows that

$$|\phi(g; x; g')| \leqslant c_3 \sum_{1 \leqslant j \leqslant p} |\phi(x; g_j)|$$

for $x \in G$ where $c_3 = c^2 c_1 c_2$. Since our hypotheses are symmetrical with respect to left and right, the assertion of the lemma is now obvious.

§ 12. Application to $C_F(G)$

For any finite subset F of $\mathcal{E}_K$, define α_F as in § 6 and let $C_F(G)$ denote the subspace of all elements in $C(G)$ of the form $\alpha_F * f * \alpha_F$ $(f \in C(G))$. It is clear that an element $f \in C(G)$ lies in $C_F(G)$ if and only if $\alpha_F * f * \alpha_F = f$. Hence $C_F(G)$ is closed in $C(G)$.

Put
$$\nu_{m,g}(f) = \sup \, (1+\sigma)^m \Xi^{-1} |gf|,$$

$$_g\nu_m(f) = \sup \, (1+\sigma)^m \Xi^{-1} |\varrho(g)f|$$

for $m \geqslant 0$, $g \in \mathfrak{G}$ and $f \in C(G)$ and let $S_1 = \{\nu_{m,g}; \, m \geqslant 0, \, g \in \mathfrak{G}\}$ and $S_2 = \{_g\nu_m; \, m \geqslant 0, \, g \in \mathfrak{G}\}$.

LEMMA 18. *Let F be a finite subset of $\mathcal{E}_K$. Then each of the three sets of seminorms S_1, S_2 and S define the same topology on $C_F(G)$.*

Consider $C(K \times K)$ as a Banach space with the norm

$$|f| = \sup_{k_1, k_2 \in K} |f(k_1, k_2)| \quad (f \in C(K \times K))$$

and let $\mu_1(k)f$ and $f\mu_2(k)$ $(k \in K)$ respectively denote the functions

$$(k_1, k_2) \to f(k^{-1}k_1, k_2) \quad \text{and} \quad (k_1, k_2) \to f(k_1, k_2 k^{-1}) \quad (k_1, k_2 \in K).$$

Then $\mu = (\mu_1, \mu_2)$ is a double representation of K on $C(K \times K)$. Let C_F be the subspace of all $f \in C(K \times K)$ such that

$$f = \int_K \alpha_F(k) \mu_1(k) f \, dk = \int_K \alpha_F(k) f \mu_2(k) \, dk.$$

Then C_F is a finite-dimensional space invariant under μ. We denote the restriction of μ on C_F by μ_F.

For any $f \in C_F(G)$, define the μ_F-spherical function f^* from G to C_F as follows. If $x \in G$, $f^*(x)$ is the function
$$(k_1, k_2) \to f(k_1^{-1} x k_2^{-1}) \quad (k_1, k_2 \in K)$$
in C_F. It is clear that

$$|f^*(g_1; x; g_2)| = \sup_{k_1, k_2 \in K} |f(g_1^{k_1}; k_1 x k_2^{-1}; g_2^{k_2})|$$

for $g_1, g_2 \in \mathfrak{G}$ and $x \in G$. Therefore if we apply Lemma 17 with $V = C_F$, Lemma 18 follows immediately.

§ 13. Density of $C_c^\infty(G)$ in $C(G)$

Now we come to the proof of Theorem 2. We have to show that $C_c^\infty(G)$ is dense in $C(G)$. Fix a finite subset F of $\mathcal{E}_K$. In view of Lemma 16, it would be enough to verify the following result.

LEMMA 19. $C_c^\infty(G) \cap C_F(G)$ *is dense in* $C_F(G)$.

For any $t > 0$, let G_t denote the open set consisting of all $x \in G$ with $\sigma(x) < t$. Also let ξ_t denote the characteristic function of G_t. Fix $a > 0$ and an element $\alpha \in C_c^\infty(G_a)$ such that $\alpha(k_1 x k_2) = \alpha(x)$ $(k_1, k_2 \in K;\ x \in G)$ and

$$\int_G \alpha(x)\,dx = 1.$$

Put
$$u_t = (1 - \xi_t) * \alpha = 1 - \xi_t * \alpha,$$

where the star denotes convolution on G as usual. It is clear thas $u_t \in C_c^\infty(G)$.

LEMMA 20. *We have*
$$u_t(x) = \begin{cases} 0 \ \textit{if } \sigma(x) \leqslant t - a, \\ 1 \ \textit{if } \sigma(x) \geqslant t + a, \end{cases}$$

and
$$|u_t(x;g)| \leqslant \int_{\bar{G}} |\alpha(y;g)|\,dy \quad (x \in G)$$

for $g \in \mathfrak{G}$.

It is clear that
$$u_t(x) = \int_{G_a} (1 - \xi_t(xy^{-1}))\,\alpha(y)\,dy$$

and if we fix $y \in G_a$, it follows from Lemma 10 that

$$\xi_t(xy^{-1}) = \begin{cases} 0 \ \text{if } \sigma(x) \geqslant t + a, \\ 1 \ \text{if } \sigma(x) \leqslant t - a. \end{cases}$$

This gives the first statement of the lemma. Now fix $g \in \mathfrak{G}$. Since g is left-invariant, we have

$$gu_t = (1 - \xi_t) * g\alpha.$$

Therefore
$$u_t(x;g) = \int (1 - \xi_t(xy^{-1}))\,\alpha(y;g)\,dy$$

and this implies the desired inequality.

Now fix $f \in C_F(G)$ and put

$$f_t = (1 - u_t)f = (\xi_t * \alpha)f.$$

Since α and ξ_t are both invariant under left and right translations of K, it is obvious that $f_t \in C_c^\infty(G) \cap C_F(G)$. We now claim that $f_t \to f$ in $C_F(G)$ as $t \to +\infty$. Fix $g \in \mathfrak{G}$. Then

$$g(f - f_t) = g(u_t f) = \sum_{1 \leqslant i \leqslant p} g_i' u_t \cdot g_i f,$$

where g_i, g_i' are suitable elements in $\mathfrak{G}$. Moreover,

$$f(x) - f_t(x) = u_t(x)f(x) = f(x)$$

if $\sigma(x) \geqslant t + a$ and

$$(1 + \sigma(x))^m \Xi(x)^{-1}|f(x; g)| \leqslant (1 + t)^{-1} \nu_{m+1, g}(f)$$

for $\sigma(x) \geqslant t$ and $m \geqslant 0$ in the notation of § 12. Hence

$$(1 + \sigma(x))^m \Xi(x)^{-1}|f(x; g) - f_t(x; g)| \leqslant (1 + t)^{-1} \nu_{m+1, g}(f)$$

for $\sigma(x) \geqslant t + a$. Now suppose $\sigma(x) < t + a$. By Lemma 20, $f(x) - f_t(x) = 0$ if $\sigma(x) \leqslant t - a$. So let us assume that $t - a < \sigma(x) < t + a$. Then if $t > a$, we conclude from Lemma 20 that

$$(1 + \sigma(x))^m \Xi(x)^{-1}|f(x; g) - f_t(x; g)| \leqslant \sum_{1 \leqslant i \leqslant p} c_i(1 + \sigma(x))^m \Xi(x)^{-1}|f(x; g_i)|$$

$$\leqslant \sum_i c_i(1 + t - a)^{-1} \nu_{m+1, g_i}(f),$$

where

$$c_i = \int_G |\alpha(y; g_i')| \, dy \quad (1 \leqslant i \leqslant p).$$

This shows that $\nu_{m, g}(f - f_t) \to 0$ as $t \to +\infty$ and therefore by Lemma 18, f_t converges to f in $C_F(G)$. This proves Lemma 19 and therefore also Theorem 2.

§ 14. An inequality

Let $\mathfrak{h} = \theta(\mathfrak{h})$ be a Cartan subalgebra of $\mathfrak{g}$ and $\mathfrak{m}$ and $\bar{M}$ the centralizers of $\mathfrak{h} \cap \mathfrak{p}$ in $\mathfrak{g}$ and G respectively. Let M denote the connected component of 1 in $\bar{M}$.

Fix compatible orders on the spaces of real-valued linear functions on $\mathfrak{h} \cap \mathfrak{p}$ and $\mathfrak{h} \cap \mathfrak{p} + (-1)^{\frac{1}{2}}\mathfrak{h} \cap \mathfrak{k}$ and let P be the set of all positive roots of $(\mathfrak{g}, \mathfrak{h})$. Let P_M be the subset of those roots in P which vanish identically on $\mathfrak{h} \cap \mathfrak{p}$ and P_+ the complement of P_M in P. Put

$$\mathfrak{n}_\mathfrak{h} = \mathfrak{g} \cap \left(\sum_{\alpha \in P_+} \mathbb{C}X_\alpha \right)$$

in the notation of (4 (n), § 4]. Then $\mathfrak{n_h}$ is a nilpotent subalgebra of $\mathfrak{g}$ and $\tilde{M}$ normalizes $\mathfrak{n_h}$. Put

$$d(m) = |\det (\mathrm{Ad}\,(m))_{\mathfrak{n_h}}|^{\frac{1}{2}} \quad (m \in \tilde{M})$$

and let $N_\mathfrak{h}$ denote the analytic subgroup of G corresponding to $\mathfrak{n_h}$.

Extend $\mathfrak{h} \cap \mathfrak{p}$ to a maximal abelian subspace $\mathfrak{a}$ of $\mathfrak{p}$ and introduce an order on the dual of $\mathfrak{a}$ which is compatible with that chosen above on the dual of $\mathfrak{h} \cap \mathfrak{p}$. Let $\mathfrak{g} = \mathfrak{k} + \mathfrak{a} + \mathfrak{n}$ be the corresponding Iwasawa decomposition. Then $\mathfrak{n_h} \subset \mathfrak{n}$. Put

$$\varrho(H) = \tfrac{1}{2}\,\mathrm{tr}\,(\mathrm{ad}\,H)_\mathfrak{n} \quad (H \in \mathfrak{a})$$

and define $\mathfrak{a}^+$ and $A^+ = \exp \mathfrak{a}^+$ as usual (see [4 (q), § 21]). Then by [4 (j), Theorem 3], we can choose a number $d \geqslant 0$ such that

$$\sup_{h \in A^+} (1 + \sigma(h))^{-d}\, e^{\varrho(\log h)}\, \Xi(h) < \infty.$$

Let Ξ_M denote the function on M corresponding to Ξ if we replace (G, K) by $(M, K \cap M)$.

LEMMA 21. *Let dn denote the Haar measure of $N_\mathfrak{h}$ and fix $r > r' \geqslant 0$. Then we can choose a number $c > 0$ such that*

$$d(m) \int_{N_\mathfrak{h}} (1 + \sigma(mn))^{-(r+2d)}\, \Xi(mn)\, dn \leqslant c(1 + \sigma(m))^{-r'}\, \Xi_M(m)$$

for $m \in M$. Moreover, there exists a number $c_0 \geqslant 1$ such that

$$1 + \sigma(m) \leqslant c_0(1 + \sigma(mn)) \quad (m \in M,\ n \in N_\mathfrak{h}).$$

It is clear that for a fixed [1] $m \in M$, $dn^m = d(m)^2 dn$. Therefore since $mn = n^m \cdot m$, we get the following result immediately.

COROLLARY. *In the above notation, we also have*

$$d(m)^{-1} \int_{N_\mathfrak{h}} (1 + \sigma(nm))^{-(r+2d)}\, \Xi(nm)\, dn \leqslant c(1 + \sigma(m))^{-r'}\, \Xi_M(m)$$

for $m \in M$.

We shall give a proof of Lemma 21 in § 44.

---

[1] $y^x = xyx^{-1}$ $(x, y \in G)$ as usual.

§ 15. The mapping of $C(G)$ into $C(\tilde{M})$

Let us keep to the above notation. Then $M \cap K$ has finite index in $\tilde{M} \cap K$ and therefore $\tilde{M}/M$ is finite. Hence the space $C(\tilde{M})$ is well defined (see § 9). Let Ω be a compact subset of G. Then it follows from Lemma 10, [4 (q), Lemma 32] and § 14 that, for any $f \in C(G)$, the integrals

$$\int_{N_{\mathfrak{h}}} |f(xn)| \, dn, \quad \int_{N_{\mathfrak{h}}} |f(nx)| \, dn$$

converge uniformly for $x \in \Omega$ (see also Corollary 2 of Lemma 90). Put

$$u_f(m) = d(m) \int_{N_{\mathfrak{h}}} f(mn) \, dn = d(m^{-1}) \int_{N_{\mathfrak{h}}} f(nm) \, dn \quad (m \in M),$$

and [1] $\mathfrak{M} = \mathfrak{S}(\mathfrak{m}_c)$. It is easy to verify (see [4 (q), § 10]) that there exist two automorphisms $\mu \to \mu'$ and $\mu \to {}^{\backprime}\mu$ of $\mathfrak{M}$ such that

$$\mu' = d^{-1} \mu \circ d, \quad {}^{\backprime}\mu = d\mu \circ d^{-1} \quad (\mu \in \mathfrak{M}).$$

Then it follows from what we have said above that $u_f \in C^\infty(M)$ and

$$u_f(\mu_1; m) = d(m) \int_{N_{\mathfrak{h}}} f(\mu_1'; mn) \, dn,$$

$$u_f(m; \mu_2) = d(m^{-1}) \int f(nm; {}^{\backprime}\mu_2) \, dn = d(m) \int f(mn; {}^{\backprime}\mu_2) \, dn$$

for $\mu_1, \mu_2 \in \mathfrak{M}$. Since ${}^{\backprime}\mu_2 f \in C(G)$, we conclude that

$$u_f(\mu_1; m; \mu_2) = d(m) \int f(\mu_1'; mn; {}^{\backprime}\mu_2) \, dn.$$

In view of Lemma 21, the following result is now obvious.

LEMMA 22. $f \to u_f$ *is a continuous linear mapping of* $C(G)$ *into* $C(M)$.

Put
$$\tilde{f}(x) = \int_K f(x^k) \, dk \quad (x \in G)$$

for $f \in C(G)$. Then it follows from Lemma 13 that $f \to \tilde{f}$ is a continuous endomorphism of $C(G)$. Now define

$$g_f(m) = d(m) \int_{N_{\mathfrak{h}}} \tilde{f}(mn) \, dn \quad (m \in \tilde{M}).$$

---

[1] We use here the notation of [4 (m), p. 280].

THEOREM 3. $f \to g_f$ *is a continuous linear mapping of* $C(G)$ *into* $C(\tilde{M})$.

This is obvious from Lemma 22.

As usual let $\mathfrak{Z}$ denote the center of $\mathfrak{G}$ and $\mu = \mu_{\mathfrak{g}/\mathfrak{m}}$ the homomorphism of [4 (o), § 12]. Since $C_c^\infty(G)$ is dense in $C(G)$, we get the following result from the corollary of [4 (q), Lemma 14].

COROLLARY. *Let* $z \in \mathfrak{Z}$ *and* $f \in C(G)$. *Then*

$$g_{zf} = \mu(z) g_f.$$

§ 16. Proof of Theorem 4

Let $I(G)$ be the space of all continuous functions f on G which are *bi*-invariant under K, and $I^+(G)$ the subset of all real $f \geqslant 0$. Put $I_c^\infty(G) = I(G) \cap C_c^\infty(G)$ and

$$I_c^+(G) = C_c(G) \cap I^+(G).$$

Let $\mathfrak{Q}$ be the centralizer of $\mathfrak{k}$ in $\mathfrak{G}$ and μ a seminorm [1] on $I_c^\infty(G)$ satisfying the following two conditions.

1) There exist elements $q_1, \ldots, q_p \in \mathfrak{Q}$ such that

$$\mu(f) \leqslant \sum_{1 \leqslant i \leqslant p} \int_G |q_i f| \, \Xi \, dx \quad (f \in I_c^\infty(G)).$$

2) If f_1, f_2 are two elements in $I_c^\infty(G)$ such that $f_1 \geqslant f_2 \geqslant 0$, then $\mu(f_1) \geqslant \mu(f_2)$.

For any $\phi \in I^+(G)$, put
$$\mu(\phi) = \sup \mu(f),$$

where f runs over all functions in $I^+(G) \cap I_c^\infty(G)$ such that $\phi \geqslant f$. Fix a number $r \geqslant 0$ as in Lemma 11.

THEOREM 4. *Under the above conditions*

$$\mu(\Xi(1+\sigma)^{-r}) < \infty.$$

Fix $\alpha \in I^+(G) \cap I_c^\infty(G)$. Then $\alpha * f \in I_c^\infty(G)$ for $f \in I_c^+(G)$ and

$$\mu(\alpha * f) \leqslant \sum_{1 \leqslant i \leqslant p} \int |q_i \alpha * f| \, \Xi \, dx$$

---

[1] Here we take the space $I_c^\infty(G)$ without topology.

since $q_i(\alpha * f) = q_i(f * \alpha) = f * q_i \alpha = q_i \alpha * f$, the convolution being abelian in the present case (see [5 (a), Theorem 4.1, p. 408]). Now

$$\int |q_i \alpha * f| \,\Xi\, dx \leqslant \int |\alpha(y; q_i)| \, f(y^{-1} x) \,\Xi(x)\, dy\, dx.$$

Therefore if $V = \mathrm{Supp}\,\alpha$ and $\qquad c_1 = \sum_i \sup |q_i \alpha|.$

we get

$$\mu(\alpha * f) \leqslant c_1 \int_{V \times G} f(x) \,\Xi(yx)\, dy\, dx.$$

Since V is compact, we conclude from [4 (q), Lemma 32] that there exists a number $c_2 \geqslant 0$ such that

$$\mu(\alpha * f) \leqslant c_2 \int_G f \Xi\, dx$$

for all $f \in I_c^+(G)$. It follows without difficulty from condition 2) on μ that the same inequality continues to hold for $f \in I^+(G)$.

Now take $f = \Xi(1 + \sigma)^{-r}$. Then $f \in I^+(G)$. Moreover we know that $\Xi > 0$ everywhere and (see[5 (a), p. 399])

$$\int_K \Xi(xky)\, dk = \Xi(x) \Xi(y) \quad (x, y \in G).$$

Therefore if we choose $\alpha \in I^+(G) \cap I_c^\infty(G)$ such that

$$\int \alpha \Xi\, dx = 1,$$

it is obvious that $\alpha * \Xi = \Xi$. Then

$$(\alpha * f)(x) = \int \alpha(y) \,\Xi(y^{-1} x)\, (1 + \sigma(y^{-1} x))^{-r}\, dy.$$

Now $\qquad\qquad 1 + \sigma(y^{-1} x) \leqslant (1 + \sigma(y))\,(1 + \sigma(x))$

from Lemma 10. Hence $\qquad (\alpha * f)(x) \geqslant c_0^{-1} f(x)$

where $\qquad\qquad\qquad c_0 = \sup_{y \in V} (1 + \sigma(y))^r$

and $V = \mathrm{Supp}\,\alpha$ as before. Therefore

$$\mu(f) \leqslant c_0 \mu(\alpha * f) \leqslant c_0 c_2 \int f \Xi\, dx.$$

Since $\qquad\qquad\qquad \int f \Xi\, dx = \int \Xi^2 (1 + \sigma)^{-r}\, dx < \infty,$

we get the assertion of the theorem.

§ 17. Convergence of certain integrals

We shall now derive some consequences of Theorem 4. Assume that G is acceptable (see [4 (o), § 18]) and let A be a Cartan subgroup of G. We use the notation of [4 (o), § 23].

THEOREM 5. *Fix r as in § 16. Then*

$$\sup_{a \in A'} |\Delta(a)| \int_{G^*} \Xi(a^{x^*})\,(1 + \sigma(a^{x^*}))^{-r}\,dx^* < \infty.$$

Let $\mathfrak{h}$ be the Lie algebra of A. Put

$$\mu(f) = \sup_{a \in A'} |F_f(a)|,\, \nu(f) = \int_A |\Delta_M(a)\,F_f(a)|\,da \quad (f \in I_c^\infty(G))$$

in the notation of [4 (o), Theorem 3]. Then it follows from [4 (q), Theorem 4] that there exists a number $c \geqslant 0$ such that

$$\nu(f) \leqslant c \int_G |f|\,\Xi\,dx \quad (f \in I_c^\infty(G)).$$

Moreover, by [4 (o), Theorem 3], we can select $z_1, \ldots, z_p \in \mathfrak{Z}$ such that

$$\mu(f) \leqslant \sum_{1 \leqslant i \leqslant p} \nu(z_i f) \quad (f \in I_c^\infty(G)).$$

Hence it is obvious that μ satisfies the two conditions of § 16. Moreover, it follows from the elementary properties of an integral that if $\phi \in I^+(G)$ and $a \in A'$, then

$$\int_{G^*} \phi(a^{x^*})\,dx^* = \sup_f \int_{G^*} f(a^{x^*})\,dx^*,$$

where f runs over all elements in $I^+(G) \cap I_c^\infty(G)$ such that $f \leqslant \phi$. Therefore the assertion of Theorem 5 is now an immediate consequence of Theorem 4.

Let γ be a semisimple element in G and G_γ the centralizer of γ in G. Then G_γ is unimodular and therefore the factor space $\bar{G} = G/G_\gamma$ has an invariant measure $d\bar{x}$. Let $x \to \bar{x}$ denote the projection of G on $\bar{G}$ and put

$$\gamma^{\bar{x}} = \gamma^x = x\gamma x^{-1} \quad (x \in G).$$

THEOREM 6. $$\int_{G/G_\gamma} \Xi(\gamma^{\bar{x}})\,(1 + \sigma(\gamma^{\bar{x}}))^{-r}\,d\bar{x} < \infty.$$

Let $\mathfrak{z}$ be the centralizer of γ in $\mathfrak{g}$. Since γ is semisimple, $\mathfrak{z}$ is reductive in $\mathfrak{g}$ and rank $\mathfrak{z} =$ rank $\mathfrak{g}$. Let $\mathfrak{h}$ be a Cartan subalgebra of $\mathfrak{z}$ which is fundamental in $\mathfrak{z}$ (see [4 (n),

§ 11]) and A the Cartan subgroup of G corresponding to $\mathfrak{h}$. Then $\gamma \in A$. As usual let P and $P_{\mathfrak{z}}$ denote the sets of positive roots of $(\mathfrak{g}, \mathfrak{h})$ and $(\mathfrak{z}, \mathfrak{h})$ respectively and $P_{\mathfrak{g}/\mathfrak{z}}$ the complement of $P_{\mathfrak{z}}$ in P. Put

$$\varpi_{\mathfrak{z}} = \prod_{\alpha \in P_{\mathfrak{z}}} H_\alpha$$

in the notation of [4 (n), § 4].

LEMMA 23.(¹) *There exists a number $c \neq 0$ such that*

$$F_f(\gamma; \varpi_{\mathfrak{z}}) = c \int_{G/G_\gamma} f(\gamma^{\bar{x}})\, d\bar{x}$$

for all $f \in C_c^\infty(G)$.

We observe that in view of [4 (o), Lemma 40], the left side has a well-defined meaning. Moreover, since γ is semisimple, the orbit γ^G is closed (see [1, § 10.1]) and therefore [1, § 5.1] the integral on the right is also well defined.

Normalize the invariant measure dy^* on G_γ/A_0 in such a way that $dx^* = d\bar{x}\, dy^*$. Let U be an open, connected neighborhood of 1 in A such that $\det (\mathrm{Ad}\,(a) - 1)_{\mathfrak{g}/\mathfrak{z}} \neq 0$ for $a \in U$. Put $U' = U \cap (\gamma^{-1}A')$. Then an element $a \in U$ lies in U' if and only if $\det (\mathrm{Ad}\,(\gamma a) - 1)_{\mathfrak{z}/\mathfrak{h}} \neq 0$. Moreover, we may assume (see [4 (i), Theorem 1]) that U has the following property. For any compact set Ω in G, there exists a compact subset C of $\bar{G}$ such that $xUx^{-1} \cap \Omega = \emptyset$ $(x \in G)$ unless $\bar{x} \in C$.

Fix $f \in C_c^\infty(G)$ and select C as above corresponding to $\Omega = \mathrm{Supp}\, f$. Then if $a \in U'$,

$$\int_{G^*} f((\gamma a)^{x^*})\, dx^* = \int_C d\bar{x} \int_{G_\gamma^*} f(x(\gamma a)^{y^*} x^{-1})\, dy^*.$$

Let $G_\gamma^{\,0}$ denote the connected component of 1 in G_γ and Z the center of G. Then $ZG_\gamma^{\,0}$ has finite index in G_γ (see [4 (h), Lemma 15]). Let N denote this index and choose y_i $(1 \leqslant i \leqslant N)$ in G_γ such that

$$G_\gamma = \bigcup_{1 \leqslant i \leqslant N} y_i Z G_\gamma^{\,0}.$$

Define
$$g_x(y) = \sum_{1 \leqslant i \leqslant N} f(x \gamma y_i y y_i^{-1} x^{-1}) \quad (y \in G_\gamma^{\,0})$$

for $x \in G$. Then it is clear that

$$\int_{G^*} f((\gamma a)^{x^*})\, dx^* = \int_C d\bar{x} \int_{(G_\gamma^{\,0})_*} g_x(a^{y^*})\, dy^*$$

for $a \in U'$.

Choose an open and convex neighborhood V of zero in $\mathfrak{h}$ such that $\exp V \subset U$ and $|\alpha(H)| < 1$ for $\alpha \in P$ and $H \in V$. Let V' denote the set of all points $H \in V$ where

---

(¹) Cf. Langlands [6, p. 114].

3 − 662900. *Acta mathematica.* 116. Imprimé le 10 juin 1966.

$$\pi_{\delta}(H) = \prod_{\alpha \in P_{\delta}} \alpha(H) \neq 0.$$

Then $\exp V' \subset U'$ and

$$\Delta(\gamma \exp H) = \xi_{\varrho}(\gamma) \prod_{\alpha \in P_{\mathfrak{g}/\delta}} (e^{\alpha(H)/2} - \xi_{\alpha}(\gamma)^{-1} e^{-\alpha(H)/2}) \Delta_{\delta}(H)$$

for $H \in V$. Here

$$\Delta_{\delta}(H) = \prod_{\alpha \in P_{\delta}} (e^{\alpha(H)/2} - e^{-\alpha(H)/2}).$$

Let D denote the differential operator on $\mathfrak{h}$ given by

$$D = \xi_{\varrho}(\gamma) \, \partial(\varpi_{\delta}) \circ \prod_{\alpha \in P_{\mathfrak{g}/\delta}} (e^{\alpha/2} - \xi_{\alpha}(\gamma)^{-1} e^{-\alpha/2}).$$

As usual D_0 denotes the local expression of D at the origin (see [4 (f), p. 90]).

LEMMA 24.
$$D_0 = \xi_{\varrho}(\gamma) \prod_{\alpha \in P_{\mathfrak{g}/\delta}} (1 - \xi_{\alpha}(\gamma)^{-1}) \, \partial(\varpi_{\delta}).$$

Fix $q \in S(\mathfrak{h}_c)$ such that

$$\partial(q) = D_0 - \xi_{\varrho}(\gamma) \prod_{\alpha \in P_{\mathfrak{g}/\delta}} (1 - \xi_{\alpha}(\gamma)^{-1}) \, \partial(\varpi_{\delta}).$$

If p is the number of roots in P_{δ}, it is clear that $d^0 \varpi_{\delta} = p$ and $d^0 q < p$. On the other hand, it is easy to see that $D_0^{s_\alpha} = -D_0$ for any $\alpha \in P_{\delta}$. Therefore ϖ_{δ} divides q in $S(\mathfrak{h}_c)$ (see [4 (f), Lemma 10]) and this shows that $q = 0$.

For any function $g \in C_c^{\infty}(G_{\gamma}{}^0)$, define

$$\phi_g(H) = \Delta_{\delta}(H) \int_{(G_{\gamma}{}^*)^*} f((\exp H)^{y^*}) dy^* \quad (H \in V').$$

Then by [4 (n) Theorem 3], and [4 (i), Lemma 19], there exists a number $c_0 \neq 0$ such that

$$\lim_{H \to 0} \phi_g(H; \partial(\varpi_{\delta})) = c_0 g(1) \quad (H \in V')$$

for every $g \in C_c^{\infty}(G_{\gamma}{}^0)$. Hence it follows from Lemma 24 that

$$F_f(\gamma; \varpi_{\delta}) = c_1 \int_C g_x(1) \, d\bar{x} = c_1 N \int_{G/G_{\gamma}} f(\gamma^{\bar{x}}) \, d\bar{x},$$

where
$$c_1 = c_0 \, \varepsilon_R(\gamma) \, \xi_{\varrho}(\gamma) \prod_{\alpha \in P_{\mathfrak{g}/\delta}} (1 - \xi_{\alpha}(\gamma)^{-1}).$$

This proves Lemma 23.

Now we come to the proof of Theorem 6. Put

$$\mu(f) = \left| \int_{G/G_\gamma} f(\gamma^{\bar{x}}) \, d\bar{x} \right| = \left| c^{-1} F_f(\gamma; \varpi_\mathfrak{h}) \right| \quad (f \in I_c^\infty(G))$$

and define $\nu(f)$ as in the proof of Theorem 5. By [4 (o), Theorem 3], we can choose $z_1, \dots, z_p \in \mathfrak{Z}$ such that

$$\mu(f) \leqslant \sum_{1 \leqslant i \leqslant p} \nu(z_i f) \quad (f \in I_c^\infty(G)).$$

This shows (see the proof of Theorem 5) that μ fulfills the two conditions of § 16. Moreover, it is clear that

$$\int_{G/G_\gamma} \phi(\gamma^{\bar{x}}) \, d\bar{x} = \sup_f \int_{G/G_\gamma} f(\gamma^{\bar{x}}) \, d\bar{x} \quad (\phi \in I^+(G)),$$

where f runs over all functions in $I^+(G) \cap I_c^\infty(G)$ such that $f \leqslant \phi$. Therefore Theorem 6 follows from Theorem 4.

§ 18. The mapping $f \to F_f$

We return to the notation of § 15 and define the function $D_\mathfrak{h}$ on G as in [4 (q), Lemma 35]. Also we recall that S is the set of all continuous seminorms on $C(G)$.

LEMMA 25. *Put*

$$\nu_1(f) = \int_G |f| \, |D_\mathfrak{h}|^{-\frac{1}{2}} \, dx \quad (f \in C(G)).$$

Then $\nu_1 \in S$.

Fix r as in Lemma 11 and put

$$\nu(f) = \sup |f| \, \Xi^{-1} (1 + \sigma)^r \quad (f \in C(G)).$$

Then it is clear that

$$\nu_1(f) \leqslant \nu(f) \int_G \Xi (1 + \sigma)^{-r} |D_\mathfrak{h}|^{-\frac{1}{2}} \, dx$$

and therefore our assertion follows from Lemma 11 and [4 (q), Lemma 35].

Now assume that G is acceptable and let $A = A_\mathfrak{h}$ be the Cartan subgroup of G corresponding to $\mathfrak{h}$. Since K is compact, A has only a finite number of connected components. Define $A'(I)$ as in [4 (o), § 22]. Then the space $C(A'(I))$ is well defined (see § 9). For any $f \in C_c^\infty(G)$, define the function $F_f \in C^\infty(A'(I))$ as in [4 (o), § 22].

LEMMA 26. *Let* $S(A'(I))$ *be the set of all continuous seminorms on* $C(A'(I))$. *Then* $F_f \in C(A'(I))$ *and for a given* $\nu_0 \in S(A'(I))$, *we can choose* $\nu \in S$ *such that*

$$\nu_0(F_f) \leqslant \nu(f) \quad (f \in C_c^\infty(G)).$$

We use induction on dim $\mathfrak{g}$. Let $\mathfrak{c}$ be the center of $\mathfrak{g}$ and first assume that $\mathfrak{h} \cap \mathfrak{p} \not\subset \mathfrak{c}$. Then dim $\mathfrak{m} <$ dim $\mathfrak{g}$ and the induction hypothesis is applicable to M and therefore our assertion follows immediately from Theorem 3 and [4 (o), Lemma 52] (see also [4 (q), § 10]). Hence we may suppose that $\mathfrak{h} \cap \mathfrak{p} = \mathfrak{c}_\mathfrak{p}$ where $\mathfrak{c}_\mathfrak{p} = \mathfrak{c} \cap \mathfrak{p}$ as before. Let us assume further that $\mathfrak{c}_\mathfrak{p} \neq \{0\}$ and put $\mathfrak{g}_1 = \mathfrak{k} + [\mathfrak{k}, \mathfrak{p}]$. Then $\mathfrak{g}$ is the direct sum of $\mathfrak{c}_\mathfrak{p}$ and $\mathfrak{g}_1$ and G is the direct product of the corresponding subgroups $C_\mathfrak{p}$ and G_1. Put[1] $\mathfrak{C}_\mathfrak{p} = \mathfrak{S}(\mathfrak{c}_{\mathfrak{p}c})$ and $\mathfrak{G}_1 = \mathfrak{S}(\mathfrak{g}_{1c})$. Then we may assume, without loss of generality, that

$$v_0(g) = \sup_{h \in A'(I)} (1 + \sigma(h))^r |g(h; \gamma u)| \quad (g \in C(A'(I)))$$

for some $r \geqslant 0$, $\gamma \in \mathfrak{C}_\mathfrak{p}$ and $u \in \mathfrak{S}(\mathfrak{h}_{1c})$ where $\mathfrak{h}_1 = \mathfrak{h} \cap \mathfrak{g}_1$. For any $f \in C(G)$ and $c \in C_\mathfrak{p}$, let f_c denote the function $x \to f(cx; \gamma)$ $(x \in G_1)$ on G_1. Put $A_1'(I) = G_1 \cap A'(I)$ and let F_g $(g \in C_c^\infty(G_1))$ denote the function on $A_1'(I)$ corresponding to [4 (o), § 22]. Then it is obvious that

$$(1 + \sigma(ch))^r |F_f(ch; \gamma u)| \leqslant (1 + \sigma(c))^r (1 + \sigma(h))^r |F_{f_c}(h; u)|$$

for $c \in C_\mathfrak{p}$ and $h \in A_1'(I)$. Since dim $\mathfrak{g}_1 <$ dim $\mathfrak{g}$, we can, by induction hypothesis, choose a continuous seminorm v_1 on $C(G_1)$ such that

$$(1 + \sigma(h))^r |F_g(h; u)| \leqslant v_1(g)$$

for $g \in C_c^\infty(G_1)$ and $h \in A_1'(I)$. Then it follows that

$$v_0(F_f) \leqslant \sup_{c \in C_\mathfrak{p}} (1 + \sigma(c))^r v_1(f_c) \quad (f \in C_c^\infty(G)).$$

Now put $\qquad\qquad\qquad v(f) = \sup_{c \in C_\mathfrak{p}} (1 + \sigma(c))^r v_1(f_c) \quad (f \in C(G)).$

Since $\Xi(cx) = \Xi(x)$ and $\sigma(cx) \geqslant \max(\sigma(c), \sigma(x))$ (see § 7) for $c \in C_\mathfrak{p}$ and $x \in G_1$, it is easy to verify that $v \in S$.

So now we may suppose that $\mathfrak{c}_\mathfrak{p} = \mathfrak{h} \cap \mathfrak{p} = \{0\}$ and therefore $\mathfrak{h} \subset \mathfrak{k}$. Put

$$v_1(f) = \int_G |f| \, |D_\mathfrak{h}|^{-\frac{1}{2}} \, dx \quad (f \in C(G)).$$

Then $v_1 \in S$ by Lemma 25. Therefore since $A \subset K$ in the present case, our assertion follows from [4 (o), Theorem 3]. This completes the proof of Lemma 26.

Since $C_c^\infty(G)$ is dense in $C(G)$ (Theorem 2) and $C(A'(I))$ is complete, it is clear that that $f \to F_f$ can be extended uniquely to a continuous mapping of $C(G)$ into $C(A'(I))$. Thus, for every $f \in C(G)$, we get a function $F_f \in C(A'(I))$.

---

[1] We use here the notation of [4 (m), p. 280].

LEMMA 27. *Let $f \in C(G)$. Then*

$$F_f(a) = \varepsilon_R(a)\,\Delta(a)\int_{G^*} f(a^{x^*})\,dx^* \quad (a \in A')$$

in the notation of [4 (o), § 23].

It is obvious from Theorem 5 that the integral on the right is well defined. Now choose a sequence $f_j \in C_c^\infty(G)$ $(j \geqslant 1)$ such that $f_j \to f$ in $C(G)$ and put $\phi_j = f - f_j$. Then, in view of the definition of F_f, it would be enough to verify that

$$\sup_{A'} |\Delta(a)|\int_{G^*} |\phi_j(a^{x^*})|\,dx^* \to 0.$$

But since $\phi_j \to 0$ in $C(G)$, this is obvious from Theorem 5.

Let B be another Cartan subgroup of G conjugate to A. Fix $x \in G$ such that $B = A^x$. Then the isomorphism $a \to a^x$ defines a linear bijection of $C(A'(I))$ on a subspace $C(B'(I))$ of $C^\infty(B'(I))$. We topologize $C(B'(I))$ so as to make this bijection a homeomorphism. It is easy to verify that this topology is independent of the choice of x.

Now let us drop the condition that $\mathfrak{h} = \theta(\mathfrak{h})$ and define

$$F_f(a) = \varepsilon_R(a)\,\Delta(a)\int_{G^*} f(a^{x^*})\,dx^* \quad (f \in C(G), a \in A'(I)).$$

It follows from Theorem 5 that this integral exists. Since $\mathfrak{h}$ is conjugate to some Cartan subalgebra which is stable under θ, it is obvious from Lemmas 26 and 27 that $f \to F_f$ is a continuous mapping of $C(G)$ into $C(A'(I))$.

LEMMA 28. $$F_f(\gamma; \varpi_{\mathfrak{h}}) = c\int_{G/G_\gamma} f(\gamma^{\bar{x}})\,d\bar{x}$$

for $f \in C(G)$ in the notation of Lemma 23.

It follows from Theorem 6 that the integral on the right is well defined. The rest of the argument is similar to that given above for Lemma 27.

Let us now return to the notation of § 15. If we replace G by M, we get the corresponding mapping $g \to F_g^M$ of $C(M)$ into $C(A_0'(I))$ where $A_0 = A \cap M$ and $A_0'(I) = A_0 \cap A'(I)$. Define $Z_A = Z(A)$ as in [4 (q), § 12]. Then the following result is obvious from [4 (o), Lemma 52] (see also [4 (q), § 10]), Theorem 3 and Lemma 26.

LEMMA 29. *For any $a \in Z_A$, put*

$$g_{f,a}(m) = g_f(am) \quad (m \in M, f \in C(G))$$

in the notation of Theorem 3. Then there exists a number $c>0$ such that

$$F_f(ah) = c\xi_\varrho(a)\, F_{\vartheta_{f,\bullet}}{}^M(h) \quad (a \in Z_A,\, f \in C(G))$$

for $h \in A_0'(I)$.

§ 19.　A criterion for an invariant eigendistribution to be tempered

Let Θ be a distribution on G. Then Θ is said to be $\mathfrak{Z}$-finite, if the space of all distributions of the form $z\Theta$ $(z \in \mathfrak{Z})$ has finite dimension. We recall (see [4 (o), Theorem 2]) that an invariant and $\mathfrak{Z}$-finite distribution is actually a locally summable function which is analytic on the regular set G'.

Define D as in [4 (o), § 28] and let $A_i\,(1 \leqslant i \leqslant r)$ be a complete set of Cartan subgroups of G, no two of which are conjugate in G. As usual put $A_i' = A_i \cap G'$.

THEOREM 7. *Let Θ be an invariant and $\mathfrak{Z}$-finite distribution on G. Then Θ is tempered if and only if there exists a number $s \geqslant 0$ such that*

$$\sup_{a \in A_i'} (1 + \sigma(a))^{-s}\, |D(a)|^{\frac{1}{2}}\, |\Theta(a)| < \infty \quad (1 \leqslant i \leqslant r).$$

Let $\mathfrak{a}_i$ be the Lie algebra of A_i. In view of Lemma 10, we can obviously assume that $\theta(\mathfrak{a}_i) = \mathfrak{a}_i$. Let us now use the notation of [4 (p), Lemma 63]. Then

$$\int_G f\, dx = \sum_{1 \leqslant i \leqslant r} c_i \int_{A_i} \varepsilon_{i,R}\, \mathrm{conj}\, \Delta_i \cdot F_{f,i}\, d_i a \quad (f \in C_c^\infty(G)),$$

where $F_{f,i} = F_f$ and $\varepsilon_{i,R} = \varepsilon_R$ for $A = A_i$ (see [4 (o), § 22]). Now choose $c \geqslant 0$ such that

$$|D(a)|^{\frac{1}{2}}\, |\Theta(a)| \leqslant c(1 + \sigma(a))^s \quad (a \in A_i',\, 1 \leqslant i \leqslant r).$$

Since $|D(a)| = |\Delta_i(a)|^2$ $(a \in A_i)$, it is clear that

$$|\Theta(f)| = \left| \int_G \Theta f\, dx \right| \leqslant c \sum_{1 \leqslant i \leqslant r} c_i \int_{A_i} (1 + \sigma(a))^s\, |F_{f,i}(a)|\, d_i a$$

for $f \in C_c^\infty(G)$. Hence it follows immediately from Lemma 26, that Θ is tempered.

Before proving the converse, we shall derive the following consequence of the theorem.

COROLLARY. *Suppose Θ is tempered. Then, in the above notation,*

$$\Theta(f) = \sum_{1 \leqslant i \leqslant r} c_i \int_{A_i} \varepsilon_{i,R}\, \Phi_i\, F_{f,i}\, d_i a \quad (f \in C(G)),$$

where
$$\Phi_i(a) = \Theta(a) \cdot \operatorname{conj} \Delta_i(a) \quad (a \in A_i').$$

Fix $f \in C(G)$ and choose a variable element $\alpha \in C_c^\infty(G)$, which converges to f in $C(G)$. Then
$$\Theta(f) = \lim_{\alpha \to f} \Theta(\alpha).$$

But
$$\Theta(\alpha) = \int \Theta \alpha \, dx = \sum_{1 \leqslant i \leqslant r} c_i \int_{A_i} \varepsilon_{i,R} \Phi_i F_{\alpha,i} \, d_i a$$

and so our assertion is an immediate consequence of Lemma 26 and the above theorem.

Now in order to prove the second part of Theorem 7, we need some preparation. Define S and S_1 as in Lemma 18.

LEMMA 30. *Let Θ be a tempered and invariant distribution on G. Then Θ is continuous in the topology defined on $C(G)$ by S_1.*

Fix a function $\alpha \in C_c^\infty(G)$, such that $\int \alpha \, dx = 1$ and, for any $f \in C_c^\infty(G)$, put
$$f_0(x) = \int_G \alpha(y) f(x^y) \, dy \quad (x \in G).$$

Then
$$f_0(z_1 x z_2) = \int_G \alpha(y z_1^{-1}) f(y x z_2 z_1 y^{-1}) \, dy \quad (z_1, z_2 \in G).$$

Fix $g', g'' \in \mathfrak{G}$. Then it is clear that we can select $g_1, \ldots, g_p \in \mathfrak{G}$ such that
$$|f_0(g'; x; g'')| \leqslant \sum_{1 \leqslant i \leqslant p} \sup_{y \in \Omega} |f(yxy^{-1}; g_i)| \quad (x \in G)$$

for all $f \in C_c^\infty(G)$. Here $\Omega = \operatorname{Supp} \alpha$. Now fix $m \geqslant 0$. Then by Lemma 10 and [4 (q), Lemma 32], we can choose $c \geqslant 0$ such that
$$\sup_G |f_0(g'; x; g'')| \, \Xi(x)^{-1}(1 + \sigma(x))^m \leqslant c \sum_{1 \leqslant i \leqslant p} \sup_G |f(x; g_i)| \, \Xi(x)^{-1}(1 + \sigma(x))^m.$$

This shows that, for a given $\nu \in S$, there exists a number $c \geqslant 0$ and a finite set $(\nu_1, \ldots, \nu_p)$ of elements in S_1, such that
$$\nu(f_0) \leqslant c \sum_{1 \leqslant i \leqslant p} \nu_i(f) \quad (f \in C_c^\infty(G)).$$

On the other hand, since Θ is tempered, we can choose $\nu \in S$ such that
$$|\Theta(f)| \leqslant \nu(f) \quad (f \in C_c^\infty(G)).$$

Moreover,
$$\Theta(f) = \Theta(f_0),$$
since Θ is invariant. Hence

$$|\Theta(f)| = |\Theta(f_0)| \leqslant \nu(f_0) \leqslant c \sum_{1 \leqslant i \leqslant p} \nu_i(f) \quad (f \in C_c^{\infty}(G))$$

and this proves our assertion.

Now fix a Cartan subalgebra $\mathfrak{h} = \theta(\mathfrak{h})$ of $\mathfrak{g}$ and let A be the corresponding Cartan subgroup of G. Let A_0 be the center of A and $\tilde{A}$ the normalizer of A in G. Put $W_A = \tilde{A}/A_0$ and let $x \to x^*$ denote the natural projection of G on $G^* = G/A_0$. Then W_A is a finite group and a^s and $x^* s$ ($s \in W_A$, $a \in A$, $x^* \in G^*$) are defined as usual (see [4 (o), §20]). If β is any function on A', we denote by β^s the function $a \to \beta(a^{s^{-1}})$ ($a \in A'$, $s \in W_A$).

Put $G_A = (A')^G$ as usual and normalize the measures dx, da and dx^* in accordance with Lemma 91. Fix a function $\alpha^* \in C_c^{\infty}(G^*)$ such that $\alpha^*(x^* s) = \alpha^*(x^*)$ ($x^* \in G^*$, $s \in W_A$) and

$$\int_{G^*} \alpha^* \, dx^* = 1.$$

Then, for any $\beta \in C_c^{\infty}(A')$, we define $f_\beta \in C_c^{\infty}(G_A)$ as follows.

$$f_\beta(a^x) = \alpha^*(x^*) \Delta(a)^{-1} \sum_{s \in W_A} \varepsilon(s) \beta(a^s) \quad (a \in A', \ x \in G).$$

Here $\varepsilon(s)$ has the usual meaning so that $\Delta^s = \varepsilon(s) \Delta$. Then it follows from Lemma 91 that

$$\Theta(f_\beta) = \int_A \Phi \beta \, da \quad (\beta \in C_c^{\infty}(A')),$$

where $\Phi(a) = \Theta(a) \cdot \text{conj } \Delta(a)$ ($a \in A'$).

Let Q be the set of all roots of $(\mathfrak{g}, \mathfrak{h})$. Then for each $\alpha \in Q$, we have the character ξ_α of A (see [4 (o), §18]). Let η_α denote the function $(1 - \xi_\alpha^{-1})^{-1}$ on A' and let $\mathcal{R}$ be the ring of analytic functions on A' generated over $\mathbb{C}$ by 1 and η_α ($\alpha \in Q$). It is obvious that $\mathcal{R}$ is stable under W_A. Moreover, one finds directly by differentiation that

$$H\eta_\alpha = \alpha(H) \eta_\alpha(1 - \eta_\alpha) \quad (\alpha \in Q, \ H \in \mathfrak{h}).$$

This shows that $\mathcal{R}$ is also stable under the differential operators in $\mathfrak{S}(\mathfrak{h}_c)$.

Fix a connected component A^+ of A'. Since K is compact, A' has only a finite number of connected components (see [4 (e), Lemma 9]) and so, in order to complete the proof of Theorem 7, it would be enough to show that there exist numbers c, $m \geqslant 0$ such that

$$|\Phi(a)| \leqslant c(1 + \sigma(a))^m \quad (a \in A^+).$$

Put $\mathfrak{h}_1 = \mathfrak{h} \cap \mathfrak{k}$, $\mathfrak{h}_2 = \mathfrak{h} \cap \mathfrak{p}$, $A_1 = A \cap K$, $A_1^0 = \exp \mathfrak{h}_1$ and $A_2 = \exp \mathfrak{h}_2$. Then A_1^0 is the connected component of 1 in A_1. For any $a \in A$, let a_i denote the component of a in A_i ($i = 1, 2$) so that $a = a_1 a_2$. Let $\mathfrak{h}_2'$ be the set of all $H \in \mathfrak{h}_2$ such that $\alpha(H) \neq 0$ for any root α of $(\mathfrak{g}, \mathfrak{h})$ which is not identically zero on $\mathfrak{h}_2$. Then $\mathfrak{h}_2'$ has only a finite number of connected components. Fix a connected component $\mathfrak{h}_2^+$ of $\mathfrak{h}_2'$ such that A^+ meets $A_1 A_2^+$. (Here $A_2^+ = \exp \mathfrak{h}_2^+$.) It would be sufficient to prove the required inequality for $a \in A^{++} = A^+ \cap (A_1 A_2^+)$.

Fix an element $b \exp H_0$ $(b \in A_1,\ H_0 \in \mathfrak{h}_2^+)$ in A^{++} and let $\mathfrak{z}$ denote the centralizer of bA_1^0 in $\mathfrak{g}$. Let $Q_\mathfrak{z}$ be the set of all roots of $(\mathfrak{z}, \mathfrak{h})$. Then every root in $Q_\mathfrak{z}$ is real (see [4 (n), § 4]) and $\alpha(H_0) \neq 0$ for $\alpha \in Q_\mathfrak{z}$.

Let $\mathfrak{F}$ and $\mathfrak{F}_2$ be the spaces of real linear functions on $(-1)^{\frac{1}{2}}\mathfrak{h}_1 + \mathfrak{h}_2$ and $\mathfrak{h}_2$ respectively. Introduce compatible orders in $\mathfrak{F}$ and $\mathfrak{F}_2$ such that an element $\lambda \in \mathfrak{F}_2$ is positive whenever $\lambda(H_0) > 0$. Let P be the set of all positive roots of $(\mathfrak{g}, \mathfrak{h})$ under this order and put $P_\mathfrak{z} = P \cap Q_\mathfrak{z}$. We may use this order for the definition of Δ and ξ_ϱ (see [4 (o), § 19]). Then

$$\Delta^{-1} = \xi_\varrho^{-1}\eta,$$

where $\eta = \prod_{\alpha \in P} \eta_\alpha \in \mathcal{R}$.

Fix a compact set $C = C^{-1}$ in G such that Supp $\alpha^* \subset C^*$.

LEMMA 31. *For any $g \in \mathfrak{G}$, we can select $u_i \in \mathfrak{S}(\mathfrak{h}_c)$ and $\eta_i \in \mathcal{R}$ $(1 \leq i \leq p)$ such that*

$$\left| f_\beta(a^x; g) \right| \leq \left| \xi_\varrho(a) \right|^{-1} \sum_{1 \leq i \leq p} \sum_{s \in W_A} \left| \eta_i(a) \beta^s(a; u_i) \right|$$

for $\beta \in C_c^\infty(A')$, $a \in A'$ and $x \in C$.

Assuming this lemma, we shall first finish the proof of Theorem 7. By Lemma 10 and [4 (q), Lemma 32], we can choose $c \geq 1$ such that

$$\Xi(y^x) \leq c\Xi(y),\ 1 + \sigma(y^x) \leq c(1 + \sigma(y)) \quad (x \in C,\ y \in G).$$

Since $C = C^{-1}$, this implies that

$$\left| f_\beta(a^x; g) \right| \Xi(a^x)^{-1} (1 + \sigma(a^x))^m$$

$$\leq c^{m+1} \left| f_\beta(a^x; g) \right| \Xi(a)^{-1}(1 + \sigma(a))^m$$

$$\leq c^{m+1} \left| \xi_\varrho(a) \right|^{-1} \Xi(a)^{-1}(1 + \sigma(a))^m \sum_{1 \leq i \leq p} \sum_{s \in W_A} \left| \eta_i(a) \beta^s(a; u_i) \right|$$

for $\beta \in C_c^\infty(A')$, $a \in A^{++}$, $x \in C$ and $m \geq 0$. Put $A^0 = (A^{++})^{W_A}$. If $\beta \in C_c^\infty(A^0)$, it is clear that Supp $f_\beta \subset (A^{++})^C$. Hence

$$\sup_{x \in G} \left| f_\beta(x; g) \right| \Xi(x)^{-1}(1 + \sigma(x))^m$$

$$\leq c^{m+1} \sup_{a \in A^{++}} \left| \xi_\varrho(a) \Xi(a) \right|^{-1}(1 + \sigma(a))^m \sum_{1 \leq i \leq p} \sum_{s \in W_A} \left| \eta_i(a) \beta^s(a; u_i) \right|,$$

for $\beta \in C_c^\infty(A^0)$ and $m \geq 0$.

It is clear that $A^{++} \subset bA_1^0 A_2^+$. Extend $\mathfrak{h}_2$ to a maximal abelian subspace $\mathfrak{a}$ of $\mathfrak{p}$ and define an order on the dual of $\mathfrak{a}$, which is compatible with the one chosen on $\mathfrak{F}_2$. It is clear that

$$|\xi_\varrho(a)| \, \Xi(a) = \Xi(a_2) e^{\varrho(\log a_2)} \quad (a \in A).$$

Since $a_2 \in A_2^+$ for $a \in A^{++}$, it follows easily from [4 (j), Lemma 36] that

$$|\xi_\varrho(a)| \, \Xi(a) \geqslant 1$$

for $a \in A^{++}$. Therefore since $\mathcal{R}$ is stable under W_A, we obtain the following result from Lemma 30.

LEMMA 32. *We can choose* $m \geqslant 0$, $\eta_i \in \mathcal{R}$ *and* $u_i \in \mathfrak{S}(\mathfrak{h}_c)$ $(1 \leqslant i \leqslant p)$ *such that*

$$\left| \int_A \beta \Phi \, da \right| \leqslant \sum_{1 \leqslant i \leqslant p} \sup_{a \in A'} (1 + \sigma(a))^m |\eta_i(a) \beta(a; u_i)|$$

for $\beta \in C_c^\infty(A^0)$.

Moreover, we have the following lemma.

LEMMA 33. *Let* $\alpha \in Q$. *Then*

$$|\eta_\alpha^{\cdot}(a) \alpha(\log a_2)| \leqslant 1 + |\alpha(\log a_2)| \quad (a \in A').$$

Since $|c_1 - c_2| \geqslant ||c_1| - |c_2||$, for two complex numbers c_1, c_2, it is clear that

$$|\eta_\alpha(a) \alpha(\log a_2)| \leqslant |\alpha(\log a_2)| \, |1 - e^{-\alpha(\log a_2)}|^{-1}.$$

On the other hand, $\qquad\qquad t(1 - e^{-t})^{-1} = t + t(e^t - 1)^{-1}$

and $\qquad\qquad\qquad\qquad t(e^t - 1)^{-1} \leqslant 1 \quad (t \geqslant 0)$.

Hence $\qquad\qquad\qquad |t| \, |1 - e^{-t}|^{-1} \leqslant 1 + |t| \quad (t \in \mathbf{R})$

and this implies our assertion.

Fix a non-empty, open and connected subset U of bA_1^0 such that UA_2 meets A^{++} and ξ_α $(\alpha \in P)$ never takes the value 1 on U unless $\alpha \in P_\mathfrak{z}$. Clearly this is possible. Then $UA_2^+ \subset A^{++} \subset bA_1^0 A_2^+ \subset A'(R)$, where $A'(R)$ is defined as in [4 (o), § 19]. Also we know that Φ extends to an analytic function on $A'(R)$ (see [4 (o), Lemma 31] and [4 (p), Lemma 64]). For $\delta \in C_c^\infty(U)$ and $\gamma \in C_c^\infty(\mathfrak{h}_2^+)$, define the function $\delta \times \gamma$ in $C_c^\infty(A^{++})$ by

$$(\delta \times \gamma)(a) = \delta(a_1) \gamma(\log a_2) \quad (a \in A).$$

Let da_1 denote the Haar measure on A_1 and dH the Euclidean measure on $\mathfrak{h}_2$. We normalize them in such a way that

$$da = da_1 \, dH$$

for $a = a_1 \exp H \in A$. Put

$$\Phi_\delta(H) = \int_{A_1} \Phi(a_1 \exp H) \, \delta(a_1) \, da_1 \quad (H \in \mathfrak{h}_2{}^+).$$

Then it is clear that
$$\int_A (\delta \times \gamma) \cdot \Phi \, da = \int_{\mathfrak{h}_2{}^+} \Phi_\delta \gamma \, dH.$$

Now fix $\delta \in C_c^\infty(U)$ and put $V = \mathrm{Supp}\,\delta$. Then V is a compact subset of U. Let Q_I be the set of all imaginary roots (see [4 (n), § 4]) of $(\mathfrak{g}, \mathfrak{h})$. Then if $\alpha \in Q_I$, it follows from the definition of U, that η_α remains bounded on V. Let P' denote the complement of $P_I = P \cap Q_I$ in P. Put

$$q(H) = \prod_{\alpha \in P'} \alpha(H) \quad (H \in \mathfrak{h}_2).$$

Then q is a polynomial function on $\mathfrak{h}_2$, which is not identically zero, and we conclude easily from Lemmas 32 and 33 that there exists an integer $m \geq 0$ such that the distribution

$$T_\delta : \gamma \to \int_{\mathfrak{h}_2{}^+} \Phi_\delta q^m \gamma \, dH \quad (\gamma \in C_c^\infty(\mathfrak{h}_2{}^+))$$

on $\mathfrak{h}_2{}^+$ is tempered.

On the other hand, since $bA_1{}^0 A_2{}^+ \subset A'(R)$ and Φ is analytic on $A'(R)$, we conclude from [4 (e), Theorem 2], [4 (p), Lemma 64], and the $\mathfrak{Z}$-finiteness of Θ, that

$$\Phi(ba_1 \exp H) = \sum_{1 \leq i \leq N} \xi_i(a_1) \sum_{1 \leq j \leq n} p_{ij}(H) \, e^{\lambda_j(H)} \quad (a_1 \in A_1{}^0, \; H \in \mathfrak{h}_2{}^+),$$

where ξ_i $(1 \leq i \leq N)$ are distinct characters of $A_1{}^0$, λ_j $(1 \leq j \leq n)$ distinct linear functions and p_{ij} polynomial functions on $\mathfrak{h}_{2c}$.

LEMMA 34. *Fix j $(1 \leq j \leq n)$. Then $p_{ij} = 0$ $(1 \leq i \leq N)$ unless* [1] *$\Re\lambda_j(H) \leq 0$ for all $H \in \mathfrak{h}_2{}^+$.*

For otherwise suppose that $\Re\lambda_j(H) > 0$ for some $H \in \mathfrak{h}_2{}^+$. Then it follows from [4 (p), Lemma 15] that

$$\sum_{1 \leq i \leq N} p_{ij} \int_{A_1} \xi_i(a_1) \, \delta(b^{-1} a_1) \, da_1 = 0.$$

Since this holds for every $\delta \in C_c^\infty(U)$, we conclude from [4 (h), Lemma 20] that $p_{ij} = 0$ $(1 \leq i \leq N)$.

It is now obvious from Lemma 34 that we can choose numbers c, $m \geq 0$ such that

$$|\Phi(a)| \leq c(1 + \sigma(a))^m \quad (a \in A^{++})$$

and this proves Theorem 7.

---

[1] $\Re c$ denotes the real part of a complex number c.

We have still to prove Lemma 31. Put $\mathfrak{q}=[\mathfrak{g}, \mathfrak{h}]$ and for any $x \in G$, define the linear mapping Γ_x of $\mathfrak{G} \otimes \mathfrak{G}$ into $\mathfrak{G}$ as in [4 (o), § 2]. Then if $a \in A'$, Γ_a defines a bijection of $\mathfrak{S}(\mathfrak{q}_c) \otimes \mathfrak{S}(\mathfrak{h}_c)$ onto $\mathfrak{G}$ (see [4 (o), Lemma 10]). Let γ_a denote the inverse of this mapping.

LEMMA 35. *Fix $g \in \mathfrak{G}$. Then we can choose $q_i \in \mathfrak{S}(\mathfrak{q}_c)$, $u_i \in \mathfrak{S}(\mathfrak{h}_c)$ and $\eta_i \in \mathcal{R}$ $(1 \leqslant i \leqslant p)$ such that*

$$\gamma_a(g) = \sum_{1 \leqslant i \leqslant p} \eta_i(a) (q_i \otimes u_i) \quad (a \in A').$$

We use the notation of [4 (o), § 2] and put $d_X = L_X - R_X$ $(X \in \mathfrak{g}_c)$. Then d_X is a derivation of $\mathfrak{G}$. Define X_α $(\alpha \in Q)$ as in [4 (n), § 4]. Then

$$\sigma_a(X_\alpha) = \xi_\alpha(a)^{-1} L_{X_\alpha} - R_{X_\alpha} = -(1 - \xi_\alpha(a)^{-1}) L_{X_\alpha} + d_{X_\alpha} \quad (a \in A)$$

and therefore $\qquad \eta_\alpha(a) \sigma_a(X_a) = -(L_{X_\alpha} - \eta_\alpha(a) d_{X_\alpha}) \quad (a \in A').$

This shows that

$$(-1)^r \eta_{\alpha_1}(a) \ldots \eta_{a_r}(a) \Gamma_a(X_{\alpha_1} X_{\alpha_2} \ldots X_{\alpha_r} \otimes u) = (L_{X_{\alpha_1}} - \eta_{\alpha_1}(a) d_{X_{\alpha_1}}) \ldots (L_{X_{\alpha_r}} - \eta_{\alpha_r}(a) d_{X_{\alpha_r}}) u$$

for $a \in A'$, $u \in \mathfrak{S}(\mathfrak{h}_c)$ and $\alpha_1, \ldots, \alpha_r \in Q$. The assertion of the lemma now follows by an easy induction on the degree of g.

For any $f \in C^\infty(G)$, put $f(x:a) = f(a^x) (x \in G, a \in A)$.

COROLLARY. *Let $f \in C^\infty(G)$. Then*

$$f(a^x; g^x) = \sum_{1 \leqslant i \leqslant p} \eta_i(a) f(x; q_i : a; u_i) \quad (x \in G, a \in A')$$

in the above notation.

This is obvious from [4 (o), Lemma 4].

We are now ready to prove Lemma 31. Let $g_1, \ldots, g_N$ be a base for the vector space over C spanned by g^x $(x \in G)$. Then

$$g^{x^{-1}} = \sum_{1 \leqslant i \leqslant N} c_i'(x) g_i \quad (x \in G),$$

where c_i' are analyti functions on G. Hence

$$f(a^x; g) = \sum_{1 \leqslant i \leqslant N} c_i'(x) f(a^x; g_i^x) \quad (x \in G, a \in A)$$

for $f \in C^\infty(G)$. Therefore it follows from the corollary of Lemma 35, that we can choose analytic functions c_i on G and $q_i \in \mathfrak{S}(\mathfrak{q}_c)$, $u_i \in \mathfrak{S}(\mathfrak{h}_c)$, $\eta_i \in \mathcal{R}$ $(1 \leqslant i \leqslant p)$ such that

$$f(a^x; g) = \sum_{1 \leqslant i \leqslant p} c_i(x)\, \eta_i(a)\, f(x; q_i : a; u_i) \quad (x \in G,\ a \in A')$$

for any $f \in C^\infty(G)$. Put $\alpha(x) = \alpha^*(x^*)$ and $\alpha_i(x) = c_i(x)\alpha(x; q_i)$ $(x \in G)$. Then it is clear that

$$f_\beta(a^x; g) = \sum_{1 \leqslant i \leqslant p} \alpha_i(x)\, \eta_i(a) \sum_{s \in W_A} \varepsilon(s)\, \beta^s(a; u_i \circ \Delta^{-1}) \quad (x \in C,\ a \in A')$$

for $\beta \in C_c^\infty(A')$. Let $u \to {}^\backprime u$ denote the automorphism of $\mathfrak{S}(\mathfrak{h}_c)$ such that ${}^\backprime H = H - \varrho(H)$ $(H \in \mathfrak{h}_c)$. Then since $\Delta^{-1} = \xi_\varrho^{-1}\eta$, it is clear that

$$u \circ \Delta^{-1} = \xi_\varrho^{-1}\, {}^\backprime u \circ \eta \quad (u \in \mathfrak{S}(\mathfrak{h}_c)),$$

as a differential operator on A'. Therefore since $\mathcal{R}$ is stable under both $\mathfrak{S}(\mathfrak{h}_c)$ and W_A, Lemma 31 follows immediately from the compactness of C.

§ 20. Proof of Theorem 8

Now suppose rank $G = \operatorname{rank} K$ and fix a Cartan subgroup B of G such that $B \subset K$. Define L and Θ_λ $(\lambda \in L)$ as in [4 (q), § 8]. Then it follows from Theorem 7 and [4 (p), Lemma 52] that Θ_λ is tempered. Similarly we conclude from [4 (p), § 29] that

$$f \to \sum_{\lambda \in L} |\varpi(\lambda)\, \Theta_\lambda(f)| \quad (f \in C(G))$$

is a continuous mapping of $C(G)$ into C. Therefore

$$f \to \sum_{\lambda \in L} \varpi(\lambda)\, \Theta_\lambda(f) \quad (f \in C(G))$$

is a tempered distribution.

Now we use the notation of [4 (q), § 14]. It follows from [4 (q), Lemma 27 and Cor. 1 of Lemma 57] that T_A is a tempered distribution on $C(\tilde{A}_R)$. Define

$$\phi_f(h) = F_f(h; \varpi_0) \quad (h \in \tilde{A}_R)$$

for $f \in C(G)$. Then it follows from § 18 that $f \to \phi_f$ is a continuous mapping of $C(G)$ into $C(\tilde{A}_R)$ and therefore

$$f \to T_A(\phi_f) \quad (f \in C(G))$$

is a tempered distribution on G. Hence we obtain the following result from [4 (q), Theorem 3].

THEOREM 8. *Let* $f \in C(G)$. *Then*

$$cf(1) = (-1)^q \sum_{\lambda \in L} \varpi(\lambda)\, \Theta_\lambda(f) + \sum_{2 \leqslant i \leqslant r} T_i(\phi_{f,i})$$

in the notation of [4 (q), *Theorem 3*].

Let $\mathcal{I}(G)$ denote the subspace of all $f \in C(G)$ which are invariant under left and right translations of K. It follows from Lemma 19 that $I_c^\infty(G) = \mathcal{I}(G) \cap C_c^\infty(G)$ is dense in $\mathcal{I}(G)$. Moreover if we take into account Lemma 18 and [4 (j), Theorem 3 and Lemma 36], it follows without difficulty that the space $\mathcal{I}(G)$, including its topology, is the same as $I(G)$ of [4(k), § 12].

COROLLARY. *Let $f \in \mathcal{I}(G)$. Then, if G is not compact,*

$$cf(1) = \sum_{2 \leqslant i \leqslant r} T_i(\phi_{f,i}).$$

This is an immediate consequence of [4 (q), Theorem 6].

§ 21. Proof of an earlier conjecture

We now drop the assumption that rank $G =$ rank K. Let $\mathfrak{a}$ be a maximal abelian subspace of $\mathfrak{p}$. Define A, N, $\mathfrak{n}$ and σ as in [4 (q), § 16] and fix a Haar measure dn on N. Put

$$\Phi_f(a) = e^{\sigma(\log a)} \int_N f(an)\,dn \quad (a \in A)$$

for $f \in \mathcal{I}(G)$. Then we know from Theorem 3 that $f \to \Phi_f$ is a continuous mapping of $\mathcal{I}(G)$ into $C(A)$.

LEMMA 36. *Let f be an element in $\mathcal{I}(G)$ such that $\Phi_f = 0$. Then $f = 0$.*

We shall prove this by induction on dim G. However we first verify the following result.

LEMMA 37. *It is sufficient to show that, under the conditions of Lemma 36, $f(1) = 0$.*

Fix an element $\alpha \in I_c^\infty(G)$ and consider $f_0 = \alpha * f$. Then

$$f_0 = \int_G \alpha(x)\,l(x)f\,dx \in \mathcal{I}(G)$$

from the results of § 10. Let da denote the Haar measure on A. Then we can assume that

$$dx = e^{2\,\sigma(\log a)}\,dk\,da\,dn \quad (x = kan,\ k \in K,\ a \in K,\ n \in N)$$

and therefore

$$f_0(y) = \int_G \alpha(x^{-1})\,f(xy)\,dx = \int e^{2\sigma(\log a)}\,\alpha(n^{-1}a^{-1})\,f(any)\,da\,dn \quad (y \in G).$$

Hence it follows without difficulty that

$$\Phi_{f_0}(a_0) = \int_A \Phi_\alpha(a)\,\Phi_f(a^{-1}a_0)\,da = 0 \quad (a_0 \in A)$$

and so we conclude that
$$\int \alpha(x^{-1})f(x)\,dx = f_0(1) = 0.$$

Since this holds for every $\alpha \in I_c^\infty(G)$, it is obvious that $f = 0$.

We shall now undertake some preparation for the proof of Lemma 36. Let $B = \theta(B)$ be a Cartan subgroup of G and $\mathfrak{b}$ its Lie algebra. For any $f \in C(G)$, define F_f corresponding to B as in § 18. Let $\mathfrak{m}$ be the centralizer of $\mathfrak{b} \cap \mathfrak{p}$ in $\mathfrak{g}$ and M the analytic subgroup of G corresponding to $\mathfrak{m}$. Define the differential operator ϖ on B as in [4, (o), § 22].

LEMMA 38. *The statements of* [4 (o), Lemma 52] *remain true for* $f \in C(G)$. *Moreover, if* $\mathfrak{b}$ *is fundamental in* $\mathfrak{g}$, *there exists a positive number* c *such that*

$$cf(1) = (-1)^q F_f(1; \varpi)$$

for $f \in C(G)$. *Here* $q = \frac{1}{2}(\dim G/K - \operatorname{rank} G + \operatorname{rank} K)$.

The first part follows from Theorem 3, its corollary and Lemma 29. The second is a consequence of [4 (o), § 22], [4 (i), p. 759] and [4 (n), Lemmas 17 and 18].

Since $\mathfrak{a}$ is maximal abelian in $\mathfrak{p}$, we can choose $k \in K$ such that $(\mathfrak{b} \cap \mathfrak{p})^k \subset \mathfrak{a}$. Hence we may assume that $\mathfrak{b} \cap \mathfrak{p} \subset \mathfrak{a}$. Put $\mathfrak{n}_1 = \mathfrak{m} \cap \mathfrak{n}$, $N_1 = M \cap N$, $K_1 = M \cap K$ and

$$\sigma_1(H) = \operatorname{tr} (\operatorname{ad} H)_{\mathfrak{n}_1} \quad (H \in \mathfrak{a}).$$

For any $f \in \mathcal{J}(G)$, define u_f as in Lemma 22 (for $\mathfrak{h} = \mathfrak{b}$). Then it is clear that u_f is bi-invariant under K_1 and therefore $u_f \in \mathcal{J}(M)$. The following lemma is a simple consequence of the definition of u_f.

LEMMA 39. *The Haar measure* dn_1 *on* N_1 *can be so normalized that*

$$\Phi_f(a) = e^{\sigma_1(\log a)} \int_{N_1} u_f(an_1)\,dn_1 \quad (a \in A)$$

for $f \in \mathcal{J}(G)$.

Now we come to the proof of Lemma 36. We may obviously assume that G is not compact. Fix $f \in \mathcal{J}(G)$ such that $\Phi_f = 0$ and first suppose that $\mathfrak{m} \neq \mathfrak{g}$. Then the induction hypothesis is applicable to M and therefore we conclude from Lemma 39 that $u_f = 0$. But then $F_f = 0$ from Lemma 29.

In view of Lemma 37, it would be sufficient to verify that $f(1) = 0$. We now use the notation of the proof of Lemma 26. Let f_1 be the restriction of f on G_1. If $\mathfrak{c}_\mathfrak{p} \neq \{0\}$, our induction hypothesis is applicable to G_1 and therefore $f(1) = f_1(1) = 0$. So we may assume that $c_p = \{0\}$. Choose $\mathfrak{b}$ so that it is fundamental in $\mathfrak{g}$. If rank $\mathfrak{g} >$ rank $\mathfrak{k}$, it is clear that $\mathfrak{b} \cap \mathfrak{p} \neq \{0\}$. Therefore since $c_p = \{0\}$, it follows that $\mathfrak{m} \neq \mathfrak{g}$. But as we have seen above, this implies that $F_f = 0$ and therefore again $f(1) = 0$ from Lemma 38.

So we may now suppose that rank $\mathfrak{k} =$ rank $\mathfrak{g}$. Then $\mathfrak{b} \cap \mathfrak{p} \not\subset c$, if $\mathfrak{b}$ is not fundamental in $\mathfrak{g}$, and therefore $F_f = 0$. Hence it follows immediately from the corollary of Theorem 8 that $f(1) = 0$. This completes the proof of Lemma 36.

In [4 (k), § 16] the proof of the Plancherel formula for G/K was reduced to two conjectures. The first of these has been verified by Gindikin and Karpelevič [3] (see also [5 (b), § 3]) while Lemma 36 proves the second. Hence in particular, [4 (k), Corollary 2, p. 611] holds for all $f \in \mathcal{J}(G)$.

§ 22. Proof of Lemma 40 (first part)

We return to the notation of § 18 except that we write $B = A_{\mathfrak{h}}$. Let dh denote the Haar measure on B. Fix a finite subset F of $\mathcal{E}_K$ and define α_F as in § 12. Put $_FC(G) = \alpha_F * C(G)$. It is obvious that $_FC(G)$ consists exactly of those $f \in C(G)$ for which $\alpha_F * f = f$.

LEMMA 40. *Given $r_0 \geqslant 0$ and $u \in \mathfrak{S}(\mathfrak{h}_c)$, we can choose $r \geqslant 0$ and a finite set of elements $g_1, ..., g_p$ in $\mathfrak{G}$ such that*

$$\int_B (1 + \sigma(h))^{r_0} |F_f(h; u)| \, dh \leqslant \sum_{1 \leqslant i \leqslant p} \int_G |g_i f| \, \Xi (1 + \sigma)^r \, dx$$

for all $f \in {}_FC(G)$.

First suppose $\mathfrak{h} \cap \mathfrak{p} = \{0\}$. Then $B \subset K$ and

$$\int_B |F_f(h; u)| \, dh \leqslant \sup_{h \in B'} |F_f(h; u)|,$$

provided the total measure of B is 1. Let ϕ_B denote the characteristic function of $G_B = (B')^G$. Then it follows from [4 (o), Lemma 41 and Theorem 3] and Lemma 25 that there exist elements $z_1, ..., z_p \in \mathfrak{B}$ such that

$$\sup_{h \in B'} |F_f(h; u)| \leqslant \sum_{1 \leqslant i \leqslant p} \int_G |z_i f| \, \phi_B \, dx \quad (f \in C(G)).$$

Now suppose $g \in {}_FC(G)$. Then $g = \alpha_F * g$ and therefore

$$|g(x)| \leqslant |\alpha_F|_\infty \, g_1(x) \quad (x \in G),$$

where $|\alpha_F|_\infty = \sup |\alpha_F|$ and $\qquad g_1(x) = \int_K |g(kx)| \, dk.$

Therefore
$$\int |g| \phi_B \, dx \leqslant |\alpha_F|_\infty \int g_1 \phi_B \, dx \leqslant c \int |g| \, \Xi \, dx \quad (g \in {}_F C(G))$$

from [4 (q), Theorem 5], where c is a positive number independent of g. Hence

$$\sup_{h \in B'} |F_f(h; u)| \leqslant c \sum_{1 \leqslant i \leqslant p} \int |z_i f| \, \Xi \, dx \quad (f \in {}_F C(G))$$

and this implies our assertion in this case.

Now, in order to prove the lemma in general, we use induction on $\dim G$. Let us keep to the notation of the proof of Lemma 26 and first assume that $c_\mathfrak{p} \neq \{0\}$. We can obviously suppose that $u = \gamma u_1$ where $\gamma \in \mathfrak{C}_\mathfrak{p}$ and $u_1 \in \mathfrak{S}(\mathfrak{h}_{1c})$. For $f \in C(G)$ and $c \in C_\mathfrak{p}$, let f_c denote the function on G_1 given by

$$f_c(x) = (1 + \sigma(c))^{r \bullet} f(cx; \gamma) \quad (x \in G_1).$$

Then $f_c \in C(G_1)$. Moreover, since $K \subset G_1$, it is obvious that $f_c \in {}_F C(G_1)$ if $f \in {}_F C(G)$. Finally

$$(1 + \sigma(ch))^{r \bullet} |F_f(ch; u)| \leqslant (1 + \sigma(h))^{r \bullet} |F_{f_c}(h; u_1)|$$

for $c \in C_\mathfrak{p}$ and $h \in B_1' = B' \cap G_1$. Therefore our assertion follows immediately by applying the induction hypothesis to G_1 and observing that

$$\max (\sigma(c), \sigma(x)) \leqslant \sigma(cx) \quad (c \in C_\mathfrak{p}, x \in G_1).$$

So we may now suppose that $c_\mathfrak{p} = \{0\}$ and $\mathfrak{h} \cap \mathfrak{p} \neq \{0\}$. Then the induction hypothesis is applicable to M.

§ 23. Proof of Lemma 40 (second part)

Let dm denote the Haar measure on M. Define Ξ_M, u_f ($f \in C(G)$) and Z_B as in §§ 14, 15 and Lemma 29 respectively. For $b \in Z_B$ and $f \in C(G)$, put

$$f_b(x) = \int_K f(kbxk^{-1}) \, dk \quad (x \in G).$$

Since $Z_B \subset K$, it is obvious that if $f \in {}_F C(G)$, the same holds for f_b.

LEMMA 41. *Fix $r \geqslant 0$ and $\zeta \in \mathfrak{S}(\mathfrak{m}_c)$. Then we can choose $v_1, \dots, v_p$ in $\mathfrak{S}$ such that*

$$\int_M |\zeta u_{f_b}| \, \Xi_M (1 + \sigma)^r \, dm \leqslant \sum_{1 \leqslant i \leqslant p} \int_G |v_i f| \, \Xi (1 + \sigma)^r \, dx$$

for $f \in {}_F C(G)$ and $b \in Z_B$.

4 – 662900. *Acta mathematica.* 116. Imprimé le 10 juin 1966.

Put $N_2 = N_b$ in the notation of § 14. We have seen in § 15 that $\zeta u_g = u_{\zeta g}$ $(g \in C(G))$ where $`\zeta = d\zeta \circ d^{-1}$. Fix $b \in Z_B$. Then if $f \in {}_F C(G)$, it is obvious that $f' = {}`\zeta f_b$ is also in ${}_F C(G)$ and

$$|f'(x)| \leqslant |\alpha_F|_\infty \int_K |f'(kx)|\, dk = |\alpha_F|_\infty \int_{K \times K} |f(k_1 x k_2^{-1}; {}`\zeta^{k_1})|\, dk_1\, dk_2.$$

Let $\zeta_1, \ldots, \zeta_p$ be a base for the subspace of $\mathfrak{G}$ spanned by $`\zeta^k$ $(k \in K)$. Then

$$`\zeta^k = \sum_{1 \leqslant i \leqslant p} a_i(k)\, \zeta_i \quad (k \in K),$$

where a_i are continuous functions on K. Put

$$c_1 = \max_i \ \sup |a_i|$$

and $c_2 = |\alpha_F|_\infty c_1$. Then it is clear that

$$|f'(x)| \leqslant c_2 \sum_i \int_{K \times K} |f_i(k_1 x k_2)|\, dk_1\, dk_2,$$

where $f_i = \zeta_i f$.

On the other hand, it is clear that

$$\int |\zeta u_{f_b}|\, \Xi_M (1 + \sigma)^r\, dm \leqslant \int_{M \times N_2} |f'(mn)|\, d(m)\, \Xi_M(m)\, (1 + \sigma(m))^r\, dm\, dn.$$

By Lemma 21, there exists a number $c_0 \geqslant 1$ such that

$$1 + \sigma(m) \leqslant c_0 (1 + \sigma(mn)) \quad (m \in M,\ n \in N_2).$$

Put $c_3 = c_0^{\,r} c_2$. Then

$$\int |\zeta u_{f_b}|\, \Xi_M (1 + \sigma)^r\, dm \leqslant c_3 \int f_0(x)\, e^{-\varrho(H_2(x))}\, \Xi_M(\mu(x))\, (1 + \sigma(x))^r\, dx$$

in the notation of § 42, where

$$f_0(x) = \sum_{1 \leqslant i \leqslant p} \int_{K \times K} |f_i(k_1 x k_2)|\, dk_1\, dk_2.$$

But we know from the corollary of Lemma 84 that

$$\int_K e^{-\varrho(H_2(xk))}\, \Xi_M(\mu(xk))\, dk = \Xi(x) \quad (x \in G).$$

Hence it is clear that

$$\int |\zeta u_{f_b}| \, \Xi_M (1+\sigma)^r \, dm \leqslant c_3 \sum_i \int |\zeta_i f| \, \Xi (1+\sigma)^r \, dx$$

and this implies the assertion of the lemma.

We can now finish the proof of Lemma 40. Let F_1 be the set of all irreducible classes of $K_1 = K \cap M$, which occur in the reduction, with respect to K_1, of some element of F. It is clear from Lemma 22 that $u_f \in_{F_1} C(M)$ if $f \in_F C(G)$. Since the induction hypothesis is applicable to M, we can choose $\zeta_1, ..., \zeta_q$ in $\mathfrak{S}(\mathfrak{m}_c)$ and $r \geqslant 0$ such that

$$\int_{B_0} |F_g^M(h; u)| \, (1+\sigma(h))^{r \cdot} dh \leqslant \sum_{1 \leqslant i \leqslant p} \int_M |\zeta_i g| \, \Xi_M (1+\sigma)^r \, dm$$

for $g \in_{F_1} C(M)$. (Here $B_0 = B \cap M$ and F_g^M is defined as in Lemma 29.) Therefore the required result follows immediately from Lemmas 29 and 41.

§ 24. Proof of Theorem 9

Let Θ be an invariant and $\mathfrak{Z}$-finite distribution on G. Fix $\mathfrak{b} \in \mathcal{E}_K$ and let $\Theta_\mathfrak{b}$ denote the corresponding Fourier component of Θ (see [4 (q), § 17]). Then we know from [4 (q), Lemma 33] that $\Theta_\mathfrak{b}$ is an analytic function on G.

THEOREM 9. *Suppose Θ is tempered. Then we can choose c, $m \geqslant 0$ such that*

$$|\Theta_\mathfrak{b}(x)| \leqslant c \Xi(x)(1+\sigma(x))^m \quad (x \in G).$$

It follows from Theorem 1 that we can choose $\beta' \in C_c^\infty(G)$ such that $\Theta_\mathfrak{b} = \Theta_\mathfrak{b} * \beta'$. Put $\beta(x) = \beta'(x^{-1})$. Then

$$\Theta_\mathfrak{b}(f) = \Theta_\mathfrak{b}(f * \beta) \quad (f \in C_c^\infty(G)).$$

On the other hand, from Theorem 7 and its corollary, we can choose numbers c_0, $s \geqslant 0$ such that

$$|\Theta(g)| \leqslant c_0 \sum_{1 \leqslant i \leqslant r} \int_{A_i} (1+\sigma(a))^s |F_{g,i}(a)| \, d_i a \quad (g \in C_c^\infty(G)).$$

Therefore from Lemma 40, there exist elements $v_1, ..., v_p \in \mathfrak{G}$ and $m \geqslant 0$ such that

$$|\Theta_\mathfrak{b}(g)| = |\Theta(g')| \leqslant \sum_{1 \leqslant i \leqslant p} \int |v_i g'| \, \Xi (1+\sigma)^m \, dx \leqslant d(\mathfrak{b})^2 \sum_i \int |v_i g| \, \Xi (1+\sigma)^m \, dx.$$

Here $g' = \alpha_\mathfrak{b} * g$ and we have made use of the fact that

$$\Theta_\mathfrak{b}(g) = \Theta(g * \alpha_\mathfrak{b}) = \Theta(\alpha_\mathfrak{b} * g)$$

which follows from the invariance of Θ. Therefore if we put $\beta_i = v_i\beta$, we get

$$|\Theta_b(f)| = |\Theta_b(f * \beta)| \leqslant \sum_i d(\mathfrak{b})^2 \int |f * \beta_i| \, \Xi(1+\sigma)^m dx \quad (f \in C_c^\infty(G)).$$

Now put $\Omega = \operatorname{Supp} \beta$ and $\qquad c_1 = d(\mathfrak{b})^2 \sum_i \sup |\beta_i|.$

Then $\qquad |\Theta_b(f)| \leqslant c_1 \int_\Omega (1+\sigma(y))^m dy \int_G |f(x)| \, \Xi(xy)(1+\sigma(x))^m dx.$

But we can choose (see [4 (q), Lemma 32]) $c_2 \geqslant 0$ such that

$$\Xi(xy) \leqslant c_2 \Xi(x) \quad (x \in G, \, y \in \Omega).$$

Then $\qquad c = c_1 c_2 \int_\Omega (1+\sigma(y))^m dy < \infty$

and $\qquad |\Theta_b(f)| \leqslant c \int_G |f| \, \Xi(1+\sigma)^m dx$

for $f \in C_c^\infty(G)$. The assertion of Theorem 9 is now obvious.

COROLLARY. *Let $f \in C(G)$. Then*

$$\Theta_b(f) = \int_G \Theta_b f \, dx.$$

Let α be a variable element in $C_c^\infty(G)$ which converges to f in $C(G)$. Then

$$\Theta_b(f) = \lim \Theta_b(\alpha) = \lim \int \Theta_b \alpha \, dx = \int \Theta_b f \, dx$$

from Lemma 11.

§ 25. Application to tempered representations

Let π be a representation of G on a locally convex space $\mathfrak{H}$. We say that an element $\phi \in \mathfrak{H}$ is K-finite (under π), if the space spanned by the vectors $\pi(k)\phi$ $(k \in K)$ has finite dimension. Now suppose $\mathfrak{H}$ is a Hilbert space and π is unitary and irreducible. Let Θ_π denote the character of π (see [4 (b), § 5] and [8]). We say that π is tempered if Θ_π is tempered.

THEOREM 10. *Let π be an irreducible unitary representation of G on a Hilbert space $\mathfrak{H}$, which is tempered. Then there exists a number $m \geqslant 0$ with the following property. For any two K-finite vectors $\phi, \psi \in \mathfrak{H}$, we can choose a constant $c \geqslant 0$ such that*

$$\left|(\phi, \pi(x)\psi)\right| \leqslant c\Xi(x)(1+\sigma(x))^m \quad (x \in G).$$

Let Θ denote the character of π. Then Θ is an invariant eigendistribution of $\mathfrak{Z}$ on G. Define $E_\mathfrak{b}$ ($\mathfrak{b} \in \mathcal{E}_K$) as in § 2 and put $\mathfrak{H}_\mathfrak{b} = E_\mathfrak{b}\mathfrak{H}$. Choose an orthonormal base ψ_i ($i \in J$) for $\mathfrak{H}$ such that every ψ_i lies in $\mathfrak{H}_\mathfrak{b}$ for some $\mathfrak{b}$. Let $J_\mathfrak{b}$ be the set of all i such that $\psi_i \in \mathfrak{H}_\mathfrak{b}$. Then [1]

$$[J_\mathfrak{b}] = \dim \mathfrak{H}_\mathfrak{b} \leqslant Nd(\mathfrak{b})^2 \quad (d \in \mathcal{E}_K),$$

where N is a positive integer independent of $\mathfrak{b}$ (see [4 (b), Theorem 4]). It follows from the definition of Θ that

$$\Theta(f) = \sum_{i \in J} \int f(x)\,(\psi_i,\, \pi(x)\,\psi_i)\,dx \quad (f \in C_c^\infty(G)),$$

and therefore it is clear that

$$\Theta_\mathfrak{b}(x) = \operatorname{tr}\,(E_\mathfrak{b}\pi(x)\,E_\mathfrak{b}) \quad (x \in G)$$

for $\mathfrak{b} \in \mathcal{E}_K$. Now fix $\mathfrak{b}_0 \in \mathcal{E}_K$ such that $\mathfrak{H}_{\mathfrak{b}_0} \neq \{0\}$. Then, by Theorem 9, we can choose numbers c_0, $m \geqslant 0$ such that

$$\left|\Theta_{\mathfrak{b}_0}(x)\right| \leqslant c_0\Xi(x)(1+\sigma(x))^m \quad (x \in G).$$

Let ϕ and ψ be two K-finite elements in $\mathfrak{H}$. For any finite subset F of $\mathcal{E}_K$, put

$$E_F = \sum_{\mathfrak{b} \in F} E_\mathfrak{b}$$

and $\mathfrak{H}_F = E_F\mathfrak{H}$. Also define α_F as in § 6. We can obviously choose F so large that $\mathfrak{b}_0 \in F$ and ϕ, ψ lie in $\mathfrak{H}_F$. Let $\mathcal{L}_F(G)$ be the space of all functions $f \in C_c^\infty(G)$ such that $f = \alpha_F * f * \alpha_F$. Then $\mathcal{L}_F(G)$ is an algebra under convolution and it is clear that $\mathfrak{H}_F$ is stable under $\pi(f)$ for $f \in \mathcal{L}_F(G)$. Let $\pi_F(f)$ denote the restriction of $\pi(f)$ on $\mathfrak{H}_F$. Then π_F is a representation of $\mathcal{L}_F(G)$ on $\mathfrak{H}_F$. We claim that this representation is irreducible. Fix an element $\psi_0 \neq 0$ in $\mathfrak{H}_F$. Since $\mathfrak{H}$ is irreducible under π, elements of the form $\pi(f)\psi_0$ ($f \in C_c^\infty(G)$) are dense in $\mathfrak{H}$. Therefore, since

$$E_F\pi(f)\psi_0 = \pi(\alpha_F * f * \alpha_F)\psi_0$$

and $\dim \mathfrak{H}_F < \infty$, it is clear that $\mathfrak{H}_F = \pi(\mathcal{L}_F(G))\psi_0$ and this shows that π_F is irreducible. Hence, by the Burnside theorem, we can choose $\alpha, \beta \in \mathcal{L}_F(G)$ such that

$$\phi = \pi(\alpha)\psi_i, \ \psi = \pi(\beta)\psi_i \quad (i \in J_{\mathfrak{b}_0}).$$

Therefore $\qquad (\phi, \pi(x)\psi) = (\psi_i, \pi(\check{\alpha})\pi(x)\pi(\beta)\psi_i) \quad (x \in G,\ i \in J_{\mathfrak{b}_0}),$

where $\check{\alpha}(y) = \operatorname{conj} \alpha(y^{-1})$ ($y \in G$). This shows that

---

$$(\phi, \pi(x)\,\psi).\ \dim \mathfrak{H}_{b_\bullet} = \mathrm{tr}\ (E_{b_\bullet}\pi(\dot\alpha)\,\pi(x)\,\pi(\beta)\,E_{b_\bullet}) = \int_{G\times G} \tilde\alpha(y)\,\Theta_{b_\bullet}\,(yxz)\,\beta(z)\,dy\,dz,$$

and the required result now follows immediately from Lemma 10 and [4 (q), Lemma 32], if we observe that $\mathfrak{H}_{b_\bullet} \neq \{0\}$.

Part II. Spherical functions and differential equations

§ 26. Two key lemmas and their first reduction

Let V be a (complex) Hilbert space of finite dimension and $\mu = (\mu_1, \mu_2)$ a continuous and unitary double representation[1] of K on V. Define $c_\mathfrak{p}$ as in § 7.

LEMMA 42. *Let $\phi \neq 0$ be a C^∞ μ-spherical function (see § 11) from G to V such that:*

1) *The space of functions of the form $z\phi$ ($z \in \mathfrak{Z}$) has finite dimension.*
2) *There exist numbers c, $r \geqslant 0$ such that*

$$|\phi(x)| \leqslant c\,\Xi(x)\,(1+\sigma(x))^r \quad (x \in G).$$

Put
$$\|\phi\|_t = \left\{ \int_{\sigma(x)\leqslant t} |\phi(x)|^2 dx \right\}^{\frac{1}{2}} \quad (t \geqslant 0).$$

Then there exists a unique integer $\nu \geqslant 0$ such that

$$0 < \liminf_{t\to\infty} t^{-\nu/2}\|\phi\|_t \leqslant \limsup_{t\to\infty} t^{-\nu/2}\|\phi\|_t < \infty.$$

Moreover, $\nu \geqslant \dim c_\mathfrak{p}$ and, for any $\varepsilon > 0$, we can choose t_0, $\delta > 0$ such that

$$t_1^{-\nu}\{\|\phi\|_{t_2}^2 - \|\phi\|_{t_1}^2\} \leqslant \varepsilon$$

for $t_0 \leqslant t_1 \leqslant t_2 \leqslant (1+\delta)t_1$.

Fix a maximal abelian subspace $\mathfrak{a}_\mathfrak{p}$ of $\mathfrak{p}$ and let $A_\mathfrak{p} = \exp \mathfrak{a}_\mathfrak{p}$ be the corresponding subgroup of G. Introduce an order in the space of (real) linear functions on $\mathfrak{a}_\mathfrak{p}$ and let $\mathfrak{g} = \mathfrak{k} + \mathfrak{a}_\mathfrak{p} + \mathfrak{n}$ be the corresponding Iwasawa decomposition of $\mathfrak{g}$. As usual put

$$\varrho(H) = \tfrac{1}{2}\,\mathrm{tr}\,(\mathrm{ad}\,H)_\mathfrak{n} \quad (H \in \mathfrak{a}_\mathfrak{p})$$

and let $\mathfrak{a}_\mathfrak{p}^+$ be the set of all $H \in \mathfrak{a}_\mathfrak{p}$ where $\alpha(H) \geqslant 0$ for every positive root α of $(\mathfrak{g}, \mathfrak{a}_\mathfrak{p})$. We recall (see § 7) that $\mathfrak{p}$ is a Hilbert space with respect to the norm $\|X\|$ ($X \in \mathfrak{p}$). Consider the set S^+ of all points $H \in \mathfrak{a}_\mathfrak{p}^+$ with $\|H\| = 1$ and put $A_\mathfrak{p}^+ = \exp \mathfrak{a}_\mathfrak{p}^+$.

---

[1] This means that V is a left K-module under μ_1 and a right K-module under μ_2. Moreover, the operations of K on the left, commute with those on the right.

LEMMA 43. *We keep to the notation of Lemma 42. Then the following three conditions are mutually equivalent.*

1) $\nu = 0$.

2) $\lim\limits_{t \to +\infty} e^{t\varrho(H)} \phi(h \exp tH) = 0$ *for* $H \in S^+$ *and* $h \in A_{\mathfrak{p}}{}^+$.

3) $\phi \in C(G) \otimes V$.

It is convenient to prove the above two lemmas together. We shall call ν the index of ϕ. The uniqueness of ν is obvious from its definition. For the rest we use induction on $\dim \mathfrak{g}$. If $\mathfrak{p} = \{0\}$, G is compact and all our assertions are true trivially with $\nu = 0$. So we may suppose that $\dim \mathfrak{p} \geqslant 1$. First assume that $\mathfrak{c}_{\mathfrak{p}} \neq \{0\}$.

Fix an element $H_0 \in \mathfrak{c}_{\mathfrak{p}}$ with $\|H_0\| = 1$ and let $\mathfrak{g}_1 = \mathfrak{k} + \mathfrak{p}_1$ where $\mathfrak{p}_1$ is the orthogonal complement (see § 7) of $\mathbf{R} H_0$ in $\mathfrak{p}$. Then $\mathfrak{g}_1$ is an ideal in $\mathfrak{g}$. Let G_1 be the analytic subgroup of G corresponding to $\mathfrak{g}_1$. Then the mapping $(t, y) \to \exp tH_0 \cdot y$ $(t \in \mathbf{R}, y \in G_1)$ defines an analytic diffeomorphism of $\mathbf{R} \times G_1$ onto G. Put $\phi(t:y) = \phi(\exp tH_0 \cdot y)$. Since H_0 lies in the center of $\mathfrak{g}$, we can, in view of condition 1) of Lemma 42, choose complex numbers c_i $(0 \leqslant i \leqslant m)$ such that $c_0 = 1$ and

$$\sum_{0 \leqslant i \leqslant m} c_i H_0{}^{m-i} \phi = 0.$$

Therefore it is clear that

$$\phi(t:y) = \sum_{1 \leqslant i \leqslant p} e^{(-1)^{\frac{1}{2}} \lambda_i t} \phi_i(t:y) \quad (t \in \mathbf{R}, y \in G_1),$$

where $\lambda_1, \ldots, \lambda_p$ are distinct complex numbers,

$$\phi_i(t:y) = \sum_{0 \leqslant j \leqslant d} t^j \phi_{ij}(y) \quad (1 \leqslant i \leqslant p)$$

and ϕ_{ij} are[1] C^∞ functions from G_1 to V. We may assume that $\phi_i \neq 0$. Then it follows (see [4 (j), § 15]) from condition 2) of Lemma 42 that $\lambda_1, \ldots, \lambda_p$ are all real. Now $K \subset G_1$ and $\dim G_1 < \dim G$. Therefore it is easy to see that, if $\phi_{ij} \neq 0$, Lemmas 42 and 43 are applicable to ϕ_{ij} by induction hypothesis. Let dy denote the Haar measure on G_1 and ν_{ij} the index of the function ϕ_{ij} on G_1. Moreover we put $\nu_{ij} = -\infty$ if $\phi_{ij} = 0$. Now let

$$\nu = 1 + \max_{i, j} (2j + \nu_{ij}).$$

We shall prove that ν is the index of ϕ.

We may obviously assume that $dx = dt\, dy$. Then for $T \geqslant 0$,

---

[1] Here we make use of the fact that the functions $f_{ij}(t) = t^j e^{(-1)^{\frac{1}{2}} \lambda_i t} (1 \leqslant i \leqslant p, j \geqslant 0)$ are linearly independent over $\mathbf{C}$.

$$\|\phi\|_T \leqslant \left\{\int_0^T dt \int_{\sigma(y)\leqslant T} |\phi(t:y)|^2 \, dy\right\}^{\frac{1}{2}} \leqslant \sum_{i,j} \left\{\int_0^T t^{2j} dt \int_{\sigma(y)\leqslant T} |\phi_{ij}(y)|^2 \, dy\right\}^{\frac{1}{2}}$$

by the triangle inequality. Now fix i, j such that $\phi_{ij} \neq 0$. Then by the definition of ν_{ij}, we can choose a number $b_{ij} > 0$ such that

$$\int_{\sigma(y)\leqslant T} |\phi_{ij}(y)|^2 \, dy \leqslant b_{ij}(1+T)^{\nu_{ij}}$$

for all $T \geqslant 0$. Therefore since $\nu \geqslant 1 + 2j + \nu_{ij}$, it is clear that

$$\limsup_{T\to\infty} T^{-\nu/2}\|\phi\|_T < \infty.$$

On the other hand, in order to show that

$$\liminf_{T\to\infty} T^{-\nu/2}\|\phi\|_T > 0,$$

it would be sufficient to obtain the following result.

LEMMA 44. *Fix $\delta > 0$. Then*

$$\liminf_{T\to\infty} T^{-\nu} \int_{\substack{t^2+\sigma(y)^2\leqslant T^2\\ \sigma(y)\leqslant\delta t}} |\phi(t:y)|^2 \, dt\, dy > 0.$$

We may obviously assume that $\delta \leqslant 1$. For $T \geqslant 0$, let $J(T)$ denote the interval $T/2 \leqslant t \leqslant T/\sqrt{2}$ and $G_1(T)$ the set of all points $y \in G_1$ with $\sigma(y) \leqslant T$. If $t \in J(T)$ and $y \in G_1(\delta T/2)$, it is obvious that $t^2 + \sigma(y)^2 \leqslant T^2$ and $\sigma(y) \leqslant \delta t$. Hence it would be enough to show that

$$\liminf_{T\to\infty} T^{-\nu} I(T) > 0,$$

where
$$I(T) = \int_{J(T)} dt \int_{G_1(\delta T/2)} |\phi(t:y)|^2 \, dy.$$

Let $\langle v_1, v_2\rangle$ $(v_1, v_2 \in V)$ denote the scalar product in V and put

$$I_{ij}(T) = \int_{J(T)} dt \int_{G_1(\delta T/2)} e^{(-1)^{\frac{1}{2}}(\lambda_j - \lambda_i)t} \langle \phi_i(t:y), \phi_j(t:y)\rangle \, dy$$

and $I_i(T) = I_{ii}(T)$ $(1 \leqslant i, j \leqslant p)$. Then it is clear that[1]

$$I(T) = \sum_{1\leqslant i\leqslant p} I_i(T) + 2\Re \sum_{1\leqslant i<j\leqslant p} I_{ij}(T).$$

Fix m, n $(0 \leqslant m, n \leqslant d)$ and put

---

[1] $\Re c$ denotes the real part of a complex number c.

$$I_{im,jn}(T) = \int_{J(T)} dt \int_{G_1(\delta T/2)} e^{(-1)^{\frac{1}{2}}(\lambda_j-\lambda_i)t} t^{m+n} \langle \phi_{im}(y), \phi_{jn}(y) \rangle \, dy$$

for $1 \leqslant i, j \leqslant p$. Then if $i \neq j$, there exists a constant $a(im, jn) \geqslant 0$ such that

$$\left| \int_{J(T)} t^{m+n} e^{(-1)^{\frac{1}{2}}(\lambda_j-\lambda_i)t} \, dt \right| \leqslant a(im, jn)(1+T)^{m+n} \quad (T \geqslant 0).$$

(This follows by integrating by parts and using induction on $m+n$.) Hence it is clear that

$$|I_{im,jn}(T)| \leqslant a(im, jn)(1+T)^{m+n} \| \phi_{im} \|_{\delta T/2} \| \phi_{jn} \|_{\delta T/2},$$

where
$$\| \phi_{kl} \|_t^2 = \int_{G_1(t)} |\phi_{kl}(y)|^2 \, dy \quad (t \geqslant 0)$$

for $1 \leqslant k \leqslant p$, $1 \leqslant l \leqslant d$. Therefore if $\phi_{im} \neq 0$, $\phi_{jn} \neq 0$, we get

$$|I_{im,jn}(T)| \leqslant b(im, jn)(1+T)^{m+n+(\nu_{im}+\nu_{jn})/2} \leqslant b(im, jn)(1+T)^{\nu-1} \quad (T \geqslant 0),$$

where $b(im, jn)$ is a positive number independent of T. Therefore it would be enough to verify that

$$\liminf_{T \to \infty} T^{-\nu} \sum_{1 \leqslant i \leqslant p} I_i(T) > 0.$$

Fix i such that $\nu_{ij} + 2j + 1 = \nu$ for some j. Then it would suffice to show that

$$\liminf_{T \to \infty} T^{-\nu} I_i(T) > 0.$$

Let Q be the set of all j $(0 \leqslant j \leqslant d)$ such that $\nu_{ij} + 2j + 1 = \nu$. Then if $j \notin Q$, we have $\nu_{ij} + 2j + 1 < \nu$. Hence

$$T^{-\nu} \int_{J(T)} t^{2j} \, dt \int_{G_1(\delta T/2)} |\phi_{ij}(y)|^2 \, dy \leqslant T^{\nu_{ij}+2j+1-\nu} T^{-\nu_{ij}} \| \phi_{ij} \|_T^2 \to 0$$

as $T \to \infty$. Put

$$\phi_i^0(t:y) = \sum_{j \in Q} t^j \phi_{ij}(y) \quad \text{and} \quad I_i^0(T) = \int_{J(T)} dt \int_{G_1(\delta T/2)} |\phi_i^0(t:y)|^2 \, dy.$$

Then it would obviously be enough to prove that

$$\liminf_{T \to \infty} T^{-\nu} I_i^0(T) > 0.$$

Now fix $\varepsilon (0 < \varepsilon \leqslant \delta/2)$ and put

$$I(\varepsilon:T) = \int_{J(T)} dt \int_{G_1(\varepsilon T)} |\phi_i^0(t:y)|^2 \, dy.$$

Then $I(\varepsilon:T) \leqslant I_i^0(T)$ and so it is sufficient to verify that

$$\liminf_{T \to \infty} T^{-\nu} I(\varepsilon:T) > 0.$$

Put
$$I_{mn}(\varepsilon:T) = \int_{J(T)} dt \int_{G_1(\varepsilon T)} t^{m+n} \langle \phi_{im}(y), \, \phi_{in}(y) \rangle \, dy$$

and
$$I_m(\varepsilon:T) = I_{mm}(\varepsilon:T) \quad (m, n \in Q).$$

Then
$$I(\varepsilon:T) = \sum_{j \in Q} I_j(\varepsilon:T) + 2\Re \sum_{\substack{m, n \in Q \\ m < n}} I_{mn}(\varepsilon:T).$$

Moreover, since $\phi_{ij} \neq 0$ for $j \in Q$, we can, by induction hypothesis, choose positive numbers a, b such that

$$\|\phi_{ij}\|_T \leqslant b(1+T)^{\nu_{ij}/2} \quad (T \geqslant 0, \, j \in Q)$$

and
$$\|\phi_{ij}\|_T \geqslant aT^{\nu_{ij}/2} \quad (j \in Q)$$

for T sufficiently large. Then

$$|I_{mn}(\varepsilon:T)| \leqslant b^2 T^{m+n+1}(1+\varepsilon T)^{(\nu_{im}+\nu_{in})/2} \quad (m, n \in Q).$$

Hence
$$I(\varepsilon:T) \geqslant \sum_{j \in Q} c_j T^{2j+1}(\varepsilon T)^{\nu_{ij}} - 2\sum_{\substack{m, n \in Q \\ m < n}} b^2 T^{m+n+1}(1+\varepsilon T)^{(\nu_{im}+\nu_{in})/2},$$

for large T, where
$$c_j = a^2(2^{-(j+1)} - 2^{-(2j+1)})/(2j+1) > 0.$$

Let $k = \max_{j \in Q} j$. Then $\nu_{ij} - \nu_{ik} = 2(k-j)$ and therefore

$$I(\varepsilon:T) \geqslant T^{\nu} \varepsilon^{\nu_{ik}} \Big\{ \sum_{j \in Q} c_j \varepsilon^{2(k-j)} - 2\sum_{\substack{m, n \in Q \\ m > n}} b^2 \varepsilon^{2k-m-n}(1+(\varepsilon T)^{-1})^{(\nu_{im}+\nu_{in})/2} \Big\}.$$

This shows that

$$\liminf_{T \to \infty} T^{-\nu} I(\varepsilon:T) \geqslant \varepsilon^{\nu_{ik}} \Big\{ c_k + \sum_{j \in Q'} c_j \varepsilon^{2(k-j)} - 2\sum_{\substack{m, n \in Q \\ m > n}} b^2 \varepsilon^{2k-m-n} \Big\},$$

where Q' is the set of all $j \neq k$ in Q. Now $c_k > 0$ and $2k - m - n \geqslant 1$ since $m < n$. Hence

$$\liminf_{T \to \infty} T^{-\nu} I(\varepsilon:T) > 0$$

if ε is sufficiently small. This proves Lemma 44.

If $l = \dim \mathfrak{c}_\mathfrak{p}$, it is obvious thas $\dim(\mathfrak{c}_\mathfrak{p} \cap \mathfrak{g}_1) = l-1$. Hence, by induction hypothesis, $\nu_{ij} \geqslant l-1$ whenever $\phi_{ij} \neq 0$. Therefore

$$\nu = \max_{i,j}(\nu_{ij} + 2j + 1) \geqslant l = \dim \mathfrak{c}_\mathfrak{p}.$$

Now we come to the last assertion of Lemma 42. By the triangle inequality,

$$\{\|\phi\|_{T_2} - \|\phi\|_{T_1}^2\}^{\frac{1}{2}} \leqslant \sum_{1 \leqslant i \leqslant p} \{\|\phi_i\|_{T_2}^2 - \|\phi_i\|_{T_1}^2\}^{\frac{1}{2}}$$

for $T_2 \geqslant T_1 \geqslant 0$. Here

$$\phi_i(\exp tH_0 \cdot y) = \phi_i(t:y) = e^{(-1)^{\frac{1}{2}}\lambda_i t} \sum_{0 \leqslant j \leqslant d} t^j \phi_{ij}(y) \quad (t \in \mathbf{R}, \ y \in G_1)$$

as before. Put $\nu_i = 1 + \max_j (2j + \nu_{ij})$. Then $\nu = \max_i \nu_i$ and so it would clearly be enough to consider the case when $p = 1$. Hence we may assume that

$$\phi(t:y) = e^{(-1)^{\frac{1}{2}}\lambda t} \sum_{0 \leqslant j \leqslant d} t^j \phi_j(y),$$

where λ is real. Then again, by the triangle inequality, we have

$$\{\|\phi\|_{T_2}^2 - \|\phi\|_{T_1}^2\}^{\frac{1}{2}} \leqslant \sum_{0 \leqslant j \leqslant d} \left\{ \int_{T_1^2 \leqslant t^2 + \sigma(y)^2 \leqslant T_2^2} t^{2j} |\phi_j(y)|^2 \, dt \, dy \right\}^{\frac{1}{2}}$$

and so it would be sufficient to consider the case when

$$\phi(t:y) = e^{(-1)^{\frac{1}{2}}\lambda t} t^j \phi_1(y).$$

Then $\nu = 2j + 1 + \nu_1$ where ν_1 is the index of the function ϕ_1 on G_1.

Now suppose $0 \leqslant T_1 \leqslant T_2$. Then

$$\|\phi\|_{T_2}^2 - \|\phi\|_{T_1}^2 = 2 I_1(T_1 : T_2) + 2 I_2(T_1 : T_2),$$

where

$$I_1(T_1 : T_2) = \int_0^{T_1} t^{2j} \, dt \int_{(T_1^2 - t^2) \leqslant \sigma(y)^2 \leqslant (T_2^2 - t^2)} |\phi_1(y)|^2 \, dy$$

and

$$I_2(T_1 : T_2) = \int_{T_1}^{T_2} t^{2j} \, dt \int_{\sigma(y)^2 \leqslant T_2^2 - t^2} |\phi_1(y)|^2 \, dy.$$

Fix $\varepsilon > 0$ and select a small number $\varepsilon_1 > 0$. Then, by induction hypothesis, we can choose $t_0 \geqslant 1$ and $\delta_1 (0 < \delta_1 < 1)$ with the property that

$$\int_{t_1 \leqslant \sigma(y) \leqslant t_2} |\phi_1(y)|^2 \, dy \leqslant \varepsilon_1 t_1^{\nu_1}$$

if $t_0 \leqslant t_1 \leqslant t_2 \leqslant (1+\delta_1) t_1$. Fix a small positive number $\delta (0 < \delta < 1)$ and put $\gamma = 1 - \delta^{\frac{1}{4}}$. Then if $0 \leqslant t \leqslant \gamma T_1$ and $0 < T_1 \leqslant T_2 \leqslant (1+\delta) T_1$, we have

$$(T_2{}^2 - t^2)^{\frac{1}{2}} (T_1{}^2 - t^2)^{-\frac{1}{2}} - 1 \leqslant (T_2{}^2 - t^2)(T_1{}^2 - t^2)^{-1} - 1$$
$$= (T_2{}^2 - T_1{}^2)(T_1{}^2 - t^2)^{-1} \leqslant \{(1+\delta)^2 - 1\}(1-\gamma^2)^{-1} \leqslant 3\delta^{\frac{1}{4}} \leqslant \delta_1,$$

provided δ is sufficiently small. Moreover,

$$(T_1{}^2 - t^2)^{\frac{1}{2}} \geqslant T_1(1-\gamma^2)^{\frac{1}{2}} \geqslant T_1 \delta^{\frac{1}{4}} \geqslant t_0.$$

if $T_1 \geqslant T_0 = \delta^{-\frac{1}{4}} t_0$. Therefore if $T_0 \leqslant T_1 \leqslant T_2 \leqslant (1+\delta) T_1$ and $0 \leqslant t \leqslant \gamma T_1$, we have

$$\int_{(T_1{}^2 - t^2) \leqslant \sigma(y)^2 \leqslant (T_2{}^2 - t^2)} |\phi_1(y)|^2 \, dy \leqslant \varepsilon_1 (T_1{}^2 - t^2)^{\nu_1/2}.$$

But then since $\nu = \nu_1 + 2j + 1$, we conclude that

$$\int_0^{\gamma T_1} t^{2j} \, dt \int_{(T_1{}^2 - t^2) \leqslant \sigma(y)^2 \leqslant (T_2{}^2 - t^2)} |\phi_1(y)|^2 \, dy \leqslant \varepsilon_1 T_1{}^\nu.$$

On the other hand, we can select $b > 0$ such that

$$\int_{G_1(t)} |\phi_1(y)|^2 \, dy \leqslant b(1+t)^{\nu_1}$$

for $t \geqslant 0$. Therefore if $T_0 \leqslant T_1 \leqslant T_2 \leqslant (1+\delta) T_1$, we have

$$\int_{\gamma T_1}^{T_1} t^{2j} \, dt \int_{(T_1{}^2 - t^2) \leqslant \sigma(y)^2 \leqslant (T_2{}^2 - t^2)} |\phi_1(y)|^2 \, dy \leqslant b \int_{\gamma T_1}^{T_1} t^{2j}(1+T_2)^{\nu_1} \, dt \leqslant 3^{\nu_1} b T_1{}^\nu \delta^{\frac{1}{4}}$$

since $1 + T_2 \leqslant 1 + (1+\delta) T_1 \leqslant 3 T_1$. This proves that

$$I_1(T_1 : T_2) \leqslant T_1{}^\nu (3^{\nu_1} b \delta^{\frac{1}{4}} + \varepsilon_1)$$

if $T_0 \leqslant T_1 \leqslant T_2 \leqslant (1+\delta) T_1$. On the other hand, it is obvious that

$$I_2(T_1 : T_2) \leqslant T_2{}^{2j}(T_2 - T_1) b (1+T_2)^{\nu_1} \leqslant 3^\nu b T_1{}^\nu \delta.$$

Therefore

$$\|\phi\|_{T_2}{}^2 - \|\phi\|_{T_1}{}^2 = 2 I_1(T_1 : T_2) + 2 I_2(T_1 : T_2) \leqslant 2(\varepsilon_1 + 3^{\nu_1} b \delta^{\frac{1}{4}} + 3^\nu b \delta) T_1{}^\nu \leqslant \varepsilon T_1{}^\nu,$$

if ε_1 and δ are chosen sufficiently small. This completes the proof of all the statements of Lemma 42 in case $\dim \mathfrak{c}_\mathfrak{p} \geqslant 1$.

Now we come to Lemma 43. Since $\nu \geqslant \dim \mathfrak{c}_\mathfrak{p} \geqslant 1$, condition 1) of Lemma 43 cannot be fulfilled in the present case. Since it is obvious that 3) implies 2), it would be enough to verify that 2) is never satisfied in our case. But since $\varrho(H_0) = 0$, it is clear from [4 (j), § 15] that 2) implies that $\phi = 0$, giving a contradiction. This proves Lemma 43 when $\dim \mathfrak{c}_\mathfrak{p} \geqslant 1$.

We state the following result, which has been proved above, as a lemma for future reference.

LEMMA 45. *Given $\varepsilon > 0$, we can choose $\delta > 0$ and $T_0 > 0$ such that*

$$\int_{T_1{}^2 \leqslant t^2 + \sigma(y)^2 \leqslant T_2{}^2} |\phi(t:y)|^2 dt\, dy \leqslant \varepsilon T_1{}^\nu$$

for $T_0 \leqslant T_1 \leqslant T_2 \leqslant (1+\delta)\, T_1$.

§ 27. The differential equation for Φ

In order to complete the proofs of Lemmas 42 and 43, we may now assume that $\mathfrak{c}_\mathfrak{p} = \{0\}$ and $\mathfrak{p} \neq \{0\}$. Fix an element $H_0 \in S^+$ and let $\mathfrak{m}$ be the centralizer of H_0 in $\mathfrak{g}$ and $\mathfrak{l}$ the centralizer of $\mathfrak{m}$ in $\mathfrak{a}_\mathfrak{p}$. Since $\mathfrak{a}_\mathfrak{p} \subset \mathfrak{m}$ and $\mathfrak{a}_\mathfrak{p}$ is maximal abelian in $\mathfrak{p}$, $\mathfrak{l}$ is also the centralizer of $\mathfrak{m}$ in $\mathfrak{p}$. Moreover, $\dim \mathfrak{m} < \dim \mathfrak{g}$.

Put $\mathfrak{M} = \mathfrak{S}(\mathfrak{m}_c)$ and let $\mathfrak{Z}_\mathfrak{m}$ denote the center of $\mathfrak{M}$ and μ_0 the homomorphism (see [4 (o), § 12]) of $\mathfrak{Z}$ into $\mathfrak{Z}_1 = \mathfrak{Z}_\mathfrak{m}$. Define $\mathfrak{U}_1 = \mathfrak{Z}_1 \mu_0(\mathfrak{U})$ where $\mathfrak{U}$ is the ideal consisting of all $u \in \mathfrak{Z}$ such that $u\phi = 0$. Then it follows from condition 1) of Lemma 42 and [4 (o), Lemma 21] that $\dim \mathfrak{Z}/\mathfrak{U}$ and $\dim \mathfrak{Z}_1/\mathfrak{U}_1$ are both finite. Let $\zeta \to \zeta^*$ denote the natural projection of $\mathfrak{Z}_1$ on $\mathfrak{Z}_1{}^* = \mathfrak{Z}_1/\mathfrak{U}_1$. We regard $\mathfrak{Z}_1{}^*$ as a $\mathfrak{Z}_1$-module in the usual way so that $z\zeta^* = (z\zeta)^*$ $(z, \zeta \in \mathfrak{Z}_1)$. Let $\mathfrak{Z}_1{}^{**}$ be the vector space dual to $\mathfrak{Z}_1{}^*$. Then since $\mathfrak{Z}_1$ is abelian, $\mathfrak{Z}_1{}^{**}$ is also a left $\mathfrak{Z}_1$-module by duality. Put $\mathbf{V} = V \otimes \mathfrak{Z}_1{}^{**}$ and let $\mathbf{\Gamma}$ denote the representation of $\mathfrak{Z}_1$ on $\mathbf{V}$ defined by

$$\mathbf{\Gamma}(z)(v \otimes \zeta^{**}) = v \otimes z\zeta^{**} \quad (z \in \mathfrak{Z}_1,\ v \in V,\ \zeta^{**} \in \mathfrak{Z}_1{}^{**}).$$

Moreover, by making K act trivially on $\mathfrak{Z}_1{}^{**}$, we can regard $\mathbf{V}$ as a double K-module. Note that $\mathbf{\Gamma}(z)$ $(z \in \mathfrak{Z}_1)$ commutes with the operations of K on $\mathbf{V}$.

Fix elements $\eta_1 = 1,\ \eta_2, \ldots, \eta_p$ in $\mathfrak{Z}_1$ such that $\eta_1{}^*, \ldots, \eta_p{}^*$ is a base for $\mathfrak{Z}_1{}^*$. Let $\eta_i{}^{**}$ $(1 \leqslant i \leqslant p)$ be the dual base for $\mathfrak{Z}_1{}^{**}$. We regard $\mathfrak{Z}_1{}^{**}$ as a Hilbert space with $(\eta_1{}^{**}, \ldots, \eta_p{}^{**})$ as an orthonormal base. Then $\mathbf{V}$ also becomes a Hilbert space.

Let Σ be the set of all positive roots of $(\mathfrak{g}, \mathfrak{a}_\mathfrak{p})$, Σ_1 the subset of those $\alpha \in \Sigma$ which vanish identically on $\mathfrak{l}$ and Σ_2 the complement of Σ_1 in Σ. For any $\alpha \in \Sigma$, let $\mathfrak{g}_\alpha$ denote the set of all $X \in \mathfrak{g}$ such that $[H, X] = \alpha(H)X$ for $H \in \mathfrak{a}_\mathfrak{p}$. Put

$$\mathfrak{n}_i = \sum_{\alpha \in \Sigma_i} \mathfrak{g}_\alpha \quad (i = 1, 2)$$

and $\mathfrak{n} = \mathfrak{n}_1 + \mathfrak{n}_2$. Let M be the analytic subgroup of G corresponding to $\mathfrak{m}$. Then M normalizes $\mathfrak{n}_2$. Put

$$d(m) = \left| \det \left(\mathrm{Ad}\,(m) \right)_{\mathfrak{n}_2} \right|^{\frac{1}{2}} \quad (m \in M)$$

and $\zeta' = d^{-1} \zeta \circ d$ for $\zeta \in \mathfrak{Z}_1$. It is easy to verify (see the Appendix, § 45) that $\zeta \to \zeta'$ is an automorphism of $\mathfrak{Z}_1$. Now define

$$\Phi(m) = \sum_{1 \leqslant i \leqslant p} \phi_i(m) \otimes \eta_i^{**} \quad (m \in M),$$

where
$$\phi_i(m) = d(m) \phi(m; \eta_i').$$

Then it is obvious that Φ is a spherical function from M to $\mathbf{V}$ corresponding to the double representation of $K_1 = M \cap K$ on $\mathbf{V}$ which we denote by $\mu = (\mu_1, \mu_2)$.

For any $\zeta \in \mathfrak{Z}_1$, there exist unique complex numbers c_{ij} such that

$$u_i(\zeta) = \zeta \eta_i - \sum_{1 \leqslant j \leqslant p} c_{ij} \eta_j \in \mathfrak{U}_1 \quad (1 \leqslant i \leqslant p).$$

LEMMA 46. *Let $\zeta \in \mathfrak{Z}_1$. Then*

$$\Phi(m; \zeta) - \Gamma(\zeta) \Phi(m) = \sum_{1 \leqslant i \leqslant p} d(m) \phi(m; u_i(\zeta)') \otimes \eta_i^{**}$$

for $m \in M$.

Fix ζ and define c_{ij} as above. Then

$$\Phi(m; \zeta) = \sum_i \phi_i(m; \zeta) \otimes \eta_i^{**} \quad \text{and} \quad \zeta' \eta_i' = \sum_j c_{ij} \eta_j' + u_i',$$

where $u_i = u_i(\zeta)$. Therefore since

$$\phi_i(m; \zeta) = d(m) \phi(m; \zeta' \eta_i'),$$

our assertion is obvious.

Put
$$\Psi_\zeta(m) = \sum_{1 \leqslant i \leqslant p} d(m) \phi(m; u_i(\zeta)') \otimes \eta_i^{**} \quad (\zeta \in \mathfrak{Z}_1, \ m \in M).$$

Then the following corollary is merely a restatement of the above lemma.

COROLLARY 1. $\Phi(m; \zeta) = \Gamma(\zeta) \Phi(m) + \Psi_\zeta(m)$ *for $m \in M$ and $\zeta \in \mathfrak{Z}_1$.*

Now $\mathfrak{l}$ lies in the center of $\mathfrak{m}$. Hence if $H \in \mathfrak{l}$, we conclude from the above corollary that

$$d(e^{-t\Gamma(H)} \Phi(m \exp tH))/dt = e^{-t\Gamma(H)} \Psi_H(m \exp tH)$$

for $t \in \mathbf{R}$ and $m \in M$. Therefore the following result is now obvious.

COROLLARY 2. $\Phi(m \exp TH) = e^{T\Gamma(H)}\Phi(m) + \int_0^T e^{(T-t)\Gamma(H)}\Psi_H^\circ(m \exp tH)\,dt$ for $m \in M$, $H \in \mathfrak{l}$ and $T \in \mathbf{R}$.

§ 28. Some estimates for Φ and Ψ_ζ

Let $\Xi_M = \Xi_1$ be the function on M which corresponds to Ξ when (G, K) is replaced by (M, K_1).

LEMMA 47. *There exist numbers* c_0, $r_0 \geqslant 0$ *such that*

$$d(\exp H)\Xi(\exp H) \leqslant c_0 \Xi_1(\exp H)(1 + \|H\|)^{r_0}$$

for $H \in \mathfrak{a}_\mathfrak{p}^+$.

We shall give a proof of this lemma in § 45.

Put $M^+ = K_1 A_\mathfrak{p}^+ K_1$. Since K_1 lies in the kernel of the homomorphism d, the following corollary is obvious.

COROLLARY. $d(m)\Xi(m) \leqslant c_0 \Xi_1(m)(1 + \sigma(m))^{r_0}$ $(m \in M^+)$.

Fix $r \geqslant 0$ as in condition 2) of Lemma 42.

LEMMA 48. *Given* g_1, $g_2 \in \mathfrak{G}$, *we can choose a number* $c(g_1, g_2) \geqslant 0$ *such that*

$$|\phi(g_1; x; g_2)| \leqslant c(g_1, g_2)\Xi(x)(1 + \sigma(x))^r \quad (x \in G).$$

In view of Lemma 17, it is enough to consider the case $g_1 = 1$. By Theorem 1, we can choose $\alpha \in C_c^\infty(G)$ such that $\phi * \alpha = \phi$. Then $g\phi = \phi * (g\alpha)$ $(g \in \mathfrak{G})$ and so our assertion follows immediately from Lemma 10 and [4 (q), Lemma 32].

Put $r_1 = r + r_0$.

COROLLARY. *For any* $v \in \mathfrak{M}$, *we can choose a number* $c(v) \geqslant 0$ *such that*

$$|\Phi(m; v)| \leqslant c(v)\Xi_1(m)(1 + \sigma(m))^{r_1} \quad (m \in M^+).$$

This is obvious from Lemma 48 and the corollary of Lemma 47.

For $\alpha \in \Sigma$, define $\mathfrak{g}_\alpha$ as in § 27.

LEMMA 49. *Fix* $g \in \mathfrak{G}$, $\alpha \in \Sigma$ *and* $X \in \mathfrak{g}_\alpha$. *Then we can choose* $c_1 \geqslant 0$ *such that*

$$|d(h)\phi(h; \theta(X)g)| \leqslant c_1 e^{-\alpha(\log h)}\Xi_1(h)(1 + \sigma(h))^{r_1}$$

for $h \in A_\mathfrak{p}^+$.

Since $\phi(h; \theta(X)g) = e^{-\alpha(\log h)}\phi(\theta(X); h; g)$, this follows from Lemmas 47 and 48.

COROLLARY. *Fix* $\zeta \in \mathfrak{Z}_1$ *and* $v \in \mathfrak{M}$. *Then we can choose* $c \geqslant 0$ *such that*

$$|\Psi_\zeta(m \exp H; v)| \leqslant c e^{-\beta(H)}\Xi_1(m)(1 + \sigma(m))^{r_1}(1 + \|H\|)^{r_1}$$

for $H \in \mathfrak{l}^+ = \mathfrak{l} \cap \mathfrak{a}_{\mathfrak{p}}^+$ and $m \in M^+$. Here

$$\beta(H) = \min_{\alpha \in \Sigma_2} \alpha(H).$$

We know that (see the Appendix, § 45)

$$z - \mu_0(z)' \in \theta(\mathfrak{n}_2)\,\mathfrak{G}.$$

Fix $g \in \mathfrak{G}$, $u \in \mathfrak{U}$ and put $g_1 = -(u - \mu_0(u)')g \in \theta(\mathfrak{n}_2)\,\mathfrak{G}$. Since $u\phi = 0$, it is clear that

$$\phi(m;\, \mu_0(u)'g) = \phi(m;\, g_1).$$

Define the automorphism $v \to v' = d^{-1}v \circ d$ on $\mathfrak{M}$ as in the Appendix, § 45. Fix v and ζ as above. Then

$$\Psi'_\zeta(m;\, v) = \sum_{1 \leqslant i \leqslant p} d(m)\,\phi(m;\, v'\,u_i(\zeta)') \otimes \eta_i^{**}.$$

Since $u_i(\zeta) \in \mathfrak{U}_1 \subset \mathfrak{Z}_1$, v commutes with $u_i(\zeta)$ and therefore

$$vu_i(\zeta) = u_i(\zeta)v \in \mu_0(\mathfrak{U})\,\mathfrak{M}.$$

Hence, in view of the remark above, we can choose $g_i \in \theta(\mathfrak{n}_2)\,\mathfrak{G}$ such that

$$\phi(m;\, v'\,u_i(\zeta)') = \phi(m;\, g_i) \quad (m \in M).$$

Since $M^+ = K_1 A_{\mathfrak{p}}^{*+} K_1$, our assertion now follows easily from Lemma 49.

§ 29. The function Θ

Let Q be the set of all eigenvalues of $\Gamma(H_0)$. Then

$$\mathbf{V} = \sum_{\lambda \in Q} \mathbf{V}_\lambda,$$

where the sum is direct and $\mathbf{V}_\lambda$ is the subspace consisting of all $v \in \mathbf{V}$ such that $(\Gamma(H_0) - \lambda)^m v = 0$ for some integer $m \geqslant 0$. Let E_λ denote the projection of $\mathbf{V}$ on $\mathbf{V}_\lambda$ corresponding to the above sum. We divide Q into three disjoint sets Q^+, Q^0 and Q^-. An element $\lambda \in Q$ lies in Q^+, Q^0 or Q^- according as[1] $\Re\lambda > 0$, $\Re\lambda = 0$ or $\Re\lambda < 0$. Put

$$E^\pm = \sum_{\lambda \in Q^\pm} E_\lambda, \quad E^0 = \sum_{\lambda \in Q^0} E_\lambda$$

and
$$\mathbf{V}^\pm = E^\pm \mathbf{V}, \quad \mathbf{V}^0 = E^0 \mathbf{V}.$$

Since $\mathfrak{Z}_1$ is abelian, $\Gamma(\zeta)\ (\zeta \in \mathfrak{Z}_1)$ commutes with the projections E_λ.

---

[1] $\Re c$ denotes the real part of a complex number c.

Put $Q' = Q^+ \cup Q^-$ and let [1]

$$4\,\varepsilon_0 = \min_{\lambda \in Q'} |\Re\,\lambda|.$$

Then $\varepsilon_0 > 0$. Put
$$\beta(H) = \min_{\alpha \in \Sigma_1} \alpha(H) \quad (H \in \mathfrak{l})$$

as before, so that $\beta(H_0) > 0$. Fix a number $\varepsilon (0 < \varepsilon \leqslant \varepsilon_0)$ and an open and relatively compact neighborhood Ω of H_0 in $\mathfrak{l}$. By selecting them sufficiently small, we can arrange that $\beta(H_0) \geqslant 5\,\varepsilon$ and[2]

$$\beta(H) \geqslant 4\,\varepsilon, \ |\Gamma(H) - \Gamma(H_0)| < \varepsilon/2$$

for $H \in \Omega$.

LEMMA 50. *We can choose a number $c \geqslant 0$ such that*

$$\left| e^{-t\Gamma(H)} E^+ \right| + \left| e^{t\Gamma(H)} E^- \right| \leqslant c e^{-2\,\varepsilon_0\,t} \quad (t \geqslant 0$$

and
$$\left| e^{t\Gamma(H)} E^0 \right| \leqslant c e^{\varepsilon|t|} \quad (t \in \mathbf{R})$$

for $H \in \Omega$.

Fix $\lambda \in Q$. Then

$$\Gamma(H)\,E_\lambda = \{\lambda + (\Gamma(H) - \Gamma(H_0)) + (\Gamma(H_0) - \lambda)\}\,E_\lambda$$

and therefore
$$\left| e^{t\Gamma(H)} E_\lambda \right| \leqslant \exp\{t\Re\lambda + \varepsilon|t|/2\}\left| e^{t(\Gamma(H_0) - \lambda)} E_\lambda \right|$$

for $H \in \Omega$ and $t \in \mathbf{R}$. Since $(\Gamma(H_0) - \lambda)\,E_\lambda$ is nilpotent, our assertions follow without difficulty.

Put
$$\Phi^\pm(m) = E^\pm\,\Phi(m), \ \Phi^0(m) = E^0\,\Phi(m) \quad (m \in M).$$

Then by Corollary 2 of Lemma 46,

$$\Phi^-(m \exp TH) = e^{T\Gamma(H)}\Phi^-(m) + \int_0^T e^{(T-t)\Gamma(H)} E^- \Psi_H(m \exp tH)\,dt$$

for $H \in \Omega$ and $T \geqslant 0$. Therefore if $v \in \mathfrak{M}$, we conclude from Lemma 50 that

$$\left| \Phi^-(m \exp TH; v) \right| \leqslant c e^{-2\,\varepsilon_0 T}\left| \Phi^-(m; v) \right| + c \int_0^T e^{-2(T-t)\varepsilon_0}\left| \Psi_H(m \exp tH; v) \right| dt.$$

Put $f(t) = \Psi_H(m \exp tH; v)$ for fixed m, H, v. Then

<hr>

[1] We define $\varepsilon_0 = 1$, in case Q' is empty.

[2] If T is a bounded linear operator on a Banach space $\mathfrak{B}$, $|T| = \sup_{|b| \leqslant 1} |Tb|$ $(b \in \mathfrak{B})$, as usual.

5 - 662900. *Acta mathematica.* 116. Imprimé le 10 juin 1966.

$$\int_0^T e^{-2(T-t)\varepsilon_0}\,|f(t)|\,dt \le e^{-T\varepsilon_0}\int_0^{T/2}|f(t)|\,dt + \int_{T/2}^T |f(t)|\,dt$$

$$\le e^{-T\varepsilon_0}\int_0^\infty |f(t)|\,dt + \int_{T/2}^\infty |f(t)|\,dt.$$

On the other hand, since Ψ_H depends linearly on H and Ω is relatively compact, we have the following result from the corollary of Lemma 49.

LEMMA 51. *For a given $v \in \mathfrak{M}$, there exists a number $c(v) \ge 0$ such that*

$$|\Psi_H(m\exp tH;\,v)| \le c(v)\,e^{-3\varepsilon t}\Xi_1(m)\,(1+\sigma(m))^{r_1}$$

for $m \in M^+$, $H \in \Omega$ and $t \ge 0$.

Therefore in view of the corollary of Lemma 48, we obtain the following lemma.

LEMMA 52. *For any $v \in \mathfrak{M}$, we can select a number $c^-(v) \ge 0$ such that*

$$|\Phi^-(m\exp TH;\,v)| \le c^-(v)\,e^{-\varepsilon T}\Xi_1(m)\,(1+\sigma(m))^{r_1}$$

for $m \in M^+$, $H \in \Omega$ and $T \ge 0$.

Now we come to Φ^+. Again, by Corollary 2 of Lemma 46, we have

$$\Phi^+(m;\,v) = e^{-T\Gamma(H)}\Phi^+(m\exp TH;\,v) - \int_0^T e^{-t\Gamma(H)}E^+\Psi_H(m\exp tH;\,v)\,dt.$$

Fix $H \in \Omega$, $m \in M^+$ and let T tend to $+\infty$. Then it is clear from Lemmas 50 and 51 and the corollary of Lemma 48, that

$$\Phi^+(m;\,v) = -\int_0^\infty e^{-t\Gamma(H)}E^+\Psi_H(m\exp tH;\,v)\,dt$$

and therefore

$$\Phi^+(m\exp TH;\,v) = -\int_T^\infty e^{-(t-T)\Gamma(H)}E^+\Psi_H(m\exp tH;\,v)\,dt.$$

In view of Lemmas 50 and 51, this gives the following result.

LEMMA 53. *For any $v \in \mathfrak{M}$, there exists a number $c^+(v) \ge 0$ such that*

$$|\Phi^+(m\exp TH;\,v)| \le c^+(v)\,e^{-\varepsilon T}\Xi_1(m)(1+\sigma(m))^{r_1}$$

for $m \in M^+$, $H \in \Omega$ and $T \ge 0$.

We shall now consider Φ^0. But first we need some preparation.

LEMMA 54. *Let C be a compact subset of M. Then we can choose $T_0 \geqslant 0$ such that $m \exp TH \in M^+$ for $m \in C$, $H \in \Omega$ and $T \geqslant T_0$.*

We may obviously assume that $K_1 C K_1 = C$. Let C_1 be the set of all $h \in C \cap A_\mathfrak{p}$ such that $\alpha(\log h) \geqslant 0$ for all $\alpha \in \Sigma_1$. Then C_1 is also compact and $C = K_1 C_1 K_1$. Now choose $T_0 \geqslant 0$ such that

$$\alpha(\log h) + 4\varepsilon T_0 \geqslant 0$$

for all $h \in C_1$ and $\alpha \in \Sigma_2$. Then it is clear that $C_1 \exp TH \subset A_\mathfrak{p}^+$ and therefore $C \exp TH \subset M^+$ for $T \geqslant T_0$ and $H \in \Omega$.

Now fix $v \in \mathfrak{M}$ and $H \in \Omega$. Then we conclude from Lemmas 50, 51 and 54 that the integral

$$\int_0^\infty \left| e^{-t\Gamma(H)} E^0 \Psi_H(m \exp tH; v) \right| dt$$

converges uniformly for $m \in C$. Put

$$\Theta_H(m) = \Phi^0(m) + \int_0^\infty e^{-t\Gamma(H)} E^0 \Psi_H(m \exp tH) \, dt \quad (m \in M).$$

Then it is clear that Θ_H is a C^∞ function on M and

$$\Theta_H(m; v) = \Phi^0(m; v) + \int_0^\infty e^{-t\Gamma(H)} E^0 \Psi_H(m \exp tH; v) \, dt.$$

Moreover, it follows from Corollary 2 of Lemma 46 that

$$\Theta_H(m; v) = \lim_{T \to +\infty} e^{-T\Gamma(H)} \Phi^0(m \exp TH; v).$$

So, in particular, the following result is obvious.

LEMMA 55. $\Theta_H(m \exp tH) = e^{t\Gamma(H)} \Theta_H(m)$ $(m \in M,\ H \in \Omega,\ t \in \mathbf{R})$.

We now claim that Θ_H is actually independent of H. Fix $H_1, H_2 \in \Omega$ and $m \in M$ and choose $T_0 \geqslant 0$ such that $m \exp tH \in M^+$ for $t \geqslant T_0$ and $H \in \Omega$. Put $m_2 = m \exp T_2 H_2 \, (T_2 \geqslant T_0)$. Then by Corollary 2 of Lemma 46,

$$e^{-\Gamma(T_1 H_1)} \Phi^0(m_2 \exp T_1 H_1) = \Phi^0(m_2) + \int_0^{T_1} e^{-t\Gamma(H_1)} E^0 \Psi_{H_1}(m_2 \exp tH_1) \, dt,$$

and therefore

$$e^{-\Gamma(T_1 H_1 + T_2 H_2)} \Phi^0(m \exp (T_1 H_1 + T_2 H_2)) - e^{-\Gamma(T_2 H_2)} \Phi^0(m \exp T_2 H_2)$$

$$= \int_0^{T_1} e^{-\Gamma(t_1 H_1 + T_2 H_2)} E^0 \Psi_{H_1}(m \exp (t_1 H_1 + T_2 H_2)) \, dt_1$$

for $T_1 \geqslant 0$. But it follows from Lemma 50 and the corollary of Lemma 49 that there exists a number $c \geqslant 0$ such that

$$\left| e^{-\Gamma(t_1 H_1 + T_2 H_2)} E^0 \Psi_{H_1}(m \exp (t_1 H_1 + T_2 H_2)) \right| \leqslant c e^{-s(t_1 + T_2 - T_0)}$$

for $t_1 \geqslant 0$ and $T_2 \geqslant T_0$. Therefore by making T_1, T_2 tend to $+\infty$, we get

$$\Theta_{H_2}(m) = \lim_{T_1, T_2 \to +\infty} e^{-\Gamma(T_1 H_1 + T_2 H_2)} \Phi^0 (m \exp (T_1 H_2 + T_2 H_2)).$$

Since the right side is symmetrical in H_1, H_2, we conclude that $\Theta_{H_1}(m) = \Theta_{H_2}(m)$.

Hence we may now write Θ instead of Θ_H.

LEMMA 56. $\Theta(m \exp H) = e^{\Gamma(H)} \Theta(m)$ $(m \in M, \ H \in \mathfrak{l})$.

Since Ω is open in $\mathfrak{l}$, every $H \in \mathfrak{l}$ can be written in the form $H = \Sigma_{1 \leqslant i \leqslant q} t_i H_i$ $(t_i \in \mathbf{R}, H_i \in \Omega)$. Our assertion therefore follows from Lemma 55.

LEMMA 57. *For any $v \in \mathfrak{M}$, we can choose a number $c^0(v) \geqslant 0$ such that*

$$\left| e^{-t\Gamma(H)} \Phi^0 (m \exp tH; v) - \Theta(m; v) \right| \leqslant c^0(v) \Xi_1(m) (1 + \sigma(m))^{r_1} e^{-2st}$$

for $m \in M^+$, $t \geqslant 0$ and $H \in \Omega$.

Since $\qquad \Theta(m; v) - e^{-T\Gamma(H)} \Phi^0 (m \exp TH; v) = \int_T^\infty e^{-t\Gamma(H)} E^0 \Psi_H(m \exp tH; v) \, dt,$

our assertion follows immediately from Lemmas 50 and 51.

COROLLARY. *For any $v \in \mathfrak{M}$, there exists a number $c(v) \geqslant 0$ such that*

$$\left| \Phi(m \exp tH; v) - \Theta(m \exp tH; v) \right| \leqslant c(v) \Xi_1(m) (1 + \sigma(m))^{r_1} e^{-st}$$

for $m \in M^+$, $H \in \Omega$ and $t \geqslant 0$.

For,

$$\left| \Phi(m \exp tH; v) - \Theta(m \exp tH; v) \right|$$

$$\leqslant \left| \Phi^+(m \exp tH; v) \right| + \left| \Phi^-(m \exp tH; v) \right| + \left| e^{t\Gamma(H)} E^0 \right| \left| e^{-t\Gamma(H)} \Phi^0(m \exp tH; v) - \Theta(m; v) \right|,$$

and so our assertion follows immediately from Lemmas 50, 52, 53 and 57.

LEMMA 58. *Let $k_1, k_2 \in K_1$, $m \in M$ and $\zeta \in \mathfrak{Z}_1$. Then*

$$\Theta(k_1 m k_2) = \mu_1(k_1) \Theta(m) \mu_2(k_2), \ \Theta(m; \zeta) = \Gamma(\zeta) \Theta(m).$$

Moreover, there exist numbers c_2, $r_2 \geqslant 0$ such that

$$|\Theta(m)| \leqslant c_2 \Xi_1(m)(1+\sigma(m))^{r_1} \quad (m \in M).$$

Fix $H \in \Omega$. Since $\Gamma(H)$ commutes with the operations of K on $\mathbf{V}$ (see § 27), the first assertion follows from the relation

$$\Theta(m) = \lim_{t \to +\infty} e^{-t\Gamma(H)} \Phi^0(m \exp tH).$$

Moreover, E^0, $\Gamma(H)$ and $\Gamma(\zeta)$ commute with each other and

$$\Phi(m \exp tH; \zeta) = \Gamma(\zeta)\Phi(m \exp tH) + \Psi_\zeta(m \exp tH)$$

from Corollary 1 of Lemma 46. Hence

$$\Theta(m; \zeta) = \lim_{t \to +\infty} e^{-t\Gamma(H)} \Phi^0(m \exp tH; \zeta) = \Gamma(\zeta)\Theta(m) + \lim_{t \to +\infty} e^{-t\Gamma(H)} E^0 \Psi_\zeta(m \exp tH).$$

But from Lemma 50 and the corollary of Lemma 49, the limit on the right is zero. Therefor $\zeta\Theta = \Gamma(\zeta)\Theta$.

Now put $v = 1$ and $t = 0$ in Lemma 57. Then we conclude from the corollary of Lemma 48 that there exists a number $c_1 \geqslant 0$ such that

$$|\Theta(m)| \leqslant c_1 \Xi_1(m)(1+\sigma(m))^{r_1} \quad (m \in M^+).$$

On the other hand, we can obviously choose a number $\nu_0 \geqslant 0$ such that

$$|\alpha(\log h)| \leqslant \nu_0 \sigma(h) \quad (\alpha \in \Sigma_2, \; h \in A_\mathfrak{p}).$$

Put $\nu = \max(1, \nu_0/4\varepsilon)$. Then it is clear (see the proof of Lemma 54) that m $\exp tH \in M^+$ ($m \in M$, $H \in \Omega$) provided $t \geqslant \nu\sigma(m)$. Now fix $m \in M$ and put $t_0 = \nu\sigma(m)$ and $m_0 = m \exp t_0 H_0$. Then $m_0 \in M^+$ and

$$\Theta(m) = e^{-t_0 \Gamma(H_0)} \Theta(m_0).$$

Therefore $$|\Theta(m)| \leqslant c_1 |e^{-t_0 \Gamma(H_0)} E^0| \Xi_1(m_0)(1+\sigma(m_0))^{r_1}.$$

But $\Xi_1(m_0) = \Xi_1(m)$, $\sigma(m_0) \leqslant \sigma(m) + t_0 = (\nu+1)\sigma(m)$ and $\Gamma(H_0) E^0$ has only pure imaginary eigenvalues. Therefore (see the proof of Lemma 50), the last statement of the lemma follows immediately.

For any linear function λ on $\mathfrak{l}_c$, let $\mathbf{V}(\lambda)$ denote the subspace of all $v \in \mathbf{V}$ such that

$$(\Gamma(H) - \lambda(H))^N v = 0 \quad (H \in \mathfrak{l}_c),$$

where $N = \dim \mathbf{V}$. Let $^0\mathbf{V}$ denote the sum

$$\sum_{\Re\lambda=0} \mathbf{V}(\lambda),$$

where λ runs over those linear functions which take only pure imaginary values on $\mathfrak{l}$.

LEMMA 59. $\Theta(m)\in{}^0\mathbf{V}$ $(m\in M)$.

Since $\Theta(m\exp tH)=e^{t\Gamma(H)}\Theta(m)$ $(H\in\Omega,\ t\in\mathbf{R})$, this is obvious from the last statement of Lemma 58.

§ 30. Application of the induction hypothesis to θ

We recall (see § 7) that $\mathfrak{p}$ is a Hilbert space under the norm $\|X\|$ $(X\in\mathfrak{p})$. Put $\mathfrak{a}_2=\mathfrak{l}$ and let $\mathfrak{a}_1$ be the orthogonal complement of $\mathfrak{a}_2$ in $\mathfrak{a}_\mathfrak{p}$. It is clear that Σ_1 is the set of all positive roots of $(\mathfrak{m},\mathfrak{a}_\mathfrak{p})$ and $\mathfrak{l}$ is the set of those $H\in\mathfrak{a}_\mathfrak{p}$ where $\alpha(H)=0$ $(\alpha\in\Sigma_1)$. Hence $\mathfrak{a}_1\subset[\mathfrak{m},\mathfrak{m}]$. Let $\mathfrak{a}_1{}^+$ be the set of all $H\in\mathfrak{a}_1$ where $\alpha(H)\geqslant0$ $(\alpha\in\Sigma_1)$. Put $A_1{}^+=\exp\mathfrak{a}_1{}^+$ and fix a number $N\geqslant1$ such that if $h_1\in A_1{}^+$ and $t\geqslant N\sigma(h_1)$, then $h_1\exp tH\in A_\mathfrak{p}{}^+$ for $H\in\Omega$. This is possible (see the proof of Lemma 58).

LEMMA 60. *We can choose a number $c'\geqslant0$ such that*

$$\left|\Phi(h_1\exp tH)-\Theta(h_1\exp tH)\right|\leqslant c'e^{-s(t-N\sigma(h_1))}\Xi_1(h_1)(1+\sigma(h_1))^{r_1}$$

for $h_1\in A_1{}^+$, $H\in\Omega$ and $t\geqslant N\sigma(h_1)$.

Put $h_0=h_1\exp t_0H$ where $t_0=N\sigma(h_1)$. Then $\Xi_1(h_1)=\Xi_1(h_0)$ and $\sigma(h_0)\leqslant\sigma(h_1)+N\sigma(h_1)\|H\|$ Since Ω is relatively compact, our assertion follows from the corollary of Lemma 57.

Define $\theta_j(m)\in V$ by

$$\Theta(m)=\sum_{1\leqslant j\leqslant p}\theta_j(m)\otimes\eta_j{}^{**}\quad(m\in M)$$

and put $\theta=\theta_1$.

COROLLARY. *Under the conditions of Lemma 60,*

$$\left|d(h_1\exp tH)\phi(h_1\exp tH)-\theta(h_1\exp tH)\right|\leqslant c'e^{-s(t-N\sigma(h_1))}\Xi_1(h_1)(1+\sigma(h_1))^{r_1}.$$

This is obvious since $\eta_1=1$.

Let $\mathfrak{b}$ be the orthogonal complement of H_0 in $\mathfrak{a}_\mathfrak{p}$ and $\mathfrak{b}^+$ the set of all $H\in\mathfrak{b}$ where $\alpha(H)\geqslant0$ for every $\alpha\in\Sigma_1$. Let E and E_i denote the orthogonal projections of $\mathfrak{a}_\mathfrak{p}$ on $\mathfrak{b}$ and $\mathfrak{a}_i$ $(i=1,2)$ respectively. Fix a number $\delta>0$ and let U be the set of all $H\in S^+$ such that

$$\|EH\|\leqslant\delta\langle H_0,H\rangle,$$

the scalar product being defined as in § 7. We assume that δ is so small that $\delta N\leqslant\tfrac{1}{2}$,

$E_2 U \subset \Omega$ and $\alpha(H) \geqslant 3\varepsilon$ for $\alpha \in \Sigma_2$ and $H \in U$. Clearly this is possible. Then U consists of all elements of the form $cH_0 + H_1$ where $H_1 \in \mathfrak{h}^+$, $c^2 + \|H_1\|^2 = 1$ and $\|H_1\| \leqslant \delta c$ $(c > 0)$, provided δ is chosen sufficiently small. Therefore, in particular, $(1 + \delta^2)^{-\frac{1}{2}} \leqslant c \leqslant 1$. For any $T \geqslant 0$, let $U(T)$ denote the set of all elements tH with $H \in U$ and $0 \leqslant t \leqslant T$.

Put $\varrho(H) = \frac{1}{2} \operatorname{tr}(\operatorname{ad} H)_{\mathfrak{n}}$ and $\varrho_i(H) = \frac{1}{2} \operatorname{tr}(\operatorname{ad} H)_{\mathfrak{n}_i}$ $(H \in \mathfrak{a}_\mathfrak{p}, \ i = 1, 2)$ in the notation of § 27. Then $\varrho = \varrho_1 + \varrho_2$ and we claim that $\varrho_i(H) = \varrho(E_i H)$ $(H \in \mathfrak{a}_\mathfrak{p})$. Since I centralizes $\mathfrak{n}_1 = \mathfrak{m} \cap \mathfrak{n}$, $\varrho_1 = 0$ on $\mathfrak{a}_2 = I$. Moreover $\mathfrak{m}$ normalizes $\mathfrak{n}_2$ and $\mathfrak{a}_1 \subset [\mathfrak{m}, \mathfrak{m}]$. Therefore $\varrho_2 = 0$ on $\mathfrak{a}_1$ and clearly this implies our assertion. We also note that $d(\exp H) = e^{\varrho_2(H)}$ $(H \in \mathfrak{a}_\mathfrak{p})$.

LEMMA 61. *There exists a number $c \geqslant 0$ with the following property. Suppose $\|H_1\| \leqslant \delta t$ $(H_1 \in \mathfrak{h}^+, \ t \geqslant 0)$. Then*

$$\left| e^{\varrho(tH_0 + E_1 H_1)} \phi(\exp(tH_0 + H_1)) - \theta(\exp(tH_0 + H_1)) \right| \leqslant c e^{-\varepsilon t/3} e^{-\varrho_1(H_1)}.$$

If $\alpha \in \Sigma_1$, it is clear that $\alpha(E_1 H_1) = \alpha(H_1) \geqslant 0$. Hence $E_1 H_1 \in \mathfrak{a}_1^+$. Moreover $N \|E_1 H_1\| \leqslant N \|H_1\| \leqslant N \delta t \leqslant t/2$. Therefore it follows without difficulty from the corollary of Lemma 60 that the left side is majorized by

$$c' e^{-\varepsilon t/2} \Xi_1(h_1)(1 + \sigma(h_1))^{r_1},$$

where $h_1 = \exp E_1 H_1$. Therefore our assertion follows by applying [4 (j), Theorem 3] to Ξ_1 and observing that $\sigma(h_1) = \|E_1 H_1\| \leqslant \delta t$.

As in § 7, put

$$D(h) = \prod_{\alpha \in \Sigma} \left\{ e^{\alpha(\log h)} - e^{-\alpha(\log h)} \right\}^{m_\alpha} \quad (h \in A_\mathfrak{p}),$$

where m_α is the multiplicity of α. Then $D(h) \geqslant 0$ for $h \in A_\mathfrak{p}^+$. Put

$$F_U(T) = \int_{\log h \,\in\, U(T)} |\phi(h)|^2 D(h) \, dh \quad (T \geqslant 0),$$

where dh is the Haar measure on A. Also let us recall that $\phi \neq 0$ and it is analytic (see [4 (q), Lemma 33]).

LEMMA 62. *There exists a unique integer $\nu \geqslant 0$ with the following property. We can select numbers a, $b (0 < a \leqslant b < \infty)$ such that*

$$a T^\nu \leqslant F_U(T) \leqslant b T^\nu$$

for all $T \geqslant 1$. $\nu = 0$ if and only if $\theta = 0$.

The uniqueness of ν is obvious from its definition. So we have only to verify its existence.

Let $U_0(T)$ denote the subset of all $H \in \mathfrak{a}_\mathfrak{p}$ of the form $H = tH_0 + H_1$ where $0 \leqslant t \leqslant T$, $H_1 \in \mathfrak{b}^+$ and $\|H_1\| \leqslant \delta t$. Then

$$U(T) \subset U_0(T) \subset U(T(1+\delta^2)^{\frac{1}{2}}).$$

Put
$$F_0(T) = \int_{\log h \, \in \, U_0(T)} |\phi(h)|^2 D(h)\, dh.$$

Then
$$F_0(T(1+\delta^2)^{-\frac{1}{2}}) \leqslant F(T) \leqslant F_0(T),$$

where $F = F_U$. Therefore it would be enough to prove the existence of an integer $\nu \geqslant 0$ such that

$$0 < \liminf_{T \to \infty} T^{-\nu} F_0(T) \leqslant \limsup_{T \to \infty} T^{-\nu} F_0(T) < \infty.$$

For any $c \geqslant 0$, let $\mathfrak{b}^+(c)$ denote the set of all $H \in \mathfrak{b}^+$ with $\|H\| \leqslant c$. Then if $\psi(H) = \phi(\exp H)$ $(H \in \mathfrak{a}_\mathfrak{p})$, we have

$$F_0(T) = \int_0^T dt \int_{\mathfrak{b}^+(\delta t)} |e^{\varrho(tH_0 + E_2 H_1)} \psi(tH_0 + H_1)|^2 D_1(H_1) \prod_{\alpha \in \Sigma_2} (1 - e^{-2\alpha(tH_0 + H_1)})^{m_\alpha}\, dH_1,$$

where
$$D_1(H_1) = \prod_{\alpha \in \Sigma_1} (e^{\alpha(H_1)} - e^{-\alpha(H_1)})^{m_\alpha} \quad (H_1 \in \mathfrak{b})$$

and dH_1 is the (suitably normalized) Euclidean measure on $\mathfrak{b}$. Choose $T_0 \geqslant 0$ so large that $2\alpha(tH_0 + H_1) \geqslant \log 2$ for $\alpha \in \Sigma_2$, $H_1 \in \mathfrak{b}^+(\delta t)$ and $t \geqslant T_0$, and put

$$J(T) = \int_{T_0}^T dt \int_{\mathfrak{b}^+(\delta t)} |e^{\varrho(tH_0 + E_2 H_1)} \psi(tH_0 + H_1)|^2 D_1(H_1)\, dH_1$$

for $T \geqslant T_0$. Then if $q = \dim \mathfrak{n}_2$, it is clear that

$$2^{-q} J(T) \leqslant F_0(T) - F_0(T_0) \leqslant J(T) \quad (T \geqslant T_0).$$

Also put $\psi_\infty(H) = \theta(\exp H)$ $(H \in \mathfrak{a}_\mathfrak{p})$ and

$$J_\infty(T) = \int_{T_0}^T dt \int_{\mathfrak{b}^+(\delta t)} |\psi_\infty(tH_0 + H_1)|^2 D_1(H_1)\, dH_1.$$

Then, by the triangle inequality, we have

$$|J(T)^{\frac{1}{2}} - J_\infty(T)^{\frac{1}{2}}|^2 \leqslant \int_{T_0}^T dt \int_{\mathfrak{b}^+(\delta t)} |e^{\varrho(tH_0 + E_2 H_1)} \psi(tH_0 + H_1) - \psi_\infty(tH_0 + H_1)|^2 D_1(H_1)\, dH_1.$$

Now apply Lemma 61 and observe that

$$e^{-2\varrho_1(H_1)} D_1(H_1) \leqslant 1 \quad (H_1 \in \mathfrak{b}^+).$$

Then it follows easily that

$$|J(T)^{\frac{1}{2}} - J_\infty(T)^{\frac{1}{2}}| \leqslant c \quad (T \geqslant T_0),$$

where c is a fixed positive number.

Now first suppose $\theta = 0$. Then $\psi_\infty = 0$ and therefore $J_\infty(T) = 0$. Hence $J(T() \leqslant c^2$ for all $T \geqslant T_0$ and it is clear that we can take $\nu = 0$.

So let us now suppose that $\theta \neq 0$. Then it is obvious from Lemma 58 that all the assumptions of Lemma 42 are fulfilled if we replace (G, K, ϕ) by (M, K_1, θ). Since $\dim \mathfrak{m} < \dim \mathfrak{g}$ and $H_0 \in \mathfrak{l} \subset \mathfrak{m} \cap \mathfrak{p}$, it follows from the induction hypothesis that the index ν_∞ of θ (on M) is positive. Hence it follows from Lemma 44 and [4 (d), Lemma 38] that

$$0 < \liminf_{T \to \infty} T^{-\nu_\infty} J_\infty(T) \leqslant \limsup_{T \to \infty} T^{-\nu_\infty} J_\infty(T) < \infty.$$

Since $\nu_\infty \geqslant 1$, it is clear that similar inequalities hold if we replace $J_\infty(T)$ by $J(T)$. Hence we can take $\nu = \nu_\infty$.

LEMMA 63. *Define ν as in Lemma 62. Then for any $\varepsilon_1 > 0$, we can choose $T_0 > 0$ and $\delta_1 > 0$ such that*
$$T_1^{-\nu}\{F_U(T_2) - F_U(T_1)\} \leqslant \varepsilon_1$$
for $T_0 \leqslant T_1 \leqslant T_2 \leqslant (1 + \delta_1) T_1$.

Put $F = F_U$ and let $0 \leqslant T_1 \leqslant T_2$. Then

$$F(T_2) - F(T_1) \leqslant \int_{\mathfrak{a}(T_1, T_2)} |e^{\varrho(tH_0 + E_2 H_1)} \psi(tH_0 + H_1)|^2 D_1(H_1)\, dt\, dH_1,$$

where $\mathfrak{a}(T_1, T_2)$ is the set of all points in $\mathfrak{a}_\mathfrak{p}$ of the form $tH_0 + H_1$ with $t \geqslant 0$, $H_1 \in \mathfrak{b}^+(\delta t)$ and $T_1^2 \leqslant t^2 + \|H_1\|^2 \leqslant T_2^2$. Therefore

$$\{F(T_2) - F(T_1)\}^{\frac{1}{2}} \leqslant I_1(T_1, T_2)^{\frac{1}{2}} + I_2(T_1, T_2)^{\frac{1}{2}},$$

where
$$I_1(T_1, T_2) = \int_{\mathfrak{a}(T_1, T_2)} |\psi_\infty(tH_0 + H_1)|^2 D_1(H_1)\, dt\, dH_1$$

and

$$I_2(T_1, T_2) = \int_{\mathfrak{a}(T_1, T_2)} |e^{\varrho(tH_0 + E_2 H_1)} \psi(tH_0 + H_1) - \psi_\infty(tH_0 + H_1)|^2 D_1(H_1)\, dt\, dH_1.$$

If we apply Lemma 61 and observe, as before, that $e^{-2\varrho_1(H_1)} D_1(H_1) \leqslant 1$ for $H_1 \in \mathfrak{b}^+$, it follows easily that

$$I_2(T_1, T_2) \leqslant c_2 \, e^{-\varepsilon' T_1} \quad (T_2 \geqslant T_1 \geqslant 0,$$

where c_2 is a fixed positive number and $\varepsilon' = (1+\delta^2)^{-\frac{1}{2}} \cdot \varepsilon/2$.

Now first suppose $\theta = 0$. Then $\nu = 0$ and $I_1(T_1, T_2) = 0$. Hence

$$F(T_2) - F(T_1) \leqslant I_2(T_1, T_2) \leqslant c_2 \, e^{-\varepsilon' T_1}$$

and the statement of Lemma 63 follows immediately. So let us assume that $\theta \neq 0$. Then $\nu = \nu_\infty \geqslant 1$, as we have seen above, and therefore the required assertion is a consequence of Lemma 45.

§ 31. Completion of the proofs of Lemmas 42 and 43

We shall now finish the proofs of Lemmas 42 and 43. For any open subset U of S^+ define

$$F_U(T) = \int_{\log h \,\in\, U(T)} |\phi(h)|^2 D(h) \, dh \quad (T \geqslant 0),$$

where $U(T)$ is the set of all elements in $\mathfrak{a}_\mathfrak{p}^+$ of the form tH $(0 \leqslant t \leqslant T, H \in U)$. Since S^+ is compact, we can choose open sets U_i in S^+ and integers $\nu_i \geqslant 0$ $(1 \leqslant i \leqslant q)$ such that $S^+ = \bigcup_{1 \leqslant i \leqslant q} U_i$ and the statements of Lemmas 62 and 63 hold for (U_i, ν_i) in place of (U, ν). Put $F_i = F_{U_i}$ and $\nu = \max_i \nu_i$. It is clear (see [4 (d), Lemma 38] that

$$\max_i F_i(T) \leqslant \|\phi\|_T^2 \leqslant \sum_{1 \leqslant i \leqslant q} F_i(T),$$

if the measure dh is suitably normalized. So it is obvious that

$$0 < \liminf_{T \to \infty} T^{-\nu/2} \|\phi\|_T \leqslant \limsup_{T \to \infty} T^{-\nu/2} \|\phi\|_T < \infty.$$

Moreover, $\qquad \|\phi\|_{T_2}^2 - \|\phi\|_{T_1}^2 \leqslant \sum_{1 \leqslant i \leqslant q} \{F_i(T_2) - F_i(T_1)\} \quad (0 \leqslant T_1 \leqslant T_2).$

Therefore the last assertion of Lemma 42 follows immediately from Lemma 63.

Now we come to the proof of Lemma 43. First assume that $\nu = 0$ and fix $H_0 \in S^+$. Then $\nu_i = 0$ $(1 \leqslant i \leqslant q)$ in the above proof. We may suppose that $H_0 \in U_1$. Define U as in § 30 for H_0 and let $A_\mathfrak{p}(U)$ denote the set of all $h \in A_\mathfrak{p}$ of the form $h = \exp tH$ $(t \geqslant 0, H \in U)$. We may obviously assume that $U \subset U_1$. Then it follows from Lemma 62 that $\theta = 0$ and therefore

$$|e^{\varrho(tH)} \phi(\exp tH)| \leqslant c e^{-\varepsilon' t} \quad (H \in U, t \geqslant 0)$$

from Lemma 61 where $\varepsilon' = (1+\delta^2)^{-\frac{1}{2}} \cdot \varepsilon/3$. Hence it is clear (see [4 (j), Lemma 36]) that

$$\sup_{h \in A_{\mathfrak{p}}(U)} |\phi(h)| \, \Xi(h)^{-1} (1 + \sigma(h))^r < \infty$$

for every $r \geqslant 0$. Since $G = K A_{\mathfrak{p}}^+ K$ and S^+ is compact, this means that

$$\sup_{x \in G} |\phi(x)| \, \Xi(x)^{-1} (1 + \sigma(x))^r < \infty$$

for any $r \geqslant 0$. But then it follows easily (see the proof of Lemma 48) that $\phi \in C(G) \otimes V$. Thus 1) implies 3) in Lemma 43. In view of Lemma 11 and [4 (j), Theorem 3], it is obvious that 3) implies both 1) and 2). Hence it remains to prove that 2) implies 1).

So suppose 2) holds. Fix $H_0 \in S^+$ and use the notation of § 30. Then it follows from Lemma 61 that

$$\lim_{t \to +\infty} \theta(\exp(tH_0 + H_1)) = 0 \quad (H_1 \in \mathfrak{b}^+).$$

Now fix $H_1 \in \mathfrak{b}^+$ and put $f(t) = \theta \, (\exp(tH_0 + H_1)) \; (t \in \mathbf{R}).)$. We have seen in § 29 that

$$\Theta(\exp(tH_0 + H_1)) = e^{t\Gamma(H_0)} \Theta(\exp H_1)$$

and all eigenvalues of $\Gamma(H_0) E^0$ are pure imaginary. Hence it is clear that

$$f(t) = \sum_{1 \leqslant i \leqslant r} p_i(t) \, e^{(-1)^{\frac{1}{2}} \lambda_i t} \quad (t \in \mathbf{R}),$$

where $\lambda_1, ..., \lambda_r$ are distinct real numbers and p_i are polynomial functions from C to V. Since $f(t) \to 0$ as $t \to +\infty$, we conclude (see [4 (j), § 15]) that $f = 0$. Since $\mathfrak{b}^+ + \mathbf{R} H_0$ is open in $\mathfrak{a}_{\mathfrak{p}}$ and θ is analytic (see [4 (q), Lemma 33]), it follows that $\theta = 0$ and therefore $\nu = 0$ in Lemma 62. This being true for every $H_0 \in S^+$, we conclude (see the proof of Lemma 42 given above) that the index of ϕ is zero. This shows that 2) implies 1) and so the proof of Lemma 43 is now complete.

Part III. Applications to harmonic analysis on G

§ 32. Lemma 64 and its consequences

Let $\mathfrak{a} = \theta(\mathfrak{a})$ be a Cartan subalgebra of $\mathfrak{g}$ and A the corresponding Cartan subgroup of G. For $f \in C(G)$, define $F_f \in C(A'(I))$ as in § 18. Let J be the algebra of all invariants of W $(\mathfrak{g}/\mathfrak{a})$ in $\mathfrak{A} = \mathfrak{S}(\mathfrak{a}_c)$. Then we have a canonical isomorphism γ of $\mathfrak{Z}$ onto J (see [4 (e), Lemma 19]). Moreover, since $C_c^\infty(G)$ is dense in $C(G)$, it follows (see § 18 and [4 (o), § 22]) that

$$F_{zf} = \gamma(z) F_f \quad (z \in \mathfrak{Z}, \, f \in C(G)).$$

Let T be a distribution on G. We recall that T is said to be $\mathfrak{Z}$-finite, if the space of distributions of the form zT $(z \in \mathfrak{Z})$ has finite dimension. In particular, we can speak of a locally summable function being $\mathfrak{Z}$-finite.

LEMMA 64. *Let f be a $\mathfrak{Z}$-finite function in $C(G)$. Then $F_f = 0$ unless $\mathfrak{a} \cap \mathfrak{p} = \{0\}$.*

Put $\mathfrak{a}_1 = \mathfrak{a} \cap \mathfrak{k}$, $\mathfrak{a}_2 = \mathfrak{a} \cap \mathfrak{p}$, $A_1 = A \cap K$ and $A_2 = \exp \mathfrak{a}_2$ and suppose that $\mathfrak{a}_2 \neq \{0\}$. Fix a point $a \in A'(I)$ and let $a = a_1 a_2$ $(a_i \in A_i,\ i = 1, 2)$. Select an open and connected neighborhood $\mathfrak{a}_1{}^0$ of zero in $\mathfrak{a}_1$ such that

$$a_1 \exp \mathfrak{a}_1{}^0 \cdot A_2 \subset A'(I).$$

This is clearly possible. Put $\mathfrak{a}^0 = \mathfrak{a}_1{}^0 + \mathfrak{a}_2$ and $A^0 = a_1 \exp \mathfrak{a}^0$.

Let $\mathfrak{U}$ be the set of all $u \in \mathfrak{Z}$ such that $uf = 0$. Then $\mathfrak{U}$ is an ideal in $\mathfrak{Z}$ of finite codimension. Since $\mathfrak{A}$ is a finite module over J, it follows that $\mathfrak{B} = \mathfrak{A}\gamma(\mathfrak{U})$ has finite codimension in $\mathfrak{A}$. Moreover $\gamma(u) F_f = F_{uf} = 0$ $(u \in \mathfrak{U})$ and therefore $vF_f = 0$ for $v \in \mathfrak{B}$. Since $\mathfrak{a}^0$ is connected, we conclude from [4 (e), p. 131] that

$$F_f(a_1 \exp H) = \sum_{1 \leqslant i \leqslant r} p_i(H) e^{\lambda_i(H)} \quad (H \in \mathfrak{a}^0),$$

where λ_i are linear functions and p_i plynomial functions on $\mathfrak{a}_c$ $(1 \leqslant i \leqslant r)$. Put

$$g(H) = F_f(a_1 \exp H) \quad (H \in \mathfrak{a}_2).$$

Since $F_f \in C(A'(I))$, it is clear that

$$\sup_{\mathfrak{a}_2} |g(H)| (1 + \|H\|)^m < \infty$$

for any $m \geqslant 0$. Since $\mathfrak{a}_2 \neq \{0\}$, we conclude from [4 (j), § 15] that $g = 0$. This shows that $F_f = 0$ on $a_1 A_2$ and therefore $F_f(a) = 0$.

COROLLARY 1. *Suppose $f \neq 0$ in the above lemma. Then* rank $G =$ rank K.

For let us otherwise assume that rank $G >$ rank K. Choose $\mathfrak{a}$ so that it is fundamental in $\mathfrak{g}$. Then $\mathfrak{a} \cap \mathfrak{p} \neq \{0\}$ and therefore $F_f = 0$. But then it follows from Lemma 38 that $f(1) = 0$. Now fix $x \in G$ and put $f_x = r(x)f$ in the notation of § 10. Then f_x is also a $\mathfrak{Z}$-finite function in $C(G)$ and $f(x) = f_x(1) = 0$, from the above proof. This shows that $f = 0$, giving a contradiction.

COROLLARY 2. *Suppose* rank $G =$ rank K *and f is a $\mathfrak{Z}$-finite function in $C(G)$. Then*

$$cf(1) = (-1)^q \sum_{\lambda \in L} \varpi(\lambda)\, \Theta_\lambda(f)$$

in the notation of Theorem 8.

It is clear from Lemma 64 that $\phi_{f,i}=0$ $(2 \leqslant i \leqslant r)$ in Theorem 8. Hence our assertion is obvious.

An element γ of G is called elliptic if it is contained in some compact Cartan subgroup.

COROLLARY 3. *Suppose γ is a semisimple element of G which is not elliptic. Then if f is a $\mathfrak{Z}$-finite function in $C(G)$, we can conclude that*

$$\int_{G/G_\gamma} f(\gamma^{\bar{x}}) \, d\bar{x} = 0$$

in the notation of Lemma 28.

Define $\mathfrak{z}$ and $\mathfrak{h}$ as in Theorem 6. By replacing γ by γ^y for some $y \in G$, we may assume that $\theta(\mathfrak{h})=\mathfrak{h}$. Let A be the Cartan subgroup of G corresponding to $\mathfrak{h}$. Then $\gamma \in A$. Since γ is not elliptic, we conclude that A is not compact and therefore $\mathfrak{h} \cap \mathfrak{p} \neq \{0\}$. Our assertion now follows from Lemmas 28 and 64.

§ 33. Proof of a conjecture of Selberg

Let f be a measurable function on G. We say that[1] f is K-finite, if the left and right translates of f, under K, span a finite-dimensional space.

LEMMA 65. *Let f be a function in $C^\infty(G) \cap L_2(G)$ which is K-finite as well as $\mathfrak{Z}$-finite. Then there exist numbers c, $r \geqslant 0$ such that*

$$|f(x)| \leqslant c\Xi(x)(1+\sigma(x))^r \quad (x \in G).$$

We shall give a proof of this lemma in § 38.

COROLLARY 1. *Let f be a function in $L_2(G)$ which is both K-finite and $\mathfrak{Z}$-finite. Then $f \in C(G)$.*

We regard $\mathfrak{H}=L_2(K \times K)$ as a Hilbert space in the usual way and define a unitary double representation $\mu^0 = (\mu_1, \mu_2)$ of K on $\mathfrak{H}$ as follows (cf. § 12). If $u \in \mathfrak{H}$ and $k \in K$, then the functions $u_1 = \mu_1(k)u$ and $u_2 = u\mu_2(k)$ are given by

$$u_1(k_1, k_2) = u_1(k^{-1}k, k_2), \; u_2(k_1, k_2) = u(k_1, k_2 k^{-1}) \quad (k_1, k_2 \in K).$$

It follows from [4 (q), Lemma 33] that f is analytic. For any $x \in G$, let $\phi(x)$ denote the function

---

$$(k_1, k_2) \to f(k_1^{-1} x k_2^{-1}) \quad (k_1, k_2 \in K)$$

in $\mathfrak{H}$. Then it is clear that $\phi(k_1 x k_2) = \mu_1(k_1)\phi(x)\mu_2(k_2)$ $(k_1, k_2 \in K)$. Let V be the subspace of $\mathfrak{H}$ spanned by $\phi(x)$ for all $x \in G$. Then V is stable under μ^0 and $\dim V < \infty$ since f is K-finite. Let μ denote the restriction of μ^0 on V. Then ϕ is a C^∞ μ-spherical function from G to V and it is clear from Lemma 65 that Lemmas 42 and 43 are applicable to ϕ, provided $f \neq 0$. Since $f \in L_2(G)$, we conclude that the index of ϕ is zero and therefore $\phi \in C(G) \otimes V$ from Lemma 43. Obviously this implies that $f \in C(G)$.

COROLLARY 2. *Suppose* $f \neq 0$ *in Corollary* 1. *Then* rank $G =$ rank K.

This follows immediately from Corollary 1 of Lemma 64.

If we combine Corollary 3 of Lemma 64 with Corollary 1 of Lemma 65, we get the following theorem.

THEOREM 11. *Suppose* γ *is a semisimple element of* G, *which is not elliptic, and* f *a function in* $L_2(G)$, *which is both K-finite and $\mathfrak{Z}$-finite. Then* $f \in C(G)$ *and, in the notation of Lemma* 28, *the integral*

$$\int_{G/G_\gamma} f(\gamma^{\bar{x}})\, d\bar{x}$$

exists and its value is zero.

This theorem represents, essentially, a conjecture of Selberg [9, p. 70]. I understand that R. P. Langlands had obtained a similar but somewhat weaker result, a few years ago.

§ 34. The behaviour of certain eigenfunctions at infinity

We now return to the notation of § 27. Extend $\mathfrak{a}_\mathfrak{p}$ to a Cartan subalgebra $\mathfrak{a}$ of $\mathfrak{g}$. Define $\varpi = \varpi^\mathfrak{a}$, $W = W(\mathfrak{g}/\mathfrak{a})$ and $W_1 = W(\mathfrak{m}/\mathfrak{a})$ as usual (see [4 (p), § 12]) and, for a given linear function λ on $\mathfrak{a}_c$, put

$$\chi_\lambda(z) = \chi_\lambda{}^\mathfrak{a}(p_z) \quad (z \in \mathfrak{Z})$$

in the notation of [4 (p), § 12] and [4 (o), § 14]. Let $\mathfrak{U}_\lambda$ denote the kernel of χ_λ in $\mathfrak{Z}$ and put $\mathfrak{U}_{1\lambda} = \mathfrak{Z}_1 \mu_0(\mathfrak{U}_\lambda)$, $\mathfrak{Z}_{1\lambda}{}^* = \mathfrak{Z}_1/\mathfrak{U}_{1\lambda}$. (Here $\mu_0 = \mu_{\mathfrak{g}/\mathfrak{m}}$ as in § 27.) Let σ_λ denote the natural representation of $\mathfrak{Z}_1$ on $\mathfrak{Z}_{1\lambda}{}^*$.

Let $r = [W : W_1]$ and select elements $s_1 = 1, s_2, \dots, s_r$ in W such that $W = \bigcup_{1 \leqslant i \leqslant r} W_1 s_i$. Consider the subalgebras J and J_1 of all invariants of W and W_1 respectively in $\mathfrak{S}(\mathfrak{a}_c)$. Then we have the canonical isomorphisms $\gamma : \mathfrak{Z} \to J$ and $\gamma_1 : \mathfrak{Z}_1 \to J_1$ (see [4(o), § 12]) and $\gamma = \gamma_1 \circ \mu_0$. We identify $\mathfrak{S}(\mathfrak{a}_c)$ with $S = S(\mathfrak{a}_c)$ as usual and denote by $\gamma_1(\zeta : \lambda)$ $(\zeta \in \mathfrak{Z}_1)$ the value at λ of the element $\gamma_1(\zeta) \in S$.

LEMMA 66. dim $\mathfrak{Z}_{1\lambda}{}^{*}=r$ *and, if* $\varpi(\lambda)\neq 0$, *we can choose a base* $(v_1, \ldots, v_r)$ *for* $\mathfrak{Z}_{1\lambda}{}^{*}$ *such that*

$$\sigma_\lambda(\zeta)v_i = \gamma_1(\zeta:s_i\lambda)v_i \quad (\zeta\in\mathfrak{Z}_1,\ 1\leqslant i\leqslant r).$$

Since W and W_1 are both generated by reflexions, the results of [4 (j), § 3] are applicable. Therefore by taking into account the isomorphisms γ and γ_1, our assertions follow immediately from Lemmas 13 and 15 of [4 (j)].

Let μ and V have the same meaning as in Lemma 42.

LEMMA 67. *Let* λ *be a linear function on* $\mathfrak{a}_c$ *and* ϕ *a* C^∞ μ-*spherical function from* G *to* V. *Suppose the following conditions are fulfilled:*

1) rank $G=$ rank K.
2) λ *takes only real values on* $\mathfrak{a}\cap\mathfrak{p}+(-1)^{i}\mathfrak{a}\cap\mathfrak{k}$ *and* $\varpi(\lambda)\neq 0$.
3) $z\phi=\chi_\lambda(z)\phi \quad (z\in\mathfrak{Z})$.
4) *There exist numbers* c, $s\geqslant 0$ *such that*

$$|\phi(x)| \leqslant c\Xi(x)(1+\sigma(x))^{s} \quad (x\in G).$$

Then $\phi\in C(G)\otimes V$.

We may obviously assume that $\phi\neq 0$ and G is not compact so that $\mathfrak{a}_\mathfrak{p}\neq\{0\}$. Then, in view of Lemma 43, it would be enough to verify that

$$\lim_{t\to+\infty} e^{t\varrho(H)}\phi(h\exp tH)=0$$

for $H\in S^+$ and $h\in A_\mathfrak{p}{}^+$. For any $H_0\in S^+$, let $\mathfrak{m}_{H_0}$ denote the centralizer of H_0 in $\mathfrak{g}$. Suppose the above condition does not hold. Then we can choose $H_0\in S^+$ such that:

1) For some $h\in A_\mathfrak{p}{}^+$, $e^{t\varrho(H_0)}\phi(h\exp tH_0)$ does not tend to zero as $t\to+\infty$.
2) dim $\mathfrak{m}_{H_0}$ is minimum possible consistent with condition 1).

Put $\mathfrak{m}=\mathfrak{m}_{H_0}$, $\mathfrak{m}_1=\mathfrak{m}\cap\mathfrak{k}+[\mathfrak{m},\mathfrak{m}]\cap\mathfrak{p}$ and let $\mathfrak{l}$ be the centralizer of $\mathfrak{m}$ in $\mathfrak{p}$. Then $\mathfrak{m}_1\cap\mathfrak{p}$ is the orthogonal complement of $\mathfrak{l}$ in $\mathfrak{m}\cap\mathfrak{p}$. Let M_1 be the analytic subgroup of G corresponding to $\mathfrak{m}_1$. We now use the notation of §§ 27–30 for this particular H_0. (Note that $\mathfrak{c}_\mathfrak{p}=\{0\}$ in the present case since rank $\mathfrak{g}=$ rank $\mathfrak{k}$.) Define $\Theta(m)$ and $\theta(m)$ $(m\in M)$ as in §§ 29, 30. We know from Lemma 66 that the representation Γ of $\mathfrak{Z}_1$ is semisimple. Moreover it is clear from condition 2) of Lemma 67 that $s_i\lambda$ takes only real values on $\mathfrak{a}_\mathfrak{p}$ and therefore also on $\mathfrak{l}$. Hence we conclude from Lemma 59 that

$$\Theta(m\exp H) = \Theta(m) \quad (m\in M, H\in\mathfrak{l}).$$

This implies, in particular, that

$$\theta(m \exp H) = \theta(m) \quad (m \in M,\ H \in \mathfrak{l}).$$

Hence it follows from Lemma 61 and the definition of H_0, that $\theta \neq 0$.

Fix an element $H \neq 0$ in $\mathfrak{a}_1^+$. Then if $h \in A_1^+$, we claim that

$$e^{t\varrho_1(H)} \theta(h \exp tH) \to 0$$

as $t \to + \infty$. Put $H_1 = c_1(H_0 + cH)$, where c is a small positive number and $c_1 = \|H_0 + cH\|^{-1}$. Then $H_1 \in S^+$ and it is obvious that

$$\dim \mathfrak{m}_{H_1} < \dim \mathfrak{m}_{H_0}.$$

Hence we conclude from definition of H_0 that

$$e^{t\varrho(H_1)} \phi(h_0 \exp tH_1) \to 0 \quad (h_0 \in A_\mathfrak{p}^+)$$

as $t \to + \infty$. Define U as in § 30 and let U_0 denote the interior of U in S^+. Then, by choosing c sufficiently small, we can assume that $H_1 \in U_0$ and therefore

$$\lim_{t \to +\infty} \left| e^{t\varrho(H_1)} \phi(h \exp tH_1) - e^{\varrho_1(tH_1)} \theta(h \exp tH_1) \right| = 0 \quad (h \in A_1^+)$$

from Lemma 61. Fix $h \in A_1^+$. Since $\alpha(H_1) > 0$ $(\alpha \in \Sigma_2)$, we can choose $t_0 \geqslant 0$ such that $h_0 = h \exp t_0 H_1 \in A_\mathfrak{p}^+$. Hence it follows from what we have seen above that

$$e^{t\varrho(H_1)} \phi(h \exp tH_1) \to 0$$

as $t \to + \infty$. Put $c_2 = c_1 c$. Then $\varrho_1(H_1) = c_2 \varrho_1(H)$ and $H_1 = c_1 H_0 + c_2 H$. Therefore since $H_0 \in \mathfrak{l}$, we conclude that

$$e^{t\varrho_1(H)} \theta(h \exp tH) \to 0$$

and this proves our assertion.

Let θ_1 denote the restriction of θ on M_1. It is clear that $\theta_1 \neq 0$ and we conclude from Lemmas 43 and 58 that $\theta_1 \in C(M_1) \otimes V$. But then rank $\mathfrak{m}_1 = $ rank $(\mathfrak{m}_1 \cap \mathfrak{k})$ from Corollary 1 of Lemma 64.

Fix a Cartan subalgebra $\mathfrak{c}$ of $\mathfrak{m}_1 \cap \mathfrak{k}$. Then $\mathfrak{h} = \mathfrak{l} + \mathfrak{c}$ is a Cartan subalgebra of $\mathfrak{m}$ and therefore also of $\mathfrak{g}$. Since rank $\mathfrak{g} = $ rank $\mathfrak{k}$ and $H_0 \in \mathfrak{l}$, $\mathfrak{h}$ cannot be fundamental in $\mathfrak{g}$. Hence there exists a root β of $(\mathfrak{g}, \mathfrak{h})$ such that[1] $H_\beta \in \mathfrak{h} \cap \mathfrak{p} = \mathfrak{l}$ (see [4 (g), Lemma 33]). Let M_c denote the (connected) complex adjoint group of $\mathfrak{m}_c$. We can choose $y \in M_c$ such that $\mathfrak{h}_c^y = \mathfrak{a}_c$. Then $\alpha = \beta^y$ is a root of $(\mathfrak{g}, \mathfrak{a})$ and $H_\alpha = (H_\beta)^y = H_\beta \in \mathfrak{l}$.

Now we know that $\Theta \neq 0$ and by Lemma 59 $\Theta(m) \in {}^0\mathbf{V}$ $(m \in M)$. Therefore it follows from Lemma 66 and the definition of Γ, that there exists an element $s \in W$ such that $s\lambda = 0$ on $\mathfrak{l}$. But then $s\lambda(H_\alpha) = 0$ and therefore $\varpi(\lambda) = 0$, contrary to our hypothesis. This proves the lemma.

---

[1] Here H_β has the usual meaning (see [4 (n), § 4]).

§ 35. Eigenfunctions of $\mathfrak{Z}$ in $C(G)$

Let us now assume that rank $G=$ rank K and use the notation of § 20. Let L' be the set of all $\lambda \in L$ where $\varpi(\lambda) \neq 0$. We denote by χ_λ $(\lambda \in L')$ the corresponding homomorphism (see [4 (p), § 29]) of $\mathfrak{Z}$ into $\mathbf{C}$ so that $z\Theta_\lambda = \chi_\lambda(z)\Theta_\lambda$ $(z \in \mathfrak{Z})$. Consider the space $C_\lambda(G)$ of all functions $f \in C(G)$ such that $zf = \chi_\lambda(z)f$ $(z \in \mathfrak{Z})$. Let $\mathfrak{H}_\lambda$ denote the closure of $C_\lambda(G)$ in $L_2(G)$ and $\mathfrak{H}$ the smallest closed subspace of $L_2(G)$ containing $\bigcup_{\lambda \in L'} \mathfrak{H}_\lambda$.

It is obvious from the definition of Θ_λ (see [4 (p), Theorem 3]) that

$$\Theta_\lambda(x^{-1}) = \operatorname{conj} \Theta_\lambda(x) = (-1)^m \Theta_{-\lambda}(x) \quad (\lambda \in L', \ x \in G'),$$

where $m = \frac{1}{2}$ (dim $\mathfrak{g}$ − rank $\mathfrak{g}$). Hence it follows that

$$\chi_\lambda(z^*) = \operatorname{conj} \chi_\lambda(\eta(z)) = \chi_{-\lambda}(z) \quad (z \in \mathfrak{Z}),$$

where z^* denotes the adjoint of the differential operator z and η the conjugation of $\mathfrak{g}_c$ with respect to $\mathfrak{g}$.

LEMMA 68. *Let f be any eigenfunction of $\mathfrak{Z}$ in $C(G)$. Then $f \in C_\lambda(G)$ for some $\lambda \in L'$.*

We may obviously suppose that $f \neq 0$. Let χ be the homomorphism of $\mathfrak{Z}$ into $\mathbf{C}$ such that $zf = \chi(z)f$ $(z \in \mathfrak{Z})$. We have to show that $\chi = \chi_\lambda$ for some $\lambda \in L'$. Suppose this is false. Fix $\lambda \in L$ and consider $\Theta_\lambda(f)$. Then

$$\chi(z)\Theta_\lambda(f) = \Theta_\lambda(zf) = \chi_\lambda(z^*)\Theta_\lambda(f) = \chi_{-\lambda}(z)\Theta_\lambda(f) \quad (z \in \mathfrak{Z}).$$

Since $\chi \neq \chi_{-\lambda}$, we conclude that $\Theta_\lambda(f) = 0$. In view of Corollary 2 of Lemma 64, this implies that $f(1) = 0$.

Now fix $x \in G$ and put $f_x = r(x)f$ in the notation of § 10. Then the above proof is applicable to f_x and therefore $f(x) = f_x(1) = 0$. This shows that $f = 0$ and so we get a contradiction. Hence the lemma.

COROLLARY. *Let ϕ be an element in $L_2(G)$ which is an eigendistribution of $\mathfrak{Z}$. Then $\phi \in \mathfrak{H}_\lambda$ for some $\lambda \in L'$.*

We may again assume that $\phi \neq 0$. Let l and r respectively denote the left- and right-regular representations of G on $L_2(G)$ and ν the usual norm on $L_2(G)$. For $\alpha, \beta \in C(K)$, define

$$\alpha * \phi * \beta = \int_{K \times K} \alpha(k_1)\beta(k_2)\, l(k_1)\, r(k_2^{-1})\, \phi\, dk_1\, dk_2$$

as usual. Let $\mathcal{L}$ denote the space of all K-finite functions in $C(K)$. Since $\mathcal{L}$ is dense in $C(K)$

6 − 662900. *Acta mathematica.* 116. Imprimé le 10 juin 1966.

in the norm $|\alpha|_\infty = \sup|\alpha|$ $(\alpha \in C(K))$, it follows easily (see § 3) that, for any $\varepsilon > 0$, we can choose $\alpha, \beta \in \mathcal{L}$ such that

$$\nu(\phi - \alpha * \phi * \beta) < \varepsilon.$$

Put $\psi = \alpha * \phi * \beta$ and suppose $\varepsilon < \nu(\phi)$. Then it is clear that $\psi \neq 0$ and $z\psi = \chi(z)\psi$ $(z \in \mathfrak{Z})$. Hence we conclude from Lemma 68 and Corollary 1 of Lemma 65 that $\psi \in C_\lambda(G)$ for some $\lambda \in L'$. Therefore, in particular, $\chi = \chi_\lambda$. Since the space $C_\lambda(G)$ depends only on χ_λ (and not on λ), this shows that $\phi \in \mathrm{Cl}(C_\lambda(G)) = \mathfrak{H}_\lambda$.

§ 36. The role of the distributions Θ_λ in the harmonic analysis on G

For any $\mathfrak{b} \in \mathcal{E}_K$, let $\Theta_{\lambda,\mathfrak{b}}$ $(\lambda \in L')$ denote the corresponding Fourier component of Θ_λ (see [4 (q), § 17]).

THEOREM 12. $\Theta_{\lambda,\mathfrak{b}} \in C_\lambda(G)$ for $\lambda \in L'$ and $\mathfrak{b} \in \mathcal{E}_K$.

This follows from Theorem 9 and Lemma 67 (see also the proof of Corollary 1 of Lemma 65).

COROLLARY 1. $C_\lambda(G) \neq \{0\}$ for $\lambda \in L'$.

Since $\Theta_\lambda \neq 0$, we conclude from Lemma 9 that $\Theta_{\lambda,\mathfrak{b}} \neq 0$ for some $\mathfrak{b} \in \mathcal{E}_K$. This implies our assertion.

Fix $\lambda_0 \in L'$ and let $L(\lambda_0)$ denote the set of all $\lambda \in L$ of the form $\lambda = s\lambda_0$ $(s \in W = W(\mathfrak{g}/\mathfrak{h})$ in the notation of [4 (p)]). Let E_{λ_0} denote the orthogonal projection of $L_2(G)$ on $\mathfrak{H}_{\lambda_0}$ and define

$$(f, g) = \int_G (\mathrm{conj}\, f)g\, dx \quad (f, g \in L_2(G))$$

as usual.

Let $\mathcal{L}(G)$ denote the space of all K-finite functions in $C_c^\infty(G)$.

COROLLARY 2. Let $\gamma \in \mathcal{L}(G)$ and $\lambda_0 \in L'$. Then $E_{\lambda_0}\gamma \in C_{\lambda_0}(G)$ and

$$\Theta_\lambda(E_{\lambda_0}\gamma) = \begin{cases} \Theta_\lambda(\gamma), & \text{if } \lambda \in L(-\lambda_0), \\ 0 & \text{otherwise,} \end{cases}$$

for $\lambda \in L'$.

It is obvious that E_{λ_0} commutes with the translations of G and therefore $f = E_{\lambda_0}\gamma$ is K-finite. Hence we conclude from Corollary 1 of Lemma 65 that $f \in C_{\lambda_0}(G)$. Therefore (see the proof of Lemma 68), $\Theta_\lambda(f) = 0$ $(\lambda \in L')$ unless $\lambda \in L(-\lambda_0)$. Now fix $\lambda \in L(-\lambda_0)$. Then $\mathrm{conj}\, \Theta_{\lambda,\mathfrak{b}} \in C_\lambda(G)$ from Theorem 12 and therefore

$$\Theta_{\lambda,\mathfrak{b}}(\gamma) = (\operatorname{conj}\Theta_{\lambda,\mathfrak{b}}, \gamma) = (\operatorname{conj}\Theta_{\lambda,\mathfrak{b}}, E_{\lambda_0}\gamma) = \Theta_{\lambda,\mathfrak{b}}(f) \quad (\mathfrak{b}\in\mathcal{E}_K)$$

from the corollary of Theorem 9. Therefore, since γ and f are both K-finite, we have

$$\Theta_\lambda(\gamma) = \sum_{\mathfrak{b}} \Theta_{\lambda,\mathfrak{b}}(\gamma) = \sum_{\mathfrak{b}} \Theta_{\lambda,\mathfrak{b}}(f) = \Theta_\lambda(f).$$

LEMMA 69. *Fix $\lambda_0\in L'$ and define c and q as in Theorem 8. Then*

$$(-1)^q \sum_{\lambda\in L(-\lambda_0)} \varpi(\lambda)\,\Theta_\lambda(\ddot{\alpha}\!*\!f) = c(\alpha, E_{\lambda_0}f)$$

for $\alpha\in C_c^\infty(G)$ and $f\in C(G)$. Here $\ddot{\alpha}(x)=\operatorname{conj}\alpha(x^{-1})$ $(x\in G)$.

Since $\ddot{\alpha}\!*\!f\in C(G)$ (see § 10), the left side is defined. Fix $\alpha_0, \beta\in \mathcal{L}(G)$ and put $g = E_{\lambda_0}\beta$. Then $\ddot{\alpha}_0\!*\!g = E_{\lambda_0}(\ddot{\alpha}_0\!*\!\beta)$. Now apply Corollary 2 of Lemma 64 to $\ddot{\alpha}_0\!*\!g$, taking into account Corollary 2 of Theorem 12 with $\gamma = \ddot{\alpha}_0\!*\!\beta$. Then we get

$$c(\alpha_0, E_{\lambda_0}\beta) = (-1)^q \sum_{\lambda\in L(-\lambda_0)} \varpi(\lambda)\,\Theta_\lambda(\ddot{\alpha}_0\!*\!\beta).$$

On the other hand, $\mathcal{L}(G)$ is dense both in $C_c^\infty(G)$ and $C(G)$, by Lemmas 9, 16 and 19. Moreover, convergence in either one of these spaces implies convergence in $L_2(G)$ (see Lemma 11). Finally, if α_0 and β are two variable elements of $\mathcal{L}(G)$, which converge to α and f in $C_c^\infty(G)$ and $C(G)$ respectively, then it is obvious from § 10 that $\ddot{\alpha}_0\!*\!\beta$ tends to $\ddot{\alpha}\!*\!f$ in $C(G)$. Therefore the statement of Lemma 69 now follows immediately.

Define the representation r of G on $C(G)$ as in § 10.

COROLLARY 1. *Let $f\in C(G)$. Then $E_{\lambda_0}f$ is a continuous function on G given by*

$$E_{\lambda_0}f(x) = c^{-1}(-1)^q \sum_{\lambda\in L(-\lambda_0)} \varpi(\lambda)\,\Theta_\lambda(r(x)f) \quad (x\in G).$$

It is obvious that the right side is continuous in x and the equality follows from Lemma 69, if we observe that $\Theta_\lambda(\ddot{\alpha}\!*\!f) = \Theta_\lambda(f\!*\!\ddot{\alpha})$ $(\alpha\in C_c^\infty(G))$, in view of the invariance of Θ_λ.

Let E denote the orthogonal projection of $L_2(G)$ on $\mathfrak{H}$.

COROLLARY 2. $c(\alpha, Ef) = (-1)^q \sum_{\lambda\in L} \varpi(\lambda)\Theta_\lambda(\ddot{\alpha}\!*\!f)$ *for $\alpha\in C_c^\infty(G)$ and $f\in C(G)$.*

This is obvious from Lemma 69.

COROLLARY 3. *For $f\in C(G)$, put*

$$f^\natural(x) = c^{-1}(-1)^q \sum_{\lambda\in L} \varpi(\lambda)\,\Theta_\lambda(r(x)f) \quad (x\in G).$$

Then $f^\natural$ is a continuous function on G and $f^\natural = Ef$.

Put
$$T(f) = c^{-1}(-1)^q \sum_{\lambda \in L} \varpi(\lambda) \Theta_\lambda(f) \quad (f \in C(G)).$$

We have seen in § 20 that T is a tempered distribution. Hence it is obvious that $f^\natural$ is continuous. The rest follows from Corollary 2, if we take into account the fact that

$$T(\tilde\alpha \ast f) = T(f \ast \tilde\alpha) = \int f^\natural(x)\,\mathrm{conj}\,\alpha(x)\cdot dx.$$

Let $W(\lambda)$ $(\lambda \in L')$ be the set of all $s \in W$ such that $s\lambda \in L$. Put [1]

$$\Theta_\lambda^* = [W(\lambda)]^{-1} \sum_{s \in W(\lambda)} \varepsilon(s)\,\Theta_{s\lambda} \quad (\lambda \in L').$$

COROLLARY 4. *For any $\lambda \in L'$, the distribution $(-1)^q \varpi(\lambda) \Theta_\lambda^*$ is of positive type.*

This is obvious from Lemma 69 since $c(\alpha, E_\lambda \alpha) \geqslant 0$ for $\alpha \in C_c^\infty(G)$.

LEMMA 70. *Fix $\mathfrak{b} \in \mathcal{E}_K$. Then there exist only a finite number of $\lambda \in L'$ such that $\Theta_{\lambda,\mathfrak{b}} \neq 0$.*

Let $\mathfrak{c}$ be the center and $\mathfrak{g}_1$ the derived algebra of $\mathfrak{g}$. Fix a quadratic form Q on $\mathfrak{g}$ such that 1) Q is negative-definite on $\mathfrak{c}$, 2) $\mathfrak{c}$ and $\mathfrak{g}_1$ are mutually orthogonal under Q and 3) $Q(X) = \mathrm{tr}\,(\mathrm{ad}\,X)^2$ for $X \in \mathfrak{g}_1$. Then Q is negative-definite on $\mathfrak{k}$, positive-definite on $\mathfrak{p}$ and it is invariant under G. Moreover, $\mathfrak{k}$ and $\mathfrak{p}$ are orthogonal under Q. Fix bases $(Y_1, ..., Y_p)$ and $(Z_1, ..., Z_q)$ for $\mathfrak{p}$ and $\mathfrak{k}$ which are orthonormal with respect to Q and $-Q$ respectively and put

$$\omega_\mathfrak{p} = Y_1^2 + ... + Y_p^2, \quad \omega_\mathfrak{k} = -(Z_1^2 + ... + Z_q^2)$$

in $\mathfrak{G}$. Then
$$\omega = \omega_\mathfrak{p} + \omega_\mathfrak{k} \in \mathfrak{Z}$$

Let γ denote the canonical isomorphism of $\mathfrak{Z}$ into [2] $S(\mathfrak{b}_c) = \mathfrak{S}(\mathfrak{b}_c)$ (see [4 (e), Lemma 19]) and $\gamma(z : \mu)$ $(z \in \mathfrak{Z}, \mu \in \mathfrak{F})$ the value of the polynomial function $\gamma(z)$ at μ. Put

$$\varrho = \tfrac{1}{2} \sum_{\alpha \in P} \alpha,$$

where P is the set of all positive roots of $(\mathfrak{g}, \mathfrak{b})$. Then (see [4 (e), p. 144])

$$|\mu|^2 = \gamma(\omega : \mu) + \tfrac{1}{2} \sum_{\alpha \in P} \varrho(H_\alpha) \quad (\mu \in \mathfrak{F})$$

is a positive-definite quadratic form on $\mathfrak{F}$ and $\gamma(\omega : \varrho) = 0$, so that

$$\gamma(\omega : \mu) = |\mu|^2 - |\varrho|^2.$$

Moreover, $\chi_\lambda(\omega) = \gamma(\omega : \lambda) = |\lambda|^2 - |\varrho|^2$ for $\lambda \in L'$.

---

[1] See the foot-note on p. 3.
[2] Here the notation is the same as in [4 (p)].

Let $\Re$ be the subalgebra of $\mathfrak{G}$ generated by $(1, \mathfrak{k}_c)$ and $\mathfrak{Z}_K$ the center of $\Re$. Then $\omega_{\mathfrak{k}} \in \mathfrak{Z}_K$ and

$$\omega_{\mathfrak{k}} \, \alpha_{\mathfrak{b}} = \chi_{\mathfrak{b}}(\omega_{\mathfrak{k}}) \, \alpha_{\mathfrak{b}},$$

where $\chi_{\mathfrak{b}}(\omega_{\mathfrak{k}})$ is a number $\geqslant 0$ (see § 3). Therefore since L is a lattice in $\mathfrak{F}$, it would be sufficient to prove the following lemma.

LEMMA 71. *Suppose λ and $\mathfrak{b}$ are two elements in L' and $\mathcal{E}_K$ respectively such that $\Theta_{\lambda, \mathfrak{b}} \neq 0$. Then*

$$|\lambda|^2 \leqslant \chi_{\mathfrak{b}}(\omega_{\mathfrak{k}}) + |\varrho|^2.$$

We know from Theorem 12 that $f = \Theta_{\lambda, \mathfrak{b}} \in C_\lambda(G) \subset L_2(G)$. Let r denote the right-regular representation of G on $L_2(G)$, V the smallest closed subspace of $L_2(G)$ containing f which is invariant under r, and π the restriction of r on V. Since convergence in $C(G)$ implies convergence in $L_2(G)$, it follows from Lemma 15 that f is differentiable under π and

$$\pi(\omega) f = \chi_\lambda(\omega) f, \qquad \pi(\omega_{\mathfrak{k}}) f = \chi_{\mathfrak{b}}(\omega_{\mathfrak{k}}) f.$$

Hence $\qquad\qquad \{|\lambda|^2 - |\varrho|^2\} \|f\|^2 = (f, \pi(\omega) f) = \chi_{\mathfrak{b}}(\omega_{\mathfrak{k}}) \|f\|^2 + (f, \pi(\omega_{\mathfrak{p}}) f),$

where $\| \ \|$ denotes the usual norm in $L_2(G)$. But since π is unitary and f is differentiable under π, it is obvious that

$$(f, \pi(\omega_{\mathfrak{p}}) f) = - \sum_{1 \leqslant i \leqslant p} \|\pi(Y_i) f\|^2 \leqslant 0.$$

Hence our assertion follows immediately from the fact that $\|f\| > 0$.

LEMMA 72. *Let f be a K-finite function in $C(G)$. Then $E_\lambda f \in C_\lambda(G)$ ($\lambda \in L'$) and $E_\lambda f = 0$ for all $\lambda \in L'$ except a finite number. Hence $Ef \in C(G)$ and it is both K-finite and $\mathfrak{Z}$-finite.*

Since E_λ commutes with the translations of G, it is clear that $E_\lambda f$ is K-finite and therefore, by Corollary 1 of Lemma 65, it lies in $C_\lambda(G)$. Now select a finite subset F of $\mathcal{E}_K$ such that $f = \alpha_F * f * \alpha_F$ in the notation of § 12. Replacing F by $F \cup F^*$, we may assume that $F = F^*$. (Recall that $\mathfrak{b}^*$ is the class in $\mathcal{E}_K$ contragradient to $\mathfrak{b}$.) Put

$$f_\lambda(x) = \Theta_\lambda(r(x) f) \quad (x \in G)$$

for $\lambda \in L'$. Then it is obvious that

$$f_\lambda(x) = \Theta_{\lambda, F}(r(x) f),$$

where $\qquad\qquad\qquad \Theta_{\lambda, F} = \sum_{\mathfrak{b} \in F} \Theta_{\lambda, \mathfrak{b}}.$

Hence we conclude from Lemma 70 that $f_\lambda=0$ for all $\lambda\in L'$ except a finite number, and therefore the assertions of the lemma follow from Corollaries 1 and 3 of Lemma 69.

Define $B'=B\cap G'$ as in § 19.

LEMMA 73. *Fix* $\lambda\in L'$ *and let* Θ *be a tempered and invariant distribution on* G *such that* $z^*=\chi_\lambda(z)\Theta$ $(z\in\mathfrak{Z})$. *Then* $\Theta_\mathfrak{b}\in C_\lambda(G)$ *for* $\mathfrak{b}\in\mathcal{E}_K$. *Moreover, in order to show that* $\Theta=0$, *it is sufficient to verify either one of the following two conditions.*

1) $\Theta(f)=0$ *for every* K-*finite function* $f\in C_{-\lambda}(G)$.
2) $\Theta=0$ *pointwise*(1) *on* B'.

It follows from Theorem 9 and Lemma 67 that $\Theta_\mathfrak{b}\in C_\lambda(G)$ $(\mathfrak{b}\in\mathcal{E}_K)$ and therefore $f=\mathrm{conj}\,\Theta_\mathfrak{b}\in C_{-\lambda}(G)$ (see § 35). Then

$$\Theta_\mathfrak{b}(f)=\int_G|\Theta_\mathfrak{b}(x)|^2\,dx$$

from the corollary of Theorem 9. Therefore $\Theta_\mathfrak{b}=0$ under condition 1). In view of Lemma 9, this implies that $\Theta=0$.

On the other hand, by Lemma 64 and the corollary of Theorem 7, 2) implies 1) and so the lemma is proved.

Define W_G as in [4 (p), Theorem 3]. Then L is stable under W_G and $\Theta_{s\lambda}=\varepsilon(s)\Theta_\lambda$ $(s\in W_G,\ \lambda\in L')$.

LEMMA 74. *Let* Θ *be an invariant eigendistribution of* $\mathfrak{Z}$ *on* G, *which is tempered. Suppose there exists an element* $\mathfrak{b}\in\mathcal{E}_K$ *such that* $\Theta_\mathfrak{b}\neq0$ *and* $\Theta_\mathfrak{b}\in L_2(G)$. *Then we can choose* $\lambda\in L'$ *such that* $z\Theta=\chi_\lambda(z)\Theta$ $(z\in\mathfrak{Z})$. *Moreover, for any such* λ, *there exist unique complex numbers* c_s $(s\in W(\lambda))$ *such that* $c_{ts}=c_s$ $(t\in W_G)$ *and*

$$\Theta=[W_G]^{-1}\sum_{s\in \overline{W}(\lambda)}\varepsilon(s)\,c_s\Theta_{s\lambda}.$$

The first statement follows from the corollary of Lemma 68. Now put

$$\Phi(b)=\Delta(b)\Theta(b)\quad(b\in B'),$$

where Δ has the usual meaning (see [4 (p), Theorem 3]). Then it follows from [4 (o), Lemma 31] that Φ extends to an analytic function on B. Moreover, we know from [4 (e), Theorem 2] that

$$\gamma(z)\Phi=\chi_\lambda(z)\Phi\quad(z\in\mathfrak{Z}),$$

---

(1) Here we have to make use of Theorem 2 of [4 (o)].

where $\gamma(z) \in \mathfrak{S}(\mathfrak{b}_c)$ is to be regarded as a differential operator on B. Finally it is obvious that $\Phi(b^s) = \varepsilon(s)\Phi(b)$ for $s \in W_G$ and $b \in B$. Therefore (see [4 (f), p. 102] and [4 (p), Theorem 3]), we can choose unique complex numbers c_s $(s \in W(\lambda))$ such that $c_{ts} = c_s$ $(t \in W_G)$ and the distribution

$$\Theta' = \Theta - [W_G]^{-1} \sum_{s \in W(\lambda)} \varepsilon(s) c_s \Theta_{s\lambda}$$

vanishes pointwise on B'. It is clear (see § 20) that Θ' is tempered and $z\Theta' = \chi_\lambda(z)\Theta'$ $(z \in \mathfrak{Z})$. Therefore $\Theta' = 0$ from Lemma 73.

LEMMA 75. *Let $\mathfrak{b}_0$ denote the class of the trivial representation of K. Then $\Theta_{\lambda, \mathfrak{b}_0} = 0$ for $\lambda \in L'$.*

Fix $\lambda \in L'$ and put $\Phi_\lambda = \Theta_{\lambda, \mathfrak{b}_0}$. Then from Theorem 9, there exist numbers $c, m \geqslant 0$ such that

$$|\Phi_\lambda| \leqslant c\Xi(1+\sigma)^m.$$

This is the analogue of [4 (q), Lemma 43]. By making use of the corollary of [4 (q), Lemma 47], we prove in the same way as in [4 (q), § 20] that $\Phi_\lambda = 0$.

§ 37. The discrete series for G

Let G be a locally compact unimodular group satisfying the second axiom of countability. Fix a Haar measure dx on G. By a unitary representation of G, we mean a representation of G on a Hilbert space, which is unitary. Let $\mathcal{E}$ be the set of all equivalence classes of irreducible unitary representations of G.

Let π be an irreducible unitary representation of G on a Hilbert space $\mathfrak{H}$. We say that π is *square-integrable* if any one of the following two mutually equivalent conditions holds (see [7, p. 640]).

(1)   There exist nonzero elements ϕ, ψ in $\mathfrak{H}$ such that

$$\int_G |(\phi, \pi(x)\,\psi)|^2\,dx < \infty.$$

(2)   There exists a closed subspace V of $L_2(G)$ stable under the right-regular representation r of G on $L_2(G)$, such that π is equivalent to the restriction of r on V.

It is known (see [7, p. 640]) that, if π is square-integrable, there exists a number $d(\pi) > 0$ such that

$$\int_G |(\phi, \pi(x)\,\psi)|^2\,dx = d(\pi)^{-1}\,\|\phi\|^2\,\|\psi\|^2 \quad (\phi, \psi \in \mathfrak{H}),$$

where $\|\phi\|$ is the norm of ϕ in $\mathfrak{H}$. We shall call $d(\pi)$ the *formal degree* of π (see [4 (d), § 3]). It is obvious that square-integrability, as well as the formal degree, are invariant under equivalence. We call a class $\omega \in \mathcal{E}$ *discrete* if every representation $\pi \in \omega$ is square-integrable and put $d(\omega) = d(\pi)$.

Let $\mathcal{E}_d$ denote the set of all discrete classes in $\mathcal{E}$. Then $\mathcal{E}_d$ is called the discrete series of G.

Now let us return to the case when G and K are defined as in § 7. For any $w \in \mathcal{E}$, let Θ_ω denote the character (see [4 (b), § 5]) and χ_ω the infinitesimal character of ω so that $z\Theta_\omega = \chi_\omega(z)\Theta_\omega$ $(z \in \mathfrak{Z})$.

LEMMA 76. *Let* $\omega \in \mathcal{E}_d$. *Then* Θ_ω *is tempered and* $\Theta_{\omega,\mathfrak{b}} \in L_2(G)$ *for* $\mathfrak{b} \in \mathcal{E}_K$.

Fix $\pi \in \omega$ and let $\mathfrak{H}$ be the representation space of π. We now use the notation of § 25 and put
$$\phi_i(x) = (\psi_i, \pi(x)\psi_i) \quad (x \in G, i \in J).$$

Then
$$\Theta_\omega(f) = \sum_{i \in J} \int f\phi_i \, dx \quad (f \in C_c^\infty(G)),$$

the series being absolutely convergent (see [4 (b), p. 243]). Moreover, ϕ_i is analytic from [4 (q), Lemma 33]. Fix an integer $m \geqslant 0$ as in Lemma 7. Then
$$\int f\phi_i \, dx = c(\mathfrak{b})^{-m} \int f \cdot \Omega^m \phi_i \, dx = c(\mathfrak{b})^{-m} \int \Omega^m f \cdot \phi_i \, dx \quad (i \in J_\mathfrak{b}),$$

in the notation of Lemma 6. Hence, by the Schwartz inequality, we get
$$\left| \int f\phi_i \, dx \right| \leqslant c(\mathfrak{b})^{-m} \| \Omega^m f \| \, d(\pi)^{-\frac{1}{2}} \quad (i \in J_\mathfrak{b}),$$

where $\| \ \|$ denotes the usual norm in $L_2(G)$. This shows that
$$|\Theta_\omega(f)| \leqslant d(\omega)^{-\frac{1}{2}} \| \Omega^m f \| \sum_{\mathfrak{b} \in \mathcal{E}_K} c(\mathfrak{b})^{-m} \dim \mathfrak{H}_\mathfrak{b} \quad (f \in C_c^\infty(G)).$$

But
$$\sum_\mathfrak{b} c(\mathfrak{b})^{-m} \dim \mathfrak{H}_\mathfrak{b} \leqslant N \sum_\mathfrak{b} c(\mathfrak{b})^{-m} d(\mathfrak{b})^2 < \infty$$

from Lemma 7. Since
$$\nu(g) = \| \Omega^m g \| \quad (g \in C(G))$$

is a continuous seminorm on $C(G)$ (see Lemma 11), we conclude that Θ_ω is tempered. Moreover, (see § 25),
$$\Theta_{\omega,\mathfrak{b}} = \sum_{i \in J_\mathfrak{b}} \phi_i \quad (\mathfrak{b} \in \mathcal{E}_K)$$

and therefore $\Theta_{\omega,\mathfrak{b}} \in L_2(G)$.

§ 38. Proof of Lemma 65

As before, let r denote the right-regular representation of G on $L_2(G)$.

LEMMA 77. *Suppose $f \neq 0$ is a $\mathfrak{Z}$-finite function in $L_2(G)$ such that the right translates of f, under K, span a finite-dimensional space. Let V be the smallest closed subspace of $L_2(G)$ containing f, which is stable under r. Then*

$$V = \sum_{1 \leqslant i \leqslant p} U_i,$$

where U_i are mutually orthogonal closed subspaces of V, which are invariant and irreducible under r.

Let π denote the restriction of r on V. For any finite subset F of $\mathcal{E}_K$, define E_F as in § 25 and put $V_F = E_F V$. It is clear that F can be so chosen that $f \in V_F$. Define V^∞ as in Lemma 4. We know from [4 (q), Lemma 33] that f is analytic and therefore, by Theorem 1, $f = f \ast \alpha$ for some $\alpha \in C_c^\infty(G)$. This shows (Lemma 2) that $f \in V^\infty$. Moreover, a simple argument (see § 8) shows that $W = \pi(\mathfrak{G})f$ is dense in V. Finally we conclude from [4 (a), Theorem 1] that

$$W = \sum_{b \in \mathcal{E}_K} E_b W$$

and $W_F = E_F W$ has finite dimension. Since W_F is dense in V_F, it follows that $W_F = V_F$.

Let $U \neq \{0\}$ be any closed subspace of V stable under π. We claim that $E_F U \neq \{0\}$. For otherwise suppose $E_F U = \{0\}$. Let U' denote the orthogonal complement of U in V. Since π is unitary, U' is also stable under π and $f \in V_F \subset U'$. But this implies that $V \subset U'$ and therefore $U = \{0\}$, contradicting our hypothesis.

Let U_i $(1 \leqslant i \leqslant p)$ be a finite set of mutually orthogonal, closed, nonzero subspaces of V, which are stable under π. Then since $U_i \cap V_F \neq \{0\}$, we conclude that $p \leqslant \dim V_F < \infty$. Therefore the required result follows immediately by assuming that p has the largest possible value.

Now we come to the proof of Lemma 65. We may assume that $f \neq 0$. Define V and U_i $(1 \leqslant i \leqslant p)$ as in Lemma 77 and put $f_i = E_i f$, where E_i is the orthogonal projection of V on U_i. It is obvious that f_i is $\mathfrak{Z}$-finite as well as K-finite. Hence by [4 (q), Lemma 33], it is analytic. Moreover, $f = f_1 + ... + f_p$. Therefore it would be sufficient to prove Lemma 65 for each f_i. Thus we may assume that V is irreducible under π so that π is square-integrable. By Theorem 1, there exists an element $\alpha \in C_c^\infty(G)$ such that $f = \alpha \ast f$. Since f is K-finite, we can obviously assume that α is also K-finite. But then

$$f(x) = (\tilde{\alpha}, \pi(x)f) = (E\tilde{\alpha}, \pi(x)f) \quad (x \in G)$$

where $\tilde{\alpha}$ is defined as in Lemma 69 and E is the orthogonal projection of $L_2(G)$ on V. The required inequality is now an immediate consequence of Theorem 10 and Lemma 76.

§ 39. The existence of the discrete series

Henceforward we assume, for convenience, that G is acceptable.

Theorem 13. *G has a discrete series if and only if* rank $G=$rank K.

Suppose $\mathcal{E}_d \neq \varnothing$. Fix $\omega \in \mathcal{E}_d$ and choose $\mathfrak{b} \in \mathcal{E}_K$ such that $f = \Theta_{\omega,\mathfrak{b}} \neq 0$. Then it follows from Lemma 76 and Corollary 2 of Lemma 65 that rank $G=$rank K.

Conversely suppose rank $G=$rank K. Fix $\lambda \in L'$ and choose $\mathfrak{b} \in \mathcal{E}_K$ such that $\Theta_{\lambda,\mathfrak{b}} \neq 0$. Then by Theorem 12, the function $f = \Theta_{\lambda,\mathfrak{b}}$ satisfies the hypotheses of Lemma 77. Let π_i denote the restriction of r on U_i ($1 \leqslant i \leqslant p$), in the notation of Lemma 77. Then π_i is square-integrable and therefore $\mathcal{E}_d \neq \varnothing$.

§ 40. The characters of the discrete series

In view of Theorem 13, we shall now assume that rank $G=$rank K and use the notation of §§ 36, 37.

For any $\lambda \in L'$, let $\mathcal{E}_d(\lambda)$ denote the set of all $\omega \in \mathcal{E}_d$ such that $\chi_\omega = \chi_\lambda$. The following result is an immediate consequence of Lemmas 74 and 76.

Lemma 78. $\mathcal{E}_d = \bigcup_{\lambda \in L'} \mathcal{E}_d(\lambda)$. *Moreover, for any* $\omega \in \mathcal{E}_d(\lambda)$ ($\lambda \in L'$), *there exist unique complex numbers* $c_s(\omega)$ ($s \in W(\lambda)$) *such that* $c_{ts}(\omega) = c_s(\omega)$ ($t \in W_G$) *and*

$$\Theta_\omega = [W_G]^{-1} \sum_{s \in W(\lambda)} \varepsilon(s)\, c_s(\omega)\, \Theta_{s\lambda}.$$

For any $\omega \in \mathcal{E}$, define the analytic function Φ_ω on B (see § 36) by

$$\Phi_\omega(b) = \Delta(b)\, \Theta_\omega(b) \quad (b \in B')$$

and put

$$F_f(b) = \Delta(b) \int_G f(b^x)\, dx \quad (b \in B')$$

for $f \in C(G)$ as in § 18. Put $G_B = (B')^G$ and let db denote the normalized Haar measure on B. Then G_B is open in G and (see Lemma 91)

$$\int_G \alpha(x)\, dx = (-1)^m [W_G]^{-1} \int_B \Delta F_\alpha\, db \quad (\alpha \in C_c^\infty(G_B)),$$

where $m = \tfrac{1}{2} \dim G/B$.

Lemma 79. *Let f be a $\mathfrak{Z}$-finite function in* $C(G)$. *Then*

$$\Theta_\omega(f) = (-1)^m [W_G]^{-1} \int_B F_f \Phi_\omega\, db \quad (\omega \in \mathcal{E}_d).$$

This follows immediately from Lemmas 64 and 76 and the corollary of Theorem 7.

LEMMA 80. *Let π be a square-integrable representation of G on $\mathfrak{H}$ and ω its class in $\mathcal{E}_d$. Fix two K-finite elements ϕ, $\psi \in \mathfrak{H}$ and put*

$$f(x) = (\phi, \pi(x)\,\psi) \quad (x \in G).$$

Then
$$F_f = d(\omega)^{-1}(\phi, \psi)\,\Phi_\omega$$

on B'.

First observe that $f \in C(G)$ from Corollary 1 of Lemma 65 and therefore F_f is defined. Now fix $\alpha \in C_c^\infty(G)$ and consider the operator

$$\pi(\alpha) = \int \alpha(x)\,\pi(x)\,dx.$$

Then $\Theta_\omega(\alpha) = \operatorname{tr} \pi(\alpha)$ and the argument of [4 (d), 576] shows that

$$\int_G dx \int_G \alpha(y)\,f(y^x)\,dy = \int_G (\phi, \pi(x)\,\pi(\alpha)\,\pi(x^{-1})\,\psi)\,dx = d(\omega)^{-1}(\phi, \psi)\,\Theta_\omega(\alpha).$$

Now suppose $\alpha \in C_c^\infty(G_B)$. Then we claim that

$$\int |\alpha(y)\,f(y^x)|\,dx\,dy < \infty.$$

For we can choose $\alpha_0 \in C_c^\infty(G_B)$ such that $\alpha_0 \geqslant |\alpha|$. Then

$$\int |\alpha(y)\,f(y^x)|\,dx\,dy \leqslant \int \alpha_0(y)\,|f(y^x)|\,dx\,dy = [W_G]^{-1} \int_B |F_{\alpha_0}(b)| \left\{|\Delta(b)| \int_G |f(b^y)|\,dy\right\} db.$$

Since $|F_{\alpha_0}|$ is bounded on B' (Lemma 26), our assertion follows from Theorem 5.

Therefore we conclude from Fubini's theorem that

$$d(\omega)^{-1}(\phi, \psi)\,\Theta_\omega(\alpha) = \int \alpha(y)\,dy \int f(y^x)\,dx \quad (\alpha \in C_c^\infty(G_B)).$$

But it is clear that

$$\int \alpha(y)\,dy \int f(y^x)\,dx = (-1)^m [W_G]^{-1} \int_B F_\alpha F_f\,db$$

and
$$\Theta_\omega(\alpha) = \int \alpha\,\Theta_\omega\,dx = (-1)^m [W_G]^{-1} \int_B F_\alpha \Phi_\omega\,db.$$

This shows that

$$\int F_\alpha F_f\, db = d(\omega)^{-1}(\phi,\psi)\int F_\alpha \Phi_\omega\, db \quad (\alpha \in C_c^\infty(G_B)).$$

On the other hand, it is easy to verify (see [4 (o), § 20]) that there exists a C^∞ function u on G_B such that

$$\Delta(b)\,u(b^x) = F_f(b) - d(\omega)^{-1}(\phi,\psi)\,\Phi_\omega(b) \quad (b \in B',\ x \in G).$$

Then if follows from the above result that

$$\int \alpha\, u\, dx = 0 \quad (\alpha \in C_c^\infty(G_B))$$

and therefore $u=0$. This implies the assertion of the lemma.

For any $\omega \in \mathcal{E}_d$, we define a subspace $\mathfrak{H}_\omega$ of $L_2(G)$ as follows. Fix $\pi \in \omega$ and let U be the representation space of π. Then $\mathfrak{H}_\omega$ is the smallest closed subspace of $L_2(G)$ containing all functions f of the form

$$f(x) = (\phi, \pi(x)\psi) \quad (x \in G),$$

where $\phi,\ \psi \in U$. It is clear that this definition is independent of the particular choice of π and $\mathfrak{H}_\omega$ is stable under both left and right translations of G. Put $C_\omega(G) = \mathfrak{H}_\omega \cap C(G)$. Then it follows from Lemma 11 that $C_\omega(G)$ is closed in $C(G)$.

THEOREM 14. $C_\omega(G)$ *is dense in* $\mathfrak{H}_\omega$ *and*

$$F_f = d(\omega)^{-1} f(1)\Phi_\omega$$

for $f \in C_\omega(G)$ *and* $\omega \in \mathcal{E}_d$.

Choose an orthonormal base ψ_i $(i \in J)$ for U as in § 25 so that ψ_i $(i \in J_b)$ is a base for U_b $(b \in \mathcal{E}_K)$ and put

$$f_{ij}(x) = (\psi_i, \pi(x)\psi_j) \quad (x \in G).$$

Then it follows from Corollary 1 of Lemma 65 that $f_{ij} \in C_\omega(G)$. This shows that $C_\omega(G)$ is dense in $\mathfrak{H}_\omega$.

Let V be the set of all $f \in C_\omega(G)$ such that

$$F_f = d(\omega)^{-1} f(1)\Phi_\omega.$$

Then it is clear that V is a closed subspace[1] of $C_\omega(G)$. Hence it would be enough to show that V is dense in $C_\omega(G)$.

---

[1] The topology of $C_\omega(G)$ is the one inherited from $C(G)$.

Fix $f_0 \in C_\omega(G)$. Then in view of Lemma 16, it would be enough to prove that $f = \alpha_{b_1} * f_0 * \alpha_{b_2} \in V$ for b_1, $b_2 \in \mathcal{E}_K$. Since J_b is a finite set for every $b \in \mathcal{E}_K$, it is clear that f is a finite linear combination of f_{ij} $(i, j \in J)$ and therefore $f \in V$ from Lemma 80. This proves the theorem.

Remark. The above proof shows that f_{ij} $(i, j \in J)$ span a dense subspace of $C_\omega(G)$. Moreover, $f_{ii}(1) = 1$ for $i \in J$.

For any $\omega \in \mathcal{E}$, let ω^* denote the class contragredient to ω. It is clear that $\Theta_{\omega^*} = \mathrm{conj}\ \Theta_\omega$ as functions on G, and ω^* is discrete whenever ω is discrete.

LEMMA 81. *Let ω, ω' be two elements in $\mathcal{E}_d$. Then*

$$\Theta_\omega(f) = \begin{cases} d(\omega)^{-1} f(1) & \text{if}\ \omega' = \omega^*, \\ 0 & \text{otherwise,} \end{cases}$$

for $f \in C_{\omega'}(G)$.

We keep to the above notation. Then it follows easily from Lemma 19 that

$$\Theta_\omega(f) = \sum_i \int f f_{ii}\, dx = \sum_i (\mathrm{conj}\ f_{ii},\ f)$$

for any K-finite function f in $C_{\omega'}(G)$. Now conj $f_{ii} \in \mathfrak{H}_{\omega^*}$ and if $\omega^* \neq \omega'$, we conclude from the Schur orthogonality relations [4 (d), Theorem 1] that $\mathfrak{H}_{\omega^*}$ is orthogonal to $\mathfrak{H}_{\omega'}$ and therefore $\Theta_\omega(f) = 0$. Since K-finite functions are dense in $C_{\omega'}(G)$ by Lemma 16, we get the required assertion in this case.

Now suppose $\omega^* = \omega'$ and $f = \mathrm{conj}\ f_{ij}$ $(i, j \in J)$. Then it follows again from the Schur orthogonality relations that

$$\Theta_\omega(f) = d(\omega)^{-1} f(1).$$

But we have seen above that conj f_{ij} $(i, j \in J)$ span a dense subspace of $C_{\omega^*}(G)$ and so the assertion of the lemma is now obvious.

As before, let db denote the normalized Haar measure of B.

COROLLARY 1.

$$\int_B (\mathrm{conj}\ \Phi_\omega)\Phi_{\omega'}\, db = \begin{cases} [W_G] & \text{if}\ \omega = \omega', \\ 0 & \text{otherwise.} \end{cases}$$

Fix $f \in C_\omega(G)$ such that $f(1) \neq 0$ (see the remark after Theorem 14) and put $g = \mathrm{conj}\ f$. Then

$$\Theta_{\omega'}(g) = (-1)^m [W_G]^{-1} \int_B F_g \Phi_{\omega'}\, db$$

from Lemma 79. But $\qquad F_g = (-1)^m \operatorname{conj} F_f$

and therefore our assertion follows from Theorem 14 and Lemma 81.

Define the number $c > 0$ by the relation (see Lemma 38)

$$F_f(1; \omega) = (-1)^q cf(1) \quad (f \in \mathcal{C}(G)).$$

Then c has the same value as in Theorem 8 (see [4 (q), § 15]). Fix $\lambda \in L'$ and for any $\omega \in \mathcal{E}_d(\lambda)$, define $c_s(\omega) \, (s \in W(\lambda))$ as in Lemma 78. Then

$$\Theta_\omega = \sum_{s \in W_G \backslash W(\lambda)} \varepsilon(s) \, c_s(\omega) \, \Theta_{s\lambda},$$

where the sum is over a complete system of representatives.

COROLLARY 2. *Let* $\omega \in \mathcal{E}_d(\lambda)$. *Then*

$$d(\omega) = (-1)^q c^{-1} \varpi(\lambda) \sum_{s \in W(\lambda)} c_s(\omega)$$

and $\qquad\qquad\qquad \sum_{s \in W_G \backslash W(\lambda)} |c_s(\omega)|^2 = 1.$

We know from Theorem 14 that

$$(-1)^q cf(1) = F_f(1; \varpi) = d(\omega)^{-1} f(1) \Phi_\omega(1; \varpi) \quad (f \in \mathcal{C}_\omega(G)),$$

and the first relation is an immediate consequence of this fact. The second follows by putting $\omega' = \omega$ in Corollary 1 above.

Let $\mathfrak{T}(\lambda) \, (\lambda \in L')$ be the space of all tempered and invariant distributions Θ on G such that $z\Theta = \chi_\lambda(z) \Theta (z \in \mathfrak{Z})$.

THEOREM 15. *Fix* $\lambda \in L'$. *Then* $\Theta_\omega \, (\omega \in \mathcal{E}_d(\lambda))$ *form a base for* $\mathfrak{T}(\lambda)$ *over* $\mathbf{C}$ *and*

$$[\mathcal{E}_d(\lambda)] = \dim \mathfrak{T}(\lambda) = [W(\lambda)][W_G]^{-1}.$$

Moreover, $\qquad\qquad \sum_{\omega \in \mathcal{E}_d(\lambda)} d(\omega) \Theta_\omega = (-1)^q c^{-1} \sum_{s \in W(\lambda)} \varpi(s\lambda) \Theta_{s\lambda},$

where c and q have the same meaning as in Theorem 8.

We know from Lemma 76 that $\Theta_\omega \, (\omega \in \mathcal{E}_d(\lambda))$ lie in $\mathfrak{T}(\lambda)$ and from Lemma 81 that they are linearly independent. Now fix $\Theta \in \mathfrak{T}(\lambda)$. We have to show that Θ is a linear combination of $\Theta_\omega \, (\omega \in \mathcal{E}_d(\lambda))$. Define analytic functions Φ and Φ_0 on B as follows.

$$\Phi(b) = \Delta(b) \Theta(b) \quad (b \in B')$$

and
$$\Phi_0 = \Phi - \sum_{\omega \in \mathcal{E}_d(\lambda)} c(\omega) \Phi_\omega,$$

where
$$c(\omega) = [W_G]^{-1} \int_B \Phi \operatorname{conj} \Phi_\omega \, db.$$

Then it is clear from Corollary 1 of Lemma 81 that Φ_0 is orthogonal to Φ_ω ($\omega \in \mathcal{E}_d(\lambda)$) in $L_2(B)$. Put
$$\Theta_0 = \Theta - \sum_{\omega \in \mathcal{E}_d(\lambda)} c(\omega) \Theta_\omega.$$

We claim that $\Theta_0 = 0$. In view of Lemma 73, it would be enough to verify that $\Theta_0(f) = 0$ for any K-finite function $f \in C_{-\lambda}(G)$. We may obviously assume that $f \neq 0$. Define V and U_i ($1 \leqslant i \leqslant p$) as in Lemma 77 and let E_i denote the orthogonal projection of V on U_i. Put $f_i = E_i f$ ($1 \leqslant i \leqslant p$). Then it follows from Corollary 1 of Lemma 65 that $f_i \in C_{-\lambda}(G)$. Therefore since
$$\Theta_0(f) = \sum_{1 \leqslant i \leqslant q} \Theta_0(f_i),$$

it would be enough to consider the case when V is irreducible under r.

Let π denote the restriction of r on V and ω the class in $\mathcal{E}_d$ such that $\pi \in \omega^*$. Then $\omega \in \mathcal{E}_d(\lambda)$ and, as we have seen in § 38, there exists an element $\alpha \in C_c^\infty(G)$ such that

$$f(x) = (\alpha, \pi(x)f) = (E\alpha, \pi(x)f) \quad (x \in G).$$

(Here E denotes the orthogonal projection of $L_2(G)$ on V.) This shows that $f \in C_{\omega^*}(G)$. On the other hand, it follows from Lemma 64 and the corollary of Theorem 7, that

$$\Theta_0(f) = (-1)^m [W_G]^{-1} \int_B F_f \Phi_0 \, db.$$

Therefore we conclude from Theorem 14 and the definition of Φ_0 that $\Theta_0(f) = 0$.

Let $s_1, s_2, \ldots, s_p$ be a complete set of representatives of $W_G \backslash W(\lambda)$ in $W(\lambda)$. Then by Lemma 74, the distributions $\Theta_{s_i\lambda}$ ($1 \leqslant i \leqslant \lambda$) also form a base for $\mathfrak{T}(\lambda)$ and therefore

$$[W(\lambda)][W_G]^{-1} = p = \dim \mathfrak{T}(\lambda) = [\mathcal{E}_d(\lambda)].$$

Now put
$$\Theta = \sum_{\omega \in \mathcal{E}_d(\lambda)} d(\omega) \Theta_\omega - (-1)^q c^{-1} \sum_{s \in W(\lambda)} \varpi(s\lambda) \Theta_{s\lambda}.$$

We have to show that $\Theta = 0$. In view of Lemma 73, it would be enough to verify that $\Theta(f) = 0$ for any K-finite function $f \in C_{-\lambda}(G)$. By the argument given above, we are reduced to the case when $f \in C_{\omega^*}(G)$ for some $\omega \in \mathcal{E}_d(\lambda)$. But then $\Theta(f) = 0$ from Lemma 81 and Corollary 1 of Lemma 69. This completes the proof of Theorem 15.

§ 41. Explicit determination of these characters

Put $\varepsilon(\lambda) = \mathrm{sign}\ \varpi(\lambda)$ for $\lambda \in L'$.

THEOREM 16.[1] *For any $\lambda \in L'$, there exists a unique element $\omega(\lambda) \in \mathcal{E}_d$ such that*
$\Theta_{\omega(\lambda)} = (-1)^q \varepsilon(\lambda) \Theta_\lambda$. *The mapping $\lambda \to \omega(\lambda)$ of L' into $\mathcal{E}_d$ is surjective and*

$$d(\omega(\lambda)) = c^{-1}[W_G]\,|\varpi(\lambda)|$$

in the notation of Theorem 15. Finally $\omega(\lambda_1) = \omega(\lambda_2)\,(\lambda_1,\ \lambda_2 \in L')$, if and only if $\lambda_1,\ \lambda_2$ are conjugate under W_G.

We begin by proving the surjectivity first. Fix $\omega \in \mathcal{E}_d$. Then by Lemma 78, Φ_ω is a finite linear combination of the characters of B. Introduce an order on $\mathfrak{F}$ and let λ be the highest element in L such that

$$c_0 = \int_B \Phi_\omega \operatorname{conj} \xi_\lambda\, db \neq 0.$$

(As before, ξ_λ has the same meaning as in [4 (p), § 24].) Then $\lambda \in L'$.

LEMMA 82. $\Theta_\omega = (-1)^q \varepsilon(\lambda) \Theta_\lambda$.

For the proof of this lemma, we may, by going over to a finite covering group of G, assume that K is also acceptable (see [4 (o), § 18]). Let P be the set of all positive roots of $(\mathfrak{g}, \mathfrak{b})$ and P_0, P_+ respectively the sets of all compact and singular roots in P (see [4 (n), § 4]). Put

$$\varrho = \tfrac{1}{2}\sum_{\alpha \in P} \alpha, \qquad \varrho_0 = \tfrac{1}{2}\sum_{\alpha \in P_0} \alpha, \qquad \varrho_+ = \tfrac{1}{2}\sum_{\alpha \in P_+} \alpha.$$

Then ϱ, ϱ_0, ϱ_+ are all in L and $\varrho = \varrho_0 + \varrho_+$. Hence we can define two analytic functions Δ_0 and Δ_+ as follows.

$$\Delta_0(\exp H) = \prod_{\alpha \in P_0} (e^{\alpha(H)/2} - e^{-\alpha(H)/2}), \qquad \Delta_+(\exp H) = \prod_{\alpha \in P_+} (e^{\alpha(H)/2} - e^{-\alpha(H)/2}) \quad (H \in \mathfrak{b}),$$

so that $\Delta = \Delta_0 \Delta_+$. It is clear that

$$\Delta_0(b^s) = \varepsilon(s) \Delta_0(b), \quad \Delta_+(b^s) = \Delta_+(b) \quad (b \in B)$$

for $s \in W_G$.

Let db and dk denote the normalized Haar measures on B and K respectively. Fix a function $\alpha \in C_c^\infty(G)$ such that α is invariant under right translations by K and

$$\int \alpha(x)\, dx = 1.$$

---

[1] Cf. [4 (c), p. 40] and [4(d), Theorem 4].

For any $\beta \in C_c^{\infty}(B')$, we can define $f_\beta \in C_c^{\infty}(G_B)$ (see [4 (o), § 20]) by

$$f_\beta(b^x) = \alpha(x)\,\Delta(b)^{-1} \sum_{s \in W_G} \varepsilon(s)\,\beta(b^s) \quad (x \in G,\ b \in B').$$

Similarly define $g_\beta \in C^{\infty}(K)$ by

$$g_\beta(b^k) = \Delta_+(b)\,\Delta_0(b)^{-1} \sum_{s \in W_G} \varepsilon(s)\,\beta(b^s) \quad (k \in K,\ b \in B').$$

Fix $\pi \in \omega$ and let U be the representation space of π. Define $E_{\mathfrak{b}}$ and $U_{\mathfrak{b}}\,(\mathfrak{b} \in \mathcal{E}_K)$ as usual and put

$$\phi_{\mathfrak{b}}(x) = \operatorname{tr}(E_{\mathfrak{b}}\,\pi(x)\,E_{\mathfrak{b}}) \quad (x \in G).$$

Then $\phi_{\mathfrak{b}}(k) = n(\mathfrak{b})\,\chi_{\mathfrak{b}}(k)\ (k \in K)$, where $n(\mathfrak{b})$ is a nonnegative integer and $\chi_{\mathfrak{b}}$ is the character of $\mathfrak{b}$. Moreover, as we have seen in § 25, there exists an integer $N \geqslant 1$ such that $n(\mathfrak{b}) \leqslant N d(\mathfrak{b})\ (\mathfrak{b} \in \mathcal{E}_K)$.

Put $m = \frac{1}{2} \dim G/B$, $m_0 = \frac{1}{2} \dim K/B$ and

$$T_\beta = \int_K g_\beta(k)\,\pi(k)\,dk, \quad \pi(f_\beta) = \int_G f_\beta(x)\,\pi(x)\,dx \quad (\beta \in C_c^{\infty}(B')).$$

Then $m = m_0 + q$ and it is clear that

$$\pi(f_\beta) = (-1)^m [W_G]^{-1} \int_B \Delta(b) \sum_{s \in W_G} \varepsilon(s)\,\beta(b^s)\,db \int_G \alpha(x)\,\pi(b^x)\,dx$$

$$= (-1)^m \int_B \Delta(b)\,\beta(b)\,db \int_G \alpha(x)\,\pi(b^x)\,dx$$

since $\alpha(xk) = \alpha(x)\ (k \in K)$. Similarly

$$T_\beta = (-1)^{m_0}[W_G]^{-1} \int_B \Delta_0(b)^2\,g_\beta(b)\,db \int_K \pi(b^k)\,dk = (-1)^{m_0} \int_B \Delta(b)\,\beta(b)\,db \int_K \pi(b^k)\,dk.$$

Therefore if follows that

$$\pi(f_\beta) = (-1)^q \int_G \alpha(x)\,\pi(x)\,T_\beta\,\pi(x^{-1})\,dx.$$

Now, by [4 (c), Lemma 24], the operator T_β is summable and therefore it follows easily that

$$\Theta_\omega(f_\beta) = \operatorname{tr}\pi(f_\beta) = (-1)^q \operatorname{tr}T_\beta = (-1)^q \sum_{\mathfrak{b} \in \mathcal{E}_K} n(\mathfrak{b}) \int_K g_\beta(k)\,\chi_{\mathfrak{b}}(k)\,dk.$$

On the other hand, $\qquad \Theta_\omega(f_\beta) = \int \Theta_\omega f_\beta\,dx = (-1)^m \int_B \Phi_\omega \beta\,db.$

7 − 662900. *Acta mathematica*. 116. Imprimé le 14 juin 1966.

Therefore we have obtained the following result.

LEMMA 83. *Put* $\eta_{\mathfrak{b}}(b) = \Delta_0(b)\,\chi_{\mathfrak{b}}(b)$ $(\mathfrak{b} \in \mathcal{E}_K,\ b \in B)$. *Then*

$$\int_B \Phi_\omega \beta\, db = \sum_{\mathfrak{b} \in \mathcal{E}_K} n(\mathfrak{b}) \int_B \Delta_+ \beta\, \eta_{\mathfrak{b}}\, db$$

for $\beta \in C_c^\infty(B')$, *the series being absolutely convergent.*

Define $\mathfrak{K}$ and Ω as in the proof of Lemma 5 and let $\mathfrak{Z}_K$ denote the center of $\mathfrak{K}$. Then $\Omega \in \mathfrak{Z}_K$. Let u be the image of Ω in $\mathfrak{S}(\mathfrak{b}_c)$ under the canonical isomorphism (see [4 (e), Lemma 19]) of $\mathfrak{Z}_K$ into $\mathfrak{S}(\mathfrak{b}_c)$. Then it follows from [4 (e), Theorem 2] that

$$u\eta_{\mathfrak{b}} = c(\mathfrak{b})\,\eta_{\mathfrak{b}} \quad (\mathfrak{b} \in \mathcal{E}_K)$$

in the notation of Lemma 6. Hence

$$\int \beta\eta_{\mathfrak{b}}\,db = c(\mathfrak{b})^{-p} \int u^{*p}\beta \cdot \eta_{\mathfrak{b}}\,db \quad (\beta \in C^\infty(B)),$$

where u^* is the adjoint of the differential operator u and p any positive integer. It follows from Weyl's formula for $\chi_{\mathfrak{b}}$ that $|\eta_{\mathfrak{b}}| \leqslant [W_G]$. Therefore

$$\sum_{\mathfrak{b} \in \mathcal{E}_K} n(\mathfrak{b}) \left| \int \beta\eta_{\mathfrak{b}}\,db \right| \leqslant N[W_G] \sup |u^{*p}\beta| \sum_{\mathfrak{b} \in \mathcal{E}_K} c(\mathfrak{b})^{-p}\, d(\mathfrak{b}),$$

and so we conclude from Lemma 7 that there exists a distribution S_0 on B such that

$$S_0(\beta) = \sum_{\mathfrak{b} \in \mathcal{E}_K} n(\mathfrak{b}) \int_B \beta\eta_{\mathfrak{b}}\,db \quad (\beta \in C^\infty(B)).$$

Put
$$S = \Phi_\omega - \Delta_+ S_0.$$

Then it follows from Lemma 83 that $S = 0$ on B'. Therefore since B is compact, we can choose (see [4 (m), Lemma 21]) an integer $p \geqslant 0$ such that $\Delta^p S = 0$. This means that

$$\int \Phi_\omega \Delta^p \beta\, db = \sum_{\mathfrak{b} \in \mathcal{E}_K} n(\mathfrak{b}) \int \beta \Delta^p \Delta_+ \eta_{\mathfrak{b}}\, db$$

for $\beta \in C^\infty(B)$. Now put $\beta = \mathrm{conj}\ \xi_{\lambda+p\varrho}$. Then it is clear that the left side is equal to c_0. Moreover, we know from Weyl's formula that $\Delta^p\Delta_+\eta_{\mathfrak{b}}$ is a finite linear combination of characters of B with coefficients in $\mathbf{Z}$. Therefore, since $n(\mathfrak{b})$ is an integer, we conclude that $c_0 \in \mathbf{Z}$. Since $c_0 \neq 0$, this shows that $|c_0| \geqslant 1$.

On the other hand, let $s_1=1$, s_2, ..., s_r, be a complete set of representatives in $W(\lambda)$ for $W_G\backslash W(\lambda)$. Then, by Lemma 78 there exist unique complex numbers c_i such that

$$\Theta_\omega = \sum_{1\leqslant i\leqslant r} \varepsilon(s_i)\, c_i\, \Theta_{s_i\lambda}.$$

It is obvious that $c_1=c_0$ and we know from Corollary 2 of Lemma 81 that

$$\sum_{1\leqslant i\leqslant r} |c_i|^2 = 1.$$

Therefore $c_1=c_0=\pm 1$ and $c_i=0$ for $i\geqslant 2$. This shows that

$$\Theta_\omega = c_0\Theta_\lambda.$$

By Theorem 15, there exist exactly r distinct elements $\omega_1=\omega$, ω_2, ..., ω_r in $\mathcal{E}_d(\lambda)$. For each i, we can, by the above proof, choose $s_i\in W(\lambda)$ ($s_1=1$) and a number $c_i=\pm 1$ such that

$$\Theta_{\omega_i}=c_i\Theta_{s_i\lambda} \quad (1\leqslant i\leqslant r).$$

Then it follows from the linear independence of Θ_{ω_i} (Lemma 81) that $s_1, s_2, ..., s_r$ form a complete system of representatives of $W_G\backslash W(\lambda)$. Therefore

$$\sum_i d(\omega_i)\Theta_{\omega_i} = (-1)^q c^{-1}[W_G]\sum_i c_i\varpi(s_i\lambda)\Theta_{\omega_i}$$

from Theorem 15 and this shows that

$$d(\omega_i) = (-1)^q c^{-1}[W_G]|\varpi(\lambda)|c_i\varepsilon(s_i\lambda) \quad (1\leqslant i\leqslant r).$$

But $d(\omega_i)>0$ and so we conclude that $c_i=(-1)^q\varepsilon(s_i\lambda)$ and

$$d(\omega_i) = c^{-1}[W_G]|\varpi(\lambda)|.$$

Hence in particular $c_0=c_1=(-1)^q\varepsilon(\lambda)$ and this proves Lemma 82.

We now come to Theorem 16. Since a class $\omega\in\mathcal{E}$ is completely determined by its character (see [4 (b), p. 250]), the uniqueness of $\omega(\lambda)$ ($\lambda\in L'$) is obvious. Moreover since $\Theta_{s\lambda}=\varepsilon(s)\Theta_\lambda$ ($s\in W_G$) from the definition of Θ_λ (see [4 (p), Theorem 3]), it follows from the linear independence of the characters, that $\omega(\lambda_1)=\omega(\lambda_2)$ ($\lambda_1, \lambda_2\in L'$) if and only if λ_1, λ_2 are conjugate under W_G. Now fix $\lambda\in L'$ and let $r=\dim \mathfrak{T}(\lambda)$. Then by Theorem 15, there are exactly r distinct elements $\omega_1, \omega_2, ..., \omega_r$ in $\mathcal{E}_d(\lambda)$. Moreover, from the above proof, we can choose a complete set of representatives $(s_1, s_2, ..., s_r)$ for $W_G\backslash W(\lambda)$ such that

$$\Theta_{\omega_i} = (-1)^q\varepsilon(s_i\lambda)\Theta_{s_i\lambda} \quad (1\leqslant i\leqslant r).$$

We may assume that $s_1 \in W_G$ and therefore

$$\Theta_{\omega_1} = (-1)^q \varepsilon(s_1\lambda)\,\Theta_{s_1\lambda} = (-1)^q \varepsilon(\lambda)\,\Theta_\lambda.$$

This shows that $\omega_1 = \omega(\lambda)$. We have already seen that $d(\omega_1) = c^{-1}[W_G]\,|\omega(\lambda)|$ and therefore the proof of Theorem 16 is now complete.

Theorem 16 shows that

$$c^{-1}(-1)^q \sum_{\lambda \in L} \varpi(\lambda)\,\Theta_\lambda = \sum_{\omega \in \mathcal{E}_d} d(\omega)\,\Theta_\omega$$

and we know (see [4 (d), § 5]) that this distribution represents the contribution of the discrete series to the Plancherel formula of G.

Part IV. Some inequalities and their consequences

§ 42. Proof of the inequalities

Let us use the notation of § 14 and put $\mathfrak{a}_2 = \mathfrak{h} \cap \mathfrak{p}$, $\mathfrak{n}_2 = \mathfrak{n}_\mathfrak{h}$, $\mathfrak{a}_1 = \mathfrak{a} \cap [\mathfrak{m}, \mathfrak{m}]$, $\mathfrak{k}_1 = \mathfrak{k} \cap \mathfrak{m}$, $\mathfrak{n}_1 = \mathfrak{n} \cap \mathfrak{m}$ and $\mathfrak{m}_1 = \mathfrak{k}_1 + \mathfrak{a}_1 + \mathfrak{n}_1$. We denote the analytic subgroup of G corresponding to a subalgebra of $\mathfrak{g}$ by the corresponding capital latin letter e.g. A_2 and N correspond to $\mathfrak{a}_2$ and $\mathfrak{n}$ respectively. Then $G = KAN$ and $M_1 = K_1 A_1 N_1$ are the Iwasawa decompositions of G and M_1 respectively. For any $x \in G$, let $\varkappa(x)$ and $H(x)$ denote the unique elements $k \in K$ and $H \in \mathfrak{a}$ respectively, such that $x = k \exp H \cdot n$ $(n \in N)$. Let $H_i(x)$ denote the component of $H(x)$ in $\mathfrak{a}_i$ $(i = 1, 2)$ so that $H(x) = H_1(x) + H_2(x)$.

We fix orders in the duals of the real vector spaces $\mathfrak{a}_2$, $\mathfrak{a}$ and $\mathfrak{h}^* = \mathfrak{h} \cap \mathfrak{p} + (-1)^{\frac{1}{2}} \mathfrak{h} \cap \mathfrak{k}$ and assume that they are compatible for the pairs $(\mathfrak{a}, \mathfrak{a}_2)$ and $(\mathfrak{h}^*, \mathfrak{a}_2)$. Let P denote the set of positive roots of $(\mathfrak{g}, \mathfrak{h})$ and Σ the set of positive roots of $(\mathfrak{g}, \mathfrak{a})$. Let P_1 and Σ_1 be the sets of those elements in P and Σ respectively, whose restrictions on $\mathfrak{a}_2$ are zero. We denote by P_2 and Σ_2 the complements of P_1 and Σ_1 in P and Σ respectively.

Put $M_{1\mathfrak{p}} = \exp(\mathfrak{m}_1 \cap \mathfrak{p})$ so that $G = K M_{1\mathfrak{p}} A_2 N_2$ (see [4 (g), Lemma 11]). Fix $x \in G$ and let $x = kman$ $(k \in K,\ m \in M_{1\mathfrak{p}},\ a \in A_2,\ n \in N_2)$. Then k, m, a, n are uniquely determined. Put $\mu(x) = m$. Since M_1 and A_2 commute, it is easy to verify that

$$\varkappa(x) = k\varkappa(m), \quad H_2(x) = \log a, \quad H_1(x) = H(m).$$

Define ϱ and $\Xi_1 = \Xi_M$ as in § 14. Since $\mathfrak{m}$ normalizes $\mathfrak{n}_2$, it is clear that $\operatorname{tr}(\operatorname{ad} X)_{\mathfrak{n}_2} = 0$ for $X \in [\mathfrak{m}, \mathfrak{m}]$. Therefore since $\mathfrak{n} = \mathfrak{n}_1 + \mathfrak{n}_2$, we conclude that

$$\varrho(H) = \tfrac{1}{2} \operatorname{tr}(\operatorname{ad} H)_{\mathfrak{n}_1} \quad (H \in \mathfrak{a}_1).$$

LEMMA 84. *Let $d_1 k$ denote the normalized Haar measure on K_1. Then*

$$\int_{K_1} e^{-\varrho(H(xk))} d_1 k = e^{-\varrho(H_1(x))} \Xi_1(\mu(x)) \quad (x \in G).$$

Since K_1 normalizes N_2, it is easy to see that

$$H(xk) = H_1(xk) + H_2(xk) = H(\mu(xk)) + H_2(x) = H(\mu(x) k) + H_2(x) \quad (k \in K_1).$$

Hence our assertion follows from the fact (see [4 (q), Lemma 31]) that

$$\Xi_1(m) = \int_{K_1} e^{-\varrho(H(mk))} d_1 k \quad (m \in M_1).$$

As usual let dk denote the normalized Haar measure on K. Then the following result is an immediate consequence of Lemma 84 and [4 (q), Lemma 31].

COROLLARY.

$$\int_K e^{-\varrho(H_2(xk))} \Xi_1(\mu(xk)) \, dk = \Xi(x) \quad (x \in G).$$

Put $\bar{N} = \theta(N)$, $\bar{N}_i = \theta(N_i)$ $(i = 1, 2)$ and

$$\beta(H) = \inf_{\alpha \in \Sigma_2} \alpha(H) \quad (H \in \mathfrak{a}).$$

LEMMA 85. $\varrho(H_2(\bar{n})) \geqslant 0$ *and* $\varrho(H(\bar{n})) \geqslant 0$ *for* $\bar{n} \in \bar{N}$. *Moreover, if* $a \in A_2$ *and* $\beta(\log a) \geqslant 0$, *we have*

$$\exp \varrho(H_2(\bar{n}^a)) \leqslant 1 + \exp \left\{ -\tfrac{1}{2}\beta(\log a) + \varrho(H_2(\bar{n})) \right\} \quad (\bar{n} \in \bar{N})$$

and
$$\exp \varrho(H(\bar{n}^a)) \leqslant 1 + \exp \left\{ -\beta(\log a) + \varrho(H(\bar{n})) \right\} \quad (\bar{n} \in \bar{N}_2).$$

Let Z be the center of G. Then, for the purpose of this lemma, we can obviously replace G by G/Z. Hence we may agree to subscribe to the assumptions and conventions of [4 (j), p. 244].

For any linear function λ on $\mathfrak{h}_c$, define $H_\lambda \in \mathfrak{h}_c$ as usual by the condition

$$\mathrm{tr} \, (\mathrm{ad} \, H \, \mathrm{ad} \, H_\lambda) = \lambda(H) \quad (H \in \mathfrak{h}_c).$$

Also put $\langle \lambda_1, \lambda_2 \rangle = \lambda_1(H_{\lambda_2})$ for two such functions λ_1, λ_2. Let J_P denote the set of all λ such that $2\langle \lambda, \alpha \rangle / \langle \alpha, \alpha \rangle$ is a nonnegative integer for every $\alpha \in P$. Then for every $\lambda \in J_P$, we have an irreducible representation π_λ of G on a finite-dimensional (complex) Hilbert space V_λ with the highest weight λ (with respect to $\mathfrak{h}$). We denote the corresponding representation of $\mathfrak{G}$ also by π_λ. Let v_λ denote a unit vector in V_λ belonging to the highest weight λ.

LEMMA 86. *Fix $\lambda \in J_P$ and let U be the subspace consisting of all $v \in V_\lambda$ such that $\pi_\lambda(H)v = \lambda(H)v$ for all $H \in \mathfrak{a}_2$. Then U is invariant and irreducible under $\pi_\lambda(\mathfrak{m})$. Moreover, $\dim U = 1$ if and only if $\langle \lambda, \alpha \rangle = 0$ for $\alpha \in P_1$.*

We write V and π for V_λ and π_λ respectively. It is clear that

$$\mathfrak{g} = \theta(\mathfrak{n}_2) + \mathfrak{m} + \mathfrak{n}_2$$

and therefore

$$\mathfrak{G} = \theta(\mathfrak{N}_2)\mathfrak{M}\mathfrak{N}_2,$$

where[1] $\mathfrak{M} = \mathfrak{S}(\mathfrak{m}_c)$ and $\mathfrak{N}_2 = \mathfrak{S}((\mathfrak{n}_2)_c)$.

For any $\alpha \in P$ define X_α, $X_{-\alpha}$ as in [4 (n), § 4] and put

$$\mathfrak{l}_\alpha = CH_\alpha + CX_\alpha + CX_{-\alpha}.$$

Fix $u \neq 0$ in U. Then if $\alpha \in P_2$, it is clear that

$$\pi(HX_\alpha)u = (\lambda(H) + \alpha(H))\pi(X_\alpha)u \quad (H \in \mathfrak{a}_2).$$

Since λ is the highest weight of π, we conclude from the definition of our order, that $\pi(X_\alpha)u = 0$. Hence

$$V = \pi(\mathfrak{G})u = \pi(\theta(\mathfrak{N}_2)\mathfrak{M})u.$$

But then it is obvious from the definition of U that $U = \pi(\mathfrak{M})u$. This proves that U is invariant and irreducible under $\pi(\mathfrak{m})$.

Now fix $\alpha \in P$ and observe that $\pi(X_\alpha)v_\lambda = 0$. Hence by considering the subalgebra $\mathfrak{l}_\alpha$, it follows (see [4 (m), Lemma 25]) that $\pi(X_{-\alpha})v_\lambda = 0$ if and only if $\langle \lambda, \alpha \rangle = 0$. On the other hand, $\mathfrak{m}' = [\mathfrak{m}, \mathfrak{m}]$ is clearly generated, as a Lie algebra, by $\theta(\mathfrak{n}_1) + \mathfrak{n}_1$. Hence $\pi(\mathfrak{m}')v_\lambda = \{0\}$ if and only if $\langle \lambda, \alpha \rangle = 0$ for all $\alpha \in P_1$. Since U is irreducible under $\pi(\mathfrak{m})$, the second assertion of the lemma is now obvious

LEMMA 87. *Fix $\Lambda \in J_p$. Then $\Lambda(H_2(\bar{n})) \geq 0$ and*

$$\exp \Lambda(H_2(\bar{n}^a)) \leqslant 1 + \exp\left\{-\tfrac{1}{2}\beta(\log a) + \Lambda(H_2(\bar{n}))\right\}$$

for $\bar{n} \in \bar{N}$ and $a \in A_2$ provided $\beta(\log a) \geq 0$.

Put $\lambda = \Lambda - \theta\Lambda$ so that

$$\langle \lambda, \alpha \rangle = \langle \Lambda, \alpha \rangle - \langle \Lambda, \theta\alpha \rangle \quad (\alpha \in P).$$

Obviously this is zero if $\alpha \in P_1$. On the other hand, $-\theta\alpha \in P_2$ whenever $\alpha \in P_2$ and therefore,

---

[1] We use here the notation of [4 (m), p. 280].

since[1] $|\theta\alpha|^2 = |\alpha|^2$, it follows that $\lambda \in J_p$. Put $\pi = \pi_\lambda$, $V = V_\lambda$, $v = v_\lambda$ and observe that $\lambda = 2\Lambda$ on $\mathfrak{h} \cap \mathfrak{p} = \mathfrak{a}_2$ and $\lambda = 0$ on $\mathfrak{h} \cap \mathfrak{k}$. Hence dim $U = 1$ and $\pi(\mathfrak{m}')U = \{0\}$ from Lemma 86. On the other hand, it is easy to see that $\mathfrak{m}_1 \cap \mathfrak{p} \subset \mathfrak{m}'$ so that $\pi(m)v = v$ for $m \in M_{1\mathfrak{p}}$. Therefore it is obvious that

$$|\pi(x)v| = e^{\lambda(H_2(x))} = e^{2\Lambda(H_2(x))} \quad (x \in G).$$

Let E denote the orthogonal projection of V on U. Then if $X \in \theta(\mathfrak{n}_2)$, it is obvious that $E\pi(X^r)v = 0$ for $r \geq 1$. Moreover, $\theta(\mathfrak{n}_1) \subset \mathfrak{m}'$ and therefore $\pi(\theta(\mathfrak{n}_1))v = \{0\}$. On the other hand, $\mathfrak{n}_2$ is an ideal in $\mathfrak{n}$ and therefore $\bar{N} = \bar{N}_2 \bar{N}_1$. Hence $E\pi(\bar{n})v = v$ and this shows that $|\pi(\bar{n})v| \geq |v| = 1$ $(\bar{n} \in \bar{N})$. Hence

$$\Lambda(H_2(\bar{n})) \geq 0 \quad (\bar{n} \in \bar{N}).$$

Put $E' = 1 - E$ and let $\|T\|$ denote the Hilbert–Schmidt norm of a linear transformation T in V. Since $U = Cv$, it is clear that

$$\exp 4\Lambda(H_2(\bar{n}^a)) = \|\pi(\bar{n}^a)E\|^2 = \|E\pi(\bar{n}^a)E\|^2 + \|E'\pi(\bar{n}^a)E\|^2 = 1 + \|E'\pi(\bar{n}^a)E\|^2$$

since $E\pi(\bar{n}^a)E = E$ as we saw above. On the other hand, we have seen during the proof of Lemma 86 that

$$V = \pi(\theta(\mathfrak{N}_2)\mathfrak{M})v = \pi(\theta(\mathfrak{N}_2))v.$$

Therefore every weight of π, other than λ, is of the form $\lambda - \sigma$ with

$$\sigma = \sum_{1 \leq i \leq r} \alpha_i,$$

$\alpha_i \in P_2$ and $r \geq 1$. Let $\lambda = \lambda_0 > \lambda_1 > \ldots \lambda_p$ be all the weights of π and V_i the subspace of V consisting of all vectors belonging to the weight λ_i $(0 \leq i \leq p)$. Since $\mathfrak{h} = \theta(\mathfrak{h})$, V is the orthogonal sum of V_i $(0 \leq i \leq p)$. Put $\sigma_i = \lambda - \lambda_i$ and let E_i denote the orthogonal projection of V on V_i. Then it is clear that $E' = E_1 + \ldots + E_p$ and therefore

$$E'\pi(\bar{n}^a)E = \sum_{1 \leq i \leq p} e^{-\sigma_i(\log a)} E_i\pi(\bar{n})E.$$

On the other hand, since $\beta(\log a) \geq 0$, it is obvious that

$$\sigma_i(\log a) \geq \inf_{\alpha \in P_2} \alpha(\log a) = \beta(\log a).$$

Therefore

$$\|E'\pi(\bar{n}^a)E\|^2 \leq e^{-2\beta(\log a)} \sum_{1 \leq i \leq p} \|E_i\pi(\bar{n})E\|^2$$

$$\leq e^{-2\beta(\log a)}|\pi(\bar{n})v|^2 = \exp\{-2\beta(\log a) + 4\Lambda(H_2(\bar{n}))\},$$

and the assertion of the lemma is now obvious.

---

[1] As usual, $|\alpha|^2 = \langle \alpha, \alpha \rangle$.

Let π be an irreducible finite-dimensional representation of G on V and Λ the highest weight of π with respect to $\mathfrak{a}$. Let U be the subspace of those vectors $u \in V$ for which $\pi(H)u = \Lambda(H)u$ ($H \in \mathfrak{a}_2$). We denote by E the orthogonal projection of V on U.

LEMMA 88. *Fix $a \in A_2$ such that $\beta(\log a) \geqslant 0$. Then*

$$\exp \Lambda(H(\bar{n}^a)) \leqslant 1 + \exp \{ -\beta(\log a) + \Lambda(H(\bar{n})) \}$$

and $$\|\pi(\bar{n}^a) E\| \leqslant \|E\| + e^{-\beta(\log a)} \|\pi(\bar{n}) E\|$$

for $\bar{n} \in \bar{N}_2$. Moreover, $\Lambda(H(\bar{n})) \geqslant 0$ for $\bar{n} \in \bar{N}$.

Let $\Lambda = \Lambda_1 > \Lambda_2 > ... > \Lambda_p$ be all the weights of π with respect to $\mathfrak{a}$. It follows from the definition of our orders that there exists an integer $q \geqslant 1$ such that $\sigma_i = \Lambda - \Lambda_i$ is zero on $\mathfrak{a}_2$ for $i \leqslant q$ while $\sigma_i \neq 0$ on $\mathfrak{a}_2$ for $i > q$. Let E_i denote the orthogonal projection of V on the space V_i consisting of all vectors belonging to the weight Λ_i. Then $E = E_1 + ... + E_q$ and

$$E' = 1 - E = \sum_{q < i \leqslant p} E_i.$$

Fix $i > q$. Then it is clear (see the proof of Lemma 86) that

$$\sigma_i = \sum_{\alpha \in \Sigma} r(\alpha) \alpha,$$

where $r(\alpha)$ are nonnegative integers and $r(\alpha) \geqslant 1$ for some $\alpha \in \Sigma_2$. Hence if v is a unit vector in V_1, it is clear that

$$|\pi(\bar{n}^a) v|^2 = 1 + |E' \pi(\bar{n}^a) v|^2 \leqslant 1 + e^{-2\beta(\log a)} |\pi(\bar{n}) v|^2 \quad (\bar{n} \in \bar{N}_2)$$

and from this the first inequality follows immediately. Moreover,

$$\|\pi(\bar{n}^a) E\|^2 = \|E\|^2 + \|E' \pi(\bar{n}^a) E\|^2 \leqslant \|E\|^2 + e^{-2\beta(\log a)} \|E' \pi(\bar{n}) E\|^2$$
$$\leqslant \|E\|^2 + e^{-2\beta(\log a)} \|\pi(\bar{n}) E\|^2 \quad (\bar{n} \in \bar{N}_2)$$

and this gives the second inequality. The last statement of Lemma 88 has already been proved in [4 (j), Lemma 43].

Now if we take $\Lambda = \frac{1}{2}\Sigma_{\alpha \in P} \alpha$ in Lemma 87, and choose π, in Lemma 88, such that its highest weight, with respect to $\mathfrak{a}$, is ϱ, then we get Lemma 85 immediately.[1]

---

[1] The second inequality of Lemma 88 has been proved for later applications.

§ 43. Applications of the above inequalities

Define the number $d \geqslant 0$ as in § 14 and observe that $\varrho(H(\bar{n})) \geqslant 0$ for $\bar{n} \in \bar{N}$ from Lemma 85. Let $d_i \bar{n}$ denote the Haar measure on $\bar{N}_i$ ($i = 1, 2$).

LEMMA 89.[1] *For any $\varepsilon > 0$,*

$$\int_{\bar{N}_2} e^{-\varrho(H(\bar{n}))} \{1 + \varrho(H(\bar{n}))\}^{-(d+\varepsilon)} d_2 \bar{n} < \infty.$$

We can choose $c > 0$ (see [4 (j), Theorem 3]) such that

$$e^{\varrho(\log a)} \Xi(a) \leqslant c(1 + \sigma(a))^d \quad (a \in A).$$

On the other hand, by [4 (j), Cor. 2, p. 289],

$$e^{\varrho(\log a)} \Xi(a) = \int_{\bar{N}} \exp\left\{-\varrho(H(\bar{n}^a)) - \varrho(H(\bar{n}))\right\} d\bar{n} \quad (a \in A),$$

where $d\bar{n}$ is the (suitably normalized) Haar measure on $\bar{N}$. We may assume that $d\bar{n} = d_2 \bar{n}_2 \cdot d_1 \bar{n}_1$ for $\bar{n} = \bar{n}_2 \bar{n}_1$ ($\bar{n}_i \in \bar{N}_i$, $i = 1, 2$). Now

$$\bar{n} = \bar{n}_2 \bar{n}_1 \in \bar{n}_2 k^{-1} \exp H(\bar{n}_1) \cdot N_1,$$

where $k = \varkappa(\bar{n}_1)^{-1} \in K_1$. Since M normalizes $\theta(\mathfrak{n}_2)$, we get

$$H(\bar{n}) = H(\bar{n}_2{}^k) + H(\bar{n}_1).$$

We may normalize $d_1 \bar{n}_1$ in such a way (see [4 (j), Lemma 44]) that

$$\int_{\bar{N}_1} e^{-2\varrho(H(\bar{n}_1))} d_1 \bar{n}_1 = 1.$$

Then, since $d_2 \bar{n}_2{}^k / d_2 \bar{n}_2 = 1$ ($k \in K_1$) and A_2 commutes with $\bar{N}_1$, we conclude that

$$e^{\varrho(\log a)} \Xi(a) = \int_{\bar{N}_2} \exp\left\{-\varrho(H(\bar{n}^a)) - \varrho(H(\bar{n}))\right\} d_2 \bar{n} \leqslant c(1 + \sigma(a))^d \quad (a \in A_2).$$

Put $a_t = \exp tH$ ($t \in \mathbf{R}$) where H is an element in $\mathfrak{a}_2$ such that

$$b = \beta(H) > 0.$$

Then if $a = a_t$ ($t \geqslant 0$), it follows from Lemma 85 that

$$\exp \varrho(H(\bar{n}^a)) \leqslant 1 + \exp\left\{\varrho(H(\bar{n})) - bt\right\} \quad (\bar{n} \in \bar{N}_2)$$

and therefore

---

[1] Cf. [4 (j), Lemma 45].

$$\int_{\bar{N}_2} e^{-\varrho(H(\bar{n}))}\{1 + e^{\varrho(H(\bar{n}))-bt}\}^{-1} d_2\bar{n} \leqslant c(1+\sigma(a_t))^d \leqslant c'(1+t)^d$$

for $t \geqslant 0$. Here c' is a positive constant independent of t. Let $\bar{N}_{2,r}$ denote the set of all $\bar{n} \in \bar{N}_2$ with $\varrho(H(\bar{n})) \leqslant 2^r$ and put $t = 2^r b^{-1}$. Then

$$e^{\varrho(H(\bar{n}))-bt} \leqslant 1$$

for $\bar{n} \in \bar{N}_{2,r}$ and therefore

$$\int_{\bar{N}_{2,r}} e^{-\varrho(H(\bar{n}))} d_2\bar{n} \leqslant 2c'(1+b^{-1}2^r)^d \leqslant c_1 2^{rd} \quad (r \geqslant 0),$$

where c_1 is a positive number independent of r. Let $\bar{N}_2(r)$ denote the complement of $\bar{N}_{2,r-1}$ in $\bar{N}_{2,r}$ $(r \geqslant 1)$. Then

$$\int_{\bar{N}_2(r)} e^{-\varrho(H(\bar{n}))}\{1 + \varrho(H(\bar{n}))\}^{-(d+\varepsilon)} d_2\bar{n} \leqslant c_1 2^{rd-(r-1)(d+\varepsilon)} = c_1 2^{d-(r-1)\varepsilon} \quad (r \geqslant 1).$$

Since $\bar{N}_{2,0}$ is compact [4 (j), Lemma 40], we get the required result from the convergence of the series $\sum_{r \geqslant 0} 2^{-r\varepsilon}$.

LEMMA 90. *Put* $\bar{n} = \theta(n^{-1})$ *for* $n \in N$. *Then there exists a number* $c \geqslant 1$ *such that*

$$1 + \max\,(\sigma(h), \varrho(H(\bar{n}))) \leqslant c(1+\sigma(hn))$$

and $$\Xi(hn) \leqslant c(1+\sigma(hn))^d \exp\{-\varrho(\log h) - \varrho(H(\bar{n}))\}$$

for $h \in A$ *and* $n \in N$.

It is clear that, for the proof of this lemma, we may assume, as in § 42, that the conditions of [4 (j), p. 244] hold. Fix an irreducible finite-dimensional representation π of G with the highest weight ϱ with respect to $\mathfrak{a}$. Define A^+ as in § 14. Then $G = KA^+K$ and therefore $hn = k_1 h' k_2$ $(k_1, k_2 \in K;\ h' \in A^+)$. Then

$$\bar{n}h = \theta(hn)^{-1} = k_2^{-1}h'k_1^{-1}$$

and therefore $\|\pi(hn)\| = \|\pi(h')\| = \|\pi(\bar{n}h)\|$. Hence if p is the degree of π, we get

$$e^{2\varrho(\log h')} \leqslant \|\pi(h')\|^2 \leqslant p e^{2\varrho(\log h')}.$$

Let $\mathfrak{w}$ be the Weyl group of $\mathfrak{g}$ with respect to $\mathfrak{a}$ (see [4 (j), p. 249]). Fix $s \in \mathfrak{w}$ and choose $k \in K$ such that $\mathrm{Ad}\,(k)H = sH$ for all $H \in \mathfrak{a}$. Then (see [1, § 7.4]) $\bar{n}^k = \bar{n}_1 n_2$ where

$$\bar{n}_1 \in \bar{N} \cap \bar{N}^k, \quad n_2 \in N \cap \bar{N}^k.$$

Hence
$$\|\pi(h'\| = \|\pi(\bar{n}^k h^s)\| = \|\pi(\bar{n}_1 h^s n_2')\|,$$

where $n_2' = (h^s)^{-1} n_2 h^s$. Therefore

$$\|\pi(h')\|^2 \geqslant |\pi(\bar{n}_1 h^s)\,\psi|^2 \geqslant e^{2\varrho(\log h^s)},$$

where ψ is a unit vector belonging to the highest weight ϱ. This shows that

$$\varrho(\log h') + \tfrac{1}{2} \log p \geqslant \max_{s\in\mathfrak{w}} \varrho(\log h^s) \geqslant |\varrho(\log h)|$$

(see [4 (j), p. 281]). On the other hand, we can obviously choose c_1, $c_2 > 0$ such that

$$c_1 \sigma(h_1) \leqslant \varrho(\log h_1) \leqslant c_2 \sigma(h_1) \quad (h_1 \in A^+).$$

Then
$$c_2 \sigma(h') + \tfrac{1}{2} \log p \geqslant \max_{s\in\mathfrak{w}} \varrho(\log h^s) \geqslant c_1 \sigma(h).$$

Since $\sigma(h') = \sigma(hn)$, this shows that we can choose $c_3 > 0$ such that

$$c_3(1 + \sigma(hn)) \geqslant 1 + \sigma(h) \quad (h \in A,\ n \in N).$$

Moreover, we know [4 (k), Lemma 42] that

$$\varrho(\log h') \geqslant \varrho(\log h) + \varrho(H(\bar{n})).$$

Since
$$\varrho(\log h') + \tfrac{1}{2} \log p \geqslant -\varrho(\log h)$$

by our result above, we conclude that

$$2\varrho(\log h') + \tfrac{1}{2} \log p \geqslant \varrho(H(\bar{n})).$$

Hence we can choose $c_4 > 0$ such that

$$1 + \varrho(H(\bar{n})) \leqslant c_4(1 + \sigma(hn)) \quad (h \in A,\ n \in N).$$

Now select $c_5 > 0$ such that

$$\Xi(h_1) \leqslant c_5\, e^{-\varrho(\log h_1)} (1 + \sigma(h_1))^d \quad (h_1 \in A^+).$$

Then

$$\Xi(hn) = \Xi(h') \leqslant c_5\, e^{-\varrho(\log h')} (1 + \sigma(h'))^d \leqslant c_5 (1 + \sigma(hn))^d \exp\{-\varrho(\log h) - \varrho(H(\bar{n}))\}.$$

This proves Lemma 90.

Put $\varrho_i(H) = \mathrm{tr}\,(\mathrm{ad}\,H)_{\mathfrak{n}_i}$ $(H \in \mathfrak{a},\ i = 1, 2)$ so that $\varrho = \varrho_1 + \varrho_2$.

COROLLARY 1. *Suppose r_1, r_2 are two numbers $\geqslant 0$ and $r = r_1 + r_2$. Then*

$$e^{\varrho_\mathbf{2}(\log h)}\,\Xi\,(hn)\,(1 + \sigma(hn))^{-(r+d)} \leqslant c^{r+1}\,e^{-\varrho_\mathbf{1}(\log h)}\,(1 + \sigma(h))^{-r_\mathbf{1}}\,e^{-\varrho(H(\bar{n}))}\,(1 + \varrho(H(\bar{n})))^{-r_\mathbf{1}}$$

for $h \in A$ and $n \in N$.

For, in the above notation, we have

$$\Xi\,(hn)\,(1 + \sigma(hn))^{-(r+d)} = \Xi\,(h')\,(1 + \sigma(h'))^{-(r+d)} \leqslant c\,e^{-\varrho(\log h')}\,(1 + \sigma(h'))^{-r}.$$

But $$(1 + \sigma(h'))^{-r} \leqslant c^r(1 + \sigma(h))^{-r_\mathbf{1}}\,(1 + \varrho(H(\bar{n})))^{-r_\mathbf{2}}$$

and $$\varrho(\log h') \geqslant \varrho(\log h) + \varrho(H(\bar{n})).$$

Hence our assertion is obvious.

Let $d_2 n$ denote the Haar measure on N_2.

COROLLARY 2. *Let Ω be a compact set in G. Then if $r > 2d$, the integral*

$$\int_{N_\mathbf{2}} \Xi\,(xn)\,(1 + \sigma(xn))^{-r}\,d_2 n$$

converges uniformly for $x \in \Omega$.

Let $x = k_0 h n_0$ ($k_0 \in K$, $h \in A$, $n_0 \in N$). Then $\Xi(xn) = \Xi(hn_0 n)$ and $\sigma(xn) = \sigma(hn_0 n)$. Now let $\bar{n}_0 = \bar{n}_2 \bar{n}_1$ where $\bar{n}_i \in \bar{N}_i$ ($i = 1, 2$). Then h and $\bar{n}_i$ remain bounded[1] and

$$H(\theta(n_0 n)^{-1}) = H(\bar{n}\bar{n}_2\bar{n}_1) = H(\bar{n}\bar{n}_2 k^{-1}) + H(\bar{n}_1) = H((\bar{n}\bar{n}_2)^k) + H(\bar{n}_1) \quad (n \in N_2),$$

where $k = \varkappa(\bar{n}_1)^{-1} \in K_1$. Fix a compact set U in $\bar{N}_2$ such that $\bar{n}_2$ stays within U. Since $r - d > d$, we can, by Lemma 89, choose, for a given $\varepsilon > 0$, a compact set V_0 in $\bar{N}_2$ such that

$$\int_{{}^c V_\mathbf{0}} e^{-\varrho(H(\bar{n}))} \{1 + \varrho(H(\bar{n}))\}^{-r+d}\,d_2\bar{n} \leqslant \varepsilon,$$

where ${}^c V_0$ denotes the complement of V_0 in $\bar{N}_2$. Put $V = V_0^{K_1} U^{-1}$. Then if $\bar{n} \in {}^c V$, it is clear that $(\bar{n}\bar{n}_2)^k \in {}^c V_0$. Moreover, since h remains bounded, we can choose c_1 such that $c^{r+1}\,e^{-\varrho(\log h)} \leqslant c_1$. Then we conclude from Corollary 1 above that

$$\Xi\,(hn)\,(1 + \sigma(hn))^{-r} \leqslant c_1\,e^{-\varrho(H(\bar{n}))}\,(1 + \varrho(H(\bar{n})))^{-r+d} \quad (n \in \bar{N}_2).$$

We may obviously assume that $d_2 n = d_2\bar{n}$ under the mapping $n \to \bar{n}$. Therefore since $\varrho(H(\bar{n}_1)) \geqslant 0$ by Lemma 85, it is clear that

---

[1] This means that they stay within compact sets as x varies in Ω.

$$\int_{CV} \Xi(hn_0 n)\,(1+\sigma(hn_0 n))^{-r}\,d_2 n \leqslant c_1 \int_{CV_0} e^{-\varrho(H(\bar{n}))}\{1+\varrho(H(\bar{n}))\}^{-r+d}\,d_2 \bar{n} \leqslant c_1 \varepsilon.$$

This proves our assertion.

§ 44. Proof of Lemma 21

We now come to the proof of Lemma 21. Put $A_1^{+}=\exp \mathfrak{a}_1^{+}$, $\mathfrak{a}_1^{+}$ being the set of all $H \in \mathfrak{a}_1$ where $\alpha(H) \geqslant 0$ $(\alpha \in \Sigma_1)$. Then $M_1 = K_1 A_1^{+} K_1$ and $M = M_1 A_2 = K_1 (A_1^{+} A_2) K_1$. Therefore it is obviously enough to consider the case when $m = h = h_1 h_2$ $(h_1 \in A_1^{+}, h_2 \in A_2)$. Put $r_1 = r'$ and $r_2 = d + r - r'$. Then it follows from Lemma 89 and Corollary 1 of Lemma 90 that

$$e^{\varrho_1(\log h)} \int_{N_2} \Xi(hn)\,(1+\sigma(hn))^{-(r+2d)}\,d_2 n \leqslant c_1\, e^{-\varrho_1(\log h)}\,(1+\sigma(h))^{-r'},$$

where c_1 is a positive number independent of h. Since

$$e^{\varrho_1(\log h)} \Xi_1(h) = e^{\varrho_1(\log h_1)} \Xi_1(h_1) \geqslant 1$$

from [4 (j), Lemma 36], the first statement of Lemma 21 is now obvious. The second is an immediate consequence of Lemma 90 and the relation $M = K_1 A K_1$.

§ 45. Appendix

We now use the notation of §§ 27, 28. Put $\varrho_i(H) = \frac{1}{2} \operatorname{tr}\,(\operatorname{ad} H)_{\mathfrak{n}_i}$ $(H \in \mathfrak{a}_\mathfrak{p}, i=1,2)$ so that $\varrho = \varrho_1 + \varrho_2$. Let M_1 and A_2 be the analytic subgroups of G corresponding to $\mathfrak{m}_1 = \mathfrak{k} + \mathfrak{p} \cap [\mathfrak{m}, \mathfrak{m}]$ and $\mathfrak{a}_2 = \mathfrak{l}$ respectively. Then $M = M_1 A_2$ and $d(ma) = e^{\varrho(\log a)}$ $(m \in M_1, a \in A_2)$. Hence it follows without difficulty that

$$d^{-1} X \circ d = X' \quad (X \in \mathfrak{m}),$$

where $X \rightarrow X'$ is the isomorphism of $\mathfrak{m}$ into $\mathfrak{M}$ given by $H' = H + \varrho(H)$, $Y' = Y$ $(H \in \mathfrak{l}, Y \in \mathfrak{m}_1)$. This gives rise to an automorphism $v \rightarrow v'$ of $\mathfrak{M}$ which preserves $\mathfrak{Z}_1$.

Now let $H \in \mathfrak{a}_\mathfrak{p}^{+}$. Then

$$d(\exp H)\,\Xi(\exp H) = e^{\varrho_2(H)}\,\Xi(H) = e^{-\varrho_1(H)}\,e^{\varrho(H)}\,\Xi(H).$$

The assertion of Lemma 47 now follows immediately if we apply [4 (j), Theorem 3] and observe [4 (j), Lemma 36] that

$$1 \leqslant e^{\varrho_1(H)}\,\Xi_1(\exp H).$$

Since $\mathfrak{g} = \theta(\mathfrak{n}_2) + \mathfrak{m} + \mathfrak{n}_2$, it is clear that

$$\mathfrak{G} = \theta(\mathfrak{N}_2)\,\mathfrak{M}\mathfrak{N}_2,$$

where $\mathfrak{N}_2 = \mathfrak{S}(\mathfrak{n}_{2c})$. We know (see the proof of the corollary of Lemma 13 of [4 (q)]) that

$$z - \mu_0(z)' \in \mathfrak{G}\mathfrak{n}_2 \quad (z \in \mathfrak{Z}).$$

Put $u = z - \mu_0(z)'$. Then u commutes with $\mathfrak{l}$ and since

$$\mathfrak{G} = \mathfrak{M}\mathfrak{N}_2 + \theta(\mathfrak{n}_2)\,\mathfrak{G},$$

it is obvious that $u \in \theta\,(\mathfrak{n}_2)\,\mathfrak{G}\mathfrak{n}_2$. This is the result needed in § 28.

Now suppose $\mathfrak{g}$ and G are defined as in § 7. Let A be a Cartan subgroup of G, A_0 the center of A and $\tilde{A}$ the normalizer of A in G. Put $W_A = \tilde{A}/A_0$. Then A_0 is open in A and W_A is a finite group (see [4 (o), § 20]). We denote by $x \to x^*$ the natural projection of G on $G^* = G/A_0$.

Let $\mathfrak{h}$ be the Lie algebra of A and dx, da the Haar measures on G and A respectively. Put $G_A = (A')^G$ as usual (see [4 (o), § 20]).

LEMMA 91. *Let dx^* be the invariant measure on G^* such that*

$$\int_G f(x)\,dx = \int_{G^*} dx^* \int_{A_0} f(xa)\,da \quad (f \in C_c(G)).$$

Then $\qquad \displaystyle \int_{G_A} f(x)\,dx = [W_A]^{-1} \int_A \nu(a)\,da \int_{G^*} f(a^{x^*})\,dx^* \quad (f \in C_c(G_A))$

in the notation of [4 (o), § 22], where

$$\nu(a) = \left| \det\,(\mathrm{Ad}\,(a^{-1}) - 1)_{\mathfrak{g}/\mathfrak{h}} \right|.$$

Let ϕ denote the mapping $(x^*, a) \to a^{x^*}$ of $G^* \times A'$ onto G_A. Then we know (see [4 (o), § 20]) that ϕ is regular and $\phi^{-1}(x)$ $(x \in G_A)$ contains exactly $[W_A]$ points in $G^* \times A'$. Hence our result follows from a simple computation which gives the functional determinant of this mapping.

References

[1]. BOREL, A. & HARISH-CHANDRA, Arithmetic subgroups of algebraic groups. *Ann. of Math.*, 75 (1962), 485–535.

[2]. BRUHAT, F., Sur les représentations induites des groupes de Lie. *Bull. Soc. Math. France*, 84 (1956), 97–205.

[3]. GINDIKIN, S. G. & KARPELEVIČ, F. I., Plancherel measure of Riemannian symmetric spaces of nonpositive curvature. *Soviet Math.*, 3 (1962), 962–965.

[4]. HARISH-CHANDRA, (a) Representations of a semisimple Lie group on a Banach space, I. *Trans. Amer. Math. Soc.*, 75 (1953), 185–243.

(b) Representations of semisimple Lie groups, III. *Trans. Amer. Math. Soc.*, 76 (1954), 234–253.

(c) Representations of semisimple Lie groups, V. *Amer. J. Math.*, 78 (1956), 1–41.

(d) Representations of semisimple Lie groups, VI. *Amer. J. Math.*, 78 (1956), 564–628.

(e) The characters of semisimple Lie groups. *Trans. Amer. Math. Soc..*, 83 (1956), 98–163.

(f) Differential operators on a semisimple Lie algebra. *Amer. J. Math.*, 79 (1957), 87–120.

(g) Fourier transforms on a semisimple Lie algebra, I. *Amer. J. Math.*, 79 (1957), 193–257.

(h) Fourier transforms on a semisimple Lie algebra, II. *Amer. J. Math.*, 79 (1957), 653–686.

(i) A formula for semisimple Lie groups. *Amer. J. Math.*, 79 (1957), 733–760.

(j) Spherical functions on a semisimple Lie group, I. *Amer. J. Math.*, 80 (1958), 241–310.

(k) Spherical functions on a semisimple Lie group, II. *Amer. J. Math.*, 80 (1958), 553–613.

(l) Invariant eigendistributions on semisimple Lie groups. *Bull. Amer. Math. Soc.*, 69 (1963), 117–123.

(m) Invariant distributions on Lie algebras. *Amer. J. Math.*, 86 (1964), 271–309.

(n) Some results on an invariant integral on a semisimple Lie algebra. *Ann. of Math.*, 80 (1964), 551–593.

(o) Invariant eigendistributions on a semisimple Lie group. *Trans. Amer. Math. Soc.*, 119 (1965), 457–508.

(p) Discrete series for semisimple Lie groups, I. *Acta Math.*, 113 (1965), 241–318.

(q) Two theorems on semisimple Lie groups. *Ann. of Math.*, 83 (1966), 74–128.

[5]. HELGASON, S., (a) *Differential geometry and symmetric spaces.* Academic Press, New York, 1962.

(b) Fundamental solutions of invariant differential operators on symmetric spaces. *Amer. J. Math.*, 86 (1964), 565–601.

[6]. LANGLANDS, R. P., The dimension of spaces of automorphic forms. *Amer. J. Math.*, 85 (1963), 99–125.

[7]. MACKEY, G. W., Infinite-dimensional group representations. *Bull. Amer. Math. Soc.*, 69 (1963), 628–686.

[8]. SEGAL, I. E., Hypermaximality of certain operators on Lie groups. *Proc. Amer. Math. Soc.*, 3 (1952), 13–15.

[9]. SELBERG, A., Harmonic analysis and discontinuous groups in weakly symmetric Riemannian spaces with applications to Dirichlet series. *J. Indian Math. Soc.*, 20 (1956), 47–87.

[10]. WEYL, H., Theorie der Darstellung kontinuierlicher halbeinfacher Gruppen durch lineare Transformationen, III. *Math. Z.*, 24 (1926), 377–395.

Received October 22, 1965

Reprinted from
Acta. Math.
116 (1966), 1–111

HARMONIC ANALYSIS
ON SEMISIMPLE LIE GROUPS

Harish-Chandra

INSTITUTE FOR ADVANCED STUDY, PRINCETON, NEW JERSEY

Let G be a locally compact group and let dx denote its left-invariant Haar measure. We assume that 1) G is unimodular, i.e., dx is both left and right invariant, and 2) G is separable, i.e. it satisfies the second axiom of countability. A representation π of G on a Hilbert space $\mathfrak{H}$ is a mapping that assigns to every $x \epsilon G$ a bounded linear operator $\pi(x)$ on $\mathfrak{H}$ such that 1) $\pi(1) = I$, $\pi(xy) = \pi(x)\pi(y)$, and 2) the mapping $(x, \psi) \to \pi(x)\psi$ of $G \times \mathfrak{H}$ into $\mathfrak{H}$ is continuous. π is called unitary if $\pi(x)$ is unitary for every x. Equivalence of two representations is defined in the usual way. Finally, π is called irreducible if no closed subspace of $\mathfrak{H}$, other than $\{0\}$ and $\mathfrak{H}$, is stable under $\pi(x)$ for all $x \epsilon G$.

Consider $L_1(G)$. Then $L_1(G)$ is an algebra under convolution given by

$$(f * g)(x) = \int f(y)g(y^{-1}x)\, dy.$$

π being a unitary representation of G, we get a representation of $L_1(G)$ by setting

$$\pi(f) = \int f(x)\pi(x)\, dx.$$

Then $\pi(f * g) = \pi(f)\pi(g) \qquad [f, g \in L_1(G)]$.

Now suppose G is a connected semisimple Lie group and π an irreducible unitary representation of G on a Hilbert space $\mathfrak{H}$. Then $\mathfrak{H}$ is separable. Select an orthonormal base ψ_1, ψ_2, ψ_3, $\cdots$ for $\mathfrak{H}$ and consider the series

$$\sum_i (\psi_i, \pi(x)\psi_i).$$

35

This is a series of continuous functions on G which does not, in general, converge pointwise. But it does converge in the space of distributions on G. More precisely, this means the following. Let $C_c^\infty(G)$ denote the space of all complex-valued C^∞ functions on G with compact support. Then for any $f \in C_c^\infty(G)$, the series

$$\sum_i \int f(x)(\psi_i, \pi(x)\psi_i)\, dx$$

converges absolutely. Let $\Theta_\pi(f)$ denote its sum. Then Θ_π is a linear functional on $C_c^\infty(G)$ which satisfies certain continuity properties, and therefore it is a distribution in the sense of Laurent Schwartz.

Let D be any differential operator on G. Then we denote by D^* its adjoint with respect to the Haar measure, so that

$$\int Df \cdot g\, dx = \int f D^* g\, dx \qquad [f, g \in C_c^\infty(G)].$$

If T is a distribution on G, then the mapping $f \to T(D^* f)$ $[f \in C_c^\infty(G)]$ is also a distribution that we denote by DT.

Let $\mathfrak{Z}$ be the algebra of all differential operators on G which commute with both left and right translations of G. Then $\mathfrak{Z}$ is abelian. Now G operates on itself by inner automorphisms: $y^x = xyx^{-1}$. Hence it operates also on the space of functions: $f^x(y) = f(y^{x^{-1}})$, and on the space of distributions: $T^x(f) = T(f^{x^{-1}})$. A distribution T is called invariant if $T^x = T (x \in G)$. Moreover, it is called an eigendistribution of $\mathfrak{Z}$ if $zT = \chi(z)T$ $(z \in \mathfrak{Z})$, where χ is a homomorphism of $\mathfrak{Z}$ into $\mathbf{C}$. Now the following two facts about Θ_π are fairly obvious:

1) Θ_π is independent of the choice of the orthonormal base $\psi_1, \psi_2, \cdots$. Also, it depends only on the equivalence class ω of π. Hence we may denote it by Θ_ω.

2) Θ_ω is an invariant eigendistribution of $\mathfrak{Z}$.

Let t be an indeterminate, and consider the polynomial $\det[t + 1 - \mathrm{Ad}(x)]$. It is a polynomial in t whose coefficients are analytic functions on G. Let l be the least integer such that the coefficient of t^l is not identically zero. Then l is called the rank of G. Let $D(x)$ denote the coefficient of t^l. An element $x \in G$ is called singular or regular according as $D(x) = 0$ or not. Let G' be the set of all regular elements in G. Then G' is an open and dense subset of G whose complement is of measure zero.

Theorem. *Let Θ be an invariant eigendistribution of $\mathfrak{Z}$ on G. Then Θ is a function which is analytic on G'.*

This means that there exists a locally summable function F on G which is analytic on G' such that

$$\Theta(f) = \int fF \, dx$$

for all $f \in C_c^\infty(G)$.

Let $\mathscr{E}$ be the set of all equivalence classes of irreducible unitary representations of G. For every $\omega \in \mathscr{E}$, we have the distribution Θ_ω, which we call the character of ω. It is a fact that $\Theta_{\omega_1} = \Theta_{\omega_2}$ if and only if $\omega_1 = \omega_2$ $(\omega_1, \omega_2 \in \mathscr{E})$.

Now the main problem of harmonic analysis is the following. Given an invariant distribution T on G, express T as a "linear combination" of the irreducible characters $\Theta_\omega(\omega \in \mathscr{E})$. More precisely, this means the following. For any $f \in C_c^\infty(G)$, define

$$\hat{f}(\omega) = \Theta_\omega(f) \qquad (\omega \in \mathscr{E}).$$

We think of $\hat{f}$ as the Fourier transform of f. Now given an invariant distribution T on G, we would like to find a "distribution" $\hat{T}$ on $\mathscr{E}$ such that $T(f) = \hat{T}(\hat{f})$ for all $f \in C_c^\infty(G)$. (By a "distribution," here we mean some sort of linear functional on a suitable space of functions on $\mathscr{E}$.) Then one thinks of $\hat{T}$ as the Fourier transform of T.

In view of what is known already for the real line **R**, it seems clear that one can hope to define $\hat{T}$ only if T is "tempered," i.e., its behavior at infinity is suitably restricted. So the problem is now to define the Fourier transform of an invariant and "tempered" distribution.

Let us consider some examples of such a transform:

1) *Plancherel formula.* The Plancherel theorem says that there exists a unique positive measure μ on $\mathscr{E}$ such that

$$\delta(f) = f(1) = \int_{\mathscr{E}} \hat{f} \, d\mu.$$

So we may say that $\hat{\delta} = \mu$. Call an element $\omega \in \mathscr{E}$ exceptional if its character Θ_ω is not tempered. By the exceptional series, one means the set $\mathscr{E}_{\mathrm{ex}}$ of all exceptional classes. It seems to be a general fact (I have a proof, which, however, has not yet been written down in all details) that $\mu(\mathscr{E}_{\mathrm{ex}}) = 0$.

2) Fix an element $\gamma \in G$, and let G_γ denote the centralizer of γ in G. Then it is not difficult to show that G_γ is unimodular, and therefore $G^* = G/G_\gamma$ has an invariant measure dx^*.

Consider the integral

$$J(f) = \int_{G/G_\gamma} f(\gamma^{x^*})\, dx^* \qquad [f \in C_c^\infty(G)]$$

where $\gamma^{x^*} = x\gamma x^{-1}$ ($x \in G$). If γ is semisimple [i.e., $\mathrm{Ad}(\gamma)$ is semisimple], one can show that the integral is well defined, and, in fact, J is an invariant tempered distribution on G. (It seems likely that this remains true even if γ is not semisimple.) The problem is to find $\hat{J}$. Note that $J = \delta$ if $\gamma = 1$. The importance of this problem was pointed out to me by Langlands, who was led to it by the Selberg trace formula.

3) *Selberg trace formula.* Here the situation has not yet been investigated precisely, and so our discussion is only formal. Let Γ be a discrete subgroup of G such that the total measure of G/Γ is finite. Then we have the obvious representation π of G on $L_2(G/\Gamma)$. For $f \in C_c^\infty(G)$, let us consider $\mathrm{tr}\,\pi(f)$. Under suitable conditions (e.g., if G/Γ is compact), this trace exists and defines an invariant distribution

$$T(f) = \mathrm{tr}\,\pi(f) \qquad [f \in C_c^\infty(G)].$$

It is not obvious (and, in fact, does not seem to be true, in general) that T is tempered. However, ignoring these complications, the main problem is to find $\hat{T}$.

Example 3 is related to example 2, for, in the computation of $\mathrm{tr}\,\pi(f)$, integrals of the form

$$\int_{G/G_\gamma} f(\gamma^{x^*})\, dx^* \qquad (\gamma \in \Gamma)$$

occur.

The Plancherel measure of G has been computed in some special cases, and the following picture emerges. Assume for simplicity that $G \subset G_c$, and G_c is simply connected. Let $A_1, \ldots, A_r$ be a complete set of nonconjugate Cartan subgroups of G, and let $\mathscr{E}_0$ be the set of all $\omega \in \mathscr{E}$ which are not exceptional. (Sometimes $\mathscr{E}_0$ is called the principal series.) Let $\hat{A}_i$ be the character group of A_i. Then $\mathscr{E}_0$ seems to be in close correspondence with the disjoint union $\hat{A}_1 \cup \hat{A}_2 \cup \cdots \cup \hat{A}_r$.

This means that for each $\hat{a} \in \hat{A}_i$ we have an invariant and tempered eigendistribution $\Theta_{\hat{a}}$ of $\mathfrak{Z}$ on G. Put

$$\hat{f}_i(\hat{a}) = \Theta_{\hat{a}}(f) \qquad [f \in C_c^\infty(G),\, \hat{a} \in \hat{A}_i].$$

Thus the "Fourier transform" of f consists of r functions $(\hat{f}_1, \ldots, \hat{f}_r)$. Now $\hat{A}_i$ is an abelian Lie group. So one can speak without difficulty of "rapidly decreasing" functions or tempered distributions on it. Then f_i is rapidly decreasing on $\hat{A}_i$. Moreover, if T is an invariant and tempered distribution on G, its Fourier transform consists of r tempered distributions $(\hat{T}_1, ..., \hat{T}_r)$ ($\hat{T}_i$ a tempered distribution on $\hat{A}_i$) such that

$$T(f) = \sum_{1 \leqslant i \leqslant r} \hat{T}_i(f_i).$$

Perhaps an example will help here. Let $G = SL(2, \mathbf{R})$,

$$H = \begin{pmatrix} 1 & 0 \\ 0 & -1 \end{pmatrix}, \qquad X = \begin{pmatrix} 0 & 1 \\ 0 & 0 \end{pmatrix}, \qquad Y = \begin{pmatrix} 0 & 0 \\ 1 & 0 \end{pmatrix},$$

$$\mathfrak{a} = \mathbf{R}H, \qquad \mathfrak{b} = \mathbf{R}(X - Y), \qquad k_\theta = \exp \theta(X - Y) = \begin{pmatrix} \cos \theta & \sin \theta \\ -\sin \theta & \cos \theta \end{pmatrix}.$$

Then we have the corresponding Cartan subgroups

$$A = \{1, k_\pi\} \times A^0, \qquad B = \exp \mathfrak{b}$$

where $A^0 = \exp \mathfrak{a}$ is the connected component of 1 in A. Now $\hat{A} \simeq \{1, -1\} \times \mathbf{R}$, $\hat{B} \simeq \mathbf{Z}$. For $\lambda \in \mathbf{R}$, we get two irreducible characters $T_\lambda^\pm$ (with the single exception that T_0^- is not irreducible). Similarly, for every $m \in \mathbf{Z}$ ($m \neq 0$), we get an irreducible character Θ_m. For $m = 0$, we get two irreducible characters Θ_0^+, Θ_0^-. Then $T_0^- = \Theta_0^+ + \Theta_0^-$. Let $\omega = \frac{1}{2}H^2 + XY + YX$ and $\Omega = 2\omega + 1$. (Here we regard H, X, Y as left-invariant differential operators on G.) Then $\Omega \in \mathfrak{z}$ and $\mathfrak{z} = \mathbf{C}[\Omega]$. Moreover,

$$\Omega T_\lambda^\pm = -\lambda^2 T_\lambda^\pm \qquad (\lambda \in \mathbf{R})$$

$$\Omega \Theta_m = m^2 \Theta_m \qquad (m \in \mathbf{Z}, m \neq 0)$$

$$\Omega \Theta_0^\pm = 0$$

All of these characters are distinct except that

$$T_\lambda^\pm = T_{-\lambda}^\pm \qquad (\lambda \in \mathbf{R}).$$

These are all the characters of the principal (i.e., nonexceptional) series. Θ_m ($m \in \mathbf{Z}, m \neq 0$) are all the characters of the discrete series. The Plancherel formula reads as follows:

$$8\pi f(1) = \sum_{-\infty < m < \infty} |m| \, \Theta_m(f) + \int_0^\infty \frac{\lambda}{2} \tanh\left(\frac{\pi\lambda}{2}\right) \cdot T_\lambda^+(f) \, d\lambda$$

$$+ \int_0^\infty \frac{\lambda}{2} \coth\left(\frac{\pi\lambda}{2}\right) \cdot T_\lambda^-(f) \, d\lambda$$

for $f \in C_c^\infty(G)$. Put

$$F_f(\theta) = -(e^{i\theta} - e^{i\theta}) \int_{G/B} f(k_\theta x^*) \, dx^*$$

where $\theta \not\equiv 0 \bmod \pi$. Then

$$\frac{\pi}{i}[F_f(\theta) - F_f(-\theta)] = \sum_m \Theta_m(f) \sin |m|\theta$$

$$- \frac{1}{2}\int_0^\infty T_\lambda^+(f) \cdot \frac{\cosh[(\pi/2) - \theta]\lambda}{\cosh(\pi\lambda/2)} \, d\lambda - \frac{1}{2}\int_0^\infty T_\lambda^-(f) \cdot \frac{\sinh[(\pi/2) - \theta]\lambda}{\sinh(\pi\lambda/2)} \, d\lambda$$

for $0 < \theta < \pi$. Similarly,

$$\pi[F_f(\theta) + F_f(-\theta)] = -\Theta_0(f) - \sum_{m \neq 0} (\operatorname{sign} m) \cos m\theta \cdot \Theta_m(f)$$

for all θ. Here $\Theta_0 = \frac{1}{2}(\Theta_0^+ - \Theta_0^-)$.

Similar results are true for an arbitrary semisimple Lie group. There seem to be strong indications that a corresponding theory for p-adic semisimple groups should be possible. For example, it seems likely that the characters of irreducible unitary representations always exist and are functions. Even the general form of these functions seems to be very similar to what we find in the real case. Actually, the only case that has been studied in some detail so far is that of $SL(2)$.

Since there are no differential operators in the p-adic case, it is clear that the method of proofs there will have to be totally different.

Characters of Semi-Simple Lie Groups

HARISH-CHANDRA

INSTITUTE FOR ADVANCED STUDY
Princeton, New Jersey

Let C be the circle group, that is, the multiplicative group of all complex numbers c with $|c| = 1$. Then for every integer n, we have the character χ_n of C given by $\chi_n(c) = c^n$, and these are the only (irreducible) characters of C. Moreover, the main result in the theory of Fourier series asserts that

$$f(1) = \sum_{-\infty < n < \infty} \int_0^{2\pi} f(e^{i\theta})e^{in\theta}\frac{d\theta}{2\pi}$$

for any smooth function f on C. Now $dc = d\theta/2\pi$ may be regarded as the normalized Haar measure on C, so that the total measure of C is 1. Then this equation becomes

$$\delta(f) = \sum_n \int f\chi_n\, dc$$

where δ is the Dirac measure on C concentrated at 1. We may restate this equation in the form

$$\delta = \sum_n \chi_n$$

in the sense of the Schwartz theory of distributions. This relation is called the *Plancherel formula* for C.

Now let G be a compact topological group and $\mathscr{E}$ the set of all equivalence classes of finite-dimensional irreducible unitary representations of G. If π is such a representation, we define

$$\Theta_\pi(x) = \operatorname{Tr} \pi(x) \qquad (x \in G)$$

The function Θ_π is called the *character* of π. It depends only on the

137

class ω of π in $\mathscr{E}$. Hence, we may denote it by Θ_ω. Then the mapping $\omega \to \Theta_\omega$ is one-to-one, and $d(\omega) = \Theta_\omega(1)$ is the degree of any representation in ω. In this case, the Plancherel formula states that

$$\delta = \sum_{\omega \in \mathscr{E}} d(\omega)\Theta_\omega$$

When G is Abelian, $d(\omega) = 1$ for all ω, and this reduces to

$$\delta = \sum_{\omega \in \mathscr{E}} \Theta_\omega$$

This is just what we had above for $G = C$.

However, when G is not compact, it need not have any finite-dimensional *unitary* representation at all, except the trivial one whose kernel is the whole of G. Therefore, we are forced to consider infinite-dimensional representations.

Let G be a connected semi-simple Lie group. (For our purposes, *semi-simple* means that every normal Abelian subgroup of G is finite.) A representation π of G on a Hilbert space $\mathscr{H}$ is a mapping which assigns to every $x \in G$ a unitary operator $\pi(x)$ on $\mathscr{H}$ such that (a) $\pi(xy) = \pi(x)\pi(y)$, $\pi(1) = I$, (b) the mapping $(x, \psi) \to \pi(x)\psi$ of $G \times \mathscr{H}$ into $\mathscr{H}$ is continuous. π is called *irreducible* if no closed subspace of $\mathscr{H}$, other than $\{0\}$ and $\mathscr{H}$, is stable under $\pi(x)$ for all $x \in G$.

Now fix an irreducible (unitary) representation π of G. We wish to define

$$\mathrm{Tr}\ \pi(x) \qquad (x \in G)$$

Let us choose an orthonormal base ψ_i $(i = 1, 2, \ldots)$ in $\mathscr{H}$. We would like to define

$$\mathrm{Tr}\ \pi(x) = \sum_{i \geq 1} [\psi_i, \pi(x)\psi_i]$$

However, the difficulty is that the series does not converge in general. For example, if we take $x = 1$, we get

$$\sum_{i \geq 1} |\psi_i|^2 = \sum_{i \geq 1} 1 = \infty$$

provided $\dim \mathscr{H} = \infty$. Now put $\phi_i(x) = [\psi_i, \pi(x)\psi_i]$. Then ϕ_i is a continuous function on G and

$$\sum_i \phi_i$$

regarded as a series of functions, does converge in the space of distributions.

Let $C_c^\infty(G)$ denote the space of all indefinitely differentiable functions f which vanish outside some compact subset of G. Also, let dx denote the Haar measure on G. (Apart from a constant factor, dx is uniquely determined by the condition that

$$\int f(yx)\,dx = \int f(x)\,dx \qquad (y \in G)$$

for any continuous function f on G with compact support.) Then one can show that

$$\sum_{i \geq 1} \left| \int f(x)[\psi_i, \pi(x)\psi_i]\,dx \right| < \infty$$

for any $f \in C_c^\infty(G)$. Put

$$\Theta_\pi(f) = \sum_{i \geq 1} \int f(x)[\psi_i, \pi(x)\psi_i]\,dx$$

Then Θ_π is a linear function on the vector space $C_c^\infty(G)$. Such functions, provided they satisfy some mild continuity conditions, have been named "distributions" by Schwartz. One can verify that Θ_π is, in fact, a distribution, and we shall call it the character of π.

We now enumerate some properties of Θ_π. Let $\mathscr{E}$ be the set of all equivalence classes of irreducible unitary representations of G.

1. *Θ_π is independent of the choice of the orthonormal base. Moreover it depends only on the class ω of π in $\mathscr{E}$. Hence, we may denote it by Θ_ω. The mapping $\omega \to \Theta_\omega$ is one-to-one.*

Write $y^x = xyx^{-1}$ $(x, y \in G)$. For any function f on G define f^x to be the function given by

$$f^x(y) = f(x^{-1}yx) = f(y^{x^{-1}})$$

f is called invariant if $f^x = f$ for all $x \in G$. Similarly let T be a distribution on G. Then T^x is the distribution given by

$$T^x(f) = T(f^{x^{-1}}) \qquad [f \in C_c^\infty(G)]$$

Again we say that T is invariant if $T^x = T$ for all $x \in G$. Then the second property of Θ_ω may be stated as follows.

2. *For any $\omega \in \mathscr{E}$, the distribution Θ_ω is invariant.* This is an obvious generalization of the well-known fact that $\operatorname{Tr}\pi(y^x) = \operatorname{Tr}\pi(y)$ for any finite-dimensional representation π.

Now we come to the third and the most important property of Θ_ω. Let D be a differential operator on G. Then its adjoint D^* is the unique differential operator such that

$$\int f D^* g\,dx = \int Df g\,dx$$

for all $f, g \in C_c^\infty(G)$. If T is a distribution, then the distribution DT is defined by

$$(DT)(f) = T(D^*f) \qquad [f \in C_c^\infty(G)]$$

Fix $x \in G$ and, for any $f \in C_c^\infty(G)$, define the function f_x by $f_x(y) = f(yx)$. We say that f_x is obtained from f by right translation by x. Moreover D is said to commute with right translations if

$$(Df)_x = Df_x$$

For all $f \in C_c^\infty(G)$ and $x \in G$. Similarly one can speak about commuting with left translations.

Let $\mathfrak{Z}$ be the algebra of all differential operators on G which commute with both left and right translations. It is known that $\mathfrak{Z}$ is Abelian.

3. Θ_ω *is an eigendistribution of* $\mathfrak{Z}$. This means that for every $z \in \mathfrak{Z}$, there exists a complex number $\chi(z)$ such that $z\Theta_\omega = \chi(z)\Theta_\omega$.

Let $\mathscr{G}$ be the Lie algebra of G and σ the adjoint representation of G on $\mathscr{G}$. (In case G is a matrix group, $\mathscr{G}$ also consists of matrices and $\sigma(x)X = xXx^{-1}$ for $x \in G$ and $X \in \mathscr{G}$.) Let t be an indeterminate. Then

$$\det [t + 1 - \sigma(x)]$$

is a polynomial in t whose coefficients are analytic functions on G. Moreover the highest coefficient is 1. Let l be the least integer $\geqslant 0$ such that the coefficient $D(x)$ of t^l is not identically zero. A point x in G is called singular or regular according as $D(x)$ is or is not equal to zero. Let G' denote the set of all regular elements. Then G' is open and dense in G and the singular set has Haar measure zero.

THEOREM: *Let* Θ *be an invariant eigendistribution of* $\mathfrak{Z}$ *on* G. *Then there exists a locally summable function* F *on* G *which is analytic on* G' *such that*

$$\Theta(f) = \int F\, dx$$

for all $f \in C_c^\infty(G)$. *Moreover,* F *is uniquely determined on* G' *by these properties.*

Now let Θ_π be the character of an irreducible unitary representation of G. Then the above theorem is applicable to Θ_π. Let F_π denote the corresponding function on G. Then

$$\sum_i \int f(x)(\psi_i, \pi(x)\psi_i)\, dx = \Theta_\pi(f) = \int f(x) F_\pi(x)\, dx$$

for $f \in C_c^\infty(G)$. Hence, it seems reasonable to say that

$$\operatorname{Tr} \pi(x) = F_\pi(x) \qquad (x \in G)$$

Of course, in general, the function F_π will have singularities on the singular set. Nevertheless, it will be convenient to define

$$\Theta_\pi(x) = F_\pi(x) \qquad (x \in G')$$

Let us now take an example. Consider the group G of all 2×2 real matrices

$$\begin{pmatrix} a & b \\ c & d \end{pmatrix}$$

with determinant 1 so that $ad - bc = 1$. There are two series of irreducible unitary representations in this case, called the *continuous* and the *discrete series*, respectively.

1. The continuous series is parameterized by a real number λ and for each λ there are two characters T_λ^+, T_λ^-.
2. The discrete series is parameterized by an integer $m \neq 0$ and we denote the corresponding character by Θ_m.

All these characters are distinct and irreducible, except that

$$T_\lambda^+ = T_{-\lambda}^+$$

and T_0^- decomposes into two irreducible characters. (We have ignored the exceptional series here, since it does not occur in the Plancherel formula.) Moreover,

$$8\pi f(1) = \sum_{m \neq 0} |m| \Theta_m(f) + \int_0^\infty \frac{\lambda}{2} \tanh \frac{\pi\lambda}{2} T_\lambda^+(f)\, d\lambda$$

$$+ \int_0^\infty \frac{\lambda}{2} \coth \frac{\pi\lambda}{2} T_\lambda^-(f)\, d\lambda$$

for all $f \in C_c^\infty(G)$. This is the Plancherel formula for G.

Put

$$h_t = \begin{pmatrix} e^t & 0 \\ 0 & e^{-t} \end{pmatrix} (t \in R) \qquad \gamma = \begin{pmatrix} -1 & 0 \\ 0 & -1 \end{pmatrix}$$

and let A be the group of all diagonal matrices in G. Then $A = A_0 \cup \gamma A_0$, where A_0 is the subgroup consisting of all h_t. Similarly, define

$$k_\theta = \begin{pmatrix} \cos\theta & \sin\theta \\ -\sin\theta & \cos\theta \end{pmatrix} \qquad (\theta \in R)$$

and let B be the subgroup consisting of all k_θ. Note that $\gamma = k_\pi$.

Let G' be the regular set of G as before. Then if

$$x = \begin{pmatrix} a & b \\ c & d \end{pmatrix}$$

is an element of G, it is easy to verify that $x \in G'$ if and only if $|\mathrm{Tr}\, x| = |a + d| \neq 2$. Put $A' = A \cap G'$, $B' = B \cap G'$ and

$$G_A = \bigcup_{x \in G} xA'x^{-1} \qquad G_B = \bigcup_{x \in G} xB'x^{-1}$$

Then G' is the disjoint union of G_A and G_B.

Now for each integer $m \geqslant 1$, G has (apart from equivalence) exactly one irreducible (nonunitary) representation σ_m of degree m. Its character is given by

$$\mathrm{Tr}\, \sigma_m(h_t) = \frac{e^{mt} - e^{-mt}}{e^t - e^{-t}} = (-1)^{n-1}\, \mathrm{Tr}\, \sigma_m(\gamma h_t)$$

$$\mathrm{Tr}\, \sigma_m(k_\theta) = \frac{e^{im\theta^*} - e^{-im\theta}}{e^{i\theta} - e^{-i\theta}}$$

On the other hand, the character $\Theta_m(m \neq 0)$ of the discrete may be described as follows.

$$\Theta_m(h_t) = \frac{e^{-|mt|}}{|e^t - e^{-t}|} = (-1)^{n-1}\Theta_m(\gamma h_t)$$

$$\Theta_m(k_\theta) = -(\mathrm{sign}\, m)\frac{e^{im\theta}}{(e^{i\theta} - e^{-i\theta})}$$

There is a remarkable resemblance between the two formulas.

Similarly,

$$T_\lambda^+(h_t) = T_\lambda^-(h_t) = \frac{e^{i\lambda t} + e^{-i\lambda t}}{|e^t - e^{-t}|}$$

$$= T_\lambda^+(\gamma h_t) = -T_\lambda^-(\gamma h_t)$$

and

$$T_\lambda^+(k_\theta) = T_\lambda^-(k_\theta) = 0$$

for the characters of the continuous series. Finally,

$$T_0^- = \Theta_0^+ + \Theta_0^-$$

where $\Theta_0^\pm$ are irreducible characters given by

$$\Theta_0^\pm(h_t) = |e^t - e^{-t}|^{-1} = -\Theta_0^\mp(\gamma h_t)$$

$$\Theta_0^\pm(k_\theta) = \mp(e^{i\theta} - e^{-i\theta})^{-1}$$

Reprinted from
Symposia on Theoretical Physics
4 (1967), 137–142

HARMONIC ANALYSIS ON SEMISIMPLE LIE GROUPS

HARISH-CHANDRA

1. Introduction

The theory of semisimple Lie groups has, in recent years, become the meeting ground of several different branches of mathematics—differential geometry, topology, algebraic geometry, arithmetic and analysis. In this lecture I wish to speak about some recent progress in Fourier analysis on such groups. The results are far from complete. Although the case of real groups is beginning to be fairly well understood, our knowledge of the p-adic groups is still very rudimentary. Nevertheless there appears to be a deep-seated analogy between these two cases. In my opinion, one of the major tasks confronting us, is to try to discover and comprehend the reasons for this similarity. Once local Fourier analysis is well understood, one would have to globalize the problem by going over to the adèle group. It is in this global setting, which seems to provide the right frame-work for the understanding of the work of Hecke and Siegel, that the deeper connections between Fourier analysis and arithmetic are likely to emerge. This is indeed a big project which may take several decades to complete. All that one can say at present, is that this promises to be an extraordinarily rich and fruitful field.

2. The discrete series

Let G be a locally compact, separable and unimodular group. A unitary representation π of G on a Hilbert space $\mathfrak{H}$ is a mapping π which assigns to every $x \in G$ a unitary operator $\pi(x)$ on $\mathfrak{H}$ such that :

$$(1) \qquad \pi(xy^{-1}) = \pi(x)\,\pi(y)^{-1} \qquad (x,\, y \in G),$$

(2) The mapping $(x, \psi) \to \pi(x)\,\psi$ of $G \times \mathfrak{H}$ into $\mathfrak{H}$ is continuous.

π is said to be irreducible if $\mathfrak{H} \neq \{0\}$ and no closed subspace of $\mathfrak{H}$, other than $\{0\}$ and $\mathfrak{H}$ itself, is stable under $\pi(x)$ for all $x \in G$. The equivalence of two representations is defined as usual. Let $\mathscr{E}$ be the set of all equivalence classes of irreducible unitary representations of G.

Fix a Haar measure dx on G and let r denote the right-regular representation of G on $L_2(G)$. A class $\omega \in \mathscr{E}$ is called discrete, if there exists a closed, invariant and irreducible subspace $\mathfrak{H}$ of $L_2(G)$ such that the restriction of r on $\mathfrak{H}$ lies in ω. Let $\mathscr{E}_d$ denote the set of all discrete classes. Then $\mathscr{E}_d$ is called the discrete series for G. Let π be an irreducible unitary representation of G on a Hilbert space $\mathfrak{H}$ and ω the class of π. Then ω is discrete if and only if

$$\int_G |(\phi, \pi(x)\psi)|^2 \, dx < \infty$$

for all ϕ, $\psi \in \mathfrak{H}$. Moreover if $\omega \in \mathscr{E}_d$, there exists a number $d(\omega) > 0$, called the formal degree of ω (or π), such that

$$\int_G |(\phi, \pi(x)\psi)|^2 \, dx = |\phi|^2 |\psi|^2 d(\omega)^{-1} \quad (\phi, \psi \in \mathfrak{H}).$$

Now suppose G is a connected (real) semisimple Lie group with finite center. Then the following theorem gives a criterion for the existence of the discrete series.

T h e o r e m 1. *In order that $\mathscr{E}_d$ should not be empty, it is necessary and sufficient that G should have a compact Cartan subgroup.*

It is believed that a p-adic semisimple group always has a compact Cartan subgroup. Therefore, by analogy, we should expect it to have a discrete series. There are some indications that this is indeed so.

3. Characters

Let $\ell = \operatorname{rank} G$ and D the coefficient of t^ℓ in $\det(t + 1 - \operatorname{Ad}(x))$ $(x \in G)$, where t is an indeterminate. Then $D = D(x)$ is an analytic function which is not identically zero. Let G' be the set of all $x \in G$ where $D(x) \neq 0$. Then G' is an open and dense subset of G whose complement has measure zero.

Let u be a differential operator on G. Its adjoint u^* is the differential operator given by the relation

$$\int_G uf \cdot g \, dx = \int_G f \cdot u^* g \, dx \qquad (f, g \in C_c^\infty(G)).$$

Let T be a distribution on G. Then $f \to T(u^* f)$ $(f \in C_c^\infty(G))$ is also a distribution which we denote by uT. A locally summable function F on G defines a distribution T_F by the rule

$$T_F(f) = \int fF \, dx \qquad (f \in C_c^\infty(G)).$$

We say that a given distribution T is a function, if there exists a locally summable function F such that $T = T_F$. Then F is unique up to a set of measure zero and it is convenient to write $T = F$.

Fix a maximal compact subgroup K of G and let $\mathfrak{B}$ denote the algebra of all differential operators on G which commute with both left and right translations of G. A distribution T on G is said to be $\mathfrak{Z}$-finite if the space of all distributions of the form zT $(z \in \mathfrak{Z})$ has finite dimension. Similarly it is called K-finite if the left and right translates of T under K span a vector space of finite dimension. It is easy to see that if T is both $\mathfrak{Z}$-finite and K-finite, it satisfies an elliptic differential equation and so it is an analytic function. We say that T is invariant if it is left fixed by all inner automorphisms of G.

T h e o r e m 2. *Let Θ be an invariant and $\mathfrak{Z}$-finite distribution on G. Then Θ is a function which is analytic on G'.*

Let π be an irreducible unitary representation of G and ω the class of π. For any $f \in C_c^\infty(G)$, define

$$\pi(f) = \int f(x)\,\pi(x)\,dx.$$

Then it can be shown that $\pi(f)$ is an operator of the trace class and there exists a distribution Θ_ω on G such that

$$\Theta_\omega(f) = \operatorname{tr} \pi(f) \qquad (f \in C_c^\infty(G)).$$

Θ_ω is called the character of ω (or π). It is easy to show that Θ_ω is an invariant eigendistribution of $\mathfrak{Z}$. Therefore by Theorem 2, it is a function which is analytic on G'.

In the $\mathfrak{p}$-adic case one defines $C_c^\infty(G)$ to be the space of all locally constant functions with compact support. Then it seems plausible that $\pi(f)$ is still of the trace class. Put

$$\Theta_\omega(f) = \operatorname{tr} \pi(f) \qquad (f \in C_c^\infty(G)).$$

Then again Θ_ω should turn out to be a locally summable function on G which is locally constant on the regular set G'.

4. The Selberg principle

For any $\gamma \in G$, let G_γ denote the centralizer of γ in G. It is not difficult to show that G_γ is unimodular and therefore the factor space $\bar{G} = G/G_\gamma$ has an invariant measure $d\bar{x}$. Put

$$\gamma^{\bar{x}} = x\gamma x^{-1} \qquad (x \in \bar{G}),$$

where $x \to \bar{x}$ is the projection of G on $\bar{G}$. An element $\gamma \in G$ is said to be elliptic, if it is contained in some compact Cartan subgroup of G.

T h e o r e m 3. (The Selberg Principle.) *Let γ be a semisimple element of G and f a K-finite and $\mathfrak{Z}$-finite function in $L_2(G)$. Then* *the integral*

$$\int\limits_{G/G_\gamma} f(\gamma^{\bar{x}})\, d\bar{x}$$

is well defined and, if γ is not elliptic, its value is zero.

Let Γ be a discrete subgroup of G such that G/Γ is compact. Then G operates on G/Γ and so we get a representation λ of G on $L_2\,(G/\Gamma)$. It is easy to see that λ decomposes into a direct sum of irreducible representations. Moreover the multiplicity $m\,(\omega)$ of each class $\omega \in \mathscr{E}$, is finite in λ. Let π be an irreducible unitary representation on $\mathfrak{H}$ lying in a given class ω. We say that π (or ω) is integrable if

$$\int |(\phi,\, \pi\,(x)\,\psi)|\, dx < \infty$$

for any two K-finite vectors $\phi,\, \psi \in \mathfrak{H}$. (A vector $\phi \in \mathfrak{H}$ is called K-finite, if the space spanned by $\pi(k)\,\phi$ $(k \in K)$, is finite-dimensional.) The Selberg principle allows us to obtain an explicit formula for the multiplicity $m\,(\omega)$ corresponding to any integrable class ω.

I believe that the Selberg principle holds also for the $\mathfrak{p}$-adic groups after a slight reformulation. Let K be any open and compact subgroup of G and π a representation on $\mathfrak{H}$ of the discrete class. Take

$$f\,(x) = (\phi,\, \pi\,(x)\,\psi) \qquad (x \in G)$$

where ϕ and ψ are two K-finite vectors in $\mathfrak{H}$. Then the statement of Theorem 3 should remain true for f.

5. Formula for the characters

We return to the case when G is real and suppose that B is a Cartan subgroup of G contained in K. Let $\mathfrak{g}$, $\mathfrak{b}$ be the Lie algebras of G and B respectively and G_c the simply connected complex analytic group corresponding to the complexification $\mathfrak{g}_c$ of $\mathfrak{g}$. Let us assume that G is the real analytic subgroup of G_c corresponding to $\mathfrak{g}$. Let P be the set of all positive roots of $(\mathfrak{g}, \mathfrak{b})$ under some fixed order. For any $\alpha \in P$, we denote by H_α the element in $\mathfrak{b}_c$ such that

$$\operatorname{tr}\,(\operatorname{ad} H \operatorname{ad} H_\alpha) = \alpha\,(H) \qquad (H \in \mathfrak{b}).$$

Then

$$\tilde{\omega} = \prod_{\alpha \in P} H_\alpha$$

can be considered as a differential operator on B. There exists an analytic function Δ on B such that

$$\Delta\,(\exp H) = \prod_{\alpha \in P} (e^{\alpha(H)/2} - e^{-\alpha(H)/2}) \qquad (H \in \mathfrak{b}).$$

Let $\mathfrak{F}$ be the space of all real-valued linear functions on $(-1)^{1/2}\,\mathfrak{b}$ and L the lattice of those $\lambda \in \mathfrak{F}$ for which there exists a character ξ_λ of B given by $\xi_\lambda (\exp H) = e^{\lambda\,(H)} (H \in \mathfrak{b})$. It is clear that $\widetilde{\omega}$ can also be regarded as a polynomial function on $\mathfrak{F}$ and

$$\widetilde{\omega}\xi_\lambda = \widetilde{\omega}\,(\lambda)\,\xi_\lambda \qquad (\lambda \in L).$$

Put $W_G = \widetilde{B}/B$ where $\widetilde{B}$ is the normalizer of B in G. Then W_G is a subgroup of the Weyl group W of $(\mathfrak{g}, \mathfrak{b})$. Put $B' = B \cap G'$ and let L' denote the set of all $\lambda \in L$ such that $\widetilde{\omega}\,(\lambda) \neq 0$.

T h e o r e m 4. *For any $\lambda \in L'$, there exists exactly one invariant eigendistribution Θ_λ of $\mathfrak{Z}$ on G such that*

1) $$\sup_{x\in G'} |\,D\,(x)\,|^{1/2}\,|\,\Theta_\lambda\,(x)\,| < \infty,$$

2) $$\Delta\,(b)\,\Theta_\lambda\,(b) = \sum_{s\in W_G} \varepsilon\,(s)\,\xi_{s\lambda}\,(b) \qquad (b\in B').$$

Here D has the same meaning as in § 3 and we have to bear in mind Theorem 2. Moreover the character ε of W is defined as usual.

Put $\varepsilon\,(\lambda) = \mathrm{sign}\,\widetilde{\omega}\,(\lambda)\ (\lambda \in L')$ and $q = \dfrac{1}{2}\,\dim G/K$. Then q is an integer.

T h e o r e m 5. *For each $\lambda \in L'$, there exists a unique class $\omega\,(\lambda) \in \mathscr{E}_d$ such that $\Theta_{\omega\,(\lambda)} = (-1)^q\,\varepsilon\,(\lambda)\,\Theta_\lambda$. The mapping $\lambda \to \omega\,(\lambda)$ of L' into $\mathscr{E}_d$ is surjective and $\omega\,(\lambda_1) = \omega\,(\lambda_2)\ (\lambda_1, \lambda_2 \in L')$ if and only if $\lambda_2 = s\lambda_1$ for some $s \in W_G$.*

Define

$$F_f\,(b) = \Delta\,(b)\int_G f\,(xbx^{-1})\,dx \qquad (b\in B')$$

for $f \in C_c^\infty\,(G)$. Then there exists a number $c > 0$ such that

$$\lim_{b\to 1} (\widetilde{\omega}F_f)\,(b) = (-1)^q\,cf\,(1)$$

for all $f \in C_c^\infty(G)$. Moreover

$$d\,(\omega\,(\lambda)) = c^{-1}\,[W_G]\,|\,\widetilde{\omega}\,(\lambda)\,| \qquad (\lambda \in L')$$

where $[W_G]$ is the order of W_G and $d(\omega(\lambda))$ the formal degree of $\omega\,(\lambda)$ (see § 2). It is possible to compute c explicitly for a suitable normalization of dx.

T h e o r e m 6. *Let f be a K-finite and $\mathfrak{Z}$-finite function in $L_2\,(G)$. Then the integral*

$$\int_G f\Theta_\lambda\,dx \qquad (\lambda \in L')$$

*is well defined. Let Θ_λ (f) denote its value. Then Θ_λ (f) = 0 for all
$\lambda \in L'$ except a finite number and*

$$(-1)^q \, cf \, (1) = \sum_{\lambda \in L'} \omega \, (\lambda) \, \Theta_\lambda \, (f).$$

We observe that f is automatically analytic (see § 3) and therefore f (1) is well defined. The Selberg principle enters in an essential way in the proofs of Theorems 5 and 6.

For $\mathfrak{p}$-adic groups the number of conjugacy classes of compact Cartan subgroups is, in general, more than 1. Also there does not seem to be any analogue of the algebra $\mathfrak{Z}$. Therefore the problem of determining all the characters of the discrete series appears to be much more difficult in this case.

6. Concluding remarks

There are some striking similarities between the real and the $\mathfrak{p}$-adic groups. Let K be a maximal compact subgroup of G in the real case and any open compact subgroup in the $\mathfrak{p}$-adic case. Let $\mathscr{E}_K$ be the set of all equivalence classes of irreducible unitary representations of K. For a given irreducible unitary representation π of G on $\mathfrak{H}$, let m ($\mathfrak{d}$) ($\mathfrak{d} \in \mathscr{E}_K$) denote the multiplicity of the class $\mathfrak{d}$ in the restriction of π on K. In the real case it is known that m ($\mathfrak{d}$) is finite. The same is believed to be true in the $\mathfrak{p}$-adic case but so far no general proof for this has been found.

The function $|\, D \,|^{-1/2}$ is locally summable on G in the real case. I believe that this fact is also true in the $\mathfrak{p}$-adic case, provided we interpret the absolute value in the $\mathfrak{p}$-adic sense.

In the real case there is a simple connection between the asymptotic behaviour of the elementary spherical functions and the Plancherel measure for the space G/K. The same appears to be true also in the $\mathfrak{p}$-adic case. (For $G = SL$ (2), this follows from the results of Mautner.)

The algebra $\mathfrak{Z}$ of bi-invariant differential operators plays a very important role in the real case. However, as we have already observed, there does not seem to exist any $\mathfrak{p}$-adic analogue of $\mathfrak{Z}$. Nevertheless it appears likely that all the final results of Fourier analysis continue to hold, after some slight reformulation, for $\mathfrak{p}$-adic groups. The unravelling of this mystery would, in my opinion, be an important achievement.

The Institute for Advanced Study,
Princeton, N.J., USA

Reprinted from
*Proceedings of the International
Congress of Mathematicians,*
(1966), 89–94

Automorphic Forms on Semisimple Lie Groups

HARISH-CHANDRA

Introduction

Let G be a connected semisimple Lie group with finite center and Γ a closed
subgroup of G , which we assume to be unimodular. Let μ denote the invariant measure
on G/Γ. We have the obvious representation λ of G on $L_2(G/Γ) = L_2(G/Γ,μ)$. The
main problem here is to carry out, as explicitly as possible, the reduction of λ. Let
K be a maximal compact subgroup of G. In case Γ = K , this problem has been solved.
Similarly if Γ = {1}, it has been more or less solved. (It is then essentially the
same problem as the explicit determination of the Plancherel formula.)

Let $\mathcal{E}$ be the set of all equivalence classes of irreducible unitary representations of G. Fix $ω ∈ \mathcal{E}$. We say that an element $f ∈ L_2(G/Γ)$ is of type ω, if the
smallest closed invariant subspace V_f of $L_2(G/Γ)$ containing f , is irreducible
under λ and the corresponding representation $λ_f$ lies in ω. Let $\mathcal{H}_ω$ be the
smallest closed subspace of $L_2(G/Γ)$ containing all elements of type ω. Then $\mathcal{H}_ω$ can
be written as an orthogonal sum of irreducible subspaces. Let m(ω) denote the number
of irreducible subspaces in this sum. Then m(ω) is called the multiplicity of ω in
λ. It is an important problem to compute m(ω).

ω is said to occur discretely in λ, if m(ω) > 0. For example, if G is not
compact and Γ = K, m(ω) = 0 for all ω and therefore λ has no discrete components.
If Γ = {1}, the discrete components occur if and only if G has a compact Cartan subgroup. Moreover one then knows which discrete components occur. Finally m(ω) = ∞ for
every discrete component ω, provided G is not compact.

A class $ω ∈ \mathcal{E}$ is said to be L_p (p = 1, 2) if some matrix coefficient of ω
lies in $L_p(G)$. If Γ is a discrete subgroup of G such that G/Γ is compact, then
m(ω) < ∞ for every ω and $L_2(G/Γ) = Cl(\underset{ω}{Σ} \mathcal{H}_ω)$. Moreover if ω is an L_1-class, there
exists an explicit formula for m(ω). It is important to generalize this formula in two
directions. Firstly we should drop the assumption that ω is L_1 and, secondly, we
should relax the condition that G/Γ be compact. However no general results in this
direction have been obtained so far.

From now on we assume that Γ is a discrete subgroup of G such that G/Γ has
finite invariant measure. For G = SL(2,ℝ) and Γ = SL(2,ℤ) (and perhaps also for
some other Γ), the problem of the reduction of λ was solved by Selberg. The case

667

G = SL(n,R), Γ = SL(n,Z) has been considered by Gelfand and Piatetsky-Shapiro. Finally the case when G is any real algebraic semisimple group defined over $\mathbb{Q}$ and Γ any arithmetic subgroup of G, has been studied by Langlands and he has obtained the deepest and the most general results so far. In this case it is possible to show that $m(\omega) < \infty$. However the problem of computing $m(\omega)$ remains largely untouched.

There is another way of approaching the problem of the reduction of $L_2(G/\Gamma)$. Let Let $\mathcal{Z}$ be the algebra of all differential operators on G which commute with both left and right translations. Let $\mathcal{Y}$ be the Lie algebra of G and $\mathcal{U}$ the universal enveloping algebra of $\mathcal{Y}_c$. Then $\mathcal{U}$ may be identified with the algebra of all left-invariant differential operators on G and $\mathcal{Z}$ is the center of $\mathcal{U}$. Put $\mathcal{H} = L_2(G/\Gamma)$ and let $\mathcal{H}_\infty$ denote the subspace of all $f \in \mathcal{H}$ such that the mapping $x \to \lambda(x)f$ of G into $\mathcal{H}$ is of class C^∞. Then one gets in a natural way a representation of $\mathcal{U}$ on $\mathcal{H}_\infty$, which we denote by λ_∞. For any differential operator D on G, let D^* denote its adjoint so that

$$\int Df.gdx = \int f.D^*gdx \qquad\qquad (f,g \in C_c^\infty(G)) \ .$$

(dx is the Haar measure on G.) Then $\mathcal{U}^* = \mathcal{U}$ and $\mathcal{Z}^* = \mathcal{Z}$. Let η denote the conjugation of $\mathcal{Y}_c$ with respect to $\mathcal{Y}$. We say that an element $z \in \mathcal{Z}$ is hermitian if $\eta(z^*) = z$.

Lemma. _If $z \in \mathcal{Z}$ is hermitian, then $\lambda_\infty(z)$ is an essentially self-adjoint operator in the sense of Hilbert-space theory._

Now $\mathcal{H}_\infty$ is stable under $\lambda(x)$ $(x \in G)$ and $\lambda_\infty(z)$ $(z \in \mathcal{Z})$ commutes with $\lambda(x)$. Hence we may regard our reduction problem, at least as a first approximation, as that of obtaining a simultaneous spectral decomposition for all the operators $\lambda_\infty(z)$ $(z \in \mathcal{Z})$. So, roughly speaking, it becomes an eigenfunction expansion problem. Given $f \in L_2(G/\Gamma)$, we have to express it as a linear combination of eigenfunctions of $\mathcal{Z}$ on G/Γ. Fix a maximal compact subgroup K of G. There is no essential loss of generality in assuming that f is K-finite on the left. Then the eigenfunctions that we are looking for, may also be assumed to be left K-finite. It is not difficult to show that such eigenfunctions are in fact analytic.

Thus it is natural to introduce the following definition. Let σ be a unitary representation of K on a finite-dimensional complex Hilbert space V and χ a character of $\mathcal{Z}$ i.e. a homomorphism of $\mathcal{Z}$ into C such that $\chi(1) = 1$. By an automorphic form of type (σ,χ), we mean a C^∞ function $f : G/\Gamma \to V$ such that $f(kx) = \sigma(k)f(x)$ $(k \in K, x \in G)$ and $zf = \chi(z)f$ $(z \in \mathcal{Z})$. However one finds that some of these eigenfunctions have nothing to do with harmonic analysis on G/Γ, since they grow much too fast at infinity. Therefore we impose, in addition, a mild growth condition on f , so as to exclude these extraneous functions.

VI

Let $\mathscr{A}(G/\Gamma,\sigma,\chi)$ denote the space of all automorphic forms of type (σ,χ). Then the first result is that dim $\mathscr{A}(G/\Gamma,\sigma,\chi) < \infty$ (provided Γ satisfies certain reasonable conditions). For example this is true if $G = SL(n,R)$, $\Gamma = SL(n,Z)$. In the case of $G = Sp(n,R)$, $\Gamma = Sp(n,Z)$ and holomorphic forms, this result was first proved by Siegel.

Now a few words about how to construct such eigenfunctions. When G/Γ is compact, no general method of constructing them explicitly is known (except when G has a discrete series). However their existence is assured from the fact that $L_2(G/\Gamma)$ reduces, in this case, into a discrete orthogonal sum of irreducible subspaces. Every such subspace then consists of eigenfunctions.

So let us now consider the case when G/Γ is not compact e.g. $G = SL(n,R)$, $\Gamma = SL(n,Z)$. Then we have the Iwasawa decomposition $G = KAN$, where $K = SO(n)$, $A =$ the group of diagonal matrices in G with all diagonal elements > 0, and $N = \left\{ \left(\begin{smallmatrix} 1 & * \\ 0 & 1 \end{smallmatrix} \right) \right\}$. Let $x = kan$ ($k \in K$, $a \in A$, $n \in N$). We write $k(x) = k$ and $H(x) = \log a \in \mathcal{U} =$ Lie algebra of A. Then

$$dx = e^{2\varrho(\log a)} dk\,da\,dn$$

where ϱ is a certain linear function on $\mathcal{U}$. Let σ be a unitary representation of K on a finite-dimensional space V as above. Then for any linear function λ on $\mathcal{U}_c$, the function

$$x \longrightarrow \sigma(k(x))e^{(\lambda-\varrho)(H(x))} \qquad\qquad (x \in G)$$

from G to the space $\mathcal{E}(V)$ of endomorphisms of V, is an eigenfunction of $\mathcal{Z}$. Let χ_λ denote the corresponding character of $\mathcal{Z}$. Then $\chi_{s\lambda} = \chi_\lambda$ for $s \in W$ (= Weyl group of $(\mathcal{Y},\mathcal{U})$). This function may be considered as a function on G/N. Therefore if the series

$$E_\lambda(x) = \sum_{\gamma \in \Gamma/\Gamma \cap N} \sigma(k(x\gamma))e^{(\lambda-\varrho)(H(x\gamma))}$$

converges, it gives an eigenfunction of $\mathcal{Z}$ on G/Γ. Now the series does converge for suitable λ. But the corresponding χ_λ is such that it is clearly not the infinitesimal character of an irreducible unitary representation of G. Therefore we are forced to study $E_\lambda(x)$ as a function of λ and show that, by analytic continuation, it can be extended to a meromorphic function of λ on $\mathcal{U}_c^*$ (= dual of $\mathcal{U}_c$).

For simplicity, let us consider the case when $\sigma = 1$ and $V = C$. Normalize the Haar measure dn on N in such a way that $N/N \cap \Gamma$ has measure 1. Then

$$e^{\varrho(\log a)} \int_{N/N \cap \Gamma} E_\lambda(an)\,dn = \sum_{s \in W} c(s\!:\!\lambda)e^{s\lambda(\log a)} \qquad\qquad (a \in A) \ ,$$

where $c(s\!:\!\lambda)$ are meromorphic functions on $\mathcal{U}_c^*$ and $c(1\!:\!\lambda) = 1$. Moreover we have the functional equations

$$c(st:\lambda) = c(s:t\lambda)c(t:\lambda)$$

and

$$E_\lambda = c(s:\lambda)E_{s\lambda} \qquad\qquad (s,t \in W).$$

Since E_λ is actually defined by means of a Dirichlet series, there is a certain obvious analogy with the ζ-functions and L-series of number theory. In particular the functions $c(s:\lambda)$ seem to have a product formula. On the other hand there is also a strong analogy with the theory of elementary spherical functions. If we put

$$\emptyset_\lambda(x) = \int_K e^{(\lambda-\varrho)(H(xk))}dk \qquad\qquad (x \in G) \ ,$$

then it is known that $\emptyset_{s\lambda} = \emptyset_\lambda$ $(s \in W)$.

For $G = SL(n,R)$ and $\Gamma = SL(n,Z)$, Selberg had obtained the analytic continuation of all the Eisenstein series (and not just those mentioned above). Langlands has now done this in the general case. Actually he does not confine himself to the arithmetic case but makes a certain set of assumptions on (G,Γ). In view of the recent work of two Russians, Vinberg and Makarov, where they construct non-arithmetic discrete groups Γ such that G/Γ has finite measure, it seems conceivable that the assumptions of Langlands are more general. However they are rather unwieldly and so we shall confine ourselves to the arithmetic case.

Reprinted from
Lecture Notes in Mathematics, No. **62**
Springer-Verlag, Berlin-Heidelberg-New York, 1968